The Flavonoids:
Advances in Research

The Flavonoids: Advances in Research

Edited by

J. B. HARBORNE
Professor, Department of Botany, University of Reading, UK

and

T. J. MABRY
Professor, Department of Botany, University of Texas at Austin, USA

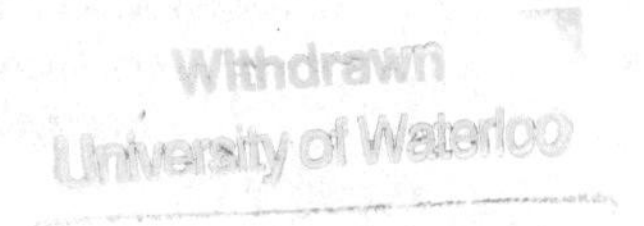

LONDON NEW YORK
Chapman and Hall

First published 1982 by
Chapman and Hall Ltd
11 New Fetter Lane, London EC4P 4EE
Published in the USA by
Chapman and Hall
733 Third Avenue, New York NY 10017

Printed in Great Britain at the
University Press, Cambridge

ISBN 0 412 22480 1

British Library Cataloguing in Publication Data

The Flavonoids: advances in research
1. Flavonoids
I. Harborne, J. B. II. Mabry, T. J.
514'.7 QP925.F5

ISBN 0-412-22480-1

Contents

Contributors

E. Besson (deceased)
Laboratoire de Chimie Biologie, Université de Lyon, Villeurbanne, France

B. A. Bohm
Department of Botany, University of British Columbia, Vancouver, Canada

M. L. Bouillant
Laboratoire de Chimie Biologie, Université de Lyon, Villeurbanne, France

J. Chopin
Laboratoire de Chimie Biologie, Université de Lyon, Villeurbanne, France

P. M. Dewick
Department of Pharmacy, University Park, Nottingham, UK

J. Ebel
Biologisches Institut II, Albert-Ludwigs Universität, Freiburg, West Germany

H. Geiger
Institut für Chemie, Universität Hohenhein, Stuttgart, West Germany

L. A. Griffiths
Department of Biochemistry, University of Birmingham, UK

K. Halbrock
Biologisches Institut II, Albert-Ludwigs Universität, Freiburg, West Germany

J. B. Harborne
Plant Science Laboratories, University of Reading, UK

E. Haslam
Department of Chemistry, The University, Sheffield, UK

K. Hostettmann
École de Pharmacie, Université de Lausanne, Switzerland

M. Hostettmann
École de Pharmacie, Université de Lausanne, Switzerland

G. Hrazdina
Department of Food Science, Cornell University, New York State Agricultural Experimental Station, USA

T. J. Mabry
Department of Botany, The University of Texas at Austin, USA

K. R. Markham
Chemistry Division, DSIR, Petone, New Zealand

V. Mohan Chari
Institut für Pharmazeutische Biologie der Universität Munchen, West Germany

C. Quinn
School of Botany, University of New South Wales, Australia

Christine A. Williams
Plant Science Laboratories, University of Reading, UK

E. Wollenweber
Institut für Botanik, Technische Hochschule, Darmstadt, West Germany

Preface

The flavonoid pigments, one of the most numerous and widespread groups of natural constituents, are of importance and interest to a wide variety of physical and biological scientists and work on their chemistry, occurrence, natural distribution and biological function continues unabated. In 1975, a monograph covering their chemistry and biochemistry was published by Chapman and Hall under our editorship entitled *The Flavonoids*. The considerable success of this publication indicated that it filled an important place in the scientific literature with its comprehensive coverage of these fascinating and versatile plant substances.

The present volume is intended to update that earlier work and provide a detailed review of progress in the flavonoid field during the years 1975 to 1980. Although cross references are made to *The Flavonoids*, this supplement is entirely self-contained and where necessary, tabular data from the earlier volume are included and expanded here.

The choice of topics in *Recent Advances* has been dictated by the developments that have occurred in flavonoid research since 1975, so that not all subjects covered in *The Flavonoids* are reviewed again here. A major advance in flavonoid separation has been the application of high performance liquid chromatography (HPLC) and this is reviewed *inter alia* in the opening chapter on separation techniques. An equally important development in the spectral analysis of flavonoids has been the measurement of carbon-13 NMR spectra and this subject is authoritatively discussed in Chapter 2 and is also illustrated with the spectra of 125 representative flavonoids.

With the continuing emphasis on the isolation and characterization of flavonoid structures from novel plant sources, a most important prerequisite for all flavonoid workers is to have listings of known substances and the plants from which they have been obtained. Very remarkably, the number of known structures among some classes of flavonoid has more than doubled during the six years under review. In this respect, the burden of the reviewer covering new isoflavonoids has been particularly onerous. The central section of this book, therefore, is given over to a series of chapters covering in detail all the major classes of flavonoid, with extensive tabulation and description of new substances. A novel feature introduced into all these chapters is the provision of check lists of all known compounds. This allows the reader to see at a glance which new

structures have been described since 1975 and which compounds are recorded in the earlier literature.

One further chapter in *Recent Advances* covers flavonoid biosynthesis. The major development here has been the isolation and description of the enzymes of biosynthesis. The final chapter on the mammalian metabolism of flavonoids covers an important area of research which was only briefly referred to in *The Flavonoids*. Here it receives a proper, extensive and up-to-date treatment.

As editors, we are particularly grateful to our contributors who have met all our deadlines and allowed us to produce this supplement on schedule. The literature coverage is comprehensive to the end of 1980 and there are also some 1981 papers included. To our readers, we ask for comments, criticisms and suggestions, since we hope to produce a second supplement within the next five years. One of us (T.J.M.) would like to thank Dr Lin Yong-long, Mr Richard Pfeil and Mr Dan Leisure for editorial and technical assistance. We are both grateful to our publishers for their continued support and interest in this endeavour.

Jeffrey B. Harborne
Reading

Tom J. Mabry
Austin

CHAPTER ONE

Isolation Techniques for Flavonoids

KURT HOSTETTMANN and MARYSE HOSTETTMANN

1.1 INTRODUCTION

Within the last five years, numerous new chromatographic techniques have become available for the flavonoid chemist. They not only reduce the separation time, but simplify the isolation of previously unknown or unstable constituents from crude plant extracts or other complex biological sources. High-performance liquid chromatography (HPLC) and droplet counter-current chromatography (DCCC) are major innovations and are complementary to each other. Whereas HPLC has been extensively used for analytical separations, DCCC is an ideal method for the isolation of flavonoids on a preparative scale. Centrifugal thin-layer chromatography and rotation locular counter-current chromatography are other new promising techniques, but at this date, little is known about their application in the field of flavonoids. However, there is no single method of separation capable of solving all isolation problems. Classical techniques such as column chromatography and thin-layer chromatography are still very useful.

In the present chapter, emphasis is given to recently developed separation techniques. It is not an exhaustive survey of the literature, but illustrations are given of the large variety of separations already achieved.

1.2 THIN-LAYER CHROMATOGRAPHY (TLC)

Thin-layer chromatography (TLC) remains an important method for the detection and separation of flavonoids in crude plant extracts. Since the spray reagents for the detection of flavonoids were reviewed by Markham (1975), some new chromogenic reagents have been reported. Saleh (1976) observed that sodium ethoxide (0.05 M in ethanol) can be employed for differentiating between the isomeric 3′- and 4′-methyl ethers of quercetin, isorhamnetin and tamarixetin. Alkaline chloramine-T reagent has been recommended as an ideal reagent for the detection of different phenolic compounds on thin-layer plates (Bajaj and Arora, 1980). The detection of flavanones and 3-hydroxyflavanones on TLC plates has been discussed by Schmidtlein and Herrmann (1976), who used diazobenzene sulphonic acid reagent or sodium borohydride-aluminium chloride reagent.

Suitable adsorbents and solvent systems for the TLC of flavonoids were mentioned in the review of Markham (1975). Some systems devised more recently are presented here. Flavone aglycones have been separated on silica gel with benzene–pyridine–ammonia (80:20:1 drop) by Voirin and Jay (1977) and with benzene–pyridine–formic acid (72:18:10) by Adzet and Martinez-Verges (1980), whereas benzene–acetic acid (45:4) or methylene chloride–acetic acid–water (2:1:1) was found to be ideal for flavanone aglycones (Schmidtlein and Herrmann, 1976). Flavone *O*-glycosides, flavone *C*-glycosides and flavonol *O*-glycosides have been separated recently on silica gel with the following solvent systems: butan-l-ol–acetic acid–water (3:1:1) (Kery *et al.*, 1977), formic acid–ethyl acetate–water (9:1:1) (Kwasniewski, 1978), chloroform–ethyl acetate–acetone (5:1:4) (Chopin *et al.*, 1978), ethyl acetate–pyridine–water–methanol (16:4:2:1) (Markham and Porter, 1979), ethanol–pyridine–water–methanol (80:12:10:5) (Bouillant *et al.*, 1979) and chloroform–methanol–water (65:45:12) (Tschesche *et al.*, 1980). For the separation of various flavanone *O*-glycosides, Ghosal *et al.* (1978) employed chloroform–acetic acid (100:4), benzene–acetic acid (100:4) or chloroform–acetic acid–methanol (90:5:5).

Solvent systems suitable for the separation of flavonoid aglycones on polyamide are the following: benzene–methyl ethyl ketone–methanol (4:3:3), (Ulubelen *et al.*, 1980), benzene–petrol–methyl ethyl ketone–methanol (60:26:7:7) (Mues *et al.*, 1979) or (30:60:5:5) (Jay *et al.*, 1979). Glycosides have been separated with water–ethanol–methyl ethyl ketone–acetyl acetone (65:15:15:5) (Teuber and Herrmann, 1978), water–butan-l-ol–acetone–acetic acid (16:2:2:1) (Parker *et al.*, 1979), nitromethane–methanol (3 : 4) (Geiger and de Groot-Pfleiderer, 1979), methanol–water–acetic acid (90:5:5) (Hostettmann-Kaldas and Jacot-Guillarmod, 1978) and water-methyl ethyl ketone–methanol–2,4-pentanedione (13:3:3:1) (Theodor *et al.*, 1980).

For TLC on cellulose, the classical solvent systems 5–40 % acetic acid, butan-l-ol–acetic acid–water (4:1:5) or 2-methylpropan-2-ol–acetic acid–water (3:1:1) are still often used. For flavone aglycones, benzene–acetic acid–water (125:72:3) or chloroform–acetic acid–water (10:9:1) is recommended (Markham and Wallace, 1980; Markham and Porter, 1979).

1.3 HIGH-PERFORMANCE THIN-LAYER CHROMATOGRAPHY (HPTLC)

Recent research has shown that high-performance thin-layer chromatography (HPTLC) can provide much better separations than TLC of complex flavonoid mixtures (Zlatkis and Kaiser, 1977). HPTLC is a development of TLC carried out using small particles, usually about 5 μm in diameter, all of a closely similar size. However, Guiochon *et al.* (1978) have shown that the use of very fine particles is not necessarily the best method, since the mobile phase flows slowly and diffusion becomes an important factor. Rapid separations can be achieved horizontally by linear, circular and anti-circular developments (Jänchen and Schmutz, 1979;

Kadar-Pauncz, 1979; Kaiser, 1978). Thus, HPTLC can be of interest for chemotaxonomic studies of flavonoids as it requires very small samples (nanogram amounts). An application has been reported by Hiermann and Kartnig (1977) who separated numerous flavonoids on HPTLC silica gel plates, followed by quantitative determination *in situ*. The solvent system for the aglycones was benzene–ethyl acetate–formic acid (40:10:5), whereas acetone–methyl ethyl ketone–formic acid (50:35:5) was used for the glycosides. The latter solvent system was also used for the identification of three flavone glycosides in the leaves of *Digitalis lanata* (Hiermann, 1978).

Chemically bonded silica gel, which is a common packing material for HPLC columns (see Section 1.6.2), has not yet been extensively employed in HPTLC, but it appears that the material of widest application is the octadecyl bonded reverse-phase silica gel or RP-18. In certain cases, RP-2 and RP-8 materials have also been applied to the separation of flavonoids (Becker *et al.*, 1979; Vanhaelen and Vanhaelen-Fastré, 1980).

Methanol–water mixtures are indicated for flavonoid aglycones but require the addition of an acid (formic acid or acetic acid) in order to avoid tailing. Lipophilic stationary phases are not suitable for the separation of highly polar glycosides since they require a high percentage of water in the mobile phase which causes difficulties in development (Hiermann, 1979).

Table 1.1 Solvent systems for reverse-phase HPTLC of flavonoid aglycones

Solvent system	*Reference*
Methanol–water–formic acid (28:10:5)	Hiermann (1979)
Methanol–water–formic acid (28:4:5)	Hiermann (1979)
Methanol–water–acetic acid (70:28:2)	Becker *et al.* (1979)
Ethanol–water (55:44)	Vanhaelen and Vanhaelen-Fastré (1980)

Reverse-phase HPTLC is indicated for the analysis of crude plant extracts since the lipophilic plant constituents remain at the application point, whereas the more polar compounds such as flavonoid aglycones are mobile (Becker *et al.*, 1979).

1.4. CENTRIFUGAL THIN-LAYER CHROMATOGRAPHY

Among the commonly used preparative-scale separation techniques, centrifugal chromatography has so far played only a minor role. Numerous attempts have been made to increase the speed of separation by acceleration of the flow rate of the mobile phase using centrifugal force (Deyl *et al.*, 1964; Heftmann *et al.*, 1972). Recently, commercially available equipment has been applied to the separation of several classes of phenolic compounds. In one of these methods, the TLC plate

(rotor) is not horizontal but inclined, and this allows the more efficient collection of the eluate. Its principle of operation is simple. The mixture to be separated is applied as a solution near the centre of a rotor, which is coated with a thin layer of adsorbent (layer thickness 1–4 mm). Elution with a solvent produces concentric rings of the components, which are spun off in turn from the edge of the rotor, together with solvent. A collection system brings each eluate to a single output tube. UV-absorbing compounds can be observed directly through a quartz lid during the separation.

The efficiency of this method was demonstrated by Derguini *et al.* (1979), who separated 100 mg amounts of *cis*/*trans* isomeric esters on silica with *n*-hexane–ether (99:1). The recovery of pure esters was 90% whereas it was only 80% in preparative liquid chromatography. Recently, Hostettmann *et al.* (1980) and Hostettmann-Kaldas *et al.* (1981) applied centrifugal TLC to the isolation of xanthone aglycones from *Gentiana* species. From a crude chloroform extract (400 mg) of *Gentiana detonsa* (Gentianaceae) 7 mg of 1-hydroxy-3,7,8-trimethoxyxanthone and 12 mg of 1,7-dihydroxy-3,8-dimethoxyxanthone were isolated within 30 min. As flavones are structurally related to xanthones and possess a similar chromatographic behaviour, preparative centrifugal TLC should be an ideal method for their isolation. This very rapid technique can handle up to 1 g of sample and can replace preparative TLC (where bands have to be scraped off in order to recover the sample), and, in some instances, column chromatography. However, its application is limited by the relatively low resolution and the restriction in the choice of stationary phases. The procedure described by Gilpin and Sisco (1976) for converting silica gel layers into octadecyl reverse-phase layers may greatly increase the versatility of centrifugal TLC and make it more suitable for the separation of glycosides.

1.5. COLUMN CHROMATOGRAPHY

Column chromatography remains a very useful technique for preliminary purification or for preparative scale separation of large quantities of flavonoids from crude plant extracts. It has been discussed in depth by Markham (1975). Polyamide is the most widely used packing material. Aglycones have been separated with chloroform–methanol–methyl ethyl ketone–acetone (40:20:5:1) (Ulubelen *et al.*, 1980) or with benzene–petrol–methyl ethyl ketone–methanol (60:26:3.5:3.5) Mues *et al.*, 1979). Numerous glycosides have been isolated from crude extracts with methanol–water mixtures (Parker *et al.*, 1979; Ulubelen *et al.*, 1980; Marston *et al.*, 1976; Bouillant *et al.*, 1979; Geiger and de Groot-Pfleiderer, 1979). Flavonol glycosides have been resolved by Geiger *et al.* (1978) using water with increasing amounts of acetone.

On silica gel, flavone aglycones have been separated with chloroform–methanol mixtures (Karl *et al.*, 1977), whereas for glycosides, it is necessary to add water to such systems. Good separations were obtained with chloro-

form–methanol–water in the proportions (80:20:1) (Higuchi and Donnelly, 1978) or in the proportions (65:20:2) and (80:18:2) (Tschesche *et al.*, 1980). The last authors also employed ethyl acetate–acetone–water (25:5:1). Homoisoflavonoids have been separated on silicic acid with chloroform–acetone mixtures by Tada *et al.* (1980). The solvent systems used for cellulose TLC (see Section 1.2) are suitable for column chromatography on the same adsorbent.

Sephadex gel is an excellent stationary phase and its advantages in the isolation of flavonoids have already been discussed (Markham, 1975). Table 1.2 summarizes some of the gel types and the corresponding solvent systems which are most widely used.

Table 1.2 Column chromatography of flavonoids on Sephadex

Type	*Solvent system*	*Flavonoids*	*Reference*
G-10	Water	Flavonols	Cabrera and Juliani (1979)
G-10	Water → water–methanol (6:4)	6-Hydroxyflavones	Ulubelen *et al.* (1980)
G-25	Water–acetone (8:2 → 6:4)	Flavonoid glycosides	Parker *et al.* (1979)
LH-20	Methanol–water (3:7)		
	Methanol → methanol–water (8:2)		Mues *et al.* (1979)
			Higuchi and Donnelly (1978)
	Chloroform–methanol (9:1)		Karl *et al.* (1977)
	Methanol–water (3:1)		Geiger and de Groot-Pfleiderer (1979)
	Acetone–methanol–water (2:1:1)		Geiger *et al.* (1978)
	Methanol, ethanol	Homoisoflavonoids	Tada *et al.* (1980)
		Flavone *C*-glycosides	Kaldas *et al.* (1975)

1.6. HIGH-PERFORMANCE LIQUID CHROMATOGRAPHY (HPLC)

High-performance liquid chromatography (HPLC) has proven to be one of the most useful techniques available today for separating complex mixtures of organic substances. The range of compounds that have been successfully separated by HPLC continues to expand at a very rapid rate. Within the last five years, HPLC has been applied to highly complex separations in the field of natural products chemistry. There has also been a dramatic increase in the scope and versatility of this technique, due mainly to the availability of the necessary instrumentation and to improvements in the efficiency and nature of column packing materials. An excellent and very useful recent review is that of Kingston (1979). Some selected examples of HPLC separations have been presented by Adams and Nakanishi (1979), while Van Sumere *et al.* (1979) have reviewed the application of HPLC to the separation of plant phenolics. Useful information on the theory and on the relationship of theory to practice in HPLC has been published (Snyder and Kirkland, 1974; Engelhardt, 1979) and will not be discussed here.

It is noteworthy that in the flavonoid field HPLC has been mainly used as an analytical technique, e.g. for quantitative determination of plant constituents, for checking the purity of natural samples, and for chemotaxonomic comparisons. For the isolation of flavonoids on a preparative scale, the full potential of HPLC has not yet been exploited. Without doubt, by using larger reverse-phase columns with high efficiencies, HPLC will become a method of choice for the isolation of previously unknown flavonoid structures.

1.6.1 Separations on silica gel columns

This column packing material has not been used often, but can be recommended for the separation of non-polar or weakly polar flavonoid aglycones. HPLC using LiChrosorb Si 60 as adsorbent and a mixture of heptane–propan-2-ol (60:40) as eluent was found to be a very efficient method for the separation of the major polymethoxylated flavones present in citrus fruits (Bianchini and Gaydou, 1980). Tangeretin, tetra-*O*-methylscutellarein, nobiletin and sinensetin have been separated and identified in crude extracts of tangerine and orange peels within 25 min. Complex mixtures of isoflavone aglycones have been resolved on a μPorasil column using different gradient systems (Carlson and Dolphin, 1980). On the other hand, isocratic chromatography of a *Glycine max* hydrolysate with [dichloromethane–ethanol–acetic acid (97:3:0.2)]–hexane (8:2) resulted in a rapid separation of the two isoflavones present, daidzein and genistein. The isocratic run does not require solvent re-equilibration for each sample and thus increases the rate of analysis. Silica gel has also been used by Ward and Pelter (1974) for the separation of isoflavones with hexane–tetrahydrofuran (2:1) and biflavonoids with isopropyl ether containing 8 % methanol. Recently, a range of flavonoid acetates has been separated and determined on 5-μm LiChrosorb Si 60, using four solvent systems by Galensa and Herrmann (1980) (see Table 1.3). Benzene–acetonitrile was the most suitable mobile phase, especially at 45° C With this solvent system, the flavonoids were readily detected by absorption measurements at 285 nm, and flavanones could therefore also be determined. Benzene–acetonitrile (85:40) separated the flavone glycosides very well, whereas flavonol and flavanone glycosides showed very short retention times with this solvent mixture. In this work, flavonoid glycosides were acetylated prior to HPLC separation and an excellent resolution was obtained. By using reverse-phase column packing materials (see below), however derivatization can be avoided.

1.6.2 Separations on reverse-phase columns

Soon after the introduction of HPLC, numerous new types of stationary phases appeared on the market. The most important of these are the so-called chemically bonded stationary phases which are prepared by bonding various organosilane molecules (e.g. octadecyltrichlorosilane, octyltrichlorosilane or phenyltrichloro-

silane) to the hydroxyl groups of a silica type surface. One of the most important factors affecting the performance of bonded-phase columns is the amount of organic material deposited on the silica surface (Snyder and Kirkland, 1974). Since the trichlorosilanes polymerize readily in the presence of moisture, the amount of material bonded to the surface may vary unless the amount of water present is standardized. In practice, numerous types of packed columns possessing a high degree of reproducibility are commercially available. But simple procedures for preparing octadecylsilyl bonded stationary phase have been described (Kingston and Gerhart, 1976; Gilpin *et al.*, 1975) and may be needed, for example, when packing a wide column for preparative-scale separations. Wehrli *et al.* (1978) estimated that 80% of current separations are done on octadecylsilyl bonded phase columns, commonly called C_{18} columns (see Table 1.3).

When using the above column material, the stationary phase is less polar than the mobile phase and the procedure is called reverse-phase chromatography. Thus, highly polar solutes possess shorter retention times than less polar solutes. This is of prime importance in the separation of complex mixtures of flavonoids: glycosides will be eluted first, followed by the aglycones, generally in order of decreasing polarities. For aglycones possessing only a few free hydroxyl groups or fully methoxylated aglycones, it is obvious that reverse-phase chromatography is not necessarily indicated, since this class of flavonoid can easily be separated by adsorption chromatography on silica gel (Bianchini and Gaydou, 1980; Carlson and Dolphin, 1980). Nevertheless, several authors have obtained excellent separations of non-polar aglycones on octadecylsilyl or octylsilyl bonded columns (Rouseff and Ting, 1979; Ting *et al.*, 1979). A practical application of the reverse-phase separation technique can be seen in an analysis of isoflavone aglycones in soybeans (West *et al.*, 1978). In this case, the positional isomers 5,7,4′-trihydroxyisoflavone, genistein, and 6,7,4′-trihydroxyisoflavone were separated with water–acetonitrile (4:1) on a Partisil-10-ODS column.

Acetonitrile–water mixtures or methanol–water containing small amounts of acetic acid are commonly used solvent systems for flavonoid glycosides. These mobile phases are suitable for use with UV detection and can easily be employed in gradient systems for complex separations (Niemann and Koerselmann-Kooy, 1977; Strack and Krause, 1978; Quercia *et al.*, 1978). In many cases, the acetic acid in the mobile phase has been replaced by very small amounts of phosphoric acid (Williams *et al.*, 1978; Court, 1977; Niemann, 1977; Niemann and Van Brederode, 1978; Seitz and Wingard, 1978). Depending on the compounds to be separated, it may be advantageous to replace C_{18}-chemically bonded phases by C_8 reverse-phase material. For example, isovitexin arabinoside and vitexin rhamnoside were poorly resolved on a LiChrosorb RP-18 column, whereas these two *C*-glycosides were clearly separated on a RP-8 column with a methanol–acetic acid–water solvent system (Strack *et al.*, 1979). This system has been applied to the analysis of flavone glycosides in crude extracts of *Avena sativa* leaves. Strack and Krause (1978) compared the selectivities of RP-8 and RP-18 columns and obtained, in

Table 1.3 HPLC of flavonoids

Separated flavonoids	*Column*	*Mobile phase*	*Reference*
Anthocyanidins	μ-Bondapak C_{18}	Methanol–acetic acid–water (20:5:75)	Adamovics and Stermitz (1976)
	μ-Bondapak C_{18}	Methanol–acetic acid–water (19:10:71)	Wilkinson *et al.* (1977)
Anthocyanidin glycosides	μ-Bondapak C_{18}		Williams *et al.* (1978)
3-Glucosides		Acetic acid–water (15:85)	
3,5-Diglucosides		0.1 % H_3PO_4 in acetic acid–water (10:90)	
p-Coumarylglucosides		Methanol–acetic acid–water (20:15:65)	
Anthocyanins	Pellidon	Chloroform–methanol (87:13)	Manley and Shubiak (1975)
Methoxylated flavones	Zorbax C_8	Tetrahydrofuran–acetonitrile–water (22:6:72)	Rouseff and Ting (1979)
	Micropak C_{18}	Acetonitrile–water (40:60)	Ting *et al.* (1979)
	Zorbax C_8	Tetrahydrofuran–water (25:75)	
	Lichrosorb Si 60	Heptane–ethanol (75:25)	Bianchini and Gaydou (1980)
Flavones and flavonols			
Aglycones	μ-Bondapak alkylphenyl	Ethanol–acetic acid–water (47.5:5:47.5)	Vanhaelen and Vanhaelen-Fastré (1980)
	μ-Bondapak C_{18}	Methanol–acetic acid–water (30:5:60)	Wulf and Nagel (1976)
	μ-Bondapak C_{18}	40 % methanol in phosphate buffer	Court (1977)
	Zipax HCP (hydrocarbon polymer)	Phosphate buffer–ethanol–ethyl acetate (100:1:0.1)	Okuda *et al.* (1979)
Flavones and flavonols			
Glycosides	Zorbax ODS	45 % → 100 % methanol containing 0.1 % acetic acid	Niemann and Koerselmann-Kooy (1977)
	Vydac ODS	Methanol–phosphoric acid–water (15:0.5:85)	
	Lichrosorb RP8	Acetonitrile–water (1:9 → 9:1)	
	μ-Bondapak C_{18}	Methanol–acetic acid–water (30:5:60)	Wulf and Nagel (1976)
	μ-Bondapak C_{18}	Methanol–acetic acid–water (30:5:70)	Adamovics and Stermitz (1976)
	μ-Bondapak C_{18}	40 % methanol in phosphate buffer	Court (1977)
	Zorbax ODS	20 % → 100 % ethanol containing 0.1 % phosphoric acid	Niemann (1977)
	Zypax HCP (hydrocarbon polymer)	Phosphate buffer–ethanol–ethyl acetate (100:1:0.1)	Okuda *et al.* (1979)

Flavone C-glycosides	Lichrosorb RP-8	(A) Methanol–acetic acid–water (5:5:90)	Strack *et al.* (1979)
	Lichrosorb RP-18	(B) Methanol–acetic acid–water (90:5:5)	
		Linear gradient from 0 → 35 % B	
	Lichrosorb RP-8	Linear gradient from 0 → 50 % B	Strack *et al.* (1979)
	Zorbax ODS	Concave gradient of methanol in water: 15 % → 95 % methanol	Quercia *et al.* (1978)
O-Glycosides of glycoflavones	Lichrosorb NH_2	acetonitrile–water (1:9 → 9:1)	Becker *et al.* (1977)
	Zorbax ODS	Ethanol–water with 0.1M-H_3PO_4 20 % → 100 %	Niemann and van Brederode (1978)
Flavanone aglycones	μ-Bondapak C_{18}	Butanol–acetic acid–methanol–water (5:2:25:6)	Hardin and Stutte (1980)
	μ-Bondapak C_{18}	10 % → 100 % methanol in phosphate buffer	Seitz and Wingard (1978)
Flavanone glycosides	μ-Bondapak C_{18}	Methanol–acetic acid–water (30:5:60)	Wulf and Nagel (1976)
	μ-Bondapak C_{18}	Acetonitrile–water (80:20)	Fisher and Wheaton (1976)
	μ-Bondapak C_{18}	Acetonitrile–water (80:20)	Fisher (1978)
	μ-Bondapak C_{18}	10 % → 100 % methanol in phosphate buffer	Seitz and Wingard (1978)
	Zorbax ODS	45 % → 100 % methanol containing 0.1 % acetic acid	Niemann and Koerselmann-Kooy (1977)
Isoflavone aglycones	Partisil-10-ODS	Water–acetonitrile (4:1)	West *et al.* (1978)
	μ-Porasil	Dichloromethane–ethanol–acetic acid (97:3:0.2)–hexanes (12:88) or (8:2)	Carlson and Dolphin (1980)
Isoflavone glycosides	μ-Bondapak C_{18}	32 % → 90 % methanol	Ohta *et al.* (1980)
Flavonoid acetates	Lichrosorb Si60	Benzene–acetonitrile (85:20)	Galensa and Herrmann (1980)
Flavone, flavonol, flavanone aglycones and glycosides		Benzene–acetone (90:15)	
		Benzene–ethanol (80:0.8)	
		Iso-octane–ethanol–acetonitrile (70:16:5.5)	
Flavonolignans	μ-Bondapak C_{18}	Methanol–acetic acid–water (40:5:60)	Tittel and Wagner (1977)
	Lichrosorb RP-18	Methanol–acetic acid–water (40:5:60)	Tittel and Wagner (1978)
Biflavonoids	Merckosorb S-160	Isopropyl ether containing 8 % methanol	Ward and Pelter (1974)

general, better resolutions of flavone *C*-glycosides on RP-8 reverse-phase material.

A hydrocarbon polymer column (Zipax HCP) coupled with a phosphate buffer–ethanol–ethyl acetate solvent system was used for the analysis of hydrolysable tannins and related polyphenols in *Geranium thunbergii* (Okuda *et al.*, 1979). Becker *et al.* (1977) employed an amino- chemically bonded silica gel column (LiChrosorb NH_2) eluted with acetonitrile–water for the separation of isomeric *O*-glycosides of glycoflavones. An amino- chemically bonded phase gave excellent separations of xanthones with three and four hydroxyl groups (Hostettmann and McNair, 1976). The same authors separated less polar xanthones (i.e. those with a single hydroxyl or those lacking free hydroxyl groups) on a more polar nitrile-chemically bonded phase (Micropak CN) with *n*-hexane–chloroform. Xanthone glycosides have been separated isocratically in the reverse-phase mode on C_{18}-chemically bonded silica gel with methanol–water mixtures (Pettei and Hostettmann, 1978). This system allowed the separation of isomeric diglycosides and monoglycosides as well as of polar xanthone aglycones. As flavones are structurally closely related to xanthones and show a similar chromatographic behaviour, the HPLC methods discussed above are suited for flavonoid separations. Although reverse-phase HPLC is extensively used in many areas of natural products chemistry and biochemistry (Montgomery *et al.*, 1977), excellent results can often be obtained by combining its use with other procedures (Nelson, 1977).

HPLC has proven to be one of the most powerful methods for the analysis of flavonoids. It is especially indicated for the quantitative determination of plant phenolics and will in the future certainly become more popular as a method for their isolation on a semi-preparative or preparative scale.

1.7 DROPLET COUNTER-CURRENT CHROMATOGRAPHY (DCCC)

A simple and very efficient all-liquid separation technique called droplet counter-current chromatography (DCCC) has been developed by Tanimura *et al.* (1970). DCCC is less cumbersome than counter-current distribution and has a higher resolving power. It is carried out by passing droplets of a mobile phase through columns of surrounding stationary liquid phase. The mobile phase may be either heavier or lighter than the stationary phase; when lighter, the mobile phase is delivered at the bottom of the columns (ascending mode) and, when heavier, through the top (descending mode). Recently, apparatus for DCCC has become commercially available and the method has already been applied to the isolation of many natural products. It is particularly indicated for the preparative-scale separation of polar compounds, such as glycosides (Hostettmann *et al.*, 1979a). A general account of DCCC has been given by Hostettmann (1980) who discussed the advantages and limitations of the method.

The absence of a solid support in DCCC is the main advantage of the technique

and is of great interest in flavonoid separation. The acidic hydroxyl groups in polyphenols often cause irreversible adsorption of the solute on to the solid stationary phase during column chromatographic procedures. This can be partially avoided by adding an acid to the mobile phase and is useful for analytical work. However, when developing a preparative-scale separation, the use of acid as one of the components of the mobile phase is to be avoided. The risk of sample deterioration, i.e. by hydrolysis of glucosidic links, is increased during the long operation and the recovery of the sample from the relatively large volume of eluate. Acid is not necessary for DCCC separation. Best results have been obtained with chloroform–methanol–water mixtures in different proportions and with chloroform–methanol–propanol–water or chloroform–methanol–butanol–water mixtures. The selected solvent system must, of course, form two layers. For flavonoids and related compounds bearing few hydroxyl groups and one sugar moiety or aglycones possessing several free hydroxyl groups, the separation can be achieved by using the less polar layer as mobile phase, whereas the more polar layer is suited for more polar monoglycosides or diglycosides. For example, from a crude fraction of *Tecoma stans* (Bignoniaceae), Hostettmann *et al.* (1979a) isolated within 6 hours a pure glycoside identified as quercetin-3-*O*-glucoside using the upper layer of chloroform–methanol–water (7:13:8) as mobile phase. The same solvent system has been used by Takeda *et al.* (1977) for the isolation of rutin and various isoflavone glycosides from *Sophora japonica* (Leguminosae). In this case, the glycosides were eluted in order of decreasing polarity: rutin, biochanin A 7-*O*-gentiobioside, biochanin A 7-*O*-xylosylglucoside, biochanin A 7-*O*-glucoside and irisolidone 7-*O*-glucoside.

Flavonoid separations carried out by DCCC are summarized in Table 1.4. It is noteworthy that the technique is ideal for the isolation of various glycosides, including *C*-glycosylflavonoids. A remarkable separation of two closely related flavonol monoacetylrhamnosides has been reported by Tanaka *et al.* (1978). These authors isolated 3″-*O*-acetylquercitrin and 4″-*O*-acetylquercitrin from the aerial parts of *Pteris grandifolia* (Pteridaceae) using chloroform–methanol–water (4:4:3) in the ascending mode (mobile phase: upper layer). A base-line separation of both isomers was obtained. Separations of isomeric acylated flavonoid glycosides are usually very difficult to achieve, by any other means. Thus, DCCC opens new possibilities and will be of great help in the future for difficult glycosidic isolations. Although, DCCC is more suited for glycoside separations, some examples of aglycone isolations have been reported (Hostettmann *et al.*, 1979b; Tanaka *et al.*, 1978).

Although DCCC is an excellent and highly promising technique for the preparative-scale separation of flavonoids, it possesses limitations arising from the fact that the efficiency of the method depends entirely upon droplet formation. By using columns with 2 mm internal diameter, the choice of solvent systems is limited since the droplets formed must have a smaller size than the internal diameter of the column. In most of the DCCC separations achieved up to now, chloroform–methanol–water or chloroform–methanol–propanol–

Table 1.4 Flavonoids isolated by DCCC

Flavonoids	*Solvent system*	*Plant source*	*Reference*
C-glycosylflavonoids			
Iso-orientin	Chloroform–methanol–propanol–water (5:6:1:4)	*Gentiana strictiflora*	Hostettmann-Kaldas *et al.* (1981)
Iso-orientin	Chloroform–methanol–butanol–water (10:10:1:6)	*Lespedeza cuneata*	Numata *et al.* (1980) Numata *et al.* (1979)
Isovitexin			
Lucenin-2			
Vicenin-2			
Swertisin	Chloroform–methanol–water (7:13:8)	*Zizyphus jujuba*	Tanimura *et al.* (1975)
Vitexin	Chloroform–methanol–water (5:6:4)	*Lindsaea ensifolia*	Satake *et al.* (1978)
Flavonol aglycones and glycosides			
Quercetin 3-*O*-glucoside	Chloroform–methanol-water (7:13:8)	*Tecoma stans*	Hostettmann *et al.* (1979a)
Quercetin 3-*O*-rutinoside	Chloroform–methanol–water (7:13:8)	*Sophora japonica*	Takeda *et al.* (1977)
3″-*O*-acetylquercitrin	Chloroform–methanol–water (4:4:3)	*Pteris grandifolia*	Tanaka *et al.* (1978)
4″-*O*-acetylquercitrin			
Quercetin 3-*O*-glucoside	Chloroform–methanol–butanol–water (10:10:1:6)	*Polygonum nodosum*	Isobe *et al.* (1980)
Kaempferol			
Kaempferol 3-*O*-glucoside			
Kaempferol 3-*O*-glucoside-2″-gallate			
Isoflavone glycosides			
Biochanin A 7-*O*-glucoside	Chloroform–methanol–water (7:13:8)	*Sophora japonica*	Takeda *et al.* (1977)
Biochanin A 7-*O*-gentiobioside			
Biochanin A 7-*O*-xylosylglucoside			
Irisolidone 7-*O*-glucoside			

water mixtures were used as solvent systems (Hostettmann, 1980). Recently, 40 cm columns with 3.4 mm internal diameters have become commercially available. It is likely that they will allow the use of many other solvent systems, e.g. butanol–pyridine–water, ethyl acetate–ethanol–water, etc. This naturally will greatly increase the versatility of the technique and make it suitable for the isolation of highly polar flavonoids, such as tri- or tetra-glycosides. However, an increase in the internal diameter of the columns reduces the resolving power. However, in the case of difficult separations, up to 1000 columns can be interconnected.

1.8 GAS–LIQUID CHROMATOGRAPHY (GLC)

Direct GLC analysis is restricted to the few thermally stable naturally occurring aglycones which are sufficiently volatile, e.g. the polymethoxylated flavones in the fruit peel of *Citrus reticulata* (Iinuma *et al.* (1980)) in such a GLC analysis found a correlation between flavone structure and the relative retention time. Retention times of flavones having one methoxyl in ring B were shorter than those of flavones with two methoxyls. The retention time did not change with the introduction of a methoxyl into the 8-position. When the mode of substitution in ring B was the same, the retention times of 5,6,7-trisubstituted flavones were shorter than those of 5,7,8-trisubstituted flavones. The stationary phase used was 1 % OV-17 and the separations were monitored with a flame ionization detector (FID).

Derivatization of flavonoids prior to GLC analysis is usually needed in order to increase their volatility. Different silylation procedures have been reviewed by Vande Casteele *et al.* (1976). These authors reported an excellent separation of naturally occurring benzoic acid, cinnamic acid, coumarin, flavone and iso-flavone derivatives. By using *NN*-bis-(trimethylsilyl) trifluoroacetamide (BSTFA) as a silylating agent, a glass column packed with Chromosorb WAW DMCS coated with 1.5 % SE-30 + 1.5 % SE-52 and a temperature program from 80 to 300° C, about 36 of the above mentioned substances were separated in a single run. In this study, preparative GLC was also used. After hydrolysis of the collected trimethylsilyl (TMS) derivatives, it was established by TLC that the substances could be recovered unchanged in yields between 81 and 85 %.

GLC of plant glycosides as the TMS derivatives should ideally be combined with mass spectrometry (Martinelli, 1980). Three flavone glycosides (luteolin 4′-*O*-glucoside, luteolin 7-*O*-glucoside and luteolin 7-*O*-rhamnoglucoside) have been identified in artichoke extracts (*Cynara scolymus*) by Bombardelli *et al.* (1977a). The column used was filled with 0.5 % OV-101 liquid phase loaded on silanized Chromosorb WHP. The same stationary phase was employed in a gas–liquid chromatography–mass spectrometry (GLC–MS) investigation of numerous anthocyanidins (Bombardelli *et al.*, 1976) and anthocyanins (Bombardelli *et al.*, 1977b). Treatment of anthocyanins with trimethylchlorosilane (TMCS) and

hexamethyldisilazane (HMDS) yielded nitrogen-containing derivatives which, after injection into the gas chromatograph, produced quinoline-like compounds. The latter were easily separated and showed fragmentation patterns which were useful for identification purposes. Attempts are currently being made to employ capillary GLC columns for the analysis of highly complex mixtures of anthocyanins. GLC–MS has also been applied to the determination of new flavonoid triglycosides from the leaves of *Cerbera manghas* (Sakushima *et al.*, 1980). Nine flavonoid aglycones (as silylated derivatives) have been separated on a column packed with OV-17 by Vanhaelen and Vanhaelen-Fastré (1980). These authors reported that the temperature and gas flow rate required to elute flavonoids are inconsistent with the use of capillary columns.

Although interesting results have been obtained by GLC, this technique is not commonly used for the separation of flavonoids because of the disadvantages associated with the need to derivatize the sample. The application of GLC to the separation of plant phenolics in general has been reviewed by Van Sumere *et al.* (1979).

1.9 MISCELLANEOUS TECHNIQUES

1.9.1 Paper electrophoresis

This technique has only limited application in the field of flavonoid isolation and separation (Markham, 1975). It is, however, important for the separation of flavonoids, when they occur in conjugation with inorganic salts as sulphates (Harborne, 1977). Measurement of electrophoretic mobility is an important criterion in the structural identification of such flavonoid conjugates. Disulphates are very clearly separated from monosulphates by this means. Harborne (1977) reported that at pH 2.2 flavonol sulphates generally move further than flavone sulphates, and that flavonoid sulphates containing sugar residues are usually more mobile on electrophoretograms than the corresponding conjugates lacking sugar substitution. The presence of a sulphate group in 6-methoxyflavonoids from *Brickellia californica* has been confirmed by high-voltage electrophoresis (1.5 kV) on paper for 1.5 h at pH 1.9 (Mues *et al.*, 1979). The same conditions were used by Ulubelen *et al.* (1980) for the isolation and identification of quercetagetin 3-methyl ether 7-sulphate from the leaves of *Neurolaena oaxacana*.

Cabrera and Juliani (1976, 1977, 1979) isolated various flavonols with a high degree of sulphation from the genus *Flaveria* by paper electrophoresis at pH 2.2. The electrophoretic mobilities of some of these sulphates are given in Table 1.5.

1.9.2 Paper chromatography

Most separations are achieved with acetic acid at different concentrations, with butan-1-ol–acetic acid–water or with 2-methylpropan-2-ol–acetic acid–water.

Table 1.5 Electrophoretic mobilities of some flavonol sulphates (Cabrera and Juliani, 1979)

Flavonol sulphate	*Mobility**
Isorhamnetin 3-sulphate	1.17
Quercetin 3-sulphate	1.00
Quercetin 4′-sulphate	0.39
Quercetin 3,4′-disulphate	4.80
Quercetin 3,7,4′-trisulphate	7.0
Quercetin 3-acetyl-7,3′,4-trisulphate	7.1
Quercetin 3,7,3′,4′-tetrasulphate	8.2

* Relative to quercetin 3-sulphate, run at pH 2.2 (formate–acetate buffer) for 5 h at 10 V/cm.

Some other solvents have been used by Saito (1976), namely butan-1-ol–ethanol–water (5:1:4), butan-1-ol–pyridine–water (30:20:15), propan-2-ol–water (6:4), ethyl acetate–pyridine–water (2:1:2) or phenol saturated with water. These systems are also indicated for preparative paper chromatography. Harborne (1977) has described the separation of the recently discovered flavonoid sulphates on paper. Some flavonoid sulphates can be separated on paper with water as solvent (Cabrera and Juliani, 1979).

1.9.3 Counter-current distribution

This liquid–liquid partition technique, invented by Craig (1944), has rarely been used for flavonoid seprations. Since it can handle large sample size, it is of interest in the preliminary separation of crude extracts. Nine kaempferol glycosides have been isolated from *Equisetum telmateja* by Craig counter-current distribution with a methyl ethyl ketone–water solvent system (Geiger *et al.*, 1978). However, pure compounds could only be obtained by combining the Craig distribution technique with column chromatography on Sephadex LH-20.

1.9.4 Rotation locular counter-current chromatography

A highly interesting partition technique called rotation locular counter-current chromatography (RLCC) has been invented by Ito and Bowman (1970). This method employs a column prepared by placing multiple centrally perforated partitions across the diameter of a tubular column which divide the space into multiple cells called locules. In each locule, the liquids form an interface while the solute partitioning is promoted by stirring of each phase induced by either rotation or gyration of the column. The column itself, inclined at an angle of 30° from the horizontal, is filled with the lower phase. The upper phase (mobile phase) is continuously introduced through the bottom of the column and then displaces the lower phase in each locule down to the level of the hole which leads

to the next locule, the process being continued throughout the column. Consequently, solute introduced into the column is subjected to a multistep partition process, promoted by the rotation of the column and collected at the outlet of the system. In practice, a modified system is used in which multiple column units, interconnected by fine Teflon tubes, are mounted lengthwise on a rotating shaft to produce a similar effect (Ito and Bowman, 1971). Depending on the compounds to be separated, the mobile phase may be either the upper or the lower layer.

Recently, an RLCC instrument became commercially available. Preliminary assays have shown that this new technique is suitable for flavonoid separations (Hostettmann, 1981). It is a complementary partition technique to droplet counter-current chromatography (no limitation in the choice of solvent system) and to counter-current distribution (less cumbersome and higher resolution). It will certainly gain more importance in the near future for the preparative-scale separation of polar natural products.

REFERENCES

Adamovics, J. and Stermitz, F. R. (1976), *J. Chromatogr.* **129**, 464.

Adams, M. A. and Nakanishi, K. (1979), *J. Liquid Chromatogr.* **2**, 1097.

Adzet, T. and Martinez-Verges, F. (1980), *Pl. méd. Phytothér.* **14**, 8.

Bajaj, K. L. and Arora, Y. K. (1980), *J. Chromatogr.* **196**, 309.

Becker, H., Wilking, G. and Hostettmann, K. (1977), *J. Chromatogr.* **136**, 174.

Becker, H., Exner, J. and Bingler, T. (1979), *J. Chromatogr.* **172**, 420.

Bianchini, J. P. and Gaydou, E. M. (1980), *J. Chromatogr.* **190**, 233.

Bombardelli, E., Bonati, A., Gabetta, B., Martinelli, E. M., Mustich, G. and Danieli, B. (1976), *J. Chromatogr.* **120**, 115.

Bombardelli, E., Gabetta, B. and Martinelli, E. M. (1977a), *Fitoterapia* **48**, 143.

Bombardelli, E., Bonati, A., Gabetta, B., Martinelli, E. M. and Mustich, G. (1977b), *J. Chromatogr.* **139**, 111.

Bouillant, M. L., Ferreres de Arce, F., Favre-Bonvin, J., Chopin, J., Zoll, A. and Mathieu, G. (1979), *Phytochemistry* **18**, 1043.

Cabrera, J. L. and Juliani, H. R. (1976), *Lloydia* **39**, 253.

Cabrera, J. L. and Juliani, H. R. (1977), *Phytochemistry* **16**, 400.

Cabrera, J. L. and Juliani, H. R. (1979), *Phytochemistry* **18**, 510.

Carlson, R. E. and Dolphin, D. (1980), *J. Chromatogr.* **198**, 193.

Chopin, J., Bouillant, M. L., Ramachandran, A. G., Ramesh, P. and Mabry, T. J. (1978), *Phytochemistry* **18**, 1043.

Court, W. A. (1977), *J. Chromatogr.* **130**, 287.

Craig, L. C. (1944), *J. Biol. Chem.* **155**, 519.

Derguini, F., Balogh-Nair, V. and Nakanishi, K. (1979), *Tetrahedron Lett.*, 4899.

Deyl, Z., Rosmus, J. and Pavlicek, M. (1964), *Chromatogr. Rev.* **6**, 19.

Engelhardt, H. (1979), *High Performance Liquid Chromatography*, Springer-Verlag, Heidelberg.

Fisher, J. F. (1978), *J. Agric. Food Chem.* **26**, 1459.

Fisher, J. F. and Wheaton, T. A. (1976), *J. Agric. Food Chem.* **24**, 899.

Galensa, R. and Herrmann, K. (1980), *J. Chromatogr.* **189**, 217.

Geiger, H. and de Groot-Pfleiderer, W. (1979), *Phytochemistry* **18**, 1709.

Geiger, H., Lang, U., Britsch, E., Mabry, T. J., Suhrschücker, U., Vander-Velde, G. and Waldrum, H. (1978), *Phytochemistry* **17**, 336.

Ghosal, S., Jaiswal, D. J. and Biswas, K. (1978), *Phytochemistry* **17**, 2119.
Gilpin, R. K. and Sisco, W. R. (1976), *J. Chromatogr.* **124**, 257.
Gilpin, R. K., Korpi, J. A. and Janicki, C. A. (1975), *Anal. Chem.* **47**, 1498.
Guiochon, G., Siouffi, A., Engelhardt, H. and Halasz, I. (1978), *J. Chromatogr. Sci.* **16**, 152.
Harborne, J. B. (1977). In *Progress in Phytochemistry* (eds. L. Reinhold, J. B. Harborne and T. Swain), Pergamon Press, Oxford, pp. 189–208.
Hardin, J. M. and Stutte, C. A. (1980), *Anal. Biochem.* **102**, 171.
Heftmann, E., Krochta, J. M., Farkas, D. F. and Schwimmer, S. (1972), *J. Chromatogr.* **66**, 365.
Hiermann, A. (1978), *Planta Med.* **34**, 443.
Hiermann, A. (1979), *J. Chromatogr.* **174**, 478.
Hiermann, A. and Kartnig, Th. (1977), *J. Chromatogr.* **140**, 322.
Higuchi, R. and Donnelly, D. M. X. (1978), *Phytochemistry* **17**, 787.
Hostettmann, K. (1980), *Planta Med.* **39**, 1.
Hostettmann, K. and Hostettmann, M. (1981), GIT, *Suppl. Chromatogr.* 22.
Hostettmann, K. and McNair, H. M. (1976), *J. Chromatogr.* **116**, 201.
Hostettmann, K., Hostettmann-Kaldas, M. and Nakanishi, K. (1979a), *J. Chromatogr.* **170**, 355.
Hostettmann, K., Hostettmann-Kaldas, M. and Sticher, O. (1979b), *Helv. Chim. Acta* **62**, 2079.
Hostettmann, K., Hostettmann-Kaldas, M. and Sticher O. (1980), *J. Chromatogr.* **202**, 154.
Hostettmann-Kaldas, M. and Jacot-Guillarmod, A. (1978), *Phytochemistry* **17**, 2083.
Hostettmann-Kaldas, M., Hostettmann, K. and Sticher, O. (1981), *Phytochemistry*, **20**, 443.
Iinuma, M., Matsuura, S., Kurogochi, K. and Tanaka, T. (1980), *Chem. Pharm. Bull.* **28**, 717.
Isobe, T., Ito, N. and Noda, Y. (1980), *Phytochemistry* **19**, 1877.
Ito, Y. and Bowman, R. L. (1970), *J. Chromatogr. Sci.* **8**, 315.
Ito, Y. and Bowman, R. L. (1971), *Anal. Chem.* **43**, 69A.
Jänchen, D. and Schmutz, H. R. (1979), *J. High Resol. Chromatogr.* **2**, 133.
Jay, M., Wollenweber, E. and Favre-Bonvin, J. (1979), *Phytochemistry* **18**, 153.
Kadar-Pauncz, J. (1979), *J. Chromatogr.* **170**, 203.
Kaiser, R. E. (1978), *J. High Resol. Chromatogr.* **1**, 164.
Kaldas, M., Hostettmann, K. and Jacot-Guillarmod, A. (1975), *Helv. Chim. Acta* **58**, 2188.
Karl, Ch., Pedersen, P. A. and Schwarz, C. (1977), *Phytochemistry* **16**, 1117.
Kery, A., Verzar-Petri, G. and Incze, J. (1977), *Acta Pharmaceutica Hung.* **47**, 11.
Kingston, D. G. I. (1979), *J. Natural Prod.* **42**, 237.
Kingston, D. G. I. and Gerhart, B. B. (1976), *J. Chromatogr.* **116**, 182.
Kwasniewski, V. (1978), *Dtsch. Apotheker Ztg.* **118**, 1049.
Manley, C. H. and Shubiak P. (1975), *J. Can. Inst. Food Sci. Technol.* **8**, 35.
Markham, K. R. (1975). In *The Flavonoids* (eds. J. B. Harborne, T. J. Mabry and H. Mabry), Chapman and Hall, London, pp. 1–44.
Markham, K. R. and Porter, L. J. (1979), *Phytochemistry* **18**, 611.
Markham, K. R. and Wallace, J. W. (1980), *Phytochemistry* **19**, 415.
Marston, A., Hostettmann, K. and Jacot-Guillarmod, A. (1976), *Helv. Chim. Acta* **59**, 2596.
Martinelli, E. M. (1980), *Eur. J. Mass Spectrom.* **1**, 33.
Montgomery, J. A., Johnston, T. P., Thomas, H. J., Piper, J. R. and Temple, C. (1977), *Adv. Chromatogr.* **15**, 169.
Mues, R., Timmermann, B. N., Ohno, N. and Mabry, T. J. (1979), *Phytochemistry* **18**, 1379.
Nelson, J. A. (1977), *Adv. Chromatogr.* **15**, 273.
Niemann, G. J. (1977), *Z. Naturforsch.* **32c**, 1015.
Niemann, G. J. and Koerselmann-Kooy, J. W. (1977), *Planta Med.* **31**, 297.
Niemann, G. J. and Van Brederode, J. (1978), *J. Chromatogr.* **152**, 523.
Numata, A., Hokimoto, K., Shimada, A., Yamaguchi, H. and Takaishi, K. (1979), *Chem. Pharm. Bull.* **27**, 602.
Numata, A., Hokimoto, K. and Yamaguchi, H. (1980), *Chem. Pharm. Bull.* **28**, 964.
Ohta, N., Kuwata, G., Akahori, H. and Watanabe, T. (1980), *Agric. Biol. Chem.* **44**, 469.
Okuda, T., Mori, K., Seno, K. and Hatano, T. (1979), *J. Chromatogr.* **171**, 313.

Parker, W. H., Maze, J. and McLachlan, D. G. (1979), *Phytochemistry* **18**, 508.
Pettei, M. J. and Hostettmann, K. (1978), *J. Chromatogr.* **154**, 106.
Quercia, V., Turchetto, L., Pierini, N., Cuozzo, V. and Percaccio, G. (1978), *J. Chromatogr.* **161**, 396.
Rouseff, R. L. and Ting, S. V. (1979), *J. Chromatogr.* **176**, 75.
Saito, K. (1976), *Planta Med.* **30**, 349.
Sakushima, A., Hisada, S., Ogihara, Y. and Nishibe, S. (1980), *Chem. Pharm. Bull.* **28**, 1219.
Saleh, N. A. M. (1976), *J. Chromatogr.* **124**, 174.
Satake, T., Murakami, T., Saiki, Y. and Chen, C.-M. (1978), *Chem. Pharm. Bull.* **26**, 2600.
Schmidtlein, H. and Herrmann, K. (1976), *J. Chromatogr.* **123**, 385.
Seitz, C. T. and Wingard, R. E. (1978), *J. Agric. Food Chem.* **26**, 278.
Snyder, L. R. and Kirkland, J. J. (1974), *Introduction to Modern Liquid Chromatography*, John Wiley & Sons, New York.
Strack, D. and Krause, J. (1978), *J. Chromatogr.* **156**, 359.
Strack, D., Fuisting, K. and Popovici, G. (1979), *J. Chromatogr.* **176**, 270.
Tada, A., Kasai, R., Saitoh, T. and Shoji, J. (1980), *Chem. Pharm. Bull.* **28**, 1477.
Takeda, T., Ishiguro, I., Masegi, M. and Ogihara, Y. (1977), *Phytochemistry* **16**, 619.
Tanaka, N., Murakami, T., Saiki, Y., Chen, C. -M. and Gomez, L. D. (1978), *Chem. Pharm. Bull.* **26**, 3580.
Tanimura, T., Pisano, J. J., Ito, Y. and Bowman, R. L. (1970), *Science* **169**, 54.
Tanimura, T., Otsuka, H. and Ogihara, Y. (1975), *Kagaku no Ryoiki* **29**, 43.
Teuber, H. and Herrmann, K. (1978), *Z. Lebensm.-Unters. Forsch.* **167**, 101.
Theodor, R., Zinsmeister, H. D., Mues, R. and Markham, K. R. (1980), *Phytochemistry* **19**, 1695.
Ting, S. B., Rouseff, R. L., Dougherby, M. H. and Attaway, J. A. (1979), *J. Food Sci.* **44**, 69.
Tittel, G. and Wagner, H. (1977), *J. Chromatogr.* **135**, 499.
Tittel, G. and Wagner, H. (1978), *J. Chromatogr.* **153**, 227.
Tschesche, R., Braun, T. M. and von Sassen, W. (1980), *Phytochemistry* **19**, 1825.
Ulubelen, A., Kerr, K. M. and Mabry, T. J. (1980), *Phytochemistry* **19**, 1761.
Vande Casteele, K., De Pooter, H. and Van Sumere, C. F. (1976), *J. Chromatogr.* **121**, 49.
Vanhaelen, M. and Vanhaelen-Fastré, R. (1980), *J. Chromatogr.* **187**, 255.
Van Sumere, C. F., Van Brussel, W., Vande Casteele, K. and Van Rompaey, L. (1979). In *Biochemistry of Plant Phenolics, Recent Advances in Phytochemistry, Vol. 12* (eds. T. Swain, J. B. Harborne and C. F. Van Sumere), Plenum Press, New York, pp. 1–28.
Voirin, B. and Jay, M. (1977), *Phytochemistry* **16**, 2043.
Ward, R. S. and Pelter, A. (1974), *J. Chromatogr. Sci.* **12**, 570.
Wehrli, A., Hildenbrand, J. C., Keller, H. P., Stampfli, R. and Frei, R. W. (1978), *J. Chromatogr.* **149**, 199.
West, L. G., Birac, P. M. and Pratt, D. E. (1978), *J. Chromatogr.* **150**, 266.
Wilkinson, M., Sweeny, J. G. and Iacobucci, G. A. (1977), *J. Chromatogr.* **132**, 349.
Williams, M., Hrazdina, G., Wilkinson, M., Sweeny, J. G. and Iacobucci, G. A. (1978), *J. Chromatogr.* **155**, 389.
Wulf, L. W. and Nagel, C. W. (1976), *J. Chromatogr.* **116**, 271.
Zlatkis, A. and Kaiser, R. E. (eds) (1977), *High Performance Thin-layer Chromatography, J. Chromatography Library, Vol. 9*, Elsevier, Amsterdam.

CHAPTER TWO

Carbon-13 NMR Spectroscopy of Flavonoids

KEN R. MARKHAM and V. MOHAN CHARI
(with TOM J. MABRY, Section 2.5)

2.1 INTRODUCTION

Carbon-13 with a natural abundance of only 1.1% and a smaller magnetic moment than ^{1}H (and hence a weaker NMR signal) was for many years a nucleus relatively inaccessible to NMR spectroscopists. However, with the advent in recent years of pulsed NMR and Fourier Transform (FT) analysis, ^{13}C-NMR spectroscopy has become more readily available. As a result the past five years have seen a dramatic increase in the number of publications on, or involving, ^{13}C-NMR spectroscopy of flavonoids. For example major compilations include those of Ternai and Markham (1976), Markham and Ternai (1976), Markham *et al.* (1978), Wagner *et al.* (1976), Chari *et al.* (1977a), Pelter *et al.* (1976) and Wenkert and Gottlieb (1977). The increasing availability and sensitivity of modern FT-NMR spectrometers is assurance that this powerful technique will become a major tool for the structure elucidation of flavonoids and their glycosides.

It should be stressed, however, that ^{13}C-NMR is in no sense superseding ^{1}H-NMR, but is complementary to it. Information gained relates to the carbon 'backbone' of the molecule while ^{1}H-NMR gives information about the structural environment of each proton. Further, it is possible to obtain a good ^{1}H-NMR spectrum on a sample that is much too small for carbon-13 analysis. Modern FT-NMR techniques now permit measurement of proton spectra on samples down to a few micrograms whereas ^{13}C spectra still require in excess of 5 mg of sample if they are to be measured without excessive expenditure of instrument time.

It is not the intention in this chapter to discuss instrumental or theoretical aspects of this technique. For this type of information the reader is referred to Breitmaier *et al.* (1971) and to the now standard texts of Levy and Nelson (1972) and Stothers (1972). The present chapter will be devoted largely to comment on the interpretation of spectra and to a discussion of the general usefulness of the technique in solving flavonoid structural problems. Illustrations of spectra have been avoided, since Section 2.5 is devoted entirely to their pictorial display.

2.1.1 Sample size and solvents

As mentioned above, sample size can be reduced to a few milligrams with modern FT-NMR techniques. However, the smaller the sample, the more instrument time is required to produce a spectrum of reasonable intensity. Thus in practice sample sizes in the range 10–50 mg (ca. 1/10 mmol) are preferred.

The two key factors in solvent selection are, as in ^{1}H-NMR, sample solubility and the position of solvent (carbon) resonances. Solvents commonly found suitable include deuterochloroform ($CDCl_3$), dioxan, hexadeuteroacetone [$(CD_3)_2CO$], hexadeuterodimethylsulphoxide (DMSO-d_6), deuterium oxide (D_2O) and combinations of these. A deuterium lock for the instrument is, in most of these, provided by the solvent. Resonances are normally expressed in parts per million (p.p.m.) downfield from tetramethylsilane (TMS) which may or may not be added as reference since solvent carbon signals usually suffice for this purpose. As a guide to solvent selection, chemical shifts reported in the literature for some common solvents (relative to TMS) are listed below:

Cyclohexane	27.7 p.p.m.
$(CD_3)_2CO$	29.2 p.p.m. (middle signal)
DMSO-d_6	39.5 p.p.m. (middle signal)
Dioxan	66.5 p.p.m.
$CDCl_3$	77.2 p.p.m. (middle signal)
Benzene	128.7 p.p.m.
Pyridine-d_5	123.4, 135.3, 149.8 p.p.m.
CS_2	192.8 p.p.m.

In mixed solvents these values may be altered slightly owing to solvent effects. Varying the solvent may also have important effects on ^{13}C chemical shifts. Thus Pelter *et al.* (1976) report solvent shifts of up to 0.5 p.p.m. (in flavonoid spectra) for individual carbon resonances when the solvent is changed from $CDCl_3$ to DMSO-d_6. Again, pyridine, owing to its basicity and polarity, is known to shift signals of phenolic carbons as well as those of nearby carbons. As early as 1964 Maciel and James (1964) noted significant variation of the chemical shifts in phenolic carbon resonances with changing solvent and examples are presented in Table 2.1. Even the detection of certain coupling constants may be affected by small changes in solvent composition, e.g. by the presence of H_2O in DMSO-d_6 (Chang, 1978).

2.1.2 The spectrum

Carbon-13 resonances (for flavonoids) occur over a range of 0–200 p.p.m. downfield from TMS compared with a range of only 0–10 p.p.m. for ^{1}H resonances. For this reason ^{13}C-NMR spectra are much more highly resolved than are ^{1}H-NMR spectra. In proton-decoupled spectra (see later) each carbon atom is represented by one line and its chemical shift is determined primarily by the electron density at that carbon atom (for flavonoids see e.g. Kalabin *et al.*, 1977). Thus the carbon resonances at lowest field are generally those of carbonyl

Table 2.1 Carbon-13 chemical shifts of phenol in different solvents*

Solvent	*Phenolic carbon*	*Ortho-carbon*	*Meta-carbon*	*Para-carbon*
Cyclohexane	155.6	116.1	130.5	120.8
Benzene	156.6			
Dioxan	157.6	115.6	129.8	119.3
Acetone	157.9	116.0	130.5	120.5
DMSO	158.5	115.5	130.2	119.8

* The original spectra measured by Maciel and James (1964) relative to benzene have been converted to the TMS standard by using 128.7 p.p.m. as the chemical shift of benzene relative to TMS.

carbons and oxygenated aromatic carbons, whereas those at highest field will represent non-oxygenated aliphatic carbons. As a general guide approximate chemical shift ranges for carbon types encountered in flavonoids are outlined in Table 2.2. For more detailed data see the appropriate section later in this chapter.

Table 2.2 Carbon-13 chemical-shift ranges for various carbon types encountered in flavonoids

Carbon type	*Approximate chemical-shift range (p.p.m. from TMS)*
Carbonyl (4-keto, acyl)	210–170
Aromatic and olefinic:	
(a) Oxygenated	165–155 (no *o*/*p*-oxygenation); 150–130 (with *o*/*p*-oxygenation); ca. 135 (C -3-OH)
(b) Non-oxygenated	135–125 (no *o*/*p*-oxygenation); 125–90 (with *o*/*p*-oxygenation); 102–110 (C – 3).
Aliphatic:	
(a) Oxygenated (sugars)	105–55
(b) Non-oxygenated (C-2, 3 flavanones)	80–40 (epicatechin C-4, 28 p.p.m.)
Methylenedioxy	ca. 100
Aromatic – OCH_3	55–60 (59–64 for *ortho*-disubstituted)
Aromatic C-$\underline{C}H_3$, $CO\underline{C}H_3$	ca. 20
Aliphatic C-$\underline{C}H_3$	ca. 17

2.1.3 ^{13}C–^{1}H and ^{13}C–^{13}C coupling and the nuclear Overhauser effect

In principle ^{13}C resonances couple with both ^{1}H and other ^{13}C nuclei in the molecule. But in practice ^{13}C–^{13}C coupling is not normally observed, since the natural abundance of carbon-13 at 1.1% is so low that the chances of two ^{13}C atoms being adjacent to one another in the same molecule are remote (1 in 10 000). ^{13}C–^{1}H coupling, however, is observed and can be most useful in signal assignment problems. With complex molecules, however, the ^{13}C–^{1}H coupling

complicates the spectrum considerably by producing multiple signals for most carbons. Thus, '^{1}H-decoupled' or 'noise-decoupled' spectra are favoured for initial analysis. For such spectra, irradiation of the sample at all ^{1}H frequencies simultaneously (= radiofrequency noise) during the measurement of the spectrum is carried out. This results in a spectrum in which each carbon is represented by one resonance signal. An added advantage of this decoupling is the signal enhancement achieved (of up to 3 times) due to the nuclear Overhauser effect (n.O.e). This enhancement is due to the fact that the irradiated protons relax the ^{13}C nuclei more efficiently, thereby favourably perturbing the population ratio of ^{13}C nuclei in the excited and ground states. The enhancement may be retained in ^{1}H-coupled spectra by use of techniques such as 'gated' decoupling and 'off-resonance' decoupling.

2.1.4 Integration

In contrast to the value of integration in ^{1}H-NMR spectroscopy, integration of ^{13}C-NMR spectra is largely uninformative unless special precautions are taken. As a result, integrations are rarely reported. The Fourier Transform operation utilizes an average carbon relaxation time (T_1), but in fact T_1 values for individual carbon atoms differ widely, depending on the environment of the carbon, and this results in non-quantitative representation of resonance lines. For example, in the spectrum of rutin (Section 2.5) it can be seen that the one-carbon signal at 104.2 p.p.m. (C-10) is markedly less intense than the one-carbon signal at 115.3 p.p.m. (C-2′). This is due to the longer relaxation time of the quaternary carbon.

2.1.5 Substituent effects – general

The introduction of a new substituent into a molecule will normally cause changes in the chemical shifts of nearby ^{13}C atoms. Such changes are referred to as 'substituent effects' and a knowledge of the extent of these effects is of immense predictive value. In particular, from reference spectra of flavonoids of known structure, e.g. apigenin, it is possible to predict with some accuracy the spectrum of an unavailable flavonoid, e.g. 6-carboxyapigenin, by application of the appropriate substituent effects. Substituent effects on the carbon resonances in substituted benzenes are obviously of prime importance for the interpretation of flavonoid spectra and a number of compilations of these data derived from studies of simple aromatics have appeared in the literature. Table 2.3 represents a selection of substituent effects from the compilations of Stothers (1972) and Levy and Nelson (1972) which are most relevant to the flavonoid field.

The effects listed in Table 2.3 are to a large extent additive and so more than one substituent can be taken into consideration in any calculation. However, the inaccuracies inherent in data obtained from simple molecules and applied to complex molecules are also additive, and the more substituents allowed for, the

Table 2.3 Substituent effects on the carbon-13 chemical shifts of benzene*

Substituent	*C-1*	*Ortho*	*Meta*	*Para*
OH	+26.9	−12.7	+1.4	− 7.3
OCH_3	+31.4	−14.4	+1.0	− 7.7
O^-†	+39.6	− 8.2	+1.9	−13.6
$OCOCH_3$†	+23.0	− 6.4	+1.3	− 2.3
O-phenyl†	+29.2	− 9.4	+1.6	− 5.1
F	+34.8	−12.9	+1.4	− 4.5
Cl	+ 6.2	+ 0.4	+1.3	− 1.9
Br	− 5.5	+ 3.4	+1.7	− 1.6
Phenyl	+13.1	− 1.1	+0.4	− 1.2
CH_3	+ 8.9	+ 0.7	−0.1	− 2.9
CH_2OH†	+12.3	− 1.4	−1.4	− 1.4
$CH{=}CH_2$†	+ 9.5	− 2.0	+2.0	− 0.5
CO_2H	+ 2.1	+ 1.5	0.0	+ 5.1
$COCH_3$	+ 9.1	+ 0.1	0.0	+ 4.2

* Shifts in Hertz units. Positive shifts are downfield.
† Data from Stothers (1972), other data from Levy and Nelson (1972).

greater the overall errors will be. In general terms the additivity principles established for simple aromatic systems have been found to be a reliable guide to the interpretation of flavonoid spectra; see e.g. Markham and Ternai (1976), Wagner *et al.* (1976), Kingsbury and Looker (1975), Pelter *et al.* (1976) and Chari *et al.* (1977a). The main exceptions noted have been in rings A and C in situations where the new substituent might be expected to have a modified influence: e.g. the introduction of hydroxyls at C-3 and C-5 (where they would be hydrogen-bonded to the 4-keto function) and the introduction of a methoxyl *ortho* to an existing methoxyl (where steric interaction would be expected). In the case of the introduction of a 3-hydroxyl, C-3 was found to shift downfield (i.e. deshielded by a larger than expected 33 p.p.m.) and the '*ortho*'-effect on C-2 was also large at − 17 p.p.m. The introduction of a 5-hydroxyl causes a downfield shift of about 31 p.p.m. in the C-5 signal and the '*meta*'-related C-4 resonance is deshielded by about 6 p.p.m. The *para*-related C-8 shifts upfield by an unexpectedly large 11 p.p.m. (Ternai and Markham, 1976). In one further example of non-adherence to established substitution effects, Pelter *et al.* (1976) noted that the shifts induced by the introduction of a methoxyl group into the A ring did not correlate well, although B-ring substitution gave good agreement except when *ortho*- substituents were involved.

2.1.6 Substituent effects of acetylation and methylation

Acetylation of free phenolic hydroxyl groups produces marked changes in the ^{13}C-NMR of flavonoids (Pelter *et al.*, 1976) and can be used to detect the location

of hydroxyl groups on the aromatic rings. On acetylation, the signal of the hydroxylated carbon moves upfield by 6.6–15.6 p.p.m., the *ortho*- and *para*-carbon signals are moved downfield by 4.1–12.1 and 2–7.9 p.p.m. respectively, and the *meta*- carbon signals are only slightly affected (+0.9 to −4.3 p.p.m.).

Methylation of free phenolic hydroxyl groups generally produces smaller, opposite and rather variable effects (Markham *et al.*, 1978; Chari *et al.*, 1977a). The signal of the hydroxylated carbon shifts downfield by 1.0–4.7 p.p.m. and the signal of the *ortho*-related carbon, upfield by 0.8–3.6 p.p.m. Methylation of the 5-hydroxyl in apigenin, however, produces an upfield shift of 1.6 p.p.m. in C-5 and shifts of −3.7, +4.3 and −6.1 p.p.m. in the resonances of C-2,-3 and -4 respectively (Wagner *et al.*, 1976), the anomalous behaviour presumably being due to the disruption of H-bonding with the 4-keto function [cf. 2-hydroxyacetophenone which is also anomalous (Kobayashi *et al.*, 1976)].

2.2 FLAVONOID AGLYCONES

2.2.1 General

Although ^{13}C-NMR is not normally the method of choice for distinguishing this class of a flavonoid, it may be of use in special situations. The different types of aglycone are not distinguishable on the basis of the aromatic carbon resonances alone, but chemical shifts for the central three-carbon unit are often quite distinctive. This is exemplified by the data in Table 2.4, most of which have been taken from Pelter *et al.* (1976).

Table 2.4 Carbon-13 resonances for ring C in flavonoids*

	C-2	C-3	>C=O†
Chalcones	136.9–145.4(d)‡	116.6–128.1(d)‡	188.6–194.6(s)
Flavanones	75.0– 80.3(d)	42.8– 44.6(t)	189.5–195.5(s)
Flavones	160.5–165(s)	103–111.8(d)	176.3–184(s)
Flavonols	145–150	136–139	172–177
Isoflavones	149.8–155.4(d)	122.3–125.9(s)	174.5–181(s)
Aurones	146.1–147.7(s)	111.6–111.9(d) (=CH−)	182.5–182.7(s)
Pterocarpans	66.4– 66.5(t) (C−6)	39.5– 40.2(d) (C−6a)	78.4– 78.5(d) (C−11a)

* Data for $CDCl_3$ or $CDCl_3$/DMSO-d_6 solutions according to Pelter *et al.* (1976) except for flavonols, DMSO-d_6 (Markham *et al.*, 1978; Ternai and Markham, 1976).

† Shifts affected by presence or absence of H-bonded hydroxyls. Data for flavones and isoflavones have been altered to include 5-OH derivatives not studied by Pelter *et al.* (1976). 4-OH aurones have not been studied.

‡ For chalcones, C-2 and C-3 represent the C-β and C-α, respectively.

(s) = singlet, (d) = doublet, (t) = triplet

2.2.2 ^{13}C–^{1}H coupling and signal assignments

The earliest report of a ^{13}C-NMR study of a flavonoid aglycone appears to be that of Joseph-Nathan *et al.* (1974) in which a complete assignment of the spectrum of flavone itself was made. These assignments were based on a study of the proton noise-decoupled spectra of 5,6,7,8-tetradeuteroflavone, 2′,3′,4′,5′,6′-pentadeuteroflavone and 4′-bromoflavone and provided a sound base for the later interpretation of more complex spectra via substituent effect and ^{13}C–^{1}H coupling data. At about the same time, Kingsbury and Looker (1975) published a study of simple methoxylated flavones and made most of their assignments using the known substituent effect of a methoxyl group and observed ^{13}C–^{1}H coupling. These methods of signal assignment have proven generally reliable with flavonoids and have been used extensively in all recent compilations of flavonoid ^{13}C-NMR spectra.

^{13}C-^{1}H coupling data are especially valuable for confirmation of tentative assignments based on standard substitutent effect shift calculations (see Introduction). For example, the presence of a directly attached proton will give rise to a coupling of 155–170 Hz while a *meta*-related proton will produce a highly diagnostic coupling of 4.5–9 Hz. *Ortho*- and *para*- related protons are generally recognized by a 1–4 Hz coupling, although this coupling is unreliable and is not always observed.

Examples of commonly encountered ^{13}C–^{1}H couplings in oxygenated flavonoid nuclei are listed in Table 2.5 and are derived largely from the data of Chang (1978), Ternai and Markham (1976) and Markham and Ternai (1976). The figures quoted are based on a limited number of samples and as such must be considered only as guides to expected coupling constants. For a more detailed study of flavonoid ^{13}C–^{1}H coupling constants (especially long-range), the reader is referred to the work of Chang (1978). Many of the interactions referred to in Table 2.5 are exemplified in the ^{1}H-coupled ^{13}C-NMR spectrum of luteolin (Fig. 2.1).

Coupling data have been used to good effect in the distinction of C-6 from C-8 signals, and C-5 from C-9 signals, in the spectra of 5,7-dihydroxy-flavones, -flavonols, -isoflavones and -flavanones (Wehrli, 1975; Chang, 1976, 1978; Wagner *et al.*, 1976; Lallemand and Duteil, 1977). In particular, coupling between the 5-hydroxyl proton and each of the above carbons can be observed for certain compounds in *dry* DMSO solutions. The degree of coupling identifies each carbon and demonstrates that C-5 resonates downfield from C-9 and that C-6 resonates downfield from C-8 (the reverse of the order for H-6 and H-8 resonances in the ^{1}H-NMR; see Mabry *et al.*, 1970).

Assignments for C-6 and C-8 in flavones and flavonols have also been made on the basis of heteronuclear decoupling experiments. Thus the selective decoupling of H-6 and H-8 led to the unambiguous assignment of the respective carbon resonances (Wagner *et al.*, 1976). Similarly, low-power irradiation of H-6 and H-8 have also led to definite assignments for C-9 and C-5, in the two compound types,

Table 2.5 ^{13}C-^{1}H coupling in oxygenated flavonoid nuclei

*Coupled atoms**	*Type location*	*Coupling constant (J) (Hz)*
CH-aromatic	A and B ring	155–170
H-*ortho*/C-aromatic	A and B ring	1.5–4
H-*meta*/C-aromatic	A and B ring	4.5–9
H-*para*/C-aromatic	A and B ring	1–3
H-3/C-3	Flavones	155–170
H-3/C-2	Flavones	ca. 5
H-2′/C-2	Flavones, Flavonols Flavanones, Isoflavones (C-3/H-2′)	ca. 3–4.5
H-2/C-2	Isoflavones	ca. 200
H-3/C-3	Flavanones	ca. 130
H-2/C-2	Flavanones	ca. 150
5-OH/C-6	Flavanones, Isoflavones (dry DMSO) Flavanone	ca. 7.0 (anti)
5-OH/C-10	(dry DMSO)	ca. 4.7 (syn)
5-OH/C-7		ca. 1.5
OCH_2O	Methylenedioxy	ca. 175
OCH_3	Methoxyl	ca. 145

* Underlining is used where necessary to indicate atoms involved.

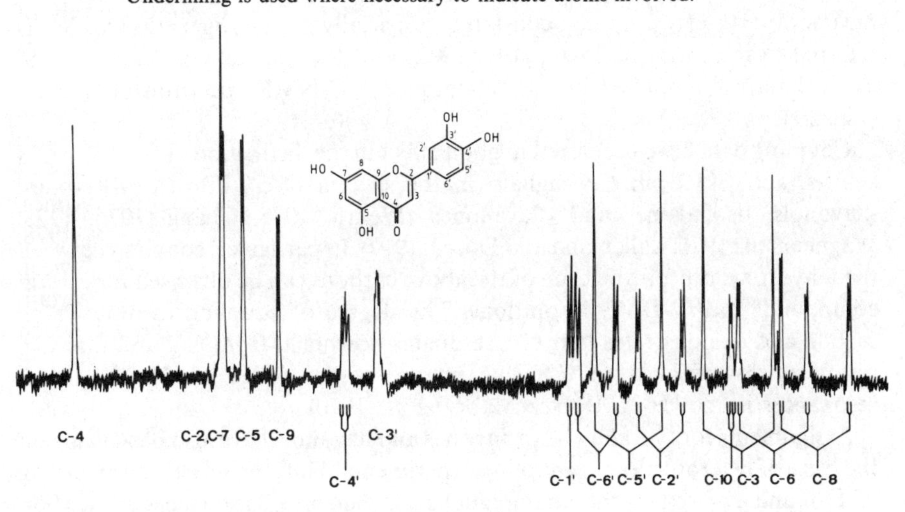

Fig. 2.1 ^{1}H-coupled ^{13}C-NMR spectrum of luteolin.

taking into consideration their respective $^2J_{CH}$ interactions (Chari *et al.*, 1977a). Thus, not only were early tentative assignments based on 'substituent effects' reversed, but a valuable method was established for the distinction of 6- from 8-*C*-alkylated flavonoids.

2.2.3 The effect of 6- and 8-substitution

From the data presented in Table 2.6, it is evident that substitution of C-6 or C-8, in 5,7-dihydroxyflavonoids, does not markedly alter the chemical shift of the unsubstituted aromatic methine carbon atom. The resonances for C-6 and C-8 in such compounds are distinct and, as their shift ranges do not overlap, can be unambiguously identified. Substitution at C-6 or C-8 can thus readily be determined (Chari *et al.*, 1978a). This also enables differentiation of 5,6,7- and 5,7,8-trioxygenation patterns and of 6-*C*- and 8-*C*-glycoflavones (see Section 2.4).

Table 2.6 Chemical shifts of C-6 and C-8 in some 5,7-dihydroxyflavonoids in DMSO-d_6

Compound	*C-6*	*C-8*	*Reference*
1. 5,7-Dihydroxyflavanone	96.1	95.1	Wagner *et al.* (1976)
2. 6-*C*-Methyl-5,7-dihydroxyflavanone	102.1	94.7	Chari *et al.* (1977a)
3. 8-*C*-Methyl-5,7-dihydroxyflavanone	95.7	101.9	Chari *et al.* (1977a)
4. 6-*C*-Prenyl-5,7,3′,4′-tetrahydroxyflavanone	108.6	95.0	Yakushijin *et al.* (1980)
5. 5,7,4′-Trihydroxyflavone	98.8	94.0	Wagner *et al.* (1976) Markham and Ternai (1976)
6. 5,6,7,3′,4′-Pentahydroxyflavone	140.4	93.6	Chari *et al.* (1977a)
7. 6-*C*-Glucosyl-5,7,4′-trihydroxyflavone	108.8	94.2	Chari *et al.* (1978b)
8. 8-*C*-Glucosyl-5,7,4′-trihydroxyflavone	98.9	104.2	Chari *et al.* (1978b)

2.2.4 Flavones and flavonols

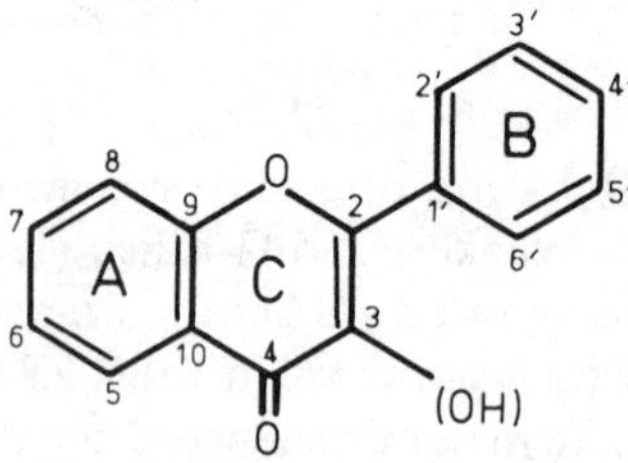

(2.1)

Numerous workers have published compilations of flavone and flavonol aglycone spectra and these include: Kingsbury and Looker (1975), Ternai and Markham (1976), Wagner *et al.* (1976), Pelter *et al.* (1976), Kalabin *et al.* (1977), Wenkert and Gottlieb (1977), Calvert *et al.* (1979) and Iinuma *et al.* (1980). In

addition, prenylated flavones have been studied by Wenkert and Gottlieb (1977), Chari *et al.* (1978b) and Waterman and Khalid (1980), and several 6-oxygenated flavones by Panichpol and Waterman (1978) and Iinuma *et al.* (1980). A selection of spectra from the work of Wagner *et al.* (1976), Ternai and Markham (1976), and Markham *et al.* (1978) is presented in Table 2.7 to highlight the effects of differing hydroxylation and methoxylation patterns on the spectra. Not evident from these data is the effect of 2′-oxygenation on the chemical shift of C-3 in flavones. In compounds such as wightin, echioidin and echioidinin which all possess 2′-oxygenation, C-3 appears at ca. 109–110 p.p.m., considerably downfield from its position in other flavones. Also not evident is the observation (Iinuma *et al.*, 1980) that the presence of 6-oxygenation reverses the relative positions of the C-5 and C-9 signals as defined for 5,7-dihydroxyflavones (above).

Apart from effects due to changes in the oxygenation pattern, the most notable differences evident between spectra listed in Table 2.7 are those observed in the C-ring resonances. The carbonyl carbon, C-4, for example resonates at around 175–178 p.p.m. when the carbonyl is not hydrogen-bonded, but in the presence of hydrogen-bonding to a 5-hydroxyl group it moves downfield to about 182 p.p.m. When a 3-hydroxyl is present as well as a 5-hydroxyl the resonance returns to about 176 p.p.m., but with the 3-hydroxyl alone the resonance appears at about 171–173 p.p.m. The marked effects brought about by the presence of 5- and 3-hydroxyl groups on both A- and C-ring carbon resonances were discussed earlier in Section 2.1.5.

2.2.5 Isoflavones

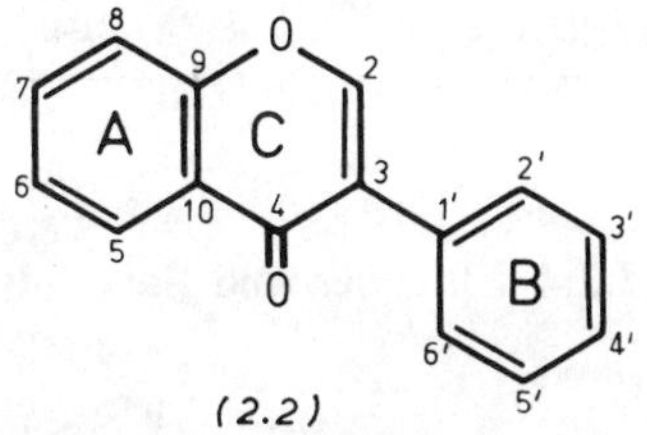

(2.2)

Reports on the ^{13}C-NMR spectra of isoflavones include those of Pelter *et al.* (1976), Pelter *et al.* (1978), Wenkert and Gottlieb (1977) and Jha *et al.* (1980). Crombie *et al.* (1975) have reported studies on a range of natural rotenoids. A selection of isoflavone spectra is presented in Table 2.8, and it is evident that only in the C-ring signals are there major differences from the spectra of flavones and flavonols. As a consequence of the different B-ring linkage site, C-2 in isoflavones resonates at higher field than in flavones and C-3 at lower field. Additionally, the C-2 signal is readily identified by its large ^{13}C–^{1}H coupling constant of about 200 Hz. A further feature of the coupled spectrum which is distinctive for isoflavones is the appearance of C-4 as a doublet ($J = 6.5$ Hz) due to a $^{3}J_{CH}$ interaction with H-2.

Table 2.7 A selection of flavone and flavonol ^{13}C-NMR spectra (solvent, DMSO-d_6)

Flavones		*C-2*	*C-3*	*C-4*	*C-5*	*C-6*	*C-7*	*C-8*	*C-9*	*C-10*	*C-1′*	*C-2′*	*C-3′*	*C-4′*	*C-5′*	*C-6′*
Flavone	**	162.1	106.7	176.7	125.1	124.5	133.8	118.2	155.3	123.1	130.8	126.0	128.7	131.4	128.7	126.0
5-OH flavone	*	164.1	105.6	182.9	159.8	110.8	135.6	107.2	155.9	110.1	130.5	126.4	128.9	132.0	128.9	126.4
5,4′-OH flavone	*	165.5	103.9	183.5	160.0	111.4	136.2	108.0	156.4	110.5	121.9	129.2	116.7	161.7	116.7	129.2
5,7-OH flavone		163.0	105.0	181.6	161.5	98.9	164.3	94.0	157.3	103.9	131.7	126.0	128.0	130.6	128.0	126.0
Apigenin		164.1	102.8	181.8	161.5	98.8	163.7	94.0	157.3	103.7	121.3	128.4	116.0	161.1	116.0	128.4
Acacetin		163.9	103.9	182.3	162.2	99.4	164.8	94.3	157.9	104.4	123.5	128.4	114.8	162.8	114.8	128.4
5-Me apigenin		160.4†	107.1	175.7	159.9†	96.4	162.4	95.2	159.0†	106.0	121.5	127.7	115.8	160.6	115.8	127.7
7-Me apigenin		164.6	103.4	182.3	161.8	98.2	165.6	92.9	157.7	105.0	121.6	128.8	116.3	161.8	116.3	128.8
Luteolin		164.5†	103.3	182.2	162.1	99.2	164.7†	94.2	157.9	104.2	122.1	113.8	146.2	150.1	116.4	119.3††
Chrysoeriol		163.7†	103.8	181.8	161.6	98.8	164.2†	94.0	157.4	103.3	121.7	110.2	150.8	148.0	115.8	120.4††
Diosmetin		163.6	104.0	181.8	161.7	99.0	164.4	94.0	157.5	103.7	123.3	113.1	146.9	151.2	112.1	118.7††
Tricetin	*	164.2	103.2	181.6	161.6	99.0	164.2	93.9	157.5	104.0	120.9	106.0	146.5	137.9	146.5	106.0
Flavonols																
Kaempferol		146.8	135.6	175.9	160.7	98.2	163.9	93.5	156.2	103.1	121.7	129.5	115.4	159.2	115.4	129.5
Quercetin		146.9	135.8	175.9	160.8	98.3	164.0	93.5	156.2	103.1	122.1	115.2	145.1	147.7	115.7	120.1
5-Me quercetin		142.0	137.1	171.1	160.6	96.0	162.6	94.8	158.1	105.2	122.4	114.6	145.1	147.1	115.7	119.3
7-Me quercetin		147.3	136.0	175.9	160.4	97.4	164.9	91.9	156.0	103.7	121.9	115.2	145.0	147.8	115.6	120.1
4′,7-Me quercetin		146.7	136.4	176.0	160.4	97.4	164.9	91.8	156.0	104.0	123.4	114.8	146.2	149.4	111.7	119.8
Morin	*	149.6	136.5	176.7	161.3	99.6	164.1	95.0	156.9	105.0	111.2	158.1	105.0	161.0	109.3	132.3
Myricetin (3-*O*-glyc)	*	–	–	–	161.2	98.6	164.0	93.3	156.2	104.0	120.0	108.8	145.3	136.6	145.2	108.8

† Assignments may be reversed.
†† Assignments for C-1′ and C-6′ reversed according to Iinuma *et al.* (1980).
* Spectra from Ternai and Markham (1976) or Markham *et al.* (1978). All others from Wagner *et al.* (1976).
** Spectrum from Wenkert and Gottlieb (1977).

Table 2.8 A selection of isoflavone ^{13}C-NMR spectra

Isoflavones	*	*C-2*	*C-3*	*C-4*	*C-5*	*C-6*	*C-7*	*C-8*	*C-9*	*C-10*	*C-1′*	*C-2′*	*C-3′*	*C-4′*	*C-5′*	*C-6′*
7-OMe	P	152.4	125.1	175.3	127.6	114.4	163.8	100.0	157.7	118.3	127.9	128.3	128.8	131.8	128.8	128.3
2′,7-OMe	P	155.8	122.7	175.5	127.9	114.4	164.0	100.3	158.1	118.6	121.1	157.7	111.4	131.8	120.6	129.7
4′-OH, 7-OMe	P	151.8	124.4	175.2	127.0	114.3	163.5	100.0	157.2	117.9	122.4	129.8	115.2	157.5	115.2	129.8
7,3′,4′-OMe	W	151.8	124.3†	175.3	127.2	114.2	163.5	99.8	157.4	118.0	124.4†	112.3	148.4‡	148.7‡	110.9	120.6
6,7,4′-OMe	W	151.5	124.2†	175.2	104.7	147.4	154.1	99.3	151.9	117.6	124.1†	129.8	113.7	159.2	113.7	129.8
5,7,4′-OH	C	153.6	121.4†	180.2	157.6	98.6	164.3	93.7	157.6	104.6	122.4†	130.0	115.2	162.1	115.2	130.0
5-OH,3′,4′-(O-CH_2-O)	M	153.7	123.4†	174.5	126.9	115.6	161.4	103.9	156.9	118.7	125.6†	109.3	147.0	147.0	107.9	122.2
5,7-OH, 6,2′,3′,4′-OMe	W	154.5	119.3	180.6	152.6	130.9	156.3	93.6	153.0	105.5	110.4	151.6	97.8	149.6	142.4	115.0

* Source of spectrum: P = Pelter *et al.* (1976), Solvent $CDCl_3$ and $CDCl_3$/DMSO-d_6 for 4′-OH,7-OMe.
W = Wenkert and Gottieb (1977); Solvent $CDCl_3$ and $CDCl_3$/MeOH for 5,7-OH, 6,2′3′4′-OMe.
C = Chari *et al.* (1977), Solvent DMSO-d_6.
M = Markham and Ternai (1976), Solvent DMSO-d_6,

† or ‡ Assignments may be reversed.

2.2.6 Chalcones

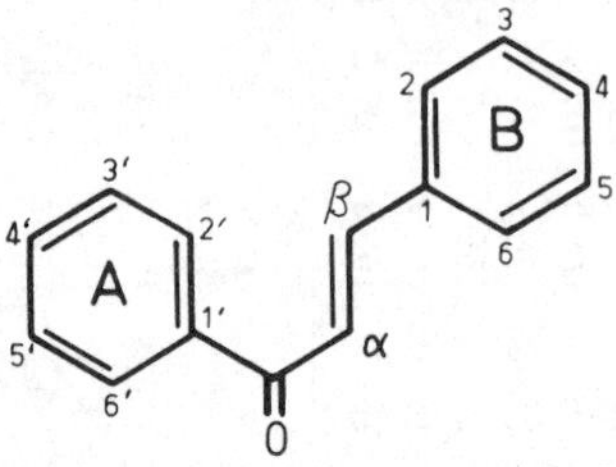

(2.3)

The ^{13}C-NMR spectrum of chalcone itself was originally assigned by Radeglia *et al.* (1976) and since that time the spectra of a number of oxygenated (Pelter *et al.*, 1976; Markham and Ternai, 1976; Solaniova *et al.*, 1976) and other (Solaniova *et al.*, 1976) derivatives have appeared in the literature. Some of the former are presented in Table 2.9. Of interest primarily are the chemical shifts of the α, β and carbonyl carbons. The carbonyl carbon at 188–195 p.p.m. is, as expected, shifted downfield by the presence of a 2′-hydroxyl due to hydrogen-bonding. C-α and C-β give rise to signals in the ranges 116–128 and 136–146 p.p.m. respectively, and according to Pelter *et al.* (1976) can be identified by their characteristic appearance as a six-line multiplet in the off-resonance-decoupled spectrum. Both are affected by changes in the B-ring substitution and C-α also by the presence or absence of a 2′-hydroxyl group.

2.2.7. Aurones

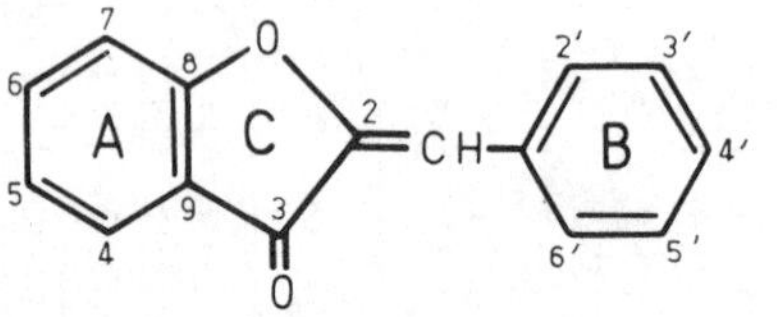

(2.4)

Very little has been published on the ^{13}C-NMR of aurones, the major contributions known to the authors being those of Pelter *et al.* (1976, 1979). The aurones studied are all synthetic and several of the reported spectra are presented in Table 2.9. The data in the table refer to the (natural product type) *Z*-isomers which were shown by Pelter *et al.* (1979) to be readily distinguishable from the equivalent *E*-isomers by the chemical shift of the exocyclic olefinic carbon. In *E*-isomers this resonance occurs at about 10 p.p.m. downfield from its position in *Z*-isomers. The carbonyl carbon resonates in the range 181–186 p.p.m. in the aurones studied, but all of these lack the hydrogen-bonding 4-hydroxyl.

Table 2.9 A selection of chalcone and aurone ^{13}C-NMR spectra

Chalcones	*	*C-1*	*C-2*	*C-3*	*C-4*	*C-5*	*C-6*	*C-β*	*C-α*	*C=0*	*C-1′*	*C-2′*	*C-3′*	*C-4′*	*C-5′*	*C-6′*
Chalcone	R	134.7	128.3	128.3	130.3	128.3	128.3	144.4	121.7	189.8	138.0	128.3	128.7	132.6	128.7	128.3
2′-OH	P	134.5	128.9	128.6	130.8	128.6	128.9	145.3	118.5	193.6	119.9	163.6	118.8	136.3	119.9	129.6
2′-OH, 4′-OMe	P	134.6	128.8	128.4	130.5	128.4	128.8	144.2	120.2	191.6	113.9	166.5	101.0	166.0	107.6	131.1
2′-OH 4-OMe	P	127.4	130.6	114.6	162.1	114.6	130.6	145.4	118.6	193.7	120.2	163.6	118.8	136.2	117.7	129.6
2, 2′-OH 4′-OMe	P	122.0	157.7	116.7	131.8	120.4	130.0	141.1	119.8	192.8	114.4	166.5	101.1	166.1	107.4	131.5
2′,3-OH 4′-OMe	P	135.8	115.1	157.7	118.2	129.8	120.0	144.5	119.8	191.7	113.9	166.3	101.0	165.9	107.3	131.3
3-OH 2′,4,4′-OMe	P	127.8	114.9	146.9	150.5	111.7	122.2	144.7	118.1	191.8	114.0	166.1	101.0	165.8	107.2	131.0
Aurones		*C-1′*	*C-2′*	*C-3′*	*C-4′*	*C-5′*	*C-6′*	*=CH–*	*C-2*	*C-3*	*C-4*	*C-5*	*C-6*	*C-7*	*C-8*	*C-9*
4′-OMe	P	125.0	133.4	114.4	161.0	114.4	133.4	112.8	145.8	184.3	124.4	123.2	136.4	113.3	165.7	121.9
6-OMe	P	132.3	131.1	128.7	129.4	128.7	131.1	111.6	147.7	182.7	125.6	112.0	167.2	96.5	168.3	114.7
4′-OH,6-OMe	P	123.4	133.2	116.1	159.3	116.1	133.2	111.9	146.1	182.5	125.3	112.8	166.9	96.5	167.8	114.9
2′-OH,6-OMe	P	119.0	157.2	115.6	131.1	119.3	130.9	105.9	146.8	181.8	125.0	112.0	164.9	96.6	167.7	114.4

* Source of spectrum: R = Radeglia *et al.* (1976). Solvent $CDCl_3$.
P = Pelter *et al.* (1976) – chalcones; (1979) – aurones. Solvent $CDCl_3$, except for 2′,4-OH,4′-OMe chalcone, 4′-OH,6-OMe and 2′-OH,6-OMe aurones, $CDCl_3/DMSO-d_6$.

2.2.8. Biflavonoids

Among the naturally occurring biflavonoids, those possessing a C–C linkage to at least one of the A rings (C-6 or C-8) predominate. Distinguishing the C-6 from the C-8 signal is thus of considerable significance in structure studies. The chemical shift difference between C-6 and C-8 in 5,7-dihydroxyflavonoids is small in flavanones and 3-hydroxyflavanones (ca. 1 p.p.m.) and larger in flavones, isoflavones and flavonols (ca. 4.8 p.p.m.). However, they may be distinguished unambiguously by selective heteronuclear decoupling (Wagner *et al.*, 1976) and on the basis of proton-coupled spectra (Wehrli, 1975; Markham and Ternai, 1976).

Examples of the range of C–C linkage types studied by ^{13}C-NMR and involving apigenin and/or naringenin are presented in Table 2.10. Interflavonoid linkages between rings A and A, A and B, and A and C are exemplified.

Table 2.10 Comparison of chemical shifts in the region 90–110 p.p.m. in the ^{13}C-NMR spectra of C–C-linked biflavonoids involving ring A

Carbon atom		*C-3*	*C-6*	*C-8*	*C-10*
Apigenin		102.8	98.8	94.0	103.7
Cupressuflavone	I	102.8	99.0	98.7	104.3
[(I-8, II-8) biapigenin]	II	102.8	99.0	98.7	104.3
Agathisflavone	I	103.1	103.6	93.7	103.8
[(I-6, II-8) biapigenin]	II	102.8	98.9	98.7	104.0
Robustaflavone	I	102.9	99.0	94.0	103.9
[(I-3′, II-6) biapigenin]	II	102.9	103.5	93.9	103.9
Amentoflavone	I	103.2	98.9	94.2	104.0
[(I-3′, II-8) biapigenin]	II	102.8	99.1	104.1	104.0
Naringenin		42.0	95.9	95.0	101.8
Succedaneaflavone	I	42.2	101.1	94.7	101.9
[(I-6, II-6) binaringenin]	II	42.2	101.1	94.7	101.9
Volkensiflavone	I	48.2	96.4	95.3	101.7
[Naringeninyl (I-3, II-8) apigenin]	II	102.8	98.5	100.6	103.6

These data were taken from Chari *et al.* (1977c). The spectra were run in DMSO-d_6 at 80°C.

In the spectrum of cupressuflavone, apigeninyl (I-8, II-8)* apigenin, the quaternary C-8 resonance appears at 98.7 p.p.m. whereas the chemical shift of C-6 is at 99.0 p.p.m. The only other resonances, each for two carbons, in the region 93.0–104.0 p.p.m. are those at 102.8 p.p.m. (I-3, II-3) and 104.3 p.p.m. (I-10, II-10). From a comparison with the chemical shifts of the corresponding carbon atoms in apigenin, it is evident that the C-8 signal experiences a downfield shift of 4.7 p.p.m. and that the others are largely unaffected. The spectrum of agathisflavone [apigeninyl (I-8, II-6) apigenin] shows eight distinct resonances in this region but only the signals for I-8 and II-6 are shifted downfield relative to the

* For nomenclature see Locksley (1973).

signals of C-8 and C-6 in the spectrum of apigenin. The spectrum of the naturally occurring (II-8, I-5′) biluteolin (Osterdahl, 1979) is characterized by the appearance of two C-6 signals (I-6, II-6) at 99.3 and 99.2 p.p.m. as well as that of I-8 at 94.3 p.p.m. In addition, the signal for the quaternary I-5′ appears at 120.6 p.p.m. and is 4.5 p.p.m. downfield relative to that of II-5′. Another example, 5-methoxybilobetin [apigeninyl (I-8, II-5′)-3′,4′-di-*O*-methyl-luteolin] has also recently been studied (Joly *et al.*, 1980). The only example of linkage involving the rings A and C is that of I-8, II-3, e.g. volkensiflavone. Owing to the restricted rotation about the C–C interflavonoid linkage, the ^{13}C-NMR spectrum of volkensiflavone at room temperature shows multiple resonances as well as considerable broadening of signals. Other (I-3, II-8)-linked biflavonoids that have been investigated by ^{13}C-NMR spectroscopy are GB-1 (naringeninyl-naringenin), morelloflavone [naringeninyl (I-3, II-8) luteolin; Duddeck *et al.*, 1978], kolaflavanone [naringeninyl (I-3, II-8) taxifolin 4′-*O*-methyl ether; Cotterill *et al.*, 1978] and manniflavanone [eriodictyolyl (I-3, II-8) taxifolin; Crichton and Waterman, 1979].

The evidence above suggests that an inspection of the chemical shift values of the signals in the region 90.0–105 p.p.m. can give a good indication of the position of the linkage in a C–C-linked biflavanoid provided that the component monoflavonoids are known. Additional confirmation can be gained by running the off-resonance and proton-coupled spectra.

2.2.9 Flavanones (2,3-dihydroflavones)

Published ^{13}C chemical-shift data for flavanones include those of Wehrli (1975), Wagner *et al.* (1976), Markham and Ternai (1976), Panichpol and Waterman (1978) and Wenkert and Gottlieb (1977). Prenylated flavanones have been studied by Komatsu *et al.* (1978), van Zyl *et al.* (1979) and Yakushijin *et al.* (1980). ^{13}C-NMR spectral investigations on naturally occurring *C*-benzylated flavanones have been reported by Hufford and Lasswell (1978) and on their methyl ethers by El-Sohly *et al.* (1979).

Increased polarization of the carbonyl group in flavanones as compared with that in flavones, results in a ca. 14 p.p.m. downfield shift of the C-4 resonance of the former relative to the latter. The carbonyl resonance in 5-hydroxylated flavanones occurs in the region 195 $\pm$2 p.p.m. The C-3 methylene carbon signal is characteristic and is at 42$\pm$0.5 p.p.m. whereas the C-2 oxymethine carbon resonates at 78$\pm$0.3 p.p.m. for such compounds. Further consequences of the reduction of the 2,3 double bond in flavones, as exemplified by naringenin, are downfield shifts of the C-5 (2 p.p.m.), C-7 (3 p.p.m.) and C-9 (5.5 p.p.m.) resonances as compared with the corresponding carbon signals in apigenin. The C-6 signal, however, shifts upfield by about 2 p.p.m. possibly owing to increased electron donation to this carbon by the C-9 ring oxygen. Loss of conjugation between the rings B and C results in a downfield shift of the C-1′ resonance by 8.6 p.p.m. with an upfield shift of 3.3 p.p.m. for C-4′ as compared with apigenin.

The chemical-shift data for the carbon atoms in some representative flavanones are presented in Table 2.11.

2.2.10 2,3-Dihydroflavonols

Hydroxyl substitution of C-3 in flavanones leads to 30.5 and 4.8 p.p.m. downfield shifts of the signals for the carbon atoms C-3 and C-2 respectively as exemplified by taxifolin (Table 2.11; Markham and Ternai, 1976). In contrast, the C-4 resonance is only slightly influenced (+0.9 p.p.m.). In the case of dihydromorin (Table 2.11) the C-2 signal shifts upfield by 4.8 p.p.m. relative to C-2 in taxifolin, presumably owing to a γ-type of interaction between H-2 and the 2′-OH (Wenkert and Gottlieb, 1977). There has been a report that *O*-glucosidation of the 3-OH in (2*R*, 3*R*)-taxifolin leads to a 7.8 p.p.m. upfield shift of the C-4 resonance (Sakurai and Okumura, 1978).

2.2.11 Isoflavanones (2,3-dihydroisoflavones)

The ^{13}C-NMR spectra of two (methoxylated) isoflavanones have been reported by Wenkert and Gottlieb (1977). They are characterized by C-2 oxymethylene resonances at 71.7 p.p.m. and 71.5 p.p.m. and C-3 methine carbon signals at 50.9 p.p.m. and 51.5 p.p.m., so affording a clear differentiation from the isomeric flavanones. Other examples of oxygenated isoflavanones investigated are the mono- and di-isoprenylated 5,7,4′-trihydroxy-2′-methoxyisoflavanones (Komatsu *et al.*, 1978). In both cases, however, interaction of the C-2′ substituent with H-3 shifts the C-3 resonances upfield to 44.6 p.p.m. and 44.7 p.p.m. respectively (cf. dihydromorin). Selected examples are presented in Table 2.11.

2.2.12 Miscellaneous flavonoid conjugates

^{13}C-NMR spectroscopy has been used in the structure elucidation of several flavonoid conjugates. Reports in the literature on the various types are listed briefly in this section.

Empirical use of chemical-shift data and model compounds has enabled the structure elucidation of a C_{30} proanthocyanidin (Jacques *et al.*, 1974; Schilling *et al.*, 1973). Schilling (1975) elegantly demonstrated by ^{13}C-NMR spectroscopy that there exists an isomer of the above mentioned compound in which the position of the interflavonoid linkage was different. The two structure types were characterized by deuteration studies on model compounds. ^{13}C-NMR spectroscopy was used by Fletcher *et al.* (1977) to establish the absolute stereochemistry of 2,3-*cis*-procyanidins at C-4. Additionally characteristic spectral differences were noted between the configurational isomers of procyanidin dimers and their derivatives. By a combination of ^{13}C-NMR, chiroptical and chemical degradative evidence it has been demonstrated that polymeric proanthocyanidins consist exclusively of repeating flavan-3-ol units with C-4–C-6 or C-4–C-8 interflavonoid

Table 2.11 A selection of ^{13}C-NMR spectra of dihydroflavonoids

Flavonoid	*Carbon number*															†
	C-2	*C-3*	*C-4*	*C-5*	*C-6*	*C-7*	*C-8*	*C-9*	*C-10*	*C-1′*	*C-2′*	*C-3′*	*C-4′*	*C-5′*	*C-6′*	
Flavanone	78.7	43.5	191.1	128.3	121.2	136.0	117.8	160.8	120.5	138.7	126.3	128.3	126.1	128.3	126.3	W
5,7-Dihydroxyflavanone (Pinocembrin)	78.4	42.2	195.8	163.6	96.1	166.6	95.1	162.7	101.9	138.0	126.5	128.5	128.5	128.5	126.5	Wa
5,7,4′-Trihydroxyflavanone (Naringenin)	78.4	42.0	196.2	163.6	95.9	166.7	95.0	162.9	101.8	128.9	128.2	115.2	157.8	115.2	128.2	Wa
5,7,3′,4′-Tetrahydroxy-flavanone (Eriodictyol)	78.3	42.2	196.2	163.4	95.7	166.6	94.8	162.8	101.7	129.4	114.2	145.1	145.6	115.3	117.8	Wa
5,7,3′-Trihydroxy-4′-methoxyflavanone. (Hesperetin)	78.5	42.1	196.2	163.8	96.2	166.9	95.4	163.0	102.1	131.4	114.3	146.7	148.1	112.1	118.0	Wa
5,7,4′-Trihydroxy-6-prenylflavanone (Sophoraflavanone B)	78.3	42.0	197.1	161.6	107.1	164.7	95.4	160.1	*	129.5	128.3	115.4	157.9	115.4	128.3	K
5,7,8-Trimethoxyflavanone	79.0	45.6	189.2	156.2	89.5	158.7	131.0	157.8	106.3	138.9	125.9	128.7	128.4	128.7	125.9	P
3,5,7,3′,4′-Pentahydroxy-flavanone(Taxifolin)	83.1	71.7	197.1	163.3	96.1	166.8	95.1	162.5	100.6	128.1	115.3	144.9	145.7	115.3	119.2	M
3,5,7,2′,4′-Pentahydroxy-flavanone(Dihydromorin)	78.3	70.9	198.4	163.7	96.5	167.1	95.5	163.1	100.9	114.2	159.0	103.0	157.5	107.1	130.3	M
7,4′-Dimethoxyiso-flavanone	71.7	50.9	190.6	129.1	109.8	165.6	100.4	163.1	114.6	127.1	129.3	114.0	158.7	114.0	129.3	M
7,3′,4′-Trimethoxyiso-flavanone	71.5	51.5	190.4	129.0	109.7	165.6	100.3	163.0	114.4	127.5	111.6	148.2	148.7	111.2	120.2	M
5,7,4′-Trihydroxy-2′-methoxy-3-prenyl isoflavanone (Sophoraisoflavanone A)	70.7	44.6	198.0	164.2	·96.1	167.0	95.0	163.5	102.4	121.3	156.3	118.9	157.9	111.4	127.3	K

* Signal not observed

† Source of spectrum:
W = Wenkert and Gottlieb (1977), DMSO-d_6, $CDCl_3$.
Wa = Wagner *et al.* (1976), DMSO-d_6.
K = Komatsu *et al.* (1978), DMSO-d_6.
P = Panichpol and Waterman (1978), $CDCl_3$.
M = Markham and Ternai (1976), DMSO-d_6.

linkages (Czochanska *et al.*, 1979, 1980). Likewise some natural trimeric proanthocyanidins have recently been shown to possess mixed C-4–C-6 and C-4–C-8 interflavonoid linkages (Hemmingway *et al.*, 1981). Chemical-shift data on six rotenoids, which are essentially prenylated isoflavonoids with an oxymethylene bridge between C-2′ and C-2 have been published by Crombie *et al.* (1975). Chalmers *et al.* (1977) have reported on the ^{13}C-NMR spectral data of pterocarpans. ^{13}C-NMR spectroscopy has played a key role in the structure elucidation of silychristin, a flavonolignan, for which two alternative formulations were advanced (Chari *et al.*, 1977a; Wagner *et al.*, 1978). Further reports on the carbon chemical-shift data for flavonolignans are those of Parthasarathy *et al.* (1979), Cooper *et al.* (1977) and Merlini *et al.* (1980). A number of reports on the ^{13}C-NMR of prenylated and chromeno-flavonoids have appeared in the literature. They include Wenkert and Gottlieb (1977), Chari *et al.* (1978a), Nomura *et al.* (1978), Nomura and Fukai (1978, 1979, 1980), Komatsu *et al.* (1978), Van Zyl *et al.* (1979), Yakushijin *et al.* (1980) and Oshima *et al.* (1980a,b).

2.3 FLAVONOID *O*-GLYCOSIDES

Because sugar carbon resonances occur largely well clear of the resonances of the flavonoid nucleus carbons, ^{13}C-NMR offers a convenient non-destructive method of studying the sugar moieties of flavonoid glycosides. *O*-Glycoside spectra are often best measured at higher than ambient temperature as this not only sharpens the sugar signals but also eliminates conformational equilibria (through hindered rotation in branched oligosaccharides).

2.3.1 Monosaccharides

The sugars commonly occurring in glycosidic form are all readily distinguishable from one another by ^{13}C-NMR spectroscopy. This is clearly seen from the selection of spectra presented in Table 2.12. These data were obtained by Gorin and Mazurek (1975) using D_2O solutions of the sugars. Solvent effects with different solvents, however, may alter the chemical shifts markedly, and this is well exemplified by the spectra reported by Tori *et al.* (1978) for C_5D_5N solutions which differ significantly from those listed in Table 2.12. Different levels of water in DMSO-d_6 solutions are also known to affect the chemical shifts of sugar carbons. It is evident from the data in Table 2.12 that pyranoses having the same hydroxyl configurations (i.e. homomorphous pyranoses) show similar patterns of carbon resonances. Thus, C-1,-2,-3 and -4 of xylose exhibit chemical shifts close to those of C-1,-2,-3 and -4 of glucose, and similarly, the arabinose spectra are related to those of galactose. Also evident from the table is the potential usefulness of ^{13}C-NMR spectra for the identification of the C-1 configuration in sugars (and glycosides).

Table 2.12 ^{13}C-NMR spectra* of selected monosaccharides in D_2O solvent

Sugar		*C-1*	*C-2*	*C-3*	*C-4*	*C-5*	*C-6*
D-Glucopyranose	α	92.7	72.1	73.4	70.4	72.1	61.3
	β	96.5	74.8	76.4	70.3	76.6	61.5
D-Galactopyranose	α	93.6	69.8	70.6	70.6	71.7	62.5
	β	97.7	73.3	74.2	70.1	76.3	62.3
D-Mannopyranose	α	95.0	71.7	71.3	68.0	73.4	62.1
	β	94.6	72.3	74.1	67.8	77.2	62.1
D-Xylopyranose	α	93.3	72.5	73.9	70.4	62.1	—
	β	97.6	75.1	76.9	70.3	66.3	—
L- (or D) Arabinopyranose	α	97.8	73.0	73.5	69.6	67.5	—
	β	93.7	69.6	69.8	69.8	63.6	—
L-Rhamnopyranose	α	95.0	71.9	71.1	73.3	69.4	18.0
	β	94.6	72.4	73.8	72.9	73.1	18.0

* Chemical shifts expressed in p.p.m. from TMS.

Furanose sugars are readily distinguished from their pyranose isomers as is demonstrated by the comparative data on methyl glycopyranosides and methyl glycofuranosides presented in Table 2.13. The data are selected from the published work of Gorin and Mazurek (1975) and (for glycofuranosides) from Ritchie *et al.* (1975).

Table 2.13 ^{13}C-NMR spectra* of selected methyl glycopyranoside/furanoside pairs in D_2O solvent

Methyl glycoside	*C-1*	*C-2*	*C-3*	*C-4*	*C-5*	*C-6*
β-D-Glucopyranoside	104.3	74.2	76.9	70.8	76.9	61.9
β-D-Glucofuranoside	110.0	80.6	75.8	82.3	70.7	64.7
β-D-Galactopyranoside	104.9	71.8	73.9	69.8	76.2	62.1
β-D-Galactofuranoside	109.9	81.3	78.4	84.7	71.7	63.6
β-D-Xylopyranoside	105.1	74.0	76.9	70.4	66.3	—
β-D-Xylofuranoside	109.7	81.0	76.0	83.6	62.2	—
α-L-Arabinopyranoside	105.1	71.8	73.4	69.4	67.3	—
α-L-Arabinofuranoside	109.2	81.8	77.5	84.9	62.4	—

* Chemical shifts expressed in p.p.m. from TMS.

2.3.2 Mono-*O*-glycosides

In *O*-glycosides carbon-1 of the monosaccharide is linked via a hemiacetal bond to the aglycone moiety. This produces a shift in the C-1 resonance to lower field. In methyl glycosides this shift is of the order of 7–8 p.p.m. (cf. data in Tables 2.12 and 2.13), and apart from a much smaller shift of the C-2 resonance the rest of the spectrum is largely unaffected. In flavonoid glycosides the extent of the downfield shift of the C-1 signal depends very much on the environment of the phenolic

hydroxyl group (Markham *et al.*, 1978), and, although it is generally 4–6 p.p.m., the precise variation appears to be of some diagnostic value (see Table 2.14). The presence of oxygen substituents in both *ortho* positions appears to shift the C-1 sugar signal downfield to about 107 p.p.m. (Osazawa and Takino, 1979) and this also could be of diagnostic value.

Table 2.14 Chemical shifts (p.p.m.) of sugar C-1 resonances in the ^{13}C-NMR spectra of flavone and flavonol glycosides

*Glucosides**	*Shift*	*Rhamnosides**	*Shift*
(Glucose)†	96.5	(Rhamnose)†	95.0
7-*O*-Glucoside	100.4	7-*O*-Rhamnoside	98.9
2′-*O*-Gluctoside	100.5		
3-*O*-Glucoside	101.4	3-*O*-Rhamnoside	101.9
4′-*O*-Glucoside	102.2		
3′-*O*-Glucoside	102.4		
5-*O*-Glucoside	104.3		

* Solvent: DMSO-d_6.
† Solvent: D_2O.

Other sugar carbon resonances are little affected by glycosylation (see Table 2.15), and so in principle the ^{13}C-NMR spectrum of a flavonoid monoglycoside can be used to identify both the sugar and the configuration of its linkage from data such as those listed in Tables 2.12 and 2.13 (provided that the same solvent is used for sugar and glycoside).

More diagnostic than the glycosylation effect on the sugar carbon signals is the effect on the signals of the flavonoid nucleus itself. In general terms the carbon at the site of glycosylation is shifted to a higher field following glycosylation, whereas the *ortho*- and *para*-related carbons shift downfield. Thus, the effect of glycosylation of the 7-hydroxyl on the C-7 signal (see Table 2.15) is a 1.4 ppm upfield shift and this is accompanied by downfield shifts of about 1 p.p.m. in the *ortho*-related C-6 and C-8 signals and 1.7 p.p.m. in the *para*-related C-10 signal (Markham *et al.*, 1978). Shifts qualitatively similar but quantitatively different have been observed for glycosylation at sites such as 3′ and 4′ in flavones and flavonols (Markham *et al.*, 1978; Österdahl, 1979). Significantly, the most reliable indicator of glycosylation appears to be the downfield shift of the *para*-related carbon signal which is invariably larger than the other shifts and generally in the range 1.7–4 p.p.m. There is some evidence that glycosylation with rhamnose produces shifts of a different magnitude from those above (Markham *et al.*, 1978), but this has yet to be confirmed with a significant number of reference compounds.

Glycosylation of the 3- and 5-hydroxyls (Markham *et al.*, 1978; Chari *et al.*, 1977a), as expected, produces unusual effects (see Table 2.15). With the

Table 2.15 Sugar carbon resonances* in the ^{13}C-NMR spectra of a selection of flavonoid-*O*-glycosides

Glycoside†	‡	*C-1*	*C-2*	*C-3*	*C-4*	*C-5*	*C-6*	*Reference*
A-7-*O*-glc	7	100.2	73.3	76.6[a]	69.8	77.4[a]	60.9	Chari *et al.* (1977c)
A-7-*O*-glc(2-1)apiose	7	99.7	76.8	76.6[a]	70.2	77.2[a]	60.9	Markham *et al.* (1978)
	2″	109.0	76.5[a]	79.1	74.0	64.4	–	
Ac-7-*O*-glc(6-1)rha	7	101.3	73.7	77.0[a]	70.4	76.4[a]	66.7	Chari *et al.* (1977c)
	6″	100.3	71.5	71.0	73.0	69.0	17.8	
Ac-7-*O*-glc(6-1)rha(2‴-OMB)	6″	97.8	73.4[a]	68.4	72.8[a]	69.0	17.8	
Ac-7-*O*-glc(6-1)rha(3‴-OMB)	6″	100.3	69.2[a]	72.3	69.5	69.0[a]	17.8	
Is-4′-OMe-7-*O*-glc(2-1)	7	102.5	82.5	75.5	69.2	77.2	60.6	Chari *et al.* (1981)
-allose(6‴-OAc)	2″	100.0	70.7	71.5	66.9	71.5	63.6	
G-5-*O*-glc(6-1)xyl	5	104.3[a]	74.0[b]	76.5[c]	70.8	76.4	69.1	Garcia-Granados and Saenz de Buruaga (1980)
	6″	104.4[a]	73.7[b]	76.8[c]	70.1	65.8		
K-3-*O*-glc	3	101.4	74.2	76.5[a]	70.1	77.2[a]	61.0	Markham *et al.* (1978)
K-3-*O*-glc(2-1)glc	3	98.6	82.0	76.6[a]	70.0[b]	76.6[a]	61.0[c]	
	2″	103.6	74.3	76.7[a]	70.5[b]	76.7[a]	61.4[c]	
K-3-*O*-glc(6-1)rha	3	101.5	74.2	76.5[a]	70.1	75.8[a]	66.9	
	6″	100.6	70.3[b]	70.7[b]	72.0	68.1	17.4	
K-7-*O*-glc-3-*O*-glc	7	100.3	73.3	76.6[a]	70.0[b]	77.3[a]	61.0	
	3	101.3	74.4	76.6[a]	70.2[b]	77.4[a]	61.0	
K-7-*O*-glc-3-*O*-glc(6‴-OAc)	7	100.3	73.2	76.5[a]	70.0	77.1[a]	60.9	
	3	101.3	74.1	76.4	70.0	74.1	62.8	
K-7-*O*-glc-3-*O*-glc(2-1)	7	100.4	73.2	76.7[a]	70.0[b]	77.0[a]	61.2[c]	
-glc(2‴-*O*-sinapoyl)	3	97.7	80.0	76.7[a]	70.4[b]	77.0[a]	61.0[c]	
	2″	99.1	74.6[d]	74.1[d]	70.7[b]	76.1[a]	61.0[c]	

K-3-*O*-gal(6-1)rha(4‴-OAc)-(3-1)glc	3	102.3	71.1	72.0	68.0	73.3	66.5	Guinaudeau *et al.* (1981)
	6″	100.5	69.7	77.2	73.3	66.0	17.3	
	3‴	103.9	73.0	76.7	69.6	76.3	60.7	
K-3-*O*-allopyranoside	3	99.9	71.6	71.6	67.2	75.1	61.3	Okuyama *et al.* (1978)
Q-3-*O*-gal	3	102.3	71.3	73.4	68.0	75.8	60.8	
Q-3-*O*-gal(6″-*O*-galloyl)	3	102.4	71.2	73.1	68.0	72.7	62.3	
Q-3-*O*-rha	3	101.9	70.4[a]	70.6[a]	71.5	70.1[a]	17.3	Markham *et al.* (1978)
Q-3-*O*-ara(pyranose)	3	101.8	71.7	70.8	65.9	64.1	–	
Q-3-*O*-ara(furanose)	3	108.1	82.1	77.2	86.2	61.0	–	

* In p.p.m. from TMS, DMSO-d_6 solvent.

† Abbreviations: A, apigenin; Ac, acacetin; K, kaempferol; Q, quercetin; G, genkwanin; Is, isoscutellarein; OAc, acetyl; MB, 2-methylbutyryl, glc, glucose; rha, rhamnose; xyl, xylose; ara, arabinose.

‡ Glycosylation site.

[a, b, c, d] Assignments bearing the same superscript in any one spectrum may be reversed.

3-hydroxyl, although the upfield shift of the C-3 signal is of the expected order (ca. 2 p.p.m.), the downfield shift of the '*ortho*-related' C-2 signal at 9.2 p.p.m. is especially pronounced. Glycosylation of the 5-hydroxyl in luteolin has a marked effect on the resonances of all A- and C-ring carbons (see Fig. 2.2), presumably owing to the disruption of 5-hydroxy-4-keto hydrogen-bonding.

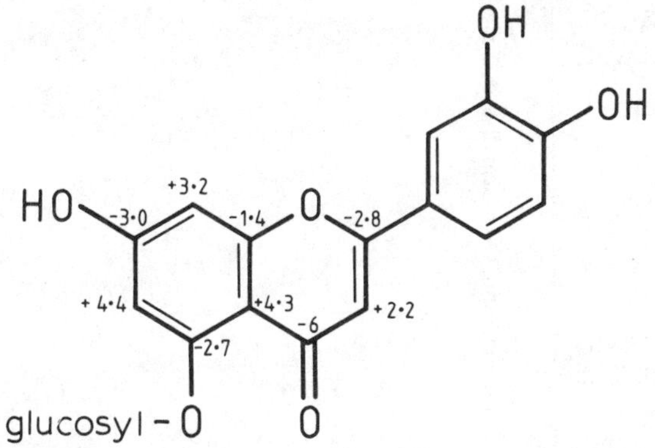

Fig. 2.2 Signal shifts observed on glycosylation of the 5-hydroxyl of luteolin (Markham *et al.*, 1978).

2.3.3 *O*-Diglycosides

The site at which a second sugar is attached to the sugar of a flavonoid mono-*O*-glycoside is readily determined by ^{13}C-NMR spectroscopy, and this is perhaps the most significant information contained in the spectrum, as it is difficult to obtain it by other methods (see Chapter 5, Section 5.3). It has been established that glycosylation of sugar hydroxyls produces a sizeable downfield shift in the resonance of the hydroxylated carbon and upfield shifts in the resonances of adjacent carbons (Yamoaka *et al.*, 1971; Colson *et al.*, 1974). Likewise, this has been observed in studies of flavonoid *O*-diglycosides (e.g. Markham *et al.*, 1978; Österdahl, 1979a).

Rhamnosylglucosides

A well-studied example of the usefulness of ^{13}C-NMR spectroscopy in this field is the distinction of flavonoid neohesperidosides [rhamnosyl(1–2)glucosides], rutinosides [rhamnosyl(1–6)glucosides] and rhamnosyl(1–4)glucosides, all of which occur naturally and are otherwise difficult to distinguish. Reports by Österdahl (1978a) and Markham *et al.* (1978) indicate that rutinosides show a 4.5–6.0 p.p.m. downfield shift of the C-6 signal, rhamnosyl(1–4)glucosides a similar downfield shift of the C-4 and neohesperidosides a 3–4 p.p.m. downfield

shift of the C-2. The upfield shifts of adjacent carbons seem to be less consistent. It is of interest that substitution of a glucoside with apiose at the 2-position has been shown to have an effect similar to that of rhamnose (see Table 2.15).

Glucosylglucosides

Shifts qualitatively similar to those reported for rhamnosylation of glucosides have been reported for glucosylation of glucosides (Markham *et al.*, 1978; Österdahl, 1979a), although as yet only flavonoid sophorosides, gluco(1–2) glucosides, appear to have been studied. In sophorosides a much larger downfield shift (8 p.p.m.) of the glycosylated carbon is observed than in neohesperidosides. This is accompanied by a 2.9 p.p.m. upfield shift of the C-1 signal and is in accord with data previously reported for disaccharides of this type (Yamoaka *et al.*, 1971; Colson *et al.*, 1974). It is to be expected that other linkage types, too, would parallel the equivalent disaccharides in terms of carbon-13 shifts.

Others

For apiosylglucosides, xylosylglucosides and allosylglucosides see Table 2.15.

2.3.4 Polyglycosides

Data such as the above have also been used with some success in defining the glycoside structures of flavonoid tri- (Markham *et al.*, 1978; Österdahl, 1979a; Guinaudeau *et al.*, 1981) and tetra-glycosides (Österdahl, 1978a, 1979a,b) involving mainly glucose and rhamnose. Assignments for polyglycosides are generally achieved by the 'best fit' method, but the reader is referred to the original work for further details.

Sugar sequencing

The use of T_1 data for sugar sequencing has recently been reported by Guinaudeau *et al.* (1981). The principle of the technique, as reported earlier in connection with studies of the linear oligosaccharide chain of a cardiac glycoside (Bock and Pederson, 1974), is that the average NT_1 values for sugar carbons increase with increasing distance from the aglycone moiety. Thus in the compound studied, faralatroside {kaempferol-3-*O*-[β-D-glucopyranosyl-(1–3)-4''''-*O*-acetyl-α-L-rhamnopyranosyl-(1–6)-β-D-galactopyranoside]} the glucose was demonstrated to be terminal by the relatively larger NT_1 values of its carbon resonances, and in particular C-3 and C-5 (see Table 2.16), for which assignments were unambiguous.

Table 2.16 Relaxation times of the saccharide carbon atoms in faralatroside

Carbon	*Relaxation time (chemical shift*, p.p.m.)*		
	Galactose	*Rhamnose*	*Glucose*
C-1	0.21 (103.7)	0.25 (100.7)	0.32 (104.3)
C-2	0.21 (71.7)	0.25 (70.4)	0.34 (73.4)
C-3	0.22 (72.4†)	0.24 (78.2)	0.35 (76.5†)
C-4	0.18 (68.6)	0.23 (73.7)	0.36 (70.0)
C-5	0.21 (73.9†)	0.25 (66.4)	0.34 (76.3†)
C-6	0.12 (67.3)	0.59 (16.7)	0.15 (61.2)

* Solvent: C_5D_5N.
† Assignments in any one sugar are reversible.

The chemical shifts of the sugar anomeric carbons may also be of use in sugar sequencing since the downfield shift experienced by C-1 on glycosylation of a phenol (4–6 p.p.m.) is consistently less than the 7–8 p.p.m. associated with glycosylation of a sugar (see Table 2.15). Thus in a flavonoid-*O*-diglucoside, C-1 of the first glucose will normally resonate in the range 100–102.5 p.p.m. whereas C-1 of the terminal glucose will resonate at about 104 p.p.m. Likewise with rhamnose, the C-1 resonance occurs at about 98.8 p.p.m. in 7-*O*-rhamnosides but in rutinosides and neohesperidosides it occurs at about 100.6 p.p.m.

2.3.5 Acylated glycosides

The site of acylation in the sugar moiety of flavonoid glycosides is readily determined from the ^{13}C-NMR spectrum and is evidenced by a downfield shift of the acylated carbon and upfield shifts of the adjacent carbons. It has been shown (Vignon and Vottero, 1976) that as the electron-withdrawing power of the acylating function increases, so also does the extent of the downfield shift of the acylated carbon. This principle also applies to the spectra of flavonoid glycosides and accordingly different acyl functions will exert quantitatively different effects.

Acetates have been studied by a number of workers (e.g. Markham *et al.*, 1978; Guinaudeau *et al.*, 1981; Chari *et al.*, 1980; Markham and Wallace, 1980; Redaelli *et al.*, 1980), and it has been established that acetylation at C-6 of a sugar is evidenced by a downfield shift of 1.8–3 p.p.m. in the C-6 signal and an upfield shift of about 3.4 p.p.m. in the C-5 signal (see Table 2.15). The acetate carbons resonate at about 21 p.p.m. ($\underline{C}H_3CO$) and 171 p.p.m. ($CH_3\underline{C}O$).

Acylation with 2-methylbutyric acid at sugar C-2 and C-3 sites has been shown (Chari *et al.*, 1977b) to result in similar downfield shifts to acetylation (Table 2.15), and adjacent carbons exhibit 1.2–2.6 p.p.m. upfield shifts in their resonances. Visible portions of the 2-methylbutyryl spectrum included C-3, C-4, C-5 and the carbonyl, at 26.5, 11.4, 16.4 and 175.5 p.p.m. respectively.

Glycoside benzoates studied by ^{13}C-NMR include gallates (Markham *et al.*, 1978; Isobe *et al.*, 1979) and benzoates themselves (Markham and Wallace, 1980). Here, downfield shifts of the acylated carbon are variable in the range 0–3 p.p.m. More reliable indicators of the acylation site appear to be the upfield shifts of adjacent carbons by about 3 p.p.m. The galloyl carbons resonate at 165 ($\underline{C}O_2$), 145.4 (C-3,5), 138.4 (C-4), 119.5 (C-1) and 108.8 (C-2, 6) p.p.m.

With sinapates and *p*-coumarates (Rahman *et al.*, 1978; Chari *et al.*, 1978a,b; Markham *et al.*, 1978; Ansari *et al.*, 1979) the downfield shift of the acylated carbon is in the range 2–3 p.p.m. if acylation is at C-6, but is often not observed when acylation is at C-2 (see also Table 2.15). However, the concomitant 2–4 p.p.m. upfield shifts of the adjacent carbons do seem to be reliable. *p*-Coumaroyl carbons resonate at 166.4 ($\underline{C}O_2$), 144.9 (C-β), 130.3 (C-2,6), 125 (C-1), 115.7 (C-3,5) and 113.9 (C-α) p.p.m., and sinapoyl carbons at 165.4 ($\underline{C}O_2$), 145.4 (C-3,5), 138.4 (C-4), 119.5 (C-1) and 108.8 (C-2,6) p.p.m.

2.4 FLAVONOID *C*-GLYCOSIDES

Flavonoid-C-glycosides, in which the sugar residue is a hexosyl or pentosyl moiety, may be considered as 1-arylated-1, 5-anhydro-hexitols or- pentitols. In contrast to *O*-glycosides, the signal for C-1 in *C*-glycosides appears in the same region as do the other oxymethine carbon resonances, owing to the presence of only one oxygen substituent on this carbon. Thus with the exception of the 6-desoxy derivatives, e.g. rhamnosyl, all carbons of the glycosyl moiety resonate between 60.0 and 82.0 p.p.m.

2.4.1 Model compounds: 1-arylated-1,5-anhydroalditols

Table 2.17 lists the chemical shifts for the carbon atoms of the sugar residue of some synthetic 1-arylated-1,5-anhydroalditols (Chari *et al.*, 1981). The assignments were made on the basis of chemical-shift data for the 1,5-anhydroalditols themselves (Que and Gray, 1974) as well as by selective decoupling and off-resonance experiments. In the spectrum of 1-phenyl-1,5-anhydroglucitol, the C-1 resonance is easily identified by selective decoupling of the H-1 and appears at 77.1 p.p.m. This ca. 30 p.p.m. upfield shift of the signal for C-1 as compared with the chemical shift of the anomeric carbon in alkyl and phenolic glycosides is due to the lack of the oxygen substituent on that carbon atom. The shifts induced by the phenyl substituent at C-1, as compared with 1,5-anhydroglucitol itself, are + 7.4 p.p.m. for C-1 and + 3.8 p.p.m. for C-2. Values for the other carbon atoms remain essentially unchanged. When the aryl substituent is 2′,4′,6′-trimethoxyphenyl the C-1 resonance appears at 73.3 p.p.m. and that of C-2 at 71.0 p.p.m. This 3.8 p.p.m. upfield shift of the C-1 signal from its position in the 1-phenyl analogue can be rationalized as being due to the γ-effect of the 2′-methoxyl substituent. Steric interaction of the same group with the axial H-2 results in a

Table 2.17 Chemical shifts of sugar carbons in 1-aryl-1, 5-anhydroalditols*

Model Compound	*C-1*	*C-2*	*C-3*	*C-4*	*C-5*	*C-6*	*References*
1,5-Anhydroglucitol*	69.7	70.2	78.3	70.6	81.2	61.4	Que and Gray (1974)
1-Phenyl-1,5-anhydroglucitol	77.1	74.0	79.7	69.7	81.6	60.9	Chari *et al.* (1981)
1-(2′,4′,6′-Trimethoxyphenyl)-β-D-1,5-anhydroglucitol	73.3	71.0	79.2	70.7	80.9	61.9	Chari *et al.* (1981).
1,5-Anhydrogalactitol*	70.0	67.3	75.0	70.0	80.3	62.2	Que and Gray (1974)
1-(2′,4′,6′-Trimethoxyphenyl)-β-D-1,5-anhydrogalactitol	74.0	68.6	76.2	69.5	79.3	61.0	Chari *et al.* (1981)
1-(2′,4′,6′-Trimethoxyphenyl)-β-D-1,5-anhydroxylitol	74.3	70.7	79.6	70.2	70.2	—	Chari *et al.* (1981)
1-(2′,4′,6′-Trimethoxyphenyl)-β-D-1,5-anhydroarabitol	73.2	67.1	74.4	68.2	69.0	—	Chari *et al.* (1981)

* Solvent: DMSO-d_6. Ref. DMSO-d_6 39.6 p.p.m. at 80°C.

† These values were taken from Que and Gray (1974) and converted using $\delta_{CS_2} - \delta_{TMS} = 193.5$ p.p.m. The chemical shifts were determined in D_2O solution.

− 3.1 p.p.m. shift of the C-2 signal. The same considerations also apply to the galactitol analogue. The assignments for C-1, and C-2 for the xylitol and arabitol analogues (homomorphous with glucitol and galactitol), were made by selective proton decoupling, whereas the C-5 of the arabinose compound was confirmed by an off-resonance experiment. In all cases, the $^1J_{CH}$ values for the oxymethine carbons were of similar magnitude being 143 ± 3.0 Hz. Hence an assignment of the C-1 signal on the basis of the $^1J_{CH}$ value alone, which can be done with *O*-glycosides, is not possible for *C*-glycosides. The only significant effect of the sugar moiety on the chemical shifts of the phenolic residue was a 16.0 p.p.m. downfield shift of the resonance of the glycosylated carbon.

2.4.2 Flavonoid mono- and di-*C*-glycosides

The data in Table 2.17 serve as a basis for the signal assignments for the sugar carbons in the ^{13}C-NMR spectra of 6-*C* and 8-*C* as well as 6,8-di-*C*-glycosides of 5,7-dihydroxyflavonoids and their derivatives. The assignments for iso-orientin were determined by selective decoupling experiments at highfield (270 MHz ^{1}H and 67.88 MHz for ^{13}C), (Chari *et al.*, 1978a) and necessitated revision of previous assignments (Österdahl, 1978b; Hostettmann and Guillarmod, 1976; Miura *et al.*, 1979). From the data in Table 2.18 it is evident that the position of linkage of the *C*-glycosyl unit to the 5,7-dihydroxyflavonoid aglycone does not affect the sugar carbon resonances (cf. the pairs, vitexin and isovitexin; orientin and iso-orientin). The position of linkage can, however, be inferred from the chemical shift of the remaining aromatic methine of ring A, since the only significant effect of *C*-glucosylation on the aglycone spectrum is a downfield shift of ca. 9.8 p.p.m. of the signal for the glycosylated carbon atom. Small upfield shifts of the *ortho* and *para* carbon resonances are also observed but the *meta* carbon resonance is unaffected.

Substitution of the 7-OH of the aglycone by methylation or glycosylation introduces restriction of rotation on the *C*-glucosyl moiety leading to the presence of conformers in solution at room temperature. This can be of use in structure elucidation. Such behaviour is exhibited, for example, by 7,3′,4′-tri-*O*-methyl-iso-orientin and saponarin as well as their derivatives. Multiple signals as well as peak broadening can be observed in their spectra at ambient temperature (Chari *et al.*, 1981).

Assignments analogous to those for *C*-glucosides have also been made for 8-galactosylapigenin (Chari *et al.*, 1980). In this case the C-5 assignment was facilitated by comparison of the shift values with those of the naturally occurring 6″-*O*-acetyl derivative.

When both C-6 and C-8 are glycosylated with the same sugar, either both sets of sugar carbon signals are observed in the region 60.0 p.p.m. to 82.0 p.p.m., or coincidence of some signals may occur. Thus the spectrum of 6,8-di-*C*-glucosylchrysoeriol exhibits 11 signals for the two hexosyl units, (Table 2.18) while the spectra of apigenin and tricetin 6,8-di-*C*-*β*-D-glucopyranosides

Table 2.18 Chemical shifts of sugar carbons in flavonoid C- and C-O-glycosides and derivatives*

Compound	*Sugar linkage*	*Carbon chemical shifts* *C-1*	*C-2*	*C-3*	*C-4*	*C-5*	*C-6*	*Reference*
Vitexin (8-*C*-glucosylapigenin)	8-C-glc	73.9	71.4	78.8	70.8	81.4	61.5	
Isovitexin (6-*C*-glucosylapigenin)	6-C-glc	73.4	70.7	79.0	70.7	81.3	61.6	
Orientin (8-*C*-glucosyl-luteolin)	8-C-glc	74.6	75.0	79.5	71.0	81.4	60.2	
Isoorientin (6-*C*-glucosyl-luteolin)	6-C-glc	72.9	70.4	78.8	70.1	81.3	61.3	
7,3′,4′-Tri-*O*-methylisoorientin	6-C-glc	72.6	70.8	78.9	70.0	81.0	61.7	Chari *et al.* (1980)
8-*C*-Galactosylapigenin	8-C-gal	73.9	68.5	75.4	69.1	80.5	61.3	Chari *et al.* (1980)
8-*C*-(6″-*O*-acetylgalactosyl)apigenin	8-C-gal	73.8	68.2	75.0	69.4	76.7	64.8	Chari *et al.* (1980)
Molludistin (7-*O*-methyl-8-*C*-arabinosyl-apigenin)	8-C-ara	74.1	67.8	74.9	68.9	71.0	–	Besson *et al.* (1979)
Saponarin (7-*O*-β-glucosylisovitexin)	6-C-glc	73.0	70.5	79.2	70.1	81.0	61.1	Wagner *et al.* (1979)
	7-O-glc	101.7	73.8	76.2	71.0	77.3	61.1	
Schaftoside (6-*C*-glucosyl-8-*C*-arabinosylapigenin)	6-C-glc	73.7	71.1	78.7	70.2	81.5	61.0	Chari, unpublished
	8-C-ara	74.6	68.9	75.1	69.2	71.1	–	
6-*C*-Arabinosyl-8-*C*-galactosylapigenin	6-C-ara	74.4	69.1	74.4	69.1	70.3	–	Chari, unpublished
	8-C-gal	74.4	69.1	75.4	69.1	80.1	61.3	
6-*C*-Galactosyl-8-*C*-arabinosylapigenin	6-C-gal	73.5	69.4	74.8	68.3	79.0	60.7	Chari, unpublished
	8-C-ara	74.4	68.2	74.5	68.9	71.0	–	
6,8-Di-*C*-glucosylchrysoeriol	6-C-glc	73.4	70.8	78.1	69.6	80.8	60.5	Mues *et al.* (1980)
	8-C-glc	74.0	71.1	78.3	69.8	81.1	60.5	
Vicenin-1 (6-*C*-xylosyl-8-*C*-glucosylapigenin)	6-C-xyl	74.6	70.3	78.5	70.0	70.0	–	Chopin, unpublished
	8-C-glc	72.1	70.3	78.6	70.0	81.6	61.2	
Violanthin (6-*C*-glucosyl-8-*C*-rhamnosylapigenin)	6-C-glc	73.5	70.6	79.2	70.2	81.4	61.6	Chari, unpublished
	8-C-α-rha	77.3	75.0	75.0	72.2	72.2	18.1	
Linoside B (2″-*O*-α-L-rhamnosyl 7,3′,4′-tri-*O*-methylisoorientin)	6-C-glc	71.6	75.4	79.4	70.2	80.5	61.5	Chari, unpublished
	2″-O-α-rha	101.1	70.2	70.7	70.7	67.6	17.0	
Linoside A (6″-*O*-acetyl-linoside B)	6-C-glc	71.6	75.5	79.3	70.3	77.5	63.8	Chari, unpublished
	2″-O-α-rha	100.3	70.3	70.8	70.3	67.8	17.2	
2″-*O*-α-L-rhamnosylvitexin	8-C-glc	71.4	75.0	79.8	70.3	81.7	61.0	Chari, unpublished
	2″-O-α-rha	100.2	70.3	71.4	70.3	68.1	17.6	

Compound	Sugar							Reference
4‴-*O*-acetyl-2″-*O*-α-L-rhamnosylvitexin	8-C-glc	71.6	74.0	79.8	70.1	81.8	61.0	Chari, unpublished
	2″-O-α-rha	99.2	70.5	67.6	73.5	65.3	17.0	
Adonivernith (2″-*O*-β-D-xylosylvitexin)	8-C-glc	71.6	80.8	78.5	70.5	82.0	61.4	Chari, unpublished
	2″-O-β-xyl	102.4	73.7	75.9	69.4	65.5	–	
Spinosin (2″-*O*-β-D-glucosyl-7-*O*-methylisovitexin)	6-C-glu	70.7	80.3	78.2	70.3	81.0	61.3	Woo *et al.* (1979)
	2″-O-β-glc	104.6	74.3	75.9	69.6	76.1	60.5	
6‴-*O*-Feruloylspinosin	6-C-glc	71.2	81.4	78.8	70.7	81.4	61.9	Woo *et al.* (1980)
	2″-O-β-glc	105.1	74.6	76.5	69.4	73.6	62.6	
6″-*O*-Arabinosyliso-orientin	6-C-glc	72.4	70.6	78.7	70.3	79.7	68.7	Hostettmann and Guillarmod (1976)
	6″-O-ara	103.7	70.3	73.1	67.2	64.7	–	
2″-*O*-β-glucosyl-7-*O*-β-galactosylisovitexin	6-C-glc	71.4	81.0	78.4	70.3	81.0	60.9	Wagner *et al.* (1979)
	2″-O-β-glc	104.8	74.8	76.1	70.3	76.6	60.9	
	7-O-β-gal	102.0	71.4	73.2	68.5	76.0	61.3	
2″-*O*-α-rhamnosyl-7-*O*-β-galactosyl-isovitexin	6-C-glc	71.9	79.7	78.8	70.5	80.8	60.4	Wagner *et al.* (1979)
	2″-O-α-rha	100.3	70.5	70.8	71.9	68.1	17.3	
	7-O-β-gal	102.6	71.4	72.9	68.1	75.9	60.4	

* Solvent: DMSO-d_6.

The assignments in this table represent those determined by Chari, unpublished. Previous assignments were either not made or were incorrect.

(Österdahl, 1978b; Theodor *et al.*, 1980) exhibit 12. When the two glycosyl moieties are different as in schaftoside (6-*C*-glucosyl-8-*C*-arabinosylapigenin) it is not possible by ^{13}C-NMR to determine the position of linkage of each sugar residue. However, this can often be done by mass spectrometry of the permethyl ether derivative (see Chapter 8).

2.4.3 *O*-Glycosides of *C*-glycosylflavonoids

Glycosylation of the existing *C*-glycosyl moiety in naturally occurring flavonoid *C*-glycosides has been reported so far only at the C-2 and C-6 positions (see Table 2.18). The resulting downfield shift of the C-2 resonance is ca. 10 p.p.m. where the second sugar is β-glucose (e.g. spinosin) or β-xylose (e.g. adonivernith), and the corresponding shift for α-rhamnosylation is ca. 5.0 p.p.m. (e.g. linoside A and B). However, in the presence of additional 7-*O*-glycosylation, the rhamnosylation shift is about 9 p.p.m. (see Table 2.18; Wagner *et al.*, 1979). Glycosylation at C-2 also results in an upfield shift of ca. 2.0 p.p.m. in the C-1 resonance of the *C*-glycosyl moiety. This parallels the effects observed with the corresponding flavonoid (2-*O*-glycosyl-*O*-glycosides) (see Section 2.3.3). Interestingly the spectrum of 2″-*O*-β-xylosylvitexin shows the presence of more than one conformer in solution at room temperature but that of 2″-*O*-α-rhamnosylvitexin does not.

Arabinosylation of the primary hydroxyl at C-6 results in a 7.5 p.p.m. downfield shift of the signal of the glycosylated carbon (Table 2.18) and it is probable that the effects of glycosylation at C-6 by other sugars would parallel those observed with *O*-glycosides (see Section 2.3.3). The identity of the C-6 signal can be readily established by its distinctive proton-coupling.

Similar solvent effects are encountered with *C*-glycosides as for *O*-glycosides. Examination of the chemical shift values in Table 2.18 reveals variations ($\pm$0.8 p.p.m.) in chemical shifts for corresponding carbon atoms in the same glycosyl moieties in different compounds. This is presumably due to the varying moisture content of the sample or the DMSO-d_6 solvents used by different workers.

2.4.4 Acylated *C*-glycosides

^{13}C-NMR spectroscopy is the easiest and only general method for the location of an acyl group when it is linked to the *C*-glycosyl moiety. The effects observed are very similar to those previously discussed for acylated *O*-glycosides. For example, in the case of linoside A (6″-*O*-acetyl-2″-*O*-α-L-rhamnosyl-7,3′,4′-tri-*O*-methyliso-orientin; see Table 2.18) the C-5″ signal of the *C*-glucosyl unit is shifted upfield by 3.0 p.p.m. to 77.5 p.p.m. compared with the C-5″ resonance in linoside B, the deacetylated product (Chari *et al.*, 1981). This established that

C-6″ bears the acetyl group in linoside A. Analogously, the position of the acetyl group in 6″-*O*-acetyl-8-β-D-galactosylapigenin has been established by ^{13}C-NMR spectroscopy (Chari *et al.*, 1980). The location of the feruloyl group on the C-6‴ of the *O*-glucosyl (rather than the *C*-glucosyl) moiety in 6‴-*O*-feruloylspinosin is clearly demonstrated by the appearance of the C-5″ resonance of the *C*-glucose moiety at 81.4 p.p.m., which is very similar to that of C-5″ in spinosin itself (Woo *et al.*, 1979, 1980).

2.5 C-13 NMR SPECTRA OF FLAVONOIDS

Tom J. Mabry, Ken R. Markham and V. Mohan Chari

Spectra 1–125 follow the index of oxidation patterns commencing on the next page.

Index of ^{13}C-NMR spectra of flavonoids

Spectrum no.	Flavonoid	Oxidation pattern 2	3	4	5	6	7	8	2′	3′	4′	5	6
	Flavones												
1	5-Hydroxyflavone				OH								
2	Chrysin				OH		OH						
3	γ,γ-Dimethyl-5-hydroxychromenoflavone				OH	prenyl	OR						
4	5,4′-Dihydroxyflavone				OH						OH		
5	3′,4′-Dihydroxyflavone									OH	OH		
6	3′,4′-Dimethoxyflavone									OMe	OMe		
7	5-Hydroxy-3′,4′-dimethoxyflavone				OH					OMe	OMe		
8	Apigenin				OH		OH				OH		
9	Apigenin 7-*O*-β-D-glucoside				OH		O-Glc				OH		
10	Apigenin 7-*O*-neohesperidoside				OH		O-Neohesp				OH		
11	Apigenin 7-*O*-apiosylglucoside				OH		O-Api-Glc				OH		
12	Vitexin				OH		OH	C-Glc			OH		
13	2″-*O*-Rhamnosylvitexin				OH		OH	C-Neohesp			OH		
14	Saponarin				OH	C-Glc	O-Glc				OH		
15	Schaftoside				OH	C-Glc	OH	C-Ara			OH		
16	Violanthin				OH	C-Glc	OH	C-Rha			OH		
17	Thevetiaflavone				OMe		OH				OH		
18	Genkwanin				OH		OMe				OH		
19	Spinosin				OH	C-Soph	OMe				OH		
20	Acacetin				OH		OH				OMe		
21	Apigenin trimethyl ether				OMe		OMe				OMe		
22	Scutellarein 7-*O*-glucoside				OH	OH	O-Glc				OH		
23	Pectolinarigenin				OH	OMe	OH				OMe		
24	Salvigenin				OH	OMe	OMe				OMe		
25	Isoscutellarein 4′-methyl ether 7-*O*-[2″-*O*-(6‴-acetyl)-β-allopyranosyl]-β-gluconyranoside				OH		O-(6‴-acetyl)-OH Allo (1-2)Glc				OMe		
26	Luteolin				OH		OH			OH	OH		

27 Luteolin 5-*O*-glucoside		O-Glc		OH			OH	OH	
28 Luteolin 7-*O*-glucoside		OH		O-Glc			OH	OH	
29 Luteolin 3′-*O*-glucoside		OH		OH			O-Glc	OH	
30 Isoorientin		OH	C-Glc	OH			OH	OH	
31 Adonivernith		OH		OH	C-Xyl(1-2)Glc		OH	OH	
32 Diosmetin		OH		OH			OH	OMe	
33 Chrysoeriol		OH		OH			OMe	OH	
34 Chrysoeriol 7-*O*-(2″-*β*-allopyranosyl)-*β*-glucopyranoside		OH		O-Allo(1-2) Glc			OMe	OH	
35 Hydnocarpin		OH		OH			OR	OR	
36 6-Hydroxyluteolin 6,7,3′-trimethyl ether		OH	OMe	OMe			OMe	OH	
37 6-Hydroxyluteolin pentamethyl ether		OMe	OMe	OMe			OMe	OMe	
38 Tricetin		OH		OH			OH	OH	OH
39 Isoaffinetin		OH	C-Glc	OH			OH	OH	OH
40 Tricetin 3′,4′,5′-trimethyl ether		OH		OH			OMe	OMe	OMe
41 Wightin		OH		OMe	OMe	OMe	OH		
Flavonols									
42 Galangin	OH	OH		OH					
43 3′-4′-Dihydroxyflavonol	OH						OH	OH	
44 3′,4′-Dimethoxyflavonol	OH						OMe	OMe	
45 Kaempferol	OH	OH		OH				OH	
46 Kaempferol 3-*O*-glucoside	O-Glc	OH		OH				OH	
47 Tiliroside	O-(6″-*p*-coumaroyl)Glc	OH		OH				OH	
48 Kaempferol 3-*O*-sophoroside	O-Soph	OH		OH				OH	
49 Kaempferol 3-*O*-rutinoside	O-Rut	OH		OH				OH	
50 Kaempferol 7-*O*-glucoside	OH	OH		O-Glc				OH	
51 Kaempferol 7-*O*-neohesperidoside	OH	OH		O-Neohesp				OH	
52 Kaempferol 3,7-di-*O*-glucoside	O-Glc	OH		O-Glc				OH	
53 Kaempferol 3-*O*-(6‴-acetyl) glucoside 7-*O*-glucoside	O-(6‴-acetyl) Glc	OH		O-Glc				OH	

Index of ^{13}C-NMR spectra of flavonoids (*Contd.*)

Spectrum no.	*Flavonoid*	*Oxidation pattern* 2	3	4	5	6	7	8	2′	3′	4′	5	6
54	Kaempferol 3-*O*-sophoroside 7-*O*-glucoside		O-Soph		OH		O-Glc				OH		
55	Kaempferol 3-*O*-(2‴-sinapyl) sophoroside 7-*O*-glucoside		2‴-sinapylsoph		OH		O-Glc				OH		
56	Kaempferol 3-*O*-glucoside 7-*O*-rhamnoside		O-Glc		OH		O-Rha				OH		
57	Kaempferol 3-*O*-(6″′-acetyl) glucoside 7-*O*-rhamnoside		O-(6″′-acetyl)Glc		OH		O-Rha				OH		
58	Kaempferol 3-*O*-rutinoside 7-*O*-rhamnoside		O-Rut		OH		O-Rha				OH		
59	Kaempferol 4′-methyl ether		OH		OH		OH				OMe		
60	Quercetin		OH		OH		OH			OH	OH		
61	Quercetin 3-*O*-glucoside		O-Glc		OH		OH			OH	OH		
62	Quercetin 3-*O*-rutinoside		O-Rut		OH		OH			OH	OH		
63	Quercetin 3-*O*-galactoside		O-Gal		OH		OH			OH	OH		
64	Quercetin 3-*O*-(6″-galloyl) galactoside		O-(6″-galloyl)Glc		OH		OH			OH	OH		
65	Quercetin 3-*O*-rhamnoside		O-Rha		OH		OH			OH	OH		
66	Quercetin 3-*O*-arabinopyranoside		O-Ara		OH		OH			OH	OH		
67	Quercetin 3-*O*-arabinofuranoside		O-Ara		OH		OH			OH	OH		
68	Quercetin 7-*O*-glucoside		OH		OH		O-Glc			OH	OH		
69	Quercetin 4′-*O*-glucoside		OH		OH		OH			OH	O-Glc		
70	Azaleatin		OH		OMe		OH			OH	OH		
71	Rhamnetin		OH		OH		OMe			OH	OH		
72	Isorhamnetin		OH		OH		OH			OMe	OH		
73	Isorhamnetin 3-*O*-rutinoside		O-Rut		OH		OH			OMe	OH		
74	Tamarixetin 7-*O*-rutinoside		OH		OH		O-Rut			OH	OMe		
75	Quercetin 3,4′-dimethyl ether		OMe		OH		OH			OH	OMe		
76	Quercetin 7,4′-dimethyl ether		OH		OH		OMe			OH	OMe		
77	7,8,3′4′-Tetrahydroxyflavonol		OH				OH	OH		OH	OH		

78	Morin		OH		OH		OH	OH		OH	
79	Patuletin		OH		OH	OMe	OH		OH	OH	
80	Myricetin 3-*O*-galactoside		O-Gal		OH		OH		OH	OH	OH
81	Myricetin 3-*O*-rhamnoside		O-Rha		OH		OH		OH	OH	OH
	Isoflavones										
82	5,7-Dihydroxyisoflavone				OH		OH				
83	Formononetin						OH			OMe	
84	Genistein				OH		OH			OH	
85	Wighteone				OH	prenyl	OH			OH	
86	Pseudobaptisin						O-Rut		O	CH_2-O-	
87	2′-Hydroxyformononetin						OH	OH		OMe	
88	Luteone				OH	prenyl	OH	OH		OH	
89	2′-Hydroxygenistein				OH		OH	OH		OH	
90	Licoisoflavone A				OH		OH	OH	prenyl	OH	
91	Licoisoflavone B				OH		OH	OH	prenyl	OR	
92	Angustone A				OH	prenyl	OH	OH	prenyl	OH	
93	Angustone B				OH	prenyl	OH	OH	prenyl	OR	
94	Angustone C				OH	prenyl	OR	OH	prenyl	OH	
	Chalcones										
95	2,2′-Dihydroxychalcone	OH						OH			
96	Isoliquiritigenin			OH				OH		OH	
	Aurones										
97	Sulfuretin					OH			OH	OH	
	Dihydroflavones										
98	5,7-Dimethyl-4′-methoxyflavanone				Me		Me			OMe	
99	Pinocembrin				OH		OH				
100	Naringenin				OH		OH			OH	
101	Naringenin 7-*O*-(6″-*p*-coumaroyl)glucoside				OH		O-(6″-*p*-coumaroyl) Glc			OH	
102	Naringin				OH		O-Neohesp			OH	
103	Naringenin 4′-methyl ether				OH		OH			OMe	
104	Eriodictyol				OH		OH		OH	OH	
105	Homoeriodictyol				OH		OH		OMe	OH	
106	Hesperetin				OH		OH		OH	OMe	

Index of ^{13}C-NMR spectra of flavonoids (*Contd.*)

Spec-trum no.	*Flavonoid*	*Oxidation pattern* 2	3	4	5	6	7	8	2′	3′	4′	5	6
107	Hesperetin 7-*O*-rutinoside (Hesperidin)				OH		O-Rut			OMe	OH		
	Dihydroflavonols												
108	(+)-Dihydroquercetin (Taxifolin)		OH		OH		OH			OH	OH		
109	(+)-Dihydroquercetin 5,7,3′4′-tetramethyl ether		OH		OMe		OMe			OMe	OMe		
	Dihydroisoflavone												
110	Sophoraisoflavanone A				OH		OH		OMe	prenyl	OH		
	Biflavones												
111	Robustaflavone				$(OH)_2$	(6→3′)	$(OH)_2$			(3′→6)	$(OH)_2$		
112	Cupressuflavone				$(OH)_2$		$(OH)_2$	(8→8)			$(OH)_2$		
113	Cupressuflavone hexamethyl ether				$(OMe)_2$		$(OMe)_2$	(8→8)			$(OMe)_2$		
114	Volkensiflavone		(3→8)		$(OH)_2$		$(OH)_2$	(8→3)			$(OH)_2$		
	Biflavanones												
115	Succedaneaflavanone				$(OH)_2$	(6→6)	$(OH)_2$				$(OH)_2$		
	Catechin/tannins												
116	(−)-Epicatechin		OH		OH		OH			OH	OH		
117	Procyanidin B5		$(OH)_2$	(4→6)	$(OH)_2$	(6→4)	$(OH)_2$			$(OH)_2$	$(OH)_2$		
118	Procyanidin B2		$(OH)_2$	(4→8)	$(OH)_2$		$(OH)_2$	(8→4)		$(OH)_2$	$(OH)_2$		
119	Procyanidin trimer		$(OH)_3$	$(4→8)_2$	$(OH)_3$		$(OH)_3$	$(8→4)_2$		$(OH)_3$	$(OH)_3$		
120	Procyanidin polymer		$(OH)_n$	$(4→8)_{n-1}$	$(OH)_n$		$(OH)_n$	$(8→4)_{n-1}$		$(OH)_n$	$(OH)_n$		
	Miscellaneous												
121	Protogenkwanin 4′-*O*-glucoside*												
122	Cassiachromone*												
123	Aloesin*												
124	Silydianin*												
125	Silymonin (3-Deoxysilydianin)*												

* See spectra pages for formulae.

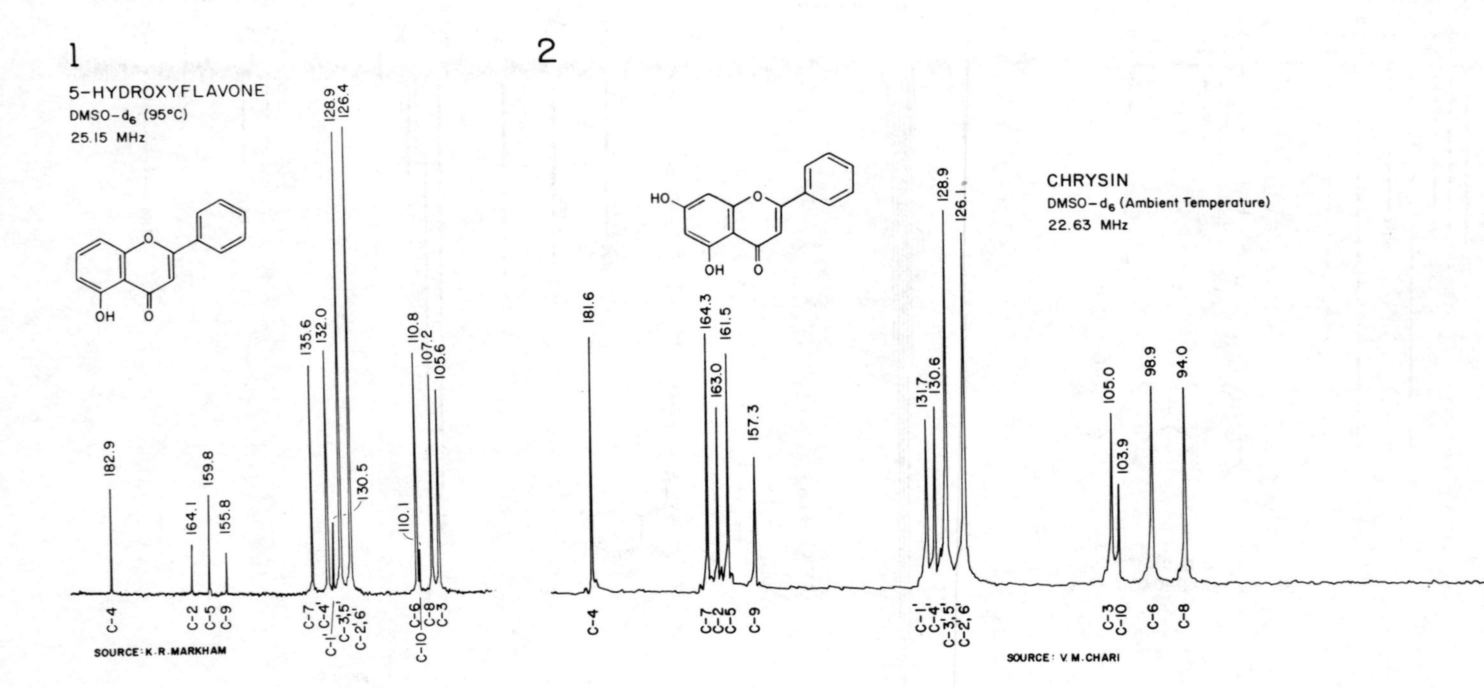

1
5-HYDROXYFLAVONE
DMSO-d6 (95°C)
25.15 MHz
OH
O
182.9
164.1
159.8
155.8
135.6
132.0
128.9
126.4
130.5
110.8
107.2
105.6
110.1
C-4
C-2
C-5
C-9
C-7
C-1'
C-4'
C-3',5'
C-2',6'
C-10
C-6
C-8
C-3
SOURCE: K.R. MARKHAM
2
CHRYSIN
DMSO-d6 (Ambient Temperature)
22.63 MHz
HO
OH
O
181.6
164.3
163.0
161.5
157.3
131.7
130.6
128.9
126.1
105.0
103.9
98.9
94.0
C-4
C-7
C-2
C-5
C-9
C-1'
C-4'
C-3',5'
C-2',6'
C-3
C-10
C-6
C-8
SOURCE: V.M. CHARI

3

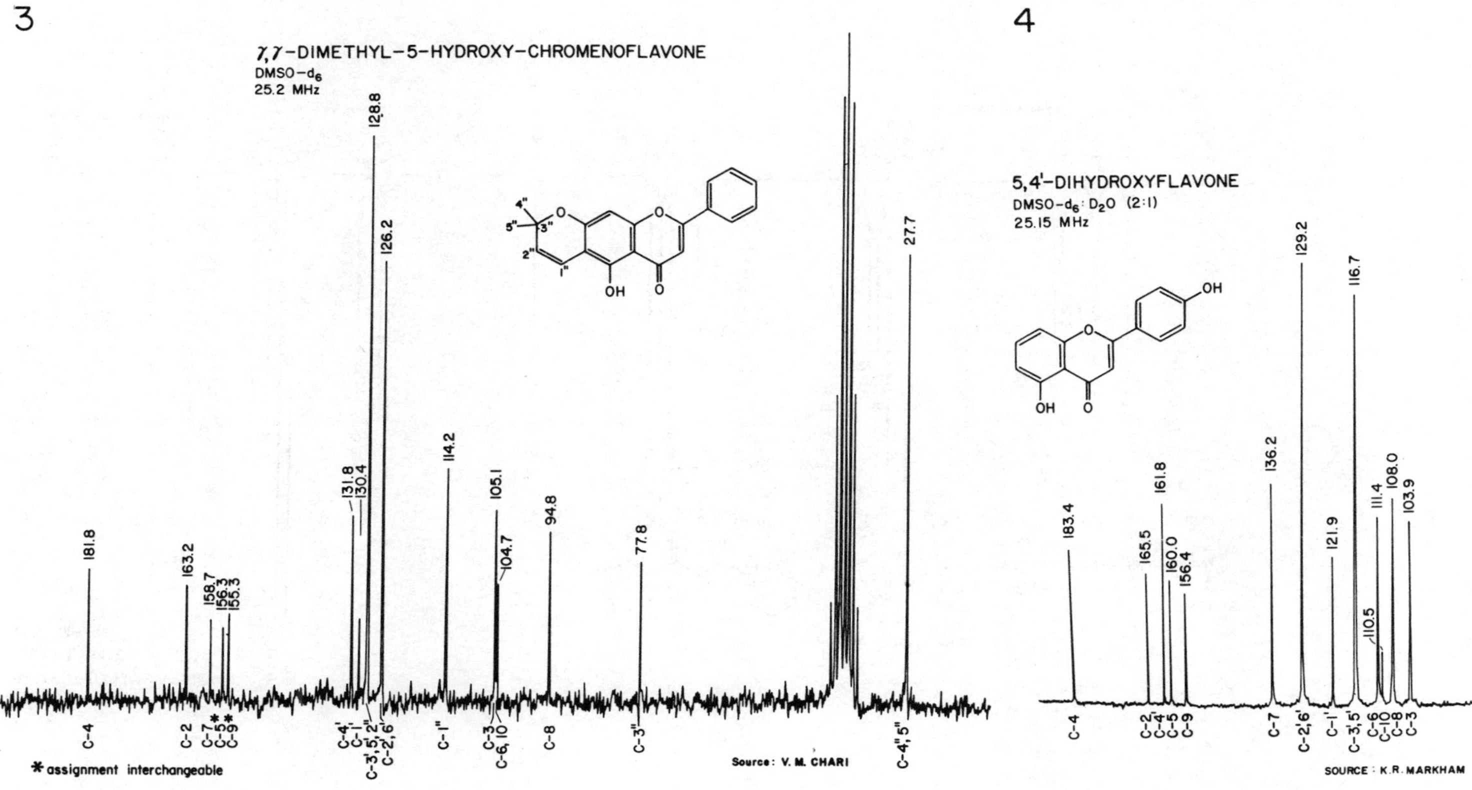
7,7-DIMETHYL-5-HYDROXY-CHROMENOFLAVONE
DMSO-d_6
25.2 MHz
181.8
163.2
158.7
156.3
155.3
131.8
130.4
128.8
126.2
114.2
105.1
104.7
94.8
77.8
27.7
C-4
C-2
C-7*
C-5*
C-9*
C-4'
C-1'
C-3',5',2"
C-2',6'
C-1"
C-3
C-6,10
C-8
C-3"
C-4",5"
* assignment interchangeable
Source: V. M. CHARI
4
5,4'-DIHYDROXYFLAVONE
DMSO-d_6 : D_2O (2:1)
25.15 MHz
OH
183.4
165.5
161.8
160.0
156.4
136.2
129.2
121.9
116.7
111.4
110.5
108.0
103.9
C-4
C-2
C-4'
C-5
C-9
C-7
C-2',6'
C-1'
C-3',5'
C-6
C-10
C-8
C-3
SOURCE: K.R. MARKHAM

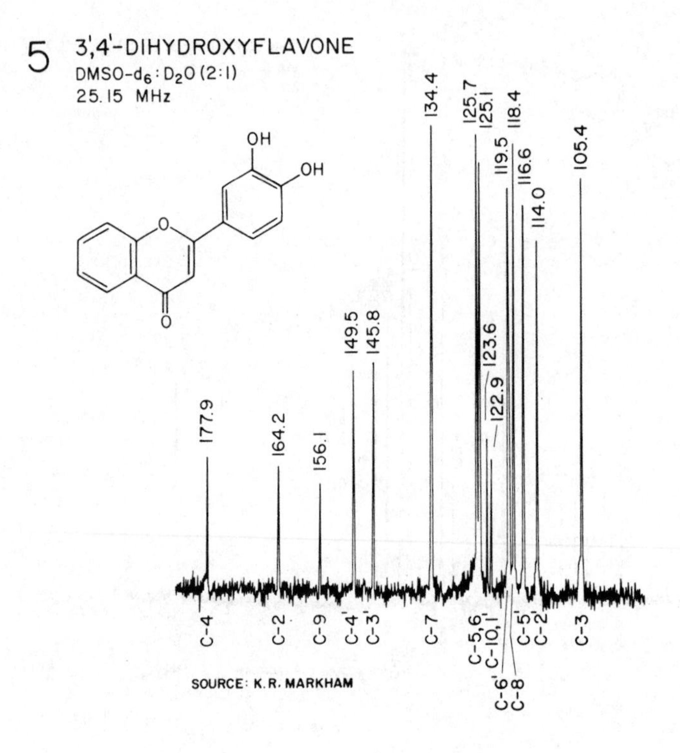

5
3',4'-DIHYDROXYFLAVONE
DMSO-d6 : D2O (2:1)
25.15 MHz
OH
OH
O
O
177.9
164.2
156.1
149.5
145.8
134.4
125.7
125.1
123.6
122.9
119.5
118.4
116.6
114.0
105.4
C-4
C-2
C-9
C-4'
C-3'
C-7
C-5,6
C-10,1'
C-6'
C-8
C-5'
C-2'
C-3
SOURCE: K.R. MARKHAM

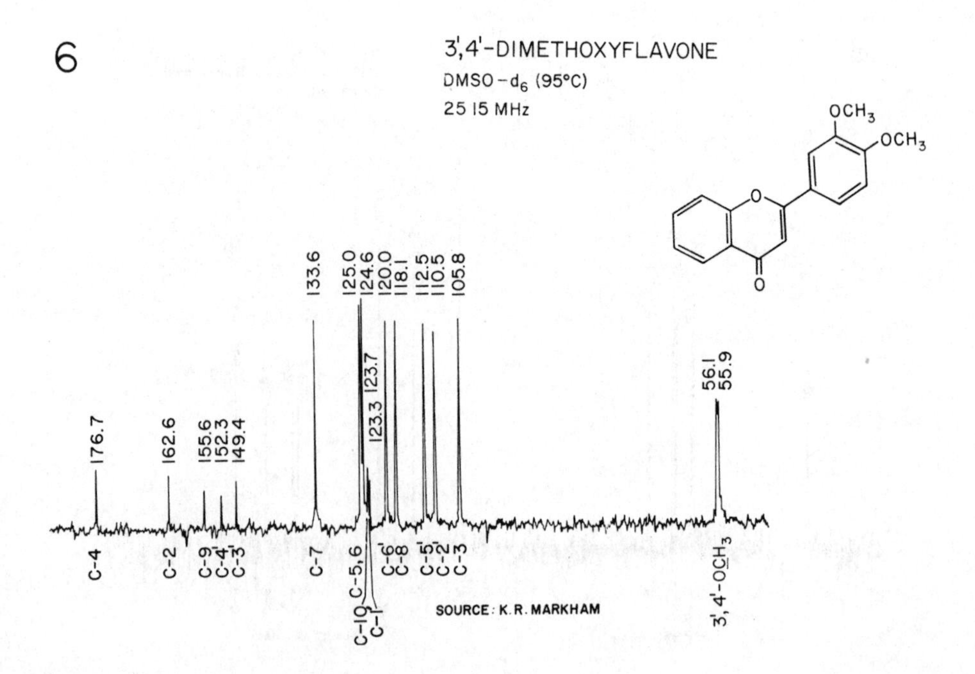

6
3',4'-DIMETHOXYFLAVONE
DMSO-d6 (95°C)
25.15 MHz
OCH3
OCH3
O
O
176.7
162.6
155.6
152.3
149.4
133.6
125.0
124.6
123.7
123.3
120.0
118.1
112.5
110.5
105.8
56.1
55.9
C-4
C-2
C-9
C-4'
C-3'
C-7
C-5,6
C-10
C-1'
C-6'
C-8
C-5'
C-2'
C-3
3',4'-OCH3
SOURCE: K.R. MARKHAM

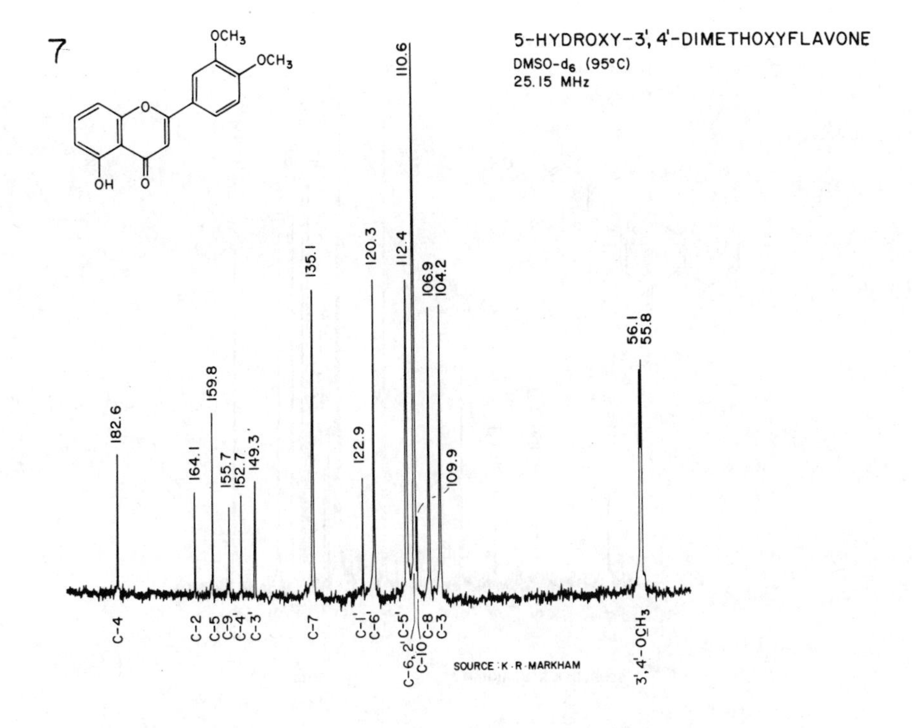
7
5-HYDROXY-3', 4'-DIMETHOXYFLAVONE
DMSO-d6 (95°C)
25.15 MHz
182.6
164.1
159.8
155.7
152.7
149.3
135.1
122.9
120.3
112.4
110.6
109.9
106.9
104.2
56.1
55.8
C-4
C-2
C-5
C-9
C-4'
C-3'
C-7
C-1'
C-6'
C-5'
C-6, 2'
C-10
C-8
C-3
3', 4'-OCH3
SOURCE : K · R · MARKHAM

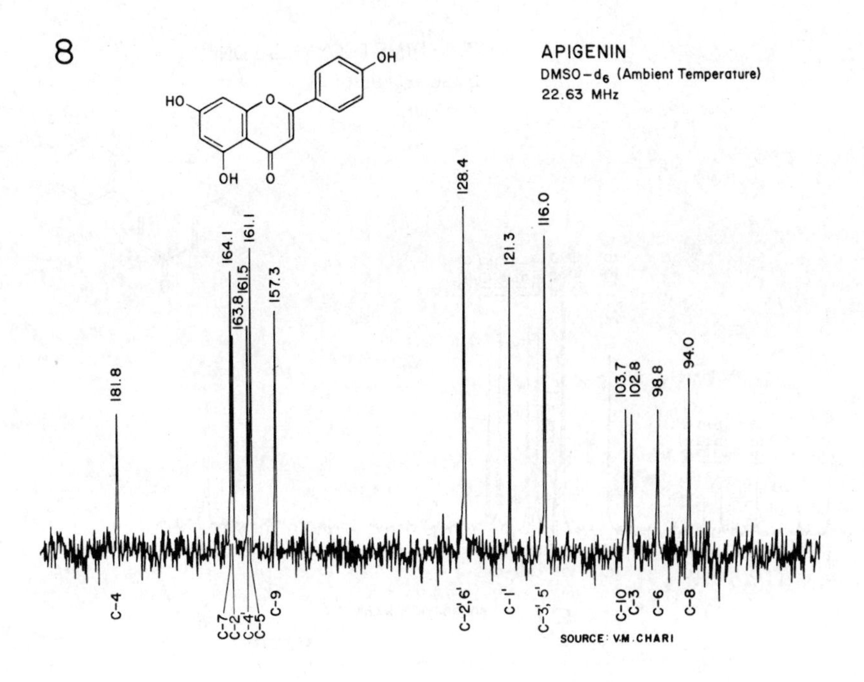
8
APIGENIN
DMSO-d6 (Ambient Temperature)
22.63 MHz
181.8
164.1
163.8
161.5
161.1
157.3
128.4
121.3
116.0
103.7
102.8
98.8
94.0
C-4
C-7
C-2
C-4'
C-5
C-9
C-2', 6'
C-1'
C-3', 5'
C-10
C-3
C-6
C-8
SOURCE: V.M. CHARI

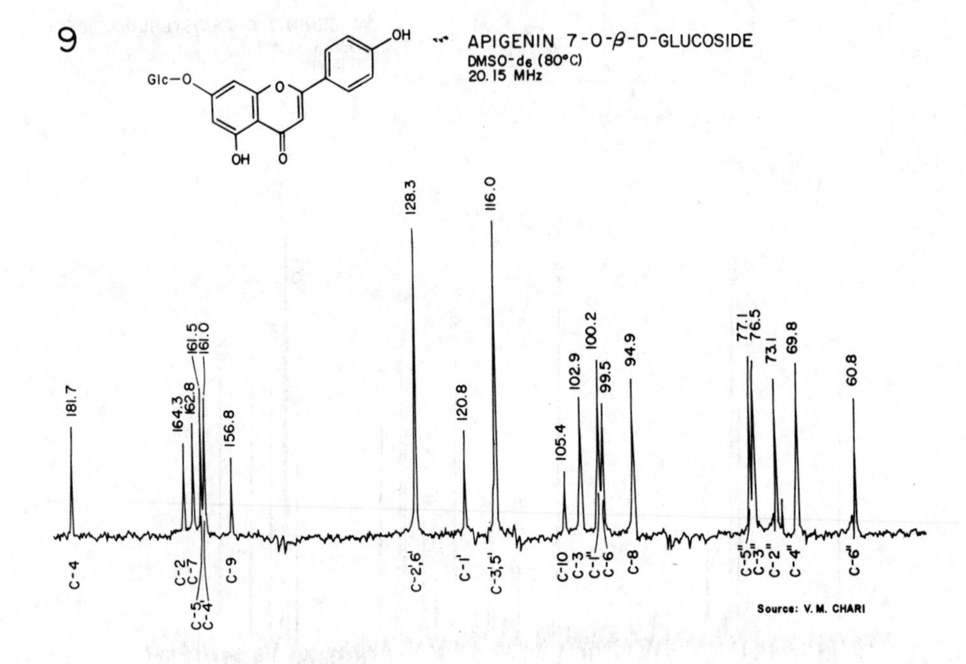
9
APIGENIN 7-O-β-D-GLUCOSIDE
DMSO-d6 (80°C)
20.15 MHz
Glc—O
OH
O
181.7
164.3
162.8
161.5
161.0
156.8
128.3
120.8
116.0
105.4
102.9
100.2
99.5
94.9
77.1
76.5
73.1
69.8
60.8
C-4
C-2
C-7
C-5
C-4'
C-9
C-2',6'
C-1'
C-3',5'
C-10
C-3
C-1''
C-6
C-8
C-5''
C-3''
C-2''
C-4''
C-6''
Source: V.M. CHARI

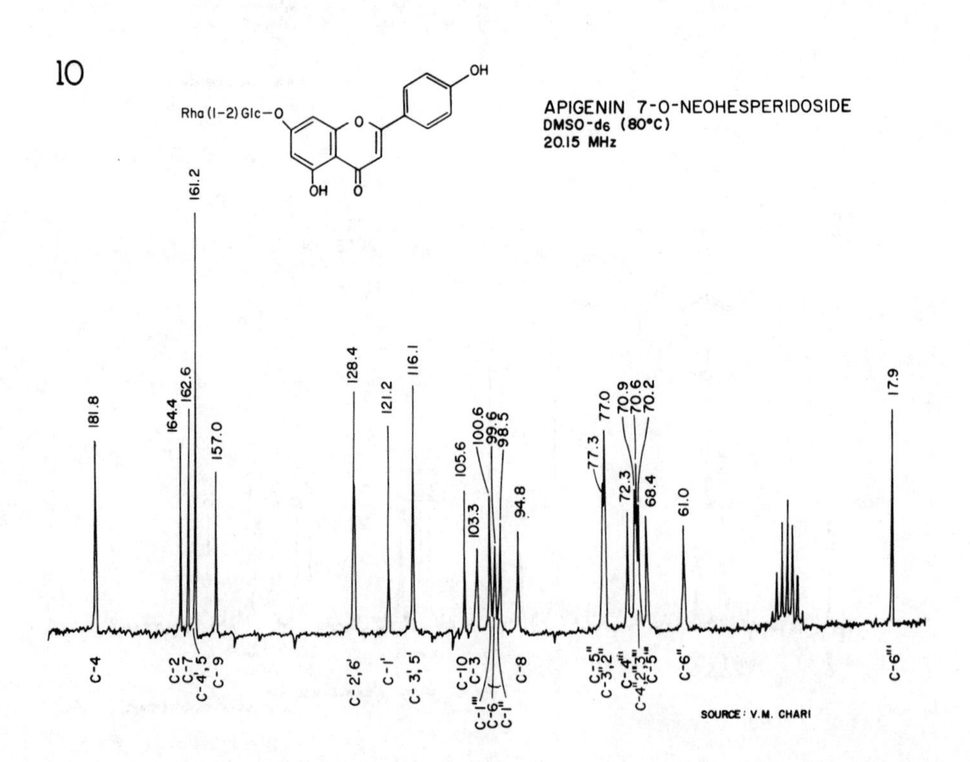
10
APIGENIN 7-O-NEOHESPERIDOSIDE
DMSO-d6 (80°C)
20.15 MHz
Rha (1-2) Glc—O
OH
O
181.8
164.4
162.6
161.2
157.0
128.4
121.2
116.1
105.6
103.3
100.6
99.6
98.5
94.8
77.3
77.0
72.3
70.9
70.6
70.2
68.4
61.0
17.9
C-4
C-2
C-7
C-4',5
C-9
C-2',6'
C-1'
C-3',5'
C-10
C-3
C-1'''
C-6
C-1''
C-8
C-5''
C-3'',2''
C-4''
C-4'''
C-2'''
C-3'''
C-5'''
C-6''
C-6'''
SOURCE: V.M. CHARI

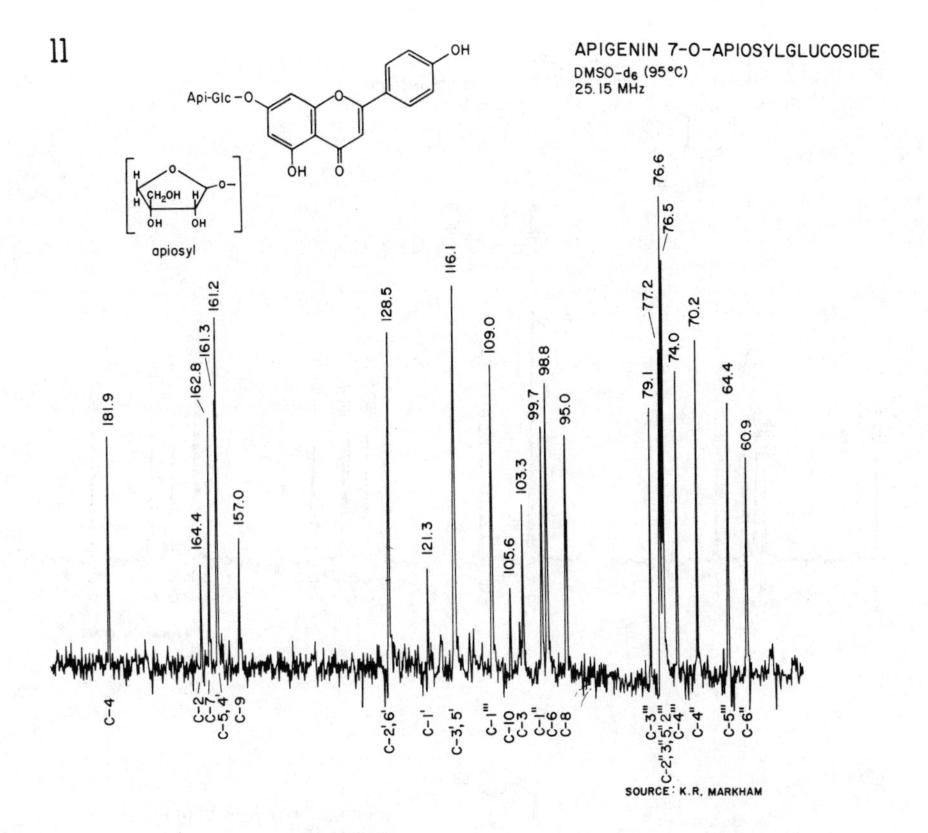
11
APIGENIN 7-O-APIOSYLGLUCOSIDE
DMSO-d_6 (95°C)
25.15 MHz
Api-Glc-O
OH
OH
O
apiosyl
181.9
164.4
162.8
161.3
161.2
157.0
128.5
121.3
116.1
109.0
105.6
103.3
99.7
98.8
95.0
79.1
77.2
76.6
76.5
74.0
70.2
64.4
60.9
C-4
C-2
C-7
C-5, 4'
C-9
C-2, 6'
C-1'
C-3', 5'
C-1'''
C-10
C-3
C-1''
C-6
C-8
C-3'''
C-2'', 3'', 5'', 2'''
C-4'''
C-4''
C-5'''
C-6''
SOURCE: K.R. MARKHAM

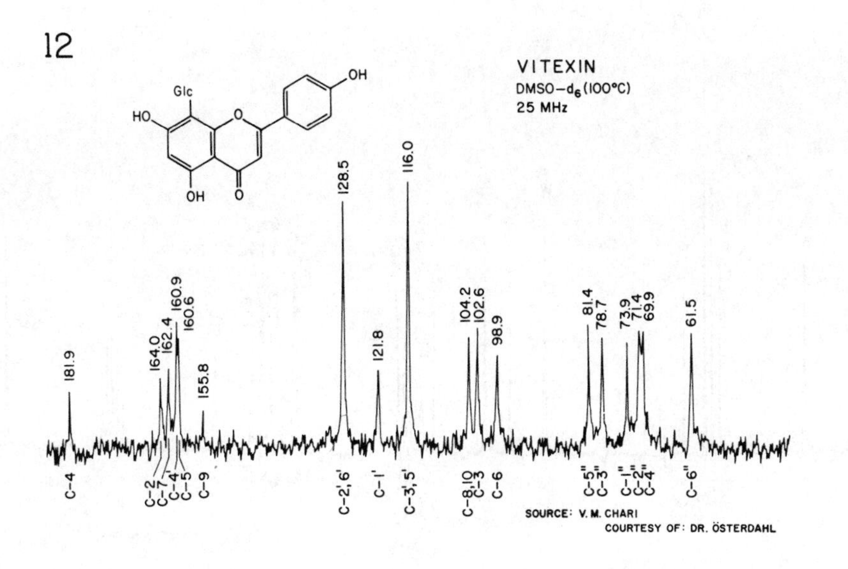
12
VITEXIN
DMSO-d_6 (100°C)
25 MHz
Glc
HO
OH
OH
O
181.9
164.0
162.4
160.9
160.6
155.8
128.5
121.8
116.0
104.2
102.6
98.9
81.4
78.7
73.9
71.4
69.9
61.5
C-4
C-2
C-7
C-4'
C-5
C-9
C-2', 6'
C-1'
C-3', 5'
C-8, 10
C-3
C-6
C-5''
C-3''
C-1''
C-2''
C-4''
C-6''
SOURCE: V. M. CHARI
COURTESY OF: DR. ÖSTERDAHL

13

2''-O-RHAMNOSYLVITEXIN

DMSO-d_6 (Ambient Temperature)
22.6 MHz

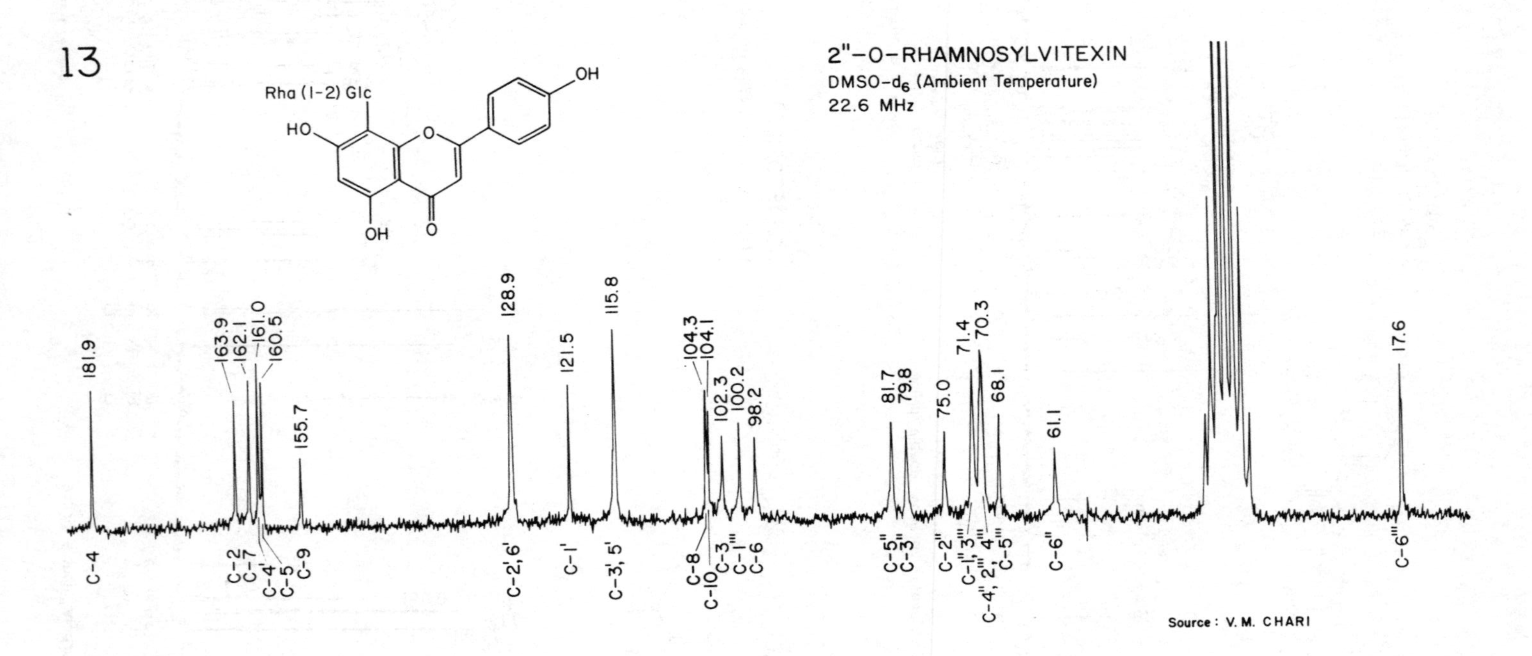

Source : V. M. CHARI

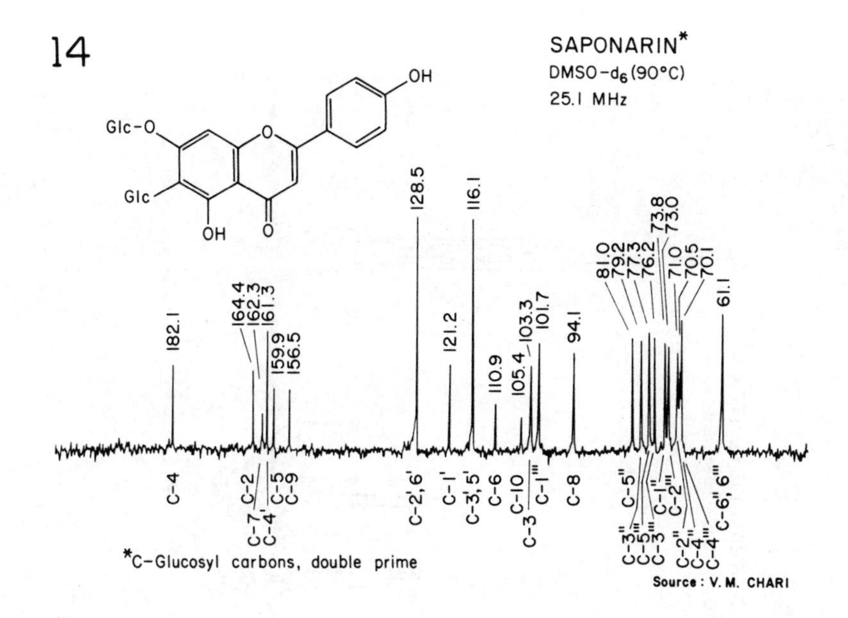
14
SAPONARIN*
DMSO-d_6 (90°C)
25.1 MHz
Glc-O
Glc
OH
O
182.1
164.4
162.3
161.3
159.9
156.5
128.5
121.2
116.1
110.9
105.4
103.3
101.7
94.1
81.0
79.2
77.3
76.2
73.8
73.0
71.0
70.5
70.1
61.1
C-4
C-2
C-7'
C-4'
C-5
C-9
C-2',6'
C-1'
C-3',5'
C-6
C-10
C-3
C-1'''
C-8
C-5''
C-3''
C-5'''
C-3'''
C-1''
C-2'''
C-2''
C-4''
C-4'''
C-6'', 6'''
*C-Glucosyl carbons, double prime
Source: V.M. CHARI

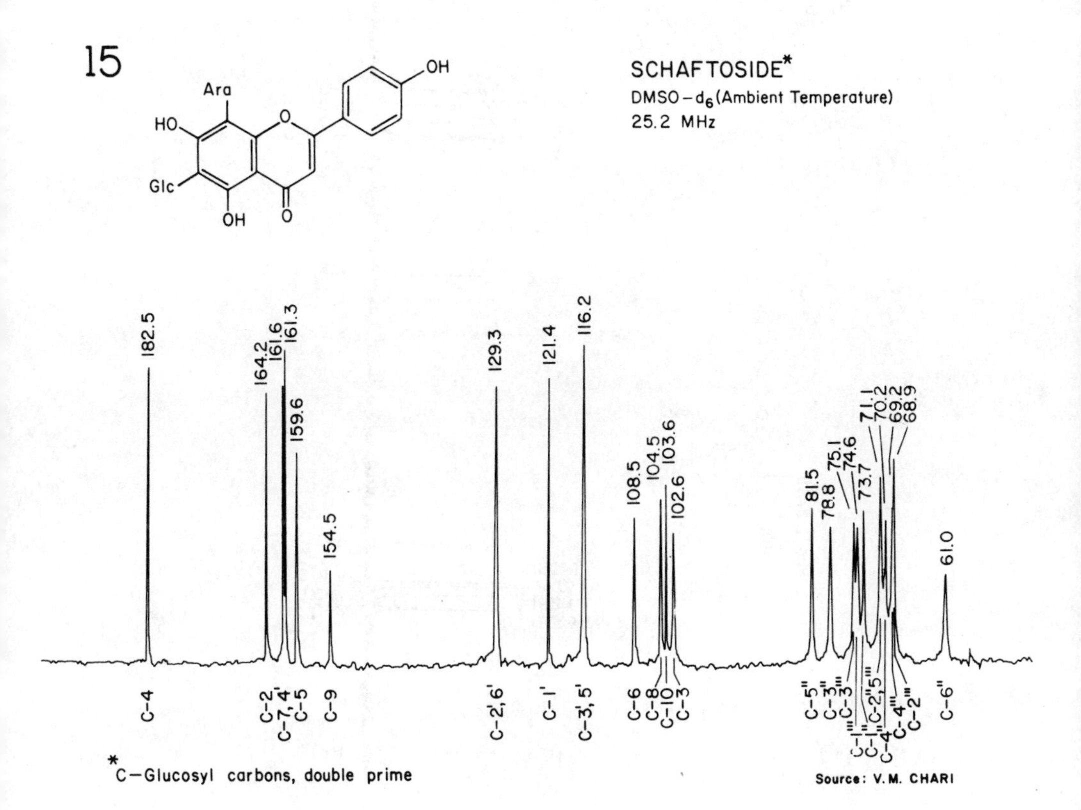
15
SCHAFTOSIDE*
DMSO-d_6 (Ambient Temperature)
25.2 MHz
Ara
HO
Glc
OH
O
OH
182.5
164.2
161.6
161.3
159.6
154.5
129.3
121.4
116.2
108.5
104.5
103.6
102.6
81.5
78.8
75.1
74.6
73.7
71.1
70.2
69.2
68.9
61.0
C-4
C-2
C-7,4'
C-5
C-9
C-2',6'
C-1'
C-3',5'
C-6
C-8
C-10
C-3
C-5''
C-3''
C-3'''
C-1'''
C-1''
C-2'',5'''
C-4''
C-4'''
C-2'''
C-6''
*C-Glucosyl carbons, double prime
Source: V.M. CHARI

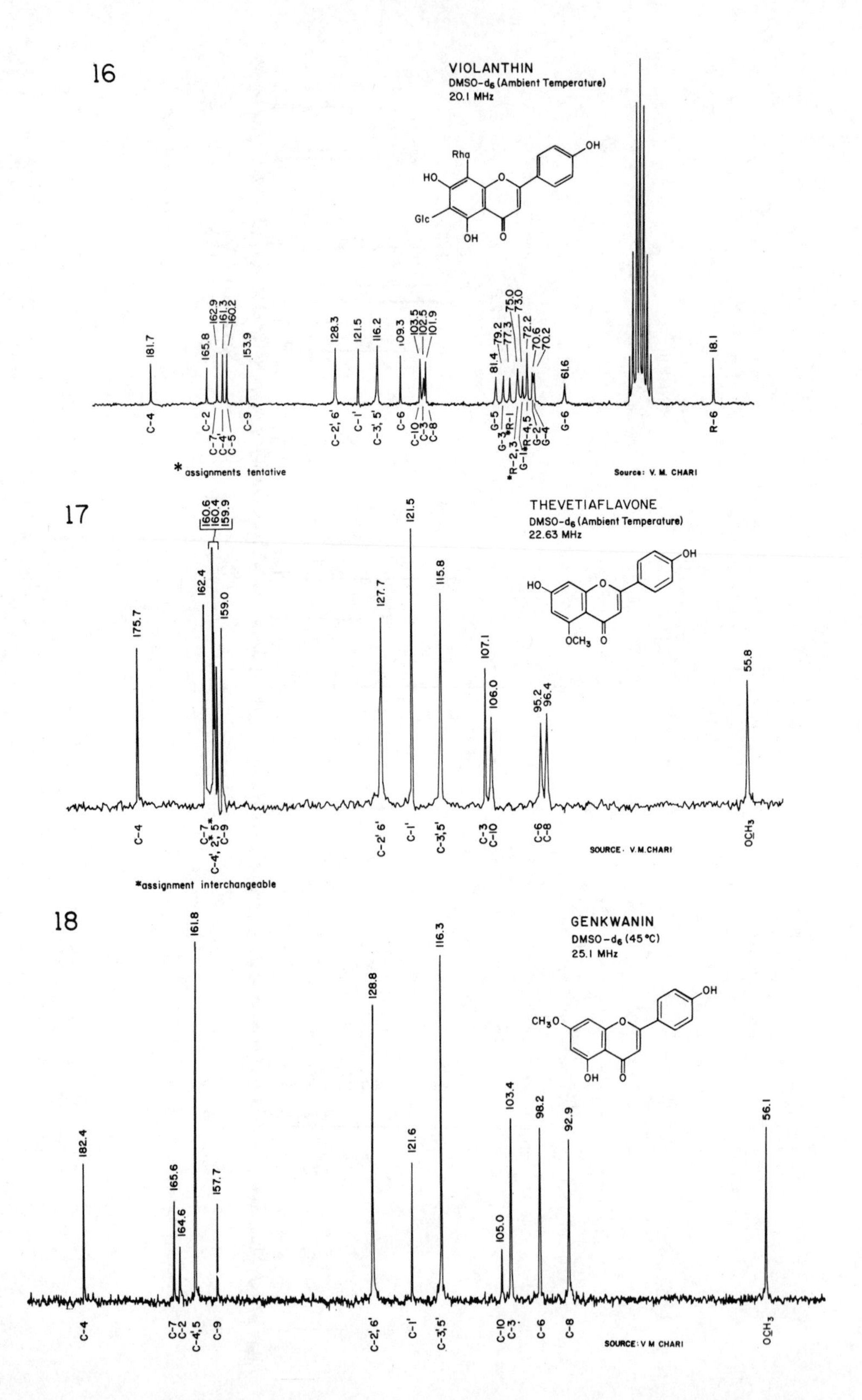
16
VIOLANTHIN
DMSO-d_6 (Ambient Temperature)
20.1 MHz
Rha
HO
Glc
OH
O
OH
181.7
165.8
162.9
161.3
160.2
153.9
128.3
121.5
116.2
109.3
103.5
102.5
101.9
81.4
79.2
77.3
75.0
73.0
72.2
70.6
70.2
61.6
18.1
C-4
C-2
C-7
C-4'
C-5
C-9
C-2', 6'
C-1'
C-3', 5'
C-6
C-10
C-3
C-8
G-5
G-3
*R-1
*R-2,3
G-1
*R-4,5
G-2
G-4
G-6
R-6
* assignments tentative
Source: V. M. CHARI
17
THEVETIAFLAVONE
DMSO-d_6 (Ambient Temperature)
22.63 MHz
HO
OH
OCH_3
O
175.7
162.4
160.6
160.4
159.9
159.0
127.7
121.5
115.8
107.1
106.0
95.2
96.4
55.8
C-4
C-7
C-4', 2* 5*
C-9
C-2' 6'
C-1'
C-3',5'
C-3
C-10
C-6
C-8
OCH_3
SOURCE · V.M.CHARI
*assignment interchangeable
18
GENKWANIN
DMSO-d_6 (45 °C)
25.1 MHz
CH_3O
OH
OH
O
182.4
165.6
164.6
161.8
157.7
128.8
121.6
116.3
105.0
103.4
98.2
92.9
56.1
C-4
C-7
C-2
C-4',5
C-9
C-2',6'
C-1'
C-3',5'
C-10
C-3
C-6
C-8
OCH_3
SOURCE: V M CHARI

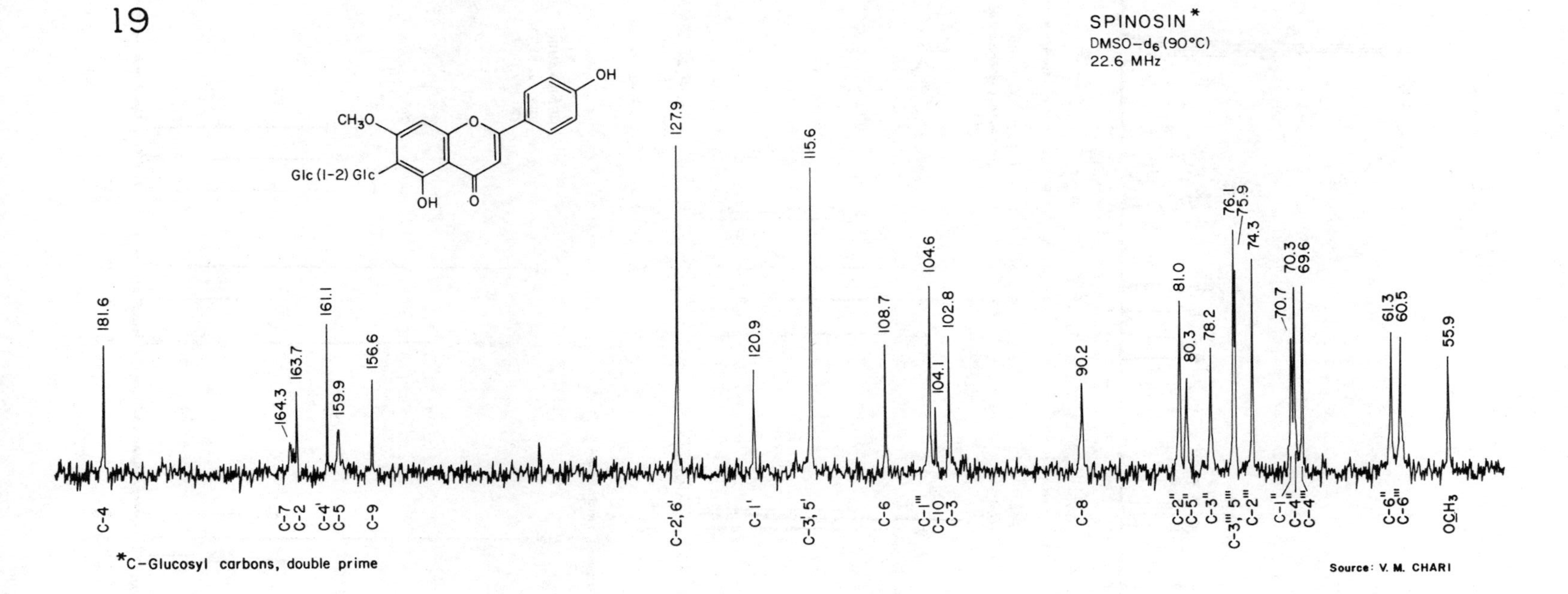
19
SPINOSIN*
DMSO-d6 (90°C)
22.6 MHz
CH3O
Glc (1-2) Glc
OH
O
OH
181.6
164.3
163.7
161.1
159.9
156.6
127.9
120.9
115.6
108.7
104.6
104.1
102.8
90.2
81.0
80.3
78.2
76.1
75.9
74.3
70.7
70.3
69.6
61.3
60.5
55.9
C-4
C-7
C-2
C-4'
C-5
C-9
C-2',6'
C-1'
C-3',5'
C-6
C-1'''
C-10
C-3
C-8
C-2''
C-5''
C-3''
C-3''', 5'''
C-2'''
C-1''
C-4''
C-4'''
C-6''
C-6'''
OCH3
*C-Glucosyl carbons, double prime
Source: V. M. CHARI

20

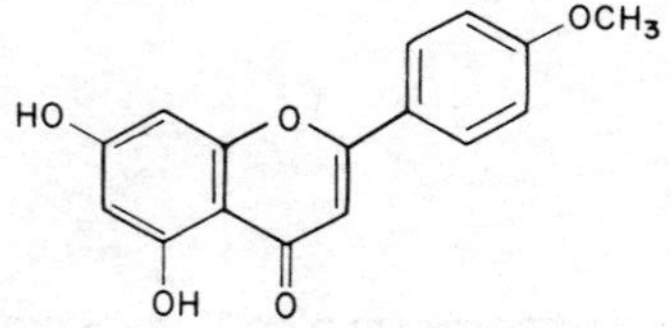

ACACETIN

DMSO-d_6 + $CDCl_3$ (Ambient Temperature)

25.2 MHz

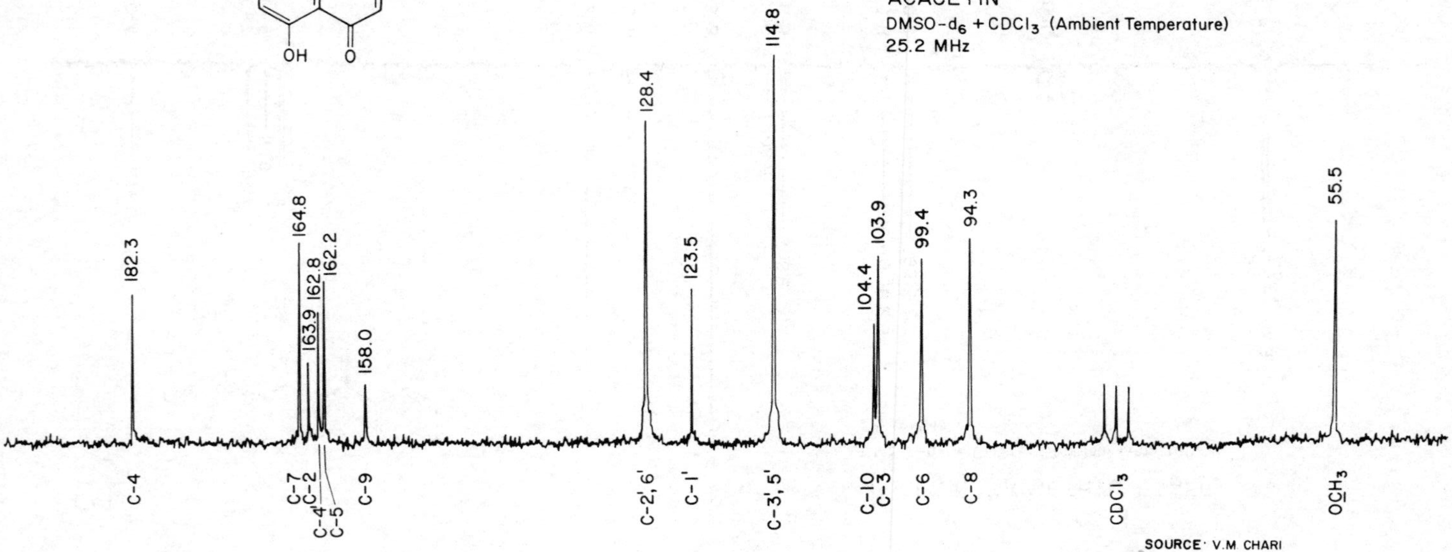

HO OH O OCH_3

182.3 164.8 163.9 162.8 162.2 158.0 128.4 123.5 114.8 104.4 103.9 99.4 94.3 55.5

C-4 C-7 C-2 C-4' C-5 C-9 C-2',6' C-1' C-3',5' C-10 C-3 C-6 C-8 $CDCl_3$ OCH_3

SOURCE: V.M CHARI

21

APIGENIN TRIMETHYL ETHER

DMSO-d_6 + $CDCl_3$ (Ambient Temperature)

25.1 MHz

SOURCE · V M CHARI

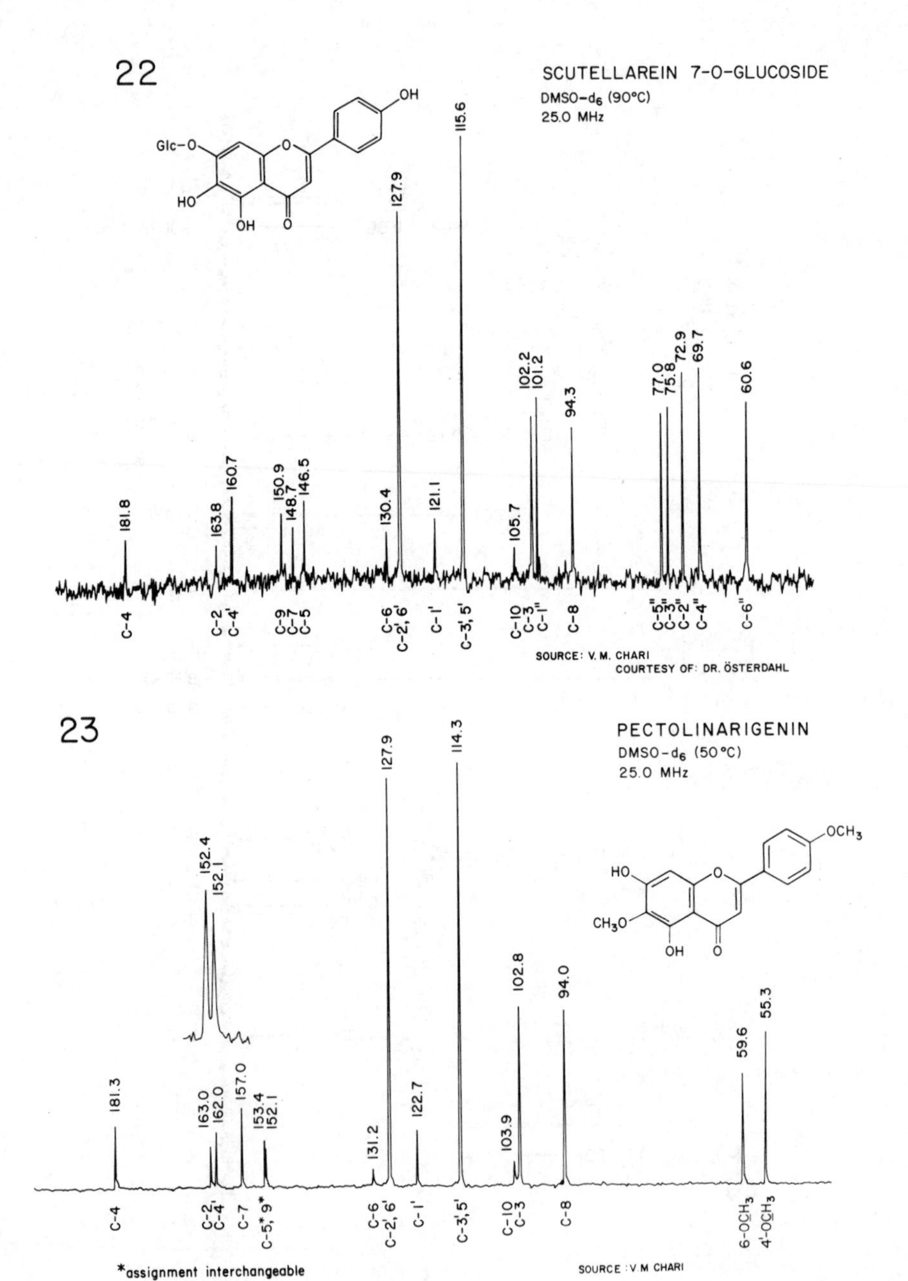

22
SCUTELLAREIN 7-O-GLUCOSIDE
DMSO-d6 (90°C)
25.0 MHz
Glc–O
HO
OH
O
OH
181.8
163.8
160.7
150.9
148.7
146.5
130.4
127.9
121.1
115.6
105.7
102.2
101.2
94.3
77.0
75.8
72.9
69.7
60.6
C-4
C-2
C-4'
C-9
C-7
C-5
C-6
C-2', 6'
C-1'
C-3', 5'
C-10
C-3
C-1''
C-8
C-5''
C-3''
C-2''
C-4''
C-6''
SOURCE: V. M. CHARI
COURTESY OF: DR. ÖSTERDAHL
23
PECTOLINARIGENIN
DMSO-d6 (50°C)
25.0 MHz
OCH3
HO
CH3O
OH
O
152.4
152.1
181.3
163.0
162.0
157.0
153.4
152.1
131.2
127.9
122.7
114.3
103.9
102.8
94.0
59.6
55.3
C-4
C-2, C-4'
C-7
C-5,* 9*
C-6
C-2', 6'
C-1'
C-3', 5'
C-10
C-3
C-8
6-OCH3
4'-OCH3
*assignment interchangeable
SOURCE : V.M CHARI

24

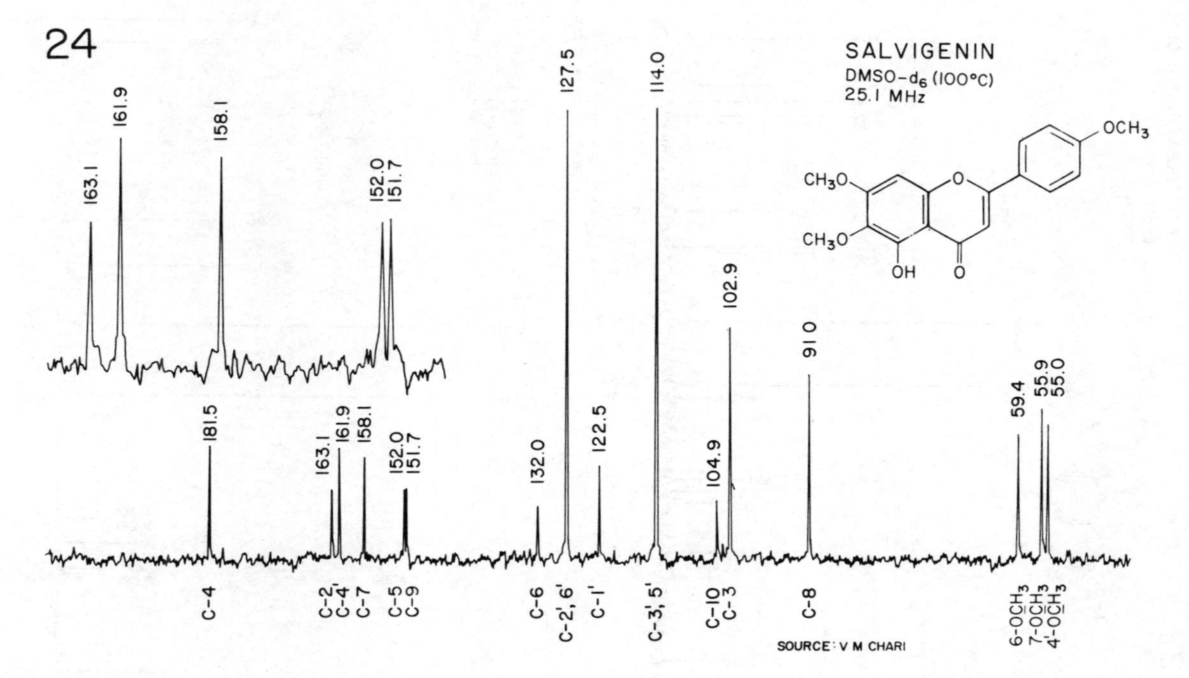
SALVIGENIN
DMSO-d6 (100°C)
25.1 MHz
OCH3
CH3O
CH3O
OH
O
163.1
161.9
158.1
152.0
151.7
181.5
163.1
161.9
158.1
152.0
151.7
132.0
127.5
122.5
114.0
104.9
102.9
91.0
59.4
55.9
55.0
C-4
C-2
C-4'
C-7
C-5
C-9
C-6
C-2', 6'
C-1'
C-3', 5'
C-10
C-3
C-8
6-OCH3
7-OCH3
4'-OCH3
SOURCE: V M CHARI

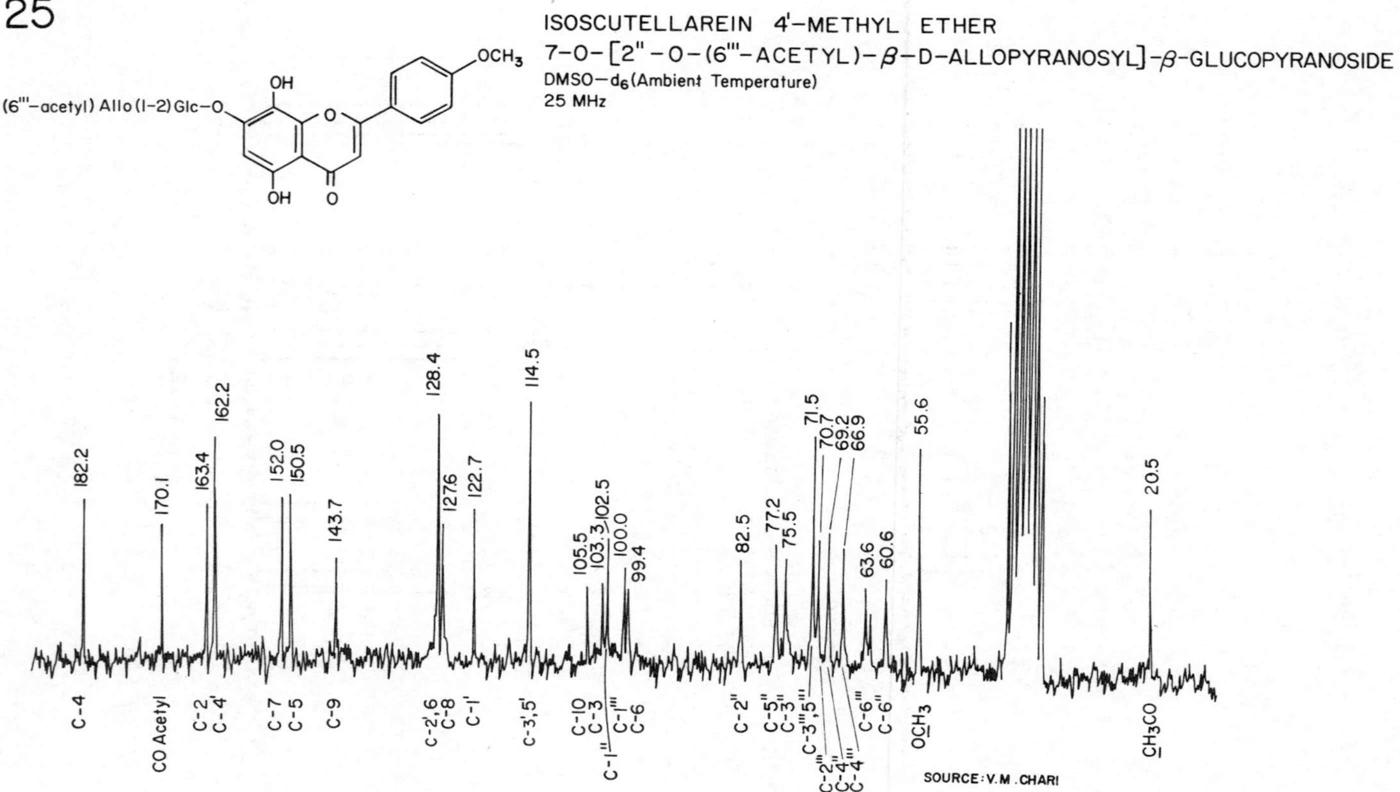
25
ISOSCUTELLAREIN 4'-METHYL ETHER
7-O-[2''-O-(6'''-ACETYL)-β-D-ALLOPYRANOSYL]-β-GLUCOPYRANOSIDE
DMSO-d6 (Ambient Temperature)
25 MHz
(6'''-acetyl) Allo (1-2) Glc-O
OH
OH
O
OCH3
182.2
170.1
163.4
162.2
152.0
150.5
143.7
128.4
127.6
122.7
114.5
105.5
103.3
102.5
100.0
99.4
82.5
77.2
75.5
71.5
70.7
69.2
66.9
63.6
60.6
55.6
20.5
C-4
CO Acetyl
C-2
C-4'
C-7
C-5
C-9
C-2',6'
C-8
C-1'
C-3',5'
C-10
C-3
C-1''
C-1'''
C-6
C-2''
C-5''
C-3''
C-3''',5'''
C-2'''
C-4''
C-4'''
C-6'''
C-6''
OCH3
CH3CO
SOURCE: V.M. CHARI

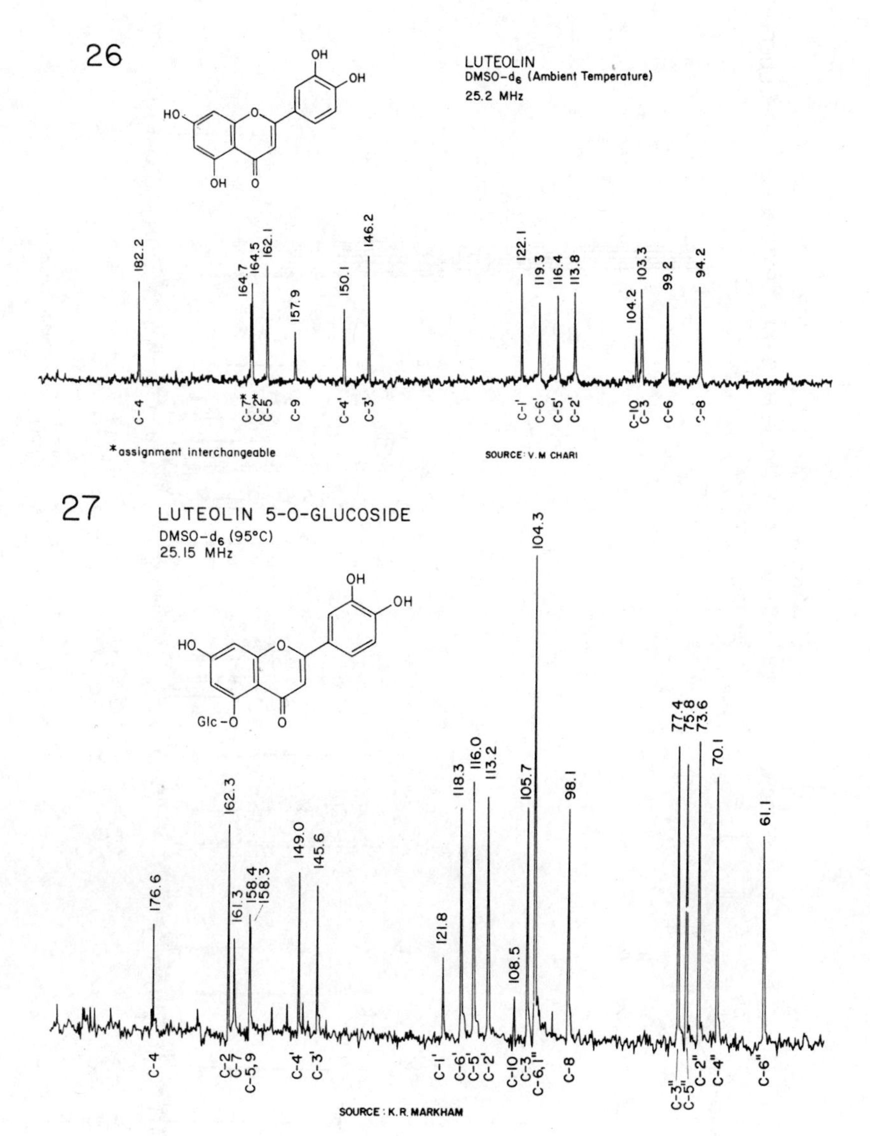
26
LUTEOLIN
DMSO-d_6 (Ambient Temperature)
25.2 MHz
OH
OH
HO
O
OH
O
182.2
164.7
164.5
162.1
157.9
150.1
146.2
122.1
119.3
116.4
113.8
104.2
103.3
99.2
94.2
C-4
C-7*
C-2*
C-5
C-9
C-4'
C-3'
C-1'
C-6'
C-5'
C-2'
C-10
C-3
C-6
C-8
*assignment interchangeable
SOURCE: V.M CHARI
27
LUTEOLIN 5-O-GLUCOSIDE
DMSO-d_6 (95°C)
25.15 MHz
OH
OH
HO
O
Glc-O
O
176.6
162.3
161.3
158.4
158.3
149.0
145.6
121.8
118.3
116.0
113.2
108.5
105.7
104.3
98.1
77.4
75.8
73.6
70.1
61.1
C-4
C-2
C-7
C-5,9
C-4'
C-3'
C-1'
C-6'
C-5'
C-2'
C-10
C-3
C-6,1"
C-8
C-3"
C-5"
C-2"
C-4"
C-6"
SOURCE: K.R. MARKHAM

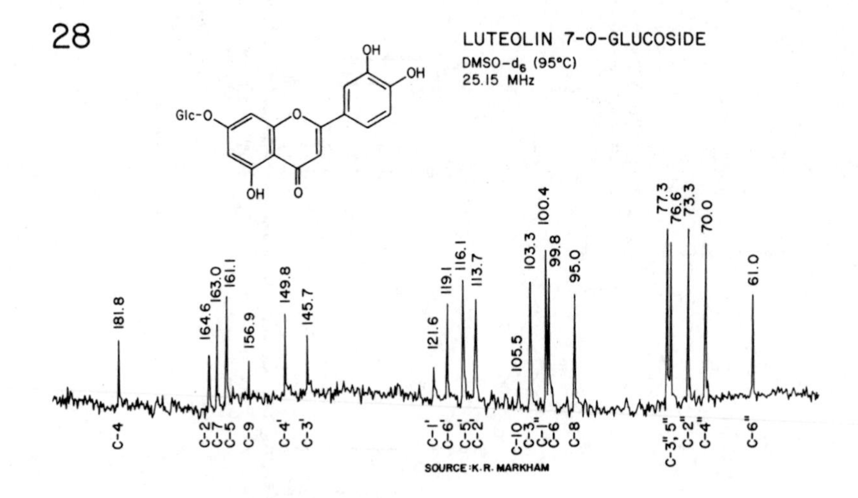
28
LUTEOLIN 7-O-GLUCOSIDE
DMSO-d6 (95°C)
25.15 MHz
SOURCE: K.R. MARKHAM

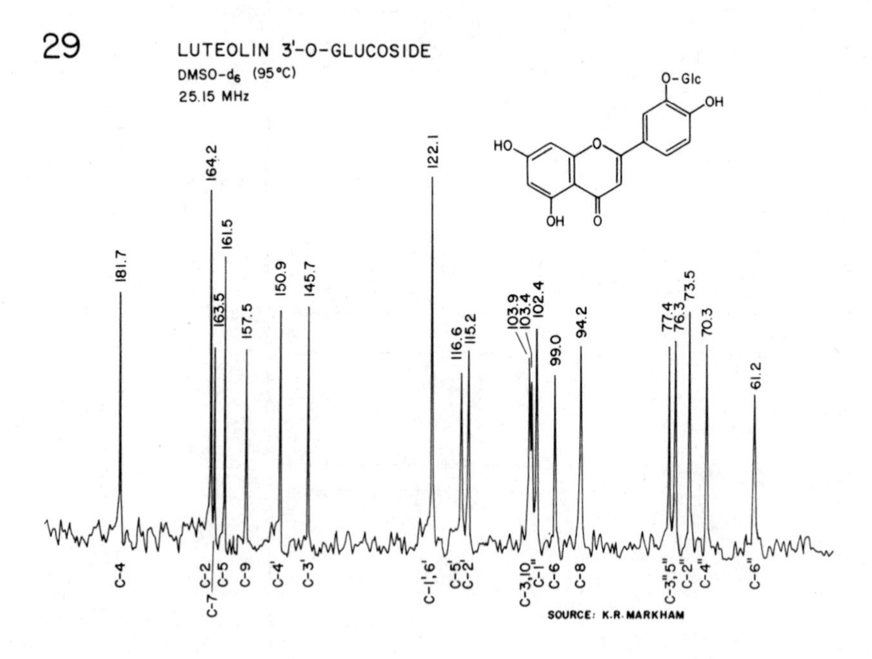
29
LUTEOLIN 3'-O-GLUCOSIDE
DMSO-d6 (95°C)
25.15 MHz
SOURCE: K.R. MARKHAM

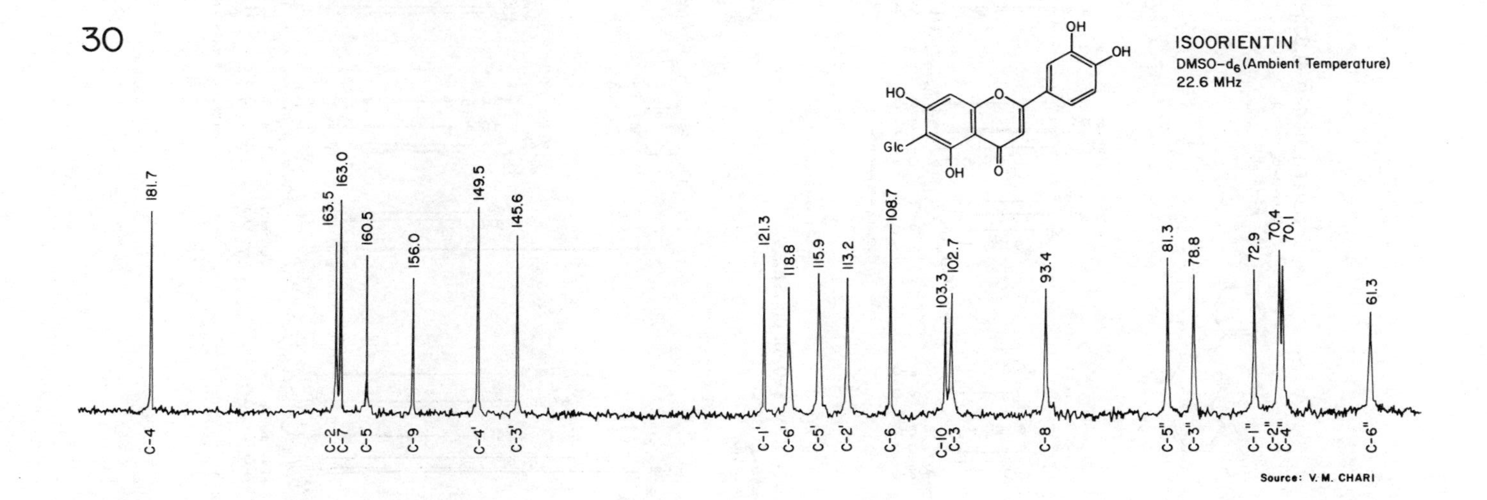
30
ISOORIENTIN
DMSO-d6 (Ambient Temperature)
22.6 MHz
HO
OH
OH
Glc
OH
O
181.7
163.5
163.0
160.5
156.0
149.5
145.6
121.3
118.8
115.9
113.2
108.7
103.3
102.7
93.4
81.3
78.8
72.9
70.4
70.1
61.3
C-4
C-2
C-7
C-5
C-9
C-4'
C-3'
C-1'
C-6'
C-5'
C-2'
C-6
C-10
C-3
C-8
C-5"
C-3"
C-1"
C-2"
C-4"
C-6"
Source: V. M. CHARI

31

ADONIVERNITH

DMSO-d_6 (90°C)
20.0 MHz

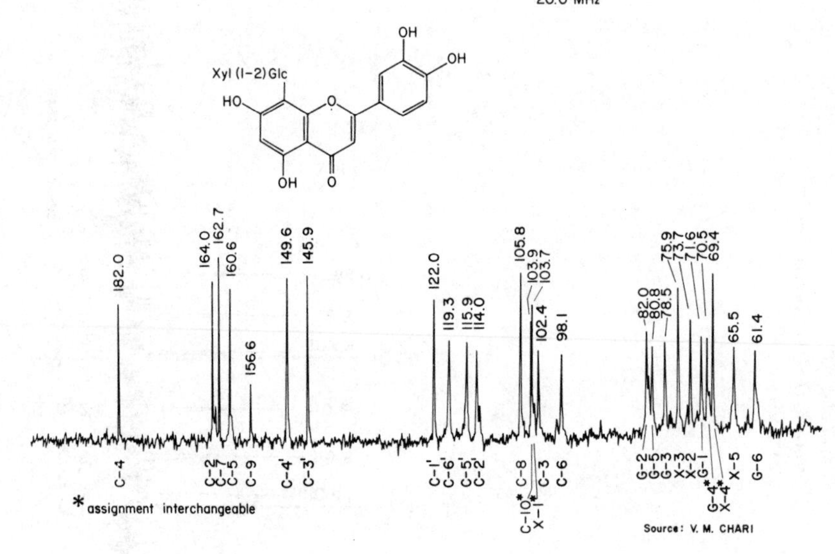

32

DIOSMETIN

DMSO-d_6 (Ambient Temperature)

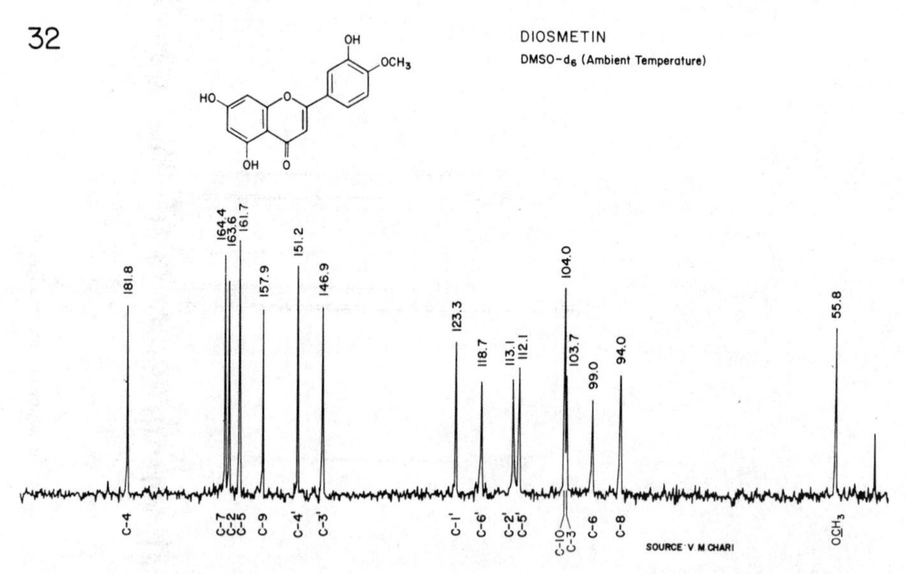

33

CHRYSOERIOL

DMSO-d_6 (Ambient Temperature)

22.63 MHz

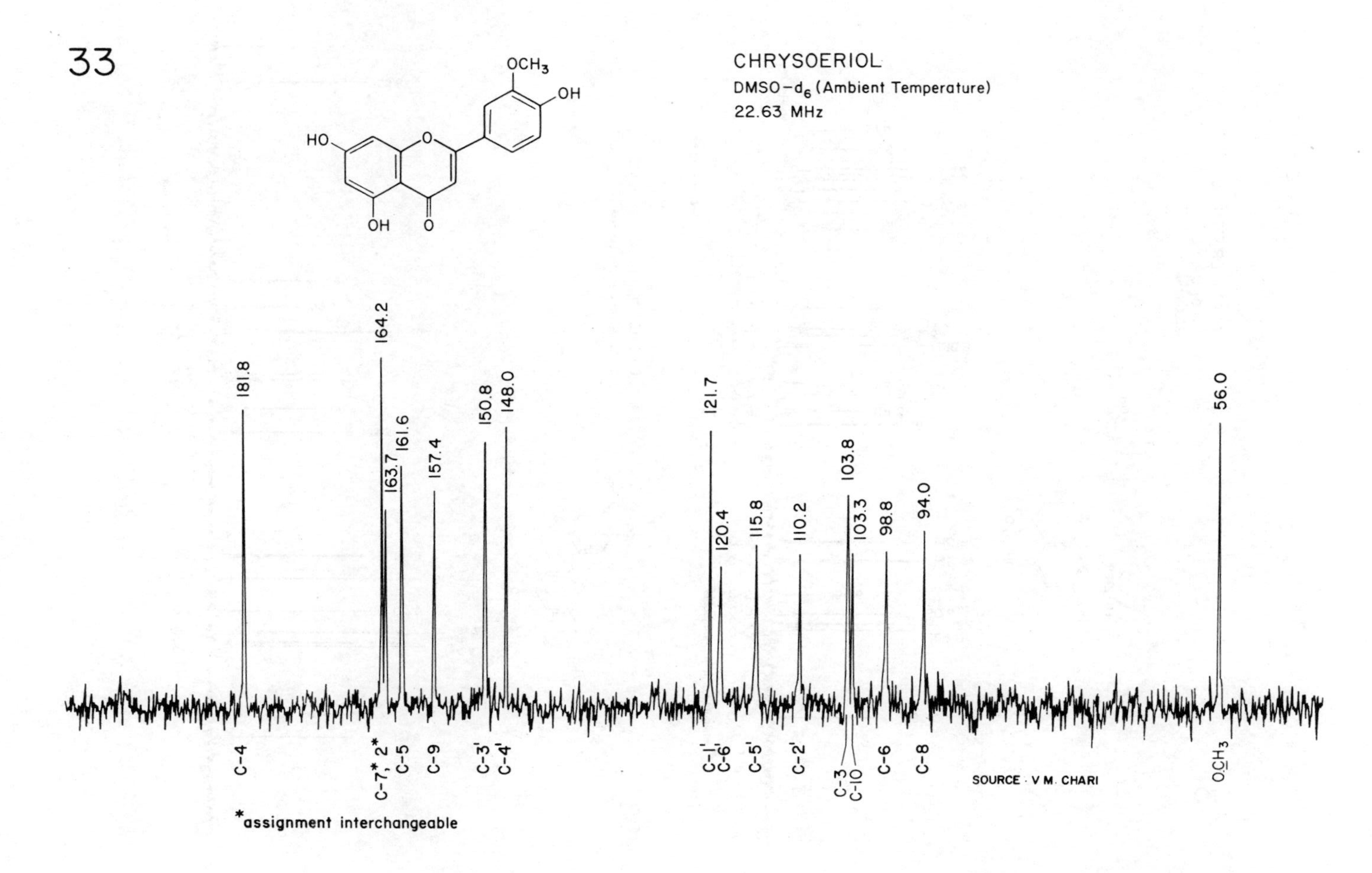

34

CHRYSOERIOL 7-O-(2"-β-allopyranosyl)-β-glucopyranoside
DMSO-d_6 (80°)
20.15 MHz

Allo(1-2) Glc–O
OCH_3
OH
OH
O

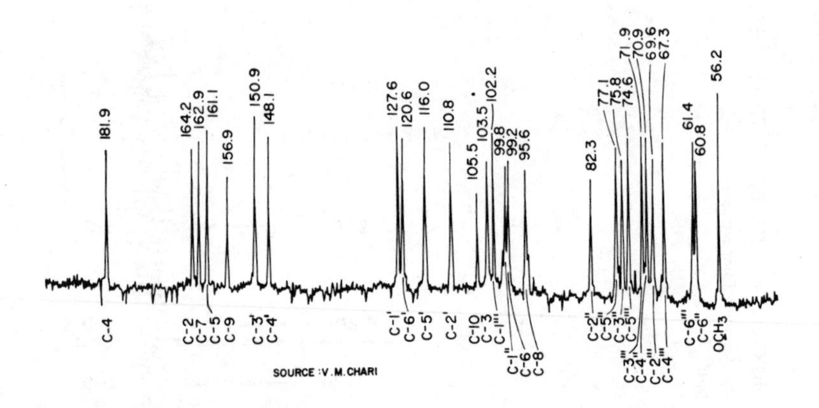

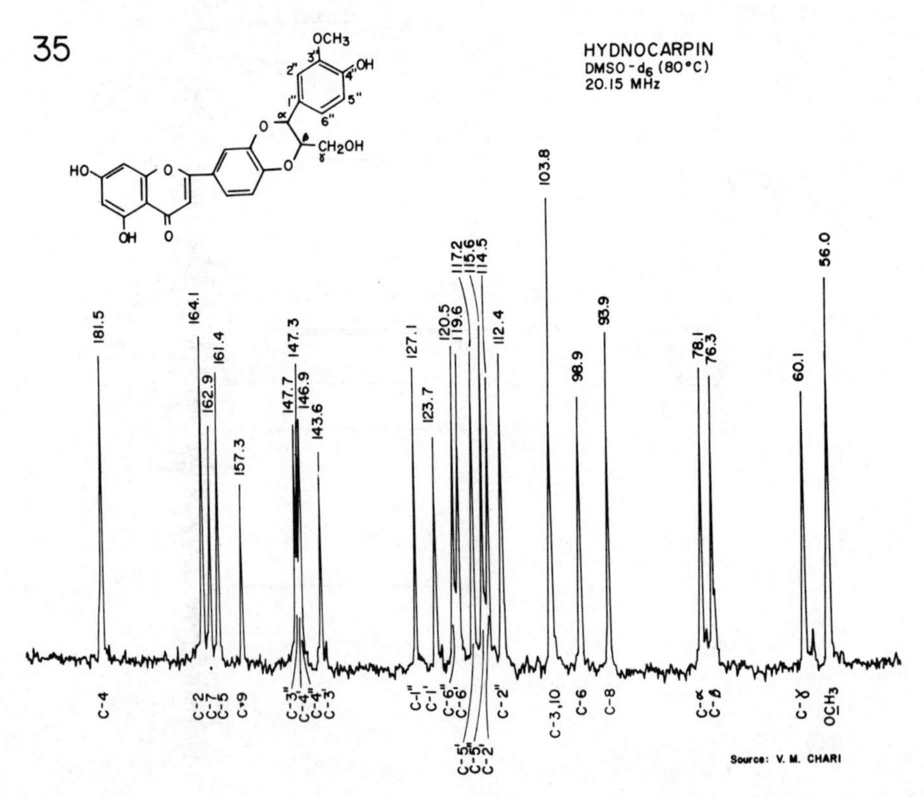

36

6-HYDROXYLUTEOLIN 6,7,3'-TRIMETHYL ETHER

DMSO-d_6 (Ambient Temperature)

22.63 MHz

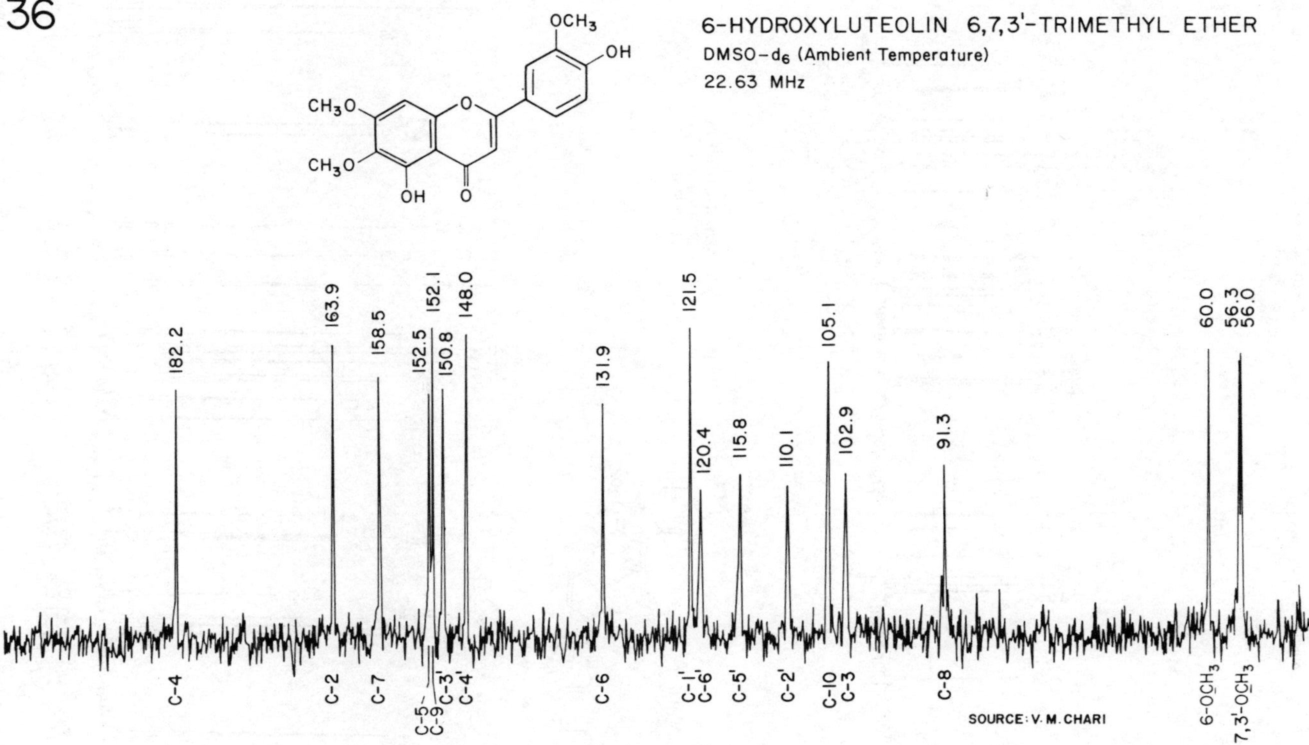

6-HYDROXYLUTEOLIN PENTAMETHYL ETHER

$CDCl_3$ (Ambient Temperature)
22 63 MHz

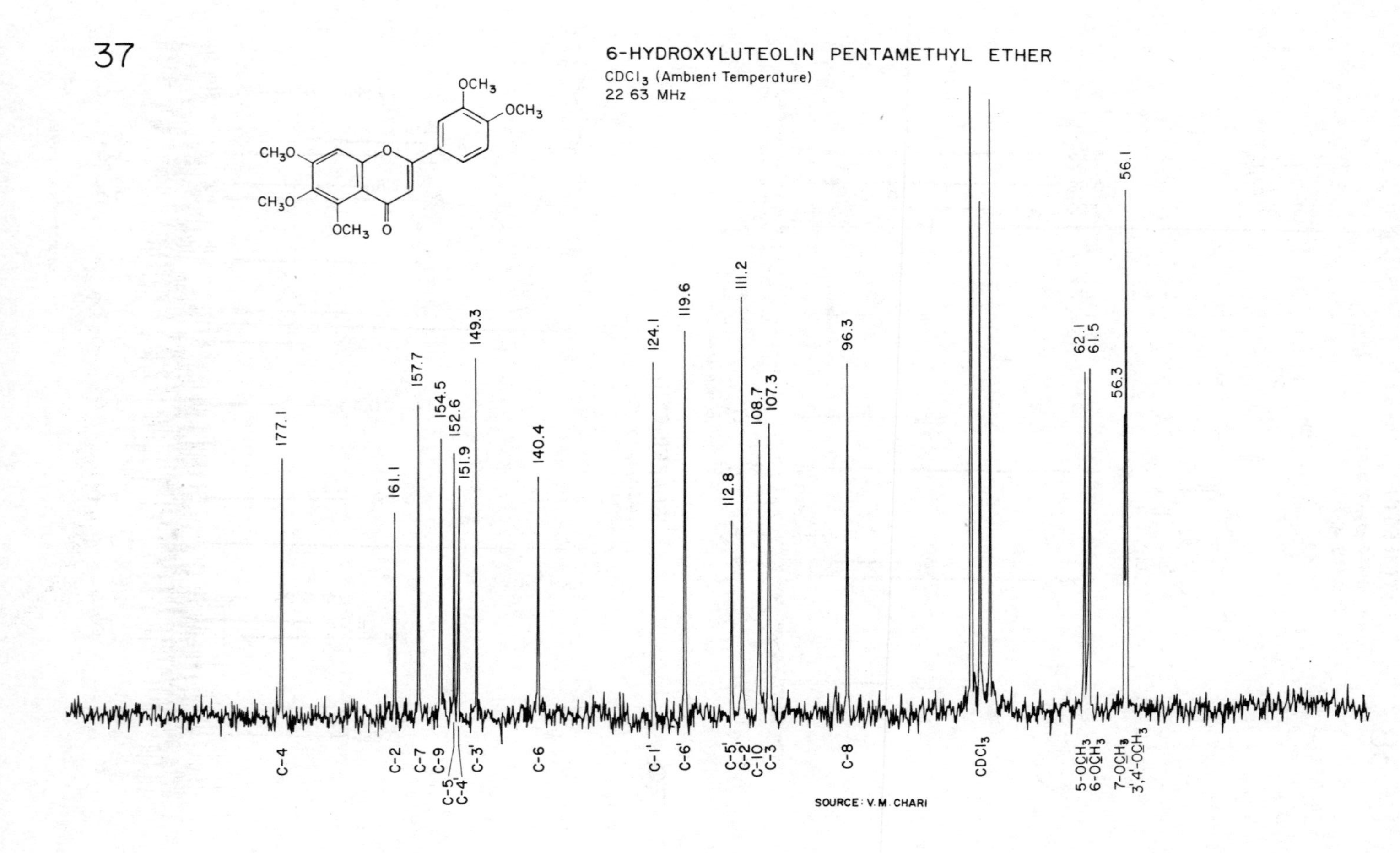

SOURCE: V.M. CHARI

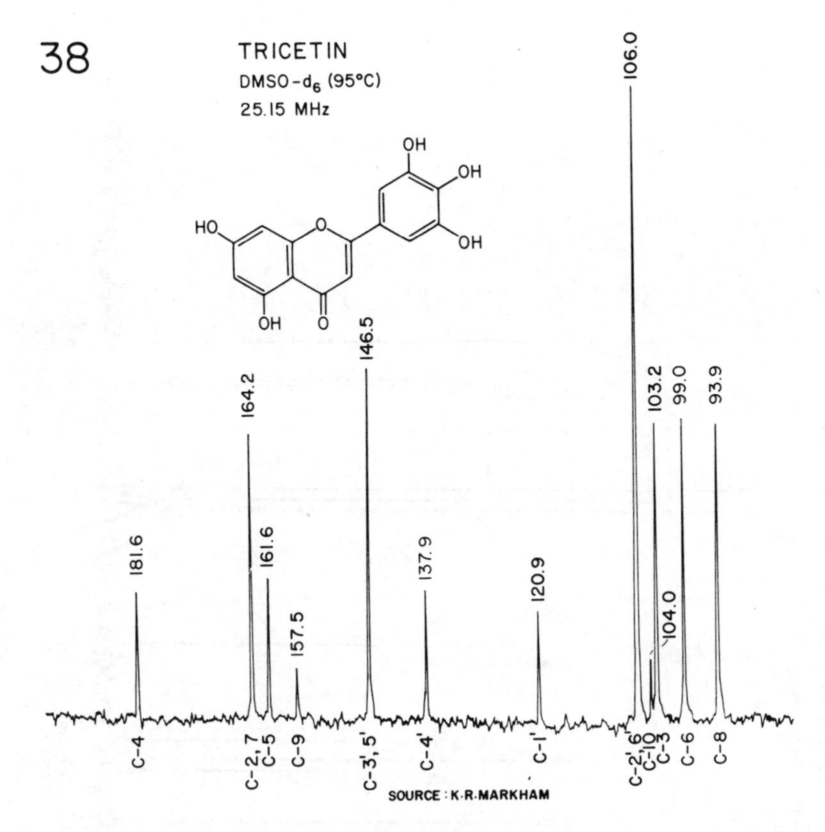
38
TRICETIN
DMSO-d_6 (95°C)
25.15 MHz
HO
OH
OH
OH
OH
O
O
181.6
164.2
161.6
157.5
146.5
137.9
120.9
106.0
104.0
103.2
99.0
93.9
C-4
C-2,7
C-5
C-9
C-3',5'
C-4'
C-1'
C-2',6'
C-10
C-3
C-6
C-8
SOURCE: K.R.MARKHAM

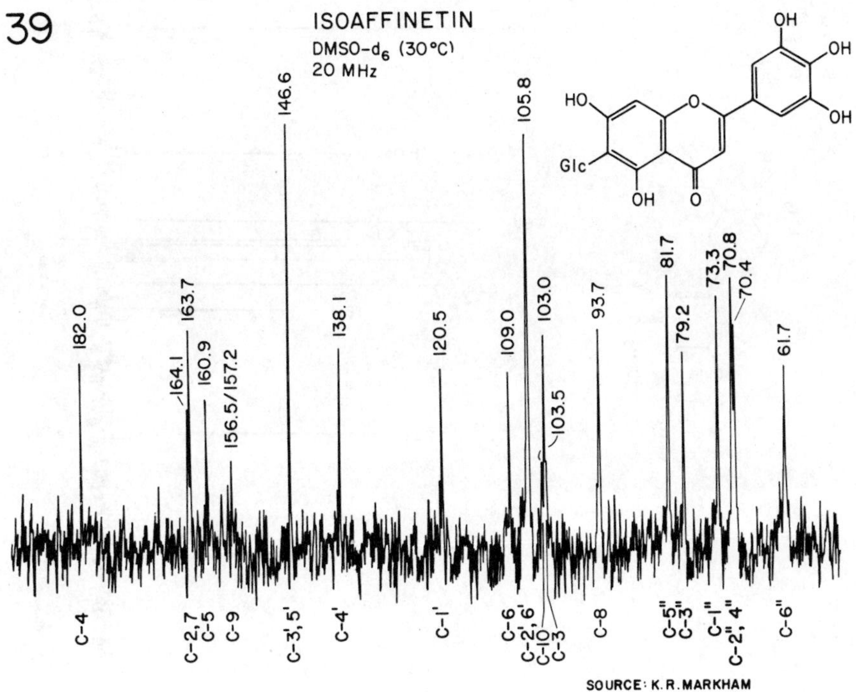
39
ISOAFFINETIN
DMSO-d_6 (30°C)
20 MHz
HO
Glc
OH
O
O
OH
OH
OH
182.0
164.1
163.7
160.9
156.5/157.2
146.6
138.1
120.5
109.0
105.8
103.5
103.0
93.7
81.7
79.2
73.3
70.8
70.4
61.7
C-4
C-2,7
C-5
C-9
C-3',5'
C-4'
C-1'
C-6
C-2',6'
C-10
C-3
C-8
C-5''
C-3''
C-1''
C-2'', 4''
C-6''
SOURCE: K. R. MARKHAM

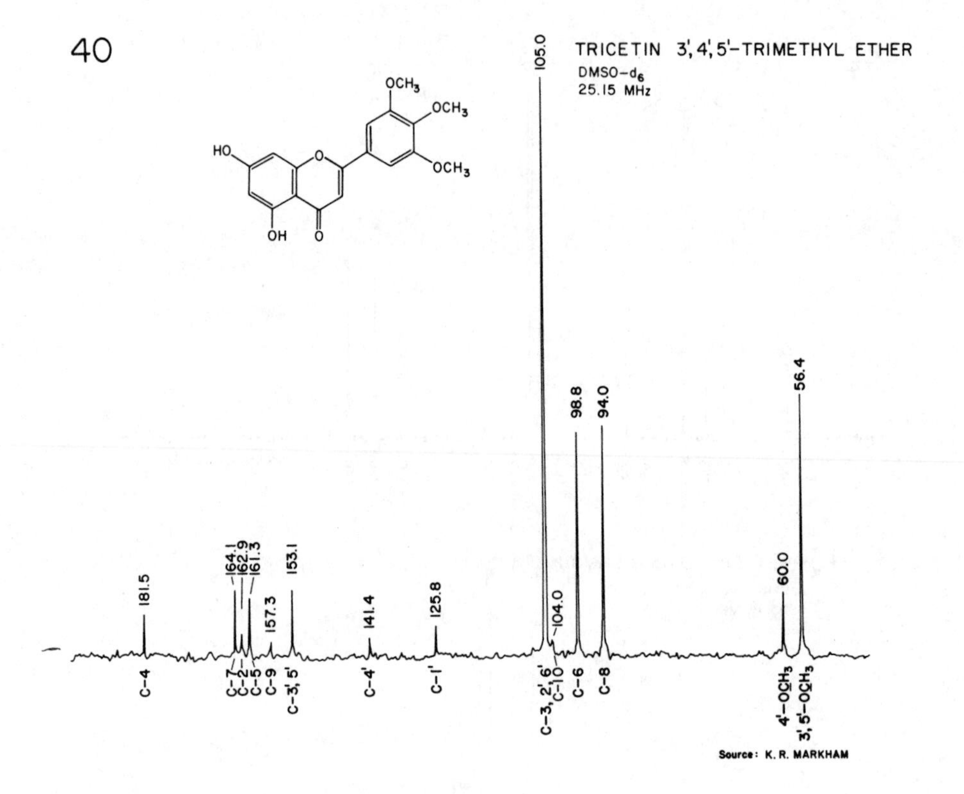
40
TRICETIN 3',4',5'-TRIMETHYL ETHER
DMSO-d6
25.15 MHz
181.5
164.1
162.9
161.3
157.3
153.1
141.4
125.8
105.0
104.0
98.8
94.0
60.0
56.4
C-4
C-7
C-2
C-5
C-9
C-3',5'
C-4'
C-1'
C-3, 2',6'
C-10
C-6
C-8
4'-OCH3
3',5'-OCH3
Source: K. R. MARKHAM

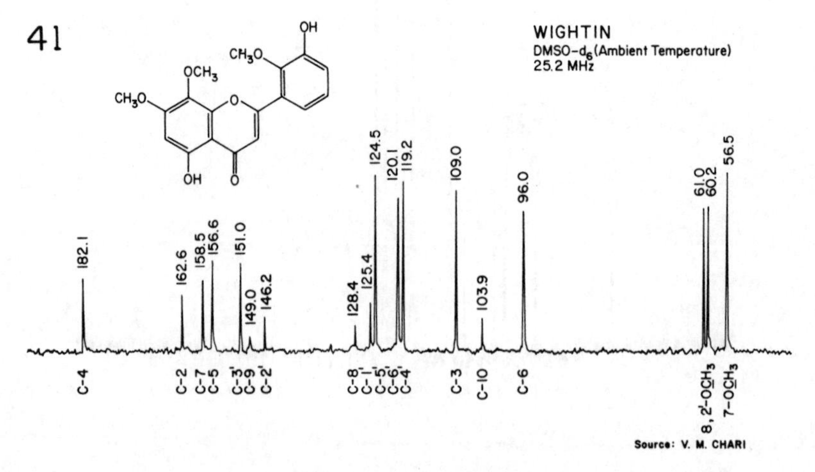
41
WIGHTIN
DMSO-d6 (Ambient Temperature)
25.2 MHz
182.1
162.6
158.5
156.6
151.0
149.0
146.2
128.4
125.4
124.5
120.1
119.2
109.0
103.9
96.0
61.0
60.2
56.5
C-4
C-2
C-7
C-5
C-3'
C-9
C-2'
C-8
C-1'
C-5'
C-6'
C-4'
C-3
C-10
C-6
8,2'-OCH3
7-OCH3
Source: V. M. CHARI

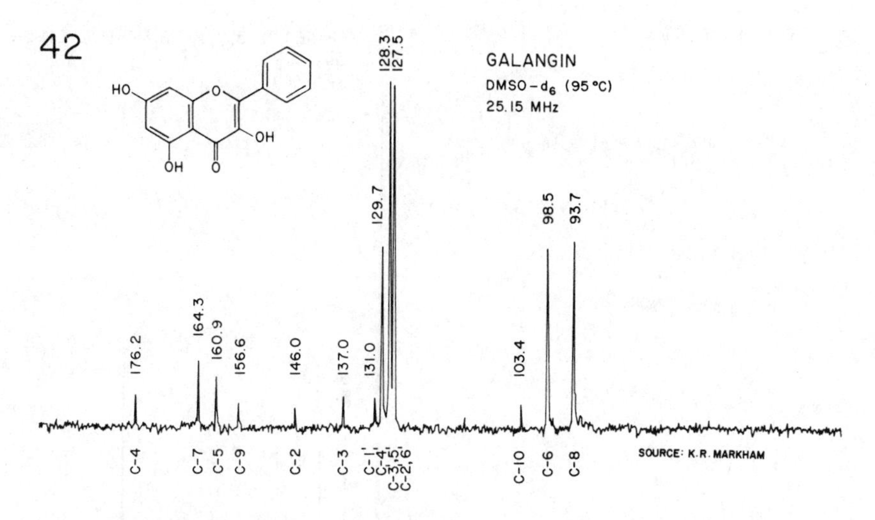

42
GALANGIN
DMSO-d_6 (95 °C)
25.15 MHz
HO
OH
OH
O
176.2
164.3
160.9
156.6
146.0
137.0
131.0
129.7
128.3
127.5
103.4
98.5
93.7
C-4
C-7
C-5
C-9
C-2
C-3
C-1'
C-4'
C-3',5'
C-2',6'
C-10
C-6
C-8
SOURCE: K.R. MARKHAM

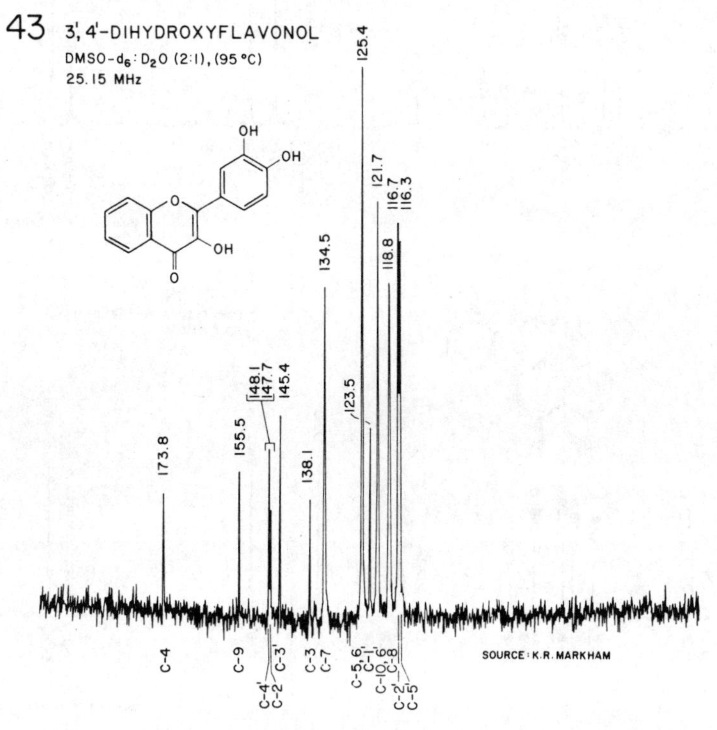

43
3',4'-DIHYDROXYFLAVONOL
DMSO-d_6 : D_2O (2:1), (95 °C)
25.15 MHz
OH
OH
OH
O
173.8
155.5
148.1
147.7
145.4
138.1
134.5
125.4
123.5
121.7
118.8
116.7
116.3
C-4
C-9
C-4'
C-2
C-3'
C-3
C-7
C-5,6
C-1'
C-10,6'
C-8
C-2'
C-5'
SOURCE: K.R. MARKHAM

44

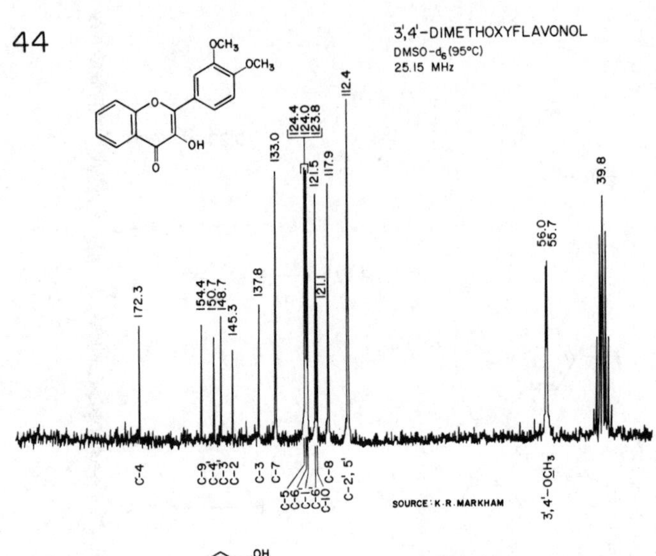
3',4'-DIMETHOXYFLAVONOL
DMSO-d6 (95°C)
25.15 MHz
OCH3
OCH3
OH
O
172.3
154.4
150.7
148.7
145.3
137.8
133.0
124.4
124.0
123.8
121.5
121.1
117.9
112.4
56.0
55.7
39.8
C-4
C-9
C-4'
C-3'
C-2
C-3
C-7
C-5
C-6'
C-1'
C-6
C-10
C-8
C-2', 5'
3', 4'-OCH3
SOURCE: K.R. MARKHAM

45

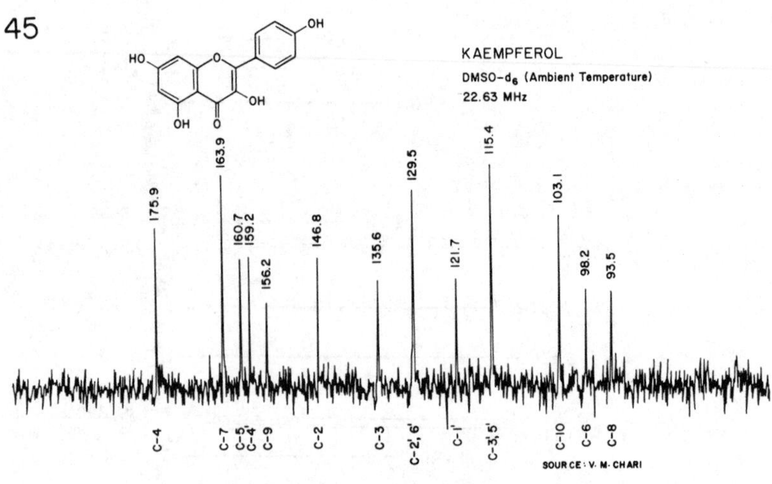
KAEMPFEROL
DMSO-d6 (Ambient Temperature)
22.63 MHz
HO
OH
OH
OH
O
175.9
163.9
160.7
159.2
156.2
146.8
135.6
129.5
121.7
115.4
103.1
98.2
93.5
C-4
C-7
C-5
C-4'
C-9
C-2
C-3
C-2', 6'
C-1'
C-3', 5'
C-10
C-6
C-8
SOURCE: V. M. CHARI

46

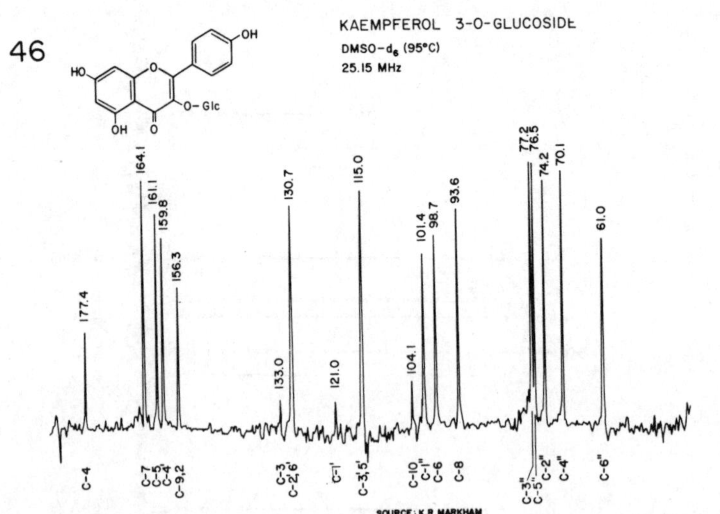
KAEMPFEROL 3-O-GLUCOSIDE
DMSO-d6 (95°C)
25.15 MHz
HO
OH
OH
O
O-Glc
177.4
164.1
161.1
159.8
156.3
133.0
130.7
121.0
115.0
104.1
101.4
98.7
93.6
77.2
76.5
74.2
70.1
61.0
C-4
C-7
C-5
C-4'
C-9,2
C-3
C-2',6'
C-1'
C-3',5'
C-10
C-1''
C-6
C-8
C-3''
C-5''
C-2''
C-4''
C-6''
SOURCE: K.R. MARKHAM

47

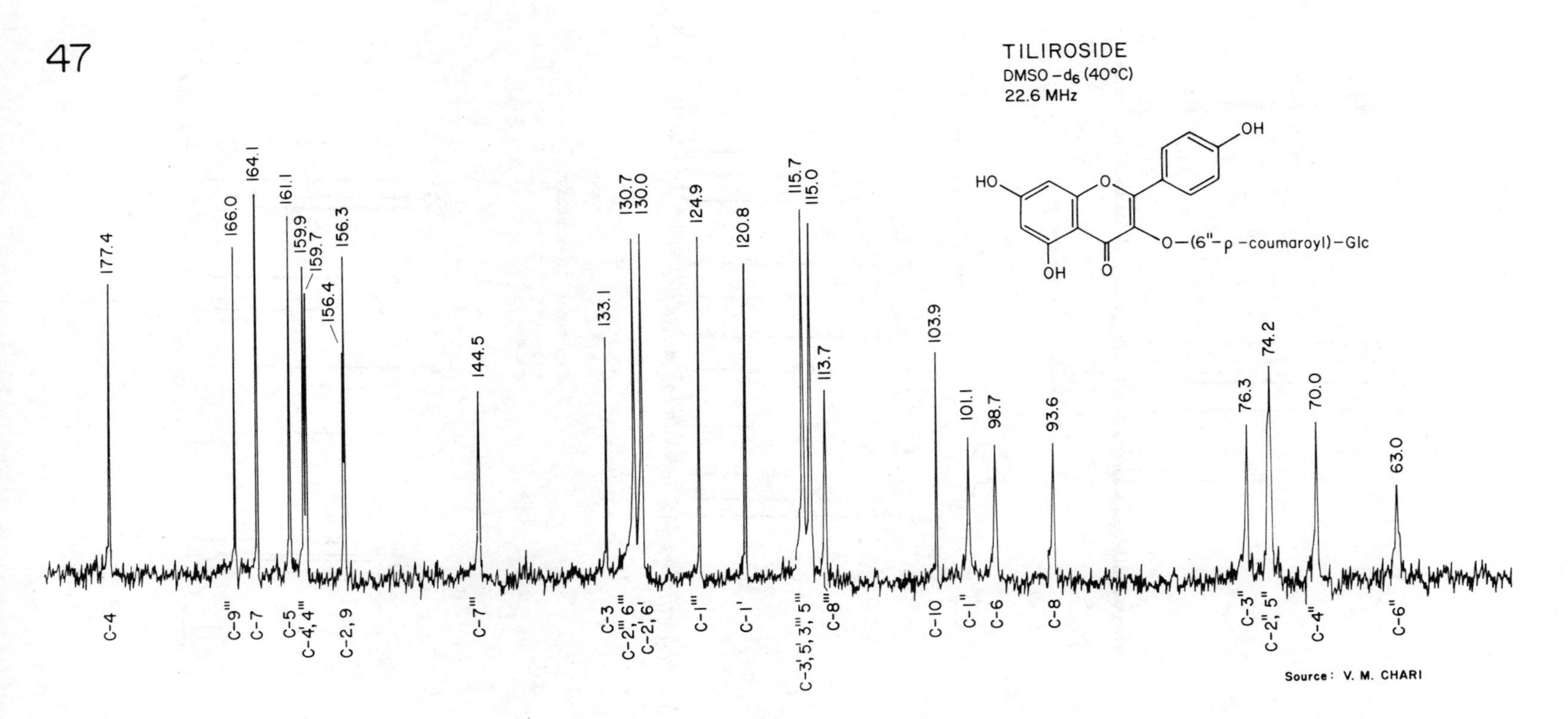

Source: V. M. CHARI

KAEMPFEROL 3-O-SOPHOROSIDE

DMSO-d_6 (95°C)
25.15 MHz

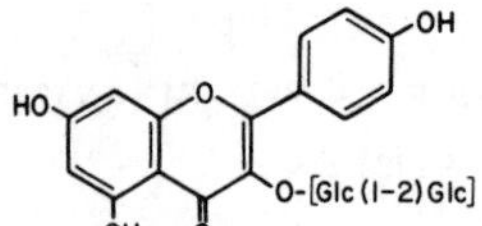

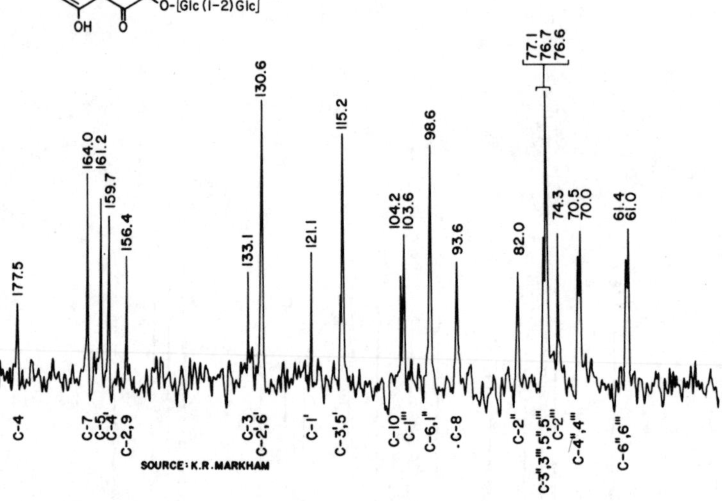

49

KAEMPFEROL 3-O-RUTINOSIDE

DMSO-d_6 (95°C)
25.15 MHz

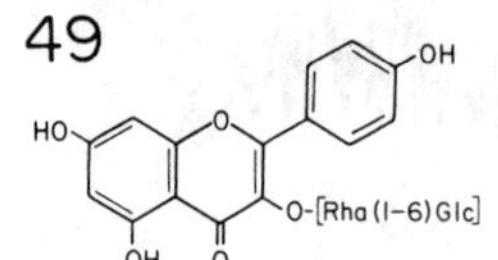

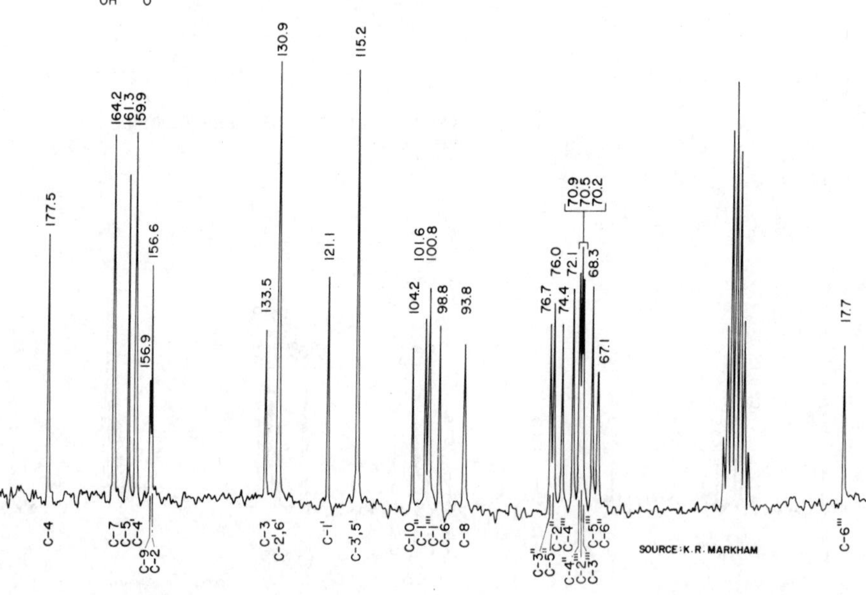

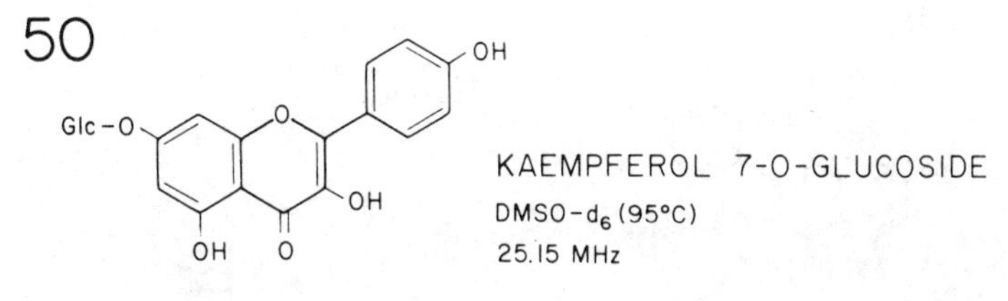
50
Glc-O
OH
OH
OH
O
O
KAEMPFEROL 7-O-GLUCOSIDE
DMSO-d_6 (95°C)
25.15 MHz

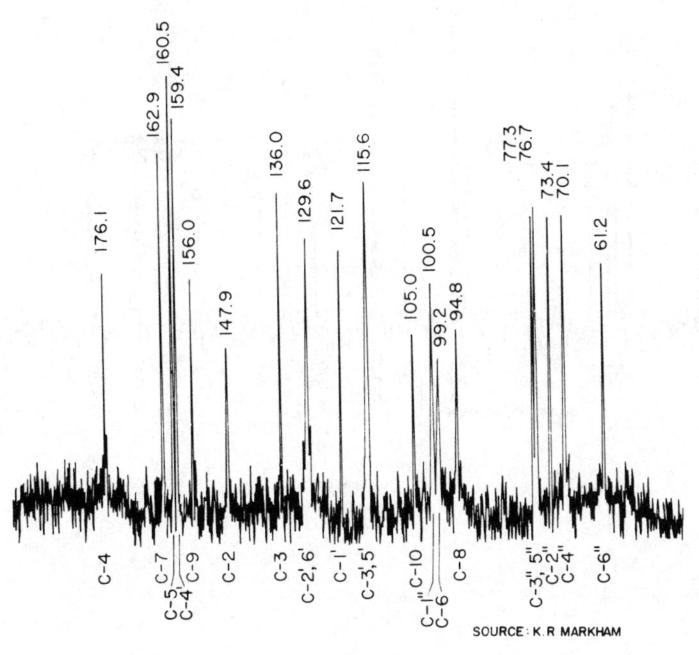
176.1
162.9
160.5
159.4
156.0
147.9
136.0
129.6
121.7
115.6
105.0
100.5
99.2
94.8
77.3
76.7
73.4
70.1
61.2
C-4
C-7
C-5
C-4'
C-9
C-2
C-3
C-2', 6'
C-1'
C-3', 5'
C-10
C-1''
C-6
C-8
C-3'', 5''
C-2''
C-4''
C-6''
SOURCE: K.R MARKHAM

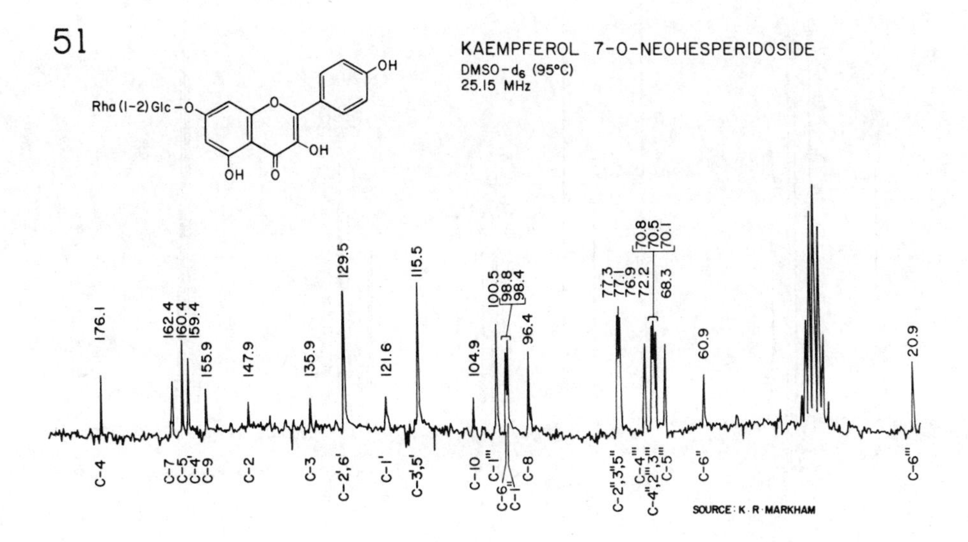
51
KAEMPFEROL 7-O-NEOHESPERIDOSIDE
DMSO-d_6 (95°C)
25.15 MHz
Rha (1-2) Glc-O
OH
OH
OH
O
O
176.1
162.4
160.4
159.4
155.9
147.9
135.9
129.5
121.6
115.5
104.9
100.5
98.8
98.4
96.4
77.3
77.1
76.9
72.2
70.8
70.5
70.1
68.3
60.9
20.9
C-4
C-7
C-5
C-4'
C-9
C-2
C-3
C-2',6'
C-1'
C-3',5'
C-10
C-1'''
C-6
C-1''
C-8
C-2'',3'',5''
C-4'''
C-4'',2''',3'''
C-5'''
C-6''
C-6'''
SOURCE: K·R·MARKHAM

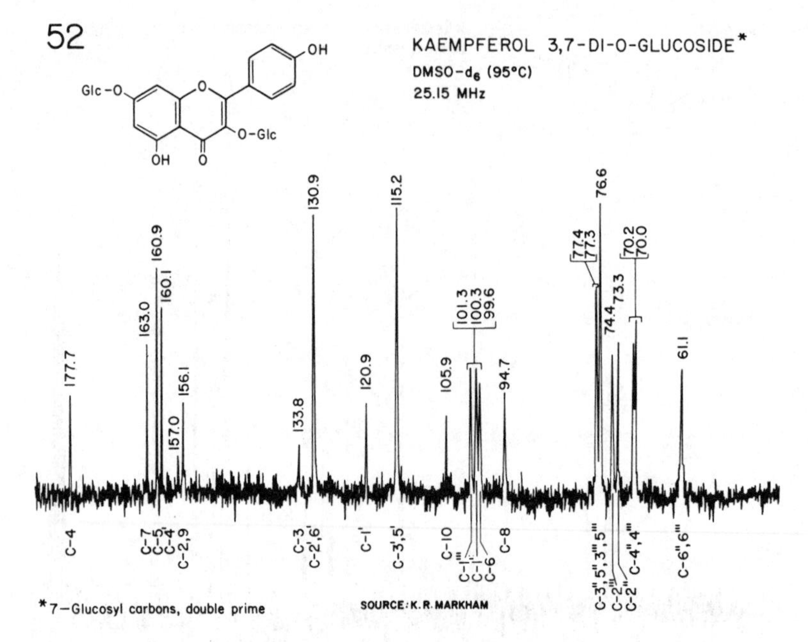
52
KAEMPFEROL 3,7-DI-O-GLUCOSIDE*
DMSO-d6 (95°C)
25.15 MHz
Glc-O
OH
O-Glc
OH
O
177.7
163.0
160.9
160.1
157.0
156.1
133.8
130.9
120.9
115.2
105.9
101.3
100.3
99.6
94.7
77.4
77.3
76.6
74.4
73.3
70.2
70.0
61.1
C-4
C-7
C-5
C-4'
C-2,9
C-3
C-2',6'
C-1'
C-3',5'
C-10
C-1'''
C-1''
C-6
C-8
C-3'',5'',3''',5'''
C-2'''
C-2''
C-4'',4'''
C-6'',6'''
*7-Glucosyl carbons, double prime
SOURCE: K.R. MARKHAM

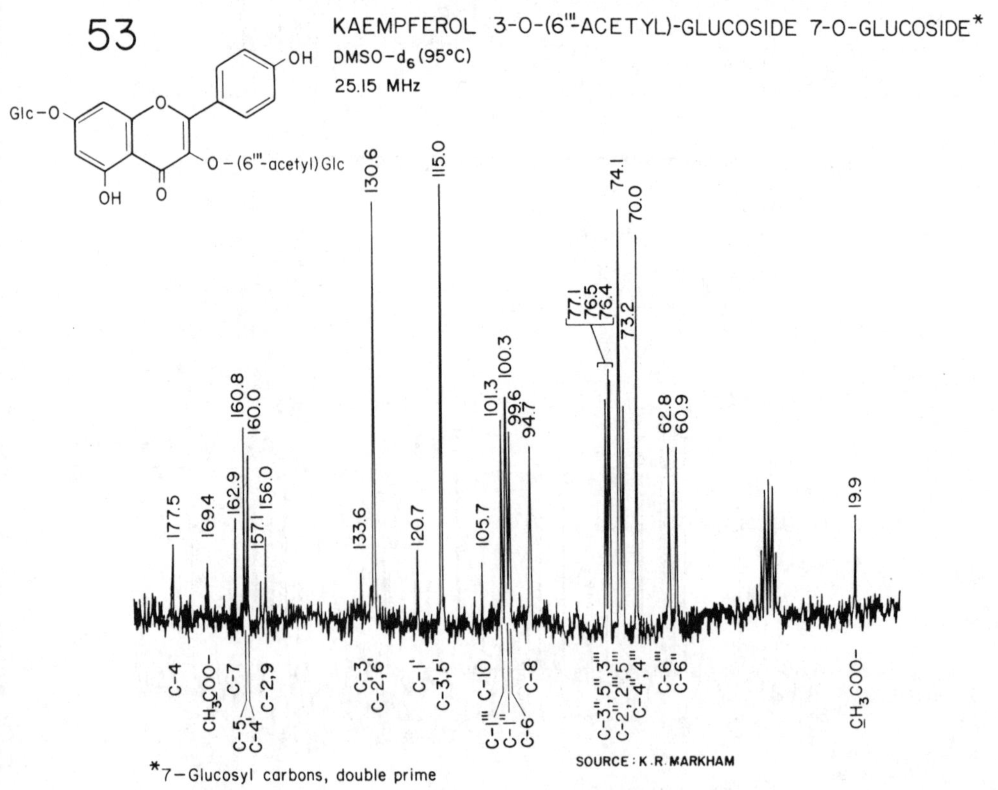
53
KAEMPFEROL 3-O-(6'''-ACETYL)-GLUCOSIDE 7-O-GLUCOSIDE*
DMSO-d6 (95°C)
25.15 MHz
Glc-O
OH
O-(6'''-acetyl) Glc
OH
O
177.5
169.4
162.9
160.8
160.0
157.1
156.0
133.6
130.6
120.7
115.0
105.7
101.3
100.3
99.6
94.7
77.1
76.5
76.4
74.1
73.2
70.0
62.8
60.9
19.9
C-4
CH3COO-
C-7
C-5
C-4'
C-2,9
C-3
C-2',6'
C-1'
C-3',5'
C-10
C-1'''
C-1''
C-6
C-8
C-3'',5'',3'''
C-2'',2''',5'''
C-4'',4'''
C-6'''
C-6''
CH3COO-
*7-Glucosyl carbons, double prime
SOURCE: K.R. MARKHAM

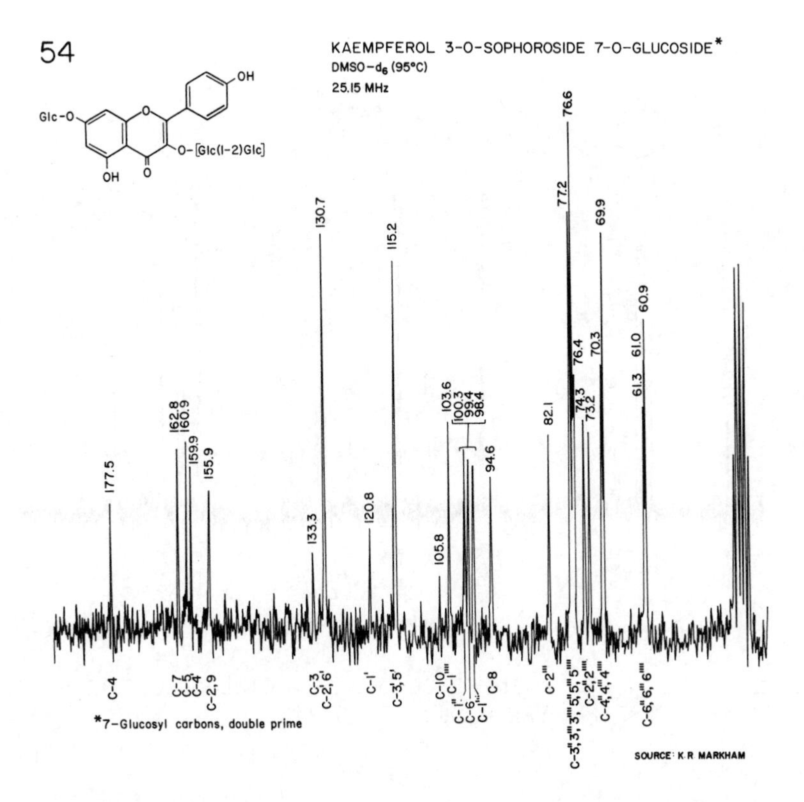

55 KAEMPFEROL 3-O-(2''''-SINAPYL)-SOPHOROSIDE 7-O-GLUCOSIDE*

DMSO-d_6 (95°C)

25.15 MHz

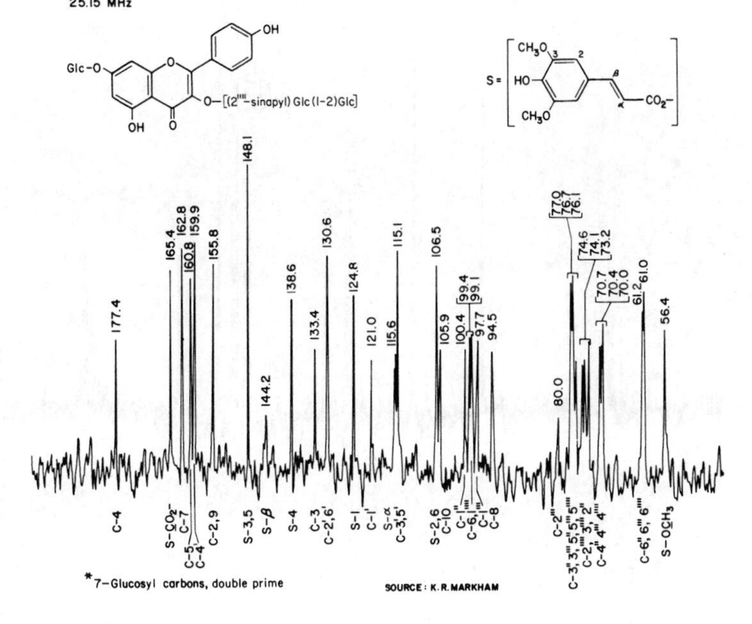

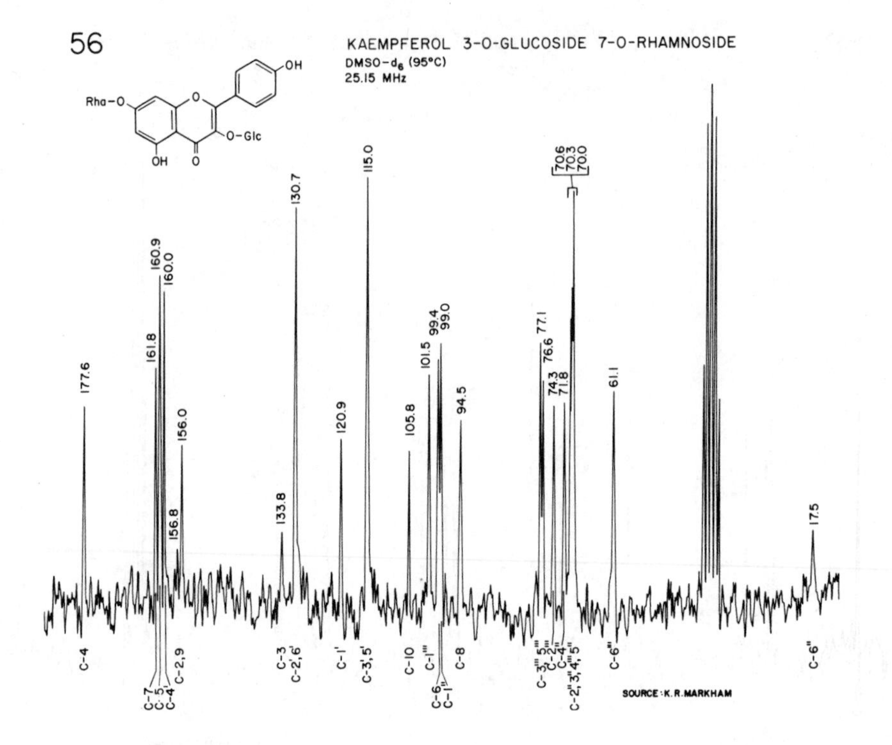

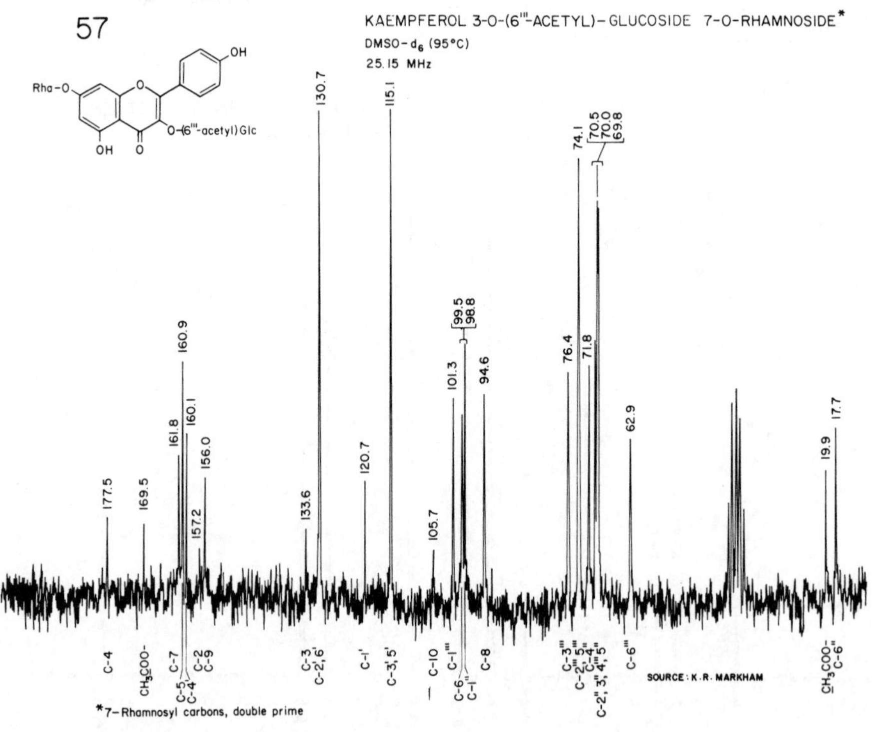

*7-Rhamnosyl carbons, double prime

58

KAEMPFEROL 3-O-RUTINOSIDE 7-O-RHAMNOSIDE*

DMSO-d_6 (95°C)
25.15 MHz

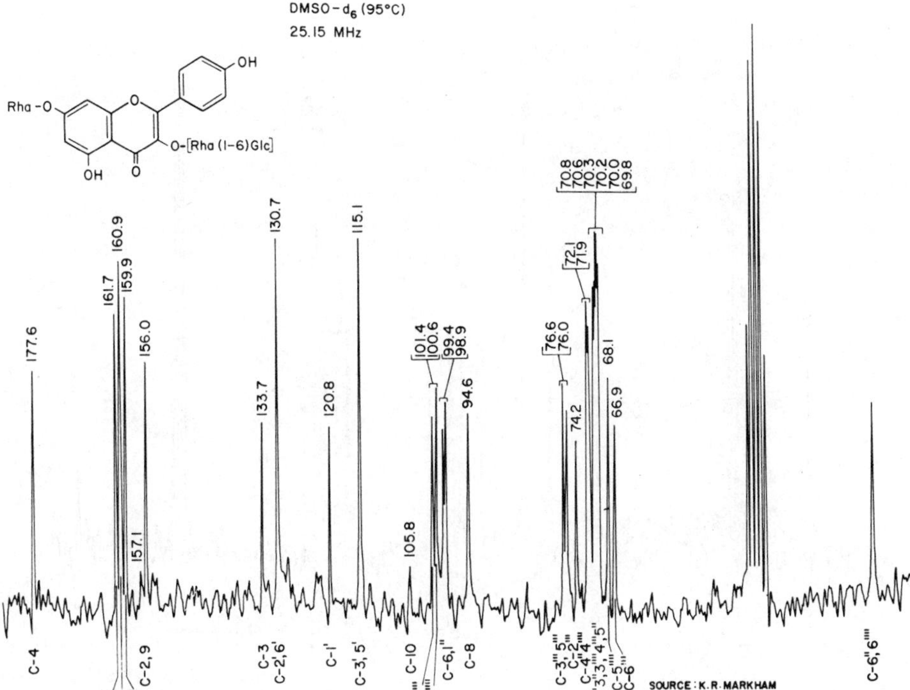

*7-Rhamnosyl carbons, double prime

59

KAEMPFEROL 4'-METHYL ETHER

DMSO-d_6 (95°C)
25.15 MHz

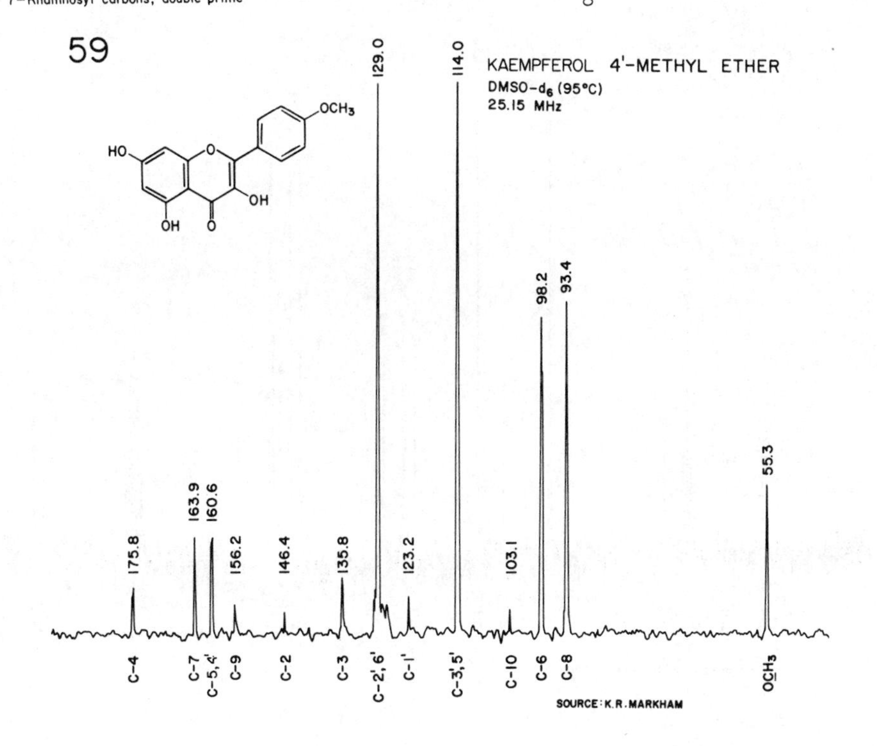

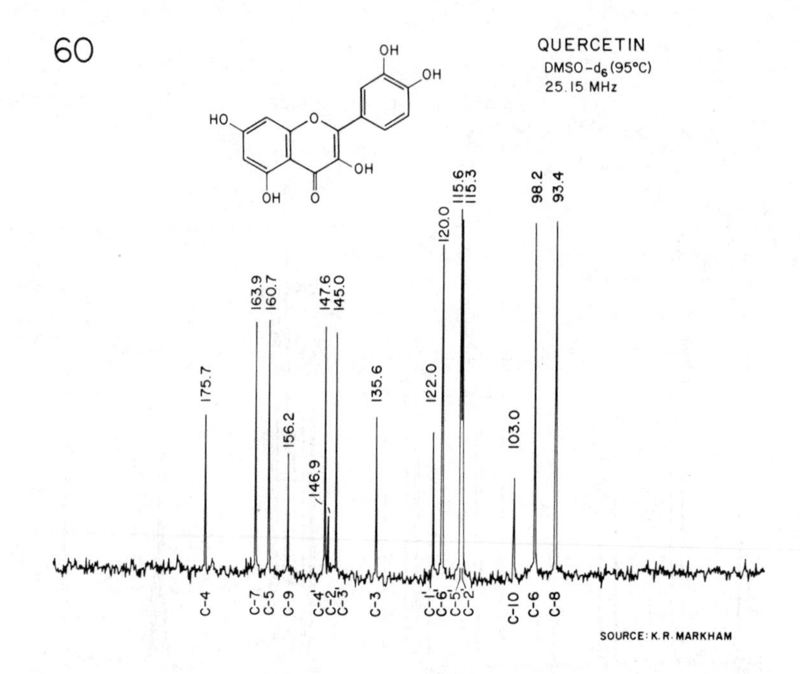
60
QUERCETIN
DMSO-d6 (95°C)
25.15 MHz
OH
OH
HO
OH
OH
O
175.7
163.9
160.7
156.2
146.9
147.6
145.0
135.6
122.0
120.0
115.6
115.3
103.0
98.2
93.4
C-4
C-7
C-5
C-9
C-4'
C-2
C-3'
C-3
C-1'
C-6'
C-5'
C-2'
C-10
C-6
C-8
SOURCE: K.R. MARKHAM

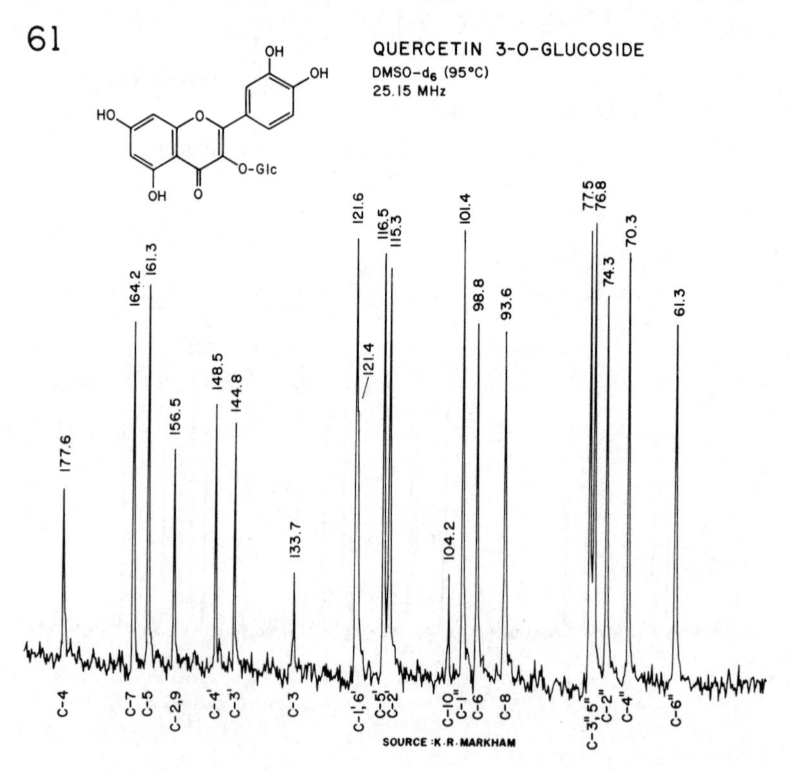
61
QUERCETIN 3-O-GLUCOSIDE
DMSO-d6 (95°C)
25.15 MHz
OH
OH
HO
O-Glc
OH
O
177.6
164.2
161.3
156.5
148.5
144.8
133.7
121.6
121.4
116.5
115.3
104.2
101.4
98.8
93.6
77.5
76.8
74.3
70.3
61.3
C-4
C-7
C-5
C-2,9
C-4'
C-3'
C-3
C-1',6'
C-5'
C-2'
C-10
C-1"
C-6
C-8
C-3",5"
C-2"
C-4"
C-6"
SOURCE: K.R. MARKHAM

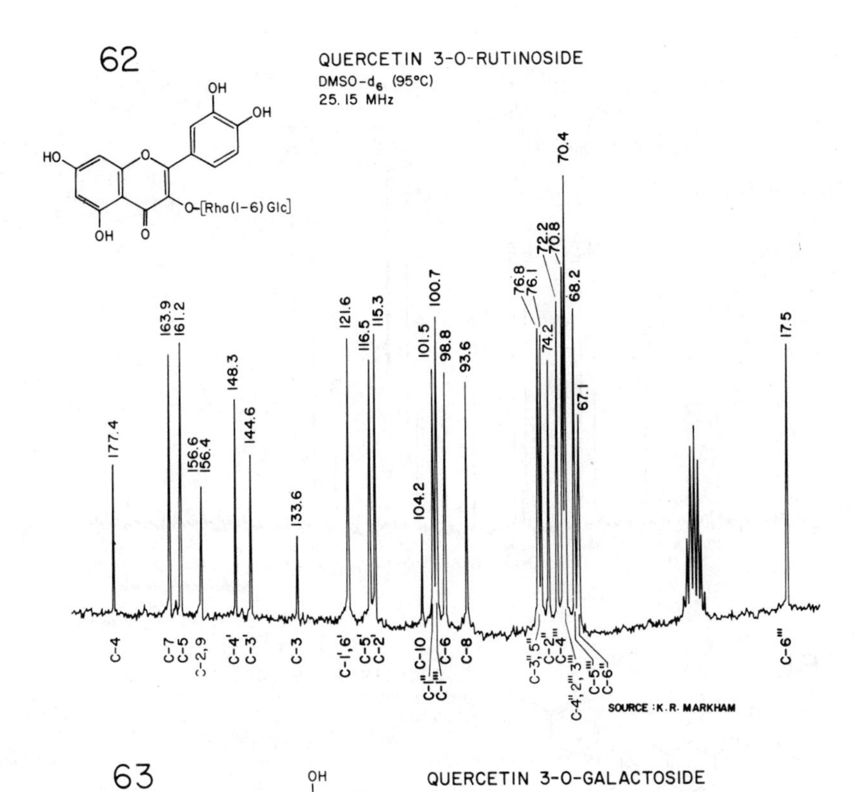

62
QUERCETIN 3-O-RUTINOSIDE
DMSO-d6 (95°C)
25.15 MHz
O-[Rha(1-6) Glc]
177.4
163.9
161.2
156.6
156.4
148.3
144.6
133.6
121.6
116.5
115.3
104.2
101.5
100.7
98.8
93.6
76.8
76.1
74.2
72.2
70.8
70.4
68.2
67.1
17.5
C-4
C-7
C-5
C-2,9
C-4'
C-3'
C-3
C-1',6'
C-5'
C-2'
C-10
C-1''
C-1'''
C-6
C-8
C-3'',5''
C-2'''
C-4'''
C-4'',2''',3'''
C-5'''
C-6''
C-6'''
SOURCE : K.R. MARKHAM

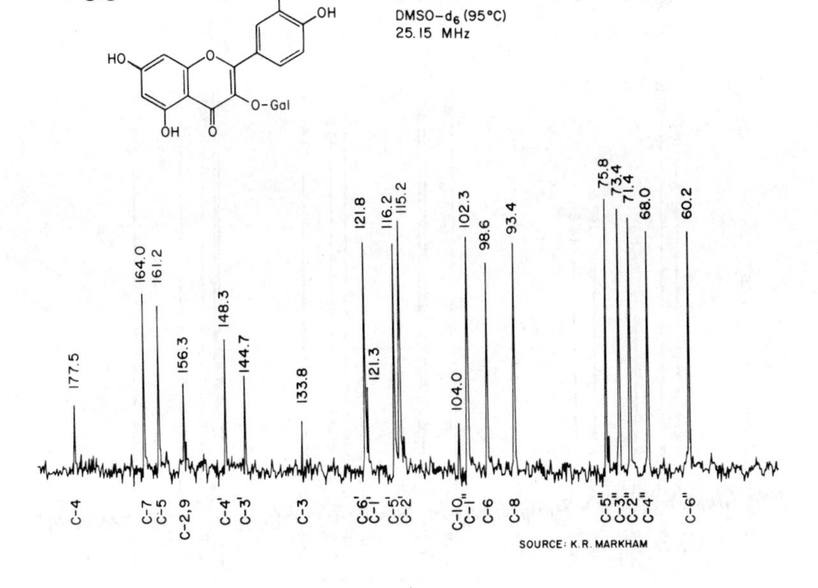

63
QUERCETIN 3-O-GALACTOSIDE
DMSO-d6 (95°C)
25.15 MHz
O-Gal
177.5
164.0
161.2
156.3
148.3
144.7
133.8
121.8
121.3
116.2
115.2
104.0
102.3
98.6
93.4
75.8
73.4
71.4
68.0
60.2
C-4
C-7
C-5
C-2,9
C-4'
C-3'
C-3
C-6'
C-1'
C-5'
C-2'
C-10
C-1''
C-6
C-8
C-5''
C-3''
C-2''
C-4''
C-6''
SOURCE: K.R. MARKHAM

64

QUERCETIN 3-O-(6''-GALLOYL)-GALACTOSIDE
DMSO-d_6 (95°C)
25.15 MHz

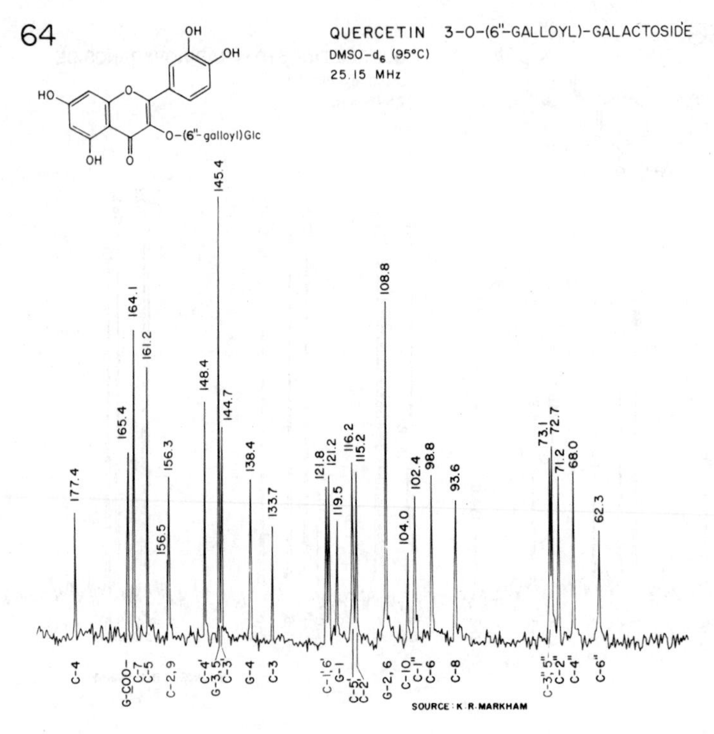

65

QUERCETIN 3-O-RHAMNOSIDE
DMSO-d_6 (95°C)
25.15 MHz

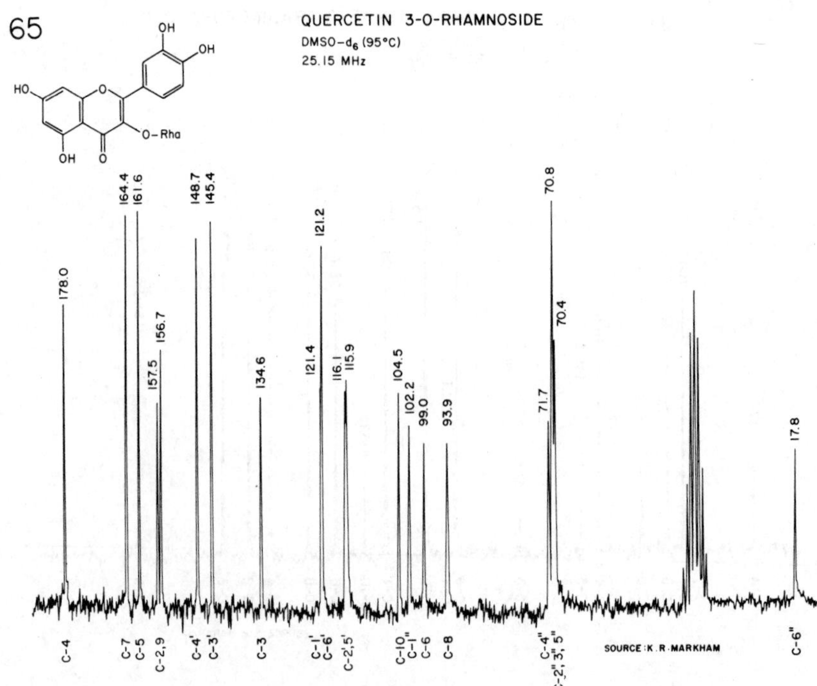

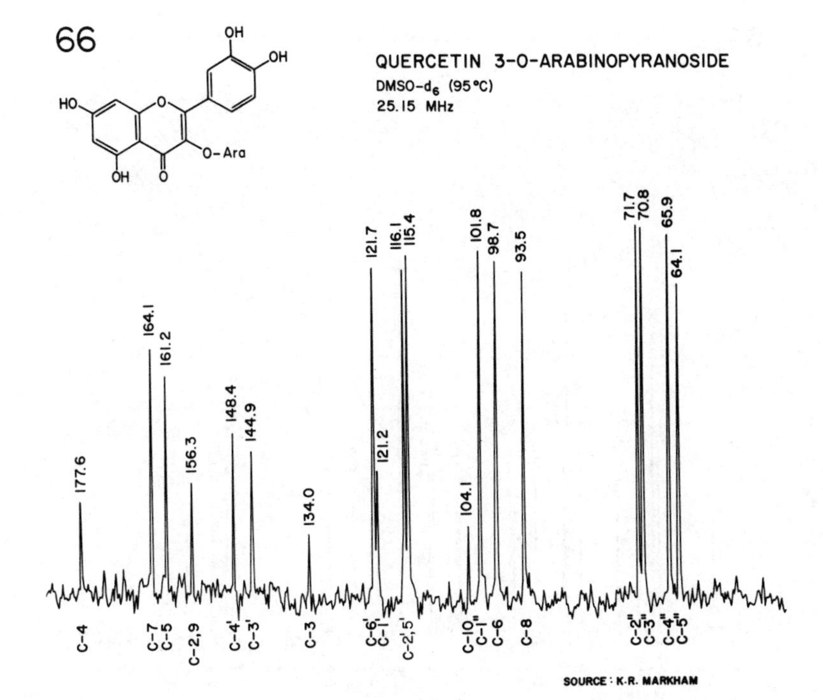

66
OH
OH
HO
O
O-Ara
OH
O
QUERCETIN 3-O-ARABINOPYRANOSIDE
DMSO-d_6 (95°C)
25.15 MHz
177.6
164.1
161.2
156.3
148.4
144.9
134.0
121.7
121.2
116.1
115.4
104.1
101.8
98.7
93.5
71.7
70.8
65.9
64.1
C-4
C-7
C-5
C-2,9
C-4'
C-3'
C-3
C-6'
C-1'
C-2',5'
C-10
C-1''
C-6
C-8
C-2''
C-3''
C-4''
C-5''
SOURCE: K.R. MARKHAM

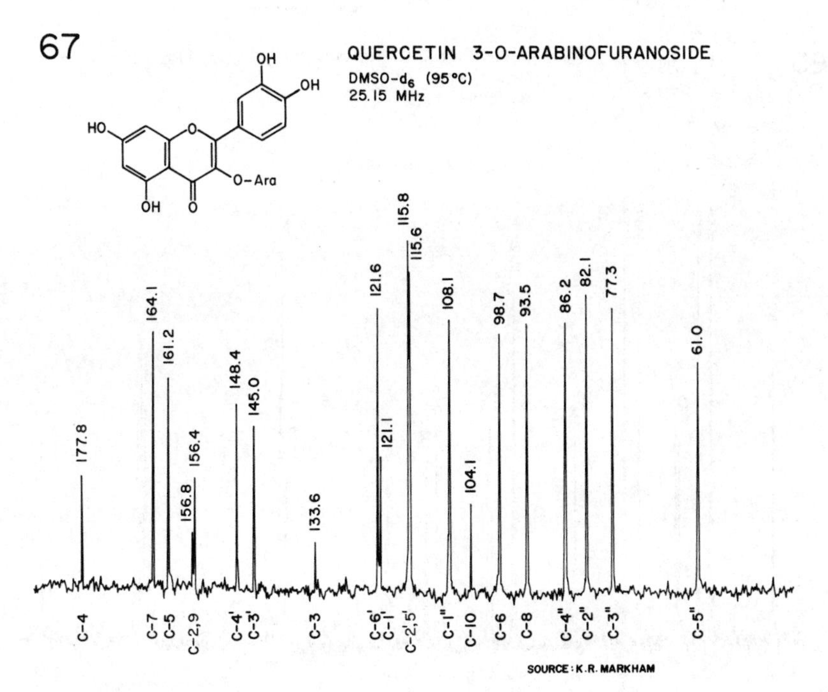

67
OH
OH
HO
O
O-Ara
OH
O
QUERCETIN 3-O-ARABINOFURANOSIDE
DMSO-d_6 (95°C)
25.15 MHz
177.8
164.1
161.2
156.8
156.4
148.4
145.0
133.6
121.6
121.1
115.8
115.6
108.1
104.1
98.7
93.5
86.2
82.1
77.3
61.0
C-4
C-7
C-5
C-2,9
C-4'
C-3'
C-3
C-6'
C-1'
C-2',5'
C-1''
C-10
C-6
C-8
C-4''
C-2''
C-3''
C-5''
SOURCE: K.R. MARKHAM

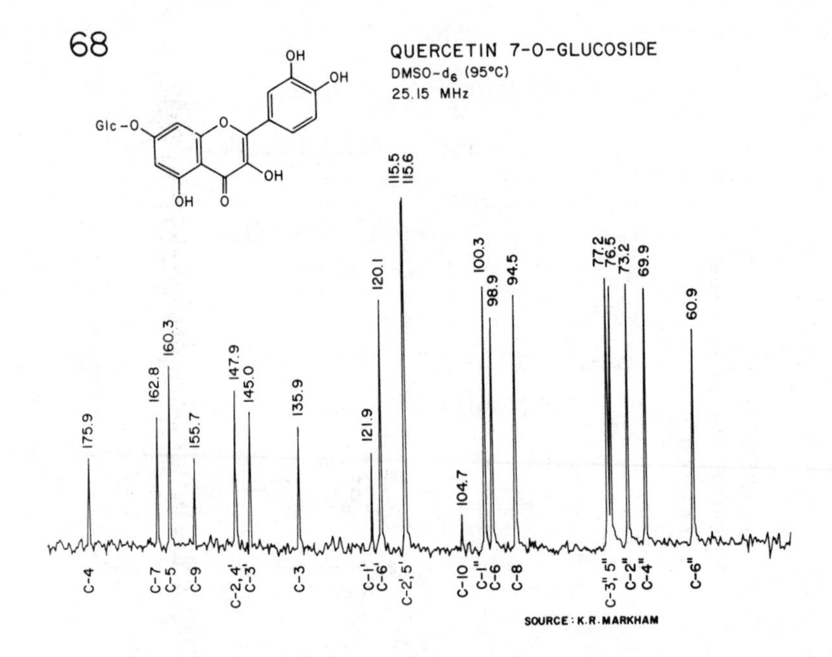
68
QUERCETIN 7-O-GLUCOSIDE
DMSO-d_6 (95°C)
25.15 MHz
OH
OH
Glc-O
OH
OH
O
175.9
162.8
160.3
155.7
147.9
145.0
135.9
121.9
120.1
115.5
115.6
104.7
100.3
98.9
94.5
77.2
76.5
73.2
69.9
60.9
C-4
C-7
C-5
C-9
C-2,4'
C-3'
C-3
C-1'
C-6'
C-2',5'
C-10
C-1''
C-6
C-8
C-3'',5''
C-2''
C-4''
C-6''
SOURCE: K.R.MARKHAM

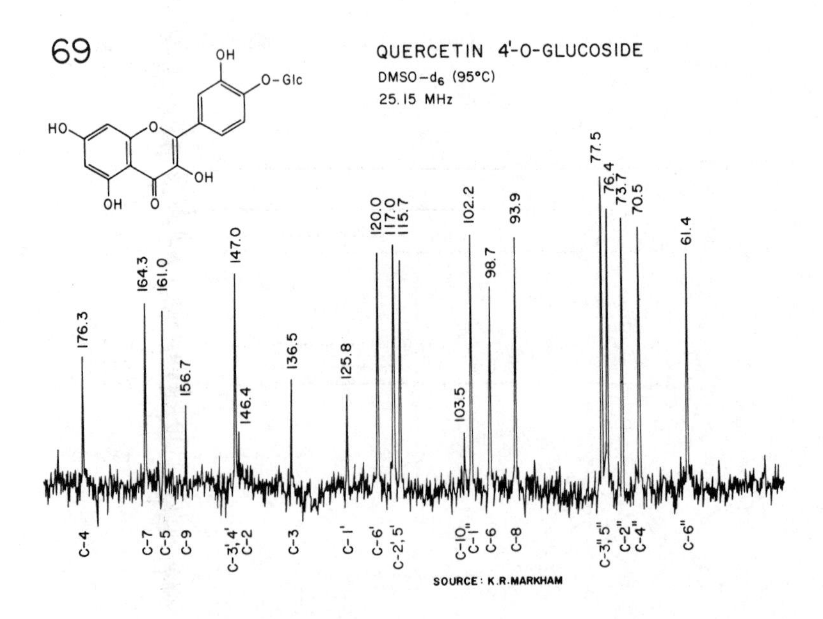
69
QUERCETIN 4'-O-GLUCOSIDE
DMSO-d_6 (95°C)
25.15 MHz
OH
O-Glc
HO
OH
OH
O
176.3
164.3
161.0
156.7
147.0
146.4
136.5
125.8
120.0
117.0
115.7
103.5
102.2
98.7
93.9
77.5
76.4
73.7
70.5
61.4
C-4
C-7
C-5
C-9
C-3',4'
C-2
C-3
C-1'
C-6'
C-2',5'
C-10
C-1''
C-6
C-8
C-3'',5''
C-2''
C-4''
C-6''
SOURCE: K.R.MARKHAM

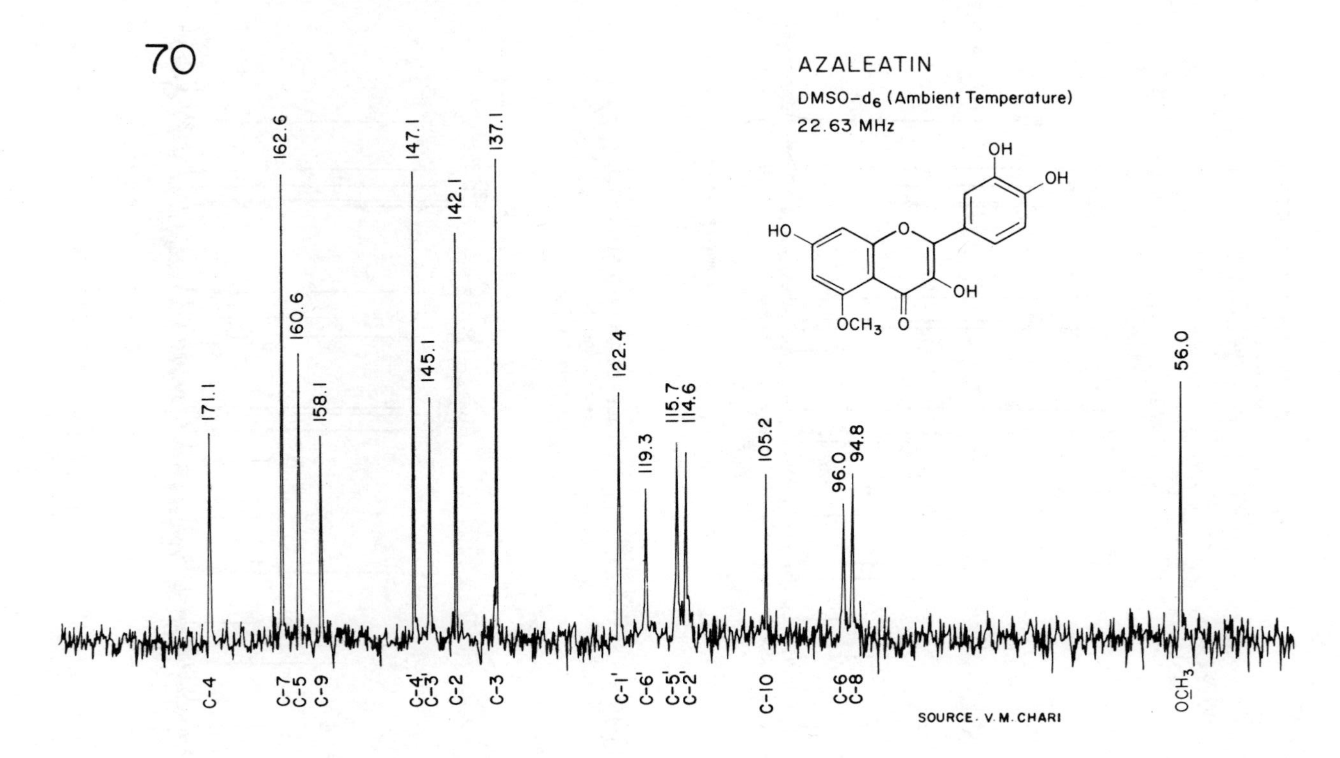
70
AZALEATIN
DMSO-d6 (Ambient Temperature)
22.63 MHz
HO
OH
OH
OH
OCH3
O
171.1
162.6
160.6
158.1
147.1
145.1
142.1
137.1
122.4
119.3
115.7
114.6
105.2
96.0
94.8
56.0
C-4
C-7
C-5
C-9
C-4'
C-3'
C-2
C-3
C-1'
C-6'
C-5'
C-2'
C-10
C-6
C-8
OCH3
SOURCE. V.M.CHARI

71

RHAMNETIN

DMSO-d_6 (Ambient Temperature)

22.63 MHz

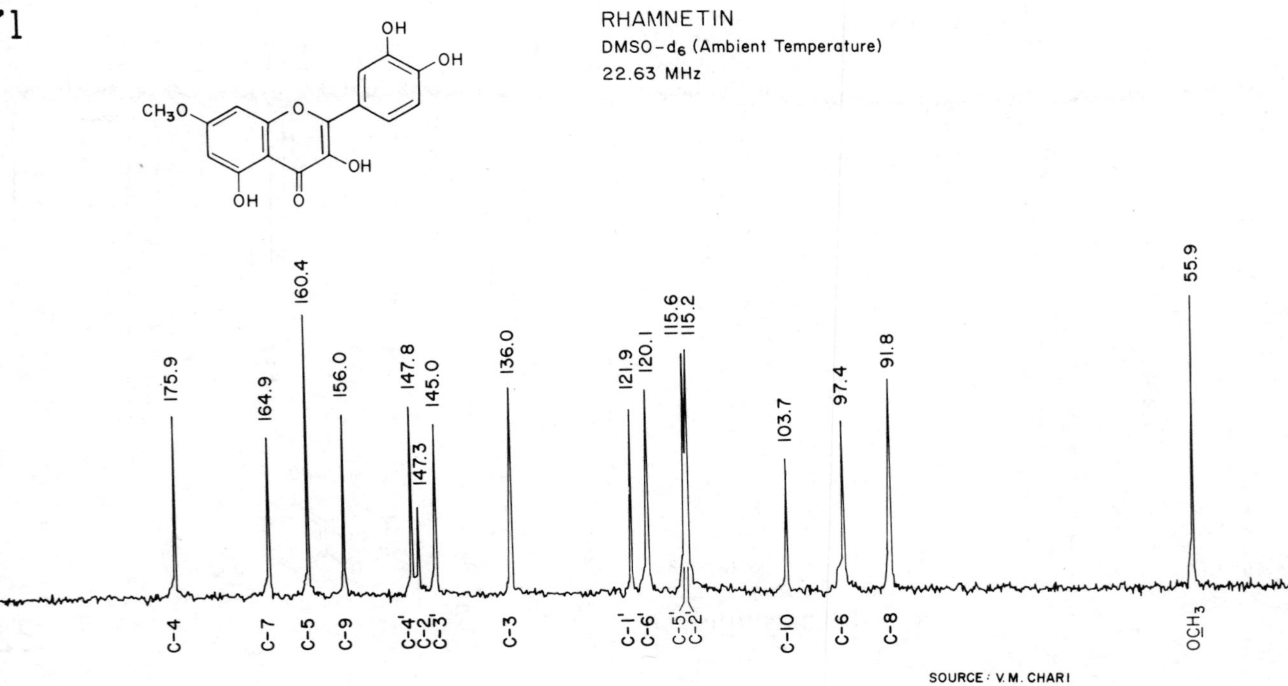

SOURCE: V.M. CHARI

72

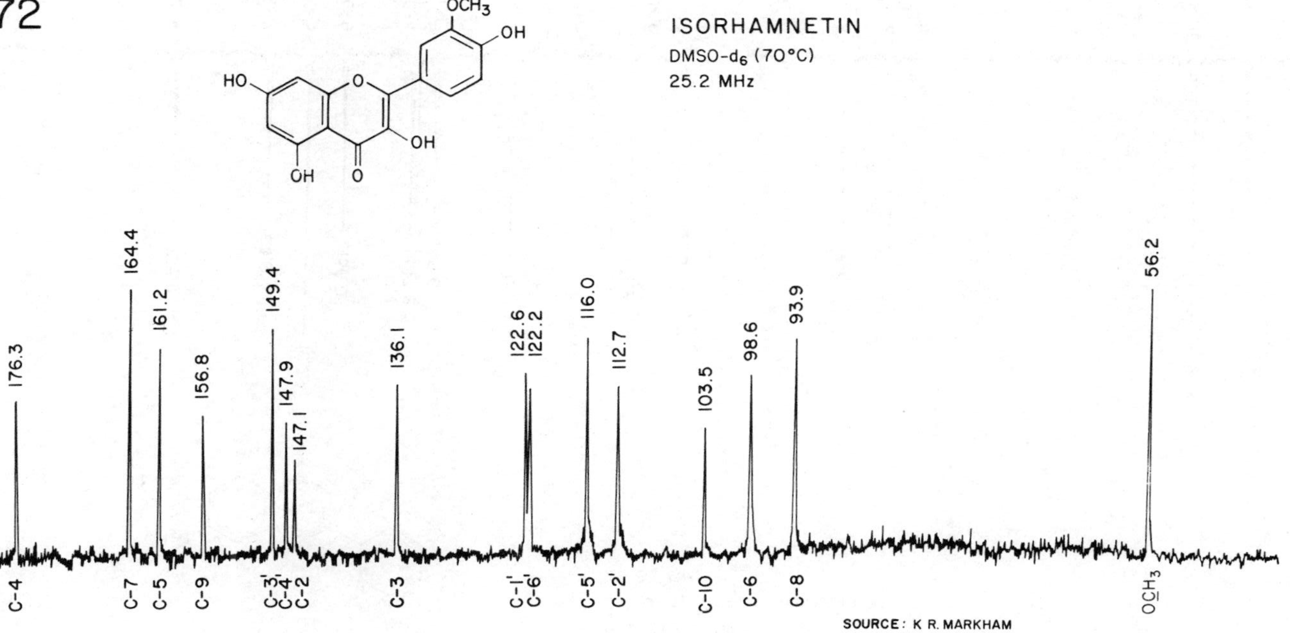
ISORHAMNETIN
DMSO-d6 (70°C)
25.2 MHz
OCH3
OH
HO
O
OH
OH
O
176.3
164.4
161.2
156.8
149.4
147.9
147.1
136.1
122.6
122.2
116.0
112.7
103.5
98.6
93.9
56.2
C-4
C-7
C-5
C-9
C-3'
C-4'
C-2
C-3
C-1'
C-6'
C-5'
C-2'
C-10
C-6
C-8
OCH3
SOURCE: K R. MARKHAM

73

ISORHAMNETIN 3-O-RUTINOSIDE

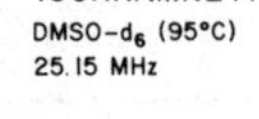

DMSO-d_6 (95°C)
25.15 MHz

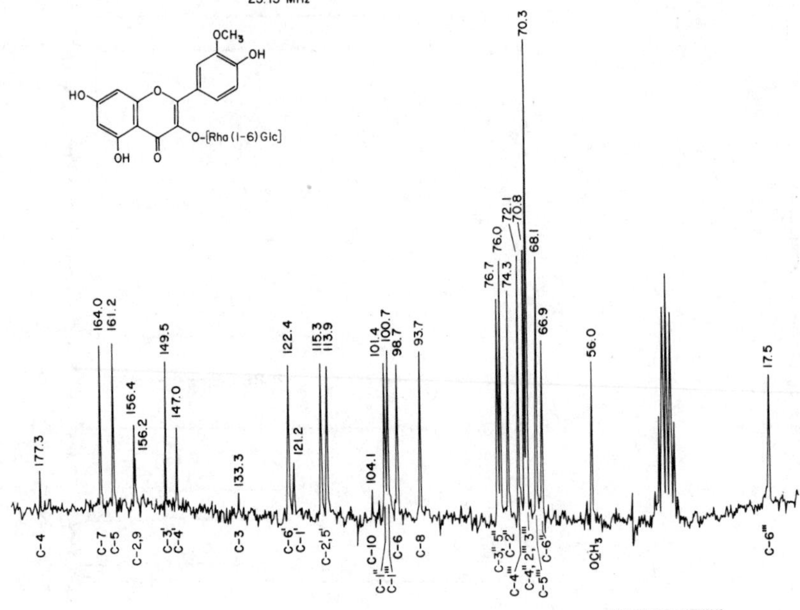

74

TAMARIXETIN 7-O-RUTINOSIDE

DMSO-d_6 (95°C)
25.15 MHz

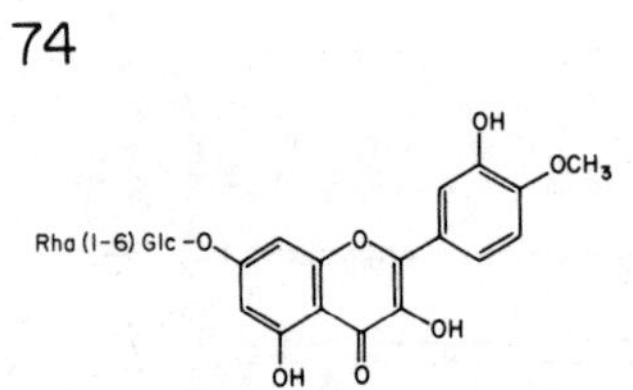

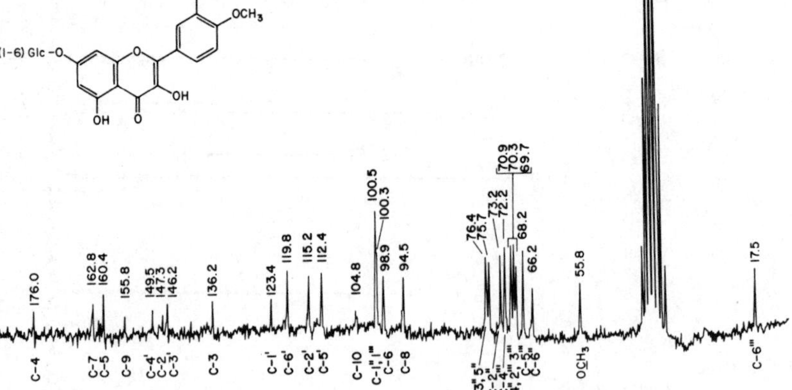

75

QUERCETIN 3,4'-DIMETHYL ETHER

DMSO-d_6 (Ambient Temperature)

22.63 MHz

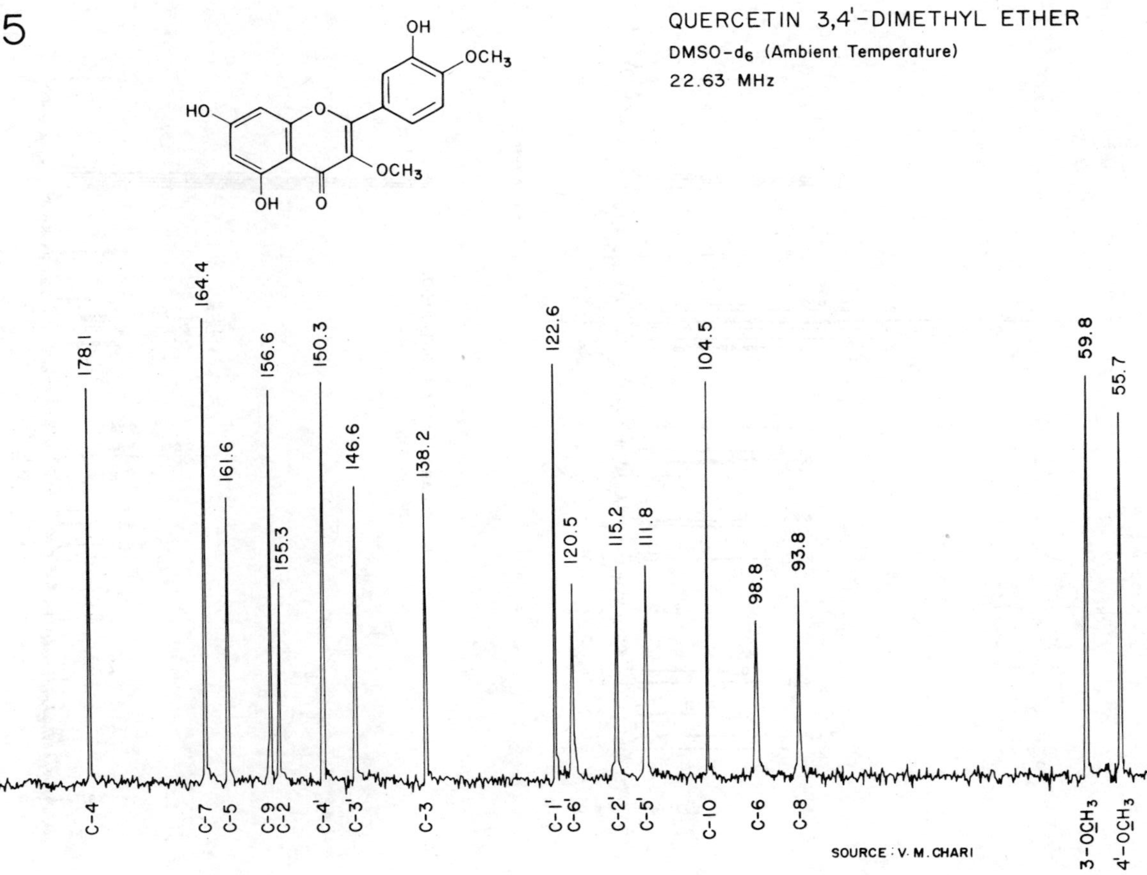

76

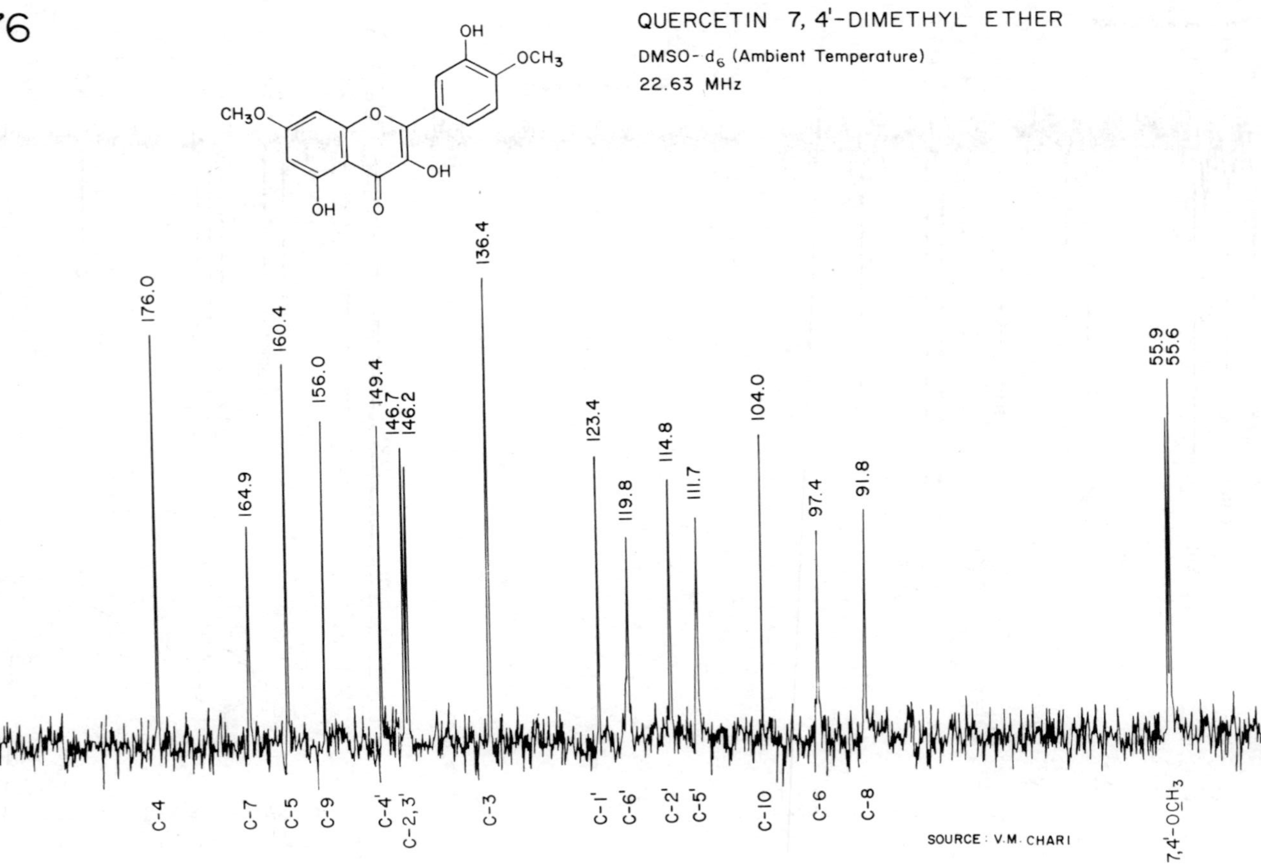
QUERCETIN 7, 4'-DIMETHYL ETHER
DMSO-d_6 (Ambient Temperature)
22.63 MHz
OH
OCH_3
CH_3O
O
OH
OH
O
176.0
164.9
160.4
156.0
149.4
146.7
146.2
136.4
123.4
119.8
114.8
111.7
104.0
97.4
91.8
55.9
55.6
C-4
C-7
C-5
C-9
C-4'
C-2,3'
C-3
C-1'
C-6'
C-2'
C-5'
C-10
C-6
C-8
SOURCE: V.M. CHARI
7,4'-OCH_3

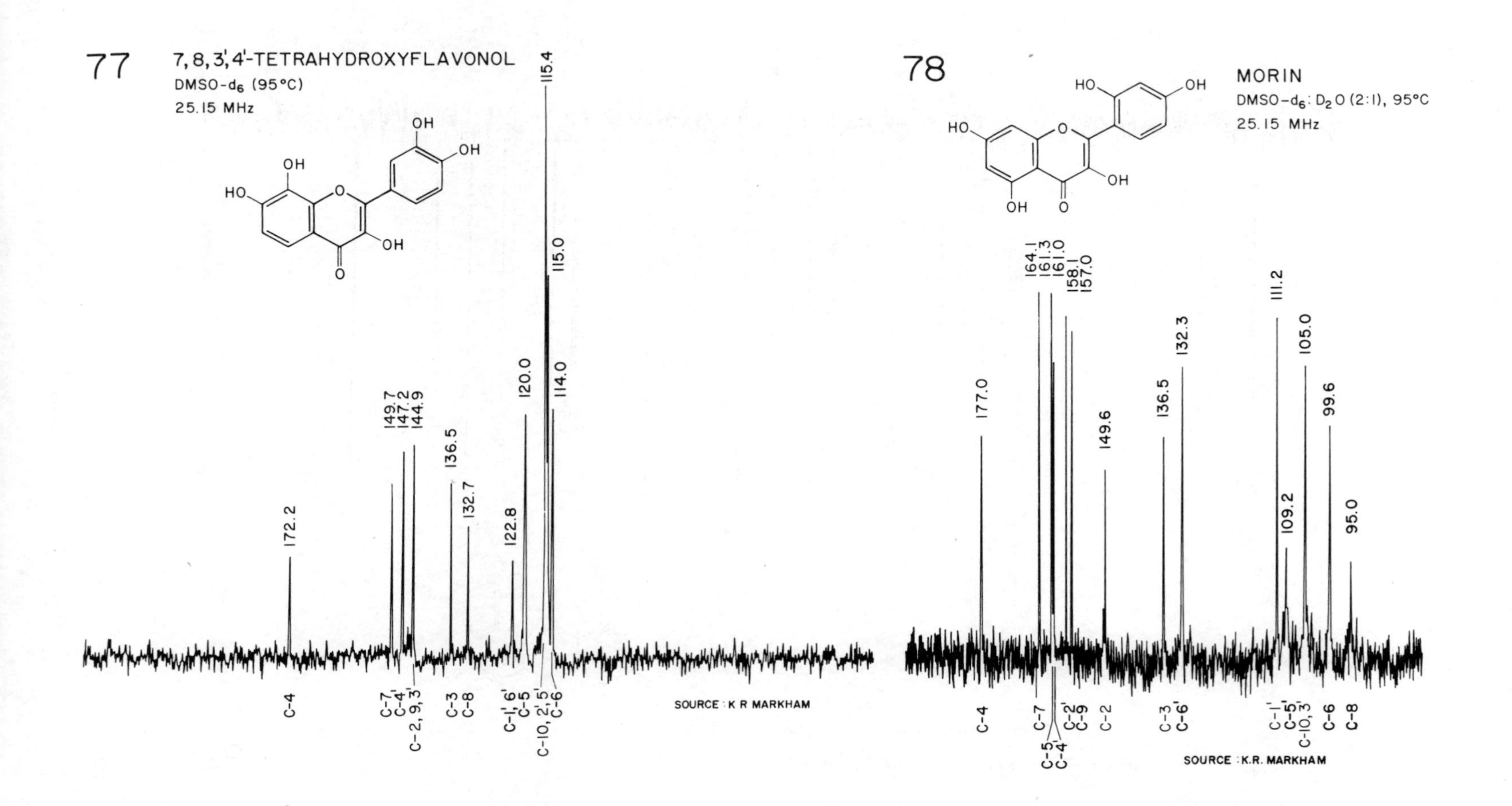
77
7,8,3',4'-TETRAHYDROXYFLAVONOL
DMSO-d_6 (95°C)
25.15 MHz
OH
OH
OH
HO
O
OH
O
172.2
149.7
147.2
144.9
136.5
132.7
122.8
120.0
115.4
115.0
114.0
C-4
C-7
C-4'
C-2, 9, 3'
C-3
C-8
C-1', 6'
C-5
C-10, 2', 5'
C-6
SOURCE : K R MARKHAM
78
MORIN
DMSO-d_6 : D_2O (2:1), 95°C
25.15 MHz
HO
OH
HO
O
OH
OH
O
177.0
164.1
161.3
161.0
158.1
157.0
149.6
136.5
132.3
111.2
109.2
105.0
99.6
95.0
C-4
C-7
C-5'
C-4'
C-2'
C-9
C-2
C-3
C-6'
C-1'
C-5'
C-10, 3'
C-6
C-8
SOURCE : K.R. MARKHAM

79

PATULETIN

DMSO-d_6 (Ambient Temperature)

20 MHz

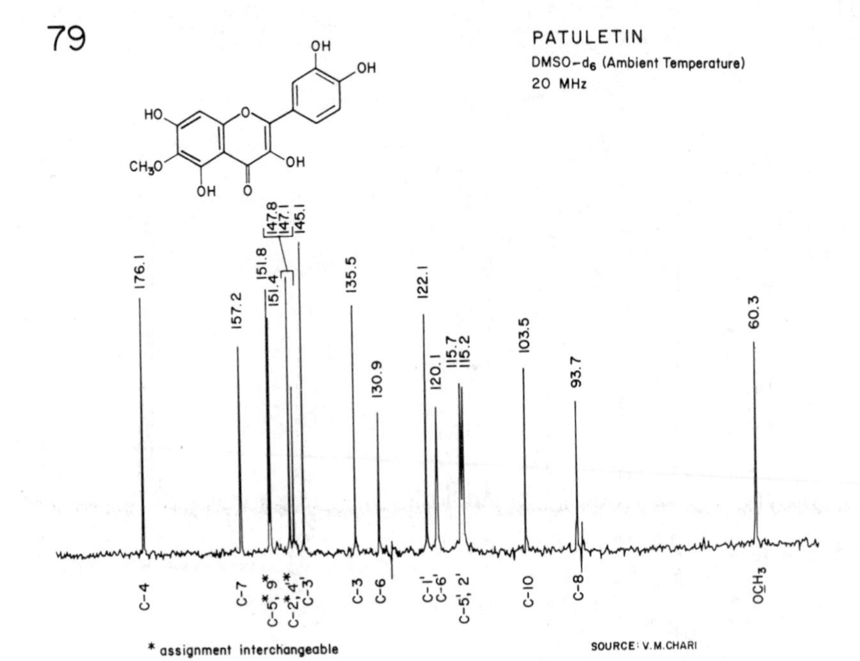

* assignment interchangeable

SOURCE: V.M.CHARI

80

MYRICETIN 3-O-GALACTOSIDE

DMSO-d_6 (95.°C)

25.15 MHz

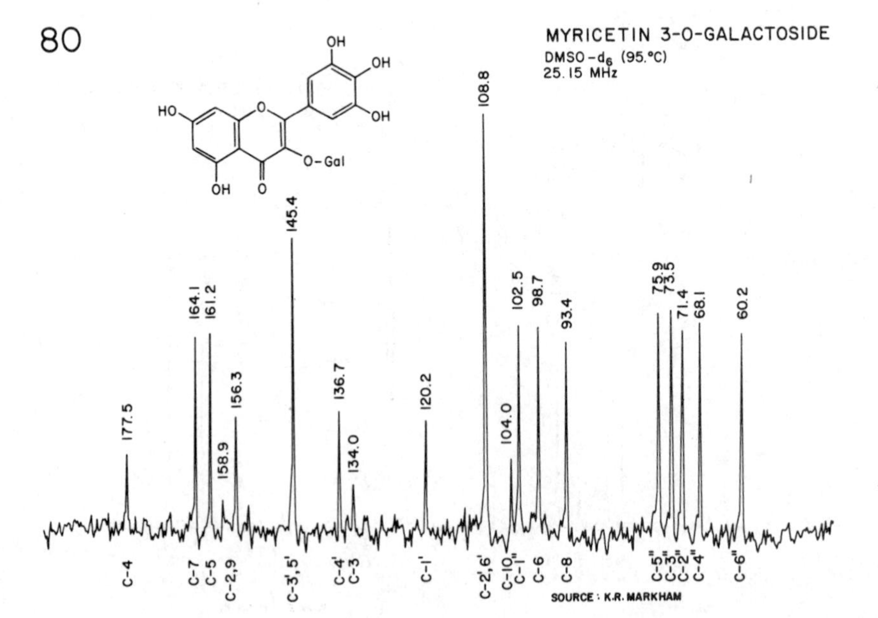

SOURCE: K.R. MARKHAM

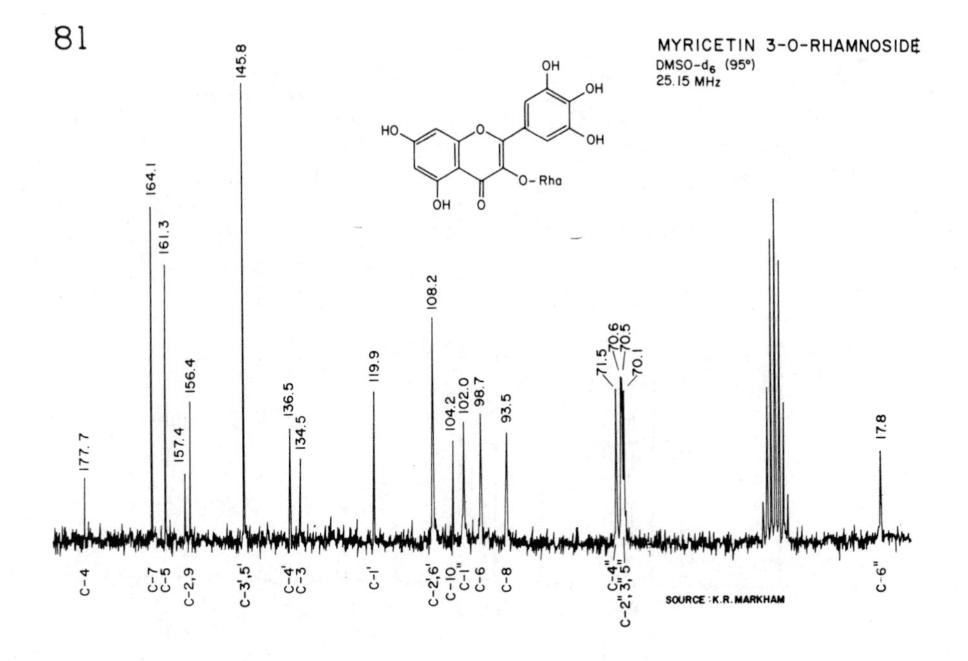
81
MYRICETIN 3-O-RHAMNOSIDE
DMSO-d6 (95°)
25.15 MHz
HO
OH
OH
OH
O-Rha
OH
O
177.7
164.1
161.3
157.4
156.4
145.8
136.5
134.5
119.9
108.2
104.2
102.0
98.7
93.5
71.5
70.6
70.5
70.1
17.8
C-4
C-7
C-5
C-2,9
C-3',5'
C-4'
C-3
C-1'
C-2',6'
C-10
C-1"
C-6
C-8
C-4"
C-2",3",5"
C-6"
SOURCE: K.R. MARKHAM

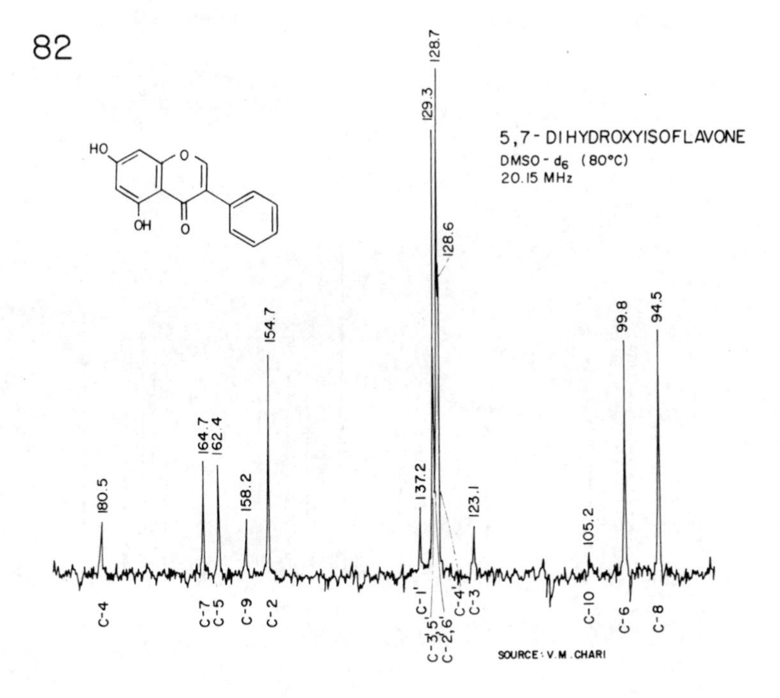
82
5,7-DIHYDROXYISOFLAVONE
DMSO-d6 (80°C)
20.15 MHz
HO
O
OH
O
180.5
164.7
162.4
158.2
154.7
137.2
129.3
128.7
128.6
123.1
105.2
99.8
94.5
C-4
C-7
C-5
C-9
C-2
C-1'
C-3',5'
C-2',6'
C-4'
C-3
C-10
C-6
C-8
SOURCE: V.M. CHARI

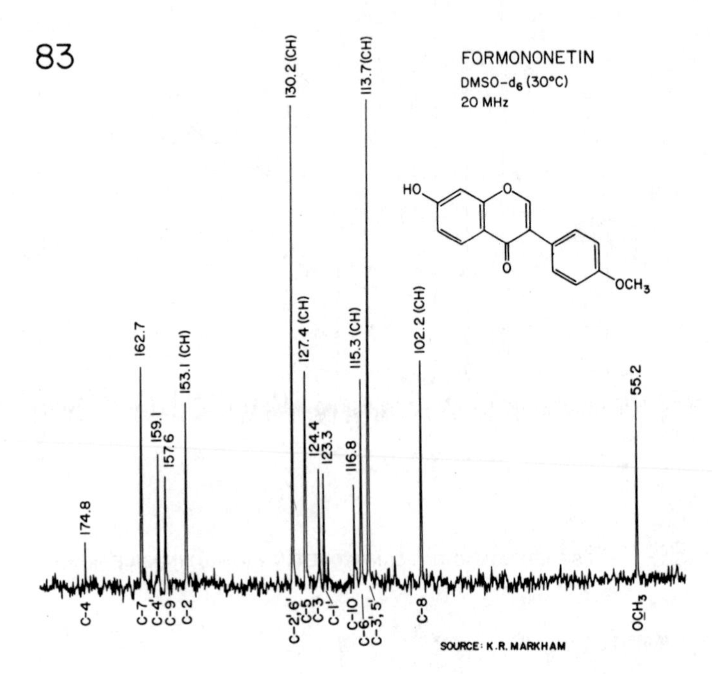

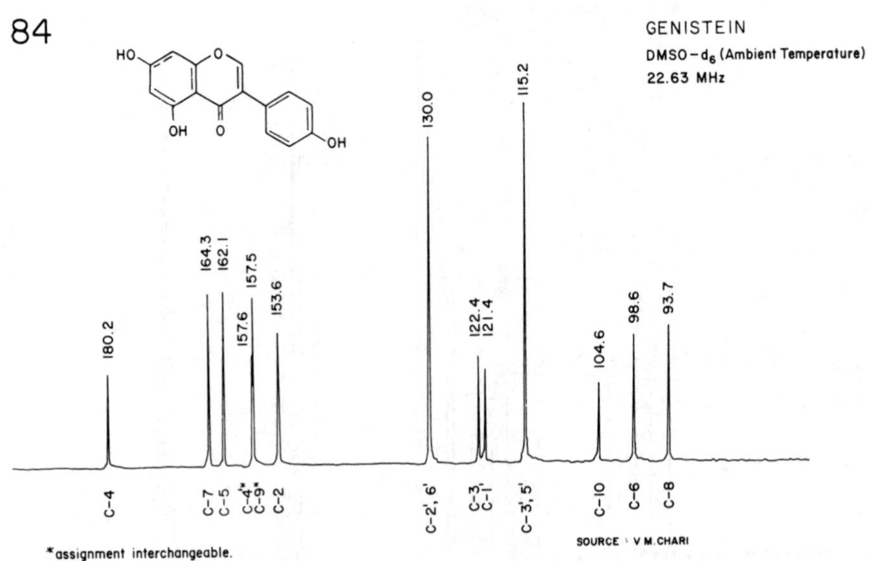

*assignment interchangeable.

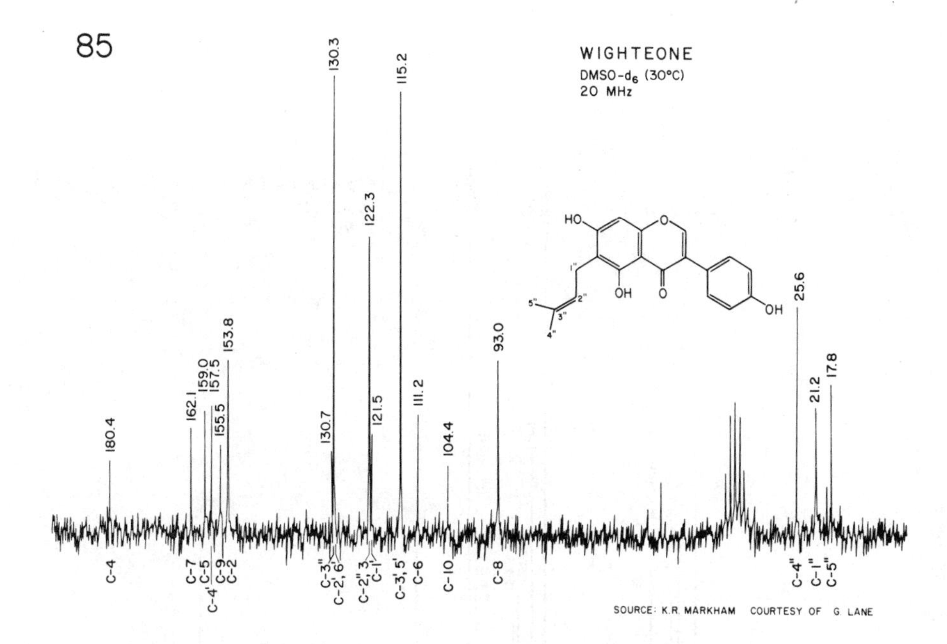

86 PSEUDOBAPTISIN (FORMONONETIN 7-O-RUTINOSIDE Impurity*)
DMSO-d_6 (95°C)
25.15 MHz

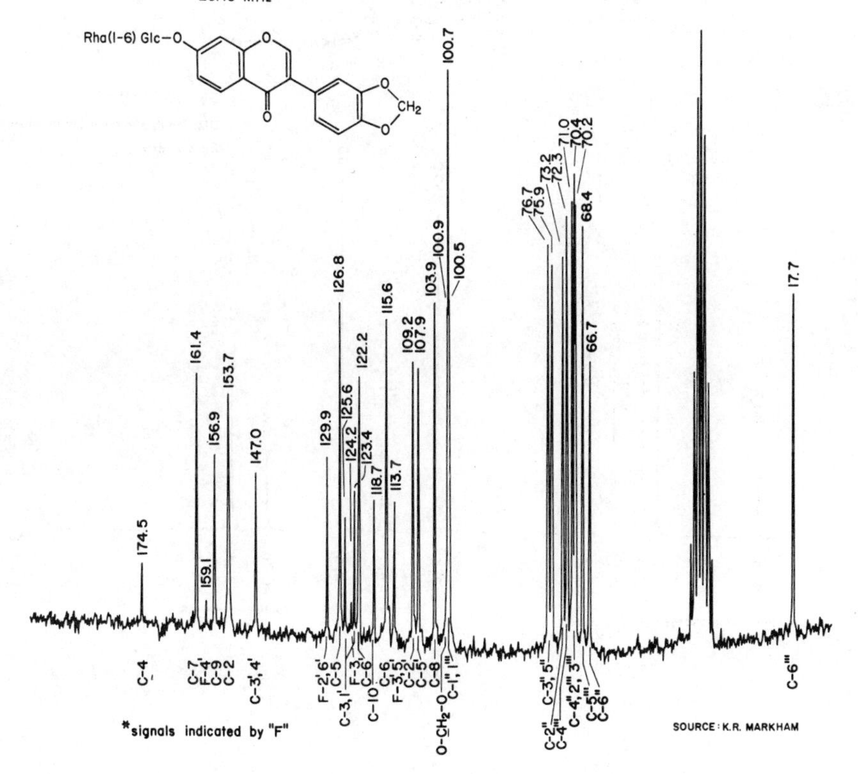

87

2' - HYDROXYFORMONONETIN

DMSO - d_6 (30°C)
20 MHz

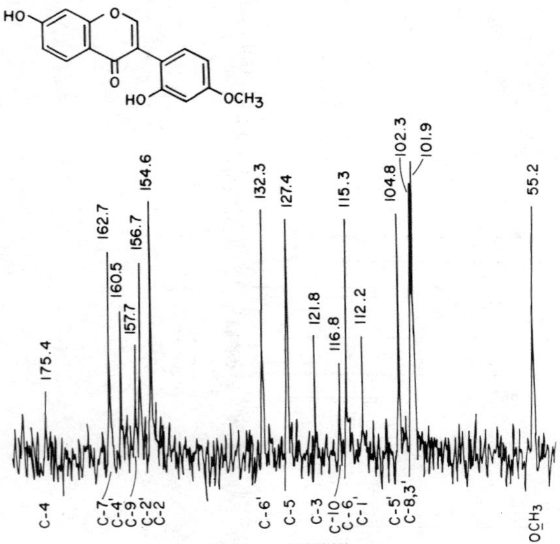

175.4 C-4
162.7 C-7
160.5 C-4'
157.7 C-9
156.7 C-2'
154.6 C-2
132.3 C-6'
127.4 C-5
121.8 C-3
116.8 C-10
115.3 C-6
112.2 C-1'
104.8 C-5'
102.3 C-8,3'
101.9
55.2 OCH_3

SOURCE: K.R. MARKHAM
COURTESY OF G. LANE

88

LUTEONE

DMSO-d_6 (30°C)
20 MHz

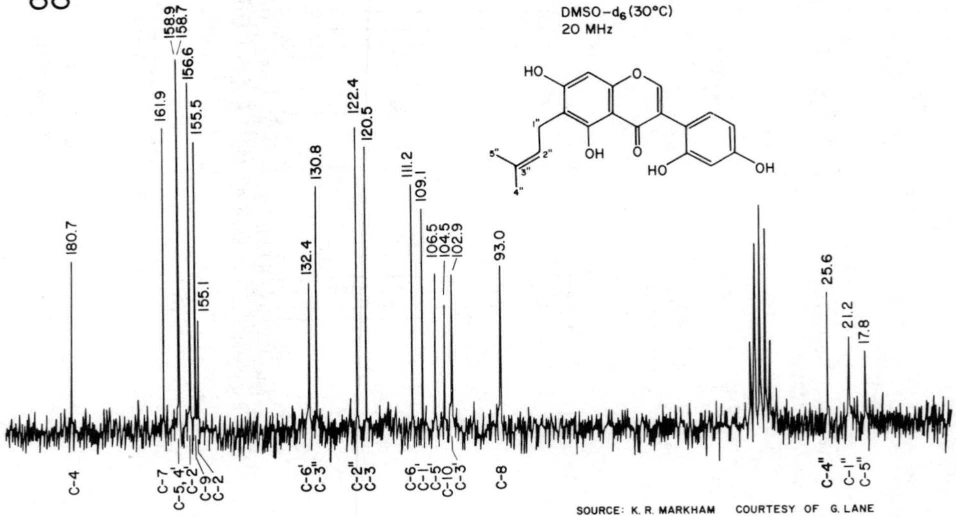

180.7 C-4
161.9 C-7
158.9 C-5,4'
158.7
156.6 C-2'
155.5 C-9
155.1 C-2
132.4 C-6'
130.8 C-3''
122.4 C-2''
120.5 C-3
111.2 C-6
109.1 C-1'
106.5 C-5'
104.5 C-10
102.9 C-3'
93.0 C-8
25.6 C-4''
21.2 C-1''
17.8 C-5''

SOURCE: K. R. MARKHAM COURTESY OF G. LANE

89 2'-HYDROXYGENISTEIN

DMSO-d_6 (30°C)
20 MHz

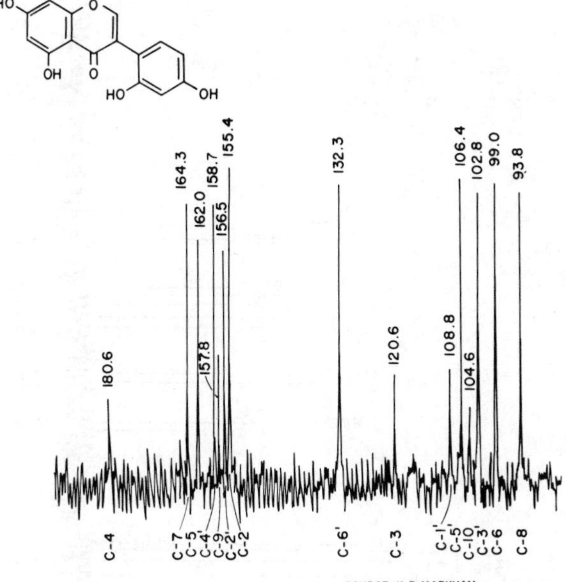

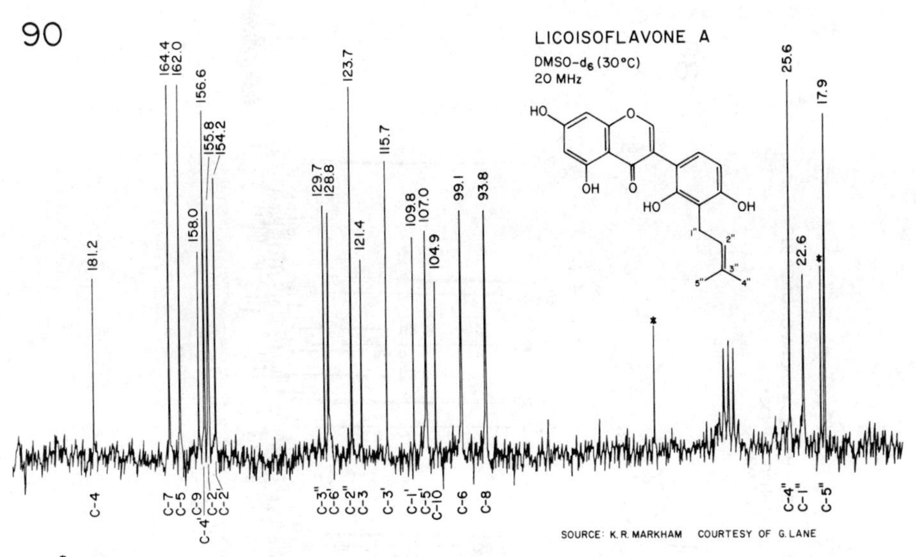

*Ethanol Peaks

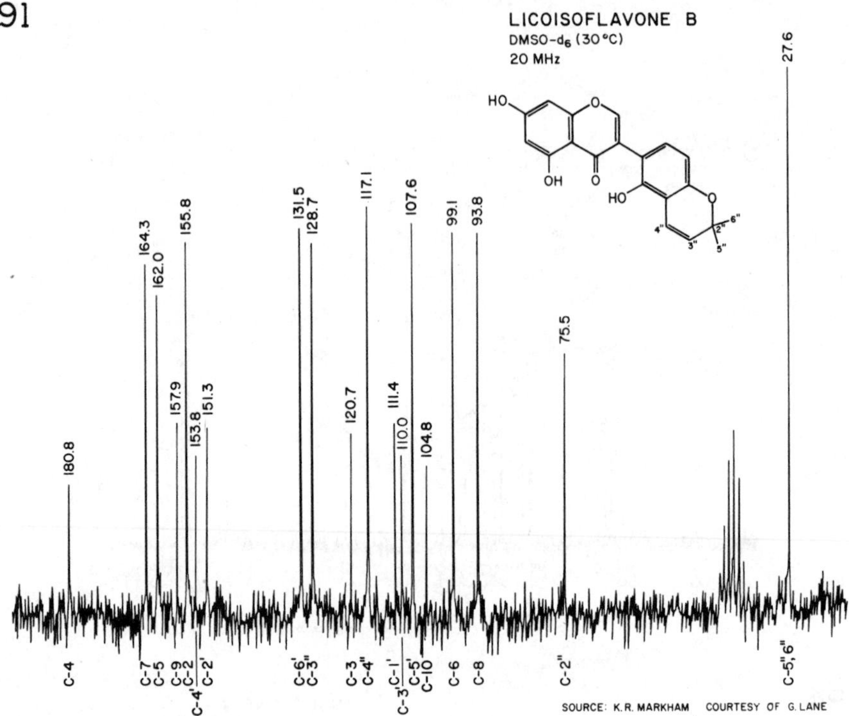
LICOISOFLAVONE B
DMSO-d6 (30°C)
20 MHz
180.8
164.3
162.0
157.9
155.8
153.8
151.3
131.5
128.7
120.7
117.1
111.4
110.0
107.6
104.8
99.1
93.8
75.5
27.6
C-4
C-7
C-5
C-9
C-2
C-4'
C-2'
C-6'
C-3"
C-3
C-4"
C-1'
C-3'
C-5'
C-10
C-6
C-8
C-2"
C-5",6"
SOURCE: K.R. MARKHAM COURTESY OF G. LANE

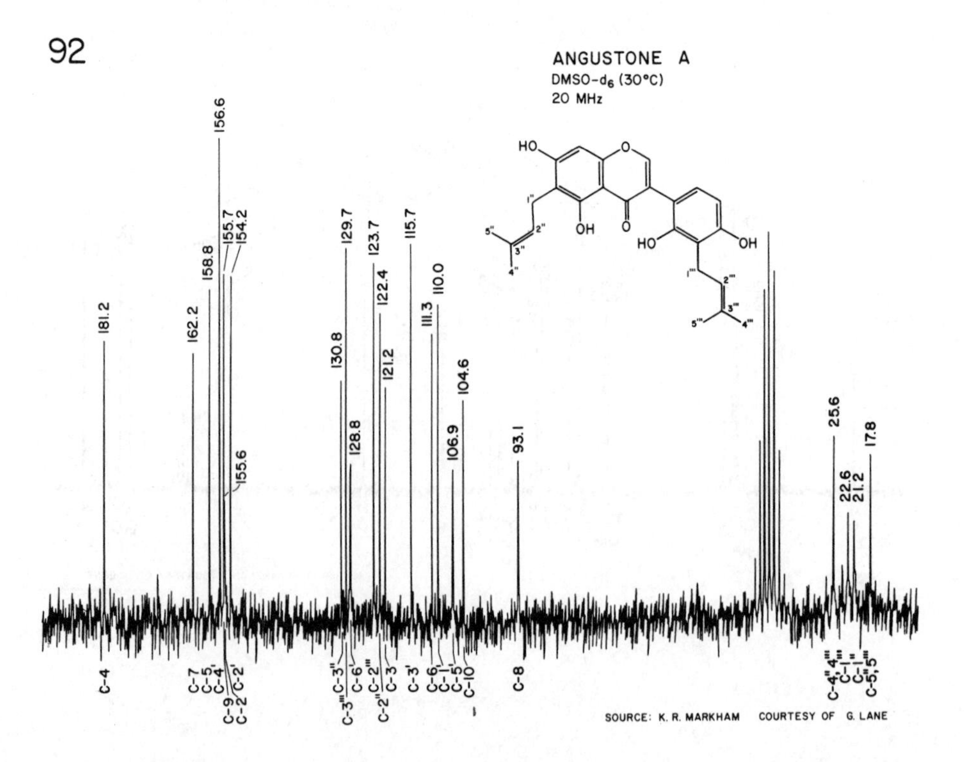
92
ANGUSTONE A
DMSO-d6 (30°C)
20 MHz
181.2
162.2
158.8
156.6
155.7
154.2
155.6
130.8
129.7
128.8
123.7
122.4
121.2
115.7
111.3
110.0
106.9
104.6
93.1
25.6
22.6
21.2
17.8
C-4
C-7
C-5
C-4'
C-9
C-2
C-2'
C-3'"
C-3"
C-6'
C-2'"
C-2"
C-3
C-3'
C-6
C-1
C-5
C-10
C-8
C-4",4'"
C-1'"
C-1"
C-5",5'"
SOURCE: K.R. MARKHAM COURTESY OF G. LANE

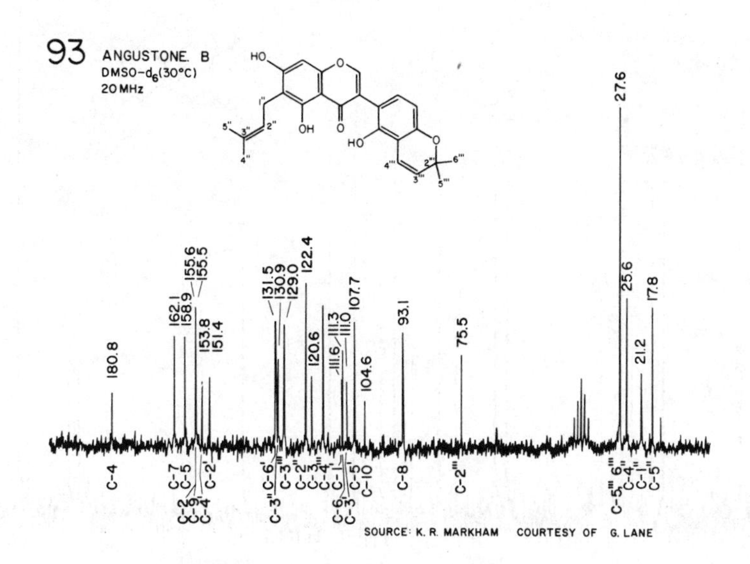
93
ANGUSTONE B
DMSO-d_6 (30°C)
20 MHz
HO
O
1''
2''
3''
4''
5''
OH
O
HO
O
2'''
3'''
4'''
5'''
6'''
180.8
162.1
158.9
155.6
155.5
153.8
151.4
131.5
130.9
129.0
122.4
120.6
111.6
111.3
111.0
107.7
104.6
93.1
75.5
27.6
25.6
21.2
17.8
C-4
C-7
C-5
C-2
C-9
C-4'
C-2'
C-3''
C-6'
C-3'''
C-2''
C-3'
C-4'''
C-1'
C-6
C-3
C-5'
C-10
C-8
C-2'''
C-5''', 6'''
C-4''
C-1''
C-5''
SOURCE: K. R. MARKHAM COURTESY OF G. LANE

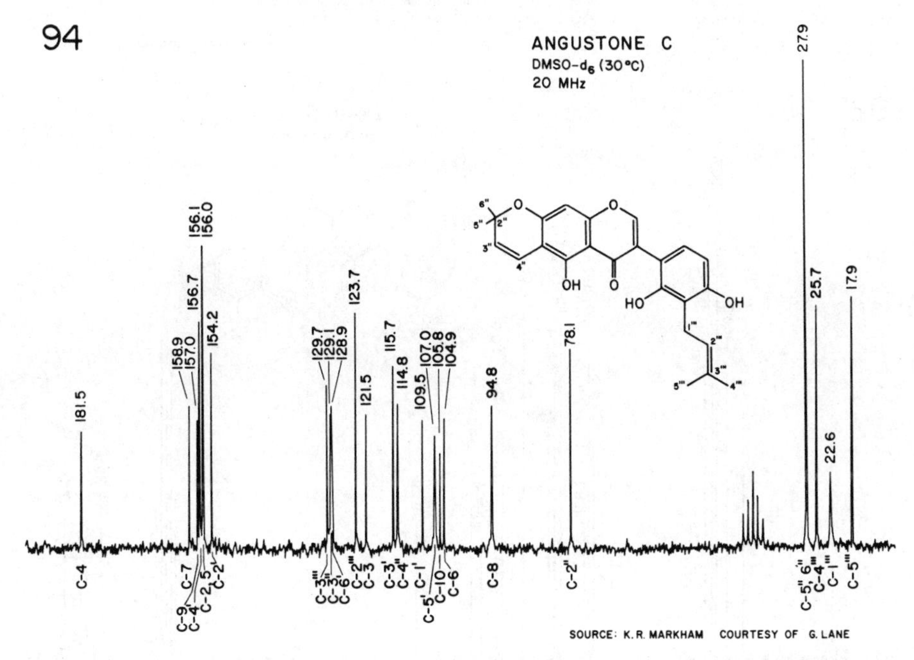
94
ANGUSTONE C
DMSO-d_6 (30°C)
20 MHz
O
O
2''
3''
4''
5''
6''
OH
O
HO
OH
1'''
2'''
3'''
4'''
5'''
181.5
158.9
157.0
156.7
156.1
156.0
154.2
129.7
129.1
128.9
123.7
121.5
115.7
114.8
109.5
107.0
105.8
104.9
94.8
78.1
27.9
25.7
22.6
17.9
C-4
C-9
C-4
C-7
C-2, 5
C-2'
C-3'''
C-3''
C-6'
C-2'''
C-3
C-3'
C-4''
C-1'
C-5'
C-10
C-6
C-8
C-2''
C-5'', 6''
C-4'''
C-1'''
C-5'''
SOURCE: K. R. MARKHAM COURTESY OF G. LANE

95

2,2'-DIHYDROXYCHALCONE

DMSO-d_6 D_2O (7:2) (95°C)
25.15 MHz

96

ISOLIQUIRITIGENIN

DMSO-d_6 (95°C)
25.15 MHz

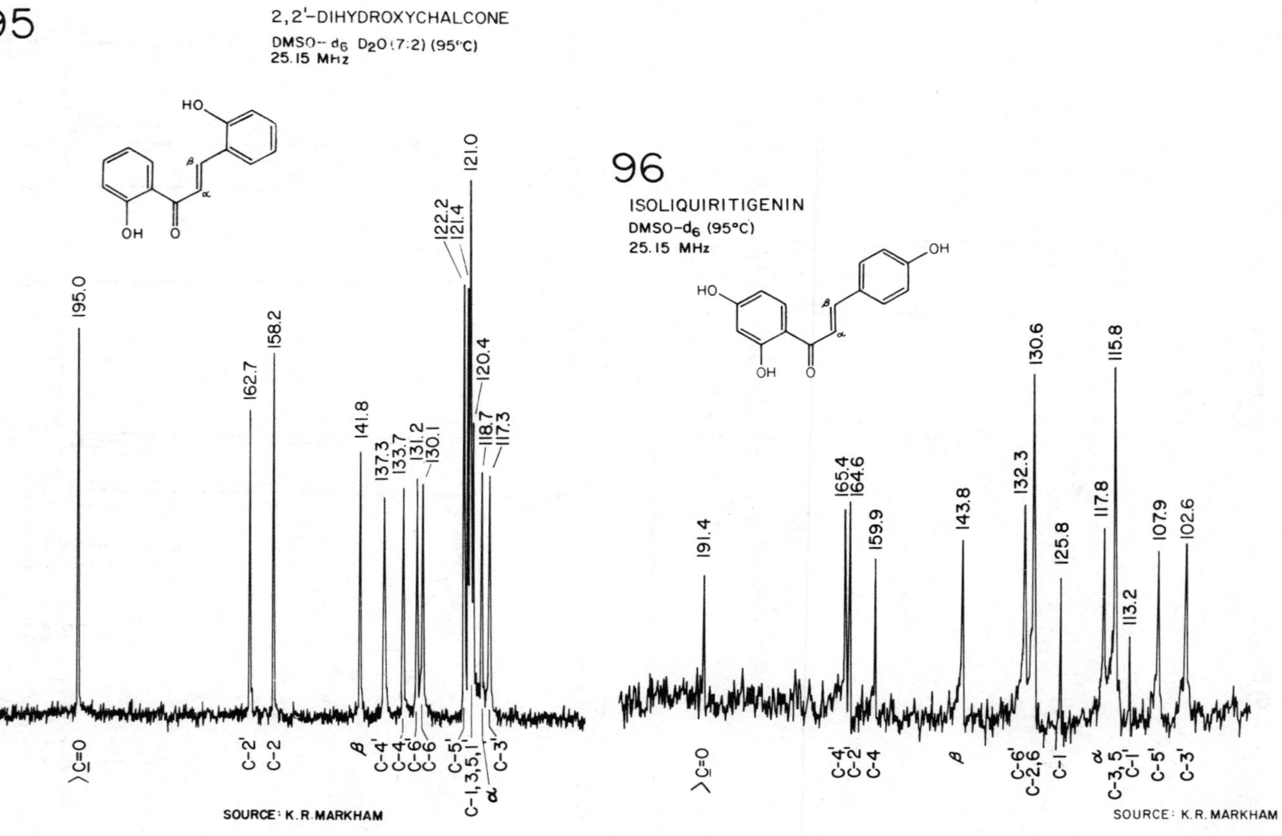

97

SULFURETIN
DMSO-d_6 (60° C)
20.15 MHz

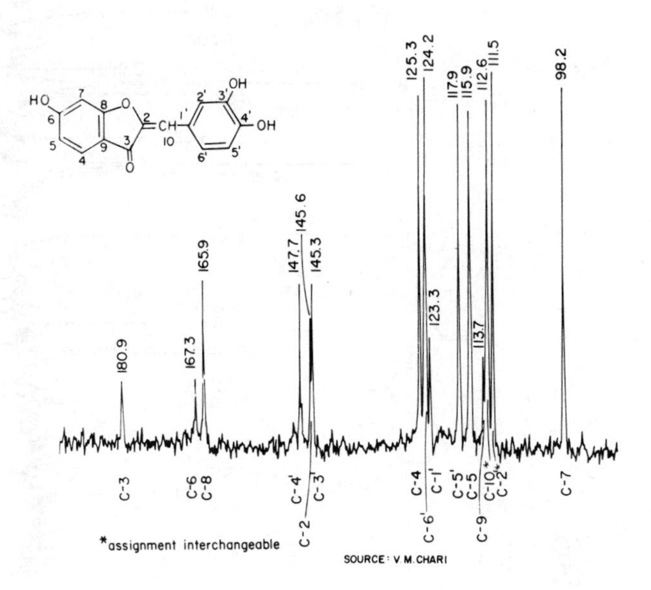

98

5,7-DIMETHYL 4'-METHOXYFLAVANONE
DMSO-d_6 (95°C)
25.15 MHz

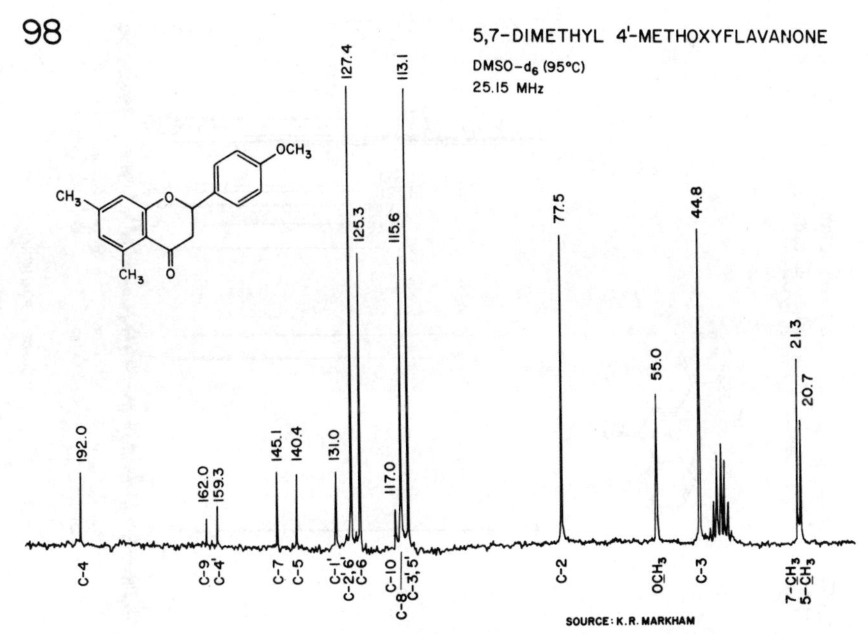

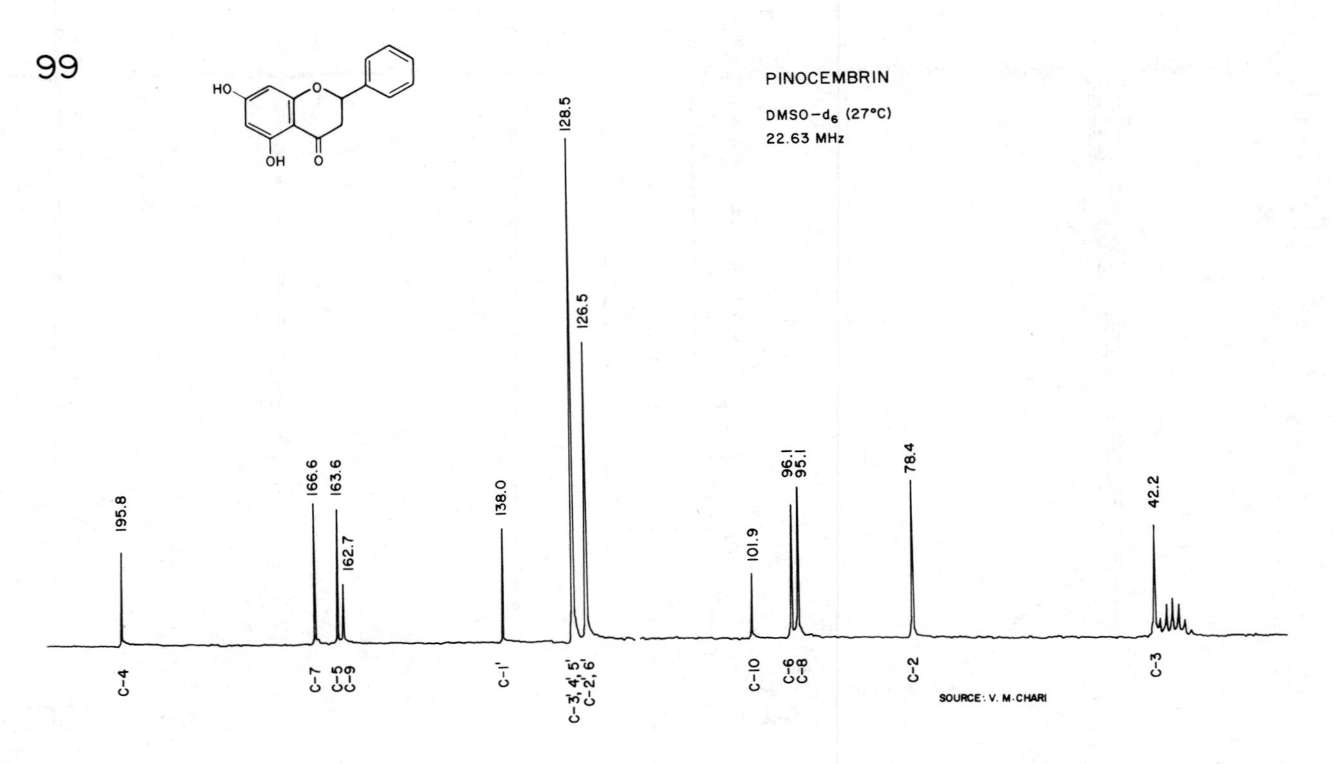
99
HO
O
OH
O
PINOCEMBRIN
DMSO-d_6 (27°C)
22.63 MHz
195.8
166.6
163.6
162.7
138.0
128.5
126.5
101.9
96.1
95.1
78.4
42.2
C-4
C-7
C-5
C-9
C-1'
C-3', 4', 5'
C-2', 6'
C-10
C-6
C-8
C-2
C-3
SOURCE: V. M. CHARI

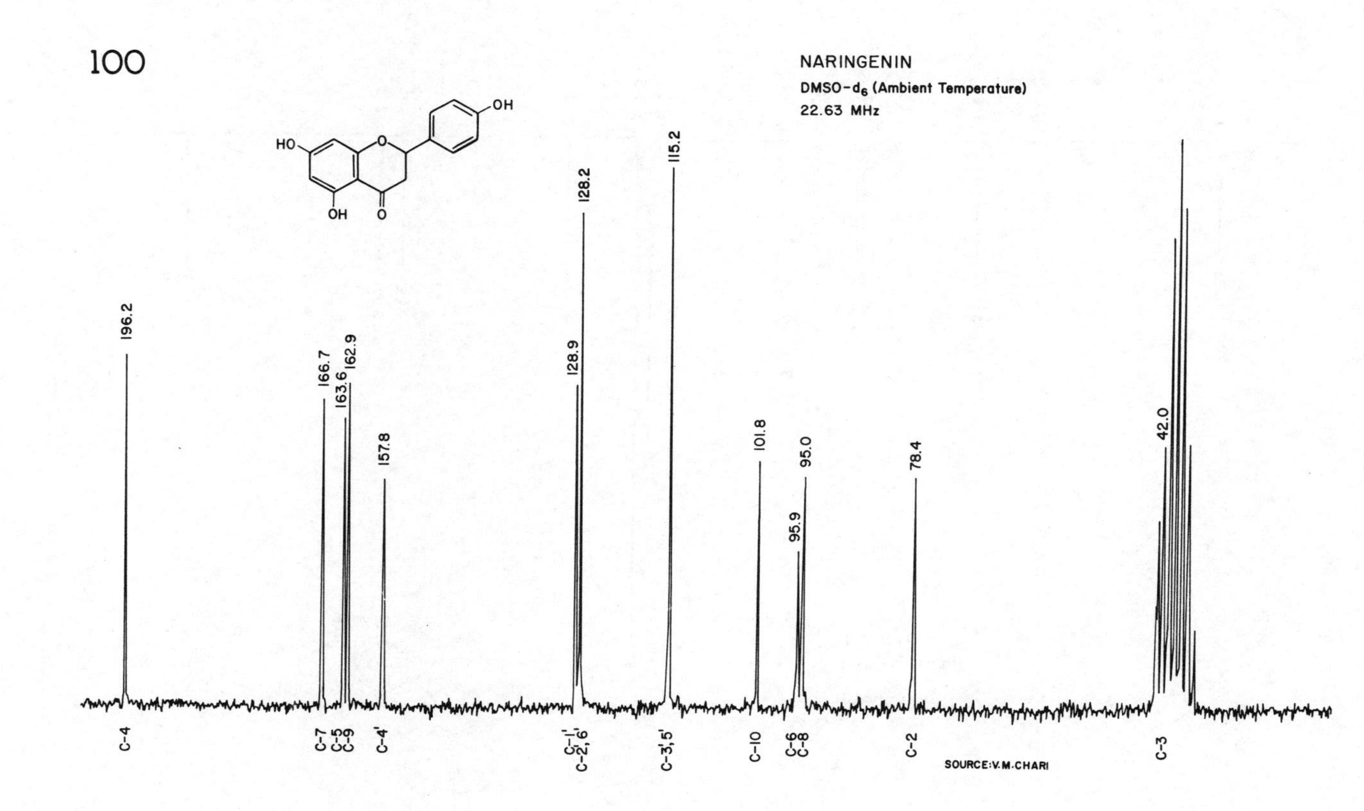
100
NARINGENIN
DMSO-d6 (Ambient Temperature)
22.63 MHz
HO
OH
O
OH
O
196.2
166.7
163.6
162.9
157.8
128.9
128.2
115.2
101.8
95.9
95.0
78.4
42.0
C-4
C-7
C-5
C-9
C-4'
C-1'
C-2,6'
C-3,5'
C-10
C-6
C-8
C-2
C-3
SOURCE: V.M.CHARI

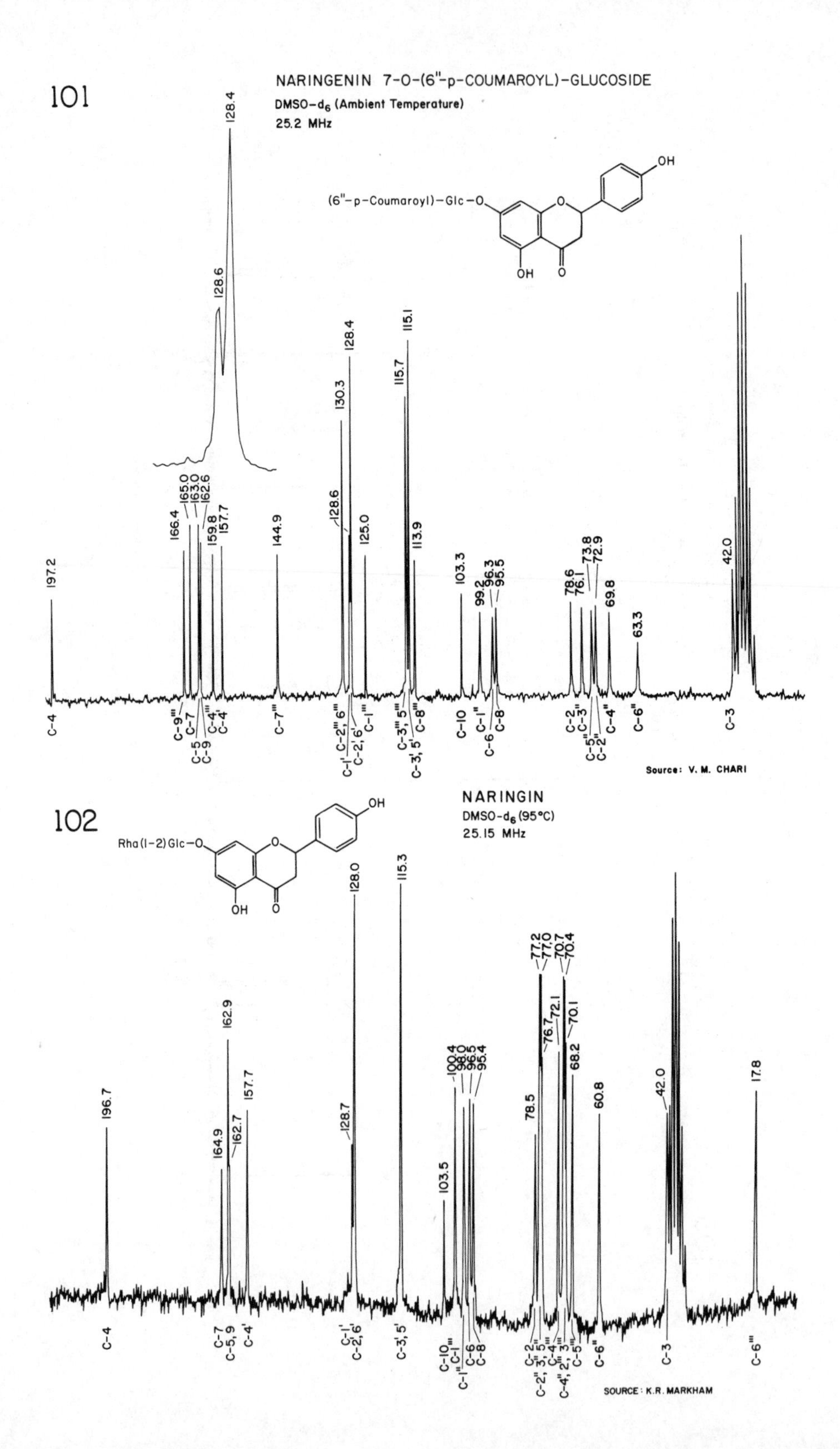
101
NARINGENIN 7-O-(6"-p-COUMAROYL)-GLUCOSIDE
DMSO-d6 (Ambient Temperature)
25.2 MHz
(6"-p-Coumaroyl)-Glc-O
OH
OH
O
Source: V. M. CHARI
102
NARINGIN
DMSO-d6 (95°C)
25.15 MHz
Rha(1-2)Glc-O
OH
OH
O
SOURCE: K.R. MARKHAM

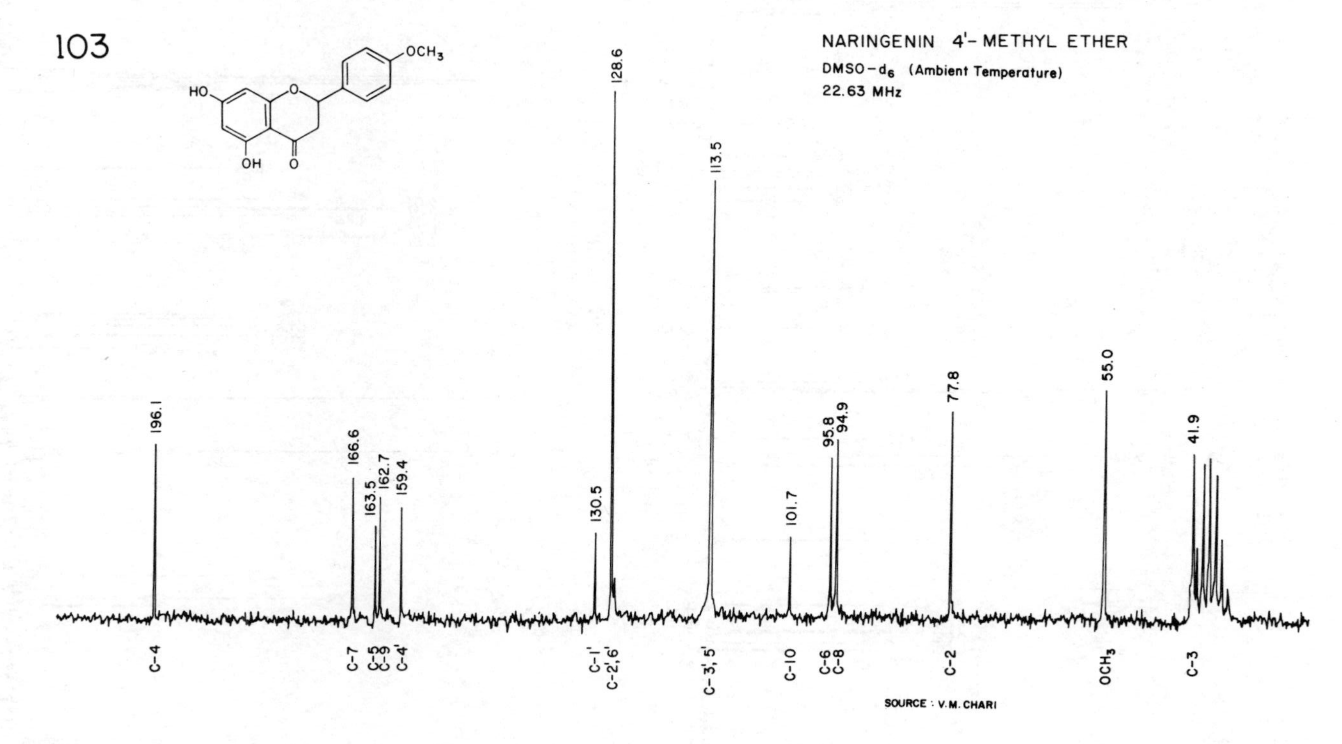
103
NARINGENIN 4'- METHYL ETHER
DMSO-d_6 (Ambient Temperature)
22.63 MHz
HO
OH
O
OCH_3
196.1
166.6
163.5
162.7
159.4
130.5
128.6
113.5
101.7
95.8
94.9
77.8
55.0
41.9
C-4
C-7
C-5
C-9
C-4'
C-1'
C-2',6'
C-3',5'
C-10
C-6
C-8
C-2
OCH_3
C-3
SOURCE : V.M. CHARI

104

ERIODICTYOL

DMSO-d_6 (95 °C)

25.15 MHz

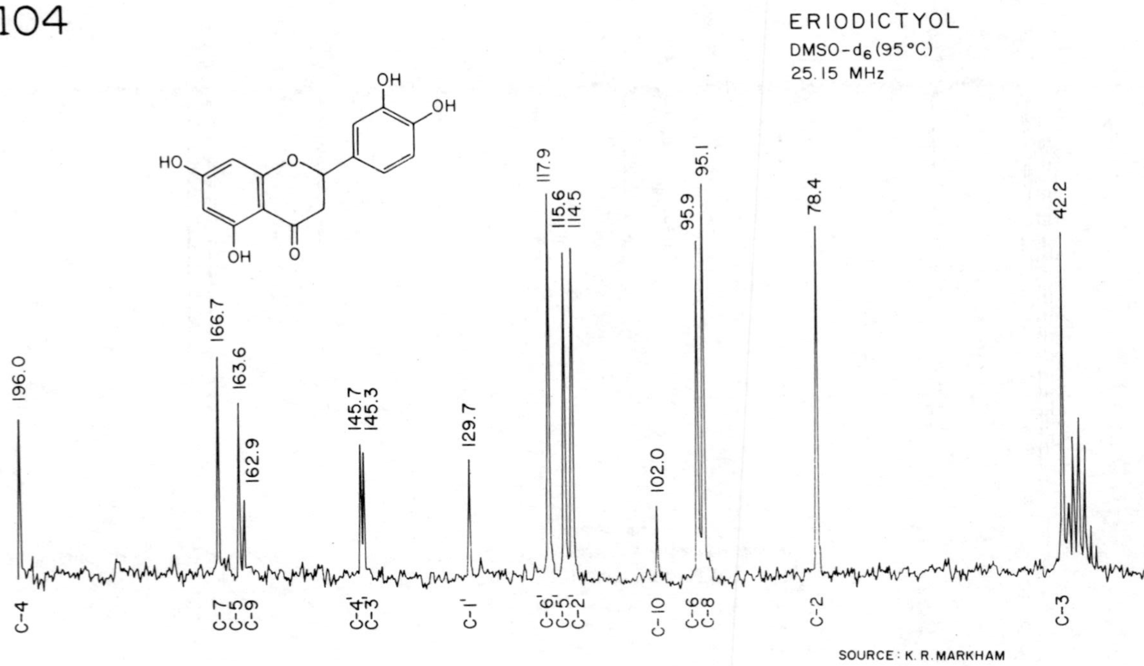

SOURCE: K. R. MARKHAM

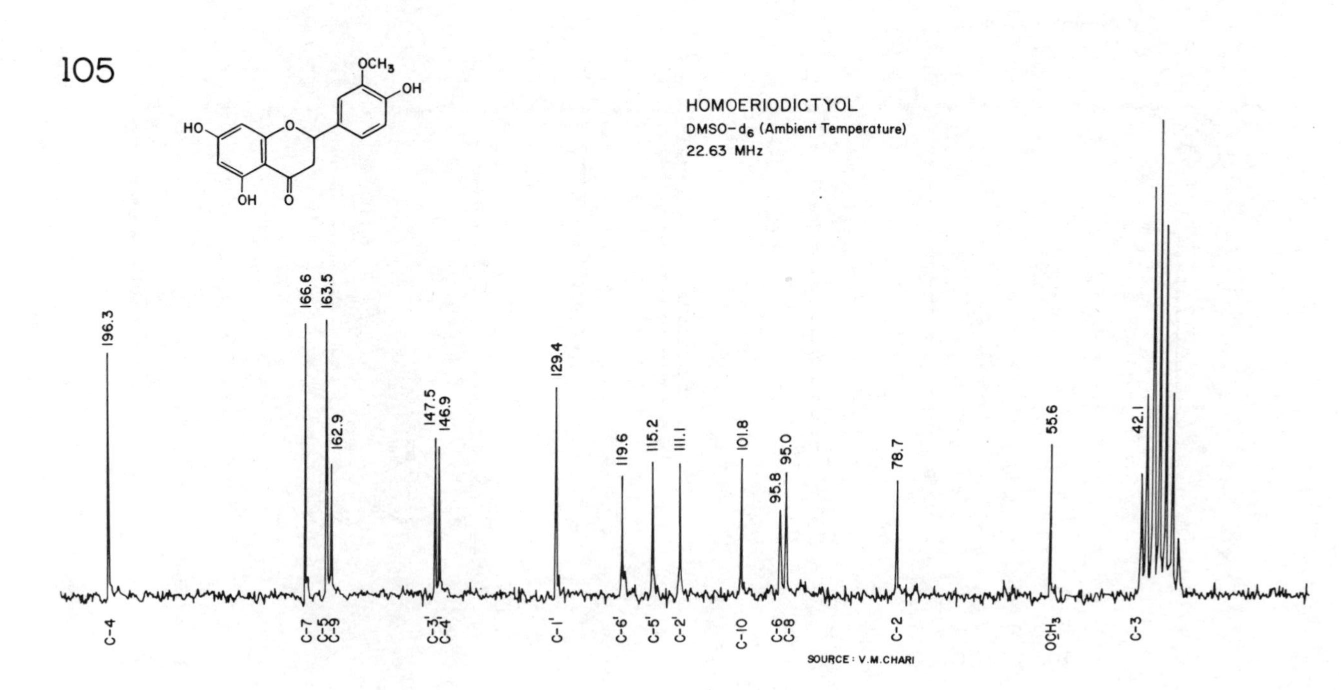
105
OCH3
OH
HO
O
OH
O
HOMOERIODICTYOL
DMSO-d6 (Ambient Temperature)
22.63 MHz
196.3
166.6
163.5
162.9
147.5
146.9
129.4
119.6
115.2
111.1
101.8
95.8
95.0
78.7
55.6
42.1
C-4
C-7
C-5
C-9
C-3'
C-4'
C-1'
C-6'
C-5'
C-2'
C-10
C-6
C-8
C-2
OCH3
C-3
SOURCE: V.M.CHARI

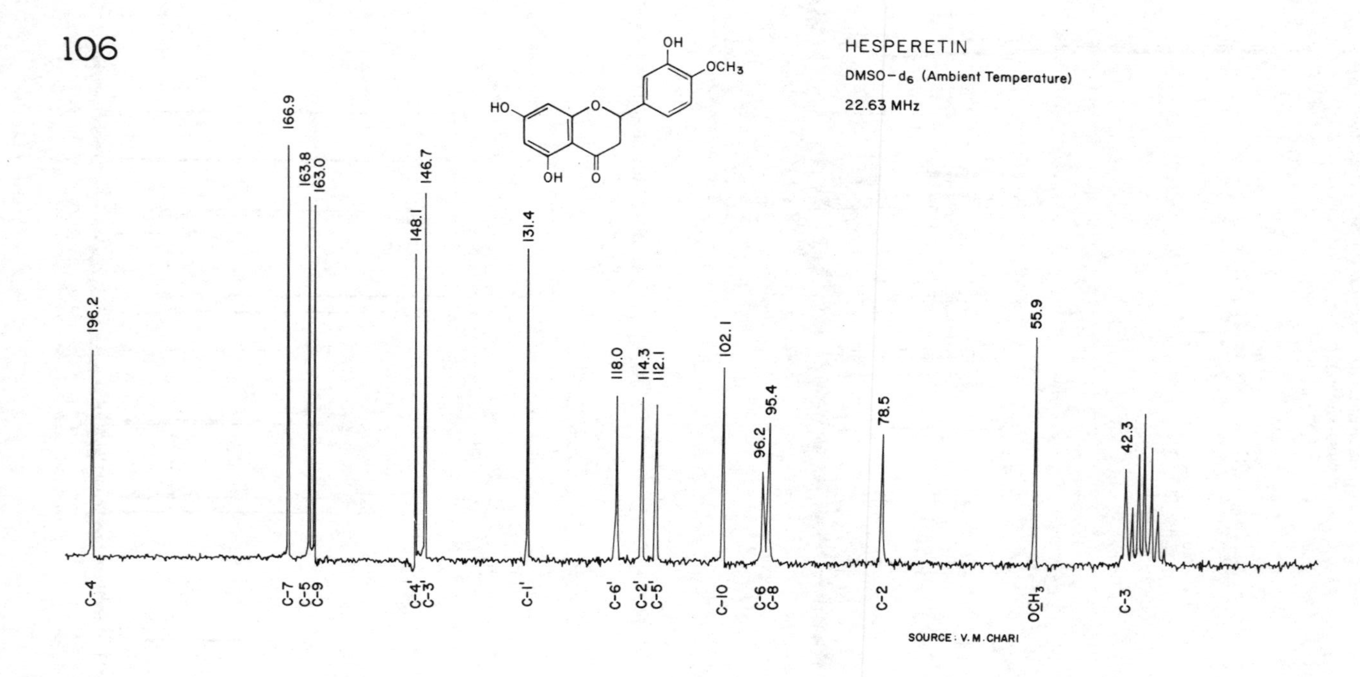
106
HESPERETIN
DMSO-d6 (Ambient Temperature)
22.63 MHz
HO
OH
O
OCH3
196.2
166.9
163.8
163.0
148.1
146.7
131.4
118.0
114.3
112.1
102.1
96.2
95.4
78.5
55.9
42.3
C-4
C-7
C-5
C-9
C-4'
C-3'
C-1'
C-6'
C-2'
C-5'
C-10
C-6
C-8
C-2
OCH3
C-3
SOURCE: V. M. CHARI

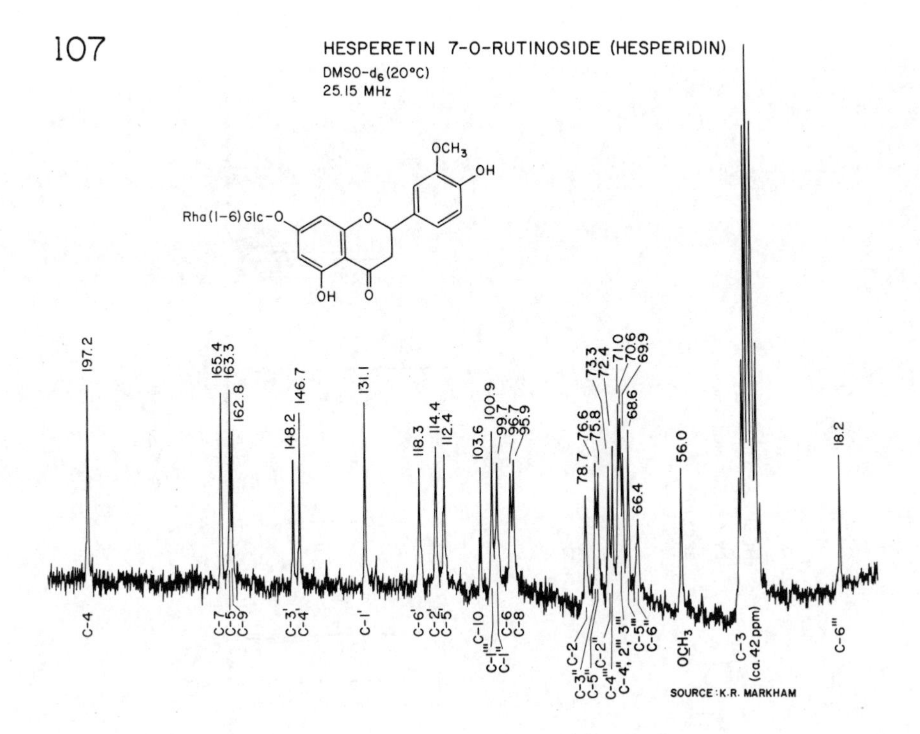
107
HESPERETIN 7-O-RUTINOSIDE (HESPERIDIN)
DMSO-d6 (20°C)
25.15 MHz
OCH3
OH
Rha(1-6)Glc-O
OH
O
197.2
165.4
163.3
162.8
148.2
146.7
131.1
118.3
114.4
112.4
103.6
100.9
99.7
96.7
95.9
78.7
76.6
75.8
73.3
72.4
71.0
70.6
69.9
68.6
66.4
56.0
18.2
C-4
C-7
C-5
C-9
C-3'
C-4'
C-1'
C-6'
C-2'
C-5'
C-10
C-1'''
C-1''
C-6
C-8
C-3''
C-5''
C-2
C-2''
C-4'''
C-4'', 2''', 3'''
C-5'''
C-6''
OCH3
C-3
(ca. 42 ppm)
C-6'''
SOURCE: K.R. MARKHAM

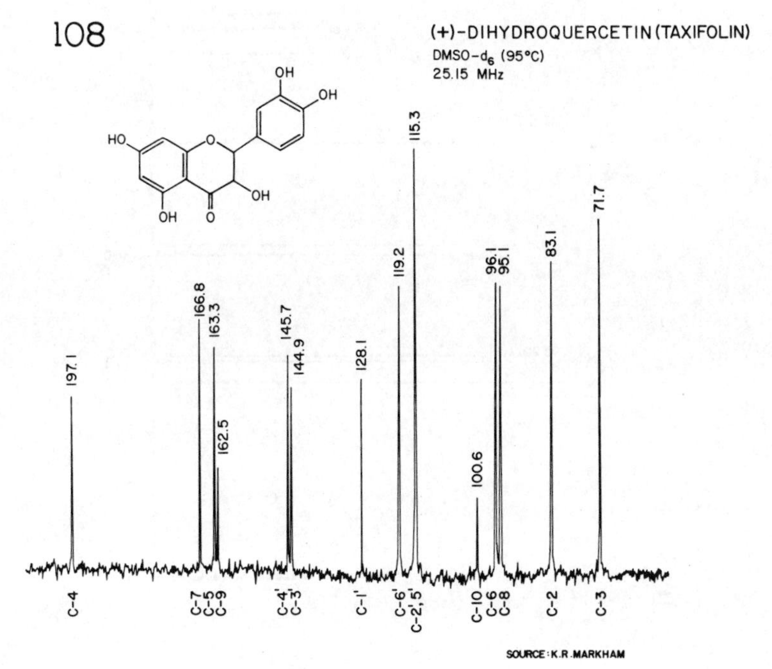
108
(+)-DIHYDROQUERCETIN (TAXIFOLIN)
DMSO-d6 (95°C)
25.15 MHz
OH
OH
HO
OH
OH
O
197.1
166.8
163.3
162.5
145.7
144.9
128.1
119.2
115.3
100.6
96.1
95.1
83.1
71.7
C-4
C-7
C-5
C-9
C-4'
C-3'
C-1'
C-6'
C-2',5'
C-10
C-6
C-8
C-2
C-3
SOURCE: K.R. MARKHAM

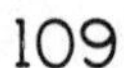

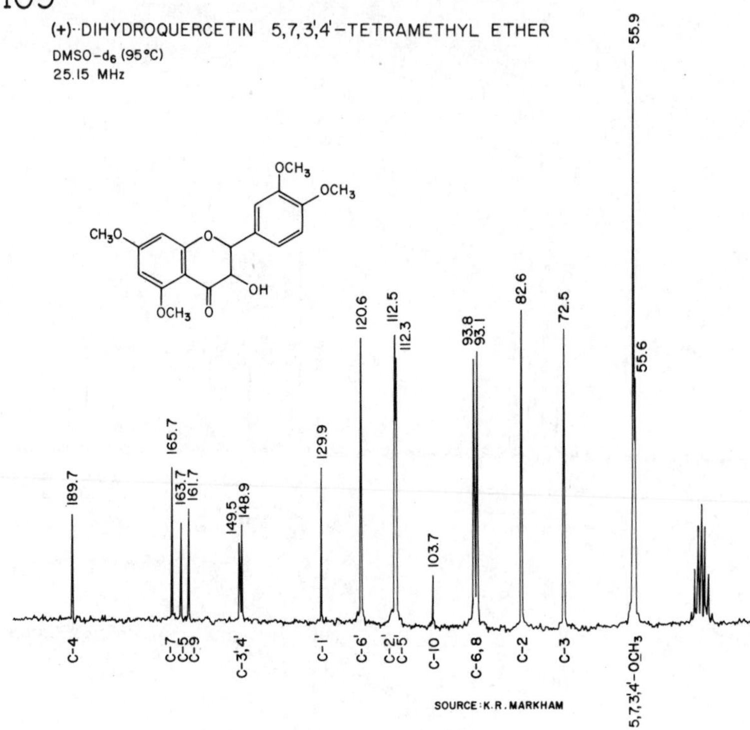

(+)-DIHYDROQUERCETIN 5,7,3',4'-TETRAMETHYL ETHER
DMSO-d6 (95°C)
25.15 MHz
OCH3
OCH3
CH3O
O
OH
OCH3
O
189.7
165.7
163.7
161.7
149.5
148.9
129.9
120.6
112.5
112.3
103.7
93.8
93.1
82.6
72.5
55.9
55.6
C-4
C-7
C-5
C-9
C-3',4'
C-1'
C-6'
C-2'
C-5'
C-10
C-6, 8
C-2
C-3
5,7,3',4'-OCH3
SOURCE: K.R. MARKHAM

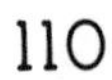

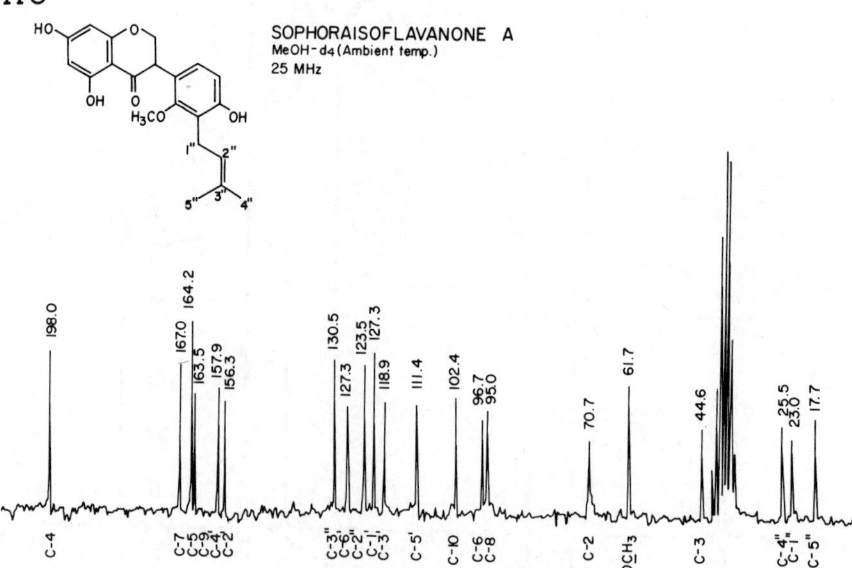

SOPHORAISOFLAVANONE A
MeOH-d4 (Ambient temp.)
25 MHz
HO
O
OH
O
H3CO
OH
1"
2"
3"
4"
5"
198.0
167.0
164.2
163.5
157.9
156.3
130.5
127.3
123.5
127.3
118.9
111.4
102.4
96.7
95.0
70.7
61.7
44.6
25.5
23.0
17.7
C-4
C-7
C-5
C-9
C-4'
C-2'
C-3"
C-6'
C-2"
C-1'
C-3'
C-5'
C-10
C-6
C-8
C-2
OCH3
C-3
C-4"
C-1"
C-5"
SOURCE: V.M.CHARI COURTESY OF KOMATSU

111

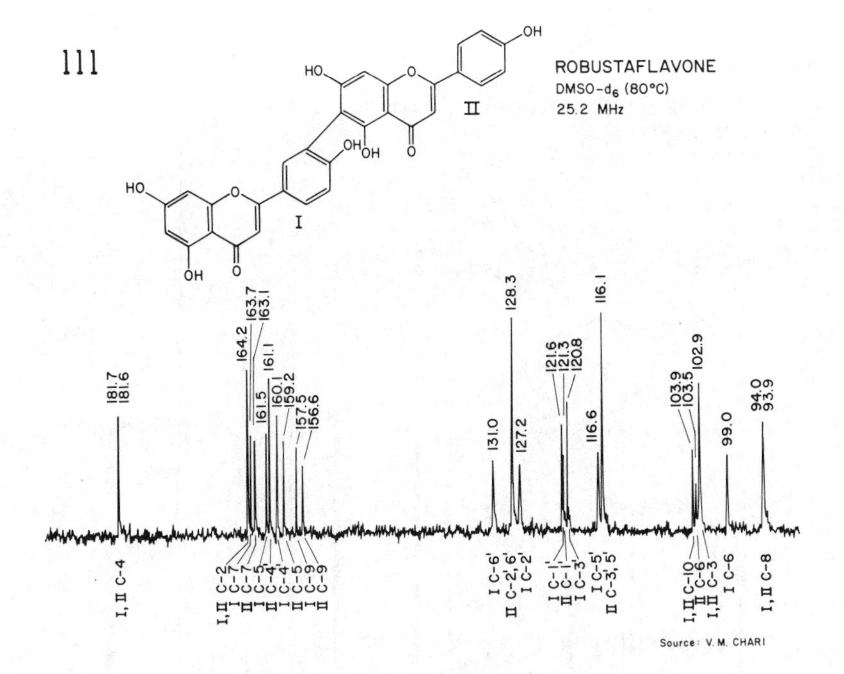

ROBUSTAFLAVONE
DMSO-d6 (80°C)
25.2 MHz
HO
OH
O
I
II
181.7
181.6
164.2
163.7
163.1
161.5
161.1
160.1
159.2
157.5
156.6
131.0
128.3
127.2
121.6
121.3
120.8
116.6
116.1
103.9
103.5
102.9
99.0
94.0
93.9
I, II C-4
I, II C-2
I C-7
II C-7
I C-5
II C-4'
I C-4'
II C-5
I C-9
II C-9
I C-6'
II C-2',6'
I C-2'
I C-1'
II C-1'
I C-3'
I C-5'
II C-3',5'
I, II C-10
II C-6
I, II C-3
I C-6
I, II C-8
Source: V.M. CHARI

112

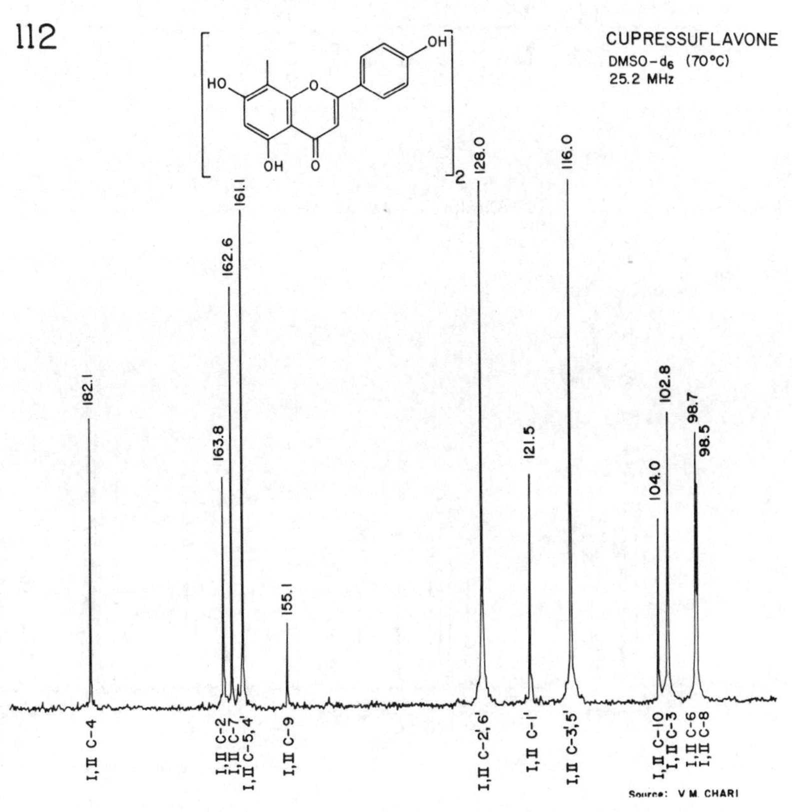

CUPRESSUFLAVONE
DMSO-d6 (70°C)
25.2 MHz
HO
OH
O
2
182.1
163.8
162.6
161.1
155.1
128.0
121.5
116.0
104.0
102.8
98.7
98.5
I, II C-4
I, II C-2
I, II C-7
I, II C-5,4'
I, II C-9
I, II C-2',6'
I, II C-1'
I, II C-3',5'
I, II C-10
I, II C-3
I, II C-6
I, II C-8
Source: V.M. CHARI

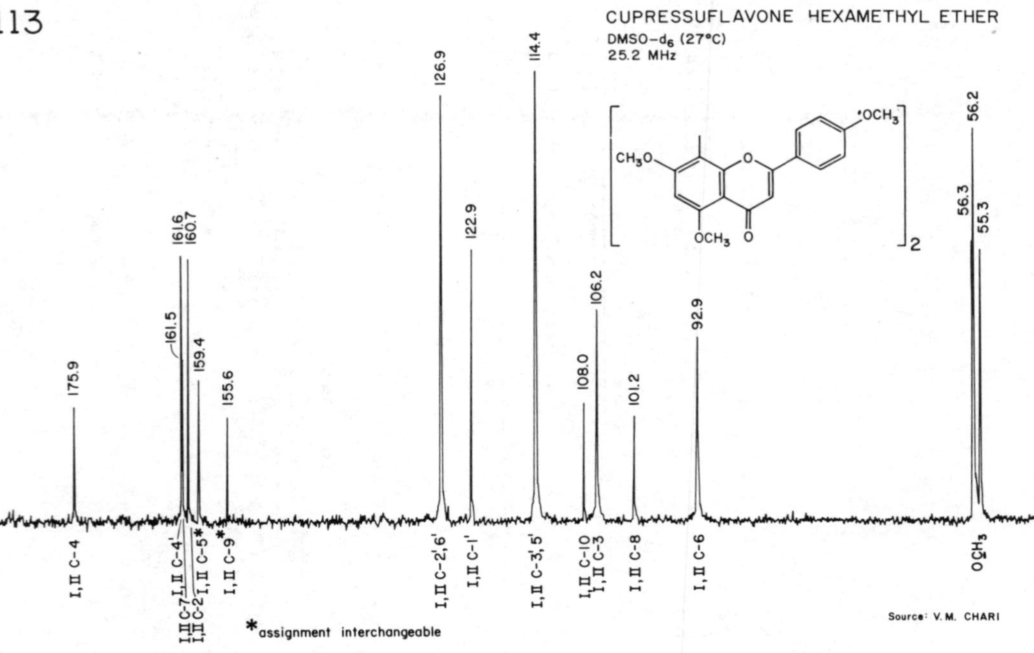
113
CUPRESSUFLAVONE HEXAMETHYL ETHER
DMSO-d_6 (27°C)
25.2 MHz
CH_3O
OCH_3
*OCH_3
O
2
175.9
161.6
160.7
161.5
159.4
155.6
126.9
122.9
114.4
108.0
106.2
101.2
92.9
56.2
56.3
55.3
I,II C-4
I,II C-7
I,II C-2
I,II C-4'
I,II C-5*
I,II C-9*
I,II C-2',6'
I,II C-1'
I,II C-3',5'
I,II C-10
I,II C-3
I,II C-8
I,II C-6
OCH_3
*assignment interchangeable
Source: V.M. CHARI
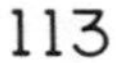

114

VOLKENSIFLAVONE

DMSO-d_6 (100°C)

25.2 MHz

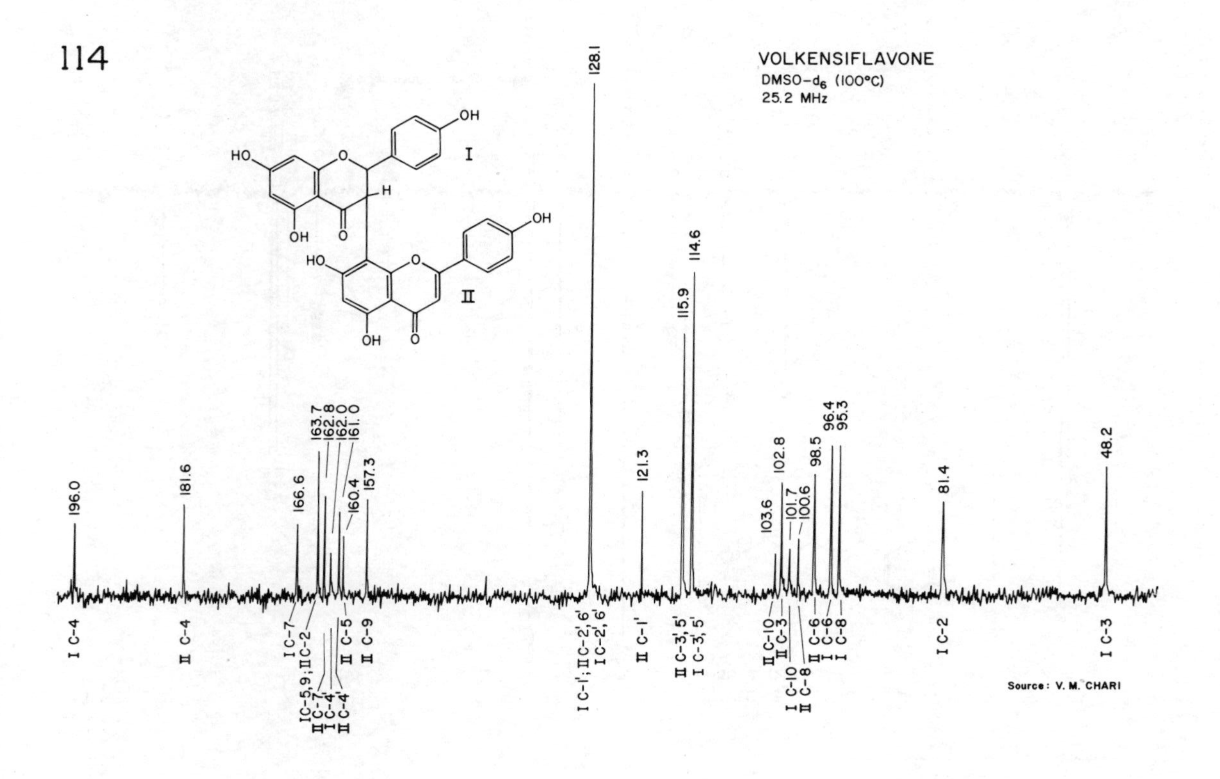

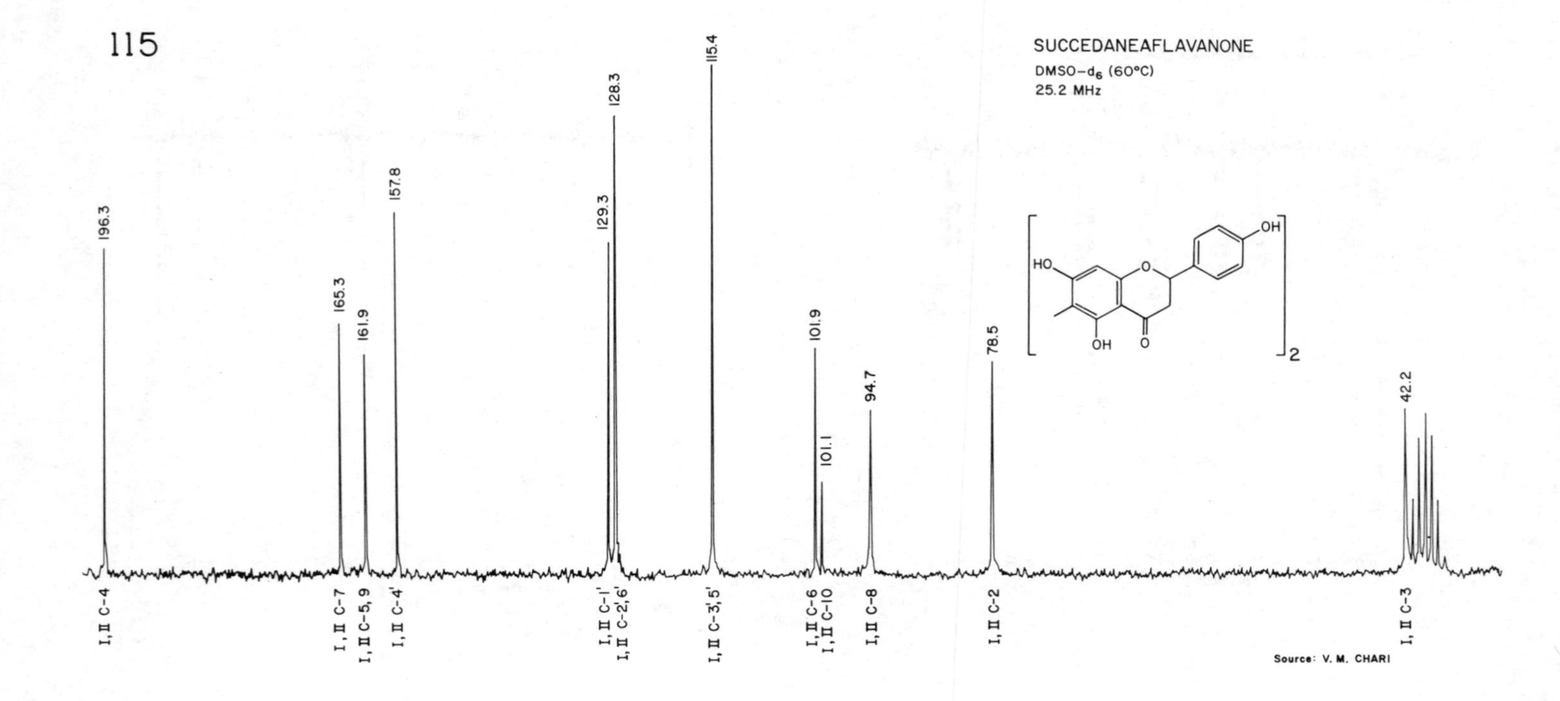
SUCCEDANEAFLAVANONE
DMSO-d_6 (60°C)
25.2 MHz
HO
OH
O
OH
O
2
196.3
165.3
161.9
157.8
129.3
128.3
115.4
101.9
101.1
94.7
78.5
42.2
I, II C-4
I, II C-7
I, II C-5, 9
I, II C-4'
I, II C-1'
I, II C-2',6'
I, II C-3',5'
I, II C-6
I, II C-10
I, II C-8
I, II C-2
I, II C-3
Source: V. M. CHARI

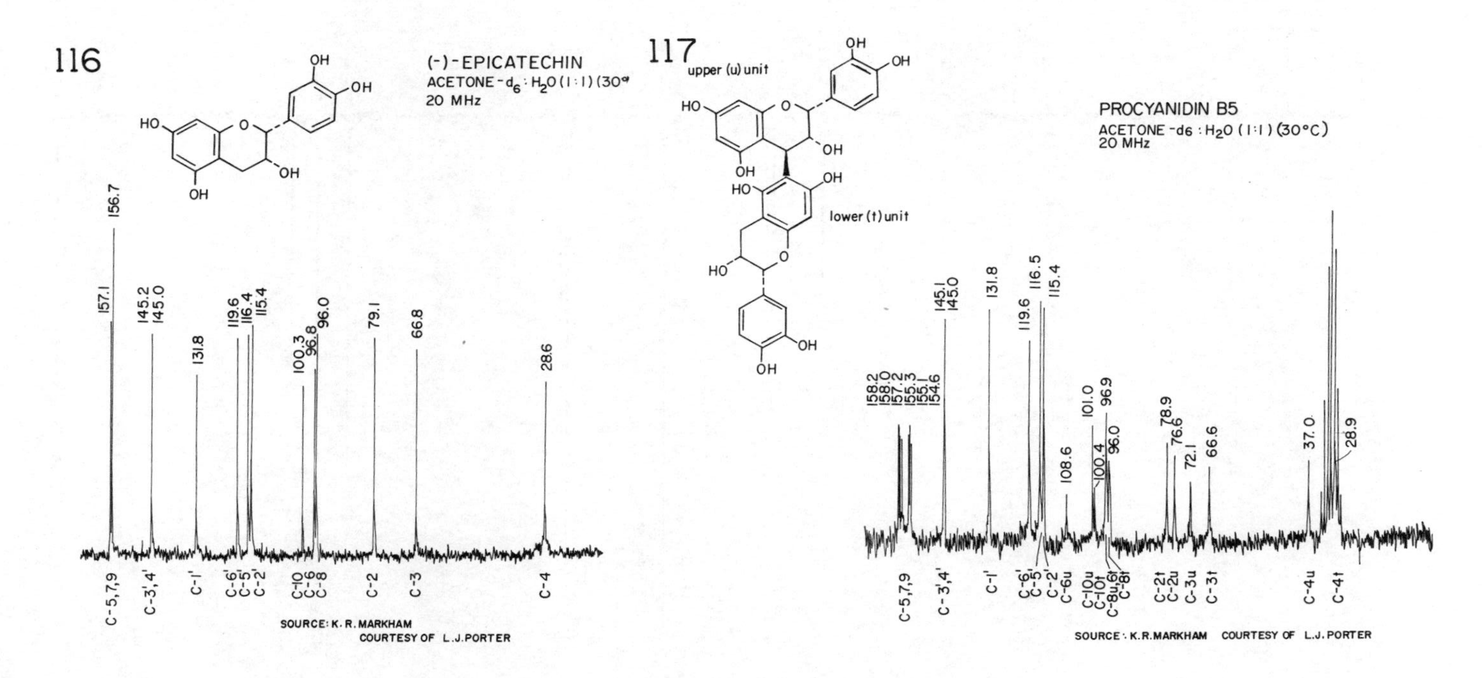
116
(-)-EPICATECHIN
ACETONE-d6 : H2O (1:1) (30°
20 MHz
HO
OH
OH
OH
OH
156.7
157.1
145.2
145.0
131.8
119.6
116.4
115.4
100.3
96.8
96.0
79.1
66.8
28.6
C-5,7,9
C-3',4'
C-1'
C-6'
C-5'
C-2'
C-10
C-6
C-8
C-2
C-3
C-4
SOURCE: K. R. MARKHAM
COURTESY OF L.J.PORTER
117
upper (u) unit
lower (t) unit
PROCYANIDIN B5
ACETONE-d6 : H2O (1:1) (30°C)
20 MHz
158.2
158.0
157.2
155.3
155.1
154.6
145.1
145.0
131.8
119.6
116.5
115.4
108.6
101.0
100.4
96.9
96.0
78.9
76.6
72.1
66.6
37.0
28.9
C-5,7,9
C-3',4'
C-1'
C-6'
C-5'
C-2'
C-6u
C-10u
C-10t
C-8u,6t
C-8t
C-2t
C-2u
C-3u
C-3t
C-4u
C-4t
SOURCE: K.R.MARKHAM COURTESY OF L.J.PORTER

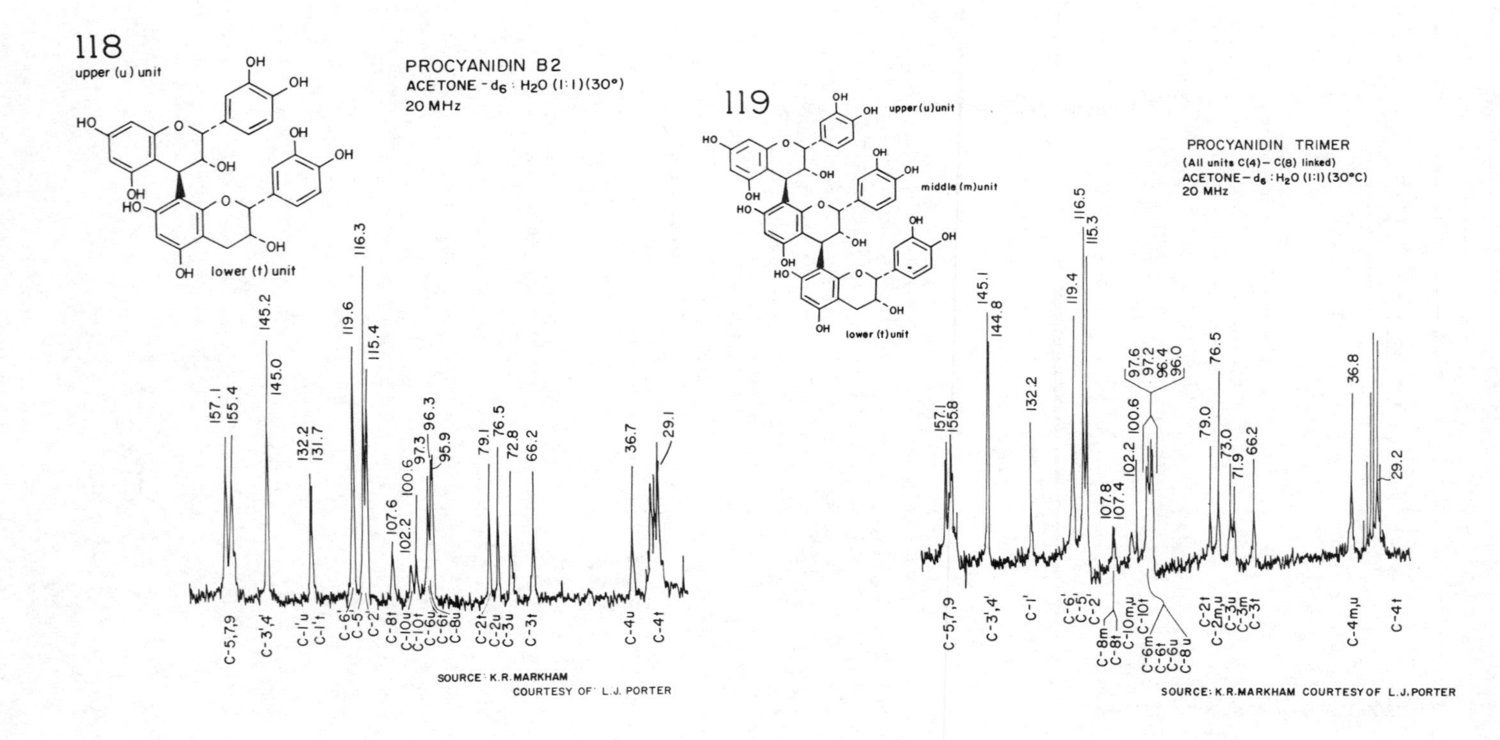
118
upper (u) unit
lower (t) unit
PROCYANIDIN B2
ACETONE - d6 : H2O (1:1)(30°)
20 MHz
SOURCE: K.R. MARKHAM
COURTESY OF L.J. PORTER
119
upper (u) unit
middle (m) unit
lower (t) unit
PROCYANIDIN TRIMER
(All units C(4) – C(8) linked)
ACETONE – d6 : H2O (1:1) (30°C)
20 MHz
SOURCE: K.R. MARKHAM COURTESY OF L.J. PORTER

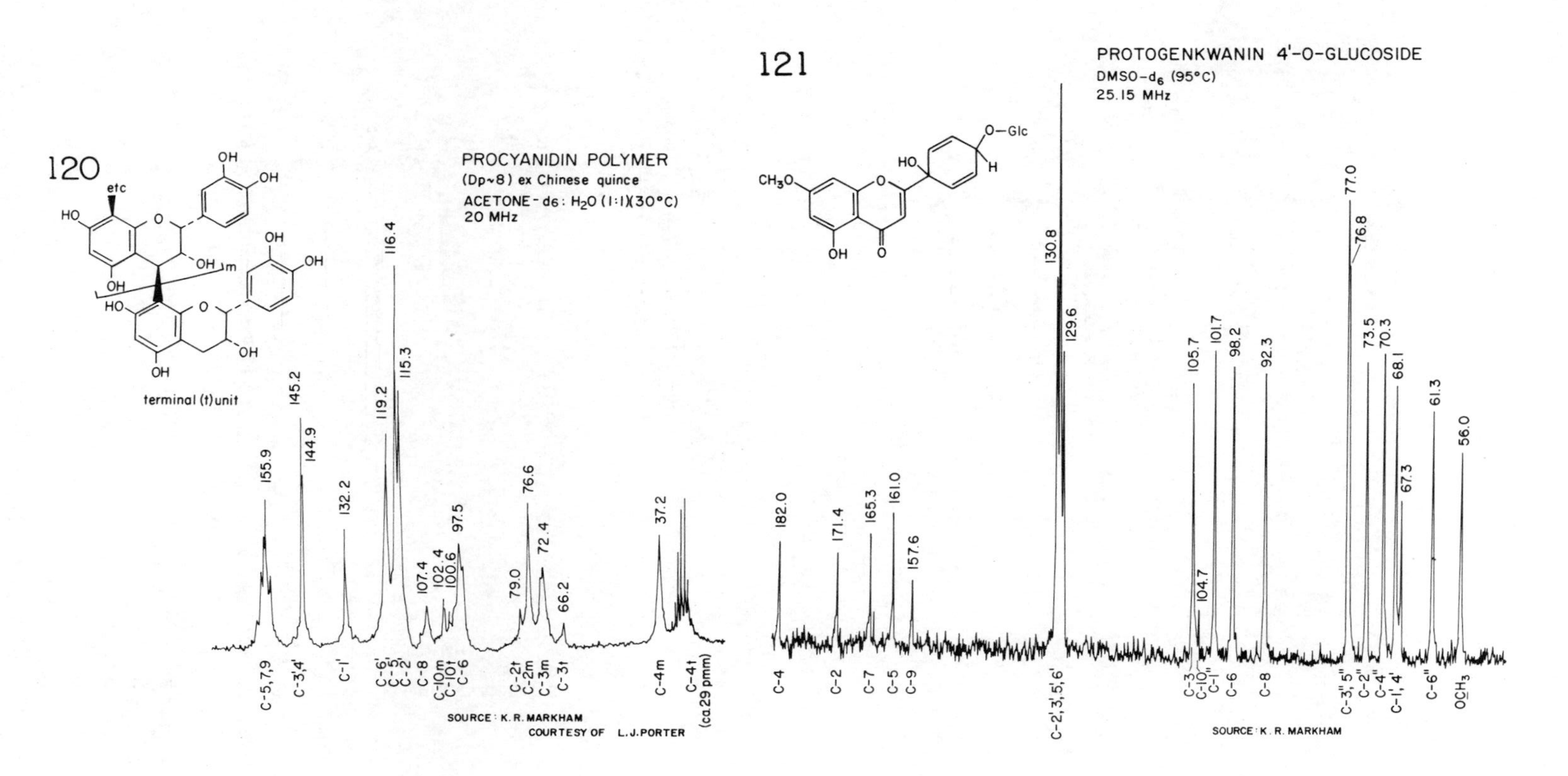
120
PROCYANIDIN POLYMER
(Dp~8) ex Chinese quince
ACETONE-d6: H2O (1:1)(30°C)
20 MHz
terminal (t)unit
155.9
145.2
144.9
132.2
119.2
116.4
115.3
107.4
102.4
100.6
97.5
79.0
76.6
72.4
66.2
37.2
C-5,7,9
C-3',4'
C-1'
C-6'
C-5'
C-2'
C-8
C-10m
C-10t
C-6
C-2t
C-2m
C-3m
C-3t
C-4m
C-4t
(ca.29 pmm)
SOURCE: K.R. MARKHAM
COURTESY OF L.J.PORTER
121
PROTOGENKWANIN 4'-O-GLUCOSIDE
DMSO-d6 (95°C)
25.15 MHz
182.0
171.4
165.3
161.0
157.6
130.8
129.6
105.7
104.7
101.7
98.2
92.3
77.0
76.8
73.5
70.3
68.1
67.3
61.3
56.0
C-4
C-2
C-7
C-5
C-9
C-2',3',5',6'
C-3
C-10
C-1''
C-6
C-8
C-3'',5''
C-2''
C-4''
C-1',4'
C-6''
OCH3
SOURCE: K.R. MARKHAM

122

CASSIACHROMONE

DMSO-d_6 (Ambient Temperature)
20.0 MHz

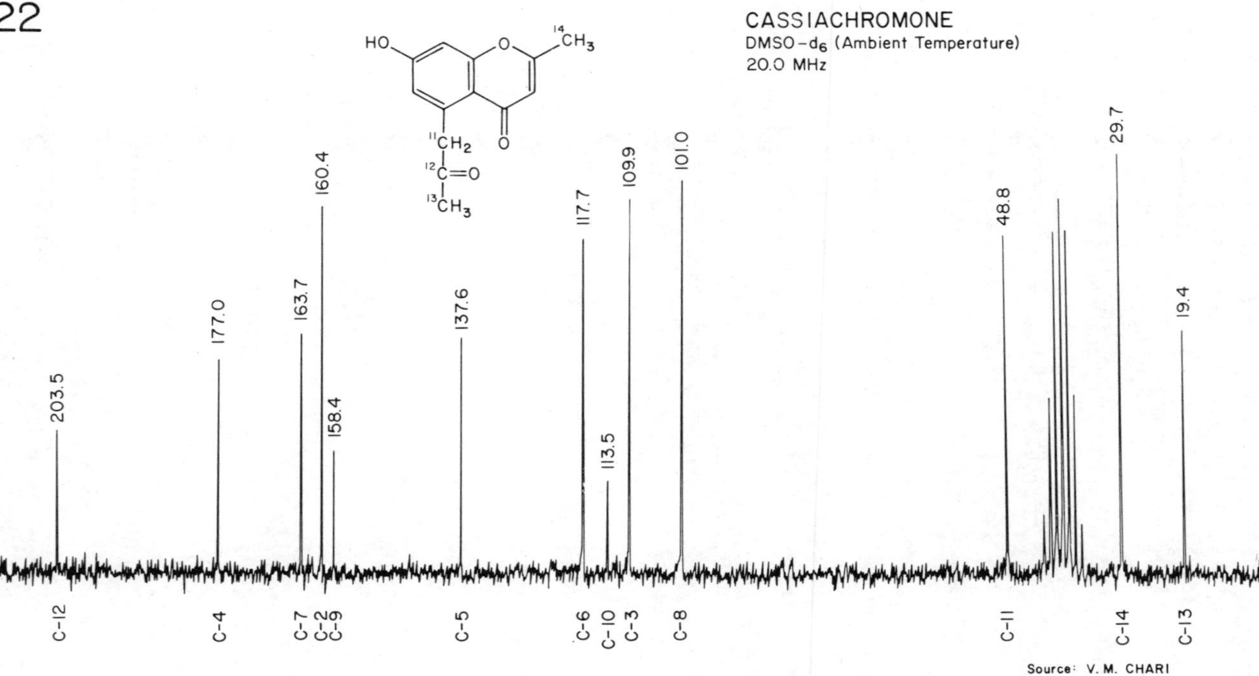

Source: V. M. CHARI

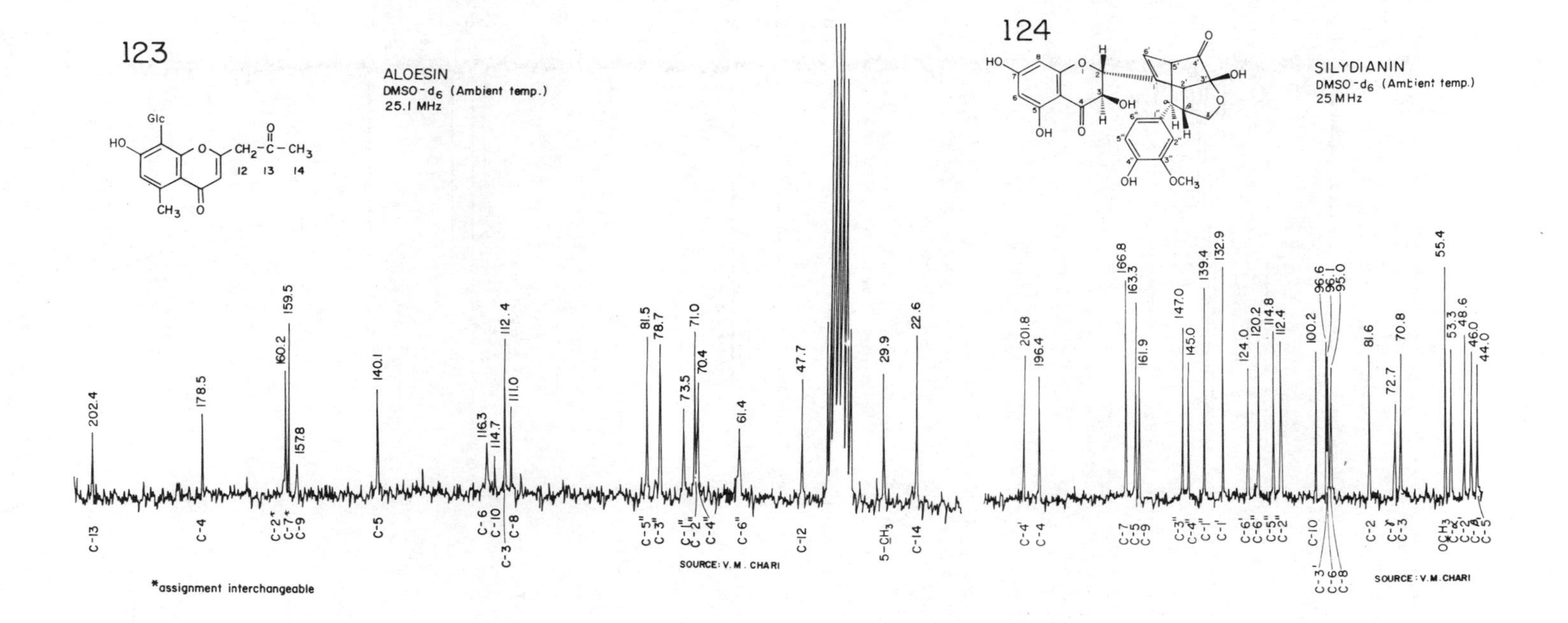
123
ALOESIN
DMSO-d6 (Ambient temp.)
25.1 MHz
*assignment interchangeable
SOURCE: V. M. CHARI
124
SILYDIANIN
DMSO-d6 (Ambient temp.)
25 MHz
SOURCE: V. M. CHARI

125

SILYMONIN (3-DEOXYSILYDIANIN)

DMSO-d_6 (80°C)
20.15 MHz

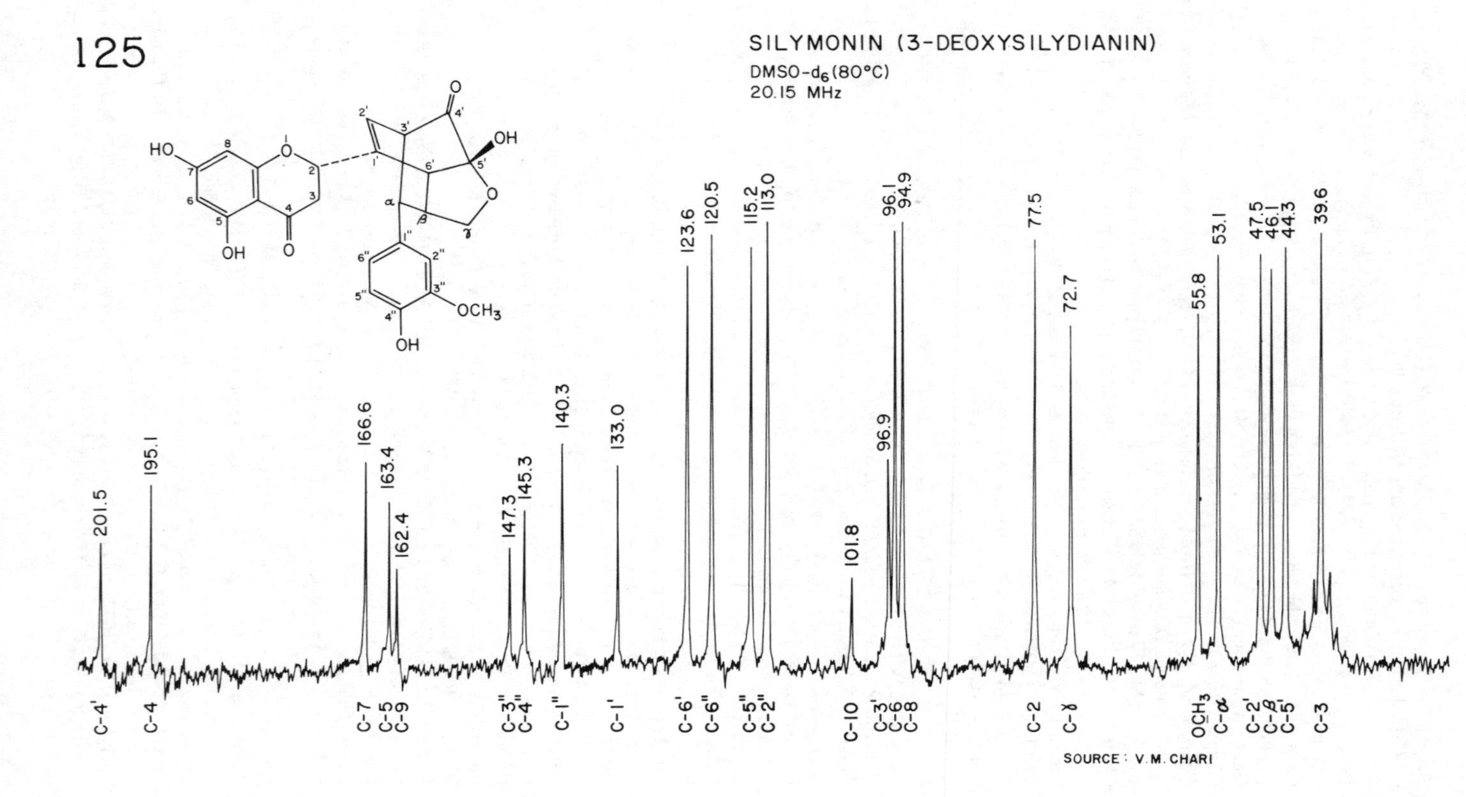

SOURCE: V.M. CHARI

REFERENCES

Ansari, F. R., Ansari, W. H., Rahman, W., Seligmann, O., Chari, V. M., Wagner, H. and Osterdahl, B. G. (1979), *Planta Med.* **36**, 196.

Besson, E., Dombris, A., Raynaud, J. and Chopin, J. (1979), *Phytochemistry* **18**, 1899.

Bock, K. and Pederson, C. (1974), *J. Chem. Soc. Perkin II* 293.

Breitmaier, E., Jung, G. and Voelter, W. (1971), *Angew Chem. Int. Edit.* **10**, 673.

Calvert, D. J., Cambie, R. C. and Davis, B. R. (1979), *Org. Magn. Reson.* **12**, 583.

Chalmers, A. A., Rall, G. J. H. and Oberholzer, M. E. (1977), *Tetrahedron* **33**, 1735.

Chang, Ching-Jer (1976), *J. Org. Chem.* **41**, 1883.

Chang, Ching-Jer (1978), *Lloydia* **41**, 17.

Chari, V. M., Wagner, H. and Neszmelyi, A. (1977a), *Proceedings of the 5th Hungarian Bioflavonoid Symposium, Matrafured, Hungary*, p. 49.

Chari, V. M., Wagner, H. and Thies, P. (1977b), *Phytochemistry* **16**, 1110.

Chari, V. M., Ilyas, M., Wagner, H., Neszmelyi, A., Chen, F-C., Chen, L-K., Lin, Y.-C. and Lin, Y.-M. (1977c) *Phytochemistry* **16**, 1273.

Chari, V. M., Ahmad, S. and Osterdahl, B-G (1978a), *Z. Naturforsch.* **33b**, 1547.

Chari, V. M., Jordan, M. and Wagner, H. (1978b), *Planta Med.* **34**, 93.

Chari, V. M., Harborne, J. B. and Williams, C. A. (1980), *Phytochemistry* **19**, 983.

Chari, V. M., Grayer-Barkmeijer, R. J., Harborne, J. and Österdahl, B. G. (1981), *Phytochemistry* **20**, 1977.

Colson, P., Jennings, H. J. and Smith, I. C. P. (1974), *J. Amer. Chem. Soc.* **96**, 8081.

Cooper, R., Gottlieb, H. and Lavie, D. (1977), *Isr. J. Chem.* **16**, 12.

Cotterill, P. J., Scheinmann, F. and Stenhouse, I. A. (1978), *J. Chem. Soc., Perkin I* 532.

Crichton, E. G. and Waterman, P. G. (1979), *Phytochemistry* **18**, 1553.

Crombie, L., Kilbee, G. W. and Whiting, D. A. (1975), *J. Chem. Soc., Perkin I* 1497.

Czochanska, Z., Foo, L. Y., Newman, R., Porter, L. J., Thomas, W. A. and Jones, W. T. (1979), *J.C.S. Chem. Commun.* 375.

Czochanska, Z., Foo, L. Y., Newman, R. and Porter, L. J. (1980), *J. Chem. Soc. Perkin I* 2278.

Duddeck, H., Snatzke, G. and Yemul, S. S. (1978), *Phytochemistry* **17**, 1369.

El-Sohly, Lasswell, W. L. and Hufford, C. D. (1979), *Lloydia*, 264.

Fletcher, A. C., Porter, L. J., Haslam, E. and Gupta, R. K. (1977), *J. Chem. Soc., Perkin I* 1628.

Garcia-Granados, A. and Saenz de Buruaga, J. M. (1980), *Org. Magn. Reson.* **13**, 462.

Gorin, P. A. J. and Mazurek, M. (1975), *Canad. J. Chem.* **53**, 1212.

Guinaudeau, H., Seligmann, O., Wagner, H. and Neszmelyi, A. (1981), *Phytochemistry* **20**, 1113.

Hemmingway, R. W., Foo, L. Y. and Porter, L. J. (1981), Private communication.

Hostettmann, K. and Guillarmod, A. J. (1976), *Helv. Chem. Acta* **59**, 1584.

Hufford, C. D. and Lasswell, W. L. (1978), *Lloydia* **41**, 151.

Iinuma, M., Matsuura, S. and Kusuda, K. (1980), *Chem. Pharm. Bull.* **28**, 708.

Isobe, T., Fukushige, T. and Noda, Y. (1979), *Chem. Lett.* 27.

Jacques, D., Haslam, E., Bedford, G. R. and Greatbanks, D. (1974), *J. Chem. Soc., Perkin I* 2663.

Jha, H. C., Zilliken, F. and Breitmaier E. (1980), *Canad. J. Chem.* **58**, 1211.

Joly, M., Hagg-Berrurier, M. and Anton, R. (1980), *Phytochemistry* **19**, 1999.

Joseph-Nathan, P., Mares, J. and Hernandez, Ma. C. (1974), *J. Magn. Reson.* **16**, 447.

Kalabin, G. A., Pogodaeva, N. N., Tyukavkina, N. A. and Kushnarev, D. F. (1977) *Khim. Prir. Soed.* No. 4, 513 (Engl. Transl. p. 429).

Kingsbury, C. A. and Looker, J. H. (1975), *J. Org. Chem.* **40**, 1120.

Kobayashi, M., Terni, Y., Tori, K. and Tsuji, N. (1976), *Tetrahedron Lett.* no. 8, 619.

Komatsu, M., Yokoe, I. and Shirataki, Y. (1978), *Chem. Pharm. Bull.* **26**, 3863.

Lallemand, J. Y. and Duteil, M. (1977), *Org. Magn. Reson.* **9**, 179.

Levy, G. C. and Nelson, G. L. (1972), *Carbon-13 Nuclear Magnetic Resonance for Organic Chemists*, Wiley, New York.

Locksley, H. D. (1973), in *Zechmeister's Progress in the Chemistry of Organic Natural Products*, Vol. 30. (eds. W. Herz, H. Grisebach and G. W. Kirby), Springer-Verlag, Vienna and New York.

Mabry, T. J., Markham, K. R. and Thomas, M. B. (1970), *The Systematic Identification of Flavonoids*, Springer-Verlag, Berlin.
Maciel, G. E. and James, R. V. (1964), *J. Amer. Chem. Soc.* **86**, 3893.
Markham, K. R. and Ternai, B. (1976), *Tetrahedron* **32**, 2607.
Markham, K. R. and Wallace, J. W. (1980), *Phytochemistry* **19**, 415.
Markham, K. R., Ternai, B., Stanley, R., Geiger, H. and Mabry, T. J. (1978), *Tetrahedron* **34**, 1389.
Merlini, L., Zanarotti, A., Pelter, A., Rochefort, M. P. and Hänsel, R. (1980), *J. Chem. Soc. Perkin I* 775.
Miura, I., Hostettmann, K. and Nakanishi, K. (1979), *Nouv. J. Chim.* **2**, 653.
Nomura, T. and Fukai, T. (1978), *Heterocycles* **9**, 1295.
Nomura, T. and Fukai, T. (1979), *Heterocycles* **12**, 1289.
Nomura, T. and Fukai, T. (1980), *Chem. Pharm. Bull.* **28**, 2548.
Nomura, T., Fukai, T. and Katayanagi, M. (1978), *Heterocycles* **9**, 745.
Okuyama, T., Hosoyama, K., Hiraga, Y., Kurono, G. and Takemoto, K. (1978), *Chem. Pharm. Bull.* **26**, 3071.
Oshima, Y., Konno, C., Hikino, H. and Matsushita, K. (1980a), *Tetrahedron Lett.* 3381.
Oshima, Y., Konno, C., Hikino, H. and Matsushita, K. (1980b), *Heterocycles* **14**, 1287.
Osazawa, T. and Takino, Y. (1979), *Agric. Biol. Chem.* **43**, 1173.
Österdahl, B. G. (1978a), *Acta Chem. Scand.* **B32**, 714.
Österdahl, B. G. (1978b), *Acta Chem. Scand.* **B32**, 93.
Österdahl, B. G. (1979a), *Acta Universitatis Upsaliensis (Abstracts of Uppsala Dissertations from the Faculty of Science)*, No. 516.
Österdahl, B. G. (1979b), *Acta Chem. Scand.* **B33**, 119.
Panichpol, K. and Waterman, P. G. (1978), *Phytochemistry* **17**, 1363.
Parthasarathy, M. R., Ranganathan, K. R. and Sharma, D. K. (1979), *Phytochemistry* 506.
Pelter, A., Ward, R. S. and Gray, T. I. (1976), *J. Chem. Soc., Perkin I* 2475.
Pelter, A., Ward, R. S. and Bass, R. J. (1978), *J. Chem. Soc., Perkin I* 666.
Pelter, A., Ward, R. S. and Heller, H. G. (1979), *J. Chem. Soc., Perkin I* 328.
Que, L. and Gray, G. R. (1974), *Biochemistry* **13**, 146.
Radeglia, R., Altenburg, W., Orlow, W. D. and Reinhardt, M. (1976), *Z. Chem.* **9**, 363.
Rahman, W., Ishratullah, K., Wagner, H., Seligmann, O., Chari, V. M. and Österdahl, B. G. (1978), *Phytochemistry* **17**, 1064.
Redaelli, C., Formentini, L. and Santaniello, E. (1980), *Phytochemistry* **19**, 985.
Ritchie, R. G. S., Cyr, N., Korsch, B., Koch, H. J. and Perlin, A. S. (1975), *Canad. J. Chem.* **53**, 1424.
Sakurai, A. and Okumura, Y. (1978), *Chem. Lett.* 259.
Schilling, G. (1975), *Liebigs Ann. Chem.* 1822.
Schilling, G., Weinges, K., Müller, O. and Mayer, W. (1973), *Liebigs Ann. Chem.* 147.
Solaniova, E., Toma, S. and Gronowitz, S. (1976) *Org. Magn. Reson.* **8**, 439.
Stothers, J. B. (1972), *Carbon-13 NMR Spectroscopy, Volume 24 of Organic Chemistry, A Series of Monographs*, Academic Press, New York.
Ternai, B. and Markham, K. R. (1976), *Tetrahedron* **32**, 565.
Theodor, R., Zinsmeister, H. D., Mues, R. and Markham, K. R. (1980), *Phytochemistry* **19**, 1695.
Tori, K., Seo, S., Yoshimura, Y., Arita, H. and Tomita, Y. (1978), *Tetrahedron Lett.* **2**, 179.
Van Zyl, J. J., Rall, G. J. H. and Roux, D. G. (1979), *J. Chem. Res.* 1301.
Vignon, M. R. and Vottero, Ph. J. A. (1976), *Tetrahedron Lett.* **28**, 2445.
Wagner, H., Chari, V. M. and Sonnenbichler, J. (1976), *Tetrahedron Lett.* **21**, 1799.
Wagner, H., Chari, V. M., Seitz, M. and Riess-Maurer, I. (1978), *Tetrahedron Lett.* 381.
Wagner, H., Obermeier, G., Seligmann, O. and Chari, V. M. (1979), *Phytochemistry* **18**, 907.
Waterman, P. G. and Khalid, S. A. (1980), *Phytochemistry* **19**, 909.
Wehrli, F. W. (1975), *J. C. S. Chem. Commun.* 663.
Wenkert, E. and Gottlieb, H. E. (1977), *Phytochemistry* **16**, 1811.
Woo, W. S., Kang, S. S., Wagner, H., Chari, V. M., Seligmann, O. and Obermeier, G. (1979), *Phytochemistry* **18**, 353.

Woo, W. S., Kang, S. S., Wagner, H., Seligmann, O. and Chari, V. M. (1980), in press.
Yakushijin, K., Shibayama, K., Murata, H. and Furukawa, H. (1980), *Heterocycles* **14**, 397.
Yamoaka, N., Usui, T., Matsuda, K., Tuzimura, K., Sugiyama, H. and Seto, S. (1971), *Tetrahedron Lett.* **23**, 2047.

CHAPTER THREE

Anthocyanins

GEZA HRAZDINA

3.1 INTRODUCTION

Anthocyanin research in the past 5 years has remained dynamic. Progress in the field can best be summarized under twelve headings. These are (1) Further investigations into the occurrence of anthocyanins in plants; (2) Taxonomic role of anthocyanins; (3) Anthocyanins as genetic markers; (4) Anthocyanins as stress markers; (5) Food-related aspects of anthocyanins; (6) Progress in analytical techniques; (7) Progress in chemistry; (8) Subcellular status of anthocyanins; (9) Photocontrol of anthocyanin biosynthesis; (10) Enzymology of anthocyanin biosynthesis; (11) Effect of physiologically active chemicals on anthocyanin biosynthesis; (12) Physiological role of anthocyanins. This chapter has been divided along these lines in the present review of the past 5 years' progress.

All reports on anthocyanins which have appeared in Biological Abstracts and Chemical Abstracts between 1974 and 1980 are considered here.

3.2. OCCURRENCE OF ANTHOCYANINS

Investigations on the occurrence of anthocyanins in the past 5 years included hitherto unreported families, and detailed studies on previously reported genera. Major reviews appeared on the anthocyanin profiles of the Compositae (Harborne, 1977), Polemoniaceae (Harborne and Smith, 1978) and Orchidaceae (Arditti and Fisch, 1977); therefore these are not treated here. The results of surveying various monocotyledonous families for their anthocyanin patterns are summarized in Stirton and Harborne (1980). Other reports on occurrence of anthocyanins in plants are collected together in Table 3.1. Anthocyanins are not commonly found in marine organisms; the report of pelargonidin 3-glucoside and cyanidin 3-glucoside, their *p*-coumaryl derivatives and cyanidin 3,5-diglucoside (Garcia-Jalon *et al.*, 1974) in *Urginea maritima* is therefore of special note.

While the majority of the plants investigated contained anthocyanins previously identified in other species, a number of new anthocyanins were reported. These are cyanidin 3-xylosylglucosylgalactoside and its ferulyl derivative

identified in the black carrot (Harborne, 1976b), and a malonyl derivative of cyanidin 3-sophoroside-5-glucoside in *Brassica* (Hrazdina *et al.*, 1977) and of cyanidin 3-glucoside in *Spirodela* (Krause and Strack, 1979). The latter two pigments are especially interesting because they indicate that in addition to the hydroxycinnamic acid derivatives (the precursors of flavonoid and lignin biosynthesis) malonate, a precursor of both flavonoid and fatty acid biosynthesis, can also be utilized as an acylating agent. It is possible that the acylating reactions have controlling or regulating effects in the biosynthetic processes in the cell. Hitherto unreported anthocyanins are also pelargonidin 3-di (*p*-hydroxybenzoyl) rutinoside-7-glucoside in *Campanula* flowers (Asen *et al.*, 1979) and caffeylferulyl-*p*-coumaryldelphinidin 3-rutinoside-5,3′,5′-triglucoside isolated from *Lobelia erinus* (Yoshitama, 1977). The latter compound is unique among the flavonoids because of its exceptional glycosylation pattern, particularly the presence of sugars attached to the B-ring hydroxyls.

3.3 TAXONOMIC ROLE OF ANTHOCYANINS

The evolution of anthocyanins in plants has been well studied in the past. From these studies it appears that pelargonidin- and delphinidin-type pigments occur more frequently in advanced plants, replacing cyanidin which is considered to be the simplest and most primitive pigment (Harborne, 1976a). Since biosynthetically pelargonidin is a simpler pigment than cyanidin (seee Chapter 11), its production in replacement of cyanidin in the flowers of certain highly advanced plant families (e.g. Polemoniaceae; Harborne and Smith, 1978) must be by loss mutation. Correlations between the flower anthocyanins and pollination ecology is further discussed by these authors.

Against the background, anthocyanins have been used successfully by a number of authors to characterize families, genera, species or cultivars. Giannasi's work (1975) on the chemotaxonomy of 21 taxa of *Dahlia* indigenous to Mexico, Central America and Northern South America showed that this genus is made up of two major chemical lines, one with sections *Dahlia* and *Pseudodendron*, the other representing section *Entomophyllon*. These chemotaxonomic findings confirm earlier calssifications based on morphological and taxonomic studies. A study by Monties *et al.* (1976) underlines the importance of quantitative studies in plant taxonomic investigations. Quantitative varietal differences were found in maize in the overall content and relative proportions of anthocyanins in the sheath of the first leaf of 10-day-old plants. Similar differences exist in the amounts of anthocyanins and other flavonoids in the petals of red or white carnations. Anthocyanins have also been used as chemotaxonomic markers in the analysis of synthetic hybrids and backcrosses of *Camelia japonica* and *C. saluenensis* (Parks and Kondo, 1974) and in investigations on species belonging to the Saxifragaceae, Rosaceae and Caprifoliaceae (Sobolevskaya and Demina, 1971, Publ. 1973).

Table 3.1 Occurrence of anthocyanins in plant orders, families, genera and species

Orders, family, genus, species	*Organ examined*	*Pigments present*	*Reference*
PTEROPHYTA			
SALVINIACEAE			
Azolla mexicana	Fronds	Lu 5-glucoside	Holst (1977)
A. filiculoides			Pieterse *et al.* (1977)
A. caroliniana			
ANGIOSPERMAE			
MONOCOTYLEDONEAE			
AMARYLLIDACEAE			
Allium cepa	Bulb	Cy 3-glucoside Cy, 3-laminaribioside	Du *et al.* (1974a)
Clivia nobilis	Fruit	Pg 3-rutinose; Pg glucoside; Cy 3-rutinoside, Cy glucoside	Ishikura and Sugahara (1979)
ARACEAE			
Anthurium andreanum	Flower	Pg, Cy 3-rhamnosylglucosides	Iwata *et al.* (1979)
DIOSCOREACEAE			
Dioscorea tryphida	Tuber	Pn, Mv 3,5-diglucoside, Mv 3,5-diglucosideferulate	Carreno-Diaz and Grau (1977)
GRAMINEAE			
Saccharum officinarum	Epidermis	Pn 3-galactoside	Misra and Dubey (1974)
Zea mays	Seed	Cy 3-glucoside	Nakatani *et al.* (1979)
HYDROCHARITACEAE			
Egeria densa	Leaves	5-*O*-Me-Cy 3-glucoside	Momose *et al.* (1977)
IRIDACEAE			
Iris ensata	Flowers	Dp, Pt, Mv glucosides	Ishikura and Yamamoto (1978)
I. spontanea	Flower	Dp, Pt, Mv glycosides	
LILIACEAE			
Disporum sessile	Fruit	Cy 3-glucoside, Cy glycoside	Ishikura and Sugahara (1979)
Liriope platyphylla	Fruit	Dp 3-glucoside, Dp, Pt 3-rutinosides	Ishikura and Sugahara (1979)
Majanthemum bifolium	Fruit	Cy 3-glucoside, 3-rhamnoglucoside	Ishikura (1975c)
MARANTACEAE			
Thaumatococcus danielli	Flower	Pg glucoside, Cy, Dp 3-rutinoside	Adesina and Harborne (1978).
Ophiopogon japonicus	Fruit	Dp 3-rutinoside, Pt glycoside	Ishikura and Sugahara (1979)
Smilacina japonica	Fruit	Cy 3-rutinoside, Pn glycoside	Ishikura and Sugahara (1979)

Table 3.1 (*Contd.*)

Orders, family, genus, species	*Organ examined*	*Pigments present*	*Reference*
Smilax china	Fruit	Cy 3-glucoside, 3-rhamnoglucoside	Ishikura (1975c)
S. riparie	Fruit	Cy 3-glucoside, 3-rutinoside	Ishikura and Sugahara (1979)
Spirodela polyrrhiza		Cy 3-malonylglucoside	Krause and Strack (1979)
Urginea maritima	Whole plant	Pg, Cy 3-glucosides, Pg, Cy 3-*p*-coumarylglucosides	Garcia-Jalon *et al.* (1974)
ORCHIDACEAE			
Cymbidium sp.	Petal	Cy 3-glucoside, 3-diglucoside	Sugiyama *et al.* (1977)
Orchis sp.	Petal	Cy 3-glucoside, 3-diglucoside, 3,5-diglucoside	Uphoff (1979)
ZINGIBERACEAE			
Renealmia regmelliana	Fruit	Pt 3-glucoside, Dp, Pt, Mv 3,5-diglucosides	Gibaja-Oviedo (1978)
DICOTYLEDONEAE			
ACANTHACEAE			
Justicia procumbense	Flower	Pn 3-glucoside	Tiwari *et al.* (1978)
Peristropha bicalyculata	Flower	Pt 3-rhamnoglucoside	Tiwari *et al.* (1978)
Ruellia sp.	Flower	Pg, Mv 3,5-diglucosides	Bloom (1976)
ANACARDIACEAE			
Schinus molle	Fruit	Cy 3-galactoside, 3-rutinoside, Pn 3-glucoside	Aziz-Ur-Rahman *et al.* (1974)
APOCYNACEAE			
Catharanthus roseus	Tissue culture	Hirs, Pt, Mv glucosides	Krueger and Carew (1975); Carew and Krueger (1976)
AQUIFOLIACEAE			
Ilex buergeri	Fruit	Pg, Cy 3-xylosylglucoside	Ishikura (1975a)
I. crenata cv. Fukasawa	Fruit	Cy, 3-glucoside, 3-sambubioside	Ishikura and Sugahara (1979)
I. geniculata	Fruit	Pg, Cy 3-xylosylglucosides	Ishikura and Sugahara (1979)
I. geniculata cv. *glabra* Ossuyama	Fruit	Pg, Cy 3-xylosylglucosides	Ishikura and Sugahara (1979)
I. kiusiana	Fruit	Cy 3-glucoside, Pg, Cy 3-xylosylglucosides	Ishikura (1975a)
I. macropoda	Fruit	Cy 3-xylosylglucoside	Ishikura (1975a)
I. micrococca	Fruit	Cy 3-glucoside	Ishikura (1975a)
I. nipponica	Fruit	Pg, Cy 3-xylosylglucosides	Ishikura (1975a)
I. sugeroki cv. *brevipedunculata*	Fruit	Cy 3-glucoside, 3-xylosylglucoside	Ishikura (1975a)

ARALIACEAE			
Acanthopanax divaricatus	Fruit	Dp 3-xylosylgalactoside	
Aralia elata	Fruit	Cy 3-xylosylgalactoside	Ishikura (1975b)
A. canescens			
Dendropanax trifidus	Fruit	Cy 3-xylosylgalactoside	
ASCLEPIADACEAE			
Calatropis procera	Flower	Cy 3-rhamnoglucoside	Tiwari *et al.* (1978)
Periploca gracea	Stem	Cy, Pn glucosides	Melin (1975)
BALSAMINACEAE			
Impatiens spp.	root	Cy 3-glycosides	Thakur and Nozzolillo (1978)
BEGONIACEAE			
Begonia sp.	Flower	Cy 3-(2^G-xylosylrutinoside)	Langhammer and Grandet (1974)
BERBERIDACEAE			
Berberis vulgaris	Leaves	Cy, Pn, Dp 3-glucosides	Semkina (1975)
B. fortunei	Fruit	Cy, Dp, Pt, Mv 3-glucosides, Pn-glucoside, Cy, Dp Pt, Mv 3-rutinosides	Ishikura and Sugahara (1979)
BORAGINACEAE			
Anchusa italica	Flower	Cy, Dp-glycosides	
A. officinalis	Flower	Cy, Dp-glycosides	Delorme *et al.* (1977)
Borago officinalis	Flower	Cy, Dp-glycosides	
Cerinthe minor	Flower	Dp glycosides	Ishikura (1974)
Cynoglossum amabile	Flower	Dp 3,5-diglucoside	Delorme *et al.* (1977)
C. cheirfolium	Flower	Dp glycosides	Delorme *et al.* (1977)
Echium amoneum	Flower	Cy, Dp-glycosides	
E. vulgare	Flower	Dp-glycosides	
Heliotropium peruvianum	Flower	Dp-glycosides	
Lithospermum purpureo-	Flower	Cy, Dp-glycosides	
Lycopsis arvensis	Flower	Cy, Dp-glycosides	Ishikura (1974)
Myosotis alpestris	Flower	Dp 3,5-diglucoside	Delorme *et al.* (1977)
		Cy, Dp, Mv glucosides	
M. silvatica	Flower	Cy, Dp, Mv glucosides	
M. versicolor	Flower	Cy, Dp, Mv glucosides	
Pulmonaria officinalis	Flower	Cy, Dp-glycosides	

Table 3.1 (*Contd.*)

Orders, family, genus, species	*Organ examined*	*Pigments present*	*Reference*
P. visianii	Flower	Dp, Pt, Mv glycosides	Coassini-Lokar and Poldini (1978)
Symphytum asperum	Flower	Dp, Mv 3,5-diglucosides	Ishikura (1974)
S. officinale	Flower	Cy, Dp, Mv glycosides	Delorme *et al.* (1977)
Tournefortia heliotropium	Flower	Cy, Mv glycosides	
Trigonatis peduncularis	Flower	Dp 3,5-diglucoside	Ishikura (1974)
BUXACEAE			
Buxus microphylla	Fruit	Cy 3-glucoside, 3-rhamnoglucoside	Ishikura (1975c)
CAMPANULACEAE			
Campanula sp. cv. Rose	Flower	Pg 3-di (*p*-hydroxybenzoyl)-rutinoside-7-glucoside	Asen *et al.* (1979)
Lobelia erinus	Flower	Caffeyl-ferulyl-*p*-coumaryl-Dp-3-rutinoside-5,3′,5′triglucoside	Yoshitama (1977)
CAPRIFOLIACEAE			
Lonicera hypoglauca	Fruit	Cy 3-glucoside, 3,5-diglucoside	Ishikura and Sugahara (1979)
L. japonica	Fruit	Cy, Pn 3,5-diglucoside	
Sambucus ebulus	Fruit	Cy 3-sambubioside, 3-sambubioside-5-glucoside	Chiarlo *et al.* (1978)
S. nigra	Fruit	Cy 3 (2-1-xylosylglucoside) Sambicyanin	Reichel and Reichwald (1977)
Viburnum awabuki	Fruit	Cy-3-xylosylglucoside	
V. dilatatum	Fruit	Cy 3-xylosylglucoside	Ishikura (1975c)
V. erosum	Fruit	Cy 3-xylosylglucoside	
V. furcatum	Fruit	Cy 3-xylosylglucoside	
V. japonicum	Fruit	Cy-3-xylosylglucoside	Ishikura (1975c)
V. lautata	Fruit	Cy 3-glucoside	Guichard *et al.* (1976)
V. phlebotrichum	Fruit	Cy 3-xylosylglucoside	Ishikura (1975c)
V. sargentii	Fruit	Cy 3-xylosylglucoside	
V. tinus	Fruit	Cy 3-glucoside, 3-sambubioside	Godeau *et al.* (1979)
V. urceolatum	Fruit	Cy 3-glucoside	Ishikura (1975b)
V. wrightii	Fruit	Cy 3-xylosylglucoside	

CARYOPHYLLACEAE			
Silene dioica	Flower	Pg, Cy 3-glucosides, 3,5-diglucosides, 3-rutinosides 3 rutinoside-5-glucosides, Pg 3(4-*O*-*p*-coumaryl-rhamnoglucoside), Cy 3-(6-*O*-caffeylglucoside)	Kamsteeg *et al.* (1978)
CELASTRACEAE			
Eunymus fortunei	Fruit	Cy 3-glucoside	Ishikura (1975a)
E. oxyphyllus	Fruit	Cy 3-glucoside	
E. sieboldianus	Fruit	Cy 3-glucoside	
COMBRETACEAE			
Lumnitzera littorea	Flower	Pg, Cy, Pn 3-glucosides, Pg, Pn 3,5-diglucosides	Lowry (1976b)
Quisqualis indica	Flower	Cy 3-glucoside	
Terminalia catappa	Fruit	Cy 3-glucoside	
COMPOSITAE			
Callistephus chinesis	Flower	Pg, Cy, Dp 3-glucosides, 3-acylglucosides, Pg, Cy, Dp 3,5-diglucosides, acylated 3,5-diglucosides Pg, Cy, Dp 3-glucosides, 3,5-diglucoside, Pg, Cy, 3-galactosides, Pg, Cy-3-rhamnoglucosides, 5-glucosides, 2^G-glucosylrutinosides	Forkmann (1977b) Guzewski (1977)
Centaurea lugdunensis	Flower	Cy-3-glucoside, 3,5-diglucoside	Kamanzi and Raynaud (1977)
C. montana	Flower	Cy-3-glucoside, 3,5-diglucoside	
Chrysanthemum sp.	Flower	Cy 3-glucoside; caffeylglucoside; 3-diglucoside; 3-caffeyldiglucoside	Sugiyama and Takano (1974a,b)
Cosmos caudatus	Flower	Cy 3-glucoside, 3-rahmnoglucoside	Samata *et al.* (1977)
C. sulphureus	Flower	Cy 3-glucoside, 3-rhamnoglucoside	
Cynara scolymus	Flower	Cy glycoside	Foury and Aubert (1977)
Senecio cruentus	Whole plant	Dp 3,7,3′-dicaffeyltriglucoside	Yoshitama *et al.* (1975)
CORNACEAE			
Cornus alba	Fruit	Cy 3-glucoside; 3-rutinoside, Pt 3-rutinoside, Cy 3-galactoside, Cy-3-arabinoside	Du *et al.* (1975b)
C. alternifolia	Fruit	Cy 3-glucoside, 3-rutinoside, Pt 3-rutinoside, Cy 3-galactoside, Cy 3-arabinoside	
C. brachypoda	Fruit	Cy glucoside, Dp glucoside	Ishikura and Sugahara (1979)
C. canadensis	Fruit	Pg, Cy 3-robinobiosides	Du *et al.* (1974b)
C. controversa	Fruit	Cy, Dp, Mv glucosides	Ishikura and Sugahara (1979)

Table 3.1 (*Contd.*)

Orders, family, genus, species	*Organ examined*	*Pigments present*	*Reference*
C. florida	Fruit	Cy 3-glucoside	Ishikura (1975c)
C. kousa	Fruit	Cy 3-glucoside	
CRUCIFEREAE			
Brassica juncea	Leaves	Pn, 3-glucoside, 3-galactoside	Park (1979a,b)
Brassica oleracea cv. Red Danish	Seedlings	Cy 3-malonylsophoroside-5-glucoside, 3-*p*-coumarylsophoroside-5-glucoside, 3-di-*p*-coumarylsophoroside-5-glucoside, 3-ferulylsophoroside-5-glucoside, 3-diferulylsophoroside-5-glucoside, 3-sinapylsophoroside-5-glucoside, 3-disinapylsophoroside-5-glucoside	Hrazdina *et al.* (1977)
EPACRIDACEAE			
Acrotricha serrulata	Flower, bark	Cy-3-arabinoside, 3-galactoside, Dp 3-arabinoside	Jarman and Crowden (1974)
Archeria comberi	Petals, leaves, bark	Cy 3-arabinoside, 3-galactoside	
A. eriocarpa			
A. hirtella			
A. serpyllifolia			
Astroloma humifusum	Leaves	Cy 3-arabinoside, 3-galactoside	
Brachyloma daphnoides	Leaves	Cy 3-arabinoside, 3-galactoside	
Cyathodes spp.	Leaves, flowers	Cy 3-arabinoside, 3-galactoside, 3-rhamnogalactoside	
	Fruit, bark, stem	Dp 3-galactoside	
Dracophyllum spp.	Leaf, flower	Cy 3-arabinoside, 3-galactoside	
Epacris spp.	Leaves, flowers	Cy 3-arabinoside, 3-galactoside	
Leucopogon spp.	Flower	Cy 3-arabinoside, 3-galactoside, 3-rhamnogalactoside	
Lissantha spp.	Leaf, flower fruit	Cy 3-arabinoside, 3-galactoside	
Monotoca spp.	Leaf, flower	Cy 3-arabinoside, 3-galactoside, 3-rhamnogalactoside	
Pentachondra spp.	Leaf, flower	Cy, Dp 3-arabinosides, 3-galactosides, Cy-3-rhamnogalactosides	
Richea spp.	Leaves	Cy 3-arabinoside, 3-galactoside	
Sprengelia incarnata	Leaf, flower	Cy 3-arabinoside, 3-galactoside	
Styphelia spp.		Cy 3-arabinoside, 3-galactoside	

Trochocarpa spp.		Cy, Dp 3-arabinosides, 3-galactosides, 3-glucosides, 3-rhamnogalactosides	Jarman and Crowden (1974)
Woollsia pangens	Flower	Cy 3-arabinoside, 3-galactoside	
ERICACEAE			
Andromeda polyfolia	Leaves	Cy, Mv glucosides	Tamas *et al.* (1975)
Cyanococcus sp.	Fruit	Cy, Pn, Dp, Pt, Mv galactosides and arabinosides Cy, Pn, glycosides; Cy-3-xylosylglucoside; Dp 3-glycoside; Dp 5-glycoside, Dp 3-rhamnoglycoside	Ballinger *et al.* (1979)
Erica, 39 sp	Flower	Pg, Cy, Pn, Dp, Pt, Mv glucosides, arabinosides and biosides	Crowden and Jarman (1976)
Rhododendron simsii	Flower	Cy, Pn, Mv glycosides	Deloose (1974)
Vaccinium bracteatum	Fruit	Cy, Dp, Pt glycosides	Ishikura and Sugahara (1979)
V. corymbosum	Leaf	Cy, Dp, Pt, Mv 3-glucosides	Pourrat *et al.* (1978)
V. floribundum	Fruit	Cy-glycosides	Martinod *et al.* (1975)
V. myrtillis	Fruit	Cy 3-xylosylglucoside	Ballinger *et al.* (1979)
V. smallii	Fruit	Dp, Pt, Mv glycosides	Ishikura and Sugahara (1979)
Vaccinium stamineum	Fruit	Cy, Pn glycosides	Ballinger *et al.* (1979)
V. vitis-idae		Dp 3-glucoside, Dp 5-glucoside, 3-rhamnoglucoside	
Vaccinium sp.	Fruit	Cy 3-glucoside, 3-rhamnoglucoside, 3-rhamnoglucosyl-glucoside	Fikiin and Chau (1977)
FOUQUIERACEAE			
Fouquieria sp.	Flowers, stem	Pg, Cy, Pn, Dp, Pt, 3-glucosides	Scogin, (1977)
FRANKENIACEAE			
Frankenia grandiflora	Flower	Dp 3-glucoside, Dp 3-galactoside, Pn 3-galactoside	Scogin (1979)
FUMARIACEAE			
Corydalis cava	Flower	Cy, Pn-glucosides	Coassini-Lokar and Poldini (1978)
C. solida	Flower	Cy, Pn-glucosides	
GENTIANACEAE			
Tripterospermum japonicum	Fruit	Cy 3-rutinoside, Cy glycoside	Ishikura and Sugahara (1979)
GERANIACEAE			
Erodium cicutarium	Flower	Cy, Pn, Mv 3-glucosides, Cy 5-glucoside, Cy 3-rutinoside Mv 3,5-diglucoside	Medrano *et al.* (1978)
E. malacoides	Flower	Mv 3-glucoside, 3,5-diglucoside	

Table 3.1 (*Contd.*)

Orders, family, genus, species	*Organ examined*	*Pigments present*	*Reference*
HYDROPHYLLACEAE			
Hydrophyllum macrophyllum	Flower	Cy 3,5-diglucoside	Beckmann (1979)
H. occidentale	Flower	Cy 3,5-diglucoside	
H. tenupes	Flower	Cy 3,5-diglucoside	
H. virginianum	Flower	Cy 3,5-diglucoside	
LARDIZABALACEAE			
Akebia quinata	Flower	Cy 3-coumarylglucoside, 3-*p*-coumarylxylosylglucoside	Ishikura (1975c)
A. trifoliata		3-xylosylglucoside	Ishikura and Nagamizo (1976)
Stauntonia hexaphylla	Flower	Cy 3-*p*-coumarylglucoside, 3-*p*-coumarylxylosyl-glucoside Cy 3-xylosylglucoside	Ishikura (1975c)
LAURACEAE			
Cinnamomum camphora	Fruit	Cy 3-rhamnoglucoside; Cy 3-glucoside	Ishikura (1975c)
C. japonicum			
Lindera glauca	Fruit	Cy 3-glucoside, 3-rutinoside	Ishikura and Sugahara (1979)
Litsea japonica	Fruit	Cy 3-glucoside, 3-glucosylrutinoside	Ishikura (1975c)
Machilus thumbergii	Fruit	Cy, Pn 3-glucosides, 3-rutinosides	Ishikura and Sugahara, (1979)
Neolitsea sericae	Fruit	Cy 3-rhamnoglucoside	Ishikura (1975c)
LECYTHIDACEAE			
Couropita quinensis	Flower	Cy, Dp 3-glucosides	Lowry, (1976a)
LEGUMINOSAE			
Albizia julibrissin	Flower	Cy 3-glucoside	Ishikura *et al.* (1978)
Anthyllis alpestris	Flower	Cy, Pn glycosides	
A. iberica	Flower	Cy, Dp glycosides	
A. maura	Flower	Cy, Pn, Dp-glycosides	
A. montana ssp.	Flower	Cy, Pn, Dp, Mv-glycosides	
A. praepropera	Flower	Cy, Dp glycosides	
A. pyrenaica	Flower	Cy, Pn, Dp glycosides	
Anthyllis vulneraria	Leaf	Cy, 3-galactoside, Pn, Dp 3-galactosides	Sterk *et al.* (1977)
	Flower	Cy, Pn, Dp glycosides	

Astralagus sinicus	Flower	Cy 3-xylosylglucoside, Dp diglucoside	Ishikura *et al*, (1978)
Bauhinia purpurea	Flower	Pg 3-glucoside, 3-triglucoside	Tiwari *et al*, 1978)
Canavalia gladiata	Flower	Pt 3,5-diglucoside, Mv 3-glucoside; Mv 3,5-diglucoside	Ishikura *et al*, (1978)
C. lineata		Pt 3,5-diglucoside, Mv 3-glucoside; Mv 3,5-diglucoside	
Cassia javanica	Flower	Pn 3-rhamnoside	Singh *et al*. (1979)
Cercis chinensis	Flower	Pt 3-glucoside; Mv 3-glucoside, Mv 3-diglucoside; Mv 3,5-diglucoside	Ishikura (1978)
Clitoria ternata	Flower	Dp, Pt, Mv 3-glucoside	Srivastava and Pande (1977)
Desmodium adhesivum	Leaves	Cy 3-glucoside, 3,5-diglucoside	Martinod *et al*. (1978)
Dumosia truncata	Fruit	Cy 3-glucoside, Mv 3,5-diglucoside, 3-rutinoside-5-glucoside, Mv glycoside	Ishikura and Sugahara (1979)
Erythrina suberosa	Flower	Pg, Cy Dp 3,5-diglucosides	Verma *et al*. (1977)
Indigofera pseudotinctorium	Flower	Pt, 3-glucoside, Mv 3,5-diglucoside	Ishikura *et al*. (1978)
Lathyrus sativus	?	Pg, Cy 3,5-diglucosides	Sarkar and Banerjee (1977)
Lespedeza bicolor	Flower	Mv 3-glucoside, 3,5-diglucoside	
L. homoloba	Flower	Mv 3-glucoside, 3,5-diglucoside	Ishikura *et al*. (1978)
L. penduliflora	Flower	Pt, Mv 3-glucosides, Mv 3,5-diglucoside	
Mucuna macrocarpa	Flower	Dp, Pt, Mv 3-glucoside	
Pisum sativum	Stem	Pg, Cy, Dp 3-glucosides, 3-oligosides	Nozzolillo (1978)
Pueraria lobata	Flower	Dp 3-glucoside, 3-diglucoside, Mv 3,5-diglucoside	Yokoyama (1976)
Rhynchosia acuminatifolia	Fruit	Cy 3-glucoside	Ishikura and Sugahara (1979)
Saraca indica	Flower	Pg, Cy 3,5-diglucosides	Lakshmi and Chauhan (1976)
Trifolium pratense	Flower	Pn 3-glucoside, 3-diglucoside, 3-xylosylglucoside, Dp, Pt, Mv 3-rhamnoglucosides	
Vicia angustifolia	Flower	Pt, Mv 3-glucosides, Mv 3,5-diglucoside	
V. cracca	Flower	Pt, Mv 3,5-diglucosides, 3-rhamnosylglucosides	
V. deflexa	Flower	Dp, Pt 3-rhamnoside-5-glucoside, Mv 3,5-diglucoside 3-rhamnoside-5-glucoside	Ishikura *et al*. (1978)
V. pseudo-orobus	Flower	Dp diglucoside, 3-rhamnoside-5-glucoside, Pt, Mv 3-rhamnoside-5-glucoside	
V. unijuga	Flower	Dp, Pt, Mv 3,5-diglucoside, 3-rhamnoside-5-glucoside Dp 3-triglucoside	
Wisteria brachybotrys	Flower	Dp, 3,5-diglucoside; Pt 3-rhamnoside-5-glucosides	Ishikura (1978)
W. floribunda	Flower	Dp, 3,5-diglucoside; Pt 3,5-diglucoside Mv 3,5-diglucoside	

Table 3.1 (*Contd.*)

Orders, family, genus, species	*Organ examined*	*Pigments present*	*Reference*
LINACEAE			
Linum usitatissimum	Flower	Pg, Cy, Dp 3-glucosylrutinosides, Cy, Dp 3-triglucosides	Dubois and Harborne (1975)
LYTHRACEAE			
Lagerstromia indica	Flower	Dp, Pt, Mv 3-glucosides	Egolf and Santamour (1975)
L. speciosa	Flower	Pn, Mv 3,5-diglucosides	Lowry, (1976a)
Woodfordia fruticosa	Flower	Pg, 3,5-diglucoside	Nair *et al.* (1976)
MAGNOLIACEAE			
Michelia compressa	Fruit	Cy 3-glucoside, 3-rutinoside	Ishikura and Sugahara (1979)
MALVACEAE			
Abelmoschus spp.	Flower	Cy 3-glucoside, 3-rutinoside, 3-sambubioside, 3-sophoroside, Dp 3-sambubioside	Lowry (1976b)
Hibiscus syriacus	Flower	Dp, Pt, Mv 3-glucosides	Egolf and Santamour (1975)
Hibiscus spp.	Flower	Cy, 3-glucoside, 3-rutinoside, 3-sophoroside, 3-sambubioside Dp 3-sambubioside	Lowry (1976b)
Malva silvestris	Flower	Dp, Mv 3-glucosides, Mv 3,5-diglucoside	Rakhimkhanov *et al.* (1975)
Thespesia spp.	Flower	Cy 3-glucoside, 3-rutinoside, 3-sophoroside, Cy, Dp, 3-sambubioside	Scogin (1979)
MARTYNIACEAE			
Martynia annua	Flower	Cy 3-galactoside, Pg 3,5-diglucoside	Tiwari *et al.* (1978)
MELASTOMATACEAE			
Clidemia hirta	Fruit	Acylated Dp 3,5-diglucoside	
Dissochaeta bractata	Fruit	Dp 3-(*p*-coumarylglucoside)-5-glucoside	
D. celebica	Fruit	Dp, Mv 3-acylglucoside-5-glucoside	
Dissotis rotundifolia	Flower	Mv 3,5-diglucoside	
Driessenia glandulifera	Flower	Cy diglucoside acylated	
Marumia nemorosa	Flower	Mv 3-acylglucoside-5-glucoside	Lowry (1976a)
M. reticulata	Flower	Mv 3-acylglucoside-5-glucoside	
Medinilla hasseltii	Fruit	Pg 3-acylglucoside-5-glucoside	
M. heteroanthera	Flower	Pg 3-acylglucoside-5-glucoside	
M. schortechinii	Flower	Pg 3-p-coumarylglucoside-5-glucoside	

Melastoma malabathricum	Flower	Mv 3,5-diglucoside	Lowry (1976a)
M. sanguineum	Flower	Cy, Mv 3,5-diglucosides	
Memecylon amplexicaule	Fruit	Cy 3,5-diglucoside	
M. caeruleum	Fruit	Mv 3,5-diglucoside	
Oritrephes grandiflora	Flower	Cy 3-glucoside	
Osheckia perakensis	Flower	Mv 3-(acylglucoside)-5-glucoside	
Oxyspora hispida	Inflorescence	Mv 3-(*p*-coumarylrutinoside)-5-glucoside	
Phyllagathis elliptica	Flower	Acylated Mv glucoside	
P. rotundifolia	Flower	Mv 3-acylglucoside-5-glucoside	
P. scortechinii	Flower	Acylated Pn glucoside	
P. tuberculata	Flower	Mv 3-*p*-coumarylglucoside-5-glucoside	
Pternandra coerulescens	Fruit	Acylated Mv glycoside	
Sarcopyramus nepalensis	Flower	Mv 3,5-diglucoside, acylated glycoside	
Sonerila argentata	Flower	Acylated Mv diglycoside	
S. erecta	Flower	Acylated Pn, Mv glycosides	
S. heterostemon	Flower	Acylated Mv glycosides	
S. rudis	Flower	Acylated Mv glycosides	
S. prostrata	Flower	Acylated Mv glycosides	
Tibouchina semidecandra	Flower	Mv 3-(*p*-coumarylglucoside)-5-glucoside	
MENISPERMACEAE			
Cocculus trilobus	Fruit	Cy 3-glucoside	Ishikura (1975)
MORACEAE			
Ficus carica	Fruit epidermis	Pg 3-rhamnoglucoside, Cy 3-glucoside, 3-rhamnoglucoside; 3,5-diglucoside	Puech *et al.* (1975)
F. erecta	Fruit	Cy 3-glucoside 3-rhamnoglucoside	Ishikura (1975c)
F. nipponica	Fruit	Cy 3-glucoside, 3-rutinoside, 3,5-diglucoside	Ishikura and Sugahara (1979)
F. pumila	Fruit	Cy 3-glucoside	
F. wrightiana	Fruit	Cy 3-glucoside, 3-rhamnoglucoside	Ishikura (1975)
Morus alba	Fruit	Cy 3-glucoside, 3-rhamnoglucoside	Ishikura (1975)
MYRICACEAE			
Myrica rubra	Fruit	Cy 3-glucoside	Ishikura and Sugahara (1979)
MYRTACEAE			
Barringtonia macrostahya	Flower	Cy, Dp 3-sambubioside	Lowry (1976a)
B. racemosa			

Table 3.1 (*Contd.*)

Orders, family, genus, species	*Organ examined*	*Pigments present*	*Reference*
Beaufortia squarrosa	Flower	Pg, Pn 3,5-diglucosides	Lowry (1976a)
Callistemon lanceolatus	Flower	Cy 3-glucoside	
C. phoenicious	Flower	Cy 3-glucoside	
Calothamnus gilesii	Flower	Cy 3,5-diglucoside	
C. oldfieldii		Pn, Cy 3-glucosides; Cy-3,5-diglucoside	
C. quadrificus		Cy 3-glucoside	
C. torulosa		Cy 3-glucoside	
Chamelaucium uncinatum	Flower	Dp 3-glucoside; Dp glucosides	
Darwinia citriodora	Flower	Pg, Cy 3,5-diglucosides	
D. oldfieldii	Flower	Cy, 3,5-diglucosides	
Eucalyptus caesia	Flower	Cy 3-glucoside	
E. macrocarpa	Flower	Cy glucoside	
Eucalyptus sp.	Leaves, bark	Cy 3-glucoside, 3,5-diglucoside, Dp glycosides, galactosides	Sharma and Crowden (1974)
Eugenia aquea	Fruit	Cy 3-glucoside	Lowry, (1976a)
E. jambolana	Fruit	Dp, Pt 3-gentiobioside, Mv 3-laminaribioside	Jain and Seshadri (1975)
E. malaccensis	Flower	Mv 3,5-diglucoside	Lowry (1976a)
Feijoa sellowiana	Flower	Cy 3-glucoside	
Hypocalymma robustum	Flower	Mv, 3,5-diglucoside	
Kunzea baxterii	Flower	Cy, Dp 3-glucosides	
K. jocunda	Flower	Mv 3,5-diglucoside	
Leptospermum erubescence	Flower	Mv 3,5-diglucoside	
L. flavescens	Flower	Cy 3-glucoside	
L. scoparium	Flower	Cy, Pn 3-glucosides	
L. seriaceum	Flower	Mv 3,5-diglucoside	
Lophomyrtus bullata	Fruit	Cy 3-glucoside	
L. obcardata	Fruit	Cy 3-glucoside	
Melaleuca nesophila	Flower	Mv 3,5-diglucoside	
M. steedmani	Flower	Cy 3,5-diglucoside	
Meterosideros excelsa	Flower	Cy, Pn, Mv 3-glucosides	

M. fulgens	Flower	Cy, Dp 3-glucosides	
M. kermadecensis	Flower	Pn(?) Cy, Pn 3-glucosides, Pn, Mv 3,5-diglucosides	
Neomyrtus pedunculata	Fruit	Cy 3-glucoside	
Pileanthus pendunculatus	Flower	Dp 3-glucoside, Dp-rhamnoside	
Psidium guajavica	Fruit	Cy 3-glucoside	
Rhodomyrtus tomentosa	Flower	Mv 3-glucoside	
Scholtzia capitata	Flower	Cy 3-glucoside, 3,5-diglucoside	Lowry (1976a)
S. parvifolia	Flower	Cy 3-glucoside, 3,5-diglucoside	
Thryptomene denticulata	Flower	Cy 3-glucoside, 3,5-diglucoside	
T. maisonervii	Flower	Cy 3-glucoside, Dp 3-rhamnoside	
Verticordia monadelpha	Flower	Cy, Dp, 3,5-diglucoside	
V. picta	Flower	Cy, 3,5-diglucoside, Cy glycoside	
OLEACEAE			
Ligustrum amurense	Fruit	Cy 3-glucoside, 3-rutinoside, Mv 3-rutinoside-5-glucoside	
L. lucidum	Fruit	Cy 3-glucoside, 3-rutinoside, Mv 3-rutinoside-5-glucoside	
L. sinense	Fruit	Cy 3-glucoside, 3-rutinoside, Mv 3-rutinoside-5-glucoside	
L. vulgare	Fruit	Cy 3-glucoside, 3-rutinoside, Mv 3-rutinoside-5-glucoside	Medrano *et al.* (1977)
L. ibota	Fruit	Cy glycosides	
L. ovalifolium	Fruit	Cy 3-caffeylglucoside, 3-caffeylrutinoside	
L. obtusifolium	Fruit	Cy 3-caffeylglucoside, 3-caffeylrutinoside	Ishikura and Sugahara (1979)
Olea europaea	Fruit	Cy 3-glucoside	Vazquez-Roncero *et al.* (1974)
Osmanthus fragrans	Fruit	Cy 3-rhamnoglucoside	Ishikura (1975c)
O. ilicifolius	Fruit	Cy 3-glucosylrhamnoside	
ONAGRACEAE			
Fuchsia spp.	Flowers	Pg, Cy, Pn, Dp, Pt, Mv 3-glucosides, 3,5-diglucosides	Crowden *et al.* (1977)
PAPAVERACEAE			
Papaver bracteatum	Flower	Pg, Cy 3-glucosides, Cy 3-sophoroside	Yaklich and Gentner (1974)
PASSIFLORACEAE			
Passiflora edulis	Fruit	Cy 3-glucoside	Ishikura and Sugahara (1979)
P. quadrangularis	Flower	Dp, Pt, Mv 3-glucosides, 3,5-diglucosides	Billot (1974)
POLYGONACEAE			
Polygonum japonica	Leaf	Cy, Dp, Mv glycosides, Cy 3,5-caffeylglucosylarabinoside, Cy 3,5-diglucoside	
P. pentjutini	Leaf	Cy, Dp, Mv glycosides, Cy 3,5-caffeylglucosylarabinoside Cy 3,5-diglucoside	Kuznetsova (1979)

Table 3.1 (*Contd.*)

Orders, family, genus, species	*Organ examined*	*Pigments present*	*Reference*
P. perfoliatum	Fruit	Mv 3,5-diglucoside	Ishikura and Sugahara (1979)
P. sadralinense	Leaf	Cy, Dp, Mv glycosides, Cy 3,5-caffeylglucosylarabinoside Cy 3,5-diglucoside	Kuznetsova (1979)
P. tomentosum	Flower	Pg 3-glucoside	Tiwari *et al.* (1978)
P. weyridzii	Leaf	Cy, Dp, Mv glycosides, Cy 3,5-caffeylglucosylarabinoside Cy 3,5-diglucoside	Kuznetsova (1979)
PUNICACEAE			
Punica granatum	Fruit	Cy, Dp, Pt 3-glucoside, 3,5-diglucosides, Cy, Dp Mv glucosides, Pg 3,5-diglucosides	Du *et al.* (1975b) Abdurazakova and Gabrazova (1972); Lowry (1976a)
RANUNCULACEAE			
Clematis terniflora	Fruit	Cy 3-glucoside	Ishikura (1975c)
Paeonia japonica	Fruit	Pn 3-glucoside 3,5-diglucoside	Ishikura and Sugahara (1979)
RHIZOPHORACEAE			
Gynotroches axillaris	Fruit	Cy 3-glucoside	Lowry (1976a)
ROSACEAE			
Cotoneaster melanocarpus	Fruit	Cy 3-glucoside	Demina (1974)
C. uniflora	Fruit	Cy 3-glucoside	
Duchesnea chrysantha	Fruit	Cy, Pn 3-glucosides, Pn 3-rutinoside	Ishikura and Sugahara (1979)
Fragaria ananassa	Fruit	Pg, Cy 3-glucosides	Ishikura (1975c)
F. vesca	Fruit	Pg, Cy 3-glucosides	Fikiin and Chau (1977)
Laurocerasus officinalis	Fruit	Cy, Pn 3-arabinosides	Tsiklauri (1975)
Malus sp.	Fruit epidermis	Cy 3-arabinoside, 3-galactoside, 3-arabinoside	Posokhlyarova (1975)
Photini glabra	Fruit	Cy 3-glucoside, 3-rhamnoglucoside	} Ishikura (1975c)
Pourthiaea villosa	Fruit	Cy 3-glucoside	}
Prunus cerasus	Fruit	Cy 3-glucoside, Cy, Pn 3-rhamnoglucosides	Posello and Bonzini (1977b) Rahman *et al.* (1975) Takeda and Sano (1978)
P. jamasakura	Fruit	Cy, 3-glucoside, 3-rhamnoglucoside	} Ishikura (1975c)
P. mume	Fruit	Cy 3-glucoside, 3-rhamnoglucoside	}
P. persica	Fruit	Cy 3-glucoside, 3-rhamnoglucoside	}
P. salicina	Fruit	Cy 3-glucoside, 3-rhamnoglucoside	}

P. sargentii	Fruit	Pg, Cy 3-glucosides, 3-rutinosides, Cy 3-diglucoside	Du *et al.* (1975)
P. tomentosa	Fruit	Pg, Cy 3-rhamnoglucosides	Ishikura (1975c)
P. yedoensis	Fruit	Cy 3-glucoside, 3-rhamnoglucoside	Ishikura (1975c)
Raphiolepis umbellata	Fruit	Cy 3-glucoside	Ishikura (1975c)
Rasselia multiflora	Flower	Pn 3-glucoside	Tiwari *et al.* (1978)
Rosa plathyacantha	Flower	Cy 3-glucoside, 3,5-diglucoside	Chumbalov *et al.* (1976)
R. spinosissima	Fruit	Cy, Dp glucosides	Plekhanova *et al.* (1978)
Rubus crataegifolius	Fruit	Pg glucosides	Torre and Barritt (1977)
R. hirsutus	Fruit	Pelargonidin-3-glucoside	Ishikura and Sugahara (1979)
R. idaeus	Fruit	Cy 3-glucosides, Cy 3-sophorosides, Cy 3-rutinoside Cy 3-glucosylrutinoside, Cy 3-sambubioside, Cy 3-xylosylrutinoside	Barritt and Torre (1975)
R. illecebrosus	Fruit	Pg glucosides	Torre and Barritt (1977)
R. morifolius	Fruit	Pg glucosides	Torre and Barritt (1977)
R. parviflorus	Fruit	Pg glucosides	Torre and Barritt (1977)
Sorbus commixta	Flower	Cy 3-glucoside	Ishikura (1975c)
Spiraea cantoniensis	Fruit	Cy 3-glucoside	Ishikura (1975c)
RUBIACEAE			
Cinchona ledgerana	Leaves	Cy 3-arabinoside, galactoside, 7-arabinoside	Paris and Jacquemin (1975)
C. succirubra	Leaves		Posokhlyarova (1975)
Damnacanthus indicus	Fruit	Pg 3-glycoside	Ishikura and Sugahara (1979)
D. macrophyllus	Fruit	Pg 3-glycoside, 3-rutinoside	Ishikura and Sugahara (1979)
Lasianthus japonicus	Flower	Pg 3-rutinoside, Pg-glycosides	Ishikura and Sugahara (1979)
Mitchella undulata	Fruit	Pg 3-rutinoside, Pg-glycosides	Ishikura and Sugahara (1979)
Rubia akane	Fruit	Cy 3-rutinoside	Ishikura and Sugahara (1979)
RUTACEAE			
Skimmia japonica	Fruit	Pg 3-glucoside, 3-rhamnoglucoside	Ishikura (1975c)
Zanthoxylum piperitum	Fruit	Cy 3-rhamnoglucoside	Ishikura (1975c)
SABIACEAE			
Meliosma tenuis	Fruit	Cy, Dp 3-rutinosides, Dp 3-glucoside	Ishikura and Sugahara (1979)
SAPINDACEAE			
Litchi chinensis	Fruit	Pg, Cy 3-glucosides, Pg 3,5-diglucoside, Cy 3-galactoside	Prasad and Jha (1978)
SAXIFRAGACEAE			
Ribes nigrum	Fruit	Cy, Dp 3-glucosides, 3-rutinosides, 3-sophoroside, Pg 3-rutinoside	Lous *et al.* (1975)

Table 3.1 (*Contd.*)

Orders, family, genus, species	*Organ examined*	*Pigments present*	*Reference*
Ribes spp.	Fruit	Cy 3-sophoroside, 3-(2^G-glucosyl)-rutinoside, 3-sambubioside	Oydvin (1975)
Ribes spp.	Fruit	Cy 3-glucoside, 3-glucosylrhamnoside, 3-glucosylxyloside, 3-diglucoside	Nilsson and Trajkovski (1977)
SCROPHULARIACEAE			
Penstemon centranthifolius	Flower	Pg 3,5-diglucoside	
P. grinnellii	Flower	Cy, Dp 3,5-diglucosides	Harborne and Smith (1978)
P. spectabilis	Flower	Cy, Dp 3,5-diglucosides	
SOLANACEAE:			
Cyphomandra betacea	Fruit	Pg, Cy, Dp 3-glucosides; Pg, Cy, Dp 3-rutinosides	Wrolstad and Heatherbell (1974)
STERCULIACEAE			
Brachychiton acerifolia	Flower	Cy 3-glucoside, Dp 3-glucoside, Pg 3-glucoside	
B. discolor	Flower	Cy 3-glucoside, Cy 3-arabinoside	
B. populneum	Flower	Cy 3-glucoside, Cy 3-arabinoside	
Cheirostemon platinoides	Flower	Cy 3-glucoside	
Dombeya burgessia	Flower	Cy 3-glucoside	
D. cayeuxii	Flower	Cy 3-glucoside	
D. burgessia cv. Seminole	Flower	Cy 3-glucoside	Scogin (1979)
Firmiana platanifolia	Flower	Cy 3-glucoside	
Fremonita californica	Flower	Cy 3-glucoside, Dp 3-glucoside	
F. mexicana	Flower	Cy 3-glucoside, Dp 3-glucoside	
F. californica hybrid	Flower	Cy 3-glucoside, Dp 3-glucoside	
Guichenotia sarotes	Flower	Cy 3-glucoside, Dp 3-glucoside	
Theobroma cacao	Flower	Cy 3-galactoside, Cy 3-arabinoside	
SYMPLOCACEAE			
Symplocos chinensis	Fruit	Dp 3-xylosylglucoside	Ishikura (1975c)
S. coreana	Fruit	Dp 3-xylosylglucoside, 3-diglucoside-5-glucoside	
S. lucida	Fruit	Dp 3-glucoside, 3-sambubioside	Ishikura and Sugahara (1979)
S. myrtacea	Fruit	Dp 3-xylosylglucoside	Ishikura (1975c)

TAMARICACEAE			
Tamarix tetranda	Flower	Cy 3-glucoside, Cy 3,5-diglucoside	Scogin (1977)
THEACEAE:			
Camellia japonica	Fruit	Cy 3-glucoside	Ishikura and Sugahara (1979)
Cleyera japonica	Fruit	Cy 3-glucoside; Cy 3-rhamnoglucoside	Ishikura (1975c)
Eurya emarginata	Fruit	Cy 3-glucoside; 3-rutinoside, Cy glucosides	Ishikura and Sugahara (1979)
Ternstroemia japonica	Fruit	Cy 3-glucoside	Ishikura (1975c)
THYMELAECEAE			
Daphne kiusiana	Fruit	Cy 3-glucoside	Ishikura (1975c)
UMBELLIFERAE			
Angelica decursiva	Fruit	Cy 3-sambubioside, 3-rutinoside	Ishikura and Sugahara (1979)
Conium maculatum	Stems	Cy 3-ferulylglucosylgalactoside	Harborne (1976b)
Daucus carota ssp *sativa*	Root	Cy 3-lathyroside, 3-xylosylglucosylgalactoside 3-ferulylxylosylglucosylgalactoside	
D. carota ssp *maritimum*	Stem	Cy 3-glucosylgalactoside	
Foeniculum vulgare	Stem	Cy 3-sinapylglucosylgalactoside	
VERBENACEAE			
Callicarpa spp.	Fruit	Cy glucoside, Pn glycoside	Ishikura (1975c)
VITACEAE			
Parthenocissus tricuspidatum	Fruit	Cy, Pn, Dp, Pt, Mv 3-glucosides, Cy, Mv 3-rutinosides 3-rutinosides-5-glucosides, Cy glycosides	Ishikura and Sugahara (1979)
Vitis amurensis	Fruit	Pn, Dp, Pt, Mv 3,5-diglucosides, Dp, Mv 3-*p*-coumaroylglucosides Pn, Dp, 3-*p*-coumarylglucoside-5-glucoside	Akuta *et al.* (1977a,b)
V. flexuosa	Fruit	Dp, Pt, Mv 3-glucosides, Mv 3,5-diglucoside, 3-sophoroside-5-glucoside, Mv acylglucoside	Ishikura and Ito (1976)
V. vinifera cvs	Fruit	Cy, Pn, Dp, Pt, Mv 3-glucosides, 3-*p*-coumarylglucosides, 3,5 diglucosides, 3-*p*-coumarylglucoside-5-glucosides, Mv 3,5-acetyldiglucoside	Diaz *et al.* (1975a,b); Fong *et al.* (1974); Hrazdina (1975); Hrazdina and Franzesa (1974); Malikov (1971); Matsudomi *et al.* (1977); Ohta *et al.* (1978)

One group of plants, the order Centrospermae, are unique among the angiosperms in not synthesizing anthocyanins but producing the alkaloid-like betalains. However, there are two families within this order, the Caryophyllaceae and Molluginaceae, which have retained the ability to produce anthocyanins (Mabry, 1976). It is the classification and evolutionary status of these two families that causes considerable controversy among both chemists and taxonomists (Harborne, 1976a). Other families thought to be somewhat related morphologically to the Centrospermae, namely the Plumbaginaceae, Polygonaceae, Primulaceae (Nowicke, 1975; Nowicke and Skvarla, 1977) and Vivianiaceae (Behnke and Mabry, 1977) have recently been excluded from this order because of their ability to synthesize anthocyanins.

While most chemotaxonomic investigations of anthocyanins are based on paper chromatography, a recent development has been the use of high-pressure liquid chromatography (see Section 6). For example, Asen (1979) obtained a rapid and accurate separation of cyanidin 3-galactoside, the 3-glucosides and rutinosides of pelargonidin and cyanidin present in poinsettia cultivars. Because of the accuracy and reproducibility of such separations, the method can be adapted for resolving difficulties in the description of new cultivars protected by plant patent law.

3.4. ANTHOCYANINS AS GENETIC MARKERS

Flower anthocyanins have been consciously used as genetic markers since Mendel's 'Experiments on hybridization of plants' using pea plants appeared in 1855. While other members of the Leguminosae have been used since for genetic investigations (Arkatov *et al.* 1976; Picard, 1976; Nozzolillo, 1978, 1979), experiments with plant species from other families have provided most information on genetic control of anthocyanin production.

Certain subspecies of the genus *Rosa* have been found to be new genetic sources of peonin and provide a new combination of anthocyanins (Marshall, 1975). *Petunia hybrida* has a special place among flowers for plant biochemists, since the development of Hess's 'cinnamic acid starter hypothesis' (Hess, 1968) was based on work with this plant. Genetic experiments designed to clarify whether the substitution pattern in anthocyanin is established at the C_9 or at the C_{15} level suggest the involvement of two polymeric genes with independent inheritance (Wiering and DeVlaming, 1977). The action of these genes results in the conversion of petunidin glycosides into malvidin glycosides, suggesting that modification in the B ring of anthocyanins (and possibly flavonoids in general) takes place at the C_{15} level. Similar data have accrued for the 28-chromosome 'Siberian Iris', which is able to convert delphinidin into malvidin (Vaughn and Lyerla, 1978). Genetic manipulations have also revealed the presence of unstable systems affecting anthocyanin production in *Petunia* flowers. The cause of these unstable systems is at the *an2-1* allele, having a regulatory effect on anthocyanin

biosynthesis (Cornu, 1977; Farcy and Cornu, 1979). Genetic manipulations affecting anthocyanin production were also carried out on bluebells (Stickland and Harrison, 1977), on *Centaurea phrygia* (Szweykowski and Bobowicz, 1976), on rhododendrons (Heursel and Horn, 1977) and on *Antirrhinum majus* (Gilbert and Richards, 1976; Sastry, 1976; Linnert, 1978).

Among crop plants, members of the Gramineae have received the closest attention. Barley was investigated for mutations affecting anthocyanin and flavonoid synthesis (Babayan *et al.*, 1977; Jende-Strid, 1978; Von Wettstein *et al.*, 1978) and for the nature of the linkage controlling anthocyanin pigmentation (Hayashi *et al.*, 1977). In rye, anthocyanin synthesis, waxy covering, ligula and spike development were shown to be under independent gene action (Surikov and Romanova, 1978). This plant was also shown to have three pairs of dominant gene suppressors involved in its anthocyanin-synthesizing mechanism (Ruebenbauer and Ruebenbauer, 1978). While genetic manipulation can provide unstable anthocyanin-producing systems, anthocyanin inheritance is nevertheless relatively constant in most tetraploid and hexaploid wheats (Raut *et al.*, 1978).

Investigations in maize on the location of *Dotted* (a gene controlling anthocyanin development in the stalks, leaves and endosperm), usually located in the terminal knob of chromosome 9, showed that the location of this gene is not permanent, but it can change position with chromosomal complements (Doerschug, 1976). Similar investigations by Dooner and Kermicle (1976) and Chen and Coe (1977) established the presence of the anthocyanin-pigmenting factor on a chromosomal segment identical with the standard duplicated sequence. Instability of the colour-producing system was found to be caused by segment duplication in a displaced position on the chromosome. This extra segment can easily be lost, resulting in diminished colour-producing ability of the plant (Gavazzi *et al.*, 1977). In addition to the position affecting phenomena for induced pigmentation, the size of the incorporated genetic material can also play a role, as has been shown for bacterial insertion sequences (Gonella and Peterson, 1978). While test crosses of maize often provide unstable systems, several stable germinal derivatives with varied phenotype expressions were also obtained (Reddy and Peterson, 1976). Controlling elements of anthocyanin synthesis in maize must act early in the flavonoid biosynthetic path, since not only the production of anthocyanins, but also that of quercetin was inhibited (Reddy and Peterson, 1977, 1978) in colourless tissues.

There is insufficient information available presently even to guess the number of genes involved in the control of anthocyanin biosynthesis in maize, or any other plant. The number of controlling genes is almost certainly greater than the number of enzymes involved, since it has been shown (Dooner and Nelson, 1977a,b, 1979) that three or more genes are involved in the control of production and activity of UDPG: flavonol-3-*O*-glucosyltransferase, an enzyme involved in the later stages of anthocyanin (and other flavonoid) biosynthesis.

Genetic experiments on crop plants involving anthocyanins are designed

mainly to improve the quality or marketability of the produce. Major efforts in this direction have been undertaken with Brussels sprouts (Ockendon, 1977), on intergeneric hybrids of *Raphanus sativus* × *Brassica oleracea, Raphanus sativus* × *Brassica caulorapa* (Bubarova, 1974), in turnips (Hoshi, 1975), green cabbage (Bagett, 1978), grapes (Golodriga and Dubovenko, 1977), globe artichoke, *Cynara scolymus* (Foury and Aubert, 1977), *Capsicum* cultivars (Daskalov, 1973; Pochard, 1977), pears (Tuz, 1978), *Nicotiana* cultivars (Sand, 1976), sweet potato (Ayyangar and Sampathkumar, 1978), *Solanum* species (Rao, 1978), red clover (Broda, 1979), *Medicago* hybrids (Lesins *et al.*, 1976), cotton (Konoplya *et al.*, 1978) and apples (Williams *et al.*, 1979). While all the above experiments were carried out with whole plants, anthocyanins have also been used in cytological identifications of colony formation in intergenic somatic hybrid cells (Gosch and Reinert, 1978).

3.5 ANTHOCYANINS AS STRESS MARKERS

Some plant species, acyanic when grown under normal conditions, produce anthocyanins in the presence of high amounts of sugars, or when deficient in certain metals. The appearance of autumn coloration (e.g. reddening of leaves) is thought to be caused by high sugar production during the photosynthetic period and lower metabolic rates and/or reduced transport during the cool nights, resulting in a steady increase in the concentration of sugars in the photosynthetic tissues. Some plants develop red spots around the infection site after attack by pathogens. Since these phenomena do not occur during the normal development of the plants, such discoloration is obviously a sign of stress.

Irradiation of plants by UV light apparently affects anthocyanin production. While tissue cultures of *Haplopappus gracilis* grown in the dark can be induced to form anthocyanins by UV radiation (Wellmann *et al.*, 1976), detached leaves of *Zebrina pendula*, when irradiated on their lower surface with UV light and placed in the dark become water-logged and necrotic within 5 days, losing anthocyanins in their lower epidermis (Witztum, 1978a, b). While the combined effect of light and low quanta of ionizing radiation seems to enhance anthocyanin production in seedlings of *Lactuca sativa, Brassica oleracea, Secale cereale* and *Raphanus raphanistrum* (Kuzin and Vagabova, 1978), plants developed from haploid protoplasts of *Datura innoxia* after treatment with ionizing radiation lose their ability to synthesize anthocyanins (Schieder, 1976). In 4-day-old seedlings of buckwheat, *Fagopyrum esculentum*, raised within the first year after irradiation of the seeds with γ-rays (3,6 and 12 R total dose), slight but significant increase in the content of rutin and leucoanthocyanidins (but not anthocyanins and flavone glycosides) were found in the cotyledons (Margna and Vainjarv, 1976). During this time hypocotyls accumulated mainly anthocyanins and leucoanthocyanidins.

The effect of water stress on plants in also significant and results in changes in

the formation of flavonoid compounds. *Quercus* species, when exposed to water stress, show higher amounts of anthocyanins than plants grown under normal conditions (Spyropoulos and Mavrommatis, 1978). Exposure to higher temperature during a period of water stress further increases the anthocyanin concentration in these plants (Spyropoulos and Lambiris, 1979). In desert plants the appearance of anthocyanins during the dehydration process has been correlated with survivability in some species (Gaff, 1977).

The simultaneous effect of mechanical damage, thermal shock and deprivation of water seems to be rather complex. One hour after slightly bruising the leaf blades of *Coleus blumei*, levels of flavonoid and phenolic compounds containing an *o*-dihydroxy phenylpropanoid moiety decreased in the entire leaf blade, while the level of flavonoid containing only a monohydroxyl in the B ring increased. After 5 h exposure to 35°C, a slight reduction in the anthocyanin content was observed with a simultaneous increase in caffeic acid derivatives (Tronchet, 1975). Exposure to low temperatures (0–5 °C) resulted in an increase in anthocyanin content and a decrease in the content of cinnamic acids. Water deprivation caused a decrease of apigenin glycoside content in the leaves. Roses reacted similarly (Plant *et al.*, 1979).

Deficiency of important minerals, such as phosphorus, also results in an increase in anthocyanin content in plants. Thus phosphorus-deficient maize plants accumulate larger quantities of cyanidin 3-rhamnoside (Bhatla and Plant, 1977). Infection of tomato plants with *Mycoplasma* reportedly results in phosphorus-deficient plants containing anthocyanins, while healthy plants do not accumulate these pigments (Rodeia and Borges, 1974). Administration of nitrogen to plants seems to have the reverse effect on anthocyanin production. Isolated buckwheat cotyledons and hypocotyls produce lower amounts of anthocyanins and flavonol glycosides (Margna *et al.*, 1974a) when incubated in a 1% NH_4NO_3 solution. The effect on intact seedlings is much less pronounced.

Exposure to ozone of leaves of *Brassica pekinensis*, *Zelkova serrata* and *Prunus yedoensis* resulted in an increased formation of anthocyanins (Nouchi and Oodaira, 1974), while fumigation of *Coleus blumei* leaves with HF destroyed their anthocyanin pigments (Lamprecht and Powell, 1977). The latter is not surprising since HF is among the strongest known acids.

Some plants respond with anthocyanin and general phenolic compound production when invaded by pathogenic organisms. Resistant and hypersensitive varieties of corn plants respond more rapidly to infection by *Colletrotrichum graminicola* with accumulations of anthocyanins than susceptible varieties (Hammerschmidt and Nicholson, 1977). 'Scar skin' and 'dapple apple', diseases whose cause has not been satisfactorily identified, result in a decreased anthocyanin production in both Red Delicious and Hyslop Crab apples (Huang and Agrios, 1979). Infection of cowberry plants by *Marsonina potentillae fragariae* and *Exobasidium vaccinii* also causes the production of less anthocyanin in the leaves (Radaitiene, 1977). By contrast, anthocyanin-containing

cultivars of rice plants respond by producing higher amounts of anthocyanins when infected with rice tungro virus (Rao *et al.*, 1979).

Embryos of *Ilex aquifolium*, when excised and cultured on a medium lacking hormones and 'optional' constituents, produce anthocyanins and fail to develop to mature plants upon illumination (Hu *et al.*, 1978). Similar anthocyanin accumulation was observed in emasculated *Cymbydium* flowers (Van Staden, 1979), pointing to metabolic disturbances in the plants following this manipulation.

3.6. FOOD-RELATED ASPECTS OF ANTHOCYANINS

Because of recent restrictions in the use of possibly harmful synthetic colourants in foods, anthocyanins as safe natural pigments have considerable potential in the food industry. There have been many reports in this area. Patents concerning anthocyanins describe basically minor modifications in their synthesis from precursors (Wakihira *et al.*, 1976) or their isolation from raw materials (Mitsuoka and Nishi, 1974; Nakamura *et al.*, 1975; Finedoc-Sica, 1976; Urushido, 1977; Abutalybov *et al.*, 1977; Kikuchi *et al.*, 1977; Minoda *et al.*, 1978; Asen *et al.*, 1978; VanPraag and Duijf, 1978; Wallin, 1979; Josse and Klaeui, 1979) and their recovery from industrial by-products (Inagami and Koga, 1975; Philip, 1976; Okajima *et al.*, 1976; Mirabel and Meiller, 1978). Large efforts spent on finding plant-material sources for anthocyanin isolation are summarized in reviews (Lowry and Chew, 1974; Francis, 1975, 1977; Shrikhande, 1976; Lancrenon, 1978a,b). The plants investigated include red cabbage (Shewfelt and Ahmed, 1977), roselle (Du and Francis, 1974), miracle fruit (Buckmire and Francis, 1976, 1978), garlic (Du and Francis, 1976), apple skins (Posokhlyarova, 1975), leaves of *Acalypha* (Billot, 1975), sour cherries (Shrikhande and Francis, 1974a,b) and olives (Maestro-Duran and Vazquez-Roncero, 1976; Fantozzi and Montedoro, 1978). Work has also continued on the characterization of anthocyanins in economically important plants. Thus the anthocyanin composition of currants (Dankanits, 1976), eggplant (Dikii *et al.*, 1978), taro (Chen *et al.*, 1977), strawberries (Selvaraj *et al.*, 1976), plums (Astegiano and Ciolfi, 1974), Ukranian garlic (Borukh and Demkevich, 1976), whortleberries and bilberries (Borukh, 1974), cowberries (Bandzaitiene and Butkus, 1975) and cranberries (Schmid, 1977) was investigated.

Because anthocyanins appear in fruits near maturity, they have been used in monitoring maturation (Christensen and Graslund, 1979; Gasanov, 1975; Bisson and Ribereau-Gayon, 1978; Carbonneau *et al.*, 1978). Owing to the direct relationship between anthocyanin content and maturity in certain fruits, anthocyanin content is often used as a criterion for quality (Cantarelli, 1975; Kushman and Ballinger, 1975a,b; Mosorinski, 1975a,b; Andreotti *et al.*, 1976; Deubert, 1978; Bourzeix *et al.*, 1977).

Unlike the synthetic red pigments used previously by the food industry, anthocyanins are not stable and readily undergo changes during storage and processing of the raw material. These changes result in loss of colour, browning of the product and formation of precipitates in liquids. Because all these changes affect quality, investigations on pigment changes during storage (Posello and Bonzini, 1977a, b) and processing (Do *et al.*, 1976) were undertaken in order to obtain a better understanding of the events taking place. Methods were devised and suggested to enable simple quality control (Watade and Abbott, 1975; Johnson *et al.*, 1976; Little, 1977), and attempts undertaken to improve simple chromatographic separation of the pigments (Tanner and Zuerrer, 1976). Investigations on the degradation of anthocyanins in model systems (Tanchev and Ioncheva, 1974, 1975, 1976; Shrikhande and Francis, 1974a,b) and in food products (Fang-Yung and Kuznetsova, 1974; Markakis, 1974; Sistrunk and Cash, 1974; Skalski and Sistrunk, 1974; Sakanishi and Fukuzumi, 1975; Mosorinski, 1975a,b; Bourzeix and Heredia, 1976; Du *et al.*, 1975a,b,c; Fang-Yung and Kuznetsova, 1976; Flora, 1976, 1978; Sistrunk, 1976; Calvi and Francis, 1978; Tanchev, 1978, Abers and Wrolstad, 1979; Kim *et al.*, 1979; Manabe *et al.*, 1979; Yoon and Lee, 1979), while giving some hints on the general nature of the reactions, failed to provide the answers sought because of the complexity of the systems. The reactions of anthocyanins in food products have been reviewed (Hrazdina, 1974).

The basic problem with the use of anthocyanins in food products is their colour sensitivity towards changes in pH. While at low pH values (pH = 2) the pigments are present as the red-coloured flavylium salts, at higher pH their colour fades or changes to purple/blue. This colour change is caused by structural transformations as has already been described (Timberlake and Bridle, 1975). Therefore anthocyanins can presently be used only in those food products where the pH is relatively low. Even here a relatively large amount of pigment has to be added to compensate for the pH-dependent colour loss. To improve the colour stability of anthocyanin preparations in foods, it is possible to add di- and tri-valent cations (Starr and Francis, 1974; Gabor, 1977; Kuusi *et al.*, 1977); this, however, also changes the colour towards blue and has limited applications at best.

Problems associated with the pigments were reviewed by Lancrenon (1978a,b). A recently reported condensation product of anthocyanins with catechins in the presence of acetaldehyde has potential value as a food colouring agent, because its colour is resistant to pH-variation (Timberlake and Bridle, 1977). The colour of the condensation product is ca. 30–40 nm towards the blue but when mixed with anthocyanin, it could provide the major characteristics required by the industry. There seem to be current limitations concerning the use of this preparation, since only ca. 10% of the available anthocyanin reacts with catechin and acetaldehyde. Investigation of the reaction products showed the presence of at least six compounds with similar spectral and chromatographic

behaviour, the exact structures of which have not yet been determined (L. J. Chen, unpublished data).

Notwithstanding the above limitations of anthocyanin preparations as food colourants, a number of investigators are promoting their use in certain food products (Clydesdale *et al.*, 1978; Kolesnik *et al.*, 1978; Main *et al.*, 1978; Shewfelt and Ahmed, 1978; Yoon *et al.*, 1978; Clydesdale *et al.*, 1979a,b,c; Ferenczi and Kerenyi, 1979). An anthocyanin preparation isolated from cherries has also been used to produce maraschino-type cherries (McClellan and Cash, 1979). While the presence of anthocyanins is desired in most cases to add colour to the food products, in others their presence can cause problems during processing (Darmaniyan and Dudkin, 1976; Moy *et al.*, 1977).

Another related area where anthocyanins play an important role is the production of red wines. Investigations have covered the anthocyanin composition of grapes (Ballinger *et al.*, 1974; Bourzeix *et al.*, 1974, 1975; Philip, 1974a,b; Sakellarides and Luh, 1974; Kantarev, 1975; Diaz *et al.*, 1976; Cash *et al.*, 1976; Bourzeix, 1978; Ohta and Osajima, 1978a,b; Symonds and Cantagrel, 1978; Ohta *et al.*, 1979a), correlation between quality and pigment parameters (Somers and Evans, 1974, 1977; Jackson *et al.*, 1978; Timberlake *et al.*, 1978), and changes of colour and anthocyanins during the vinification process (Erdoss and Fodor, 1976; Gadzhiev and Vlasova, 1976; Getov *et al.*, 1976; Glories and Augustin, 1976; Kapustina, 1976; Samvelyan, 1976; Tikhonova and Lebedeva, 1976; Timberlake and Bridle, 1976a,b; Dancic, 1977; Glories, 1978a,b,c; Margheri, 1978; Morgan and Mitjavila, 1978; Maksimova, 1979; Mandzhukov, 1979; Odarchenko, 1979; Somers and Evans, 1979).

During the vinification process a relatively large amount of pomace is produced in a winery. To reduce the cost of its disposal and eliminate possible water pollution problems, investigations were carried out on the extraction of anthocyanins from the pomace (Philip, 1974a,b; Saquet, 1974; Champemont *et al.*, 1975; Bernou *et al.*, 1976; Peri and Bonini, 1976; Bakos *et al.*, 1977; Mandzhukov, 1977). If the pomace contains sufficient quantities of anthocyanins, this process could prove to be an economic way to produce anthocyanin preparations for the food industry.

3.7 PROGRESS IN THE ANALYTICAL TECHNIQUES OF ANTHOCYANINS

Paper and thin-layer chromatography are still the traditional tools of investigators working with anthocyanins. In addition, some more elaborate techniques using conventional column chromatography are being adopted by an increasing number of investigators. Recent developments in column support materials for high-pressure liquid chromatography make this technique the most powerful in the analysis of natural products. Some of the above techniques have been reviewed recently (Glories, 1976; Saquet, 1976; Hrazdina, 1979).

3.7.1 Paper and thin-layer chromatography

Progress using these traditional techniques has consisted mainly in developing new solvent systems for sharper separations (Krishnamurty and Krishnaswami, 1975; Gombkoto, 1977; Vacarri and Pifferi, 1978) and adaptation of the existing system for semiquantitative determination of anthocyanins (Ohta *et al.*, 1979b). Both of these methods, in spite of their general use, have drawbacks. Paper chromatography is cumbersome and time consuming when analysing large numbers of samples. Thin-layer chromatography is not as reproducible as paper chromatography and requires the use of reference compounds in most cases.

3.7.2 Column chromatography

Development of column chromatographic methods for the separation of anthocyanins was necessitated by the need for pure compounds in large amounts for chemical and biochemical investigations or as reference substances for qualitative or quantitative analysis. Polyamide powders show good retention of anthocyanins when aqueous plant extracts are percolated through the columns, and permit the removal of sugars, organic acids, soluble polysaccharides, proteins and cations by simply washing with water. While certain anthocyanin classes can be resolved on polyamide columns (Strack and Mansell, 1975), in most cases the resolution is mediocre and one must turn to thin-layer or paper chromatography for further separations and purification. Insoluble polyvinylpyrrolidone (PVP, Polyclar AT) can separate the individual anthocyanins within specific groups, e.g. 3-monosides, 3-biosides, 3,5-diglucosides and their acyl derivatives (Hrazdina, 1970). While single columns of either polyamide or polyvinylpyrrolidone do not give complete separation of anthocyanins, a combination of both is usually sufficient to resolve up to 20 different pigments (Hrazdina and Franzese, 1974a, b; Hrazdina, 1975). These methods are being increasingly adapted for anthocyanin isolation (Brouillard and Delaporte, 1978; Ohta *et al.*, 1979a), or in a combination of silica gel, ion exchange, molecular sieve chromatography, ultrafiltration and electrophoretic separations (Lin and Hilton, 1980). While the combination of polyamide–polyvinylpyrrolidone columns works well in general, certain plant species can contain pigment mixtures where purification of the individual anthocyanin requires further separation on paper chromatograms (Hrazdina *et al.*, 1977). Ion-exchage chromatography usually does not separate individual anthocyanins; it, however, can be used for preliminary separation of the pigments from sugars, polysaccharides and proteins, and to remove trace amounts of metals from anthocyanin preparations (Pifferi and Vaccari, 1980). Satisfactory separations of anthocyanins and other phenolic compounds have also been achieved by chromatography on a poly (ethylene glycol) dimethacrylate gel column (Goldstein, 1976).

3.7.3 Gas chromatography

Anthocyanin chemistry has benefited little from developments of gas chromatographic techniques because of their non-volatility and general instability. It is the structure and properties of anthocyanins that makes their separation difficult. Derivatization of anthocyanins for gas chromatographic (GC) or gas chromatographic–mass spectrometric (GC–MS) analysis is strongly hampered by the structural transformations which these pigments undergo. At pH of 0.5–1.0 only the red-coloured flavylium salt is present, but this pH range is unfortunately not suitable for derivatization. As the pH is raised, the flavylium salt undergoes structural transformations (see following section), where the colourless carbinol base and the purple-coloured anhydrobases are formed. At pHs above 7 the anthocyanins also undergo ring fission to produce chalcones. Therefore, under the conditions best suited for derivatization (e.g. alkaline), at least four different structural transformation forms of these pigments are present. This is the reason why it is practically impossible to obtain homogeneous derivatives of anthocyanins. In spite of these serious limitations, some reports have appeared on the GC and GC–MS analysis of anthocyanins (Bombardelli *et al.*, 1977; Drawert and Leupold, 1976a,b; Lanzarini *et al.*, 1977).

3.7.4 High-pressure liquid chromatography (HPLC)

Recent developments in both instrumentation and column support material such as Lichrosorb, Zorbax, Bondapack and μ-Bondapack treated with octadecyltrichlorosilane have opened up new possibilities for the separation of plant products using this technique. The adoption of variable-wavelength spectrophotometers as detectors for HPLC made possible the monitoring of each group of compounds at their specific absorption maxima. With these techniques, anthocyanins can be selectively monitored at their visible maxima without interference from other materials.

The first HPLC separation of anthocyanins was reported by Manley and Shubiak (1975), who used Pellidone as column support material. This technique, which resolved mixtures of the 3-glucosides of peonidin, petunidin and malvidin, did not prove readily useful, because the pigments had to be separated first from sugars and organic acids and required prior concentration on polyvinylpyrrolidone. A report by Wulf and Nagel (1976) on the successful separation of phenolic acids and flavonoids on a reverse-phase column support material soon led to methods for separating anthocyanidins (Adamovics and Stermitz, 1976; Wilkinson *et al.*, 1977). These methods permitted the identification of anthocyanidins in plant extract hydrolysates, but did not permit the analysis of anthocyanin composition.

An almost simultaneous development in this methodology using reverse-phase Lichrosorb (Wulf and Nagel, 1978) and μ-Bondapack column support materials (Williams *et al.*, 1978) enabled the separation of complex anthocyanin mixtures.

Since anthocyanins can be detected selectively in the 520 nm region, where few other compounds absorb, these methods do not require preliminary treatment of the samples. Centrifuged or ultrafiltered samples free from any particulate material can be directly injected on to the columns and separation of 20 or more anthocyanins accomplished in 90–120 min.

These methods have been successfully used in the identification of plant cultivars (Stewart *et al.*, 1979) and food-research-related work (Camire and Clydesdale, 1979). The methods were originally developed for qualitative analysis of plant extracts. However, simple modification by using standard curves obtained with pure compounds enables their use for quantitative determination of the individual anthocyanins and other flavonoids in complex mixtures (Nagel and Wulf, 1979; Moskowitz and Hrazdina, 1980). The technique is very sensitive. Using the above method (Williams *et al.*, 1978), the qualitative and quantitative composition of anthocyanins, flavonoid glycosides and hydroxycinnamic acid esters could be determined in vacuoles isolated from grape berry subepidermal tissues (Moskowitz and Hrazdina, 1980).

3.8 PROGRESS IN THE CHEMISTRY OF ANTHOCYANINS

Because of the complexity of reactions and the large number of steps involved in the direct synthesis of anthocyanins, it is often more convenient to obtain them by conversion of more readily available flavonoids. In this manner a number of 3-deoxyanthocyanidins were produced by benzoquinone oxidation of the corresponding flavan-4-ol derivatives (Sweeny and Iacobucci, 1977a,b). Flavan-4-ols are rather conveniently prepared by $NaBH_4$ reduction of flavones. Treatment of the flavan-4-ol with benzoquinone results in the formation of flav-3-ene intermediates of which further abstraction of the C_2 allylic hydrogen produces the 3-deoxyanthocyanidin derivatives. However, not all compounds behave in a similar manner. Flavan-3,4-diols, prepared by $NaBH_4$ reduction of tetramethyldihydroquercetin, showed no improved yield under conditions when the amount of the benzoquinone was varied or chloranil was used. A reason for this may be the known tendency of flavan-3,4-diols to undergo condensation and polymerization reactions in the presence of strong acids. Using a modification of the above method, apigeninidin and 4′-*O*-methyl-luteolinidin were synthesized from naringenin and hesperetin respectively (Sweeny and Iacobucci, 1977a,b).

The main reason why the chemical properties of anthocyanins have been little investigated is the difficulty encountered in their isolation, and their general instability, especially in aqueous solution. From earlier work (see Timberlake and Bridle, 1975) with synthetic flavylium salts, some natural anthocyanidins and anthocyanins, it was established that these reactions in solution involve the formation of anhydrobases (*3.2*), carbinol base (*3.3*) and chalcone (*3.4*) in acidic media with the former, and anhydrobases and carbinol base with the latter two compounds in addition to the flavylium salt (*3.1*). The decolourization of an

anthocyanin solution in the pH region 4–5 is thought to occur via the anhydrobase (*3.2*) by hydration. Investigations by Abe *et al.* (1977) suggest that there are differences even among the anthocyanidin 3,5-diglucosides in the formation of their anhydrobases. Anhydrobases of pelargonidin, cyanidin and peonidin 3,5-diglucosides reportedly have the 7-keto structure, while that of malvidin 3,5-diglucoside is present in the 4′-keto structure.

Recent investigations by Brouillard (Brouillard and Dubois, 1977; Brouillard and Delaporte, 1977; Brouillard *et al.*, 1978, 1979; Brouillard and El Hage Chanine, 1980) suggest that structural transformations do not proceed on the above paths. On the basis of kinetic and thermodynamic experiments these authors suggest, that in aqueous acidic solutions there is an acid–base equilibrium between flavylium salt (*3.1*) and the anhydrobase (*3.2*), and that the formation of the colourless carbinol base (*3.3*) is due solely to the hydroxylation of the flavylium salt (*3.1*). The carbinol base (*3.3*) forms a tautomeric equilibrium with the chalcone (*3.4*) (Fig. 3.1). The majority of investigations have been carried out with commercial anthocyanin preparations under the assumption that there is only one anhydrobase structure of the anthocyanins, which is blue, and that the colourless, chalcone absorbs in the 340 nm region of the spectrum. Spectroscopic evidence, however, shows that in a mildly acidic medium there are two purple anhydrobase species absorbing at 534 nm and 556 nm respectively, and these ionize in the pH region of 6–6.5, producing the blue-coloured ionized anhydrobase, which absorbs at 610 nm (L. J. Chen, unpublished data). The chalcone form of anthocyanins absorbs at 396 nm, and this absorption band does not appear below pH 6. Both the purple and blue colours fade and turn colourless or yellowish (from the ionized anhydrobase) producing the carbinol base or ionized carbinol base upon storage, and the latter undergoes ring fission to the chalcone (L. J. Chen, unpublished).

From the above data and observations, the following transformation mechanism can be suggested: in aqueous acidic media there is an acid–base equilibrium between the flavylium salt (*3.1*) and the quinoidal anhydrobases (*3.2*), with increasing stability of the flavylium salt under pH 4–4.5 as the pH decreases. The stability of the anhydrobases increases with increasing pH up to 6–6.5. Nucleophilic attack of water on the flavylium salt (*3.1*) produces the carbinol base (*3.3*), which can also be produced by hydration of the anhydrobases (*3.2*). The chalcone form appears above pH 6 and is produced from the carbinol base, or ionized carbinol base under these conditions.

These structural transformation reactions are mainly responsible for the fact that NMR and MS spectral methods are of limited value in the structural investigations on anthocyanins. Similar transformations also occur during oxidation. Thus under strongly acidic conditions (pH 1–4) anthocyanidin 3,5-diglucosides and their *p*-coumaryl derivatives, when oxidized with H_2O_2, undergo a Bayer–Villiger reaction to produce malvone-type compounds. However, when the same reaction is carried out under low acidity to neutral conditions (e.g. with anhydrobases), the B ring of the pigments becomes lost and

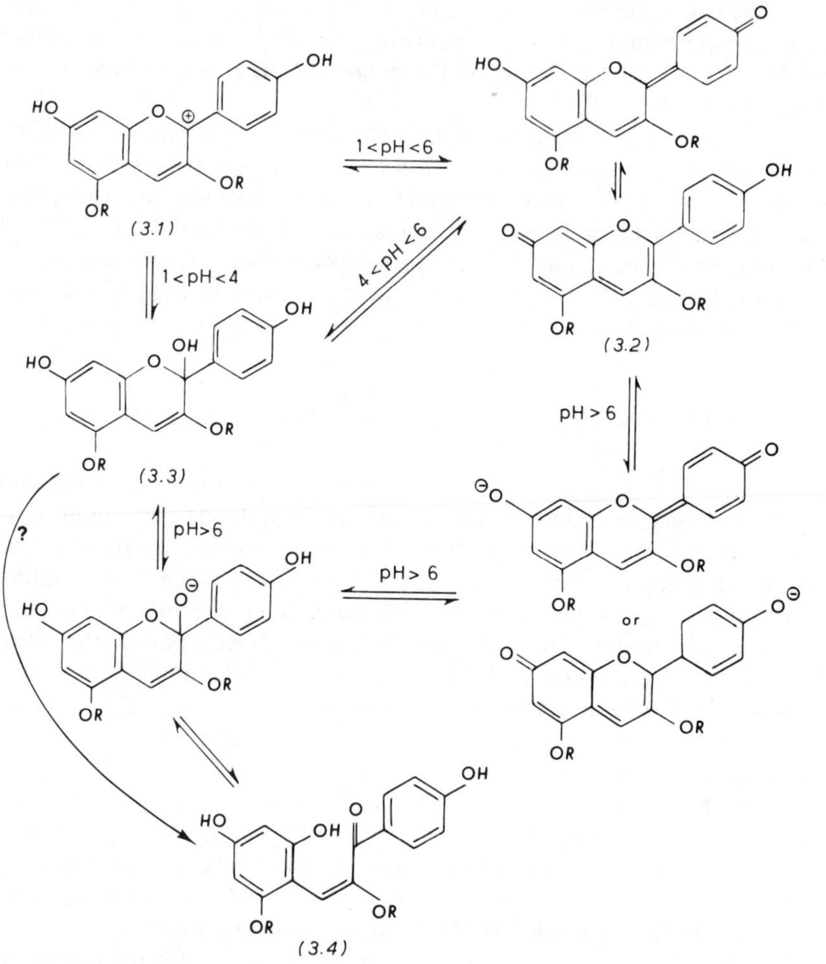

Fig. 3.1 Structural transformation reactions of anthocyanins.

the end products of the reaction are coumarin derivatives (Hrazdina and Franzese, 1974a,b). The yield on these unique oxidation products is rather low and changes significantly when the pH is varied. However, reaction products in higher yields could possibly be produced at other pH values also, since it has been reported that high concentrations of neutral salts have a stabilizing effect on both flavylium salts and anhydrobases (Goto *et al.*, 1976).

Another interesting reaction of anthocyanins is their condensation with catechins and catechin-type compounds in the absence and presence of acetaldehyde (Timberlake and Bridle, 1976a,b). While in the absence of acetaldehyde the condensation reaction is very slow, and xanthylium salts seem

to be the end products, in the presence of acetaldehyde a number of intensely coloured compounds are produced. Owing to limitations in available analytical methods, the exact structure of the condensation products have not been determined.

Photochemical investigations on an anthocyanin mixture indicate stability of compounds in the pH 2–4 range (Riva *et al.*, 1974). Investigations on the stability of anthocyanin in wines show that the compounds undergo continuous changes, but the intermediates formed are not known (Nagel and Wulf, 1979). The chemical problems related to structure identification of anthocyanins were reviewed by Ueno and Saito (1976), those pertinent to food production by Hrazdina (1974).

3.9 SUBCELLULAR STATUS OF ANTHOCYANINS

Anthocyanins accumulate in the vacuoles of epidermal or subepidermal tissues in plants, and thus are responsible for the pink, red, purple and blue colours present in flowers and fruits. This is true of all angiosperm families, except for members of the Centrospermae (see Section 3.3). Several different factors are involved in plant colour expression, the most important being the pH of the cell sap (vacuole), the nature of the aglycone, the extent of glycosylation, the concentration of anthocyanin(s) and other flavonoids in the vacuole, and metal-complexing (Timberlake and Bridle, 1975).

The role of metals in the colour expression in some plants was further confirmed by Takeda (1977), and by Takeda and Hayashi (1977), who found that commelinin, the pigment of *Commelina communis*, consists of awobanin, flavocommelin and magnesium in the ratio 2 : 2 : 1. Manganese, cobalt, nickel, zinc and cadmium can replace magnesium in the complex without significant colour changes. Iron can also form complexes with the anhydrobases of anthocyanins (Kim and Ahn, 1978); its role in colour expression, however, seems to be limited at best because of the scarcity of iron-accumulating flowering plants. The role of pH in colour expression has been further confirmed by measurements in the vacuolar sap (Stewart *et al.*, 1975). These showed that the pH value of the vacuolar environment can vary widely from 2.5 in *Begonia* species to 7.5, as in the Morning Glory cv. Heavenly Blue. In these two extreme examples the formation of the coloured species is due to the flavylium salt (red) at 2.3 and the ionized anhydrobase (blue) at 7.5. Anhydrobases of anthocyanins start to ionize around pH 6. In most plants, however, the pH of the vacuolar environment is in the range of 4–5, where anthocyanins are expected to be present in the colourless carbinol base form. The intense red–purple colouration in the epidermal tissue of flowers where the vacuolar pH is in this region is an unmistakable sign of another structural factor, the existence of pigment–co-pigment complexing.

Recent work (Sugiyama and Takano, 1974a,b; Asen, 1975, 1976; Asen *et al.*,

1975, 1977; Ishikura and Shimizu, 1975; Saleh and Ishak, 1976; Yazaki, 1976; Maekawa and Nakamura, 1977; DeLoose, 1978; Yoshitama, 1978) has confirmed that anthocyanins in most flowers are present as anhydrobase complexes with other flavonoids. While acylated anthocyanins in general show a much lower response to co-pigmentation reactions than do the 3,5-diglucosides (Scheffeldt and Hrazdina, 1978; Williams and Hrazdina, 1979), in one instance it was reported (Hoshino *et al.*, 1980) that acyl groups have a stabilizing effect on complexes formed between acylated anthocyanins and flavone *C*-glycosides. The complex between anthocyanin and flavone (or flavonol) is a rather loose association and is pH-dependent, therefore hydrogen-bond formation was originally suggested as the mechanism (Asen *et al.*, 1972; Scheffeldt and Hrazdina, 1978; Williams and Hrazdina, 1979). Systematic investigations with a large number of differently substituted flavonoids suggest that *in addition* to hydrogen-bond formation between the carbonyl group of the anthocyanin anhydrobases and aromatic hydroxyl groups of the complexing flavonoids, other factors are also important. These are unsaturation at C_2–C_3 in the flavonoid co-pigment, electrostatic forces and configurational or steric effects between the two molecules (Chen and Hrazdina, 1981). In addition to flavonoids, pectin or protopectin has also been reported to form complexes with anthocyanins (Yasuda, 1976). How this association can take place in the living tissue is not quite clear, since anthocyanins are accumulated in the vacuole, and pectin or protopectin is a component of the middle lamella.

While it is generally true that anthocyanins are present as complexes with other flavonoids in the epidermal tissues of flowers, this may not necessarily be the case in other plant tissues. In vacuoles isolated from grape berry subepidermal tissues, anthocyanins were found to be present as the flavylium salts, not associated with flavonoids or other compounds (Moskowitz and Hrazdina, 1980). Eggplant epidermal tissue, on the other hand, seems to contain the stabilized anhydrobase form of these compounds (Nothmann *et al.*, 1976).

Anthocyanins and flavonoid glycosides are known to accumulate in the vacuole of plant cells since this has been shown with isolated vacuole preparations (Wagner and Siegelman, 1975; Hrazdina *et al.*, 1978; Wagner, 1979). Therefore a report on their presence in chloroplasts (Saunders and McClure, 1976) requires further confirmation.

3.10 PHOTOCONTROL OF ANTHOCYANIN BIOSYNTHESIS

The synthesis of anthocyanins is, as are most processes in plants, light-dependent. The light promotion of anthocyanin biosynthesis is mediated by at least two photoreactions: *1*, a low-energy, red/far-red (R/FR) reversible, phytochrome-controlled reaction, and *2*, a 'high-energy' reaction, referred to as high irradiance response (HIR), most effective in the blue and far-red spectral regions. Recent experiments have been designed to prove directly or indirectly

these reactions in carnations (Maekawa, 1975), apples (Bishop and Klein, 1975), grapes (Kliewer, 1977), tea plants (Koretskaya and Zaprometov, 1976), water-pepper (Asahira and Masuda, 1977) and/or to establish the quantitative light requirements (Dzhaparidze, 1974; Axelsson *et al.*, 1979). Detailed investigations on the effect of light on anthocyanin synthesis and of some enzymes involved in mustard seedlings (Blaauw-Jansen, 1974; Mohr *et al.*, 1974, 1979; Drumm and Mohr, 1974; Drumm *et al.*, 1975; Steinitz *et al.*, 1976, 1979; Steinitz and Bergfeld, 1977; Buehler *et al.*, 1978; Kinnersley and Davies, 1976, 1977), red cabbage (Pecket and Hathout-Bassim, 1974 a,b; Mancinelli and Rabino, 1975; Mancinelli *et al.*, 1975; Rabino *et al.*, 1977; Mancinelli and Walsh, 1979), buckwheat (Tokhver, 1976; Tokhver and Margna, 1978; Kefeli and Amrhein, 1977a,b) and *Spirodela* (Mancinelli, 1977) showed or reconfirmed the involvement of phytochrome in the above reactions.

Similar investigations with sorghum (Drumm and Mohr, 1978) and maize seedlings (Duke and Naylor, 1974, 1976a,b; Duke *et al.*, 1976; Styles and Ceska, 1977; McCormic, 1978; Tselas *et al.*, 1979) indicated that a HIR system is responsible for anthocyanin production in these plants. Thus HIR of anthocyanin accumulation has been usually interpreted either in terms of phytochrome, or of another, as yet unknown, photoreceptor. The above investigations now suggest the presence and function of phytochrome in these two plants and seem to rule out the controversial contribution of photosynthesis to the HIR anthocyanin accumulation.

There is, however, a serious flaw in attempting to correlate the activity of PAL (phenylalanine-ammonia-lyase) directly to the synthesis of anthocyanins. While PAL is an important enzyme in anthocyanin biosynthesis, its reaction product, cinnamic acid, and its hydroxylated and activated form (see Chapter 11) do not enter the flavonoid path exclusively. The activated hydroxycinnamic acids also enter in transformations leading to the biosynthesis of lignin and to glucose, quinic and tartaric acid derivatives, and are also the major acylating components of complex flavonoids and anthocyanins. Therefore only a fraction of the enzymic transformation products of phenylalanine is converted directly into anthocyanins. This is why investigations on the activity of this enzyme do not permit a direct correlation with anthocyanin production. To establish the true effect of phytochrome on the biosynthesis of anthocyanins, an enzyme, such as a specific UDP-glucose: anthocyanidin 3-*O*-glucosyltransferase, or other enzymes located toward the end of the biosynthetic path (see Chapter 11), should be investigated. An observation in the maize system (Duke and Naylor, 1976) that P_{fr} controls an enzymic reaction other than PAL supports this proposition.

Reports on the direct effect of UV light on anthocyanin formation (Balazs and Toth, 1974; Wellmann, 1975; Wellman *et al.*, 1976: Lindoo and Caldwell, 1978) are suggestive of a specific UV photoreceptor system in some plants. While such a system cannot be excluded presently, it is also possible that anthocyanin production upon UV irradiation is a photo-independent stress response.

Similarly, the effect of 'penetrant carriers' in relation to photocontrol (Dumortier and Vendrig, 1978) could also be due to the chemical stress caused by the compounds used.

3.11 BIOSYNTHESIS OF ANTHOCYANINS

Since the biosynthesis of all flavonoids is covered in detail in Chapter 11, only certain aspects of anthocyanin biosynthesis are described here. Phenylalanine ammonia lyase (PAL), the first enzyme on the phenylpropanoid path, has been further investigated (Duke and Naylor, 1976; Melin *et al.*, 1977; Faragher and Chalmers, 1977; Heinzmann and Seitz, 1977; Tan, 1979, 1980) in relationship to anthocyanin production. As mentioned in the previous section (3.10) it is difficult to correlate PAL activity directly with anthocyanin production because of the involvement of its product(s) in other pathways. In rye and radish seedlings, no direct relationship between the above-discussed parameters could be established (Margna *et al.*, 1974b,c). Feeding experiments with phenylalanine indicate that exogenous material does not equilibrate with the endogenous pool of this precursor. It also appears that enzymic complexes responsible for the synthesis of different flavonoids are not in an equal position regarding their capability of consuming common precursors (Margna and Margna, 1978). A closer relationship between phenylpropanoid accumulation and anthocyanin biosynthesis (Braun and Seitz, 1975) may exist in tissue cultures, where lignification reactions play only a minor role. The solubility of precursors used in feeding experiments may also play an important role. Solubility properties of phenylalanine and cinnamic acid may explain the preferential uptake of the former and its incorporation into anthocyanins in *Petunia* (Steiner, 1975, 1977).

Investigations on genetically blocked mutants of *Callistephus* (Kuhn *et al.*, 1978), *Petunia* (Forkmann and Kuhn, 1979) and in tapetum sections of *Tulipa* (Suetfeld and Wiermann, 1980) indicated that chalcones are the first C_{15} intermediates in the pathway. Isolation of a chalcone synthase from *Tulipa* anther tapetum sections, producing naringenin chalcone at the interphase of an acidic biphasic system, seems to confirm this. This would also explain the necessity of the chalcone flavanone isomerase in the biosynthetic path. The presence of β-mercaptoethanol in the incubation assay, which is known to cause the premature release of 'short chain release products' (Hrazdina *et al.*, 1976), and the strong acidity of the medium (pH 4–4.8) cannot presently exclude the possibility that the chalcone is prematurely released owing to the stress effect of β-mercaptoethanol or the low pH on the 'flavanone synthase'. Previous investigations with *Haplopappus gracilis* tissue culture (Wellman *et al.*, 1976), petals of *Hippeastrum* and *Tulipa* (Hrazdina *et al.*, 1978) red cabbage seedlings (Hrazdina and Creasy, 1979) and *Pisum, Phaseolus, Brassica* and *Spinacia* seedlings (Hrazdina *et al.*, 1980) indicated 'flavanone synthase' as the enzyme responsible for the establishment of the C_{15} flavonoid skeleton. Further

transformation of the naringenin so produced to eriodictyol by a microsomal preparation and of dihydrokaempferol to dihydroquercetin from *Haplopappus gracilis* cultures (Fritsch, 1974; Fritsch and Grisebach, 1975) suggest that in some plants the establishment of the hydroxylations (or generally substitution) pattern of the B ring in flavonoids occurs *after* establishment of the C_{15} skeleton. Supporting this view are sequential and genetically controlled investigations on anthocyanin biosynthesis in *Antirrhinum majus* (Harrison and Stickland, 1974, 1978; Stickland and Harrison, 1974), *Petunia* (Kho, 1978; Tabak *et al.*, 1978), *Matthiola* (Forkmann, 1977a,b), *Pisum* (Statham and Crowden, 1974) and *Silene dioica* (Kamsteeg *et al.*, 1976), and the substrate-specificity experiments of hydroxycinnamate:CoA-ligase from anthocyanin-containing and anthocyanin-free carrot cell cultures (Heinzmann *et al.*, 1977). Contrary to these data are the results of Endress (1974a,b), whose chalcone-feeding experiments indicate preferential incorporation of hydroxylated and methoxylated chalcone glucosides into corresponding anthocyanins.

The major difference between the biosynthetic path of anthocyanins and that of flavone and flavonol glycosides is the reduction of the 4-carbonyl in the former, followed by water abstraction. The only information available on these parts of the biosynthetic path is indirect, obtained from feeding experiments with dihydroquercetin and/or its glycoside (Kho *et al.*, 1975, 1977; Kho, 1979; Stickland and Harrison, 1977). These experiments uniformly show the incorporation of this compound into cyanidin and delphinidin. Naringenin, naringin, kaempferol and quercetin were not converted. From these data the anthocyanin path *via* dihydroflavanol→ flavan-3, 4-diol→ flav-3-ene→ anthocyanidin (anhydrobase) seems to emerge as the mechanism in plants. The exact biosynthetic mechanisms can however only be established by isolation and identification of every enzyme in the pathway.

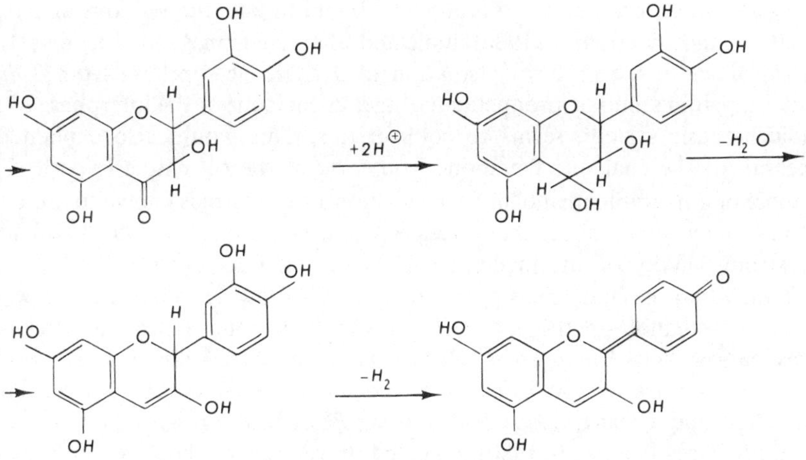

Fig. 3.2 Schematic biosynthetic paths of anthocyanidin from naringenin.

Direct methylation experiments with anthocyanins have not been carried out yet. Data concerning this are available from a genetically designed experiment indicating the conversion of petunidin glycosides into malvidin glycosides (Wiering and DeVlaming, 1977), suggesting that methylation here, similarly to other flavonoid compounds, is among the last steps in biosynthesis. Glycosylation and acylation are the last steps in flavonoid, hence also anthocyanin, biosynthesis. Experiments with *Silene dioica* (Kamsteeg *et al.*, 1979), maize (Dooner and Nelson, 1977a,b) and *Impatiens* (Strack and Mansell, 1979) confirm this (see Chapter 11) and also the *de novo* synthesis of some of the enzymes involved. In tissue cultures of *Daucus carota* the size of the cell aggregates seems to be one of the determining factors for an increased activation of anthocyanin biosynthesis (Seitz, 1976; Kinnersley and Dougall, 1980). In intact plants the accumulation of sugars seems to coincide with the activation of this pathway (Creasy, 1974; Ishikura, 1977; Maekawa and Nakamura, 1978).

Recently, there has been speculation about the localization of enzymic activities; it has been suggested (Fritsch and Grisebach, 1975), that some of the enzymic activities associated with anthocyanin (flavonoid) biosynthesis may occur on *or* in the tonoplast, or in the vacuolar sap. There were also suggestions that anthocyanins (and other flavonoids) can be produced in chloroplasts (Ranjeva *et al.*, 1977; McClure, 1975; Banerji and Sharma, 1979). Experiments with protoplasts, vacuoles and tonoplasts isolated from *Hippeastrum* and *Tulipa* (Hrazdina *et al.*, 1978) and with intact chloroplast preparations obtained from *Pisum, Phaseolus, Brassica* and *Spinacia* cultivars (Hrazdina *et al.*, 1980) showed that none of the above subcellular sites is the location of the anthocyanin (and flavonoid) biosynthesis. It is thought presently that the overall conversion of phenylalanine and malonyl-CoA to anthocyanins and other flavonoids occurs near the endoplasmic reticulum (ER) by cytoplasmic and ER-bound enzymes acting in close proximity to each other. The anthocyanins and other flavonoids produced are then transported by an unknown transport mechanism to the central vacuole of the cells, where they accumulate.

3.12 EFFECT OF PHYSIOLOGICALLY ACTIVE CHEMICALS ON THE BIOSYNTHESIS OF ANTHOCYANINS

The use of synthetic growth regulators, herbicides and other physiologically active compounds in agriculturally important plants, or in tissue cultures, can either stimulate (Table 3.2), or inhibit (Table 3.3) anthocyanin synthesis in the plants. While it is not possible at present to generalize about the chemicals which produce these effects, it seems that kinetin-type compounds seem to promote and auxin analogues inhibit anthocyanin synthesis.

The biochemical action of these compounds is not clear, with two exceptions. These exceptions are the direct action of α-amino-oxy-β-phenylpropionic acid and glyphosate (*N*-phosphomethylglycine). The action of α-amino-oxy-β-

Table 3.2 Chemicals promoting anthocyanin synthesis

Compound	*Plant*	*Reference*
6-Benzylaminopurine	*Saintpaulia ionantha*	Jungnickel (1976)
Benzyladenine	*Helianthus* sp.	Servettaz *et al.* (1975a,b)
CGA-15281 (?) R-27969 (?)	*Vaccinium macrocarpon*	Devlin and Demoranville (1978)
2-Chloroethyltrimethyl ammonium chloride	*Primula obconica*	Abou-Zied and Bakry (1978)
PRB-8 [α-chloro-β-(3-chloro-*O*-tolyl) propionitrile]	*Malus pumila*	Clijsters (1974)
NN-Dimethylformamide	*Phaseolus aureus*	Dumortier and Vendrig (1978)
Dimethyl sulphoxide		
2,2-Dimethylhydrazide	*Primula obconica*	Abou-Zied and Bakry (1978)
Di-1-*p*-menthene (Wilt-pruf)	*Malus* sp.	Creasy (1976)
Cytokinin	*Olea europaea*	Shulman and Lavee (1976)
EDTA	*Brassica oleracea*	Hathout-Bassim and Pecket (1975)
	Spirodela oligorhiza	Elliot (1977)
Ethylene	*Sinapis alba*	Buehler *et al.* (1978)
	Sorghum vulgare	Craker (1975)
Ethephon (2-chloroethylphosphonic acid)	*Ficus carica*	Puech *et al.* (1976)
	Vitis vinifera	Steenkamp *et al.* (1977)
	Vaccinium sp.	Shawa (1979)
Kinetin	*Brassica oleracea*	Pecket and Hathout-Bassim (1974a,b)
α-Naphthaleneacetic acid	*Saintpaulia ionantha*	Jungnickel (1976)

Table 3.3 Chemicals inhibiting anthocyanin synthesis

Compound	*Plant*	*Reference*
Abscisic acid	*Zea mays*	Pilet and Takahashi (1979)
	Raphanus sativus	Guruprasad and Laloraya (1980)
Actinomycin D	*Helianthus annuus, Sinapis alba*	Gerhardt (1974)
Amino-oxyacetate	*Fagopyrum esculentum*	Amrhein (1979)
N-(phosphomethyl)glycine (Glyphosate)	*Glycine max*	Hoagland *et al.* (1979)
	Zea mays	Duke and Hoagland (1978)
	Fagopyrum esculentum	Amrhein *et al.* (1980)
α-Amino-oxy-β-phenylpropionic acid	*Ipomea tricolor, Catharanthus roseus, Brassica* sp.	Amrhein and Hollaender (1979)
Cycloheximide	*Elodea densa*	Witztum (1978a,b)
Ethylene	*Solanum tuberosum*	Mingo-Castel *et al.* (1976)
Ethylenediamine-di-*O*-hydroxyphenyl acetic acid (EDDHA)	*Spirodela punctata*	Scharfetter *et al.* (1978)
Ethyl methanesulphonate	*Amaranthus tricolor*	Behera and Patnaik (1975)
Gibberellic acid	*Lycopersicon* sp.	Aung and Bryan (1974)
	Daucus carota	Seitz and Heinzmann (1975)
Hydroxylamine	*Fagopyrum esculentum*	Margna *et al.* (1978)
Indoleacetic acid, auxin analogues	*Fagopyrum esculentum*	Amrhein *et al.* (1976)
Malformin	*Phaseolus vulgaris*	Curtis and John (1975)
Metyrapone	*Phaseolus aureus*	Geuns (1977)
Paraquat (1,1′-dimethyl-4,4′-bipyridinium ion)	*Glycine max*	Hoagland (1980)

phenylpropionic acid has been shown to be competitive inhibition of phenylalanine ammonia-lyase (Amrhein and Hollaender, 1979). It is presently not known if the inhibition is based purely on lack of substrate for further enzymic transformation, or if it is caused by the absence of induction or activation of further enzymes in the path.

Feeding experiments with glyphosate in soybean seedlings suggested that the inhibition of anthocyanin synthesis is caused by the inhibitory effect of the substance at some point on aromatic amino acid biosynthesis (Duke *et al.*, 1979) and a concomitant increase in PAL activity that resulted in a depletion of the phenylalanine pool (Duke *et al.*, 1980). Recent investigations on the effect of this compound in buckwheat show that its direct action is the inhibition of shikimate conversion to chorismate (Amrhein *et al.*, 1980). A diminished production of phenylalanine, and thus depletion of substrate for the phenylpropanoid and flavonoid pathways results. Since phenylalanine is also an essential substrate for protein biosynthesis, it would be interesting to learn if the decreased production of this amino acid affects only the secondary metabolic path, or if it also interferes with the synthesis of protein in these plants.

3.13 PHYSIOLOGICAL ROLE OF ANTHOCYANINS

Physiological properties have been attributed to anthocyanins in both plants and animals. In plants the presence of these compounds is associated with resistance to pathogens in species of *Brassica* (Weisaeth, 1976), sunflower (Burlov and Kostyuk, 1976), pea seedlings (Kraft, 1977; Muehlbauer and Kraft, 1978), maize (Hammerschmidt and Nicholson, 1977) and roots of avocado trees (VanLelyveld and Bester, 1978). Anthocyanins have also been implicated in effecting the growth of *Diplodia maydis* in liquid cultures (Larson and Bussard, 1979), as rhizobium specific markers in lupin nodules (Caradus and Silvester, 1979); as enhancers of photosynthesis in leaves of tropical rain-forest plants (Lee *et al.*, 1979); or regulators of photosynthesis in some woody plant species grown in the Far East (Chernyshev, 1975). In some palm species they are thought to provide protection against herbivores by causing the leaves to appear brown (thus seemingly dry) in combination with chlorophyll (Stone, 1979). Their presence in the seeds or seed coats of some desert plants is thought to prevent germination (Makhamadzhanov, 1975).

All the evidence presented above is circumstantial at best, and no direct role of anthocyanins has been shown. Our present knowledge on their physiological properties does not permit firm conclusions. The protective effect of anthocyanins in plants against herbivores seems to lack support, since the compounds do not have any pronounced taste even at high concentrations. The physiological effect of anthocyanins in mammals, if any, seems to be related to the prevention of capillary fragility. This phenomenon was recently reviewed by Pourrat (1977) and Proserpio (1977). In addition to investigations on their stability in

pharmaceutical preparations (Bonati and Crippa, 1978), their effect in rats (Lietti and Forni, 1975, 1976) and mechanism of action has been studied (Mian *et al.*, 1977; Robert *et al.*, 1977; Lagrue *et al.*, 1979; Detre and Jellinek, 1979). It has also been suggested that anthocyanins provide protection against UV radiation (Kano and Miyakoshi, 1975, 1976).

ACKNOWLEDGEMENTS

The author thanks Mrs Ruth Bowers and Mrs Grace Parsons for their help with the manuscript.

REFERENCES

Abdurazakova, S. K. and Gabrazova, L. R. (1972), *Chem. Abstr.* **83**, 122569.
Abe, K., Sakaino, Y., Kakinuma, J. and Kakisawa, H. (1977), *Chem. Abstr.* **87**, 153403.
Abers, J. E. and Wrolstad, R. E. (1979), *J. Food Sci.* **44**, 75.
Abou-Zied, E. N. and Bakry, M. Y. (1978), *Acta Hortic.* **9**, 175.
Abutalybov, M. G., Aslanov, S. M., Gorchieva, S. E., Novruzov, E. N. and Farkhadova, M. T. (1977), *U.S.S.R. Patent No.* 574 455
Adamovics, J. and Stermitz, F. R. (1976), *J. Chromatogr.* **129**, 464.
Adesina, S. K. and Harborne, J. B. (1978), *Planta Med.* **34**, 323.
Akuta, S., Ohta, H., Osajima, Y., Matsudomi, N. and Kobayashi, K. (1977a), *Chem. Abstr.* **91**, 171711.
Akuta, S., Ohta, H., Sakane, Y. and Yutaka, O. (1977b), *Chem. Abstr.* **91**, 207400.
Amrhein, N. (1979). *Phytochemistry* **18**, 585.
Amrhein, N. and Hollaender, H. (1979), *Planta* **144**, 385.
Amrhein, N., Goedeke, K. H. and Kefeli, V. I. (1976), *Ber. Dtsch. Bot. Ges.* **89**, 247.
Amrhein, N., Deus, B., Gehrke, P., Hollaender, H., Schulz, A. and Steinruecken, H. C. (1980), *Plant Physiol.* **65**(S), 97.
Andreotti, R.,Tomasicchio, M. and Macchiavelli, L. (1976), *Ind. Conserve* **51**, 193.
Arditti, J. and Fisch, M. H. (1977), *Orchid Biol.* **1**, 117.
Arkatov, V. V., Andreev, V. S. and Ratkin, A. V. (1976), Genetica **12**, 30.
Asahira, T. and Masuda, M. (1977), *J. Jpn. Soc. Hort. Sci.* **46**, 225.
Asen, S. (1975), *Acta Hortic.* **41**, 57.
Asen, S. (1976), *Acta Hortic.* **63**, 217.
Asen, S. (1979), *J. Am. Soc. Hortic. Sci.* **104**, 223.
Asen, S., Stewart, R. N. and Norris, K. H. (1972), *Phytochemistry* **11**, 1139.
Asen, S., Stewart, R. N. and Norris, K. H. (1975), *Phytochemistry* **14**, 2677.
Asen, S., Stewart, R. N. and Norris, K. H. (1977), *Phytochemistry* **16**, 1118.
Asen, S., Stewart, R. N. and Norris, K. H. (1978), *U. S. Patent Appl.* 910 152.
Asen, S., Stewart, R. N. and Norris, K. H. (1979), *Phytochemistry* **18**, 1251.
Astegiano, V. and Ciolfi, G. (1974), *Riv. Vitic. Enol.* **27**, 473.
Aung, L. H. and Bryan, H. H. (1974), *Plant Growth Subst., Proc. Int. Conf. 8th, 1973*, p. 52.
Axelsson, L., Klockare, B. and Sundquist, C. (1979), *Physiol. Plant.* **45**, 387.
Ayyangar, K. R. and Sampathkumar, R. (1978), *Indian J. Genet. Plant Breed.* **38**, 262.
Aziz-Ur-Rahman, T. M. A. and Frontera, M. A. (1974), *An. Asoc. Quim. Argent.* **62**, 169.
Babayan, R. S., Airapetyan, R. B. and Saakyan, M. A. (1977), *Genetika* **13**, 973.
Bagett, J. R. (1978), *Euphytica* **27**, 593.

Bakos, A., Kállai, M. and Szövényi, E. (1977), *Borgazdasag.* **25**, 140.
Balázs, E. and Tóth, A. (1974), *Chem. Abstr.* **83**, 093943.
Ballinger, W. E., Maness, E. P., Nesbitt, W. B. and Caroll, D. E. (1974), *J. Food Sci.* **38**, 909.
Ballinger, W. E., Galletta, G. J. and Maness, E. P. (1979), *J. Am. Soc. Hortic. Sci.* **104**, 554.
Bandzaitiene, Z. and Butkus, V. (1975), *Chem. Abstr.* **84**, 14629.
Banerji, D. and Sharma, V. (1979), *Phytochemistry* **18**, 1767.
Barritt, B. H. and Torre, L. C. (1975), *J. Am. Soc. Hortic. Sci.* **100**, 98.
Beckmann, R. (1979), *Am. J. Bot.* **66**, 1053.
Behera, B. and Patnaik, S. N. (1975), *Curr. Sci.*, **44**, 319.
Behnke, H-D. and Mabry, T. J. (1977), *Plant Syst. Evol.* **126**, 371.
Bernou, J., Borzeix, M., Touzel, M., DuBreil de Pontbriand, P. and Heredia, H. (1976), *Ann. Falsif. Expert. Chim.* **69**, 153.
Bhatla, S. C. and Pant, R. C. (1977), *Curr. Sci.* **46**, 700.
Billot, J. (1974), *Phytochemistry* **13**, 2886.
Billot, J. (1975), *Physiol. Veg.* **13**, 407.
Bishop, R. C. and Klein, R. M. (1975), *HortScience* **10**, 126.
Bisson, J. and Ribereau-Gayon, P. (1978), *Am. Technol. Agric.* **27**, 827.
Blaauw-Jansen, G. (1974), *Acta Bot. Neerl.* **23**, 513.
Bloom, M. (1976), *Am. J. Bot.* **63**, 399.
Bombardelli, E., Bonati, A., Gabetta, B., Martinelli, E. M. and Mustich, G. (1977), *J. Chromatogr.* **139**, 111.
Bonati, A. and Crippa, F. (1978), *Fitoterapia* **49**, 10.
Borukh, I. F. (1974), *Chem. Abstr.* **82**, 196066.
Borukh, I. F. and Demkevich, L. I. (1976), *Chem. Abstr.* **85**, 119571.
Bourzeix, M. (1978), *Bull. Liaison, Groupe Polyphenols.* **8**, 459.
Bourzeix, M. and Heredia, N. (1976), *C. R. Seances Acad. Agric. Fr.* **62**, 750.
Bourzeix, M., Dubernet, M. O. and Heredia, N. (1974), *Bull. Liaison, Groupe Polyphenols*, **5**, 16.
Bourzeix, M., Dubernet, M. O. and Heredia, N. (1975), *Ind. Aliment. Agric.* **92**, 1057.
Bourzeix, M., Heredia, N., Meriaux, S., Rollin, H. and Rutten, P. (1977), *C. R. Hebd. Seances, Acad. Sci., Ser. D.* **284**, 365.
Braun, G. and Seitz, U. (1975), *Biochem. Physiol. Pflanz.* **168**, 93.
Broda, Z. (1979), *Genet. Pol.* **20**, 75.
Brouillard, R. and Delaporte, B. (1977), *J. Am. Chem. Soc.* **99**, 8461.
Brouillard, R. and Delaporte, B. (1978). In *Protons and Ions Involved in Fast Dynamic Phenomena*, Elsevier, Amsterdam, pp. 403–412.
Brouillard, R. and Dubois, J. E. (1977), *J. Am. Chem. Soc.* **99**, 1359.
Brouillard, R. and El Hage Chanine, J. M. (1980), *J. Am. Chem. Soc.* **102**, 5375.
Brouillard, R., Delaporte, B. and Dubois, J. E. (1978), *J. Am. Chem. Soc.* **100**, 6202.
Brouillard, R., Delaporte, B., El Hage Chanine, J. M. and Dubois, J. E. (1979), *J. Chim. Phys.* **76**, 273.
Bubarova, M. (1974), *Chem. Abstr.* **82**, 14088.
Buckmire, R. E. and Francis, F. J. (1976), *J. Food Sci.* **41**, 1363.
Buckmire, R. E. and Francis, F. J. (1978), *J. Food Sci.* **43**, 908.
Buehler, B., Drumm, H. and Mohr, H. (1978), *Planta* **142**, 109.
Burlov, V. V. and Kostyuk, S. V. (1976), *Genetika* **12**, 44.
Calvi, J. P. and Francis, F. J. (1978), *J. Food Sci.* **43**, 1448.
Camire, A. L. and Clydesdale, F. M. (1979), *J. Food Sci.* **44**, 926.
Cantarelli, C. (1975), *Conv. Int. Polifenoli (Relaz. Comun.)* 337.
Caradus, J. R. and Silvester, W. B. (1979), *Plant and Soil* **51**, 437.
Carbonneau, A., Casteran, P. and Leclair, P. (1978), *Ann. Amelior. Plants (Paris)* **28**, 195.
Carew, D. P. and Krueger, R. J. (1976), *Phytochemistry* **15**, 442.
Carreno-Diaz, R. and Grau, N. (1977), *J. Food Sci.* **42**, 615.
Cash, J. N., Sistrunk, W. A. and Stutte, C. A. (1976), *J. Food Sci.* **41**, 1398.

Champemont, R., Pech, J. C., Fallot, J., Julien, H. and Fourmaux, L. P. (1975), *Ind. Aliment. Agric.* **92**, 115.
Chen, H. T., Kao-Jao, T. H. C. and Nakayama, T. O. M. (1977), *J. Food Sci.* **43**, 19.
Chen, L. J. and Hrazdina, G. (1981), *Phytochemistry* **20**, 297.
Chen, S-M. and Coe, E. H. (1977), *Biochem. Genet.* **15**, 333.
Chernyshev, V. D. (1975), *Lesovedenie* **6**, 63.
Chiarlo, B., Cajelli, E. and Piazzai, G. (1978), *Fitoterapia* **49**, 99.
Christensen, P. E. and Graslund, J. (1979), *Tidssk. Planteavl.* **83**, 95.
Chumbalov, T. K., Nurgalieva, G. M. and Beisekova, K. D. (1976), *Chem. Abstr.* **91**, 16701.
Clijsters, H. (1974), *Acta Hortic.* **2**, 309.
Clydesdale, F. M., Main, J. H., Francis, F. J. and Damon, R. A. (1978), *J. Food Sci.* **43**, 1687.
Clydesdale, F. M., Main, J. H. and Francis, F. J. (1979a), *J. Food Prot.* **42**, 196.
Clydesdale, F. M., Main, J. H. and Francis, F. J. (1979b), *J. Food Prot.* **42**, 204.
Clydesdale, F. M., Main, J. H., Francis, F. J. and Hayes, K. M. (1979c), *J. Food Prot.* **42**, 225.
Coassini-Lokar, L. and Poldini, L. (1978), *G. Bot. Ital.* **112**, 327.
Cornu, A. (1977), *Mutat. Res.* **42**, 235.
Craker, L. E. (1975), *Proc. Northeast Weed Sci. Soc.* **29**, 177.
Creasy, L. L. (1974), *Phytochemistry* **13**, 1391.
Creasy, L. L. (1976), *HortScience* **11**, 251.
Crowden, R. K. and Jarman, S. J. (1976), *Phytochemistry* **15**, 1796.
Crowden, R. K., Wright, J. and Harborne, J. B. (1977), *Phytochemistry* **16**, 400.
Curtis, R. W. and John, W. W. (1975), *Plant and Cell Physiol.* **16**, 719.
Dancic, M. (1977) *Chem. Abstr.* **87**, 100601.
Dankanits, E. (1976), *Élelmiszervizsgálati Közl.* **22**, 48.
Darmaniyan, E. B. and Dudkin, M. S. (1976), *Chem. Abstr.* **86**, 54065.
Daskalov, S. (1973), *Chem. Abstr.* **85**, 59771.
DeLoose, R. (1974), *Meded. Fac. Landbowwet. Rijksuniv. Gent.* **39**, 1369.
DeLoose, R. (1978), *Scientia Hortic.* **9**, 285.
Delorme, P., Jay, M. and Ferry, S. (1977), *Plant Med. Phytother.* **11**, 5.
Demina, T. G. (1974), *Chem. Abstr.* **82**, 40688.
Detre, Z. R. and Jellinek, H. (1979), *Front. Matrix Biol.* **7**, 201.
Deubert, K. H. (1978), *J. Agric. Food Chem.* **26**, 1452.
Devlin, R. M. and Demoranville, I. E. (1978), *Proc. Northeast Weed Sci. Soc.* **32**, 108.
Diaz, L. S., Ferrero, J. H., Gasque, F. and Lafuente, B. (1975a), *Rev. Agroquim. Tecnol. Aliment.* **15**, 530.
Diaz, L. S., Gasque, F. and Lafuente, B. (1975b), *Rev. Agroquim. Tecnol. Aliment.* **15**, 408.
Diaz, L. S., Gasque, F. and Lafuente, B. (1976), *Rev. Agroquim. Tecnol. Aliment.* **16**, 509.
Dikii, S. P., Anikeenko, A. P. and Studentsova, L. I. (1978), *Chem. Abstr.* **88**, 166759.
Do, J. Y., Potewiratananond, S., Salunkhe, D. K. and Rahman, A. R. (1976), *J. Food Technol.* **11**, 265.
Doerschug, E. B. (1976), *Theor. Appl. Genet.* **48**, 119.
Dooner, H. K. and Kermicle, J. L. (1976), *Genetics* **82**, 309.
Dooner, H. K. and Nelson, O. E. (1977a), *Proc. Natl. Acad. Sci. U.S.A.* **74**, 5623.
Dooner, H. K. and Nelson, O. E. (1977b), *Biochem. Genet.* **15**, 509.
Dooner, H. K. and Nelson, O. E. (1979), *Genetics* **91**, 309.
Drawert, F. and Leupold, G. (1976a), *Z. Lebensm-Unters. Forsch.* **162**, 401.
Drawert, F. and Leupold. G. (1976b), *Chromatographia* **9**, 605.
Drumm, H. and Mohr, H. (1974), *Photochem. Photobiol.* **20**, 151.
Drumm, H. and Mohr, H. (1978), *Photochem. Photobiol.* **27**, 241.
Drumm, H., Wildermann, A. and Mohr, H. (1975), *Photochem. Photobiol.* **21**, 269.
Du, C. T. and Francis, F. J. (1974), *J. Food Sci.* **38**, 810.
Du, C. T. and Francis, F. J. (1976), *J. Food. Sci.* **40**, 1101.
Du, C. T., Wang, P. L. and Francis, F. J. (1974a), *J. Food. Sci.* **39**, 1265.

Du, C. T., Wang, P. L. and Francis, F. J. (1974b), *Phytochemistry* **13**, 2002.
Du, C. T., Wang, P. L. and Francis, F. J. (1975a), *J. Food Sci.* **40**, 417.
Du, C. T., Wang, P. L. and Francis, F. J. (1975b), *J. Food. Sci.* **40**, 1142.
Du, C. T., Wang, P. L. and Francis, F. J. (1975c), *HortScience* **10**, 36.
Dubois, J. A. and Harborne, J. B. (1975), *Phytochemistry* **14**, 2491.
Duke, S. O. and Hoagland, R. E. (1978), *Plant Sci. Lett.* **11**, 185.
Duke, S. O. and Naylor, A. W. (1974), *Plant Sci. Lett.* **2**, 289.
Duke, S. O. and Naylor, A. W. (1976a), *Physiol. Plant.* **37**, 62.
Duke, S. O. and Naylor, A. W. (1976b), *Plant Sci. Lett.* **6**, 361.
Duke, S. O., Fox, S. B. and Naylor, A. W. (1976), *Plant Physiol.* **57**, 192.
Duke, S. O., Hoagland, R. E. and Elmore, C. D. (1979), *Physiol. Plant.* **46**, 307.
Duke, S. O., Hoagland, R. E. and Elmore, C. D. (1980), *Plant Physiol.* **65**, 17.
Dumortier, F. and Vendrig, J. (1978), *Z. Pflanzenphysiol.* **87**, 313.
Dzhaparidze, I. (1974), *Chem. Abstr.* **84**, 102392.
Egolf, D. R. and Santamour, F. S., Jr. (1975), *HortScience* **10**, 223.
Elliott, D. C. (1977), *Aust. J. Plant Physiol.* **4**, 39.
Endress, R. (1974a), *Phytochemistry* **13**, 421.
Endress, R. (1974b), *Phytochemistry* **13**, 599.
Erdoss, T. and Fodor, J. (1976), *Chem. Abstr.* **87**, 165952.
Fang-Yung, A. F. and Kuznetsova, N. A. (1974), *Chem. Abstr.* **82**, 15477.
Fang-Yung, A. F. and Kuznetsova, N. A. (1976), *Chem. Abstr.* **84**, 178454.
Fantozzi, P. and Montedoro, G. (1978), *Ind. Aliment. Agric.* **95**, 1335.
Faragher, J. D. and Chalmers, D. J. (1977), *Aust. J. Plant Physiol.* **4**, 133.
Farcy, E. and Cornu, A. (1979), *Theor. Appl. Genet.* **55**, 273.
Ferenczi, S. and Kernyi, Z. (1979), *Elelmez. Ip.* **33**, 137.
Fikiin, A. and Le Gnoc Chau (1977), *Chem. Abstr.* **88**, 150776.
Finedoc-Sica, S. A. (1976), *Fr. Demande* 2 299 385.
Flora, L. F. (1976), *J. Food Sci.* **41**, 1312.
Flora, L. F. (1978), *J. Food Sci.* **43**, 1819.
Fong, R. A., Webb, A. D. and Kepner, R. E. (1974), *Phytochemistry* **13**, 1001.
Forkmann, G. (1977a), *Planta* **137**, 159.
Forkmann, G. (1977b), *Phytochemistry* **16**, 299.
Forkmann, G. and Kuhn, B. (1979), *Planta* **144**, 189.
Foury, C. and Aubert, S. (1977), *Ann. Amelior. Plant*, **27**, 603.
Francis, F. J. (1975), *Food Technol.* **29**, 52, 54.
Francis, F. J. (1977), In *Curr. Aspects Food Color.*, 19.
Fritsch, H. (1974), *Bull. Liaison, Groupe Polyphenol.* **5**, 2.
Fritsch, H. and Grisebach, H. (1975), *Phytochemistry* **14**, 2437.
Gabor, E. (1977), *Flavonoids and Bioflavonoids, Proc. Hung. Bioflavonoid Symp. 5th*, 219.
Gadzhiev, D. M. and Vlasova, O. K. (1976), *Chem. Abstr.* **86**, 41756.
Gaff. D. F. (1977), *Oecologia (Berl.)* **31**, 95.
Garcia-Jalon, J., Vega, F. A., Fernandez, M. and Ygartua, P. (1974), *Cienc. Ind. Farm.* **6**, 181.
Gasanov, T. G. (1975), *Chem. Abstr.* **86**, 186051.
Gavazzi, G., Sandri, M., Anzani, G. and Ghidoni, A. (1977), *Heredity* **38**, 349.
Gerhardt, B. (1974), *Z. Pflanzenphysiol.* **74**, 14.
Getov, G., Furstov, K. and Numoloshanu, J. (1976), *Chem. Abstr.* **85**, 107430.
Geuns, J. M. C. (1977), *Biochem. Physiol. Pflanz.* **171**, 435.
Giannasi, D. E. (1975), *Mem. N.Y. Bot. Gard.* **26**, 125.
Gibaja-Oviedo, S. (1978), *Bol. Soc. Quim. Peru* **44**, 67.
Gilbert, R. I. and Richards, D. B. (1976), *Genetica (The Hague)* **46**, 211.
Glories, Y. (1976), *Groupe Polyphenol.* 12.
Glories, Y. (1978a), *Ann. Technol. Agric.* **27**, 253.
Glories, Y. (1978b), *Ann. Nutr. Aliment.* **32**, 1163.

Glories, Y. (1978c), *Bull. Liaison, Groupe Polyphenol.* **8**, 468.
Glories, Y. and Augustin, M. (1976), *Connaiss. Vigne. Vin* **10**, 51.
Godeau, R. P., Pelissier, Y. and Fouraste, J. (1979). *Plant Med. Phytother.* **13**, 37.
Goldstein, G. (1976), *J. Chromatogr.* **129**, 466.
Golodriga, P. Y. and Dubovenko, N. P. (1977), *Chem. Abstr.* **90**, 183309.
Gombkótö, G. (1977), *Chem. Abstr.* **90**, 36326.
Gonella, J. A. and Peterson, P. A. (1978), *Mol. Gen. Genet.* **167**, 29.
Gosch, G. and Reinert, J. (1978), *Protoplasma* **96**, 23.
Goto, T., Hoshino, T. and Ohba, M. (1976), *Agric. Biol. Chem.* **40**, 1593.
Guichard, J. P., Regerat, F. and Pourrat, H. (1976), *Plant Med. Phytother.* **10**, 105.
Guruprasad, K. N. and Laloraya, M. M. (1980), *Plant Sci. Lett.* **19**, 73.
Guzewski, W. A. (1977), *Acta Hortic.* **63**, 229.
Hammerschmidt, R. and Nicholson, R. L. (1977), *Phytopathology* **67**, 247, 251.
Harborne, J. B. (1976a), *Nova Acta Leopold., Suppl.* **7**, 563.
Harborne, J. B. (1976b), *Biochem. Syst. Ecol.* **4**, 31.
Harborne, J. B. (1977). In *The Biology and Chemistry of the Compositae* (V. H. Heywood, J. B. Harborne, and B. L. Turner, eds.), vol. 1, Academic Press, London, pp. 359–384.
Harborne, J. B. and Smith, D. M. (1978), *Biochem. Syst. Ecol.* **6**, 127.
Harrison, B. J. and Stickland, R. G. (1974), *Heredity* **33**, 112.
Harrison, B. J. and Stickland, R. G. (1978), *Heredity* **40**, 127.
Hathout-Bassim, T. A. and Pecket, R. C. (1975), *Phytochemistry* **14**, 731.
Hayashi, J., Takahashi, R. and Moriya, J. (1977), *Nogaku Kenkyu*, **56**, 167.
Heinzmann, U. and Seitz, U. (1977), *Planta* **135**, 63.
Heinzmann, U., Seitz, U. and Seitz, U. (1977), *Planta* **135**, 313.
Hess, D. (1968), *Biochemische Genetik*, Springer Verlag, Berlin, pp. 86–106.
Heursel, J. and Horn, W. (1977), *Z. Pflanzenzuecht.* **79**, 238.
Hoagland, R. (1980), *Plant Physiol.* **65(S)**, 98.
Hoagland, R. E., Duke, S. O. and Elmore, C. D. (1979), *Physiol. Plant* **46**, 357.
Holst, R. (1977). *Am. Fern. J.* **67**, 99.
Hoshi, T. (1975), *Bot. Mag.* **88**, 249.
Hoshino. T., Matsumoto, U. and Goto, T. (1980), *Phytochemistry* **19**, 663.
Hrazdina, G. (1970), *J. Agric. Food Chem.* **17**, 243.
Hrazdina, G. (1974), *Lebensm.-Wiss. Technol.* **7**, 193.
Hrazdina, G. (1975), *Lebensm.-Wiss. Technol.* **8**, 111.
Hrazdina, G. (1979). In *Liquid Chromatographic Analysis of Food and Beverages* (ed. G. Charalambous), Academic Press, New York, pp. 141–159.
Hrazdina, G. and Creasy, L. L. (1979), *Phytochemistry* **18**, 581.
Hrazdina, G. and Franzese, A. J. (1974a), *Phytochemistry* **13**, 225.
Hrazdina, G. and Franzese, A. J. (1974b), *Phytochemistry* **13**, 231.
Hrazdina, G., Kreuzaler, F., Hahlbrock, K. and Grisebach, H. (1976), *Arch. Biochem. Biophys.* **175**, 392.
Hrazdina, G., Iredale, H. andMattick, L. R. (1977), *Phytochemistry* **16**, 297.
Hrazdina, G., Wagner, G. J. and Siegelman, H. W. (1978), *Phytochemistry* **17**, 53.
Hrazdina, G., Alscher-Herman, R. and Kish, V. M. (1980), *Phytochemistry* **19**, 1355.
Hu, C. Y., Ochs, J. D. and Mancini, F. M. (1978), *Z. Pflanzenphysiol.* **89**, 41.
Huang, M. C. and Agrios, G. N. (1979), *Phytopathology* **69**, 35.
Inagami, K. and Koga, T. (1975), *Chem Abstr.* **84**, 29455.
Ishikura, N. (1974), *Kumamoto J. Sci. Biol.* **12**, 17.
Ishikura, N. (1975a), *Phytochemistry* **14**, 743.
Ishikura, N. (1975b), *Phytochemistry* **14**, 1439.
Ishikura, N. (1975c), *Bot. Mag. Tokyo* **88**, 41.
Ishikura, N. (1977), *Bot. Mag. Tokyo* **89**, 251.
Ishikura, N. (1978), *Plant and Cell Physiol.* **19**, 887.

Ishikura, N. and Ito, S. (1976), *Kumamoto J. Sci. Biol.* **13**, 7.
Ishikura, N. and Nagamizo, N. (1976), *Phytochemistry* **15**, 442.
Ishikura, N. Shimizu, M. (1975), *Kumamoto J. Sci. Biol.* **12**, 41.
Ishikura, N. and Sugahara, K. (1979), *Bot. Mag. Tokyo* **92**, 157.
Ishikura, N. and Yamamoto, E. (1978), *Kumamoto J. Sci. Biol.* **14**, 9.
Ishikura, N., Ito, S. and Shibata, M. (1978), *Bot. Mag. Tokyo* **91**, 25.
Iwata, R. Y., Tang, C. S. and Kamemoto, H. (1979), *J. Am. Soc. Hortic. Sci.* **104**, 464.
Jackson, M. G., Timberlake, C. F., Bridle, P. and Vallis, L. (1978), *J. Sci. Food Agric.* **29**, 715.
Jain, M. C. and Seshadri, T. R. (1975), *Indian J. Chem.* **13**, 20.
Jarman, S. J. and Crowden, R. K. (1974), *Phytochemistry* **13**, 743.
Jende-Strid, B. (1978), *Carlsberg Res. Commun.* **43**, 265.
Johnson, L. E., Clydesdale, F. M. and Francis, F. J. (1976), *J. Food Sci.* **41**, 74.
Josse, R. and Klaeui, H. (1979), *Ger. Offen.* 2 901 071.
Jungnickel, F. (1976), *Biochem. Physiol. Pflanz.* **170**, 457.
Kamanzi, K. and Raynaud, J. (1977), *Plant Med. Phytother.* **11**, 289.
Kamsteeg, J., VanBrederode, J. and Van Nigtevecht, G. (1976), *Phytochemistry* **15**, 1917.
Kamsteeg, J., VanBrederode, J., Kuppers, F. J. E. M. and Van Nigtevecht, G. (1978), *Z. Naturforsch.* **33C**, 475.
Kamsteeg, J., VanBrederode, J. and Van Nigtevecht, G. (1979), *Phytochemistry* **18**, 659.
Kano, E. and Miyakoshi, J. (1975), *J. Radiat. Res.* **16**, 91.
Kano, E. and Miyakoshi, J. (1976), *J. Radiat. Res.* **17**, 55.
Kantarev, J. (1975), *Chem. Abstr.* **84**, 71469.
Kapustina, V. V. (1976), *Sadovod. Vinograd. Vinodel. Mold.* **31**, 24.
Kefeli, V. I. and Amrhein. N. G. (1977a), *Sov. Plant Physiol.* **24**, 91.
Kefeli, V. I. and Amrhein, N. G. (1977b), *Fiziol. Rast.* **24**, 118.
Kho, K. F. F. (1978), *Phytochemistry* **17**, 245.
Kho, K. F. F. (1979), *Pharm. Weekbl.* **114**, 325.
Kho, K. F. F., Bennink, G. J. H. and Wiering, H. (1975), *Planta* **127**, 271.
Kho, K. F. F., Bolsman-Louwen, A. C., Wuik, J. C. and Bennink, G. J. H. (1977), *Planta* **135**, 109.
Kikuchi, K., Chiba, A., Miyake, K., Nakai, T. and Tokuda, M. (1977), *Chem. Abstr.* **88**, 150953.
Kim, H.-S. and Ahn, S.-Y. (1978), *Chem. Abstr.* **90**, 70796.
Kim, H.-S., Lee, S.-J. and Yoon, T.-H. (1979), *Chem. Abstr.* **91**, 209597.
Kinnersley, A. M. and Davies, P. J. (1976), *Plant Physiol.* **58**, 777.
Kinnersley, A. M. and Davies, P. J. (1977), *Plant Physiol.* **60**, 175.
Kinnersley, A. M. and Dougall, D. K. (1980), *Planta* **149**, 200.
Kliewer, W. M. (1977), *Am. J. Enol. Viticult.* **28**, 96.
Kolesnik, A. A., Ogneva, O. K. and Gogua, O. V. (1978), *Chem. Abstr* **91**, 18314.
Konoplya, S. P., Fursov. V. N., Druzhkov, A. A. and Nuryeva, G. N. (1978) *Genetika* **9**, 154.
Koretskaya, T. F. and Zaprometov, M. N. (1976), Sov. Plant Physiol. **22**, 1975.
Kraft, J. M. (1977), *Phytopathology* **67**, 1057.
Krause, J. and Strack, D. (1979), *Z. Pflanzenphysiol.* **95**, 183.
Krishnamurty, H. G. and Krishnaswami, L. (1975), *J. Chromatogr.* **114**, 286.
Krueger, R. J. and Carew, D. P. (1975), *Lloydia*, **38**, 542.
Kuhn, B., Forkmann, G. and Seyffert. W. (1978), *Planta* **138**, 199.
Kushman, L. J. and Ballinger, W. E. (1975a), *J. Am. Soc. Hortic. Sci.* **100**, 561.
Kushman, L. J. and Ballinger, W. E. (1975b), *J. Am. Soc. Hortic. Sci.* **100**, 564.
Kuusi, T., Pyysalo, H. and Pippuri, A. (1977), *Z. Lebensm. Unters. Forsch.* **163**, 196.
Kuzin, A. M. and Vagabova, M. E. (1978), *Radiobiologiya* **18**, 242.
Kuznetsova, Z. P. (1979), *Chem. Abstr.* **91**, 171655.
Lagrue, G., Robert, A. M., Miskulin, M., Robert, L., Pinaudeau, Y., Hirbec, G. and Kamalodine, T. (1979), *Front. Matrix Biol.* **7**, 324.
Lakshmi, V. and Chauhan, J. S. (1976), *Q. J. Crude Drug Res.* **14**, 65.
Lamprecht, W. O., Jr. and Powell, R. D. (1977), *Econ. Bot.* **31**, 148.

Lancrenon, X. (1978a). *Process Biochem.* **13**, 16.
Lancrenon, X. (1978b), *Ind. Aliment. Agric.* **95**, 965.
Langhammer, L. and Grandet, M. (1974), *Planta Med.* **26**, 260.
Lanzarini, G., Morselli, L. and Pifferi, P. (1977), *J. Chromatogr.* **130**, 261.
Larson, R. and Bussard, J. B. (1979), *Z. Pflanzenkr. Pflanzenschutz* **86**, 247.
Lee, D. W., Lowry, J. B. and Stone, B. C. (1979), *Biotropica* **11**, 70.
Lesins, K., Sadasivaiah, R. S. and Singh, S. M. (1976). *Can. J. Genet. Cytol.* **18**, 345.
Lietti, A. and Forni, G. P. (1975), *Conv. Int. Polifenoli*, 117.
Lietti, A. and Forni, G. (1976), *Arzneim.-Forsch.* **26**, 832.
Lin, R. I. and Hilton, B. W. (1980), *J. Food Sci.* **45**, 297.
Lindoo, S. J. and Caldwell, M. M. (1978), *Plant Physiol.* **61**, 278.
Linnert, G. (1978), *Biol. Zentralbl.* **97**, 513.
Little, A. (1977), *J. Food Sci.* **42**, 1570.
Lous, J., Majoie, B., Moriniere, J. L. and Wulfert, E. (1975), *Ann. Pharm. Fr.* **33**, 393.
Lowry, J. B. (1976a), *Phytochemistry* **15**, 513.
Lowry, J. B. (1976b), *Phytochemistry* **15**, 1395.
Lowry, J. B. and Chew, L. (1974), *Econ. Bot.* **29**, 61.
Mabry, T. J. (1976), *Plant Syst. Evol.* **126**, 79.
Maekawa, S. (1975), *J. Jpn. Soc. Hortic. Sci.* **43**, 443.
Maekawa, S. and Nakamura, N. (1977), *Chem. Abstr.* **86**, 185909.
Maekawa, S. and Nakamura, N. (1978), *Chem. Abstr.* **89**, 39476.
Maestro-Duran, R. and Vazquez-Roncero, A. (1976), *Grasas Aceites* **27**, 237.
Main, J. H., Clydesdale, F. M. and Francis, F. J. (1978), *J. Food Sci.* **43**, 1693.
Makhamadzhanov, I. (1975), *Chem. Abstr.* **86**, 136377.
Maksimova, S. Y. (1979), *Chem. Abstr.* **92**, 56769.
Malikov, V. M. (1971), *Chem. Abstr.* **83**, 75374.
Manabe, M., Nakamichi, K., Shingai, R. and Tarutani, T. (1979), *Chem. Abstr.* **91**, 73331.
Mancinelli, A. L. (1977), *Plant Physiol.* **59(S)**, 49.
Mancinelli, A. L. and Rabino, I. (1975), *Plant Physiol.* **56**, 351.
Mancinelli, A. L. and Walsh, L. (1979), *Plant Physiol.* **63**, 841.
Mancinelli, A. L., Yang, C.-P. H., Lindquist, P., Anderson, O. R. and Rabino, I. (1975), *Plant Physiol.* **55**, 251.
Mandzhukov, B. (1977), *Chem. Abstr.* **88**, 72940.
Mandzhukov, B. (1979), *Chem. Abstr.* **91**, 138813.
Manley, C. H. and Shubiak, P. (1975), *Can. Inst. Food Sci. Technol. J.* **8**, 35.
Margheri, G. (1978), *Vignevini* **5**, 29.
Margna, U. and Margna, E. (1978), *Biochem. Physiol. Planz.* **173**, 347.
Margna, U. and Vainjarv, T. (1976), *Environ. Exp. Bot.* **16**, 201.
Margna, U., Laanest, L., Margna, E., Otter, M. and Vainjarv. T. (1974a), *Chem. Abstr.* **82**, 72078.
Margna, U., Laanest, L., Margna, E., Otter, M. and Vainjarv, T. (1974b), *Eesti NSV Tead. Akad. Toim. Biol.* **23**, 19.
Margna, U., Laanest, L., Margna, E., and Vainyarv, T. (1974c), *Eesti NSV Tead. Akad. Toim. Biol.* **23**, 221.
Margna, U., Vainjarv, T. and Laanest, L. (1978), *Chem. Abstr.* **90**, 51515.
Markakis, P. (1974), *CRC Crit. Rev. Food Technol.* **4**, 437.
Marshall, H. H. (1975), *J. Am. Soc. Hortic.* **100**, 336.
Martinod, P., Hidalgo, J., Guevara, C. and Pazmino, C. (1975), *Polytecnica* **3**, 151.
Martinod, P., Hidalgo, J., Guevara, C., Arteaga, M. and Lucero, M. (1978), *Polytecnica* **4**, 74.
Matsudomi, N., Yamamura, M., Kobayashi, K., Ohta, H. and Akuta, S. (1977), *Chem. Abstr.* **91**, 207395.
McClellan, M. R. and Cash, J. N. (1979), *J. Food Sci.* **44**, 483.
McClure, J. W. (1975). In *The Flavonoids* (eds. J. B. Harborne, T. J. Mabry, and H. Mabry), Chapman and Hall, London, p. 1005.

McCormic, S. (1978), *Biochem. Genet.* **16**, 777.
Medrano, M. A., Tomas, M. R. and Frontera, M. A. (1977). *An. Asoc. Quim. Argent.* **65**, 59.
Medrano, M. A., Frontera, M. A. and Tomas, M. A. (1978), *An. Asoc. Quim. Argent.* **66**, 107.
Melin, D. (1975), *Phytochemistry* **14**, 2363.
Melin, C., Moulet, A. M., Dupin, J. F. and Hartmann, C. (1977), *Phytochemistry* **16**, 75.
Mian, E., Curri, S. B., Lietti, A. and Bombardelli, E. (1977), *Minerva Med.* **68**, 3565.
Mingo-Castel, A. M., Smith, O. E. and Kumamoto, J. (1976), *Plant Physiol.* **57**, 480.
Minoda, Y., Kodama, T. and Kurahashi, O. (1978), *Chem. Abstr.* **89**, 194144.
Mirabel, B. and Meiller, F. (1978), *Ger. Offen.* 2 802 789.
Misra, K. and Dubey, R. C. (1974), *Curr. Sci.* **43**, 544.
Mitsuoka, S. and Nishi, T. (1974), *Chem. Abstr.* **82**, 136033.
Mohr, H., Drumm, H. and Kasemir, H. (1974), *Ber. Dtsch. Bot. Ges.* **87**, 49.
Mohr, H., Drumm, H., Schmidt, R. and Steinitz, B. (1979), *Planta* **146**, 369.
Momose, T., Abe, K. and Yoshitama, K. (1977), *Phytochemistry* **16**, 1321.
Monties, B., Chesneaux, M.-T. and Hutin, C. (1976), *Ann. Amelior. Plant.* **26**, 495.
Morgan, P. and Mitjavila, S. (1978), Ann. Nutr. Aliment. **32**, 1171.
Moskowitz, A. H. and Hrazdina, G. (1980), *Plant Physiol.* **65(S)**, 97.
Mosorinski, N. (1975a), *Chem. Abstr.* **83**, 130229.
Mosorinski, N. (1975b), *Chem. Abstr.* **83**, 191553.
Moy, J. H., Wang, N. T. S. and Nakayama, T. O. M. (1977), *J. Food Sci.* **42**, 917.
Muehlbauer, F. J. and Kraft, J. M. (1978), *Crop Sci.* **18**, 321.
Nagel, C. W. and Wulf, L. W. (1979), *Am. J. Enol. Vitic.* **30**, 11.
Nair, A. G. R., Kotiyal, J. P., Ramesh, P. and Subramanian, S. S. (1976), *Indian J. Pharm.* **38**, 110.
Nakamura, M., Hosokawa, M., Yajima, J. and Hayashi, K. (1975), *Chem. Abstr.* **83**, 176995.
Nakatani, N., Fukuda, H. and Fuwa, H. (1979), *Agric. Biol. Chem.* **43**, 389.
Nilsson, F. and Trajkovski, V. (1977), *Lantbrukshogskolans meddelande Seria A*, NR **282**, Uppsala.
Nothmann, J., Rylski, I. and Spiegelman, M. (1976), *Scientia Hortic.* **4**, 191.
Nouchi, J. and Oodaira, T. (1974), *Chem. Abstr.* **87**, 63585.
Nowicke, J. W. (1975), *Grana* **15**, 51.
Nowicke, J. W. and Skvarla, J. J. (1977), *Smithson. Contrib. Bot.* **37**, 1.
Nozzolillo, C. (1978), *Can. J. Bot.* **56**, 2890.
Nozzolillo, C. (1979), *Can. J. Bot.* **57**, 2557.
Ockendon, D. J. (1977), *Heredity* **39**, 149.
Odarchenko, V. Ya. (1979), *Chem. Abstr.* **91**, 122118.
Ohta, H. and Osajima, Y. (1978a), *Chem. Abstr.* **91**, 106850.
Ohta, H. and Osajima, Y. (1978b), *Chem. Abstr.* **91**, 106851.
Ohta, H., Saburo, A., Shiraishi, S. and Osajima, Y. (1978), *Chem. Abstr.* **91**, 189779.
Ohta, H., Akuta, S., Okamoto, T. and Osajima, Y. (1979a), *Chem. Abstr.* **90**, 202307.
Ohta, H., Akuta, S. and Osajima, Y. (1979b), *Chem. Abstr.* **91**, 209444.
Ohta, H., Watanabe, H. and Osajima, Y. (1979c), *Chem. Abstr.* **91**, 89805.
Okajima, K., Kuriki, T., Oue, K. and Matsuda, S. (1976), *Chem. Abstr.* **86**, 44359.
Oydvin, J. (1975), *Hortic. Res.* **14**, 1.
Paris, R. R. and Jaquemin, H. (1975), *Ann. Pharm. Fr.* **33**, 73.
Park, K. H. (1979a), *Chem. Abstr.* **92**, 57035.
Park, K. H. (1979b), *Chem. Abstr.* **92**, 57036.
Parks, C. R. and Kondo, K. (1974), *Brittonia* **26**, 321.
Pecket, R. C. and Hathout-Bassim, T. A. (1974a), *Phytochemistry* **13**, 1395.
Pecket, R. C. and Hathout-Bassim, T. A. (1974b), *Phytochemistry* **13**, 815.
Peri, C. and Bonini, V. (1976), *J. Food Technol.* **11**, 283.
Philip, T. (1974a), *J. Food Sci.* **39**, 449.
Philip, T. (1974b), *J. Food Sci.* **39**, 859.
Philip, T. (1976), *U.S. Patent* 3, 963, 700
Picard, J. (1976), *Ann. Amelior. Plant* (*Paris*) **26**, 101.

Pieterse, A. H., Delange, L. and Van Vliet, J. P. (1977), *Acta Bot. Neerl.* **26**, 433.
Pifferi, P. G. and Vaccari, A. (1980), *Lebensm.-Wiss. Technol.* **13**, 85.
Pilet, P. E. and Takahashi, P. (1979), *Plant Sci. Lett.* **17**, 1.
Plaut, Z., Zieslin, N., Grawa, A. and Gazit, M. (1979), *Scientia Hortic.* **11**, 183.
Plekhanova, T. I., Bandyukova, V. A. and Bairamkulova, F. K. (1978), *Kim. Prir. Soedin.* 398.
Pochard, E. (1977), *Ann. Amelior, Plant (Paris)* **27**, 255.
Posello, A. and Bonzini, C. (1977a), *Ann. 1st Sper. Valorizzanione Tecnol. Prod. Agric. Milano* **8**, 47.
Posello, A. and Bonzini, C. (1977b), *Confructa* **22**, 170.
Posokhlyarova, N. S. (1975), *Chem. Abstr.* **83**, 175642.
Pourrat, H. (1977), *Plant. Med. Phytother.* **1977**, 11.
Pourrat, H., Guichard, J. P., Pourrat, A. and Lamaison, J. L. (1978), *Plant. Med. Phytother.* **12**, 212.
Prasad, U. S. and Jha, O. P. (1978), *Plant Biochem. J.* **5**, 44.
Proserpio, G. (1977), *Riv. Ital. Essenza, Profumi, Piante Off., Aromi, Saponi, Cosmet. Aerosol* **59**, 669.
Puech, A. A., Rebeiz, C. A., Catlin, P. B. and Crane, J. C. (1975), *J. Food Sci.* **40**, 775.
Puech, A. A., Rebeiz, C. A. and Crane, J. C. (1976), *Plant Physiol.* **57**, 504.
Rabino, I., Mancinelli, A. L. and Kuzmanoff, K. M. (1977), *Plant Physiol.* **59**, 569.
Radaitiene, D. (1977), *Chem. Abstr.* **88**, 3215.
Rahman, A. U., Frontera, M. A. and Tomas, M. A. (1975), *An. Asoc. Quim. Argent.* **63**, 91.
Rakhimkhanov, Z. B., Ismailov, A. I., Karimdzhanov, A. K. and Dzhuraeva, F. K. (1975), *Khim. Prir. Soedin.* **2**, 255.
Ranjeva, R., Alibert, G. and Boudet, A. M. (1977), *Plant Sci. Lett.* **10**, 225, 235.
Rao, B. L. S., Gosh, A. and John, V. T. (1979), *Phytopathol. Z.* **94**, 367.
Rao, G. R. (1978), *Acta Bot. Indica* **6**, 41.
Raut, V. M., Kulkarni, V. S., Patil, V. P. and Deodikar, G. B. (1978), *Indian J. Genet. Plant Breed.* **38**, 199.
Reddy, A. R. and Peterson, P. A. (1976), *Theor. Appl. Genet.* **48**, 269.
Reddy, A. R. and Peterson, P. A. (1977), *Can. J. Genet. Cytol.* **19**, 111.
Reddy, A. R. and Peterson, P. A. (1978), *Can. J. Genet. Cytol.* **20**, 337.
Reichel, L. and Reichwald, W. (1977), *Pharmacia* **32**, 40.
Riva, A., Vaccari, A. and Pifferi, P. G. (1974), *Ann. Chim. (Rome)* **64**, 429.
Robert, A. R., Godeau, G., Moati, F. and Miskulin, M. (1977), *J. Med.* **8**, 321.
Rodeia, N. and Borges, M. L. (1974), *Port. Acta Biol. Ser. A* **13**, 72.
Ruebenbauer, T. and Ruebenbauer, K. (1978), *Genet. Pol.* **18**, 193.
Sakanishi, Y. and Fukuzumi, H. (1975), *Chem. Abstr.* **84**, 161936.
Sakellarides, H. and Luh, B. S. (1974), *J. Food Sci.* **39**, 329.
Saleh, N. A. M. and Ishak, M. (1976), *Phytochemistry* **15**, 835.
Samata, Y., Inazu, K. and Takahashi, K. (1977), *Chem. Abstr.* **88**, 117766.
Samvelyan, A. M. (1976), *Chem. Abstr.* **85**, 157902.
Sand, S. E. (1976), *Genetics* **83**, 719.
Saquet, H. (1974), *Ind. Aliment. Agric.* **91**, 1149.
Saquet, H. (1976), *Doc. Prep. J. Etud Prog. Recent Methods Anal. Qual. Quant. Struct. Polyphenols Assem. Gen. Groupe – Groupe Polyphenols* 13.
Sarkar, J. K. and Banerjee, P. K. (1977), *Curr. Sci.* **46**, 62.
Sastry, G. R. K. (1976), *Heredity* **36**, 315.
Saunders, J. A. and McClure, J. W. (1976), *Phytochemistry* **15**, 809.
Scharfetter, E., Rottenburg, T. and Kandeler, R. (1978), *Z. Pflanzenphysiol.* **87**, 445.
Scheffeldt, P. and Hrazdina, G. (1978), *J. Food Sci.* **43**, 517.
Schieder, O. (1976), *Mol. Gen. Genet.* **149**, 251.
Schmid, P. (1977), *Acta Hortic.* **61**, 241.
Scogin, R. (1977), *Biochem. Syst. Ecol.* **5**, 265.
Scogin, R. (1979), *Biochem. Syst. Ecol.* **7**, 35.
Seitz, U. (1976), *Nova Acta Leopold. Suppl.* **7**, 89.
Seitz, U. and Heinzmann, U. (1975), *Planta Med.* **30**, 66.

Selvaraj, Y., Divakar, N. G., Suresh, E. R., Iyer, C. P. A. and Subramanyam, M. D. (1976), *J. Food Sci. Technol.* **13**, 195.
Semkina, L. (1975), *Chem. Abstr.* **84**, 147646.
Servettaz, O., Castelli, D. and Longo, C. P. (1975a), *Plant Sci. Lett.* **4**, 361.
Servettaz, O., Castelli, D. and Longo, C. P. (1975b), *G. Bot. Ital.* **109**, 179.
Sharma, P. J. and Crowden, R. K. (1974), *Aust. J. Bot.* **22**, 623.
Shawa, A. Y. (1979), *Hortic. Sci.* **14**, 168.
Shewfelt, R. L. and Ahmed, E. M. (1977), *Food Prod. Dev.* **11**, 52.
Shewfelt, R. L. and Ahmed, E. M. (1978), *J. Food Sci.* **43**, 435.
Shrikhande, A. J. (1976), *CRC Crit. Rev. Food Sci.* **7**, 193.
Shrikhande, A. J. and Francis, F. J. (1974a), *J. Food Sci.* **38**, 646.
Shrikhande, A. J. and Francis, F. J. (1974b), *J. Food Sci.* **39**, 904.
Shulman, Y. and Lavee, S. (1976), *Plant Physiol.* **57**, 490.
Singh, J., Tiwari, A. R. and Tiwari, R. D. (1979), *J. Indian Chem. Soc.* **56**, 746.
Sistrunk, W. A. (1976), *Arkansas Farm Res.* **25**, 12.
Sistrunk, W. A. and Cash, J. N. (1974), *J. Food Sci.* **39**, 1120.
Skalski, C. and Sistrunk, W. A. (1974), *J. Food Sci.* **38**, 1060.
Sobolevskaya, K. A. and Demina, T. G. (1971), *Fenolnye Soedin, Ikh. Fiziol, Svoistva, Mater. Vses. Simp. Fenolnym Soedin. 2nd 79. Chem. Abstr.* **82**, 1930.
Somers, T. C. and Evans, M. E. (1974), *J. Sci. Food Agric.* **25**, 1369.
Somers, T. C. and Evans, M. E. (1977), *J. Sci. Food Agric.* **28**, 279.
Somers, T. C. and Evans, M. E. (1979), *J. Sci. Food Agric.* **30**, 623.
Spyropoulos, C. G. and Lambiris, M. P. (1979), *Ann. Bot.* **44**, 215.
Spyropoulos, C. and Mavrommatis, M. (1978), *Exp. Bot.* **29**, 473.
Srivastava, B. K. and Pande, C. S. (1977), *Planta Med.* **32**, 138.
Starr, M. S. and Francis, F. J. (1974), *J. Food Sci.* **38**, 1043.
Statham, C. M. and Crowden, R. K. (1974), *Phytochemistry* **13**, 1835.
Steenkamp, J., Blommaert, K. L. J. and Jooste, J. H. (1977), *Agroplantae* **9**, 51.
Steiner, A. M. (1975), *Phytochemistry* **14**, 1993.
Steiner, A. M. (1977), *Phytochemistry* **16**, 1703.
Steinitz, B. and Bergfeld, R. (1977), *Planta* **133**, 229.
Steinitz, B., Drumm, H. and Mohr, H. (1976), *Planta* **130**, 23.
Steinitz, B., Schaefer, E., Drumm, H. and Mohr, H. (1979), *Plant Cell Environ.* **2**, 159.
Sterk, A. A., Devlaming, P. and Bolsman-Louwen, A. C. (1977), *Acta Bot. Neerl.* **26**, 349.
Stewart, R. N., Norris, K. H. and Asen, S. (1975), *Phytochemistry* **14**, 937.
Stewart, R. N., Asen, S., Massie, D. R. and Norris, K. H. (1979), *Biochem. Syst. Ecol.* **7**, 281.
Stickland, R. G., and Harrison, B. J. (1974), *Heredity* **33**, 108.
Stickland, R. G. and Harrison, B. J. (1977), *Heredity* **39**, 327.
Stirton, J. Z. and Harborne, S. B. (1980), *Biochem. Syst. Ecol.* **8**, 285.
Stone, B. C. (1979), *Biotropica* **11**, 26.
Strack, D. and Mansell, R. L. (1975), *J. Chromatogr.* **109**, 325.
Strack, D. and Mansell, R. L. (1979), *Z. Pflanzenphysiol.* **91**, 63.
Styles, E. D. and Ceska, O. (1977), *Can. J. Genet. Cytol.* **19**, 289.
Suetfeld, R. and Wiermann, R. (1980), *Arch. Biochem. Biophys.* **201**, 64.
Sugiyama, A., Kinoshita, M., Kako, S., Ohno, H. and Sakakibara, K. (1977), *Chem. Abstr.* **88**, 71485.
Sugiyama, A. and Takano, T. (1974a), *Chem. Abstr.* **82**, 70288.
Sugiyama, A. and Takano, T. (1974b), *Chem. Abstr.* **83**, 144528.
Surikov, I. M. and Romanova, N. P. (1978), *Genetika* **14**, 396.
Sweeny, J. G. and Iacobucci, G. A. (1977a), *Tetrahedron* **33**, 2923.
Sweeny, J. G. and Iacobucci, G. A. (1977b), *Tetrahedron* **33**, 2927.
Symonds, P. and Cantagrel, R. (1978), *Bull. Liaison, Groupe Polyphenols* **8**, 379.

Szweykowski, J. and Bobowicz, M. A. (1976), *Bull. Soc. Amis Sci. Lett. Poznan Ser. D. Sci. Biol.* **16**, 29.

Tabak, A. J. H., Meyer, H. and Bennink, G. J. H. (1978), *Planta* **139**, 67.

Takeda, K. (1977), *Proc. Jpn. Acad. Ser. B* **53**, 257.

Takeda, K. and Hayashi, K. (1977), *Proc. Jpn. Acad. Ser. B* **53**, 1.

Takeda, K. and Sano, S. (1978), *Chem. Abstr.* **90**, 200314.

Tamas, M., Lazar-Keul, G. and Soran, V. (1975), *Chem. Abstr.* **84**, 56644.

Tan, S. C. (1979), *J. Am. Soc. Hortic. Sci.* **104**, 581.

Tan, S. C. (1980), *Aust. J. Plant Physiol.* **7**, 159.

Tanchev, S. (1978), *Chem. Abstr.* **91**, 37701.

Tanchev, S. S. and Ioncheva, N. (1974), *Nahrung* **18**, 747.

Tanchev, S. S. and Ioncheva, N. (1975), *Nahrung* **20**, 629.

Tanchev, S. and Ioncheva, N. (1976), *Nahrung* **20**, 889.

Tanner, H. and Zuerrer, H. (1976), *Schweiz. Z. Obst-Weinbau* **112**, 111.

Thakur, M. and Nozzolillo, C. (1978), *Can. J. Bot.* **56**, 2898.

Tikhonova, N. P. and Lebedeva, T. V. (1976), *Chem. Abstr.* **86**, 15163.

Timberlake, C. F. and Bridle, P. (1975). In *The Flavonoids* (eds. J. B. Harborne, T. J. Mabry and H. Mabry), Chapman and Hall, London, pp. 214–266.

Timberlake, C. F. and Bridle, P. (1976a), *Vitis* **15**, 37.

Timberlake, C. F. and Bridle, P. (1976b), *Am. J. Enol. Viticult.* **27**, 97.

Timberlake, C. F. and Bridle, P. (1977), *J. Sci. Food Agric.* **28**, 539.

Timberlake, C., Bridle, P., Jackson, M. G. and Vallis, L. (1978), *Ann. Nutr. Aliment.* **32**, 1095.

Tiwari, K. P., Masood, M., Rathore, Y. K. S. and Minocha, P. K. (1978), *Vijnaja Parishad Anusandhan Patrika* **21**, 177.

Tokhver, A. (1976), *Chem. Abstr.* **87**, 16465.

Tokhver, A. and Margna, E. R. (1978), *Chem. Abstr.* **89**, 87198.

Torre, L. C. and Barritt, B. H. (1977), *J. Food Sci.* **42**, 488.

Tronchet, J. (1975), *Ann. Sci. Univ. Besancon Bot.* **3**, 17.

Tselas, S. K., Georghiou, K. C. and Thanos, C. A. (1979), *Plant Sci. Lett.* **16**, 81.

Tsiklauri, G. Ch. (1975), *Chem. Abstr.* **83**, 190350.

Tuz, A. S. (1978), *Genetika* **14**, 223.

Ueno, K. and Saito, N. (1976), *Chem. Abstr.* **90**, 24758.

Uphoff, W. (1979), *Experientia* **35**, 1013.

Urushido, H. (1977), *Chem. Abstr.* **87**, 199506.

Vaccari, A. and Pifferi, P. G. (1978), *Chromatographia* **11**, 193.

VanLelyveld, L. J. and Bester, A. I. J. (1978), *Phytopathol. Z.* **92**, 270.

VanPraag, M. and Duijf, G. (1978), *US Patent* 4 118 516.

Van Staden, J. (1979), *Scientia Hortic.* **10**, 277.

Vaughn, K. C. and Lyerla, T. A. (1978), *Theor. Appl. Genet.* **51**, 247.

Vazquez-Roncero, A., Graciani-Constante, E. and Maestro-Duran, R. (1974), *Grasas Aceites* **25**, 269.

Verma, Y. S., Saxena, V. K. and Nigam, S. S. (1977), *Proc. Natl. Acad. Sci., India, Sect. A* **47**, 71.

Von Wettstein, D., Jende-Strid, B., Ahrenst-Larsen, B. and Sorensen, J. A. (1978), *Carlsberg Res. Commun.* **42**, 341.

Wagner, G. J. (1979), *Plant Physiol.* **64**, 88.

Wagner, G. J. and Siegelman, H. W. (1975), *Science* **190**, 1298.

Wakihira, K., Kikumoto, S., Nozaki, O. and Minami, M. (1976), *Chem. Abstr.* **86**, 121704.

Wallin, B. K. (1979), *Ger. Offen.* 2 839 02

Watade, A. and Abbott, J. (1975), *J. Food Sci.* **40**, 1278.

Weisaeth, G. (1976), *Qual. Plant. Plant Foods Hum. Nutr.* **26**, 167.

Wellmann, E. (1975), *Planta Med.* **1975(S)**, 107.

Wellmann, E., Hrazdina, G. and Grisebach, H. (1976), *Phytochemistry* **15**, 913.

Wiering, H. and DeVlaming, P. (1977), *Z. Pflanzenzuecht.* **78**, 113.
Wilkinson, M., Sweeny, J. G. and Iacobucci, G. A. (1977), *J. Chromatogr.* **132**, 349.
Williams, M. and Hrazdina, G. (1979), *J. Food Sci.* **44**, 66.
Williams, M., Hrazdina, G., Wilkinson, M., Sweeny, J. G. and Iacobucci, G. A. (1978), *J. Chromatogr.* **155**, 389.
Williams, R. R., Church, R. M. and Wood, D. E. S. (1979), *J. Hortic. Sci.* **54**, 75.
Witztum, A. (1978a), *Bot. Gaz.* **139**, 53.
Witztum, A. (1978b), *Bot. Gaz.* **139**, 295.
Wrolstad, R. E. and Heatherbell, D. A. (1974), *J. Sci. Food Agric.* **25**, 1221.
Wulf, L. W. and Nagel, C. W. (1976), *J. Chromatogr.* **116**, 271.
Wulf, L. W. and Nagel, C. W. (1978), *Am. J. Enol. Vitic.* **29**, 42.
Yaklich, R. W. and Gentner, W. A. (1974), *Physiol. Plant* **31**, 326.
Yasuda, H. (1976), *Cytologia* **41**, 487.
Yazaki, Y. (1976), *Bot. Mag.* **89**, 45.
Yokoyama, Y. (1976), *Chem. Abstr.* **85**, 189220.
Yoon, T. H. and Lee, S.-J. (1979), *Chem. Abstr.* **91**, 191411.
Yoon, T. H., Lee, S.-J. and Kim, K.-S. (1978), *Chem. Abstr.* **90**, 36311.
Yoshitama, K. (1977), *Phytochemistry* **16**, 1857.
Yoshitama, K. 1978), *Bot. Mag.* **91**, 207.
Yoshitama, K., Hayashi, K., Abe, K. and Kakisawa, H. (1975), *Bot. Mag.* **88**, 213.

CHECK LIST OF ALL KNOWN ANTHOCYANIDINS AND ANTHOCYANINS

1 Apigeninidin (5,7,4′-tri-OH flavylium)
2 5-glucoside (gesnerin)
3 Adientum 1 [apigeninidin (?) glycoside]
4 Dryopteris 1 [apigeninidin (?) glycoside]

5 Luteolinidin (5,7,3′,4′-tetra-OH flavylium)
6 5-glucoside
7 5-diglucoside
8 Adientum 2 [luteolinidin (?) glycoside]
9 Dryopteris 2 [luteolinidin (?) glucoside]
10 Pteris 1 [luteolinidin (?) glycoside]
11 Pteris 2 [luteolinidin (?) glycoside]

12 Tricetinidin (5,7,3′,4′,5′-penta-OH flavylium)
13 5-(?) glucoside
14 3,5-diglucoside

15 Pelargonidin (3,5,7,4′-tetra-OH flavylium)
16 3-galactoside
17 3-glucoside
18 3-rhamnoside
19 5-glucoside
20 7-glucoside
21 3-gentiobioside
22 3-lathyroside
23 3-sambubioside
24 3-sophoroside
25 3-glucosylxyloside
26 3-rhamnosylgalactoside
27 3-rhamnoside-5-glucoside
28 3-robinobioside*
29 3-rutinoside
30 3-gentiotrioside
31 3-(2^{G}-glucosylrutinoside)
32 3-galactoside-5-glucoside
33 3,5-diglucoside
34 3,7-diglucoside
35 3-rutinoside-5-glucoside
36 3-sambubioside-5-glucoside
37 3-sophoroside-5-glucoside
38 3-sophoroside-7-glucoside
39 3-(4‴-*p*-coumarylrhamnoglucoside)*
40 3,5-diglucoside malonyl ester
41 3,5-diglucoside-*p*-coumaric acid ester (monardein)
42 3,5-diglucoside-caffeic acid ester (salvianin)
43 3-(*p*-coumarylglucoside)-5-glucoside
44 3-(caffeylglucoside)-5-glucoside
45 3-(*p*-coumarylrutinoside)-5-glucoside
46 3-(*p*-coumarylsophoroside)-5-glucoside
47 3-(dicaffeyldiglucoside)-5-glucoside
48 3-(ferulylsophoroside)-5-glucoside

49 3-(*p*-coumarylcaffeyldiglucoside)-5-glucoside
50 3-(*p*-coumarylferulylsambubioside)-5-glucoside
51 3-(dicaffeyldiglucoside)-5-glucoside
52 3-(*p*-coumaryldicaffeyldiglucoside)-5-glucoside
53 3-(di-*p*-hydroxybenzoylrutinoside)-7-glucoside*
54 3-diglucoside-5-glucoside-*p*-coumaric acid ester (raphanusin A)
55 3-diglucoside-5-glucoside-ferulic acid ester (raphanusin B)
56 3-rhamnoglucoside-5-glucoside-*p*-coumaric acid ester (pelanin)
57 3-xylosylglucoside-5-glucoside-*p*-coumaric-ferulic acid ester (matthiolanin)

58 Cyanidin (3,5,7,3′,4′-penta-OH flavylium)
59 3-arabinoside
60 3-galactoside
61 3-glucoside
62 3-rhamnoside
63 5-glucoside
64 4′-glucoside
65 3-arabinosylgalactoside
66 3-arabinosylglucoside
67 3-gentiobioside
68 3-lathyroside
69 3-robinobioside*
70 3-rutinoside
71 3-sambubioside
72 3-sophoroside
73 3-xylosylarabinoside
74 3-gentiotrioside
75 3-(2^G-glucosylrutinoside)
76 3-rhamnosyldiglucoside
77 3-xylosylglucosylgalactoside*
78 3-(2^G-xylosylrutinoside)
79 3-arabinoside-5-glucoside
80 3-galactoside-5-glucoside
81 3-rhamnoside-5-glucoside
82 3,5-diglucoside
83 3,7-diglucoside
84 3-glucoside-7-rhamnoside
85 3,4′-diglucoside
86 3-rutinoside-5-glucoside
87 3-sambubioside-5-glucoside
88 3-sophoroside-5-glucoside
89 3-rhamnosylglucoside-7-xyloside
90 3-acetylglucoside
91 3-*p*-coumarylglucoside (hyacinthin)
92 3-caffeylglucoside
93 3-*p*-coumarylgentiobioside
94 3-caffeylgentiobioside
95 3-*p*-coumarylxylosylglucoside*
96 3-caffeylglucosylarabinoside*
97 3-ferulylglucosylgalactoside*
98 3-sinapylglucosylgalactoside*
99 3-caffeylrhamnosylglucoside
100 3-caffeylrhamnosyldiglucoside
101 3-ferulylxylosylglucosylgalactoside*
102 3,5-diglucoside malonyl ester
103 3,5-diglucoside-*p*-coumaric acid ester (perillanin)
104 3-(6-*p*-coumarylglucoside)-5-glucoside
105 3-(*p*-coumarylrutinoside)-5-glucoside (cyananin)
106 3-caffeylglucoside-5-glucoside
107 3-(*p*-coumarylsophoroside)-5-glucoside
108 3-malonylsophoroside-5-glucoside
109 3-caffeylsophoroside-5-glucoside
110 3-ferulylsophoroside-5-glucoside
111 3-sinapylsophoroside-5-glucoside
112 3-(di-*p*-coumaryl)sophoroside-5-glucoside
113 3-(di-ferulyl)sophoroside-5-glucoside
114 3-(di-sinapyl)sophoroside-5-glucoside
115 3-(*p*-coumarylcaffeyl)-sophoroside-5-glucoside

116 Peonidin (3,5,7,4′-tetra-OH, 3′OMe flavylium)
117 3-arabinoside
118 3-galactoside
119 3-glucoside
120 3-rhamnoside
121 5-glucoside
122 3-gentiobioside
123 3-lathyroside
124 3-rutinoside
125 3-galactoside-5-glucoside
126 3,5-diglucoside
127 3-rhamnoside-5-glucoside
128 3-rutinoside-5-glucoside
129 3-gentiotrioside
130 3-glucosylrhamnosylglucoside (?)
131 3-acetylglucoside
132 3-*p*-coumarylglucoside
133 3-caffeylglucoside
134 3-*p*-coumarylgentiobioside
135 3-caffeylgentiobioside
136 3-(*p*-coumarylglucoside)-5-glucoside
137 3-(*p*-coumarylrutinoside)-5-glucoside (peonanin)

138 3-(*p*-coumarylsophoroside)-5-glucoside
139 3-(di-*p*-coumaryl)glucoside
140 3-(di-caffeylsophoroside)-5-glucoside
141 3-(*p*-coumarylcaffeylsophoroside)-5-glucoside

142 Delphinidin(3,5,7,3′,4′,5′-hexa-OH flavylium)
143 3-arabinoside
144 3-galactoside
145 3-glucoside
146 3-rhamnoside
147 3-rutinoside
148 3-sambubioside
149 7-galactoside
150 7-glucoside
151 3-glucosylglucoside
152 3-lathyroside
153 3-rhamnosylgalactoside
154 3,5-diglucoside
155 3-rhamnoside-5-glucoside
156 3-rutinoside-5-glucoside
157 3-acetylglucoside
158 3-*p*-coumarylglucoside
159 3-caffeylglucoside
160 3-(di-*p*-coumaryl)glucoside
161 3,5-diglucoside malonyl ester
162 3-(*p*-coumarylglucoside)-5-glucoside (awobanin)
163 3-(*p*-coumarylrutinoside)-5-glucoside (delphanin)
164 3-(*p*-coumarylsophoroside)-5-glucoside
165 3-caffeylglucoside-5-glucoside
166 3-(di-caffeylrutinoside)-5-glucoside
167 3-rutinoside-5,3′,5′-triglucoside-caffeylferulyl-*p*-coumaryl ester*

168 Petunidin (3,5,7,3′,4′-penta-OH-5′-OMe flavylium)
169 3-arabinoside
170 3-galactoside
171 3-glucoside
172 3-rhamnoside
173 3-rutinoside
174 5-glucoside
175 3-gentiobioside
176 3-sophoroside
177 3,5-diglucoside
178 3-rhamnoside-5-glucoside
179 3-rutinoside-5-glucoside
180 3-gentiotrioside
181 3-acetylglucoside
182 3-(*p*-coumarylglucoside)
183 3-caffeylglucoside

* Newly reported during 1975–1980.

184 3-(dicaffeyl)-glucoside
185 3-(*p*-coumarylglucoside)-5-glucoside
186 3-(*p*-coumarylrutinoside)-5-glucoside (petanin)
187 3-(di-*p*-coumaryl)rutinoside (guineesin)

188 Malvidin (3,5,7,4′-tetra-OH, 3′,5′-diOMe flavylium)
189 3-arabinoside
190 3-galactoside
191 3-glucoside
192 3-rhamnoside
193 5-glucoside
194 3-gentiobioside
195 3-laminaribioside*
196 3-rutinoside
197 3,5-diglucoside
198 3-rhamnoside-5-glucoside
199 3-rutinoside-5-glucoside
200 3-gentiotrioside
201 3-arabinosylglucoside-5-glucoside
202 3-sophoroside-5-glucoside*
203 3-acetylglucoside
204 3-(*p*-coumarylglucoside)
205 3-caffeylglucoside
206 3-(di-*p*-coumaryl)-glucoside
207 3,5-diglucoside acetyl ester
208 3,5-diglucoside-*p*-coumaric acid ester (tibouchinin)
209 3-(*p*-coumarylglucoside)-5-glucoside
210 3-caffeylglucoside-5-glucoside
211 3-(*p*-coumarylrutinoside)-5-glucoside (negretein)

212 Aurantinidin (3,5,6,7,4′-penta-OH flavylium)
213 3-sophoroside
214 3,5-diglucoside

215 Rosinidin (3,5,4′-tri-OH-7,3′-di-OMe flavylium)
216 3,5-diglucoside

217 Pulchellidin (3,7,3′,4′,5′-penta-OH-5-OMe flavylium)
218 3-glucoside

219 Europinidin (3,7,4′,5′-tetra-OH-5,3′-di-OMe flavylium)
220 3-glucoside
221 3-galactoside

222 Capensinidin (3,7,4′-tri-OH-5,3′,5′-tri-OMe flavylium)
223 3-rhamnoside

224 Hirsutidin (3,5,4′-tri-OH-7,3′,5′-tri-OMe flavylium)
225 3,5-diglucoside

CHAPTER FOUR

Flavones and Flavonols

ECKHARD WOLLENWEBER

4.1 FLAVONOIDS WITH HYDROXY AND/OR METHOXY SUBSTITUTION

Most reports on the identification of known flavonoids in new plant sources as well as those on novel flavonoids naturally deal with the two most abundant classes, the flavones and flavonols. For this chapter we have listed all references to the occurrence of flavone and flavonol aglycones which have appeared from 1975 to 1980, and also include such earlier references as were overlooked in the relevant chapters of 'The Flavonoids' (1975). Emphasis is given to the natural distribution of these compounds. The data are collected in Tables 4.1 and 4.2 which cite substances, plant sources and references. Since 1974, some 50 flavones and 60 flavonols with common substitution patterns have been described as novel compounds. Many are simply new methyl ethers of known compounds. The structures are arranged throughout the Tables by number and arrangement of substituents in ring A, followed by ring B. Wherever possible, the plant organs from which the compounds have been isolated are cited. In cases where authors have extracted the 'whole plant' or 'aerial parts', it does not necessarily follow that the compound is present in all parts, since these substances are often specifically located in certain organs only. In all cases where free aglycones are encountered, especially highly methylated flavonoids, it is likely that they occur externally, or are associated with secretory structures and are correlated with the production of other lipophilic secondary compounds, mainly of terpenoid origin. The occurrence and distribution of free flavonoid aglycones in nature has recently been reviewed (Wollenweber and Dietz, 1981).

Except for very common compounds, we have attempted to cite all plants from which flavonoids have been reported. In order to avoid excessive references, in some cases only the most recent literature is quoted and original reports are omitted. Plant sources of such widespread compounds as apigenin, luteolin or kaempferol, are given only in those cases where it is obvious that they occur externally. Those compounds which have as yet only been found in glycosidic combinations are also included in order to provide a complete list of all known flavones and flavonols. However, references are given only for those which have been reported for the first time since 1975. The natural occurrence of flavones and flavonols in bound form is covered separately in Chapter 5.

Table 4.1 List of flavones reported between 1975 and 1980 and their plant sources

Substitution	*Trivial name*	*Plant source*	*References*
	Flavone		Synthesis: Kasahara *et al.* (1974)
			Synthesis: Ashihara *et al.* (1977)
5-OH	Primuletin		
2′-OH			
(5,6-di-OH)			
6-Me			
5,6-di-Me			
5,7-di-OH (G)	Chrysin	*Oroxylum indicum* (Bign.) stem bark	Subramanian and Nair (1972a)
		Escallonia, 3 spp. (Sax.), leaf	Wollenweber (1981a)
		Flourensia resinosa (Ast.), leaf	Wollenweber (unpublished)
7-Me	Tectochrysin	*Andrographis serpyllifolia*, (Acanth.), leaf	Govindachari *et al.* (1968)
		Flourensia resinosa (Ast.), leaf	Wollenweber (unpublished)
5,7-di-Me*		*Helichrysum herbaceum* (Ast.), aerial parts	Bohlmann *et al.* (1979)
5,8-di-OH	Primetin		
5,2′-di-OH			
7,4′-di-OH (G)			
3′,4′-di-OH (G)		*Primula* spp. (Prim.), farinose exudate	Wollenweber (1974b)
5,6,7-tri-OH (G)	Baicalein		
6-Me (G)	Oroxylon (oroxylin A)		
7-Me (G only)	Negletein		
6,7-di-Me*		*Popowia cauliflora* (Annon.), stem	Panichpol and Waterman (1978)
5,6,7-tri-Me		*Callicarpa japonica* (Verb.)	Hosozawa *et al.* (1972)
		Kickxia lanigera (Scroph.), whole plant	Pinar (1973)
		Colebrookia oppositifolia, leaf	Ahmad *et al.* (1974)
		Popowia cauliflora (*Annon.*), *stem*	Panichpol and Waterman (1978)
		synthesis:	Matsuura *et al.* (1978)
5,7,8-tri-OH (G)	Norwogonin		
8-Me	Wogonin	*Scutellaria rivularis* (Scroph.), root	Chou (1978)
7,8-di-Me*		*Scutellaria rivularis* (Scroph.), root	Chou (1978)
		Gnaphalium pellitum (Ast.), inflorescence	Escarria *et al.* (1977)
		Andrographis paniculata (Acanth), tissue culture	Jalal *et al.* (1979)

5,7,8-tri-Me*		*Helichrysum mundii* (Ast.), aerial parts	Bohlmann *et al.* (1978)
		Actinodaphne madraspatana (Laurac.), heartwood	Adinarayana and Gunasekar (1979)
(5,6,2′-tri-OH) 5,6,2′-tri-Me			
(5,6,3′-tri-OH) 5,6,3′-tri-Me			
(5.7,2′-tri-OH) 7-Me (G)	Echioidinin		
5,7,4′-tri-OH (G)	Apigenin		
		° *Salvia glutinosa* (Lab.), aerial parts	Wollenweber (1974a)
		Larrea divaricata, L. tridentata (Zyg.), leaf-resin	Sakakibara *et al.* (1976)
		Cheilanthes spp., *Notholaena* spp. (Polypod.), farinose exudate on fronds	Wollenweber (1976c)
5-Me(G)	Thevetiaflavone	*Saccharum officinarum* (Poac.), flower	Misra and Mishra (1979)
7-Me (G)	Genkwanin	° *Salvia glutinosa* (Lab.), aerial parts	Wollenweber (1974a)
		Alnus, Betula, Oostrya (Bet.), bud excretions	Wollenweber (1975b)
		Artemisia capillaris (Ast.), aerial parts	Komiya *et al.* (1976)
		Larrea divaricata, L.tridentata (Zyg.), leaf-resin	Sakakibara *et al.* (1976)
		Cheilanthes spp., *Notholaena* spp.	Wollenweber (1976b)
		Pityrogramma spp. (Polypod.) farinose exudate	Wollenweber and Dietz (1980)
4′-Me (G)	Acacetin	*Alnus* and *Betula* spp.	Wollenweber (1975b)
		Ostrya japonica (Bet.) bud exretions	
		Eupatorium tinifolium (Ast.) aerial parts	Moreno *et al.* (1980)
		Notholaena spp. (Polypod.), farinose exudate on fronds	Wollenweber (1981b)
		Cheilanthes spp.	Wollenweber (unpublished)
5,7-di-Me (G only)*		(see sect. 4.5)	
7,4′-di-Me (G)		°*Beyeria* spp. (Euphorb.), leaf, terminal branches	Dawson *et al.* (1965)
		°*Acacia ixiophylla* (Leg.), leaf	Clark-Lewis and Dainis (1968)
		°*Piper peepuloides* (Piper.), fruit	Dhar *et al.* (1970)
		Alnus japonica, Betula spp. (Bet.), bud excretions	Wollenweber (1975b)
		Cheilanthes spp., *Notholaena* spp., (Polypod.), farinose exudate on fronds	Wollenweber (1976b)
		Baccharis crispa (Ast.), aerial parts	Bandoni *et al.* (1978)

Table 4.1 (*Contd.*)

Substitution	*Trivial name*	*Plant source*	*References*
		Sideritis gomerae (Lam.), aerial parts	González *et al.* (1978)
		Eupatorium tinifolium (Ast.) aerial parts	Moreno *et al.* (1980)
		Elaegia utilis (Rub.), bud excretion	Biftu and Stevenson (1978)
		Pityrogramma tartarea (Polypod.), farinose exudate on fronds	Wollenweber and Dietz (1980)
5,7,4′-tri-Me*		*Citrus reticulata* (Rut.), fruit peel	Iinuma *et al.* (1980)
		synthesis:	Matsuura *et al.* (1978)
5,8,2′-tri-OH			
7,3′,4′-tri-OH (G)		*Juncus trifidus, Luzula purpurea* (Juncac.), stem, leaf	Williams and Harborne (1975)
		Salvertia convallarioides, Vochysia cinnamonea, V. tucanorum (Vochys.), leaf	Lopes *et al.* (1979)
3′-Me (G)	Geraldone	*Salvertia convallariodes Vochysia cinnamonea* (Vochys.), leaf	Lopes *et al.* (1979)
7,4′-di-Me	Tithonine		
(5,6,7,8-tetra-OH)			
6,7-di-Me*		*Helichrysum herbaceum* (Ast.), aerial parts	Bohlmann *et al.* (1979)
6,8-di-Me*		*Pityrogramma triangularis* var. *triangularis* (Polypod.), farinose exudate on fronds	Dietz *et al.* (1981)
5,6,7-tri-Me*		*Helichrysum herbaceum* (Ast.), aerial parts	Bohlmann *et al.* (1979)
6,7,8-tri-Me	Alnetin		
5,6,7,8-tetra-Me		*Helichrysum herbaceum* (Ast.), aerial parts	Bohlmann *et al.* (1979)
5,6,7,4′-tetra-OH (G)	Scutellarein	*Oroxylum indicum* (Bign.), leaf	Subramanian and Nair (1972a)
		Clerodendron phlomides, Durante repens (Verb.), leaf	Subramanian and Nair (1972b)
		Scoparia dulcis (Scroph.), leaf	Ramesh *et al.* (1979)
6-Me (G)	Hispidulin, Dinatin	°*Iva nevadensis* (Ast.), whole plant	Farkas *et al.* (1966)
		°*Iva hayesiana* (Ast.), aerial parts	Herz and Sudarsanam (1970)
		°*Salvia officinalis* (Lam.), leaf	Brieskorn and Biechele (1971)
		°*Gaillardia* spp., *Helenium* spp. *Hymenoxis* spp.	Carman *et al.* (1972)

		Plummera ambigens P. floribunda, Ratibida columnaris (Ast.), leaf	
		°*Digitalis ferruginea* (Scroph.), leaf	Imre *et al.* (1973)
		Digitalis schischkinii (Scroph.), leaf	Imre *et al.* (1977)
		Betula verrucosa (Bet.), bud excretion	Popravko *et al.* (1979)
		Santolina chamaecyparissus (Ast.), aerial parts	Becchi and Carrier (1980)
		synthesis:	Ahluwalia *et al.* (1976)
7-Me (G)	Sorbifolin	*Pleocarphus revolutus* (Ast.), leaf, stem	Silva *et al.* (1977)
		Adenocalymma alliaceum (Bign.) flower	Rao and Rao (1980)
		Anvillea garcini (Ast.), leaf	Ulubelen *et al.* (1979a)
		Scoparia dulcis (Scroph.), leaf	Ramesh *et al.* (1979)
4′-Me (G)		*Clerodendron inerme* (Verb.), leaf	Vendantham *et al.* (1977)
6,7-di-Me(G)	Cirsimaritin (scrophulein)	°*Helichrysum viscosum* var. *bracteatum* (Ast.), whole plant	Geissman *et al.* (1967)
		°*Teucrium pelium* (Lam.), aerial parts	Brieskorn and Biechele (1969)
		°*Flourensia cernua* (Ast.), plant heads	Rao *et al.* (1970)
		°*Salvia officinalis* (Lam.), leaf	Brieskorn and Biechele (1971)
		Salvia tomentosa (Lam.), leaf	Ulubelen *et al.* (1979b)
		°*Helenium* spp. (Ast.)	Bierner (1973)
		Coleus aromaticus (Lam.), leaf	Misra *et al.* (1976)
		Coleus amboinicus (Lam.), leaf	Brieskorn and Riedel (1977)
		Artemisia capillaris (Ast.), aerial parts	Komiya *et al.* (1976)
		Sideritis dasygnaphala (Lam.), aerial parts	González *et al.* (1978)
		Brickellia californica (Ast.), leaf	Mues *et al.* (1979)
		Digitalis thapsi (Scroph.), leaf	Pascual Teresa *et al.* (1980)
		synthesis:	Ahluwalia *et al.* (1976)
6,4′-di-Me (G)	Pectolinarigenin	°*Hymenoxis rusbyi* (Ast.), aerial parts	Herz *et al.* (1970)
		°*Clerodendron phlomides, Durante repens* (Verb.), leaf	Subramanian and Nair (1972b)
		Clerodendron inerme, (Verb.), leaf	Vendantham *et al.* (1977)
		Alnus spp., *Betula* spp. (Bet.), bud excretions	Wollenweber (1975b)
		Cassia reniger (Leg.), stem, bark	Tiwari and Bajpai (1977)
		Digitalis schischkinii (Scroph.), leaf	Imre *et al.* (1977)
		Sideritis dasygnaphala (Lam.), aerial parts	González *et al.* (1978)
		Plummera ambigua, P. floribunda (Ast.), aerial parts	Bierner, (1978)

Table 4.1 (*Contd.*)

Substitution	*Trivial name*	*Plant source*	*References*
		Brickellia californica (Ast.), leaf	Mues *et al.* (1979)
		Brickellia dentata (Ast.), leaf	Ulubelen *et al.* (1980a)
		Santolina chamaecyparissus (Ast.), aerial parts	Becchi and Carrier (1980)
		Milletia pachycarpa (Leg.), aerial parts	Singhal *et al.* (1980)
7,4′-di-Me (G)			
6,7,4′-tri-Me (G)	Salvigenin	°*Ageratine gilbertii* (Ast.) aerial parts	Herz and Gibaja (1972)
		°*Psathyrothes annua* (Ast.), leaf	Finnegan *et al.* (1972)
		°*Eupatorium odoratum* (Ast.), whole plant	Talapatra *et al.* (1974)
		Alnus spp., *Betula nigra* bud excretions	Wollenweber (1975b)
		Salvia aethiopis (Lam.), whole plant	Ulubelen and Uygur (1976)
		Coleus amboinicus (Lam.), leaf	Brieskorn and Riedel (1977)
		Phyllarthron madagascariensis (Bignon.), leaf	Tillequin *et al.* (1977)
		Sideritis dasygnaphala (Lam.), aerial parts	Gonzalez *et al.* (1978)
		Brickellia dentata (Ast.), leaf	Ulubelen *et al.* (1980a)
		Ocimum canum (Lam.), leaf and flower	Xaasan *et al.* (1980)
		Milletia pachycarpa (Leg.), aerial parts	Singhal *et al.* (1980)
		Kickxia spuria (Scroph.), aerial parts	Tóth *et al.* (1980)
5,6,7,4′-tetra-Me		°*Callicarpa japonica* (Verb.)	Hosozawa *et al.* (1972)
		°*Kickxia lanigera* (Scroph.), whole plant	Pinar (1973)
		°*Colebrookia oppositifolia* (Verb.), leaf	Ahmad *et al.* (1974)
		Chromolaena odorata (Ast.), (= Eupat. od.) aerial parts	Barua *et al.* (1978)
		Citrus cult. 'calamondin' (Rut.), fruit peel	Tatum and Berry (1978)
		Kickxia spuria (Scroph.), aerial parts	Tóth *et al.* (1980)
		synthesis:	Matsuura *et al.* (1978)
5,7,8,2′-tetra-OH (G)			
7,8,-di-Me*	Skullcapflavone I (see Section 4.5)	*Scutellaria baicalensis* (Lam.)	Takido *et al.* (1979)
		Andrographis paniculata (Acanth.), tissue culture	Jalal *et al.* (1979)
		synthesis:	Iinuma and Matsuura (1979)

7,8,2′-tri-Me*		*Andrographis paniculata* (Acant.), tissue culture	Jalal *et al.* (1979)
5,7,8,4′-tetra-OH (G)	Isoscutellarein	synthesis:	Farkas *et al.* (1974)
7-Me*	Salvitin	*Salvia plebeja* (Lam.), flower	H. C. Gupta *et al.* (1975) (incl. synthesis)
8-Me*		*Gardenia gummifera, G. lucida* (Rut.), dikamali gum	S. R. Gupta *et al.* (1975)
4′-Me	Takakin	*Takakia ceratophylla, T. lepidozioides* (Hepat.)	Markham and Porter (1979)
7,8-di-Me*		*Citrus reticulatum* (Rut.), fruit peel	Iinuma *et al.* (1980)
8,4′-di-Me*	Cirsitakaogenin	*Cirsium japonicum* var. *takoense* (Ast.), leaf	Lin *et al.* (1978)
7,8,4′-tri-Me*		*Citrus reticulata* (Rut.), fruit peel	Iinuma *et al.* (1980)
5,7,8,4′-tetra-Me		*Citrus* cult. 'calamondin' (Rut.), fruit peel	Tatum and Berry (1978)
		Citrus reticulata (Rut.)	Iinuma *et al.* (1980)
(5,6,2′,6′-tetra-OH)			
6,2′,6′-tri-Me	Zapotinin		
5,6,2′,6′-tetra-Me	Zapotin		
(5,6,3′,5′-tetra-OH)			
5,6,3′,5′-tetra-Me	Cerosillin		
5,7,2′,4′-tetra-OH	Norartocarpetin		
7-Me	Artocarpetin		
5,7,2′,4′-tetra-Me*		*Terminalia arjuna* (Combret.), fruit	Nagar *et al.* (1979)
5,7,3′,4′-tetra-OH (G)	Luteolin	*Vernonia pectoralis* (Ast.), aerial parts	Bézanger-Beauquesne *et al.* (1975)
		Ceratodon purpureus (Ditrich.), sporophyt	Vandekerkhove (1978)
		Notholaena californica (Polypod.), farinose exudate on fronds	Wollenweber (1977b)
5-Me (G)*		*Juncus* spp., *Luzula* spp. (Juncac.), leaf, stem	Williams and Harborne (1975)
7-Me (G)		*Notholaena californica* (Polypod.), farinose exudate	Wollenweber (1977b)
3′-Me (G)	Chrysoeriol	*Coleus amboinicus* (Lam.), leaf	Brieskorn and Riedel (1977)
		Notholaena californica (Polypod.), farinose exudate on fronds	Wollenweber (1981b)
4′-Me (G)	Diosmetin	*Brickellia laciniata* (Ast.), leaf	Timmermann *et al.* (1979a)
		Stemodia viscosa (Scroph.), leaf	Ramesh *et al.* (1979)
5,3′-di-Me*		*Medicago × varia* (Leg.), seed	Gehring and Geiger (1980)
7,3′-di-Me (G) velutin		*Vernonia flexuosa* (Ast.), aerial parts	Kisiel (1975)
		Larrea divaricata, L. tridentata (Zyg.), leaf resin	Sakakibara *et al.* (1976)
		Notholaena californica (Polypod.), farinose exudate on fronds	Wollenweber (1981b)

Table 4.1 (*Contd.*)

Substitution	*Trivial name*	*Plant source*	*References*
7,4′-di-Me (G) pilloin		*Baccharis crispa* (Ast.), aerial parts	Bandoni *et al.* (1978)
		Lychnophora affinis (Ast.), leaf, stem	Le Quesne *et al.* (1976)
		Notholaena californica (Polypod.), farinose exudate	Wollenweber (1981b)
		synthesis:	Jain and Sharma (1973)
3′,4′-di-Me (*G only*)*		*Rhynchosia beddomei* (Leg.), leaf	Adinarayana *et al.* (1980a)
7,3′,4′-tri-Me		*Betula nigra* (Bet.), bud excretion	Wollenweber (unpublished)
		Salvia aethiopis (Lam.), whole plant	Ulubelen and Uygur (1976)
		Turnera diffusa (Turner.), aerial parts	Dominguez and Hinojosa (1976)
		Lychnophora affinis (Ast.), leaf, stem	Le Quesne *et al.* (1976)
5,7,3′,4′-tetra-Me*		°*Merillia caloxylon* (Rut.), fruit	Fraser and Lewis (1974)
(6,7,3′,4′-OH)			
6,4′-di-Me	Abrectorin	*Abrus precatorius* (Leg.), seed kernel	Bhardwaj *et al.* (1980a)
(5,6,7,8,4′-penta-OH)			
6,8-di-Me	Desmethoxy-sudachitin	°*Hymenoxis scaposa* (Ast.), leaf	Thomas and Mabry (1968)
		Gardenia gummifera, *G. lucida* (Rub.), ‘dikamali gum’	S. R. Gupta *et al.* (1975)
6,7,8-tri-Me	Xanthomicrol	*Sideritis dasygnaphala* (Lam.), aerial parts	González *et al.* (1978)
6,8,4′-tri-Me	Nevadensin	°*Gardenia gummifera*, *G. lucida* (Rub.), ‘dikamali gum’	Krishnamurti *et al.* (1972)
		Helianthus pumilis (Ast.), whole plant	Herz and de Groote (1977)
		Ocimum canum (Lab.), leaf and flower	Xaasan *et al.* (1980)
5,6,8,4′-tetra-Me*		*Helichrysum herbaceum* (Ast.), aerial parts	Bohlmann *et al.* (1979)
6,7,8,4′-tetra-Me	Gardenin B, desmethyltangeretin	*Brickellia squarrosa* (Ast.),	Rodriguez *et al.* (1974)
5,6,7,8,4′-penta-Me	Tangeretin (ponkanetin)	synthesis:	Matsuura *et al.* (1978)
(5,7,8,2′,6′-penta-OH)			
7,8,6′-tri-Me	Rivularin	*Scutellaria rivularis* (Scroph.), root	Chou (1978)
5,6,7,3′,4′-penta-OH (G)	6-Hydroxyluteolin		
6-Me (G)	Nepetin (eupafolin)	°*Eupatorium cuneifolium* (Ast.), leaf	Kupchan *et al.* (1969)
		°*Eupatorium subhastatum* (Ast.), leaf	Ferraro and Coussio (1973)
		°*Salvia officinalis* (Lam.), leaf	Brieskorn and Biechele (1971)

		°*Helenium* spp. (Ast.)	Bierner (1973)
		Artemisia tridentata spp. *roseyana* (Ast.)	Brown *et al.* (1975)
		Digitalis schischkinii (Scroph.), leaf	Imre *et al.* (1977)
		Brickellia californica (Ast.), leaf	Mues *et al.* (1979)
		Anvillea garcini (Ast.), leaf	Ulubelen *et al.* (1979a)
		Brickellia dentata (Ast.), leaf	Ulubelen *et al.* (1980a,b)
		Santolina chamaecyparissus (Ast.), aerial parts	Becchi and Carrier (1980)
		synthesis:	Midge and Rama Rao (1975)
7-Me (G)	Pedalitin	*Eupatorium inulaefolium* var. *suaveolens* (Ast.), aerial parts	Ferraro *et al.*(1977)
3′-Me (G)	Nodifloretin (batatifolin)	*Neurolaena oaxacana* (Ast.), aerial parts	Ulubelen *et al.* (1980b)
6,7-di-Me (G)	Cirsiliol	*Brickellia californica* (Ast.), leaf	Mues *et al.* (1979)
		Brickellia dentata (Ast.), leaf	Ulubelen *et al.* (1980a)
6,3′-di-Me	Jaceosidin	°*Helichrysum viscosum* var. *bracteatum* (Ast.), aerial parts	Geissman *et al.* (1967)
		°*Helenium alternifolium* (Ast.), aerial parts	Herz *et al.* (1968)
		°*Artemisia dracunculoides* (Ast.), aerial parts	Herz *et al.* (1970)
		Plummera ambigens P. floribunda (Ast.)	Bierner (1978)
		Salvia tomentosa (Lam.)	Ulubelen *et al.* (1979b)
		Santolina chamaecyparissus (Ast.), aerial parts	Becchi and Carrier (1980)
6,4′-di-Me (G)	Desmethoxy-centaureidin	*Brickellia californica* (Ast.), leaf	Mues *et al.* (1979)
		Brickellia laciniata (Ast.), leaf	Timmermann *et al.* (1979a)
7,3′-di-Me (*G only*)*		*Citharexylum subserratum* (Verb.), leaf	Mathuram *et al.* (1976)
7,4′-di-Me		*Eupatorium inulaefolium* var. *suaveolens* (Ast.), aerial parts	Ferraro *et al.* (1977)
6,7,3′-tri-Me (G)	Cirsilineol (fastigenin, anisomelin)	*Ajania fastigiata* (Ast.)	Chumbalov *et al.* (1974)
		Artemisia capillaris (Ast.)	Komiya *et al.* (1976)
		Artemisia monosperma (Ast.)	Khafagy *et al.* (1979)
		Sideritis mugronensis (Lam.), aerial parts	B. Rodriguez (1977)
		Anisomela malabarica (Lam.), aerial parts	Devi *et al.* (1979b)
		Salvia tomentosa (Lam.), aerial parts	Ulubelen *et al.* (1979b)
6,7,4′-tri-Me	Eupatorin	°*Orthosiphon stamineus* (Lam.), aerial parts	Matsuura *et al.* (1973)
		°*Salvia lavanduloides* (Lam.)	J. Rodriguez *et al.* (1974)

Table 4.1 (*Contd.*)

Substitution	*Trivial name*	*Plant source*	*References*
		Centaurea pseudomaculosa (Ast.), aerial parts	Zapesochnaya *et al.* (1977)
		Eupatorium altissimum (Ast.), aerial parts	Dobberstein *et al.* (1977)
		Otostegia limbata (Lam.), stem	Ikram *et al.* (1979)
		Brickellia dentata, B. laciniata (Ast.), leaf	Timmermann *et al.* (1979a)
		synthesis:	Wagner *et al.* (1974)
6,3′,4′-tri-Me	Eupalitin	*Sideritis gomera* aerial parts	González *et al.* (1978)
		synthesis:	Gupta *et al.* (1979)
5,6,7,4′-tetra-Me		°*Orthosiphon stamineus* (Lam.)	Matsuura *et al.* (1973)
6,7,3′,4′-tetra-Me		°*Merillia caloxylon* (Rut.), root	Fraser and Lewis (1974)
		°*Salvia lavanduloides* (Lam.), whole plant	Rodriguez *et al.* (1974)
		Salvia tomentosa (Lam.), leaf	Ulubelen *et al.* (1979b)
		Sideritis mugronensis (Lam.), aerial parts	B. Rodriguez (1977)
		Sideritis gomerae (Lam.)	González *et al.* (1978)
		Otostegia limbata (Ast.)	Ikram *et al.* (1979)
		Eupatorium altissium (Ast.), whole plant	Dobberstein *et al.* (1977)
		Citrus reticulata (Rut.), fruit peel	Iinuma *et al.* (1980)
		synthesis:	Komiya *et al.* (1976)
5,6,7,3′,4′-penta-Me	Sinensetin	°*Orthosiphon stamineus* (Lam.)	Matsuura *et al.* (1973)
		Chromolaena odorata (Ast.) (= *Eupatorium od.*), aerial parts	Barua *et al.* (1978)
		Kickxia spuria (Scroph.), aerial parts	Tóth *et al.* (1980)
		Citrus cult. 'calamondin' (Rut.), fruit peel	Tatum and Berry (1978)
		Citrus reticulata (Rut.), fruit peel	Iinuma *et al.* (1980)
(5,7,8,2′,3′-penta-OH)	Norwightin	synthesis:	Matsuura *et al.* (1978)
7,8,2′-tri′-Me	Wightin		
7,8,2′,3′-tetra-Me			
5,7,8,3′,4′-penta-OH (G)	Hypolaetin		
8-Me	Onopordin	*Gardenia gummifera, G. lucida* (Rub.), 'dikamali gum'	Chhabra *et al.* (1976a)
		synthesis:	Chhabra *et al.* (1978)

Compound	Trivial name	Source	Reference
3′-Me (*G only*)*		*Solanum grayi* (Solan.)	Whalen and Mabry (1979)
7,3′-di-Me*		*Solanum grayi* (Solan.), aerial parts	Whalen and Mabry (1979)
8,3′-di-Me			
7,8,3′,4′-tetra-Me		*Lychnophora affinis* (Ast.), leaf, stem	Le Quesne *et al.* (1976) (includes synthesis)
		synthesis:	Devi *et al.* (1979a)
5,7,8,3′,4′-penta-Me	Isosinensetin	*Fortunella japonica* (Rut.), leaf, branches	Talapatra *et al.* (1974)
		Citrus cult. 'calamondin' (Rut.), fruit peel	Tatum and Berry (1978)
		Citrus reticulata (Rut.), fruit peel	Iinuma *et al.* (1980)
5,7,2′,4′,5′-penta-OH (*G only*)*	Isoetin (hieracin)	*Isoetes duriuei I. delilei* (Isoet.), 'fronds'	Voirin *et al.* (1975) (includ. synthesis)
		Hieracium pilosella (Ast.), flower	Shelyuto *et al.* (1977)
		18 Asteraceae, Cichorieae, leaf (flower in 3 spp.)	Harborne (1978)
5,7,3′,4′,5′-penta-OH (*G only*)	Tricetin	synthesis:	Gaydou and Bianchini (1978)
5′-Me (*G only*)*	Selgin	*Huperzia selago* (Lycop.),	Voirin *et al.* (1976) (incl. synthesis)
		Isoetes lacustris, I. velata (Isoet.)	Voirin and Jay (1978)
3′,5′-di-Me (G)	Tricin	°*Eperna bijuga* (Leg.)	Braz-Filho *et al.* (1973)
		°*Schefflera roxburgii* (Araliac.), root	Desai *et al.* (1973)
		Avena sativa (Poac.), aerial parts	Popovici *et al.* (1977)
3′,4′,5′-tri-Me*		°*Bromus pauciflora, Hordeum vulgare* var. *hexastichon* (Poac.), stem, leaf	Kaneta and Sugiyama (1973)
7,3′,4′,5′-tetra-Me	Corymbosin	synthesis:	Gaydou and Bianchini (1978)
5,7,3′,4′,5′-penta-Me		*Merillia caloxylon* (Rut.), root	Fraser and Lewis (1974)
(6,7,3′,4′,5′-penta-OH)			
6,3′,4′5′-tetra-Me*	Prosogerin D	*Prosopis spicigera* (Leg.), seed	Bhardwaj *et al.* (1980b)
6,7,3′,4′,5′-penta-Me*	Prosogerin C	*Prosopis spicigera* (Leg.)	Bhardwaj *et al.* (1978a)
(5,6,7,8,2′,6′-hexa-OH)			
6,7,8,6′-tetra-Me*	Skullcapflavone II	*Scutellaria baicalensis* (Lam.)	Iinuma and Matsuura (1979)
		synthesis:	Iinuma and Matsuura (1979)
(5,6,7,8,3′,4′-hexa-OH)			
6,8-di-Me (G)*		*Gardenia gummifera, G. lucida* (Rub.), 'dikamali gum'	Chhabra *et al.* (1977a)
3′,4′-di-Me		°*Citrus aurantium* (Rut.), fruit peel	Gripenberg (1962)
6,7,8-tri-Me*		*Sideritis leucantha* (Lam.), whole plant	Tomas *et al.* (1979)
6,8,3′-tri-Me (G)	Sudachitin (majoranin)	°*Majorana hortensis* (Lam.), leaf	Subramanian *et al.* (1972)
		Citrus reticulata (Rut.), fruit peel	Iinuma *et al.* (1980)

Table 4.1 (*Contd.*)

Substitution	*Trivial name*	*Plant source*	*References*
6,8,4′-tri-Me (G)	Acerosin	*Gardenia gummifera*, *G. lucida* (Rub.), 'dikamali gum'	S. R. Gupta *et al.* (1975)
6,7,8,3′-tetra-Me*		*Sideritis mugronensis* (Lam.), aerial parts	B. Rodriguez (1977)
		Sideritis leucantha (Lam.), whole plant	Tomas *et al.* (1979)
		Sideritis flavovirens (Lam.), aerial parts	Escamilla and Rodriguez (1980)
6,7,8,4′-tetra-Me*	Gardenin D	*Sideritis mugronensis* (Lam.), aerial parts	B. Rodriguez (1977)
6,8,3′,4′-tetra-Me	Hymenoxin	°*Hymenoxis linearifolia* (Ast.), aerial parts	Herz *et al.* (1970)
		°*Helianthus angustifolium* (Ast.), aerial parts	Waddell (1973)
5,6,7,8,3′-penta-Me*		*Citrus reticulata* (Rut.), fruit peel	Iinuma *et al.* (1980)
6,7,8,3′,4′-penta-Me	5-Desmethoxynobiletin	°*Citrus* cult. 'tangerine' (Rut.), fruit peel	Tatum and Berry (1972)
		Sideritis mugronensis (Lam.)	B. Rodriguez (1977)
5,6,7,8,3′,4′-hexa-Me	Nobiletin	*Citrus* cult. 'calamondin' (Rut.), fruit peel	Tatum and Berry (1978)
(5,6,7,2′,4′,5′-hexa-OH)			
6,2′,4′,5′-tetra-Me*	Tabularin	*Chukrasia tabularis* (Meliac.), leaf	Purushotaman *et al.* (1977)
5,6,7,3′,4′,5′-hexa-OH			
6,3,′,5′-tri-Me*		*Conoclinium coelestinum* (Ast.), herbac. parts	Herz *et al.* (1980)
		Artemisia frigida (Ast.), leaf and stem	Liu and Mabry (1980)
6,4′,5′-tri-Me*		*Artemisia frigida* (Ast.), leaf and stem	Liu and Mabry (1981)
6,7,3′,4′,5′-penta-Me*		°*Merillia caloxylon* (Rut.), fruit	Fraser and Lewis (1974)
(5,7,8,2′,3′,4′-hexa-OH)			
7,8,2′,3′,4′-penta-Me	Serpyllin		
(5,7,8,3′,4′,5′-hexa-OH)			
8-Me*		*Gardenia gummifera*, *G. lucida* (Rut.), 'dikamali'	Chhabra *et al.* (1976a)
		synthesis:	Chhabra *et al.* (1978)
8,4′-di-Me*		*Gardenia gummifera*, *G. lucida* (Rut.), 'dikamali gum'	Chhabra *et al.* (1977a)
8,3′,4′-tri-Me*		*Gardenia gummifera G. lucida* (Rut.)	Gupta *et al.* (1975)
7,8,4′,5′-tetra-Me*		*Lychnophora affinis* (Ast.), leaf, stem	Le Quesne *et al.* (1976)
8,3′,4′,5′-tetra-Me*		*Gardenia gummifera*, *Gardenia lucida* (Rut.) 'dikamali gum'	Gupta *et al.* (1975)

(5,6,7,8,2′,4′,5′-hepta-OH)			
6,7,8,5′-tetra-Me*	Agecorynin-D	*Ageratum corymbosum* (Ast.), aerial parts	Quijano *et al.* (1980)
5,6,7,8,2′,4′,5′-hepta-Me*	Agecorynin-C	*Ageratum corymbosum* (Ast.), aerial parts	Quijano *et al.* (1980)
(5,6,7,8,3′,4′,5′-hepta-OH)			
6,7,8,4′-tetra-Me	Gardenin E	*Gardenia gummifera*, *G. lucida* (Rut.), 'dikamali gum'	S. R. Gupta *et al.* (1975)
6,8,3′,4′-tetra-Me	Scaposin		
8,3′,4′,5′-tetra-Me*(G)	Trimethoxywogonin	*Gardenia gummifera*, *G. lucida* (Rut.), 'dikamali gum'	S. R. Gupta *et al.* (1975)
		Vitex negundo (Verb.), stem bark	Subramanian and Misra (1979)
6,7,8,4′,5′-penta-Me	Gardenin C		
6,7,8,3′,4′,5′-hexa-Me	Gardenin A	*Gardenia gummifera*, *G. lucida* (Rut.), 'dikamali gum'	S. R. Gupta *et al.* (1975)
5,6,7,8,3′,4′,5′-hepta-Me*		*Eupatorium coelestinum* (Ast.), aerial parts	Le-Van and Pham (1979)
		Ageratum conyzoides (Ast.), aerial parts	Adesogan and Okunade (1979)
		Ageratum corymbosum (Ast.) aerial parts	Quijano *et al.* (1980)

Key: G = occurs in glycosidic combination; * = compound newly reported during 1975–1980; ° = earlier report, not included in 'The Flavonoids'.

Table 4.2 List of flavonols reported between 1975 and 1980 and their plant sources

Substitution	*Trivial name*	*Plant source*	*References*
3,5,7-tri-OH (G)	Galangin	°*Helichrysum arenarium* (Ast.), seeds	Vrkoč *et al.* (1973)
		°*Alnus pendula* (Bet.), male flowers	Suga *et al.* (1972)
		Alnus viridis (Bet.), bud excretion	Wollenweber (1975b)
		°*Comptonia peregrina* (Myr.), leaf	Lau-Cam and Chan (1973)
		Populus spp. (Salic.), bud excretion	Wollenweber (1975a)
		Escallonia, 4 spp. (Sax.), on leaf	Wollenweber (1981a)
		Adiantum sulphureum, *Cheilanthes kaulfussii*, *Pellaea longimucronata* (Polypod.), farinose exudate on fronds	Wollenweber (1979)
		Pityrogramma chrysoconia, *Pityrogramma triangularis* var. *maxonii* (Polypod.), farinose exudate on fronds	Wollenweber and Dietz (1980)
3-Me (G)		*Ericameria diffusa* (Ast.)	Urbatsch *et al.* (1976)
		Populus spp. (Salic.), bud excretion	Wollenweber (1975a)
		Escallonia, 3 spp. (Sax.), on leaf	Wollenweber (1981a)
		Cheilanthes kaulfussii, *Notholaena* spp. *(Polypod.), farinose exudate on fronds*	Wollenweber (1979)
7-Me (G)	Izalpinin	*Populus* spp. (Salic.), bud excretion	Wollenweber, (1975a)
		Aniba riparia (Laurac.), trunk wood	Fernandes *et al.* (1978)
		Escallonia, 3 spp. (Sax.), leaf	Wollenweber (1981a)
		Adiantum sulphureum (Polypod.), farinose exudate on frond	Wollenweber (1976b)
		Pityrogramma chrysoconia (Polypod.), farinose exudate on frond	Wollenweber and Dietz (1980)
		Notholaena sulphurea (Polypod.), farinose exudate on frond	Wollenweber (1981b)
3,7-di-Me*		*Escallonia langlayensis*, *E. illit.* × *illit.* (Sax.), on leaf	Wollenweber (1981a)
		Cheilanthes kaulfussii, *Notholaena* spp. (Polypod.)	Wollenweber (1979)
3,5,7-tri-Me*		*Aniba riparia* (Laurac.), wood	Franca *et al.* (1976)
		Gomphrena martiana (Amar.), whole plant	Buschi *et al.* (1980)
3,7,4′-tri-OH (*G only*)	5-Deoxykaempferol	Leguminosae, tribe Loteae	Jay *et al.* (1980)
(3,5,6,7-tetra-OH)	(6-Hydroxygalangin)		
6-Me	Alnusin	synthesis:	Goudard and Chopin (1976)

3,6-di-Me*		*Helichrysum chrysargyrum*, *H. heterolasium* (Ast.), aerial parts	Bohlmann and Abraham (1979)
3,7-di-Me*		*Helichrysum chrysargyrum* (Ast.)	Bohlmann *et al.* (1979)
		Gnaphalium wrightii (Ast.)	Bohlmann (personal communication)
5,7-di-Me*		*Gomphrena martiana* (Amar.), whole plant	Buschi *et al.* (1980)
3,6,7-tri-Me	Alnustin	synthesis:	Goudard and Chopin (1976)
3,5,6,7-tetra-Me*		*Gomphrena martiana* (Amar.), whole plant	Buschi *et al.* (1980)
(3,5,7,8-tetra-OH)	(8-Hydroxygalangin)		
7-Me*		*Notholaena galapagensis*, *N. neglecta* (Polypod.), (as ester with acetic and with butyric acid), farinose exudate on fronds	Wollenweber *et al.* (1978)
3,7-di-Me	Isognaphalin		
3,8-di-Me	Gnaphalin	synthesis:	Goudard and Chopin (1976) Gupta *et al.* (1979)
7,8-di-Me*		*Gnaphalium obtusifolium* (Ast.), aerial parts	Bohlmann (personal communication)
3,7,8-tri-Me	Methylgnaphalin	*Gnaphalium obtusifolium* (Ast.), aerial parts	Bohlmann (personal communication).
3,5,7,2′-tetra-OH (G)	Datiscetin		
2′-Me	Ptaeroxylol	(see Section 4.5)	
7-Me	Datin	°*Datisca cannabina* (Datisc.)	Pangarova and Zapesochnaya (1974)
3,5,7,4′-tetra-OH (G)	Kaempferol	*Larrea tridentata* (Zyg.) leaf	Sakakibara *et al.* (1975)
		Larrea nitida (Zyg.) leaf	Timmermann *et al.* (1979b)
		Aesculus spp. (Hipp.), *Betula* spp. (Bet.), *Populus* spp. (Salic.), bud excretion	Wollenweber (1981a) Wollenweber (1978)
		Cheilanthes spp., *Notholaena* spp., *Pityrogramma* spp. (Polypod.), farinose exudate on fronds	
3-Me (G)	Isokaempferide	*Chrysothamnus viscidiflorus* (Ast.), leaf	Urbatsch *et al.* (1975)
		Ericameria diffusa (Ast.), leaf	Urbatsch *et al.* (1976)
		Flourensia ilicifolia (Ast.), leaf	Dillon and Mabry (1977)
		Didiera madagascariensis, *D. trollii* (Didier.), aerial parts	Rabesa and Lebreton (1977)
		Tetragonotheca ludoviciana, *T. texana* (Ast.) leaf	Bacon *et al.* (1978)

Table 4.2 (*Contd.*)

Substitution	*Trivial name*	*Plant source*	*References*
		Prosopis spp. (Leg.), leaf	Bragg *et al.* (1978)
		Eupatorium tinifolium (Ast.), aerial parts	Moreno *et al.* (1980)
		Parthenium spp. (Ast.), leaf	Mears (1980)
		Betula nigra (Bet.), *Aesculus* spp., (Hipp.), *Populus* spp. (*Salic.*), bud excretion	Wollenweber (1981a)
		Cheilanthes spp., *Notholaena* spp. (Polypod.), farinose exudate on frond	Wollenweber (1976c)
		Pityrogramma triangularis var. *triangularis*	Smith (1980)
5-Me (G)			
7-Me (G)	Rhamnocitrin	*Larrea tridentata* (Zyg.), leaf	Sakakibara *et al.* (1976)
		Larrea nitida (Zyg.), leaf	Timmermann *et al.* (1979b)
		Alnus spp., *Betula* spp., *Ostrya* spp. (Bet.), *Aesculus*. 4 spp. (Hipp.), *Populus*, spp (Salic.), bud excretion	Wollenweber (1981a)
		Parthenium spp. (Ast.), leaf	Mears (1980)
		Cheilanthes spp., *Notholaena* spp. (Polypod.) farinose exudate on frond	Wollenweber (1976c)
4′-Me (G)	Kaempferide	*Dillenia indica* (Dillen.), fruit, pericarp	Pavanasasivam and Sultanbawa (1975)
		Alnus spp., *Betula* spp. *Ostrya* spp. (Bet.), *Populus* spp.	Wollenweber (1981a)
		Pityrogramma triangularis (Polypod.), farinose exudate on frond	Star *et al.* (1975)
		Notholaena spp. (Polypod.), farinose exudate on frond	Wollenweber (1981b)
		Cheilanthes spp. (Polypod.), farinose exudate on frond	Wollenweber (unpublished)
3,7-di-Me	Kumatakenin		
	Jaranol	°*Ambrosia eriocentra* (Ast.), aerial parts	Herz *et al.* (1973)
		°*Salvia glutinosa* (Lam.), aerial parts	Wollenweber (1974a)
		Ericameria diffusa (Ast.), leaf	Urbatsch *et al.* (1976)
		Flourensia cernua, *F. retinophylla* (Ast.), leaf	
		Larrea nitida (Zyg.), leaf	Timmermann *et al.* (1979b)
		Digitalis thapsi (Scroph.), leaf	Pascual Teresa *et al.* (1980)

		Alnus, 3 spp., *Ostrya*, 2 spp. (Bet.), *Aesculus*, 2 spp. (Hipp.), bud excretion	Wollenweber (1981a)
		Notholaena spp. (Polypod.), farinose exudate on frond	Wollenweber (1981b)
		Cheilanthes spp., farinose exudate on frond	Wollenweber (unpublished)
3,4′-di-Me (G)	Ermanin	°*Cordia boissieri* (Borag.), flower	Dominguez *et al.* (1973)
		Betula spp., *Ostrya* spp. (Bet.), bud excretion	Wollenweber (1975b)
		Eupatorium tinifolium (Ast.), aerial parts	Moreno *et al.* (1980)
		Geranium macrorrhizum (Geran.), whole plant	Ivantcheva (1975)
		Angelonia grandiflora (Scroph.), leaf	Nair *et al.* (1978)
		Pityrogramma triangularis var. *triangularis* (Polypod.), farinose exudate on frond	Smith (1980)
		Notholaena spp. (Polypod.), farinose exudate on frond	Wollenweber (1981b)
		Cheilanthes spp. (Polypod.)	Wollenweber (unpublished)
7,4′-di-Me		*Angelonia grandiflora* (Scroph.), leaf	Nair *et al.* (1978)
		Betula spp., *Ostrya* spp. (Bet.), *Aesculus* spp. (Hipp.), *Populus* spp. (Salic.), bud excretion	Wollenweber (1981a)
		Pityrogramma triangularis (Polypod.), farinose exudate on frond	Smith (1980)
		Notholaena spp. (Polypod.), farinose exudate on frond	Wollenweber (1981b)
3,5,7-tri-Me*		*Ericameria diffusa* (Ast.)	Urbatsch *et al.* (1976)
3,7,4′-tri-Me		*Sideritis bolaeana*, *S. dasygnaphala* (Lam.), aerial parts	Gonzàlez *et al.* (1978)
		Ostrya virginiana (Bet.), *Aesculus turbinata* (Hipp.), bud excretion	Wollenweber (1981a)
		Eupatorium tinifolium (Ast.), aerial parts	Moreno *et al.* (1980)
		Pityrogramma triangularis (Polypod.), farinose exudate on frond	Smith (1980)
		Notholaena spp. (Polypod.), farinose exudate on frond	Wollenweber (1981b)
		Cheilanthes spp. (Polypod.), farinose exudate on frond	Wollenweber (unpublished)
3,5,8,4′-tetra-OH	Pratoletin	(see Section 4.5)	
3,7,8,4′-tetra-OH			
3-Me			
3,7,3′,4′-tetra-OH (G)	Fisetin		
3-Me			
3′-Me (G)	Geraldol		

Table 4.2 (*Contd.*)

Substitution	*Trivial name*	*Plant source*	*References*
4′-Me			
3,7,3′,4′-tetra-Me			
(3,5,6,7,8-penta-OH)			
6,8 di-Me			
3,6,8-tri-Me*	Araneol	*Anaphalis araneosa* (Ast.), flowering part	Ali *et al.* (1979)
		Gnaphalium elegans (Ast.), flower	Torrenegra *et al.* (1980)
6,7,8-tri-Me		°*Helichrysum graveolens* (Ast.), aerial parts	Hänsel and Cubukcu (1972)
		Artemisia ludoviciana (Ast.), stem	Dominguez and Cardenas (1975)
3,5,6,7,4′-penta-OH (*G only*)	'Galetin' (?)	(see Section 4.5)	
		Tetragonotheca texana (Ast.),	Bacon *et al.* (1978)
3-Me (G)		*Neurolaena oaxacana* (Ast.), leaf	Ulubelen *et al.* (1980b)
5-Me (G)	Vogeletin	*Neurolaena lobata* (Ast.)	Ulubelen (personal communication)
6-Me (G)	6-Methoxykaempferol	°*Artemisia arbuscula* (Ast.), whole plant	E. Rodriguez *et al.* (1972b)
		Flourensia illicifolia (Ast.), leaf	Dillon and Mabry (1977)
4′-Me (G)		*Ostrya japonica* (Bet.), bud excretion	Wollenweber (1975c)
3,6-di-Me (G)		*Alnus*, 3 spp., *Ostrya carpinifolia* (Bet.), bud excretion	Wollenweber (1975b)
		Acanthospermum glabratum (Ast.), whole plant	Saleh *et al.* (1976)
		Flourensia illicifolia (Ast.), leaf	Dillon and Mabry (1977)
		Parthenium spp. (Ast.), leaf	Mears (1980)
		Didiera trollii (Didier.), aerial parts	Rabesa and Lebreton (1977)
		Prunus cerasus (Ros.), bud excretion	Wollenweber (unpublished)
3,7-di-Me*		*Inula grandis* (Ast.), leaf	Nikinova and Nikonov (1975)
		Parthenium spp. (Ast.), leaf	Shen *et al.* (1976); Mears (1980)
		Neurolaena oaxacana (Ast.), leaf	Ulubelen *et al.* (1980b)
		Neurolaena lobata (Ast.), leaf	Ulubelen (personal communication)
		synthesis:	Wagner *et al.* (1977a)
6,7-di-Me (G)	Eupalitin	*Ostrya japonica* (Bet.), bud excretion	Wollenweber (1975b)
		Matricaria chamomilla, flowers	Exner *et al.* (1981)
		Cheilanthes argentea (Polypod.), farinose exudate on frond	Wollenweber (unpublished)

6,4′-di-Me	Betuletol	*Alnus* spp., *Betula* spp. (Bet.), bud excretion	Wollenweber (1975b)
		Alluaudia ascendens, A. procera,	Rabesa (1980)
		Decarya madagascariensis (Didier.), aerial parts	Rabesa (1980)
		synthesis:	Wagner *et al.* (1976, 1977a); Goudard and Chopin (1976)
3,6,7-tri-Me	Penduletin	°*Artemisia*, 3 spp. (Ast.), whole plant	E. Rodriguez *et al.* (1972b)
		Parthenium spp. (Ast.), leaf	Shen *et al.* (1976); Mears (1980)
		Digitalis thapsi (Scroph.), leaf	Pascual Teresa *et al.* (1980)
		Betula nigra (Bet.), *Prunus avium* (Ros.), bud excretion	Wollenweber (unpublished)
		Gardenia fosbergii (Rub.), bud exudute	Gunatilaka *et al.* (1979)
3,6,4′-tri-Me	Santin, 3-Methyl-betuletol	°*Chrysanthemum parthenium* (Ast.), leaf	J. Rodriguez *et al.* (1974)
		Alnus spp., *Betula* spp. (Bet.), bud excretion	Wollenweber (1975b)
		Alluaudia ascendens, A. procera, Decarya madagascariensis (Didier.), aerial parts	Rabesa (1980)
6,7,4′-tri-Me (G)	Mikanin	evtl. in Betulaceae, bud excretion	Wollenweber (unpublished)
3,6,7,4′-tetra-Me		*Achillea millefolium* (Ast.), flower	Falk *et al.* (1975)
		Sideritis gomerae (Lam.), aerial parts	Gonzalez *et al.* (1978)
3,5,7,8,4′-penta-OH (*G only*)	Herbacetin		
7-Me	Pollenitin	*Notholaena standleyi* (Polypod.), farinose exudate on frond	Wollenweber and Rehse (unpublished)
		N. aschenborniana (Polypod.), (as ester with acetic and butyric acid), farinose exudate on frond	Wollenweber *et al.* (1978)
8-Me (*G only*)	Sexangularetin		
3,7-di-Me		*Larrea divaricata, L. tridentata* (Zyg.), leaf resin	Sakakibara *et al.* (1975)
		synthesis:	Wagner *et al.* (1976)
3,8-di-Me			
7,8-di-Me (*G only*)*		*Rudbeckia bicolor* (Ast.), whole plant	Jauhari *et al.* (1979)
7,4′-di-Me*		*Notholaena standleyi* (Polypod.), farinose exudate on frond	Wollenweber and Rehse (unpublished)
		Notholaena affinis (Pol.), (as ester with acetic and butyric acid), farinose exudate on frond	Wollenweber *et al.* (1978)
8,4′-di-Me	Prudomestin	synthesis:	Goudard and Chopin (1976)
3,7,8-tri-Me			

Table 4.2 (*Contd.*)

Substitution	*Trivial name*	*Plant source*	*References*
3,8,3′-tri-Me		°*Cyanostegia angustifolia* (Verb.)	Ghisalberti *et al.* (1967)
3,8,4′-tri-Me		*Conyza stricta* (Ast.), whole plant	Tandon and Rastogi (1977)
7,8,4′-tri-Me	Tambulin		
3,7,8,4′-tetra-Me	Flindulatin	(see Section 4.5)	
(3,6,7,8,4′-penta-OH)			
3,6,7,8,4′-penta-Me	Auranetin		
3,5,7,2′,4-penta-OH (G)	Morin		
(3,5,7,2′,5′-penta-OH)			
3,5,2′-tri-Me*		*Inula cappa* (Ast.), aerial parts	Baruah *et al.* (1979)
3,5,7,3′,4′-penta-OH	Quercetin	widespread	
3-Me		°*Alpinia officinalis* (Zingib.), rhizome	Bleier and Chirikdjian (1972)
		°*Larrea tridentata* (Zyg.), leaf resin	Chirikdjian (1973)
		Larrea nitida (Zyg.), leaf resin	Timmermann *et al.* (1979b)
		Haplopappus beilahuen (Ast.), aerial parts	Hörhammer *et al.* (1973)
		Chrysothamnus viscidiflorus (Ast.)	Urbatsch *et al.* (1975)
		Vernonia pectoralis (Ast.), aerial parts	Bézanger-Bauquesne *et al.* (1975)
		Ericameria diffusa (Ast.), leaf	Urbatsch *et al.* (1976)
		Inula viscosa (Ast.), flower	Öksüz (1977)
		Betula nigra (Bet.), bud excretion	Wollenweber (1977a)
		Prosopis spp. (Leg.), leaf	Bragg *et al.* (1978)
		Didiera madagascariensis, *D. trollii* (Didier.), bark and spines	Rabesa (1980)
		Neurolaena oaxacana (Ast.), leaf	Ulubelen *et al.* (1980a,b)
		Notholaena californica (Polypod.), farinose exudate on frond	Wollenweber (1981b)
5-Me (G)	Azaleatin		
7-Me (G)	Rhamnetin	*Alnus maritima*, *Betula* spp. (Bet.), bud excretion	Wollenweber (1975b)
		Populus spp. (Salic.), bud excretion	Wollenweber (1975a)
		Cheilanthes spp. (Polypod.) farinose exudate on frond	Wollenweber (unpublished)

3′-Me (G)	Isorhamnetin	*Dillenia indica*, bark	Pavanasasivam and Sultanbawa (1975)
		Ericameria diffusa (Ast.), leaf	Urbatsch *et al.* (1976)
		Larrea divaricata, L. tridentata (Zyg.), leaf resin	Sakakibara *et al.* (1976)
		Alnus spp., *Betula* spp., (Bet.), *Aesculus* spp. (Hipp.), bud excretion	Wollenweber (1981a)
		Cheilanthes spp. (Polypod.), farinose exudate on frond	Wollenweber (unpublished)
4′-Me (G)	Tamarixetin		
3,5-di-Me (G)	Caryatin		
3,7-di-Me		*Ericameria diffusa* (Ast.), leaf	Urbatsch *et al.* (1976)
		Larrea divaricata, L. tridentata, (Zyg.), leaf resin	Sakakibara *et al.* (1976)
		Angelonia grandiflora (Scroph.), leaf	Nair *et al.* (1978)
		Alnus spp., *Betula* spp. (Bet.)	Wollenweber (1981a)
		Populus spp. (Salic.), bud excretion	
3,3′-di-Me		°*Melicope perspicuinervia* (Rut.), leaf	Murphy *et al.* (1974)
		Chrysothamnus viscidiflorus (Ast.), leaf	Urbatsch *et al.* (1975)
		Populus spp., (Salic.), bud excretion	Wollenweber (1975a)
		Larrea divaricata, L. tridentata (Zyg.), leaf resin	Sakakibara *et al.* (1976)
		Larrea nitida (Zyg.), leaf resin	Timmermann *et al.* (1979b)
		Betula nigra (Bet.), bud excretion	Wollenweber (1977a)
		Didiera madagascariensis (Didier.), aerial parts	Rabesa and Lebreton (1977)
		Parthenium rollinsianum (Ast.), aerial parts	E. Rodriguez (1977)
		Escallonia, 3 spp. (Sax.), on leaf	Wollenweber (1981a)
3,4′-di-Me		*Ericameria diffusa* (Ast.), leaf	Urbatsch *et al.* (1976)
		Angelonia grandiflora (Scroph.), leaf	Nair *et al.* (1978)
5,3′-di-Me*		*Medicago* × *varia* (Leg.), seed	Gehring and Geiger (1980)
7,3′-di-Me (G)	Rhamnazin	°*Artemisia pygmaea* (Ast.), aerial parts	E. Rodriguez *et al.* (1972b)
		Larrea divaricata, L. tridentata (Zyg.), leaf resin	Sakakibara *et al.* (1976)
		Alnus spp., *Betula* spp. (Bet.), *Aesculus*, 4 spp. (Hipp.), *Populus* spp. (Salic.), bud excretion	Wollenweber (1981a)
		Cheilanthes spp., *Notholaena* spp. (Polypod.), farinose exudate on frond	Wollenweber (unpublished)
7,4′-di-Me (G)	Ombuin	*Ostrya japonica* (Bet.), bud excretion	Wollenweber (1975b)
		Cassia laevigata (Leg.), leaf	Singh and Tiwari (1979)
		Angelonia grandiflora (Scroph.), leaf	Nair *et al.* (1978)

Table 4.2 (*Contd.*)

Substitution	*Trivial name*	*Plant source*	*References*
3′,4′-di-Me (G)	Dillenetin	*Dillenia indica* (Dill.), fruit, pericarp	Pavanasasivam and Sultanbawa (1975)
3,7,3′-tri-Me (G)	Pachypodol	°*Pachypodanthium confine* (Annon.), root	Cavé *et al.* (1973)
		°*Larrea tridentata* (Zyg.), leaf resin	Chirikdjian (1974)
		Betula nigra (Bet.), bud excretion	Wollenweber (1975b)
		Larrea divaricata (Zyg.), leaf resin	Sakakibara *et al.* (1976)
		Digitalis thapsi (Scroph.)	Pascual Teresa *et al.* (1980)
3,7,4′-tri-Me (G)	Ayanin	°*Apuleia leiocarpa* (Leg.), wood	Braz Filho and Gottlieb (1971)
		°*Salvia glutinosa* (Lam.), aerial parts	Wollenweber (1974a)
		Alnus spp., *Betula* spp., *Ostrya* spp. (Bet.), bud excretion	Wollenweber (1975b)
		Ericameria diffusa (Ast.), leaf	Urbatsch *et al.* (1976)
		Physalis angulata (Sol.), whole plant	Lopez *et al.* (1976)
		Distemonanthus benthamianus (Leg.), heartwood	Malan and Roux (1979)
		Escallonia, 3 spp. (Sax.)	Wollenweber (1981a)
		Notholaena spp. (Polypod.), farinose exudate on frond	Wollenweber (1981b)
		Cheilanthes spp. (Polypod.), farinose exudate on frond	Wollenweber (unpublished)
3,3′,4′-tri-Me*		*Ericameria diffusa* (Ast.), leaf	Urbatsch *et al.* (1976)
		Betula nigra (Bet.), bud exretion	Wollenweber (1977a)
7,3′,4′-tri-Me (G)		*Diospyros melanoxylon* (Ebenac.), leaf	Desai *et al.* (1976)
		Larrea divaricata, *L. tridentata* (Zyg.), leaf resin	Sakakibara *et al.* (1976)
		Larrea nitida (Zyg.)	Timmermann *et al.* (1979b)
		Alnus spp., *Betula* spp., *Ostrya* spp. (Bet.), *Aesculus* spp. (Hipp.), bud excretion	Wollenweber (1981a)
3,5,7,4′-tetra-Me			
3,5,3′,4′-tetra-Me			
3,7,3′,4′-tetra-Me	Retusin	°*Geranium macrorrhizum* (Geran.)	°Nakashima *et al.* (1973)
		°*Salvia glutinosa* (Lam.), aerial parts	Wollenweber (1974a)
		Betula nigra (Bet.), bud excretion	Wollenweber (1975b)
		Larrea divaricata, *L. tridentata* (Zyg.), leaf resin	Sakakibara *et al.* (1976)
		Distemonanthus benthamianus (Leg.), heartwood	Malan and Roux (1979)
		Parthenium spp., (Ast.) leaf	Mears (1980)
		Notholaena spp. (Polypod.), farinose exudate on frond	Wollenweber (unpublished)

3,5,8,3′,4′-penta-OH (*G only*)		(see Section 4.5)	
3,6,7,3′,4′-penta-OH*	Rhynchosin	*Rhynchosia beddomei* (Leg.), leaf	Adinarayana *et al.* (1980b)
3,7,8,3′,4′-penta-OH	Melanoxetin	*Albizia adianthifolia* (Leg.), wood	Candy *et al.* (1978)
3,7,3′,4′,5′-penta-OH	Robinetin		
(3,5,6,7,8,3′-hexa-OH)			
3,6,7,8,3′-penta-Me*	Emmaosunin	°*Gymnosperma glutinosa* (Ast.), aerial parts	Dominguez & Torre (1974)
(3,5,6,7,8,4′-hexa-OH)			
3,6,7,8-tetra-Me	Calicopterin	*Gardenia fosbergii* (Rub.) bud exudate	Gunatilaka *et al.* (1980)
5,6,7,4′-tetra-Me	Eriostemin		
3,6,8,4′-tetra-Me	Araneosol	*Anaphalis araneosa* (Ast.), flowering plant	Ali *et al.* (1979)
3,6,7,8,4′-penta-Me*		°*Calycopteris floribunda* (Combr.), leaf	Gupta *et al.* (1973)
(3,5,6,7,2′,4′-hexa-OH)			
3,6,7,2′-tetra-Me	Chrysosplenin		
3,5,6,7,3′,4′-hexa-OH (G)	Quercetagetin		
3-Me(*G only*)*		*Chrysosplenium tetrandrum* (Sax.), aerial parts	Bohm *et al.* (1977)
		Neurolaena oaxacana (Ast.), leaf	Ulubelen *et al.* (1980b)
5-Me*		*Tagetes patula* (Ast.), petals	Bhardwaj *et al.* (1980c)
6-Me (G)	Patuletin	*Tetragonotheca*, 3 spp. (Ast.), leaf	Bacon *et al.* (1978)
		Brickellia laciniata (Ast.), leaf	Timmermann *et al.* (1979a)
		Brickellia chlorolepis (Ast.), leaf	Ulubelen *et al.* (1980a)
		Brickellia cylindracea (Ast.), leaf	Timmermann (personal communication)
		Matricaria chamomilla (Ast.), flowers	Exner *et al.* (1981)
		synthesis:	Bhardwaj *et al.* (1978b)
3′-Me (*G only*)			
3,6-di-Me	Axillarin	°*Artemisia* spp. (Ast.), leaf	E. Rodriguez *et al.* (1972b)
		Chrysothamnus viscidiflorus (Ast.), leaf	Urbatsch *et al.* (1975)
		Flourensia illicifolia (Ast.), aerial parts	Dillon and Mabry (1977)
		Didiera madagascariensis (Didier.), aerial parts	Rabesa and Lebreton (1977)
		Tetragonotheca helianthoides (Ast.), leaf	Bacon *et al.* (1978)
		Didiera trollii, *Decarya madagascariensis* (Didier.), aerial parts	Rabesa (1980)
		Neurolaena oaxacana (Ast.), leaf	Ulubelen *et al.* (1980b)
		Matricaria chamomilla (Ast.), flower	Exner *et al.* (1981)
		Parthenium spp. (Ast.) leaf	Mears (1980)

Table 4.2 (*Contd.*)

Substitution	*Trivial name*	*Plant source*	*References*
3,7-di-Me (G)	Tomentin	*Parthenium* spp. (Ast.), leaf	Shen *et al.* (1976); Mears (1980)
		Neurolaena oaxacana (Ast.) leaf	Ulubelen *et al.* (1980b)
		synthesis:	Wagner *et al.* (1976, 1977b)
6,7-di-Me (G)	Eupatolitin	*Brickellia laciniata* (Ast.), leaf	Timmermann *et al.* (1979a)
		Brickellia dentata (Ast.), leaf	Ulubelen *et al.* (1980a)
		Brickellia veronicaefolia (Ast.), leaf	Roberts *et al.* (1980)
		Matricaria chamomilla (Ast.), flower	Exner *et al.* (1981)
		synthesis:	Wagner *et al.* (1974)
6,3′-di-Me (G)	Spinacetin	*Chrysothamnus viscidiflorus* (Ast.), leaf	Urbatsch *et al.* (1975)
		Brickellia californica (Ast.), leaf	Mues *et al.* (1979)
		Brickellia laciniata (Ast.), leaf	Timmermann *et al.* (1979a)
6,4′-di-Me*	Laciniatin	*Brickellia laciniata* (Ast.), leaf	Timmermann *et al.* (1979a)
3,6,7-tri-Me	Chrysosplenol-D	°*Artemisia*, 3 spp. (Ast.), aerial parts	E. Rodriguez *et al.* (1972b)
		Parthenium rollinsianum (Ast.), leaf	Shen *et al.* (1976)
		Blumea lacera (Ast.), leaf	Rao *et al.* (1977)
		Brickellia veronicaefolia (Ast.), aerial parts	Roberts *et al.* (1980)
		Matricaria chamomilla (Ast.), flower	Exner *et al.* (1981)
3,6,3′-di-Me (G)	Jaceidin	*Chrysothamnus viscidiflorus* (Ast.), leaf	Urbatsch *et al.* (1975)
		Centaurea salicifolia (Ast.), aerial parts	Zapesochnaya *et al.* (1977)
		Flourensia illicifolia (Ast.), aerial parts	Dillon and Mabry (1977)
		Didiera madagascariensis, *D. trollii* (Didier.), aerial parts	Rabesa (1980)
		Matricaria chamomilla (Ast.), flower	Exner *et al.* (1981)
		Prunus cerasus (Ros.), bud excretion	Wollenweber (unpublished)
3,6,4′-tri-Me	Centaureidin	*Alnus*, 4 spp. (Bet.), bud excretion	Wollenweber (1975b)
		Tetragonotheca repanda (Ast.), leaf	Bacon *et al.* (1978)
		Brickellia lacinata (Ast.), leaf	Timmermann *et al.* (1979a)
		Brickellia cylindracea (Ast.), leaf	Timmermann (personal communication)

3,7,3′-tri-Me	Chrysosplenol-C	*Parthenium bipinnatifidum*, *P. glomeratum* (Ast.), aerial parts	E. Rodriguez (1977)
		Chrysosplenium alternifolium, *C. opposiifolium* (Sax.), aerial parts	Jay and Voirin (1976)
3,7,4′-tri-Me	Oxyayanin-B	*Distemonanthus benthamianus* (Leg.), heartwood	Malan and Roux (1979)
6,7,3′-tri-Me (*G only*)*	Veronicafolin	*Brickellia veronicaefolia* (Ast.), leaf	Roberts *et al.* (1980)
		Brickellia chlorolepis (Ast.), leaf	Ulubelen *et al.* (1980a)
6,7,4′-tri-Me	Eupatin	*Brickellia californica* (Ast.), leaf	Mues *et al.* (1979)
		Brickellia lacinata (Ast.), leaf	Timmermann *et al.* (1979a)
		Brickellia dentata (Ast.), leaf	Ulubelen *et al.* (1980a)
		Brickellia veronicaefolia (Ast.), leaf	Roberts *et al.* (1980)
3,5,6,3′-tetra-Me*		*Chrysothamnus visidiflorus* (Ast.), leaf	Urbatsch *et al.* (1975)
3,6,7,3′-tetra-Me (G)	Chrysosplenetin	° *Lepidophyllum quadrangulare* (Ast.), leaf	Marini-Bettolo *et al.* (1957)
		Parthenium, 3 spp. (Ast.), leaf	E. Rodriguez (1977)
		Digitalis thapsi (Scroph.), leaf	Pascual Teresa *et al.* (1980)
3,6,7,4′-tetra-Me	Casticin	*Achillea millefolium* (Ast.), flowerhead	Falk *et al.* (1975)
		Parthenium, 3 spp. (Ast.), leaf	E. Rodriguez (1977)
		Brickellia chorolepis (Ast.)., leaf	Timmermann *et al.* (1979a)
		Brickellia lacinata (Ast.), leaf	Ulubelen *et al.* (1980a)
		Brickellia veronicaefolia (Ast.), leaf	Roberts *et al.* (1980)
5,6,7,4′-tetra-Me	Eupatoretin		
3,6,3′,4′-tetra-Me	Bonanzin	*Vitex trifoliata* (Verb.), (whole plant)	Nair and Subramanian (1975)
		Parthenium ligulatum (Ast.), leaf	Rodriguez (1977)
3,7,3′,4′-tetra-Me*		*Distemonanthus benthamianus* (Leg.), heartwood	Malan and Roux (1979)
3,6,7,3′,4′-penta-Me	Artemetin	*Achillea millefolium* (Ast.), flowerhead	Falk *et al.* (1975)
		Vitex trifoliata (Verb.), leaf	Nair and Subramanian (1975)
		Parthenium rollinsianum (Ast.), leaf	E. Rodriguez (1977)
		Blumea lacera (Ast.), leaf	Rao *et al.* (1977)
		Sideritis gomerae (Lam.), aerial parts	Gonzàlez *et al.* (1978)
		Brickellia cylindracea, *B. veronicifolia* (Ast.), leaf	Roberts *et al.* (1980)
		Brickellia chlorolepis (Ast.), leaf	Ulubelen *et al.* (1980a)
3,5,6,7,3′,4′-hexa-Me			
(3,5,7,8,2′,4′-hexa-OH)			
3,7,8,2′-tetra-Me*		*Notholaena affinis* (Polypod.), farinose exudate on frond	Jay *et al.* (1979b)
3,7,8,2′,4′-penta-Me*		*Notholaena affinis* (Polypod.), farinose exudate on frond	Jay *et al.* (1979b)

Table 4.2 (*Contd.*)

Substitution	*Trivial name*	*Plant source*	*References*
3,5,7,8,3′,4′-hexa-OH (G)	Gossypetin		
7-Me (*G only*)	Ranupenin	(see Section 4.5)	
		Atraphaxis purpurea (Polygon.), leaf	Chumbalov (1976)
		Eriogonum nudum (Polygon), flower	Harborne *et al.* (1978)
		Leptocarpus similis, *Staberoha cernus* (Rest.), aerial parts	Harborne (1979)
		synthesis:	Wagner *et al.* (1977c)
8-Me (*G only*)	Corniculatusin	*Lotus corniculatus* (Leg.), leaf	Jay *et al.* (1978a)
		Eriogonum nudum (Polygon.),	Harborne *et al.* (1978)
		Geraea canescens (Ast.), flower	
		synthesis:	Wagner *et al.* (1977c)
3′-Me (*G only*)		*Coronilla valentina* (Leg.), flower	Harborne, (1981)
3,7-di-Me		*Larrea divaricata*, *L. tridentata* (Zyg.), leaf resin	Sakakibara *et al.* (1976)
		synthesis:	Wagner *et al.* (1977b)
3,8-di-Me			
7,4′-di-Me (G)		*Notholaena affinis* (Polypod.), (as ester with butyric and acetic acid) farinose exudate on frond	Wollenweber *et al.* (1978)
8,3′-di-Me (G)	Limocitrin		
8,4′-di-Me (*G only*)			
3,7,8-tri-Me		°*Cyanostegia angustifolia*, (Verb.), leaf	Ghisalberti *et al.* (1967)
3,7,3′-tri Me*		*Larrea tridentata* (Zyg.), leaf resin	Sakakibara *et al.* (1976)
3,8,3′-tri-Me			
7,8,3′-tri-Me*		*Heteromma simplicifolia* (Ast.), aerial parts	Bohlmann and Fritz (1979)
3,7,8,3′-tetra-Me	Ternatin		
3,7,8,4′-tetra-Me			
3,7,3′,4′-tetra-Me			
3,8,3′,4′-tetra-Me*		*Heteromma simplicifolia* (Ast.), aerial parts	Bohlmann and Fritz (1979)
3,7,8,3′,4′-penta-Me			
3,5,7,8,3′,4′-hexa-Me			
(3,5,7,2′,3′,4′-hexa-OH)			
3,7,4′-tri-Me	Apuleidin		
		synthesis:	Kalra *et al.* (1975)

3,5,7,2′,4′,5′-hexa-OH	5′-OH-morin	*Geranium macrorrhizum* (Geran.), aerial parts	Ognyanov and Ivantcheva (1972)
		Distemonanthus benthamianus (Leg.), heartwood	Malan and Roux (1979)
3,7,4′-tri-Me	Oxyayanin-A		
3,5,7,4′-tetra-Me			
3,7,4′,5′-tetra-Me		*Distemonanthus benthamianus* (Leg.), heartwood	Malan and Roux (1979)
3,7,2′,4′,5′-penta-Me*		*Distemonanthus benthamianus* (Leg.), heartwood	Malan and Roux (1979)
3,5,7,4′,5′-penta-Me*		*Distemonanthus benthamianus* (Leg.), heartwood	Malan and Roux (1979)
3,5,7,2′,4′,5′-hexa-Me*		*Distemonanthus benthamianus* (Leg.), heartwood	Malan and Roux (1979)
(3,6,7,2′,4′,5′-hexa-OH)			
3,6,7,4′-tetra-Me	5-*O*-Demethylapulein		
3,5,7,3′,4′,5′-hexa-OH (G)	Myricetin	*Soymidia febrifuga* (Mel.), heartwood	Rao *et al.* (1979)
3-Me (*G only*)	Annulatin		
		synthesis:	Chhabra *et al.* (1976b)
5-Me (*G only*)			
7-Me (*G only*)	Europetin		
3′-Me (*G only*)*	Laricitrin	*Larix sibirica* (Pin.)	Tyukavkina *et al.* (1974)
		Cedrus atlantica (Pin.)	Niemann (1977)
		Tetragonolobus siliquosus (Leg.)	Jay *et al.* (1978b)
		Elegia galpinii, *E. parviflora* (Rest.), stem	Harborne (1979)
		Ginkgo biloba (Gink.), leaf	Geiger (1979)
4′-Me (G)	Mearnsetin	*Liatris pauciflora* (Ast.), aerial parts	Iyengar *et al.* (1976)
		Elaeocarpus lanceofolius (Elaeoc.), leaf	Ray *et al.* (1976)
		Elaeocarpus floribundus (Elaeoc.), leaf	Dasgupta *et al.* (1977)
		Alluaudia, 6 spp., *Decarya madagascariensis* (Didier.), stem bark	Rabesa (1980)
3,4′-di-Me*		*Alluaudia ascendens A. procera*, *Decarya madagascariensis* (Didier.), stem bark	Rabesa (1980)
3,5′-di-Me*		*Artemisia monosperma* (Ast.), flowerhead	Khafagy *et al.* (1979)
3′,5′-di-Me (G)	Syringetin	*Heuchera micrantha* var. *diversifolia* (Sax.), aerial parts	Wilkins and Bohm (1976)
		Elegia spp. (Rest.) stem	Harborne (1979)
3,7,4′-tri-Me			
7,3′,4′-tri-Me		synthesis:	Kumari *et al.* (1979)
3,7,3′,4′-tetra-Me		*Doliocarpus amazonicus* (Dillen.), leaf	Gurni (1979)

Table 4.2 (*Contd.*)

Substitution	*Trivial name*	*Plant source*	*References*
3,7,3′,4′,5′-penta-Me	Combretol	*Betula nigra* (Bet.), bud excretion	Wollenweber (1976a)
		Notholaena candida var. *candida* (Polypod.), farinose exudate on frond	Wollenweber (1976b)
(3,5,6,7,8,2′,4′-hepta-OH)			
3,6,7,8-tetra-Me*		*Notholaena affinis* (Polypod.), farinose exudate on frond	Jay *et al.* (1979a)
3,6,7,8,2′-penta-Me*		*Notholaena affinis* (Polypod.), farinose exudate on frond	Jay *et al.* (1979a)
3,6,7,8,4′-penta-Me*		*Notholaena affinis* (Polypod.), farinose exudate on frond	Jay *et al.* (1979a)
3,6,7,8,2′,4′-hexa-Me*		*Notholaena affinis* (Polypod.), farinose exudate on frond	Jay *et al.* (1979a)
(3,5,6,7,8,3′,4′-hepta-OH)			
3,6,8-tri-Me		*Plucha saggitalis* (Ast.)	Martino *et al.* (1976)
6,8,3′-tri-Me	Limocitrol		
6,8,4′-tri-Me	Isolimocitrol		
3,6,8,3′-tetra-Me*		*Chrysothamnus viscidiflorus* (Ast.), leaf	Urbatsch *et al.* (1975)
3,6,7,8,3′-penta-Me*		°*Calycopteris floribundus* (Combr.), leaf	Rama Rao and Varadan, (1973)
		Digitalis thapsi (Scroph.), leaf	Pascual Teresa *et al.* (1980)
3,6,7,8,3′,4′-hexa-Me			
5,6,7,8,3′,4′-hexa-Me	Natsudadain		
3,5,6,7,8,3′,4′-hepta-Me		*Citrus* cult. (Rut.), fruit peel	Tatum and Berry (1978)
(3,5,6,7,2′,3′,4′-hepta-OH)			
3,7,4′-tri-Me	Apuleisin		
3,5,6,7,2′,3′,4′-hepta-Me*		*Distemonanthus benthamianus* (Leg.), heartwood	Malan and Roux (1979)
(3,5,6,7,2′,4′,5′-hepta-OH)			
3,5,6,7,4′-penta-Me	Apulein		
3,5,7,4′,5′-penta-Me*		*Distemonanthus benthamianus* (Leg.), heartwood	Malan and Roux (1979)
3,7,2′,4′,5′-penta-Me*		*Distemonanthus benthamianus* (Leg.), heartwood	Malan and Naidoo (1980)
3,5,6,7,2′,4′5′-hepta-Me*		*Distemonanthus benthamianus* (Leg.), heartwood	Malan and Naidoo (1980)
(3,5,6,7,3′,4′,5′-hepta-OH)	(6-OH-myricetin)		
3,6-di-Me*		*Alluaudia ascendens* (Didier.), stem bark	Rabesa and Voirin (1979d)
3,4′-di-Me*		*Alluaudia ascendens* (Didier.), stem bark	Rabesa and Voirin (1979d)

3,6,3′-tri-Me*		*Alluaudia ascendens*, *A. procera*, *Decarya madagascariensis* (Didier.), aerial parts	Rabesa (1980)
3,6,4′-tri-Me*		*Alluaudia ascendens*, *A. procera*, *Decarya madagascariensis* (Didier.), aerial parts	Rabesa (1980)
6,3′,4′-tri-Me (*G only*)*		*Tillandsia usneoides* (Brom.)	Williams (1978)
3,3′,5′-tri-Me		*Gardensia fosbergii* (Rub.) bud exudate	Gunatilaka *et al.* (1979)
3,5,6,5′-tetra-Me			
3,6,7,4′-tetra-Me			
3,6,3′,4′-tetra-Me*		*Decarya madagascariensis* (Didier.), aerial parts	Rabesa (1980)
3,6,3′,5′-tetra-Me*		*Tillandsia usneoides* (Brom.)	Lewis and Mabry (1977)
5,6,3′,5′-tetra-Me			
3,7,3′,4′-tetra-Me	Apuleitrin	synthesis:	Kalra *et al.* (1977)
3,7,7,4′-tetra-Me	—	*Gardensia fosbergii* (Rub.) bud exudate	Gunatilaka *et al.* (1979)
3,5,7,3′,4′-penta-Me	Apuleirin	synthesis:	Kalra *et al.* (1977)
3,6,7,3′,4′-penta-Me		*Gardenia fosbergii* (Rub.) bud exudate	Gunatilaka *et al.* (1979)
3,6,7,3′,5′-penta-Me*	Murrayanol	*Murraya omphalocarpa* (Rut.), fruit	Wu *et al.* (1980)
6,7,3′,4′,5′-penta-Me		*Gardenia fosbergii* (Rub.) bud exudate	Gunatilaka *et al.* (1979)
3,5,6,7,3′,5′-hexa-Me		*Murraya paniculata* (Rut.), leaf	De Silva *et al.* (1980)
3,5,6,7,3′,4′,5′-hepta-Me		*Murraya omphalocarpa* (Rut.), fruit	Wu *et al.* (1980)
3,5,7,8,3′,4′,5′-hepta-OH (*G only*)	Hibiscetin		
3,7,4′-tri-Me*		*Solanum* spp. (Solan.)	Whalen and Mabry (1979)
3,7,8,4′-tetra-Me*		*Solanum* spp. (Solan.)	Whalen and Mabry (1979)
3,8,4′,5′-tetra-Me			
3,8,3′,4′,5′-penta-Me*	Conyzatin	*Conyza stricta* (Ast.), whole plant	Tandon and Rastogi (1977)
7,8,3′,4′,5′-penta-Me*		*Heteromma simplicifolia* (Ast.), aerial parts	Bohlmann and Fritz (1979)
3,5,7,8,3′,4′,5′-hepta-Me			
(3,5,6,7,8,3′,4′,5′-octa-OH)			
3,6,7,8,4′,5′-hexa-Me	Digicitrin		
3,5,6,7,3′,4′,5′-hepta-Me			
3,5,6,7,8,3′,4′,5′-octa-Me	Exoticin		

Key: G = occurs in glycosidic combination; * = compound newly reported during 1975–1980; ° = earlier report, not included in 'The flavonoids'.

General observations on the frequencies of different substitution patterns in the flavonoid skeleton have been made by Harborne (1967) and Gottlieb (1975) and are still valid. The distribution and accumulation of free aglycones in certain families and their correlation with other structural features have been reviewed elsewhere (Wollenweber and Dietz, 1981). Although a general discussion of flavones and flavonols with 'normal' substitution patterns is not therefore needed, brief comments on flavonoids with more unusual patterns (e.g. isoprenoid constituents, additional ring systems) are necessary. To our knowledge no report of important new routes for chemical synthesis of flavones and flavonols has appeared. References to the synthesis of certain individual compounds, however, are interesting enough to be included.

4.2 *C*-METHYLFLAVONOIDS

4.2.1 Flavones

The first natural *C*-methylated flavonoid to be reported was the flavanone matteucinol (6,8-di-*C*-methylnaringenin). Today *C*-methylated compounds are known among every class of flavonoid, but they are still most abundant among the flavanones (Wollenweber and Dietz, 1981). In every known case *C*-methylation takes place at C-6 and/or C-8. In 'The Flavonoids' only four *C*-methylflavones were cited, and this number has not been augmented since. However, reports of eucalyptin from the leaf wax of two *Angophora* species as well as from the leaf of *Eugenia biflora* and *Myrcia citrifolia* deserve mention. Except for the presence of strobochrysin in *Pinus* species, *C*-methylated flavones are restricted in their occurrence to plants of the Myrtaceae.

4.2.2 Flavonols

In the flavonol series, only two compounds (6-*C*-methylquercetin and 6-*C*-methylmyricetin) were known previously (Gottlieb, 1975) and these were reported from *Pinus ponderosa*. Meanwhile one new flavonol (sylpin) has been reported from *Pinus silvestris*, and two from the Californian goldenback fern. In addition, ten *C*-methylflavonols have been found in aglycone form in bark and spines of Didieraceae. This family, comprising only 11 species in four genera endemic to Madagascar, appears to produce *C*-methylflavonols as an extreme specialization. The compounds present are methyl derivatives of kaempferol and myricetin. Three further compounds, methyl derivatives of quercetin, also appear to occur in these plants (Rabesa, 1980). *C*-Methylation is regarded as a primitive character (Harborne, 1967). Taking into account that within the flavonols, it is always accompanied by *O*-methylation (except for pinoquercetin and 6-*C*-methylmyricetin) at C-3 and/or C-4′ (except pityrogrammin), Rabesa (1980) assumed that *O*-methylation at these positions also might be regarded as a

primitive character, such methylation occurring at an early stage in the formation of the C_{15} skeleton.

4.2.3 Formylflavones

A special type of *C*-methylflavone is represented by three flavones from *Unona lawii* (Table 4.3), the only formylflavones known as yet. [Two *C*-formylated chalcones have been found in *Psoralea corylifolia* (see B. K. Gupta *et al.*, 1980)]. Finally, it should be mentioned that one additional *C*-methylflavone (desmodol) is found among the pyranoflavones (Table 4.6).

4.3 METHYLENEDIOXYFLAVONOIDS

Methylenedioxyflavonoids were not treated as a separate group in 'The Flavonoids'. In the present chapter they have been listed separately to allow easier access. Completeness, however, cannot be guaranteed. Plant sources are cited for those compounds known before 1975. To our knowledge, since then six new flavones and five new flavonols with methylenedioxy groups have been reported. However, this type of substitution is more abundant among flavonols than among flavones. Only positions 6/7 and 3′/4′ are concerned; in one compound (meliternatin), all four positions are so substituted. In general, methylenedioxy substitution is a rare feature among flavones and flavonols but is more abundant with the isoflavonoids (see Chapter 10). An earlier preliminary report of a methylenedioxyflavonol from chloroplasts of spinach (Oettmeier and Heupel, 1972) has still not been confirmed by more detailed studies.

4.4 FLAVONOIDS WITH OTHER SUBSTITUENTS

4.4.1 Flavones and flavonols with isoprene substitution

Most of the compounds dealt with in this section are found in *Artocarpus*, *Morus* and *Tephrosia*, except for the flavonols with C_5 and C_{10} side chains. Venkataraman's treatment (1975) of the flavones found in these genera was appropriate. However, the more formal setting out of these substances in tabular form should allow easier access to the relevant information. For this reason plant sources are given in Table 4.5 for those compounds already covered in 'The Flavonoids'. Only for the special flavonoids of *Tephrosia* and *Pityrogramma* separate treatment is still deemed to be justified.

The majority of the flavonoids listed in Tables 4.5–4.7 bear isoprenoid substituents. Simple C_5 and C_{10} side chains occur as well as additional pyran and furan (or dihydrofuran) rings. The pyran ring is indicated in Table 4.6 by the abbreviation ODmp for the dimethylallyl unit ($Me_2C\text{-}CH{=}CH\text{-}$) linked via

Table 4.3 *C*-Methylated flavonoids

Substitution	*Trivial name*	*Plant source*	*References*
C-METHYLFLAVONES			
5,7-di-OH, 6-Me	6-Methylchrysin		
5,4′-di-OH, 7-OMe, 6,8-di-Me	Sideroxylin		
5-OH, 7,4′-di-OMe, 6-Me			
5-OH, 7,4′-di-OMe, 6,8-di-Me	Eucalyptin	°*Eugenia biflora*, *Myrcia citrifolia*, (Myrtac.), leaf	Gottlieb and Da Rocha (1972)
		°*Angophora lanceolata*, *A. subvelutina*, (Myrtac.), leaf wax	Lamberton (1964)
5,7-di-OH, 8-Me, 6-CHO	Unonal	*Unona lawii* (Anonac.), stem	Joshi and Gawad (1976)
5,7-di-OH, 6-Me, 8-CHO	Isounonal	*Unona lawii* (Anonac.), stem	Joshi and Gawad (1976)
5-OH, 7-OMe, 8-Me, 6-CHO	Unonal 7-Methyl ether	*Unona lawii* (Anonac.), stem	Joshi and Gawad (1976)
C-METHYLFLAVONOLS			
5,7-di-OH, 3-OMe, 6,8-di-Me		*Pityrogramma triangularis* var. *triangularis*, (Polypod.), farinose exudate on fronds	Dietz *et al.* (1981)
3,5,7-tri-OH, 8-OMe, 6-Me	Pityrogrammin	*Pityrogramma triangularis* var. *triangularis*, (Polypod.), farinose exudate on fronds	Star *et al.* (1975)
5,6,4′-tri-OH, 3-OMe, 8-Me	Sylpin	*Pinus silvestris*, (Pin.)	Medvedeva *et al.* (1977)
5,7,4′-tri-OH, 3-OMe, 6-Me		*Alluaudia dumosa* (Didier.), aerial parts	Rabesa and Voirin (1978)
5,7,4′-tri-OH, 3-OMe, 6,8-di-Me		*Alluaudia dumosa*, *A. humbertii* (Didier.), aerial parts	Rabesa (1980)
3,5,7,3′,4′-penta-OH, 6-Me	Pinoquercetin		
3,5,7,3′,4′-penta-OH, 6,8-di-Me		*Alluaudia humbertii* (Didier.), aerial parts	Rabesa (1980)
5,7,4′-tri-OH, 3,3′-di-OMe, 6-Me		*Alluaudia humbertii* (Didier.), aerial parts	Rabesa (1980)
3,5,7,3′,4′,5′-hexa-OH, 6-Me			
5,7,3′,4′,5′-penta-OH, 3-OMe, 6-Me	Alluaudiol	*Alluaudia dumosa* (Didier.), bark	Rabesa and Voirin (1979a)
		Alluaudia humbertii (Didier.), bark	Rabesa (1980)
5,7,3′,4′,5′-penta-OH, 3-OMe, 6,8-di-Me		*Alluaudia humbertii* (Didier.), aerial parts	Rabesa (1980)
3,5,7,3′,5′-penta-OH, 4′-OMe, 6-Me	Dumosol	*Alluaudia dumosa* (Didier.), aerial parts	Rabesa and Voirin (1979b)

3,5,7,3′,5′-penta-OH, 4′-OMe 6,8-di-Me	*Alluaudia humbertii* (Didier.), aerial parts	Rabesa and Voirin (1979c)
5,7,3′,5′-tetra-OH, 3,4′-di-OMe, 6-Me	*Alluaudia dumosa*, *A. humbertii* (Didier.), aerial parts	Rabesa and Voirin (1979b) Rabesa and Voirin (1979c)
5,7,3′,5′-tetra-OH, 3,4′-di-OMe, 6,8-di-Me	*Alluaudia dumosa*, *A. humbertii* (Didier.) aerial parts	Rabesa and Voirin (1979b)

Key: * Compound newly reported during 1975–1980; ° = earlier report, not included in *The Flavonoids.*

Table 4.4 Methylenedioxyflavonoids

Substitution	*Trivial name*	*Plant source*	*References*
FLAVONES			
5-OMe, 7,8-O_2CH_2*		*Helichrysum mundii* (Ast.), root and aerial parts	Bohlmann *et al.* (1978)
5,6-di-OMe, 3′,4′-O_2CH_2*		*Millettia ovalifolia* (Leg.), leaf	Khan and Zaman (1974)
7-OH, 6-OMe, 3′,4′-O_2CH_2*	Prosogerin-A	*Prosopis spicigera* (Leg.), flower	Bhardwaj *et al.* (1979)
5,4′-di-OH, 6,7-O_2CH_2*	Kanzakiflavone-2	*Iris unguicularis* (Ir.), rhizome	Arisawa *et al.* (1976)
		synthesis:	Bhardwaj *et al.* (1978b)
5,8-di-OH, 4′-OMe, 6,7-O_2CH_2*	Kanzakiflavone-1	*Iris unguicularis* (Ir.), rhizome	Arisawa and Morita (1976)
		synthesis:	Bhardwaj *et al.* (1978b)
5,7-di-OH, 8-di-OMe, 3′,4′-O_2CH_2	Linderoflavone A	*Lindera lucida* (Laur.), root	
5,6,7,8-tetra-OMe, 3′,4′-O_2CH_2	Linderoflavone B	*Lindera lucida* (Laur.), root	
5,6,7,8,5′-penta-OMe, 3′,4′-O_2CH_2*	Eupalestin	*Eupatorium coelestinum* (syn. *Conoclinium coel.*)(Ast.), aerial (herbaceous) parts	Le-Van and Pham (1979) Herz *et al.* (1980)
		Ageratum corymbosum (Ast.), aerial parts	Quijano *et al.* (1980)
FLAVONOLS			
3,5-di-OMe, 6,7-O_2CH_2*		*Gomphrena martiana* (Amarant.), whole plant	Buschi *et al.* (1979)
3,5-di-OMe, 6,7-O_2CH_2 3′,4′-O_2CH_2	Meliternatin	*Melicope*, 3 spp. (Rut.), wood	
3,7-di-OMe, 3′,4′-O_2CH_2	Demethoxykanungin	*Pongamia glabra* (Leg.), root and stem bark; heartwood	
3,5,4′-tri-OH, 6,7-O_2CH_2*	Gomphrenol	*Gomphrena globosa* (Amar.), leaf	Bouillant *et al.* (1978)
3,7,3′-tri-OMe, 4′,5′-O_2CH_2	Kanugin	*Pongamia glabra* (Leg.), root and stem bark; heartwood	Subramanyan *et al.* (1977)
5-OH, 3,6,7-tri-OMe, 3′,4′-O_2CH_2	Melisimplin	*Melicope simplex* (Rut.), wood	
3,5,6,7-tetra-OMe, 3′,4′-O_2CH_2	Melisimplexin	*Melicope simplex* *M. broadbentiana* (Rut.), wood	

5-OH, 3,7,8-tri-OMe, 3′,4′-O_2CH_2*		*Pelea barbigera* (Rut.) aerial parts?	Higa and Scheuer (1974)
3,5,7,8-tetra-OMe, 3′,4′-O_2CH_2	Meliternin	*Melicope ternata* (Rut.), wood	
5,3′,4′-tri-OH, 3-OMe, 6,7-O_2CH_2		*Spinacia oleracea* (Chen.), chloroplasts	(see text, Section 4.3)
5,3′,4′-tri-OH, 3-OMe, 7,8-O_2CH_2	Wharangin	*Melicope ternata* (Rut.), bark	
3,5,7-tri-OH, 6,8-di-OMe, 3′,4′-O_2CH_2*	Melinervin	*Melicope perspicuinervia* (Rut.), leaf	Murphy *et al.* (1974)
5-OH, 3,6,7,8-tetra-OMe, 3′,4′-O_2CH_2*		*Pelea barbigera* (Rut.), aerial parts?	Higa and Scheuer (1974)
3,5,6,7,8-penta-OMe, 3′,4′-O_2CH_2	Melibentin	*Melicope broadbentiana* (Rut.), wood	

Key: *Compound newly reported during 1975–1980.

Table 4.5 Flavones and flavonols with C_5, C_5-OH and C_{10} side chains†

Substitution	*Trivial name*	*Plant source*	*References*
FLAVONES			
7-OMe, 8-C_5-OH*	*trans*-Lanceolatin	*Tephrosia lanceolata*, *T. purpurea*, roots *T. apollinea* (Leg.)	Ayengar *et al.* (1973) Rao and Ranga Raju (1979) Waterman and Khalid (1980)
5,7-di-OMe, 8-C_5-OH*	*cis*-Tephrostachin	*Tephrosia polystachyoides* (Leg.)	Vleggar *et al.* (1973)
	trans-Tephrostachin	*Tephrosia bracteolata* (Leg.), seed	Khalid and Waterman (1981)
5,7-di-OMe, 8-CH=CH-C(Me)=CH_2	*trans*-Anhydro-tephrostachin	*Tephrosia bracteolata* (Leg.), seed	Khalid and Waterman (1981)
5,7,2′,4′-tetra-OH, 6-C_5	Artocarpesin	*Artocarpus* (Morac.)	
5,7,2′,4′-tetra-OH, 6-C_5-OH	Oxidihydroartocarpesin	*Artocarpus* (Morac.)	
5,7,2′,4′-tetra-OH, 3,6-di-C_5	Norartocarpin	*Artocarpus*	
5,7,2′,4′-tetra-OH, 3,6-di-C_5	Mulberrin	*Morus*	
5,2′,4′-tri-OH, 7-OMe,3-C_5*	Integrin	*Artocarpus integer* (Mor.) synthesis of trimethyl-ether	Pendse *et al.* (1976) Midge *et al.* (1977)
5,2′,4′-tri-OH, 7-OMe,3,6-di-C_5	Artocarpin	*Artocarpus*	
5,7,2′,4′-tetra-OH, 3,8-di-C_5*	Kuwanone C	*Morus alba* (Mor.), root bark	Nomura *et al.* (1977) Nomura *et al.* (1978b)
5,7,3′,4′-tetra-OH, 8-C_5		*Xanthium* (Ast.)	
7,4′,6′-tri-OH, 3-C_{10}	Rubraflavone A	*Morus*	
7,4′,6′-tri-OH, 3-C_{10}, 8-C_5	Rubraflavone B	*Morus*	
5,7,4′,6′-tetra-OH, 3-C_{10}, 6-C_5	Rubraflavone C	*Morus*	
FLAVONOLS			
3,7,4′-tri-OH, 8-C_5-OH		*Phellodendron* (Rut.)	
3,5,7,8-tetra-OH, 6-C_5*		*Platanus acerifolia* (Plat.), bud excretion	Egger *et al.* (1972)
3,5,7,4′-tetra-OH, 8-C_5	Noranhydroicaritin	*Sophora* (Leg.) *Epimedium* (Berb.)	
3,5,7,4′-tetra-OH, 8-C_5-OH	Noricaritin	*Epimedium* (Berb.)	

3,5,4′-tri-OH, 7-OMe, 8-C_5	Isoanhydroicaritin	*Sophora* (Leg.)	
3,5,7-tri-OH, 4′-OMe, 8-C_5-OH	Icaritin	*Epimedium* (Berb.)	
3,5,7,3′,4′-penta-OH, 6-C_5-OH (G only)			
5,4′-di-OH, 3,3′-di-OMe, 7-O-C_5* ‡		*Euodia glabra* (Rub.)	Fraser and Lewis (1973)
5,4′-di-OH, 3,7,3′-tri-OMe, 8-C_5		*Phebalium* (Rut.)	

Key: * = Compound newly reported during 1975–1980; G = Occurs in glycosidic combination.
† $C_5 = Me_2CH\text{-}CH{=}CH_2$ or $Me_2C{=}CH\text{-}CH_2$; $C_5\text{-}OH = Me_2C(OH)CH_2CH_2$; $C_{10} = (Me_2C = CH\text{-}CH_2)_2$.
‡ Note *O*-prenylation!

Table 4.6 Pyranoflavones and pyranoflavonols

Substitution†	*Trivial name*	*Plant source*	*References*
PYRANOFLAVONES			
5-OH,7,8-ODmp*		*Derris obtusa* (Leg.)	do Nascimento *et al.* (1976)
5-OMe, 7,8-ODmp*	Isopongaflavone	*Pongamia glabra* (Leg.), seed	Roy *et al.* (1977) (includes synthesis)
		Tephrosia bracteolata (Leg.), seed	Khalid and Waterman (1981)
5-OMe, 7,6-ODmp*	Praecansone B	*Tephrosia praecana* (Leg.), seed	Camele *et al.* (1980)
5,4′-di-OH, 7,6-ODmp*		*Pamburus missionis* (Rut.)	Dreyer and Park (1975)
5,4′-di-OH, 7,6-ODmp, 6′,3-ODmp	Cyclomulberrochromene	*Morus*	
5,4′-di-OH, 7,8-ODmp, 6′,3-ODmp*	Cyclomorusin	*Morus alba* (Mor.) root bark	Nomura *et al.* (1976) Nomura *et al.* (1978a)
5,2′,4′-tri-OH,7,6-ODmp	Cycloartocarpesin	*Artocarpus*	
5,2′,4′-tri-OH,3-C_5,7,6-ODmp	Mulberrochromene	*Morus* (Mor.)	
5,7,2′-tri-OH,3-C_5,3′,4′-ODmp*	Kuwanone B	*Morus alba* (Mor.), root bark	Nomura *et al.* (1977) Nomura *et al.* (1978b)
5,2′,4′-tri-OH,3-C_5-OH, 7,8-ODmp*	Oxydihydromorusin	*Morus alba* (Mor.), root bark	Nomura *et al.* (1977) Nomura *et al.* (1978b)
5,2′,4′-tri-OH,3-C_5, 7,8-ODmp*	Morusin	*Morus alba* root bark	Nomura *et al.* (1976) Nomura *et al.* (1978a)
5,7,4′-tri-OH,3-C_5, 2′,3′-ODmp*	Kuwanone A	*Morus alba* root bark	Nomura *et al.* (1977) Nomura *et al.* (1978b)
5,2′,4′-tri-OH,3-C_{10}, 7,6-ODmp	Rubraflavone D	*Morus*	
5,7,4′-tri-OH,6-C_5, 6′,3-ODmp	Cyclomulberrin	*Morus*	
5,4′-di-OH,7-OMe, 6-C_5, 6′,3-ODmp	Cycloartocarpin	*Artocarpus*	
5,3′,4′-tri-OH,6-CH_3, 7,8-ODmp*	Desmodol	*Desmodium caudatum* (Leg.) root and stem	Ueno *et al.* (1978)
5,3′,4′-tri-OH,8-C_5, 7,6-ODmp, 6′,3-ODmp	Cycloheterophyllin	*Artocarpus*	
5,3′,4′,6′-tetra-OH,5′-C_5, 7,6-ODmp*	Artobilichromene	*Artocarpus nobilis* (Mor.), bark	Kumar *et al.* (1977)

PYRANOFLAVONOLS			
3-OMe, 7,8-ODmp*	Karanjachromene	*Pongamia glabra* (Leg.), seed oil	Satam and Bringi (1973)
	Pongaflavone	*Pongamia glabra* (Leg.), stem	Lakshmi *et al.* (1974)
3,5-di-OMe,3',4'-O_2CH_2, 7,8-ODmp*	Pongachromene	*Pongamia glabra* (Leg.) stem bark	Subramanyam *et al.* (1977)
3,6-di-OMe, 7,8-ODmp*		*Derris obtusa* (Leg.), root bark	do Nascimento *et al.* (1976)
3,5-di-OH, 8-C_5, 7,6-ODmp	Sericetin	*Mundulea* (Leg.)	
3,5,4'-tri-OH, 7,8-ODmp	Phellamuretin	*Phellodendron japonicum* (Rut.)	Grimshaw and Lamer-Zarawska (1972)
3,5,3',4'-tetra-OH, 7,8-ODmp*		*Asclepias syriaca* (Asclep.)	Gonnet *et al.* (1973)

† For structures, see Fig. 4.1

Table 4.7 Furano-flavones and flavonols

Substitution†	*Trivial name*	*Plant source*	*References*
FURANOFLAVONES			
7,8-fur	Lanceolatin B	*Tephrosia, T. falsiformis* (Leg.)	Ghanim and Jayaraman (1979)
3′,4′-O_2CH_2, 7,8-fur*	Pongaglabrone	*Pongamia glabra* (Leg.), heartwood	Subramanyam *et al.* (1977)
5-OH, 7,8-fur*	Pongaglabol	*Pongamia glabra* (Leg.), flower	Talapatra *et al.* (1980)
5-OMe, 7,8-fur	Pinnatin	*Pongamia*	
5-OMe, 3′,4′-O_2CH_2, 7,8-fur	Gamatin	*Pongamia*	
6-OMe, 7,8-fur*		*Pongamia glabra* (Leg.), leaf	Malik *et al.* (1977)
8-OMe, 7,6-fur*		*Pongamia glabra* (Leg.), leaf	Malik *et al.* (1977)
3′-OH, 7,8-fur*		*Pongamia glabra* (Leg.), seed	Roy and Khanna (1979)
5,3′,6′-tri-OH, 4′,5′-fur-ODmp	Dihydrofurano-artibilichromene b_1 and b_2	*Artocarpus nobilis* (Mor.), bark	Kumar *et al.* (1977)
5,3′,4′-tri-OH,6′,5′-fur-ODmp	Dihydrofurano-artobilichromene a	*Artocarpus nobilis* (Mor.), bark	Kumar *et al.* (1977)
5,4′,6′-tri-OH, 3-C_5, 7,6-fur-ODmp-OH*	Mulberranol	*Morus alba* (Mor.), bark	Deshpande *et al.* (1976)
FURANOFLAVONOLS			
3-OMe, 7,8-fur	Karanjin	*Pongamia* (Leg.)	
3-OMe, 3′,4′-O_2CH_2,7,8-fur	Pongapin	*Pongamia*	
3,3′-di-OMe, 4′,5′-O_2CH_2, 7,8-fur*	3′-Methoxypongapin	*Pongamia glabra* (Leg.), leaf	Malik *et al.* (1977)

† For structures see Fig. 4.1

oxygen, the first cipher always indicating the position of the O. In Table 4.7 isoprenylfuran rings are indicated by 'fur-ODmp', while the simple furan ring is marked as 'fur' only. The latter compounds again are included here for the sake of completeness.

The terms pyranoflavonoid and furanoflavonoid have come into use in recent years. If the pyran or furan ring jointly with the aromatic ring to which they are linked form a benzochromene or a furanochromene system, these flavonoids can also be called chromenoflavonoids. This term cannot be used for those in which annulation takes place from C-6′ to C-3.

Tables 4.6 and 4.7 listing these compounds may not be absolutely complete but it is hoped they will be useful as a basic entry into this group of substances. Structures of some typical furano- and pyrano-flavones and -flavonols are given in Fig. 4.1.

The occurrence of flavones with C_5, OH-C_5 and C_{10} side chains (12 compounds) is still almost completely restricted to the Moraceae. However, in addition to the one solitary earlier example of 8-prenyl-luteolin from *Xanthium* (Asteraceae), two more 8-prenylated flavones have now been found in *Tephrosia* (Leguminosae). Flavonols with these substituents (nine compounds) occur in members of the Berberidaceae, Leguminosae, Platanaceae, Rubiaceae and Rutaceae. Two of these compounds were accidentally overlooked in Gottlieb (1975). Isoprene substitution of this type is more abundant in the flavanone series and is also encountered among chalcones and isoflavonoids. The simple 8-prenyl compounds are common in several genera of the Leguminosae.

Among the pyranoflavones and pyranoflavonols, a considerable number of new reports have appeared recently; 11 of the 18 flavones and five of the seven flavonols are new. Except for the presence of three of these compounds in the Leguminosae, all flavones are found in Moraceae. Five pyranoflavonols are from the Leguminosae, one from the Asclepiadaceae and one from the Rutaceae.

Eight of the 13 furanoflavones and furanoflavonols listed in Table 4.7 are new compounds. They are almost completely restricted to the genera *Pongamia* (Leguminosae) and *Artocarpus* (Moracae), with one compound each from *Tephrosia* (Leguminosae) and from *Morus* (Moraceae) as exceptions.

The number of flavones with a 7-membered oxepine ring have increased from two or four. Cyclointegrin (see Fig. 4.1) is the first natural flavone with an oxocin ring system.

4.4.2 *Tephrosia* flavonoids

The genus *Tephrosia* (Leguminosae, Papilionoideae) is distinguished by the production of a special group of flavonoids: 7-oxygenated or 5/7-oxygenated compounds with unsubstituted B ring and extra substituents at C-8, derived from a prenyl unit. Most are flavones, three are flavanones and one is a chalcone. Waterman and Khalid (1980) recently surveyed all such flavonoids; one new one can be added to their list, the flavanone, 2,3-dihydrosemiglabrin, or purpurin

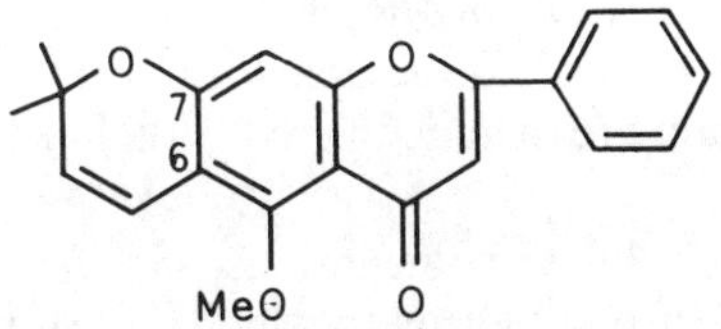

Praecansone B
(5-OMe, 7,6-ODmp)
(linear linkage)

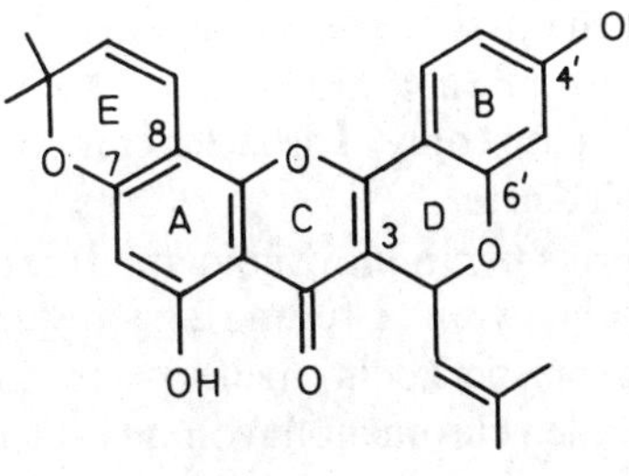

Cyclomorusin
(5,4′-OH, 7,8-ODmp, 6,3′-ODmp)
(angular linkage for AE)

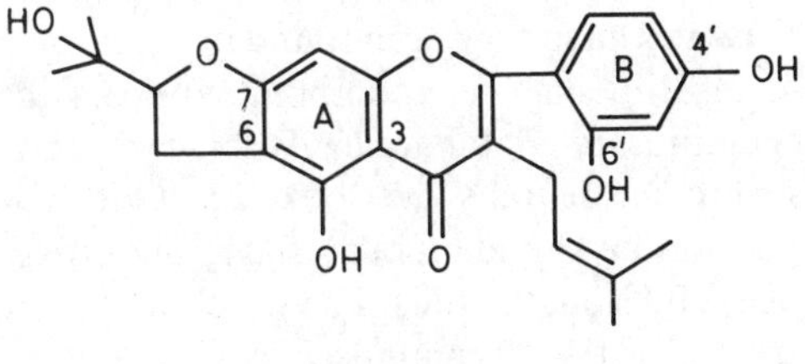

Mulberranol
(5,4′,6′-tri-OH, 3-C_5, 7,6-fur-ODmp-OH)

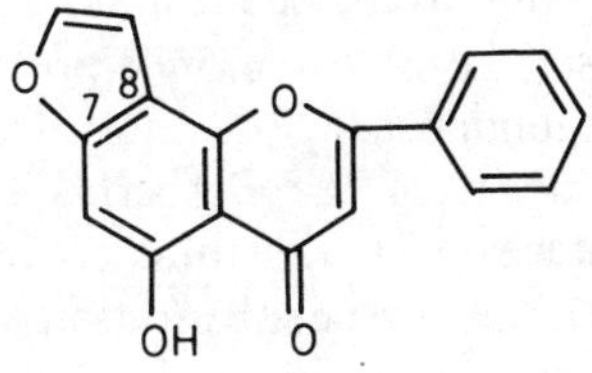

Pongaglabrol
(5-OH, 7,8-fur)

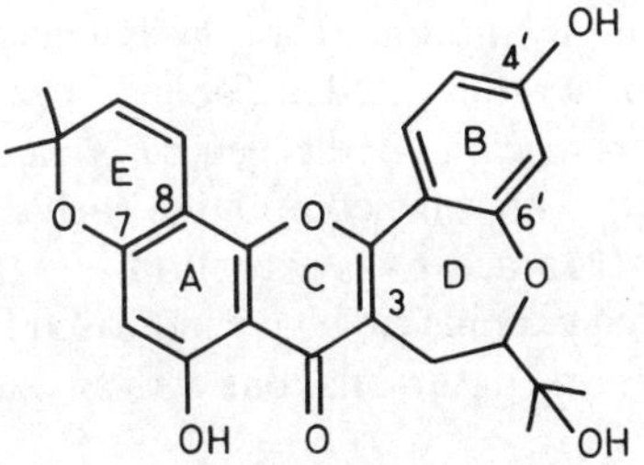

'Compound A'
(5,4′-di-OH, 7,8-ODmp, 6,3′-ox-OH)

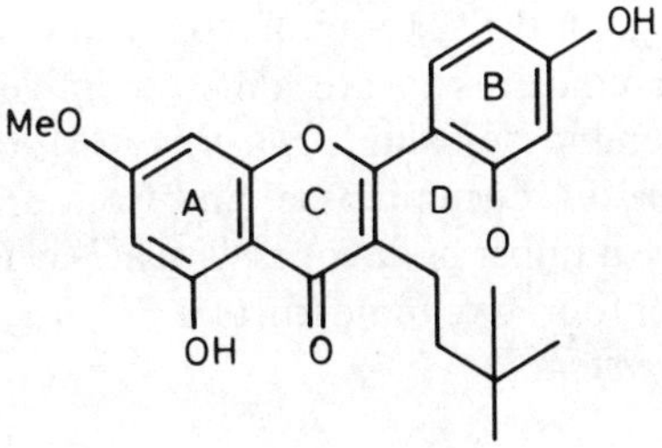

Cyclointegrin
(5,4′-OH, 7-OMe, 6′,3-oxecin)

Fig. 4.1 Structures of selected furano- and pyrano-flavones.

(R. K. Gupta *et al.*, 1980). The first reported representatives, tachrosin, stachyoidin, and tephrodin, were obtained from *Tephrosia polystachyoides*. Their structures were confirmed by synthesis (Antus *et al.*, 1977). In the

Table 4.8 Flavones with oxepine and oxocin rings

Substitution†	*Trival name*	*Plant sources*	*Reference*
5,4′-di-OH, 7,8-ODmp, 6′,3-ox-OH*	'Compound A'	*Morus alba* (Mor.) root bark	Nomura *et al.* 1976) Nomura *et al.* (1978a)
5,4′-di-OH, 7-OMe, 6′,3-ox.-OH	Oxyisocyclointegrin	*Artocarpus integer* (Mor.), heartwood	*Pendse et al.* (1976)
5,4′-di-OH, 7-OMe, 6-C_5, 6′,3-ox.-OH	Chaplashin	*Artocarpus*	
5,3′,4′-tri-OH, 6-C_5, 7,8-ODmp 6′,3-ox.	Isocycloheterophyllin	*Artocarpus*	
5,4′-di-OH, 7-OMe, 6′,3-oxocin*	Cyclointegrin	*Artocarpus integer* (Mor.), heartwood	*Pendse et al.* (1976)

† for structures see Fig. 4.1

meantime, these unusual flavonoids were also found in *T. apollinea* (apollinine, glabratephrine, pseudosemiglabrin, semiglabrin), *T. bracteolata* (the flavanone, obovatin-5-Me ether), *T. candida* (obovatin-5-Me ether), *T. multijuga* (multijugin, multijuginol), *T. obovata* (flavanones, obovatin and its 5-Me ether; the chalcone, obovatachalcone), *T. praecana* (obovatin-Me ether), *T. purpurea* (flavanone, purpurin) and *T. semiglabra* (glabratephrin, semiglabrin, semiglabrinol) (simple C-5 derivatives not cited here). From *T. polystachyoides* in addition polystachin has been reported (for references, see Waterman and Khalid, 1980). A total of 15 flavones of this type (including stereoisomers) are now known. Jain and Gupta (1978) suggest that these flavones are formed biogenetically from chrysin by prenylation at C-8, followed in many cases by a complex process of further substitution and cyclization. It may be mentioned in this context that two flavanones, lupinifolin and lupinifolinol, from *T. lupinifolia* (Smalberger *et al.*, 1974) are known which bear a $\gamma\gamma'$-dimethylallyl unit at C-8 (and an additional pyran ring), and tephrostachin and lanceolatin-A bear, respectively, *cis*- and *trans*-3-hydroxy-3-methyl-but-1-ene side chains (of prenyl origin) at C-8.

In Fig. 4.2, one complete structural formula is given at the top as an example; details of the stereochemistry are omitted. The rest of the substituents are presented in Fig. 4.2 according to Waterman and Khalid (1980).

4.4.3 *Pityrogramma* flavonoids

During extensive studies on the composition of frond exudates on gymnogrammoid ferns by Wollenweber and co-workers, several new constituents were isolated from the farinose indument of some *Pityrogramma* species. These constituents represent a novel structural type among complex flavonoids, in which a flavonoid moiety is linked with a dihydrophenylcoumarin (or

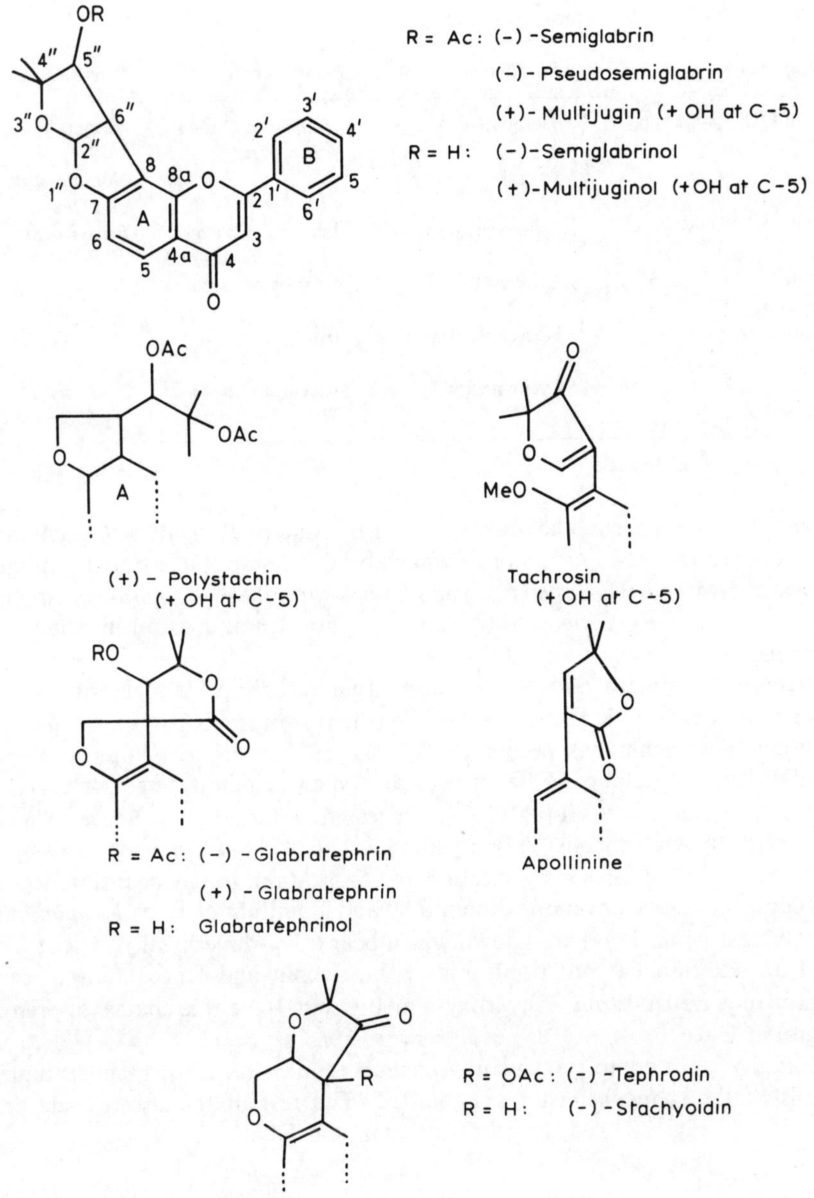

R = Ac: (–)-Semiglabrin
(–)-Pseudosemiglabrin
(+)-Multijugin (+ OH at C-5)
R = H: (–)-Semiglabrinol
(+)-Multijuginol (+OH at C-5)

(+) – Polystachin
(+ OH at C-5)

Tachrosin
(+ OH at C-5)

R = Ac: (–)-Glabratephrin
(+)-Glabratephrin
R = H: Glabratephrinol

Apollinine

R = OAc: (–)-Tephrodin
R = H: (–)-Stachyoidin

Fig. 4.2 Structures of *Tephrosia* flavonoids.

dihydroneoflavonoid) moiety via a common phloroglucinol ring. So far a dihydrochalcone (Wagner *et al.*, 1979) and three chalcones (Dietz *et al.*, 1980) have been reported as well as a flavone and a flavonol (Wagner *et al.*, 1979). The latter two compounds (Fig. 4.3) have the same substitution pattern and are

inseparable in any known chromatographic system. That they are both present is apparent from the spectral data. These compounds D-1 and D-2 contribute to the quasi genus-specific flavonoid pattern in *Pityrogramma* (Wollenweber and Dietz, 1980); they have also been labelled A and B_1/B_2, respectively, by Wagner *et al.* (1979). Closely related are two further flavones, in which no additional γ-pyrone ring is present; they bear dihydrocinnamoyl substituents. They were reported by Favre-Bonvin *et al.* (1980) as X-1 and X-2 (Fig. 4.3) and are found in some individual plants of *P. calomelanos* var. *aureoflava* only.

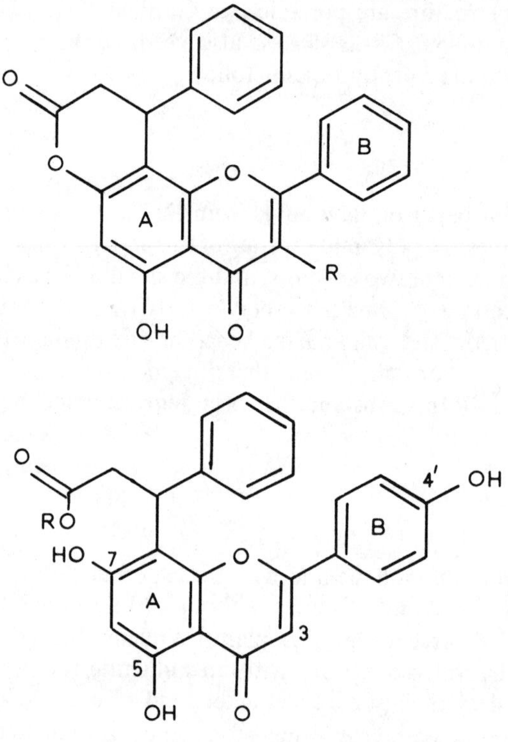

D-1 a : R = H

D-2 b : R = OH

X-1 : R = CH_3

X-2 : R = H

Fig. 4.3 Structures of *Pityrogramma* flavonoids.

4.5 REVISIONS AND PROBLEMATICAL STRUCTURES

Gottlieb (1975) included a section on revisions and problems. Some of the problems discussed there have been resolved in recent years; others still await solution. A few new topics can now be added to the list.

Uncertain structures

A variety of compounds have yet to be fully characterized. These include wharangin (Geissman, 1962, p. 430), a flavone from *Colutea arborescens*

(Geissman, 1962, p. 422), a flavone from *Acacia cavenia* (Geissman, 1962, p. 422), polycladin, atanasin (Flores and Herran, 1960), Primula F3A (possibly a gossypetin dimethyl ether), ptaeroxylol (3,5,7-trihydroxy-2′-methoxyflavone ?) and pratoletin from *Trifolium pratense*. Two unidentified flavones reported earlier from *Larrea* (Geissman, 1962, pp. 427 and 430) are probably, according to more recent papers, methyl ethers of quercetin (see Table 4.2). With regard to the structures of gardenins A to C, it is clear from more recent results that all substances from *Gardenia* have 7-*O*-methyl substitution and are flavones (see Table 4.1). The structure of sericetin from *Mundulea sericea* is still uncertain, in that both a linear or angular structure are possible (see Gottlieb, 1975). The earlier identification of galangin from *Citrus sinensis* also needs confirmation. More detailed reports on other structural problems follow.

Gossypium flavonol.

Struck and Kirk (1970), in their paper on flavonoids from *Gossypium hirsutum* flower petals, reported R_F values on polyamide for a Substance III. It has the same R_F as ombuin and rhamnazin in two solvents, and the small difference in the third is negligible. Quercetin-3′,4′-dimethyl ether in $CHCl_3$/EtOH/MEK would appear between rhamnetin and rhamnazin; hence in our opinion this possibility can be excluded. The substance must therefore be quercetin-7,4′-dimethyl ether. We do not know if this substance has been reinvestigated in the meantime.

6-Hydroxykaempferol

Harborne (1967, p. 164) failed to find a flavonoid with the expected chromatographic properties from *Galega officinalis* flowers, although Maksyutina and Litvinenko (1964) assumed its presence in this plant. Wollenweber (1974, unpublished) compared the hydrolysed extract with an authentic sample of 6-hydroxykaempferol (from demethylation of betuletol) and also could not detect it, but found only kaempferol, quercetin and isorhamnetin. 6-Hydroxykaempferol therefore appears to be absent from *Galega*. Hence the trivial name 'galetin' should not be used for this flavonol. Unfortunately, its use has already become established, see e.g. Dillon and Mabry (1977). In fact, 6-hydroxykaempferol has now been found once in nature, as a 7-*O*-glucoside in *Tetragonotheca texana* (Bacon *et al.*, 1978).

Tambulin

Chatterjee *et al.* (1976) confirmed this compound as being herbacetin-7,8,4′-trimethyl ether as originally proposed by Bose and Bose (1939), on grounds of spectroscopic investigations, chemical reactions and synthesis.

3,5,8,3′,4′-Pentahydroxyflavone

Wollenweber (1974, unpublished) in a search for galetin in *Galega* did not detect a compound which might be this flavone as had been earlier suggested by Lapinina (1965).

Apigenin 5,7-dimethyl ether

The assumed presence of this flavone as a 4′-glucoside in *Saccharum officinarum* (Dubey and Misra, 1974) still needs to be confirmed.

Ladanetin

Ladanetin and ladanein were reported as constituents of *Galeopsis ladanum* (Gritsenko and Litvinenko, 1969) and ascribed the structures: scutellarein 7-methyl ether and scutellarein 7, 4′-dimethyl ether. By synthesis of the 7-methyl ether (sorbifolin), Farkas *et al.* (1971) showed that ladanetin was not identical with this product. Although Venkataraman (1975) drew attention to this fact, ladanetin and ladanein to our knowledge still await proper identification.

Skullcapflavone I

Takido *et al.* (1975) isolated two flavones, skullcapflavone I and II, from roots of *Scutellaria baicalensis*. The structures 5,2′-dihydroxy-6,8-dimethoxyflavone and 5,2′-dihydroxy-6,7,8,6′-tetramethoxyflavone were ascribed to these compounds. Later Takido *et al.* (1979) revised the first structure because of spectral data and colour reactions and proposed that skullcapflavone I was 5,2′-dihydroxy-7,8-dimethoxyflavone. This was confirmed by synthesis of 5,2′-dihydroxy-6,8-dimethoxyflavone, 5,2′-dihydroxy-7,8-dimethoxyflavone and 5,2′-dihydroxy-6,7-dimethoxyflavone (Iinuma and Matsuura, 1979). For skullcapflavone II the earlier proposed structure has now been confirmed by synthesis (Iinuma and Matsuura, 1979).

Alnus flavone

Another flavone has been described on the basis of the UV spectra to have ‘in all likelihood’ the 5,6,8-trioxygenation pattern, which is rather unlikely biosynthetically. The structure proposed was 5,4′-dihydroxy-6,8-dimethoxyflavone, from *Alnus glutinosa* fruits (Karczewska and Krolikowska, 1974). After synthesis of the above structure, Iinuma and Matsuura (1980) found that this cannot be correct for the *Alnus* flavone. By comparison of melting points, they suggested that the natural product could be 5,4′-dihydroxy-6,7-dimethoxyflavone (cirsimaritin). Wollenweber *et al.* (1971) reported the isomeric 6,4′-dimethyl ether of scutellarein (pectolinarigenin) as a constituent of the lipophilic material on buds, male flowers (catkins) and fruits of black alder. A reinvestig-

ation (involving a direct comparison of freshly prepared extracts with scutellarein 6,7- and 6,4′-dimethyl ethers and a synthetic sample of 5,4′-dihydroxy-6,8-dimethoxyflavone) showed that the only flavone of this type present is pectolinarigenin. No trace of 5,4′-dihydroxy-6,8-dimethoxyflavone could be detected (Wollenweber unpublished).

5,4′-Dihydroxy-7,8,3′-trimethoxyflavone

Le Quesne *et al.* (1976) identified five methylated flavones in *Lychnophora affinis*, two of which were new compounds. One of these, namely the 7,8,3′-trimethyl ether of hypolaetin, was synthesized later by Bhardwaj *et al.* (1978c). However, the properties of the synthetic product were different from the natural sample. Hence the structure of the natural flavone clearly needs revision.

Luiselozindin

A flavone called luiselozindin, isolated from the essential oil of *Gymnosperma glutinosa*, was ascribed the structure of 5,7-dihydroxy-6,8,3′,4′,6′-pentamethoxyflavone (Dominguez and Torre, 1974). Chhabra *et al.* (1977b) synthesized this compound and found it to be different from the natural product. Hence the constitution of luiselozindin still needs revision.

Mikanin

In 'The Flavonoids' the name mikanin has erroneously been used for the scutellarein trimethyl ether, salvigenin (p. 272).

Tomentin

Rodriguez *et al.* (1972a) ascribed the structure of quercetagetin 3,3′-dimethyl ether to a substance 'tomentin', isolated from *Parthenium tomentosum*. Shen *et al.* (1976) then reported that the earlier statement was incorrect in that tomentin was in fact quercetagetin 3,7-dimethyl ether. This was confirmed by direct comparison with the synthetic product (Wagner *et al.*, 1976, 1977b). Unfortunately, in a review by Rodriguez (1977) on *Parthenium* constituents substance no. *38* is described as the 3,3′-dimethyl ether in the Figure of structural formulae, but as the 3,7-dimethyl ether in the Table of distribution. The latter is, clearly, the correct structure and there is no evidence now (Mabry, 1981) that the 3, 3′-dimethyl ether occurs in *Parthenium*.

Artemisia compound

The structure of quercetagetin 6,7,3′,4′-tetramethyl ether has been ascribed to a flavonol isolated from *Artemisia annua* by Djermanovic *et al.* (1975). Bhardwaj

et al. (1977), however, demonstrated that a synthetic sample of this compound had a melting point and spectral data markedly different from those reported for the natural product. The *Artemisia* compound must, therefore, have a different structure; it may be an isomeric tetramethyl ether of quercetagetin, or perhaps of gossypetin.

Ranupenin

Urschler (1968) reported a monomethyl ether of gossypetin, called ranupenin, as a glycoside in flowers of *Ranunculus repens* and ascribed it the structure of gossypetin 8-methyl ether. Wagner *et al.* (1977c) after synthesis of three monomethyl ethers of gossypetin were able to show that ranupenin is in fact gossypetin 7-methyl ether. Unfortunately later authors (Harborne *et al.*, 1978) caused trouble again by confusing these two flavonols. The plant sources of the two isomeric 7- and 8-methyl ethers of gossypetin are shown correctly in Table 4.2.

Unonia formylflavonoids

Three formylflavonoids, isolated from *Unonia lawii*, have been ascribed the structures of 5,7-dihydroxy-6-methyl-8-aldehydoflavone (unonal), 5,7-dihydroxy-8-methyl-6-aldehydoflavone (isounonal), and 5-hydroxy-7-methoxy-6-methyl-8-aldehydoflavone (unonal 7-methyl ether) by Joshi and Gawad (1976). According to B. S. Joshi (1980) they now have been revised, i.e. the positions of methyl and aldehyde groups must be exchanged. This follows from a comparison with the *C*-methylfavanones strobopinin and cryptostrobin, the structures of which have been determined by ^{13}C-NMR spectroscopy and X-ray crystal analysis of bromocryptostrobin (Joshi and Cannon, unpublished).

4.6 METHODS FOR SEPARATION AND IDENTIFICATION

4.6.1 Chromatography on polyamide

Recent advances in chromatographic techniques for flavonoids (including HPLC) are generally discussed in Chapter 1. Only certain aspects of particular importance in separating flavones and flavonols will be treated here. TLC on polyamide, for example, is especially valuable for distinguishing the many and various methyl ethers of both flavones and flavonols. In his chapter on isolation techniques in 'The Flavonoids', Markham (1975) referred to Egger's publications on polyamide TLC, especially his contribution to Stahl's textbook on TLC (Egger, 1967). Solvents for TLC of the aglycones first were based on chloroform and are cited in the literature sometimes as 'Egger's solvents'. Chloroform has been successively replaced by benzene (Wollenweber, 1970) and then by toulene, which is somewhat less toxic than benzene, since in contrast to

benzene, it can be metabolized. The 'standard solvents' currently in use are listed in Table 4.9 in order of increasing polarity. Solvents I and II (Table 4.9) are used for non-polar flavonoids, especially for highly methylated compounds (and will separate, for example, galangin 3,7-dimethyl ether from kaempferol 3,7,4'-trimethyl ether). They also are used for a first run of extracts containing much lipid material, which impedes or even prevents development with the other solvents. Solvent III is the one most often used for routine work. It is applicable to a wide range of aglycones, from chrysin 7-methyl ether to apigenin, from kaempferol 3,7,4'-trimethyl ether to kaempferol itself. It therefore is always used on unknown extracts containing lipophilic flavonoids. More polar compounds should be chromatographed in solvents IV to VI. Solvent V is used for the most common flavonols. Very polar compounds, penta- and hexa-hydorxyflavones and their monomethyl ethers are better separated in VI. VII is a useful second solvent for two-dimensional TLC. It can be helpful in the rare cases when other solvents give little or no separation, e.g. for the 3'- and 4'-methyl ethers of quercetin.

Table 4.9 Solvents for polyamide TLC

	Solvents	*Composition*
(I)	Toluene/petrol (b.p. 100–140°C)/MeCOEt/MeOH	30:90: 2:1.5
(II)	Toluene/petrol (b.p. 100–140°C)/MeCOEt/MeOH	30:60: 5:5
(III)	Toluene/petrol (b.p. 100–140°C)/MeCOEt/MeOH	60:30:10:5 (60:26: 7:7)
(IV)	Toluene/dioxan/MeOH	80:10:10
(V)	Toluene/MeCOEt/MeOH	60:25:15 (60:26:14)
(VI)	Toluene/MeOH/MeCOEt/Acetylacetone (benzene/MeCOEt/MeOH)	40:30:20:10 (40:30:30)
(VII)	HOAc/dioxan/DMF/H_2O	10:30:15:15

Previous composition with benzene is given in parentheses where the quantities have been altered.

Some of these solvents have been adopted by other laboratories (e.g. Charrière and Tissut, 1973: IV, V; Mues *et al.*, 1979: III, VI; Reichling *et al.*, 1979: III, V), sometimes with slight variations, or with benzene instead of toluene. Similar solvents have been used as eluents in column chromatography on polyamide (e.g. Sakakibara *et al.*, 1975). Usually the separations are clear, even when spots overlap. Nearly all the methyl derivatives of quercetin (except those with 5-*O*-methylation) can be distinguished on two one-dimensional chromatograms in solvents III and V (see Jay *et al.*, 1975; Wollenweber, 1975c).

For detection of flavonoids, colours in UV_{366} are much more valuable on polyamide than on paper and allow conclusions as to the substitution at C-3, C-5, presence of additional substituents at C-6 and C-8. Also the various colours produced after spraying with 'Naturstoffreagenz A' are more easily dis-

criminated on polyamide. This reagent, introduced by Neu (1956), because of its sensitivity and the variety of colours developed (see also Somaroo *et al.*, 1973) is still unsurpassed for the detection of flavonoids. Additional information comes from UV-induced changes in visual colours in the case of methyl derivatives of apigenin, luteolin, and several other compounds. The TLC plates should therefore be re-examined in daylight some hours after spraying with 'Naturstoffreagenz A' and evaluation in UV_{366}.

TLC plates are prepared by spreading a slurry of powdered polyamide (Polyamid DC-11, polyaminoundecanic acid; Macherey-Nagel, Düren) with 15% cellulose powder in MeOH on thoroughly cleaned glass plates (Egger, 1967). Thickness of the layer is 0.2 or 0.3 mm (85 g of polyamide with 15 g of cellulose yield ca. 14 plates 20 cm × 20 cm). Such 'home-made' plates give much better separations and show better UV colours than the pre-coated plates or sheets available commercially. 'Naturstoffreagenz A' (β-aminodiethyl ester of diphenylboric acid) is used in 0.5% solution in MeOH.

Preparative separations of flavones and flavonols can readily be achieved on thicker layers (0.5 mm) of polyamide, using the solvents listed in Table 4.10. Markham (1975) has noted that removal of compounds from polyamide may on occasions be difficult. In our experience, it is difficult to get rid of low-molecular-weight polyamide during elution of flavone adsorbents. This is a nuisance especially when MS and NMR studies are to follow (see next sections). To overcome this problem, plates should be run in MeOH prior to use, in order to bring most of the soluble material to the front. Then, the eluate should be thoroughly dried in a small conical bulb; the flavonoid is then taken up in benzene (or methyl ethyl ketone (MEK), but not MeOH) and pipetted off, leaving the polyamide sticking to the glass wall. If these steps are not successful, the sample should undergo a final purification on silica (TLC) or on Sephadex LH-20 (see p. 243).

Finally, some hints on the use of polyamide for column chromatographic separations may be appropriate. One of the commercially available products is Polyamid SC-6 (ε-aminopolycaprolactam; particle size 0.05–0.16 mm; Macherey-Nagel). Small particles (and especially soluble low-molecular-weight fractions) are removed by washing with MeOH. The material is stirred with excess MeOH. After standing, the supernatant is sucked off and the material dried at 50–60°C and then kept ready for use in toluene (Jay, 1978). Elution of columns is either with solvent mixtures similar to those recommended for TLC (Sakikabara *et al.*, 1975; Mues *et al.*, 1979) or with toluene (benzene) and increasing quantities of MEK and MeOH (Egger *et al.*, 1970). Polyamide can be used many times! The dried powder is suspended in water and a few ml of NH_4OH (30%) and H_2O_2 (30%) are added. After standing overnight the water is renewed several times until the supernatant remains colourless. Then the polyamide is sucked dry, washed with acetone, and ether, and after evaporation of the latter dried at 50–60°C. No loss of capacity has ever been observed (Egger, 1968; Dietz, 1980).

Table 4.10 UV spectral characteristics of flavones and flavonols

Reagent	*Substitution pattern*	*Shift of Band I (nm)*	*Remarks*
MeOH	6-OH	−8–12	Flavonols incl. 3-OMe-flavonols
		−9	Flavones
	6-OMe	−	
	8-OH	+13–16	Flavonols; +peak at ±330 nm
		−7	Flavones
$+AlCl_3$	3-OH	+60	Complex stable with H^+
	5-OH	+35–55	Complex stable with H^+
	3,5-di-OH	+50–60	Complex stable with H^+
	5,6-di-OH	+20	
	5-OH, 3,6-di-OMe	+20	
$+AlCl_3$ +HCL	o-di-OH at ring B	−30–40	Relative to spectrum with $AlCl_3$
	tri-OH at ring B	−20	Relative to spectrum with $AlCl_3$
	5-OH, 6-OH	+25–30	Relative to MeOH-spectrum; for flavones and 3-O-subst. flavonols.
	5-OH, 6-OMe	+20	
	5-OH, 8-OMe	+55–57	
+NaOMe (NaOH)	4′-OH	+40–60	Increase of intensity (or at least no decrease)
	3-OH	+50–60	Decrease of intensity
	3,4′-di-OH		Slow degeneration
	3,3′,4′-tri-OH		Rapid degeneration
	3,3′,4′,5′-tetra-OH		Rapid degeneration
	7-OH		Small additional peak or shoulder at ±330 nm
+NaOAc	7-OH	+5–20*	Minor or no shift when CH_3 or OCH_3 at C-6 and/or C-8
	5,6,7-tri-OH		Compound alkali-sensitive
	5,7,8-tri-OH		Compound alkali-sensitive
	3,3′,4′-tri-OH		Rapid decomposition
	3,4′-OH, 3′-OMe		Rapid decomposition
+NaOAc $+H_3BO_3$	o-di-OH	+12–30	at ring B
		+5–10	at 6.7 or 7.8

* This is a shift in band II, not band I.

4.6.2 Spectroscopic methods

(a) UV spectroscopy

Spectra in methanol

In general, flavones exhibit Band I between 304 and 350 nm, flavonols absorb between 352 and 385 nm. For flavonols with *O*-substitution at C-3 the general shapes of the curve as well as the ranges of Band I (328–357 nm) approach those of flavones. This is as well known as the observation that increasing oxygenation

of ring B results in a bathochromic shift of Band I. Band II appears as one peak (at about 270 nm) in compounds with monosubstituted B ring, but as two peaks or one peak (at about 258 nm) plus a shoulder (at about 272 nm) when a di- or tri-*O*-substituted B ring is present (Mabry *et al.*, 1970; Gottlieb, 1975; Hasan, 1976). The flavonoids thus have been grouped according to their B-ring substitution and UV spectra into kaempferol-types, quercetin-types etc. (Jay *et al.*, 1975). The latter authors present UV data for 23 flavones and 32 flavonols not given in Mabry *et al.* (1970). Averett (1977) has presented a table of absorption maxima for 252 flavonoids of different classes, mostly glycosides, but also a number of *O*-methylated aglycones. To this basic knowledge about the range of maxima and shape of UV curves, some additional findings may be added.

6-O-substitution

Introduction of an OH group at C-6 produces a hypsochromic shift on Band I of 8 nm in flavonols and up to 12 nm in 3-*O*-methylflavonols, but ther is no effect in the flavone series. Methylation of the OH at C-6 cancels this effect in flavonols, except if there is methylation at C-3. In such 3-methyl ethers, methylation at C-6 causes a hypsochromic effect of 9 nm on Band I. Thus 3-*O*-methylated flavonols with *O*-substitution at C-6 do not follow the usual spectral rules (Hasan, 1976).

8-O-substitution

Introduction of a OH group at C-8 in flavonols produces a bathochromic shift of 13–16 nm on Band I and an additional peak at 330 nm (Hasan, 1976). The absorption curves of these compounds thus are quite characteristic (cf. Wollenweber *et al.*, 1978; Wollenweber and Wiermann, 1979). The relatively high wavelength of the absorption maxima also means that these compounds are yellow in colour (Harborne, 1969; Wollenweber and Wiermann, 1979). In flavones, on the other hand, a hyposochromic effect of 7 nm is observed in some cases on insertion of an 8-OH (Combier *et al.*, 1975). Methylation of the OH at C-8 causes minor shifts in the same direction.

Aluminium chloride/hydrochloric acid

Mears and Mabry in 1972 reported a procedure for the UV detection of OH- and -OMe groups at C-6 in flavones and 3-*O*-substituted flavonols. They emphasized that in such compounds the bathochromic shift of Band I produced by addition of $AlCl_3$/HCl relative to the MeOH spectrum is only about 20 nm, whereas otherwise the shift is about 45 nm (or upto 57–62 nm; Hasan, 1976). This allows the discrimination of 6-*O*-substituted from 8-*O*-substituted compounds. Similarly *C*-methylation at C-6 may cause a weak $AlCl_3$ reaction of the 5-OH due to steric hindrance (Wollenweber *et al.*, 1980; Dietz *et al.*, 1981). Later, Sakakibara and Mabry (1977) after studying more compounds than Mears and

Mabry (1972) showed that the $AlCl_3/HCl$ shift on Band I of +25–30 nm, relative to the MeOH spectrum, is indicative for 5-OH flavones and 3-*O*-substituted flavonols with free OH at C-6, whereas the earlier observed shift of 20 nm is only valid for compounds *O*-methylated at C-6. The same authors (Sakakibara and Mabry, 1978) found that $AlCl_3/HCl$ produced a large shift of 55–75 nm of Band I (relative to MeOH spectrum) for flavones and 3-*O*-substituted flavonols with free OH at C-8. It may be mentioned that recent results of B. Voirin (unpublished, cited by Rabesa, 1980) will possibly require some minor corrections of these rules. Rabesa (1980) recommends that the $AlCl_3$ solution should be prepared fresh at least every month.

Sodium methoxide

Degeneration of the UV absorption peaks of flavonols on addition of NaOMe does not necessarily indicate a 3,4′-di-OH system. It can also be due to 3′,4′,5′-tri-OH substitution, e.g. in tricetin derivatives (Mues, 1975). In the flavone series, this is the only relatively reliable evidence that a 3′,4′,5′-tri-OH system is present. Occurrence of a small peak or a shoulder at +330 nm on addition of NaOMe is indicative for a free OH at C-7, or at least supports the relevant evidence from the NaOAc reaction (see below). The French workers use NaOH (1M) instead of NaOMe (Rabesa, 1980).

Sodium Acetate

This reagent for detection of 7-OH groups usually produces bathochromic shifts of band II of 8–20 nm (Mabry *et al.*, 1970). Rabesa (1980) points out that in flavones with OMe- or CH_3- groups at C-6 and/or C-8 the effect can be less than 5 nm and even zero. This has been observed also by Dietz *et al.* (1981). The adjacent substituents may reduce or even prevent the acidity of the 7-OH. Rabesa (1980) emphasizes that fused NaOAc, kept permanently in a drying oven at 120° C, should always be used.

On the basis of the publications cited above, Dietz (1980) has compiled a table of UV characteristics which is reproduced here (Table 4.10) in a slightly altered version. It should be stressed that the observations discussed above as well as those summarized in this Table must not be regarded as strict rules. They should be considered as useful hints for structural assignments, which should then be confirmed by other procedures.

(b) Mass spectroscopy

Mass spectroscopy was treated *in extenso* by Mabry and Markham (1975). The general rules for fragmentation have been confirmed repeatedly in more recent measurements. It may be mentioned that some authors still prefer the abbreviations for the fragments as introduced by Audier (1966) and adapted by Kingston (1971) to those used by Mabry and Markham. The problems of

eliminating impurities in samples before mass spectral analysis are legion. In the case of flavones and flavonols, the polyamide dissolved from the adsorbent when performing preparative TLC can cause very annoying artefact peaks. It can be eliminated by a further preparative-scale TLC on silica, preferably with a concentration zone (Wollenweber and Wiermann, 1979; Wollenweber *et al.*, 1980). A passage of the sample to be analysed via a small column of Sephadex LH-20 (eluted with MeOH) or Polyamide SC.6 (0.5 cm × 5 cm, elution with toluene, then increasing quantities of MeOH) can also be recommended (Rabesa, 1980).

Wollenweber and Dietz (1979) have prepared a valuable time-saving Table of MS parent ions. This table can be used either to find out at a glance if the molecular ion (M^+) of a substance to which one ascribed a certain structure by other methods fits the molecular formula, or it can help to determine the number of hydroxyl and/or methoxyl groups present in a flavonoid of known molecular weight. It must be kept in mind that the much less frequent *C*-methyl substitution may cause errors (one OH and one $CH_3 = OCH_3$). Differentiation between flavonoids of the same molecular ion but belonging to different classes is possible by evaluation of TLC behaviour and UV spectra and by determining the number of hydroxyl groups originally present after mass spectroscopy of the acetate (or TMS derivative).

As in UV spectroscopy, discrimination between substitution at C-6 and C-8 is a special problem which has received special attention. Kingston (1971) earlier found intense $(M-CH_3)^+$ and $(M-CH_3CO)^+$ fragments to be significant for 6- and 8-methoxyflavonoids. Nielsen and Møller (1970) also had indicated that in 6-methoxyquercetin, M^+ appears as base peak, whereas in 8-methoxyquercetin the base peak is at $M-15$. By contrast, it may be noted that in case of PM (permethyl) and PDM (perdenteromethyl) derivatives of 6-methoxyflavones and 6-methoxyflavonols, the $M-15$ peak is the base peak (Mues *et al.*, 1979). More recently Goudard *et al.* (1978) have confirmed that, in underivatized 5,7-dihydroxyflavones with methoxyl at C-6, in general M^+ is more important than $M-15$ and mostly produces the base peak, whereas in those with methoxyl at C-8 the base peak is due to $M-15$. In addition, in 6-methoxy compounds the relative intensity of $M-18$ is greater than 10%, whilst in 8-methoxy compounds it is lower. 3-*O*-Methylflavonols with methoxyl at C-6 differ from the corresponding flavonols with a free 3-hydroxyl by the relative intensity of $M-1$ being greater than 10%. According to Mues (1975) the limit is not 10% but 30%. 6-*O*-Methylflavonols exhibit a less important peak at $M-15$ than the corresponding flavonols (Goudard *et al.*, 1978). $M-13$ signals of less than 30% intensity might be indicative of underivatized 8-methoxy and/or 3-methoxyflavonols, whereas $M-43$ signals of more than 30% are indicative for underivatized 6-methoxyflavones or 6-methoxyflavonols which have a free 3-hydroxyl. These relationships are collected together in Table 4.11, from Goudard *et al.* (1978), modified by Dietz (1980). Goudard and co-workers (1979) have also reported that in underivatized 5-hydroxy-6,7-dimethoxyflavones the relative intensity of M^+ is higher (usually the base peak) than $M-15$, whereas

Table 4.11 Mass spectral-fragmentation of flavones and flavonols

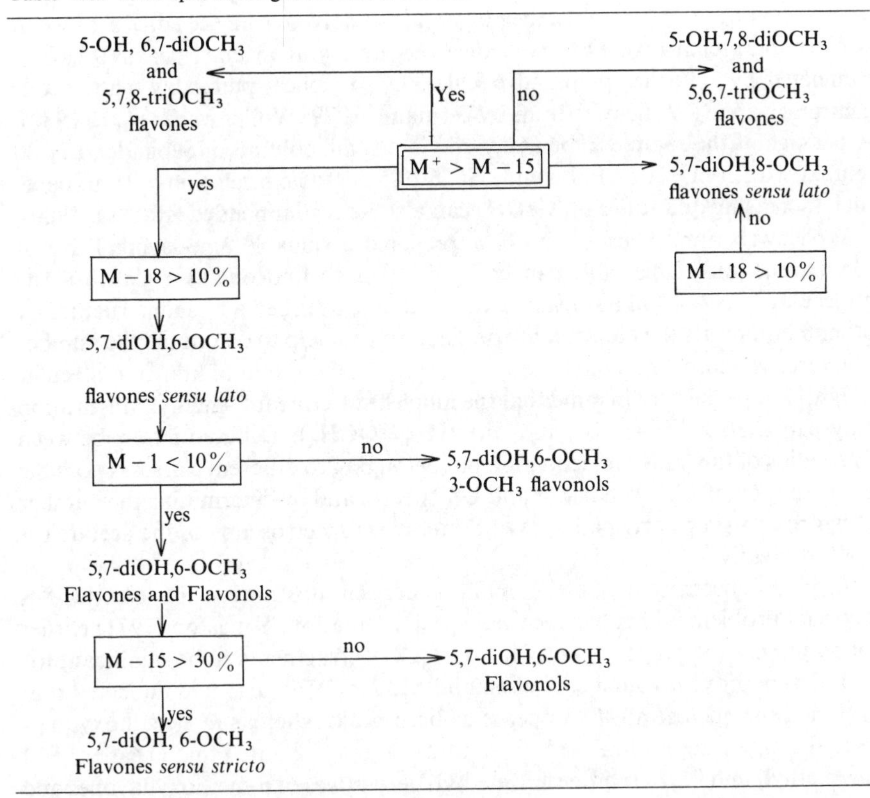

the latter fragment occurs as the base peak in 5-hydroxy-7,8-dimethoxyflavones (flavonols). For 5,6,7-trimethoxy and for 5,7,8-trimethoxy derivatives, however, the intensity relationships are reversed. It may be mentioned finally that Rama Rao *et al.* (1972) have published a paper dealing with mass spectroscopy of flavones with *C*-isoprenyl substituents, and draw attention to the significant structural information obtainable by this method.

(c) ^{1}H-NMR spectroscopy

^{1}H nuclear magnetic resonance (NMR) spectroscopy of flavones and flavonols has been treated in detail by Markham and Mabry (1975), and Chapter 2 in this book is dedicated to ^{13}C-NMR. To my knowledge, no important new results on ^{1}H-NMR spectroscopy have been published since 'The Flavonoids' appeared. Therefore I only want to mention one practical hint. Recovering the sample after measurement is often important, since a considerable amount of pure substance is then available for other purposes. Although DMSO-d_6 is the most generally useful solvent for underivatized flavonoids, it is a difficult solvent to get rid of

(Markham and Mabry, 1975). Column chromatography on polyamide, preparative TLC on silica, and preparative PC have been used for this purpose (Exner, 1978). The most convenient method for recovering aglycones, however, is passage of the sample solution over a small column (e.g. 15 cm × 3 cm; Dietz, 1980) of Sephadex LH-20. On elution with MeOH the DMSO will leave the column with the first portion (Hiermann *et al.*, 1978).

ACKNOWLEDGEMENTS

The author is very grateful to P. G. Waterman (Glasgow, UK), R. Mues (Saarbrücken, GFR) and J. Favre-Bonvin, M. Jay and B. Voirin (Lyon, France) for reviewing individual sections. The help of M. Iinuma (Gifu, Japan) and B. S. Joshi (Bombay, India) is greatly acknowledged.

REFERENCES

Adesogan, E. K. and Okunade, A. L. (1979), *Phytochemistry* **18**, 1863.
Adinarayana, D. and Gunasekar, D. (1979), *Indian J. Chem.* **18B**, 552.
Adinarayana, D., Gunasekar, D., Seligmann, O. and Wagner, H. (1980a), *Phytochemistry* **19**, 480.
Adinarayana, D., Gunasekar, D., Seligmann, O. and Wagner, H. (1980b), *Phytochemistry* **19**, 483.
Ahluwalia, V. K., Malik, B. K. and Seshadri, T. R. (1976), *Indian J. Chem.* **14B**, 592.
Ahmad, S. A., Siddiqui, S. A. and Zaman, A. (1974), *Indian J. Chem.* **12**, 1327.
Ali, E., Bagchi, D. and Pakrashi, S. C. (1979), *Phytochemistry* **18**, 356.
Antus, S., Farkas, L., Nógradi, M. and Boross, F. (1977), *J. Chem. Soc., Perkin I* **1977**, 948.
Arisawa, M. and Morita, N. (1976a), *Chem. Pharm. Bull. Jpn.* **24**, 815.
Arisawa, M., Kizu, H. and Morita, N. (1976b), *Chem. Pharm. Bull.* **24**, 1609.
Asen, S. and Horowitz, R. M. (1977), *Phytochemistry* **16**, 147.
Ashihara, Y., Nagata, Y. and Kurosawa, K. (1977), *Bull. Chem. Soc. Japan* **50**, 3298.
Audier, H. (1966), *Bull. Soc. Chim. Fr.* **1966**, 2892.
Averett, J. E. (1977), *Phytochem. Bull.* **10**, 10.
Ayengar, K. N. M., Sastry, B. R. and Rangaswamy, S. (1973), *Indian J. Chem.* **11B**, 85.
Bacon, J. D., Urbatsch, L. E., Bragg, L. H., Mabry, T. J., Neumann, P. and Jackson, D. W. (1978), *Phytochemistry* **17**, 1939.
Bandoni, A. L., Medina, J. E., Rondina, R. V. D. and Coussio, J. D. (1978), *Planta Med.* **34**, 328.
Barua, R. N., Sharma, R. P., Thyagarajan, G. and Herz, W. (1978), *Phytochemistry* **17**, 1807.
Baruah, N. C., Sharma, R. P., G. Thyagarajan, Herz, W. and Govindan, S. V. (1979), *Phytochemistry* **18**, 2003.
Becchi, M. and Carrier, M. (1980), *Planta Med.* **38**, 267.
Bennett, J. P., Gomperts, B. D. and Wollenweber, E. (1981), *Arzneim. Forsch./Drug Res.*, **31**, 433.
Bézanger-Beauquesne, L., Debray, M., Pinkas, M. and Trotin, M. F. (1975), *C. R. Hebd. Seances Acad. Sci. Ser. D* **281**, 2025.
Bhardwaj, D. K., Jain, S. C. and Sharma, G. C. (1977), *Indian J. Chem.* **15B**, 860.
Bhardwaj, D. K., Jain, S. C., Sharma, G. C. and Mehta, C. K. (1978a), *Indian J. Chem.* **16B**, 1133.
Bhardwaj, D. K., Jain, S. C., Sharma, G. C. and Singh, R. (1978b), *Indian J. Chem.* **16B**, 339.
Bhardwaj, D. K., Gupta, A. K., Jain, R. K. and Sharma, G. C. (1978c), *Curr. Sci.* **47**, 424.
Bhardwaj, D. K., Bisht, M. S., Mehta, C. K. and Sharma, G. C. (1979), *Phytochemistry* **18**, 355.
Bhardwaj, D. K., Bisht, M. S. and Mehta, C. K. (1980a), *Phytochemistry* **19**, 2040.

Bhardwaj, D. K., Bisht, M. S., Jain, R. K. and Sharma, G. C. (1980b), *Phytochemistry* **19**, 1269.
Bhardwaj, D. K., Bisht, M. S., Uain, S. C., Mahta, C. K. and Sharma, G. C. (1980c) *Phytochemistry* **19**, 713.
Bierner, M. W. (1973) *Biochem. Syst.* **1**, 55.
Bierner, M. W. (1978) *Biochem. Syst. Ecol.* **6**, 293.
Biftu, T. and Stevenson, R. (1978), *J. Chem. Soc., Perkin I*, **1978**, 360.
Blasdale, W. C. (1945), *J. Am. Chem. Soc.* **67**, 491.
Bleier, W. and Chirikdjian, J. J. (1972), *Planta Med.* **22**, 145.
Bohlmann, F. and Abraham, W.-R. (1979), *Phytochemistry* **18**, 889.
Bohlmann, F. and Fritz, U. (1979), *Phytochemistry* **18**, 1080.
Bohlmann, F., Mahanta, P. K. and Zdero, C. (1978), *Phytochemistry* **17**, 1935.
Bohlmann, F., Zdero, C. and Ziesche, J. (1979), *Phytochemistry* **18**, 1375.
Bohm, B. A., Collins, F. W. and Bose, R. (1977), *Phytochemistry* **16**, 1205.
Bose, P. K. and Bose, J. (1939), *J. Indian Chem. Soc.* **16**, 183.
Bouillant, M. L., Redolfi, P., Cantisani, A. and Chopin, J. (1978), *Phytochemistry* **17**, 2138.
Bragg, L. H., Bacon, J. D., McMillan, C. and Mabry, T. J. (1978), *Biochem. Syst. Ecol.* **6**, 113.
Braz-Filho, R. and Gottlieb, O. R. (1971), *Phytochemistry* **10**, 2433.
Braz-Filho, R., Gottlieb, O. R., Vieira Pinho, S. L., Queiroz Monte, F. J. and Da Rocha, A. I. (1973), *Phytochemistry* **12**, 1184.
Brieskorn, C. H. and Biechele, W. (1969), *Tetrahedron Lett.* **1969**, 2603.
Brieskorn, C. H. and Biechele, W. (1971), *Arch. Pharm.* **304**, 5577.
Brieskorn, C. H. and Riedel, W. (1977), *Planta Med.* **31**, 308.
Brown, D., Asplund, R. O. and McMahon, V. A. (1975), *Phytochemistry* **14**, 1083.
Brunswick, H. (1922), *Sitzber. Akad. Wien. Math.-Nat. Kl., Abt. I*, **131**, 221.
Buschi, C. A., Pomilio, A. B. and Gros, E. G. (1979), *Phytochemistry* **18**, 1249.
Buschi, C. A., Pomilio, A. B. and Gros, E. G. (1980), *Phytochemistry* **19**, 903.
Camele, G., Delle Monache, F., Delle Monache, G. and Marini-Bettolo, G. B. (1980), *Phytochemistry* **19**, 707.
Candy, H. A., Brookes, K. B., Bull, J. R., Garry, E. J. and Garry, J. M. (1978), *Phytochemistry* **17**, 1681.
Carman, N. J., Watson, T., Bierner, M. W., Averett, J., Sanderson, S., Seaman, F. C. and Mabry, T. J. (1972), *Phytochemistry* **11**, 3271.
Cavé, A., Bouquet, A. and Paris, R.-R. (1973), *C. R. Acad. Sci. D.* **276**, 1899.
Charrière, Y. and Tissut, M. (1973), *Phytochemistry* **12**, 1443.
Chatterjee, A., Malakav, D. and Ganguly, D. (1976), *Indian J. Chem.* **14B**, 233.
Chhabra, S. C., Gupta, S. R., Seshadri, T. R. and Sharma, N. D. (1976a), *Indian J. Chem.* **14B**, 651.
Chhabra, S. C., Gupta, S. R., Sharma, C. S. and Sharma, N. D. (1976b), *Indian J. Chem.* **14**, 384.
Chhabra, S. C., Gupta, S. R. and Sharma, N. D. (1977a), *Phytochemistry* **16**, 399.
Chhabra, S. C., Gupta, S. R. and Sharma, N. D. (1977b), *Indian J. Chem.* **15B**, 421.
Chhabra, S. C., Gupta, S. R. and Sharma, N. D. (1978), *Indian J. Chem.* **16B**, 1079.
Chirikdjian, J. J. (1973), *Sci. Pharm.* **41**, 206.
Chirikdjian, J. J. (1974), *Die Pharmazie* **29**, 292.
Chou, C.-J. (1978), *J. Taiwan Pharm. Assoc.* **30**, 36.
Chumbalov, T. K. (1976), *Khim. Prir. Soedin.* **1976**, 658.
Chumbalov, T. K., Mukhamed'yarova, M. M. and Omurkamzinova, V. B. (1974), *Khim. Prir. Soedin.* **1974**, 814.
Clark-Lewis, J. W. and Dainis, I. (1968), *Austr. J. Chem.* **21**, 425.
Combier, H., Jay, M., Voirin, B. and Lebreton, P. (1974), *C. R. Groupe Polyphénols* **5**.
Dasgupta, S., Dutta, S. C. and Ray, A. B. (1977), *Indian J. Chem.* **15B**, 197.
Dawson, R. M., Henrick, C. A., Jeffries, P. R. and Middleton, E. J. (1965), *Austr. J. Chem.* **18**, 1871.
Desai, H. K., Gawad, D. G., Govindachari, T. R. and Joshi, B. S. (1973) *Indian J. Chem.* **11**, 840.

Desai, H. K., Gawad, D. H., Govindachari, T. R., Joshi, B. S., Parthasarathy, P. C., Ramachandran, K. S., Ravindranath, K. R., Sidhaye, A. R. and Visnawathan, N. (1976), *Indian J. Chem.* **14B**, 473.
De Silva, L. B., de Silva, U. L. L., Mahendran, M. and Jennings, R. C. (1980), *Phytochemistry* **19**, 2794.
Deshpande, V. H., Wakharkar, P. V. and Rama Rao, A. V. (1976), *Indian J. Chem.* **14B**, 647.
Devi, G., Kapil, R. S. and Popli, S. P. (1979a), *Indian J. Chem.* **17B**, 75.
Devi, G., Kapil, R. S. and Popli, S. P. (1979b), *Indian J. Chem.* **17B**, 85.
Dhar, K., Atal, C. and Pelter, A. (1970), *Planta Med.* **19**, 332.
Dietz, V. H. (1980), Dissertation, Darmstadt.
Dietz, V. H., Wollenweber, E., Favre-Bonvin, J. and Gómez, L. D. (1980), *Z. Naturforsch.* **35c**, 36.
Dietz, V. H., Wollenweber, E., Favre-Bonvin, J. and Smith, D. M. (1981), *Phytochemistry* **20**, 1181.
Dillon, M. O. and Mabry, T. J. (1977), *Phytochemistry* **16**, 1318.
Djermanovic, M., Jokić, A., Mladenović, S. and Stefanović, M. (1975), *Phytochemistry* **14**, 1873.
Dobberstein, R. H., Tin-Wa, M., Fong, H. H. S., Crane, F. A. and Farnsworth, N. R. (1977), *J. Pharm. Sci.* **66**, 600.
Dominguez, X. A. and Cardenas, E. (1975), *Phytochemistry* **14**, 2511.
Dominguez, X. A. and Hinojosa, M. (1976), *Planta Med.* **30**, 68.
Dominguez, X. A. and Torre, B. (1974), *Phytochemistry* **13**, 1624.
Dominguez, X. A., Escarria, S. and Butruille, D. (1973), *Phytochemistry* **12**, 724.
do Nascimento, M. C., Vasconcellos Dias, R. L. de and Mors, W. B. (1976), *Phytochemistry* **15**, 1553.
Dossaji, S. and Kubo, I. (1980), *Phytochemistry* **19**, 482.
Dreyer, D. L. and Park, K.-H. (1975), *Phytochemistry* **14**, 1617.
Dubey, R. C. and Misra, K. (1974), *J. Ind. Chem. Soc.* **51**, 653.
Egger, K. (1967). In *Dünnschichtchromatographie* (ed. E. Stahl), Springer-Verlag, Berlin, Heidelberg, New York.
Egger, K. (1968), Private communication.
Egger, K., Wollenweber, E. and Tissut, M. (1970) *Z. Pflanzenphys.* **62**, 464.
Egger, K., Charrière, Y. and Tissut, M. (1972), *Z. Pflanzenphys.* **68**, 92.
Escamilla, E. M. and Rodriguez, B. (1980), *Ann. Quim.* **76**, 189.
Escarria, S., Torrenegra, R. D. and Anagarita, B. (1977), *Phytochemistry* **16**, 1618.
Exner, J. (1978), Dissertation, Heidelberg.
Exner, J., Reichling, J., Cole, T. C. H. and Becker, H. (1981), *Planta Med.* **41**, 198.
Falk, A. J., Smolenski, S. J., Bauer, L. and Bell, C. L. (1975), *J. Pharm. Sci.* **64**, 1838.
Farkas, L., Nógradi, M. Sudarsanam, V. and Herz, W. (1966), *J. Org. Chem.* **31**, 3228.
Farkas, L., Mezey-Vándor, G. and Nógradi, M. (1971), *Chem. Ber.* **104**, 2646.
Farkas, L., Mezey-Vándor, G. and Nógradi, M. (1974), *Chem. Ber.* **107**, 3878.
Favre-Bonvin, J., Jay, M., Wollenweber, E. and Dietz, V. H. (1980), *Phytochemistry* **19**, 2043.
Fernandes, J. B., Gottlieb, O. R. and Xavier, L. M. (1978), *Biochem. Syst. Ecol.* **6**, 55.
Ferraro, G. E. and Coussio, J. D. (1973), *Phytochemistry* **12**, 1825.
Ferraro, G., Martino, V. S. and Coussio, J. D. (1977), *Phytochemistry* **16**, 1618.
Finnegan, R. A., Bachman, P. L. and Knutson, D. (1972), *Lloydia (J. Nat. Prod.)* **35**, 457.
Flores, S. E. and Herran, J. (1960), *Chem. Ind.* **1960**, 291.
Franca, N. C., Gottlieb, O. R., Magalhàes, M. T., Mendes, P. H., Maia, J. G. S., Da Silva, M. L. and Gottlieb, H. E. (1976), *Phytochemistry* **15**, 572.
Fraser, A. W. and Lewis, J. R. (1973), *Phytochemistry* **12**, 1787.
Fraser, A. W. and Lewis, J. R. (1974), *Phytochemistry* **13**, 1561.
Gaydou, E. M. and Bianchini, J. P. (1978), *Bull. Soc. Chim. Fr.* **1978**, 43.
Gehring, E. and Geiger, H. (1980), *Z. Naturforsch.* **35c**, 380.
Geiger, H. (1979), *Z. Naturforsch.* **34c**, 878.
Geissman, T. A. (Ed.) (1962). *The Chemistry of Flavonoid Compounds*, Pergamon Press, Oxford.
Geissman, T. A., Mukherjee, R. and Sim, K. Y. (1967), *Phytochemistry* **6**, 1575.
Ghanim, A. and Jayaraman, I. (1979), *Ind. J. Chem.* **17B**, 648.

Ghisalberti, E. L., Jefferies, P. R. and Stacey, C. I. (1967), *Austr. J. Chem.* **20**, 1049.

Gonnet, J.-F., Kozjek, F. and Favre-Bonvin, J. (1973), *Phytochemistry* **12**, 2773.

González, A. G., Fraga, B. M., Hernandez, M. G., Larruga, F., Luis, J. G. and Ravelo, A. G. (1978), *Lloydia (J. Nat. Prod.)* **41**, 279.

Gottlieb, O. R. (1975). In *The Flavonoids* (eds. J. B. Harborne, T. J. Mabry and H. Mabry), Chapman & Hall, London.

Gottlieb, O. R. and Da Rocha, A. J. (1972), *Phytochemistry* **11**, 1183.

Goudard, M. and Chopin, J. (1976), *C. R. Hebd. Seances Acad. Sci. Ser. C*, **282**, 683.

Goudard, M., Favre-Bonvin, J., Lebreton, P. and Chopin, J. (1978), *Phytochemistry* **17**, 145.

Goudard, M., Favre-Bonvin, J., Strelisky, J., Nógradi, M. and Chopin, J. (1979), *Phytochemistry* **18**, 186.

Govindachari, T. R., Parthasarathy, P. C., Pai, B. R. and Kalyanaraman, P. S. (1968), *Tetrahedron* **24**, 7027.

Grimshaw, J. and Lamer-Zarawska, E. (1972), *Phytochemistry* **11**, 3273.

Gripenberg, J. (1962). In *The Chemistry of Flavonoid Compounds* (ed. T. A. Geissman), Pergamon Press, Oxford.

Gripenberg, J. and Silander, K. (1955), *Chem. Ind.* **1955**, 443.

Gritsenko, E. N. and Litvinenko, V. I. (1969), *Khim. Prir. Soedin.* **1969**, 55.

Gunatilaka, H. A. L., Sirimanne, S. R., Sotheeswaran, S. and Nakanishi, T. (1979), *J. Chem. Res. S.* **1979**, 216.

Gupta, B. K., Gupta, G. K., Dhar, K. L. and Atal, C. K. (1980), *Phytochemistry* **19**, 2034.

Gupta, H. C., Ayengar, K. N. N. and Rangaswamy, S. (1975), *Indian J. Chem.* **13**, 215.

Gupta, R. K., Krishnamurti, M. and Parthasarathi, J. (1980) *Phytochemistry* **19**, 1264.

Gupta, S. R., Seshadri, T. R. and Sood, G. R. (1973), *Phytochemistry* **12**, 2539.

Gupta, S. R., Seshadri, T. R., Sharma, C. S. and Sharma, N. D. (1975), *Indian J. Chem.* **13**, 785.

Gupta, S. R., Seshadri, T. R., Sharma, C. S. and Sharma, N. D. (1979), *Indian J. Chem.* **17B**, 37.

Gurni, A. A. (1979), Dissertation, Hamburg.

Hänsel, R. and Cubukcu, B. (1972), *Phytochemistry* **11**, 2632.

Harborne, J. B. (1967), *Comparative Biochemistry of the Flavonoids*, Academic Press, London and New York.

Harborne, J. B. (1968), *Phytochemistry* **7**, 1215.

Harborne, J. B. (1969), *Phytochemistry* **8**, 177.

Harborne, J. B. (1973), *Phytochemical Methods*, Chapman & Hall, London.

Harborne, J. B. (1978), *Phytochemistry* **17**, 915.

Harborne, J. B. (1979), *Phytochemistry* **18**, 1323.

Harborne, J. B. (1981) *Phytochemistry* **20**, 1117.

Harborne, J. B. and Williams, C. A. (1973), *Biochem. Syst.* **1**, 51.

Harborne, J. B., Saleh, N. A. M. and Smith, D. M. (1978), *Phytochemistry* **17**, 589.

Hasan, A. (1976) Thèse (3e cycle), Lyon.

Herz, W. and de Groote, R. (1977), *Phytochemistry* **16**, 1307.

Herz, W. and Gibaja, S. (1972), *Phytochemistry* **11**, 2625.

Herz, W. and Sudarsanam, V. (1970), *Phytochemistry* **9**, 895.

Herz, W., Gast, C. M. and Subramanian, P. S. (1968), *J. Org. Chem.* **33**, 2780.

Herz, W., Aota, K. and A. L. Hall (1970a), *J. Org. Chem.* **35**, 4117.

Herz, W., Bhat, S. V. and Santhanam, P. S. (1970b), *Phytochemistry* **9**, 891.

Herz, W., Fitzhenry, B. and Anderson, G. D. (1973), *Phytochemistry* **12**, 1181.

Herz, W., Govindan, S. V., Riess-Maurer, I., Kreil, B., Wagner, H., Farkas, L. and Strelisky, J. (1980), *Phytochemistry* **19**, 669.

Hiermann, A., Becker, H., Exner, J. and Averett, J. E. (1978), *Phytochem. Bull.* **11**, 55.

Higa, T. and Scheuer, P. J. (1974), *J. Chem. Soc.* **1974**, 1350.

Higuchi, R. and Donnelly, D. M. X. (1978), *Phytochemistry* **17**, 787.

Hörhammer, L., Wagner, H., Wilkomirsky, M. T. and Iyengar, M. A. (1973), *Phytochemistry* **12**, 2068.

Hosozawa, S., Kato, N. and Munakata, K. (1972), *Phytochemistry* **11**, 2362.
Iinuma, M. and Matsuura, S. (1979), *Yakugaku Zasshi* **99**, 657.
Iinuma, M. and Matsuura, S. (1980), *Yakugaku Zasshi* **100**, 657.
Iinuma, M., Matsuura, S., Kurogochi, K. and Tanaka, T. (1980), *Chem. Pharm. Bull.* **23**, 717.
Ikram, M., Jehangir, S., Razaq, S. and Kawano, M. (1979), *Planta Med.* **37**, 189.
Imre, S., Tulus, R. and Sengün, I. (1973), *Phytochemistry* **12**, 2317.
Imre, S., Öztunc, A. and Wagner, H. (1977), *Phytochemistry* **16**, 799.
Ivantcheva, S. (1975), Dissertation, Sofia.
Iyengar, M. A., Bhat, U. G., Katti, S. B., Wagner, H., Seligmann, O. and Herz, W. (1976), *Indian J. Chem.* **14B**, 714.
Jain, A. C. and Gupta, R. C. (1978), *Curr. Sci. (India)* **47**, 770.
Jain, A. C. and Sharma, B. N. (1973), *Phytochemistry* **12**, 1455.
Jalal, M. F., Overton, K. H. and Rycroft, D. S. (1979), *Phytochemistry* **18**, 149.
Jauhari, P. K., Sharma, S. C., Tandon, J. S. and Dhar, M. M. (1979), *Phytochemistry* **18**, 359.
Jay, M. (1978), Private communication.
Jay, M. and Voirin, B. (1976), *Phytochemistry* **15**, 517.
Jay, M., Gonnet, J.-F., Wollenweber, E. and Voirin, B. (1975), *Phytochemistry* **14**, 1605.
Jay, M., Hasan, A., Voirin, B. and Viricel, M.-R. (1978a), *Phytochemistry* **17**, 827.
Jay, M., Hasan, A., Voirin, B., Favre-Bonvin, J. and Viricel, M.-R. (1978b), *Phytochemistry* **17**, 1196.
Jay, M., Wollenweber, E. and Favre-Bonvin, J. (1979a), *Phytochemistry* **18**, 153.
Jay, M., Favre-Bonvin, J. and Wollenweber, E. (1979b), *Can. J. Chem.* **57**, 1901.
Jay, M., Voirin, B., Hasan, A., Gonnet, J.-F. and Viricel, M.-R. (1980), *Biochem. Syst. Ecol.* **8**, 127.
Joshi, B. S. (1980), Private communication.
Joshi, B. S. and Gawad, D. H. (1976), *Indian J. Chem.* **14B**, 9.
Kalra, A. J., Krishnamurty, M. and Seshadri, T. R. (1975), *Indian J. Chem.* **13**, 18.
Kalra, A. J., Krishnamurty, M. and Nath, M. (1977), *Indian J. Chem.* **15B**, 393.
Kaneta, M. and Sugiyama, N. (1973), *Agric. Biol. Chem.* **37**, 2663.
Karczewska, W. and Krolikowska, M. (1974), *Roezn. Chem. Ann. Soc. Chim. Polon.* **48**, 981.
Karl, C., Pedersen, P. A. and Müller, G. (1980), *Z. Naturforsch.* **35c**, 826.
Kasahara, A., Izumi, T. and Sohima, M. (1974), *Bull. Chem. Soc. Japan* **47**, 2526.
Khafagy, S. M., El-Ghazooly, M. G. and Metwally, A. M. (1979), *Die Pharmazie* **34**, 748.
Khalid, S. A. and Waterman, P. G. (1981), *Phytochemistry* **20**, in press.
Khan, H. and Zaman, A. (1974), *Tetrahedron* **30**, 2811.
Kingston, D. G. I. (1971), *Tetrahedron* **27**, 2691.
Kisiel, W. (1975), *Pol. J. Pharmacol. Pharm.* **27**, 339.
Komiya, T., Naruse, Y. and Oshio, H. (1976), *Yakugaku Zasshi* **96**, 855.
Krishnamurti, M., Seshadri, T. R. and Sharma, N. D. (1972), *Indian J. Chem.* **10**, 23.
Kumar, N. S., Pavanasasivam, G., Sultanbawa, M. U. S. and Mageswaran, R. (1977), *J. Chem. Soc., Perkin I* **1977**, 1243.
Kumari, D., Chhabra, S. C. and Gupta, S. R. (1979), *Indian J. Chem.* **17B**, 168.
Kupchan, S. M., Sigel, C. W., Hemingway, R. J., Knox, I. R. and Udayamurthy, M. S. (1969), *Tetrahedron* **25**, 1603.
Lakshmi, R., Srimanarayana, G. and Rao, N. V. S. (1974), *Indian J. Chem.* **12**, 8.
Lamberton, J. A. (1964), *Austr. J. Chem.* **17**, 692.
Lapinina, L. O. (1965), *Farmatsevt. Zh. (Kiev)* **20**, 57.
Lau-Cam, C. A. and Chan, H. H. (1973), *Phytochemistry* **12**, 1829.
Le Quesne, P. W., Pastore, M. P. and Raffauf, R. P. (1976), *Lloydia (J. Nat. Prod.)* **39**, 391.
Le-Van, N. and Pham, T. V. C. (1979), *Phytochemistry* **18**, 1859.
Lewis, D. S. and Mabry, T. J. (1977), *Phytochemistry* **16**, 1114.
Lin, C.-N., Arisawa, M., Shimizu, M. and Morita, N. (1978), *Chem. Pharm. Bull.* **26**, 2036.
Liu, Y.-L. and Mabry, T. J. (1981), *Phytochemistry*, **19**, 2439.

Lopes, J. L. C., Lopes, J. N. C. and Leitao Fho, H. F. (1979), *Phytochemistry* **18**, 362.
Lopez, J. A., Saenz, J. A., Slatkin, D. J., Knapp, J. E. and Schiff, P. L. (1976), *Phytochemistry* **15**, 2028.
Mabry, T. J. (1981), Private communication.
Mabry, T. J., Markham, K. R. and Thomas, M. B. (1970). *The Systematic Identification of Flavonoids*, Springer-Verlag, Berlin-Heidelberg, New York.
Maksyutina, N. P. and Litvinenko, V. I. (1964), *Dokl. Akad. Nauk. SSSR* **154**, 1123.
Malan, E. and Naidoo, S. (1980), *Phytochemistry*, **19**, 2731.
Malan, E. and Roux, D. G. (1979), *J. Chem. Soc., Perkin I* **1979**, 2696.
Malik, S. B., Sharma, P. and Seshadri, T. R. (1977), *Indian J. Chem.* **15B**, 536.
Marini-Bettolo, G. B., Chiavelli, S. and Casinovi, C. G. (1957), *Gazz. Chim. Ital.* **87**, 1185.
Markham, K. R. (1975) In *The Flavonoids*. (eds. J. B. Harborne, T. J. Mabry and H. Mabry), Chapman and Hall, London.
Markham, K. R. and Mabry, T. J. (1975), In *The Flavonoids* (eds. J. B. Harborne, T. J. Mabry and H. Mabry) Chapman and Hall, London.
Martino, V. S., Ferraro, G. E. and Coussio, J. D. (1976), *Phytochemistry* **15**, 1086.
Mathuram, S., Purushothaman, K. K., Sarada, A. and Connolly, J. D. (1976), *Phytochemistry* **15**, 838.
Matsuura, S., Iinuma, M., Ishikawa, K. and Kagei, K. (1978), *Chem. Pharm. Bull.* **26**, 305.
Matsuura, S., Kunii, T. and Iinuma, M. (1973), *J. Pharm. Soc. Jpn.* **93**, 1517.
Mears, J. A. (1980), *J. Nat. Prod.* **43**, 708.
Mears, J. A. and Mabry, T. J. (1972), *Phytochemistry* **11**, 411.
Medvedeva, S. A., Ivanova, S. Z., Tyukavkina, N. A. and Zapesochnaya, G. G. (1977), *Khim. Prir. Soedin.* **1977**, 650.
Midge, M. D. and Rama Rao, A. V. (1975), *Indian J. Chem.* **13**, 541.
Midge, M. D., Rao, A. V. R. and Venkataraman, K. (1977), *Indian J. Chem.* **15B**, 667.
Misra, K. and Mishra, C. S. (1979), *Indian J. Chem.* **18B**, 88.
Misra, T. N., Singh, R. S., Sharma, S. C. and Tandon, J. S. (1976), *J. Ind. Chem. Soc.* **53**, 1064.
Moreno, B., Delle Monache, G., Delle Monache, F. and Marini-Bettolo, G. B. (1980a), *I. Farmaco Ed. Sci.* **35**, 457.
Mues, R. (1975), Dissertation, Saarbrücken.
Mues, R., Timmermann, B. N., Ohno, N. and Mabry, T. J. (1979), *Phytochemistry* **18**, 1379.
Murphy, S. T., Ritchie, E. and Taylor, W. C. (1974), *Austr. J. Chem.* **27**, 187.
Nagar, A., Gujral, V. K. and Gupta, S. R. (1979), *Phytochemistry* **18**, 1245.
Nair, A. G. R. and Subramanian, S. S. (1975), *Curr. Sci. (India)* **44**, 214.
Nair, A. G. R., Ramesh, P., Subramanian, S. S. and Joshi, B. S. (1978), *Phytochemistry* **17**, 591.
Nakashima, R., Yoshikawa, M. and Matsuura, T. (1973), *Phytochemistry* **12**, 1502.
Neu, R. (1956), *Naturwissenschaft* **43**, 82.
Nielsen, J. G. and Møller, J. (1970), *Acta Chem. Scand.* **24**, 2665.
Niemann, G. J. (1975), *Z. Naturforsch.* **30c**, 550.
Niemann, G. J. (1977), *Z. Naturforsch.* **32c**, 1015.
Nikonova, L. P. and Nikonov, G. K. N. (1975), *Khim. Prir. Soedin.* **1975**, 96.
Nomura, T., Fukai, T. and Katayanagi, M. (1976), *Chem. Pharm. Bull.* **24**, 2898.
Nomura, T., Fukai, T. and Katayanagi, M. (1977), *Chem. Pharm. Bull.* **25**, 529.
Nomura, T., Fukai, T., Yamada, S. and Katayanagi, M. (1978a), *Chem. Pharm. Bull.* **26**, 1394.
Nomura, T., Fukai, T. and Katayanagi, M. (1978b), *Chem./Pharm. Bull.* **28**, 1453.
Oettmeier, W. and Heupel, A. (1972), *Z. Naturforsch.* **27b**, 177.
Ognyanov, I. V. and Ivantcheva, S. (1972), *Dokl. Bolg. Akad. Nauk.* **25**, 1027.
Öksüz, S. (1977), *Planta Med.* **31**, 270.
Pangarova, T. T. and Zapesochnaya, G. G. (1974), *Khim. Prir. Soedin.* **1974**, 788.
Panichpol, K. and Waterman, P. G. (1978), *Phytochemistry* **17**, 1363.
Pascual Teresa, J. D. de, Diaz, F., Sanchez, F. J., Hernandez, J. M. and Grande, M. (1980), *Planta Med.* **38**, 271.

Pavanasasivam, G. and Sultanbawa, M. U. S. (1975), *Phytochemistry* **14**, 1127.

Pendse, A. D., Pendse, R., Rao, A. V. R. and Venkataraman, K. (1976), *Indian J. Chem.* **14B**, 69.

Pinar, M. (1973), *Phytochemistry* **12**, 3014.

Popovici, G., Weissenböck, G., Bouillant, M.-L., Dellamonica, G. and Chopin, J. (1977), *Z. Pflanzenphys.* **85**, 103.

Popravko, S. A., Kononenko, G. P., Tikhomirova, V. I. and Wulfson, N. S. (1979), *Bioorgan. Khim.* **5**, 1662.

Purushotaman, K. K., Sarada, A., Saraswathi, G. and Connolly, J. D. (1977), *Phytochemistry* **16**, 398.

Quijano, L., Calderon, J. S., Gómez, F., Soria, I. E. and Rios, T. (1980), *Phytochemistry* **19**, 2439.

Rabesa, Z. A. (1980) Thèse (3ᵉcycle), Lyon.

Rabesa, Z. A. and Lebreton, P. (1977), *Comm. Coll. Plantes malgaches*

Rabesa, Z. A. and Voirin, B. (1978), *Tetrahedron Lett.* **1978**, 3717.

Rabesa, Z. A. and Voirin, B. (1979a), *Phytochemistry* **18**, 692.

Rabesa, Z. A. and Voirin, B. (1979b), *Z. Pflanzenphys.* **91**, 183.

Rabesa, Z. A. and Voirin, B. (1979c), *C. R. Hebd. Sciences Acad. Sci. Ser. C* **289**, 167.

Rabesa, Z. A. and Voirin, B. (1979d), *Phytochemistry* **18**, 360.

Rama Rao, A. V., Rathi, S. S. and Venkataraman, K. (1972), *Indian J. Chem.* **10**, 989.

Rama Rao, A. V. and Varadan, M. (1973), *Indian. J. Chem.* **11**, 403.

Ramesh, P., Nair, A. G. R. and Subramanian, S. (1979), *Curr. Sci.* **48**, 67.

Rao, C. B., Rao, T. and Muralikrishna, B. (1977), *Planta Med.* **31**, 235.

Rao, E. V. and RangaRaju, N. (1979), *Phytochemistry* **18**, 1581.

Rao, M. M. and Rao, E. V. (1980), *Curr. Sci.* **49**, 468.

Rao, M. M., Kingston, D. G. I. and Spittler, T. D. (1970), *Phytochemistry* **9**, 227.

Rao, M. M., Gupta, P. S., Krishu, E. M. and Singh, P. P. (1979), *Indian J. Chem.* **17B**, 178.

Ray, A. B., Dutta, S. C. and Dasgupta, S. (1976), *Phytochemistry* **15**, 1797.

Reichling, J., Becker, H., Exner, J. and Dräger, P.-D. (1979), *Pharmaz. Z.* **124**, 1998.

Roberts, M. F., Timmermann, B. N. and Mabry, T. J. (1980), *Phytochemistry* **19**, 127.

Rodriguez, B. (1977), *Phytochemistry* **16**, 800.

Rodriguez, E. (1977), *Biochem. Syst. Ecol.* **5**, 207.

Rodriguez, E., Carman, N. J., Chaves, P. and Mabry, T. J. (1972a), *Phytochemistry* **11**, 1507.

Rodriguez, E., Carman, N. J., Vander Velde, G., McReynholds, J. H., Mabry, T. J., Irwin, M. A. and Geissman, T. A. (1972b), *Phytochemistry* **11**, 3509.

Rodriguez, J., Tello, H., Quijano, L., Calderon, J., Gómez, F., Romo, J. and Rios, T. (1974), *Rev. Latinoam. Quim.* **5**, 41.

Roy, D. and Khanna, R. N. (1979), *Indian J. Chem.* **18B**, 525.

Roy, D., Sharma, N. N. and Khanna, R. N. (1977), *Indian J. Chem.* **15B**, 1138.

Sakakibara, M. and Mabry, T. J. (1977), *Rev. Latinoam. Quim.* **8**, 99.

Sakakibara, M. and Mabry, T. J. (1978), *Rev. Latinoam. Quim.* **9**, 92.

Sakakibara, M., Timmermann, B. N., Nakatani, N., Waldrum, H. and Mabry, T. J. (1975), *Phytochemistry* **14**, 849.

Sakakibara, M., diFeo, D., Nakatani, N., Timmermann, B. N. and Mabry, T. J. (1976), *Phytochemistry* **15**, 727.

Saleh, A. A., Cordell, G. A. and Farnsworth, N. R. (1976), *Lloydia (J. Nat. Prod.)* **39**, 456.

Satam, P. G. N. and Bringi, N. V. (1973), *Indian J. Chem.* **11**, 1188.

Shelyuto, V. L., Glyzin, V. I., Kruglova, E. P. and Smirnova, L. P. (1977), *Khim. Prir. Soedin.* **1977**, 860.

Shen, M. C., Rodriguez, E., Kerr, K. and Mabry, T. J. (1976), *Phytochemistry* **15**, 1045.

Silva, M., Wisenfeld, A., Sammes, P. G. and Taylor, T. W. (1977), *Phytochemistry* **16**, 379.

Singh, J. and Tiwari, R. D. (1979), *Phytochemistry* **18**, 2060.

Singhal, A. K., Sharma, R. P., Thyagarajan, G., Herz, W. and Govindan, S. V. (1980), *Phytochemistry* **19**, 929.

Smalberger, T. M., Vleggaar, R. and Weber, J. C. (1974), *Tetrahedron* **30**, 3927.

Smith, D. M. (1980), *Bull. Torrey Bot. Club* **107**, 134.
Somaroo, B. H., Thakur, M. L. and Grant, W. F. (1973), *J. Chromatogr.* **87**, 290.
Star, A. E., Rösler, H., Mabry, T. J. and Smith, D. M. (1975), *Phytochemistry* **14**, 2275.
Struck, R. F. and Kirk, M. C. (1970), *J. Agric. Food Chem.* **18**, 548.
Subramanian, P. M. and Misra, G. S. (1979), *J. Nat. Prod. (Lloydia)* **42**, 540.
Subramanian, S. S. and Nair, A. G. R. (1972a), *Curr. Sci.* **41**, 62.
Subramanian, S. S. and Nair, A. G. R. (1972b), *Phytochemistry* **11**, 3095.
Subramanian, S. S., Nair, A. G. R., Rodriguez, E. and Mabry, T. J. (1972), *Curr. Sci.* **41**, 202.
Subramanyam, K., Madhusudhana, J. and Rao, K. V. J. (1977), *Indian J. Chem.* **15B**, 12.
Suga, T., Iwanata, N. and Asakawa, Y. (1972), *Bull. Chem. Soc. Jpn.* **45**, 2058.
Takido, M., Aimi, M., Takahashi, S., Yamanouchi, S., Torii, H. and Dohi, S. (1975), *Yakugaku Zasshi* **95**, 108.
Takido, M., Yasukawa, K., Matsuura, S. and Iinuma, M. (1979), *Yakugaku Zasshi* **99**, 443.
Talapatra, S. K., Bhar, D. S. and Talapatra, B. (1974), *Phytochemistry* **13**, 284.
Talapatra, S. K., Mallik, A. K. and Talapatra, B. (1980), *Phytochemistry* **19**, 1199.
Tandon, S. and Rastogi, R. P. (1977), *Phytochemistry* **16**, 1455.
Tatum, J. H. and Berry, R. E. (1972), *Phytochemistry* **11**, 2283.
Tatum, J. H. and Berry, R. E. (1978), *Phytochemistry* **17**, 447.
Thomas, M. B. and Mabry, T. J. (1968), *Phytochemistry* **7**, 787.
Tillequin, F., Henri, M. E. and Paris, R. R. (1977), *Planta Med.* **31**, 76.
Timmermann, B. N., Mues, R., Mabry, T. J. and Powell, A. M. (1979a), *Phytochemistry* **18**, 1855.
Timmermann, B. N., Valesi, A. G. and Mabry, T. J. (1979b), *Rev. Latinoam. Quim.* **10**, 81.
Tiwara, R. D. and Bajpa, M. (1977), *Phytochemistry* **16**, 798.
Tomas, F., Ferreres, F. and Guirado, A. (1979), *Phytochemistry* **18**, 185.
Torrenegra, R. D., Escarria, S., Raffelsberger, B. and Achenbach, H. (1980), *Phytochemistry* **19**, 2795.
Tóth, L., Kokovay, K., Bujtás, G. and Pápay, V. (1980), *Pharmazie* **35**, 334.
Tyukavkina, N. A., Medvedeva, S. A. and Ivanova, S. Z. (1974), *Khim. Prir. Soedin.* **1974**, 157.
Ueno, A., Ikeya, Y., Fukushima, S., Noro. T., Morinaga, K. and Kuwano, H. (1978), *Chem. Pharm. Bull.* **26**, 2411.
Ulubelen, A. and Uygur, I. (1976), *Planta. Med.* **29**, 318.
Ulubelen, A., Mabry. T. J. and Aynehechi (1979a), *J. Nat. Prod. (Lloydia)* **42**, 624.
Ulubelen, A., Miski, M., Neumann, P. and Mabry, T. J. (1979b), *J. Nat. Prod. (Lloydia)* **42**, 261.
Ulubelen, A., Timmermann, B. N. and Mabry, T. J. (1980a), *Phytochemistry* **19**, 905.
Ulubelen, A., Kerr, K. M. and Mabry, T. J. (1980b), *Phytochemistry* **19**, 1761.
Urbatsch, L. E., Bacon, J. D. and Mabry, T. J. (1975), *Phytochemistry* **14**, 2279.
Urbatsch, L. E., Mabry, T. J., Miyakado, M., Ohno, N. and Yoshioka, H. (1976), *Phytochemistry* **15**, 440.
Urschler, I. (1968), *Phyton (Horn, Austria)* **13**, 15.
Vandekerkhove, O. (1978), *Z. Pflanzenphys.* **86**, 275.
Vendantham, T. N. C., Subramaniam, S. S. and Harborne, J. B. (1977), *Phytochemistry* **16**, 294.
Venkataraman, K. (1975) In *The Flavonoids.* (eds. J. B. Harborne, T. J. Mabry and H. Mabry), Chapman and Hall, London.
Vleggar, R., Smalberger, T. M. and de Waal, H. L. (1973), *J. S. Afr. Chem. Inst.* **26**, 71.
Voirin, B. and Jay, M. (1978), *Biochem. Syst. Ecol.* **6**, 99.
Voirin, B., Jay, M. and Hauteville, M. (1975), *Phytochemistry* **14**, 257.
Voirin, B., Jay, M. and Hauteville, M. (1976), *Phytochemistry* **15**, 840.
Vrkoč, J., Ubik, K. and Sedmera, P. (1973), *Phytochemistry* **12**, 2062.
Waddell, T. G. (1973), *Phytochemistry* **12**, 2061.
Wagner, H., Farkas, L., Flores, G. and Strelisky, J. (1974), *Chem. Ber.* **107**, 1049.
Wagner, H., Maurer, I., Farkas, L. and Strelisky, J. (1976), *Tetrahedron Lett.* **1976**, 67.
Wagner, H., Maurer, I., Farkas, L. and Streslisky, J. (1977a), *Tetrahedron* **33**, 1411.
Wagner, H., Maurer, I., Farkas, L. and Strelisky, J. (1977b), *Tetrahedron* **33**, 1405.

Wagner, H., Rüger, R., Maurer, G. and Farkas, L. (1977c), *Chem. Ber.* **110**, 737.
Wagner, H., Seligmann, O., Chari, M. V., Wollenweber, E., Dietz, V. H., Donnelly, D. M. X., Meegan, M. J. and O'Donell, B. (1979), *Tetrahedron Lett.* **1979**, 4269.
Wallace, J. M., *Phytochem. Bull.* **6**.
Waterman, P. G. and Khalid, S. A. (1980), *Phytochemistry* **19**, 909.
Whalen, M. D. and Mabry, T. J. (1979), *Phytochemistry* **18**, 263.
Wilkins, C. K. and Bohm, B. A. (1976), *Can. J. Bot.* **54**, 2133.
Williams, C. A. (1978), *Phytochemistry* **17**, 729.
Williams, C. A. and Harborne, J. B. (1975), *Biochem. Syst. Ecol.* **3**, 181.
Wollenweber, E. (1970), Dissertation, Heidelberg.
Wollenweber, E. (1974a), *Phytochemistry* **13**, 753.
Wollenweber, E. (1974b), *Biochem. Phys. Pflanzen* **166**, 419.
Wollenweber, E. (1975a), *Biochem. Syst. Ecol.* **3**, 35.
Wollenweber, E. (1975b), *Biochem. Syst. Ecol.* **3**, 47.
Wollenweber, E. (1975c), *Z. Pflanzenphys.* **74**, 415.
Wollenweber, E. (1976a), *Phytochemistry* **15**, 438.
Wollenweber, E. (1976b), *Phytochemistry* **15**, 2013.
Wollenweber, E. (1976c), *Ber. Dtsch. Bot. Ges.* **89**, 243.
Wollenweber, E. (1977a), *Phytochemistry* **16**, 295.
Wollenweber, E. (1977b), *Z. Pflanzenphys.* **85**, 71.
Wollenweber, E. (1978), *Am. Fern J.* **68**, 13.
Wollenweber, E. (1979), *Flora* **168**, 138.
Wollenweber, E. (1981a), In *The Biology and Chemistry of Plant Trichomes* (eds. T. J. Mabry, P. Healey and E. Rodriguez), Plenum Press, New York.
Wollenweber, E. (1981b), *Bot. J. Linn. Soc.*, in press.
Wollenweber, E. and Dietz, V. H. (1979), *Phytochem. Bull.* **12**, 48.
Wollenweber, E. and Dietz, V. H. (1980), *Biochem. Syst. Ecol.* **8**, 21.
Wollenweber, E. and Dietz, V. H. (1981), *Phytochemistry* **20**, 869.
Wollenweber, E. and Wiermann, R. (1979), *Z. Naturforsch.* **34c**, 1289.
Wollenweber, E., Bouillant, M.-J., Lebreton, P. and Egger, K. (1971), *Z. Naturforsch.* **26b**, 1188.
Wollenweber, E., Favre-Bonvin, J. and Jay, M. (1978), *Z. Naturforsch.* **33c**, 831.
Wollenweber, E., Dietz, V. H., Schillo, D. and Schilling, G. (1980), *Z. Naturforsch.* **35c**, 685.
Wu, T.-S., Tien, H.-J., Arisawa, M., Shimizu, M. and Morita, N. (1980), *Phytochemistry* **19**, 2227.
Xaasan, C. C., Xaasan Cilmi, C., Faarax, M. X., Passannanti, S., Piozzi, F. and Paternostro, M. (1980), *Phytochemistry* **19**, 2229.
Zapesochnaya, G. G., Evstratova, R. I. and Mukhametzhanov, M. N. (1977), *Khim. Prir. Soedin.* **1977**, 706.

CHECK LIST OF FLAVONES AND FLAVONOLS

I Hydroxyflavones and their methyl ethers

1 Flavone
2 Primuletin (5-OH)
3 2′-Hydroxyflavone
4 5-Hydroxy-6-methoxyflavone
5 5,6-Dimethoxyflavone
6 Chrysin (5,7-di-OH)
7 7-Methyl ether (tectochrysin)
8 5,7-Dimethyl ether
9 Primetin (5,8-di-OH)
10 5,2′-Dihydroxyflavone
11 7,4′-Dihydroxyflavone
12 3′,4′-Dihydroxyflavone
13 Baicalein (5,6,7-tri-OH)
14 6-Methyl ether (oroxylin)
15 7-Methyl ether (negletein)
16 6,7-Dimethyl ether*
17 5,6,7-Trimethyl ether
18 Norwogonin (5,7,8-tri-OH)

19 8-Methyl ether (wogonin)
20 7,8-Dimethyl ether*
21 5,7,8-Trimethyl ether*
22 5,6,2′-Trimethoxyflavone
23 5,6,3′-Trimethoxyflavone
24 5,2′-Dihydroxy-7-methoxyflavone (echiodinin)
25 Apigenin (5,7,4′-tri-OH)
26 5-Methyl ether (thevetiaflavone)
27 7-Methyl ether (genkwanin)
28 4′-Methyl ether (acacetin)
29 5,7-Dimethyl ether*
30 7,4′-Dimethyl ether
31 5,7,4′-Trimethyl ether*
32 5,8,2′-Trihydroxyflavone
33 7,3′,4′-Trihydroxyflavone
34 3′-Methyl ether (geraldone)
35 7,4′-Dimethyl ether (tithonine)
36 5,8-Dihydroxy-6,7-dimethoxyflavone
37 5,7-Dihydroxy-6,8-dimethoxyflavone
38 8-Hydroxy-5,6,7-trimethoxyflavone
39 Alnetin (5-OH,6,7,8-tri-OMe)
40 5,6,7,8-Tetramethoxyflavone
41 Scutellarein (5,6,7,4′-tetra-OH)
42 6-Methyl ether (hispidulin)
43 7-Methyl ether (sorbifolin)
44 4′-Methyl ether
45 6,7-Dimethyl ether (cirsimaritin)
46 6,4′-Dimethyl ether (pectolinarigenin)
47 7,4′-Dimethyl ether
48 6,7,4′-Trimethyl ether (salvigenin)
49 5,6,7,4′-Tetramethyl ether
50 5,7,8,2′-Tetrahydroxyflavone
51 7,8-Dimethyl ether (skullcapflavone I)*
52 7,8,2′-Trimethyl ether*
53 Isoscutellarein (5,7,8,4′-tetra-OH)
54 7-Methyl ether (salvitin)*
55 8-Methyl ether*
56 4′-Methyl ether (takakin)
56 7,8-Dimethyl ether*
57 8,4′-Dimethyl ether* (cirsitakaogenin)
58 7,8,4′-Trimethyl ether*
59 5,7,8,4′-Tetramethyl ether
60 Zapotinin (5-OH, 6,2′,6′-tri-OMe)
61 Zapotin (5,6,2′,6′-tetra-OMe)
62 Cerosillin (5,6,3′,5′-tetra-OMe)
63 Norartocarpetin (5,7,2′,4′-tetra-OH)
64 7-Methyl ether (artocarpetin)
65 5,7,2′,4′-Tetramethyl ether*
66 Luteolin (5,7,3′,4′-tetra-OH)
67 5-Methyl ether*
68 7-Methyl ether
69 3′-Methyl ether (chrysoeriol)
70 4′-Methyl ether (diosmetin)
71 5,3′-Dimethyl ether*
72 7,3′-Dimethyl ether (velutin)
73 7,4′-Dimethyl ether (pilloin)
74 3′,4′-Dimethyl ether*
75 7,3′,4′-Trimethyl ether
76 5,7,3′,4′-Tetramethyl ether*
77 Abrectorin (7,3′-di-OH, 6,4′-di-OMe)
78 Desmethoxysudachitin (5,7,4′-tri-OH, 6,8-di-OMe)
79 Xanthomicrol (5,4′-di-OH, 6,7,8-tri-OMe)
80 Nevadensin (5,7-di-OH, 6,8,4′-tri-OMe)
81 7-Hydroxy-5,6,8,4′-tetramethoxy*
82 Gardenin B (5-OH, 6,7,8,4′-tetra-OMe)
83 Tangeretin (5,6,7,8,4′-penta-OMe)
84 Rivularin (5,2′-di-OH, 7,8,6′-tri-OMe)
85 6-Hydroxyluteolin
86 6-Methyl ether (nepetin)
87 7-Methyl ether (pedalitin)
88 3′-Methyl ether (nodifloretin)
89 6,7-Dimethyl ether (cirsiliol)
90 6,3′-Dimethyl ether (jaceosidin)
91 6,4′-Dimethyl ether
92 7,3′-Dimethyl ether*
93 7,4′-Dimethly ether
94 6,7,3′-Trimethyl ether (cirsileneol)
95 6,7,4′-Trimethyl ether (eupatorin)
96 6,3′,4′-Trimethyl ether (eupalitin)
97 5,6,7,4′-Tetramethyl ether
98 6,7,3′,4′-Tetramethyl ether
99 5,6,7,3′,4′-Pentamethyl ether (sinensetin)
100 Norwightin (5,7,8,2′,3′-penta-OH)
101 7,8,2′-Trimethyl ether (wightin)
102 7,8,2′,3′-Tetramethyl ether
103 Hypolaetin (5,7,8,3′,4′-penta-OH)
104 8-Methyl ether (onopordin)
105 3′-Methyl ether*
106 7,3′-Dimethyl ether*
107 8,3′-Dimethyl ether
108 7,8,3′,4′-Tetramethyl ether
109 5,7,8,3′,4′-Pentamethyl ether (isosinensetin)
110 Isoetin (5,7,2′,4′,5′-penta-OH)* (= hieracin)
111 Tricetin (5,7,3′,4′,5′-penta-OH)
112 5′-Methyl ether (selgin)*
113 3′,5′-Dimethyl ether (tricin)
114 3′,4′,5′-Trimethyl ether*
115 7,3′,4′,5′-Tetramethyl ether (corymbosin)
116 5,7,3′,4′,5′-Pentamethyl ether

117 Prosogerin D (7-OH, 6,3′,4′,5′-tetra-OMe)
118 Prosogerin C (6,7,3′,4′,5′-penta-OMe)
119 Skullcapflavone II (5,2′-di-OH, 6,7,8,6′-tetra-OMe)*
120 5,7,3′,4′-Tetrahydroxy-6,8-dimethoxyflavone*
121 5,6,7,8′-Tetrahydroxy-3′,4′-dimethoxyflavone
122 5,3′,4′-Trihydroxy-6,7,8-trimethoxyflavone*
123 Sudachitin (5,7,4′-tri-OH, 6,8,3′-tri-OMe)
124 Acerosin (5,7,3′-tri-OH, 6,8,4′-tri-OMe)
125 5,4′-Dihydroxy-6,7,8,3′-tetramethoxyflavone*
126 Gardenin D (5,3′-di-OH, 6,7,8,4′-tetra-OMe)*
127 Hymenoxin (5,7-di-OH, 6,8,3′,4′-tetra-OMe)
128 4′-Hydroxy-5,6,7,8,3′-pentamethoxyflavone*
129 5-Desmethoxynobiletin (6,7,8,3′,4′-penta-OMe)
130 Nobiletin (5,6,7,8,3′,4′-hexa-OMe)
131 Tabularin (5,7-di-OH, 6,2′,4′,5′-tetra-OMe)*
132 5,7,4′-Trihydroxy-6,3′,5′-trimethoxyflavone*
133 5,7,3′-Trihydroxy-6,4′,5′-trimethoxyflavone*
134 5-Hydroxy-6,7,3′,4′,5′-pentamethoxyflavone*
135 5-Hydroxy-7,8,2′,3′,4′-pentamethoxyflavone
136 5,7,3′,4′,5′-Pentahydroxy-8-methoxyflavone*
137 5,7,3′,5′-Tetrahydroxy-8,4′-dimethoxyflavone*
138 5,7,5′-Trihydroxy-8,3′,4′-trimethoxyflavone*
139 5,3′-Dihydroxy-7,8,4′,5′-tetramethoxyflavone*
140 5,7-Dihydroxy-8,3′,4′,5′-tetramethoxyflavone*
141 5,2′,4′-Trihydroxy-6,7,8,5′-tetramethoxyflavone*
142 5,6,7,8,2′,4′,5′-Heptamethoxyflavone*
143 Gardenin E (5,3′,5′-tri-OH, 6,7,8,4′-tetra-OMe)
144 6,7,8,5′-Tetramethyl ether (agecorynin-O)
145 5,6,7,8,2′,4′,5′-Heptamethyl ether (agecorynin-O)
146 Scaposin (5,7,5′-tri-OH, 6,8,3′,4′-tetra-OMe)
147 5,6,7-Trihydroxy-8,3′,4′,5′-tetramethoxyflavone*
148 Gardenin C (5,3′-di-OH, 6,7,8,4′,5′-penta-OMe)
149 Gardenin A (5-OH, 6,7,8,3′,4′,5′-hexa-OMe)
150 5,6,7,8,3′,4′,5′-Heptamethoxyflavone*

II Hydroxyflavonols and their methyl ethers

1 Galangin (3,5,7-tri-OH)
2 3-Methyl ether
3 7-Methyl ether (izalpinin)
4 3,7-Dimethyl ether*
5 3,5,7-Trimethyl ether*
6 5-Deoxykaempferol (3,7,4′-tri-OH)
7 Alnusin (3,5,7,-tri-OH, 6-OMe)
8 5,7-Dihydroxy-3,6-dimethoxyflavone*
9 5,6-Dihydroxy-3,7-dimethoxyflavone*
10 3,6-Dihydroxy-5,7-dimethoxyflavone*
11 Alnustin (5-OH, 3,6,7-tri-OMe)
12 3,5,6,7-Tetramethoxyflavone*
13 3,5,8-Trihydroxy-7-methoxyflavone*
14 Isognaphalin (5,8-di-OH, 3,7-di-OMe)
15 Gnaphalin (5,7-di-OH, 3,8-di-OMe)
16 3,5-Dihydroxy-7,8-dimethoxyflavone*
17 Methylgnaphalin (5-OH, 3,7,8-tri-OMe)
18 Datiscetin (3,5,7,2′-tetra-OH)
19 7-Methyl ether (datin)
20 Kaempferol (3,5,7,4′-tetra-OH)
21 3-Methyl ether (isokaempferide)
22 5-Methyl ether
23 7-Methyl ether (rhamnocitrin)
24 4′-Methyl ether (kaempferide)
25 3,7-Dimethyl ether (kumatakenin)
26 3,4′-Dimethyl ether (ermanin)
27 7,4′-Dimethyl ether
28 3,5,7-Trimethyl ether*
29 3,7,4′-Trimethyl ether
30 Pratoletin (3,5,8,4′-tetra-OH)
31 3,7,8,4′-Tetrahydroxyflavone
32 3-Methyl ether
33 Fisetin (3,7,3′,4′-tetra-OH)
34 3-Methyl ether
35 3′-Methyl ether (geraldol)
36 4′-Methyl ether
37 3,7,3′,4′-Tetramethyl ether
38 3,5,7-Trihydroxy-6,8-dimethoxyflavone
39 Araneol (5,7-di-OH, 3,6,8-tri-OMe)

40 3,5-Dihydroxy-6,7,8-trimethoxyflavone*
41 6-Hydroxykaempferol (3,5,6,7,4′-penta-OH)
42 3-Methyl ether
43 5-Methyl ether (vogeletin)
44 6-Methyl ether
45 4′-Methyl ether
46 3,6-Dimethyl ether
47 3,7-Dimethyl ether*
48 6,7-Dimethyl ether (eupalitin)
49 6,4′-Dimethyl ether (betuletol)
50 3,6,7-trimethyl ether (penduletin)
51 3,6,4′-Trimethyl ether (santin)
52 6,7,4′-Trimethyl ether (mikanin)
53 3,6,7,4′-Tetramethyl ether
54 Herbacetin (3,5,7,8,4′-penta-OH)
55 7-Methyl ether (pollenitin)
56 8-Methyl ether (sexangularetin)
57 3,7-Dimethyl ether
58 3,8-Dimethyl ether
59 7,8-Dimethyl ether*
60 7,4′-Dimethyl ether*
61 8,4′-Dimethyl ether (prudomestin)
62 3,7,8-Trimethyl ether
63 3,8,3′-Trimethyl ether
64 3,8,4′-Trimethyl ether
65 7,8,4′-Trimethyl ether (tambulin)
66 3,7,8,4′-Tetramethyl ether (flindulatin)
67 Auranetin (3,6,7,8,4′-penta-OMe)
68 Morin (3,5,7,2′,4′-penta-OH)
69 7,5′-Dihydroxy-3,5,2′-trimethoxyflavone*
70 Quercetin (3,5,7,3′,4′-penta-OH)
71 3-Methyl ether
72 5-Methyl ether (azaleatin)
73 7-Methyl ether (rhamnetin)
74 3′-Methyl ether (isorhamnetin)
75 4′-Methyl ether (tamarixetin)
76 3,5-Dimethyl ether (caryatin)
77 3,7-Dimethyl ether
78 3,3′-Dimethyl ether
79 3,4′-Dimethyl ether
80 5,3′-Dimethyl ether*
81 7,3′-Dimethyl ether (rhamnazin)
82 7,4′-Dimethyl ether (ombuin)
83 3′,4′-Dimethyl ether (dillenetin)
84 3,7,3′-Trimethyl ether (pachypodol)
85 3,7,4′-Trimethyl ether (ayanin)
86 3,3′,4′-Trimethyl ether*
87 7,3′,4′-Trimethyl ether
88 3,5,7,4′-Tetramethyl ether
89 3,5,3′,4′-Tetramethyl ether
90 3,7,3′,4′-Tetramethyl ether (retusin)
91 3,5,8,3′,4′-Pentahydroxyflavone
92 Rhynchosin (3,6,7,3′,4′-penta-OH)
93 Melanoxetin (3,7,8,3′,4′-penta-OH)
94 Robinetin (3,7,3′,4′,5′-penta-OH)
95 Emmaosunin (5-OH,3,6,7,8,3′-penta-OMe)*
96 Calicopterin (5,4′-di-OH, 3,6,7,8-tetra-OMe)
97 Eriostemon (3,8′-di-OH, 5,6,7,4′-tetra-OMe)
98 Araneosol (5,7-di-OH, 3,6,8,4′-tetra-OMe)
99 5-Hydroxy-3,6,7,8,4′-pentamethoxyflavone*
100 Chrysosplenin (5,4′-di-OH, 3,6,7,2′-tetra-OMe)
101 Quercetagetin (3,5,6,7,3′,4′-hexa-OH)
102 3-Methyl ether*
103 5-Methyl ether*
104 6-Methyl ether (patuletin)
105 3′-Methyl ether
106 3,6-Dimethyl ether (axillarin)
107 3,7-Dimethyl ether (tomentin)
108 6,7-Dimethyl ether (eupatolitin)
109 6,3′-Dimethyl ether (spinacetin)
110 6,4′-Dimethyl ether (laciniatin)
111 3,6,7-Trimethyl ether (chrysosplenol-D)
112 3,6,3′-Trimethyl ether (jaceidin)
113 3,6,4′-Trimethyl ether (centaureidin)
114 3,7,3′-Trimethyl ether (chrysosplenol-C)
115 3,7,4′-Trimethyl ether (oxyayanin-B)
116 6,7,3′-Trimethyl ether (veronicafolin)
117 6,7,4′-Trimethyl ether (eupatin)
118 7,3′,4′-Trimethyl ether
119 3,5,6,3′-Tetramethyl ether*
120 3,6,7,3′-Tetramethyl ether (chrysosplenetin)
121 3,6,7,4′-Tetramethyl ether (casticin)
122 5,6,7,4′-Tetramethyl ether (eupatoretin)
123 3,6,3′,4′-Tetramethyl ether (bonanzin)
124 3,7,3′,4′-Tetramethyl ether*
125 3,6,7,3′,4′-Pentamethyl ether (artemetin)
126 3,5,6,7,3′,4′-Hexamethyl ether
127 5,4′-Dihydroxy-3,7,8,2′-tetramethoxyflavone*
128 5-Hydroxy-3,7,8,2′,4′-pentamethoxyflavone*
129 Gossypetin (3,5,7,8,3′, 4′-hexa-OH)
130 7-Methyl ether
131 8-Methyl ether (corniculatusin)
132 3′-Methyl ether
133 3,7-Dimethyl ether
134 3,8,-Dimethyl ether
135 7,4′,-Dimethyl ether
136 8,3′-Dimethyl ether (limocitrin)

137 8,4′-Dimethyl ether
138 3,7,8-Trimethyl ether
139 3,7,3′-Trimethyl ether*
140 3,8,3′-Trimethyl ether
141 7,8,3′-Trimethyl ether*
142 3,7,8,3′-Tetramethyl ether (ternatin)
143 3,7,8,4′-Tetramethyl ether
144 3,7,3′,4′-Tetramethyl ether
145 3,8,3′,4′-Tetramethyl ether*
146 3,7,8,3′,4′-Pentamethyl ether
147 3,5,7,8,3′,4′-Hexamethyl ether
148 Apuleidin (5,2′,3′-tri-OH, 3,7,4′-tri-OMe)
149 5′-Hydroxymorin (3,5,7,2′,4′,5′-hexa-OH)
150 3,7,4′-Trimethyl ether (oxyayanin-A)
151 3,5,7,4′-Tetramethyl ether
152 3,7,4′,5′-Tetramethyl ether
153 3,7,2′,4′,5′-Pentamethyl ether*
154 3,5,7,4′,5′-Pentamethyl ether*
155 3,5,7,2′,4′,5′-Hexamethyl ether*
156 5-Demethylapulein (2′,5′-di-OH, 3,6,7,4′-tetra-OMe)
157 Myricetin (3,5,7,3′,4′,5′-hexa-OH)
158 3-Methyl ether (annulatin)
159 5-Methyl ether
160 7-Methyl ether (europetin)
161 3′-Methyl ether (laricitrin)*
162 4′-Methyl ether (mearnsetin)
163 3,4′-Dimethyl ether*
164 3,5′-Dimethyl ether*
165 3′,5′-Dimethyl ether (syringetin)
166 3,7,4′-Trimethyl ether
167 3,3′,4′-Trimethyl ether*
168 7,3′,4′-Trimethyl ether
169 3,7,3′,4′-Tetramethyl ether
170 3,7,3′,4′,5′-Pentamethyl ether (combretol)
171 5,2′,4′-Trihydroxy-3,6,7,8-tetra-methoxyflavone*
172 5,4′-Dihydroxy-3,6,7,8,2′-penta-methoxyflavone*
173 5,2′-Dihydroxy-3,6,7,8,4′-penta-methoxyflavone*
174 5-Hydroxy-3,6,7,8,2′,4′-hexa-methoxyflavone*
175 5,7,3′,4′-Tetrahydroxy-3,6,8-tri-methoxyflavone
176 Limocitrol (3,5,7,4′-tetra-OH,6,8,3′-tri-OMe)
177 Isolimocitrol (3,5,7,3′-tetra-OH,6,8,4′-tri-OMe)
178 5,7,4′-Trihydroxy-3,6,8,3′-tetra-methoxyflavone*
179 5,4′-Dihydroxy-3,6,7,8,3′-penta-methoxyflavone*
180 5-Hydroxy-3,6,7,8,3′,4′-hexa-methoxyflavone
181 3-Hydroxy-5,6,7,8,3′,4′-hexa-methoxyflavone
182 3,5,6,7,8,3′,4′-Heptamethoxyflavone
183 Apuleisin (5,6,2′,3′-tetra-OH,3,7,4′-tri-OMe)
184 3,5,6,7,2′,3′,4′-Heptamethoxyflavone
185 Apulein (2′,5′-di-OH, 3,5,6,7,4′-penta-OMe)
186 6,2′-Dihydroxy-3,5,7,4′,5′-penta-methoxyflavone*
187 5,6-Dihydroxy-3,7,2′,4′,5′-penta-methoxyflavone*
188 3,5,6,7,2′,4′,5′,-Heptamethoxyflavone*
189 6-Hydroxymyricetin 3,6-dimethyl ether*
190 3,4′-Dimethyl ether*
191 3,6,3′-Trimethyl ether*
192 3,6,4′-Trimethyl ether*
193 6,3′,4′-Trimethyl ether*
194 3,5,6,5′-Tetramethyl ether
195 3,6,7,4′-Tetramethyl ether
196 3,6,3′,4′-Tetramethyl ether*
197 3,6,3′,5′,-Tetramethyl ether*
198 5,6,3′,5′,-Tetramethyl ether
199 3,7,3′,4′-Tetramethyl ether (apuleitrin)
200 3,5,7,3′,4′-Pentamethyl ether (apuleirin)
201 3,6,7,3′,5′-Pentamethyl ether (murrayanol)*
202 3,5,6,7,3′,5′-Hexamethyl ether
203 3,5,6,7,3′,4′,5′-Heptamethyl ether
204 Hibiscetin (3,5,7,8,3′,4′,5′-hepta-OH)
205 3,7,4′-Trimethyl ether*
206 3,7,8,4′-Tetramethyl ether*
207 3,8,4′,5′-Tetramethyl ether
208 3,8,3′,4′,5′-Pentamethyl ether (conyzatin)*
209 7,8,3′,4′,5′-Pentamethyl ether*
210 3,5,7,8,3′,4′,5′-Heptamethyl ether
211 Digicitrin (5,3′-di-OH, 3,6,7,8,4′,5′-hexa-OMe)
212 8-Hydroxy-3,5,6,7,3′,4′,5′-Hepta-methoxyflavone
213 Exoticin (3,5,6,7,8,3′,5′-octa-OMe)

III C-Methyl flavones and C-methyl flavonols

1 6-Methyl chrysin (5,7-di-OH, 6-Me)
2 Sideroxylin (5,4′-di-OH, 7-OMe, 6,8-di-Me)
3 5-Hydroxy-7,4′-dimethoxy-6-methylflavone
4 Eucalyptin (5-OH, 7,4′-di-OMe, 6,8-di-Me)

5 Unonal (5,7-di-OH, 8-Me, 6-CHO)*
6 Isounonal (5,7-di-OH, 6-Me, 8-CHO)*
7 Unonal 7-methyl ether (5-OH, 7-OMe, 8-Me, 6-CHO)*
8 5,7-Dihydroxy-3-methoxy-6,8-dimethylflavone*
9 Pityrogrammin (3,5,7-tri-OH, 8-OMe, 6-Me)*
10 Sylpin (5,6,4′-tri-OH, 3-OMe, 8-Me)*
11 5,7,4′-Trihydroxy-3-methoxy-6-methylflavone*
12 5,7,4′-Trihydroxy-3-methoxy-6,8-dimethylflavone*
13 Pinoquercetin (6-Me quercetin)
14 6,8-Dimethyl quercetin*
15 Pinoquercetin 3,3′-dimethyl ether*
16 6-Methylmyricetin
17 6-Methylmyricetin 3-methyl ether (alluaudiol)*
18 6,8-Dimethylmyricetin 3-methyl ether*
19 6-Methylmyricetin 4′-methyl ether (dumosol)*
20 6,8-Dimethylmyricetin 4′-methyl ether*
21 6-Methylmyricetin 3,4′-dimethyl ether*
22 6,8-Dimethylmyricetin 3,4′-dimethyl ether*

IV Methylenedioxyflavones and methylenedioxyflavonols

1 5-Hydroxy-7,8-methylenedioxyflavone*
2 5,6-Dimethoxy-3′,4′-methylenedioxyflavone*
3 Prosogerin-A (7-OH, 6-OMe, 3′,4′-O_2CH_2)*
4 Kanzakiflavone-2 (5,4′-di-OH, 6,7-O_2CH_2)*
5 Kanzakiflavone-1 (5,8-di-OH, 4′-OMe, 6,7-O_2CH_2)*
6 Linderoflavone A (5,7-di-OH, 6,8-di-OMe, 3′,4′-O_2CH_2)
7 Linderoflavone B (5,6,7,8-tetra-OMe, 3′,4′-O_2CH_2)
8 Eupalestin (5,6,7,8,5′-penta-OMe, 3′,4′-O_2CH_2)*
9 3,5-Dimethoxy-6,7-methylenedioxyflavone*
10 Meliternatin (3,5-di-OMe, 6,7-O_2CH_2, 3′,4′-O_2CH_2)
11 Demethoxykanungin (3,7-di-OMe, 3′,4′-O_2CH_2)
12 Gomphrenol (3,5,4′-tri-OH, 6,7-O_2CH_2)*
13 Kanugin (3,7,3′-tri-OMe, 4′,5′-O_2CH_2)
14 Melisimplin (5-OH, 3,6,7-tri-OMe, 3′,4′-O_2CH_2)
15 Melisimplexin (3,5,6,7-tetra-OMe, 3′,4′-O_2CH_2)
16 5-Hydroxy-3,7,8-trimethoxy-3′,4′-methylenedioxyflavone*
17 Meliternin (3,5,7,8-tetra-OMe, 3′,4′-O_2CH_2)
18 5,3′,4′-Trihydroxy-3-methoxy-6,7-methylenedioxyflavone
19 Wharangin (5,3′,4′-tri-OH, 3-OMe, 7,8-O_2CH_2)
20 Melinervin (3,6,7-tri-OH, 6,8-di-OMe, 3′,4′-O_2CH_2)*
21 5-Hydroxy-3,6,7,8-tetramethoxy-3′,4′-methylenedioxyflavone*
22 Melibentin (3,5,6,7,8-penta-OMe, 3′,4′-O_2CH_2)

V Isoprene-substituted flavones and flavonols

1 *Trans*-lanceolatin*
2 *Cis*-tephrostachin*
3 *Trans*-tephrostachin
4 *Trans*-anhydrotephrostachin
5 Artocarpesin
6 Oxodihydroartocarpesin
7 Norartocarpin
8 Mulberrin
9 Integrin*
10 Artocarpin
11 Kuwanone C*
12 8-Prenyl-luteolin
13 Rubraflavone A
14 Rubraflavone B
15 Rubraflavone C
16 8-Hydroxyprenyl-5-deoxykaempferol
17 6-Prenyl-3,5,7,8-tetrahydroxyflavone*
18 Noranhydroicaritin
19 Noricaritin
20 Isoanhydroicaritin
21 Icaritin
22 6-Hydroxyprenylquercetin
23 7-*O*-Prenylquercetin 3,3′-dimethyl ether*
24 8-Prenylquercetin 3,7,3′-trimethyl ether
25 7,8-Dimethylallyloxy-5-hydroxyflavone*
26 Isopongaflavone
27 Praecansone B
28 7,6-Dimethylallyloxy-5,4′-dihydroxyflavone*
29 Cyclomulberrochromene
30 Cyclomorusin*
31 Cycloartocarpesin
32 Mulberrochromene

33 Kuwanone B*
34 Hydroxydihydromorusin*
35 Morusin*
36 Kuwanone A*
37 Rubraflavone D
38 Cyclomulberrin
39 Cycloartocarpin
40 Desmodol*
41 Cycloheterophyllin
42 Artobilichromene*
43 Karanjachromene
44 Pongaflavone
45 Pongachromene*
46 7,8-Dimethylallyloxy-3,6-dimethoxyflavone*
47 Sericetin
48 Phellamuretin
49 7,8-Dimethylallyloxy-3,5,3',4'-tetrahydroxyflavone*
50 Lanceolatin B
51 Pongaglabrone*
52 Pongaglabol*
53 Pinnatin
54 Gamatin
55 6-Methoxy-7,8-furanoflavone*
56 8-Methoxy-7,8-furanoflavone*
57 3'-Hydroxy-7,8-furanoflavone*
58 Dihydrofuranoartobilichromene b*
59 Dihydrofuranoartobilichromene a*
60 Mulberranol
61 Karanjin
62 Pongapin
63 3'-Methoxypongapin
64 *Morus alba* compound A
65 Oxyisocyclointegrin
66 Chaplashin
67 Isocycloheterophyllin
68 *Artocarpus* oxocin flavone

* Structures newly reported 1975–1980

The details of isoprene substituted flavones and flavonols can be found in Tables 4.5–4.8 and Figs. 4.1–4.3; the total number of these structures may be slightly greater than 68 since the list here may not be completely exhaustive.

CHAPTER FIVE

Flavone and Flavonol Glycosides

JEFFREY B. HARBORNE and CHRISTINE A. WILLIAMS

5.1 INTRODUCTION

In 1975, some 360 flavone and flavonol glycosides were known to occur in plants (Harborne and Williams, 1975). In the intervening five years, a similar number of new compounds have been recorded, bringing the total to about 720 structures (see the check list at the end of the chapter). This is, in fact, a conservative estimate, since a variety of partly characterized glycosides, probably differing in structure from those reported in the Tables in this chapter, have been omitted from this check list. In the present chapter, therefore, we provide an account of the new substances discovered during the period 1975–1980. References to the earlier literature will be included from time to time, since a number of structural reassignments have also taken place. The term 'conjugate' might be more apt than 'glycoside', since here, as in our previous account, we are including flavones and flavonols with acyl and sulphate substituents; these usually, but not invariably, have glycosidic attachments as well.

Many of the new compounds reported have the expected structures and simply fill in gaps among known derivatives. For example, the 3-xyloside of quercetin was first described from the leaves of *Reynoutria japonica* (Polygonaceae) by Nakaoki and Morita (1956), and has been found in a number of other plants since (see Harborne, 1967). The related kaempferol and myricetin derivatives have now been discovered: kaempferol 3-xyloside occurs in *Euonymus* leaf (Celastraceae) (Ishikura and Sato, 1977); and myricetin 3-xyloside is one of a number of glycosides present in *Leptarrhena pyrolifolia* (Saxifragaceae), reported by Miller and Bohm (1979).

An increasing number of other new glycosides reported contain acyl substituents (e.g. acetyl, galloyl, *p*-coumaroyl etc.) attached to sugar hydroxyls, and a powerful tool for the location of these substituents is ^{13}C-NMR spectroscopy (see Chapter 2). The application of this technique to flavonoids has also led to the discovery of a new rare hexose, allose, in glycosidic combination. Allose occurs as such in kaempferol 3-alloside and in certain flavones in combination with glucose as a disaccharide (see Section 5.4). Allose was only recognized as a plant sugar as recently as 1973 by Perold *et al.*, who isolated some allose derivatives from the leaves of *Protea rubropilosa*; it has since also been found in glycosidic combination with iridoids (Jensen *et al.*, 1981).

A variety of new oligosaccharides has also been reported, particularly in association with kaempferol and quercetin. A notable series of four branched trisaccharides occur attached to flavonols in soya bean leaves: 2^G-rhamnosyl- and 2^G-glucosyl-rutinose, 2^G-rhamnosyl- and 2^G-glucosyl-gentiobiose (Buttery and Buzzell, 1975). Several new linear trisaccharides have also been described, e.g. rhamnosyl(1→2)rutinose. Only one tetrasaccharide (an acetylated rhamnosylsophorotriose) has yet been fully characterized as a flavonoid sugar. Since penta- and even hexa-saccharides are known to occur in glycosidic combination with triterpenoids (Chandel and Rastogi, 1980), their discovery in association with flavonoids cannot be far off.

In recent years attention has been given to the importance of glycosylation in relation to the function of flavones and flavonols in plants (Harborne, 1979a). A significant number of these compounds has been detected in plants occurring in the free state without sugar attachment (Wollenweber and Dietz, 1981). These tend to be those which have one or more *O*-methyl groups; furthermore, their occurrence is often associated with other lipophilic constituents and may be restricted to certain parts of the plant, e.g. the waxy coatings of the leaves. Free flavones and flavonols are potentially toxic to living cells and inhibit many enzymic activities (Harborne, 1979a), so that glycosylation must be an essential protective device to prevent cytoplasmic damage and to locate the flavonoids safely in the cell vacuole. While glycosylation is most frequent with the fully hydroxylated compounds such as kaempferol and quercetin, it also occurs with many of their methyl ethers. Indeed, an increasing number of novel *O*-methylated flavonols, especially those in the quercetagetin and gossypetin series, have been discovered recently. Significantly, most of these are found with at least one or two sugar residues and many of the new flavonol glycosides recorded here (see Section 5.6) are of this type.

5.2 SEPARATION AND PURIFICATION

Standard procedures based on thin-layer, paper and column chromatography are still most widely used for the separation and purification of flavone and flavonol glycosides. These techniques provide most compounds in sufficient quantity for routine structural determination. If larger amounts are desired, then other techniques may be needed (see below). It is perhaps worth emphasizing the importance of checking homogeneity of isolated compounds in as many chromatographic systems as possible. It is certainly advisable to test for purity using thin-layer chromatography on microcrystalline cellulose, polyamide and silica gel, employing several solvent systems in each case. Even so, there may be pairs of glycosides which are unresolvable in any known system. The 3-arabinosylglucoside and 3-xylosylglucoside of isorhamnetin present in *Festuca* leaf extracts are apparently inseparable, as are the related quercetin derivatives (Wong, 1980).

A number of new chromatographic techniques has been developed recently (see Chapter 1) for flavonoids and some of these will be of importance for separating glycoside mixtures. In particular, droplet counter-current (DCC) chromatography appears to offer an efficient and rapid method for separating glycosides on a preparative scale. In one trial run, a crude 130 mg fraction from the plant *Tecoma stans* (Bignoniaceae) was subjected to DCC separation using the upper phase of chloroform–methanol–water (7:13:8) to yield 27 mg of pure quercetin 3-glucoside within about 6 h (Hostettmann *et al.*, 1979).

High-performance liquid chromatography (HPLC) can also be applied on the preparative scale but up to the present it has mainly been used analytically. Its value in quantitative analysis of glycosides is nicely illustrated by recent determinations of amounts of flavonol glycosides in sepals of Poinsettia (Stewart *et al.*, 1980). By using two solvent systems, it was possible to measure individual amounts of the ten glycosides present, namely the 3-rutinosides, 3-rhamnosylgalactosides, 3-rhamnosides, 3-glucosides and 3-galactosides of kaempferol and quercetin. Difficulties were experienced in distinguishing the 3-glucoside and 3-galactoside of quercetin, but these are two glycosides which run close together in every chromatographic system. HPLC is obviously a technique that will be used more and more in the determination of the complex glycosidic patterns that are present in many angiosperm plants.

5.3 IDENTIFICATION

The most important recent development in the structural identification of flavonoids is the application of ^{13}C-NMR spectroscopy. It has been used successfully with flavone and flavonol glycosides (see Chapter 2, Section 2.3) to provide information on the nature and position of glycosidic attachment(s). The only limitation is that of sample size. It is necessary to have about 10 mg of material, preferably more, for spectral analysis. Its especial value is in revealing the sequence and mode of interglycosidic linkages of complex glycosides and in locating and identifying acyl substitution. ^{13}C-NMR spectroscopy, for example, has been used to identify the acylated trisaccharide present in two flavonols of *Colubrina faralaotra* (Rhamnaceae) as glucosyl(1→3)-4‴-acetylrhamnosyl (1→6)galactose. This sequence of the three sugars attached linearly to the 3-hydroxyl position of kaempferol and quercetin followed from relaxation time measurements (Guinaudeau *et al.*, 1981).

Even with simple flavone or flavonol monoglycosides, ^{13}C-NMR data may be of value in establishing the position of sugar attachment, the nature of the linkage and the configuration and conformation of the sugar. While all these can be determined by other methods (e.g. enzymically), the procedure has importance where controversy exists. Application of this technique should, for example, reveal finally whether any flavones or flavonols occur naturally with glucose, rhamnose or xylose attached to them, in the furanose form (see Section 5.4). It is

significant that all measurements made to date on a variety of such *O*-glycosides (see Table 2.15, Chapter 2) have shown that these three sugars always exist in the expected pyranose form.

Mass spectroscopy still has a place in the structural analysis of glycosides, since it can be applied to smaller samples (1 mg or less) than are required for NMR measurements. It is possible to detect weak molecular ion signals on permethylated or perdeuteriomethylated derivatives and thus establish the molecular weight. The fragmentation may also be quite characteristic of the glycosidic pattern present (see Harborne and Williams, 1975). According to Schels *et al.* (1977, 1978) there are advantages in making these measurements on the silyl ethers. The advantage over permethylation is that silylation occurs rapidly and there is no need to purify before analysis, as with the permethyl ethers. The molecular mass of the silylated glycoside is apparent from a weak M^+ ion (2–4% intensity of parent ion) and a stronger M^+-15 ion. In the case of silylated flavonol 3-biosides, it is possible to deduce the sugar sequence because the terminal sugar produces a very intense fragment (relative intensity > 30%) accompanied by two other related ions. Again, with flavonol 3,7-diglycosides, it is possible to deduce the positions of attachment, since the 3-*O*-sugar gives more intense signals than that in the 7-position.

The position of sugar attachment can often be determined more simply from colour reactions and UV spectral measurements. Homberg and Geiger (1980) have now shown that it can also be deduced from the fluorescence spectra. These have been determined for a representative number of flavone and flavonol glycosides adsorbed on cellulose, either alone or in the presence of various shift reagents.

Enzymic procedures for the structural determination of glycosides are very simple but are often neglected. Until recently, it has not been clear that specific enzymes can be employed for the removal of the terminal sugar or for the stepwise removal of all the sugars in complex glycosides. Buttery and Buzzell (1975), however, have shown that β-glucosidase will specifically remove terminal glucose units from flavonol triglycosides to give diglycosides in good yield. Thus kaempferol 3-(2^G-glucosylgentiobioside) by cleavage of the β1→6 linkage gives kaempferol 3-sophoroside whereas kaempferol 3-(2^G-glucosylrutinoside) by cleavage of the β1→2 linkage gives kaempferol 3-rutinoside. It is interesting that given the choice of a β1→2- and β1→6-linked glucose in the first triglucoside, β-glucosidase only hydrolyses the latter; the preference of the enzyme for a β1→6 over a β1→2 linkage is, however, expected from earlier studies (Harborne, 1965). These enzymic results, together with those from partial acidic hydrolysis, were helpful in characterizing the above kaempferol triglucoside as a new glycosidic type.

The stepwise removal of sugars from complex glycosides has also been successfully used to determine the structure of new compounds. For example, *Primula officinalis* leaves (Primulaceae) contain a quercetin 3-rhamnosyl-robinobioside (Karl *et al.*, 1981). The terminal position of the second rhamnose

was established when soyaremanase, an enzyme preparation from soya beans, selectively removed it to give quercetin 3-robinobioside. These authors were also able to recover the robinobiose [rhamnosyl(α1$\rightarrow$6)galactose] intact from this intermediate following enzymic hydrolysis with rhamnodiastase, a disaccharidase from *Rhamnus cathartica* seed.

Similarly, a kaempferol 3-(caffeoyl-laminaribioside)-7-rhamnoside from *Phyllitis scolopendrium* lost its 7-rhamnose sugar with α-rhamnosidase to give kaempferol 3-(caffeoyl-laminaribioside). This compound, on cellulase hydrolysis, lost a glucose unit to yield kaempferol 3-(caffeoylglucoside), an experiment which proved that the caffeoyl residue must be located on the sugar nearest to the aglycone moiety (Karl *et al.*, 1980). Clearly, such procedures deserve to be more widely known, since they may often simplify structural determinations. It is to be hoped that some of the more specific enzyme preparations needed will eventually become available commercially.

5.4 SUGARS OF FLAVONE AND FLAVONOL GLYCOSIDES

5.4.1 Monosaccharides

The ten monosaccharides known to occur in association with flavones and flavonols are shown in Table 5.1. Most occur both in monosaccharide combination and as components of di- and tri-saccharides attached to these substances (see Sections 5.4.2 and 5.4.3). Three of the sugars deserve further comment: allose, apiose and galacturonic acid. The hexose allose is the only new sugar reported in the last five years. It was discovered attached to kaempferol in the 3-position in the glycoside, asiaticalin, from the fern *Osmunda asiatica* (Okuyama *et al.*, 1978). It has also been found in disaccharide combination with glucose, as 2-allosylglucose, in two flavones based on chrysoeriol and isoscutellarein from *Sideritis* and *Veronica* species respectively (Chari *et al.*, 1981).

Table 5.1 Monosaccharides of flavone and flavonol glycosides

Pentoses	*Hexoses*	*Uronic acids*
D-Apiose	D-Allose	D-Galacturonic acid†
L-Arabinose*	D-Galactose	D-Glucuronic acid†
L-Rhamnose	D-Glucose	
D-Xylose	D-Mannose	

* Definitely known to occur in both pyranose and furanose forms; all other sugars (except apiose) are normally (always) in the pyranose form.
† Also reported to occur as the methyl ester.

Apiose, an unusual branched pentose, is well known to occur as a disaccharide with glucose. Indeed apigenin 7-apiosylglucoside, called apiin, from seeds of *Apium graveolens* (Umbelliferae) was one of the very first flavone glycosides to be

isolated in crystalline condition (Lindenborn, 1869). Its discovery as a monoglycoside is, however, quite novel. The 7-apioside of 6-hydroxyluteolin has just been reported in *Lepidagathis cristata* (Acanthaceae) by Ranganathan *et al.* (1980).

Until recently, galacturonic acid has only appeared quite rarely as a flavonoid sugar. However, a remarkable range of such derivatives has been uncovered in liverworts and extend the number of uronic acid derivatives considerably. Quercetin 3-galacturonide has been found for the first time, for example, in *Corsinia coriandrina* (Markham, 1980). A variety of luteolin derivatives present in *Marchantia berteroana* also deserve mention, namely the 7-galacturonide, the 3′-galacturonide, the 7,3′-digalacturonide and the 3′,4′-digalacturonide (Markham and Porter, 1975).

One other sugar, besides the ten listed in Table 5.1, may be associated with flavonoids, namely the keto sugar fructose, although the evidence for its occurrence is still not fully authenticated. The description of its presence in leaves of the tea plant *Camellia sinensis* in association with quercetin and glucose is limited by the fact that it could only be detected in processed leaves and thus might be of artifactual origin (Imperato, 1976). Fructose has also been reported in disaccharide combination with glucose attached to the flavone tricin in *Hyacinthus orientalis* (Liliaceae) (Williams, 1975) but the nature of its linkage in this glycoside is still obscure.

Of the ten sugars listed in Table 5.1, only arabinose is definitely known to occur in both the furano and pyrano forms. There have, in addition, been reports of rhamnose, glucose, galactose and xylose also occurring in furanose form when attached to flavones and flavonols, but these are exclusively from the Russian literature. Furthermore, the evidence for the presence of these furano forms is far from unequivocal and was based largely on IR-spectral and hydrolytic data. A careful examination in our laboratory of a reputed quercetin 3-α-rhamnofuranoside from horse chestnut showed that it was indistinguishable in all respects from the well known and well characterized 3-α-rhamnopyranoside (quercitrin). The same question has now been taken up by Farkas *et al.* (1976a) regarding furanose-containing forms of the glycoside robinin with similar negative conclusions.

The complete structure of robinin as kaempferol 3-*O*-(6-*O*-α-L-rhamnopyranosyl-β-D-galactopyranoside)-7-*O*-α-L-rhamnopyranoside was established by Zemplén and Bognár (1941) and confirmed by Shimokoriyama (1949). In spite of this, Maksyutina *et al.* (1965) later claimed robinin to be a mixture of four isomeric glycosides in which the 7-rhamnose moiety existed in the α- and β-furano as well as the α- and β-pyrano forms; in all these forms, the two sugars attached in the 3-position were reported to be furano. Farkas *et al.* (1976a), in a detailed rebuttal of these allegations, reconfirmed that all the sugars in robinin are in fact pyranose. Thus the kaempferol 7-rhamnoside, obtained by partial hydrolysis of robinin, was fully methylated and after hydrolysis, reduction and acetylation yielded 2,3,4-trimethylrhamnitol diacetate. The α-configuration and anomeric homogeneity of the kaempferol 7-rhamnoside was

further established by synthesis. Finally, a total synthesis of robinin itself once more confirmed the absence of furanose rings from any of its sugars.

We can only conclude from these results that all reports of furanose-containing glycosides (other than arabinosides) must be treated with caution and cannot be regarded as correct, unless supported by more conclusive evidence. We are therefore excluding these reports from our lists, although many of them were mentioned in our earlier survey (Harborne and Williams, 1975).

One final point regarding the monosaccharides needs emphasizing: the nature of the linkage to the phenolic hydroxyl. This is usually assumed to be β- for glucosides, galactosides, glucuronides and xylosides and α- for arabinosides and rhamnosides. A β-linked 3-arabinoside of quercetin is, however, well-known (Geissman, 1962) and other β-linked arabinosides may be present in plants. On the other hand, α-linked glucosides and galactosides are also possible and need to be distinguished from the much more common β-derivatives (see Harborne and Williams, 1975). Studies with appropriate hydrolytic enzymes are thus desirable in all cases when new compounds are found.

5.4.2 Disaccharides

Ten disaccharides were listed by Harborne and Williams (1975) as being fully characterized as components of flavone or flavonol glycosides. A further 11 disaccharides have now been described since 1975 and all 21 are collected together in Table 5.2. Considerable progress has thus been made in the characterization of disaccharide moieties. Nevertheless, as will be clear in later sections of this chapter, there are still many diglycosides reported in plants which probably have new disaccharides but which have not been fully characterized. Eventually, some of the gaps in Table 5.2 will be filled from further studies on such compounds.

Of the new disaccharides, the discovery of lactose in association with isorhamnetin in seeds of *Cassia multijuga* (Leguminosae) (Dubey and Gupta, 1980) is unexpected, since this disaccharide otherwise occurs in the free state in human milk. Most of the others are related to earlier reported disaccharides. Thus, the arabinosyl(α1→6)galactose, present attached to quercetin in *Hydrocotyle vulgaris* leaf (Umbelliferae) (Hiller *et al.*, 1979), is related to vicianose except that galactose replaces glucose. Again, the rhamnosyl(α1→2)galactose, in a kaempferol 3-glycoside from *Chenopodium fremontii* leaf (Chenopodiaceae) (Crawford and Mabry, 1978), is an isomer of the well-known robinobiose, which has the same two sugars in an α1→6 linkage.

The first disaccharides based on glucose and galactose have been characterized, i.e. Glc(β1→6)Gal and Gal(β1→4)Glc, as have two disaccharides based on glucose and mannose (Table 5.2). The first digalactose to be fully characterized is also reported Gal(β1→4)Gal, which is attached to a myricetin derivative from *Vitex negundo* (Verbenaceae) (Misra and Subramanian, 1980). Finally, one of the most distinctive of the new disaccharides must be Gal(α1→4)Rha where the

Table 5.2 Disaccharides of flavone and flavonol glycosides

Structure	*Trivial name*
Pentose–pentose	
4-*O*-α-L-Rhamnosyl-D-xylose*	
Pentose–hexose	
2-*O*-α-L-Rhamnosyl-D-glucose	Neohesperidose
3-*O*-α-L-Rhamnosyl-D-glucose	Rungiose
6-*O*-α-L-Rhamnosyl-D-glucose	Rutinose
2-*O*-β-D-Xylosyl-D-glucose	Sambubiose
6-*O*-α-L-Arabinosyl-D-glucose	Vicianose
2-*O*-α-L-Rhamnosyl-D-galactose*	
6-*O*-α-L-Rhamnosyl-D-galactose	Robinobiose
2-*O*-β-D-Xylosyl-D-galactose	Lathyrose
6-*O*-α-L-Arabinosyl-D-galactose*	—
Hexose–pentose	
4-*O*-β-D-Glucosyl-L-rhamnose*	—
4-*O*-β-D-Galactosyl-L-rhamnose*	—
Hexose–hexose	
2-*O*-β-D-Glucosyl-D-glucose	Sophorose
3-*O*-β-D-Glucosyl-D-glucose	Laminaribiose
6-*O*-β-D-Glucosyl-D-glucose	Gentiobiose
2-*O*-β-D-Mannosyl-D-glucose*	—
4-*O*-β-D-Glucosyl-D-mannose*	—
2-*O*-β-D-Allosyl-D-glucose*	—
4-*O*-β-D-Galactosyl-D-glucose*	Lactose
6-*O*-β-D-Glucosyl-D-galactose*	
4-*O*-β-D-Galactosyl-D-galactose*	—

* Newly discovered in diglycosidic combination, in the period 1975–1980.

normal order pentose–hexose is reversed and the rhamnose is directly attached to the aglycone with the galactose as the terminal sugar. This disaccharide occurs linked to the 3-hydroxyl of galangin in *Artocarpus lakoocha* rootbark (Moraceae) (Chauhan *et al.*, 1979a).

5.4.3 Trisaccharides

The rapid progress that has been made in the study of flavonol 3-triglycosides is apparent from the fact that seven trisaccharides (five linear, two branched) were reported as fully characterized in 1975, while a further ten (three linear, seven branched) have been described since then. All these structures are shown in Table 5.3. In all cases, the trisaccharide is attached to the 3-hydroxyl of a flavonol, either kaempferol and/or quercetin or a simple methyl ether of these two aglycones.

One reassignment of structure has occurred. This is of a triglucose attached to kaempferol and quercetin and reported in *Primula sinensis* and other *Primula*

Table 5.3 Trisaccharides of flavonol glycosides

Structure	*Trivial name*
Linear	
O-α-Rhamnosyl-(1→2)-*O*-α-rhamnosyl-(1→6)-glucose*	2′-Rhamnosylrutinose
O-β-Glucosyl-(1→2)-*O*-β-glucosyl-(1→6)-glucose*	2′-Glucosylgentiobiose
O-β-Glucosyl-(1→2)-*O*-β-glucosyl-(1→2)-glucose	Sophorotriose
O-β-Glucosyl-(1→6)-*O*-β-glucosyl-(1→4)-glucose	Sorborose
O-α-Rhamnosyl-(1→4)-*O*-α-rhamnosyl-(1→6)-galactose	Rhamninose
O-α-Rhamnosyl-(1→3)-*O*-α-rhamnosyl-(1→6)-galactose	Sugar of alaternin
O-β-Glucosyl-(1→3)-*O*-α-rhamnosyl-(1→6)-galactose*	Sugar of faraltroside†
O-β-Glucosyl-(1→6)-*O*-β-glucosyl-(1→4)-rhamnose*	—
Branched	
O-β-Apiosyl-(1→2)-*O*-[α-rhamnosyl-(1→6)-glucose]*	2G-Apiosylrutinose
O-α-Rhamnosyl-(1→2)-*O*-[α-rhamnosyl-(1→6)-glucose]*	2G-Rhamnosylrutinose
O-β-Glucosyl-(1→2)-*O*-[α-rhamnosyl-(1→6)-glucose]	2G-Glucosylrutinose
O-β-Glucosyl-(1→3)-*O*-[α-rhamnosyl-(1→2)-glucose]	3G-Glucosylneohesperidose
O-α-Rhamnosyl-(1→2)-*O*-[β-glucosyl-(1→6)-glucose]*	2G-Rhamnosylgentiobiose
O-β-Glucosyl-(1→2)-*O*-[β-glucosyl-(1→6)-glucose]*	2G-Glucosylgentiobiose
O-α-Rhamnosyl-(1→4)-*O*-[α-rhamnosyl-(1→2)-glucose]*	4G-Rhamnosylneohesperidose
O-α-Rhamnosyl-(1→2)-*O*-[α-rhamnosyl-(1→6)-galactose]*	2Gal-Rhamnosylrobinobiose
O-α-Rhamnosyl-(1→4)-*O*-[α-rhamnosyl-(1→6)-galactose]*	4Gal-Rhamnosylrobinobiose

* Newly reported 1975–1980.
† Only present with an acetyl group at the 4-hydroxyl of rhamnose.

species (Harborne and Sherratt, 1961). Since this sugar gave only gentiobiose on hydrolysis and R_M values indicated two β1→6 linkages, it was suggested that it might be gentiotriose (Harborne, 1968). More recently, both kaempferol and quercetin 3-gentiotriosides were synthesized by standard procedures (Farkas *et al.*, 1976b) and comparison with the natural kaempferol 3-triglucoside showed that the two compounds were different. Subsequently, the quercetin 3-triglucoside in *Primula officinalis* has been re-examined and the trisaccharide identified as one with both β1→2 and β1→6 links, i.e. as 2′-glucosylgentiobiose (Karl *et al.*, 1981). Since the pattern of flavonol glycosylation is probably the same throughout the genus, it is reasonable to assume that the *P. sinensis* derivatives also contain the same sugar.

Among the new trisaccharides reported (Table 5.3) are a remarkable number of branched sugars. A very rich source of kaempferol and quercetin 3-triglycosides is the soya-bean *Glycine max* (Leguminosae). From the leaf, Buttery and Buzzell (1975) have characterized four such trisaccharides in flavonol combination: 2G-glucosylrutinose, 2G-rhamnosylrutinose, 2G-glucosyl-gentiobiose and 2G-rhamnosylgentiobiose – the latter three are all new compounds. The genetic control of the synthesis of these branched triglycosides has also been examined, and their production occurs whenever plants have the ability to synthesize related pairs of diglycosides, i.e. 3-sophorosides and 3-rutinosides, 3-gentiobiosides and 3-rutinosides etc. Interestingly, the two

flavonol 3-(2G-rhamnosylrutinosides) of soya bean also occur in hops, *Humulus lupulus*, but here they are accompanied by the related linear triglycosides, the 3-(2′-rhamnosylrutinosides) (Vancraenenbraeck *et al.*, 1969).

Finally, it must be pointed out that other trisaccharides occur associated with flavonols and with flavones, which await full characterization. Many of these are indicated in later Tables in this chapter.

5.4.4 Acylated derivatives

Some 60 new acyl derivatives have been reported in the period 1975–1980. The increasing frequency of these reports reflects to some extent the introduction of ^{13}C-NMR spectroscopy for flavonoid analysis; some acyl substituents, such as acetyl, are not readily detectable by any other means. The report of 2-methylbutyric acid attached to the rhamnose moiety of acacetin 7-rutinoside in *Valeriana wallichii* represents one of the first examples of the utility of ^{13}C-NMR spectroscopy for pinpointing the nature and location of an acyl substitution (Chari *et al.*, 1977).

In toto, some seven aliphatic and eight aromatic acids have been identified as acyl substituents in flavones and flavonols (Table 5.4). The usual linkage is through one of the sugar hydroxyls and any one of the three or four free hydroxyls may be involved. The direct linkage of acyl group to a phenolic hydroxyl also occasionally occurs. Recent examples of this are the 8-*O*-acetates and 8-*O*-butyrates of several flavonols found in frond exudates of the fern genus *Notholaena* by Wollenweber *et al.* (1978a,b,c).

The most common acyl groups appear to be acetic, gallic, *p*-coumaric and ferulic acids. Monoacylation is usual, but substances with two acyl groups have been reported. Thus apigenin 4′-glucoside occurs with *p*-coumaroyl residues at the 2″ and 6″-positions of the glucose moiety, in *Lycopodium clavatum* leaf (Ansari *et al.*, 1979). Furthermore, the acyl groups may be different. Thus, kaempferol 3-glucoside has been reported as the 3″-*p*-coumarate-6″-ferulate ester in *Picea obovata* needles (Zapesochnaya *et al.*, 1978).

Several pairs of isomeric acyl derivatives are known. These may occur together in the same plant. While acyl migration might possibly occur during isolation and handling, there is no suggestion as yet that one or other of a given pair are of artifactual origin. One example is apigenin 7-glucoside, which occurs in *Matricaria chamomilla* flowers as a mixture of the 2″- and 6″-acetates (Redaelli *et al.*, 1980). Another is quercetin 3-galactoside which is present in *Tellima grandiflora* as a mixture of 2″- and 6″-gallates (Collins *et al.*, 1975).

In some cases, the number of known isomers has not yet been clarified. This is true of the mono-*p*-coumaroyl esters of kaempferol 3-glucoside. Two are of certain structure: the 6″-ester, tiliroside from *Tilea argentea* petals and recently described from *Anaphalis contorta* (Compositae) (Lin *et al.*, 1975) and from *Pinus contorta* needles (Higuchi and Donnelly, 1978); and the 3″-ester, found in needles of *Picea obovata* by Zapesochnaya *et al.* (1978). An ester (of undefined

Table 5.4 Acylating acids found in flavone and flavonol derivatives

Organic acid	*Example of occurrence*	*Reference*
Aliphatic acids		
Acetic	Apigenin 7-(2″-acetyl)glucoside	Redaelli *et al.* (1980)
Malonic	Quercetin 3-(malonyl)glucoside*	Woeldecke and Herrmann (1974)
Succinic	Kaempferol 7-(6″-succinyl)glucoside	Hiraoka and Maeda (1979)
Butyric	Herbacetin 8-butyrate	Wollenweber *et al.* (1978a)
2-Methylbutyric	Acacetin 7-[2‴-(2 methylbutyryl)rutinoside]	Chari *et al.* (1977)
Tiglic	Quercetin 7-(6″-tiglyl)glucoside	Ogawa and Ogihara (1975)
Aromatic acids		
Benzoic	Kaempferol 3-(benzoylglucoside)*	Schonsiegel *et al.* (1969)
p-Hydroxybenzoic	Kaempferol 3(*p*-hydroxybenzoylglucoside)*	Schonsiegel *et al.* (1969)
Gallic	Quercetin 3-(6″-galloyl)glucoside	Collins *et al.* (1975)
p-Coumaric	Apigenin 7-(4″-*p*-coumaroyl)glucoside	Karl *et al.* (1976)
Caffeic	Apigenin 7-glucoside-4′-caffeate	Gella *et al.* (1967)
Ferulic	Luteolin 7-glucuronide-3′-feruloylglucoside*	Markham *et al.* (1978a)
Isoferulic	Quercetin 3-(isoferuloyl)glucuronide*	El Ansari *et al.* (1976)
Sinapic	Kaempferol 3-(sinapoylsophoroside)-7-glucoside*	Stengel and Geiger (1976)

* In these examples, the position of attachment of acyl group to sugar has not been established.

linkage) is reported to be present in leaves of *Larix* and other Pinaceae by Niemann (1979), but it is not clear whether this refers to the 3″- or 6″-ester, or both. Yet another ester that has been described is tribuloside, which occurs in *Tribulus terrestris* and is known to be different from tiliroside. It cannot be the 6″-ester as originally proposed (Bhutani *et al.*, 1969) since this is the now proven structure of tiliroside, based on ^{13}C NMR studies (Pijewska and Kamecki, 1977; Chari *et al.*, 1978). The most recent work on the structure of tribuloside suggests it is, in fact, a mixture of tiliroside with the related isorhamnetin derivative (Saleh, N. A. M., private communication).

The most complex acyl derivative to date is probably an acacetin derivative from *Coptis japonica* leaves, which not only has three acetyl residues attached variously to two different monosaccharide units, but also contains the only known tetrasaccharide, namely 6^{G}-rhamnosylsophorotriose (Fujiwara *et al.*, 1976). The sugar moiety is thus glucosyl (1→ 2)glucosyl(1→ 2) [2,4-diacetyl-rhamnosyl(1→ 6)]-3-acetylglucose.

5.4.5 Sulphate conjugates

In 1975, only a relatively few sulphate conjugates had been characterized. A range of new structures has been described between 1975 and 1980 and the total number of conjugates is now 54 (Table 5.5). In both the flavone and flavonol series, the most common type is the 7- and 3-sulphate respectively. Sulphation at other hydroxyl groups is also possible and flavones with sulphation at the B-ring hydroxyls have been described from *Daucus carota* (Harborne and King, 1976), *Lachenalia unifolia* (Williams *et al.*, 1976) and *Zostera marina* (Harborne and Williams, 1976). The first known disulphated flavone, luteolin 7,3′-disulphate, was also obtained from the latter plant. While a polysulphated quercetin salt was reported earlier from *Flaveria bidentis* (Pereyra de Santiago and Juliani, 1972), several new forms, including the 3,4′-disulphate and 3,7,4′-trisulphate, have been subsequently detected in the same species (Cabrera and Juliani, 1979). Such polysulphation, however, is still rare and the normal rule is to find monosulphates only.

A significant number of glycosidesulphates has been described (Table 5.5). In some of these, the sugar and sulphate residues occupy different phenolic hydroxyls of the flavone or flavonol, but in others the sulphate is attached through one of the sugar hydroxyls. In two pigments of the latter type, the position of the sulphate has been shown to be on the 6-hydroxyl. Thus, in the kaempferol 3-α-glucosidesulphate of the fern *Asplenium filix-foemina*, 6-*O*-sulphation was established by methylation and hydrolysis to give 2,3,4-tri-*O*-methylglucose (Imperato, 1979b). Similarly, in the kampferol 3-gentiobioside sulphate present in *Asplenium fontanum*, the sulphate was located by similar methods as being at the 6-position of the terminal glucose (Imperato, 1980).

One other variation in sulphate conjugation has been detected recently, in the nature of the associated cation. While potassium is usual, a eupatin 3-sulphate

Table 5.5 Sulphate conjugates of flavones and flavonols

Type of conjugate	*Known aglycones*
Flavones	
7-Sulphate	Apigenin, luteolin, chrysoeriol, diosmetin
8-Sulphate	Hypolaetin
3′-Sulphate	Luteolin, diosmetin, tricetin
4′-Sulphate	Luteolin
7, 3′-Disulphate	Luteolin, diosmetin, tricetin
7-Glucosidesulphate	Apigenin, luteolin, chrysoeriol, tricin
7-Rutinosidesulphate	Luteolin
8-Glucoside-3′-sulphate	Hypolaetin
7-Sulphate-3′-glucoside	Luteolin
7-Sulphate-3′-rutinoside	Luteolin
Flavonols	
3-Sulphate	Rhamnocitrin, kaempferol 7, 4′-dimethyl ether, quercetin, isorhamnetin, tamarixetin, rhamnetin, rhamnazin, gossypetin, patuletin, eupatoletin, eupatin, veronicafolin
7-Sulphate	Kaempferol, isorhamnetin, patuletin, quercetagetin 3-methyl ether
3,7-Disulphate	Kaempferol, isorhamnetin
3,4′-Disulphate	Quercetin
3,5,4′-Trisulphate	Rhamnetin
3,7,4′-Trisulphate	Quercetin
3,7,3′,4′-Tetrasulphate	Quercetin
3-Glucosidesulphate	Kaempferol
3-Rhamnosidesulphate	Kaempferol, quercetin, myricetin
3-Rutinosidesulphate	Quercetin, isorhamnetin
3-Gentiobiosidesulphate	Kaempferol
3-Glucoside-7-sulphate	Patuletin
3-Glucuronide-7-sulphate	Kaempferol, quercetin, isorhamnetin
3′-Glucuronide-3,5,4′-trisulphate	Rhamnetin

present in *Brickellia californica* and *B. laciniata* has been discovered as the half calcium salt (Mues *et al.*, 1979). Interestingly, other sulphates in *Brickellia*, e.g. of patuletin, are present as the potassium salts, so that variation can clearly occur within the same plant.

5.5 NEW REPORTS OF FLAVONE GLYCOSIDES

About 130 new flavone glycosides have been reported in the period 1975–1980. These are shown in Table 5.6, together with plant source and reference. A complete check list of all known flavone glycosides appears at the end of this chapter. About twice as many compounds are now known as compared to 1975. The number of apigenin glycosides has increased from 22 to 40, luteolin from 18

Table 5.6 New flavone glycosides

Glycoside	*Source*	*Reference*
5,7-Dihydroxyflavone (Chrysin)		
7-Galactoside	*Aerva persica* roots; Amaranthaceae	Garg *et al.* (1979)
3′,4′-Dihydroxyflavone		
4′-Glucoside	*Rhus* spp. heartwood; Anacardiaceae	Young (1979)
5,7-Dihydroxy-8-methoxyflavone (Wogonin)		
7-Glucuronide	*Scutellaria galericulata*; Labiatae	Popova *et al.* (1976)
Apigenin		
7-Arabinoside	*Hieracium umbellatum* whole plant; Compositae	Guppy and Bohm (1976)
7-Xyloside	*Muscari armeniacum*; Liliaceae	Ellnain-Wojtaszek *et al.* (1978)
7-Galacturonide	*Marchantia berteroana* thallus; Hepaticae	Markham and Porter (1975)
4′-Glucuronide	*Chrysanthemum cinerarifolium*; Compositae	Rao *et al.* (1973)
7-Arabinosylglucoside	*Hieracium umbellatum* whole plant; Compositae	Guppy and Bohm (1976)
7-Sophoroside	*Glandularia bipinnatifida*; Verbenaceae	Umber (1980)
7-Rhamnosylglucuronide	*Marchantia foliacea* thallus; Hepaticae	Markham and Porter (1973)
7-Diglucuronide	*Adenocalymma alliaceum* flowers; Bignoniaceae	Rao and Rao (1980)
7-Glucuronide-4′-rhamnoside	*Conocephalum conicum* whole plant; Hepaticae	Markham *et al.* (1976)
7-Rutinoside-4′-glucoside	*Galium mollugo*; Rubiaceae	Borisov (1974)
7-Neohesperidoside-4′-glucoside	*Hedwigia ciliata*; Bryophyta	Österdahl (1979)
7-Diglucuronide-4′-glucuronide	*Conocephalum conicum* whole plant; Hepaticae	Markham *et al.* (1976)
7-Neohesperidoside-4′-sophoroside	*Hedwigia ciliata*; Bryophyta	Österdahl (1979)
7-(4″-*p*-Coumaroyl)glucoside	*Salix alba* leaf; Salicaceae	Karl *et al.* (1976)
7-(2″-Acetyl)glucoside	*Matricaria chamomilla* flowers; Compositae	Redaelli *et al.* (1980)
7-(6″-Acetyl)glucoside	*Matricaria chamomilla* flowers; Compositae	Kunde and Issac (1979)
4′-(2″,6″-Di-*p*-coumaroyl)-glucoside	*Lycopodium clavatum* leaf; Lycopodiaceae	Ansari *et al.* (1979)
7-Glucosidesulphate	*Phoenix canariensis* flowers; Palmae	Harborne *et al.* (1974)

Glycoside	*Source*	*Reference*
7-Sulphate	*Bixa orellana* leaf; Bixaceae	Harborne (1975)
Apigenin 4′-methyl ether (Acacetin)		
7-(6″-Acetyl)glucoside	*Agastache rugosa*; Labiatae	Zakharova *et al.* (1979)
7-(4‴-Acetyl)rutinoside	*Thalictrum aquilegifolium* whole plant; Ranunculaceae	Ina and Iida (1981)
7-[2‴-(2-Methyl)butyryl]rutinoside 7-[3‴-(2-Methyl)butyryl]rutinoside	*Valeriana wallichii* rhizome; Valerianaceae	Chari *et al.* (1977)
7-Glucosyl(1 → 2)glucosyl(1 → 2) [2, 4-diacetylrhamnosyl](1 → 6) 3-acetylglucoside	*Coptis japonica*; Ranunculaceae	Fujiwara *et al.* (1976)
Apigenin 5, 7-dimethyl ether		
4′-Galactoside	*Morinda citrifolia* flower; Rubiaceae	Singh and Tiwari (1976)
6-*Hydroxyapigenin* (Scutellarein)		
6-Xyloside	*Juniperus communis* fruit; Cupressaceae	Lamer-Zarawska (1980)
6-Glucuronide	*Sterculia foetida* leaf; Sterculiaceae	Nair *et al.* (1977b)
Scutellarein 6-methyl ether		
7-Rutinoside	*Oncidium excavatum* leaf; Orchidaceae	Williams (1979)
Scutellarein 7-methyl ether		
6-Rhamnosyl (1→4)xyloside	*Sorbaria sorbifolia* flowers and leaves; Rosaceae	Zaitsev *et al.* (1974)
Scutellarein 4′-methyl ether		
6-Glucoside	*Catalpa ovata*; Bignoniaceae	Okuda *et al.* (1975)
7-Glucuronide	*Clerodendron trichotomum*; Verbenaceae	Morita *et al.* (1977)
6-Sophoroside	*Catalpa ovata*; Bignoniaceae	Okuda *et al.* (1975)
Scutellarein 6, 4′-dimethyl ether (Pectolinarigenin)		
7-Glucoside	*Oncidium excavatum* leaf; Orchidaceae	Williams (1979)
7-Glucuronide 7-Glucuronide methyl ester	*Comanthosphace japonica* leaf; Labiatae	Arisawa *et al.* (1979)
7-(4‴-Acetylrutinoside)	*Linaria japonica* leaf; Scrophulariaceae	Morita *et al.* (1974)
Scutellarein 6, 7, 4′-trimethyl ether		
5-Glucoside	*Salvia virgata*; Labiatae	Ulubelen and Ayanoglu (1975)
8-*Hydroxyapigenin* (Isoscutellarein)		
8-Glucuronide	*Marchantia berteroana* whole plant; Hepaticae	Markham and Porter (1975)
4′-Mannosyl(1→ 2)-glucoside	*Stachys inflata* aerial parts; Labiatae	Komissarenko *et al.* (1978)

Table 5.6 (Contd.)

Glycoside	*Source*	*Reference*
7-(Acetyl)sophoroside	*Stachys palustris*; Labiatae	Bishara *et al.* (1976)
8-Hydroxyapigenin 4′-methyl ether (Takakin)		
8-Glucuronide 8-Xylosylglucoside	*Takakia lepidozioides* whole plant; Hepaticae	Markham and Porter (1979)
7-(6‴-Acetylallosyl)-(1→2)glucoside	*Veronica filiformis* whole plant; Scrophulariaceae	Chari *et al.* (1981)
Luteolin		
5-Galactoside	*Dracocephalum nutans*; Labiatae	Shamyrina *et al.* (1977)
7-Galacturonide	*Marchantia berteroana* thallus; Hepaticae	Markham and Porter (1975)
3′-Glucuronide	*Lunularia cruciata* thallus; Hepaticae	Markham and Porter (1974a)
3′-Galacturonide	*Marchantia berteroana* thallus; Hepaticae	Markham and Porter (1975)
4′-Arabinoside	*Hieracium umbellatum* whole plant; Compositae	Guppy and Bohm (1976)
4′-Glucuronide	*Riccia fluitans* thallus; Hepaticae	Markham *et al.* (1978a)
7-Arabinosylglucoside	*Vernonia* spp.; Compositae	Mabry *et al.* (1975)
7-Gentiobioside	*Campanula rotundifolia*; Campanulaceae	Teslov (1980)
7-Sophoroside	*Glandularia bipinnatifida*; Verbenaceae	Umber (1980)
7-Digalactoside	*Cephalaria procera*; Dipsacaceae	Ulubelen *et al.* (1978)
7-Galactosylglucoside	*Vernonia* spp.; Compositae	Mabry *et al.* (1975)
7-Glucuronide-3′-glucoside	*Riccia fluitans* thallus; Hepaticae	Markham *et al.* (1978a)
7,3′-Diglucuronide	*Marchantia polymorpha* thallus; Hepaticae	Markham and Porter (1974b)
7,3′-Digalacturonide	*Marchantia berteroana* thallus; Hepaticae	Markham and Porter (1975)
7-Galactoside-4′-glucoside	*Vernonia* spp.; Compositae	Mabry *et al.* (1975)
7-Glucuronide-3′-rhamnoside	*Riccia fluitans*; Hepaticae	Vandekerkhove (1978)
7-Glucuronide-4′-rhamnoside	*Conocephalum conicum* thallus; Hepaticae	Markham *et al.* (1976)
7,4′-Diglucuronide	*Marchantia polymorpha* thallus; Hepaticae	Markham and Porter (1974b)
3′,4′-Diglucoside	*Listera ovata* leaf; Orchidaceae	Williams (1979)
3′,4′-Diglucuronide	*Lunularia cruciata* thallus; Hepaticae	Markham and Porter (1974a)
3′,4′-Digalacturonide	*Marchantia berteroana* thallus; Hepaticae	Markham and Porter (1975)

Glycoside	*Source*	*Reference*
7-Glucosylarabinoside-4′-glucoside	*Galium mollugo*; Rubiaceae	Borisov (1974)
7-Neohesperidoside-4′-glucoside 7-Glucoside-4′-neohesperidoside	*Hedwigia ciliata*; Bryophyta	Österdahl (1979)
7-Glucuronide-3′,4′-dirhamnoside	*Conocephalum conicum* thallus; Hepaticae	Markham *et al.* (1976)
7,4′-Diglucuronide-3′-glucoside	*Riccia fluitans* thallus; Hepaticae	Markham *et al.* (1978c)
7,3′,4′-Triglucuronide	*Marchantia polymorpha* thallus; Hepaticae	Markham and Porter (1974b)
7-Neohesperidoside-4′-sophoroside	*Hedwigia ciliata*; Bryophyta	Österdahal and Lindberg (1977)
7-Glucuronide-3′-feruloylglucoside 7,4′-Diglucuronide-3′-feruloylglucoside	*Riccia fluitans* thallus; Hepaticae	Markham *et al.* (1978*c*)
7-Glucosidesulphate	*Phoenix roebelenii* leaf; Palmae	Williams *et al.* (1973)
7-Rutinosidesulphate	*Washingtonia robusta* leaf; Palmae	Williams *et al.* (1973)
7-Sulphate-3′-rutinoside	*Opsiandra maya* leaf; Palmae	Williams *et al.* (1973)
7-Sulphate	*Bixa orellana* leaf; Bixaceae	Harborne (1975)
3′-Sulphate	*Lachenalia unifolia* leaf; Liliaceae	Williams *et al.* (1976)
4′-Sulphate	*Daucus carota* leaf; Umbelliferae	Harborne and King (1976)
7,3′-Disulphate	*Zostera marina* leaf; Zosteraceae	Harborne and Williams (1976)
Luteolin 5-methyl ether		
7-Glucoside	*Juncus* spp. leaf; Juncaceae	Williams and Harborne (1975)
Luteolin 3′-methyl ether (Chrysoeriol)		
5-Glucoside	*Lycopodium scariosum* aerial parts; Lycopodiaceae	Markham and Moore (1980)
7-Xyloside	*Salvia seravschanica*; Labiatae	Smirnova *et al.* (1974)
7-Rhamnoside	*Sedum formosanum* whole plant; Crassulaceae	Chou *et al.* (1976)
4′-Glucoside	*Dianthus deltoides* aerial parts; Caryophyllaceae	Boguslavskaya *et al.* (1976)
5-Diglucoside	*Lycopodium scariosum* aerial parts; Lycopodiaceae	Markham and Moore (1980)

Table 5.6 (Contd.)

Glycoside	*Source*	*Reference*
7-Rhamnosylglucuronide	*Marchantia foliacea* thallus; Hepaticae	Markham and Porter (1973)
7-Allosyl(1→2)glucoside	*Sideritis grandiflora* whole plant; Labiatae	Rabanal *et al.* (1982)
5,4′-Diglucoside 7,4′-Diglucoside	*Lycopodium scariosum* aerial parts; Lycopodiaceae	Markham and Moore (1980)
7-Glucuronide-4′-rhamnoside	*Conocephalum conicum* thallus; Hepaticae	Markham *et al.* (1976)
7-Sophorotrioside	*Sideritis romana* aerial parts; Labiatae	Venturella *et al.* (1977)
7-Glucosidesulphate	*Juncus effusus* leaf; Juncaceae	Williams and Harborne (1975)
7-Sulphate	*Zostera marina* leaf; Zosteraceae	Harborne and Williams (1976)
Luteolin 4′-methyl ether (Diosmetin)		
7-Xylosylglucoside	*Anacyclus* spp. leaf; Compositae	Greger (1978)
7-Sophoroside	*Glandularia bipinnatifida*; Verbenaceae	Umber (1980)
7-(4^{G}-Rhamnosyl-neohesperidoside)	*Dicranum scoparium*; Dicranaceae	Österdahl (1978)
7-Sulphate	*Zostera marina* leaf; Zosteraceae	Harborne and Williams, (1976)
3′-Sulphate 7,3′-Disulphate	*Lachenalia unifolia* leaf; Liliaceae	Williams *et al.* (1976)
Luteolin 3′,4′-dimethyl ether		
7-Rhamnoside	*Linum maritimum*; Linaceae	Volk and Sinn (1968)
7-Glucuronide	*Rhynchosia beddomei* leaf; Leguminosae	Adinarayana *et al.* (1980)
6-Hydroxyluteolin		
6-Xyloside	*Juniperus communis* fruit; Cupressaceae	Lamer-Zarawska (1980)
6-Glucoside	*Stereospermum suaveolens*; Bignoniaceae	Nair (1979)
6-Glucuronide	*Sterculia colorata*; Sterculiaceae	Nair *et al.* (1976)
7-Arabinoside	*Lippia nodiflora*; Verbenaceae	Nair *et al.* (1973)
7-Xyloside	*Parahebe* spp. leaf; Scrophulariaceae	Grayer-Barkmeijer (1978)
7-Apioside	*Lepidagathis cristata*; Acanthaceae	Ranganathan *et al.* (1980)
7-Rhamnosyl(1→4)xyloside	*Pityrodia coerulea* leaf; Verbenaceae	Harborne and Stacey (1978)

Glycoside	*Source*	*Reference*
7-Arabinoside-4′-rhamnoside	*Lippia nodiflora*; Verbenaceae	Nair *et al.* (1973)
6-Hydroxyluteolin 7-methyl ether (Pedalitin)		
6-Galactoside, 6-Galactosylglucoside	*Sullivantia* spp.; Saxifragaceae	Soltis (1980)
6-Hydroxyluteolin 3′-methyl ether		
7-Diglucoside	*Sideritis leucantha*; Labiatae	Tomas (1979)
6-Hydroxyluteolin 6,3′-dimethyl ether		
7-Rhamnoside	*Cassia occidentalis* leaf; Leguminosae	Tiwari and Singh (1977)
7-(6″-Caffeoyl)glucoside	*Gnaphalium uliginosum* whole plant; Compositae	Konopleva *et al.* (1979)
6-Hydroxyluteolin 7,3′-dimethyl ether		
6-Glucoside	*Citharexylum subserratum* leaf; Verbenaceae	Mathuram *et al.* (1976)
8-*Hydroxyluteolin* (Hypolaetin)		
8-Glucuronide, 8,4′-Diglucuronide	*Marchantia berteroana* thallus; Hepaticae	Markham and Porter (1975)
8-Glucoside-3′-sulphate	*Malva sylvestris* leaf; Malvaceae	Nawwar *et al.* (1977a)
8-Sulphate	*Bixa orellana* leaf; Bixaceae	Harborne (1975)
8-Hydroxyluteolin 3′-methyl ether		
7-Glucoside	*Solanum grayi* leaf; Solanaceae	Whalen and Mabry (1979)
7-Mannosyl(1→2)glucoside	*Stachys spectabilis*; Labiatae	Derkach *et al.* (1980)
5,7,3′,4′-Tetrahydroxy-6,8-dimethoxyflavone		
7-Glucoside	*Sideritis leucantha* whole plant; Labiatae	Tomas and Ferreres (1980)
5,7,3′-Trihydroxy-6,8,4′-trimethoxyflavone (Acerosin)		
5-(6″-Acetyl)glucoside	*Vitex negundo* stem bark; Verbenaceae	Subramanian and Misra (1979)
5,7,4′-Trihydroxy-6,8,3′-trimethoxyflavone		
7-Glucoside	*Sideritis leucantha* whole plant; Labiatae	Tomas and Ferreres (1980)
Tricetin		
3′-Sulphate, 7,3′-Disulphate	*Lachenalia unifolia* leaf; Liliaceae	Williams *et al.* (1976)
Tricetin 3′-methyl ether		
7-Glucoside	*Taxodium distichum* short shoots; Taxodiaceae	Geiger and de Groot-Pfleiderer (1979)

Table 5.6 (Contd.)

Glycoside	*Source*	*Reference*
Tricetin 3′,5′-dimethyl ether (Tricin)		
4′-Glucoside 5,7-Diglucoside	*Lycopodium scariosum* aerial parts; Lycopodiaceae	Markham and Moore (1980)
7-Fructosylglucoside	*Hyacinthus orientalis* leaf; Liliaceae	Williams (1975)
7-Neohesperidoside 7-Diglucoside	*Saccharum officinarum* leaf; Gramineae	Williams *et al.* (1974)
7-Rhamnosylglucuronide	*Marchantia foliacea* thallus; Hepaticae	Markham and Porter (1973)
7-Rutinoside-4′-glucoside	*Hyacinthus orientalis* leaf; Liliaceae	Williams (1975)
7-Glucosidesulphate	*Saccharum officinarum* leaf; Gramineae	Williams *et al.* (1974)
5,7,2′,4′,5′-Pentahydroxyflavone (Isoetin)		
7-Glucoside	*Hispidella hispanica* leaf; Compositae	Harborne (1978)
2′-Arabinoside 2′-Glucoside 2′-Xylosylarabinosylglucoside	*Heywoodiella oligocephala* leaf; Compositae	Harborne (1978)
4′- or 5′-Glucoside	*Isoetes delilei*; Isoetales	Voirin *et al.* (1975)

to 53, chrysoeriol from 9 to 21 and tricin from 8 to 14. One or two reports refer to the pre-1975 literature, e.g. of luteolin 3′,4′-dimethyl ether 7-rhamnoside (Volk and Sinn, 1968), where compounds have been accidentally overlooked in our earlier listing (Harborne and Williams, 1975). Some reports represent confirmation of earlier provisional identifications. For example, the 4′-glucoside of 3′,4′-dihydroxyflavone was provisionally identified in *Primula* leaves (Harborne, 1968), but its full characterization has only recently been accomplished following its isolation from another source, in *Rhus* heartwood (Young, 1979).

Some of the more complex and unusual new glycosides reported have been isolated from lower plants. A notable example is the first apigenin derivative with two disaccharide moieties, namely the 7-neohesperidoside-4′-sophoroside from the moss *Hedwigia ciliata* (Österdahl, 1979). Among several new flavones highly substituted with glucuronic acid are apigenin 7-diglucuronide-4′-monoglucuronide from the liverwort *Conocephalum conicum* (Markham *et al.*, 1976) and luteolin 7,3′,4′-triglucuronide in *Marchantia polymorpha* (Markham and Porter, 1974b).

A number of incompletely characterized di- and tri-glycosides has been included in Table 5.6, where the structures are clearly new, but further analyses to determine interglycosidic linkages are needed in many cases. For example, a tricin 5-diglucoside was reported in *Triticum dicoccum* by Harborne and Hall (1964) and more recently in *Lycopodium scariosum* (Markham and Moore, 1980). It is

not known whether they are the same, with the same disaccharide unit, or whether they are different.

Several flavone triglycosides have been reported, but few have been fully described. Indeed, one of the few which seems to have been completely identified is chrysoeriol 7-sophorotrioside in *Sideritis romana* (Labiatae) (Venturella *et al.*, 1977). Another, diosmetin 7-(4^G-rhamnosylneohesperidoside) from the moss *Dicranum scoparium*, is unusual in having a unique branched trisaccharide as the sugar moiety (Österdahl, 1978). One other interesting one, awaiting complete analysis, is an isoetin 2′-triglycoside with the sugars xylose, arabinose and glucose reported in *Heywoodiella* (Harborne, 1978).

5.6 NEW REPORTS OF FLAVONOL GLYCOSIDES

The new flavonol glycosides reported in plants in the period 1975–1980 are shown in Table 5.7. The rate of discovery of new glycosides is apparent from the increase in the number of known kaempferol derivatives from 50 to 104, of quercetin derivatives from 62 to 112, and of myricetin derivatives from 12 to 25.

Table 5.7 New flavonol glycosides

Glycoside	*Source*	*Reference*
3,5-Dihydroxy-6-methoxyflavone		
3-Rutinoside	*Datisca cannabina* root cortex; Datiscaceae	Zapesochnaya (1969)
3,5,7-Trihydroxyflavone (Galangin)		
3-Rhamnoside	*Artocarpus lakoocha* root bark; Moraceae	Chauhan and Kumari (1979)
3-Glucoside, 7-Glucoside	*Cephalaria procera*; Dipsacaceae	Ulubelen *et al.* (1978)
3-Galactosyl(1→4)rhamnoside	*Artocarpus lakoocha* root bark; Moraceae	Chauhan *et al.* (1979a)
Galangin 3-methyl ether		
7-Glucoside	*Calluna vulgaris* flowers; Ericaceae	Stepien *et al.* (1978)
8-Hydroxygalangin 7-methyl ether		
8-Acetate, 8-Butyrate	*Notholaena* spp. farinose exudate; Filicales	Wollenweber *et al.* (1978b)
Kaempferol		
3-Xyloside	*Euonymus ciliatus* leaf; Celastraceae	Ishikura and Sato (1977)
3-Alloside	*Osmunda asiatica* trophophyll; Osmundaceae	Okuyama *et al.* (1978)

Table 5.7 (Contd.)

Glycoside	*Source*	*Reference*
7-Arabinoside 7-Xyloside	*Dryopteris* spp. frond; Dryopteridaceae	Hiraoka (1978)
7-Galactoside	*Tithymalus densus* leaf; Euphorbiaceae	Litvinenko *et al.* (1973)
4′-Rhamnoside	*Phyllanthus niruri* roots; Euphorbiaceae	Chauhan *et al.* (1977)
4′-Glucoside	*Ophiopogon jaburan* seed coat; Liliaceae	Ishikura and Hayashida (1979)
3-Arabinosyl(1→6)galactoside	*Lysichiton camtschatcense* leaf; Araceae	Williams *et al.* (1981)
3-Xylosylrhamnoside	*Woodsia polystichoides* frond; Athyriaceae	Hiraoka (1978)
3-Rhamnosylxyloside	*Euonymus alatus* leaf; Celastraceae	Ishikura *et al.* (1976)
3-Xylosylgalactoside	*Lysichiton camtschatcense* leaf; Araceae	Williams *et al.* (1981)
3-Glucosyl(1→4)rhamnoside	*Rosa multiflora* fruit; Rosaceae	Takagi *et al.* (1976b)
3-Neohesperidoside	*Humulus lupulus* cones; Cannabidaceae	Vancraenenbroeck *et al.* (1969)
3-Rhamnosyl(1→2)galactoside	*Chenopodium fremontii* leaf; Chenopodiaceae	Crawford and Mabry (1978)
3-Glucosyl(1→6)galactoside	*Eryngium planum*; Umbelliferae	Hiller *et al.* (1980)
3-Galactosylglucoside	*Heuchera* spp., whole plant; Saxifragaceae	Wells and Bohm (1980)
3-Arabinoside-7-rhamnoside 3-Rhamnoside-7-arabinoside	*Asplenium trichomanes*; Aspleniaceae	Imperato (1979a)
3-Glucoside-7-arabinoside	*Loxsoma cunninghamii* leaf; Filicales	Markham and Given (1979)
3-Glucoside-7-xyloside	*Takakia lepidozoides* whole plant; Hepaticae	Markham and Porter (1979)
3,7-Digalactoside	*Pachyphragma macrophyllum* flowers; Cruciferae	Zentsova and Dzhumyrko (1974)
3,4′-Diglucoside	*Ophiopogon jaburan* seed coat; Liliaceae	Ishikura and Hayashida (1979)
3-Glucosylxylosylarabinoside	*Antitoxicum sibiricum*; Asclepiadaceae	Shatokhina *et al.* (1974)
3-Rhamnosyl(1→2) rhamnosyl(1→6)glucoside	*Humulus lupulus* cones; Cannabidaceae	Vancraenenbroeck *et al.* (1969)
3-(2^G-Rhamnosylrutinoside) 3-(2^G-Rhamnosylgentiobioside) 3-(2^G-Glucosylgentiobioside)	*Glycine max* leaf; Leguminosae	Buttery and Buzzell (1975)
3-Rhamnosylglucosylgalactoside	*Peltophorum africanum*; Leguminosae	El Sherbeiny *et al.*(1977)
3-Rhamnosylxyloside-7-glucoside	*Euonymus ciliatus* leaf; Celastraceae	Ishikura and Sato (1977)

Glycoside	*Source*	*Reference*
3-Glucosylxyloside-7-xyloside	*Heuchera micrantha* whole plant; Saxifragaceae	Wilkins and Bohm (1976)
3-Glucosylrhamnoside-7-rhamnoside	*Oxytropis komarovii* aerial parts; Leguminosae	Baimukhambetov (1975)
3-Rutinoside-7-rhamnoside	*Equisetum silvaticum* green parts; Equisetaceae	Aly *et al.* (1975)
3-Rhamnosylgalactoside-7-arabinoside	*Dryopteris* spp. frond; Dryopteridaceae	Hiraoka (1978)
3-Glucoside-7-gentiobioside	*Nerisyrenia* spp. leaf; Cruciferae	Bacon and Mabry (1976)
3-Rutinoside-4′-diglucoside	*Prunus* spp. leaf; Rosaceae	Henning and Herrmann (1980)
3-(2″-*p*-Coumaroyl) arabinofuranoside	*Picea koraiensis* needles; Pinaceae	Ivanova *et al.*, (1978)
3-(3″-*p*-Coumaroyl)glucoside	*Picea obvata* needles; Pinaceae	Zapesochnaya *et al.* (1980)
3-(Feruloyl)glucoside	*Pseudolarix amabilis* leaf; Pinaceae	Niemann (1975)
3-(Di-*p*-coumaroyl)glucoside	*Pinus contorta* needles; Pinaceae	Higuchi and Donnelly (1978)
3-(3″-*p*-Coumaroyl-6″-feruloyl) glucoside	*Picea obvata* needles; Pinaceae	Zapesochnaya *et al.* (1978)
3-(6″-*p*-Coumaroyl)acetylglucoside	*Anaphalis contorta*; Compositae	Lin *et al.* (1976)
3-(6″-Galloyl)glucoside	*Heuchera* spp. whole plant; Saxifragaceae	Wells and Bohm (1980)
3-(2″-Galloyl)glucoside	*Polygonum nodosum* aerial parts; Polygonaceae	Isobe *et al.* (1980)
7-(6″-Succinyl)glucoside	*Cyathea contaminans* frond; Cyatheaceae, Pterophyta	Hiraoka and Maeda (1979)
3-(6‴-Acetyl)glucosyl-(1→4)rhamnoside	*Rosa multiflora* fruit; Rosaceae	Takagi *et al.* (1976b); Yamasaki *et al.* (1977)
3-(6″-Acetyl)glucoside-7-rhamnoside; 3-(6″-Acetyl)glucoside-7-glucoside	*Equisetum telmateja*; Equisetaceae	Geiger *et al.* (1978)
3-(2‴-Sinapoyl)sophoroside-7-glucoside	*Brassica napus* seed; Cruciferae	Stengel and Geiger, (1976); Markham *et al.* (1978b)
3-[Glucosyl-(1→3)-4‴-acetyl-rhamnosyl-(1→6)galactoside]	*Colubrina faralaotra* leaf; Rhamnaceae	Guinaudeau *et al.* (1981)
3-(4″-Caffeoyl)laminaribioside-7-rhamnoside	*Phyllitis scolopendrium* aerial aparts; Aspleniaceae	Karl *et al.* (1980)
3-α-(6″-Sulphate)glucoside	*Asplenium filix-foemina* frond; Aspleniaceae	Imperato (1979b)
3-Rhamnosylsulphate	*Davidsonia pruriens* leaf; Davidsoniaceae	Wilkins and Bohm (1977)
3-Glucuronide-7-sulphate	*Frankenia pulverulenta* leaf; Frankeniaceae	Harborne (1975)

Table 5.7 (Contd.)

Glycoside	*Source*	*Reference*
3-(6″-Sulphate)gentiobioside	*Asplenium fontanum*, Aspleniaceae	Imperato (1980)
7-Sulphate	*Frankenia pulverulenta* leaf; Frankeniaceae	Harborne (1975)
3,7-Disulphate	*Reaumuria mucronata* leaf; Tamaricaceae	Nawwar *et al.* (1977b)
Kaempferol 7-methyl ether (Rhamnocitrin)		
3-Glucoside, 7-Rutinoside	*Pityrogramma triangularis* frond; Polypodiaceae	Star *et al.* (1975)
3-Sulphate	*Ammi visnaga* flower; Umbelliferae	Harborne and King (1976)
Kaempferol 4′-methyl ether (Kaempferide)		
3-Glucuronide	*Cleome viscosa* root; Capparidaceae	Chauhan *et al.* (1979b)
Kaempferol 7,4′-dimethyl ether		
3-sulphate	*Tamarix aphylla* flowers; Tamaricaceae	El Ansari *et al.* (1976)
6-C-Methylkaempferol		
3-Glucoside	*Pinus contorta* needles; Pinaceae	Higuchi and Donnelly (1978)
6-Hydroxykaempferol		
7-Glucoside	*Tetragonotheca* spp. leaf; Compositae	Bacon *et al.* (1978)
6-Hydroxykaempferol 3-methyl ether		
6-Glucoside, 7-Glucoside	*Neurolaena oaxacana* leaf; Compositae	Ulubelen *et al.*(1980b)
6-Hydroxykaempferol 6-methyl ether (Eupafolin)		
3-Glucoside	*Flaveria brownii* leaf and stem; Compositae	Al-Khubaizi *et al.* (1977)
3-Galactoside	*Anvillea garcini* leaf; Compositae	Ulubelen *et al.* (1979)
7-Glucoside	*Tetragonotheca* spp. leaf; Compositae	Bacon *et al.* (1978)
3-Rutinoside	*Anvillea garcini* leaf; Compositae	Ulubelen *et al.* (1979)
6-Hydroxykaempferol 4′-methyl ether		
3,7-Dirhamnoside	*Tephrosia candida* aerial parts; Leguminosae	Sarin *et al.* (1976)
6-Hydroxykaempferol 6,7,4′-trimethyl ether (Mikanin)		
3-Galactoside	*Anvillea garcini* leaf; Compositae	Ulubelen *et al.* (1979)
8-Hydroxykaempferol (Herbacetin)		
7-Arabinoside	*Rhodiola gelida* rhizomes; Crassulaceae	Krasnov and Alekseyuk (1979)

Glycoside	*Source*	*Reference*
8-Xyloside	*Rhodiola algida*; Crassulaceae	Zapesochnaya (1978)
8-Rhamnoside	*Rhodiola litvinovii* rhizome; Crassulaceae	Krasnov (1979)
8-Glucuronide	*Melochia corchorifolia*; Sterculiaceae	Nair *et al.* (1977a)
4′-Arabinoside	*Rhodiola algida* roots; Crassulaceae	Pangarova and Zapesochnoya (1974)
4′-Glucoside	*Rhodiola gelida* rhizomes and roots; Crassulaceae	Krasnov and Alekseyuk (1979)
4′-Glucuronide	*Rhodiola algida* roots; Crassulaceae	Pangarova and Zapesochnoya (1974)
8-Rutinoside	*Fagonia arabica* whole plant; Zygophyllaceae	Saleh and El-Hadidi (1977)
8-Gentiobioside	*Calluna vulgaris* flowers; Ericaceae	Olechnowicz-Stepien *et al.* (1978)
8-Arabinoside-4′-xyloside	*Rhodiola algida*; Crassulaceae	Krasnov and Demidenko (1979)
8-(3″-Acetyl)xyloside	*Rhodiola algida*; Crassulaceae	Zapesochnaya (1978)
4′-(Acetyl)arabinoside	*Rhodiola algida* roots; Crassulaceae	Pangarova and Zapesochnoya (1974)
8-(Di-acetyl)glucoside	*Rhodiola algida*; Crassulaceae	Krasnov and Demidenko (1979)
4′-(Di-acetyl)xyloside 4′-(Tri-acetyl)xyloside	*Rhodiola algida* roots; Crassulaceae	Pangarova and Zapesochnoya (1974)
8-Acetate 8-Butyrate	*Notholaena* spp. farinose exudates; Filicales	Wollenweber *et al.* (1978a)
Herbacetin 7-methyl ether		
8-Acetate 8-Butyrate	*Notholaena* spp. farinose exudates; Filicales	Wollenweber *et al.* (1978b)
Herbacetin 8-methyl ether		
3-Glucoside	*Sorbus aucuparia* inflorescence; Rosaceae	Jerzmanowska and Kamecki (1973)
3-Rutinoside	*Fagonia arabica* whole plant; Zygophyllaceae	Saleh and El-Hadidi (1977)
Herbacetin 7,8-dimethyl ether		
3-Rhamnoside	*Rudbeckia bicolor* whole plant; Compositae	Jauhari *et al.* (1979)
Herbacetin 7,4′-dimethyl ether		
8-Acetate	*Notholaena* spp. farinose exudates; Filicales	Wollenweber *et al.* (1978a)
8-Butyrate	*Notholaena* spp. farinose exudates; Filicales	Wollenweber *et al.* (1978c)
Quercetin		
3-α-Galactoside	*Rhododendron dauricum*; Ericaceae	Liu *et al.* (1976)

Table 5.7 (Contd.)

Glycoside	*Source*	*Reference*
3-Galacturonide	*Corsinia coriandrina* thallus; Marchantia	Markham (1980)
7-Arabinoside	*Conyza dioscoridis* leaf; Compositae	Ismail *et al.* (1979)
7-β-Galactoside	*Coptis japonica*; Ranunculaceae	Fujiwara *et al.* (1976)
3′-Xyloside	*Euphorbia paralias*; Euphorbiaceae	Rizk *et al.* (1976)
3-Arabinosyl(1→6)galactoside	*Hydrocotyle vulgaris* leaf; Umbelliferae	Hiller *et al.* (1979)
3-Dixyloside 3-Xylosylglucoside	*Heuchera micrantha* whole plant; Saxifragaceae	Wilkins and Bohm (1976)
3-Glucosyl(1→4)rhamnoside	*Rosa multiflora* fruit; Rosaceae	Takagi *et al.* (1976a)
3-Robinobioside	*Crataegus pinnatifida* flowers; Rosaceae	Bykov and Glyzin (1977)
3-Galactosylglucoside	*Lasthenia* spp. whole plant; Compositae	Ornduff *et al.* (1973)
3-Glucosylmannoside	*Solanum xanthocarpum* flowers; Solanaceae	Dubey and Gupta (1978)
3-Galactosylglucuronide	*Sullivantia* spp. Saxifragaceae	Soltis (1980)
3-Glucoside-7-arabinoside	*Loxsoma cunninghamii* leaf; Filicales	Markham and Given (1979)
3-Galactoside-7-glucoside	*Verbesina encelioides* leaf and flower; Compositae	Glennie and Jain (1980)
3-(2^G^-Rhamnosylrutinoside)	*Glycine max* leaf; Leguminosae	Buttery and Buzzell (1975)
3-Rhamnosyl(1→2)rutinoside	*Humulus lupulus* cones; Cannabidaceae	Vancraenenbroeck *et al.* (1969)
3-Rhamnosylrobinobioside	*Primula officinalis* flowers; Primulaceae	Karl *et al.* (1981)
3-Glucosyl(1→6)glucosyl(1→4) rhamnoside	*Rosa multiflora* fruit; Rosaceae	Takagi *et al.* (1976a)
3-(2^G^-Rhamnosylgentiobioside)	*Glycine max* leaf; Leguminosae	Buttery and Buzzell, (1975)
3-(2^G^-Apiosylrutinoside)	*Solanum glaucophyllum* leaf; Solanaceae	Rappaportt *et al.* (1977)
3-Glucosyl(1→2)gentiobioside	*Primula officinalis* flowers; Primulaceae	Karl *et al.* (1981)
3-(2^G^-Glucosylgentiobioside)	*Glycine max* leaf; Leguminosae	Buttery and Buzzell (1975)
7-Triglucoside	*Inula helenium* inflorescence; Compositae	Kowalewska and Lutomski (1978)
3-Dixyloside-7-glucoside	*Heuchera micrantha* whole plant; Saxifragaceae	Wilkins and Bohm (1976)
3-Gentiobioside-7-glucoside	*Tribulus* spp. whole plant; Zygophyllaceae	Saleh and El-Hadidi (1977)

Glycoside	*Source*	*Reference*
3-(2^{G}-Rhamnosylrutinoside-7-glucoside)	*Cerbera manghas* leaf; Apocynaceae	Sakushima *et al.* (1980)
3-Rhamnosyldiglucoside-7-glucoside	*Coprosma repens* leaf; Rubiaceae	Wilson (1979)
3-Rutinoside-7,3′-bisglucoside 3-Rutinoside-4′-diglucoside	*Prunus* spp. leaf; Rosaceae	Henning and Herrmann (1980)
3-(6″-*p*-Coumaroyl)glucoside	*Helichrysum krausii*; Compositae	Candy *et al.* (1975)
3-(3″-*p*-Coumaroyl)glucoside	*Pinus* sp. needles; Pinaceae	Ivanova (1978)
3-Isoferuloylglucuronide	*Tamarix aphylla* flower; Tamaricaceae	El Ansari *et al.* (1976)
3-(3″-Acetyl)-α-L-arabinofuranoside	*Boehmeria tricuspis* root; Urticaceae	Takemoto and Miyase (1974)
3-(3″-Acetyl)rhamnoside 3-(4″-Acetyl)rhamnoside	*Pteris grandiflora* aerial parts; Pteridaceae	Tanaka *et al.* (1978)
3-(6″-Acetyl)glucoside	*Pinus* sp. needles; Pinaceae	Ivanova (1978)
3-(2″-Galloyl)glucoside	*Polygonum nodosum*; Polygonaceae	Isobe *et al.* (1979)
3-(6″-Galloyl)glucoside	*Tellima grandiflora* leaf; Saxifragaceae	Collins *et al.* (1975)
3-(2″-Galloyl)galactoside	*Euphorbia seguieriana*; Euphorbiaceae	Nahrstedt *et al.* (1974)
3-(Malonylglucoside)	*Lactuca sativa* leaf; Compositae	Kreuzaler and Hahlbrock (1973) Woeldecke and Hermann (1974).
3-(3″,6″-Di-*p*-coumaroyl)glucoside	*Pinus sylvestris* needles; Pinaceae	Ivanova *et al.* (1980)
3-*p*-Coumaroylsophoroside-7-rhamnoside 3-Acetylsophoroside-7-rhamnoside	*Oxytropis myriophylla*; Leguminosae	Blinova and Thuan (1977)
3-[Glucosyl-(1→3)-4‴-acetylrhamnosyl-(1→6) galactoside]	*Colubrina faralaotra* leaf; Rhamnaceae	Guinaudeau *et al.* (1981)
7-(6″-Tiglyl)glucoside	*Enkianthus nudipes* stem; Ericaceae	Ogawa and Ogihara (1975)
3-Rhamnosidesulphate	*Davidsonia pruriens* leaf; Davidsoniaceae	Wilkins and Bohm (1977)
3-Glucuronide-7-sulphate	*Frankenia pulverulenta* leaf; Frankeniaceae	Harborne (1975)
3-Rutinosidesulphate	*Arecastrum romanzoffianum* × *Butia capitata* leaf; Palmae	Williams *et al.* (1973)
3-Acetyl-7,3′,4′-trisulphate	*Flaveria bidentis*; Compositae	Cabrera and Juliani (1976)
3-Sulphate	*Ammi visnaga* leaf; Umbelliferae	Harborne and King (1976)

Table 5.7 (Contd.)

Glycoside	*Source*	*Reference*
3,4′-Disulphate 3,7,4′-Trisulphate	*Flaveria bidentis* leaf; Compositae	Cabrera and Juliani (1979)
Quercetin 5-methyl ether (Azaleatin)		
3-Arabinoside	*Carya pecan* branches; Juglandaceae	El Ansari *et al.* (1977)
5-Glucoside	*Stizolophus coronopifolius*; Compositae	Oganesyan and Dzhumyrko (1974)
3-Xylosylarabinoside 3-Rhamnosylarabinoside	*Erysimum perofskianum* Cruciferae	Matlawska and Kowalew (1976)
3-Rutinoside	*Carya pecan* branches; Juglandaceae	El Ansari *et al.* (1977)
Quercetin 7-methyl ether (Rhamnetin)		
3-Glucoside	*Thalictrum foetidum*; Ranunculaceae	Nuralieva *et al.* (1969)
3′-Glucuronide	*Tamarix aphylla*; Tamaricaceae	Saleh *et al.* (1975)
3-Rutinoside	*Suriana maritima* whole plant; Simaroubaceae	Mitchell and Geissman (1971)
3-2^{Gal}-Rhamnosylrobinobioside	*Rhamnus petiolaris* fruit; Rhamnaceae	Wagner *et al.* (1974)
3-Sulphate	*Ammi visnaga* flower; Umbelliferae	Harborne and King (1976)
3,5,4′-Trisulphate	*Tamarix aphylla*; Tamaricaceae	Saleh *et al.* (1975)
Quercetin 3′-methyl ether (Isorhamnetin)		
3-α-L-Arabinofuranoside	*Taxodium distichum* short shoot; Taxodiaceae	Geiger and de Groot-Pfleiderer (1979)
3-Xyloside	*Leptarrhena pyrolifolia* leaf; Saxifragaceae	Miller and Bohm (1979)
3-Arabinosyl(1→6)glucoside	*Papaver orientale* leaf; Papaveraceae	Sakar *et al.* (1980)
3-Xylosylglucoside	*Festuca arundinacea*; Gramineae	Wong (1980)
3-Sophoroside	*Hippophae rhamnoides*; Elaeagnaceae	Mukamedyarova and Chumbalov (1979)
3-Lactoside	*Cassia multijuga* seed; Leguminosae	Dubey and Gupta (1980)
3-Glucoside-7-xyloside	*Heuchera micrantha* whole plant; Saxifragaceae	Wilkins and Bohm (1976)
3-Xylosylrhamnosylgalactoside	*Nitraria retusa* whole plant; Zygophyllaceae	Saleh and El-Hadidi (1977)
3-Rhamnosylrutinoside	*Limnanthes douglassi* leaf; Limnanthaceae	Parker and Bohm (1975)
3-Galactosylrhamnosylgalactoside	*Nitraria retusa* whole plant; Zygophyllaceae	Saleh and El-Hadidi (1977)

Glycoside	Source	Reference
3-Arabinoside-7-rhamnoside	*Cheiranthus cheiri* flower; Cruciferae	Kowalewski *et al.* (1979)
3-Rutinoside-7-glucoside	*Limnanthes douglassi* leaf; Limnanthaceae	Parker and Bohm (1975)
3-Glucoside-7-gentiobioside	*Nerisyrenia* spp. leaf; Cruciferae	Bacon and Mabry (1976)
3-Galactoside-4′-glucoside	*Senecio subdentatus*; Compositae	Fadeeva and Nikishchenko (1979)
3-(6″-Acetyl)glucoside	*Salix viminalis* leaf; Salicaceae	Karl *et al.* (1977)
3-(6″-Acetyl)galactoside	*Pinus* sp. needles; Pinaceae	Ivanova (1978)
3-Rutinosidesulphate	*Arecastrum romanzoffianum* × *Butia capitata* leaf; Palmae	Williams *et al.* (1973)
3-Glucuronide-7-sulphate 7-Sulphate	*Frankenia pulverulenta* leaf; Frankeniaceae	Harborne (1975)
3,7-Disulphate	*Flaveria bidentis* flower; Compositae	Cabrera and Juliani (1977)
Quercetin 4′-methyl ether (Tamarixetin)		
3-Rutinoside	*Albizia amara*; Leguminosae	Sastry *et al.* (1967)
Quercetin 3,5-dimethyl ether (Caryatin)		
3′-(or 4′-)Glucoside	*Carya pecan* branches; Juglandaceae	El Ansari *et al.* (1977)
Quercetin 5,3′-dimethyl ether		
3-Glucoside	*Medicago* × *varia* seed; Leguminosae	Gehring and Geiger (1980)
Quercetin 7,3′-dimethyl ether (Rhamnazin)		
3-(4^{Gal}-Rhamnosylrobinobioside)	*Rhamnus petiolaris* Fruit; Rhamnaceae	Wagner *et al.* (1974)
Quercetin 7,4′-dimethly ether (Ombuin)		
3-Galactoside	*Cassia laevigata* pods; Leguminosae	Tiwari and Singh (1978)
3-Neohesperidoside	*Cassia laevigata* leaf; Leguminosae	Singh and Tiwari (1979)
Quercetin 3′,4′-dimethyl ether		
3-Rutinoside	*Anvillea garcini* leaf; Compositae	Ulubelen *et al.* (1979)
Quercetin 7,3′,4′-trimethyl ether		
3-Digalactoside	*Cassia laevigata* pods; Leguminosae	Tiwari and Singh (1978)
6-Hydroxyquercetin (Quercetagetin)		
3-Rhamnoside	*Juniperus macropoda* leaf; Cupressaceae	Ilyas *et al.* (1977)
3,7-Diglucoside	*Parthenium rollinsianum* whole plant; Compositae	Rodriguez (1977)

Table 5.7 (Contd.)

Glycoside	*Source*	*Reference*
Quercetagetin 3-methyl ether		
7-Glucoside 7-Sulphate	*Neurolaena oaxacana* leaf; Compositae	Ulubelen *et al.* (1980b)
Quercetagetin 6-methyl ether (Patuletin)		
3-Xyloside	*Lasthenia* spp. whole plant; Compositae	Ornduff *et al.* (1973)
3-Rhamnoside	*Lepidophorum repandum* leaf; Compositae	Harborne *et al.* (1976)
3,7-Dirhamnoside	*Kalanchoe spathulata*; Crassulaceae	Gaind *et al.* (1981)
3-Acetylglucoside	*Lagascea mollis* aerial parts; Compositae	El-Naggar and Doskotch (1979)
3-Sulphate	*Brickellia californica* leaf; Compositae	Mues *et al.* (1979)
Quercetagetin 7-methyl ether		
3-Glucoside	*Parthenium rollinsianum* whole plant; Compositae	Rodriguez (1977)
Quercetagetin 3,6-dimethyl ether		
7-Glucoside	*Artemisia taurica* flower; Compositae	Oganesyan *et al.* (1976)
Quercetagetin 3,7-dimethyl ether		
6-Glucoside	*Chrysosplenium glechomaefolium*; Compositae	Bohm and Collins (1979)
6-Galactoside	*Neurolaena oaxacana* leaf; Compositae	Ulubelen *et al.* (1980b)
4′-Glucoside	*Chrysosplenium glechomaefolium*; Compositae	Bohm and Collins (1979)
Quercetagetin 6,7-dimethyl ether (Eupatolitin)		
3-Sulphate	*Brickellia veronicaefolia* leaf; Compositae	Roberts *et al.* (1980)
Quercetagetin 6,3′-dimethyl ether (Spinacetin)		
7-Glucoside	*Lepidophorum repandum* flower; Compositae	Harborne *et al.* (1976)
3-Rutinoside	*Anvillea garcini* leaf; Compositae	Ulubelen *et al.* (1979)
Quercetagetin 3,6,7-trimethyl ether		
4′-Glucoside	*Chrysosplenium glechomaefolium*; Compositae	Bohm and Collins (1979)
Quercetagetin 3,7,4′-trimethyl ether		
6-Glucoside	*Chrysosplenium echinus*; Compositae	Bohm and Collins (1979)
3′-Glucoside	*Chrysosplenium americanum*; Compositae	Bohm and Collins (1979)

Glycoside	Source	Reference
Quercetagetin 6,7,3′-trimethyl ether (Veronicafolin)		
3-Rutinoside	*Brickellia chlorolepis* leaf; Compositae	Ulubelen *et al.* (1980a)
3-Digalactoside, 3-Sulphate	*Brickellia veronicaefolia* leaf; Compositae	Roberts *et al.* (1980)
Quercetagetin 6,7,4′-trimethyl ether (Eupatin)		
3-Sulphate	*Brickellia californica* leaf; Compositae	Mues *et al.* (1979)
Quercetagetin 3,6,7,3′,4′-pentamethyl ether (Artemetin)		
5-Glucosylrhamnoside	*Vitex negundo* stem bark; Verbenaceae	Misra and Subramanian (1980)
8-Hydroxyquercetin (Gossypetin)		
3-Sulphate	*Malva sylvestris* leaf; Malvaceae	Nawwar *et al.* (1977a)
Gossypetin 7-methyl ether		
3-Arabinoside	*Eriogonum nudum* flowers; Polygonaceae	Harborne *et al.* (1978)
3-Rhamnoside	*Atraphaxis purpurea* leaf; Polygonaceae	Chumbalov (1976)
3-Galactoside, 8-Glucoside, 3-Galactoside-8-glucoside	*Eriogonum nudum* flowers; Polygonaceae	Harborne *et al.* (1978)
3-Rhamnoside-8-acetate	*Atraphaxis purifolia* leaf; Polygonaceae	Chumbalov (1976)
Gossypetin 8-methyl ether		
3-Galactoside	*Geraea canescens* flowers; Compositae	Harborne *et al.* (1978)
Gossypetin 3′-methyl ether		
3-Rutinoside	*Coronilla valentina* petals; Leguminosae	Harborne (1981)
7-(6″-Acetyl)glucoside	*Haplophyllum perforatum*; Rutaceae	Batirov and Malikov (1980)
Gossypetin 7,4′-dimethyl ether		
8-Acetate	*Notholaena* spp. farinose exudate; Filicales	Wollenweber *et al.* (1978a)
8-Butyrate	*Notholaena* spp. farinose exudate; Filicales	Wollenweber *et al.* (1978b)
Gossypetin 8,3′-dimethyl ether		
3-Rutinoside	*Coronilla valentina* petals; Leguminosae	Harborne (1981)
Myricetin		
3-Xyloside	*Leptarrhena pyrolifolia* leaf; Saxifragaceae	Miller and Bohm (1979)
3-Dixyloside, 3-Xylosylglucoside	*Heuchera* spp. whole plant; Saxifragaceae	Wells and Bohm (1980)
3-Xylosylgalactoside	*Saxifraga* spp, whole plant; Saxifragaceae	Miller and Bohm (1980)

Table 5.7 (Contd.)

Glycoside	*Source*	*Reference*
3-Diglucoside	*Heuchera* spp. whole plant; Saxifragaceae	Wells and Bohm (1980)
3-Galactosylglucoside	*Heuchera* spp. whole plant; Saxifragaceae	Wells and Bohm (1980)
3-Rhamnosylrutinoside	*Limnanthes douglasii* leaf; Limnanthaceae	Parker and Bohm (1975)
3-Rutinoside-7-glucoside		
3-Triglucoside	*Heuchera cylindrica* whole plant; Saxifragaceae	Bohm and Wilkins (1978)
3-(6″-Galloyl)glucoside	*Sedum kamtschaticum*; Crassulaceae	Zapesochnaya and Shnyakina (1978)
3-(6″-Galloyl)galactoside		
3-(2″-Acetyl)rhamnoside	*Sedum solskianum* aerial parts; Crassulaceae	Zapesochnaya and Shnyakina, (1975)
3-Rhamnosidesulphate	*Davidsonia pruriens* leaf; Davidsoniaceae	Wilkins and Bohm (1977)
Myricetin 3′-methyl ether (Laricitrin)		
3-Rhamnoside	*Abies amabilis* needles; Pinaceae	Parker *et al.* (1979)
3-Glucoside	*Larix sibirica* needles; Pinaceae	Tyukavkina *et al.* (1974)
3-Galactoside	*Chondropetalum hookerianum* stem; Restionaceae	Harborne (1979b)
3-Rutinoside	*Limnanthes douglasii* leaf and petal; Limnanthaceae	Parker and Bohm (1975)
3-Rhamnosylrutinoside		
3-Rutinoside-7-glucoside		
3-*p*-Coumaroylglucoside	*Larix gmelinii* needles; Pinaceae	Niemann (1980)
Myricetin 4′-methyl ether		
3-Galactosyl(1→4)galactoside	*Vitex negundo* Stem bark; Verbenaceae	Misra and Subramanian (1980)
Myricetin 3′,5′-dimethyl ether (Syringetin)		
3-Rhamnoside	*Hedychium stenopetalum* leaf; Zingiberaceae	Williams and Harborne (1977)
3-Galactoside	*Philydrum lanuginosum* whole plant; Philydraceae	Bohm and Collins (1975)
3-Glucuronide		
3-Rutinoside	*Limnanthes douglasii* leaf and petal; Limnanthaceae	Parker and Bohm (1975)
3-Rhamnosylrutinoside		
3-Rutinoside-7-glucoside		
3-*p*-Coumaroylglucoside	*Larix gmelini* needles; Pinaceae	Niemann (1980)
6-Hydroxymyricetin 6,3′,5′-trimethyl ether		
3-Glucoside	*Tillandsia usneoides* leaf and stem; Bromeliaceae	Williams (1978)
6-Hydroxymyricetin 4,6,3′,5′-tetramethyl ether		
7-Glucoside	*Tillandsia usneoides* leaf and stem; Bromeliaceae	Williams (1978)

Glycoside	*Source*	*Reference*
8-Hydroxymyricetin (Hibiscetin)		
8-Glucosylxyloside	*Solanum citrullifolium* leaf; Solanaceae	Whalen and Mabry (1979)
2′-Hydroxyflavonols		
5,2′-Dihydroxy-3,7,4′-trimethoxyflavone		
2′-Glucoside	*Chrysosplenium americanum*; Compositae	Bohm and Collins (1979)
5,2′,5′-Trihydroxy-3,7,4′-trimethoxyflavone		
2′-Glucoside	*Chrysosplenium glechomaefolium*; Compositae	Bohm and Collins (1979)

A general problem in drawing up Table 5.7 has been the difficulty of relating 3-diglycosides of known structure with similar compounds in which the interglycosidic links have not been determined. In general, the assumption has been made of identity, although confirmation is desirable in all cases. For example, the kaempferol 3-glucosylgalactoside (called panasenoside) reported in *Panax ginseng* (Araliaceae) by Komatsu *et al.* (1969) may or may not be the same as the 3-glucosyl(1 → 6)galactoside recently fully characterized in *Eryngium planum* (Umbelliferae) by Hiller *et al.* (1980). The presumption that they are the same is strengthened by the facts that the two plants concerned are from closely related families and that the anthocyanin patterns in the two families are also closely similar (see Harborne, 1976).

Other examples of presumed identity are of a kaempferol 3-xylosylgalactoside in *Lysichiton camtschatcense* (Araceae) (Table 5.7) and the 3-xylosyl(1 → 6) galactoside, reported in *Laurocerasus officinale* (Rosaceae) as having the xylose moiety in the unlikely furanose form (Tsiklauri *et al.*, 1979). Again, the 3-rhamnosylglucoside-7-rhamnoside of kaempferol in *Hesperis matronalis* (Cruciferae) (Ellnain-Wojtaszek and Kowalewski, 1971) could be identical with the 3-rutinoside-7-rhamnoside reported in *Equisetum* (see Table 5.7). Also, the quercetin 3-rhamnosylgalactoside of *Lasthenia* species (Compositae) (Ornduff *et al.*, 1973) could be the same as the 3-robinobioside reported in *Crateagus* by Bykov and Glyzin (1977) (see Table 5.7). There are also problems with partly characterized flavonol triglycosides. Thus, the 3-rhamnosylrutinosides of kaempferol and quercetin reported in *Limnanthes douglasii* (Parker and Bohm, 1975) and the 3-rhamnosyldiglucoside of quercetin in *Camellia sinensis* (Roberts, 1962) may well be identical separately with various triglycosides more recently described from *Glycine* or *Humulus* (see Table 5.7).

The possibilities of isomerism among flavonol 3-diglycosides are lessened by the fact that certain glycosidic patterns are very common in plants. For example, flavonol 3-diglycosides containing rhamnose and glucose nearly always have the sugars joined in an $\alpha 1 \rightarrow 6$ linkage, as in rutinose. Thus, quercetin 3-rutinoside or

rutin is probably the most widespread of all known flavonol glycosides. The related 3-neohesperidoside, with an $\alpha 1 \rightarrow 2$ linkage, has been described but it is definitely rare in occurrence. It is interesting that a third isomer has very recently been reported in the flowers of *Crataegus pinnatifida* (Rosaceae) by Bykov and Glyzin (1977), but it is not yet clear whether the two sugars in this compound are $\alpha 1 \rightarrow 3$- or $\alpha 1 \rightarrow 4$-linked.

Several examples of isomeric flavonol 3-diglycosides occurring in the same plant have been reported, e.g. of isomeric quercetin 3-xylosylglucosides in *Heuchera micrantha* (Wilkins and Bohm, 1976), but until the nature of the isomerism is clear, it is not appropriate to include such variation in Table 5.7.

It is worth remembering here that the presence of impurities in glycoside samples can interfere both with the measurement of R_F values and also with sugar analyses. It is also important that sugar:aglycone ratios be determined wherever possible. This was not done on what appeared to be kaempferol 7,4′-diglucoside, isolated earlier from an acylated complex present in *Asplenium* (Harborne *et al.*, 1973) because of shortage of material. A recent reinterpretation of the UV spectrum and shifts (J. B. Harborne, unpublished results) now indicates that this compound must also have a glucose residue in the 3-position and is the so far undescribed 3,7,4′-triglucoside of kaempferol. A related trisubstituted kaempferol derivative containing glucose and rhamnose and with the same spectral properties has also been isolated from *Coronilla* leaves, and trisubstitution was confirmed in this case by methylation and hydrolysis to yield kaempferol 5-methyl ether. These two substances represent the first known examples of kaempferol glycosides with sugars present at three different hydroxyl groups in the flavonoid moiety; full details of their structural analyses will be published later.

One simplification to the earlier 1975 review of flavonol glycosides can now be made: the removal of foeniculin from the list. At that time, four quercetin 3-L-arabinosides were mentioned: two of known structure, the α-arabofuranoside avicularin and the α-arabopyranoside, guajaverin; and two of incompletely determined structure, the β-arabinoside polystachoside from *Polygonum polystachyum* and the arabinoside, foeniculin, from fennel *Foeniculum vulgare* (Umbelliferae). Recent studies by Jarman and Crowden (1977) of foeniculin, which occurs widely in plants of the Epacridaceae, have confirmed on the basis of enzymic and R_F data that it is different from both avicularin and polystachoside. Nevertheless, its m.p. is close to that of guajaverin and indeed, a direct R_F comparison in many solvent systems (J. B. Harborne, unpublished data) with an authentic sample of guajaverin, kindly supplied by H. Geiger, showed foeniculin to be identical with it. One final question remains to be answered with these compounds; is the sugar moiety of polystachoside in the furano- or pyrano- form?

5.7. DISTRIBUTION PATTERNS IN PLANTS

Most of the work recorded in this chapter refers to the isolation and identification of individual glycosides from single plant taxa. In addition to these detailed investigations, the flavone and flavonol glycosides present in a much wider range of plants have been surveyed over the past five years. New glycosides are often first detected during such surveys. In addition much new information on distribution patterns of known derivatives has accumulated. In most recent chemotaxonomic investigations of flavonoids, glycosidic patterns have been determined and, in many cases, significant correlations with morphological groupings have become apparent.

The discovery in liverworts of a range of interesting and unusual flavone glycosides has already been mentioned (Section 5.5). Distribution patterns of these compounds in Bryophyta have been reviewed by Markham and Porter (1978). Chemotaxonomic aspects of flavone and flavonol glycosides in ferns have also been the subject of a recent review (Cooper-Driver, 1980). Although flavones and flavonols are widespread in gymnosperms, their glycosides have not been extensively studied except in a few genera such as *Pinus* and *Larix*; recent work is summarized by Niemann (1979).

In the case of the angiosperms, so many papers have appeared on distribution patterns that only a small selection can be referred to here. At the species level, the detailed investigation of 35 flavonol glycosides variously present in 26 populations of *Anthyllis vulneraria* deserves mention (Gonnet, 1978). At the generic level, extensive surveys have been carried out in the following genera, among others: *Anacyclus* (Greger, 1978), *Hazardia* (Clark and Mabry, 1978), *Solanum* (Wietschel, 1978), *Sorbus* (Challice and Kovanda, 1978) and *Veronica* (Grayer-Barkmeijer, 1978). At the family level, reviews of known glycosides have appeared for the Celastraceae (Brüning and Wagner, 1978), Compositae (Harborne, 1977), Oleaceae (Harborne and Green, 1980) and Solanaceae (Harborne and Swain, 1979). With regard to monocotyledonous families, glycosidic patterns have been reported in a number of families, including Bromeliaceae, Cyperaceae, Gramineae and Orchidaceae; for references, see the latest paper on Araceae (Williams *et al.*, 1981).

In spite of all these investigations, our knowledge of the glycosidic complexity of flavones and flavonols is still remarkably fragmentary, and further studies cannot but uncover a yet wider range of these conjugates among vascular plants.

REFERENCES

Adinarayana, D., Gunasekar, D., Seligmann, O. and Wagner, H. (1980), *Phytochemistry* **19**, 480.
Al-Khubaizi, M. S., Mabry, T. J. and Bacon, J. (1977), *Phytochemistry* **17**, 163.
Aly, H.-F., Geiger, H., Schücker, U., Waldrum, H., Velde, G. V. and Mabry, T. J. (1975), *Phytochemistry* **14**, 1613.

Ansari, F. R., Ansari, W. H., Rahman, W., Seligmann, O., Chari, V. M., Wagner, H. and Österdahl, B. G. (1979), *Planta Med.* **36**, 196.
Arisawa, M., Fukuta, M., Shimizu, M. and Morita, N. (1979), *Chem. Pharm. Bull.* **27**, 1252.
Bacon, J. D. and Mabry, T. J. (1976), *Phytochemistry* **15**, 1087.
Bacon, J. D., Urbatsch, L. E., Bragg, L. H., Mabry, T. J., Neuman, P. and Jackson, D. W. (1978), *Phytochemistry* **17**, 1939.
Baimukhambetov, M. A. (1975), *Nek. Probl. Farm. Nauki Prackt. Mater Sézeda Farm. Kazis* **1**, 101.
Batirov, E. Kh. and Malikov, V. M. (1980), *Khim. Prir. Soedin.* **16**, 330.
Bhutani, S. P., Chibber, S. S. and Seshadri, T. R. (1969), *Phytochemistry* **8**, 299.
Bishara, S. A. R., Zinchenko, T. V. and Nikonov, G. K. (1976), *Ukr. Khim. Zh. (Russ. Ed.)* **42**, 284.
Blinova, K. F. and Thuan, B. T. (1977), *Rastit. Resur.* **13**, 466.
Boguslavskaya, L. I., Gordienko, V. G., Kovalev, I. P. and Litvinenko, V. I. (1976), *Khim. Prir. Soedin.* **12**, 391.
Bohm, B. A. and Collins, F. W. (1975), *Phytochemistry* **14**, 315.
Bohm, B. A. and Collins, F. W. (1979), *Biochem. Syst. Ecol.* **7**, 195.
Bohm, B. A. and Wilkins, C. K. (1978), *Can. J. Bot.* **56**, 1174.
Borisov, M. I. (1974), *Rast. Resur.* **10**, 66.
Brüning, R. and Wagner, H. (1978), *Phytochemistry* **17**, 1821.
Buttery, B. R. and Buzzell. R. I. (1975), *Can. J. Bot.* **53**, 219.
Bykov, V. I. and Glyzin, V. I. (1977), *Vopr. Farm. Dal'nem Vostoke* **2**, 40.
Cabrera, J. L. and Juliani, H. R. (1976), *Lloydia* **39**, 253.
Cabrera, J. L. and Juliani, H. R. (1977), *Phytochemistry* **16**, 400.
Cabrera, J. L. and Juliani, H. R. (1979), *Phytochemistry* **18**, 510.
Candy, H. A., Laing, M., Weeks, C. M. and Kruges, G. J. (1975), *Tetrahedron Lett.* 1211.
Challice, J. and Kovanda, M. (1978), *Preslia, Praha* **50**, 305.
Chandel, R. S. and Rastogi, R. P. (1980), *Phytochemistry* **19**, 1889.
Chari, V. M., Jordan, M., Wagner, H. and Thies, P. W. (1977), *Phytochemistry* **16**, 1110.
Chari, V. M., Jordan, M. and Wagner, H. (1978), *Planta Med.* **34**, 93.
Chari, V. M., Grayer, R., Harborne, J. B. and Österdahl, B. G. (1981), *Phytochemistry* **20**, 1977.
Chauhan, J. S. and Kumari, G. (1979), *Planta Med.* **37**, 86.
Chauhan, J. S., Sultan, M. and Srivastava, S. K. (1977), *Planta Med.* **32**, 217.
Chauhan, J. S., Kumari, G. and Saraswat, M. (1979a), *Ind. J. Chem.* **18**, 473.
Chauhan, J. S., Srivastava, S. K., Srivastava, S. D. (1979b), *Phytochemistry* **18**, 691.
Chou, C.-J., Wang. C.-B. and Lin, L.-C. (1976), *Phytochemistry* **15**, 1420.
Chumbalov, T. K. (1976), *Khim. Prir. Soedin.* **12**, 658.
Clark, W. D. and Mabry, T. J. (1978), *Biochem. Syst. Ecol.* **6**, 19.
Collins, F. W., Bohm, B. A. and Wilkins, C. K. (1975), *Phytochemistry* **14**, 1099.
Cooper-Driver, G. A. (1980), *Bull. Torrey Bot. Club* **107**, 116.
Crawford, D. J. and Mabry, T. J. (1978), *Biochem. Syst. Ecol.* **6**, 189.
Derkach, A. I., Komissarenko, N. F., Gordienko, V. G., Sheremet, I. P., Kovalev, I. P. and Pakaln, D. A. (1980), *Khim. Prir. Soedin.* **16**, 172.
Dubey, P. and Gupta, P. C. (1978), *Phytochemistry* **17**, 2138.
Dubey, P. and Gupta, P. C. (1980), *Planta Med.* **38**, 155.
El Ansari, M. A., Nawwar, M. A. M., El Dein, A., El Sherbeiny, A. and El Sissi, H. I. (1976), *Phytochemistry* **15**, 231.
El Ansari, M. A., Ishak, M. S., Ahmed, A. A. and Saleh, N. A. M. (1977), *Z. Naturforsch.* **32C**, 444.
Ellnain-Wojtaszek, M. and Kowalewski, Z. (1971), *Dissert. Pharm. Pharmacol.* **XXIII**, No. 6, 555.
Ellnain-Wojtaszek, M., Kowalewski, Z. and Skrzypczakowa, L. (1978), *Ann. Pharm.* (*Poznan*) **13**, 171.
El-Naggar, S. F. and Doskotch, R. W. (1979), *J. Nat. Prod.* (*Lloydia*) **42**, 126.
El Sherbeiny, A. A., El Ansari, M. A., Nawwar, M. A. M. and El Sayed, N. H. A. (1977), *Planta Med.* **32**, 165.

Fadeeva, O. V. and Nikishchenko, T. K. (1979), *Khim. Prir. Soedin.* **15**, 576.
Farkas, L., Vermes, B., Nogradi, M. and Kalman, A. (1976a), *Phytochemistry* **15**, 215.
Farkas, L., Vermes, B., Nogradi, M. and Kalman, A. (1976b), *Phytochemistry* **15**, 1184.
Fujiwara, H., Nonaka, G. and Yogi, A. (1976), *Chem. Pharm. Bull.* **24**, 407.
Gaind, K. N., Singla, A. K. and Wallace, J. W. (1981), *Phytochemistry* **20**, 530.
Garg, S. P., Bhushan, R. and Kapoor, R. C. (1979), *Indian J. Chem. Sect. B* **17**, 416.
Gehring, E. and Geiger, H. (1980), *Z. Naturforsch.* **35C**, 380.
Geiger, H. and de Groot-Pfleiderer, W. (1979), *Phytochemistry* **18**, 1709.
Geiger, H., Lang, U., Britsch, E., Mabry, T. J., Suhr-Schücker, U., Velde, G. V. and Waldrum, H. (1978), *Phytochemistry* **17**, 336.
Geissman, T. A. (ed.) (1962), *The Chemistry of Flavonoid Compounds.* Pergamon Press, Oxford.
Gella, E. V., Makarova, G. V. and Borisyuk, Y. G. (1967), *Farmatsert. Zh.* (*Kiev*) **22**, 80.
Glennie, C. W. and Jain, S. C. (1980), *Phytochemistry* **19**, 157.
Gonnet, J. F. (1978), *Phytochemistry* **17**, 1319.
Grayer-Barkmeijer, R. J. (1978), *Biochem. Syst. Ecol.* **6**, 131.
Greger, H. (1978), *Biochem. Syst. Ecol.* **6**, 11.
Guinaudeau, H., Seligmann, O., Wagner, H. and Neszmelyi, A. (1981), *Phytochemistry* **20**, 1113.
Guppy, G. A. and Bohm, B. A. (1976), *Biochem. Syst. Ecol.* **4**, 231.
Harborne, J. B. (1965), *Phytochemistry* **4**, 107.
Harborne, J. B. (1967), *The Comparative Biochemistry of Flavonoids.* Academic Press, London.
Harborne, J. B. (1968), *Phytochemistry* **7**, 1215.
Harborne, J. B. (1975), *Phytochemistry* **14**, 1331.
Harborne, J. B. (1976), *Biochem. Syst. Ecol.* **4**, 31.
Harborne, J. B. (1977), In *The Biology and Chemistry of the Compositae* (eds. V. H. Heywood, J. B. Harborne and B. L. Turner), Academic Press, London, pp. 359–384.
Harborne, J. B. (1978), *Phytochemistry* **17**, 915.
Harborne, J. B. (1979a), *Recent Adv. Phytochem.* **12**, 457.
Harborne, J. B. (1979b), *Phytochemistry* **18**, 1323.
Harborne, J. B. (1981), *Phytochemistry* **20**, 1117.
Harborne, J. B. and Green, P. S. (1980), *Bot. J. Linn. Soc.* **81**, 155.
Harborne, J. B. and Hall, E. (1964), *Phytochemistry* **3**, 421.
Harborne, J. B. and King, L. (1976), *Biochem. Syst. Ecol.* **4**, 111.
Harborne, J. B. and Sherratt, H. S. A. (1961), *Biochem. J.* **78**, 298.
Harborne, J. B. and Stacey, C. I. (1978), *Phytochemistry* **17**, 588.
Harborne, J. B. and Swain, T. (1979). In *The Biology and Taxonomy of the Solanaceae* (eds. J. G. Hawkes, R. N. Lester and A. D. Skelding), Academic Press, London, pp. 257–268.
Harborne, J. B. and Williams, C. A. (1975). In *The Flavonoids* (eds. J. B. Harborne, T. J. Mabry and H. Mabry), Chapman & Hall, London, pp. 376–441.
Harborne, J. B. and Williams, C. A. (1976), *Biochem. Syst. Ecol.* **4**, 37.
Harborne, J. B., Williams, C. A. and Smith, D. M. (1973), *Biochem. Syst.* **1**, 51.
Harborne, J. B., Williams, C. A., Greenham, J. and Moyna, P. (1974), *Phytochemistry* **13**, 1557.
Harborne, J. B., Heywood, V. H. and King, L. (1976), *Biochem. Syst. Ecol.* **4**, 1.
Harborne, J. B., Saleh, N. A. M. and Smith, D. M. (1978), *Phytochemistry* **17**, 589.
Henning, W. and Herrmann, K. (1980), *Phytochemistry* **19**, 2727.
Higuchi, R. and Donnelly, D. M. X. (1978), *Phytochemistry* **17**, 787.
Hiller, K., Voight, G., Koppel, H. and Otto, A. (1979), *Die Pharmazie* **34**, 192.
Hiller, K., Otto, A. and Grundemann, E. (1980), *Die Pharmazie* **35**, 113.
Hiraoka, A. (1978), *Biochem. Syst. Ecol.* **6**, 171.
Hiraoka, A. and Maeda, M. (1979), *Chem. Pharm. Bull.* **27**, 3130.
Homberg, H. and Geiger, H. (1980), *Phytochemistry* **19**, 2443.
Hostettmann, K., Hostettmann-Kaldas, M. and Nakanishi, K. (1979), *J. Chromatogr.* **170**, 355.
Ilyas, M., Ilyas, N. and Wagner, H. (1977), *Phytochemistry* **16**, 1456.

Imperato, F. (1976), *Phytochemistry* **15**, 439.
Imperato, F. (1979a), *Experientia* **35**, 1134.
Imperato, F. (1979b), *Chem. Ind.*, 525.
Imperato, F. (1980), *Chem. Ind.*, 540.
Ina, H. and Iida, H. (1981), *Phytochemistry* **20**, 1176.
Ishikura, N. and Hayashida, S. (1979), *Agric. Biol. Chem.* **43**, 1923.
Ishikura, N. and Sato, S. (1977), *Bot. Mag. (Tokyo)* **90**, 83.
Ishikura, N., Sato, S. and Kurosawa, K. (1976), *Phytochemistry* **15**, 1183.
Ismail, S. I., Hammouda, F. M., Rizk, A. M. and El-Kady, S. E. (1979), *Pharmazie* **34**, 112.
Isobe, T., Fukushige, T. and Noda, Y. (1979), *Chem. Lett.* **27**.
Isobe, T., Nobuo, I. and Noda, Y. (1980), *Phytochemistry* **19**, 1877.
Ivanova, S. Z. (1978), *Issled. Obl. Khim. Drev. Tezisy Dokl. Konf. Molodykh Uch.* **2**, 32 (eds. V. P. Kalwan and Z. Ird), Riga, U.S.S.R.
Ivanova, S. Z., Zapesochnaya, G. G., Medvedeva, S. A. and Tyukavkina, N. A. (1978), *Khim. Prir. Soedin.* **14**, 200.
Ivanova, S. Z., Zapesochnaya, G. G., Sheichenko, V. I., Tyukavkina, N. A. and Medvedeva, S. A. (1980), *Khim. Prir. Soedin.* **16**, 254.
Jarman, S. J. and Crowden, R. K. (1977), *Phytochemistry* **16**, 929.
Jauhari, P. K., Sharma, S. C., Tandon, J. S. and Dhar, M. M. (1979), *Phytochemistry* **18**, 359.
Jensen, S. R., Mikkelsen, C. B. and Nielsen, B. J. (1981), *Phytochemistry* **20**, 71.
Jerzmanowska, Z. and Kamecki, J. (1973), *Rocz. Chem.* **47**, 1629.
Karl, C., Müller, G. and Pedersen, P. A. (1976), *Phytochemistry* **15**, 1084.
Karl, C., Pedersen, P. A. and Schwarz, C. (1977), *Phytochemistry* **16**, 1117.
Karl, C., Pedersen, P. A. and Müller, G. (1980), *Z. Naturforsch.* **35C**, 826.
Karl, C., Müller, G. and Pedersen, P. A. (1981), *Planta Med.* **41**, 96.
Komatsu, M., Tomimori, T. and Makiguchi, Y. (1969), *Yakugaku Zasshi* **89**, 122.
Komissarenko, N. F., Derkach, A. I., Sheremet, I. P., Kovalev, I. P., Gordienko, V. G. and Pakaln, D. A. (1978), *Khim. Prir. Soedin.* **14**, 521.
Konopleva, M. M., Smirnova, L. P., Glyzin, V. T. and Shelyuto, V. L. (1979), *Khim. Prir. Soedin.* **15**, 311.
Kowalewska, K. and Lutomski, J. (1978), *Herba Pol.* **24**, 107.
Kowalewski, Z., Kortus, M. and Matlawska, I. (1979), *Herba Pol.* **25**, 183.
Krasnov, E. A. (1979), *Khim. Prir. Soedin.* **15**, 851.
Krasnov, E. A. and Alekseyuk, N. V. (1979), **15**, 860.
Krasnov, E. A. and Demidenko, L. A. (1979), *Khim. Prir. Soedin.* **15**, 404.
Kreuzaler, F. and Hahlbrock, K. (1973), *Phytochemistry* **12**, 1149.
Kunde, K. and Issac, O. (1979), *Planta Med.* **37**, 124.
Lamer-Zarawska, E. (1980), *Pol. J. Chem.* **54**, 213.
Lin, J.-H., Lin, Y.-M. and Chen, F.-C. (1975), *J. Chin. Chem. Soc. Taipei* **22**, 383.
Lin, J.-H., Lin, Y.-M. and Chen, F.-C. (1976), *J. Chin. Chem. Soc. Taipei* **23**, 57.
Lindenborn, H. (1869), Doctoral Dissertation, Würzburg, Germany.
Litvinenko, V. I., Sabirov, R. S. and Nazitov, Z. N. (1973), *Khim. Prir. Soedin.* **9**, 672.
Liu, Y. L., Fu, F. Y., Hsie, T. H. and Chang, S. L. (1976), *Hua Hsuch Hsueh Pao* **34**, 211.
Mabry, T. J., Abdel-Baset, Z., Padolina, W. G. and Jones, S. B., Jr. (1975), *Biochem. Syst. Ecol.* **2**, 185.
Maksyutina, N. P., Litvinenko, V. I. and Kovalev, I. P. (1965), *Khim. Prir. Soedin.* **2**, 388.
Markham, K. R. (1980), *Biochem. Syst. Ecol.* **8**, 11.
Markham, K. R. and Given, D. R. (1979), *Biochem. Syst. Ecol.* **7**, 91.
Markham, K. R. and Moore, N. A. (1980), *Biochem. Syst. Ecol.* **8**, 17.
Markham, K. R. and Porter, L. J. (1973), *Phytochemistry* **12**, 2007.
Markham, K. R. and Porter, L. J. (1974a), *Phytochemistry* **13**, 1553.
Markham, K. R. and Porter, L. J. (1974b), *Phytochemistry* **13**, 1937.

Markham, K. R. and Porter, L. J. (1975), *Phytochemistry* **14**, 1093.

Markham, K. R. and Porter, L. J. (1978), *Progr. Phytochem.* **5**, 181.

Markham, K. R. and Porter, L. J. (1979), *Phytochemistry* **18**, 611.

Markham, K. R., Porter, L. J., Mues, R., Zinsmeister, H. D. and Brehm, B. G. (1976), *Phytochemistry* **15**, 147.

Markham, K. R., Zinsmeister, H. D. and Mues, R. (1978a), *Phytochemistry* **17**, 1601.

Markham, K. R., Ternai, B., Stanley, R., Geiger, H. and Mabry, T. J. (1978b), *Tetrahedron* **34**, 1389.

Markham, K. R., Zinsmeister, H. D. and Mues, R. (1978c), *Phytochemistry* **17**, 1601.

Mathuram, S., Koziparambil, K., Purushothaman, K. and Sarada, A. (1976), *Phytochemistry* **15**, 838.

Matlawska, I. and Kowalew, Z. (1976), *Herba Pol.* **22**, 11.

Miller, J. M. and Bohm, B. A. (1979), *Phytochemistry* **18**, 1412.

Miller, J. M. and Bohm, B. A. (1980), *Biochem. Syst. Ecol.* **8**, 279.

Misra, G. S. and Subramanian, P. M. (1980), *Planta Med.* **38**, 155.

Mitchell, R. E. and Geissman, T. A. (1971), *Phytochemistry* **10**, 1559.

Morita, N., Shimizu, M. and Arisawa, M. (1974), *Yakugaku Zasshi* **94**, 913.

Morita, N., Arisawa, M., Ozawa, H., Chen, C.-S. and Kan, W.-S. (1977), *Yakugaku Zasshi* **97**, 976.

Mues, R., Timmermann, B. N., Ohno, N. and Mabry, T. J. (1979), *Phytochemistry* **18**, 1379.

Mukamedyarova, M. M. and Chumbalov, T. K. (1979), *Khim. Prir. Soedin.* **15**, 854.

Nahrstedt, A., Dumkow, K., Janistyn, B. and Pohl, R. (1974), *Tetrahedron Lett.* 559.

Nair, A. G. R. (1979), *Indian J. Chem. Sect. B* **18**, 188.

Nair, A. G. R., Ramachandran, R. P., Nagarajan, S. and Subramanian, S. S. (1973), *Indian J. Chem.* **11**, 1316.

Nair, A. G., Ramesh, P. and Subramanian, S. S. (1976), *Phytochemistry* **15**, 839.

Nair, A. G. R., Subramanian, S. S. and Joshi, B. S. (1977a), *Indian J. Chem.* **15B**, 1645.

Nair, A. G. R., Ramesh, P. and Subramanian, S. S. (1977b), *Curr. Sci.* **46**, 14.

Nakaoki, T. and Morita, N. (1956), *J. Pharm. Soc. Jpn.* **76**, 323.

Nawwar, M. A. M., El Sherbieny, A., El Ansari, M. A. and El Sissi, H. I. (1977a), *Phytochemistry* **16**, 145.

Nawwar, M. A. M., Ishak, M. S., El Sherbieny, A. A. and Meshaal, S. A. (1977b), *Phytochemistry* **16**, 1319.

Niemann, G. J. (1975), *Z. Naturforsch.* **30C**, 550.

Niemann, G. J. (1979), *Acta Bot. Neerl.* **28**, 73.

Niemann, G. J. (1980), *Z. Naturforsch.* **35C**, 514.

Nuralieva, Zh. S., Litvinenko, V. I. and Alibaeva, P. K. (1969), *Khim. Prir. Soedin.* **5**, 369.

Oganesyan, E. T. and Dzhumyrko, S. F. (1974), *Khim. Prir. Soedin.* **10**, 520.

Oganesyan, E. T., Smirnova, L. P., Dzhumyrko, S. F. and Kechatova, N. A. (1976), *Khim. Prir. Soedin.* **12**, 599.

Ogawa, M. and Ogihara, Y. (1975), *Yagugaku Zasshi* **95**, 655.

Okuda, T., Yoshida, T. and Ono, I. (1975), *Phytochemistry* **14**, 1654.

Okuyama, T., Hosoyama, K., Hiraga, Y., Kurono, G. and Takemoto, T. (1978), *Chem. Pharm. Bull.* **26**, 3071.

Olechnowicz-Stepien, W., Rzadkowska-Bodalska, H. and Lamer-Zarawska, E. (1978), *Pol. J. Chem.* **52**, 2167.

Ornduff, R., Bohm, B. A. and Saleh, N. A. M. (1973), *Biochem. Syst.* **1**, 147.

Österdahl, B. G. (1978), *Acta Chem. Scand.* **32**, 714.

Österdahl, B. G. (1979), *Acta Chem. Scand.* **33**, 119.

Österdahl, B. G. and Lindberg, G. (1977), *Acta Chem. Scand.* **31**, 293.

Pangarova, T. T. and Zapesochnoya, G. G. (1974), *Khim. Prir. Soedin.* **10**, 667.

Parker, W. H. and Bohm, B. A. (1975), *Phytochemistry* **14**, 553.

Parker, W. H., Maze, J. and McLachlan, D. G. (1979), *Phytochemistry* **18**, 508.

Pereyra de Santiago, O. J. and Juliani, H. R. (1972), *Experientia* **28**, 380.

Perold, G. W., Beylis, P. and Howard, A. S. (1973), *J. Chem. Soc., Perkin Trans. 1*, 643.
Pijewska, L. and Kamecki, J. (1977), *Rocz. Chim.* **51**, 291.
Popova, T. P., Pakalns, D., Chernykh, N. A., Zoz, I. G. and Litvinenko, V. I. (1976), *Rastit Resur.* **12**, 232.
Rabanal, R. A., Valverde, S., Martin-Lomas, M., Rodriguez, B. and Chari, V. M. (1982), *Phytochemistry* **21**, in press.
Ranganathan, R. M., Nagarajan, S., Mabry, T. J., Yong-Long, L. and Neuman, P. (1980), *Phytochemistry* **19**, 2505.
Rao, M. A. and Rao, E. V. (1980), *Curr. Sci.* **49**, 468.
Rao, P. R., Seshadri, T. R. and Sharma, P. (1973), *Curr. Sci.* **42**, 811.
Rappaportt, I., Giacopello, D., Seldes, A. M., Blanco, M. C. and Deulofeu, V. (1977), *Phytochemistry* **16**, 1115.
Redaelli, C., Formentini, L. and Santaniello, E. (1980), *Phytochemistry* **19**, 985.
Rizk, A. M., Youssef, A. M., Diab, M. A. and Salem, H. M. (1976), *Pharmazie* **31**, 405.
Roberts, E. A. H. (1962). In *Chemistry of the Flavonoid Compounds* (ed. T. A. Geissmann), Pergamon Press, Oxford, pp. 468–512.
Roberts, M. F., Timmermann, B. N. and Mabry, T. J. (1980), *Phytochemistry* **19**, 127.
Rodriguez, E. (1977), *Biochem. Syst. Ecol.* **5**, 207.
Sakushima, A., Nishibe, S. and Hisada, S. (1980), *Phytochemistry* **19**, 712.
Saleh, N. A. M. and El-Hadidi, M. N. (1977), *Biochem. Syst. Ecol.* **5**, 121.
Saleh, N. A. M., El Sissi, H. I. and Nawwar, M. A. M. (1975), *Phytochemistry* **14**, 312.
Sakar, M. K., Engelshowe, R. and Friedrich, H. (1980), *Planta Med.* **40**, 193.
Sarin, J. P. S., Singh, S., Garg, H. S., Khanna, N. M. and Dhar, M. M. (1976), *Phytochemistry* **15**, 232.
Sastry, C. V. R., Rukmim, C. and Row, L. R. (1967), *Indian J. Chem.* **5**, 613.
Schels, H., Zinsmeister, H. D. and Pfleger, K. (1977), *Phytochemistry* **16**, 1019.
Schels, H., Zinsmeister, H. D. and Pfleger, K. (1978), *Phytochemistry* **17**, 523.
Schonsiegel, I., Egger, K. and Keil, M. (1969), *Z. Naturforsch.* **24B**, 1213.
Shamyrina, A. A., Peshkova, V. A. and Shergina, N. I. (1977), *Khim. Prir. Soedin.* **13**, 577.
Shatokhina, R. K., Dranik, L. I. and Blinova, K. F. (1974), *Khim. Prir. Soedin.* **10**, 518.
Shimokoriyama, M. (1949), *Bot. Mag. (Tokyo)* **62**, 737.
Singh, J. and Tiwari, R. D. (1976), *J. Indian Chem. Soc.* **53**, 424.
Singh, J. and Tiwari, R. D. (1979), *Phytochemistry* **18**, 2060.
Smirnova, L. P., Glyzin, V. I. and Patudin, A. V. (1974), *Khim. Prir. Soedin.* **10**, 668.
Soltis, D. E. (1980), *Biochem. Syst. Ecol.* **8**, 149.
Star, A. E., Seigler, D. S., Mabry, T. J. and Smith, D. M. (1975), *Biochem. Syst. Ecol.* **2**, 109.
Stengel, B. and Geiger, H. (1976), *Z. Naturforsch.* **31C**, 622.
Stepien, W. O., Bodalksa, H. R. and Zarawska, E. L. (1978), *Pol. J. Chem.* **52**, 2167.
Stewart, R. N., Aseu, S., Massie, D. R. and Norris, K. H. (1980), *Biochem. Syst. Ecol.* **8**, 119.
Subramanian, P. M. and Misra, G. S. (1979), *J. Nat. Prod. (Lloydia)* **42**, 540.
Takagi, S., Yamaki, M., Masuda, K. and Kubota, M. (1976a) *Yakugaku Zasshi* **96**, 284.
Takagi, S., Yamaki, M., Masuda, K. and Kubota, M. (1976b), *Yakugaku Zasshi* **96**, 1217.
Takemoto, T. and Miyase, T. (1974), *Yakugaku Zasshi* **94**, 1597.
Tanaka, N., Murakami, T., Saiki, Y., Chen, C.-M. and Gomez, P. L. D. (1978), *Chem. Pharm. Bull.* **26**, 3580.
Teslov, L. S. (1980), *Khim. Prir. Soedin.* **16**, 334.
Tiwari, R. D. and Singh, J. (1977), *Phytochemistry* **16**, 1107.
Tiwari, R. D. and Singh, J. (1978), *Planta Med.* **34**, 319.
Tomas, F. (1979), *Rev. Agroquim. Tecnol. Aliment.* **19**, 224.
Tomas, F. and Ferreres, F. (1980), *Phytochemistry* **19**, 2039.
Tsiklauri, G. Ch., Shalashvili, A. G. and Litvinenko, V. I. (1979), *Khim. Prir. Soedin.* **15**, 98.
Tyukavkina, N. A., Medvedeva, S. A. and Ivanova, S. Z. (1974), *Khim. Prir. Soedin.* **10**, 157.

Ulubelen, A. and Ayanoglu, E. (1975), *Lloydia* **38**, 446.
Ulubelen, A., Oksuz, S., Aynehchi, Y. and Siami, A. (1978), *Lloydia* **41**, 435.
Ulubelen, A., Mabry, T. J. and Aynehchi, Y. (1979), *J. Nat. Prod.* (*Lloydia*) **42**, 624.
Ulubelen, A., Timmermann, B. N. and Mabry, T. J. (1980a), *Phytochemistry* **19**, 905.
Ulubelen, A., Kerr, K. M. and Mabry, T. J. (1980b), *Phytochemistry* **19**, 1761.
Umber, R. E. (1980), *Am. J. Bot.* **67**, 935.
Vancraenenbroeck, R., Callewaert, W., Gorissen and Lontie, R. (1969), *Proc. Eur. Brew. Conv.* **12**, 29.
Vandekerkhove, O. (1978), *Z. Pflanzenphysiol.* **86**, 217.
Venturella, P., Bellino, A. and Papucci, A. (1977), *Heterocycles* **6**, 267.
Voirin, B., Jay, M. and Hauteville, M. (1975), *Phytochemistry* **14**, 257.
Volk, O. H. and Sinn, M. (1968), *Z. Naturforsch.* **23B**, 1017.
Wagner, H., Ertan, M. and Seligmann, O. (1974), *Phytochemistry* **13**, 857.
Wells, E. F. and Bohm, B. A. (1980), *Can. J. Bot.* **58**, 1459.
Whalen, M. D. and Mabry, T. J. (1979), *Phytochemistry* **18**, 263.
Wietschel, G. (1978), Doctoral thesis, Cologne, Germany.
Wilkins, C. K. and Bohm, B. A. (1976), *Can. J. Bot.* **54**, 2133.
Wilkins, C. K. and Bohm, A. (1977), *Phytochemistry* **16**, 144.
Williams, C. A. (1975), *Biochem. Syst. Ecol.* **3**, 229.
Williams, C. A. (1978), *Phytochemistry* **17**, 729.
Williams, C. A. (1979), *Phytochemistry* **18**, 803.
Williams, C. A. and Harborne, J. B. (1975), *Biochem. Syst. Ecol.* **3**, 181.
Williams, C. A. and Harborne, J. B. (1977), *Biochem. Syst. Ecol.* **5**, 221.
Williams, C. A., Harborne, J. B. and Clifford, H. T. (1973), *Phytochemistry* **12**, 2417.
Williams, C. A., Harborne, J. B. and Smith, P. (1974), *Phytochemistry* **13**, 1141.
Williams, C. A., Harborne, J. B. and Crosby, T. S. (1976), *Phytochemistry* **15**, 349.
Williams, C. A., Harborne, J. B. and Mayo, S. J. (1981), *Phytochemistry* **20**, 217.
Wilson, R. D. (1979), *N. Z. J. Bot* **17**, 113.
Woeldecke M. and Herrmann, K. (1974), *Z. Naturforsch.* **29C**, 355.
Wollenweber, E. and Dietz, V. H. (1981), *Phytochemistry* **20**, 869.
Wollenweber, E., Favre-Bonvin, J. and Jay, M. (1978a) *Bull. Liaison Groupe Polyphenols* **8**, 341.
Wollenweber, E., Favre-Bonvin, J. and Jay, M. (1978b), *Z. Naturforsch.* **33C**, 831.
Wollenweber, E., Favre-Bonvin, J. and Lebreton, P. (1978c), *Phytochemistry* **17**, 1684.
Wong, E. (1980), *N. Z. J. Sci.* **23**, 39.
Yamasaki, K., Kasai, R., Masaki, Y., Okihara, M., Tanaka, O., Oshio, H., Takagi, S., Yamaki, M., Masuda, K., Nonaka, G., Tsuboi, M. and Nishioka, I. (1977), *Tetrahedron Lett.* **14**, 1231.
Young, D. A. (1979), *Am. J. Bot.* **66**, 502.
Zaitsev, V. G., Makarova, G. V. and Komissarenko, N. F. (1974), *Khim. Prir. Soedin.* **10**, 92.
Zakharova, O. I., Zakharov, A. M. and Glyzin, V. I. (1979), *Khim. Prir. Soedin.* **15**, 642.
Zapesochnaya, G. (1969), *Khim. Prir. Soedin.* **5**, 179.
Zapesochnaya, G. G. (1978), *Khim. Prir. Soedin.* **14**, 519.
Zapesochnaya, G. G. and Shnyakina, G. P. (1975), *Khim. Prir. Soedin.* **11**, 720.
Zapesochnaya, G. G. and Shnyakina, G. P. (1978), *Khim. Prir. Soedin.* **14**, 806.
Zapesochnaya, G. G., Ivanova, S. Z., Sheichenko, V. I., Tyukavkina, N. A. and Medvedeva, S. A. (1978), *Khim. Prir. Soedin.* **14**, 570.
Zapesochnaya, G. G., Ivanova, S. Z., Sheichenko, V. I., Tjukavkina, N. A. and Medvedeva, S. A. (1980), *Khim. Prir. Soedin.* **16**, 186.
Zemplén, G. and Bognár, R. (1941), *Ber. Dtsch. Chem. Ges.* **74**, 1483.
Zentsova, G. N. and Dzhumyrko, S. F. (1974), *Khim. Prir. Soedin.* **10**, 669.

CHECK LIST OF KNOWN FLAVONE GLYCOSIDES

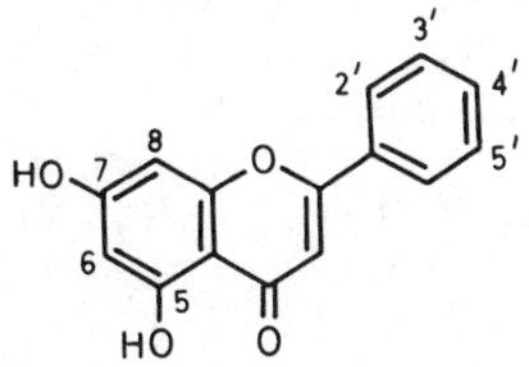

5,7-Dihydroxyflavone (Chrysin)
1 5-Glucoside (toringin)
2 7-Glucoside (aequinetin)
3 7-Galactoside*
4 7-Glucuronide
5 7-Rutinoside

7,4′-Dihydroxyflavone
6 7-Glucoside
7 7-Rutinoside

3′,4′-Dihydroxyflavone
8 4′-Glucoside*

5,6,7-Trihydroxyflavone (Baicalein)
9 6-Glucoside
10 6-Glucuronide
11 7-Rhamnoside
12 7-Glucuronide

Baicalein 6-methyl ether
13 7-Glucoside
14 7-Glucuronide

Baicalein 7-methyl ether
15 5-Glucuronide
16 5-Glucuronosylglucoside

5,7-Dihydroxy-8-methoxyflavone (Wogonin)
17 7-Glucuronide*

7,3′,4′-Trihydroxyflavone
18 7-Glucoside
19 7-Rutinoside

7,4′-Dihydroxy-3′-methoxyflavone
20 7-Glucoside

5,2′-Dihydroxy-7-methoxyflavone
21 2′-Glucoside (echioidin)

5,7,4′-Trihydroxyflavone (Apigenin)
22 5-Glucoside
23 7-Arabinoside*
24 7-Xyloside*
25 7-Rhamnoside
26 7-Glucoside (cosmosiin)
27 7-Galactoside
28 7-Glucuronide
29 7-Galacturonide*
30 7-Methylgalacturonide
31 4′-Glucoside
32 4′-Glucuronide*
33 7-Arabinosylglucoside*
34 7-Xylosyl (1→2) glucoside
35 7-Rutinoside
36 7-Neohesperidoside
37 7-Rhamosylglucuronide
38 7-Sophoroside
39 7-Apiosylglucoside (apiin)
40 7-Xylosylglucuronide
41 7-Rhamnosylglucuronide*
42 7-Diglucuronide*
43 7,4′-Diglucoside
44 7,4′-Diglucuronide
45 7-Glucuronide-4′-rhamnoside*
46 4′-Diglucoside
47 7-Rutinoside-4′-glucoside*
48 7-Neohesperidoside-4′-glucoside*
49 7-Diglucuronide-4′-glucuronide*
50 7-Neophespheridoside-4′-sophoroside*
51 7-(4″-*p*-Coumaroyl)glucoside*
52 7-(2″-Acetyl)glucoside*
53 7-(6″-Acetyl)glucoside*
54 7-(Malonyl)glucoside
55 7-Glucoside-4′-*p*-coumarate
56 7-Glucoside-4′-caffeate
57 4′-(2″,6″-Di-*p*-coumaroyl)glucoside*
58 7-(Malonyl)apiosylglucoside
59 7-Rutinoside-4′-caffeate
60 7-Glucosidesulphate*
61 7-sulphate*

Apigenin 7-methyl ether (Genkwanin)
62 5-Glucoside
63 4′-Glucoside
64 5-Xylosylglucoside

Apigenin 4′-methyl ether (Acacetin)
65 7-Glucoside (tilianine)
66 7-Glucuronide
67 7-Rutinoside (linarin)
68 7-Neohesperidoside (fortunellin)
69 7-Diglucoside
70 7-Rhamnosylgalacturonide
71 7-Glucuronosyl(1→2)glucuronide
72 7-(6″-Acetyl)glucoside*
73 7-(4‴-Acetyl)rutinoside*

74 7-[2'''-(2-Methyl)butyryl)]rutinoside*
75 7-[3'''-(2-Methyl)butyryl)]rutinoside*
76 7-Glucosyl(1→ 2)glucosyl(1→ 2) [2,4-diacetylrhamnosyl](1→ 6)-3-acetyl-glucoside*

Apigenin 5,7-dimethyl ether
77 4'-Galactoside*

Apigenin 7,4'-dimethyl ether
78 5-Xylosylglucoside

6-Hydroxyapigenin
79 5-Glucuronide
80 6-Xyloside*
81 6-Glucuronide*
82 7-Rhamnoside
83 7-Glucuronide
84 4'-Arabinoside
85 7-Xylosyl(1→ 4)rhamnoside
86 7-Rutinoside
87 7-Diglucoside
88 7,4'-Dirhamnoside
89 7-(Sinapoyl)glucuronide

Scutellarein 6-methyl ether (Hispidulin)
90 7-Glucoside (homoplantaginin)
91 7-Glucuronide
92 7-Rutinoside*

Scutellarein 7-methyl ether
93 6-Rhamnosyl(1→ 4)xyloside*

Scutellarein 4'-methyl ether
94 6-Glucoside*
95 7-Glucuronide*
96 6-Sophoroside*
97 7-[*p*-Coumaroylglucosyl (1→ 2)mannoside]

Scutellarein 6,7-dimethyl ether
98 4'-Glucoside
99 4'-Rutinoside

Scutellarein 6,4'-dimethyl ether (Pectolinarigenin)
100 7-Glucoside*
101 7-Glucuronide*
102 7-Glucuronic acid methyl ester*
103 7-Rutinoside
104 7-(4'''-Acetylrutinoside)*
105 7-Acetylrutinoside

Scutellarein 7,4'-dimethyl ether
106 6-Glucoside

Scutellarein 6,7,4'-trimethyl ether
107 5-Glucoside*

8-Hydroxyapigenin (Isoscutellarein)
108 8-Glucuronide*
109 4'-Mannosyl(1→ 2)glucoside*
110 7-(Acetyl)sophoroside*

8-Hydroxyapigenin 4'-methyl ether
111 8-Glucuronide*
112 8-Xylosylglucoside*
113 7-(6'''-Acetyl)allosyl(1→ 2)glucoside*

5,7,3',4'-Tetrahydroxyflavone (Luteolin)
114 5-Glucoside (Galuteolin)
115 5-Galactoside*
116 7-Glucoside
117 7-Galactoside
118 7-Glucuronide
119 7-Galacturonide*
120 3'-Glucoside
121 3'-Glucuronide*
122 3'-Galacturonide*
123 4'-Arabinoside*
124 4'-Glucoside
125 4'-Glucuronide*
126 7-Arabinosylglucoside*
127 7-Xylosylglucoside
128 7-Glucosylarabinoside
129 7-Rutinoside
130 7-Neohesperidoside (veronicastroside)
131 7-Gentiobioside*
132 7-Sophoroside*
133 7-Laminaribioside
134 7-Digalactoside*
135 7-Glucosylgalactoside
136 7-Galactosylglucoside*
137 7-Glucosylglucuronide
138 7-Apiosylglucoside
139 7,3'-Diglucoside
140 7-Glucuronide-3'-rhamnoside*
141 7-Glucuronide-4'-rhamnoside*
142 7-Glucuronide-3'-glucoside*
143 7,3'-Diglucuronide*
144 7,3'-Digalacturonide*
145 7-Galactoside-4'-glucoside*
146 7,4'-Diglucuronide*
147 3',4'-Diglucoside*
148 3',4'-Diglucuronide*
149 3',4'-Digalacturonide*
150 7-Glucosylarabinoside-4'-glucoside*
151 7-Rutinoside-4'-glucoside
152 7-Neohesperidoside-4'-glucoside*
153 7-Glucoside-4'-neohesperidoside*
154 7-Glucuronide-3',4'-dirhamnoside*
155 7,4'-Diglucuronide-3'-glucoside*
156 7,3',4'-Triglucuronide*

157 7-Neohesperidoside-4′-sophoroside*
158 7-*p*-Coumaroylglucoside
159 7-Glucuronide-3′-feruloylglucoside*
160 7,4′-Diglucuronide-3′-feruloylglucoside*
161 7-Glucosidesulphate*
162 7-Sulphate-3′-glucoside
163 7-Rutinosidesulphate*
164 7-Sulphate-3′-rutinoside*
165 7-Sulphate*
166 3′-Sulphate*
167 4′-Sulphate*
168 7,3′-Disulphate*

Luteolin 5-methyl ether
169 7-Glucoside*

Luteolin 7-methyl ether
170 4′-Rhamnoside
171 4′-Gentiobioside
172 5-Xylosylglucoside

Luteolin 3′-methyl ether (Chrysoeriol)
173 5-Glucoside*
174 7-Xyloside*
175 7-Rhamnoside*
176 7-Glucoside
177 7-Glucuronide
178 4′-Glucoside*
179 5-Diglucoside*
180 7-Arabinofuranosyl(1→ 2)galactoside
181 7-Rutinoside
182 7-Rhamnosylglucuronide*
183 7-Digalactoside
184 7-Apiosylglucoside
185 7-Allosyl(1→ 2)glucoside*
186 5,4′-Diglucoside*
187 7,4′-Diglucoside*
188 7-Glucuronide-4′-rhamnoside*
189 7-Sophorotrioside*
190 7-(Malonyl)glucoside
191 7-*p*-Coumaroylglucosylglucuronide
192 7-(6′′′-Acetyl)glucosyl(1→ 2)mannoside
193 7-Glucosidesulphate*
194 7-Sulphate*

Luteolin 4′-methyl ether (Diosmetin)
195 7-Glucoside
196 7-Glucuronide
197 7-α-[Glucosyl(1→ 6)arabinoside]
198 7-β-[Glucosyl(1→ 6)arabinoside]
199 7-Xylosylglucoside*
200 7-Rutinoside(diosmin)
201 7-Sophoroside*
202 7-(4^G-Rhamnosylneohesperidoside)*
203 7-Sulphate*
204 3′-Sulphate*
205 7,3′-Disulphate*

Luteolin 7,3′-dimethyl ether
206 4′-Glucoside
207 4′-Apiosylglucoside

Luteolin 7,4′-dimethyl ether
208 5-Xylosylglucoside

Luteolin 3′,4′-dimethyl ether
209 7-Rhamnoside*
210 7-Glucuronide*

Luteolin 7,3′,4′-trimethyl ether
211 5-Glucoside

6-Hydroxyluteolin
212 6-Xyloside*
213 6-Glucoside*
214 6-Glucuronide*
215 7-Arabinoside*
216 7-Xyloside*
217 7-Glucoside
218 7-Galactoside
219 7-Glucuronide
220 7-Apioside*
221 7-Rhamnosyl(1→ 4)xyloside*
222 7-Diglucoside
223 7-Arabinoside-4′-rhamnoside*

6-Methoxyluteolin
224 7-Glucoside
225 7-Rutinoside

6-Hydroxyluteolin 7-methyl ether
226 6-Glucoside (pedaliin)
227 6-Galactoside*
228 6-Galactosylglucoside*

6-Hydroxyluteolin 3′-methyl ether
229 7-Diglucoside*

6-Hydroxyluteolin 6,7-dimethyl ether (Cirsiliol)
230 4′-Glucoside

6-Hydroxyluteolin 6,3′-dimethyl ether
231 7-Rhamnoside*
232 7-Glucoside
233 7-(6′′-Caffeoyl)glucoside

6-Hydroxyluteolin 6,4′-dimethyl ether
234 7-Glucoside
235 7-Rutinoside

6-Hydroxyluteolin 7,3′-dimethyl ether
236 6-Glucoside*

6-Hydroxyluteolin 6,7,3′-trimethyl ether (Cirsilineol)
237 4′-Glucoside

8-Hydroxyluteolin (Hypolaetin)
238 7-Glucoside
239 8-Glucuronide*
240 8,4′-Diglucuronide*
241 8-Glucoside-3′-sulphate*
242 8-Sulphate*

Hypolaetin 3′-methyl ether
243 7-Glucoside*
244 7-Mannosyl(1→ 2)glucoside*

5,7,3′,4′-Tetrahydroxy-6,8-dimethoxyflavone
245 7-Glucoside*

5,7,3′-Trihydroxy-6,8,4′-trimethoxyflavone (Acerosin)
246 5-(6″-Acetyl)glucoside*

5,7,4′-Trihydroxy-6,8,3′trimethoxyflavone
247 7-Glucoside*

5,7,3′,4′,5′-Pentahydroxyflavone (tricetin)
248 7-Glucoside
249 3′-Glucoside
250 3′-Sulphate*
251 3′-Disulphate*

Tricetin 3′-methyl ether
252 7-Glucoside*

Tricetin 3′,5′-dimethyl ether (Tricin)
253 5-Glucoside
254 5-Diglucoside
255 7-Glucoside
256 7-Glucuronide
257 4′-Glucoside*
258 7-Rutinoside
259 7-Neohesperidoside*
260 7-Diglucoside*
261 7-Fructosylglucoside*
262 7-Rhamnosylglucuronide*
263 7-Diglucuronide
264 5,7-Diglucoside*
265 7-Rutinoside-4′-glucoside*
266 7-Glucosidesulphate*

5,7,2′,4′,5′-Pentahydroxyflavone (Isoetin)
267 7-Glucoside*
268 2′-Arabinoside*
269 2′-Glucoside*
270 2′-Xylosylarabinosylglucoside*
271 4′-or 5′-Glucoside*

Key to check list

In all cases except where otherwise indicated, all sugars can be assumed to be in the pyranose form and have the appropriate linkage to aglycone or between sugars (β- for glucosides, galactosides, glucuronides, xylosides, etc.; α- for rhamnosides, arabinosides). With arabinosides, β- linkages are not uncommon and are so indicated where they are known. The glycosides of a given flavone aglycone are listed in sequence: monoglycosides, diglycosides, triglycosides, tetraglycosides, acylated glycosides, sulphated derivatives. Where the di-or tri-saccharide is well known, the trivial name is used: sophorose, Glcβ1 → 2Glc; laminaribiose, Glcβ1 → 3Glc; gentiobiose, Glcβ1 → 6Glc; maltose, Glcα1 → 4Glc; rutinose, Rhaα1 → 6Glc; neohesperiodose, Rhaα1 → 2Glc; robinobiose, Rhaα1 → 6Gal; sambubiose, Xylβ1 → 2Glc; vicianose, Araα1 → 6Glc. An asterisk indicates that the compound is newly reported during the period 1975–1980.

CHECK LIST OF KNOWN FLAVONOL GLYCOSIDES

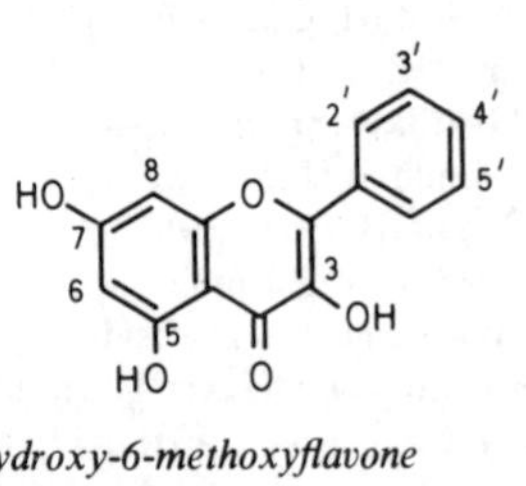

3,5-Dihydroxy-6-methoxyflavone
1 3-Rutinoside*

3,5,7-Trihydroxyflavone (Galangin)
2 3-Rhamnoside*
3 3-Glucoside*
4 7-Glucoside*
5 3-Rutinoside
6 3-Galactosyl (1→ 4) rhamnoside*

3-Methylgalangin
7 7-Glucoside*

3,7,4′-Trihydroxyflavone
8 3-Glucoside
9 7-Glucoside
10 4′-Glucoside
11 7-Rutinoside

8-Hydroxygalangin 7-methyl ether
12 8-Acetate*
13 8-Butyrate*

3,7,3′,4′-Tetrahydroxyflavone (Fisetin)
14 3-Glucoside
15 7-Glucoside
16 7-Rutinoside

3,7,4′-Trihydroxy-3′-methoxyflavone (Geraldol)
17 4′-Glucoside

3,5,7,4′-Tetrahydroxyflavone (Kaempferol)
18 3-Arabinofuranoside (juglanin)
19 3-Xyloside*
20 3-Rhamnoside (afzelin)
21 3-Glucoside (astragalin)
22 3-Galactoside (trifolin)
23 3-Glucuronide
24 3-Alloside* (asiaticalin)
25 5-Glucoside
26 7-Arabinoside*
27 7-Xyloside*
28 7-Rhamnoside
29 7-Glucoside (populnin)
30 7-Galactoside*
31 4′-Rhamnoside*
32 4′-Glucoside*
33 3-Xylosylrhamnoside*
34 3-Rhamnosylxyloside*
35 3-Arabinosyl (1→6) galactoside*
36 3-Xylosylgalactoside*
37 3-Rutinoside
38 3-Neohesperidoside*
39 3-Rhamnosyl (1→3) glucoside (rungioside)
40 3-Glucosyl (1→4) rhamnoside*
41 3-Sambubioside
42 3-Xylosylglucoside
43 3-Gentiobioside
44 3-Sophoroside
45 3-Robinobioside
46 3-Rhamnosyl (1→2) galactoside*
47 3-Glucosyl (1→6) galactoside*
48 3-Galactosylglucoside*
49 3-Apiosylglucoside
50 3-Digalactoside
51 3-Arabinoside-7-rhamnoside*
52 3-Rhamnoside-7-arabinoside*
53 3-Glucoside-7-arabinoside*
54 3-Xyloside-7-rhamnoside
55 3-Glucoside-7-xyloside*
56 3,7-Dirhamnoside
57 3,7-Diglucoside
58 3,7-Digalactoside
59 3-Glucoside-7-rhamnoside
60 3-Rhamnoside-4′-arabinoside
61 3,4′-Diglucoside*
62 3-Glucosyl-β-(1→4) arabinofuranosyl-α-(1→2) arabinopyranoside (primflasine)
63 3-Glucosylxylosylarabinoside*
64 3-Rhamnosyl (1→2) rhamnosyl (1→6) glucoside*
65 (2^G-Rhamnosylrutinoside)*
66 3-(2^G-Rhamnosylgentiobioside)*
67 3-(2^G-Glucosylrutinoside)
68 3-(2^G-Glucosylgentiobioside)*
69 3-Glucosyl (1→2) gentiobioside*
70 3-Sophorotrioside
71 3-β-Maltosyl (1→6) glucoside
72 3-Rhamnosylglucosylgalactoside*
73 3-(2′-Rhamnosyllaminaribioside)
74 3-Rhamnosylarabinoside-7-rhamnoside
75 3-Glucosylxyloside-7-xyloside*
76 3-Rhamnosylxyloside-7-glucoside*
77 3-Glucosylrhamnoside-7-rhamnoside*
78 3-Rutinoside-7-rhamnoside*
79 3-Rhamnosylgalactoside-7-arabinoside*
80 3-Robinobioside-7-rhamnoside (robinin)
81 3-Robinobioside-7-glucoside
82 3-Lathyroside-7-rhamnoside
83 3-Rutinoside-7-glucoside
84 3-Rutinoside-7-galactoside
85 3-Rutinoside-7-glucuronide
86 3-Sophoroside-7-rhamnoside
87 3-Sophoroside-7-glucoside
88 3-Glucoside-7-gentiobioside*
89 3-Sambubioside-7-glucoside
90 3-Rutinoside-4′-diglucoside*
91 3-Sophorotrioside-7-rhamnoside
92 3,7,4′-Triglucoside
93 3-(2″-p-Coumaroyl) arabinoside*
94 3-(3″-p-Coumaroyl) glucoside*
95 3-(6″-p-Coumaroyl) glucoside (tiliroside)
96 3-(6″-Feruloyl) glucoside*
97 3-Benzoylglucoside
98 3-p-Hydroxybenzoylglucoside
99 3-(2″-Galloyl) glucoside*
100 3-(6″-Galloyl) glucoside*
101 7-(6″-Succinyl) glucoside*
102 3-(Di-p-coumaroyl) glucoside*
103 3-(6″-p-Coumaroyl) acetylglucoside*
104 3-(3″-p-Coumaroyl-6″-feruloyl) glucoside*
105 3-(4″-Caffeoyl) laminaribioside-7-rhamnoside*
106 3-(Feruloylsophoroside) (petunoside)
107 3-Benzoylglucoside-7-glucoside

108 3-(*p*-Hydroxybenzoylglucoside)-7-glucoside
109 3-(6‴-Acetyl) glucosyl (1→4) rhamnoside
110 3-(6″-Acetyl) glucoside-7-rhamnoside*
111 3-(6″-Acetyl) glucoside-7-glucoside*
112 3-(*p*-Coumaroylsophorotrioside)
113 3-Glucosyl (1→3) (4‴-acetyl) rhamnosyl (1→6) galactoside
114 3-(Feruloylsophorotrioside)
115 3-(2‴-Sinapoyl) sophoroside-7-glucoside*
116 3-α-(6″-Sulphate) glucoside*
117 3-Rhamnosylsulphate*
118 3-Glucuronide-7-sulphate*
119 3-(6″-Sulphate) gentiobioside*
120 7-Sulphate*
121 3,7-Disulphate*

Kaempferol 3-methyl ether
122 7-Rutinoside

Kaempferol 5-methyl ether
123 3-Galactoside

Kaempferol 7-methyl ether (Rhamnocitrin)
124 3-Glucoside*
125 3-Glucuronide
126 3-Rutinoside*
127 3-Rhamnosyl (1→3) rhamnosyl (1→6) galactoside (alaternin)
128 4′-Rhamnosyl (1→4) rhamnosyl (1→6) galactoside (catharticin)
129 3-Sulphate*

Kaempferol 4′-methyl ether (Kaempferide)
130 3-Glucuronide*
131 3-Diglucoside

Kaempferol 3,4′-dimethyl ether
132 7-Glucoside

Kaempferol 7,4′-dimethyl ether
133 3-Sulphate

6-C-Methylkaempferol
134 3-Glucoside*

6-Hydroxykaempferol
135 7-Glucoside*

6-Hydroxykaempferol 3-methyl ether
136 6-Glucoside*
137 7-Glucoside*

6-Hydroxykaempferol 5-methyl ether
138 4′-Rhamnoside
139 3-Arabinosylrhamnoside

6-Hydroxykaempferol 6-methyl ether (Eupafolin)
140 3-Glucoside*
141 3-Galactoside*
142 7-Glucoside*
143 3-Rutinoside*

6-Hydroxykaempferol 7-methyl ether
144 6-Rhamnosyl (1→4) xyloside*

6-Hydroxykaempferol 4′-methyl ether
145 3,7-Dirhamnoside*

6-Hydroxykaempferol 3,6-dimethyl ether
146 7-Glucoside

6-Hydroxykaempferol 6,7-dimethyl ether (Eupalitin)
147 3-Rhamnoside (eupalin)
148 3-Galactoside
149 3-Diglucoside

6-Hydroxykaempferol 3,6,7-trimethyl ether
150 4′-Glucoside

6-Hydroxykaempferol 6,7,4′-trimethyl ether (Mikanin)
151 3-Galactoside*

8-Hydroxykaempferol (Herbacetin)
152 3-Glucoside
153 7-Arabinoside*
154 7-Glucoside
155 8-Xyloside*
156 8-Rhamnoside*
157 8-Glucoside
158 8-Glucuronide*
159 4′-Arabinoside*
160 4′-Glucoside*
161 4′-Glucuronide*
162 8-Rutinoside*
163 8-Gentiobioside*
164 8-Arabinoside-4′-xyloside*
165 8-(3″-Acetyl) xyloside*
166 4′-(Acetyl) arabinoside
167 8-(Diacetyl) glucoside*
168 8-Acetate*
169 8-Butyrate*
170 4′-(Diacetyl) xyloside*
171 4′-(Triacetyl) xyloside*

Herbacetin7-methyl ether
172 8-Acetate*
173 8-Butyrate*

Herbacetin 8-methyl ether
174 3-Glucoside*
175 3-Rutinoside*
176 3-Rhamnosylglucoside-7-rhamnoside

Herbacetin 7,8-dimethyl ether
177 3-Rhamnoside*

Herbacetin 7,4′-dimethyl ether
178 8-Acetate*
179 8-Butyrate*

Quercetin
180 3-α-L-Arabinofuranoside (avicularin)
181 3-α-L-Arabinopyranoside (guaijaverin)
182 3-β-L-Arabinoside (polystachoside)
183 3-Xyloside (reynoutrin)
184 3-Rhamnoside (quercitrin)
185 3-Glucoside (isoquercitrin)
186 3-β-Galactoside (hyperin)
187 3-α-Galactoside*
188 3-Glucuronide (miquelianin)
189 3-Galacturonide*
190 5-Glucoside
191 7-Arabinoside*
192 7-Rhamnoside
193 7-Glucoside (quercimeritrin)
194 7-α-Galactoside
195 7-β-Galactoside*
196 3′-Xyloside*
197 3′-Glucoside
198 4′-Glucoside (spiraeoside)
199 3-Diarabinoside
200 3-Rhamnosylarabinoside
201 3-Vicianoside
202 3-Arabinosyl (1→6) galactoside*
203 3-Galactosylarabinoside
204 3-Dixyloside*
205 3-Rhamnosylxyloside
206 3-Xylosylglucoside*
207 3-Xylosylgalactoside
208 3-Rutinoside (rutin)
209 3-Neohesperidoside
210 3-Glucosyl (1→4) rhamnoside*
211 3-Robinobioside*
212 3-Gentiobioside
213 3-Sophoroside
214 3-Galactosylglucoside*
215 3-Glucosylmannoside*
216 3-Glucosylglucuronide
217 3-Digalactoside
218 3-Galactosylglucuronide*
219 7-Rutinoside
220 3,5-Digalactoside
221 3-Arabinoside-7-glucoside
222 3-Glucoside-7-arabinoside*
223 3-Xyloside-7-glucoside
224 3-Galactoside-7-xyloside
225 3,7-Dirhamnoside
226 3-Rhamnoside-7-glucoside
227 3-Glucoside-7-rhamnoside
228 3-Galactoside-7-rhamnoside
229 3,7-Diglucoside
230 3-Galactoside-7-glucoside*
231 3,7-Diglucuronide
232 3,3′-Diglucoside
233 3,4′-Diglucoside
234 7,4′-Diglucoside
235 3-Rhamnosyl (1→2) rutinoside*
236 3-(2^G-Rhamnosylrutinoside)*
237 3-Rhamnosylrobinobioside*
238 3-(2^G-Glucosylrutinoside)
239 3-(2^G-Rhamnosylgentiobioside)*
240 3-Glucosyl (1→6) glucosyl (1→4) rhamnoside*
241 3-(2^G-Apiosylrutinoside)*
242 3-Sophorotrioside
243 3-Glucosyl (1→2) gentiobioside*
244 3-(2^G-Glucosylgentiobioside)
245 3-Dixyloside-7-glucoside*
246 3-Xylosylglucoside-7-glucoside
247 3-Rutinoside-7-glucoside
248 3-Glucoside-7-rutinoside
249 3-Rutinoside-7-galactoside
250 3-Robinobioside-7-rhamnoside
251 3-Robinobioside-7-glucoside
252 3-Rutinoside-7-glucuronide
253 3-Gentiobioside-7-glucoside*
254 3-Sophoroside-7-glucoside
255 3-Sambubioside-7-glucoside
256 3-Sambubioside-3′-glucoside
257 3-(2^G-Rhamnosylrutinoside)-7-glucoside*
258 3-Rhamnosyldiglucoside-7-glucoside*
259 3-Rutinoside-7,3′-bisglucoside*
260 3-Rutinoside-4′-diglucoside*
261 7-Triglucoside*
262 3-(6″-*p*-Coumaroyl) glucoside* (helichrysoside)
263 3-(3″-*p*-Coumaroyl) glucoside*
264 3-Isoferuloylglucuronide*
265 3-(3″-Acetyl)-α-arabinofuranoside*
266 3-(3″-Acetyl) rhamnoside*
267 3-(4″-Acetyl) rhamnoside*
268 3-(6″-Acetyl) glucoside*
269 3-(2″-Galloyl) glucoside*
270 3-(6″-Galloyl) glucoside*
271 3-(2″-Galloyl) galactoside*
272 3-(6″-Galloyl) galactoside
273 3-Malonylglucoside
274 3-(3″,6″-Di-*p*-coumaroyl) glucoside*
275 3-Glucosyl (1→3)-4‴-acetylrhamnosyl (1→6) galactoside*
276 3-*p*-Coumaroylsophoroside-7-rhamnoside*

277 3-Caffeoylsophoroside-7-glucoside
278 3-Acetylsophoroside-7-rhamnoside*
279 3-(*p*-Coumaroylsophorotrioside)
280 3-(Feruloylsophorotrioside)
281 3-Rhamnoside-7-glucoside-4′-(caffeoylgalactoside)
282 7-Acetyl-3′-glucoside
283 7-(6″-Tiglyl) glucoside*
284 3-Rhamnosidesulphate*
285 3-Glucuronide-7-sulphate*
286 3-Rutinosidesulphate*
287 3-Acetyl-7,3′,4′-trisulphate*
288 3-Sulphate
289 3,4′-Disulphate*
290 3,7,4′-Trisulphate*
291 3,7,3′,4′-Tetrasulphate

Quercetin 3-methyl ether
292 5-Glucoside*
293 7-Glucoside
294 7-Diglucoside-4′-glucoside

Quercetin 5-methyl ether (Azaleatin)
295 3-Arabinoside*
296 3-Rhamnoside (azalein)
297 3-Galactoside
298 3-Glucuronide
299 3-Xylosylarabinoside*
300 3-Rhamnosylarabinoside*
301 3-Arabinosylgalactoside
302 3-Rutinoside*
303 3-Diglucoside

Quercetin 7-methyl ether (Rhamnetin)
304 3-α-L-Arabinofuranoside
305 3-α-L-Arabinopyranoside
306 3-Glucoside*
307 3-Galactoside
308 3′-Glucuronide*
309 3-α-Diarabinoside
310 3-β-Diarabinoside
311 3-Rutinoside*
312 3-Rhamninoside
313 3-Rhamnosyl (1→2) rhamnosyl (1→6) galactoside*
314 3-Sulphate*
315 3′-Glucuronide-3,5,4′-trisulphate*

Quercetin 3′-methyl ether (Isorhamnetin)
316 3-α-L-Arabinopyranoside (distichin)
317 3-α-L-Arabinofuranoside*
318 3-Xyloside*
319 3-Glucoside
320 3-Galactoside
321 3-Glucuronide
322 5-Glucoside
323 3-Arabinosyl (1→2) rhamnoside
324 3-Arabinosyl (1→6) glucoside*
325 3-Xylosylglucoside*
326 3-Dirhamnoside
327 3-Rutinoside (narcissin)
328 3-Rhamnosylgalactoside
329 3-Sophoroside*
330 3-Gentiobioside
331 3-Lactoside*
332 3-Arabinoside-7-glucoside
333 3-Glucoside-7-arabinoside
334 3-Glucoside-7-xyloside*
335 3-Glucoside-7-rhamnoside
336 3,7-Diglucoside
337 3-Galactoside-7-glucoside
338 3-Glucoside-4′-rhamnoside
339 3,4′-Diglucoside (dactylin)
340 3-Xylosylrutinoside
341 3-Xylosylrhamnosylgalactoside*
342 3-Rhamnosylrutinoside*
343 3-Rutinosylglucoside
344 3-Glucosylrhamnosylgalactoside
345 3-Galactosylrhamnosylgalactoside*
346 3-Arabinoside-7-rhamnoside*
347 3-Rutinoside-7-glucoside*
348 3-Sophoroside-7-rhamnoside
349 3-Sophoroside-7-glucoside
350 3-Glucoside-7-gentiobioside*
351 3-Galactoside-4′-glucoside*
352 3-(6″-Acetyl) glucoside*
353 3-(6″-Acetyl) galactoside*
354 3-Rutinosidesulphate*
355 3-Glucuronide-7-sulphate*
356 3-Sulphate*
357 7-Sulphate
358 3,7-Disulphate*

Quercetin 4′-methyl ether (Tamarixetin)
359 3-Glucoside (tamarixin)
360 3-Rutinoside
361 3-Sulphate

Quercetin 3,5-dimethyl ether (Caryatin)
362 3′-(or 4′-) Glucoside*

Quercetin 5,3′-dimethyl ether
363 3-Glucoside*

Quercetin 7,3′dimethyl ether (Rhamnazin)
364 3-Glucoside
365 3-Rhamnoside
366 3-Rhamnosyl (1→4) rhamnosyl (1→6) galactoside*
367 3-Sulphate

Quercetin 7,4′-dimethyl ether (Ombuin)
368 3-Galactoside*
369 3-Rutinoside (ombuoside)
370 3-Neohesperidoside*
371 3,5-Diglucoside

Quercetin 3′,4′-Dimethyl ether
372 3-Rutinoside*

Quercetin 7,3′,4′-Trimethyl ether
373 3-Digalactoside*

6-Hydroxyquercetin (Quercetagetin)
374 3-Rhamnoside*
375 3-Glucoside (tagetiin)
376 7-Glucoside
377 3,7-Diglucoside*

Quercetagetin 3-methyl ether
378 7-Glucoside*
379 7-Sulphate*

Quercetagetin 6-methyl ether (Patuletin)
380 3-Xyloside*
381 3-Rhamnoside*
382 3-Glucoside
383 3-Galactoside
384 3-Glucuronide
385 7-Glucoside
386 3-Rutinoside
387 3-Diglucoside
388 3,7-Dirhamnoside*
389 3-Acetylglucoside*
390 3-Glucoside-7-sulphate
391 3-Sulphate*
392 7-Sulphate

Quercetagetin 7-methyl ether
393 3-Glucoside*

Quercetagetin 3′-methyl ether
394 7-Glucoside

Quercetagetin 3,6-dimethyl ether
395 7-Glucoside* (axillaroside)

Quercetagetin 3,7-dimethyl ether
396 6-Glucoside*
397 6-Galactoside*
398 4′-Glucoside*

Quercetagetin 6,7-dimethyl ether (Eupatolitin)
399 3-Rhamnoside (eupatolin)
400 3-Glucoside
401 3-Galactoside
402 3-Sulphate*

Quercetagetin 6,3′-dimethyl ether
403 7-Glucoside*
404 3-Rutinoside*

Quercetagetin 3,6,7-trimethyl ether
405 4′-Glucoside*

Quercetagetin 3,6,3′-trimethyl ether
406 7-Glucoside (jacein)

Quercetagetin 3,7,4′-trimethyl ether
407 6-Glucoside*
408 3′-Glucoside*

Quercetagetin 6,7,3′-trimethyl ether (Veronicafolin)
409 3-Rutinoside*
410 3-Digalactoside*
411 3-Sulphate*

Quercetagetin 6,7,4′-trimethyl ether (Eupatin)
412 3-Sulphate*

Quercetagetin 7,3′,4′-trimethyl ether
413 3-Rhamnoside

Quercetagetin 3,6,7,3′-tetramethyl ether
414 4′-Glucoside (chrysosplenin)

Quercetagetin 3,6,7,3′,4′-pentamethyl ether (Artemetin)
415 5-Glucosylrhamnoside*

8-Hydroxyquercetin (Gossypetin)
416 3-Glucoside
417 3-Galactoside
418 3-Glucuronide
419 7-Glucoside
420 8-Glucoside
421 3-Sulphate*
422 3-Gentiotrioside

Gossypetin 7-methyl ether
423 3-Arabinoside*
424 3-Rhamnoside*
425 3-Galactoside*
426 8-Glucoside*
427 3-Galactoside-8-glucoside*
428 3-Rhamnoside-8-acetate*

Gossypetin 8-methyl ether
429 3-Galactoside

Gossypetin 3′-methyl ether
430 3-Rutinoside*
431 7-(6″-Acetyl) glucoside*

Gossypetin 7,4′-dimethyl ether
432 8-Glucoside
433 8-Acetate*
434 8-Butyrate*

Gossypetin 8,3′-dimethyl ether
435 3-Rutinoside*

Myricetin
436 3-Arabinoside
437 3-Xyloside*
438 3-Rhamnoside (myricitrin)
439 3-Glucoside
440 3-Galactoside
441 7-Glucoside
442 3′Arabinoside
443 3′-Glucoside (cannabiscitrin)
444 3-Dixyloside*
445 3-Xylosylglucoside*
446 3-Xylosylgalactoside*
447 3-Rutinoside
448 3-Diglucoside*
449 3-Galactosylglucoside*
450 3-Digalactoside
451 3,3′-Digalactoside
452 3-Rhamnosylrutinoside*
453 3-Rutinoside-7-rhamnoside
454 3-Rutinoside-7-glucoside*
455 3-Robinobioside-7-rhamnoside
456 3-Triglucoside*
457 3-(2″-Acetyl) rhamnoside*
458 3-(6″-Galloyl) glucoside*
459 3-(6″-Galloyl) galactoside*
460 3-Rhamnosidesulphate*

Myricetin 3-methyl ether
461 3′-Glucoside

Myricetin 5-methyl ether
462 3-Rhamnoside
463 3-Galactoside

Myricetin 7-methyl ether (Europetin)
464 3-Rhamnoside

Myricetin 3′-methyl ether (Laricitrin)
465 3-Rhamnoside*
466 3-Glucoside*
467 3-Galactoside*
468 3-Rutinoside*
469 3-Rhamnosylrutinoside*
470 3-Rutinoside-7-glucoside*
471 3-*p*-Coumaroylglucoside*

Myricetin 4′-methyl ether
472 3-Rhamnoside (mearnsitrin)
473 3-Galactosyl (1→4) galactoside

Myricetin 3′,5′-dimethyl ether (Syringetin)
474 3-Rhamnoside*
475 3-Glucoside
476 3-Galactoside*
477 3-Glucuronide*
478 3-Rutinoside*
479 3-Rhamnosylrutinoside*
480 3-Rutinoside-7-glucoside*
481 3-*p*-Coumaroylglucoside*

6-Hydroxymyricetin 6,3′,5′-trimethyl ether
482 3-Glucoside*

6-Hydroxymyricetin 4,6,3′,5′-tetramethyl ether
483 7-Glucoside*

8-Hydroxymyricetin (Hibiscetin)
484 3-Glucoside
485 8-Glucosylxyloside*

3,5,7,2′-Tetrahydroxyflavone (Datiscetin)
486 3-Rutinoside

CHAPTER SIX

The Minor Flavonoids

BRUCE A. BOHM

6.1 GENERAL INTRODUCTION

As defined for the present treatment the minor flavonoids consist of chalcones, aurones (including auronols), dihydrochalcones (including β-oxygenated derivatives), flavanones and dihydroflavonols. The first three types were covered in Chapter 9 of 'The Flavonoids' while the latter two were discussed in Chapter 11. The approach adopted for the present review is essentially the same as for the above two chapters. The compounds in each class are presented and discussed in order of increasing hydroxylation of the B ring, i.e. none, one, two or three hydroxyl groups. Within each subgroup the compounds are covered roughly in order of increasing complexity of A-ring substitution. In this way readers may follow the material in this review in parallel with corresponding structural types in the earlier chapters.

Sections 6.2–6.6 of this review focus on structures but also make many references to distribution. However, in the case of compounds enjoying wide distribution only new or unusual reports are mentioned in the text. New records for families are indicated and family names are shown parenthetically following the appearance of each genus for the first time. All occurrences of flavonoids of the five classes which I have found in the literature since appearance of *The Flavonoids* are presented in tabulated form. Some references overlooked during preparation of the earlier work have also been included.

6.2 CHALCONES

6.2.1 Introduction

It may be useful to remind the reader that chalcone numbering is different from the system used for flavonoid types that possess heterocyclic rings. In chalcones the A ring, always written to the left, is given the primed numbers, while the B-ring carbons are given the unprimed numbers. The positions are shown numbered in structure (*6.1*). Bridge carbons are marked relative to the carbonyl function.

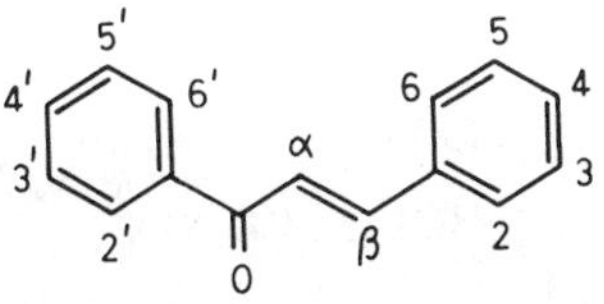

(6.1)

6.2.2 Structures of naturally occurring chalcones

Structures of naturally occurring chalcones are summarized in Tables 6.1 and 6.2. A summary of natural sources appears in Table 6.10.

(a) Chalcones lacking B-ring hydroxyls

The simplest member of this subgroup is 2′,4′-dihydroxychalcone (*6.2*) whose discovery in *Flourensia heterolepis* (Compositae) was reported by Bohlmann and Jakupovic (1979). Compounds based upon either *O*- or *C*-prenylation of the parent compound are known. The 4′-prenyl ether (*6.3*), known as 'derricidin' or 'cordoin', has been found in *Derris floribunda* (Braz Filho *et al.*, 1975b) and in *Lonchocarpus* species (De Lima *et al.*, 1973; Delle Monache *et al.*, 1978)

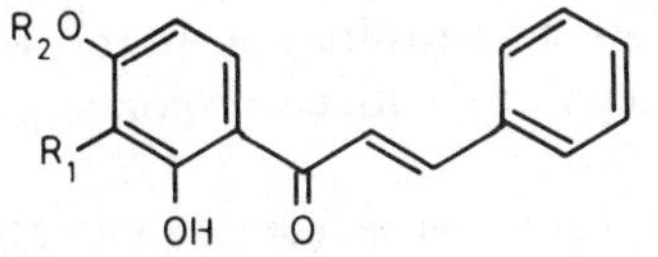

(6.2) $R_1 = R_2 = H$

(6.3) $R_1 = H, R_2 =$

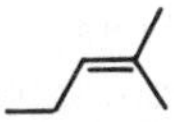

(6.4) $R_1 =$ 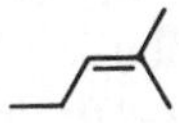 $R_2 = H$

(6.5) $R_1 =$ 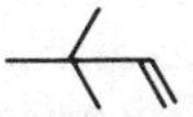 $R_2 = H$

(both Leguminosae). 2′,4′-Dihydroxy-3′-prenylchalcone (4), whose trivial names are 'isocordoin' and 'flemistrictin-A', is also known from *Derris* (Braz Filho *et al.*, 1975b), *Lonchocarpus* (De Lima *et al.*, 1973) and *Flemingia* (Rao *et al.*, 1976). A compound isomeric to structure (*6.4*) has also been isolated from *Lonchocarpus* by Delle Monache *et al.* (1974, 1978). Called 'ψ-isocordoin' by those authors, the compound is the 2′,4′-dihydroxy-3′-*C*-(-α,α-dimethylallyl) chalcone shown as structure (*6.5*).

Ring closure of the alkyl side chain of isocordoin (*6.4*) with the 4′-hydroxyl group affords the dimethylchromenochalcone 'lonchocarpin' (*6.6*) known from *Derris* (Braz Filho *et al.*, 1975b), *Pongamia glabra* (Leguminosae) (Subramanyam *et al.*, 1977) and *Lonchocarpus* (Delle Monache *et al.*, 1978). Several closely related compounds have been isolated from species of *Flemingia* (Leguminosae). Rao and co-workers (1976) isolated 'flemistrictin-C' (*6.7*) and 'flemistrictin-B' (*6.8*) from *F. stricta* while compound (*6.9*), 'flemichapparin-A', was described by Adityachaudhury and co-workers (1973) as a constituent of

Table 6.1 Structures of naturally occurring chalcones

Substituents			No. in text	Trivial name	Source*		References
OH	OCH_3	Other					
2′,4′			(*6.2*)		*Flourensia heterolepis*	Com	Bohlmann and Jakupovic (1979)
2′,4′		4′-*O*-Prenyl	(*6.3*)	Derricidin	*Derris floribunda*	Leg	Braz Filho *et al.* (1975b)
				Cordoin	*Lonchocarpus.* sp	Leg	De Lima *et al.* (1972, 1973)
2′,4′		3′-*C*-Prenyl	(*6.4*)	Isocordoin	*Lonchocarpus* sp.	Leg	De Lima *et al.* (1973)
				Flemi-strictin-A	*Flemingia stricta*	Leg	Rao *et al.* (1975a, 1976)
					Derris floribunda	Leg	Braz Filho *et al.* (1975b)
2′,4′		3′-(αα-Dimethyl-allyl)	(*6.5*)	ψ-Isocordoin	*Lonchocarpus* sp.	Leg	Delle Monache *et al.* (1974)
	2′	Furano-(2″,3″,4′,3′)	(*6.10*)	Ovalitenin-A	*Milletia ovalifolia*	Leg	Gupta and Krishnamurti (1977b)
2′		6″,6″-Dimethyl-pyrano-(2″,3″,4′,3′)	(*6.6*)	Lonchocarpin	*Derris floribunda*	Leg	Braz Filho *et al.* (1975b)
					Pongamia glabra	Leg	Subramanyam *et al.* (1977)
2′,4′	3′		(*6.11*)		*Zuccagnia punctata*	Leg	Pederiva *et al.* (1975)
2′,4′,6′			(*6.12*)		*Helichrysum acutatum*	Com	Bohlmann and Abraham (1979c)
					H. polycladum	Com	Bohlmann *et al.* (1980c)
					Populus spp.	Sal	Wollenweber and Weber (1973)
					Adiantum sulphureum	Fil	Wollenweber (1979)
2′,4′	6′		(*6.13*)		*Alnus* spp.	Bet	Wollenweber *et al.* (1974)
					Betula spp.	Bet	Wollenweber *et al.* (1974)
2′,6′	4′		(*6.14*)		*Populus* spp.	Sal	Wollenweber *et al.* (1974)
					Adiantum sulphureum	Fil	Wollenweber (1979)
					Pityrogramma chrysoconia	Fil	Wollenweber (1979)
					Onychium auratum	Fil	Ramakrishnan *et al.* (1974)
					Notholaena spp.	Fil	Wollenweber (1977a)
					Cheilanthes spp.	Fil	Wollenweber (1977a)
					Helichrysum spp.	Com	Bohlmann *et al.* (1979c, 1980c)

Table 6.1 (*Contd.*)

OH	Substituents *OCH₃*	*Other*	*No. in text*	*Trivial name*	*Source**		*References*
2′	4′,6′		(*6.15*)	Flavokawin-B	*Aniba riparia*	Lau	Fernandes *et al.* (1978)
					Pityrogramma triangularis var. *pallida*	Fil	Wollenweber *et al.* (1980b)
					Didymocarpus corchorifolia	Ges	Wollenweber *et al.* (1980)
2′,4′,6′		3′-*C*-Prenyl	(*6.16*)		*Helichrysum* spp.	Com	Bohlmann *et al.* (1979c, 1980c)
2′,6′	4′	3′-*C*-Prenyl	(*6.17*)		*H. cymosum* ssp. *calvum*	Com	Bohlmann *et al.* (1979c)
2′	4′,6′	3′-*C*-Prenyl	(*6.18*)	Ovalichalkone	*Milletia ovalifolia*	Leg	Gupta and Krishnamurti (1977a)
2′,5′		6″,6″-Dimethyl-pyrano-(2″,3″,4′,3′)	(*6.9*)	Flemi-chapparin-A	*Flemingia chappar*	Leg	Adityachaudhury *et al.* (1973)
2′	6′	6″,6″-Dimethyl-pyrano-(2″,3″,4′3′)	(*6.20*)	Ponga-chalcone-I	*Pongamia glabra*	Leg	Subramanyam *et al.* (1977)
	2′,6′,β	6″,6″-Dimethyl-pyrano-(2″,3″,4′,3′)	(*6.21*)	Praecano-sone-A	*Tephrosia praecans*	Leg	Camele *et al.* (1980)
β	2′,6′	6″,6″-Dimethyl-pyrano-(2″,3″,4′,3′)	(*6.22*)	Praecano-sone-B	*Tephrosia praecans*	Leg	Camele *et al.* (1980)
2′,4′,6′		3′-*C*-Geranyl	(*6.24*)	Cannabigerol	*Helichrysum umbraculigerum*	Com	Bohlmann and Hoffmann (1979)
	2′,6′	3′,4′-Methylene-dioxy	(*6.34*)		*H. sutherlandii*	Com	Bohlmann *et al* (1978)
2′	4′,5′,6′		(*6.35*)		*H. sutherlandii*	Com	Bohlmann *et al.* (1978)
2′,6′	4′	3′-*C*-Methyl	(*6.29*)	Triangularin	*Pityrogramma triangularis*	Fil	Star *et al.* (1978)
2′	3′,4′,6′		(*6.36*)		*Popowia cauliflora*	Ano	Panichpol and Waterman (1978)
	2′,3′,4′,6′		(*6.37*)		*Popowia cauliflora*	Ano	Panichpol and Waterman (1978)
2′	4′,6′	3′-*C*-Methyl	(*6.30*)	Aurentiacin	*Didymocarpus aurentiaca*	Ges	Adityachaudhury *et al.* (1976)
					Pityrogramma spp.	Fil	Wollenweber *et al.* (1980b)

2′,4′	6′	5′-*C*-Methyl	(*6.30a*)	Aurentiacin-A	*Didymocarpus aurentiacum*	Ges	Adityachaudhury and Daskanungo (1975)
					Dalea spp.	Leg	Dominguez *et al.* (1980)
2′,4′	3′,6′		(*6.33*)		*Polygonatum senegalense*	Lil	Maradufu and Ouma (1978)
2′,6′	3′,4′		(*6.38*)	Pashanone	*Didymocarpus pedicellata*	Ges	Agarwal *et al.* (1973)
					Onychium auratum	Fil	Ramakrishnan *et al.* (1974)
2′,4′	3′,5′,6′		(*6.39*)	Isodi-dymocarpin	*Didymocarpus pedicellata*	Ges	Bhattacharyya *et al.* (1979)
2′	3′,4′,5′,6′		(*6.40*)	Kanakugiol	*Popowia cauliflora*	Ano	Waterman and Pootakahm (1979)
2′,4′	6′	3′,5′-di-*C*-Methyl	(*6.31*)		*Myrica gale*	Myr	Malterud *et al.* (1977)
2′,4′,4			(*6.41*)	Isoliquiriti-genin	*Sophora tomentosa*	Leg	Komatsu *et al.* (1978)
					Dalbergia ecastophyllum	Leg	Matos *et al.* (1975)
					D. retusa	Leg	Manners *et al.* (1974)
					D. sericea	Leg	Parthasarathy *et al.* (1976)
					Onobrychis viciifolia	Leg	Ingham (1978)
					Medicago spp.	Leg	Ingham (1979)
					Machaerium mucronulatum	Leg	Kurosawa *et al.* (1978)
					Lespedeza cyrtobotrya	Leg	Miyase *et al.* (1980)
					Dahlia tenuicaulis	Com	Lam and Wrang (1975)
2′,4′,4		3′-*C*-Glucose	(*6.49*)		*Cladrastis platycarpa*	Leg	Ohashi *et al.* (1977)
2′,4	4′	5′-Formyl	(*6.48*)	Neobava-chalcone	*Psoralea corylifolia*	Leg	Gupta *et al.* (1975b, 1977)
4′,4	2′	5′-Formyl		Isoneobava-chalcone	*Psoralea corylifolia*	Leg	B. K. Gupta *et al.* (1980)
2′,4′,4		3′-*C*-Prenyl	(*6.42*)	4-Hydroxy-isocordoin	*Lonchocarpus* sp.	Leg	Delle Monache *et al.* (1974)
2′,4	4′	3′-*C*-Prenyl	(*6.43*)	4-Hydroxy-derricin	*Lonchocarpus* sp.	Leg	Delle Monache *et al.* (1974)
					Angelica keiskei	Umb	Kozawa *et al.* (1977)
2′,4		6″,6″-Dimethyl-pyrano-(2″,3″,4′,5′)	(*6.45*)	4-Hydroxy-loncho-carpin	*Derris floribunda*	Leg	Braz Filho *et al.* (1975b)
2′,4′,4		3′-*C*-Geranyl	(*6.44*)	Xanthoangelol	*Angelica keiskei*	Umb	Kozawa *et al.* (1977)
					Lespedeza cyrtobotrya	Leg	Miyase *et al.* (1980)

Table 6.1 (*Contd.*)

Substituents							
OH	*OCH₃*	*Other*	*No. in text*	*Trivial name*	*Source**		*References*
2′,4′,6′,4			(*6.50*)	Isosalipurpol	*Petunia hybrida*	Sol	de Vlaming and Kho (1976)
					Nothofagus antartica	Fag	Wollenweber and Wiermann (1979)
		4-*O*-Glucose	(*6.52*)		*Acacia cyanophylla*	Leg	Imperato (1978)
		2′-*O*-Glucose	(*6.51*)	Isosalipurposide	*Salix rubra*	Sal	Vinokurov (1979)
					Oenothera hookeri ssp. *venusta*	Ona	Dement and Raven (1974)
					Ruellia spp.	Aca	Bloom (1976)
2′,4′,4	6′		(*6.54*)		*Helichrysum heterolasium*	Com	Bohlmann and Abraham (1979b)
		4′-*O*-Glucose	(*6.55*)	Gnaphalin	*Gnaphalium multiceps*	Com	Maruyama *et al.* (1974); Ahluwalia and Rami (1976)
					G. affine	Com	Aritomi and Kawasaki (1974)
2′,4	4′,6′	4-*O*-Glucose	(*6.56*)		*Viscum album*	Lor	Becker *et al.* (1978)
2′,6′	4′,4				*Pterozonium brevifrons*	Fil	Wollenweber (1979)
2′	4′,6′,4		(*6.57*)		*Dahlia tenuicaulis*	Com	Lam and Wrang (1975)
2′,4	4′,5′,6′		(*6.59*)		*Chromolaena odorata*	Com	Barua *et al.* (1978)
2′	4′,5′,6′,4		(*6.60*)		*Chromolaena odorata*	Com	Barua *et al.* (1978)
2′,4	6′	3′,4′-Methylenedioxy	(*6.58*)		*Helichrysum glomeratum*	Com	Bohlmann and Suwita (1979c)
4′,4	2		(*6.62*)	Echinatin	*Glycyrrhiza echinata*	Leg	Saitoh *et al.* (1975)
4′,4	2	5-*C*-(αα-Dimethyl)-allyl	(*6.63*)	Licochalcone-A	*G. glabra*	Leg	Saitoh and Shibata (1975)
4′,3,4	2		(*6.64*)	Licochalcone-B	*G. glabra*	Leg	Saitoh and Shibata (1975)
2′,4′,3,4,α			(*6.78*)		*Peltogyne* spp.	Leg	Malan and Roux (1974)
2′,3,4		6″,6″-Dimethyl pyrano(2″,3″,4′,3′)	(*6.73*)	3,4-Dihydroxylonchocarpin	*Derris floribunda*	Leg	Braz Filho *et al.* (1975b)
2′,4	3	6″,6″-Dimethylpyrano (2″,3″,4′,3′)	(*6.74*)	Pongachalkone-II	*Pongamia glabra*	Leg	Sharma *et al.* (1973); Subramanyam *et al.* (1977)

2′		6″,6″-Dimethyl-pyrano (2″,3″,4′,3′)- and 3,4-Methyl-enedioxy	(6.75)	Glabra-chromene-II	*Pongamia glabra*	Leg	Sharma *et al.* (1973); Subramanyam *et al.* (1977)
2′	6′	6″,6″-Dimethyl-pyrano-(2″,3″,4′,3′)- and 3,4-Methyl-enedioxy	(6.76)	Glabra-chromene	*Pongamia glabra*	Leg	Subramanyam *et al.* (1977)
	2′	Furano(2″,3″,-4′,3′)-and 3,4-methylene-dioxy	(6.77)	Ovalitenin-C	*Milletia ovalifolia*	Leg	Islam *et al.* (1980)
2′,4′,3,4			(6.65)	Butein	*Machaerium mucronulatum*	Ana	Kurosawa *et al.* (1978)
					Bidens laevis	Com	Scogin and Zakar (1976)
		4′-*O*-Glucose	(6.66)	Coreopsin	*Pyrrhopappus* spp.	Com	Harborne (1977a)
					Helianthus spp.	Com	Harborne and Smith (1978)
					Viguiera laciniata	Com	Harborne and Smith (1978)
					Bidens spp.	Com	Hart (1979)
					Coreopsis bigelovii	Com	Nicholls and Bohm (1979)
		4′-*O*-Glucosyl-glucose	(6.67)		*Bidens* spp.	Com	Hart (1979)
					Coreopsis bigelovii	Com	Nicholls and Bohm (1979)
		2′,3-di-*O*-Glucose	(6.68)		*Rhus* spp.	Ana	Young (1979)
					Amphipterygium adstrigens	Ana	Young (1976)
		4′-*O*-Arabinosyl-galactose	(6.69)		*Bauhinia purpurea*	Leg	Bhartiya *et al.* (1979)
2′,4′,3	4		(6.70)		*Dahlia tenuicaulis*	Com	Lam and Wrang (1975)
2′,4′,4	3		(6.71)		*Iryanthera polyneura*	Myt	de Almeida *et al.* (1979)
2′,4′	3,4		(6.72)		*Iryanthera polyneura*	Myt	de Almeida *et al.* (1979)
2′3′,4′,3,4			(6.80)	Okanin	*Albizia adianthifolia*	Leg	Candy *et al.* (1978)
		4′-*O*-Glucose	(6.81)	Marein	*Coreopsis bigelovii*	Com	Nicholls and Bohm (1979)

Table 6.1 (*Contd.*)

Substituents			*No. in*	*Trivial*			
OH	OCH_3	*Other*	*text*	*name*	*Source**		*References*
					C. nuecensis	Com	Harborne (1977a)
					C. grandiflora	Com	Crawford and Smith (1980)
					Bidens spp.	Com	Hart (1979)
					B. laevis	Com	Scogin and Zakar (1976)
2′,4′,3,4	3′			Lanceoletin	*Coreopsis bigelovii*	Com	Nicholls and Bohm (1979)
		4′-*O*-Glucose	(*6.82*)	Lanceolin	*C. bigelovii*	Com	Nicholls and Bohm (1979)
					C. nuecensis	Com	Harborne (1977a)
					C. grandiflora	Com	Crawford and Smith (1980)
2′,3′,4′,4	3		(*6.82a*)	3-*O*-Methyl-okanin	*Bidens pilosa* s.l.	Com	Ballard (1975)
2′,3′,4′	3,4		(*6.82b*)	3,4,Di-*O*-methyl-okanin	*B. pilosa* s.l.	Com	Ballard (1975)
2′,4′,5′,3,4		4′-*O*-Glucose	(*6.83*)	Stillopsin	*Coreopsis bigelovii*	Com	Nicholls and Bohm (1979)
2′,4′	5′	3,4-Methylene-dioxy	(*6.84*)	Prosogerin-B	*Prosopis spicigera*	Leg	Bhardwaj *et al.* (1979)
2′,3	4′,6′,4		(*6.85*)		*Merillia caloxylon*	Rut	Fraser and Lewis (1974)
2′	4′,6′,3,4		(*6.86*)		*M. caloxylon*	Rut	Fraser and Lewis (1974)
2′	4′,6′	3′-*C*-Prenyl and-3,4-methyl-enedioxy	(*6.87*)	Ovali-chalkone-A	*Milletia ovalifolia*	Leg	Gupta and Krishnamurti (1979)
2′,4′	6′,2,4		(*6.88*)	Cerasin	*Prunus cerasus*	Ros	Nagarajan and Parmar (1977a)
2′	4′,6′,2,4		(*6.89*)	Cerasidin	*P. cerasus*	Ros	Nagarajan and Parmar (1977a)
					Terminalia arjuna	Cmb	Nagar *et al.* (1979)
2′	4′,5′,6′,3,4		(*6.90*)		*Chromolaena odorata*	Com	Barua *et al.* (1978)
2′,3	4′,6′,4,5		(*6.91*)		*Merrillea caloxylon*	Rut	Fraser and Lewis (1974)
2′	4′,6′,2,4,5		(*6.92*)	Rubone	*Derris robusta*	Leg	Chibber *et al* (1979)

* Key: Aca = Acanthaceae, Ana = Anacardiaceae, Ano = Anonaceae, Bet = Betulaceae, Cmb = Combretaceae, Com = Compositae, Fag = Fagaceae, Fil = Filicinae, Ges = Gesneriaceae, Lau = Lauraceae, Leg = Leguminosae, Lil = Liliaceae, Lor = Loranthaceae, Myr = Myrtaceae, Myt = Myristicaceae, Ona = Onagraceae, Ros = Rosaceae, Rut = Rutaceae, Sal = Salicaceae, Sol = Solanaceae, Umb = Umbelliferae.

Table 6.2 Structures of chalcones having complex substituents

Name	*No.*	*Source**		*References*
Flemistrictin-B	(*6.8*)	*Flemingia stricta*	Leg	Rao *et al.* (1976)
Flemistrictin-C	(*6.7*)	*F. stricta*	Leg	Rao *et al.* (1976)
4′-*O*-Methyl-4-deoxybavachromanol	(*6.19*)	*Helichrysum cymosum* spp. *calvum*	Com	Bohlmann *et al.* (1979c)
(—)-Rubranine	(*6.23*)	*Aniba rosaeodora*	Lau	de Alleluia *et al.* (1978)
5,7-Dihydroxy-8-cinnamoyl-4-phenyl-2*H*-1-benzopyran-2-ones†	(*6.26, 6.27, 6.28*)	*Pityrogramma trifoliata*	Fil	Dietz *et al.* (1980)
Ceroptin	(*6.32*)	*P. triangularis*	Fil	Star *et al.* (1975a, b)
Bavachromanol	(*6.46*)	*Psoralea corylifolia*	Leg	Suri *et al.* (1980)
Lespeol	(*6.47*)	*Lespedeza cyrtobotrya*	Leg	Miyase *et al.* (1980)

* Key: Com = Compositae, Fil = Filicinae, Lau = Lauraceae, Leg = Leguminosae.
† Trivial names for these have not been put forward.

F. chappar. Compound (*6.7*) is a hydration product of lonchocarpin, compound (*6.8*) would arise by ring closure at carbon-3″of the isopentenyl side chain (instead of at carbon-4″), and compound (*6.9*) is the result of extra hydroxylation of lonchocarpin at position 5′.

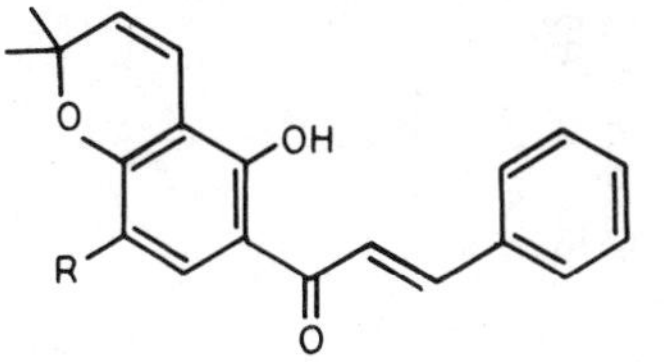

(6.6) R = H
(6.9) R = OH

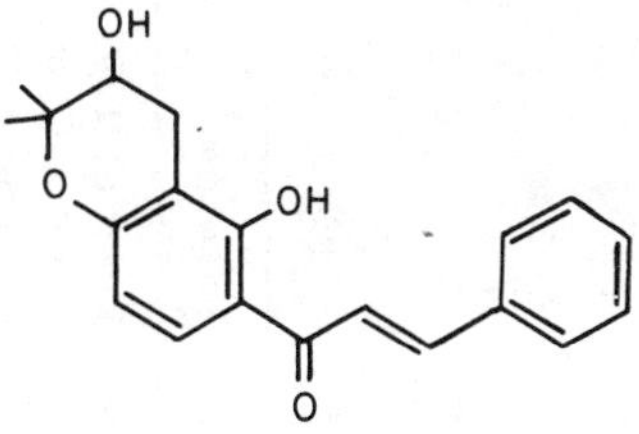

(6.7)

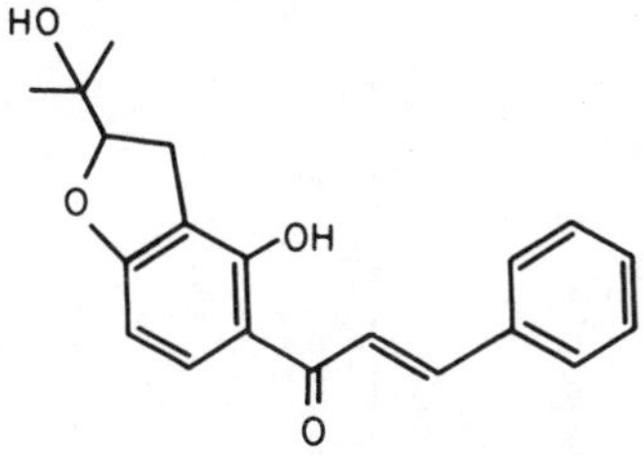

(6.8)

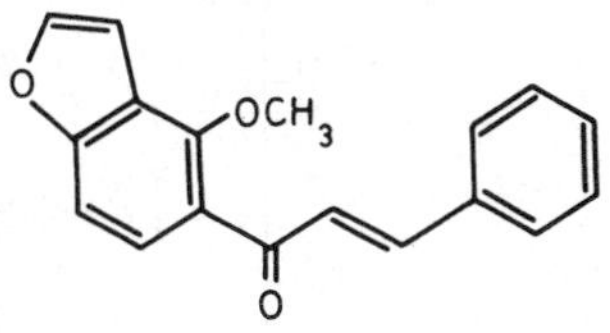

(6.10)

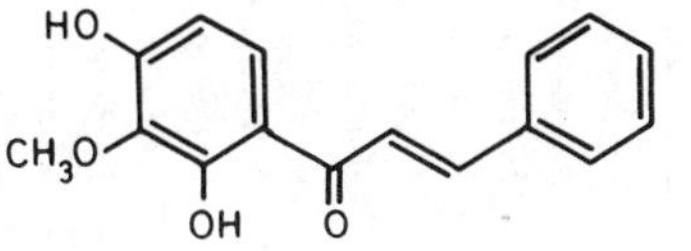

(6.11)

The furano derivative (*6.8*) represents a rare structure, at least amongst the minor flavonoids. Somewhat more common are furanochalcones such as 'ovalitenin-A' (*6.10*) which occurs in *Milletia ovalifolia* (Leguminosae) seeds along with other flavonoids having the same A-ring substitution, namely compounds 'ovalitenin-B' (*6.156*), 'ovalitenin-C' (*6.77*) and 'ovalitenone' (*6.148*) (Gupta and Krishnamurti, 1977b). Compound (*6.11*) is a simple extra-hydroxylated chalcone isolated from *Zuccagnia punctata* (Leguminosae) by Pederiva *et al.* (1975).

The next group of compounds is based upon a phloroglucinol-type A ring. The parent compound (*6.12*) occurs in *Helichrysum* (Bohlmann and Abraham, 1979c; Bohlmann *et al.*, 1980c), in the exudate of *Adiantum sulphureum* (Pteridaceae) (Wollenweber, 1979), and in the lipid fraction of *Populus* buds (Wollenweber and Weber, 1973) where it occurs with its 6′-methyl ether (*6.13*) (Wollenweber *et al.*, 1974). The 4′-methyl ether (*6.14*) occurs in *Helichrysum* (Bohlmann *et al.*, 1979c, 1980c), the oil of *Populus* buds (Wollenweber and Weber, 1973), and a variety of

ferns (Ramakrishnan *et al.*, 1974; Wollenweber 1977a, 1979; Wollenweber and Weber, 1973). 'Flavokawin-B' (*6.15*) has been found in *Helichrysum* (Bohlmann *et al.*, 1980c), *Aniba* (Lauraceae) (Fernandes *et al.*, 1978), *Didymocarpus* (Gesneriaceae) and *Pityrogramma* species (Wollenweber, unpublished).

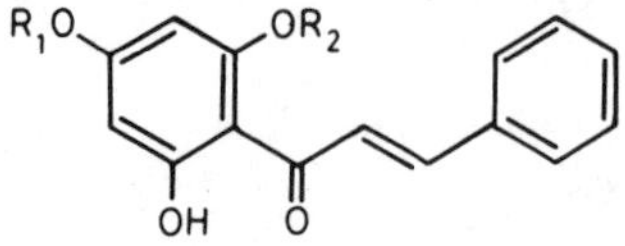

(*6.12*) $R_1 = R_2 = H$
(*6.13*) $R_1 = H, R_2 = CH_3$
(*6.14*) $R_1 = CH_3, R_2 = H$
(*6.15*) $R_1 = R_2 = CH_3$

C-Prenylation and *O*-methylation of the parent molecule yield the 3′-prenyl- and 3′-prenyl-4′-*O*-methylchalcones, (*6.16*) and (*6.17*), that Bohlmann and co-workers (1979c, 1980c) isolated from species of *Helichrysum*. Gupta and Krishnamurti (1977a) isolated 'ovalichalkone' from *Milletia ovalifolia* and showed it to be 2′-hydroxy-3′-prenyl-4′,6′-dimethoxychalcone (*6.18*).

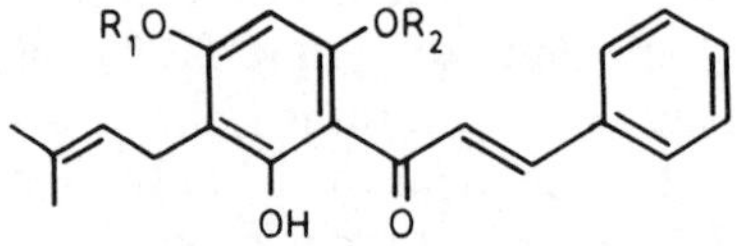

(*6.16*) $R_1 = R_2 = H$
(*6.17*) $R_1 = CH_3, R_2 = H$
(*6.18*) $R_1 = R_2 = CH_3$

Ring closure reactions in this series occur to yield such compounds as chalcone (*6.19*), isolated by Bohlmann *et al.* (1979c) from *Helichrysum cymosum* ssp. *calvum*, and 'pongachalcone-I' (*6.20*) from the heartwood of *Pongamia glabra* (Subramanyam *et al.*, 1977).

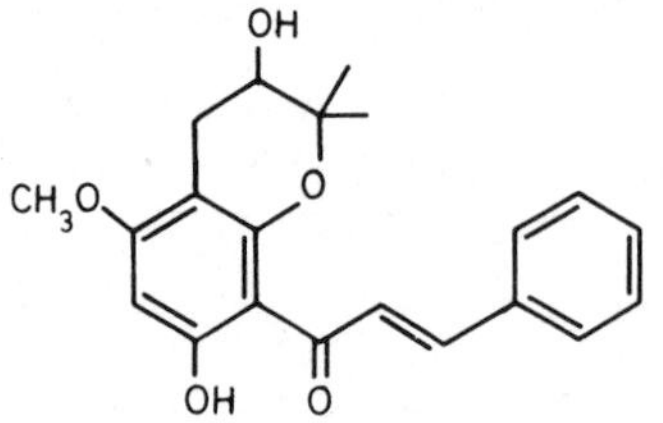

(*6.19*) 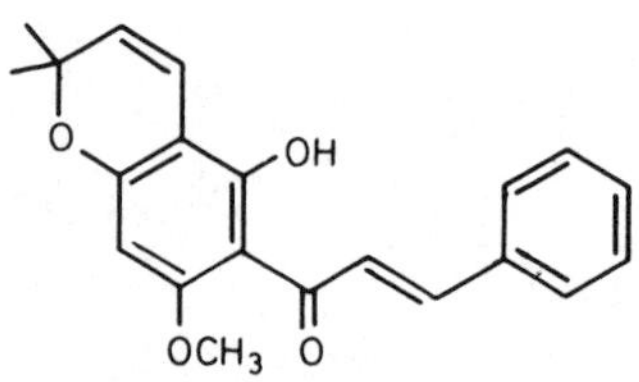 (*6.20*)

Recently, Camele and co-workers (1980) described two *β*-oxygenated chalcones from *Tephrosia praecans* (Leguminosae). 'Praecanosone-B' (*6.21*) could also be written in the tautomeric *β*-diketo form but 'praecanosone-A' (*6.22*) is locked into the chalcone structure by virtue of the *β*-methoxy group.

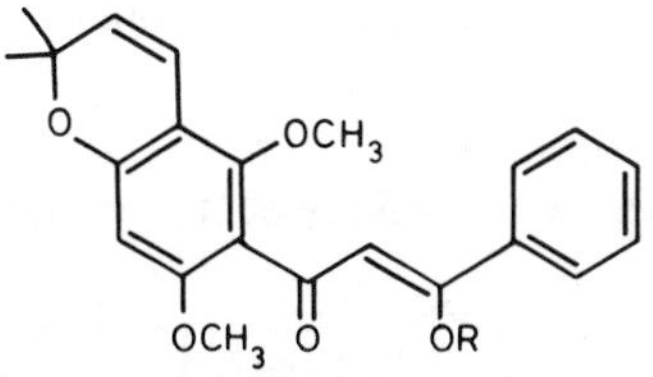

OCH3 OCH3 O OR

(6.21) R = H

(6.22) R = CH_3

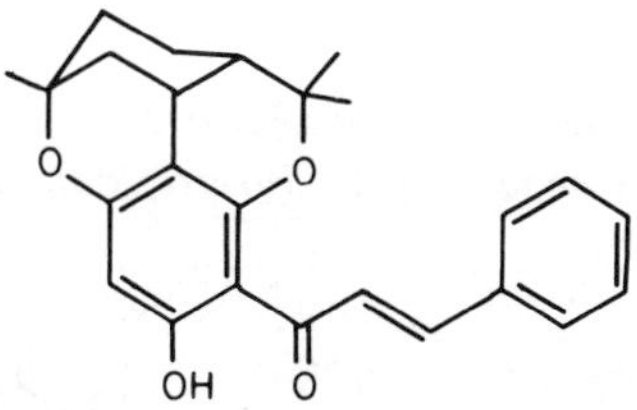

OH O

(6.23)

The structure of 'rubranine' (*6.23*) was originally proposed by Combes *et al.* (1970) for a compound that they isolated from *Aniba rosaeodora*. The existence of rubranine as a natural product was questioned by Montero and Winternitz (1973) who succeeded in synthesizing it through reaction of pinocembrin (5,7-dihydroxyflavanone) with citral in the presence of anibine, a known alkaloidal constituent of *A. rosaeodora*. The argument was joined more recently by de Alleluia and co-workers (1978), who demonstrated conclusively that (−)-rubranine is a natural product and not an artifact of isolation. Their conclusion was based upon two observations: (a) (−)-rubranine ($[\alpha]_D^{25} = -32.7°$) could be extracted from *A. rosaeodora* wood by very brief washing with hexane, conditions which minimize the opportunity for chemical transformations of the sort suggested by Montero and Winternitz (1973); and (b) citral does not accompany linalol and geraniol in *A. rosaeodora*.

Another example of the rich treasure of natural products elaborated by members of the genus *Helichrysum* is the chalcone (*6.24*) described by Bohlmann and Hoffmann (1979). It is interesting to note the co-occurrence of compound (*6.25*), a stilbene which would result from the alternative cyclization of the protoflavonoid precursor and by eventual modification of the terminal portion of the C-10 side chain.

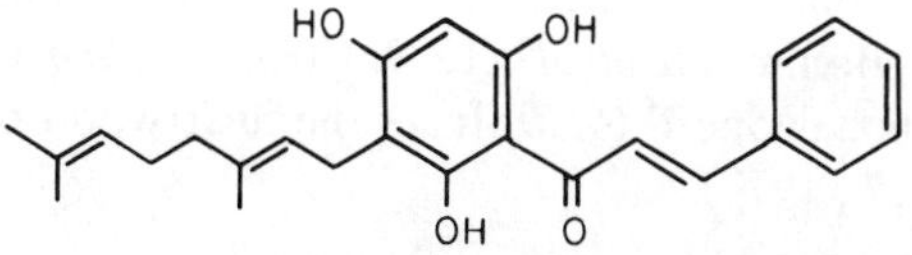

HO OH OH O

(6.24)

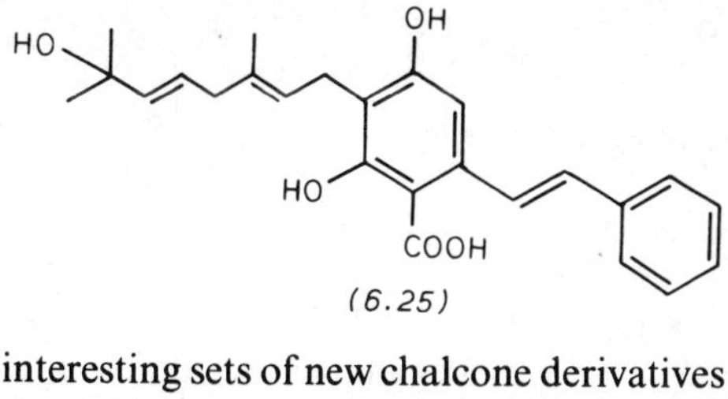

HO OH HO COOH

(6.25)

One of the most interesting sets of new chalcone derivatives was discovered by Dietz and co-workers (1980) in the exudate of *Pityrogramma trifoliata*. The three compounds, (*6.26*), (*6.27*) and (*6.28*), can be thought of as chalcones to which a

phenylpropanoid moiety is attached through esterification at the 2′-OH group and by carbon–carbon bond coupling at C-3′. Alternatively, as these authors state, these compounds may be considered 4-phenyldihydrocoumarins, structures of which are easier to visualize if they are rewritten as (*6.26a*), (*6.27a*) and (*6.28a*). Confirmation of these structures by synthesis is clearly needed. A dihydrochalcone corresponding to compound (*6.26*) has been reported from this plant as well (Wagner *et al.*, 1979; Dietz *et al.*, 1980) and a similar compound

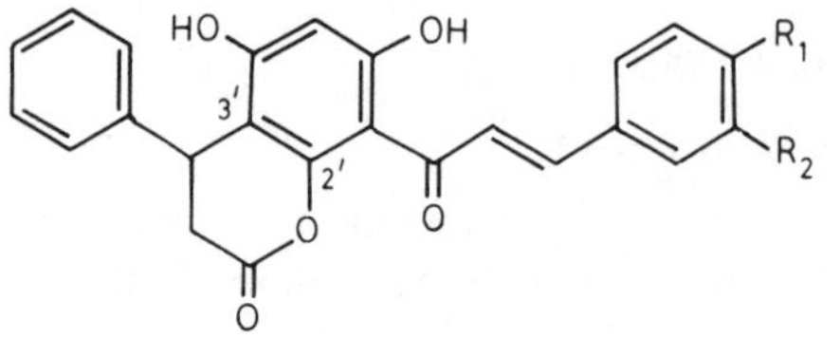

(*6.26*) $R_1 = R_2 = H$ (*6.27*) $R_1 = OH$, $R_2 = H$ (*6.28*) $R_1 = R_2 = OH$

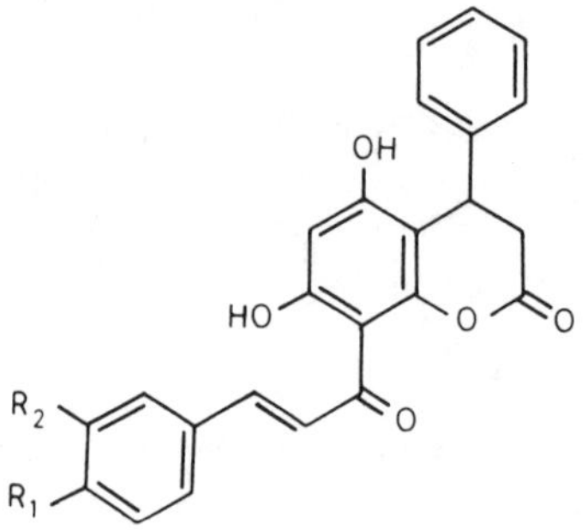

(*6.26a*) $R_1 = R_2 = H$ (*6.27a*) $R_1 = OH$, $R_2 = H$ (*6.28a*) $R_1 = R_2 = OH$

related to chalcone (*6.27*) is also indicated. Structural confirmation through unequivocal synthesis would be in order, as would biosynthetic studies, for this family of compounds.

Several *C*-methylated chalcones having the phloroglucinol A-ring pattern are also known. 'Triangularin' (*6.29*) occurs in the leaf exudate of *Pityrogramma triangularis* (Star *et al.*, 1978) while 'aurentiacin' (*6.30*) occurs in *Pityrogramma* exudates (Wollenweber *et al.*, 1980b) as well as in *Didymocarpus* (Adityachaudhury *et al.*, 1973). 'Aurentiacin-A' (*6.30a*) also occurs in *Didymocarpus* (Adityachaudhury and Daskanungo, 1975) and in *Dalea* (Leguminosae) (Dominguez *et al.*, 1980). The di-*C*-methylchalcone (*6.31*) occurs in leaves of *Myrica gale* (Myricaceae) (Malterud *et al.*, 1977). Further studies of the distribution of the known 'ceroptin' (*6.32*) in chemotypes of *Pityrogramma triangularis* have been described by Star and co-workers (1975a,b).

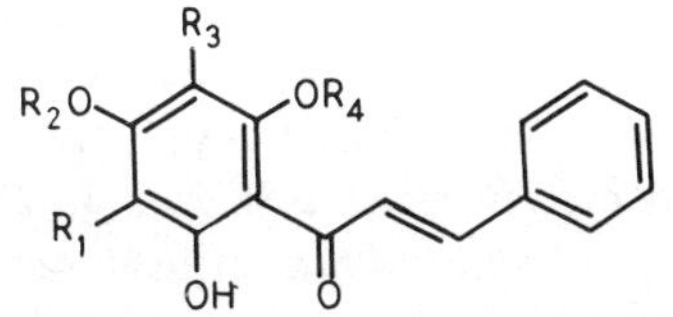

(6.29) $R_1 = R_2 = CH_3$, $R_3 = R_4 = H$

(6.30) $R_1 = R_2 = R_4 = CH_3$, $R_3 = H$

(6.30a) $R_1 = R_2 = H$, $R_3 = R_4 = CH_3$

(6.31) $R_1 = R_3 = CH_3$, $R_2 = R_4 = H$

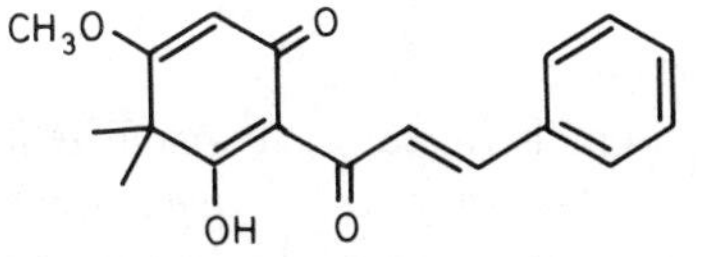

(6.32)

Chalcones having four A-ring hydroxylations occur widely. Structures and source genera are as follows: (*6.33*), *Polygonatum* (Liliaceae) (Maradufu and Ouma, 1978); (*6.34*) and (*6.35*), *Helichrysum* (Bohlmann *et al.*, 1978); (*6.36*) and (*6.37*), *Popowia* (Panichpol and Waterman, 1978); and compound (*6.38*) from both *Didymocarpus* (Agarwal *et al.*, 1973) and *Onychium auratum* (Gymnogrammataceae) (Ramakrishnan *et al.*, 1974).

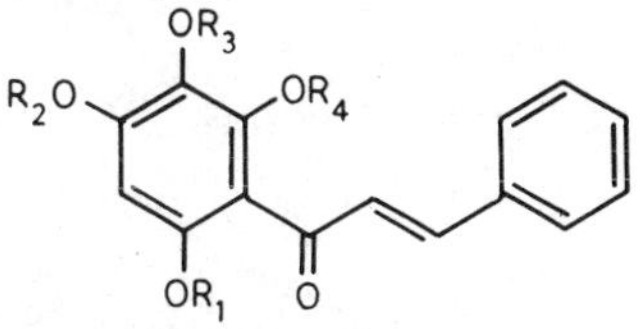

(6.33) $R_1 = R_3 = CH_3$, $R_2 = R_4 = H$

(6.34) $R_1 = R_4 = CH_3$, R_2 and R_3 = $-O-CH_2-O-$

(6.35) $R_1 = H$, $R_2 = R_3 = R_4 = CH_3$

(6.36) $R_1 = R_2 = R_3 = CH_3$, $R_4 = H$

(6.37) $R_1 = R_2 = R_3 = R_4 = CH_3$

(6.38) $R_1 = R_4 = H$, $R_2 = R_3 = CH_3$

Only two chalcones are known in this subgroup in which the A rings are completely oxygenated. Compound (*6.39*) was found in *Didymocarpus* by Bhattacharyya and co-workers (1979), and 'kanakugiol' (*6.40*) was isolated from fruit of *Popowia cauliflora* by Waterman and Pootakahm (1979).

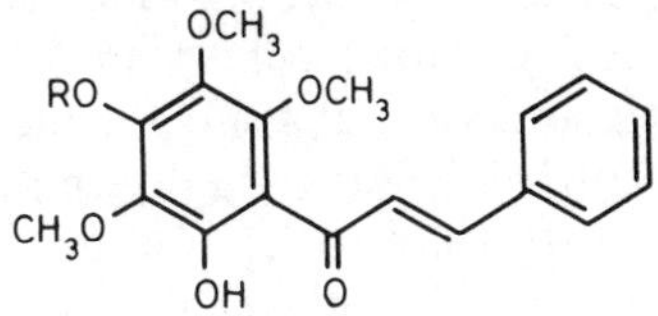

(6.39) R = H

(6.40) $R = CH_3$

(b) Chalcones having one B-ring hydroxyl

'Isoliquiritigenin' (*6.41*), the simplest chalcone encountered in this subgroup, occurs in *Dahlia* species (Lam and Wrang, 1975; see Bohm, 1975, Table 9.1) and from *Xanthorrhoea* (Xanthorraceae, *loc. cit.*, Table 9.1), but is most often encountered in members of Leguminosae. Recent work describes its occurrence in *Sophora* (Komatsu *et al.*, 1978) for the first time, and in *Dalbergia* (Matos *et al.*, 1975). Members of Leguminosae also synthesize *C*-prenylchalcones based upon isoliquiritigenin. Variations on this structural theme are seen in '4-hydroxyisocordoin' (*6.42*) and '4-hydroxyderricin' (*6.43*) which occur in *Lonchocarpus* (Delle Monache *et al.*, 1974). The chalcone (*6.43*), isolated from *Angelica keiskei* by Kozawa and co-workers (1977), accompanies the newly discovered 'xanthoangelol' (*6.44*). This represents the first report of chalcones from Umbelliferae.

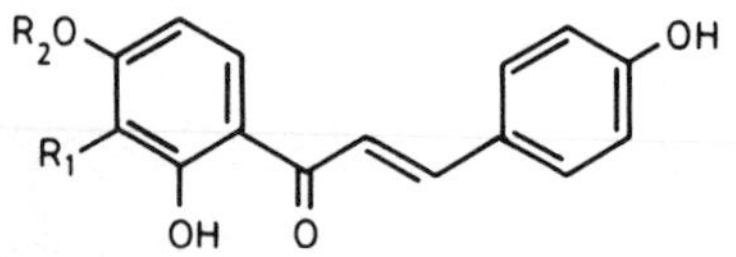

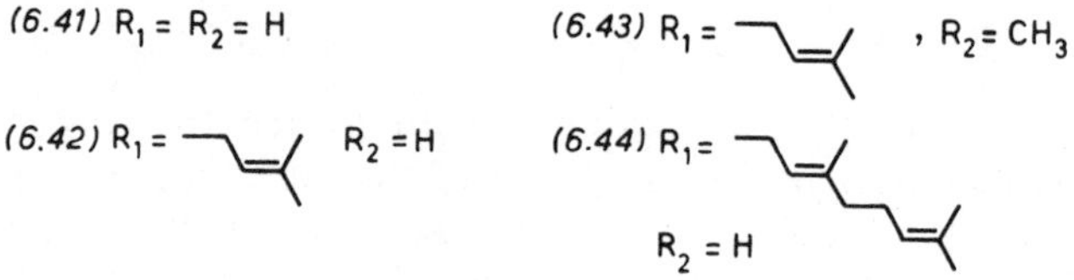

R_2O OH R_1 OH O

(*6.41*) $R_1 = R_2 = H$

(*6.42*) $R_1 =$ [prenyl] $R_2 = H$

(*6.43*) $R_1 =$ [prenyl], $R_2 = CH_3$

(*6.44*) $R_1 =$ [geranyl] $R_2 = H$

Cyclization of the prenyl group of compound (*6.42*) to the 4′-phenolic function would yield '4-hydroxylonchocarpin' (*6.45*), known as a constituent of *Derris floribunda* (Braz Filho *et al.*, 1975b), while closure at the 2′-hydroxyl function could give, after hydration, 'bavachromenol' (*6.46*) which Suri and co-workers (1980) recently isolated from seeds of *Psoralea corylifolia*. The novel compound xanthoangelol (*6.44*) also occurs in *Lespedeza cyrtobotrya* (Leguminosae) where it occurs with, and presumably could act as precursor to, another newly found chalcone, 'lespeol' (*6.47*) (Miyase *et al.*, 1980).

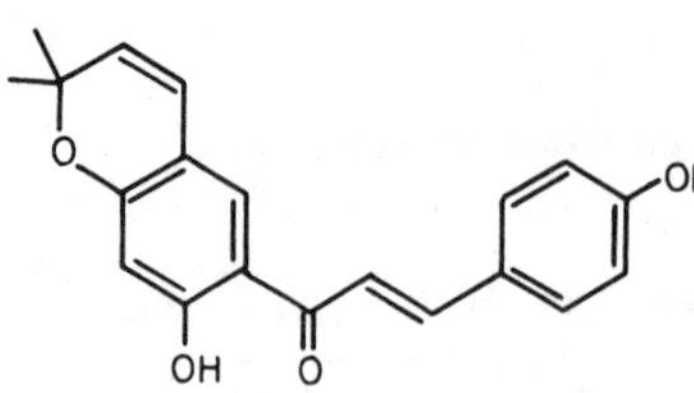

O OH OH O

(*6.45*)

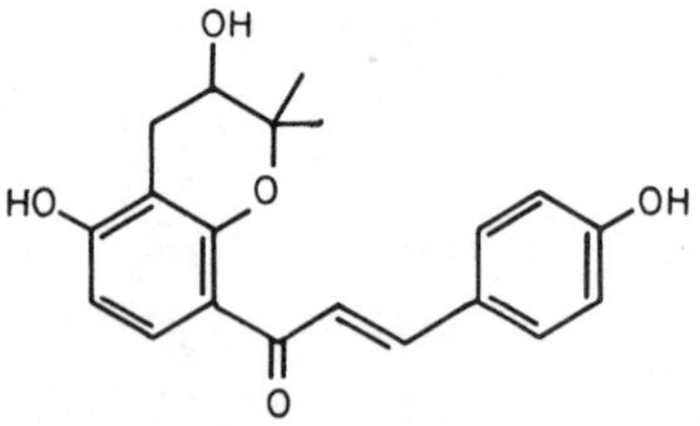

OH O HO OH O

(*6.46*)

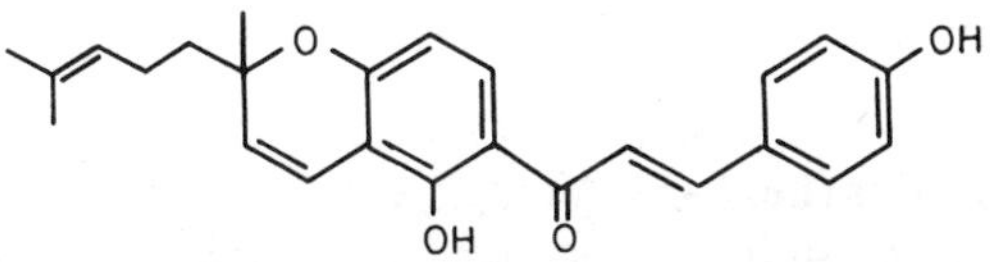

(6.47)

The last two compounds in this subsection contain unique structural features among naturally occurring chalcones. Compound (*6.48*), 'neobavachalcone', 2′,4-dihydroxy-4′-methoxy-5′-formylchalcone, isolated from *Psoralea corylifolia* (Leguminosae) and subsequently synthesized by Gupta *et al.* (1975b, 1977) is distinctive in its aldehydo substitution. 3′-*C*-(β-D-Glucopyranosyl) isoliquiritigenin (*6.49*) was isolated from the wood of *Cladrastis platycarpa* (Leguminosae) by Ohashi and co-workers (1977) and represents the first report of a naturally occurring *C*-glycosylchalcone.

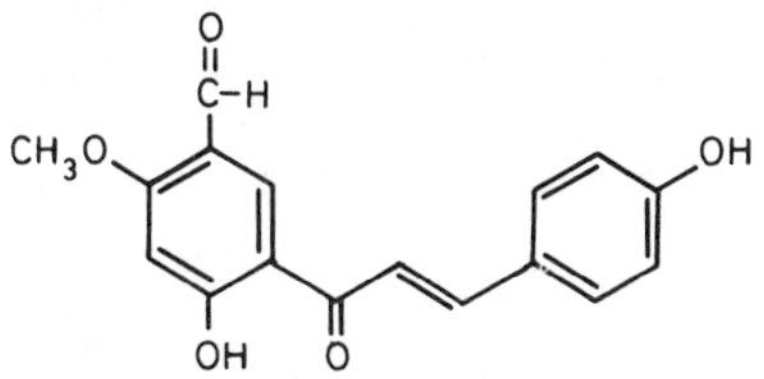

(6.48) 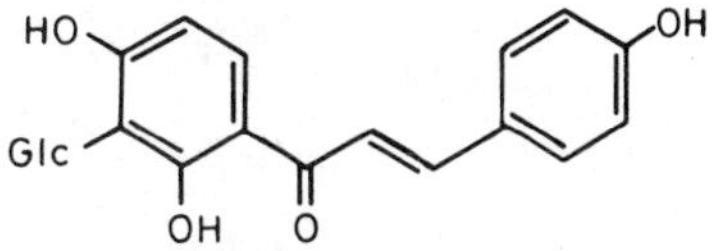 (6.49)

The only reports of 'isosalipurpol' (*6.50*) as the free phenol involve, interestingly, its occurrence in the pollen of two unrelated families, namely, *Petunia hybrida* (Solanaceae) (de Vlaming and Kho, 1976), and *Nothofagus antartica* (Fagaceae) (Wollenweber and Wiermann, 1979). Glycosides of isosalipurpol are much more common in nature and a variety of *O*-methylated derivatives is known. Glycosides include 'isosalipurposide', the 2′-*O*-glucoside (*6.51*) which occurs in *Ruellia* species (Acanthaceae) (Bloom, 1976), *Acacia*

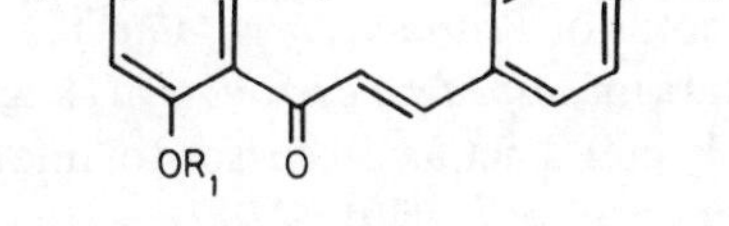

(6.50) $R_1 = R_2 = R_3 = R_4 = H$

(6.51) R_1 = Glucose, $R_2 = R_3 = R_4 = H$

(6.52) $R_1 = R_2 = R_3 = H$, R_4 = Glucose

(6.53) $R_1 = R_2 = R_3 = H$, R_4 = Glucose-X″(*p*-coumaroyl)

(6.54) $R_1 = R_2 = R_4 = H$, $R_3 = CH_3$

(6.55) $R_1 = CH_3$, R_2 = Glucose, $R_3 = R_4 = H$

(6.56) $R_1 = R_2 = CH_3$, $R_3 = H$, R_4 = Glucose

(6.57) $R_1 = R_2 = R_4 = CH_3$, $R_3 = H$

cyanophylla, where it accompanies the newly discovered 4-*O*-glucoside (*6.52*) (Imperato, 1978) and in *Salix rubra*, where it is also found as a *p*-coumaroyl derivative (*6.53*) (Vinokurov, 1979). Isosalipurpol 6′-methyl ether (*6.54*) is known as a constituent of *Helichrysum heterolasium* (Bohlmann and Abraham, 1979b), while its 4′-*O*-glucoside (*6.55*) has been found in species of *Gnaphalium* (Compositae) by Aritomi and Kawasaki (1974) and Maruyama *et al.* (1974) and synthesized by Ahluwalia and Rami (1976). The di-*O*-methylchalcone, as its glucoside (*6.56*), occurs in *Viscum album*, the European mistletoe (Becker *et al.*, 1978), and the trimethyl ether (*6.57*) is one of several flavonoids isolated from *Dahlia tenuicaulis* by Lam and Wrang (1975).

Extra hydroxylation of salipurpol followed by greater or lesser degrees of *O*-methylation would yield the following chalcones, all of which occur in the free state: (*6.58*) occurs in *Helichrysum glomeratum* (Bohlmann and Suwita, 1979c); (*6.59*) and (*6.60*) were found in *Chromolaena odorata* (Compositae) (Barua *et al.*, 1978); (*6.61*) occurs with other *O*-methylated flavonoids in *Popowia cauliflora* (Panichpol and Waterman, 1978); and (*6.61a*) plus an unidentified chalcone were found in roots of *Flemingia strobilifera* by Bhatt (1975).

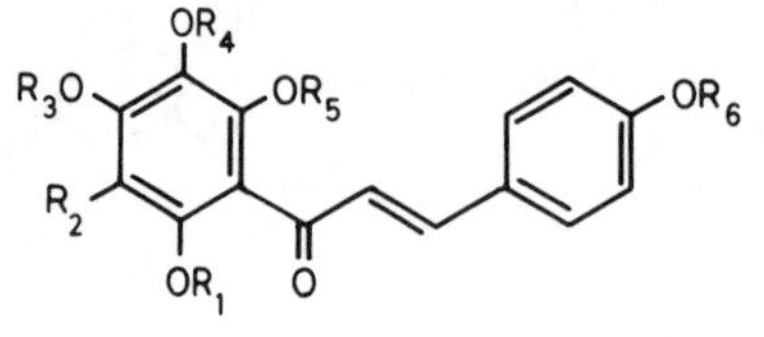

OR_4 R_3O OR_5 OR_6 R_2 OR_1 O

(*6.58*) $R_1 = CH_3$, $R_2 = R_5 = R_6 = H$, R_3 and $R_4 = -O-CH_2-O-$

(*6.59*) $R_1 = R_2 = R_6 = H$, $R_3 = R_4 = R_5 = CH_3$

(*6.60*) $R_1 = R_2 = H$, $R_3 = R_4 = R_5 = R_6 = CH_3$

(*6.61*) $R_1 = R_3 = R_4 = CH_3$, $R_2 = R_5 = R_6 = H$

(*6.61a*) $R_1 = R_4 = H$, $R_2 = OCH_3$, $R_3 = R_5 = R_6 = CH_3$

(c) Chalcones having two B-ring hydroxyls

Chemically, as opposed to biosynthetically, compounds (*6.62*) and (*6.63*) are the simplest members of this subgroup by virtue of their single A-ring hydroxyl group. [The closely related trihydroxy B-ring compound (*6.64*) will also be treated here, owing to its biosynthetic similarity to the other two compounds.] 'Echinatin' (*6.62*) occurs in cultures of *Glycyrrhiza echinata* (Saitoh *et al.*, 1975) while 'licochalcone-A' (*6.63*) and 'licochalcone-B' (*6.64*) occur in root tissue of *G. glabra* (Saitoh and Shibata, 1975). The unusual hydroxylation pattern of these compounds, with a single hydroxyl group at position-4′ and an apparent polyacetate-derived substitution pattern on ring B, prompted the name 'retro-flavonoids' for this small group. The biosynthesis of these compounds was studied by Saitoh *et al.* (1975), whose results show that the singly hydroxylated

ring in these compounds is derived from the ring of a phenylpropanoid precursor while the right-hand ring is derived from malonyl-CoA. A shift of bridge oxygen position must have occurred during their formation. The pathway suggested by Saitoh *et al.* (1975) is summarized in Fig. 6.1.

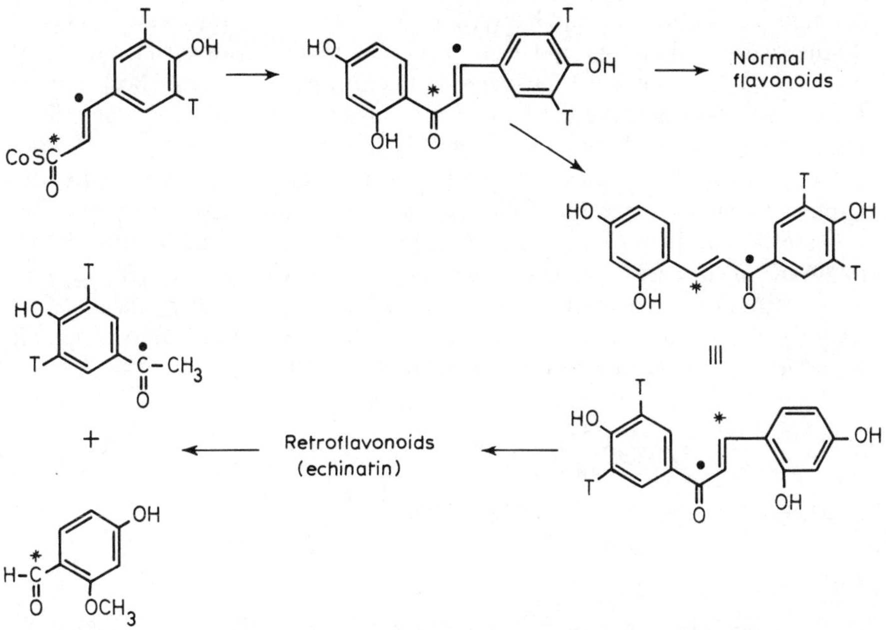

Fig. 6.1 Biosynthetic route to retroflavonoids (after Saitoh *et al.*, 1975).

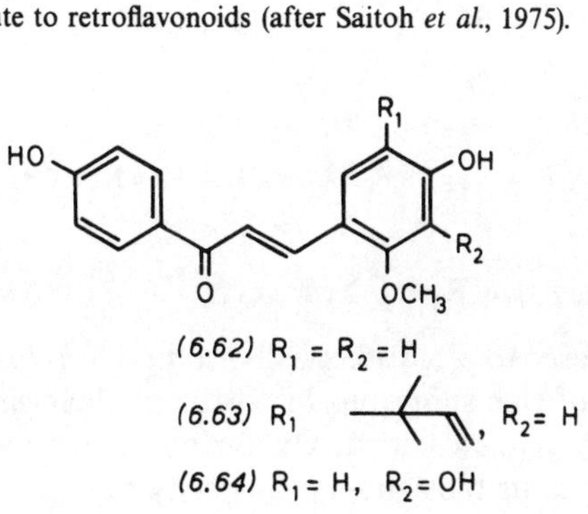

The majority of compounds in this hydroxylation subgroup are based upon a resorcinol-type A ring, seen in the parent compound butein (*6.65*). Butein itself occurs in trunkwood of *Machaerium mucronulatum* (Anacardiaceae) (Kurosawa *et al.*, 1978) and in the ligules of *Bidens laevis* (Compositae) (Scogin and Zakar, 1976). Glycosidic forms include the well-known 4′-*O*-glucoside (*6.66*) coreopsin known from several genera of Compositae, the 4′-*O*-diglucoside (*6.67*) from

Bidens (Hart, 1979) and *Coreopsis bigelovii* (Nicholls and Bohm, 1979), the 2′,3-di-*O*-glucoside (*6.68*) from species of *Rhus* (Young, 1979) and *Amphipterigium adstrigens* (Young, 1976) (both Anacardiaceae), and a novel 4′-*O*-arabinosylgalactoside (*6.69*) from *Bauhinia purpurea* (Leguminosae) (Bhartiya *et al.*, 1979).

Simple *O*-methyl derivatives of butein include the 4-*O*-methyl ether (*6.70*) from *Dahlia* (Giannasi, 1975; Lam and Wrang, 1975) and *Megalodonta* (Roberts, 1980) and the 3-*O*-methyl- (*6.71*) and 3,4-di-*O*-methyl (*6.72*) derivatives from *Iryanthera polyneura* (Myristicaceae) (de Almeida *et al.*, 1979).

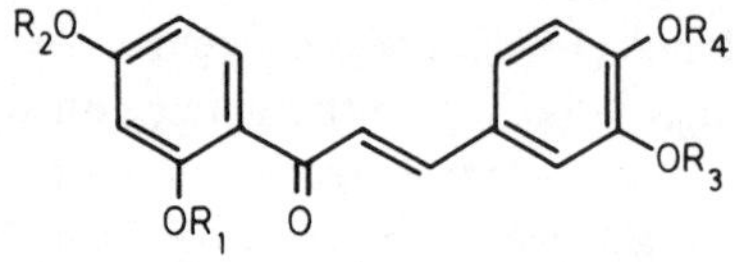

R_2O OR_4 OR_3 OR_1 O

(*6.65*) $R_1 = R_2 = R_3 = R_4 = H$

(*6.66*) $R_1 = R_3 = R_4 = H$, $R_2 = Glc$

(*6.67*) $R_1 = R_3 = R_4 = H$, $R_2 = Glc{-}O{-}Glc$

(*6.68*) $R_1 = R_3 = Glc$ $R_2 = R_4 = H$

(*6.69*) $R_1 = R_3 = R_4 = H$, $R_2 = -Gla{-}O{-}Ara$

(*6.70*) $R_1 = R_2 = R_3 = H$, $R_4 = CH_3$

(*6.71*) $R_1 = R_2 = R_4 = H$, $R_3 = CH_3$

(*6.72*) $R_1 = R_2 = H$, $R_3 = R_4 = CH_3$

The following set of compounds share a structural feature commonly seen in flavonoids from Leguminosae, the presence of the dimethylchromeno A-ring structure. 3,4-Dihydroxylonchocarpin (*6.73*), from *Derris floribunda* (Braz Filho *et al.*, 1975b) is the simplest of these. The *O*-methylated derivatives (*6.74*) and (*6.75*) occur in *Pongamia glabra* (Sharma *et al.*, 1973; Subramanyam *et al.*, 1977). The related compound (*6.76*) also occurs in *P. glabra* (Subramanyam *et al.*, 1977) along with chalcones (*6.6*) and (*6.20*) treated earlier in this chapter. 'Ovalitenin-C' (*6.77*) was isolated from *Milletia ovalifolia* (Islam *et al.*, 1980), and is a member of a series of closely related compounds having the furano A ring.

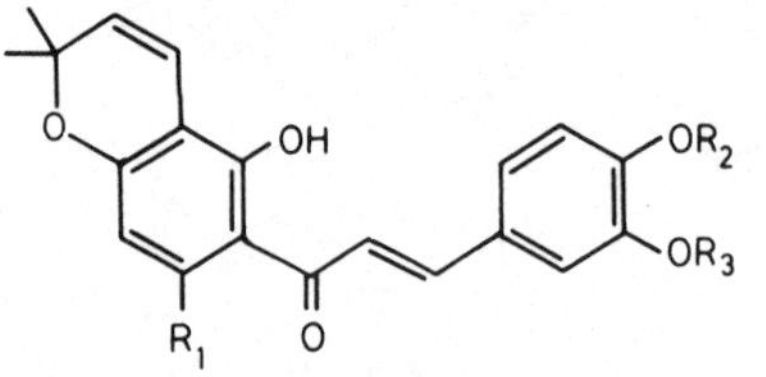

O OH OR_2 OR_3 R_1 O

(*6.73*) $R_1 = R_2 = R_3 = H$

(*6.74*) $R_1 = R_2 = H$, $R_3 = CH_3$

(*6.75*) $R_1 = H$, R_2 and $R_3 = -O{-}CH_2{-}O-$

(*6.76*) $R_1 = OCH_3$, R_2 and $R_3 = -O{-}CH_2{-}O-$

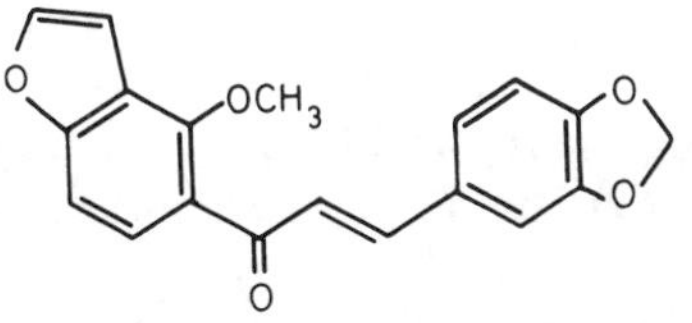

(6.77)

Extra hydroxylation of the butein skeleton at the α-carbon would yield α,2′,4′,3,4-pentahydroxychalcone (*6.78*) which has been isolated from *Peltogyne* species by Malan and Roux (1974). The occurrence of a chalcone hydroxylated at that position suggests a possible biosynthetic relationship with such compounds as the peltogynoid flavonoids represented below by peltochalcone (*6.79*). Extra hydroxylation of butein at position 3′ yields okanin (*6.80*) which is known to occur as the free phenol in wood of *Albizia adianthifolia* (Leguminosae) (Candy *et al*., 1978) and as the 4′-*O*-glucoside, marein (*6.81*), in many species of *Bidens* and *Coreopsis* (see Table 6.1 for references). *O*-Methylation of marein at position 3′ affords lanceolin (*6.82*), also known from *Coreopsis* species. 3-*O*-Methyl- (*6.82a*) and 3,4-*O*-methylokanin (*6.82b*) glucosides occur in several members of the *Bidens pilosa* group of species (Ballard, 1975). 'Stillopsidin',

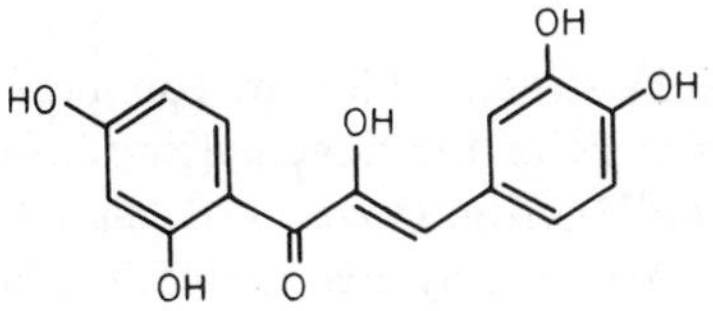

(6.78)

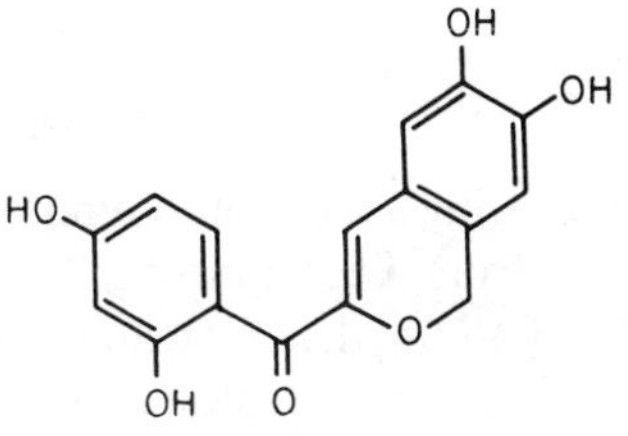

(6.79)

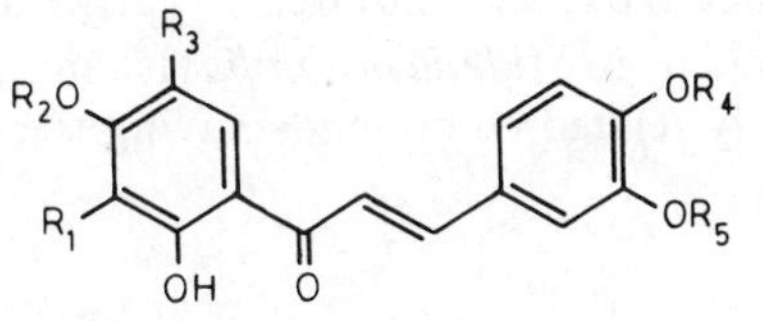

(6.80) R_1 = OH, R_2 = R_3 = R_4 = R_5 = H

(6.81) R_1 = OH, R_2 = Glc, R_3 = R_4 = R_5 = H

(6.82) R_1 = OCH_3, R_2 = Glc, R_3 = R_4 = R_5 = H

(6.82a) R_1 = OH, R_2 = R_3 = R_4 = H, R_5 = CH_3

(6.82b) R_1 = OH, R_2 = R_3 = H, R_4 = R_5 = CH_3

(6.83) R_1 = R_4 = R_5 = H, R_2 = Glc, R_3 = OH

(6.84) R_1 = R_2 = H, R_3 = OCH_3, R_4 and R_5 = -O-CH_2-O-

which is 5′-hydroxybutein, occurs as the 4′-*O*-glucoside (*6.83*) in *Coreopsis bigelovii* (Nicholls and Bohm, 1979). The 5′-hydroxybutein pattern also occurs in 'prosogerin-B' (*6.84*), a constituent of flowers of *Prosopis spicigera* (Leguminosae) (Bhardwaj *et al.*, 1979).

Only a few pentahydroxychalcones having the phloroglucinol A-ring pattern have been reported during the past few years. Two of these, (*6.85*) and (*6.86*), were isolated from *Merrillia caloxylon* (Rutaceae) by Fraser and Lewis (1974). Structure (*6.87*) was established for 'ovalichalkone-A' by Gupta and Krishnamurti (1979) who isolated it from the seeds of *Milletia ovalifolia*.

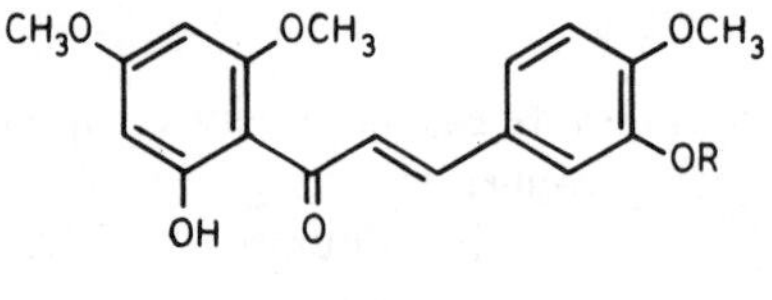

(*6.85*) R = H
(*6.86*) R = CH_3

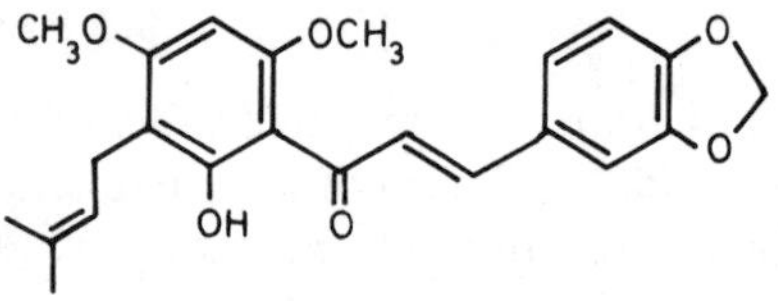

(*6.87*)

Compounds (*6.88*) and (*6.89*) exhibit the comparatively uncommon 2,4-dihydroxy B-ring substitution pattern. Both compounds occur in the heartwood of *Prunus cerasus* (Rosaceae) (Nagarajan and Parmar, 1977a,c) while (*6.89*) also occurs in *Terminalia arjuna* (Combretaceae) (Nagar *et al.*, 1979).

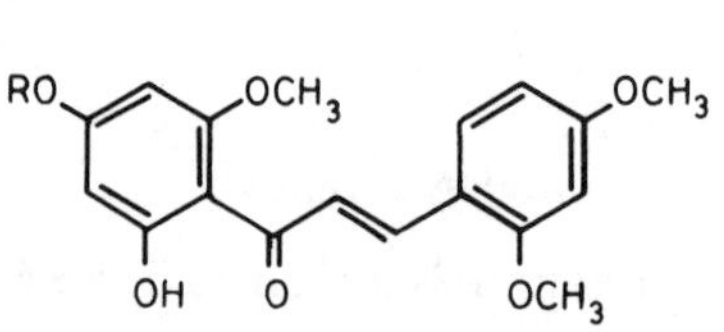

(*6.88*) R = H
(*6.89*) R = CH_3

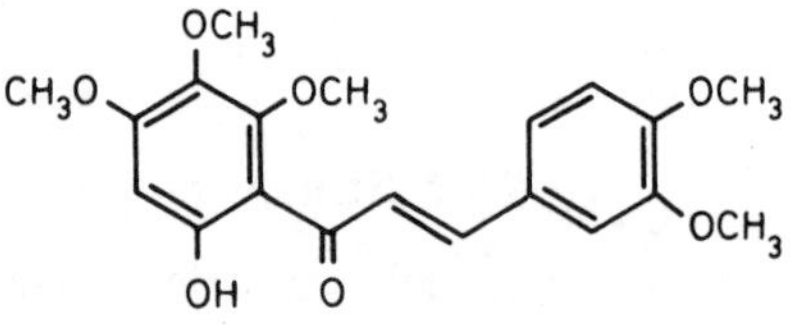

(*6.90*)

The pentamethoxychalcone (*6.90*) has been found in *Chromolaena odorata* (Compositae) by Barua *et al.* (1978) where it occurs with several other highly *O*-methylated flavonoids.

(d) Chalcones having three B-ring hydroxyls

Two new members of this group of chalcones have been discovered. Chalcone (*6.91*) was isolated from *Merrillea caloxylon* by Fraser and Lewis (1974) where it occurs with the related *O*-methyl derivatives (*6.85*) and (*6.86*). Compound (*6.92*), isolated from *Derris robusta* by Chibber *et al.* (1979), exhibits the unusual 2,4,5-trihydroxy B-ring substitution pattern.

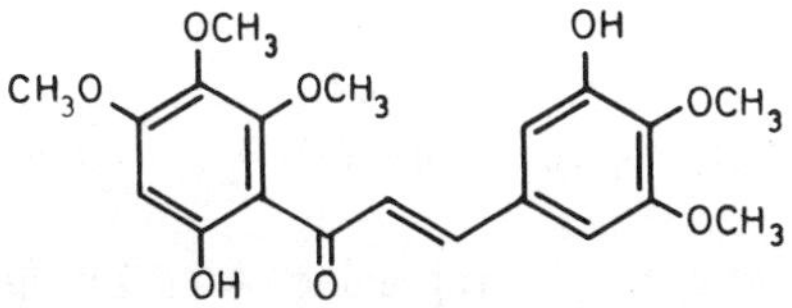

(6.91) 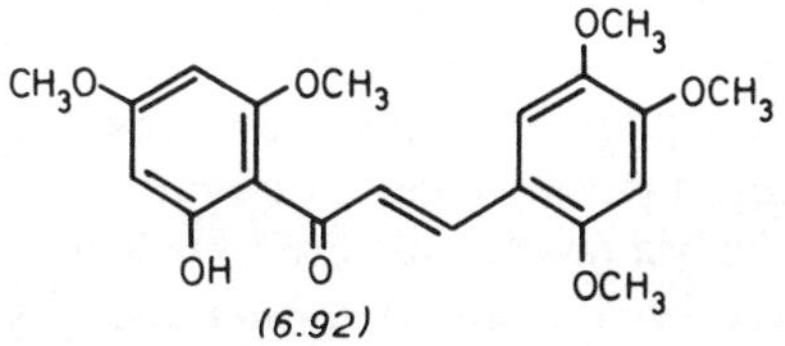 (6.92)

6.3 AURONES

6.3.1 Introduction

Aurones are hydroxylated 2-benzylidenecoumaranones, the parent compound being (*6.93*). The 'normal' numbering system applies to this group of compounds: positions on the A ring are identified by unprimed numbers and the B-ring positions by primed numbers. Note that in aurones position 4 corresponds biosynthetically to position 5 of other heterocyclic flavonoids. Included in this section are two 'auronol' derivatives, distinguished by the presence of a methoxyl group in place of the carbonyl function at position 3.

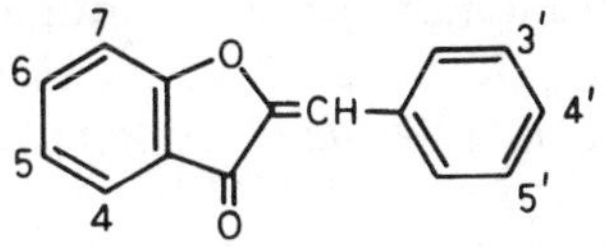

(6.93)

6.3.2 Structures of naturally occurring aurones

Structures of naturally occurring aurones and auronols appear in Table 6.3

(a) Aurones lacking B-ring hydroxyls

The first naturally occurring aurones lacking B-ring substitution were isolated from *Derris obtusa* by do Nascimento *et al.* (1976). The three compounds, (*6.94*), (*6.95*) and (6.96), were accompanied by a methylenedioxy derivative (*6.96a*), and by two auronol derivatives (*6.97*) and (*6.98*). The first suggestion of the existence of an auronol was made by Hänsel and co-workers based on a compound from *Helichrysum arenarium* (Hänsel *et al.*, 1965). Soon thereafter, however, these authors (Hänsel *et al.*, 1967) withdrew their proposal having correctly identified the compound as a flavonol. In the case of the 3-*O*-methylauronols from *Derris* do Nascimento and co-workers presented convincing evidence for their existence. In the first place, both isomeric flavonols are known and were observed to differ in all physical respects from derriobtusone-A (*6.97*) and derriobtusone-B

(*6.98*). Secondly, both flavonols were resistant to dilute refluxing HCl while the 3-*O*-methylauronols were readily demethylated under these conditions. Longer acid treatment resulted in the production of the furanocoumaranone from both compounds and benzoic acid from (*6.97*) and piperonylic acid (3,4-methylenedioxybenzoic acid) from (*6.98*).

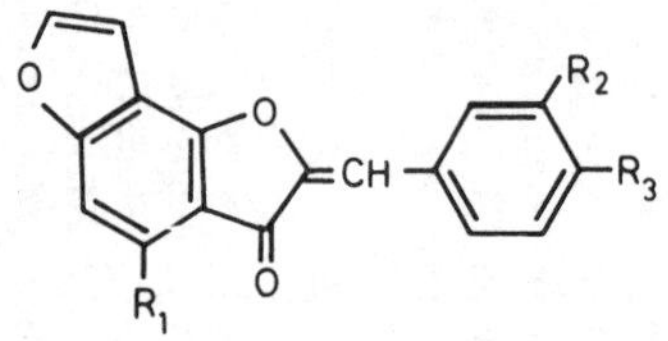

(*6.94*) $R_1 = R_2 = R_3 = H$

(*6.95*) $R_1 = OH, R_2 = R_3 = H$

(*6.96*) $R_1 = OCH_3, R_2 = R_3 = H$

(*6.96a*) $R_1 = H$, R_2 and $R_3 = -O-CH_2-O-$

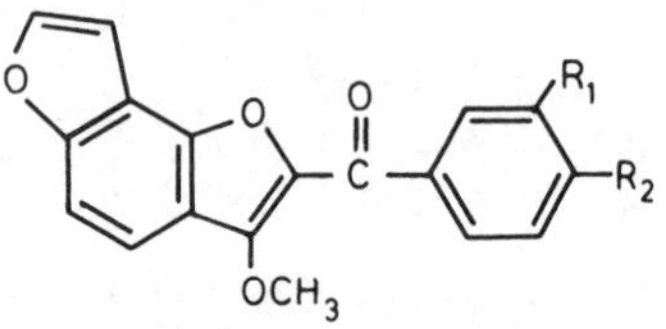

(*6.97*) $R_1 = R_2 = H$

(*6.98*) R_1 and $R_2 = -O-CH_2-O-$

(b) Aurones having one B-ring hydroxyl

Only one member of this subgroup has been reported since the 1975 review. El Sherbeiny and co-workers (1978) reported the presence of the known aurone hispidol (*6.99*) and its 6-*O*-glucoside in seeds of *Lygos raetam* (Leguminosae)

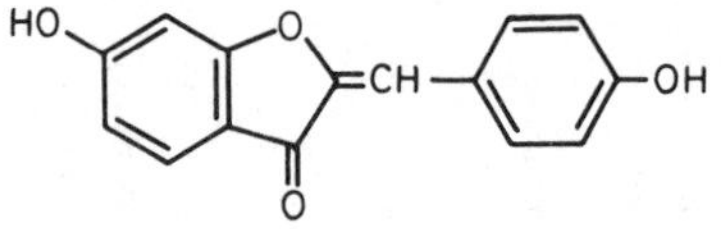

(*6.99*)

(c) Aurones having two B-ring hydroxyls

Sulfuretin (*6.100*), one of the most common aurones, has recently been reported to occur as the free phenol and as the 6,3′-di-*O*-glucoside (*6.101*) in *Rhus* species (Young, 1979) and in *Amphipterygium adstrigens* (Young, 1976), as a series of 6-mono- and di-glycosides in several *Bidens* species (Hart, 1979; Scogin and Zakar, 1976), and as the 6-*O*-glucoside (102) in several *Helianthus* species (Harborne and Smith, 1978), *Coreopsis bigelovii* (Nicholls and Bohm, 1979) and *Viguiera laciniata* (Harborne and Smith, 1978).

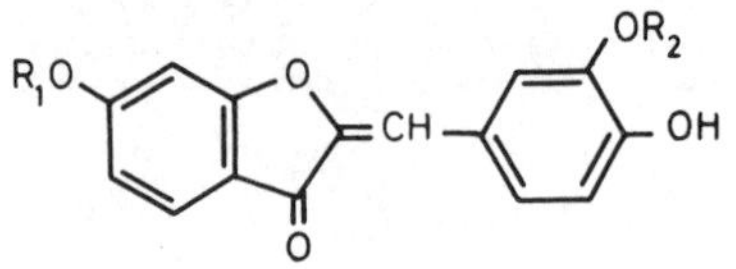

(*6.100*) $R_1 = R_2 = H$

(*6.101*) $R_1 = R_2 = Glc$

(*6.102*) $R_1 = Glc, R_2 = H$

Table 6.3 Structures of naturally occurring aurones (including auronols)

Substituents: OH	Substituents: OCH_3	Substituents: Other	No. in text	Trivial name	Source*		References
		Furano(2″,3″,6,7)	(*6.94*)		*Derris obtusa*	Leg	do Nascimento *et al.* (1976)
4		Furano(2″,3″,6,7)	(*6.95*)		*D. obtusa*	Leg	do Nascimento *et al.* (1976)
	4	Furano(2″,3″,6,7)	(*6.96*)		*D. obtusa*	Leg	do Nascimento *et al.* (1976)
	3	Furano(2″,3″,6,7); β-carbonyl	(*6.97*)	Derriobtusone-A	*D. obtusa*	Leg	do Nascimento *et al.* (1976)
6,4′			(*6.99*)	Hispidol	*Lygos raetam*	Leg	El Sherbeiny *et al.* (1978)
		6-*O*-Glucose		Hispidol glucoside	*L. raetam*	Leg	El Sherbeiny *et al.* (1978)
6,3′,4′			(*6.100*)	Sulfuretin	*Rhus* spp.	Ana	Young (1979)
					Amphipterygium adstrigens	Ana	Young (1976)
		6-*O*-Glucose	(*6.102*)	Sulfurein	*Helianthus* spp.	Com	Harborne and Smith (1978)
					Viguiera laciniata	Com	Harborne and Smith (1978)
					Bidens spp.	Com	Hart (1979)
					Coreopsis bigelovii	Com	Nicholls and Bohm (1979)
					Bidens laevis	Com	Scogin and Zakar (1976)
		6-*O*-Glucosylglucose			*Bidens* spp.	Com	Hart (1979)
		6,3′-di-*O*-Glucose	(*6.101*)		*Rhus* spp.	Ana	Young (1979)
					Amphipterygium adstrigens	Ana	Young (1976)
6,7,3′,4′		6-*O*-Glucose	(*6.103*)	Maritimein	*Coreopsis bigelovii*	Com	Nicholls and Bohm (1979)
6,3′,4′	7	6-*O*-Glucose	(*6.104*)	Leptosin	*Coreopsis grandiflora*	Com	Crawford and Smith (1980)
		Furano(2″,3″,6′,7′); 3,4-methylenedioxy	(*6.96a*)		*Derris obtusa*	Leg	do Nascimento *et al.* (1976)
	3	Furano(2″,3″,6′,7′); 3,4-methylenedioxy; β-carbonyl	(*6.98*)	Derriobtusone-B	*D. obtusa*	Leg	do Nascimento *et al.* (1976)
4,6,3′,4′			(*6.105*)	Aureusidin	*Melanorrhoea aptera*	Ana	Ogiyama *et al.* (1976)
		6-*O*-Glucose	(*6.107*)	Aureusin	*Antirrhinum orontium*	Scr	Gilbert (1975)
		6-*O*-Glucuronic acid	(*6.108*)		*Marchantia berteroana*	Hep	Markham and Porter (1978)
					M. polymorpha	Hep	Markham and Porter (1978)
					Conocephalum supra-decompositum	Hep	Markham and Porter (1978)
					Carrpos sphaerocarpos	Hep	Markham (1980)

6,3′,4′	4		(*6.106*)	Rengasin	*Melanorrhoea aptera*	Ana	Ogiyama *et al.* (1976)
					Rhus spp.	Ana	Young (1979)
					Amphipterygium adstrigens	Ana	Young (1976)
4,6,3′,4′,5′			(*6.109*)	Bracteatin	*Funaria hygrometrica*	Mus	Weitz and Ikan (1977)
		6-*O*-Glucose	(*6.110*)		*Antirrhinum orontium*	Scr	Gilbert (1975)
6,3′,4′,5′	4	6-*O*-Rhamnosylglucose	(*6.111*)	Subulin	*Amomum subulatum*	Zin	Lakshmi and Chauhan (1977)

* Key: Ana = Anacardiaceae, Com = Compositae, Hep = Hepaticae, Leg = Leguminosae, Mus = Musci, Scr = Scrophulariaceae, Zin = Zingiberaceae.

Extra hydroxylation at position 7 of sulfuretin yields maritimetin, which exists as maritimein (*6.103*) in *Coreopsis bigelovii* (Nicholls and Bohm, 1979), and in *C. grandiflora*, where it occurs with its 7-methyl ether, leptosin (*6.104*) (Crawford and Smith, 1980).

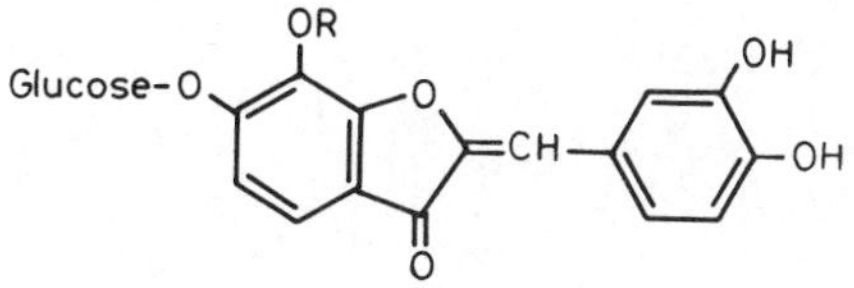

(*6.103*) R = H
(*6.104*) R = CH_3

Aureusidin, 4,6,3′,4′-tetrahydroxyaurone (*6.105*), occurs as the free phenol in heartwood of *Melanorrhoea aptera* (Anacardiaceae) along with its 4-methyl ether, rengasin (*6.106*) (Ogiyama *et al.*, 1976). Aureusin (*6.107*), the 6-*O*-glucoside of aureusidin, was found with bracteatin 6-*O*-glucoside (*6.110*) in *Antirrhinum orontium* (Scrophulariaceae) by Gilbert (1975), who discussed the observation in terms of the sectional disposition of this species. Aureusidin 6-*O*-glucuronide (*6.108*) has been found by Markham and Porter (1978) as a constituent of the antheridiophores of three liverworts: *Marchantia berteroana, M. polymorpha* and *Conocephalum supradecompositum.* The compound was not found in associated thallus tissue or in female plants and was detectable for only 4–8 weeks, which corresponds to the lifespan of the reproductive structures. Aureusidin 6-*O*-glucuronide has been found in another liverwort, *Carrpos sphaerocarpus* (Markham, 1980), as well.

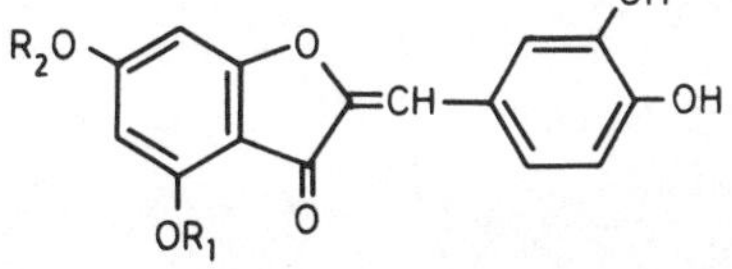

(*6.105*) $R_1 = R_2 = H$
(*6.106*) $R_1 = CH_3, R_2 = H$
(*6.107*) $R_1 = H, R_2 = Glc$
(*6.108*) $R_1 = H, R_2 = Glur$

(d) Aurones having three B-ring hydroxyls

Only two aurones having this pattern have been found in nature. Bracteatin (*6.109*) occurs as the free phenol in the sporophytes, but not in the leaves, of the moss *Funaria hygrometrica* (Weitz and Ikan, 1977) and as the 6-*O*-glucoside (*6.110*) in *Antirrhinum orontium* (Gilbert, 1975). 'Subulin', recently isolated from seeds of *Amomum subulatum* (Zingiberaceae) by Lakshmi and Chauhan (1977),

was shown to be the novel 6,3′,4′,5′-tetrahydroxy-4-methoxyaurone 6-*O*-α-L-rhamnopyranosyl-(1→4)-β-D-glucopyranoside (*6.111*).

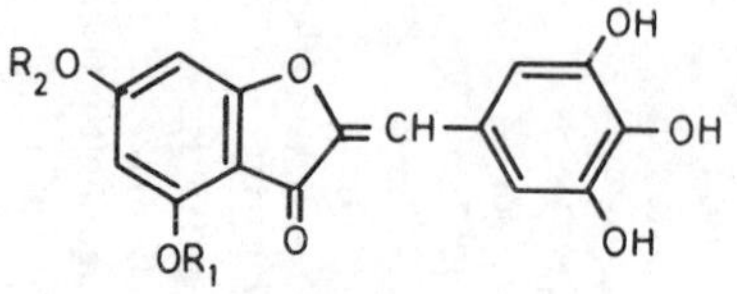

(*6.109*) $R_1 = R_2 = H$

(*6.110*) $R_1 = H, R_2 = Glc$

(*6.111*) $R_1 = CH_3, R_2 = -Glc-O-Rha$

6.4 DIHYDROCHALCONES

6.4.1 Introduction

For the identification of positions in the dihydrochalcone molecule, the numbering system is the same as that for chalcones [see structure (*6.1*)]. Included here is a group of compounds that bear oxygenation on the β-carbon and are referred to often as dibenzoylmethane derivatives in the literature.

6.4.2 Structures of naturally occurring dihydrochalcones

The structures of dihydrochalcones reported since 1974, including dibenzoylmethane derivatives, are given in Tables 6.4 and 6.5.

(a) Dihydrochalcones lacking B-ring hydroxyls

In the 1975 review only one dihydrochalcone belonging to this hydroxylation subgroup had been reported in nature, except for the existence of dihydrochalcone itself. The situation has changed markedly since that time. The simplest dihydrochalcone, in terms of hydroxylation pattern, is 2′,4′-dihydroxydihydrochalcone (*6.112*) which occurs as the 4′-(γγ-dimethyl)allyl ether (*6.113*) in *Lonchocarpus* species (Delle Monache *et al.*, 1974). A similar compound based

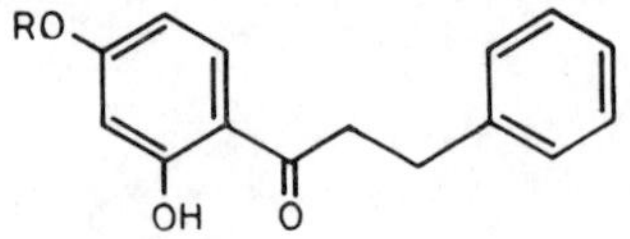

(*6.112*) R = H

(*6.113*) R =

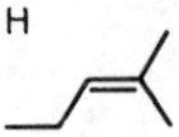

Table 6.4 Structures of naturally occurring dihydrochalcones

Substituents OH	OCH_3	Other	No. in text	Trivial name	Source*		References
2′		4′-*O*-Prenyl	(*6.113*)	Dihydrocordoin	*Lonchocarpus* sp.	Leg.	Delle Monache *et al.* (1974)
	2′,β	Furano(2″,3″,4′,3′)	(*6.156*)	Ovalitenin-B	*Milletia ovalifolia*	Leg	Gupta and Krishnamurti (1977b)
	2′	Furano(2″,3″,4′,3′); β-keto	(*6.149*)	Pongamol	*Pongamia glabra*	Leg	Talapatra *et al.* (1980)
					Tephrosia purpurea	Leg	Gupta *et al.* (1980)
2′,4′,6′			(*6.114*)		*Helichrysum tenuifolium*	Com	Bohlmann and Abraham (1979a)
		2′-*O*-Glucose	(*6.116*)		*Malus fusca*	Ros	A. H. Williams (1979)
		4′-*O*-Glucose	(*6.117*)		*Malus* cultivar	Ros	A. H. Williams (1979)
		4′-*O*-Prenyl	(*6.115*)		*Helichrysum polycladum*	Com	Bohlmann *et al.* (1980c)
2′,4′,6′		2′-*O*-Glucose; β-keto	(*6.145*)		*Malus* spp.	Ros	A. H. Williams (1979)
2′,6′	4′		(*6.118*)		*Pityrogramma* spp.	Fil	Wollenweber (1977b); Dietz *et al.* (1980)
					P. chrysoconia	Fil	Wollenweber (1977b, 1979)
					Notholaena spp.	Fil	Wollenweber (1977a)
					Adiantum sulphureum	Fil	Wollenweber (1979)
2′,4′	6′			Uvangoletin	*Uvaria angolensis*	Ano	Hufford and Oguntimein (1980)
2′	4′,6′	3′-*C*-Methyl	(*6.119*)		*Myrica gale*	Myr	Malterud *et al.* (1977)
2′,6′	4′	3′,5′-Di-*O*-methyl	(*6.120*)		*M. gale*	Myr	Anthonsen *et al.* (1971); Uyar *et al.* (1978)
2′,4′	6′	3′,5′-Di-*O*-methyl		Angoletin	*Uvaria angolensis*	Ano	Hufford and Oguntimein (1980)
	2′,6′,β	3′,4′-Methylenedioxy	(*6.157*)		*Helichrysum mundii*	Com	Bohlmann *et al.* (1978)
β	2′,6′	3′,4′-Methylenedioxy; carbonyl reduced to OH	(*6.158*, *6.159*)		*Helichrysum mundii*	Com	Bohlmann *et al.* (1978)
2′,4′,6′		3′-*C*-Methyl; 5′-formyl; β-keto	(*6.152*)		*Unona lawii*	Ano	Joshi and Gawad (1976)
2′,6′	4′	3′-*C*-Methyl; 5′-formyl; β-keto	(*6.153*)		*Unona lawii*	Ano	Joshi and Gawad (1976)
					U. angolensis	Ano	Hufford and Oguntimein (1980)
2′,4′	6′	3′-*C*-(2-Hydroxybenzyl)	(*6.131*)	Uvaretin	*Uvaria acuminata*	Ano	Cole *et al.* (1976)
					U. kiskii	Ano	Tammami *et al.* (1977)

2′,4′	6′	5′-C-(2-Hydroxybenzyl)		Isouvaretin or chaumuvarin	*U. chamae*	Ano	Okorie (1977)
					U. angolensis	Ano	Hufford and Oguntimein (1980)
2′,4′,6′	3′	5′-C-Prenyl	(*6.127*)		*Helichrysum polycladum*	Com	Bohlmann *et al* (1980c)
2′,4′,6′		3′,5′-di-C-Prenyl	(*6.128*)		*H. polycladum*	Com	Bohlmann *et al.* (1980c)
4′,6′		3′,3′,5′-tri-C-Prenyl; 2′-keto	(*6.129*)		*H. polycladum*	Com	Bohlmann *et al.* (1980c)
4′,6′	3′	3′,5′-di-C-Prenyl; 2′-keto	(*6.130*)		*H. polycladum*	Com	Bohlmann *et al.* (1980c)
					H. cymosum ssp. *calvum*	Com	Bohlmann *et al.* (1979c)
2′,4′,4				Davidogenin	*Viburnum davidii*	Cpr	Jensen *et al.* (1977)
		2′-O-Glucose	(*6.133*)	Davidioside	*Viburnum* spp.	Cpr	Jensen *et al.* (1977)
2′,4	4′		(*6.134*)	4′-O-Methyl-davidiogenin	*Viburnum* spp.	Cpr	Jensen *et al.* (1977)
2′,4′,4		β-Keto	(*6.147*)	Licodione	*Glycyrrhiza echinata*	Leg	Furuya *et al.* (1976)
2′,4′,6′,4			(*6.135*)		*Helichrysum splendidum*	Com	Bohlmann and Suwita (1979b)
		2′-O-Glucose	(*6.136*)	Phlorrhizin	*Malus* spp.	Ros	A. H. Williams (1979)
2′,4′,4	4′				*Rhododendron* spp.	Eri	Mabry *et al.* (1975); King (1977a,b)
		2′-O-Glucose	(*6.137*)	Asebotin	*Rhododendron* spp.	Eri	Mabry *et al.* (1975); King (1977a,b)
2′,4′	6′,4		(*6.138*)		*Iryanthera laevis*	Myt	Braz Filho *et al.* (1980)
2′,6′	4′,4		(*6.139*)		*Notholaena sulphurea*	Fil	Wollenweber (1977a)
2′,6′,4	4′	3′-C-Methyl	(*6.140*)		*Pityrogramma triangularis* var. *viscosa*	Fil	Wollenweber *et al.* (1979b)
2′,3′,4′,6′,4		5′-C-Geranyl	(*6.141*)		*Helichrysum monticola*	Com	Bohlmann and Zdero (1980)
2′,3′,4′,6′,4		5′-C-Neryl	(*6.142*)		*H. monticola*	Com	Bohlmann and Zdero (1980)
2′,4′,4	3′		(*6.143*)		*Iryanthera polyneura*	Myt	de Almeida *et al.* (1979)
2′,4′,6′,3,4			(*6.144*)		*Balanophora tobiracola*	Bal	Ito *et al.* (1980)
		4′-O-Glucose			*B. tobiracola*	Bal	Ito *et al.* (1980)
	2′	Furano(2″,3″,4′,3′); 3,4-methylenedioxy; β-keto	(*6.148*)	Ovalitenone	*Pongamia glabra*	Leg	Garg *et al.* (1978); Sharma and Parthasarathy (1977); Talapatra *et al.* (1980)
					Milletia ovalifolia	Leg	Gupta and Krishnamurti (1977b)
2′,4′,3,5,β	4		(*6.160*)		*Gliricidia sepium*	Leg	Manners and Jurd (1979)

* Key: Ano = Anonaceae, Bal = Balanophoraceae, Com = Compositae, Cpr = Caprifoliaceae, Eri = Ericaceae, Fil = Filicinae, Leg = Leguminosae, Myr = Myricaceae, Myt = Myristicaceae, Ros = Rosaceae.

Table 6.5 Structures of dihydrochalcones having complex substituents

*Names**	*No. in text*	*Source†*		*References*
2′,4′,4′-Tri-*C*-methylcyclopentenedione-A-ring	*(6.122)*	*Myrica gale*	Myr	Anthonsen *et al.* (1971)
3′,5′,5′-Tri-*C*-methylcyclohexanetrione-A-ring	*(6.121)*	*M. gale*	Myr	Uyar *et al.* (1978)
Cyclic 7-ring dihydrochalcone	*(6.123)*	*Radula variabilis*	Hep	Asakawa *et al.* (1978)
Chamuvaritin	*(6.132)*	*Uvaria chamae*	Ano	Okorie (1977)
5,7-Dihydroxy-8-(3-phenylpropionyl)-4-phenyl-dihydrocoumarins		*Pityrogramma* spp.	Fil	Wagner *et al.* (1979); Dietz *et al.* (1980)

* Trivial names have not been proposed for entries other than chamuvaritin.

† Key: Ano = Anonacaae, Fil = Filicinae, Hep = Hepaticeae, Myr = Myricaceae.

upon a phloroglucinol A ring (*6.115*) has been found in *Helichrysum polycladum* (Bohlmann *et al.*, 1980c) where it occurs with other prenylated flavonoids. 2′,4′,6′-Trihydroxydihydrochalcone (*6.114*), itself, occurs in *H. tenuifolium* (Bohlmann and Abraham, 1979a), while its 2′-*O*- and 4′-*O*-glucosides, (*6.116*) and (*6.117*) respectively have been reported in several taxa of *Malus* (C. A. Williams, 1979). 2′,6′-Dihydroxy-4′-methoxydihydrochalcone (*6.118*) has

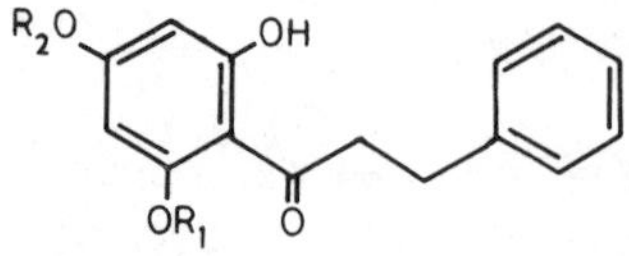

(*6.114*) $R_1 = R_2 = H$

(*6.115*) $R_1 = H$, $R_2 =$

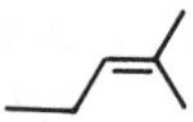

(*6.116*) $R_1 =$ Glc, $R_2 = H$

(*6.117*) $R_1 = H$, $R_2 =$ Glc

(*6.118*) $R_1 = H$, $R_2 = CH_3$

been found as a constituent of the exudate of white farinose forms of members of the following genera of ferns: *Pityrogramma* (Wollenweber, 1977b, 1979; Dietz *et al.*, 1980), *Notholaena* (Wollenweber, 1977a) and *Adiantum* (Wollenweber, 1979).

Higher levels of methylation, including *C*-methylation, are seen in several of the compounds isolated from *Myrica gale* (Myricaceae). Relatively simple members of this group include 2′,4′-dimethoxy-5′-methyl-6′-hydroxydihydrochalcone (*6.119*) (Malterud *et al.*, 1977) and 2′,6′-dihydroxy-3′, 5′-di-*C*-methyl-4′-methoxydihydrochalcone (*6.120*) (Anthonsen *et al.*, 1971; Uyar *et al.*, 1978).

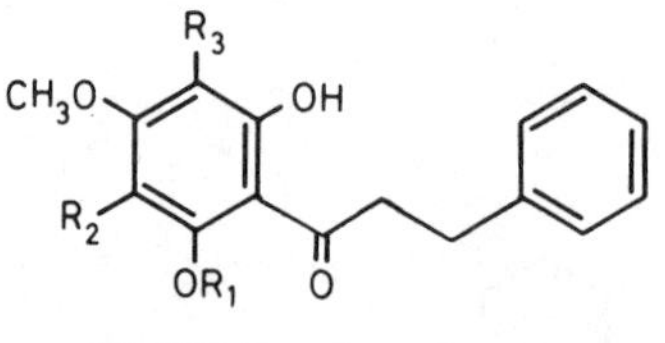

(*6.119*) $R_1 = R_3 = CH_3$, $R_2 = H$

(*6.120*) $R_1 = H$, $R_2 = R_3 = CH_3$

Two *C*-methylations at position 3′ make aromatization impossible and lead to the triketo 'dihydrochalcone' (*6.121*) discovered by Uyar and co-workers (1978). A most unusual dihydrochalcone is shown as structure (*6.122*). This compound, also from *M. gale* (Anthonsen *et al.*, 1971), is reminiscent of the contracted A-ring chalcones linderone and methyl-linderone isolated by Kiang and co-workers from *Lindera pipericarpa* (see Bohm, 1975, p. 454). In the case of the compounds from *Myrica* it would be of interest to see if the five-membered ring of (*6.122*) arises through loss of carbon-6′ from (*6.121*). There seems to be no information

available concerning the formation of contracted ring flavonoids, although the number is admittedly very small.

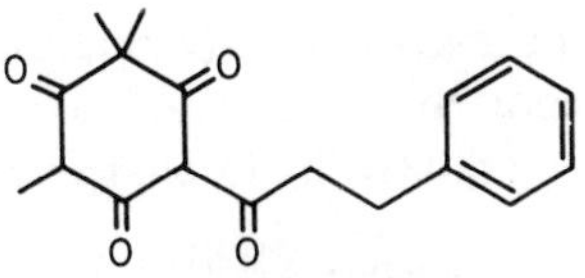

(6.121) 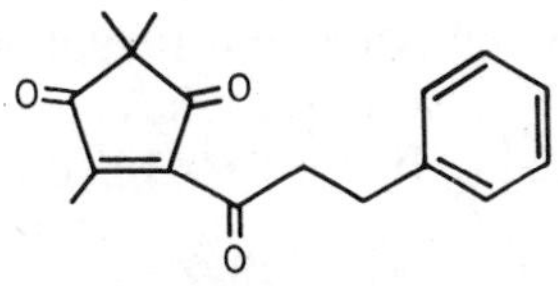 (6.122)

At the time at which the earlier review was prepared *C*-alkylated dihydrochalcones were limited to a small number of *C*-glycosyl derivatives. More recently several *C*-prenylated dihydrochalcones have been isolated from sources as varied as the liverwort *Radula variabilis* to species of *Helichrysum*. The *C*-alkylated compound from *Radula* is shown as structure (*6.123*). Asakawa and his co-workers (1978) also isolated a series of bibenzyls from *Radula*, the structures of which are shown as (*6.124*), (*6.125*) and (*6.126*) for comparison. The seven-membered ring closure is unusual and has not been reported before in any flavonoid type covered in this chapter and may be unique amongst flavonoids in general.

Recent work by Bohlmann and colleagues (1979c, 1980c) has shown a high level of *C*-alkylation in species of *Helichrysum*. Compound (*6.127*) results from

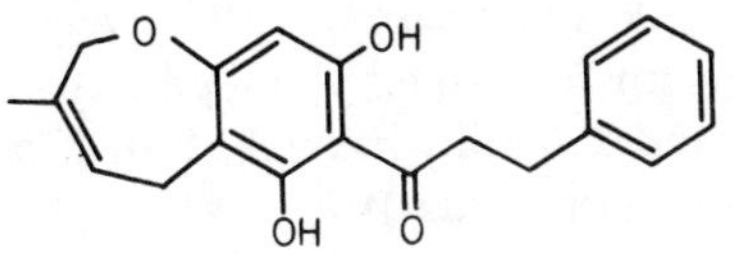

(6.123) 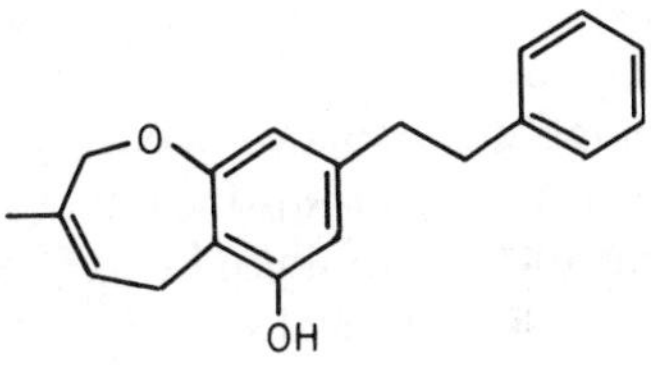 (6.124)

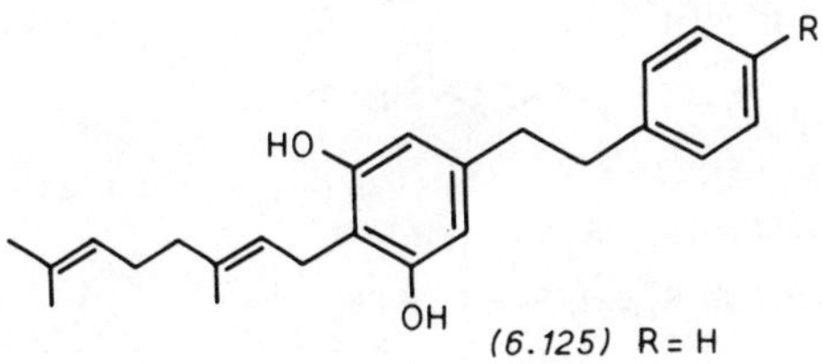

(6.125) R = H
(6.126) R = OH

C-prenylation, extra hydroxylation and *O*-methylation (not necessarily in that order) while (*6.128*) would result simply from two *C*-prenylations. Compound (*6.129*) is unique in having three prenyl substitutions on the A ring, two of which at position 3′ prevent formation of the aromatic ring. In compound (*6.130*) a prenyl group and a methoxy function occupy position 3′ which also prevents aromatization.

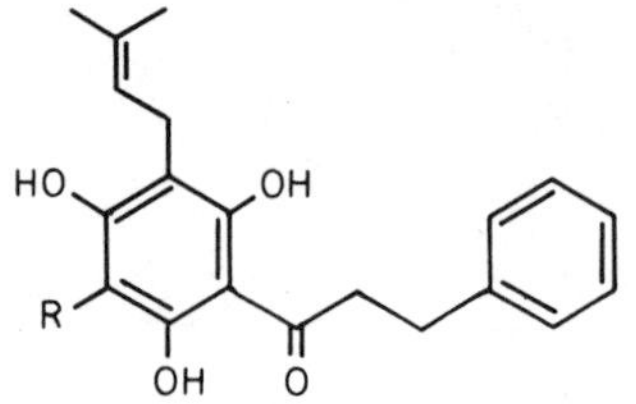

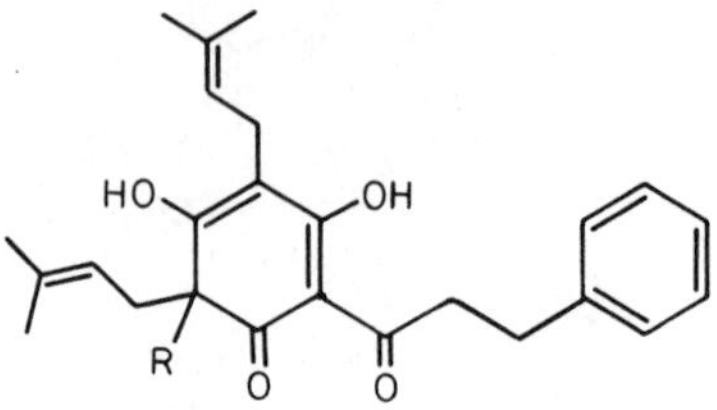

(6.127) R = OCH_3

(6.128) R = [prenyl]

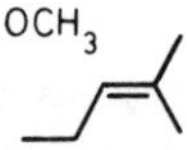

(6.129) R = [prenyl]

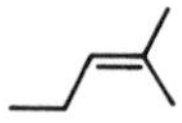

(6.130) R = OCH_3

Studies of the flavonoids of *Uvaria* (Anonaceae) have shown the presence of novel *C*-benzyldihydrochalcones in several species. Cole and co-workers (1976) established the structure of 'uvaretin', isolated from *U. acuminata*, as 2′,4′-dihydroxy-3′-(2-hydroxybenzyl)-6′-methoxydihydrochalcone (*6.131*).

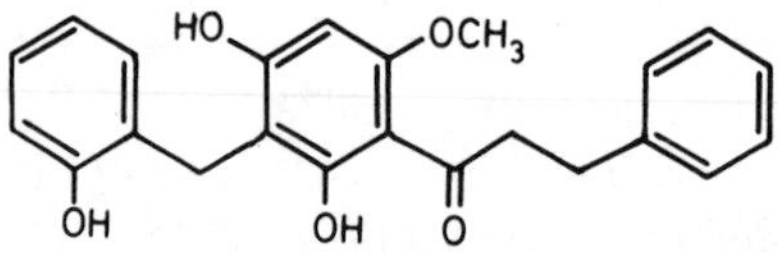

(6.131)

Uvaretin showed inhibitory activity against the P-388(3PS) lymphocytic leukaemia test system. Tammami and co-workers (1977) found uvaretin in *U. kiskii*, while Okorie (1977) isolated the 5′-(2-hydroxybenzyl) isomer of (*6.131*) from *U. chamae* and called in 'chamuvarin.' Okorie (1977) also found the 3′,5′-dibenzyl derivative 'chamuvaretin' (*6.132*) in *U. chamae*. In (*6.132*) ring closure has occurred involving the phenolic groups at the 4′ position of the flavonoid and on the 2-hydroxybenzyl substituent.

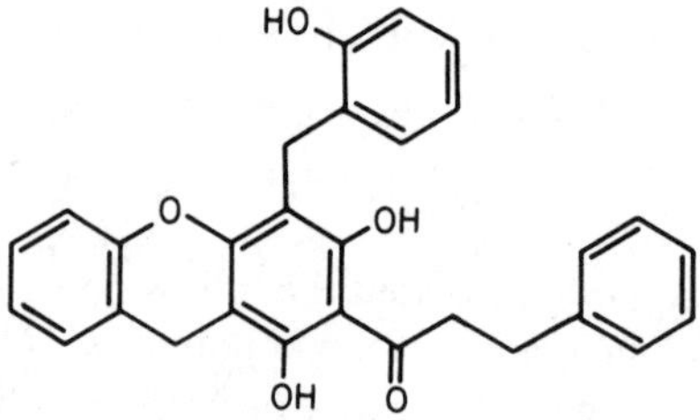

(6.132)

(b) Dihydrochalcones having one B-ring hydroxyl

Jensen and colleagues (1977) have confirmed the existence of 'davidoside' (*6.133*) in *Viburnum davidii* (Caprifoliaceae), reported its presence in *V. lantanoides*, and demonstrated that 4′-*O*-methyldavidogenin (*6.134*) is also a constituent of both species.

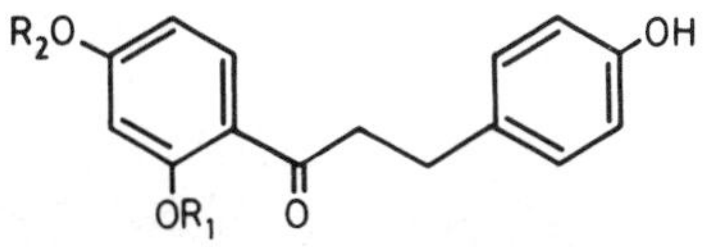

(6.133) R_1 = Glc, R_2 = H
(6.134) R_1 = H, R_2 = CH_3

4,2′,4′,6′-Tetrahydroxydihydrochalcone (*6.135*) occurs as the free phenol in *Helichrysum splendidum* (Bohlmann and Suwita, 1979b) and in the form of the well-known 2′-*O*-glucoside, 'phlorrhizin (phloridzin)' (*6.136*), in *Malus* species (A. H. Williams, 1979). Phlorrhizin 4′-methyl ether (*6.137*), commonly called 'asebotin', occurs in several species of *Rhododendron* (King, 1977a,b; Mabry *et al.*, 1975b). The 4,6′-dimethyl ether (*6.138*) occurs as a trunk-wood constituent of *Iryanthera laevis* (Myristicaceae) (Braz Filho *et al.*, 1980), while the 4,4′-dimethyl ether (*6.139*) occurs in the farina of the white form of *Notholaena sulphurea* (Wollenweber, 1977a). 4,2′,6′-Trihydroxy-3′-methyl-4′-methoxydihydrochalcone (*6.140*) also occurs as a constituent of the exudate of *Pityrogramma triangularis* var. *viscosa* (Wollenweber *et al.*, 1979b).

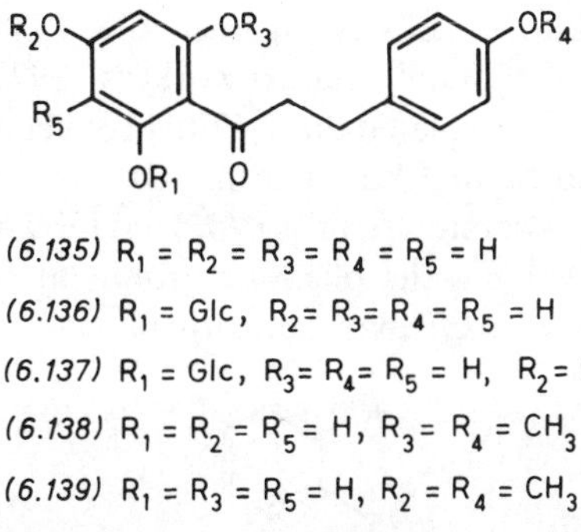

(6.135) $R_1 = R_2 = R_3 = R_4 = R_5$ = H
(6.136) R_1 = Glc, $R_2 = R_3 = R_4 = R_5$ = H
(6.137) R_1 = Glc, $R_3 = R_4 = R_5$ = H, R_2 = CH_3
(6.138) $R_1 = R_2 = R_5$ = H, $R_3 = R_4$ = CH_3
(6.139) $R_1 = R_3 = R_5$ = H, $R_2 = R_4$ = CH_3
(6.140) $R_1 = R_3 = R_4$ = H, $R_2 = R_5$ = CH_3

Bohlmann and Zdero (1980) isolated the following pair of isomeric dihydrochalcones from *Helichrysum monticola*: compound (*6.141*) is the *C*-geranyl and compound (*6.142*) is the *C*-neryl derivative of the parent dihydrochalcone of this subgroup.

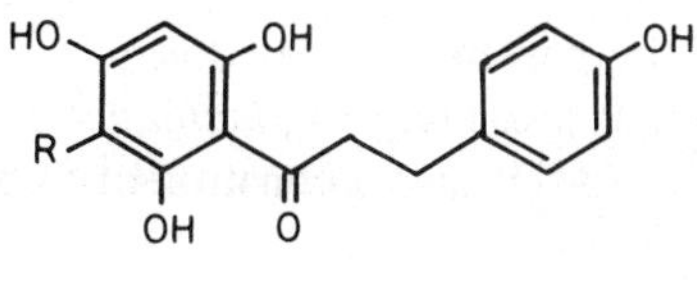

(6.141) R = 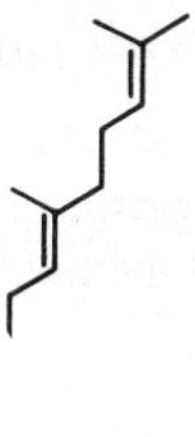
(6.142) R =

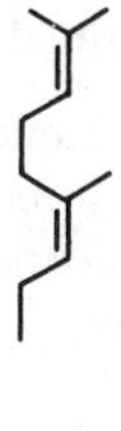

(c) Dihydrochalcones having two B-ring hydroxyls

Only two representatives of this subgroup have been reported since the 1975 review. 2′,4,4′-Trihydroxy-3-methoxydihydrochalcone (*6.143*) is known to occur in *Iryanthera polyneura* (de Almeida *et al.*, 1979) along with other flavonoids and diarylpropanoids having the same or related substitution patterns. The other member of this subgroup is the unsubstituted pentahydroxydihydrochalcone (*6.144*) which is found in *Balanophora tobiracola* (Balanophoraceae) along with its 4′-*O*-glucoside (Ito *et al.*, 1980). This represents the first report of a dihydrochalcone from Balanophoraceae.

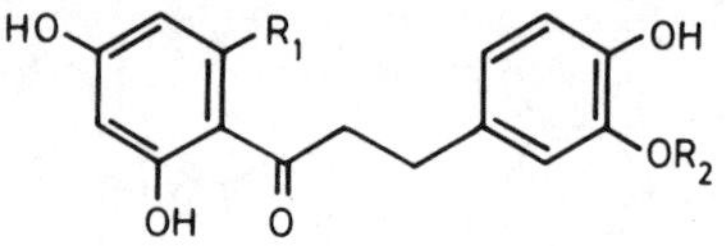

HO R1 OH OR2 OH O

(*6.143*) R_1 = H, R_2 = CH_3

(*6.144*) R_1 = OH, R_2 = H

6.4.3 Structures of naturally occurring dihydrochalcones having β-oxygenation

In 1967 A. H. Williams proposed the structure (*6.145*), 2,4,6-trihydroxydibenzoylmethane 4-glucoside, for a compound that he isolated from *Malus*. In 1972 Chadenson and co-workers suggested that the compound should in fact be written in the form of a 2-hydroxyflavanone (*6.146*). A similar structure having a 7-methoxyl function was also proposed by Chandenson *et al.* (1972) for a substance isolated from *Populus niger*. Since the time of those reports, several other groups have described the isolation of dibenzoylmethane derivatives from natural sources. In a continuation of his study of phenolic compounds in *Malus*

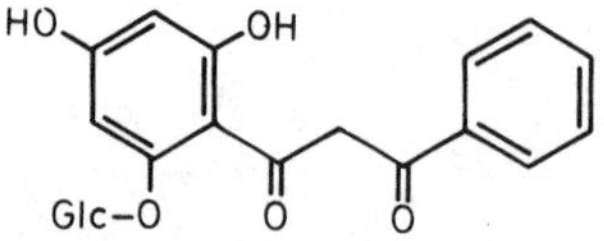

HO OH Glc–O O O

(*6.145*)

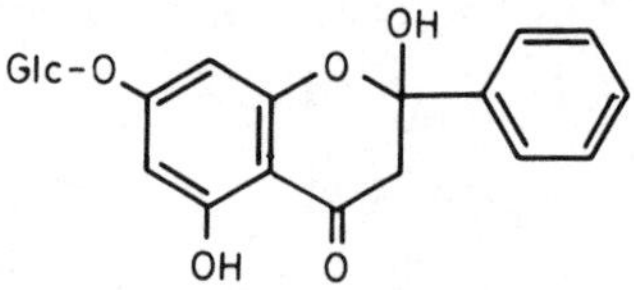

OH Glc–O O OH O

(*6.146*)

A. H. Williams (1979) reported the occurrence of (*6.145*) from six species. 'Licodione', isolated from the roots and from cultured cells of *Glycyrrhiza echinata* by Furuya *et al.* (1976), is 2,4,4′-trihydroxybenzoylmethane (*6.147*). 'Ovalitenone' (*6.148*) is known from both *Milletia ovalifolia* (Gupta and Krishnamurti, 1977b) and *Pongamia glabra* (Talapatra *et al.*, 1980; Garg *et al.*, 1978; Sharma and Parthasarthy, 1977) where it occurs with 'pongamol' (*6.149*) (Talapatra *et al.*, 1980). Traub and Geiger (1975) reported the existence of 2,5,7,3′,4′-pentahydroxyflavanone 5-*O*-glucoside in seeds of *Galega officinalis*

(Leguminosae), but represented it as being in equilibrium with the corresponding dibenzoylmethane glucoside (*6.151*).

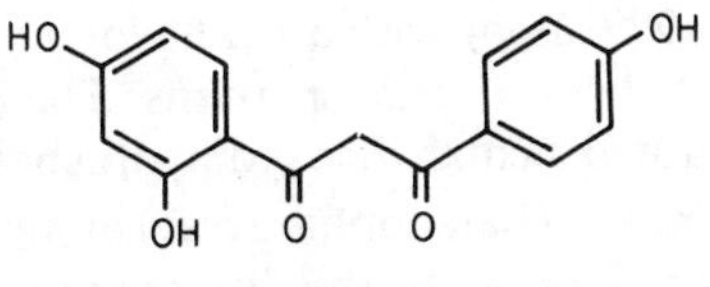

(*6.147*)

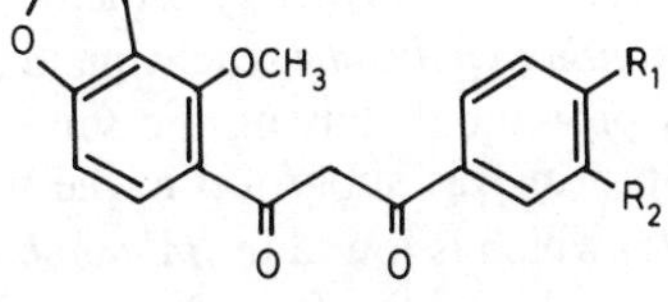

(*6.148*) R_1 and R_2 = $-O-CH_2-O-$

(*6.149*) $R_1 = R_2 = H$

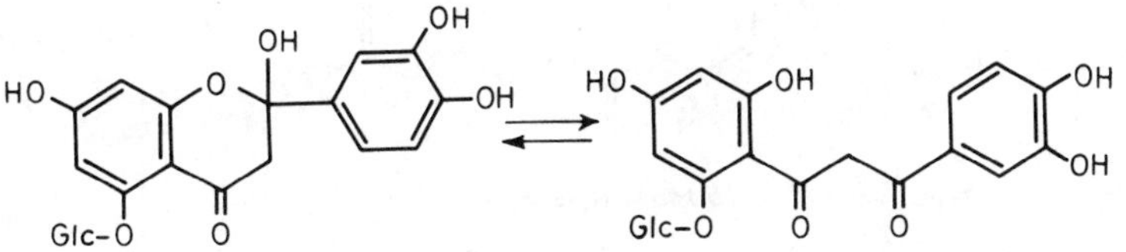

(*6.150*)

(*6.151*)

Compounds (*6.152*) and (*6.153*) were reported as constituents of *Unona lawii* (Anonaceae) by Joshi and Gawad (1976), A reinvestigation of these structures by Chopin and Hauteville in collaboration with Joshi and Gawad (Chopin *et al.*, 1978) showed that when the 4 position is occupied by a methoxyl function (*6.153*) the natural product exists in the dibenzoylmethane form, while the presence of the unsubstituted phenolic function at that position is correlated with the existence of the compound as a mixture of the two cyclic hemiketals (= 2-hydroxyflavanones) (*6.154*) and *6.155*).

The explanation for the existence of these compounds in the 2-hydroxyflavanone form would seem to be more complex than whether position 4 has a free phenolic group or whether it is substituted. While the 4-substituted compound (*6.153*) occurs in the dibenzoylmethane form, the compound from

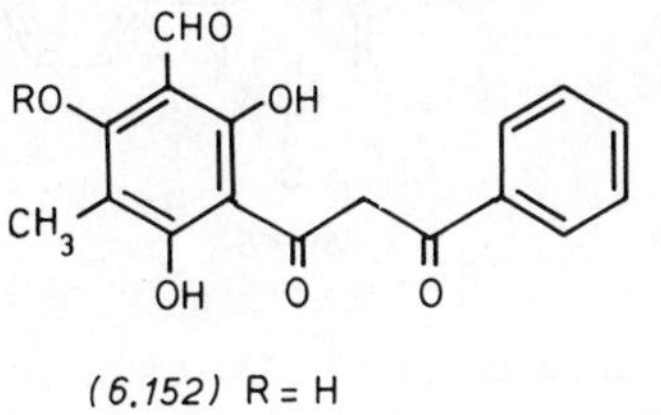

(*6.152*) R = H

(*6.153*) R = CH_3

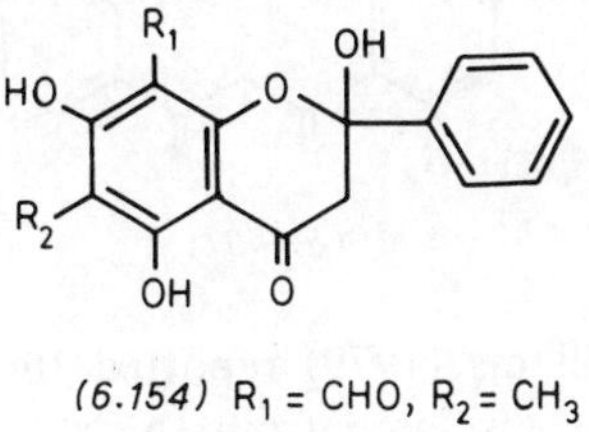

(*6.154*) R_1 = CHO, R_2 = CH_3

(*6.155*) R_1 = CH_3, R_2 = CHO

Malus, with 4-*O*-glucose, and the compound from *Populus*, with 4-*O*-methyl, appear to exist in the 2-hydroxyflavanone form. Studies with model compounds could be instructive in resolving this problem.

Three dihydrochalcones having β-hydroxyl functions have been described in the recent literature. 'Ovalitenin-B' (*6.156*) occurs in *Milletia ovalifolia* (Gupta

and Krishnamurti, 1977b) along with several other flavonoids having similar substitution patterns. It seems reasonable to suggest biosynthesis from a B-ring unsubstituted precursor which resembles pongamol (*6.149*) through reduction of the β-keto function followed by *O*-methylation. The structurally very similar (*6.157*) is known from *Helichrysum mundii* (Bohlmann *et al.*, 1978) where it occurs with the dihydroxy compounds (*6.158*) and (*6.159*). These compounds are not, strictly speaking, flavonoids, but they are sufficiently similar structurally to suggest close biosynthetic ties.

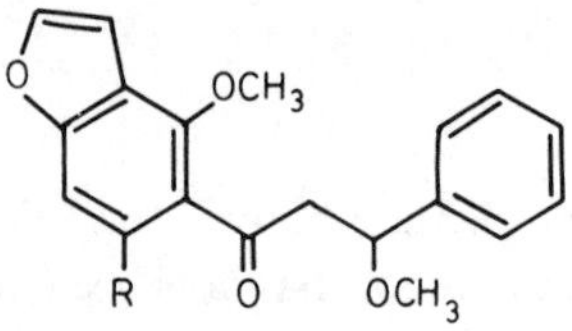

(*6.156*) R = H

(*6.157*) R = OCH_3

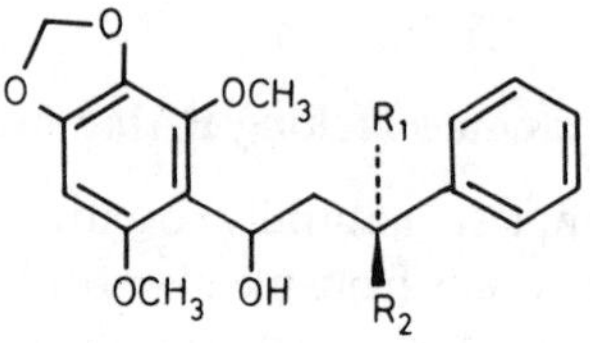

(*6.158*) R_1 = H, R_2 = OH

(*6.159*) R_1 = OH, R_2 = H

Finally, the β-hydroxy compound (*6.160*) has been reported by Manners and Jurd (1979) to occur in *Gliricidia sepium* along with the dihydroflavonol 'sepinol', described below, which is characterized by the same, rare *O*-methylation pattern, *viz.* 3′,5′-dihydroxy-4′-methoxy B ring.

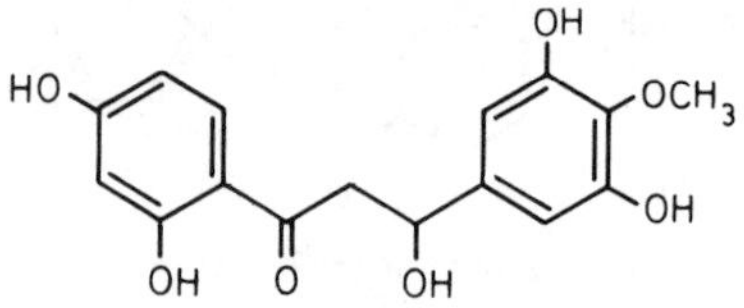

(*6.160*)

6.5 FLAVANONES

6.5.1 Introduction

The numbering system for flavanones uses primed numbers for the A ring and unprimed numbers for the B ring. Since carbon-2 of the flavanone molecule is a centre of asymmetry, two isomeric forms of each structure are possible. Evidence leading to the establishment of the (2*S*) configuration for the naturally occurring flavanones was presented in the earlier review (Bohm, 1975, pp. 594–595). Not all investigators cited in the present work determined the optical properties of flavanones with which they were working so configurations are not known for some compounds. In many cases newly discovered glycosides or methyl ethers are derived from flavanones of known configuration. In these cases I have taken the liberty of assigning these compounds the configuration of the parent compound. In other cases the (2*S*) configuration has been assigned on the basis of

reported laevorotation. An asterix above a bond (e.g. *6.165*) means that no stereochemical assignment is possible based upon the available data. Obviously, more detailed study is required for these compounds. Optical properties of flavanones for which data are available are summarized in Table 6.8.

6.5.2 Structures of naturally occuring flavanones

Structures of flavanones reported since 1974 are summarized in Tables 6.6 and 6.7.

(a) Flavanones lacking B-ring hydroxyls

The simplest naturally occurring flavanone is compound (*6.161*) known previously only from members of Leguminosae. A report has now been published describing its occurrence in *Flourensia heterolepis* (Compositae) where it occurs with its methyl ether (*6.162*) (Bohlmann and Jakupovic, 1979). Hydroxylation at position 8 followed by *O*-methylation affords compound (*6.163*) isolated from *Zuccagnia punctata* (Leguminosae) by Pederiva *et al.* (1975) while the dimethyl ether (*6.164*) was isolated from the leaves of *Tetragonia expansa* (Aizoaceae) (Kemp *et al.*, 1979).

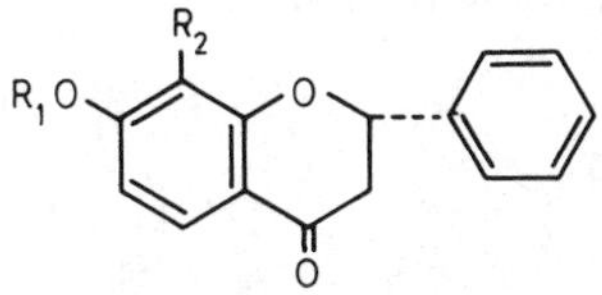

(*6.161*) $R_1 = R_2 = H$

(*6.162*) $R_1 = CH_3$, $R_2 = H$

(*6.163*) $R_1 = H$, $R_2 = OCH_3$

(*6.164*) $R_1 = CH_3$, $R_2 = OCH_3$

Prenylation of the parent compound at position 8 affords flavanone (*6.165*) while prenylation at both positions 6 and 8 gves (*6.166*); both compounds occur in *Milletia ovalifolia* where they occur with the dimethylchromenoflavanone (*6.167*) (Gupta and Krishnamurti, 1976a). 'Isolonchocarpin' (*6.168*) is the product of cyclization of the prenylated flavanone (*6.165*) and has been found in

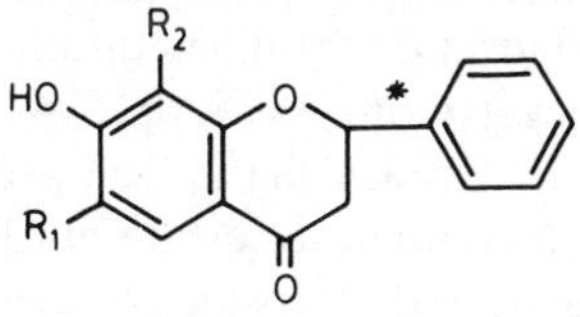

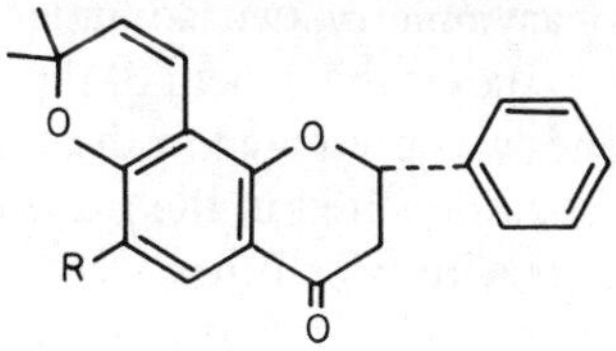

(*6.165*) $R_1 = H$, $R_2 =$ prenyl

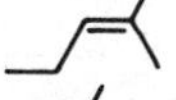

(*6.166*) $R_1 = R_2 =$ prenyl

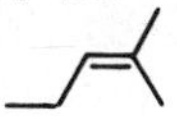

(*6.167*) $R = OCH_3$

(*6.168*) $R = H$

Table 6.6 Structures of naturally occurring flavanones

Substituents: OH	Substituents: OCH_3	Substituents: Other	No. in text	Trivial name	Source*		References
7			(*6.161*)		*Flourensia heterolepis*	Com	Bohlmann and Jakupovic (1979)
	7		(*6.162*)		*F. heterolepis*	Com	Bohlmann and Jakupovic (1979)
7	8		(*6.163*)		*Zuccagnia punctata*	Leg	Pederiva *et al.* (1975)
	7,8		(*6.164*)		*Tetragonia expansa*	Aiz	Kemp *et al.* (1979)
		6″,6″-Dimethyl-pyrano(2″,3″,7,8)	(*6.168*)	(-)-Isoloncho-carpin	*Tephrosia purpurea*	Leg	R. K. Gupta *et al.* (1980); Rao and Raju (1979)
					Pongamia glabra	Leg	Satam and Brinji (1973); Talapatra *et al.* (1980)
	6	6″,6″-Dimethyl-pyrano(2″,3″,7,8)	(*6.167*)	Ovali-chromene	*Milletia ovalifolia*	Leg	Gupta and Krishnamurti (1976b)
7		6,8-di-*C*-Prenyl	(*6.166*)		*M. ovalifolia*	Leg	Gupta and Krishnamurti (1976a)
7		8-*C*-Prenyl	(*6.165*)		*M. ovalifolia*	Leg	Gupta and Krishnamurti (1976a)
5,7			(*6.170*)	Pino-cembrin	*Chrysothamnus nauseusus*	Com	Bohlmann *et al.* (1979a)
					Glycyrrhiza glabra	Leg	Kaltaev and Nikonov (1974)
					Helichrysum calliconum	Com	Bohlmann and Abraham (1979b)
					H. oreophyllum	Com	Bohlmann *et al.* (1980c)
					H. tenuifolium	Com	Bohlmann and Abraham (1979a)
					Iryanthera polyneura	Myt	de Almeida *et al.* (1979)
					Larix dahurica	Cnf	Lapteva *et al.* (1974)
					Prunus domestica	Ros	Nagarajan and Parmar (1977b)
					Tarchonanthus trilobus	Com	Bohlmann and Suwita (1979a)
		7-*O*-Rhamnose			*Polystichum lepidocaulon*	Fil	Hiraoka (1978)
		7-*O*-Neohesperidose			*Nierembergia hippomanica*	Sol	Pomilio and Gros (1979)
7	5		(*6.171*)	Alpinetin	*Chrysothamnus nauseusus*	Com	Bohlmann *et al.* (1979a)
					Helichrysum herbaceum	Com	Bohlmann *et al.* (1979e)
					Dalea scandens var. *paucifolia*	Leg	Dominguez *et al.* (1980)

Table 6.6 (*Contd.*)

OH	*Substituents* *OCH*$_3$	*Other*	*No. in text*	*Trivial name*	*Source**		*References*
5	7		(*6.172*)	Pino-strobin	*Aniba riparia*	Lau	Fernandes *et al.* (1978)
					Helichrysum polycladum	Com	Bohlmann *et al.* (1980c)
					Larix dahurica	Cnf	Lapteva *et al.* (1974)
					Prunus cerasus	Ros	Nagarajan and Parmar (1977c)
					Agonis spathulata	Myr	Cannon and Martin (1977)
	5,7		(*6.173*)	Dimethyl-pinocembrin	*Aniba riparia*	Lau	Fernandes *et al.* (1978)
5	6,7		(*6.174*)		*Chrysothamus nauseusus*	Com	Bohlmann *et al.* (1979a)
5,7	8		(*6.175*)	Dihydro-wogonin	*Prunus cerasus*	Ros	Nagarajan and Parmar (1977c)
	5,7,8		(*6.176*)		*Popowia cauliflora*	Ano	Panichpol and Waterman (1978)
7	5,6,8		(*6.178*)	'Didymo-carpin'	*Didymocarpus pedicellata*	Ges	Bose and Adityachaudhury (1978); Garg *et al.* (1979)
	5,6,7,8		(*6.177*)	Kanakugin	*Popowia cauliflora*	Ano	Waterman and Pootakahm (1979)
5,7		6-*C*-Methyl	(*6.179*)	Strobopinin	*Pityrogramma pallida*	Fil	Wollenweber *et al.* (1979a)
5	7	6-*C*-Methyl	(*6.182*)	7-Methyl-strobopinin	*P. triangularis*	Fil	Star *et al.* (1978)
5,7		8-*C*-Methyl	(*6.180*)	(-)-Crypto-strobin	*Agonis spathulata*	Myr	Cannon and Martin (1977)
5,7		6,8-Di-*C*-methyl	(*6.181*)	(-)-Des-methoxy-matteucinol	*Agonis spathulata*	Myr	Cannon and Martin (1977)
					Dalea polyadenia	Leg	Dreyer *et al.* (1975)
					Pityrogramma pallida	Fil	Wollenweber *et al.* (1979a)
					Unona lawii	Ano	Joshi and Gawad (1974)
		7-*O*-Glucose			*Matteuccia* spp.	Fil	Hiraoka (1978)
5	7	6,8-Di-*C*-methyl	(*6.183*)		*Unona lawii*	Ano	Joshi and Gawad (1974)
5,7,2		6-(or 8)-*C*-Methyl; 8-(or 6) formyl			*Unona lawii*	Ano	Chopin *et al.* (1978)

5,7		6-Formyl; 8-*C*-Methyl	(*6.184*)	Lawinal	*Unona lawii*	Ano	Joshi and Gawad (1974)
5,7		6-*C*-Prenyl	(*6.185*)		*Chlorophora tinctoria*	Mor	Gottlieb *et al.* (1975)
					Helichrysum polycladum	Com	Bohlmann *et al.* (1980c)
					Derris floribunda	Leg	Braz Filho *et al.* (1975b)
					D. rariflora	Leg	Braz Filho *et al.* (1975a)
5	7	6-*C*-Prenyl	(*6.186*)		*D. rariflora*	Leg	Braz Filho *et al.* (1975a)
	5	6″,6″-Dimethyl-pyrano(2″,3″,7,8)	(*6.192*)		*Tephrosia praecans*	Leg	Camele *et al.* (1980)
5,7		8-*C*-Prenyl	(*6.187*)	Glabranin	*Helichrysum hypocephalum*	Com	Bohlmann and Abraham (1979d)
					Platitenia absinthifolia	?	Ubaev *et al.* (1974)
5,7		8-*C*-(β-Methyl-β-hydroxy-methylallyl)	(*6.188*)		*Helichrysum hypocephalum*	Com	Bohlmann and Abraham (1979d)
5,7		8-*C*-(β-Methyl-β-formylallyl)	(*6.189*)		*H. hypocephalum*	Com	Bohlmann and Abraham (1979d)
5,7		8-*C*-Geranyl	(*6.190*)		*H. hypocephalum*	Com	Bohlmann and Abraham (1979d)
5	7	8-*C*-Prenyl	(*6.191*)	7-Methyl-glabranin	*Tephrosia villosa*	Leg	Jayaraman *et al.* (1980)
5,7		8-*C*-(2-Hydroxybenzyl)	(*6.194*)		*Uvaria chamae*	Ano	Hufford and Laswell (1976)
5,7		6-*C*-(2-Hydroxybenzyl)	(*6.195*)		*Uvaria chamae*	Ano	Hufford and Laswell (1976)
6,8		7-*C*-(5-Hydroxy-geranyl)	(*6.197*)	Flemi-flavanone-B	*Flemingia stricta*	Leg	Sivarambabu *et al.* (1979a)
7,4′			(*6.202*)	Liquiri-tigenin	*Dalbergia* spp.	Leg.	Matos *et al.* (1975); Manners *et al.* (1974); Parthasarathy *et al.* (1976); Letcher and Shirley (1976)
					Medicago spp.	Leg	Ingham (1979)
					Myroxylon balsamum	Leg	de Oliveira *et al.* (1978)
					Onobrychis viciifolia	Leg	Ingham (1978)
					Peltogyne spp.	Leg	Malan and Roux (1974)
7,4′		8-*C*-Galactose	(*6.203*)	Aervanone	*Aerva persica*	Amr	Garg *et al.* (1980)
7,4′		6,8-Di-*C*-prenyl	(*6.204*)		*Sophora subprostrata*	Leg	Jain and Gupta (1978)
5,7,4′			(*6.205*)	Naringenin	*Baccharis alaternoides*	Com	Bohlmann *et al.* (1979b)
					Cochlospermum gillivraei	Coc	Cook and Knox (1975)
					Dillenia indica	Dil	Pavanasasivan and Sultanbawa (1975)

Table 6.6 (*Contd.*)

OH	*Substituents* *OCH₃*	*Other*	*No. in text*	*Trivial name*	*Source**		*References*
					Equisetum arvense	Equ	Syrchina *et al.* (1975, 1978)
					Larix dahurica	Cnf	Lapteva *et al.* (1974)
					Lygos raetan	Leg	El Sherbeiny *et al.* (1978)
					Matthiola incana	Cru	Forkmann (1979)
					Nothofagus antartica	Fag	Wollenweber and Wiermann (1979)
					Othonna euphorboides	Com	Bohlmann and Knoll (1978)
					Persica vulgaris	Ros	Sadykov and Sadykov (1979)
					Picea ajanensis	Cnf	Ivanova *et al.* (1975)
					Polymnia fructicosa	Com	Bohlmann and Zdero (1977)
					Prunus domestica	Ros	Nagarajan and Parmar (1977b)
					Rhynchosia beddomei	Leg	Adinarayana *et al.* (1980b)
					Senecio longilinguae	Com	Bohlmann and Zdero (1979)
					Soymida febrifuga	Mel	Rao *et al.* (1979)
					Steiractinia mollis	Com	Bohlmann *et al.* (1980b)
		5-*O*-Glucose			*Persica vulgaris*	Ros	Sadykov and Sadykov (1979)
					Salix elbursensis	Sal	Kompantsev and Shinkarenko (1975)
		7-*O*-Glucose			*Dryopteris* spp.	Fil	Hiraoka (1978)
					Matthiola incana	Cru	Forkmann (1979)
					Coriaria spp.	Cor	Bohm and Ornduff (1981)
					Cochlospermum gillivraei	Coc	Cook and Knox (1975)
		7-*O*-Glucose-6″-galloyl			*Acacia farnesiana*	Leg	El Sissi *et al.* (1974)
		7-*O*-Glucose-6″-*O*-*p*-coumaroyl			*Anacardium occidentale*	Ana	Rahman *et al.* (1978)
		7-*O*-Rhamnose			*Litsea glutinosa*	Lau	Mohan *et al.* (1978)
		?-*O*-Fructose			*Camellia sinensis*	The	Imperato (1976)
		4′-*O*-Galactose			*Cleome viscosa*	Cpr	Chauhan *et al.* (1979b)
		4′-*O*-Xylosylglucose			*C. viscosa*	Cpr	Srivastava *et al.* (1979)
		4′-*O*-Rutinose			*Bauhinia variegata*	Leg	Gupta *et al.* (1979)
		7-*O*-Rutinose		Narirutin	*Citrus* hybrids	Rut	Kamiya *et al.* (1979)

		7-*O*-Neohesperidose		Naringin	*Adiantum* spp.	Fil	Cooper-Driver and Swain (1977)
					Citrus hybrids	Rut	Kamiya *et al.* (1979)
					Clymenia polyandra	Rut	Grieve and Scora (1980)
					Fortunella japonica	Rut	Grieve and Scora (1980)
					F. margarita	Rut	Grieve and Scora (1980)
					Microcitrus inodora	Rut	Grieve and Scora (1980)
5,7,4′		6-*C*-Glucose	(*6.213*)		*Tulipa gesneriana*	Lil	Budzianowski and Skrzypczakowa (1978)
5,4′	7		(*6.206*)	Sakuranetin	*Baccharis trinervis* var. *rhexoides*	Com	Bohlmann *et al.* (1979b)
					Betula spp.	Bet	Wollenweber (1975)
					Gymnopentzia bifurcata	Com	Bohlmann and Zdero (1978)
					Notholaena fendleri	Fil	Wollenweber *et al.* (1980a)
					Polymnia fruticosa	Com	Bohlmann and Zdero (1977)
					Prunus cerasus	Ros	Nagarajan and Parmar (1977c)
5,7	4′		(*6.207*)	Isosakuranetin	*Baccharis alaternoides*	Com	Bohlmann *et al.* (1979b)
					Betula spp.	Bet	Wollenweber (1975)
					Gymnopentzia bifurcata	Com	Bohlmann and Zdero (1978)
					Notholaena fendleri	Fil	Wollenweber *et al.* (1980)
					Prunus domestica	Ros	Nagarajan and Parmar (1977b)
		5-*O*-Glucose			*Hoppea dichotoma*	Gen	Ghosal *et al.* (1978)
		7-*O*-Rutinose		Didymin	'Satsuma orange'	Rut	Bandyukova and Fishman (1979)
					Citrus hybrids	Rut	Kamiya *et al.* (1979)
		7-*O*-Neohesperidose		Poncirin	*Citrus* hybrids	Rut	Kamiya *et al.* (1979)
					Eremocitrus glauca	Rut	Grieve and Scora (1980)
					Microcitrus australasica	Rut	Grieve and Scora (1980)
					M. inodora	Rut	Grieve and Scora (1980)
7,4′	5		(*6.208*)		*Gnaphalium multiceps*	Com	Maruyama *et al.* (1974)
5	7,4′		(*6.210*)		*Betula* spp.	Bet	Wollenweber (1975)
					Dahlia tenuicaulis	Com	Lam and Wrang (1975)
					Eupatorium odoratum	Com	Arene *et al.* (1978)

Table 6.6 (*Contd.*)

Substituents *OH*	*OCH*$_3$	*Other*	*No. in text*	*Trivial name*	*Source**		*References*
4′	5,7		(*6.209*)		*Baccharis alaternoides*	Com	Bohlmann *et al.* (1979b)
		4′-*O*-Xylosyl-arabinose			*Hibiscus mutabilis*	Mal	Chauhan *et al.* (1979c)
		4′-*O*-Rhamnosyl-glucose			*Bauhinia variegata*	Leg	Gupta *et al.* (1980)
	5,7,4′		(*6.211*)		*Dahlia tenuicaulis*	Com	Lam and Wrang (1975)
5,2′	7	2′-*O*-Glucose	(*6.212*)	Haplanthin	*Haplanthus tentaculatus*	Aca	Viswanathan and Sidhaye (1976)
5,7,4′		6,8-Di-*C*-methyl	(*6.214*)	Farrerol	*Rhododendron* spp.	Eri	King (1977a,b)
		5,7-Di-*O*-glucose			*Rhododendron* spp.	Eri	King (1977a,b)
		7-*O*-Glucose			*Cyrtomium* spp.	Fil	Hiraoka (1978)
5,7	4′	6,8-Di-*C*-methyl; 7-*O*-glucose	(*6.215*)	Matteu-cinol-7-glu-coside	*Matteuccia* spp.	Fil	Hiraoka (1978)
5,7,4′		8-*C*-Prenyl	(*6.216*)		*Marshallia grandiflora*	Com	Bohlmann *et al.* (1979d)
5,4′		6-*C*-Prenyl; 6″,6″-dimethyl-pyrano-(2″,3″,7,8)	(*6.217*)	Caja-flavanone	*Cajanus cajan*	Leg	Bhanumati *et al.* (1978)
5,7,4′		6- or 8-*C*-Prenyl; 4′,7-Di-*O*-glucose			*Evoda rutaecarpa*	Rut	Grimshaw and Lamer-Zarawska (1975)
5,4′	6,7		(*6.218*)		*Cheilanthes argentea*	Fil	Wollenweber *et al.* (1980a)
5	6,7,4′		(*6.219*)		*C. argentea*	Fil	Wollenweber *et al.* (1980a)
5,4′	7,8		(*6.220*)		*C. argentea*	Fil	Wollenweber *et al.* (1980a)
5	7,8,4′		(*6.221*)		*C. argentea*	Fil	Wollenweber *et al.* (1980a)
5,6	7,8,4′		(*6.222*)		*C. argentea*	Fil	Wollenweber *et al.* (1980a)
5,4′	6,7,8		(*6.223*)		*C. argentea*	Fil	Wollenweber *et al.* (1980a)
5	6,7,8,4′		(*6.224*)		*C. argentea*	Fil	Wollenweber *et al.* (1980a)
4′	5,6,7		(*6.225*)		*Chromolaena odorata*	Com	Barua *et al.* (1978)
7,3′,4′			(*6.226*)		*Gliricidia sepium*	Leg	Jurd and Manners (1977)
					Acacia saxatilis	Leg	Fourie *et al.* (1974)
					Peltogyne spp.	Leg	Malan and Roux (1974)
					Acacia spp.	Leg	Tindale and Roux (1974)

7,8,3′,4′		7-*O*-Glucose	(*6.227*)	(2*R*)-Isookanin-7-glucoside	*Bidens tripartita*	Com	Serbin *et al.* (1975)
			(*6.228*)	(2*S*)-Isookanin-7-glucoside	*Bidens* spp.	Com	Hart (1979)
7		8-*C*-Prenyl; 3′,4′-methylenedioxy	(*6.231*)	Ovaliflavanone-C	*Milletia ovalifolia*	Leg	Islam *et al.* (1980)
7		6,8-Di-*C*-prenyl; 3′,4′-methylenedioxy	(*6.232*)	Ovaliflavanone-D	*M. ovalifolia*	Leg	Islam *et al.* (1980)
		6″,6″-Dimethylpyrano (2″,3″,7,8); 3′,4′-methylenedioxy	(*6.230*)	Ovalichromene-B	*M. ovalifolia*	Leg	Islam *et al.* (1980)
	6	6″,6″-Dimethylpyrano (2″,3″,7,8); 3′,4′-methylenedioxy	(*6.229*)	Ovalichromene-A	*M. ovalifolia*	Leg	Islam *et al.* (1980)
5,7,3′,4′			(*6.233*)	Eriodictyol	*Coleus amboinicus*	Lab	Brieskorn and Riedel (1977)
					Garcinia conrauana	Clu	Waterman and Crichton (1980)
					Lasianthus japonica	Rub	Ishikura *et al.* (1979)
		5-*O*-Rhamnose			*Cleome viscosa*	Cpr	Srivastava and Srivastava (1979a)
		7-*O*-Rhamnose			*Phyllanthus niruri*	Eup	Chauhan *et al.* (1977)
		7-*O*-Glucose			*Lasianthus japonica*	Rub	Ishikura *et al.* (1979)
					Dryopteris spp.	Fil	Hiraoka (1978)
		7-*O*-Rhamnosylglucose			*Mentha longifolia*	Lab	Bourweig and Pohl (1973)
					Dryopteris spp.	Fil	Hiraoka (1978)
		Unidentified-glycoside			*Rhododendron* spp.	Eri	King (1977a,b)
		7-*O*-Rutinose		Eriocitrin	*Citrus* hybrids	Rut	Kamiya *et al.* (1979)
		7-*O*-Neohesperidose		Neoeriocitrin	*Citrus* hybrids	Rut	Kamiya *et al.* (1979)
5,7,3′	4′		(*6.234*)	Hesperetin	*Persica vulgaris*	Ros	Sadykov and Sadykov (1979)
		5-*O*-Glucose			*P. vulgaris*	Ros	Sadykov and Sadykov (1979)
		7-*O*-Rhamnose			*Cordia obliqua*	Bor	Chauhan *et al.* (1978); Tiwari and Srivastava (1979)

Table 6.6 (*Contd.*)

Substituents *OH*	*OCH*$_3$	*Other*	*No. in text*	*Trivial name*	*Source**		*References*
		7-*O*-Rutinose		Hesperidin	*Mentha longifolia*	Lab	Bourweig and Pohl (1973)
					Vernonia brevifolia	Com	Mabry *et al.* (1975)
					Zanthoxylum avicennae	Rut	Fish *et al.* (1975a)
					Z. dipetalum	Rut	Fish *et al.* (1975b)
					Citrus hybrids	Rut	Kamiya *et al.* (1979)
					Eremocitrus glauca	Rut	Grieve and Scora (1980)
					Fortunella margarita	Rut	Grieve and Scora (1980)
					Pleiospermium sumatranum	Rut	Grieve and Scora (1980)
					Poncirus trifoliata	Rut	Grieve and Scora (1980)
		7-*O*-Neohesperidose		Neohesperidin	*Citrus* hybrids	Rut	Kamiya *et al.* (1979)
5,4′	7,3′		(*6.235*)		*Eupatorium odoratum*	Com	Arene *et al.* (1978)
5,3′	7,4′		(*6.236*)	Persicogenin	*Notholaena* spp.	Fil	Wollenweber *et al.* (1980a)
					Persica vulgaris	Ros	Sadykov and Sadykov (1979)
		5-*O*-Glucose			*P. vulgaris*	Ros	Sadykov and Sadykov (1979)
5,7	3′,4′	5-*O*-Glucose	(*6.237*)		*Hoppea dichotoma*	Gen	Ghosal *et al.* (1978)
5	7,3′,4′		(*6.238*)		*Notholaena* spp.	Fil	Wollenweber *et al.* (1980a)
5,7,3′,4′		6,8-Di-*C*-methyl	(*6.239*)	Cyrtominetin	*Cyrtomium* spp.	Fil	Hiraoka (1978)
5,7,2,3′,4′		5-*O*-Glucose			*Galega officinalis*	Leg	Traub and Geiger (1975)
5,7,2′,5′			(*6.241*)		*Inula cappa*	Com	Baruah *et al.* (1979)
	5,7,2′,4′		(*6.243*)	Arjunone	*Terminalia arjuna*	Cmb	Nagar *et al.* (1979)
7	5,2′,4′		(*6.242*)		*Prunus cerasus*	Ros	Nagarajan and Parmar (1977a)
5,3′,4′	7	8-*C*-Prenyl	(*6.240*)		*Flourensia heterolepis*	Com	Bohlmann and Jakupovic (1979)
5,7,2′	4′	6,8-Di-*C*-methyl	(*6.244*)	Flemiflava-none-A	*Flemingia stricta*	Leg	Sivarambabu *et al.* (1979a)
5,2′,4′		6″,6″-Dimethyl-pyrano(2″,3″,7,6); 8-*C*-prenyl	(*6.245*)	Flemichin-D	*F. wallichii*	Leg	Rao *et al.* (1975b)

		(enantiomer of flemichin-D)		Flemiflavanone-C	*F. stricta*	Leg	Sivarambabu *et al.* (1979a)
5	7,3′,4′,5′		(*6.248*)		*Notholaena lemmonii* var. *lemmonii*	Fil	Wollenweber *et al.* (1980a)
5	6,7,3′,4′,5′	5-*O*-Rhamnose	(*6.249*)		*Cassia renigera*	Leg	Tiwari and Bajpai (1977)

* Key: Aca = Acanthaceae, Aiz = Aizoaceae, Amr = Amaranthaceae, Ano = Anonaceae, Bet = Betulaceae, Bor = Boraginaceae, Cpr = Capparaceae, Clu = Clusiaceae, Cnf = Coniferae, Coc = Cochlospermaceae, Cmb = Combretaceae, Com = Compositae, Cor = Coriariaceae, Cru = Cruciferae, Dil = Dilleniaceae, Equ = Equisetaceae, Eri = Ericaceae, Eup = Euphorbiaceae, Fag = Fagaceae, Fil = Filicinae, Gen = Gentianaceae, Ges = Gesneriaceae, Lab = Labiatae, Lau = Lauraceae, Leg = Leguminosae, Lil = Liliaceae, Mal = Malvaceae, Mel = Meliaceae, Mor = Moraceae, Myr = Myrtaceae, Myt = Myristicaceae, Ros = Rosaceae, Rub = Rubiaceae, Rut = Rutaceae, Sal = Salicaceae, Sol = Solanaceae, The = Theaceae.

Table 6.7 Structures of flavanones having complex substituents

Name	*No. in text*	*Source**		*References*
Purpurin	(*6.169*)	*Tephrosia purpurea*	Leg	R. K. Gupta *et al.* (1980)
Uvarinol	(*6.196*)	*Uvaria chamae*	Ano	Hufford *et al.* (1979)
Louisfieserone-A	(*6.200*)	*Dalea* spp.	Leg	Dominguez *et al.* (1980)
		Indigofera suffruticosa	Leg	Dominguez *et al.* (1978a,b)
Louisfieserone-B	(*6.201*)	*Dalea* spp.	Leg	Dominguez *et al.* (1980)
Flemichin-E	(*6.246*)	*Flemingia wallichii*	Leg	Sivarambabu *et al.* (1979b)
Flemichin-A	(*6.247*)	*Flemingia wallachii*	Leg	Rao *et al.* (1974)

* Key: Ano = Anonaceae, Leg = Leguminosae.

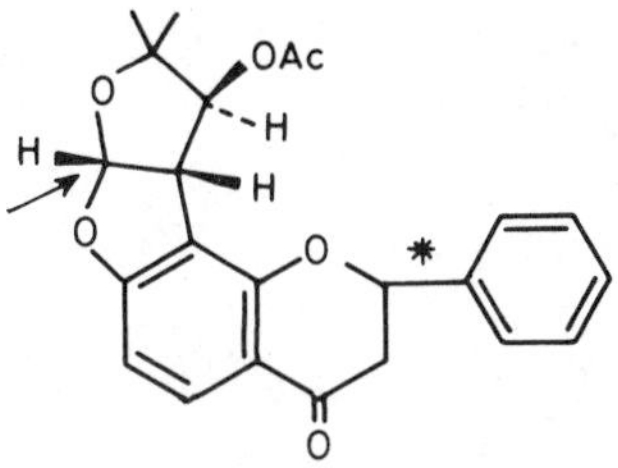

(*6.169*)

a number of leguminous genera: *Tephrosia* (Rao and Raju, 1979; Gupta *et al.*, 1980), *Pongamia* (Talapatra *et al.*, 1980; Satam and Bringi, 1973) and *Lonchocarpus* (Delle Monache *et al.*, 1978). Accompanying isolonchocarpin in *Tephrosia purpurea* is the highly unusual alkylated flavanone, 'purpurin' (*6.169*). The biosynthetic origin of purpurin could be viewed as an interaction between a normal prenyl function at C-8 and an *O*-methyl (or biochemical equivalent) at position 7 (marked with an arrow). Hydration and acetylation would complete the process.

The simplest flavanone that bears the phloroglucinol hydroxylation pattern on the A ring is pinocembrin (*6.170*). Distribution of pinocembrin and its glycosidic derivatives is summarized in Table 6.6. Pinocembrin has been reported for the first time from the following flowering plant families: Compositae, Leguminosae,

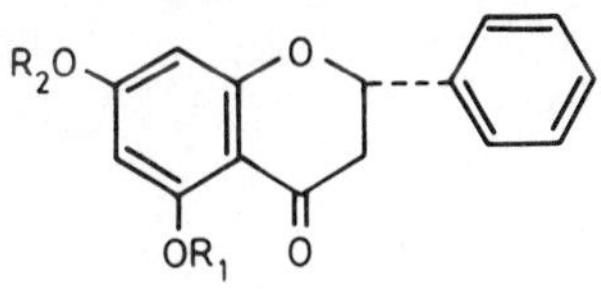

(*6.170*) $R_1 = R_2 = H$

(*6.171*) $R_1 = CH_3, R_2 = H$

(*6.172*) $R_1 = H, R_2 = CH_3$

(*6.173*) $R_1 = R_2 = CH_3$

Table 6.8 Optical properties of some flavanones

Name	No. in text	Rotation*				Sources†		References
Purpurin	(*6.169*)	−67.41	—	$CHCl_3$	25	*Tephrosia purpurea*	Leg	R. K. Gupta *et al.* (1980)
Cajaflavanone	(*6.217*)	−66.6	1.05	$CHCl_3$	25	*Cajanus cajan*	Leg	Bhanumati *et al.* (1978)
Didymocarpin‡		−12.8	0.7	$CHCl_3$	—	*Didymocarpus pedicellata*	Ges	Bose and Adityachaudhury (1978)
Eriodictyol 3′,4-dimethyl ether 5-*O*-glucoside	(*6.237*)	−40	0.5	MeOH	28	*Hoppea dichotoma*	Gen	Ghosal *et al.* (1978)
Isopentenylnaringenin 7,4′-diglucoside†		−47	0.7	MeOH–Pyridine		*Evodia rutaecarpa*	Rut	Grimshaw and Lamer-Zarawska (1975)
Ovalichromene	(*6.167*)	−90	—	—	20	*Milletia ovalifolia*	Leg	Gupta and Krishnamurti (1976b)
Ovalichromene-A	(*6.229*)	−66.3	—	—	25	*M. ovalifolia*	Leg	Gupta and Krishnamurti (1976c)
Ovalichromene-B	(*6.230*)	−91.6	—	—	25	*M. ovalifolia*	Leg	Gupta and Krishnamurti (1976c)
(-)-Salipurposide (naringenin-5-*O*-glucoside)§		−119	0.78	EtOH	20	*Salix elbursensis*	Sal	Kompantsev and Shinkarenko (1975)
Flemichin-A	(*6.247*)	−96.3	0.135	EtOH	25	*Flemingia wallichii*	Leg	Rao *et al.* (1974)
Flemichin-D	(*6.245*)	−63	0.10	EtOH	—	*F. wallichii*	Leg	Rao *et al.* (1975b)
Isolonchocarpin	(*6.168*)	−93	—	$CHCl_3$	28	*Tephrosia purpurea*	Leg	Rao and Raju (1979)
Isolonchocarpin		−120.2	0.065	$CHCl_3$	28	*Pongamia glabra*	Leg	Talapatra *et al.* (1980)
Flemiflavanone-A	(*6.244*)	+38.04	0.034	$CHCl_3$	25	*Flemingia stricta*	Leg	Sivarambabu *et al.* (1979a)
Flemiflavanone-B	(*6.197*)	−10.96	0.045	$CHCl_3$	25	*Flemingia stricta*	Leg	Sivarambabu *et al.* (1979a)
Flemiflavanone-C	(*6.245*)‖	+57.59	0.026	$CHCl_3$	25	*F. stricta*	Leg	Sivarambabu *et al.* (1979a)
Flemichin-E	(*6.246*)	−105.3	0.29	MeOH	—	*F. wallichii*	Leg	Sivarambabu *et al.* (1979b)
Kanakugin	(*6.177*)	0.00	0.66	$CHCl_3$	24	*Popowia cauliflora*	Ano	Waterman and Pootakahm (1979)

* Rotation data: rotation in degrees using sodium D-line, concentration, solvent, temperature. "—" means datum not provided by authors.

† Key: Ano = Anonaceae, Gen = Gentianaceae, Ges = Gesneriaceae, Leg = Leguminosae, Sal = Salicaceae, Rut = Rutaceae.

‡ Structure of this compound has been challenged, see text.

§ Structures of these compounds were not drawn in text.

‖ Flemiflavanone-C is thought to be an enantiomer of Flemichin-D.

Myristicaceae and Solanaceae. Pinocembrin, as the 7-rhamnoside, has also been detected in several members of Athyriaceae and Dryopteridaceae (Hiroaka, 1978). Pinocembrin 5-methyl ether, 'alpinetin' (*6.171*), has been reported from Compositae (Bohlmann *et al.*, 1979a,e) and *Dalea scandens* (Leguminosae) (Dominguez *et al.*, 1980), both of which constitute new familial records. Pinostrobin (*6.172*) and dimethylpinostrobin (*6.173*) occur in *Aniba riparia* (Lauraceae) (Fernandes *et al.*, 1978).

The dimethyl ether (*6.174*), not previously known as a natural product, has been found by Bohlmann *et al.* (1979a) in *Chrysothamnus nauseusus* (Compositae) where it occurs with alpinetin and pinocembrin. 5,7-Dihydroxy-8-methoxyflavanone (*6.175*), commonly called dihydrowogonin, occurs in *Prunus cerasus* and has been synthesized by Nagarajan and Parmar (1977c). The fully *O*-methylated flavanone (*6.176*) occurs in *Popowia cauliflora* (Panichpol and Waterman, 1978). Other constituents of *P. cauliflora* include 'kanakugin' (*6.177*) and the corresponding chalcone kanakugiol (*6.40*). In 1978 Bose and Adityachaudhury reported structure (*6.178*) for a flavanone 'didymocarpin'

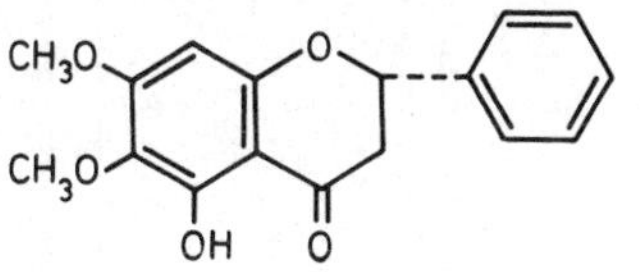

(*6.174*)

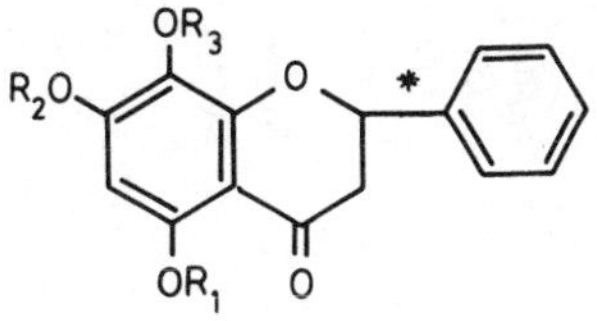

(*6.175*) $R_1 = R_2 = H$, $R_3 = CH_3$

(*6.176*) $R_1 = R_2 = R_3 = CH_3$

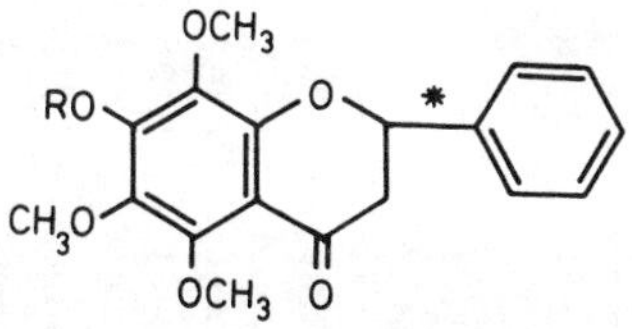

(*6.177*) $R = CH_3$

(*6.178*) $R = H$

which they isolated from *Didymocarpus pedicillata*. Garg and co-workers (1979) synthesized 7-hydroxy-5,6,8-trimethoxyflavanone but found that melting point and spectral data did not agree with those reported for didymocarpin; therefore the structure of the flavanone from *D. pedicellata* remains in question.

Several *C*-methyl derivatives of pinocembrin have also been reported from nature. Wollenweber and co-workers (1979a) found both strobopinin (*6.179*) and cryptostrobin (*6.180*) in the farina of *Pityrogramma pallida* along with the dimethylated compound desmethoxymatteucinol (*6.181*). Star and co-workers (1978) reported a trace of stobopinin 7-methyl ether (*6.182*) along with larger quantities of the corresponding chalcone (*6.29*) in the exudate of *P. triangularis*.

Desmethoxymatteucinol (*6.181*) has been reported from *Agonis spathulata* (Cannon and Martin, 1977), *Dalea polyadenia* (Dreyer *et al.*, 1975) and *Unona lawii* (Joshi and Gawad, 1974). *Unonia lawii* also contains desmethoxymatteucinol 7-methyl ether (*6.183*) and 'lawinal', 6-formyl-8-methylpinocembrin (*6.184*). A discussion of the naturally occurring forms of a pair of 2-hydroxyflavanones from *U. lawii* appears on page 348 (see structures *6.152*–*6.155*).

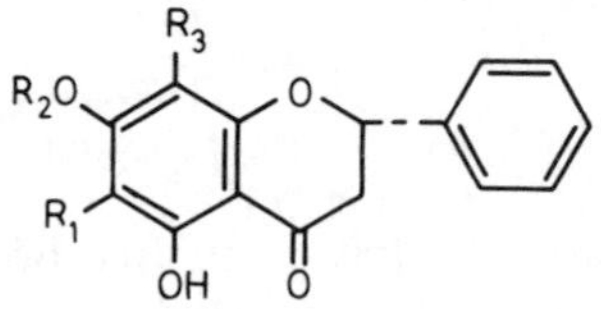

(*6.179*) $R_1 = CH_3$, $R_2 = R_3 = H$

(*6.180*) $R_1 = R_2 = H$, $R_3 = CH_3$

(*6.181*) $R_1 = R_3 = CH_3$, $R_2 = H$

(*6.182*) $R_1 = R_2 = CH_3$, $R_3 = H$

(*6.183*) $R_1 = R_2 = R_3 = CH_3$

(*6.184*) $R_1 = CHO$, $R_2 = H$, $R_3 = CH_3$

6-Prenylpinocembrin (*6.185*) occurs in *Helichrysum polycladum* (Bohlmann *et al.*, 1980c), *Chlorophora tinctoria* (Moraceae) (Gottlieb *et al.*, 1975), *Derris floribunda* (Braz Filho *et al.*, 1975b), and in *D. rariflora* where it occurs with its 7-methyl ether (*6.186*). A series of 8-prenylpinocembrin derivatives, (*6.187*), (*6.188*) and (*6.189*) and 8-geranylpinocembrin (*6.190*) occur in *Helichrysum hypocephalum* (Bohlmann and Abraham, 1979d). Compound (*6.187*), known commonly as 'glabranin', is also known from *Platitenia absinthifolia* (Fam. ?) (Ubaev *et al.*, 1974). 7-Methylglabranin (*6.191*) occurs in *Tephrosia villosa* (Jayaraman *et al.*, 1980). The dimethylchromenoflavanone (*6.192*) has been described from *T. praecans* by Camele *et al.* (1980). The structure of (*6.193*),

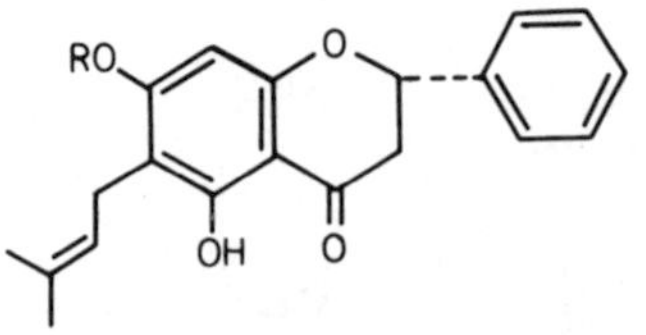

(*6.185*) R = H

(*6.186*) $R = CH_3$

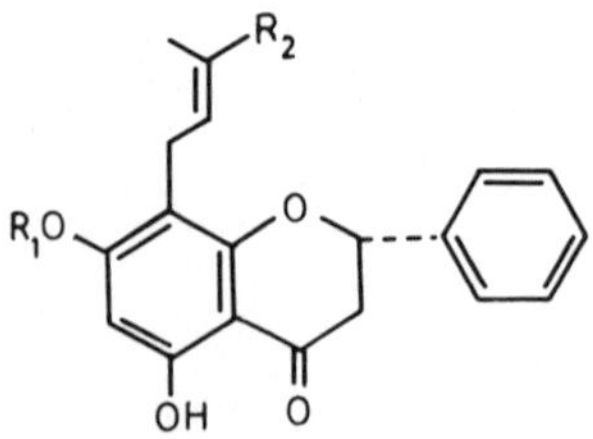

(*6.187*) $R_1 = H$, $R_2 = CH_3$

(*6.188*) $R_1 = H$, $R_2 = CH_2OH$

(*6.189*) $R_1 = H$, $R_2 = CHO$

(*6.191*) $R_1 = R_2 = CH_3$

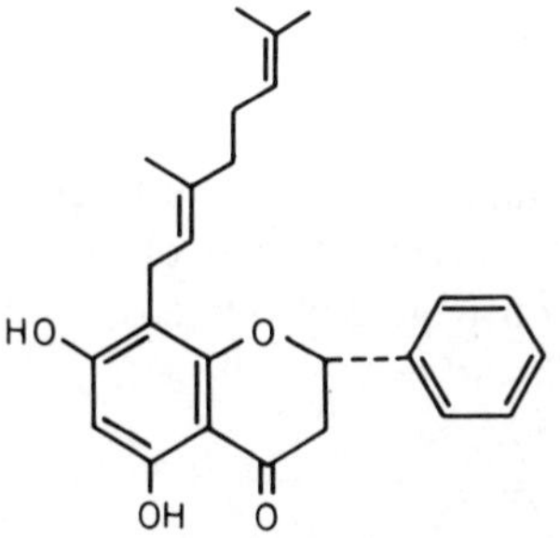

(6.190)

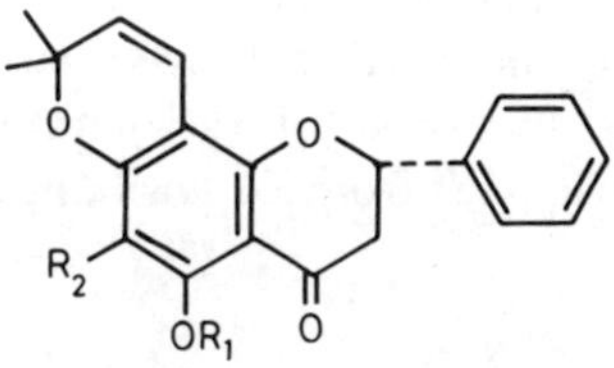

(6.192) $R_1 = CH_3, R_2 = H$

(6.193) $R_1 = H, R_2 = CH_3$

originally isolated from *Mallotus philippinensis* (see Bohm, 1975, p. 569), has been confirmed by synthesis (Manchanda and Khanna, 1977).

In a series of detailed studies by Hufford and co-workers (1976, 1978a, 1978b, 1979), several cytotoxic flavanones were isolated from *Uvaria chamae* and examined for their biological activity. 'Uvaretin' (*6.194*) or (*6.195*) and 'isouvaretin' (*6.194* or *6.195*) are the 6- and 8-alkyl derivatives of pinocembrin. The alkyl group is 2-hydroxybenzyl. The relationship between the two flavanones was established through hydrogenation which yielded a single dihydrochalcone. The most complex member of the group is the tribenzylated flavanone 'uvarinol' (*6.196*). Bark extracts of the plant were active *in vivo* against P-388 lymphocytic leukaemia and *in vitro* against cells from human carcinoma of the nasopharynx. In addition the extracts showed significant activity against *Staphlococcus aureus*, *Bacillus subtilis* and *Mycobacterium smegmatis* (Hufford and Laswell, 1978b).

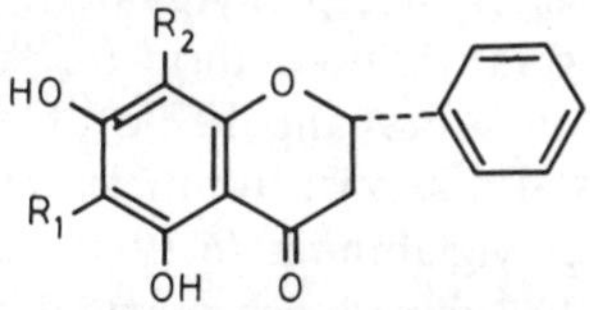

(6.194) $R_1 = H, R_2 =$ 2-hydroxybenzyl

(6.195) $R_1 =$ 2-hydroxybenzyl, $R_2 = H$

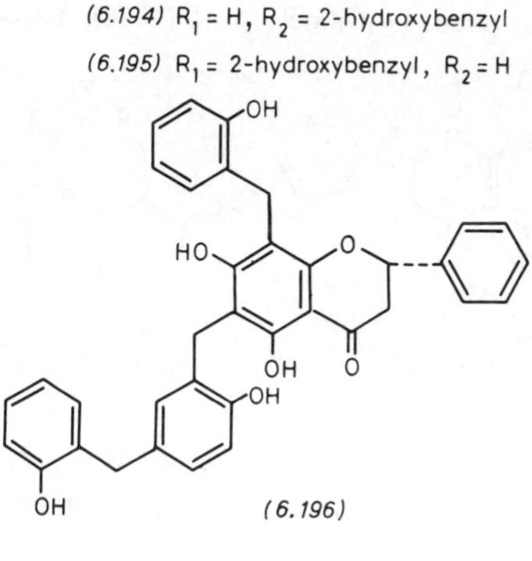

(6.196)

In 1979 Sivarambabu and co-workers described several new flavanones from *Flemingia stricta*, three of which will appear later. The structure that concerns us here is (*6.197*) proposed for 'flemiflavanone-B.' The authors considered two structures, (*6.197*) and (*6.198*), which could account for the spectral data obtained. Both of these are consistent with the presence of a single A-ring proton and the absence of a chelated hydroxyl signal in the ^{1}H-NMR spectrum. Structure (*6.197*) was chosen on the basis that the unknown compound gave a positive Gibbs test (2,6-dichlorobenzoquinone-4-chloroimide in borate buffer) to detect *para*-unsubstituted phenols. Structure (*6.197*) has such a position; structure (*6.198*) does not. Structure (*6.197*), however, poses several problems

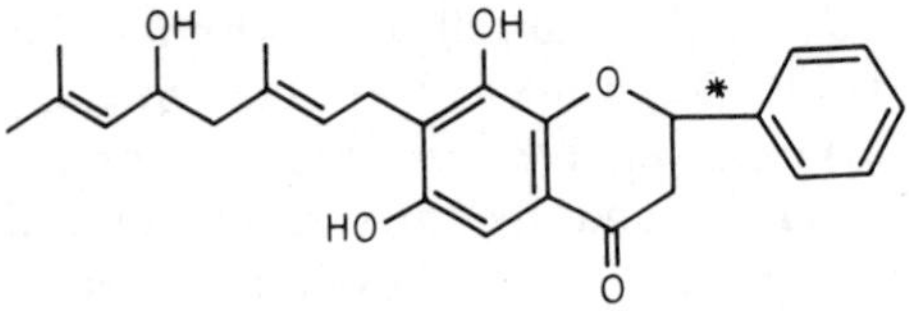

(*6.197*)

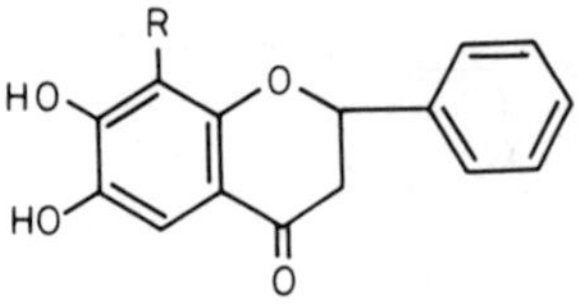

(*6.198*)

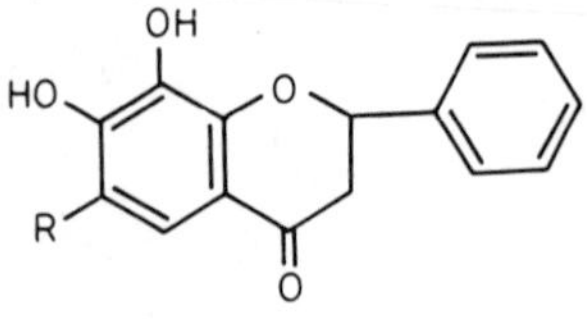

(*6.199*)

when viewed biosynthetically. In order to arrive at this structure following the normal pathway there would have to have been loss of hydroxyl groups from positions 5 and 7, oxygenation at positions 6 and 8, and alkylation at position 7. This is a tortuous pathway, although loss of oxygen from carbon-5 is a common occurrence in leguminous plants. Structure (*6.199*) is an alternative possibility which would require postulation of a less complicated pathway and would be consistent with the observed ^{1}H-NMR data and would be expected to give a positive Gibbs test. This possible structure was apparently not considered by Sivarambabu and co-workers (1979a,b).

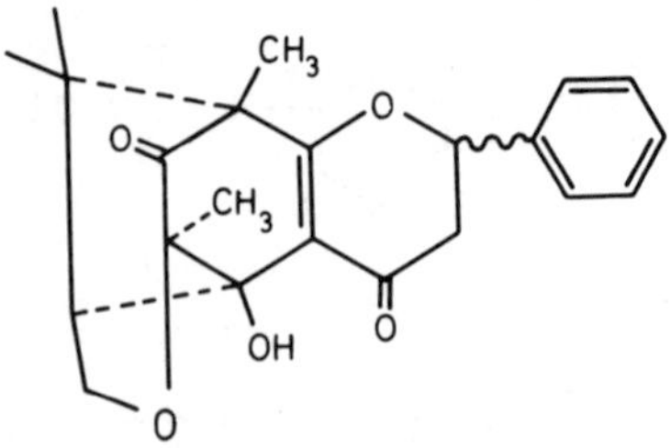

(*6.200*) α-phenyl (2*S*)

(*6.201*) β-phenyl (2*R*)

Two of the most complex A-ring alkylated flavanones to have been found in nature are compounds (*6.200*) and (*6.201*), 'louisfieserone' and 'isolouisfieserone' respectively. Both isomers have been found to occur in *Dalea* species (Dominguez *et al.*, 1980) while flavanone (*6.200*) is also known from *Indigofera suffruticosa* (Dominguez *et al.*, 1978a, 1978b).

(b) Flavanones having one B-ring hydroxyl

With the exception of an early report of liquiritigenin glycosides in *Dahlia* (Compositae) this simple flavanone (*6.202*) is otherwise known only from Leguminosae. New generic records from that family include *Onobrychis* (Ingham, 1978), *Medicago* (Ingham, 1979), *Myroxylon* (De Oliveira *et al.*, 1978) and *Peltogyne* (Malan and Roux, 1974). The finding of 8-*C*-*β*-galactosyl-liquiritigenin (*6.203*) Garg *et al.*, 1980) in *Aerva persica* represents a first report of this compound in nature and the first report of a flavanone from the Amaranthaceae. Compound (*6.204*), however, is yet another *C*-prenylated derivative of liquiritigenin from the genus *Sophora* (Jain and Gupta, 1978; see Bohm, 1975, pp. 572–573).

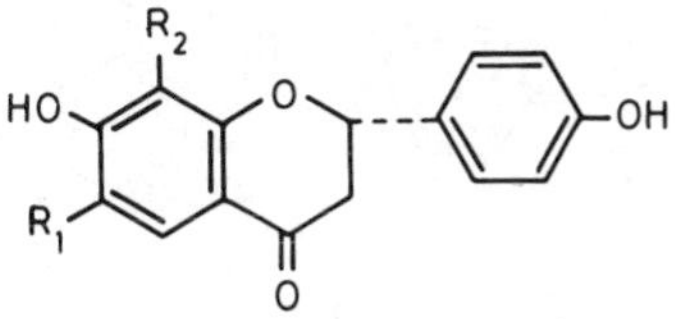

(*6.202*) $R_1 = R_2 = H$

(*6.203*) $R_1 = H$, $R_2 = Gal$

(*6.204*) $R_1 = R_2 =$ [prenyl]

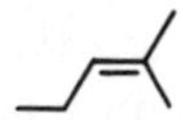

The most widely occurring of all flavanones is naringenin (*6.205*), represented in Table 6.6 by reports from 26 genera, although some of the genera had been listed in the earlier survey. New family records include: Anacardiaceae, Athyriaceae, Capparaceae, Cochlospermaceae, Coriariaceae, Dilleniaceae, Dryopteridaceae, Equisetaceae, Lauraceae, Meliaceae, Pinaceae and Theaceae (see Table 6.6 for references). Naringenin occurs as the free phenol in many

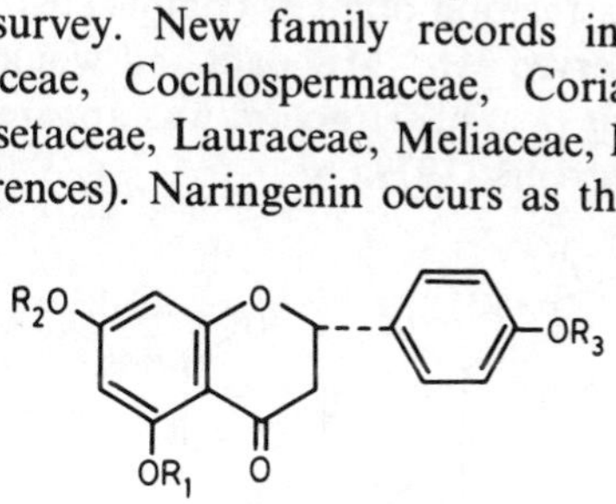

(*6.205*) $R_1 = R_2 = R_3 = H$

(*6.206*) $R_1 = R_3 = H$, $R_2 = CH_3$

(*6.207*) $R_1 = R_2 = H$, $R_3 = CH_3$

(*6.208*) $R_1 = CH_3$, $R_2 = R_3 = H$

(*6.209*) $R_1 = R_2 = CH_3$, $R_3 = H$

(*6.210*) $R_1 = H$, $R_2 = R_3 = CH_3$

(*6.211*) $R_1 = R_2 = R_3 = CH_3$

species and as a variety of glycosides in several of the others, the most noteworthy of which is the fructoside (position of attachment not determined) which Imperato (1976) discovered in leaves of *Camellia sinensis* (Theaceae).

All three mono-*O*-methyl derivatives of naringenin are known to occur naturally. Sakuranetin, the 7-methyl ether (*6.206*), and isosakuranetin, the 4′-methyl ether (*6.207*), are well-known compounds occurring either as free phenols or glycosides (Table 6.6). The remaining member of this group is 5-*O*-methyl naringenin (*6.208*) which was reported as a natural product for the first time by Maruyama *et al.* (1974), who obtained it from flowers of *Gnaphalium multiceps* (Compositae).

Naringenin-5,7-dimethyl ether (*6.209*) was first reported from nature by Bohlmann and co-workers (1979) who isolated the free phenol from *Baccharis alternoides* (Compositae). Very soon thereafter, two reports of glycosidic derivatives of this flavanone appeared. Chauhan *et al.* (1979c) obtained the 4′-β-D-xylopyranosyl-β-D-arabinopyranoside from stem tissues of *Hibiscus mutabilis* (Malvaceae). A. K. Gupta and co-workers (1980) reported the 4′-α-L-rhamno pyranosyl-β-D-glucopyranoside from stems of *Bauhinia variegata* (Leguminosae). Naringenin-7,4′-dimethyl ether (*6.210*), a well known constituent of *Betula*, has been reported as a constituent of the bud excretions of several additional *Betula* species by Wollenweber (1975), as well as from two new sources: *Eupatorium odoratum* (Compositae) (Arene *et al.*, 1978) and *Dahlia tenuicaulis* (Lam and Wrang, 1975). These last mentioned workers also reported naringenin trimethyl ether (*6.211*) from *D. tenuicaulis*, the only genus from which this flavanone has been isolated.

The flavanone (*6.212*) is a new member of the rare flavonoid group having only 2′-hydroxylation on the B ring. This member is known as 'haplanthin' and was isolated from a variety of *Haplanthus tentaculatus* (Acanthaceae) by Viswanathan and Sidhaye (1976), who also confirmed the structure by synthesis.

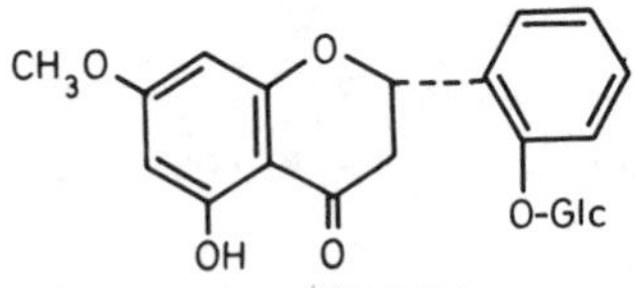

CH_3O O O-Glc OH O

(6.212)

C-Alkylation of the naringenin ring occurs in several plants to provide a variety of derivatives. The simple 6-*C*-β-D-glucopyranosyl derivative (*6.213*), reported from *Tulipa gesneriana* by Budzianowski and Skrzypczakowa (1978), is the first flavanone reported from the Liliaceae. The *C*-methylflavanone farrerol (*6.214*) and its 5,7-diglucoside occur in *Rhododendron* species belonging to subgenus *Pentanthera* (King, 1977a,b). Farrerol 7-*O*-glucoside and the closely related compound matteucinol (*6.215*) are thought to occur in several members of the Athyriaceae and Dryopteridaceae (Hiraoka, 1978), although confirmation of farrerol glucoside was lacking. 8-*C*-Prenylnaringenin (*6.216*) occurs in the

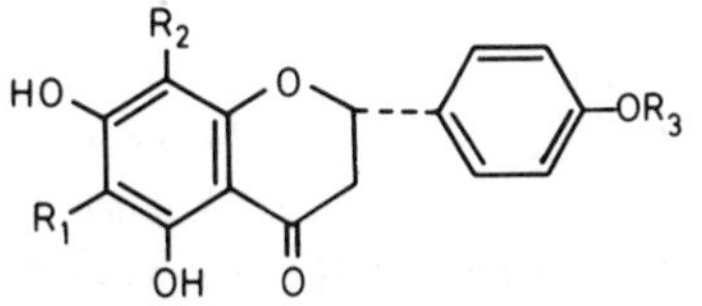

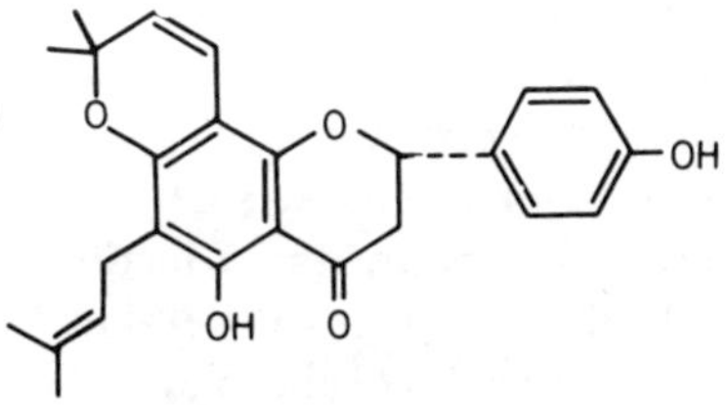

(6.213) R_1 = Glc, R_2 = R_3 = H

(6.214) R_1 = R_2 = CH_3, R_3 = H

(6.215) R_1 = R_2 = R_3 = CH_3

(6.216) R_1 = R_3 = H, R_2 =

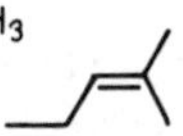

(6.217)

roots of *Marshallia grandiflora* (Compositae) (Bohlmann *et al.*, 1979d). An isopentenylnaringenin-7,4′-diglucoside has been discovered in *Evoda rutaecarpa* (Rutaceae) by Grimshaw and Lamer-Zarawska (1975), who were unable to place the alkyl group with certainty at either position 6 or 8. 'Cajaflavanone' (*6.217*) consists of naringenin having two prenyl functions one of which has been cyclized with the 7-hydroxyl group to afford the dimethylchromeno ring system. The compound was described as a natural product for the first time by Bhanumati *et al.* (1978), who isolated it from *Cajanus cajan* (Leguminosae).

Seven poly-*O*-methylated flavanones having 6-, 8- or 6,8-hydroxylation were isolated from the exudate of the fern *Cheilanthes argenta* by Wollenweber and co-workers (1980a). The compounds are shown as structures (*6.218–6.224*). If it

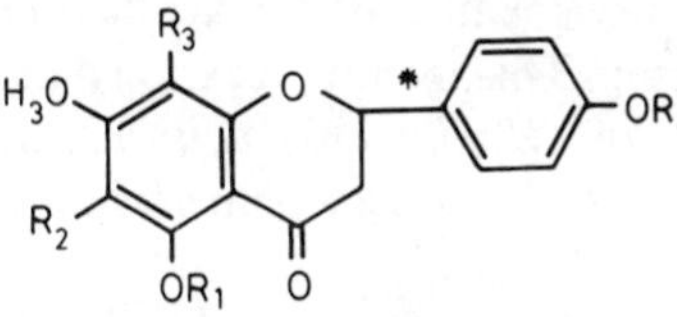

(6.218) R_1 = R_3 = R_4 = H, R_2 = OCH_3

(6.219) R_1 = R_3 = H, R_2 = OCH_3, R_4 = CH_3

(6.220) R_1 = R_2 = R_4 = H, R_3 = OCH_3

(6.221) R_1 = R_2 = H, R_3 = OCH_3, R_4 = CH_3

(6.222) R_1 = H, R_2 = OH, R_3 = OCH_3, R_4 = CH_3

(6.223) R_1 = H, R_2 = R_3 = OCH_3 R_4 = H

(6.224) R_1 = H, R_2 = R_3 = OCH_3, R_4 = CH_3

(6.225) R_1 = CH_3, R_2 = OCH_3, R_3 = R_4 = H

were possible to get the biosynthetically active tissue of this plant into a manageable form this system would offer an interesting opportunity to examine the specificities of hydroxylation and *O*-methylation reactions.

Compound (*6.225*), isolated from *Chromolaena odorata* (Compositae) by

Barua *et al.* (1978), may have been an artifact of isolation since it was obtained in optically inactive form. The corresponding 4,2′-dihydroxy-4′,5′,6′-trimethoxychalcone (*6.59*) was also isolated from the plant (Barua *et al.*, *loc. cit.*).

(c) Flavanones having two B-ring hydroxyls

The simplest flavanone of this series is (*6.226*) recorded as a constituent of *Acacia saxatilis* (Fourie *et al.*, 1974), *Peltogyne* species (Malan and Roux, 1974) and *Gliricidia sepium* (Jurd and Manners, 1977). Isoökanin is a well-known flavanone

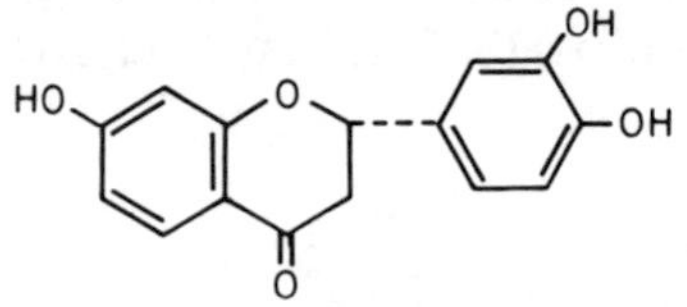

(6.226)

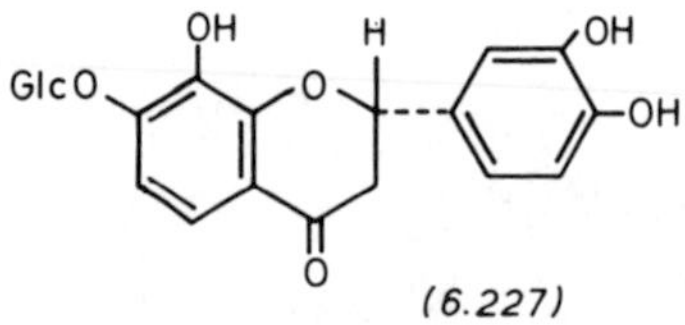

(6.227) 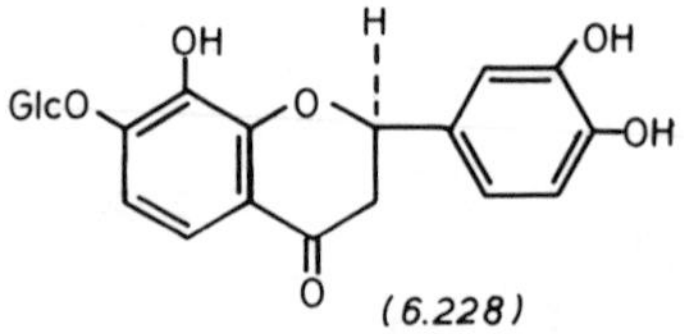 (6.228)

which was isolated as its 7-*O*-*β*-D-glucoside from several members of the *Bidens ferulaefolia* complex by Hart (1979). The majority of flavanones studied to date exhibit the (2*S*) configuration as written for compound (*6.227*). It is of interest, then, that compound (*6.228*) has also been isolated from a *Bidens* species, *B. tripartita*, by Serbin *et al.* (1975). Analysis showed that the compound has the (2*R*) configuration. Further detailed examination of flavanone stereochemistry of *Bidens* would be in order.

A complex series of *C*-alkylated flavanones, all possessing the 3′,4′-methylenedioxy function, has been studied by Gupta and his colleagues (Gupta and Krishnamurti, 1976b; Islam *et al.*, 1980). The compounds are from *Milletia ovalifolia* and are ‘ovalichromene-A’ (*6.229*), ‘ovalichromene-B’ (*6.230*), ‘ovaliflavanone-C’ (*6.231*) and ‘ovaliflavanone-D’ (*6.232*).

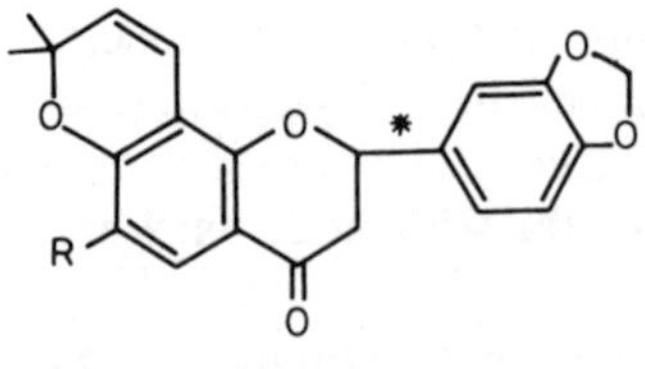

(6.229) R = OCH_3
(6.230) R = H

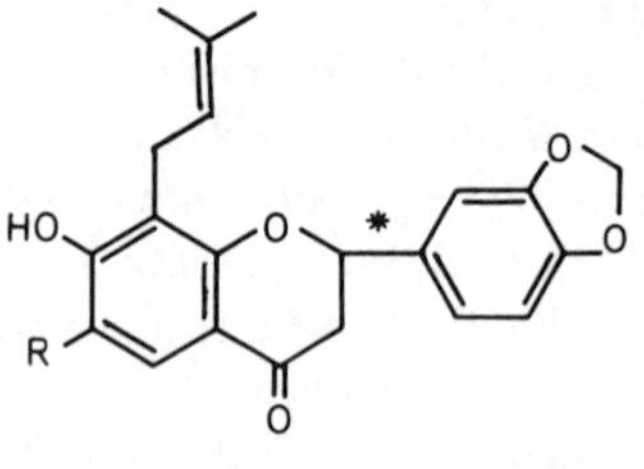

(6.231) R = H (2S)
(6.232) R =

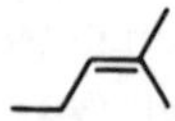

Eriodictyol (*6.233*) occurs widely in flowering plants (Table 6.6) and has recently been reported from members of the Athyriaceae and Dryopteridaceae (Hiraoka, 1978). New familial records for the compound, or its glycosidic forms, include: Capparaceae, Clusiaceae, Ericaceae, Euphorbiaceae, Labiatae and Rubiaceae (see Table 6.6 for references). Eriodictyol 4′-methyl ether, hesperitin (*6.234*), also enjoys a modest distribution with new reports from the Boraginaceae, Compositae and Labiatae (see Table 6, for references). Flavanone (*6.235*), formerly known only from *Melicope* in the Rutaceae, has recently been found in *Eupatorium odoratum* by Arene and co-workers (1978). The isomeric 7,4′-dimethyl ether (*6.236*), known as 'persicogenin', occurs as such and as the 5-glucoside in *Persica vulgaris* (Rosaceae) (Sadykov and Sadykov, 1979), and as

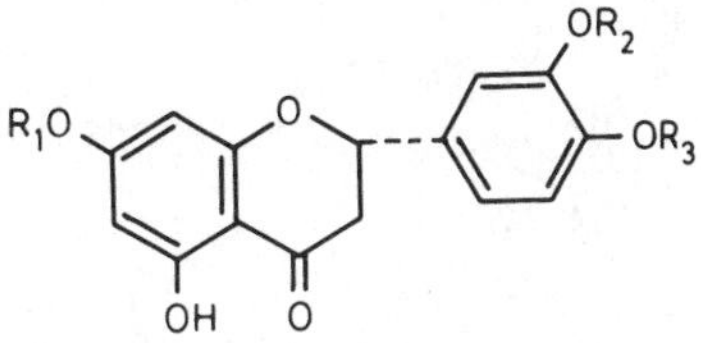

(*6.233*) $R_1 = R_2 = R_3 = H$

(*6.234*) $R_1 = R_2 = H$, $R_3 = CH_3$

(*6.235*) $R_1 = R_2 = CH_3$, $R_3 = H$

(*6.236*) $R_1 = R_3 = CH_3$, $R_2 = H$

(*6.237*) $R_1 = H$, $R_2 = R_3 = CH_3$

(*6.238*) $R_1 = R_2 = R_3 = CH_3$

the aglycone in leaf exudates of the ferns *Notholaena limitanea* var. *mexicana* and *N. fendleri* (Wollenweber *et al.*, 1980a). A third dimethyl ether was reported by Ghosal *et al.* (1978): eriodictyol 3′,4′-dimethyl ether occurs as the 5-*O*-β-glucoside (*6.237*) in *Hoppea dichotoma*, which represents a first report for this flavanone (aglycone and glycoside) and the first record for a flavanone from Gentianaceae. Eriodictyol 7,3′,4′-trimethyl ether (*6.238*) was also one of the exudate constituents found on leaves of *Notholaena* species by Wollenweber and his colleagues (1980a).

Sastry and Row (1961a,b) reported the presence of two highly *O*-methylated flavanones from the peel of *Citrus mitis*: 5,6,7,8,3′,4′-hexamethoxy- and 5-hydroxy-6,7,8,3′,4′-pentamethoxy-flavanone. Recently, Tatum and Berry (1978) examined the flavonoids of the *Citrus* cultivar 'Calamondin' and concluded that the two 'flavanones' were identical with the flavones nobiletin and 5-*O*-desmethylnobiletin.

Cyrtominetin, the 6,8-di-*C*-methyl derivative of eriodictyol (*6.239*), was detected in several species of fern (Athyriaceae and Dryopteridaceae) by Hiraoka (1978), although structural confirmation was not obtained. To date, this compound has only been reported from ferns (see Bohm, 1975, Table 11.1). An

8-prenyl derivative of eriodictyol (*6.240*) was isolated from *Flourensia heterolepis* by Bohlmann and Jakupovic (1979), and a 2-hydroxyeriodictyol derivative (*6.150*) occurs in *Galega officinalis* (Traub and Geiger, 1975).

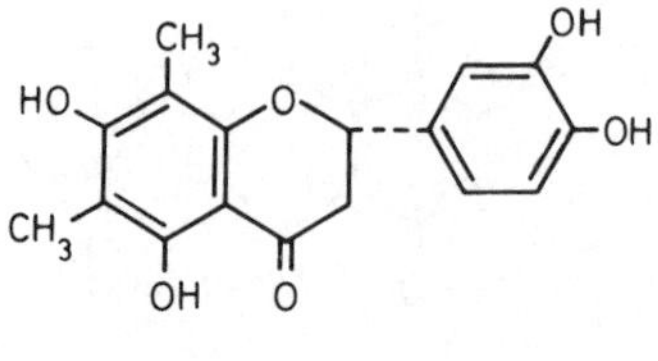

(*6.239*) 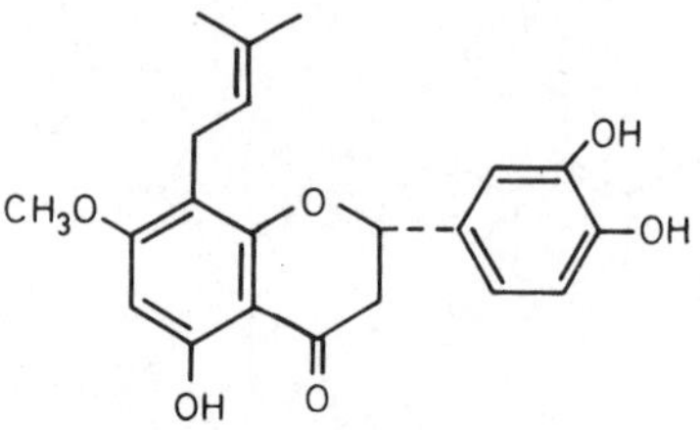 (*6.240*)

The last group of flavanones having two B-ring hydroxyls exhibit either the 2′,5′- or the 2′,4′- pattern. Only one member of the former type has been found recently. Compound (*6.241*) occurs in *Inula cappa* (Compositae) along with a dihydroflavonol (*6.269*) with the same hydroxylation pattern (Baruah *et al.*, 1979). Compound (*6.242*) was isolated from *Prunus cerasus* and its structure determined and confirmed by synthesis by Nagarajan and Parmar (1977a, 1978).

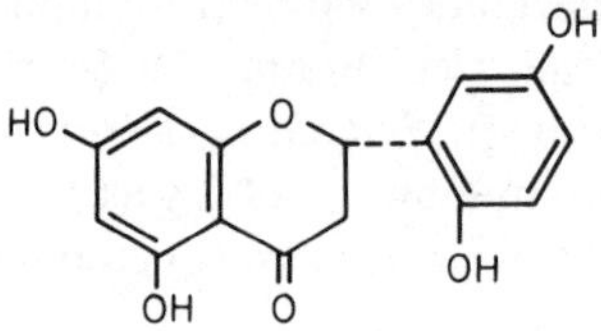

(*6.241*)

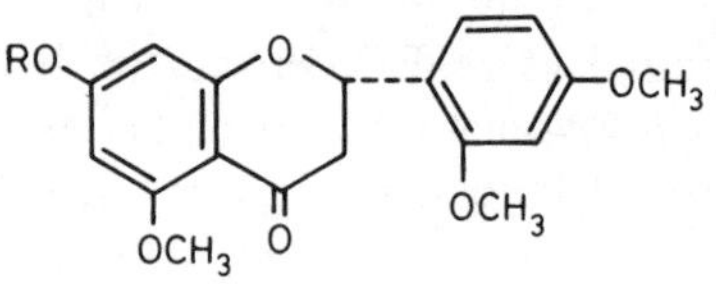

(*6.242*) R = H

(*6.243*) R = CH_3

The completely *O*-methylated compound (*6.243*), 'arjunone', has been isolated from *Terminalia arjuna* (Combretaceae) (Nagar *et al.*, 1979), where it occurs with the corresponding chalcone cerasidin (*6.89*). Five additional compounds are known that are based on the 2′,4′- pattern; all exhibit extensive *C*-alkylation and all have been found in the genus *Flemingia*. The simplest of these is 'flemiflavanone-A' (*6.244*) isolated from *F. stricta* by Sivarambabu *et al.* (1979a). 'Flemichin-D' (*6.245*), 'flemichin-E' (*6.246*) and 'flemichin-A' (*6.247*) occur

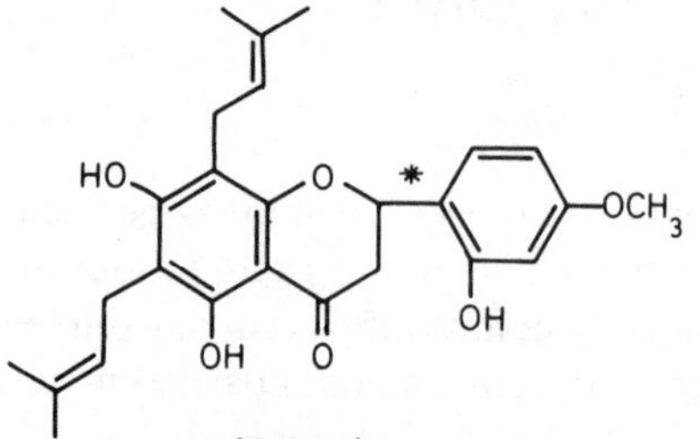

(*6.244*) 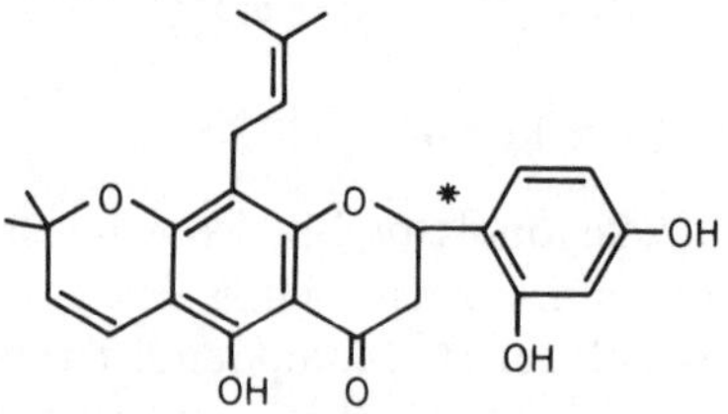 (*6.245*)

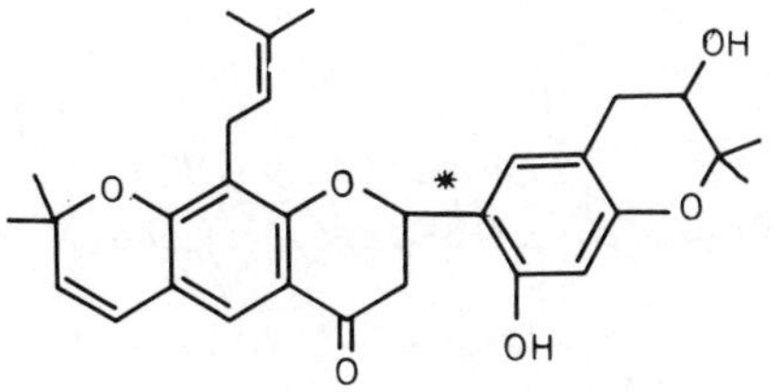

(6.246) 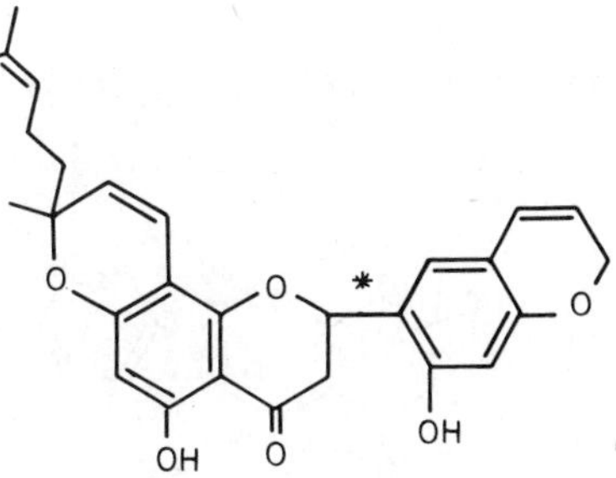 (6.247)

in *F. wallichii* (Sivarambabu *et al.*, 1979b; Rao *et al.* 1974, 1975b). 'Flemiflavanone-C' has been isolated from *F. stricta* by Sivarambabu *et al.* (1979a), who concluded on the basis of its optical rotation that it is an enantiomer of flemichin-D (*6.245*).

(d) Flavanones having three B-ring hydroxyls

Only two such compounds are known if one excludes the benzoquinone-type B-ring compounds isolated from Cyperaceae (see Bohm, 1975, p. 582). Compound (*6.248*) occurs in the leaf exudates of *Notholaena lemmonii* var. *lemmonii* (Wollenweber *et al.*, 1980a) while compound (*6.249*) was found by Tiwari and Bajpai (1977) in the stem bark of *Cassia renigera* (Leguminosae).

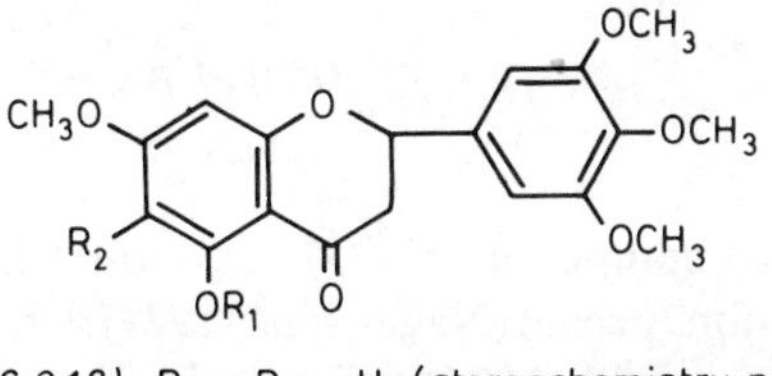

(6.248) $R_1 = R_2 = H$ (stereochemistry not determined)
(6.249) $R_1 = Rha$, $R_2 = OCH_3$, (2*S*) stereochemistry

6.6 DIHYDROFLAVONOLS

6.6.1. Introduction

Dihydroflavonols, or 3-hydroxyflavanones, are numbered in the same fashion as flavanones. Carbons-2 and -3 are asymmetric in dihydroflavonols, so four isomers are possible for each compound. The majority of naturally occurring dihydroflavonols exist in the (2*R*:3*R*) configuration, but a few compounds are known with (2*S*:3*S*) stereochemistry (see Bohm, 1975, p. 597 and examples below).

6.6.2 Structures of naturally occurring dihydroflavonols

The structures of naturally occurring dihydroflavonols are summarized in Table 6.9.

(a) Dihydroflavonols lacking B-ring hydroxyls

Compounds based upon the resorcinol A-ring pattern include 3-hydroxy-isolonchocarpin (*6.250*) from *Lonchocarpus eriocaulinalis* (Delle Monache *et al.*, 1978) and (*6.251*) from *Dalbergia ecastophyllum*. Structure (*6.251*) was originally suggested for the compound by Matos and co-workers some years ago, and has now been confirmed by synthesis (Matos *et al.*, 1975). The simplest dihydroflavonol exhibiting the phloroglucinol A-ring substitution pattern is pinobanksin (*6.252*) which is widely distributed in *Pinus*. Recent reports have described

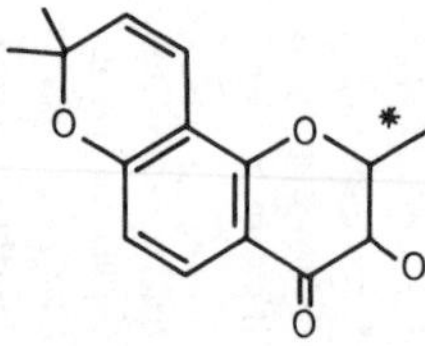

(*6.250*) *Stereochemistry unknown 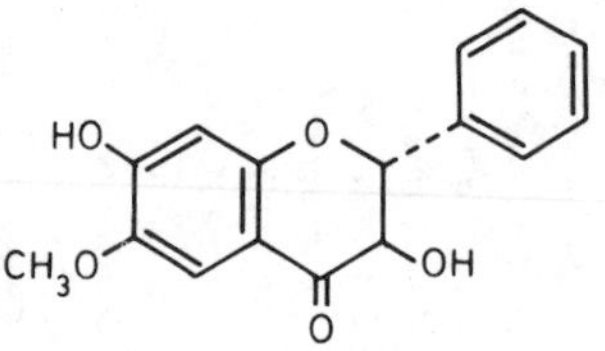 (*6.251*)

pinobanksin in aerial parts of *Helichrysum tenuifolium* (Bohlmann and Abraham, 1979a) and in the heartwood of *Larix dahurica* (Lapteva *et al.*, 1974). Pinobanksin dimethyl ether (*6.253*), a new naturally occurring dihydroflavonol, was isolated from *Aniba riparia* by Fernandes and co-workers (1978).

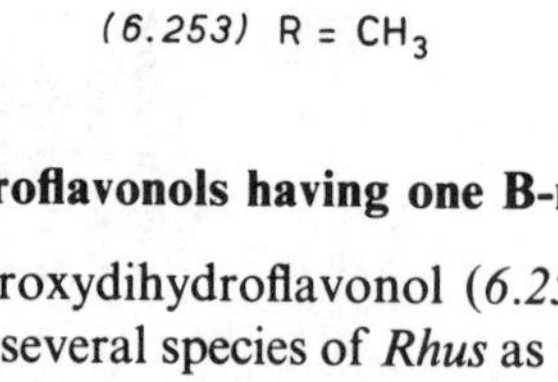

(*6.252*) R = H
(*6.253*) R = CH_3
(*6.254*) 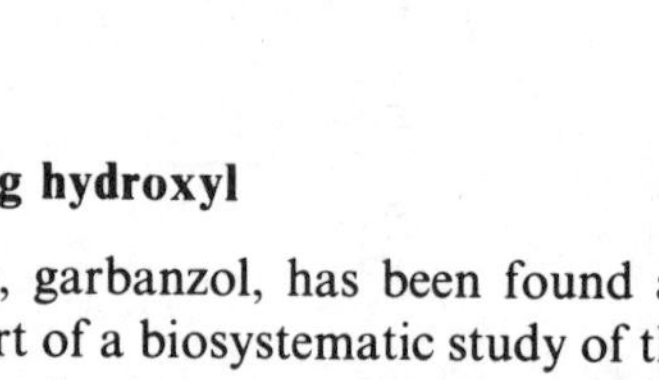

(b) Dihydroflavonols having one B-ring hydroxyl

7,4′-Dihydroxydihydroflavonol (*6.254*), garbanzol, has been found as a constituent of several species of *Rhus* as part of a biosystematic study of that genus by Young (1979). This represents the first report of this compound from Anacardiaceae.

The most widely occurring member of this subgroup is dihydrokaempferol, or aromadendrin (*6.255*). Recent reports of its natural occurrence are summarized in Table 6.9. Families from which aromadendrin has been obtained for the first

Table 6.9. Structures of naturally occurring dihydroflavonols

Substituents			No. in	Trivial			
OH	*OCH_3*	*Other*	*text*	*name*	*Source**		*References*
7	6		(*6.251*)		*Dalbergia ecastophyllum*	Leg	Matos *et al.* (1975)
5,7			(*6.252*)	Pinobanksin	*Helichrysum tenuifolium*	Com	Bohlmann and Abraham (1979a)
					Larix dahurica	Cnf	Lapteva *et al.* (1974)
	5,7		(*6.253*)		*Aniba riparia*	Lau	Fernandes *et al.* (1978)
		6″,6″-Dimethylpyrano (2″,3″,7,8)	(*6.250*)	3-Hydroxyisolonchocarpin	*Lonchocarpus eriocaulinalis*	Leg	Delle Monache *et al.* (1978)
7,4′			(*6.254*)	Garbanzol	*Rhus* spp.	Ana	Young (1979)
5,7,4′			(*6.255*)	Aromadendrin	*Cavendishia* spp.	Eri	Luteyn *et al.* (1980)
					Cudrania javanensis	Mor	S. R. Gupta *et al.* (1975a)
					Dillenia retusa	Dil	Pavanasasivan and Sultanbawa (1975)
					Equisetum arvense	Equ	Syrchina *et al.* (1975, 1978)
					Eucryphia cordifolia	Euc	Tschesche *et al.* (1979)
					Matthiola incana	Cru	Forkmann (1979)
					Morus spp.	Mor	Deshpande *et al.* (1975)
					Othonna euphorboides	Com	Bohlmann and Knoll (1978)
					Persica vulgaris	Ros	Sadykov and Sadykov (1979)
					Picea ajanensis	Cnf	Ivanova *et al.* (1975)
					Prunus spp.	Ros	Nagarajan and Parmar (1977b,c)
					Wormia triquetra	Dil	Pavanasasivan and Sultanbawa (1975)
		3-*O*-Rhamnose		Engelitin	*Eucryphia cordifolia*	Euc	Tschesche *et al.* (1979)
		3-*O*-Glucose			*Dalbergia sericea*	Leg	Dayal and Parthasarathy (1977)
		4′-*O*-Xylose			*Cleome viscosa*	Cpr	Chauhan *et al.* (1979b)
		7-*O*-Glucose			*Matthiola incana*	Cru	Forkmann (1979)
					Cudrania javanensis	Mor	Gupta *et al.* (1975a)
5,7	4′		(*6.256*)		*Prunus domestica*	Ros	Nagarajan and Parmar (1977b)
5	7,4′		(*6.257*)		*Podocarpus neriifolius*	Cnf	Rizvi and Rahman (1974)
		5-*O*-Glucose			*P. neriifolius*	Cnf	Rizvi and Rahman (1974)

5,4′	6,7				*Saxifraga caespitosa*	Sax	Miller and Bohm (1979)
5,7,4′		8-*C*-Prenyl	(*6.258*)		*Marshallia obovata*	Com	Bohlmann *et al.* (1980a)
5,4′		6″,6″-Dimethylpyrano (2″,3″,7,8)	(*6.259*)		*M. obovata*	Com	Bohlmann *et al.* (1980a)
5,4′		6″,6″-Dimethylpyrano (2″,3″,7,6); 8-*C*-prenyl	(*6.260*)		*Milletia pachycarpa*	Leg	Singhal *et al.* (1980)
5,7,4′		6-*C*-(3-Hydroxy-3-methylbutyl); 7-*O*-glucose		Phellavin	*Phellodendron* spp.	Rut.	Otryashenkova *et al.* (1977)
7,3′,4′			(*6.261*)	Fustin	*Rhus* spp.	Ana	Young (1979)
					Amphipterygium adstrigens	Ana	Young (1976)
					Acacia spp.	Leg	Fourie *et al.* (1974); Tindale and Roux (1974)
	3		(*6.262b*)	3-*O*-Methyl-2,3-*trans*-fustin	*Peltogyne* spp.	Leg	Malan and Roux (1974)
		3,7-Di-*O*-glucose			*Georgia pellucida*	Mus	Vandekerkhove (1977)
	3		(*6.262a*)	3-*O*-Methyl-2,3-*cis*-fustin	*Peltogyne* spp.	Leg	Malan and Roux (1974)
					Acacia spp.	Leg	Tindale and Roux (1974)
7,8,3′,4′			(*6.263*)	8-Hydroxyfustin	*Albizia adianthifolia*	Leg	Candy *et al.* (1978)
					Acacia saxatilis	Leg	Fourie *et al.* (1974)
5,7,3′,4′			(*6.264*)	Taxifolin	*Acacia* spp.	Leg	Tindale and Roux (1974)
					Cavendishia spp.	Eri	Luteyn *et al.* (1980)
					Coleus amboinicus	Lab	Brieskorn and Riedel (1977)
					Cudrania javanensis	Mor	Gupta *et al.* (1975a)
					Dillenia retusa	Dil	Pavanasasivan and Sultanbawa (1975)
					Eucryphia cordifolia	Euc	Tschesche *et al.* (1979)
					Abies amabilis	Cnf	Parker *et al.* (1979)
					Cedrus deodara	Cnf	Agrawal *et al.* (1980a,b)
					Larix leptolepis	Cnf	Niemann (1974)
		3-*O*-Glucose			*Aerides fieldingii*	Orc	C. A. Williams (1979)
					Zizyphus nummularia	Rhm	Srivastava and Chauhan (1977)
		3′-*O*-Glucose			*Cedrus deodara*	Cnf	Agrawal *et al.* (1980b)

Table 6.9 (*Contd.*)

Substituents			*No. in*	*Trivial*			
OH	OCH_3	*Other*	*text*	*name*	*Source**		*References*
		3-*O*-Galactose			*Rhododendron* spp.	Eri	King (1977a,b)
		7-*O*-Galactose			*Rhododendron* spp.	Eri	King (1977a,b)
		3-*O*-Rhamnose		Astilbin	*Eucryphia cordifolia*	Euc	Tschesche *et al.* (1979)
		3,5-Di-*O*-rhamnose			*Cordia obliqua*	Bor	Srivastava and Srivastava (1979b)
5,7,3′,4′	3		(*6.265*)		*Dillenia indica*	Dil	Pavanasasivan and Sultanbawa (1975)
5,7,3′,4′		6-*C*-Methyl	(*6.266*)	Cedeodarin	*Cedrus deodara*	Cnf	Agrawal *et al.* (1980a)
5,3′	7,4′	8-*C*-Prenyl	(*6.267*)	Tirumalin	*Rhynchosia cyanosperma*	Leg	Adinarayana *et al.* (1980a)
5,7,2′	5′		(*6.269*)		*Inula cappa*	Com	Baruah *et al.* (1979)
5,7,2′,4′				Dihydromorin	*Morus* spp.	Mor	Deshpande *et al.* (1975)
5,7	2′	6″,6″-Dimethyl-pyrano(2″,3″,4′,3′)	(*6.268*)	Sophoronol	*Sophora tomentosa*	Leg	Delle Monache *et al.* (1976)
7,3′,5′	4′		(*6.273*)	Sepinol	*Gliricidia sepium*	Leg	Manners and Jurd (1979)
5,7,3′,4′,5′			(*6.274*)	Ampelopsin	*Soymida febrifuga*	Mel	Rao *et al.* (1979)
					Leptarrhena pyrolifolia	Sax	Miller and Bohm (1979)
					Heuchera villosa	Sax	Wells and Bohm (1980)
					Cedrus deodara	Cnf	Agrawal *et al.* (1980a)
5,7,3′,4′,5′		6-*C*-Methyl	(*6.275*)	Cedrin	*Cedrus deodara*	Cnf	Agrawal *et al.* (1980a)
		3′-*O*-Glucose		Cedrinoside	*C. deodara*	Cnf	Agrawal *et al.* (1980a)
5,7,3′,4′	5′	4′-*O*-Rhamnose	(*6.276*)		*Diospyros peregrina*	Ebn	Chauhan *et al.* (1979a)

* Key: Ana = Anacardiaceae, Bor = Boraginaceae, Cpr = Capparaceae, Cnf = Coniferae, Com = Compositae, Cru = Cruciferae, Dil = Dilleniaceae, Equ = Equisetaceae, Eri = Ericaceae, Ebn = Ebenaceae, Euc = Eucryphiaceae, Lab = Labiatae, Lau = Lauraceae, Leg = Leguminosae, Mel = Meliaceae, Mor = Moraceae, Mus = Musci, Orc = Orchidaceae, Rhm = Rhamnaceae, Ros = Rosaceae, Rut = Rutaceae, Sax = Saxifragaceae.

time are: Capparaceae, Compositae, Cruciferae, Dilleniaceae, Equisetaceae and Eucryphiaceae (see Table 6.9 for references). Aromadendrin 4′-methyl ether

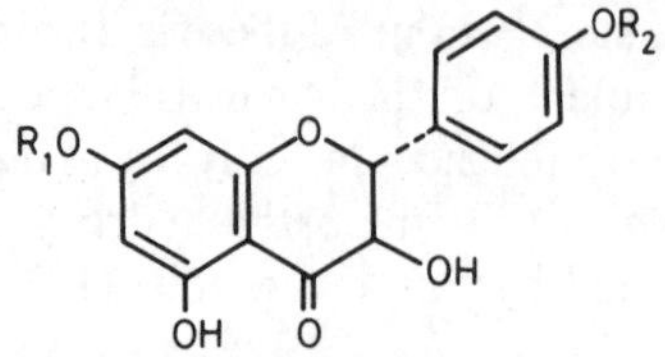

(*6.255*) $R_1 = R_2 = H$

(*6.256*) $R_1 = H, R_2 = CH_3$

(*6.257*) $R_1 = R_2 = CH_3$

(*6.256*) has been reported from *Prunus domestica* (Nagarajan and Parmar, 1977b) which is a confirmation of earlier work by Nagarajan and Seshadri (see Bohm, 1975, Table 11.8). Aromadendrin 7,4′-dimethyl ether (*6.257*) was reported to occur in *Podocarpus neriifolius* (Podocarpaceae) as the 5-*O*-glucoside (Rizvi and Rahman, 1974).

Three *C*-prenylated aromadendrin derivatives have been described in the recent literature. Bohlmann and co-workers (1980a) isolated compounds (*6.258*) and (*6.259*) from *Marshallia obovata* along with a series of flavans with identical or very similar substitution patterns. Compound (*6.259*) is clearly an intramolecular cyclization product derived from (*6.258*). A diprenylated dihydroflavonol (*6.260*) has been reported from *Millettia pachycarpa* by Singhal and co-

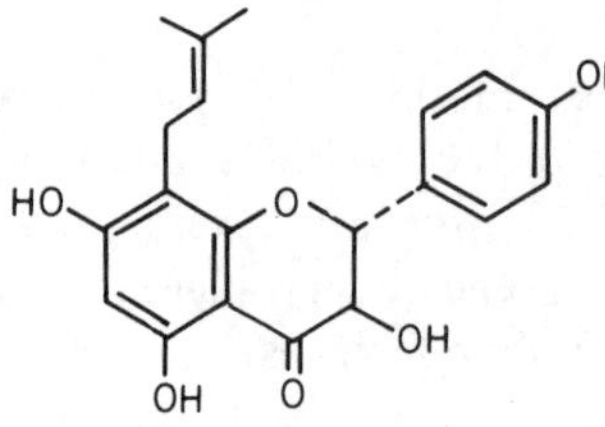

(*6.258*)

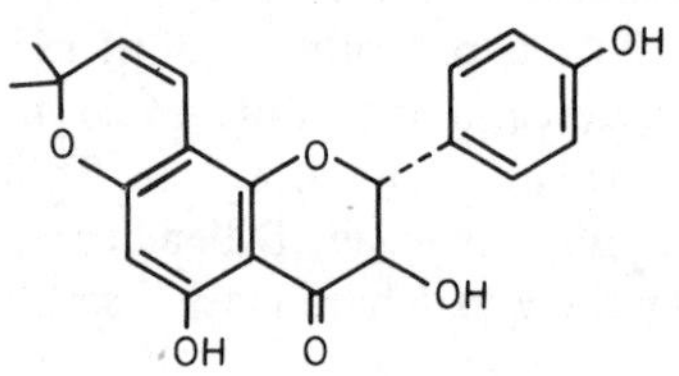

(*6.259*)

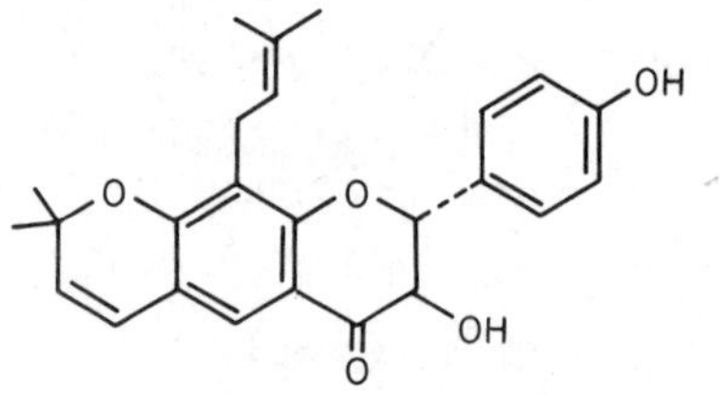

(*6.260*)

workers (1980), where it occurs with a group of similarly substituted isoflavones. These are the first reports of these dihydroflavonols as naturally occurring compounds.

(c) Dihydroflavonols having two B-ring hydroxyls

Several 5-deoxydihydroflavonols having 3′,4′-dihydroxylation have been isolated in recent years. The simplest one of these is 'fustin' (*6.261*) which has long been known as a constituent of the Anacardiaceae. Further reports of its occurrence in *Rhus* were published recently by Young (1979). A '5-deoxydihydroflavonol glycoside with B-ring *ortho*-hydroxyl groups' similar to fustin 3,7-diglycoside was reported by Vandekerkhove (1977) as a constituent of the moss *Georgia pellucida*.

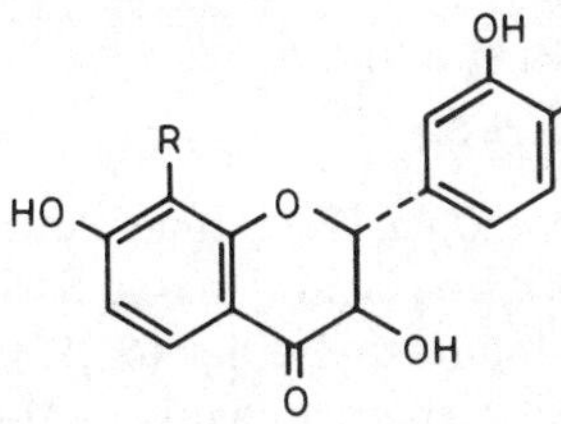

(*6.261*) R = H

(*6.263*) R = OH

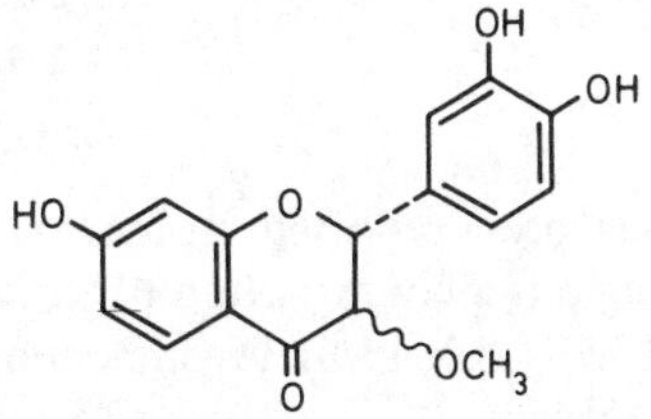

(*6.262a*) α-3-OCH_3

(*6.262b*) β-3-OCH_3

In a detailed study of the polyphenols of *Peltogyne* Malan and Roux (1974) reported finding (±)-3-*O*-methyl-2,3-*cis*-fustin (*6.262a*) and (±)-3-*O*-methyl-2,3-*trans*-fustin (*6.262b*) in both *P. pubescens* and *P. venosa*. Fourie and co-workers (1974) reported fustin (*6.261*) and 8-hydroxyfustin (*6.263*) from *Acacia saxatilis*. The latter compound has also been found to occur in *Albizia adianthifolia* (Leguminosae) wood (Candy *et al*., 1978).

The most commonly encountered member of this subgroup, however, is dihydroquercetin or taxifolin (*6.264*). New records of taxifolin and its glycosidic derivatives are summarized in Table 6.9. Of note are several new family records: Boraginaceae, Dilleniaceae, Equisetaceae, Eucryphiaceae, Labiatae, Orchidaceae and Rhamnaceae (see Table 6.9 for references).

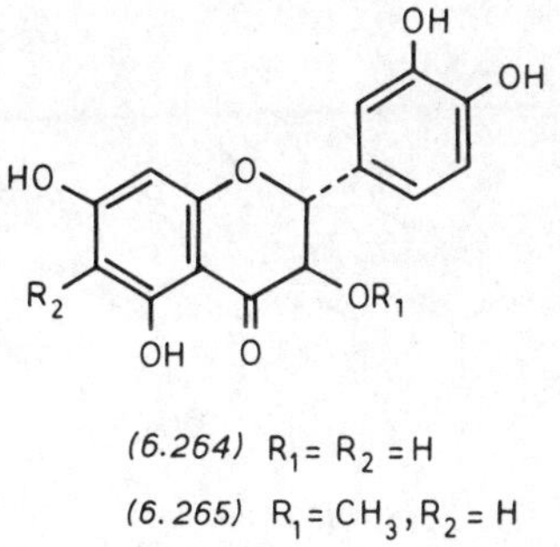

(*6.264*) $R_1 = R_2 = H$

(*6.265*) $R_1 = CH_3, R_2 = H$

(*6.266*) $R_1 = H, R_2 = CH_3$

Naturally occurring derivatives of taxifolin include 3-*O*-methyltaxifolin (*6.265*) from *Dillenia indica* (Pavanasasivan and Sultanbawa, 1975), 6-*C*-methyl

taxifolin (*6.266*), called 'cedeodarin', from *Cedrus deodara* (Agrawal *et al.*, 1980a) and (+)-(2*R*:3*R*)-8-*C*-prenyltaxifolin 7,4′-dimethyl ether (*6.267*) from *Rhynchosia cyanosperma* (Leguminosae) (Adinaryana *et al.*, 1980a).

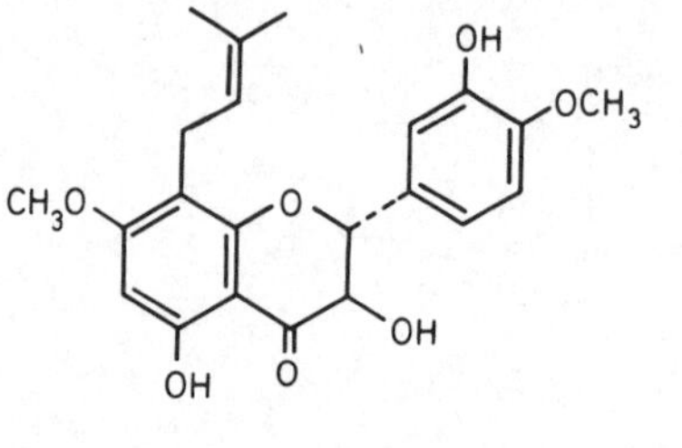

(*6.267*)

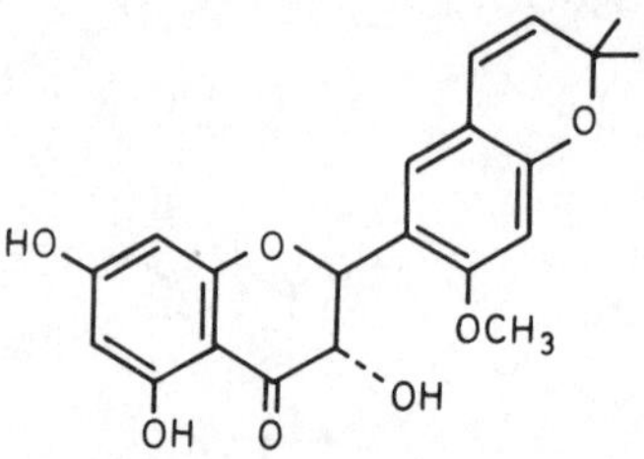

(*6.268*)

A prenylated dihydroflavonol having two unusual features was isolated from *Sophora tomentosa* by Delle Monache and co-workers (1976). 'Sophoronol' (*6.268*) possesses the uncommon 2′,4′-dihydroxylation of ring B and it exists in the rare (2*S*:3*S*) configuration. Also exhibiting an odd B-ring hydroxylation pattern is the compound whose structure is shown as (*6.269*). This compound accompanied a flavanone (*6.241*) with the same substitution pattern in *Inula cappa* (Baruah *et al.*, 1979).

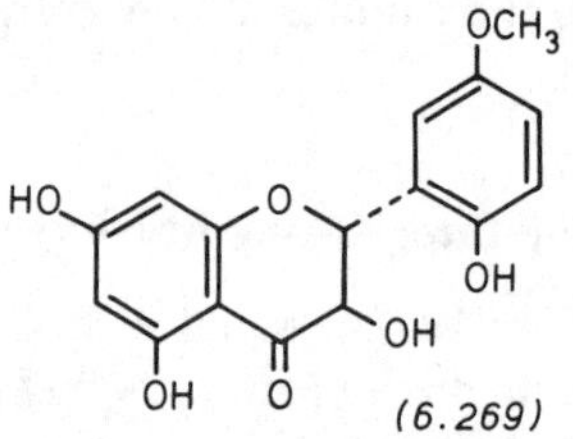

(*6.269*)

The structures of three antihepatotoxic flavonolignans isolated from *Silybum marianum* (Compositae) were discussed in the 1975 review (p. 593). Some question existed concerning the structure of one of these, however. The original structure proposed for 'silychristin' by Wagner *et al.* (1971) is shown as (*6.270*). An alternative structure (*6.271*) was proposed by Hänsel and co-workers (1976).

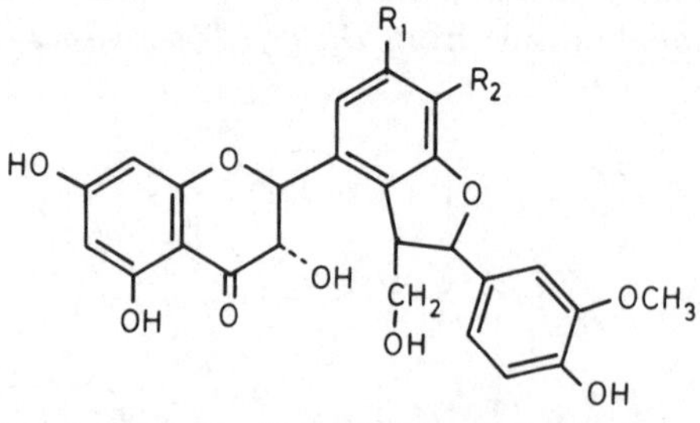

(*6.270*) R_1 = H, R_2 = OH
(*6.271*) R_1 = OH, R_2 = H

More recent work by Pelter and co-workers (1977) and by Wagner and co-workers (1978), using ^{13}C NMR of silychristin and model compounds, unequivocally support yet another structure (*6.272*) for the naturally occurring compound.

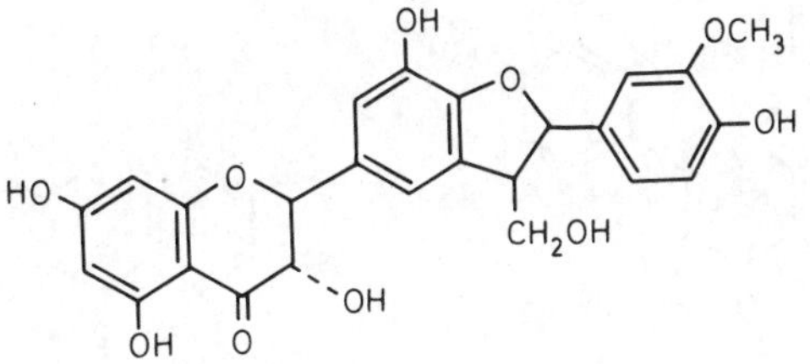

OH OCH3 OH HO O O CH2OH OH OH O

(6.272)

The correct stereochemistry of the known flavonolignans involves the (2*S*:3*S*) configuration, as shown in (*6.272*); the stereochemistry of silychristin presented in the 1978 paper by Wagner and his colleagues was in error (Wagner, 1980).

Two recent papers treat physiological activities of the *Silybum* flavonolignans. All three were shown to inhibit indoleacetic acid oxidase activity (Greimel and Koch, 1977; Becker and Schrall, 1977). Silybin, silydianin and taxifolin stimulated growth of cell cultures of *Silybum* while silychristin had no effect. All were toxic when applied to the cultures at concentrations of 10^{-2} mol. litre^{-1} (Becker and Schrall, 1977).

(d) Dihydroflavonols having three B-ring hydroxyls

'Sepinol', recently described by Manners and Jurd (1979) from *Gliricidia sepium*, is the 4′-methyl ether of dihydrorobinetin (= fustin-4′-methyl ether) (*6.273*). Dihydromyricetin (*6.274*), commonly known as ampelopsin, has been isolated from *Soymida febrifuga* (Rao *et al.*, 1979) which represents a first report for Meliaceae, and from *Cedrus deodara* (Agrawal *et al.*, 1980a) 6-*C*-Methyl-dihydromyricetin (*6.275*) and its 3′-*O*-β-D-glucopyranoside also occur in *C. deodara* (Agrawal *et al.*, 1980a). It is of interest to note that this plant also accumulates taxifolin (*6.264*) and its 6-*C*-methyl derivative (*6.266*). Chauhan and co-workers (1979a) have identified dihydromyricetin 3′-methyl ether (*6.276*) as the 4′-*O*-α-L-rhamnopyranoside from *Diospyros peregrina* (Ebenaceae).

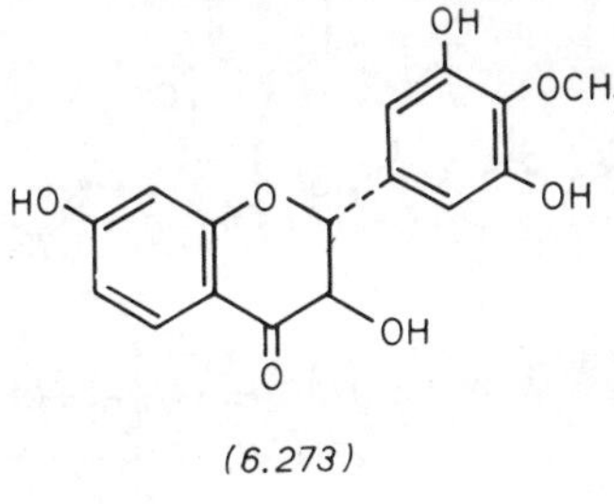

OH OCH3 HO O OH OH O

(6.273)

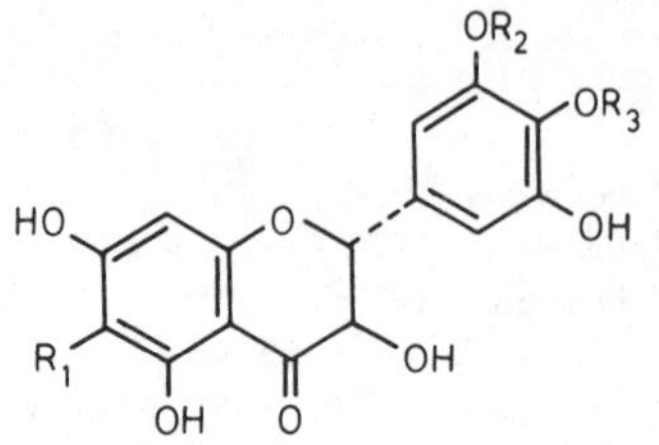

(6.274) $R_1 = R_2 = R_3 = H$

(6.275) $R_1 = CH_3, R_2 = R_3 = H$

(6.276) $R_1 = R_3 = H, R_2 = CH_3$

6.7 TAXONOMIC IMPLICATIONS OF THE DISTRIBUTION OF THE MINOR FLAVONOIDS

6.7.1 Introduction

The occurrence of the five classes of minor flavonoids are summarized in Tables 6.10 (chalcones), 6.11 (aurones), 6.12 (dihydrochalcones) 6.13 (flavanones) and 6.14 (dihydroflavonols). A few general comments on flavonoid distribution would seem in order before considering specific taxonomic problems. We are frequently required to make comparisons of flavonoid patterns when only limited data are available. For example, it is difficult to assess the significance of the paucity of minor flavonoids in the Monocotyledonae. The apparent rarity of these types may reflect a real chemical difference between the Monocotyledonae and Dicotyledonae, or there may simply be insufficient information on the monocots to justify extensive comparisons. Similarly, the amount of information available on the distribution of flavonoids (of all types) in mosses and liverworts is limited. It is important to appreciate, however, that the occurrence of flavonoids in these organisms does not simply represent a chemical novelty. Rather, it would seem that we are witnessing the early stages of accumulation of what is likely to become a taxonomically significant body of knowledge. Application of these data at higher taxonomic levels is simply not feasible yet. On the other side of the coin is the situation exemplified by the Leguminosae and Compositae with regard to minor flavonoids. These families have attracted a great deal of attention because of their size, ready availability, and, no doubt, their economic importance. The result is a very large body of information. In many cases, then, comparisons between families, one of which has been well studied and the other(s) much less so, lead only to very speculative and tentative conclusions. The discussions of the taxonomic implications of minor flavonoid distribution which follow will focus on problems at lower taxonomic levels where reasonable amounts of data are available.

Table 6.10 Summary of natural sources of chalcones by genus

FILICINAE	
Adiantum	Wollenweber (1979)
Cheilanthes	Wollenweber (1977a)
Notholaena	Wollenweber (1977a)
Onychium	Ramakrishnan *et al.* (1974)
Pityrogramma	Wollenweber (1977b, 1979); Wollenweber *et al.* (1980); Dietz *et al.* (1980); Star *et al.* (1975a,b, 1978)
Pterozonium	Wollenweber (1979)
ANGIOSPERMAE–DICOTYLEDONAE	
Acanthaceae	
Ruellia	Bloom (1976)
Anacardiaceae	
Amphipterygium	Young (1976)
Machaerium	Kurosawa *et al.* (1978)
Rhus	Young (1979)
Anonaceae	
Popowia	Panichpol and Waterman (1978)
Betulaceae	
Alnus	Wollenweber *et al.* (1974)
Betula	Wollenweber *et al.* (1974)
Combretaceae	
Terminalia	Najar *et al.* (1979)
Compositae	
Bidens	Ballard (1975); Hart (1979); Scogin and Zakar (1976)
Coreopsidineae	Crawford and Stuessy (1981), see Table 6.19
Coreopsis	Nicholls and Bohm (1979); Harborne (1977); Crawford and Smith (1980)
Chromolaena	Barua *et al.* (1978)
Dahlia	Giannasi (1975); Lam and Wrang (1975)
Flourensia	Bohlmann and Jakupovic (1979)
Gnaphalium	Maruyama *et al.* (1974); Ahluwalia and Rami (1976); Aritomi and Kawasaki (1974)
Helianthus	Harborne and Smith (1978)
Helichrysum	Bohlmann and Suwita (1979c); Bohlmann and Abraham (1979b,c); Bohlmann *et al.* (1979c, 1980c, 1978); Bohlmann and Hoffmann (1979)
Pyrrhopappus	Harborne (1977)
Simsia	Crawford and Stuessy (1981)
Viguiera	Harborne and Smith (1978)
Fagaceae	
Nothofagus	Wollenweber and Wiermann (1979)
Gesneriaceae	
Didymocarpus	Adityachaudhury *et al.* (1976); Adityachaudhury and Dasskanugo (1975); Agarwal *et al.* (1973); Bhattacharyya *et al.* (1979); Wollenweber *et al.* (1980)
Lauraceae	
Aniba	Fernandes *et al.* (1978); de Alleluia *et al.* (1978)

Leguminosae	
Acacia	Imperato (1978); Tindale and Roux (1974)
Albizia	Candy *et al.* (1978)
Bauhinia	Bhartiya *et al.* (1979)
Cladrastis	Ohashi *et al.* (1977)
Dalbergia	Matos *et al.* (1975); Manners *et al.* (1974)
Dalea	Dominguez *et al.* (1980)
Derris	Chibber *et al.* (1979); Braz Filho *et al.* (1975a,b)
Flemingia	Adithyachaudhury *et al.* (1973); Rao *et al.* (1974, 1975a,b,c, 1976); Bhatt (1975)
Glycyrrhiza	Saitoh *et al.* (1975); Saitoh and Shibata (1975)
Lespedeza	Miyase *et al.* (1980)
Lonchocarpus	Delle Monache *et al.* (1974); De Lima *et al.* (1973)
Machaerium	Kurosawa *et al.* (1978)
Medicago	Ingham (1979)
Milletia	Islam *et al.* (1980); Gupta and Krishnamurti (1977a, 1979)
Onobrychis	Ingham (1978)
Peltogyne	Malan and Roux (1974)
Pongamia	Sharma *et al.* (1973); Subramanyam *et al.* (1977)
Prosopis	Bhardwaj *et al.* (1979)
Psoralea	Gupta (1975b, 1977); Suri *et al.* (1980); B. K. Gupta *et al.* (1980)
Sophora	Komatsu *et al.* (1978)
Tephrosia	Camele *et al.* (1980)
Zuccagnia	Pederiva *et al.* (1975)
Loranthaceae	
Viscum	Becker *et al.* (1978)
Myricaceae	
Myrica	Malterud *et al.* (1977)
Myristicaceae	
Iryanthera	de Almeida *et al.* (1979)
Onagraceae	
Oenothera	Dement and Raven (1974)
Rosaceae	
Prunus	Nagarajan and Parmar (1977a,b,c)
Rutaceae	
Merrillea	Fraser and Lewis (1974)
Salicaceae	
Populus	Wollenweber *et al.* (1974); Wollenweber and Weber (1973);
Salix	Vinokurov (1979)
Solanaceae	
Petunia	de Vlaming and Kho (1976)
Umbelliferae	
Angelica	Kozawa *et al.* (1977)
MONOCOTYLEDONAE	
Liliaceae	
Polygonatum	Maradufu and Ouma (1978)

Table 6.11 Summary of natural sources of aurones by genus

BRYOPHYTA–HEPATICAE	
Carrpos	Markham (1980)
Conocephalum	Markham and Porter (1978)
Marchantia	Markham and Porter (1978)
BRYOPHYTA–MUSCI	
Funaria	Weita and Ikan (1977)
ANGIOSPERMAE–DICOTYLEDONAE	
Anacardiaceae	
Amphipterygium	Young (1976)
Melanorrhoea	Ogiyama *et al.* (1976)
Rhus	Young (1979)
Compositae	
Bidens	Hart (1979); Scogin and Zakar (1976)
Coreopsidinae	Crawford and Stuessy (1981)
Coreopsis	Nicholls and Bohm (1979); Crawford and Smith (1980)
Helianthus	Harborne and Smith (1978)
Viguiera	Harborne and Smith (1978)
Leguminosae	
Derris	do Nascimento *et al.* (1976)
Lygos	El Sherbeiny *et al.* (1978)
Scrophulariaceae	
Antirrhinum	Gilbert (1975)
MONOCOTYLEDONAE	
Zingiberaceae	
Amomum	Lakshmi and Chauhan (1977)

6.7.2 Filicineae

(a) Introduction

The flavonoid chemistry of this ancient and widespread group of plants has received comparatively little systematic study despite its size (Mabry, 1973). The last few years, however, have witnessed a growing interest in chemosystematic studies of ferns (Cooper-Driver, 1980). Several compounds belonging to the structural types treated in this chapter have been isolated from fern species. Some of the occurrences, such as the presence of naringin in six *Adiantum* species (of 58 tested) (Cooper-Driver and Swain, 1977), have little systematic value and will not be dwelt upon. Other situations, such as the flavanone distributions described by Hiraoka (1978) and the exudate chemistry of *Pityrogramma* (Wollenweber, 1978; Wollenweber and Dietz, 1980) are discussed in some detail below.

The part played by flavonoids, and a few related phenolic compounds, in fern systematics has been reviewed recently by Cooper-Driver (1980) to which the reader is referred for details. Included in that treatment is a tablulation of the distribution of flavonoid types in fern families. Flavanones and/or chalcones were reported absent from 17 families and present in only three: Pteridaceae, Dennstaedtiaceae and Dryopteridaceae (includes Athyriaceae in this treatment).

Table 6.12 Summary of natural sources of dihydrochalcones by genus

BRYOPHYTA–HEPATICAE	
Radula	Asakawa *et al.* (1978)
FILICINAE	
Adiantum	Wollenweber (1979)
Notholaena	Wollenweber (1977a)
Pityrogramma	Wollenweber (1977b, 1979); Dietz *et al.* (1980); Wagner *et al.* (1979); Wollenweber *et al.* (1979b)
ANGIOSPERMAE–DICOTYLEDONAE	
Anonaceae	
Uvaria	Hufford and Oguntimein (1980); Cole *et al.* (1976); Tammami *et al.* (1977); Okorie (1977)
Unona	Joshi and Gawad (1976)
Balanophoraceae	
Balanophora	Ito *et al.* (1980)
Compositae	
Helichrysum	Bohlmann and Abraham (1979a); Bohlmann and Suwita (1979b); Bohlmann and Zdero (1980); Bohlmann *et al.* (1978, 1979c, 1980c)
Caprifoliaceae	
Viburnum	Jensen *et al.* (1977)
Ericaceae	
Rhododendron	King (1977); Mabry *et al.* (1975)
Leguminosae	
Gliricidia	Manners and Jurd (1979)
Glycyrrhiza	Furuya *et al.* (1976)
Lonchocarpus	Delle Monache *et al.* (1974)
Milletia	Gupta and Krishnamurti (1976a,b,c, 1977a,b, 1979)
Pongamia	Talapatra *et al.* (1980); Garg *et al.* (1978); Sharma and Parthasarathy (1977)
Tephrosia	R. K. Gupta *et al.* (1980)
Myricaceae	
Myrica	Malterud *et al.* (1977); Anthonsen *et al.* (1971); Uyar *et al.* (1978)
Myristicaceae	
Iryanthera	Braz Filho *et al.* (1980); de Almeida *et al.* (1979)
Rosaceae	
Malus	A. H. Williams (1979)

The phylogenetic significance of flavonoid distribution was commented upon by Cooper-Driver (1980), who pointed out that more advanced families of ferns accumulate flavone-*C*-glycosides, flavanones, chalcones and xanthones, whereas the most ancient families accumulate only proanthocyanidins and flavonols. Some specific studies are summarized below.

(b) Athyriaceae

Hiraoka (1978) studied the flavonoid composition of 21 species representing six genera in the Athyriaceae. The flavonoids of the onocleoid ferns (*Matteuccia, Onoclea, Onocleopsis*) differ from those present in the athyreoid ferns (*Athyrium,*

Table 6.13 Summary of natural sources of flavanones by genus

Taxon	Reference
FILICINAE (Includes *Equisetum*)	
Adiantum	Cooper-Driver and Swain (1977)
Cheilanthes	Wollenweber *et al.* (1980a)
Cyrtomium	Hiraoka (1978)
Dryopteris	Hiraoka (1978)
Matteuccia	Hiraoka (1978)
Notholaena	Wollenweber *et al.* (1980a)
Pityrogramma	Wollenweber *et al.* (1979a); Star *et al.* (1978)
Polystichum	Hiraoaka (1978)
Equisetum	Syrchina *et al.* (1975, 1978)
CONIFERAE	
Larix	Lapteva *et al.* (1974)
Picea	Ivanova *et al.* (1975)
ANGIOSPERMAE–DICOTYLEDONAE	
Acanthaceae	
Haplanthus	Viswanathan and Sidhaye (1976)
Aizoaceae	
Tetragonia	Kemp *et al.* (1979)
Amaranthaceae	
Aerva	Garg *et al.* (1980)
Anacardiaceae	
Anacardium	Rahman *et al.* (1978)
Anonaceae	
Unona	Joshi and Gawad (1974); Chopin *et al.* (1978)
Uvaria	Hufford and Laswell (1976); Hufford *et al.* (1979)
Popowia	Panichpol and Waterman (1978); Waterman and Pootakahm (1979)
Betulaceae	
Betula	Wollenweber (1975)
Boraginaceae	
Cordia	Chauhan *et al.* (1978); Tiwari and Srivastava (1979)
Capparidaceae	
Cleome	Chauhan *et al.* (1979b); Srivastava *et al.* (1979) Srivastava and Srivastava (1979a)
Clusiaceae	
Garcinia	Waterman and Crichton (1980)
Cochlospermaceae	
Cochlospermum	Cook and Knox (1975)
Combretaceae	
Terminalia	Nagar *et al.* (1979)
Compositae	
Baccharis	Bohlmann *et al.* (1979b)
Bidens	Hart (1979); Serbin *et al.* (1975)
Chromolaena	Barua *et al.* (1978)
Chrysothamnus	Bohlmann *et al.* (1979a)
Dahlia	Lam and Wrang (1975)
Eupatorium	Arene *et al.* (1978)
Flourensia	Bohlmann and Jakupovic (1979)
Gnaphalium	Maruyama *et al.* (1974)
Gymnopentzia	Bohlmann and Zdero (1978)

Helichrysum	Bohlmann and Abraham (1979a,b,d); Bohlmann *et al.* (1978, 1979e, 1980c)
Inula	Baruah *et al.* (1979)
Marshallia	Bohlmann *et al.* (1979d)
Othonna	Bohlmann and Knoll (1978)
Polymnia	Bohlmann and Zdero (1977)
Senecio	Bohlmann and Zdero (1979)
Steiractinia	Bohlmann *et al.* (1980b)
Tarchonanthus	Bohlmann and Suwita (1979a)
Coriariaceae	
Coriaria	Bohm and Ornduff (1981)
Cruciferae	
Matthiola	Forkmann (1979)
Dilleniaceae	
Dillenia	Pavanasasivan and Sultanbawa (1975)
Ericaceae	
Rhododendron	King (1977a,b)
Euphorbiaceae	
Phyllanthus	Chauhan *et al.* (1977)
Fagaceae	
Nothofagus	Wollenweber and Wiermann (1979)
Gentianaceae	
Hoppea	Ghosal *et al.* (1978)
Gesneriaceae	
Didymocarpus	Bose and Adithyachaudhury (1978); Garg *et al.* (1979)
Labiatae	
Coleus	Brieskorn and Riedel (1977)
Mentha	Bourweig and Pohl (1973)
Lauraceae	
Aniba	Fernandes *et al.* (1978)
Litsea	Mohan *et al.* (1978)
Malvaceae	
Hibiscus	Chauhan *et al.* (1979c)
Meliaceae	
Soymida	Rao *et al.* (1979)
Moraceae	
Chlorophora	Gottlieb *et al.* (1975)
Myrtaceae	
Agonis	Cannon and Martin (1977)
Myristicaceae	
Iryanthera	de Almeida *et al.* (1979)
Rosaceae	
Persica	Sadykov and Sadykov (1979)
Prunus	Nagarajan and Parmar (1977a,b,c)
Rubiaceae	
Lasianthus	Ishikura *et al.* (1979)
Rutaceae	
Citrus hybrids	Kamiya *et al.* (1979)
Clymenia	Grieve and Scora (1980)
Eremocitrus	Grieve and Scora (1980)
Fortunella	Grieve and Scora (1980)
Microcitrus	Grieve and Scora (1980)
Pleiospermium	Grieve and Scora (1980)

Table 6.13 (*Contd.*)

Poncirus	Grieve and Scora (1980)
Zanthoxylum	Fish *et al.* (1975a,b)
Evoda	Grimshaw and Lamer-Zarawska (1975)
Leguminosae	
Acacia	Fourie *et al.* (1974); El Sissi *et al.* (1974); Gupta *et al.* (1980); Rao and Raju (1979)
Bauhinia	A. K. Gupta *et al.* (1979, 1980)
Cajanus	Bhanumati *et al.* (1978)
Cassia	Tiwari and Geiger (1975)
Dalbergia	Matos *et al.* (1975); Manners *et al.* (1974); Parthasarathy *et al.* (1976); Letcher and Shirley (1976)
Dalea	Dreyer *et al.* (1975); Dominguez *et al.* (1980)
Derris	Braz Filho *et al.* (1975)
Flemingia	Sivarambabu *et al.* (1979a,b); Rao *et al.* (1974, 1975b)
Galega	Traub and Geiger (1975)
Gliricidia	Jurd and Manners (1977)
Glycyrrhiza	Kaltaev and Nikonov (1974)
Indigofera	Dominguez *et al.* (1978a,b)
Lygos	El Sherbeiny *et al.* (1978)
Medicago	Ingham (1979)
Milletia	Islam *et al.* (1980); Gupta and Krishnamurti, (1976a,b)
Myroxylon	de Oliveira *et al.* (1978)
Onobrychis	Ingham (1978)
Peltogyne	Malan and Roux (1974)
Pongamia	Satam and Brinji (1973); Talapatra *et al.* (1980)
Rhynchosia	Adinarayana *et al.* (1980a,b)
Sophora	Jain and Gupta (1978)
Tephrosia	Jayaraman *et al.* (1980); Camele *et al.* (1980)
Zuccagnia	Pederiva *et al.* (1975)
Salicaceae	
Salix	Kompantsev and Shinkarenko (1975)
Solanaceae	
Nierembergia	Pomilio and Gros (1979)
Theaceae	
Camellia	Imperato (1976)
MONOCOTYLEDONAE	
Liliaceae	
Tulipa	Budzianowski and Skrzypczakowa (1978)

Table 6.14 Summary of natural sources of dihydroflavonols by genus

BRYOPHYTA–MUSCI	
Georgia	Vandekerkhove (1977)
CONIFERAE	
Abies	Parker *et al.* (1979)
Cedrus	Agrawal *et al.* (1980a,b)
Larix	Niemann (1974); Lapteva *et al.* (1974)
Picea	Ivanova *et al.* (1975)
Podocarpus	Rizvi and Rahman (1974)

ANGIOSPERMAE–DICOTYLEDONAE

Anacardiaceae	
Amphipterygium	Young (1976)
Rhus	Young (1979)
Boraginaceae	
Cordia	Srivastava and Srivastava (1979)
Capparidaceae	
Cleome	Chauhan *et al.* (1979b)
Compositae	
Helichrysum	Bohlmann and Abraham (1979a)
Inula	Barauh *et al.* (1979)
Marshallia	Bohlmann *et al.* (1980a)
Othonna	Bohlmann and Knoll (1978)
Cruciferae	
Matthiola	Forkmann (1979)
Dilleniaceae	
Dillenia	Pavanasasivan and Sultanbawa (1975)
Wormia	Pavanasasivan and Sultanbawa (1975)
Ebenaceae	
Diospyros	Chauhan *et al.* (1979a)
Ericaceae	
Cavendishia	Luteyn *et al.* (1980)
Rhododendron	King (1977a,b)
Eucryphiaceae	
Euchryphia	Tschesche *et al.* (1979)
Labiatae	
Coleus	Brieskorn and Riedl (1977)
Lauraceae	
Aniba	Fernandes *et al.* (1978)
Leguminosae	
Acacia	Fourie *et al.* (1974)
Albizia	Candy *et al.* (1978)
Dalbergia	Matos *et al.* (1975, 1976); Dayal and Parthasarathy (1977)
Gliricidia	Manners and Jurd (1979)
Lonchocarpus	Delle Monache *et al.* (1978)
Milletia	Singhal *et al.* (1980)
Peltogyne	Malan and Roux (1974)
Rhynchosia	Adinarayana *et al.* (1980a)
Sophora	Delle Monache *et al.* (1976)
Meliaceae	
Soymida	Rao *et al.* (1979)
Moraceae	
Cudrania	Gupta *et al.* (1975a)
Rhamnaceae	
Zizyphus	Srivastava and Chauhan (1977)
Rosaceae	
Persica	Sadykov and Sadykov (1979)
Prunus	Nagarajan and Parmar (1977b,c)
Rutaceae	
Phellodendron	Otryashenkova *et al.* (1977)
Saxifragaceae	
Heuchera	Wells and Bohm (1980)

Table 6.14 (*Contd.*)

Leptarrhena	Miller and Bohm (1979)
Saxifraga	Miller and Bohm (1980)
MONOCOTYLEDONAE	
Orchidaceae	
Aerides	C. A. Williams (1979)

Diplazium, Lunathyrium), a difference which parallels morphological differences between the two groups. The onocleoid ferns show a much more complex pattern of flavonol glycosides, and *Matteuccia* is further distinguished through the presence of the flavanones matteucinol 7-glucoside and desmethoxymatteucinol 7-glucoside (tentative identification).

(c) Dryopteridaceae

In the same paper Hiraoka (1978) also determined the flavonoids of 30 species from five genera of Dryopteridaceae. Flavanones figure significantly in discussions of inter- as well as infra-generic distinctions. *Cyrtomium* is clearly distinguished from other members of the family owing to the presence of the flavanones cyrtominetin and farrerol (tentative identifications). Cyrtominetin has been found only in this genus to date (see Bohm, 1975, Table 11.1, p. 568) while farrerol, also known from *Cyrtomium*, occurs in flowering plants as well (ibid., p. 568).

In support of the recognition of *Arachniodes* and *Leptorumohra* as separate genera, Hiraoka (1978) pointed out that species of the former make only flavones while the latter makes only flavonols. He did not mention that a further distinguishing compound marks separate generic status: *L. miqueliana* has been shown to possess the odd, reduced B-ring flavanone 'protofarrerol' (*6.277*) (Noro *et al.*, 1969).

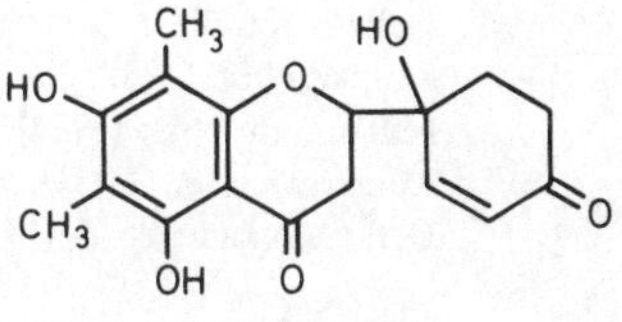

(*6.277*)

Differences between *Polysticum* and *Dryopteris* exist in the array of flavonols and flavones elaborated by each. There are differences in flavanone structure and distribution as well, but these differences are much less clear cut.

Flavonoids allow clear distinctions to be drawn between groups of species within *Dryopteris* section *Erythro-variae*, however. The first group, comprising *D. bissetiana, D. sordidipes, D. pacifica* and *D. sacrosancta*, exhibits a complex pattern of both flavones and flavonol glycosides and has naringenin 7-*O*-glucoside. The second group of species, *D. erythrosera, D. nipponensis,*

D. championii, *D. hondoensis*, *D. gymnosora* and *D. gymnophylla*, has the simpler array of flavones and flavonols and has eriodictyol 7-*O*-rhamnoglucoside. All members of both groups possess eriodictyol 7-*O*-glucoside. Hiraoka (1978) pointed out that the Japanese members of this section fall into two groups on the basis of morphological characters which parallel the flavonoid variation.

Dryopteris watanabei has been shown by cytological analysis to be a natural hybrid between *D. crassirhizoma* and *D. uniformis*. *Dryopteris uniformis* accumulates an array of flavonoids consisting of six flavonol glycosides while *D. crassirhizoma* was shown to have three flavones, two flavonol glycosides and eriodictyol 7-*O*-glucoside. The pattern of *D. watanabei* shows total additivity of the parental flavonoid phenotypes.

(d) Pteridaceae

Much intensive work on the chemistry of *Pityrogramma* has been published during the past few years. Members of this genus are commonly referred to as silver-back or gold-back ferns owing to the frequently copious production of semi-crystalline material by leaf glands. Chalcones and dihydrochalcones are major constituents of the exudates of most species while in others they occur along with a variety of methylated flavone and flavonol aglycones. Several reviews of the subject have appeared recently (Wollenweber, 1978; Star, 1980; Wollenweber and Dietz, 1980). Table 6.15 summarizes the distribution of the most commonly met chalcones and dihydrochalcones and attempts to point out some of the variation seen. A detailed discussion of the variation patterns observed is beyond the scope of this chapter and should be sought in the article by Wollenweber and Dietz (1980).

Pityrogramma triangularis is distinguished from other members of the genus studied to date by virtue of the replacement of the chalcones and dihydrochalcones seen in all other species by several flavonoids including the unusual A-ring chalcone ceroptin. The occurrence of these other flavonoids is not general throughout the species but closely follows varietal lines. These data are presented in Table 6.16. Further flavonoid variation in *P. triangularis* var. *triangularis* was found and discussed by Star and co-workers (1975a,b), who pointed out the usefulness of the vacuolar flavonoids in distinguishing ploidy levels in the chemotypes.

Although there are some exceptions, notably in *Cheilanthes* and *Notholaena*, the four main compounds whose distribution is summarized in Table 6.15, serve well as generic markers, setting *Pityrogramma* off from other Gymnogrammoid ferns (Wollenweber and Dietz, 1980). These authors point out that *P. triangularis* does not fit the pattern, but offer no suggestion as to why this taxon, with its four chemically different varieties, should be expected to be so different.

In 1852 Fée removed *P. trifoliata* from *Pityrogramma* on the basis of differences in frond morphology and placed it as the sole member of *Trismeria*. Tryon (1962) disagreed with this view and returned the taxon to *Pityrogramma*,

Table 6.15 Summary of the distribution of chalcones and dihydrochalcones in *Pityrogramma**†

Species	2′,6′-Dihydroxy-4′-methoxychalcone-4-R		2′,6′-Dihydroxy-4′-methoxydihydrochalcone-4-R		Comments
	R = H	R = OCH_3	R = H	R = OCH_3	
P. argentea	+ (9/9)‖	—	+ (4/4)	—	
P. aurentiaca	+ (4/4)	—	—	—	
P. calomelanos	+	+	+	+	Extensive intra- and inter-varietal variation*
P. chrysoconia	+ (1/7)	—	+ (5/7)	—	
P. chrysophylla	+ (8/10)	+ (6/10)	+ (3/10)	+ (3/10)	var. *marginata* has 2′,4-dihydroxy-4′,6′-dimethoxydihydrochalcone‡
P. dealbata	+ (1/9)	—	+ (8/9)	+ (2/9)	
P. dukei	—	+ (1/1)	—	—	Only one sample tested
P. lehmanii	—	—	—	+ (1/1)	Only one sample tested
P. pulchella	—	—	+ (1/1)	—	Only one sample tested
P. schizophylla†	—	+	—	+	No variation data available
P. sulphurea	+ (7/7)	+†	—	—	Has cinnamoyldihydrocoumarin ('T-1')§
P. tartarea	+	+	+	+	Extensive variation; has 2′,6′,4-trihydroxy-4′-methoxychalcone
P. triangularis	—	—	—	—	Extensive variation, see Table 6.16 for details
P. trifoliata	+ (1/27)	+ (1/27)	+ (27/27)	+ (15/27)	Has cinnamoyldihydrocoumarins ('T-1', 'T-2', 'T-3')§
P. williamsii	+ (2/3)	—	—	—	Has cinnamoyldihydrocoumarin ('T-1')§
P. trifoliata × *calomelanos*	—	—	+ (5/5)	+ (4/5)	

* Data for this Table comes from Wollenweber and Dietz (1980) with additions from other sources as indicated below.
† From Star (1980).
‡ Nilsson (1961).
§ Dietz *et al.* (1980).
‖ Number of samples positive for the compound/number of samples tested.

Table 6.16 Flavonoid variation in *Pityrogramma triangularis*

Varieties of P. triangularis	*Constituents (No.)**	*References*
var. *triangularis*	Ceroptin (*6.32*)	Star *et al.* (1975a,b)
	Galangin and kaempferol methyl ethers†	Star *et al.* (1975a,b)
	Triangularin (*6.29*)‡	Star *et al.* (1978)
var. *maxonii*	Galangin	Wollenweber and Dietz (1980)
var. *pallida*‡	Cryptostrobin (*6.180*)	Wollenweber *et al.* (1979a)
	Strobopinin (*6.179*)	Wollenweber *et al.* (1979a)
	Desmethoxymatteucinol (*6.181*)	Wollenweber *et al.* (1979a)
	Pinocembrin (*6.170*)	Wollenweber *et al.* (1979a)
	Aurentiacin (*6.30*)	Wollenweber *et al.* (1980b)
	Flavokawin-B (*6.15*)	Wollenweber *et al.* (1980b)
var. *viscosa*	2′,6′,4-Trihydroxy-4′-methoxy-5-*C*-methyldihydrochalcone (*6.140*)	Wollenweber *et al.* (1979b)

* Refer to text for structures, the numbers of which are shown.
† Flavonol derivatives.
‡ A trace of the corresponding flavanone, 7-methylstrobopinin (*6.182*), was also detected.
§ Also seen as *Pityrogramma pallida* in text and Table 6.6.

a move which is supported by the farina chemistry. *Pityrogramma trifoliata* exhibits the characteristic dihydrochalcones seen in the genus. It also can make the cinnamoyl dihydrocoumarin derivatives (T-1, T-2 and T-3 of Dietz *et al.*, 1980) in common with *P. sulphurea* and *P. williamsii*.

6.7.3 Anacardiaceae

(a) *Rhus*

Young (1979) examined the heartwood flavonoids of 23 taxa of *Rhus* as part of an investigation of infrageneric relationships. The aglycones identified were the aurones rengasin and sulfuretin, 7,4′ dihydroxydihydroflavonol, 7,3′,4′-trihydroxydihydroflavonol, 7,3′,4′-trihydroxyflavonol (fisetin) and 3′,4′-dihydroxyflavone. These compounds plus a glucoside of sulfuretin were found in all 23 taxa. Variation was seen in the distribution of glucosides of butein, fisetin and 3′,4′-dihydroxyflavone. It is this variation that sheds light on subgeneric relationships. *Rhus* subgenus *Rhus* lacks all three glucosides, *Rhus* subgenus *Lobadium* section *Terebinthifolia* has fisetin glucoside but lacks the other two compounds, section *Lobadium* has fisetin glucoside and dihydroxyflavone glucoside, while section *Styphonia*, with significant exceptions, has all three glucosides. Section *Styphonia* is subdivided into three subsections on the basis of leaf morphology. Subsections *Intermediae* and *Styphoniae* are identical in their possession of all three glucosides while some species of subsection *Compositae* lack the capacity to make the 3′,4′-dihydroxyflavone glucoside. It is of interest

that the species lacking the compound are among the more primitive members of the group. Young (1979) pointed out that the major evolutionary trend seemed to be toward accumulation of glycosides of the heartwood flavonoids. The data are summarized in Table 6.17.

Table 6.17 Distribution of selected flavonoid glycosides in *Rhus**

		Glycosides of	
	FIS	*DHF*	*BUT*
Rhus subgen. *Rhus*	–	–	–
subgen. *Lobadium*			
sect. *Terebinthifolia*	+	–	–
sect. *Lobadium*	+	+	–
sect. *Stylonia*			
subsect. *Compositae*	+	–/+	+
subsect. *Intermediae*	+	+	+
subsect. *Styphoniae*	+	+	+

* After Young (1979). Key: FIS = fisetin, DHF = 3′,4′-dihydroxyflavone, BUT = butein.

(b) Amphipterygium

In an earlier paper Young (1976) also used flavonoid chemistry to assess the relationships of the Julianaceae, a family whose position in or near the Anacardiaceae has been the subject of some discussion. Sixteen compounds were isolated from *Amphipterygium adstrigens*: seven 5-deoxyflavonoids plus rengasin from the roots, and nine 5-hydroxyflavonoids, including rengasin, and sulfuretin from the leaves. The flavonoid pattern of *Amphipterygium* is very similar to that of the Anacardiaceae, particularly to that of the tribe Rhoeae. Young concluded, on the basis of the flavonoid data as well as anatomical and morphological similarities, that the genus, and thus *Orthopterygium* by association, belonged in the Rhoeae. Differences in fruit structure and some floral characteristics, however, were taken as support for placing these genera in the subtribe Julianiinae within Rhoeae.

6.7.4 Compositae

(a) Bidens

Bidens pilosa L. was considered by Sherff (1937) to be a single species having widely variant morphological characters throughout its very extensive range. Detailed field studies coupled with an experimental breeding programme and cytological and flavonoid studies led Ballard (1975) to recognize the existence of four species (eight taxa in all). A significant contribution to Ballard's data was the taxon-specific distribution of okanin derivatives, namely, 4′-glycosides, 3′,4′-

diglycosides, 3-*O*-methyl-4′-glycosides and 3,4-di-*O*-methyl-4′-glycosides. The absence of flavonols from leaves of *B. clavatus*, along with its self-incompatibility, also helps to distinguish it from *B. odorata*, since these two species are the only diploids ($n = 12$) in the group. Ballard's data are summarized in Table 6.18. The Table includes information about *B. bigelovii* to which the *B. pilosa* complex was compared.

Table 6.18 Summary of *Bidens pilosa* complex data*

		Breed†	*Okanin derivatives*‡				
Taxon	*n*	*biol.*	*4′G*	*3′4G*	*3OM*	*34OM*	*Flav*
B. odorata							
var. *odorata*	12	comp			+		+
var. *calcicola*	12	comp	+				+
var. *oaxacensis*	12	comp		+			+
var. *chilpancingensis*	12	comp				+	+
B. clavatus	12	incomp			+		−
B. alba var. *alba*	24	incomp			+		+
var. *radiata*	24	incomp			+	+	+
B. pilosa	36	comp	+				+
B. bigelovii							
var. *bigelovii*	24	comp	+				+
var. *angustiloba*	24	comp	+				+

* From Ballard (1975).
† Breeding biology: comp = self-compatible, incomp = self-incompatible.
‡ 4′G = 4′-*O*-glucoside, 3′4G = 3′,4′di-*O*-glucoside, 3OM = 3-*O*-methyl, 34OM = 3,4-di-*O*-methyl, Flav = flavonols.

(b) Megalodonta

Megalodonta is a specialized aquatic genus closely related to *Bidens*. In a recent paper Roberts (1980) described the flavonoid chemistry of *M. beckii*, the sole member of the genus, as part of a study of the relationship between the two genera. *Megalodonta* accumulates a number of chalcones and aurones including the comparatively rare 2′,4′,3-trihydroxy-3′-methoxychalcone 4′-*O*-glucoside. This chalcone, as the free phenol, occurs in *Dahlia* (Giannasi, 1975) and in members of two sections of *Bidens: Platycarpaea* and *Heterodonta* (Roberts, 1980). The systematic significance of this distribution remains to be determined following wider sampling within *Bidens*, but a close association between the two genera is clearly supported by these phytochemical observations.

(c) Dahlia

In a detailed study of *Dahlia*, Giannasi (1975) reported the distribution of 64 glyosides and 15 aglycones in 21 taxa obtained from Mexico, Central America and northern South America. The flavonoids represented five classes of

compounds: flavones, flavonols, chalcones, aurones and anthocyanins. The chemical data are useful at several taxonomic levels. Several glycosides appear to characterize the genus: diosmetin (= luteolin 4′-*O*-methyl ether) glucosides, butein galactoside, butein-4,4′-tri- and tetra-glucosides and butein arabinosylgalactosides. At the sectional level the presence of 6-methoxyflavones clearly separates section *Entemophyllum* from sections *Dahlia* and *Pseudodendron* wherein flavonols were found. At the interspecific level individual species exhibited specific combinations of pigments, although in several cases in sections *Dahlia* and *Pseudodendron* greater use of tissue specific profiles was necessary, since total flavonoid complements were often identical.

Some distinctions at the intergeneric level can be made on the basis of localization of anthochlors. In *Dahlia*, flavones, flavonols and anthochlors occur in both leaf and floral tissue, whereas in *Cosmos*, leaves possess flavones and flavonols while anthochlors are restricted to floral tissue. It is tempting to speculate that this situation reflects specialization in the case of *Cosmos* where anthochlors are responsible for the flavonoid contribution to flower colour, whereas the even distribution of anthochlor and non-anthochlor pigments throughout aerial parts of *Dahlia* species represents the non-specialized situation. A detailed comparative study of the reproductive biology of these two genera might afford useful insight into the selective pressures associated with tissue specific distribution of anthochlors and of flavonoids in general.

(d) Coreopsidinae

Coreopsis has figured significantly in the study of the structure and distribution of anthochlor pigments. Beginning with the research of Gertz (1938) the subject has proceeded from surveys using simple colour tests to the present position where anthochlor pigments, along with a host of other characters, play an important part in helping to define the limits of the subtribe Coreopsidinae (Heliantheae) (Crawford and Stuessy, 1981). In that paper, Crawford and Stuessy reported the presence of anthochlors in 15 genera of Coreopsidinae not studied to date [their count of 16 included *Megalodonta*, work on which has now been published by Roberts (1980)]. Table 6.19 lists the genera studied and the types of compounds found.

With publication of these data 30 of 32 genera in Coreopsidinae have been tested for anthochlors (*Cyathomone* and *Sphagneticola* were not available). Of the 30 genera tested only two lacked anthochlors: *Guardiola* Cerv. *ex* H. and B (five species of 13 known) and *Venegasia* DC. (monotypic, four collections tested). The authors state that, although exact affinities within the tribe are difficult to establish, a number of characters clearly place *Guardiola* in the subtribe. They suggest that absence of anthochlors represents an evolutionary loss. The absence of anthochlors from *Venegasia* was similarly explained since the genus is closely related to *Jaumea*; both have $n = 19$, an unusual number within the subtribe. The authors suggest that loss of anthochlors could be explained through 'addition of

Table 6.19 New generic records of anthochlors in the Coreopsidinae*

Genus	*Taxa tested*†	*Anthochlors found*
Coreocarpus	3/3/10	Butein and sulfuretin glucosides
Dicranocarpus	1/1/1	Not identified
Ericentrodea	2/2/3	Butein and sulfuretin glucosides
Eringiophyllum	1/1/1	Butein and sulfuretin glucosides, okanin and maritimetin glucosides
Glossocardia	1/1/2	Not identified
Goldmanella	1/1/1	Butein and sulfuretin glucosides, okan and maritimetin glucosides
Guerreroia	1/1/1	Not identified
Henricksonia	1/1/1	Not identified
Jaumea	3/3/20	Butein and sulfuretin glucosides
Moonia	1/1/1	Not identified
Narvalina	1/1/4	Butein and sulfuretin glucosides, okanin and maritimetin glucosides
Oparanthus	1/1/2	Not identified
Petrobium	1/1/1	Not identified
Selleophytum	1/1/1	Butein and sulfuretin glucosides
Staurochlamys	1/1/1	Not identified

* After Crawford and Stuessy (1981).
† $m/n/p$ where m = number of species positive for anthochlors, n = number of species tested, p = number of species recognized in genus.

a few enzymes to the basic flavonoid biosynthetic pathway.' It is not clear to me how this could happen. Since formation of aurones appears to require an oxidase, absence of aurones would imply loss or suppression of such an enzyme. Also, since chalcones are generally considered to be obligate intermediates in flavonoid biosynthesis the absence of chalcones may only indicate that the rate of chalcone consumption exceeds the rate of formation, the result being manifested in the lack of *accumulation* of chalcone derivatives.

Crawford and Stuessy also extended the list of genera of the Heliantheae in which anthochlors do *not* occur. Also of interest is their report of two genera in Heliantheae wherein anthochlors have been found for the first time; butein and sulfuretin glycosides in *Simsia* Pers. (one species of three tested; 35 species known) and *Tithonia* Desf. (five species of 11 tested, 11 species known) (unpublished data of J. La Duke). These observations increase to four the number of genera in tribe Heliantheae subtribe Helianthinae known to accumulate anthochlors [the others being *Helianthus* L. and *Viguiera* H. B. K. (Harborne and Smith, 1978)]. The close morphological similarities of these four genera have been recognized for many years (Stuessy, 1977).

The synthesis and accumulation of anthochlors can clearly be taken as a characteristic of Coreopsidinae. Occurrence of anthochlors in other members of the Compositae, however, is not at all easily explained in evolutionary terms.

Anthochlors are known from Cardueae, Eupatorieae, Inuleae, Helenieae, Lactuceae and, as described above, Heliantheae. [The Helenieae is now considered defunct with its constitutent genera dispersed to several tribes (Turner and Powell, 1977)]. It is of interest to note that the compounds from the Cardueae, Eupatorieae and Inuleae are based upon the phloroglucinol A-ring pattern (*6.278*) while those from Heliantheae and Lactuceae are based upon the resorcinol pattern (*6.279*). Crawford and Stuessy point out that distribution of

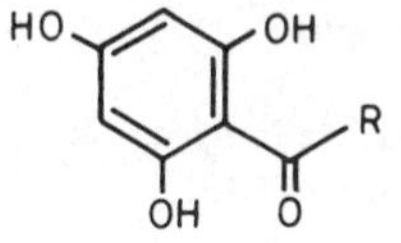

HO OH R OH O

(6.278)

HO R OH O

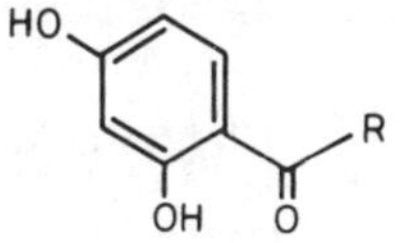

(6.279)

these structural types does not correlate well with subfamilial treatments of the Compositae (see their paper for extensive references). Parallel evolution of the two types in the subfamilies may have occurred. Until more is learned about the biosynthesis and biological function(s) of these types of compounds in their respective organisms little more can be said at this time.

In addition to work aimed at establishing relationships at generic and higher taxonomic levels studies at infrageneric levels also continue. The distribution of okanin and maritimetin proved useful in a study of *Coreopsis* section *Pseudo-agarista* in Mexico (Crawford, 1976). Chalcone–aurone glycoside pairs coupled with chromosome counts allowed a clear distinction to be drawn between section *Pseudo-agarista* and sections *Electra* and *Anathysana*. Other discussions of the significance of anthochlor distribution at the sectional level within *Coreopsis* have appeared (Crawford, 1978; Nicholls and Bohm, 1979), and Crawford and Smith (1980) showed that the anthochlors of the four varieties of *C. grandiflora* are identical despite widely different geographic ranges and chromosome numbers.

6.7.5 Ericaceae

(a) *Rhododendron* subgen. *Pentanthera*

The flavonoid complement of *Rhododendron* subgen. *Pentanthera* was shown by King (1980, and references cited therein) to include a variety of structural types: flavonols (including 5-*O*-methylflavonols), a dihydroflavonol, flavanones, dihydrochalcones and chalcones, although not all compounds were fully characterized. Despite populational variation it was possible for King to describe species-specific groups of compounds and to show that the flavonoids were useful in arranging the species into five 'alliances.' King found that the *R. canescens–R. nudiflorum–R. roseum* group (alliance-A) is characterized by the presence of the dihydrochalcone asebotin and the flavanone eriodictyol. The alliance comprising *R. speciosum–R. prunifolium–R. calendulaceum–R. bakeri* (alliance-B) can be distinquished by the presence of the flavanone farrerol and its

5,7-diglucoside. The presence of asebotin in *R. calendulaceum* suggests a possible link with alliance-A. A further link between these two alliances involves the presence of two (unidentified) chalcones in *R. prunifolium* and *R. bakeri* (alliance-B) and in three collections of *R. canescens* (alliance-A). Alliance-C, consisting of *R. alabamense* and *R. arborescens*, and alliance-D comprising *R. atlanticum*, *R. viscosum*, *R. coryi*, *R. serrulatum* and *R. oblongifolium*, are closely related and both lack the flavones and dihydrochalcones characteristic of the other alliances. Alliance-E consists of *R. austrinum* of southeastern North America, *R. occidentale* of western North America and *R. luteum* of Europe. In spite of the highly disjunct nature of this alliance the group is unique in *Pentanthera* in possessing 5-*O*-methylflavonols as major constituents.

6.7.6 Leguminosae

In the 1975 treatment of legume flavonoids it was stated (Bohm, 1975, p. 490) that, with only one exception, the anthochlors of this family were based upon the resorcinol A ring and that such 5-deoxyflavonoids (6′-deoxychalcones, 4-deoxyaurones) might be considered to characterize the Leguminosae. Since that time a good deal of work has been published concerning the flavonoid chemistry of the family with the result being that a reconsideration of that generalization is in order. In fact, shortly after the appearance of 'The Flavonoids' a paper appeared by do Nascimento and co-workers (1976) that described a number of flavonoids from *Derris obtusa* many of which had the phloroglucinol A ring (aurones, auronols, flavones and flavonols). It was pointed out by those authors that my generalization was no longer valid. A tabulation of data from the present survey shows that members of five genera make chalcones having phloroglucinol-type A rings: *Dalea, Derris, Milletia, Pongamia* and *Tephrosia*. Phloroglucinol-type A-ring aurones and auronols are found in *Derris* as mentioned above. Two genera, *Rhynchosia* and *Sophora*, make 6-oxygenated dihydroflavonols, and no less than 13 genera of legumes make 6-oxygenated flavanones: *Acacia, Bauhinia, Cajanus, Cassia, Dalea, Derris, Flemingia, Galega, Glycyrrhiza, Indigofera, Lygos, Rhynchosia* and *Tephrosia*.

6.7.7 Rutaceae

(a) *Citrus* hybrids

In a recent paper, Kamiya and co-workers (1979) studied the flavanone glycoside complement of a large number of artificial *Citrus* hybrids, in order to determine whether the product of a hybridization experiment is a true genetic hybrid or the product of nucellar embryony which is of regular occurrence in many *Citrus* species. Since nucellar embryony yields an offspring arising solely from maternal tissue the flavonoid patterns of the adult plant and the offspring would be identical. An altered pattern would indicate contribution from another

source, that is, from the male or pollen parent. In all, 94 hybrids were examined of which 53 had patterns which were different from those of the maternal plant.

(b) Aurantioideae

In another recent paper Grieve and Scora (1980) concluded that morphological specialization in subfamily Aurantioideae has been paralleled by an increase in the complexity of the flavonoid complement. Their study described the distribution of *C*- and *O*-glycosylflavones, flavonols and flavanones in members of subtribes Clauseninae and Merrilliinae of tribe Clauseneae and subtribes Triphasiinae, Citrinae and Balsamocitrinae of tribe Citrae (all *sensu* Swingle). Flavanones (naringin, poncirin, hesperidin and two unidentified compounds) were found only in Citrinae and within this subtribe they were largely restricted to members considered as true citrus trees: *Citrus, Clymenia, Eremocitrus, Fortunella, Microcitrus* and *Poncirus*. The only exceptional report involved the occurrence of hesperidin in *Pleiospermium sumatranum*, which, though still a member of the subtribe, belongs to a group of more primitive citrus trees.

6.8 NATURAL FUNCTIONS OF ANTHOCHLORS

6.8.1 Ultraviolet patterning

Significant evidence for the operation of certain flavonoid types as nectar-guide components in flowers of various plants has accumulated. The flavonoids involved in nectar-guide chemistry include 'yellow flavonols' (Harborne, 1977b, p. 369 ff.), 2′-hydroxyflavones (Harborne and Smith, 1978) and anthochlors (*vide infra*). All of these compounds exert their effect by absorbing ultraviolet light and thus appearing dark or 'bee-purple' (Proctor and Yeo, 1973, p. 168 ff.) to the eye of the pollinators. Nectar-guide compounds often occur in association with carotenoids which, in reflecting UV and yellow wavelengths, act to provide a background against which the nectar-guide is visible.

In a study of *Oenothera hookeri* ssp. *venusta* Dement and Raven (1974) observed that only carotenoids occurred in the distal portion of petals while the basal portion had carotenoids plus the chalcone isosalipurposide. The chalcone absorbs UV light and appears bee-purple to insect visitors or simply dark to the human eye. Seven other species of *Oenothera* having the chalcone in their flowers were also listed.

Scogin and Zakar (1976) observed that the uniformly golden-yellow ray florets of *Bidens laevis*, when exposed to ammonia vapours, displayed a colour change to bright orange involving the involucre, the entire disc, the basal 50–66% of the ligule, and the veins running the length of the ligule. The UV absorbance pattern, and thus the pattern which would appear bee-purple to a pollinator, was exactly

congruent with the zone of colour change caused by the ammonia. These workers went on to survey a number of *Bidens* species for pigment floral patterns. Anthochlor distribution fell into three categories: (a) uniform distribution throughout the ligule, (b) distribution restricted to basal portion of the ligule (ranging from 25 to 80% of the ligules in the samples tested) and (c) pigment absent except for some in the veins of the ligule. The distribution patterns in 36 species correlated reasonably well with sectional taxonomy of the genus. The compounds determined in *B. laevis* were marein, sulfurein, coreopsin, butein and sulfuretin. Very similar results were obtained by Scogin and his co-workers (Scogin, 1976; Scogin *et al.*, 1977) in a study of anthochlor distribution in flowers of 34 species of *Coreopsis*.

Details of pigment localization in ligules of *Coreopsis bigelovii* were reported by Harborne and Smith (1978), who found that the entire lower surface of ray florets gave a positive anthochlor test and absorbed ultraviolet light, while on the upper surface only the basal two-thirds gave a positive test. Closer examination using microtome sections showed that the upper surfaces were papillate while epidermal cells on the undersides were only slightly convex. The anthochlor test was observed to occur in papillate cells. These observations are in complete accord with the report of anthochlor-bearing papillate cells on the upper epidermis of ligules of *Lasthenia chrysostoma* (Brehm and Krell, 1975). Further discussion of ligule microstructure including reference to UV-absorption patterns can be found in a review by Baagøe (1977).

Harborne and Smith (1978) also identified anthochlors from ligules of *Viguiera laciniata* and three species of *Helianthus*, and detected anthochlors by means of their colour reaction with base, in an additional 28 species of *Helianthus* (37 tested). Nectar guides occur in some of these taxa but not in all, i.e. *Viguiera*, where the entire ligule is pigmented. In a recent study of *Megalodonta* Roberts (1980) showed that a UV-absorbing pattern does not exist in contrast to the patterns seen in *Bidens* the genus most closely related to *Megalodonta*. According to Roberts (1980) the uniformly absorbing capitulum of *Megalodonta* may be an adaptation to the aquatic habitat that has the effect of maximizing the 'target area' of the head against the reflective water surface.

6.8.2 Larval-growth inhibition

Five flavanone derivatives and two dihydroflavonol derivatives were tested, along with several other flavonoids, for their activity against the growth of larva of *Heliothis zea* (Elliger *et al.*, 1980). Eriodictyol, dihydroquercetin and dihydroquercetin 3-*O*-rhamnoside were active but naringenin, naringin, hesperetin and neohesperidin were inactive. The results were discussed in terms of the number and disposition of phenolic groups in the active molecules. The presence of *ortho*-dihydroxylation seems to be important. The oxidation level of the C ring appears not to be an important factor, nor does glycosylation appear to play a major part. No mechanism of action was postulated.

6.8.3 Allelopathy

Star (1980) tested the effects of 2′,6′-dihydroxy-4′-methoxydihydrochalcone, 2′,6′-dihydroxy-4′-methoxychalcone and 5-hydroxy-7-methoxyflavonol (izalpinin) on the germination of spores of *Pityrogramma calomelanos*. The dihydrochalcone was the most effective causing inhibition of germination at concentrations of 5×10^{-5} M and higher. The chalcone was inhibitory at 5×10^{-4} M and 5×10^{-6} M but stimulatory at 5×10^{-7} M and 5×10^{-9} M. Izalpinin was stimulatory at 5×10^{-4} and 5×10^{-3} M.

The author (Star, 1980) went on to describe observations of dispersal patterns of *P. calomelanos* plants at three sites in the Hawaiian Islands. Two features of these patterns were described in terms of the amounts of exudate substances in the soil. Distances from the parent plants to their nearest neighbours were greater than the distances between the smaller neighbours suggesting that higher concentrations of exudate near the parent plant had inhibited spore germination. At greater distances from the parent, however, amount of exudate would be lower, hence stimulation of germination could be expected. Despite the correlation between the laboratory and field observations one would like to see studies undertaken to determine actual concentrations of the putatively active compounds in the soil. Such an attempt to link cause and effect under natural conditions would be very welcome.

6.8.4 Flavonoids and disease resistance in plants

(a) Malus

The relationship between dihydrochalcones and the resistance of apples to attack by scab (*Venturia inaequalis* Cke. Wint) has investigated by Hunter (1975). Neither trilobatin (2′,4′,6′,4-tetrahydroxydihydrochalcone 4′-*O*-glucoside) nor sieboldin (2′,4′,6′,3,4-pentahydroxydihydrochalcone 4′-*O*-glucoside) conferred resistance to infection on seedlings of *Malus* crosses. He also reported a natural infection of *M. astrosanguinea* by scab despite the presence of both phlorrhizin (2′,4′,6′,4-tetrahydroxydihydrochalcone 2′-*O*-glucoside) and sieboldin. Cultures of the pathogen could use phloretin, phlorrhizin, trilobatin, sieboldin or 3-hydroxyphlorrhizin as carbon source.

(b) Medicago

Ingham (1979) tested the ability of leaves of 25 species of *Medicago* to produce phytoalexins following inoculation by *Helminthosporium carbonum*. Pterocarpans and isoflavans were detected in most species. However, diffusates of *M. sativa* cv. Du Puits and *M. lupulina* yielded the chalcone–flavanone pair isoliquiritigenin and liquiritigenin. Control leaves (uninfected) of *M. lupulina* did not appear to contain either of these compounds. Bioassays using *Cladosporium herbarum* showed minor inhibition of growth with the chalcone–flavanone mixture.

(c) Beta

Evidence has also appeared suggesting that the flavanone 'betagarin', 5,2'-dimethoxy-6,7-methylenedioxyflavanone (*6.280*), acts as a phytoalexin in sugar beet (*B. vulgaris*) infected with leaf-spot fungus *Cercospora beticola* (Johnson *et al.*, 1976; Martin, 1977).

OCH_3 O OCH_3

(*6.280*)

ACKNOWLEDGEMENTS

Preparation of this chapter and work from our laboratory cited herein were supported by operating grants from the Natural and Engineering Sciences Research Council of Canada for which I express my sincere appreciation. I should also like to thank Professor Dr H. Wagner for his comments on the flavonolignan stereochemistry problem and Drs D. Crawford, T. Stuessy and E. Wollenweber for making unpublished data available to me. Thanks go, as usual, to Dr Elijah Tannen for his unfailing friendship and support.

REFERENCES

Adinarayana, D., Gunsekar, D., Ramachandraiah, P., Seligmann, O. and Wagner, H. (1980a), *Phytochemistry* **19**, 478.

Adinarayana, D., Gunsekar, D., Seligmann, O. and Wagner, H. (1980b), *Phytochemistry* **19**, 480.

Adityachaudhury, N. and Daskanungo, P. (1975), *Plant Biochem. J.* **2**, 65.

Adityachaudhury, N., Ghosh, D., Choudhury, A. and Kirtaniya, C. L. (1973), *J. Indian Chem. Soc.* **50**, 363.

Adityachaudhury, N., Das, A. K., Choudhury, A. and Daskanungo, P. L. (1976), *Phytochemistry* **15**, 229.

Agarwal, S. C., Bhaskar, A. and Seshadri, T. R. (1973), *Indian J. Chem.* **11B**, 9.

Agrawal, P. K., Agarwal, S. K. and Rastogi, R. P. (1980a), *Phytochemistry* **19**, 893.

Agrawal, P. K., Agarwal, S. K. and Rastogi, R. P. (1980b), *Phytochemistry* **19**, 1260.

Ahluwalia, V. K. and Rami, N. (1976), *Indian J. Chem.* **14B**, 594.

Anthonsen, T., Falkenberg, I., Laake, M., Midelfart, A. and Mortensen, T. (1971), *Acta Chem. Scand.* **25**, 1929.

Arene, E. O., Pettit, G. R. and Ode, R. H. (1978), *Lloydia* **41**, 68.

Aritomi, M. and Kawasaki, T. (1974), *Chem. Pharm. Bull.* **22**, 1800.

Asakawa, Y., Toyota, M. and Takemoto, T. (1978), *Phytochemistry* **17**, 2005.

Baagøe, J. (1977). In *The Biology and Chemistry of the Compositae* (eds. V. H. Heywood, J. B. Harborne and B. L. Turner) Academic Press, New York, Vol. 1, pp. 119–139.

Ballard, R. E. (1975), *Dissertation Abstracts* **36B**, 1564.

Bandyukova, V. A. and Fishman, G. M. (1979), *Khim. Prir. Soed.* **15**, 354 (English translation).

Barua, R. N., Sharma, R. P., Thyagarajan, G. and Herz, W. (1978), *Phytochemistry* **17**, 1807.
Baruah, N. C., Sharma, R. P., Thyagarajan, G., Herz, W. and Govindan, S. V. (1979), *Phytochemistry* **18**, 2003.
Becker, H. and Schrall, R. (1977). *Z. Pflanzenphysiol.* **83**, 137.
Becker, H., Exner, J. and Schilling, G. (1978), *Z. Naturforsch.* **33**, 771.
Bhanumati, S., Chhabra, S. C., Gupta, R. S. and Krishnamoorthy (1978), *Phytochemistry* **17**, 2045.
Bhardwaj, D. K., Bisht, M. S., Mehta, C. K. and Sharma, G. C. (1979), *Phytochemistry* **18**, 355.
Bhartiya, H. P., Dubey, P., Katiyer, S. B. and Gupta, P. C. (1979), *Phytochemistry* **18**, 689.
Bhatt, S. (1975), *Indian J. Chem.* **13B**, 1105.
Bhattacharyya, A., Choudhury, A., Mitra, S. R. and Adityachaudhury, N. (1979) *Indian J. Chem.* **16B**, 176.
Bloom, M. (1976), *Am. J. Bot.* **63**, 399.
Bohlmann, F. and Abraham, W.-R. (1979a), *Phytochemistry* **18**, 839.
Bohlmann, F. and Abraham, W.-R. (1979b), *Phytochemistry* **18**, 889.
Bohlmann, F. and Abraham, W.-R. (1979c), *Phytochemistry* **18**, 1754.
Bohlmann, F. and Abraham, W.-R. (1979d), *Phytochemistry* **18**, 1851.
Bohlmann, F. and Hoffmann, E. (1979), *Phytochemistry* **18**, 1371.
Bohlmann, F. and Jakupovic, J. (1979), *Phytochemistry* **18**, 1189.
Bohlmann, F. and Knoll, K.-H. (1978), *Phytochemistry* **17**, 461.
Bohlmann, F. and Suwita, A. (1979a), *Phytochemistry* **18**, 677.
Bohlmann, F. and Suwita, A. (1979b), *Phytochemistry* **18**, 885.
Bohlmann, F. and Suwita, A. (1979c), *Phytochemistry* **18**, 2046.
Bohlmann, F. and Zdero, C. (1977), *Phytochemistry* **16**, 492.
Bohlmann, F. and Zdero, C. (1978), *Phytochemistry* **17**, 1155.
Bohlmann, F. and Zdero, C. (1979), *Phytochemistry* **18**, 125.
Bohlmann, F. and Zdero, C. (1980), *Phytochemistry* **19**, 683.
Bohlmann, F., Mahanta, P. K. and Zdero, C. (1978), *Phytochemistry* **17**, 1935.
Bohlmann, F., Dutta, L., Robinson, H. and King, R. M. (1979a), *Phytochemistry* **18**, 1889.
Bohlmann, F., Knauf, W., King, R. M. and Robinson, H. (1979b), *Phytochemistry* **18**, 1011.
Bohlmann, F., Ziesche, J. and Mahanta, P. K. (1979c), *Phytochemistry* **18**, 1033.
Bohlmann, F., Zdero, C., King, R. M. and Robinson, H. (1979d), *Phytochemistry*, **18**, 1246.
Bohlmann, F., Zdero, C. and Ziesche, J. (1979e), *Phytochemistry* **18**, 1375.
Bohlmann, F., Jakupovic, J., King, R. M. and Robinson, H. (1980a), *Phytochemistry* **19**, 1815.
Bohlmann, F., Knoll, K.-H., Robinson, H. and King, R. M. (1980b), *Phytochemistry* **19**, 971.
Bohlmann, F., Zdero, C., Abraham, W.-R., Suwita, A. and Grenz, M. (1980c), *Phytochemistry* **19**, 873.
Bohm, B. A. (1975). In *The Flavonoids* (eds. J. B. Harborne, T. J. Mabry and H. Mabry), Chapman and Hall, London.
Bohm, B. A. and Ornduff, R. (1981), *Syst. Bot.*, **6**, 15.
Bose, P. C. and Adityachaudhury (1978), *Phytochemistry* **17**, 587.
Bourweig, D. and Pohl, R. (1973) *Planta Med.* **24**, 304.
Braz Filho, R., Gottlieb, O. R. and Mourao, A. P. (1975a), *Phytochemistry* **14**, 261.
Braz Filho, R., Gottlieb, O. R., Mourao, A. P., Da Rocha, A. I. and Oliviera, F. S. (1975b), *Phytochemistry* **14**, 1454.
Braz Filho, R., Da Silva, M. S. and Gottlieb, O. R. (1980), *Phytochemistry* **19**, 1195.
Brehm, B. G. and Krell, D. (1975), *Science* **190**, 1221.
Brieskorn, C. H. and Riedel, W. (1977), *Planta Med.* **31**, 308.
Budzianowski, J. and Skrzypczakowa (1978), *Phytochemistry* **17**, 2044.
Camele, G., Delle Monache, F., Delle Monache, G. and Marini-Bettolo, G. B. (1980), *Phytochemistry* **19**, 707.
Candy, H. A., Brooks, K. B., Bull, J. R., McGarry, E. J. and McGarry, J. M. (1978) *Phytochemistry* **17**, 1681.
Cannon, J. R. and Martin, P. F. (1977), *Austr. J. Chem.* **30**, 2099.

Chadenson, M., Hauteville, M. and Chopin, J. (1972), *Compt. Rend.* **275**, 1291.
Chauhan, J. S., Sultan, M. and Srivastava, S. K. (1977), *Planta Med.* **32**, 217.
Chauhan, J. S., Srivastava, S. K. and Sultan, M. (1978), *Phytochemistry* **17**, 334.
Chauhan, J. S., Saraswat, M. and Kumari, G. (1979a), *Planta Med.* **35**, 373.
Chauhan, J. S., Srivastava, S. K. and Srivastava, S. D. (1979b), *Indian J. Chem.* **17B**, 300.
Chauhan, J. S., Vidyapati, T. J. and Gupta, A. K. (1979c), *Phytochemistry* **18**, 1766.
Chibber, S. S., Sharma, R. P. and Dutt, S. K. (1979), *Phytochemistry* **18**, 2056.
Chopin, J., Hauteville, M., Joshi, B. S. and Gawad, D. H. (1978), *Phytochemistry* **17**, 332.
Cole, J. R., Torrance, S. J., Wiedhopf, R. M., Arora, S. K. and Bates, R. B. (1976), *J. Org. Chem.* **41**, 1852.
Combes, G., Vassort, P. and Winternitz, F. (1970), *Tetrahedron* **26**, 5981.
Cook, I. F. and Knox, J. R. (1975), *Phytochemistry* **14**, 2510.
Cooper-Driver, G. (1980), *Bull. Torrey Bot. Club* **107**, 116.
Cooper-Driver, G. and Swain, T. (1977), *Bot. J. Linnean Soc.* **74**, 1.
Crawford, D. J. (1976), *Brittonia* **28**, 329.
Crawford, D. J. (1978), *Phytochemistry* **17**, 1680.
Crawford, D. J. and Smith, E. B. (1980), *Brittonia* **32**, 154.
Crawford, D. J. and Stuessy, T. (1981), *Am. J. Bot.* **68**, 107.
Dayal, R. and Parthasarathy, M. R. (1977), *Planta Med.* **31**, 245.
de Alleluia, I. B., Braz Filho, R., Gottlieb, O. R., Magalhaes, E. G. and Marques, R. (1978), *Phytochemistry* **17**, 517.
de Almeida, M. E. L., Braz Filho, R., von Bülow, M. V., Correa, J. J. L., Gottlieb, O. R., Maia, J. G. S. and da Silva, M. S. (1979), *Photochemistry* **18**, 1015.
De Lima, O. G., Marini-Bettolo, G. B., De Mello, J. F., Delle Monache, F., Coêlho, J. S. De B., Lyra, F. D. De A. and De Albuquerque, M. M. F. (1973), *Gazz. Chem. Ital.* **103**, 771.
De Lima, O. G., Marini-Bettolo, G. B., De Mello, J. F., Delle Monache, F., De A. Lyra, F. D., Machado, M. and De Albuquerque, F. (1972), *Rend. Accad. Naz. Lincei, Classe Sci. Fis. Mat. e. Nat.*, **53**, 433.
Delle Monache, G., De Mello, J. F., Delle Monache, F., Marini-Bettolo, G. B., De Lima, O. G. and Coêlho, J. S. De B. (1974), *Gazz. Chim. Ital.* **104**, 861.
Delle Monache, F., Delle Monache, G., Marini-Bettolo, G. B., De Albuquerque, M. M. F., De Mello, J. F. and De Lima, O. G. (1976), *Gazz. Chim. Ital.* **106**, 935.
Delle Monache, F., Cuca Suraez, L. E. and Marini-Bettolo, G. B. (1978), *Phytochemistry* **17**, 1812.
Dement, W. A. and Raven, P. (1974), *Nature* (*London*) **252**, 705.
De Oliveira, A. B., Iracema, M., Madruga, L. M. and Gottlieb, O. R. (1978), *Phytochemistry* **17**, 593.
Deshpande, V. H., Srinivasan, R. and Rao, A. V. R. (1975), *Indian J. Chem.* **13B**, 453.
de Vlaming, P. and Kho, K. F. F. (1976), *Phytochemistry* **15**, 348.
Dietz, V. H., Wollenweber, E., Favre-Bonvin, J. and Gómez P., L. D. (1980), *Z. Naturforsch.* **35C**, 36.
Dominguez, X. A., Martinez, C., Calero, A., Dominguez, X. A., Jr., Hinojosa, M., Zamudio, A., Watson, W. H. and Zabel, V. (1978a), *Planta Med.* **34**, 172.
Dominguez, X. A., Martinez, C., Calero, A., Dominguez, X. A., Jr., Hinojosa, M., Zamudio, A., Zabel, V., Smith, W. B. and Watson, W. H. (1978b), *Tetrahedron Lett.*, 429.
Dominguez, X. A., Franco, R., Zamudio, A., Barradas, D. M., Watson, W. H., Zabel, V. and Merijanian, A. (1980), *Phytochemistry* **19**, 1262.
do Nascimento, M. C., Dias, R. L. De V. and Mors, W. B. (1976), *Photochemistry* **15**, 1553.
Dreyer, D. L., Munderloch, K. P. and Thiessen, W. E. (1975), *Tetrahedron* **31**, 287.
Elliger, C. A., Chan, B. C. and Waiss, A. C., Jr. (1980), *Naturwissenschaft* **67**, 358.
El Sherbeiny, A. E. A., El Sissi, H. I., Nawwar, M. A. M. and El Ansari, M. A. (1978), *Planta Med.* **34**, 335.
El Sissi, H. I., Saleh, N. A. M., El Negoumy, S. I., Wagner, H., Iyengar, M. A. and Seligmann, O. (1974), *Phytochemistry* **13**, 2843.
Fée, A. (1852), *Mem. Fam. Foug.* **5**, 165.
Fernandes, J. B., Gottlieb, O. R. and Xavier, L. M. (1978), *Biochem. Syst. Ecol.* **6**, 55.

Fish, F., Gray, A. I. and Waterman, P. G. (1975a), *Phytochemistry* **14**, 841.
Fish, F., Gray, A. I. and Waterman, P. G. (1975b), *Phytochemistry* **14**, 2073.
Forkmann, G. (1979), *Phytochemistry* **18**, 1973.
Fourie, T. G., Ferreira, D. and Roux, D. G. (1974), *Phytochemistry* **13**, 2573.
Fraser, A. W. and Lewis, J. R. (1974), *Phytochemistry* **13**, 1561.
Furuya, T., Ayaba, S.-i, and Kobayashi, M. (1976) *Tetrahedron Lett.*, 2539.
Garg, G. P., Sharma, N. N. and Khanna, R. N. (1978), *Indian J. Chem.* **16B**, 658.
Garg, S. K., Gupta, S. R. and Sharma, N. D. (1979), *Indian J. Chem.* **17B**, 394.
Garg, S. P., Bhushan, R. and Kapoor, R. C. (1980), *Phytochemistry* **19**, 1265.
Gertz, O. (1938), *Kungl. Fysiog. Sallsk. Forch.* **8**, 62.
Ghosal, S., Jaiswal, D. K. and Biswas, K. (1978), *Phytochemistry* **17**, 2119.
Giannasi, D. E. (1975), *Mem. N. Y. Bot. Gard.* **26**, 1.
Gilbert, R. I. (1975), *Phytochemistry* **14**, 1461.
Gottlieb, O. R., de Lima, R. A., Mendes, P. H. and Magalhaes, M. T. (1975), *Phytochemistry* **14**, 1674.
Greimel, A. and Koch, H. (1977), *Planta Med.* **32**, 331.
Grieve, C. M. and Scora, R. W. (1980), *Syst. Bot.* **5**, 39.
Grimshaw, J. and Lamer-Zarawska, E. (1975), *Phytochemistry* **14**, 838.
Gupta, A. K., Vidyapati, T. J. and Chauhan, J. S. (1979), *Indian J. Chem.* **18B**, 85.
Gupta, A. K., Vidyapati, T. J. and Chauhan, J. S. (1980), *Planta Med.* **38**, 174.
Gupta, B. K., Gupta, G. K., Dhar, K. L. and Atal, C. K. (1980), *Phytochemistry* **19**, 2034.
Gupta, R. K. and Krishnamurti, M. (1976a), *Phytochemistry* **15**, 832.
Gupta, R. K. and Krishnamurti, M. (1976b), *Phytochemistry* **15**, 1795.
Gupta, R. K: and Krishnamurti, M. (1976c), *Phytochemistry* **15**, 2011.
Gupta, R. K. and Krishnamurti, M. (1977a), *Phytochemistry* **16**, 293.
Gupta, R. K. and Krishnamurti, M. (1977b), *Phytochemistry* **16**, 1104.
Gupta, R. K. and Krishnamurti, M. (1979), *Indian J. Chem.* **17B**, 291.
Gupta, R. K., Krishnamurti, M. and Parthasarathi, J. (1980), *Phytochemistry* **19**, 1264.
Gupta, S. R., Seshadri, T. R. and Sood, G. R. (1975a), *Indian J. Chem.* **13B**, 868.
Gupta, S. R., Seshadri, T. R. and Sood, G. R. (1975b), *Indian J. Chem.* **13B**, 632.
Gupta, S. R., Seshadri, T. R. and Sood, G. R. (1977), *Phytochemistry* **16**, 1995.
Hänsel, R., Rimpler, H. and Schwartz, R. (1965), *Tetrahedron Lett.*, 1545.
Hänsel, R., Rimpler, M. and Schwartz, R. (1967), *Tetrahedron Lett.*, 735.
Hänsel, R., Kaloga, M. and Pelter, A. (1976), *Tetrahedron Lett.*, 2241.
Harborne, J. B. (1977a), *Phytochemistry* **16**, 927.
Harborne, J. B. (1977b). In *The Biology and Chemistry of the Compositae* (eds. V. H. Heywood, J. B. Harborne and B. L. Turner), Academic Press, New York, pp. 359–384.
Harborne, J. B. and Smith, D. M. (1978), *Biochem. Syst. Ecol.* **6**, 287.
Hart, C. R. (1979), *Syst. Bot.* **4**, 130.
Hiraoka, A. (1978), *Biochem. Syst. Ecol.* **6**, 171.
Hufford, C. D. and Lasswell, W. L. (1976), *J. Org. Chem.* **41**, 1297.
Hufford, C. D. and Lasswell, W. L. (1978a), *Lloydia* **41**, 151.
Hufford, C. D. and Lasswell, W. L. (1978b), *Lloydia* **41**, 156.
Hufford, C. D. and Oguntimein, B. O. (1980), *Phytochemistry* **19**, 2036.
Hufford, C. D., Lasswell, W. L., Hirotsu, K. and Clardy, J. (1979), *J. Org. Chem.* **44**, 4709.
Hunter, L. D. (1975), *Phytochemistry* **14**, 1519.
Imperato, F. (1976), *Phytochemistry* **15**, 439.
Imperato, F. (1978), *Phytochemistry* **17**, 822.
Ingham, J. L. (1978), *Z. Naturforsch.* **33c**, 146.
Ingham, J. L. (1979), *Biochem. Syst. Ecol.* **7**, 29.
Ishikura, N., Sugahara, K. and Kurosawa, K. (1979), *Z. Naturforsch.* **34c**, 628.
Islam, A., Gupta, R. K. and Krishnamurti, M. (1980), *Phytochemistry* **19**, 1558.
Ito, K., Itoigawa, M., Haruna, M. Murata, H. and Furukawa, H. (1980), *Phytochemistry* **19**, 476.

Ivanova, S. Z., Medvedeva, S. A., Lutskii, V. I., Tyukavkina, N. A. and Zelenikina, N. D. (1975), *Khim. Prir. Soed.* **11**, 817 (English Translation).
Jain, A. C. and Gupta, R. C. (1978), *Indian J. Chem.* **16B**, 1126.
Jayaraman, I., Ghanim, A. and Khan, H. A. (1980), *Phytochemistry* **19**, 1267.
Jensen, S. R., Nielsen, B. J. and Norn, V. (1977), *Phytochemistry* **16**, 2036.
Johnson, G., Maag, D. D., Johnsen, D. K. and Thomas, R. D. (1976), *Physiol. Plant Pathol.* **8**, 225.
Joshi, B. S. and Gawad, D. H. (1974), *Indian J. Chem.* **12B**, 1033.
Joshi, B. S. and Gawad, D. H. (1976), *Indian J. Chem.* **14B**, 9.
Jurd, L. and Manners, G. (1977), *J. Agric. Food Chem.* **25**, 723.
Kaltaev, N. Sh. and Nikonov, G. K. (1974), *Khim. Prir. Soed.* **10**, 94 (English translation).
Kamiya, S., Esaki, S. and Konishi, F. (1979), *Agric. Biol. Chem.* **43**, 1529.
Kemp, M. S., Burden, R. S. and Brown, C. (1979), *Phytochemistry* **18**, 1765.
King, B. L. (1977a), *Am. J. Bot.* **64**, 350.
King, B. L. (1977b), *Syst. Bot.* **2**, 14.
King, B. L. (1980). In *Contributions toward a Classification of Rhododendron*, (ed. J. L. Luteyn), New York Botanical Garden, New York, pp. 163–185.
Komatsu, M., Yokoe, I. and Shirataki, Y. (1978), *Chem. Pharm. Bull.* **26**, 1274.
Kompantsev, V. A. and Shinkarenko, A. L. (1975), *Khim. Prir. Soed.* **11**, 682 (English translation).
Kozawa, M., Morita, N., Baba, K. and Hata, K. (1977), *Chem. Pharm. Bull.* **25**, 515.
Kurosawa, K., Ollis, W. D., Sutherland, I. O., Gottlieb, O. R. and de Oliveira, A. B. (1978), *Phytochemistry* **17**, 1389.
Lam, J. and Wrang, P. (1975), *Phytochemistry* **14**, 1621.
Lapteva, K. I., Lutskii, V. I. and Tyuvkina, N. A. (1974), *Khim. Prir. Soed.* **10**, 102 (English translation).
Letcher, R. M. and Shirley, I. M. (1976), *Phytochemistry* **15**, 353.
Lakshmi, V. and Chauhan, J. S. (1977), *Indian J. Chem.* **15**, 814.
Luteyn, J. L., Harborne, J. B. and Williams, C. A. (1980), *Brittonia* **32**, 1.
Mabry, T. J. (1973). In *Chemistry in Evolution and Systematics* (ed. T. Swain), Butterworths, London, p. 389.
Mabry, T. J., Abdel-Baset, Z., Padolina, W. G. and Jones, S. B., Jr. (1975a), *Biochem. Syst. Ecol.* **2**, 185.
Mabry, T. J., Sakakibara, M. and King, B. (1975b), *Phytochemistry* **14**, 1448.
Malan, E. and Roux, D. G. (1974), *Phytochemistry* **13**, 1575.
Malterud, K. E., Anthonsen, T. and Lorentzen, G. B. (1977), *Phytochemistry* **16**, 1805.
Manchanda, V. P. and Khanna, R. N. (1977), *Indian J. Chem.* **15B**, 660.
Manners, G. D. and Jurd, L. (1979), *Phytochemistry* **18**, 1037.
Manners, G. D., Jurd, L. and Stevens, K. L. (1974), *Phytochemistry* **13**, 292.
Maradufu, A. and Ouma, J. H. (1978), *Phytochemistry* **17**, 823.
Markham, K. R. (1980), *Biochem. Syst. Ecol.* **8**, 11.
Markham, K. R. and Porter, L. J. (1978), *Phtochemistry* **17**, 159.
Martin, S. S. (1977), *Physiol. Plant Pathol.* **11**, 297.
Maruyama, M., Hayasaka, K., Sasaki, S.-I., Hosokawa, S. and Uchiyama, H. (1974), *Phytochemistry* **13**, 286.
Matos, F. J. de A., Gottlieb, O. R. and Andrade, C. H. S. (1975), *Phytochemistry* **15**, 825.
Miller, J. M. and Bohm B. A. (1979), *Phytochemistry* **18**, 1412.
Miller, J. M. and Bohm, B. A. (1980), *Biochem. Syst. Ecol.* **8**, 279.
Miyase, T., Ueno, A., Noro, T. and Fukushima, S. (1980), *Chem. Pharm. Bull.* **28**, 1172.
Mohan, H., Sudhansty and Parthak, H. D. (1978), *Indian J. Chem.* **16B**, 244.
Montero, J. L. and Winternitz, F. (1973), *Tetrahedron* **29**, 1243.
Nagar, A., Gujral, V. K. and Gupta, S. R. (1979), *Planta Med.* **37**, 183.
Nagarajan, G. R. and Parmar, V. S. (1977a), *Phytochemistry* **16**, 1317.
Nagarajan, G. R. and Parmar, V. S. (1977b), *Planta Med.* **31**, 146.
Nagarajan, G. R. and Parmar, V. S. (1977c), *Planta Med.* **32**, 50.

Nagarajan, G. R. and Parmar, V. S. (1978), *Indian J. Chem.* **16B**, 439.
Nicholls, K. W. and Bohm, B. A. (1979), *Phytochemistry* **18**, 1076.
Niemann, G. J. (1974), *Planta Med.* **26**, 101.
Nilsson, M. (1961), *Acta Chem. Scand.* **15**, 154, 211.
Noro, T., Fukushima, S., Saiki, Y. Uemo, A. and Akahori, Y. (1969), *Yakugaku Zasshi* **89**, 851.
Ogiyama, K., Yasue, M. and Niva, G. (1976), *Phytochemistry* **15**, 2025.
Ohashi, H., Goto, M. and Imamura, H. (1977), *Phytochemistry* **16**, 1106.
Okorie, D. A. (1977), *Phytochemistry* **16**, 1591.
Otryashenkova, V. E., Glyzin, V. I. and Mashnin, A. I. (1977), *Acta Pharm. Jugosl.* **27**, 131.
Panichpol, K. and Waterman, P. G. (1978), *Phytochemistry* **17**, 1363.
Parker, W. H., Maze, J. and McLachlan, D. G. (1979), *Phytochemistry* **18**, 508.
Parthasarathy, M. R., Seshadri, T. R. and Varma, R. S. (1976), *Phytochemistry* **15**, 226.
Pavanasasivan, G. and Sultanbawa, M. U. S. (1975), *Phytochemistry* **14**, 1127.
Pederiva, R., Kavda, J. and D'Arcangelo, A. (1975), *An. Asoc. Quim. Argent.* **63**, 85.
Pelter, A. Hänsel, R. and Kaloga, M. (1977), *Tetrahedron Lett.*, 4547.
Pomilio, A. B. and Gros, E. G. (1979), *Phytochemistry* **18**, 1410.
Proctor, M. and Yeo, P. (1973), *The Pollination of Flowers*, Wm. Collins Sons & Co., London.
Rahman, W., Ishratullah, K., Wagner, H., Seligmann, O., Chari, W. M. and Österdahl, B.-G. (1978), *Phytochemistry* **17**, 1064.
Ramakrishnan, G., Banerjii, A. and Chadha, M. S. (1974), *Phytochemistry* **13**, 2317.
Rao, E. V. and Raju, N. R. (1979), *Phytochemistry* **18**, 1581.
Rao, J. M., Subrahmanyam, K. and Rao, J. V. J. (1974), *Indian J. Chem.* **12B**, 762.
Rao, R. M., Subrahmanyam, K. and Rao, K. V. J. (1975a), *Curr. Sci.* **44**, 158.
Rao, J. M., Subrahmanyam, K., Rao, K. V. J. and Rao, M. G. (1975b), *Indian J. Chem.* **13B**, 775.
Rao, J. M., Subrahmanyam, K. and Rao, K. V. J. (1975c), *Indian J. Chem.* **13B**, 1000.
Rao, J. M., Subrahmanyam, K., Rao, K. V. J. and Ramish, T. S. (1976), *Indian J. Chem.* **14B**, 339.
Rao, M. M., Gupta, P. S., Krishna, E. M. and Singh, P. P. (1979), *Indian J. Chem.* **17B**, 178.
Rizvi, S. H. M. and Rahman, W. (1974), *Phytochemistry* **13**, 2879.
Roberts, M. L. (1980), *Biochem. Syst. Ecol.* **8**, 115.
Sadykov, A. A. and Sadykov, A. S. (1979), *Khim. Prir. Soed.* **15**, 273 (English translation).
Saitoh, T. and Shibata, S. (1975), *Tetrahedron Lett.* 4461.
Saitoh, T., Shibata, S., Sankawa, U., Furuya, T. and Ayabe, S. (1975), *Tetrahedron Lett.* 4463.
Sastry, G. P. and Row, L. R. (1961a), *J. Sci. Ind. Res. India*, **20B**, 187.
Sastry, G. P. and Row, L. R. (1961b), *Tetrahedron*, **15**, 111.
Satam, P. G. N. and Bringi, N. V. (1973), *Indian J. Chem.* **11B**, 209.
Scogin, R. (1976), *Aliso* **8**, 429.
Scogin, R. and Zakar, K. (1976), *Biochem. Syst. Ecol.* **4**, 165.
Scogin, R., Young, D. A. and Jones, C. E. (1977), *Bull Torrey Bot. Club.* **104**, 155.
Serbin, A. G., Borisov, M. I., Chernobai, V. T., Kovalev, I. P. and Gordienko, V. G. (1975), *Khim. Prir. Soed.* **11**, 160 (English Translation).
Sharma, P. and Parthasarathy, M. R. (1977), *Indian J. Chem.* **15B**, 866.
Sharma, P., Seshadri, T. R. and Mukerjee, S. K. (1973), *Indian J. Chem.* **11B**, 985.
Sherff, E. E. (1937). *The Genus Bidens*. Field Museum Publications in Botany, Chicago, Vol. 16, pp. 1–709.
Singhal, A. K., Sharma, R. P., Thygarajan, G., Herz, W. and Govindan, S. V. (1980), *Phytochemistry* **19**, 929.
Sivarambabu, S., Rao, J. M. and Rao, K. V. J. (1979a), *Indian J. Chem.* **17B**, 85.
Sivarambabu, S., Rao, J. M. and Rao, K. V. J. (1979b), *Indian J. Chem.* **18B**, 388.
Srivastava, S. K. and Chauhan, J. S. (1977), *Planta Med.* **32**, 384.
Srivastava, S. K. and Srivastava, S. D. (1979a), *Indian J. Chem.* **18B**, 86.
Srivastava, S. K. and Srivastava, S. D. (1979b), *Phytochemistry* **18**, 2058.
Srivastava, S. K., Chauhan, J. S. and Srivastava, S. D. (1979), *Phytochemistry* **18**, 2057.
Star, A. E., Rösler, H., Mabry, T. J. and Smith, D. M. (1975a), *Phytochemistry* **14**, 2275.

Star, A. E., Siegler, D. S., Mabry, T. J. and Smith, D. M. (1975b), *Biochem. Syst. Ecol.* **2**, 109.
Star, A. E., Mabry, T. J. and Smith, D. M. (1978), *Phytochemistry* **17**, 586.
Star, A. E. (1980), *Bull. Torrey Bot. Club*, **107**, 146.
Stuessy, T. F. (1977), In *The Biology and Chemistry of the Compositae* (eds. V. H. Heywood, J. B. Harborne and B. L. Turner), Academic Press, New York, pp. 621–672.
Subramanyam, K., Rao, J. M. and Rao, K. V. J. (1977), *Indian J. Chem.* **15B**, 12.
Suri, J. L., Gupta, G. K., Dhar, K. L. and Atal, C. K. (1980), *Phytochemistry* **19**, 336.
Syrchina, A. I., Voronkov, M. G. and Tyukavina, N. A. (1975), *Khim. Prir. Soed.* **11**, 439 (English Translation).
Syrchina, A. I., Voronkov, M. G. and Tyukavkina, N. A. (1978), *Khim. Prir. Soed.* **14**, 685 (English Translation).
Talapatra, S. K., Mallik, A. K. and Talapatra, B. (1980), *Phytochemistry* **19**, 1199.
Tammami, B., Torrance, S. J., Fabela, F. V., Wiedhopf, R. M. and Cole, J. R. (1977), *Phytochemistry* **16**, 2040.
Tatum, J. H. and Berry, R. E. (1978), *Phytochemistry* **17**, 447.
Tindale, M. D. and Roux, D. G. (1974), *Phytochemistry* **13**, 829.
Tiwari, R. D. and Bajpai, M. (1977), *Phytochemistry* **16**, 798.
Tiwari, K. P. and Srivastava, S. S. D. (1979), *Planta Med.* **36**, 191.
Traub, A. and Geiger, H. (1975), *Z. Naturforsch.* **30c**, 823.
Tryon, R. (1962), *Contrib. Gray Herb.* **189**, 52.
Tschesche, R., Delhvi, S., Sepulveda, S. and Breitmaier, E. (1979), *Phytochemistry* **18**, 867.
Turner, B. L. and Powell, A. M. (1977). In *The Biology and Chemistry of the Compositae* (eds. V. H. Heywood, J. B. Harborne and B. L. Turner), Academic Press, New York, pp. 699–738.
Ubaev, Kh. U., Urdusheva, B. and Nikonov, G. K. (1974), *Khim. Prir. Soed.* **10**, 257 (English translation).
Uyar, T., Malterud, K. E. and Anthonsen, T. (1978), *Phytochemistry* **17**, 2011.
Vandekerkhove, O. (1977), *Z. Pflanzenphysiol.* **82**, 455.
Vinokurov, I. I. (1979), *Khim. Prir. Soed.* **15**, 355 (English Translation).
Viswanathan, N. and Sidhaye, A. R. (1976), *Indian J. Chem.* **14B**, 644.
Wagner, H. (1980), Personal communication.
Wagner, H., Seligmann, O., Hörhammer, L. and Seitz, M. (1971), *Tetrahedron Lett.*, 1895.
Wagner, H., Chari, V. M., Seitz, M. and Riess-Maurer, I. (1978), *Tetrahedron Lett.*, 381.
Wagner, H., Seligmann, O., Chari, M., Wollenweber, E., Dietz, V. H., Donnally, D. M., Meegan, M. M. and O'Donnell, B. (1979), *Tetrahedron Lett.*, 4269.
Waterman, P. G. and Crichton, E. G. (1980), *Phytochemistry* **19**, 1187.
Waterman, P. G. and Pootakahm, K. (1979), *Planta Med.* **35**, 366.
Weitz, S. and Ikan, R. (1977), *Phytochemistry* **16**, 1108.
Wells, E. F. and Bohm, B. A. (1980), *Can. J. Bot.* **58**, 1459.
Williams, A. H. (1967), *Phytochemistry* **6**, 1583.
Williams, A. H. (1979), *Phytochemistry* **18**, 1897.
Williams, C. A. (1979) *Phytochemistry* **18**, 803.
Wollenweber, E. (1975), *Biochem. Syst. Ecol.* **3**, 47.
Wollenweber, E. (1977a), *Z. Naturforsch.* **32c**, 1013.
Wollenweber, E. (1977b), *Z. Pflanzenphysiol.* **85**, 71.
Wollenweber, E. (1978), *Am. Fern J.* **68**, 13.
Wollenweber, E. (1979) *Flora* **168**, 138.
Wollenweber, E. and Dietz, V. H. (1980), *Biochem. Syst. Ecol.* **8**, 21.
Wollenweber, E. and Weber, W. (1973), *Z. Pflanzenphysiol.* **69**, 125.
Wollenweber, E. and Wiermann, R. (1979), *Z. Naturforsch.* **34c**, 1289.
Wollenweber, E., Jay, M. and Favre-Bonvin, J. (1974), *Phytochemistry* **13**, 2618.
Wollenweber, E., Dietz, V. H., MacNeill, C. D. and Schilling, G. (1979a), *Z. Pflanzenphysiol.* **94**, 241.

Wollenweber, E., Dietz, V. H., Smith, D. M. and Seigler, D. S. (1979b), *Z. Naturforsch.* **34c**, 876.
Wollenweber, E., Dietz, V. H., Schillo, D. and Schilling, G. (1980a), *Z. Naturforsch.* **35c**, 685.
Wollenweber, E., Rehse, C. and Dietz, V. H. (1980b), in press.
Young, D. A. (1976), *Syst. Bot.* **1**, 149.
Young, D. A. (1979), *Am. J. Bot.* **66**, 502.

CHECK LIST OF CHALCONES

1 4′-Methoxychalcone
2 2′,4′-Dihydroxychalcone
3 2′-Hydroxy-4′-prenyloxychalcone
4 2′,4′-Dihydroxy-3′-*C*-prenylchalcone*
5 2′-Hydroxy-3′-C-prenyl-4′-methoxychalcone*
6 2′,4′-Dihydroxy-3′- (αα-dimethylallyl)chalcone*
7 2′-Methoxyfurano(2″,3″,4′,3′)chalcone*
8 2′-Hydroxy-6″,6″-dimethylpyrano(2″,3″,4′,3′)chalcone
9 2′,4′-Dihydroxy-3′-methoxychalcone*
10 2′,4′-Dihydroxy-5′-methoxychalcone
11 2′,4′,6′-Trihydroxychalcone*
12 2′,4′-Dihydroxy-6-methoxychalcone
13 2′,6′-Dihydroxy-4′-methoxychalcone
14 2′-Hydroxy-4′,6′-dimethoxychalcone
15 2′,4′,6′-Trihydroxy-3′-*C*-prenylchalcone*
16 2′,6′-Dihydroxy-4′-methoxy-3′- *C*-prenylchalcone*
17 2′-Hydroxy-3′-*C*-prenyl-4′,6′-dimethoxychalcone*
18 2′,5′-Dihydroxy-6″,6″-dimethylpyrano(2″,3″,4′,3′)chalcone
19 2′,4′-Dihydroxy-3′-*C*-methyl-6″, 6″-dimethylpyrano(2″,3″,6′,5′) chalcone
20 2′-Hydroxy-6′-methoxy-6″,6″-dimethylpyrano(2″,3″,4′,3′) chalcone*
21 2′,6′,*β*-Trimethoxy-6″,6″-dimethylpyrano(2″,3″,4′,3′) chalcone*
22 *β*-Hydroxy-2′,6′-dimethoxy-6″,6″-dimethylpyrano(2″,3″,4′,3′) chalcone*
23 2′,4′,6′-Trihydroxy-3′-*C*-geranylchalcone*
24 2′,6′-Dimethoxy-3′,4′-methylenedioxychalcone*
25 2′-Hydroxy-4′,5′,6′-trimethoxychalcone*
26 2′,3′,4′,5′,6′-Pentamethoxychalcone
27 2′-Hydroxy-3′,6′-diketo-4′,5′-dimethoxychalcone
28 2′,5′-Dihydroxy-3′,6′-diketo-4′-methoxychalcone
29 2′,5′-Dihydroxy-3′,4′,6′-trimethoxychalcone
30 2′,4′-Dihydroxy-5′-*C*-methyl-6′-methoxychalcone
31 2′,6′-Dihydroxy-3′-*C*-methyl-4′-methoxychalcone*
32 2′-Hydroxy-3′-*C*-methyl-4′,6′-dimethoxychalcone*
33 2′,4′-Dihydroxy-3′,5′-di-*C*-methyl-6′-methoxychalcone*
34 2′-Hydroxy-3′,4′,6′-trimethoxychalcone*
35 2′,3′,4′,6′-Tetrahydroxychalcone*
36 2′,4′-Dihydroxy-3′,6′-dimethoxychalcone*
37 2′,6′-Dihydroxy-3′,4′-dimethoxychalcone*
38 2′,4′-Dihydroxy-3′,5′,6′-trimethoxychalcone*
39 2′-Hydroxy-3′,4′,5′,6′-tetramethoxychalcone*
40 2′,4′,4-Trihydroxychalcone
41 4′-*O*-Glucoside
42 4′-*O*-Glucosylglucoside
43 4′,4-Di-*O*-glucoside
44 4′-*O*-Glucosylglucoside-4-*O*-glucoside
45 4-*O*-Glucoside
46 4-*O*-Apiosylglucoside
47 4-*O*-Rhamnosylglucoside
48 2′,4′,4-Trihydroxy-3′-*C*-glucosylchalcone
49 2′,4-Dihydroxy-4′-methoxychalcone
50 2′,4-Dihydroxy-4′-methoxy-5′-formylchalcone
51 2′-Methoxy-4′,4-dihydroxy-5′-formylchalcone*
52 2′,4′,4-Trihydroxy-3′-*C*-prenylchalcone*
53 2′,4-Dihydroxy-3′-*C*-prenyl-4′-methoxychalcone*
54 2′,4-Dihydroxy-4′-methoxy-5′-*C*-prenylchalcone
55 2′,4′,4-Trihydroxy-3′-*C*-geranylchalcone*
56 2′,4′,4-Trihydroxy-3,3′,5-tri-*C*-prenylchalcone
57 2′,4-Dihydroxy-6″,6″-dimethylpyrano(2″,3″,4′,5′)chalcone*
58 2′,4′,6′,4-Tetrahydroxychalcone*
59 4-*O*-Glucoside*

60 2′-*O*-Glucoside
61 2′,4′,4-Trihydroxy-3′-*C*-prenyl-6′-methoxychalcone*
62 2′,4′,4-Trihydroxy-6′-methoxychalcone*
63 4′-*O*-Glucoside*
64 6′,2,4-Trihydroxy-4′-methoxychalcone 2′-*O*-glucoside
65 2′,4-Dihydroxy-4′,6′-dimethoxychalcone 4-*O*-glucoside*
66 2′,6′-Dihydroxy-4′,4-dimethoxychalcone
67 2′-Hydroxy-4′,6′,4-trimethoxychalcone*
68 2′,4-Dihydroxy-4′,5′,6′-trimethoxychalcone*
69 2′-Hydroxy-4′,5′,6′,4-tetramethoxychalcone*
70 2′,4-Dihydroxy-3′,4′-methylenedioxy-6′-methoxychalcone*
71 4′,4-Dihydroxy-2-methoxychalcone*
72 4′,3,4-Trihydroxy-2-methoxychalcone*
73 4′,4-Dihydroxy-2-methoxy-5-(α,α-dimethylallyl)chalcone*
74 2′,4′,3,4,α-Pentahydroxychalcone*
75 2′,3′,4′,4-Tetrahydroxychalcone
76 2′,3′,4′6′,4-Pentahydroxychalcone 6′-*O*-glucoside
77 2′,4′,4-Trihydroxy-3′,6′-diketochalcone 2′-*O*-glucoside
78 2′,5′-Dihydroxy-3′,4,4′,6′-tetramethoxychalcone
79 2′,3,4-Trihydroxy-6″,6″-dimethylpyrano(2″,3″,4′,3′)chalcone*
80 2′,4-Dihydroxy-3-methoxy-6″,6″-dimethylpyrano(2″,3″,4′,3′) chalcone*
81 2′-Hydroxy-3,4-methylenedioxy-6″,6″-dimethylpyrano(2″,3″,4′,3′) chalcone*
82 2′-Hydroxy-3,4-methylenedioxy-6′-methoxy-6″,6″-dimethylpyrano (2″,3″,4′,3′)chalcone*
83 2′-Methoxy-3,4-methylene-dioxyfurano(2″,3″,4′,3′) chalcone*
84 2′,4′,3,4-Tetrahydroxychalcone
85 3-*O*-Glucoside
86 4′-*O*-Glucoside
87 3,4′-Di-*O*-glucoside
88 2′,3-Di-*O*-glucoside
89 4′-*O*-Glucosylglucoside
90 4′-*O*-Arabinosylgalactoside
91 2,′4′,3-Trihydroxy-4-methoxychalcone*
92 2′,4′,4-Trihydroxy-3-methoxychalcone
93 4-*O*-Glucoside
94 2′,4′-Dihydroxy-3,4-dimethoxychalcone*
95 2′,3′,4′,3,4-Pentahydroxychalcone
96 4′-*O*-Glucoside
97 2′,4′,3,4,α-Pentahydroxychalcone
98 2′,4′,3,4-Tetrahydroxy-3′-methoxychalcone 4′-*O*-Glucoside
99 2′,4′,5′,3,4-Pentahydroxychalcone
100 4′-*O*-Glucoside
101 2′,3′,4′,4-Tetrahydroxy-3-methoxychalcone*
102 2′,3′,4′-Trihydroxy-3,4-dimethoxychalcone*
103 2′,4′-Dihydroxy-3,4-methylenedioxy-5′-methoxychalcone*
104 2′,4′,6′,3,4-Pentahydroxychalcone
105 2′-*O*-Glucoside
106 2′,3-Dihydroxy-4′,6′,4-trimethoxychalcone
107 2′-Hydroxy-4′,6′,3,4-tetramethoxychalcone
108 2′-Hydroxy-3,4-methylenedioxy-3′-*C*-prenyl-4′,6′-tetramethoxychalcone*
109 2′,4′-Dihydroxy-6′,2,4-trimethoxychalcone*
110 2′-Hydroxy-4′,6′,2,4-tetramethoxychalcone*
111 2′-Hydroxy-4′,5′,6′,3,4-pentamethoxychalcone*
112 2′,4′,4,2-Tetrahydroxy-3′-(2-isopropylidene-5-methylhex-4-enyl)-5′-methoxychalcone
113 2′,4′,2-Trihydroxy-3′-geranyl-5′-methoxychalcone
114 2′,4′,2,5-Tetrahydroxy-3′-geranyl-5′-methoxychalcone
115 2′,4′,3,4,5-Pentahydroxychalcone
116 2′,4′,6′,3,4,5-Hexahydroxychalcone 2′-*O*-glucoside
117 2′,3′,4′,5′,3,4,α-Heptahydroxychalcone 2′-*O*-glucoside
118 2′,3-Dihydroxy-4′,6′,4,5-tetramethoxychalcone*
119 2′-Hydroxy-4′,6′,3,4,5-pentamethoxychalcone
120 2′-Hydroxy-4′,6′,2,4,5-pentamethoxychalcone*
121 5,7-Dihydroxy-8-cinnamoyl-4-phenyl-2H-1-benzopyran-2-one*
122 5,7-Dihydroxy-8-(4-hydroxycinnamoyl)-4-phenyl-2H-1-benzopyran-2-one*
123 5,7-Dihydroxy-8-(3,4-dihydroxycinnamoyl)-4-phenyl-2H-1-benzopyran-2-one*
124 2-Cinnamoyl-3-hydroxy-4,4-dimethyl-5-methoxycyclohexadienone (= ceroptin)
125 2′,4′-Dihydroxy-3′-*C*-(2,4,6-trihydroxy-3-acetyl-5-methylbenzyl)-6″,6″-di-

methylpyrano(2″,3″,4′,5′)chalcone (= rottlerin)
126 2′,4′,4-Trihydroxy-3′-C-(2,4,6,-trihydroxy-3-acetyl-5-methylbenzyl)-6″,6″-dimethylpyrano(2″,3″,4′,5′)chalcone (= 4-hydroxyrottlerin)
127 4′,2,3,4-Tetrahydroxy-3′-C-(2,4,6-trihydroxy-3-acetyl-5-methylbenzyl)-6″,6″-dimethylpyrano(2″,3″,4′,5′) chalcone(= 3,4-dihydroxyrottlerin)
128 2′,4′-Dihydroxy-3′,3-di-C-prenyl-6″,6″-dimethylpyrano(2″,3″,4,5) chalcone
129 Cryptocaryone (I 8)†
130 Flemingin-A (I 69)†
131 Flemingin-B (I 70)†
132 Flemingin-C (I 71)†
133 Peltogynoid chalcone (I 64)†
134 (−)-Rubranine (23)‡
135 Bavachromanol (46)*‡
136 4′-O-Methyl-4-deoxybavachromanol (19)*‡
137 Flemistrictin-B (8)*‡
138 Flemistrictin-C (7)*‡
139 Lespeol (47)*‡

* New structures reported during 1975–1980.
† Refers to structure number in Bohm (1975) Chapter 9.
‡ Refers to structure number in present chapter.

CHECK LIST OF AURONES

1 Furano(2″,3″,6,7)aurone*
2 4′-Hydroxyfurano(2″,3″,6,7)aurone*
3 4′-Methoxyfurano(2″,3″,6,7)aurone*
4 6,4′-Dihydroxyaurone*
5 6-O-Glucoside
6 4,6,4′-Trihydroxyaurone 4-O-glucoside
7 6,3′,4′-Trihydroxyaurone
8 6-O-Glucoside
9 6,3′-Di-O-glucoside
10 6-O-Glucosylglucoside
11 6,7,3′,4′-Tetrahydroxyaurone
12 6-O-Glucoside
13 6,3′,4′-Trihydroxy-7-methoxyaurone
14 6-O-Glucoside
15 4,3-Methylenedioxyfurano (2″,3″,6,7)aurone*
16 4,3′,4′,6′-Tetrahydroxyaurone
17 6-O-Glucoside
18 6-O-Glucuronide*
19 6,3′,4′-Trihydroxy-4-methoxyaurone
20 4,6,3′,4′,5′-Pentahydroxyaurone*
21 4-O-Glucoside
22 6-O-Glucoside
23 6,3′,4′,5′-Tetrahydroxy-4-methoxyaurone 6-O-rhamnosylglucoside*
24 Derriobtusone (*6.97*)*†

* New structures reported during 1975–1980.
† Refers to structure in current chapter.

CHECK LIST OF DIHYDROCHALCONES

1 Dihydrochalcone
2 2′-Hydroxy-4′-prenyloxydihydrochalcone*
3 2′,β-Dimethoxyfurano (2″,3″,4′,3′)dihydrochalcone*
4 2′-Methoxyfurano(2″,3″,4′,3′)-β-ketodihydrochalcone*
5 2′,4′,6′-Trihydroxydihydrochalcone*
6 2′-O-Glucoside*
7 4′-O-Glucoside*
8 2′,6′-Dihydroxy-4′-prenyloxydihydrochalcone*
9 2′,4′,6′-Trihydroxy-β-ketodihydrochalcone 2′-O-glucoside*
10 2′,6′-Dihydroxy-4′-methoxy-dihydrochalcone
11 2′,4′-Dihydroxy-6′-methoxy-dihydrochalcone*
12 2′-Hydroxy-3′-C-methyl-4′,6′-dimethoxydihydrochalcone*
13 2′,6′-Dihydroxy-3′,5′-di-C-methyl-4′-methoxydihydrochalcone*
14 2′,4′-Dihydroxy-3′,5′-di-C-methyl-6′-

methoxydihydrochalcone*
15 2′,6′,β-Trimethoxy-3′,4′-methylenedioxydihydrochalcone*
16 1-Phenyl-3-(2,6-dimethoxy-3,4-methylenedioxyphenyl)-propan-1,3-diol*
17 2′,4′,6′-Trihydroxy-3′-*C*-methyl-5′-formyl-β-ketodihydrochalcone*
18 2′,6′-Dihydroxy-3′-*C*-methyl-4′-methoxy-5′-formyl-β-ketodihydrochalcone*
19 2′,4′-Dihydroxy-3′-*C*-(2-hydroxybenzyl)-6′-methoxydihydrochalcone*
20 2′,4′-Dihydroxy-5′-*C*-(2-hydroxybenzyl)-6′-methoxydihydrochalcone*
21 2′,4′,6′-Trihydroxy-3′-methoxy-5′-*C*-prenyldihydrochalcone*
22 2′,4′,6′-Trihydroxy-3′,5′-di-*C*-prenyldihydrochalcone*
23 4′,6′-Dihydroxy-3′,3′,5′-tri-*C*-prenyl-β-ketodihydrochalcone*
24 4′,6′-Dihydroxy-3′-methoxy-3′,5′-di-*C*-prenyl-β-ketodihydrochalcone*
25 2′,4′,4-Trihydroxydihydrochalcone*
26 2′-*O*-Glucoside
27 2′,4-Dihydroxy-4′-methoxydihydrochalcone*
28 2′,4′,4-Trihydroxy-β-ketodihydrochalcone*
29 2′,4′,6′,4-Tetrahydroxydihydrochalcone*
30 2′-*O*-Glucoside
31 2′-*O*-Rhamnoside
32 4′-*O*-Glucoside
33 2′-*O*-Xylosylglucoside
34 2′,6′,4-Trihydroxy-4′-methoxydihydrochalcone 2′-*O*-glucoside
35 2′,6′-Dihydroxy-4,4′-dimethoxydihydrochalcone
36 2′,4′-Dihydroxy-4,6′-dimethoxydihydrochalcone*
37 2′,6′,4-Trihydroxy-3′-*C*-methyl-4′-methoxydihydrochalcone*
38 2′,4′,6′,4-Tetrahydroxy-3′-*C*-glycosyldihydrochalcone
39 2′,3′,4′,6′,4-Pentahydroxy-5′-geranyldihydrochalcone*
40 2′,3′,4′,6′,4-Pentahydroxy-5′-neryldihydrochalcone*
41 2′,4′,4-Trihydroxy-3′-methoxydihydrochalcone*
42 2′,4′,6′,3,4-Pentahydroxydihydrochalcone*
43 4′-*O*-Glucoside
44 2′-Methoxy-3,4-methylenedioxyfurano(2″,3″,4′,3′)-β-ketodihydrochalcone*
45 2′,4′,6′,3,4-Pentahydroxy-3′-*C*-glucosyldihydrochalcone
46 2′,4′,3,5,β-Pentahydroxy-4-methoxydihydrochalcone*

* New structures reported during 1975–1980.

CHECK LIST OF FLAVANONES

1 7-Hydroxyflavanone
2 7-Methoxyflavanone*
3 7-Hydroxy-8-methoxyflavanone*
4 7,8-Dimethoxyflavanone*
5 7-Prenyloxyflavanone
6 6″,6″-Dimethylpyrano(2″,3″,7,8)flavanone
7 6-Methoxy-6″,6″-dimethylpyrano(2″,3″,7,8)flavanone*
8 7-Hydroxy-6,8-di-*C*-prenylflavanone*
9 7-Hydroxy-8-*C*-prenylflavanone*
10 5,7-Dihydroxyflavanone*
11 5-*O*-Glucoside
12 7-*O*-Rhamnoside
13 7-*O*-Neohesperidoside*
14 5-Methoxy-7-hydroxyflavanone
15 5-Hydroxy-7-methoxyflavanone
16 5,7-Dimethoxyflavanone
17 5,7-Dihydroxy-6-methoxyflavanone
18 5-Hydroxy-6,7-dimethoxyflavanone*
19 5,7-Dihydroxy-8-methoxyflavanone
20 7-*O*-Glucoside
21 5,7,8-Trimethoxyflavanone*
22 7-Hydroxy-5,6,8-trimethoxyflavanone* (structure has been challenged)
23 6-Hydroxy-5,7,8-trimethoxyflavanone*
24 5,6,7,8-Tetramethoxyflavanone*
25 5,7-Dihydroxy-6-*C*-methylflavanone
26 5-Hydroxy-6-*C*-methyl-7-methoxyflavanone*
27 5,7-Dihydroxy-8-*C*-methylflavanone
28 5,7-Dihydroxy-6,8-di-*C*-methylflavanone
29 7-*O*-Glucoside*

30 5-Hydroxy-6,8-di-*C*-methyl-7-methoxy-flavanone*
31 2,5,7-Trihydroxy-6-(or 8)-*C*-methyl-8-(or 6)-formylflavanone*
32 5,7-Dihydroxy-6-formyl-8-*C*-methyl-flavanone*
33 5,7-Dihydroxy-6-*C*-prenylflavanone*
34 5,7-Dihydroxy-8-*C*-prenylflavanone
35 5-Hydroxy-6-*C*-prenyl-7-methoxy-flavanone*
36 5-Methoxy-6″,6″-dimethylpyrano (2″,3″,7,8)flavanone*
37 5,7-Dihydroxy-8-*C*-(β-methyl-β-hydroxy-methylallyl)flavanone*
38 5,7-Dihydroxy-8-*C*-(β-methyl-β-formylallyl)flavanone*
39 5,7-Dihydroxy-8-*C*-geranylflavanone*
40 5-Hydroxy-7-methoxy-8-*C*-prenyl-flavanone*
41 5,7-Dihydroxy-6-*C*-(2-hydroxybenzyl) flavanone*
42 5,7-Dihydroxy-8-*C*-(2-hydroxybenzyl) flavanone
43 6,8-Dihydroxy-7-*C*-(5-hydroxygeranyl) flavanone* (structure has been challenged)
44 7,8,4′-Trihydroxyflavanone
45 7,4′-Dihydroxyflavanone
46 4′-*O*-Glucoside
47 7-*O*-Glucoside
48 4′-*O*-Rhamnosylglucoside
49 7-*O*-Glucosylglucoside
50 7,4′-Dihydroxy-8-*C*-galactosylflavanone*
51 7,4′-Dihydroxy-6,8-*C*-prenylflavanone*
52 5,7,4′-Trihydroxyflavanone
53 5-*O*-Glucoside
54 7-*O*-Glucoside
55 7-*O*-Rhamnoside*
56 4′-*O*-Glucoside
57 4′-*O*-Galactoside
58 ?-*O*-Fructoside
59 7-*O*-Glucoside-6″-*O*-galloyl*
60 7-*O*-Glucoside-6″-*p*-coumaroyl*
61 5,7-Di-*O*-glucoside
62 5-*O*-Glucosylglucoside
63 7-*O*-Rutinoside
64 7-*O*-Neohesperidoside
65 4′-*O*-Rutinoside
66 4′-*O*-Xylosylglucoside*
67 5,7,4′-Trihydroxy-6-*C*-glucosylflavanone*
68 5,4′-Dihydroxy-7-methoxyflavanone
69 5-*O*-Glucoside
70 5,7-Dihydroxy-4′-methoxyflavanone
71 5-*O*-Glucoside*
72 7-*O*-Glucoside
73 7-*O*-Rutinoside
74 7-*O*-Neohesperidoside
75 7-*O*-4′-(Rhamnosyl)glucoside
76 5-Methoxy-7,4′-dihydroxyflavanone*
77 5-Hydroxy-7,4′-dimethoxyflavanone
78 4′-Hydroxy-5,7-dimethoxyflavanone*
79 4′-*O*-Xylosylarabinoside*
80 4′-*O*-Rhamnosylglucoside*
81 5,7,4′-Trimethoxyflavanone
82 5,6,7,4′-Tetrahydroxyflavanone 5-*O*-glucoside
83 4′-Prenyloxy-5,7-dihydroxyflavanone
84 5,7,4′-Trihydroxy-6-*C*-glucosylflavanone
85 5,7,4′-Trihydroxy-8-*C*-glucosylflavanone
86 5,2′-Dihydroxy-7-methoxyflavanone 2′-*O*-glucoside*
87 5,7,4′-Trihydroxy-6-*C*-methylflavanone
88 7-*O*-Glucoside
89 5,7,4′-Trihydroxy-6,8-di-*C*-methylflavanone
90 7-*O*-Glucoside
91 5,7-Di-*O*-glucoside
92 5,4′-Dihydroxy-6,8-di-*C*-methyl-7-methoxy flavanone
93 4′-Methoxy-5,7-dihydroxy-6,8-di-*C*-methyl flavanone
94 7-*O*-Glucoside*
95 7-*O*-Glucosylglucoside
96 5,7,4′-Trihydroxy-8-*C*-prenylflavanone*
97 7,4′-Dihydroxy-5-methoxy-8-*C*-prenylflavanone
98 7,4′-Dihydroxy-8-*C*-prenylflavanone
99 7,4′-Dihydroxy-6-*C*-prenylflavanone
100 4′-Hydroxy-6-*C*-prenyl-7-methoxy-flavanone
101 5,4′-Dihydroxy-6-*C*-prenyl-6″,6″-dimethyl-pyrano(2″,3″,7,8)flavanone*
102 5,7,4′-Trihydroxy-6-(or 8)-*C*-prenyl-flavanone 7,4′-di-*O*-glucoside*
103 7,4′-Dihydroxy-8,3′,5′-tri-*C*-prenylflavanone
104 7-Hydroxy-8,5′-di-*C*-prenyl-6″,6″-dimethyl-pyrano(2″,3″,4′,3′)flavanone
105 5,4′-Dihydroxy-6,7-dimethoxyflavanone*
106 5-Hydroxy-6,7,4′-trimethoxyflavanone*
107 5,4′-Dihydroxy-7,8-dimethoxyflavanone*
108 5-Hydroxy-7,8,4′-trimethoxyflavanone*
109 5,6-Dihydroxy-7,8,4′-trimethoxyflavanone*
110 5,4′-Dihydroxy-6,7,8-trihydroxy-flavanone*

111 5-Hydroxy-6,7,8,4′-tetramethoxyflavanone*
112 4′-Hydroxy-5,6,7-trimethoxyflavanone*
113 5,4′-Dimethoxy-6,7-methylenedioxyflavanone*
114 7,3′,4′-Trihydroxyflavanone
115 3′-*O*-Glucoside
116 7-*O*-Glucoside
117 7,3′-Di-*O*-glucoside
118 6,7,3′,4′-Tetrahydroxyflavanone
119 7,8,3′,4′-Tetrahydroxyflavanone
120 7-*O*-Glucoside
121 7,3′,4′-Trihydroxy-8-methoxyflavanone
122 3′,4′-Methylenedioxy-7-hydroxy-8-*C*-prenylflavanone*
123 3′,4′-Methylenedioxy-6,8-di-*C*-prenyl-7-hydroxyflavanone*
124 3′,4′-Methylenedioxy-6″,6″-dimethylpyrano(2″,3″,7,8)flavanone*
125 3′,4′-Methylenedioxy-6-methoxy-6″,6″-dimethylpyrano(2″,3″,7,8)flavanone*
126 5,7,3′,4′-Tetrahydroxyflavanone
127 5-*O*-Rhamnoside*
128 7-*O*-Glucoside
129 7-*O*-Rhamnoside
130 5,3′-Di-*O*-glucoside
131 7-*O*-Rhamnosylglucoside
132 7-*O*-Rutinoside
133 7-*O*-Neohesperidoside
134 5,7,4′-Trihydroxy-3′-methoxyflavanone
135 5,7,3′-Trihydroxy-4′-methoxyflavanone
136 5-*O*-Glucoside
137 7-*O*-Rhamnoside*
138 7-*O*-Rutinoside
139 7-*O*-Neohesperidoside
140 5,4′-Dihydroxy-7,3′-dimethoxyflavanone
141 5,3′-Dihydroxy-7,4′-dimethoxyflavanone
142 5-*O*-Glucoside
143 5,7-Dihydroxy-3′,4′-methylenedioxyflavanone 5-*O*-glucoside*
144 5-Hydroxy-7,3′,4′-trimethoxyflavanone*
145 5-Hydroxy-7,3′-dimethoxy-4′-prenyloxyflavanone
146 5,7,3′,4′-Tetrahydroxy-6-*C*-methylflavanone 7-*O*-glucoside
147 5,7,3′,4′-Tetrahydroxy-6,8-di-*C*-methylflavanone
148 2,5,7,3′,4′-Pentahydroxyflavanone 5-*O*-glucoside*
149 5,7,2′,5′-Tetrahydroxyflavanone*
150 7-Hydroxy-5,2′,4′-trimethoxyflavanone*
151 5,7,2′,4′-Tetramethoxyflavanone*
152 5,3′,4′-Trihydroxy-7-methoxy-8-*C*-prenylflavanone*
153 5,7,2′-Trihydroxy-4′-methoxy-6,8-di-*C*-methylflavanone*
154 5,2′,4′-Trihydroxy-8-*C*-prenyl-6″,6″-dimethylpyrano(2″,3″,7,6)flavanone*
155 5-Hydroxy-6,7,8,3′,4′-pentamethoxyflavanone
156 5,6,7,8,3′,4′-Hexamethoxyflavanone
157 5-Hydroxy-7,3′,4′,5′-tetramethoxyflavanone*
158 5-Hydroxy-6,7,3′,4′,5′-pentamethoxyflavanone 5-*O*-rhamnoside*
159 Artocarpanone (I 78)†
160 Breverin, 6- or 8- (I 78b)†
161 Kurarinone (I 77)†
162 Nor-kurarinone (I 78)†
163 Remerin (I 78a)†
164 Scaberin (I 78c)†
165 Steppogenin (I 75)†
166 Stepposide (I 76)†
167 Flemichin-A (*6.247*)†*
168 Flemichin-E (*6.246*)†*
169 Louisfieserone-A (*6.200*)‡*
170 Louisfieserone-B (*6.201*)‡*
171 Purpurin (169)‡*
172 Uvarinol (196)‡*

* New structures reported during 1975–1980.
† Refers to structures in Bohm (1975) Chapter 11.
‡ Refers to structures in current chapter.

CHECK LIST OF DIHYDROFLAVONOLS

1 7-Hydroxydihydroflavonol
2 6-Methoxy-7-hydroxydihydroflavonol
3 5,7-Dihydroxydihydroflavonol
4 5-Hydroxy-7-methoxydihydroflavonol
5 3-Acetyl-5-hydroxy-7-methoxydihydroflavonol

6 5,7-Dimethoxydihydroflavonol*
7 5,7-Dihydroxy-6-methoxydihydroflavonol
8 5,7-Dihydroxy-6-*C*-methyldihydroflavonol
9 6″,6″-Dimethylpyrano(2″,3″,7,8) dihydroflavonol*
10 7,4′-Dihydroxydihydroflavonol
11 3-*O*-Glucoside
12 7,8,4′-Trihydroxydihydroflavonol
13 5,7,4′-Trihydroxydihydroflavonol
14 3-*O*-Glucoside
15 3-*O*-Rhamnoside
16 4′-*O*-Xyloside
17 7-*O*-Glucoside
18 4′-Hydroxy-7-methoxydihydroflavonol (*cis* and *trans* isomers)
19 5,7-Dihydroxy-4′-methoxydihydroflavonol
20 4′-Hydroxy-5,7-dimethoxydihydroflavonol
21 5-Hydroxy-4′,7-dimethoxydihydroflavonol
22 5-*O*-Glucoside
23 5,4′-Dihydroxy-6,7-dimethoxydihydroflavonol
24 5,7-Dihydroxy-8,4′-dimethoxydihydroflavonol 7-*O*-glucoside
25 5,4′-Dihydroxy-6-*C*-glucosyl-7-methoxydihydroflavonol
26 5,7,4′-Trihydroxy-6-(3-hydroxy-3-methylbutyl)dihydroflavonol 7-*O*-glucoside
27 5,7,4′-Trihydroxy-8-(3-hydroxy-3-methylbutyl)dihydroflavonol 7-*O*-glucoside
28 5,7,4′-Trihydroxy-8-(3-glucosyloxy-3-methylbutyl)dihydroflavonol 7-*O*-glucoside
29 5,7,4′-Trihydroxy-8-*C*-prenyldihydroflavonol*
30 5,4′-Dihydroxy-6″,6″-dimethylpyrano (2″,3″,7,8)dihydroflavonol*
31 5,4′-Dihydroxy-8-*C*-prenyl-6″,6″-dimethylpyrano(2″,3″,7,6)dihydroflavonol*
32 7,3′,4′-Trihydroxydihydroflavonol
33 3-*O*-Glucoside
34 3,7-Di-*O*-glucoside*
35 7,3′,4′-Trihydroxy-3-*O*-methyldihydroflavonol (*cis* and *trans* isomers)*
36 7,8,3′,4′-Tetrahydroxydihydroflavonol*
37 5,7,3′,4′-Tetrahydroxydihydroflavonol
38 3-*O*-Glucoside*
39 3-*O*-Galactoside*
40 3-*O*-Rhamnoside
41 3-*O*-Xyloside
42 4′-*O*-Glucoside
43 7-*O*-Galactoside*
44 3,5-Di-*O*-rhamnoside*
45 5,7,3′,4′-Tetrahydroxy-3-*O*-methyldihydroflavonol*
46 5,3′,4′-Trihydroxy-7-methoxydihydroflavonol
47 3-*O*-Glucoside
48 5,4′-(or 3′)-Dihydroxy-7,3′-(or 4′)-dimethoxydihydroflavonol
49 5,7,3′,4′-Tetrahydroxy-6-*C*-methyldihydroflavonol
50 5,6,3′,4′-Tetrahydroxy-8-*C*-methyldihydroflavonol (?)
51 5,3′-Dihydroxy-7,4′-dimethoxy-8-*C*-prenyldihydroflavonol*
52 5,7,2′-Trihydroxy-5′-methoxydihydroflavonol*
53 5,7,2′,4′-Tetrahydroxydihydroflavonol
54 5,7-Dihydroxy-2′-methoxy-6″,6″-dimethylpyrano(2″,3″,4′,3′)dihydroflavonol
55 7,3′,4′,5′-Tetrahydroxydihydroflavonol
56 7,3′,5′-Trihydroxy-4′-methoxydihydroflavonol*
57 5,7,3′,4′,5′-Pentahydroxydihydroflavonol
58 5,7,3′,4′,5′-Pentahydroxy-6-*C*-methyldihydroflavonol*
59 3′-*O*-Glucoside*
60 5,7,3′,4′-Tetrahydroxy-5′-methoxydihydroflavonol 4′-*O*-rhamnoside*
61 5,7,4′-Trihydroxy-3′,5′-dimethoxydihydroflavonol

* New structures reported during 1975–1980. To this list may be added three peltogynoids [compounds 108, 109 and 110 in Bohm (1975)] and three flavonolignans (compounds 111, 113 and 114).

CHAPTER SEVEN

Proanthocyanidins

EDWIN HASLAM

7.1 INTRODUCTION

Rapid advances in our knowledge of the proanthocyanidins were made during the last decade, and although this review necessarily concentrates upon recent aspects, this is also an appropriate moment to look at the subject in a broader context. With this viewpoint in mind facets of earlier work which are of importance – such as the solvolytic reactions of flavan-3,4-diols – and related areas of research into natural phenolic compounds are discussed.

One of the pioneers of research into the proanthocyanidins and to whom much of our present knowledge of their biochemistry is due is Dr E. C. Bate-Smith. His name will always be inextricably linked with the 'LA's' and it is a pleasure to dedicate this chapter to him on the occasion of his 80th birthday.

7.1.1 Leucoanthocyanins, leucoanthocyanidins and proanthocyanidins

What's in a name? that which we call a rose
By any other name would smell as sweet.
Shakespeare

The subtleties and puritanical rigours of chemical nomenclature should never be permitted to stand in the way of good chemistry. However, the field of proanthocyanidin chemistry has been bedevilled by the changes in nomenclature which have paralleled advances in knowledge. Confusion still arises from the use by latterday Capulets of older forms of nomenclature and in particular the word *leucoanthocyanin.* A full historical analysis of the way in which the nomenclature was derived was given in an earlier review (Haslam, 1975) and it is perhaps useful to reiterate the nomenclature introduced by Freudenberg and Weinges (1960). These workers collectively designated all the colourless substances isolated from plants which when heated with acid form anthocyanidins as *proanthocyanidins.* This is a chemical term and does not imply any biogenetic relationship. The terminology of *leucoanthocyanidin* was reserved (Weinges *et al.*, 1969) for the monomeric proanthocyanidins such as the phenolic flavan-3,4-diols and the name condensed proanthocyanidin for the various flavan-3-ol dimers and higher oligomers which, on the basis of present evidence, constitute the

overwhelming majority of proanthocyanidins found in plants. This is the nomenclature used in this review and except where it is important to develop points from a historical viewpoint the description *leucoanthocyanin* is one which should no longer be used.

7.1.2 Proanthocyanidins and vegetable tannins

Bate-Smith (1962) in a classical study of phenolic metabolism demonstrated that three classes of phenol predominate in the leaves of vascular plants. These are the esters, amides and glycosides of the various hydroxycinnamic acids and hexahydroxydiphenic acid, the glycosylated flavonols (derivatives of kaempferol, quercetin and myrecitin) and the proanthocyanidins (*7.1*) in particular those which gave cyanidin or delphinidin with acid. Bate-Smith used the pattern of distribution of these three classes of phenolic constituent as a taxonomic guide for the classification of plants and in particular he noted the correlation between the presence of leucoanthocyanins (as proanthocyanidins were known at that point in time!) and a woody habit of growth in the plant. Bate-Smith and Swain (1953) had earlier also noted the close similarity in systematic distribution between leucoanthocyanins and many members of that class of substance rather indefinitely defined in the botanical literature as 'tannins', and they were led to suggest that leucoanthocyanins were most commonly responsible for the broad range of reactions – such as the precipitation of proteins, polysaccharides, alkaloids; the impaired nutritional characteristics of some foodstuffs and their astringent taste – which have generally been attributed to the presence in plants of 'tannins'.

A curious but significant observation also made by Bate-Smith was the complete absence of 3,4,5-trihydroxycinnamic acid from plant tissues. In situations where the acid might have been predicted to occur he found hexahydroxydiphenic acid (*7.2*) and Bate-Smith suggested that this was the taxonomic equivalent of the missing acid. Later work (E. Haslam and R. K. Gupta, unpublished observations) has demonstrated the widespread occurrence in plants of (*7.2*) as esters with D-glucose and the putative biogenetic precursors of these esters the galloyl-D-glucoses. In contrast to the natural phenolic esters of the hydroxycinnamic acids, which are invariably found as mono- and bis- esters with polyols, and the esters of *p*-hydroxybenzoic and protocatechuic acid which are extremely rare, gallic and hexahydroxydiphenic acid (*7.2*) occur in a range of forms esterified to D-glucose and with molecular weights spanning the range 500 to at least 2000. A typical example is the ester (*7.3*)* found widely in the Hamamalidaceae, Paeonaceae, Aceraceae and Anacardiaceae and sporadically in the Ericaceae.

Whilst the biosynthetic implications of these observations remain to be resolved the metabolism of these esters and the various proanthocyanidins by a whole range of plants is worthy of comment and consideration. These compounds are unique to the plant kingdom and because of their polyphenolic

* Recent work has led to a revision of this structure (E. Haslam and R. K. Gupta).

(R = R′ = H .... Profisetinidin)
(R = H ; R′ = OH .... Procyanidin)
(R = R′ = OH .... Prodelphinidin)
(7.1)

(7.2)

Heterogeneity : n = 0,1,2
C-1 galloyl group
may be absent

(7.3) Tannic acid – Chinese gallotannin

character they possess the distinctive properties referred to above – such as that of precipitating proteins. The overwhelming circumstantial evidence points unmistakably to the fact that these properties derive from the accumulation, within a molecule of moderate molecular size, of a substantial number of unconjugated phenolic groups with an *ortho*-dihydroxy (catechol) or *ortho*-trihydroxy orientation within a phenyl ring. Recent experimental support for this suggestion has come from quantitative studies on the interaction of phenols with proteins (J. McManus, unpublished observations), using equilibrium dialysis and microcalorimetry. The affinity of resorcinol for proteins is weak but with catechol, pyrogallol and methyl gallate the binding to protein was considerably enhanced, both in its strength and in the number of primary and secondary sites on the protein molecule which are available for complexation. These observations give quantitative support to earlier proposals (Haslam, 1974a) that the *ortho*-dihydroxyphenolic groups in natural polyphenols are the primary points for association with protein and that isolated phenolic groups do not participate to any significant extent. A further interesting feature of the observations is that the affinity for protein of the isolated catechol group and the simple esters of gallic acid are broadly comparable. The two functionalities are found respectively as the pendant aromatic nuclei of the procyanidins (*7.1*) (R′ = OH,R = H) and of the natural galloyl esters (*7.3*) which constitute the most widely distributed polyphenols of the vegetable tannin group found in plants.

Despite their apparently disparate modes of construction these two classes of natural polyphenols display many similarities in shape and disposition of phenolic groups and coupled with their similar physical properties it is tempting to suggest that they have similar roles in plant metabolism. A word of caution is, however, appropriate. The elaboration during biosynthesis of the proanthocyanidins of the difficultly soluble and the insoluble polymeric forms (*vide infra*) does raise the question whether these metabolites have a function which is principally structural in character. Bate-Smith has repeatedly drawn attention to the empirical relationship which exists between the presence of proanthocyanidins in plants and the 'woodiness' of a particular family or species. From the phylogenetic point of view proanthocyanidins first appeared as plants developed a vascular character – in ferns, for example, proanthocyanidins are frequently found in polymeric forms. Although Bate-Smith generally discounted a connection with lignification, an examination of the proanthocyanidins to ascertain if they have a structural role would be very valuable.

In a recent series of papers Bate-Smith has amplified and collated earlier data of a semiquantitative nature and has commented on the various classes of complex polyphenol as markers of evolutionary trends in plants (Bate-Smith, 1972, 1973, 1975, 1977). Proanthocyanidins are present in the most primitive of vascular plants such as ferns and gymnosperms, but the galloyl and hexahydroxydiphenoyl esters are not. These latter two groups appear for the first time in dicotyledons and Bate-Smith was led to propose a phylogenetic order – proanthocyanidins – hexahydroxydiphenoyl esters – galloyl esters – which,

incidentally parallels their order of increasing efficiency as precipitants for proteins (Bate-Smith, 1972). Further facets of polyphenol metabolism have been revealed by other studies which are worthy of brief mention. Several plant families (e.g. Aceraceae, Ericaceae, Fagaceae and Rosaceae) retain the capacity to biosynthesize different types of complex polyphenol but one particular species may specialize in just one particular form. Thus in the Ericaceae, species of the three genera *Calluna*, *Erica* and *Rhododendron* are all rich sources of proanthocyanidins but *Arctostaphylos uva ursi* has only a minimal capacity for proanthocyanidin synthesis and combines with this a very high level of gallic acid metabolism (principally as *7.3*). This distinctive feature has been noted on several occasions.

7.2 NATURAL LEUCOANTHOCYANIDINS (FLAVAN-3,4-DIOLS)

7.2.1 Occurrence, distribution and structure

The principal flavan-3,4-diols of natural origin whose structure and stereochemistry have been fully characterized are shown in Table 7.1. The majority, it may be noted, have been isolated from the wood or bark of *Acacia* species. Extensive compilations of chemical and spectroscopic data for these flavan-3,4-diols are contained in a review by Weinges and his collaborators (Weinges *et al.*, 1969).

The question of the natural occurrence of other flavan-3,4-diols – particularly those of the leucopelargonidin, leucocyanidin and leucodelphinidin classess – remains essentially unresolved. Definitive stereochemical work has not been completed for any member of these last three groups of leucoanthocyanidins, and in most cases indeed the evidence for the actual flavan-3,4-diol structure is not wholely adequate.

The various physical and chemical methods used for the determination of structure and stereochemistry of leucoanthocyanidins have been fully described in a number of earlier publications (Weinges *et al.*, 1969; Clark-Lewis, 1968; Clark-Lewis *et al.*, 1969; Drewes, 1968). An observation of particular value for stereochemical assignments in six-membered ring systems is the upfield shift for the resonance of a carbon atom which is gauche to another carbon or heteroatom at the γ-position (Roberts *et al.*, 1970). Similar results obtain in the flavan ring and the C-2 resonance in the ^{13}C resonance spectrum of flavan 3,4-diols occurs at higher field when a quasi-axial hydroxy or acetoxy group is placed at the C-4 position. This γ-effect (4.1–4.6 p.p.m.) has been observed by several workers (Weinges *et al.*, 1973; Haslam *et al.*, 1977a) and is useful for the assignment of relative configurations in flavan-3,4-diols. The phenomenon is discussed in more detail in the assignment of relative configuration to procyanidins (*vide infra*).

The preferred conformation of the heterocyclic ring of flavan derivatives has been discussed by Whalley (1962) and by Clark-Lewis (1968), and two conformations (*7.4* and *7.5*) in which the aromatic ring attached to position 2 is in

Table 7.1 Natural leucoanthocyanidins – occurrence, structure and stereochemistry

Trivial name	*Hydroxylation pattern*	*Absolute configuration*	*Source*
(+)-Guibourtacacidin	7,4′	2*R*:3*S*:4*S*	*Guibourtia coleosperma* (Saayman and Roux, 1965)
			Acacia cultriformis (du Preez and Roux, 1970)
	7,4′	2*R*:3*S*:4*R*	*Acacia cultriformis* (du Preez and Roux, 1970)
(+)-Mollisacacidin	7,3′,4′	2*R*:3*S*:4*R*	*Acacia mearnsii* (Keppler, 1957)
			Acacia saxatilis (Fourie *et al*., 1974)
(+)-Gleditsin			*Gleditsia japonica* (Mitsuno and Yoshizaki, 1958; Clark-Lewis and Mitsuno, 1958)
			Acacia cultriformis (du Preez and Roux, 1970)
	7,3′,4′	2*R*:3*S*:4*S*	*Acacia cultriformis* (du Preez and Roux, 1970)
			Acacia saxatilis (Fourie *et al*., 1974)
			Neuratanenia amboensis (Oberholzer *et al*., 1980)
	7,3′,4′-tri-OMe		*Neuratanenia amboensis* (Oberholzer *et al*., 1980)
	3,4,7,3′,4′-penta-OMe		*Neuratanenia amboensis* (Oberholzer *et al*., 1980)
(−)-Leucofisetinidin	7,3′,4′	2*S*:3*R*:4*S*	*Schinopsis lorentzii* (Drewes and Roux, 1964)
(−)-Teracacidin	7,8,4′	2*R*:3*R*:4*R*	*Acacia orites* (Clark-Lewis *et al*., 1961)
(−)-Isoteracacidin	7,8,4′	2*R*:3*R*:4*S*	*Acacia sparsiflora*,
			Acacia obtusifolia (Clark-Lewis and Dainis, 1964, 1967)
			Acacia maidenii (Clark-Lewis and Porter, 1972)
			Acacia auriculiformis (Drewes and Roux, 1966)
	7,8,4′-tri-OMe	2*R*:3*S*:4*R*	*Acacia cultriformis* (du Preez and Roux 1970)
(−)-Melacacidin	7,8,3′,4′	2*R*:3*R*:4*R*	*Acacia melanoxylon* (King and Bottomley, 1954)
	7,3′,4′,3,8-di-OMe		*Acacia saxatilis* (Fourie *et al*., 1974)
(−)-Isomelacacidin	7,8,3′,4′	2*R*:3*R*:4*S*	*Acacia harpophylla* (Clark-Lewis and Mortimer, 1960)
			Acacia excelsa,
	7,8,3′,4′-tetra-OMe	2*R*:3*S*:4*S*	*Acacia cultriformis* (du Preez and Roux, 1970)
			Acacia saxatilis (Fourie *et al*., 1974)
(+)-Leucorobinetinidin	7,3′,4′,5′	2*R*:3*S*:4*R*	*Robinia pseudacacia* (Weinges, 1958)

the quasi-equatorial position have received particular attention. Definitive evidence of the shape of the flavan-3-ol ring system has come from X-ray analysis of 8-bromotetramethyl-(+)-catechin (*7.6a*) (Roux *et al.*, 1978a), and this shows the conformation of the puckered heterocyclic ring is in fact somewhere between the 'half-chair' and 'sofa' (*7.4* and *7.5*). Structure (*7.6b*) shows the distances out of the mean plane of the adjacent benzene ring in pm for the atoms of the heterocyclic ring of the bromo-(+)-catechin derivative. Structure (*7.7b*) shows a stereo drawing of the corresponding 8-bromo-tetramethyl-(−)-epicatechin (*7.7a*) derived from X-ray data (M. J. Begley, D. A. Whiting and E. Haslam, unpublished observations). In this instance the atoms O(1), C(8a), C(5a) and C-4 lie virtually co-planar, C-2 lies 0.004 nm, C-3 lies 0.07 nm and C-1′ lies 0.07 nm below that plane. The shape of the heterocyclic ring is in this case best described as an envelope and it is clearly much flatter than that in the (+)-catechin derivative. These X-ray structures define the shape of the flavan ring in the crystal structure of (*7.6a* and *7.7a*) and in solution conformational equilibria and solvation may well modify these shapes.

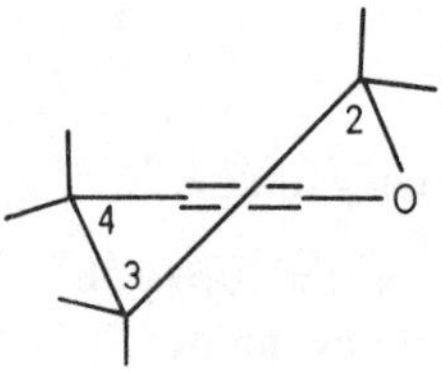

(7.4)

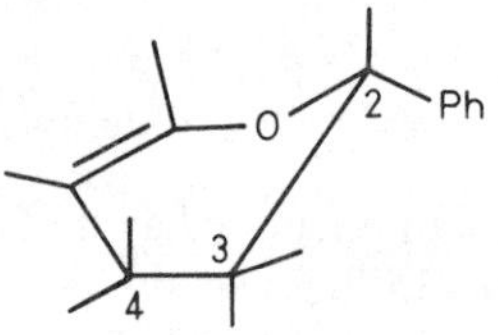

(7.5)

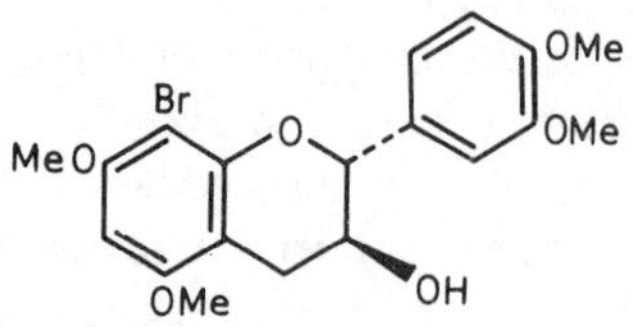

(7.6a)

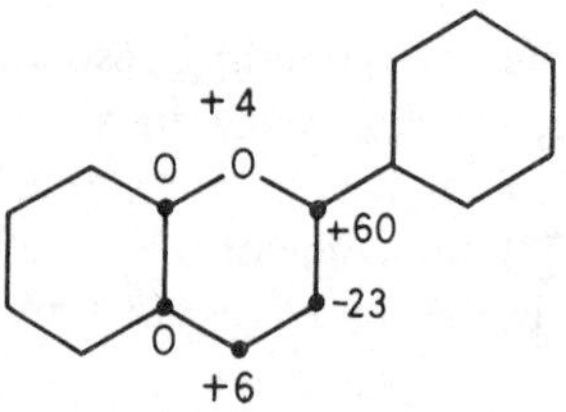

(7.6b)

Dihedral angles

C-8a-O	24°
O — C-2	-57°
C-2 — C-3	68°
C-3 — C-4	-45°
C-4 — C-5a	13°

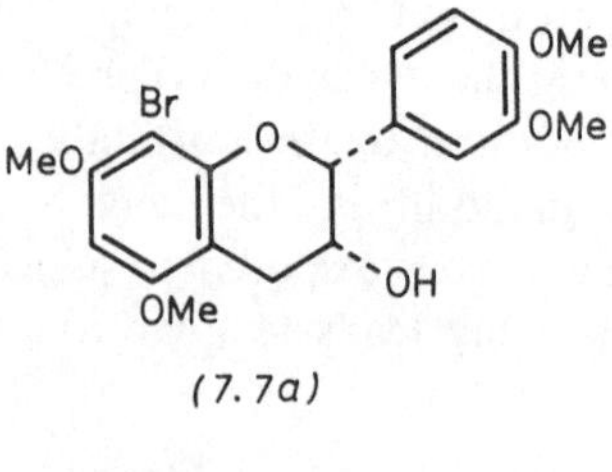

(7.7a)

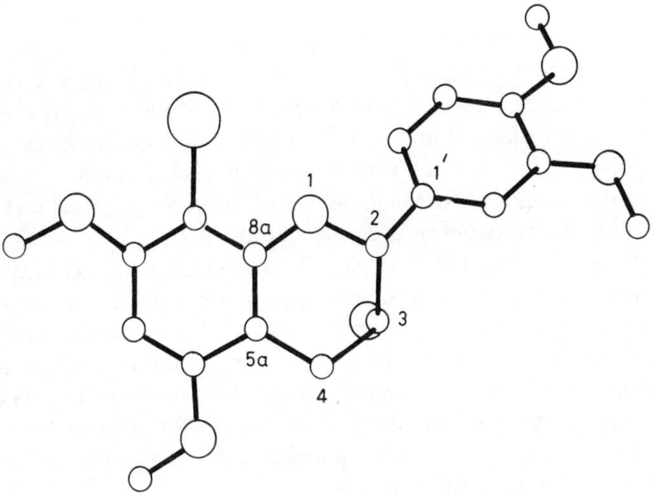

(7.7b)

7.2.2 Solvolysis of leucoanthocyanidins

The hallmark of phenolic flavan-3,4-diols is their partial (up to 50%) conversion with hot mineral acid to the corresponding anthocyanidin (Robertson, 1959; Creasey and Swain, 1965). The reaction is believed to proceed by an initial dehydration to the flav-3-en-3-ol followed by aerial oxidation to yield the anthocyanidin (*7.9*). The environment of the 4-hydroxyl group in the naturally occurring phenolic flavan-3,4-diols is such that in acidic media it is readily solvolysed to give the corresponding resonance-stabilized benzyl carbocation (*7.10*), and this property is the essence of much of the chemistry of these compounds. These solvolytic reactions were discussed at length in an earlier review (Haslam, 1975), but some of the salient features are summarized here as they are relevant to the more recent chemistry of the condensed proanthocyanidins.

The ease of ionization at the 4 position is influenced by the stereochemistry at both the 4 and 3 positions. Clark-Lewis and Mortimer (1960) first noted that isomelacacidin (*7.8*) was more readily solvolysed than its epimer at the 4 position, melacacidin (*7.12*). This was rationalized on stereochemical grounds. Thus (*7.8*) has a quasi-axial hydroxyl group at C-4 and a small deformation of the geometry of the heterocyclic ring brings the C-4 hydroxyl group into a truly axial position and this facilitates overlap of the aromatic π orbitals to assist in the development of the C-4 carbocation. In addition, the neighbouring *trans*-oriented C-3 hydroxyl group is also favourably disposed to assist in this solvolysis at the benzylic positions and probably leads to a protonated epoxide intermediate (*7.11*).

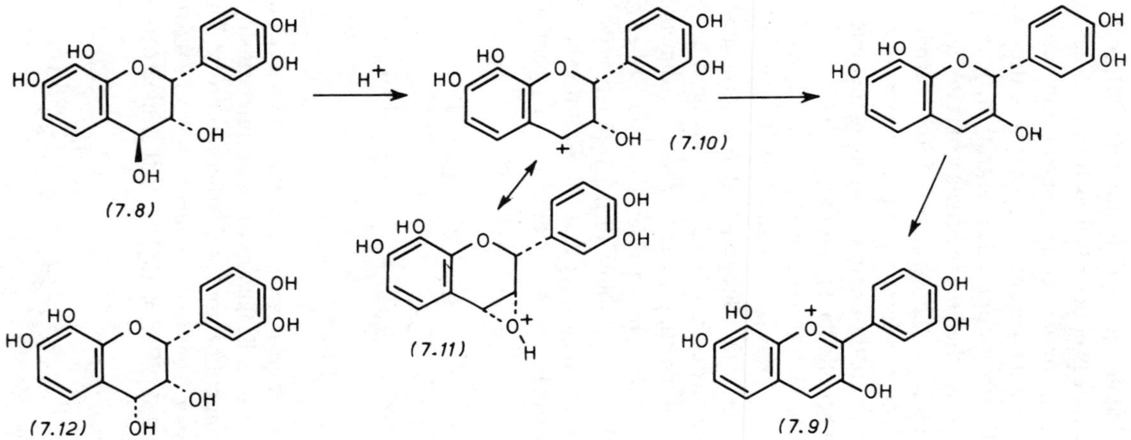

H^+

(7.8)

(7.10)

(7.11)

(7.12)

(7.9)

The carbocation, once formed, may be captured by other nucleophiles such as water, alcohols, nucleophilic phenols, such as resorcinol and phloroglucinol which result in the formation of new C–C bonds, and a variety of sulphur nucleophiles. The products which result from these reactions are often determined by the acidity of the reaction conditions. Thus if the acidity of the reaction is such that the product is not transformed back once again to the carbocation, then that product which is formed fastest will predominate. The reaction is said to be under *kinetic* control and factors which influence the approach of the external nucleophile to the carbocation will dominate the energy profile of the reaction (steric approach control). If, however, the reaction is under *thermodynamic* control, then the products which are most stable will be formed. Under these conditions the capture of the carbocation by the external nucleophile is reversible under the acidity of the reaction. These are the principal features which dominate the factors controlling the range of products formed in the solvolysis of phenolic flavan-3,4-diols.

Brown and Shaw (1974) have recently extended their observations on the solvolysis of a range of flavan-4-ols and 4-arylflavans in the presence of external sulphur nucleophiles. They showed that the carbocation formed from these substrates in acidic media may be captured by both benzene thiol and benzene sulphinic acid and interestingly showed that 5,7-dimethoxyflavan-4β-ol (*7.13*) reacts with benzene thiol – in the absence of acid – to give the thioether (*7.14*) in 88 % yield. A study of the stability of the 4-thioether derivatives of flavans lent strong support to the suggestion that the 4-α-thioether derivatives are formed under kinetically rather than thermodynamically controlled S_N1 reactions.

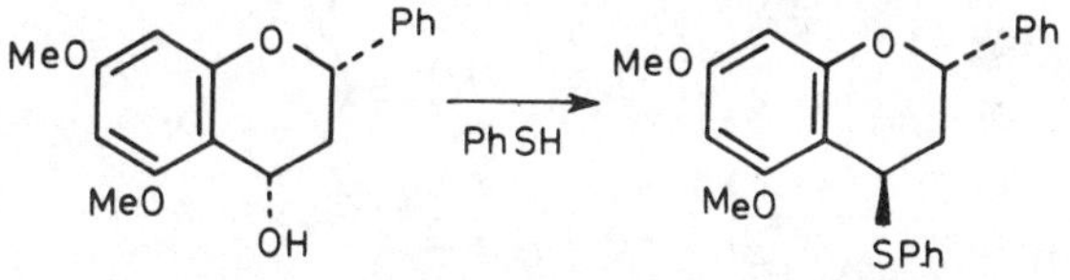

MeO, MeO, O, Ph, OH — PhSH → MeO, MeO, O, Ph, SPh

(7.13) (7.14)

Brown and Shaw further suggested that this different lability of groups attached to the 4 position of phenolic flavans towards different sulphur nucleophiles, with and without acid, might be used as a structural probe. Thus a reaction in which the '*condensed tannin*' from *Calluna vulgaris* was refluxed in aqueous ethanol with benzene thiol yielded, after methylation, the thioether (*7.15*). This result they

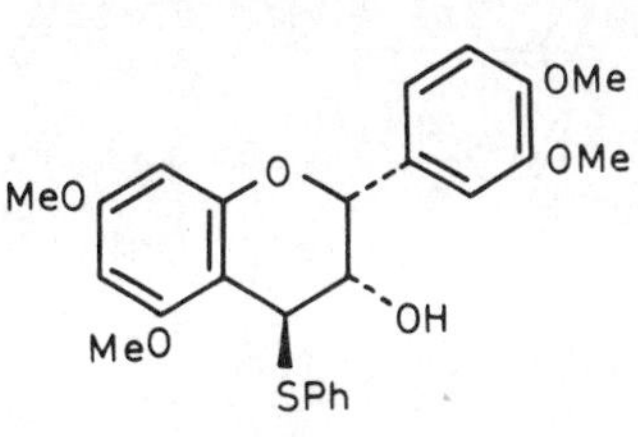

MeO, MeO, O, OMe, OMe, OH, SPh

(7.15)

interpreted as showing the presence in the tannin of interflavan bonds comprising C–O linkages at the 4 position.

7.3 CONDENSED PROANTHOCYANIDINS

7.3.1 Procyanidins

Oligomeric* condensed procyanidins are most commonly responsible for the distinctive reactions associated with many plants which have been attributed by biochemists and botanists alike to '*condensed tannins*'. Studies of these compounds – the procyanidins – has shed considerable light on their chemistry and biochemistry. As such it forms a significant contribution to an understanding of the class of '*condensed tannins*' as a whole and hence to a solution to one of the classical unsolved problems of natural product chemistry.

Fruit-bearing plants have proved to be particularly rich sources of oligomeric condensed procyanidins, and studies have concentrated upon the four major naturally occuring dimeric* procyanidins (B-1, B-2, B-3 and B-4) which have been isolated from the fruit, fruit pods, seeds and seed shells of a diverse range of plants. The procyanidins occur free and unglycosylated and almost invariably are found with one or both of the flavan-3-ols, (+)-catechin or (−)-epicatechin (Fig. *7.1*). These patterns of occurrence of the procyanidins are most effectively revealed by two-dimensional paper chromatography as a distinctive 'fingerprint' which may be used as a taxonomic guide (Fig. *7.2*).

Although many plants contain mixtures of the procyanidins B-1, B-2, B-3 and B-4 and (+)-catechin and (−)-epicatechin, several simpler and highly characteristic 'fingerprints' have been discerned. Probably the most commonly encountered is that found, for example, in hawthorn (*Crataegus monogyna*). Co-occurring with (−)-epicatechin is one major procyanidin dimer [B-2, formally two (−)-epicatechin units C-4 to C-8-linked], a minor procyanidin dimer [B-5, two (−)-epicatechin units C-4 to C-6-linked], a major trimer C-1 and various higher oligomers (Fig. 7.2). Much less widely found is a second 'fingerprint' (present for example in the catkins of the male Sallow, *Salix caprea*) where (+)-catechin, a major procyanidin dimer [B-3, two (+)-catechin units C-4 to C-8-linked], a minor procyanidin dimer [B-6, two (+)-catechin units C-4 to C-6-linked], a trimer C-2 and various higher oligomers occur together (Fig. 7.2). Two further categories, which are the most significant from a biosynthetic point of view, are again not widely found. Thus in various *Rubus* sp. (−)-epicatechin, procyanidin B-4 and associated oligomers are found, and in the cereal sorghum (*Sorghum vulgare*) the seed coat contains (+)-catechin, procyanidin B-1 and various higher oligomers. The two flavan-3-ol units which formally compose the structures of procyanidins B-1 and B-4 are of opposite absolute stereochemistry

* This is the most convenient terminology to employ in this field, since the procyanidins are formally built up of numbers of flavan-3-ol units.

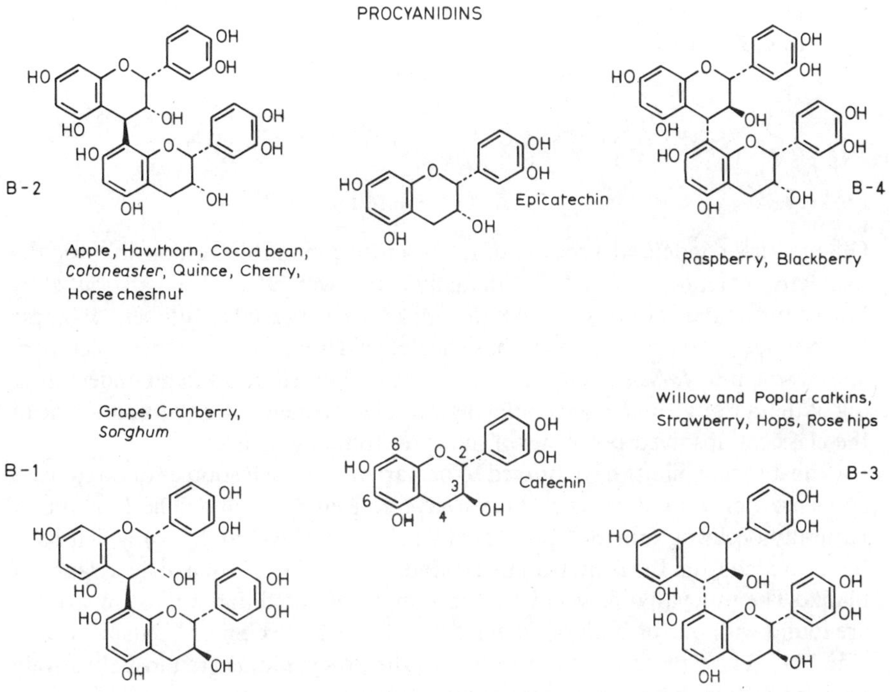

Fig. 7.1 The principal naturally occurring procyanidin dimers: plant sources (Weinges *et al.*, 1968a,b; Thompson *et al.*, 1972).

at C-3, but in the plants noted they co-occur with only *one* of the two possible flavan-3-ols (Haslam, 1977).

The procyanidins may be isolated as their deca-acetates (Weinges *et al.*, 1968a) or in their free phenolic forms by chromatography on Sephadex LH-20 (Thompson *et al.*, 1972). The yields in some cases may be quite impressive. Thus 1 kg of fresh horse chestnut shells (*Aesculus hippocastanum* or *A. carnea*) yields upwards of 1 g of procyanidin B-2. Weinges and his colleagues (Weinges *et al.*, 1968a) successfully assigned the structures and some of the stereochemical details of the four procyanidins (B-1, B-2, B-3 and B-4) on the basis principally of an examination of the ^{1}H-NMR spectra of the deca-acetates. Analysis of the ^{1}H-NMR evidence coupled with the presumption that the two 'halves' of the procyanidin dimers were related in absolute stereochemistry at C-2, 2′ and C-3,3′ to the natural flavan-3-ols [(+)-catechin and (−)-epicatechin] permitted Weinges to define the absolute stereochemistry at C-4 as 4 *S* in procyanidins B-3 and B-4. The 'lower' flavan-3-ol unit in these two dimeric procyanidins thus occupies a quasi-equatorial position at C-4 on the heterocyclic ring of the upper (+)-catechin unit.

Later work illustrated some of the important facets of the chemistry of the procyanidins. Most distinctive are the reactions based upon the acid-catalysed

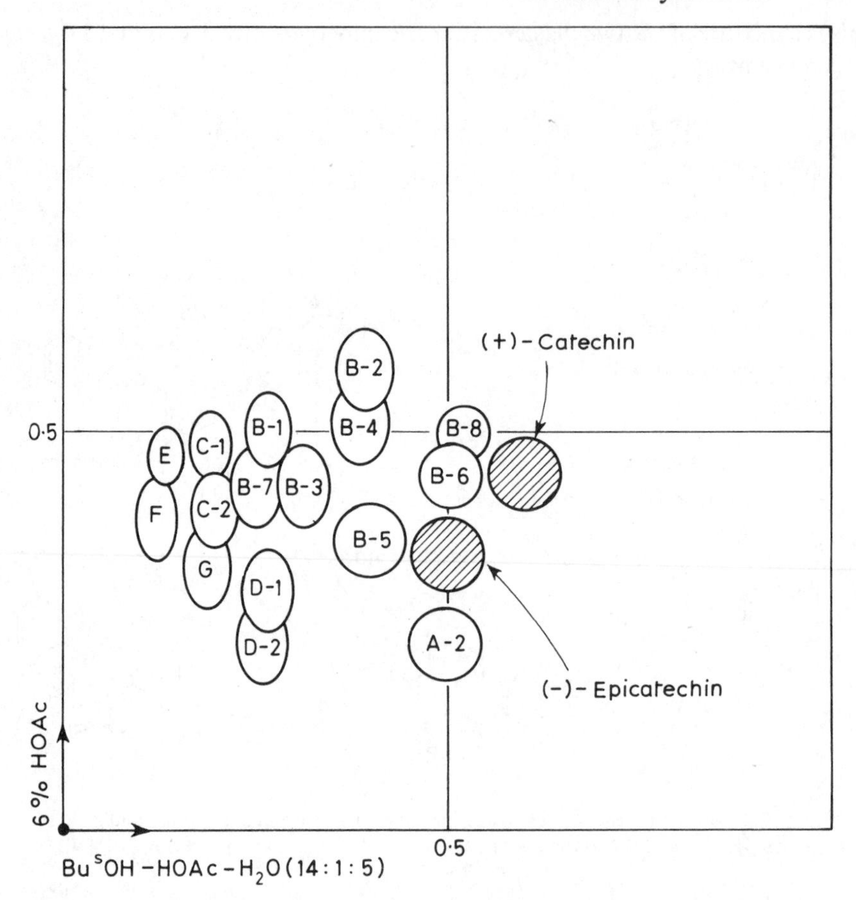

Fig. 7.2 Paper chromatography of the plant procyanidins (Thompson *et al.*, 1972).

solvolysis of the C–C interflavan bond (Fig. 7.3). Thus the procyanidin dimers are degraded by acid to give initially the familiar benzylic carbocation from the 'upper half' of the molecule and the appropriate flavan-3-ol, (+)-catechin or (−) epicatechin from the 'lower half' of the dimer. The carbocation, as is the case with the leucoanthocyanidins, is normally rapidly converted by proton loss and oxidation to give cyanidin, but, under appropriate conditions, it may be intercepted by various nucleophiles. Thus in the presence of toluene-α-thiol the flavanyl-4-thioether (*7.16*) is formed and with phloroglucinol the product is the 4-arylflavan (*7.17*). The thioether-forming reaction occurs in the presence of acetic acid and is a kinetically controlled reaction. It has proved of particular value for structural identification purposes. Capture of the flavan-4-yl carbocation (*7.18*) with 2,3-*cis* [(−)-epicatechin] stereochemistry, (from procyanidins B-1 and B-2) gives almost exclusively one thioether (*7.16*)with 3,4-*trans* stereochemistry. On the other hand, the alternative flavanyl carbocation with 2,3-*trans*[(+)-catechin] stereochemistry derived from procyanidins B-3 and B-4

gives a mixture of the two diastereoisomeric thioethers with 3,4-*cis* and 3,4-*trans* stereochemistry.

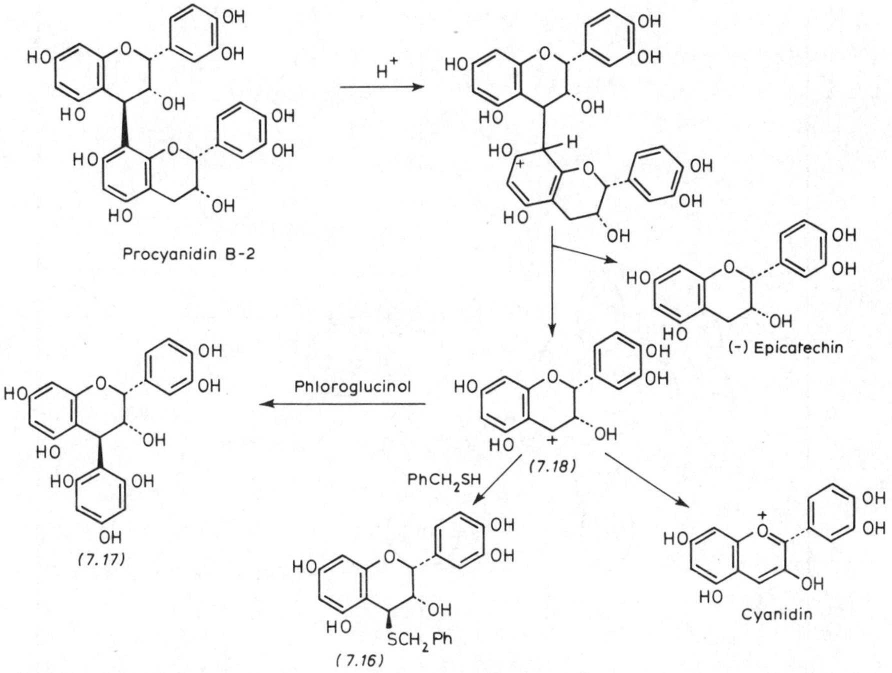

Fig. 7.3 Solvolytic reaction of plant procyanidins: reactions of procyanidin B-2 (Thompson *et al.*, 1972; Haslam *et al.*, 1977a).

The absolute stereochemistry in procyanidins B-1 and B-2 was determined as 4*R* in later work by the Sheffield group (Haslam *et al.*, 1977a) using ^{13}C-NMR. The C–C interflavan bond in procyanidins B-1 and B-2 thus occupies a quasi-axial position at C-4 on the 'upper' flavan-3-ol heterocyclic ring. These observations were based on the 'γ effect' in the ^{13}C-NMR spectra of the carbon atoms of the heterocyclic ring of the flavan-3-ol system. When an aryl group is substituted at C-4 in flavan its effect on the ^{13}C chemical shift at C-2 is dependent on its orientation (Fig. 7.4). Thus in 2,4-*cis*-4-phenylflavan (*7.19*) (4-phenyl group quasi-equatorial) the effect of the substitution of the phenyl group at C-4 on the resonance at C-2 is minimal (+0.4 p.p.m); however, in 2,4-*trans*-4-phenylflavan (*7.20*), (4-phenyl group quasi-axial) a characteristic upfield shift (γ effect, −4.5 p.p.m.) was observed on the ^{13}C chemical shift of C-2. Similar γ-effects were observed in the resonances at C-2 in the 'upper' flavan-3-ol unit of procyanidin B-1 and of procyanidin B-2 (*7.22*) when compared to the C-2 resonance of the parent flavan-3-ol (−)-epicatechin (*7.21*). These observations demonstrated that the configuration at C-4 in the 'upper' flavan-3-ol unit of these

procyanidin dimers is 4*R* and this places the 'lower' flavan-3-ol unit in a quasi-axial position at C-4. In contrast no γ-effects were seen in the ^{13}C-NMR spectra of the procyanidin dimers B-3 (*7.24*) and B-4 when compared to the parent flavan-3-ol (+)-catechin (*7.23*). It is thus noteworthy that the major natural procyanidin dimers (B-1, B-2, B-3 and B-4) all possess a *trans* orientation of the hydroxyl group at C-3 and the flavan substituent at C-4 in the 'upper' flavan-3-ol unit.

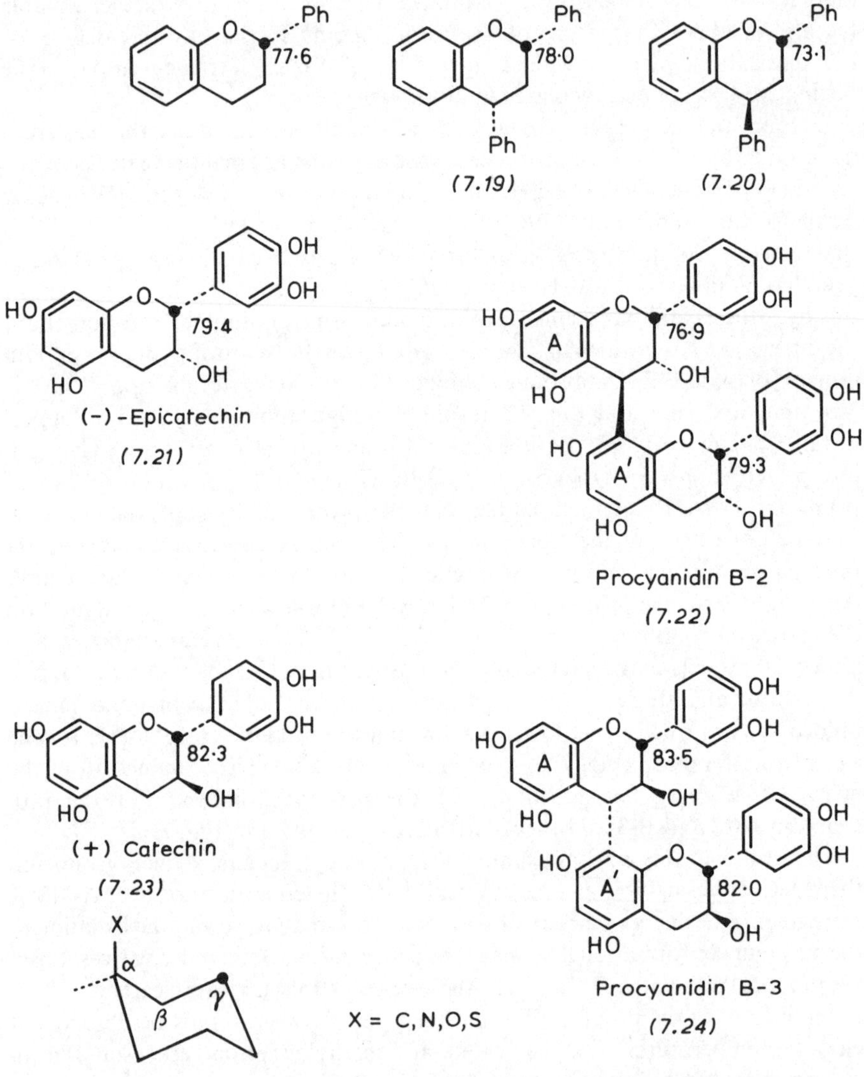

(7.19)

(7.20)

(−)-Epicatechin (7.21)

Procyanidin B-2 (7.22)

(+) Catechin (7.23)

X = C, N, O, S

Procyanidin B-3 (7.24)

Fig. 7.4. ^{13}C-NMR analysis of procyanidin stereochemistry: the γ-effect in flavan derivatives and some procyanidins (Haslam *et al.*, 1977a). $^{13}C-\delta$ values (p.p.m. from TMS).

These stereochemical features of the natural procyanidins were confirmed by circular dichroism (CD) measurements (Scopes *et al.*, 1979). The principal feature of the CD spectra is a very large positive or negative couplet at short wavelength (200–220 nm), and this possibly originates from the two aryl chromophores linked by a benzylic carbon atom (i.e. rings A and A′ in procyanidin B-2). Procyanidins B-1 and B-2 with 4*R* stereochemistry and related models all show a strong positive couplet whilst procyanidins B-3 and B-4 and models with 4*S* stereochemistry exhibit a negative couplet. A further notable feature of the CD spectra of the procyanidins is that as the degree of polymerization increases (with the formation of trimers, tetramers etc.) then the distinctive CD couplet increases in amplitude.

Despite the successful application of NMR methods to the structure determination of the various dimeric procyanidins, anomalous features of the spectra (Thompson *et al.*, 1972) led to speculation by several workers (Weinges *et al.*, 1970; Jurd and Lundin, 1968; du Preez *et al.*, 1971; Thompson *et al.*, 1972) that restricted rotation about the interflavan bond existed in this group of natural products. A detailed study (Haslam *et al.*, 1977a) has revealed the correctness of this hypothesis and has defined two different forms of hindered rotation about the interflavan bond which may be associated with the two groups of procyanidin dimers [B-1 and B-2, 4*R* absolute configuration, (−)-epicatechin 'upper' flavan-3-ol unit] and [B-3 and B-4, 4*S* absolute configuration, (+)-catechin 'upper' flavan-3-ol unit] (Fig. 7.5). For the dimers B-1 and B-2 the observations have been interpreted (Fig. 7.5 – procyanidin B-2) in terms of restricted rotation caused primarily by steric interactions between the proton at C-2 and the aromatic π system of ring A of the 'upper' flavan unit, and the substituents *ortho* to the inter-flavan linkage on the phloroglucinol ring of the 'lower' flavan unit. Alternatively for the procyanidin dimers B-3 and B-4 with the 4*S* configuration the restricted rotation around the interflavan bond has several analogies, and molecular models show that the oxygen substituents at C-3 and C-5 in the 'upper' flavan unit and those in the *ortho* position to the interflavan bond on the phloroglucinol ring in the 'lower' flavan unit are primarily responsible for the steric interference (Fig. 7.5 – procyanidin B-3). Typical values for the $\Delta G_{rot}^{\neq}$ (free energy of activation for the rotational process) for the procyanidins B-2 and B-3 and their derivatives are shown in Fig. 7.5.

The results of this work thus imply that, although dimers of the B group and their derivatives exhibit atropisomerism under the conditions of the ^{1}H-NMR experiment, the energy barriers observed are too small to permit the isolation of the different conformational forms of the procyanidins or their derivatives. From the practical point of view, however, these observations indicate clearly that first-order spectra of the procyanidins and their derivatives can only be observed at elevated temperatures – for the acetate and methyl ether derivatives at 180° in pentadeuterionitrobenzene and for the free phenols at 60–90° in hexadeuteriodimethylsulphoxide.

Although the ^{1}H NMR experiments show that for each procyanidin a number

of conformations about the interflavan bond are accessible, examination of molecular models suggests that one preferred conformation probably exists for each of the free phenolic forms of the procyanidins. These are shown in Fig. 7.5 for procyanidin B-2 and procyanidin B-3. An interesting feature of these two conformations is that inspected from different view points (as indicated by the arrows) they bear an almost object-to-mirror-image relationship; the structures are quasi-enantiomeric.

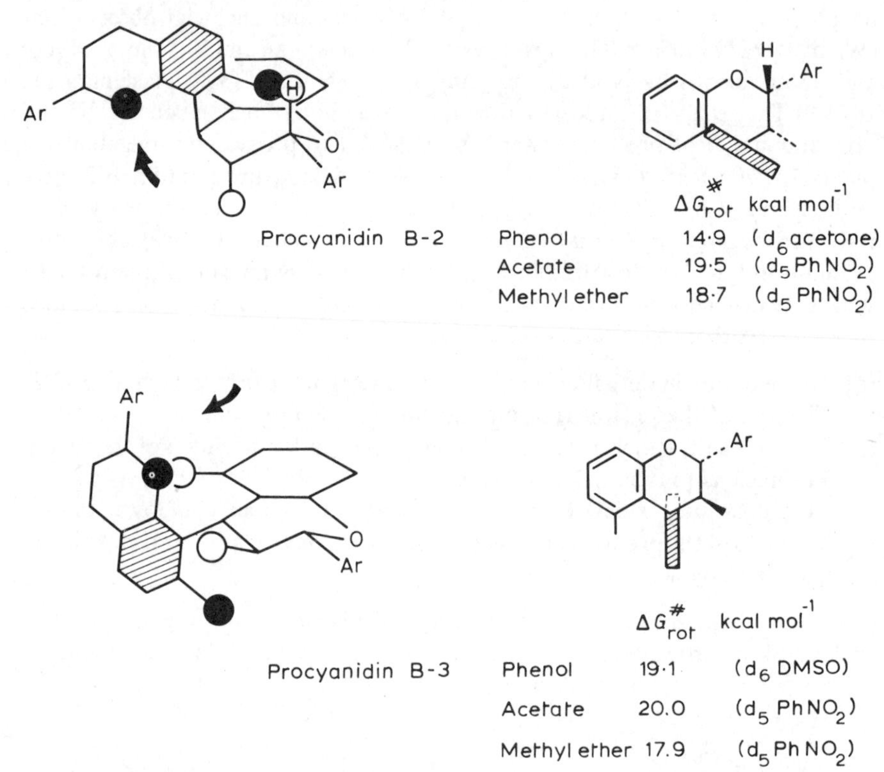

Fig. 7.5 Restricted rotation in procyanidin dimers: preferred conformations of procyanidins (Haslam *et al.*, 1977a). Rotational barriers in procyanidin dimers.

Elaboration of the oligomeric forms of procyanidins by the addition of further flavan-3-ol units [(−)-epicatechin to B-1 and B-2 and (+)-catechin to B-3 and B-4], bearing in mind the type of conformational restraint about the interflavan bond which has been demonstrated, leads to two helical structures. The central core of these linear polymers is composed of rings A and C of the flavan repeat unit and ring B – the *ortho* dihydroxyphenyl ring – projects laterally from this core. Significantly perhaps those formed from units related to (−)-epicatechin (e.g. the extension of the procyanidin B-1 or B-2) are *left-hand* helices, whilst those of the type related to procyanidin B-3 or B-4 formed by the addition of

(+)-catechin units are *right-hand* helices. Although the biological significance of these observations is not clear, they are a further interesting example of the economy of nature. Thus different plants have evolved the means of synthesis of isomeric oligomers and polymers with opposite helicities by the simple expedient of a change in stereochemistry of the hydroxyl group at C-3 in the carbocation precursor (*vide infra*).

Ideas on the biosynthesis of the plant proanthocyanidins and associated flavan-3-ols are based upon a range of structural and chemical observations (Weinges *et al.*, 1968a,b; Thompson *et al.*, 1972; Geissman and Yoshimura, 1966) and upon various biosynthetic experiments (Opie *et al.*, 1977; Haslam *et al.*, 1977b). The results of these experiments, in which variously labelled (^{14}C and ^{3}H) cinnamic acid precursors were administered to procyanidin-metabolizing plants [*A. hippocastanum*, *A. carnea* – (−)-epicatechin, procyanidin B-2, proanthocyanidin A-2; *Rubus fruticosus*, *R. idaeus* – (−)-epicatechin, procyanidin B-4; *Salix caprea*, *S. inorata*, *Chamaecyparis lawsoniana* – (+)-catechin, procyanidin B-3] are summarized in Fig. 7.6. The most significant of the results – which are in general accord with ideas on the biosynthesis of flavonoids outlined by earlier workers (Grisebach, 1968) – are:

(a) The retention in both flavan units in the dimers of the proton H_a at C-2, C-2′.
(b) The loss of the proton H_c from the cinnamic acid precursor.
(c) The NIH shift which accompanies the hydroxylation of ring B of the flavan-3-ol nucleus (40–50% loss of H_b).
(d) The differential extent of labelling obtained in the 'upper' and 'lower' flavan-3-ol units of the procyanidin dimers (dependent on the individual experiment but generally a ratio of 1:2 to 5).

The most straightforward interpretation of this last result (d) is that the two 'halves' of the procyanidin dimers are derived from different metabolic sources.

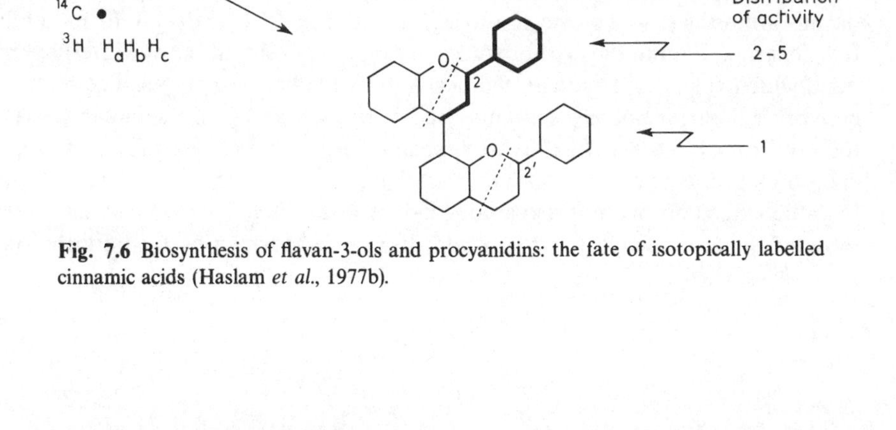

Fig. 7.6 Biosynthesis of flavan-3-ols and procyanidins: the fate of isotopically labelled cinnamic acids (Haslam *et al.*, 1977b).

The other results have been incorporated along with (d) in a general scheme of procyanidin and flavan-3-ol biosynthesis which is illustrated in Fig 7.7 (Haslam *et al.*, 1977b). The biosynthesis of the $C_6.C_3.C_6$ carbon skeleton of the flavonoids is thought to involve the chalcone as the first formed intermediate (Fig. 7.7). However, the sequence of chemical changes (e.g. oxidation, reduction etc.) in the C-3 fragment which lead to the production of the individual flavonoids nevertheless remains poorly defined. Assuming the chalcone–dihydroflavone pair to be key intermediates on the biosynthetic pathway, several mechanistically plausible schemes can lead to the flavan-3-ols. Fig. 7.7, route A shows one such pathway and route B is a possible alternative scheme of biogenesis in which the $C_6.C_3$ precursor is the α-keto acid, phenylpyruvic acid. A key intermediate in both schemes is the flav-3-en-3-ol and it is postulated that a two-stage stereospecific reduction of this compound (proton transfer, followed by hydride addition) might lead to the two stereoisomeric flavan-3-ols (+)-catechin and (−)-epicatechin. It has been argued that this reaction also provides a basis for the biogenesis of the procyanidins. If the supply of biological reductant (say NADPH) is rate-limiting, the intermediate C-4 carbocations (3*S* or 3*R* absolute stereochemistry) might then escape from the active site of the enzyme and react with the reduced product [the flavan-3-ol, (+)-catechin or (−)-epicatechin] to produce dimers, trimers, tetramers and higher oligomers (Haslam *et al.*, 1977b; Haslam, 1977). The hypothesis has been elaborated in greater detail (Haslam, 1977) to account for the four major procyanidin 'fingerprints' (*vide supra*) and in particular to account for the derivation of procyanidin dimers, such as B-1 and B-4, in which the component 'halves' have opposite absolute stereochemistry at C-3.

The various distinctive patterns of procyanidins found in plants are therefore thought to be derived directly by reaction of one or both of the parent flavan-3-ols [(+)-catechin, (−)-epicatechin] utilizing their nucleophilic character at C-8 or C-6 with one or both of the carbocations (related in stereochemistry at C-2 and C-3). Support for this theory has been obtained from experiments *in vitro* in which the postulated carbocation intermediate(s) are released in solution in the presence of the appropriate flavan-3-ol metabolite (Haslam, 1974b). The simplest procedure is to use a procyanidin dimer as a source of the appropriate carbocation. Thus procyanidin B-2 in acidic media releases (Fig. 7.3) the carbocation related in absolute stereochemistry at C-3 to (−)-epicatechin, and similarly procyanidin B-3 gives the analogous carbocation to the 2,3-*trans* stereochemistry of (+)-catechin. Correct adjustment of the acidity enables a biomimetic synthesis to be set up. Each of the four possible reactions can be reproduced in the laboratory and the products, both qualitatively and quantitatively, match those found in particular plants (Fig. 7.1). The biomimetic syntheses closely resemble some of the solvolytic reactions discussed earlier for phenolic flavan-3,4-diols (Section 7.2.2). The experiments *in vitro* are clearly under thermodynamic control and a point therefore of some interest is that in each "biomimetic situation' the major dimer which is produced is that in which the carbocation at C-4 captures the flavan-3-ol with the formation of a new C–C bond *trans* to the C-3 hydroxyl group of the carbocation.

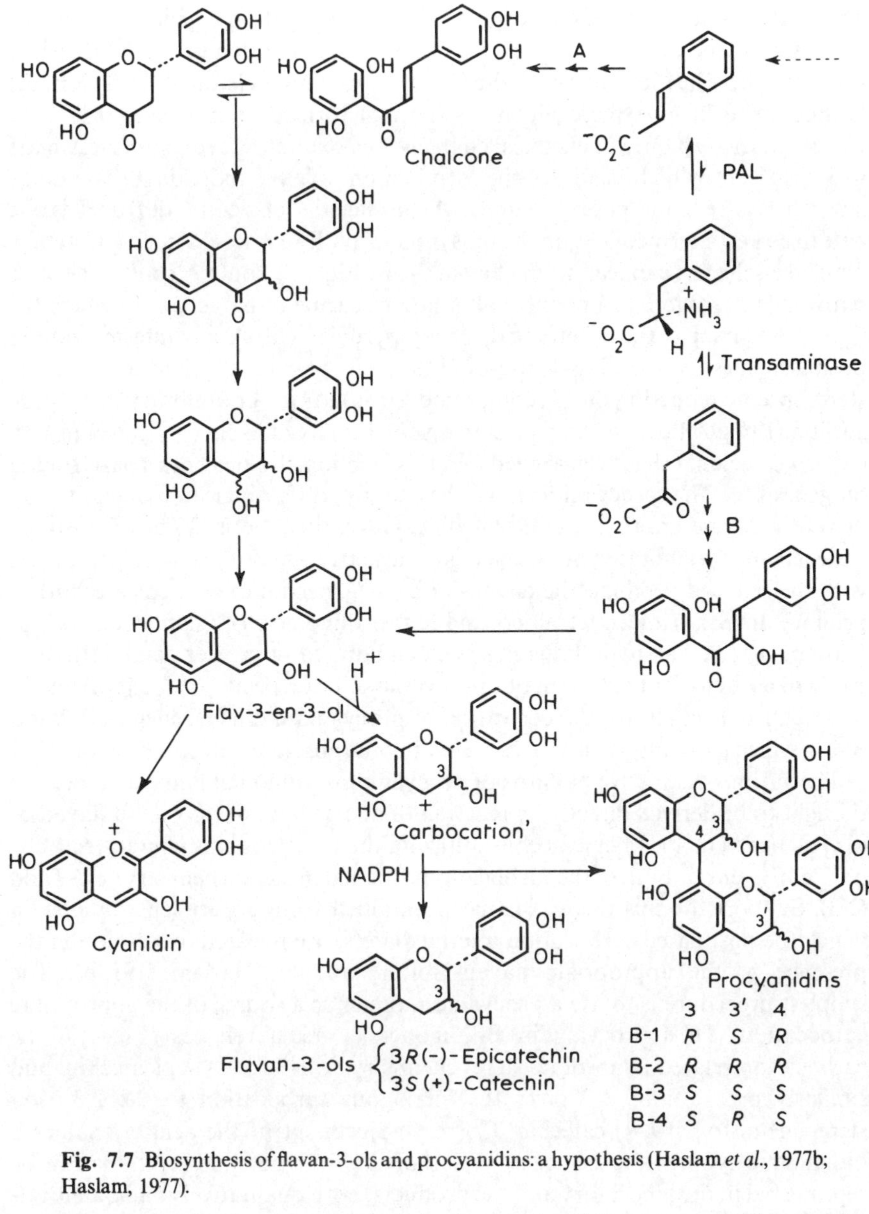

Fig. 7.7 Biosynthesis of flavan-3-ols and procyanidins: a hypothesis (Haslam *et al.*, 1977b; Haslam, 1977).

The status of the flavan-3,4-diols in procyanidin biosynthesis remains something of an enigma, since to date *no* flavan-3,4-diol with the, 5,7,3′,4′-tetrahydroxy pattern of phenolic hydroxyl groups has been found in the vegetative tissues of a procyanidin-metabolizing plant, and attempts to obtain experimental evidence to support their suggested role have so far proved negative (Haslam *et al.*, 1977b).

7.3.2 Prodelphinidins

In contrast to the ubiquitous procyanidins, little is yet known in detail of the chemistry of the prodelphinidins, although Bate-Smith in his earlier surveys (Bate-Smith, 1962) showed that a wide range of 'woody' plants yield delphinidin (usually with cyanidin) when fruit or leaf extracts are heated with butanol and hydrochloric acid. The limited amount of work reported on these substances does, however, suggest a very similar chemistry to the procyanidins and an occurrence which follows the pattern of occurrence of (+)-gallocatechin and (−)-epigallocatechin in plants. Thus Porter and Foo (1978) have isolated a prodelphinidin (*7.25*), (−)-epigallocatechin and (+)-gallocatechin as their polyacetates from the leaves of *Ribes sanguineum* and a further prodelphinidin (*7.26*), with a mixed B-ring oxidation pattern from *Salix caprea*. Likewise E. Haslam and R. K. Gupta (unpublished observations) have obtained (+)-catechin, (+)-gallocatechin and the prodelphinidin (*7.27*) from *Pinus sylvestris*. The structures of these prodelphinidins were established by methods analogous to those outlined above for the procyanidins (NMR, degradation in acid in the presence of a thiol or phloroglucinol). These limited observations support the suggestion of a relationship between prodelphinidin metabolism and that of the gallocatechins. However, the occurrence of the dimers (*7.26*) and (*7.27*) is noteworthy and has important biosynthetic implications.

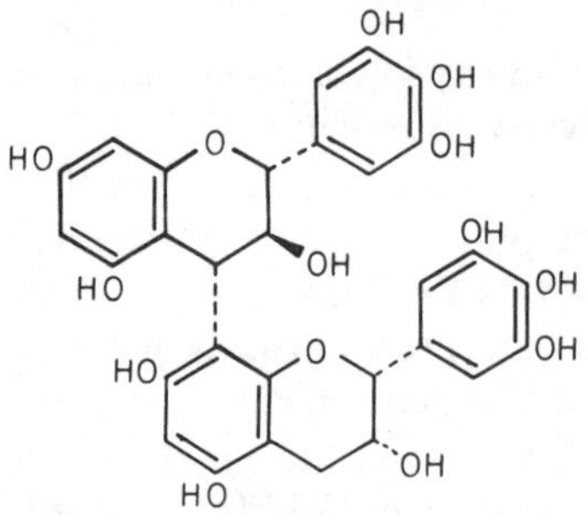

(*7.25*)

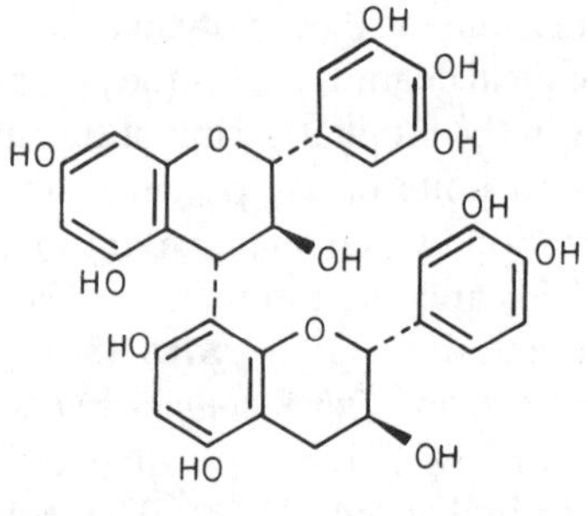

(*7.26*)

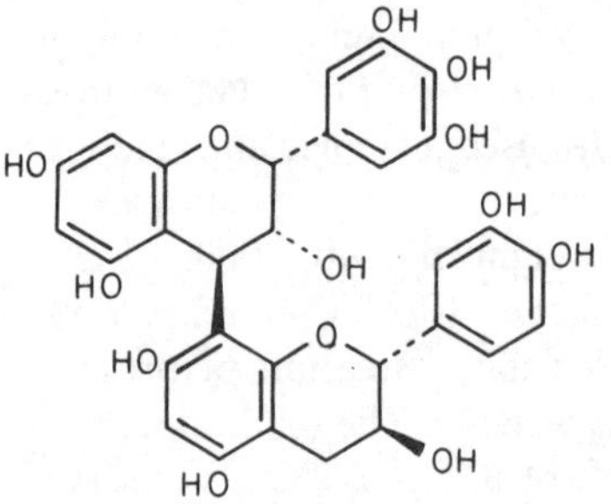

(*7.27*)

7.3.3 Polymeric procyanidins and prodelphinidins

In any plant tissue where proanthocyanidin synthesis occurs there is invariably found a range of proanthocyanidins – from the monomeric flavan-3-ols (catechins and gallocatechins) to the polymers (*7.1*). For each tissue the balance between these molecular forms is probably determined by the corresponding balance between the metabolic flux to the flav-3-en-3-ol (Fig. 7.7) and the rate of supply of the biological reductant NADPH. Tissues in which the flux is low and the NADPH supply is high will contain a range of proanthocyanidins of varying molecular size but those tissues in which the flux is high and the supply of NADPH is low will conversely contain predominantly the higher oligomeric forms – for example the seed coat of the cereal sorghum (Gupta and Haslam, 1978). This facet of proanthocyanidin metabolism was noted very early by Sir Robert and Lady Robinson (Robinson and Robinson, 1935) who recognized the presence in plants of leucoanthocyanins (proanthocyanidins!) of quite different solubilities. With increasing degrees of polymerization the proanthocyanidins become more difficult to solubilize in aqueous and alcoholic media, and it has been suggested (Porter *et al.*, 1979) that those which can be coaxed into solution may have molecular weights up to 7000 – corresponding to the accumulation of up to 20 flavan-3-ol units (1, $n = 18$) in the polymer chain. Currently available techniques enable such polymers to be isolated and purified in an undegraded form (Jones *et al.*, 1976), and very recent work has concentrated upon structural examinations of these polymers.

Chemical degradation of the polymers occurs by solvolysis in acidic media and results in the familiar fission of the interflavan C–C bonds (Fig. 7.3). The chain-extension units of the polymer are released as the corresponding flavan-4-yl carbocation which may be trapped with phloroglucinol or toluene α-thiol (Fig. 7.3), and the terminal unit is released as the corresponding flavan-3-ol. Porter and Foo (1978) thus showed that the predominantly prodelphinidin polymers from *Ribes sanguineum* consisted largely of chain-extension units corresponding to the (+)-gallocatechin structure linked to a (−)-epigallocatechin terminal group (*7.28*). The polymer was quantitatively degraded in the presence of toluene α-thiol (Porter *et al.*, 1979) and the products (*7.29* and *7.30*) were separated and analysed by GLC as their trimethylsilyl ethers. In this way the ratio of chain extension to terminal units in the polymer structure was estimated as 21:1, implying a molecular weight of ~ 6600 and an average structure (*7.28*). Gupta and Haslam (1978) used a similar approach for the determination of the average structure of the procyanidin polymer from the seed coat of sorghum as (*7.31*) and a molecular weight of 1700–2000.

Porter and his colleagues (Porter *et al.*, 1979) have described a novel spectroscopic method for the estimation of proanthocyanidin polymer compositions and molecular weights. The method uses ^{13}C-NMR as the analytical probe and makes use of the different signals observed for carbon atoms in the heterocyclic rings of procyanidin dimers (Haslam *et al.*, 1977a). Thus the ^{13}C signal for C-3 of the terminal flavan-3-ol unit of a proanthocyanidin oligomer

usually occurs at 65–66 p.p.m., quite separate from the C-3 signal of flavan-3-ol chain-extension units which occur at 72–73 p.p.m. (*7.32*). The mean molecular weight of the *Ribes* polymer was calculated from the ratio of the areas of these two signals and gave a chain extension:chain terminal unit ratio of 18:1 (i.e. a mean molecular weight of 5900). Estimation of molecular weight by ^{13}C-NMR spectroscopy is controlled by the observable limit of the C-3 signal from the terminal unit and this was suggested to mean a limit of about 8000 to the molecular weight of the polymer under examination. The ratio of procyanidin to prodelphinidin chain-extension units in the polymer from *R. sanguineum* was similarly determined by an examination of the aryl C atom region of the ^{13}C-NMR spectrum (145–150 p.p.m) and a value of ~ 1:4 was obtained, in good agreement with other measurements of this ratio.

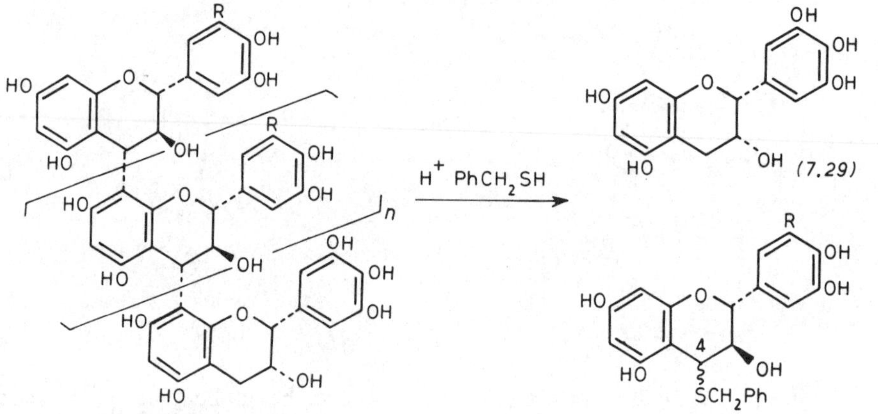

H$^+$ PhCH$_2$SH

(n = 17-20, R = H or OH, ratio ~1 : 4)
(*7.28*)

(*7.29*)

(R=H or OH, 4*R* and 4*S* stereochemistry)
(*7.30*)

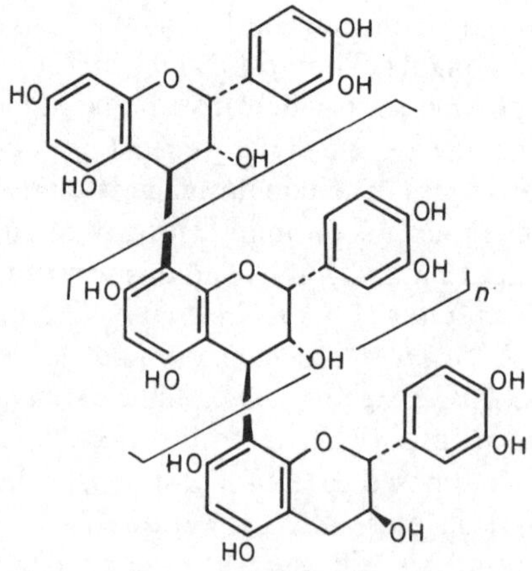

(n = 4-5)
(*7.31*)

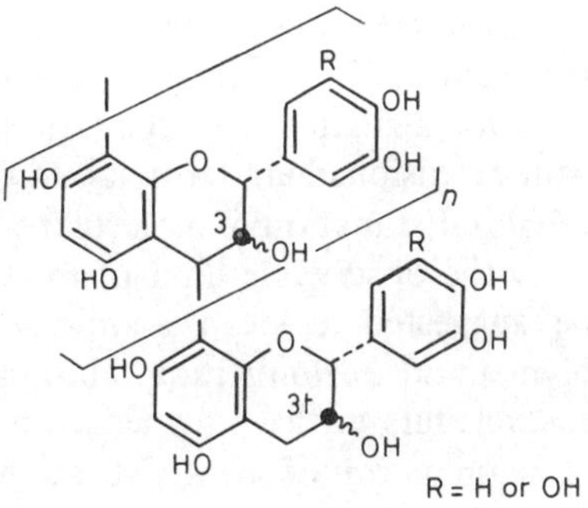

R = H or OH

(^{13}C-NMR δ values: C-3t...65-66 p.p.m., C-3 ... 72-73 p.p.m.)

(7.32)

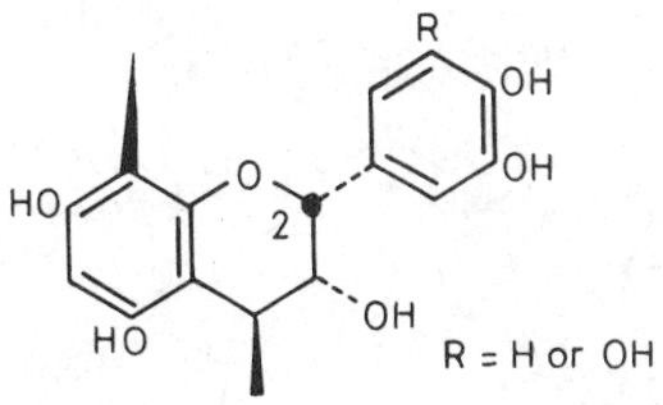

R = H or OH

(^{13}C-NMR δ value: C-2 ... ~77 p.p.m.)

(7.33)

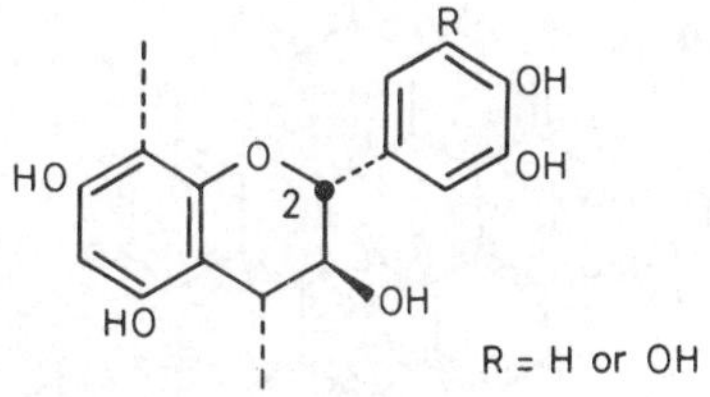

R = H or OH

(^{13}C-NMR δ value: C-2 ... ~83 p.p.m.)

(7.34)

Porter and his collaborators have also outlined the application of ^{13}C and optical rotary dispersion (ORD) measurements for the determination of polymer structures containing chain-extension units of mixed stereochemistry, i.e. those with 2,3-*cis* [(−)-epicatechin or (−)-epigallocatechin] stereochemistry and 2,3-*trans* [(+)-catechin or (+)-gallocatechin] stereochemistry (Porter *et al.*, 1979, 1980). Earlier surveys of plants (Thompson *et al.*, 1972) suggest that these types of proanthocyanidin polymers are widely distributed in the plant kingdom. The ^{13}C-NMR method again makes use of the γ-effect (*vide supra*: Sections 7.2.1 and 7.3.1) in the ^{13}C-NMR spectra of certain flavan derivatives. The interflavan bonds between flavan-3-ol chain-extension units with 2,3-*cis* stereochemistry occupy quasi-axial positions at C-4 in these units and this substitution leads to a γ-effect on the C-2 signal in such units (*7.33*, C-2 signal ~ 77 p.p.m.). This effect is not present in units with 2,3-*trans*-stereochemistry and the corresponding signal is now quite distinct (*7.34*) (C-2 signal ~ 83 p.p.m.). Estimation of the ratio of the areas under each signal thus yields the ratio of chain-extension units in the polymer with 2,3-*cis* as opposed to 2,3-*trans* stereochemistry. Table 7.2 gives a summary and comparison of results for various proanthocyanidin polymers determined by Porter and his colleagues (Porter *et al.*, 1979; Porter and Foo, 1980) by the methods outlined.

Table 7.2 Composition of proanthocyanidin polymers (taken from Porter *et al.*, 1979).

Proanthocyanidin	$[\alpha]_{578}$	*Mole fraction of 2,3-cis-isomer* From rotation	From ^{13}C-NMR	*Ratio, prodelphinidin:procyanidin* Acid degradation	Thiol degradation	^{13}C-NMR	*Mean Mol. wt.* ^{13}C-NMR
Sainfoin*†	+345°	0.91	0.87	81:19	83:17	81:19	‖
*Lotus**†	+255°	0.83	0.75	77:23	80:20	82:18	‖
*Ribes**†	−692°	0	0	87:13	93:7	91:9	5900
Siebel*‡	+436°	0.99	1.0	20:80	22:78	10:90	‖
Beaujolais*‡	+390°	0.95	0.91	20:80	18:82	13:87	4900
*Cyathea**†	+455°	1.00	1.0	41:59	38:62	40:60	‖
*Dicksonia**†	+425°	0.98	1.0	18:82	5:95	13:87	2900
*Pinus radiata**§	+217°	0.80	0.76	50:50	49:51	48:52	1800
Rose hips*‡	+111°	0.70	0.72	10:90	13:87	13:87	6400

* Polymer. † Isolated from leaves. ‡ Isolated from unripe fruit.
§ Isolated from phloem. ‖ C_3(t) Signal not observed.

Electrophilic substitution reactions of the (+)-catechin and (−)-epicatechin nucleus occur much more readily at the C-8 than the C-6 position, and all the proanthocyanidin biosynthetic (and hence structural) arguments are based on this assumption – namely that the proanthocyanidin C–C-bond-forming reaction occurs predominantly *via* C-4 of one flavan unit (the carbocation) and C-8 of the next unit. On this basis the proanthocyanidin polymers probably have essentially linear structures with C-4 to C-8 interflavan bonds. However, simply on a statistical basis the formation of some C-4 to C-6 linkages might be expected to be present in the polymers and hence that they may well possess a degree of branching within them. Roux and his collaborators (Roux and Hundt, 1978), on the basis of model experiments, have suggested degradative bromination and ^{1}H-NMR as methods to determine the position (C-6 or C-8) of the interflavan linkage. The ^{1}H-NMR method utilizes the position of the chemical shift of the singlet proton (C-6 or C-8)- in the (C-8 or C-6)-substituted flavan-3-ol nucleus. A number of C-8- and C-6-substituted derivatives of (−)-epicatechin and (+)-catechin tetramethyl ethers were prepared by unequivocal methods. It was observed that in these derivatives the residual A-ring proton absorbs in one of two narrow ranges dependent on its location at C-6 or C-8 – δ (8H) falls in the range 6.32–6.47 and δ (6H) at 6.10–6.22. There are some problems applying this method to natural procyanidins and prodelphinidins because of the hindered rotation about the C–C interflavan bond at ambient temperatures (*vide supra*, Section 7.3.1, Fig. 7.5). However, the chemical shift of the singlet A-ring proton in procyanidin B-2 octamethyl ether at 170° in d_5 nitrobenzene is at δ 6.08, and *if* the extrapolation can be made, this confirms the assignment of the C-8 to C-4 linkage in this dimer (Haslam, 1977; Haslam *et al.*, 1977a).

7.3.4 Miscellaneous proanthocyanidins

The heartwood and bark of trees are often rich sources of proanthocyanidins, often of an unusual type, and Roux and his collaborators have continued their

notable researches into the profisetinidins, prorobinetinidins and proguibourtinidins which they have found principally in trees of the *Acacia* family (Roux, 1972). They have favoured a scheme of biosynthesis for many of these proanthocyanidins which is based on the derivation of the appropriate phenolic flavan-4-yl carbocation from the associated flavan-3,4-diol which is frequently found alongside the proanthocyanidins in the wood. In recent work they have optimized the conditions for biomimetic synthesis of proanthocyanidins derived from (+)-mollisacacidin (*7.35*) in black wattle (*Acacia mearnsii*). The flavan-3,4-diol was thus condensed at ambient temperatures (20°C) in 0.1 M-HCl with appropriate nucleophilic substrates. Phloroglucinol was thus stereoselectively captured to give (*7.36*) whilst with resorcinol both the 3,4-*cis* and 3,4-*trans* products were obtained (Roux *et al.*, 1978b). Under similar conditions (−)-teracacidin (*7.37*) gave with both phloroglucinol and resorcinol a product with the same (4*R*) absolute stereochemistry at C-4 (*7.38*). Roux and his colleagues attributed these differences to varying degrees of steric-approach control in the C–C-bond-forming reaction and in particular to the influence of the orientation of the C-3 hydroxyl group. The differences in absolute stereochemistry at C-4 in these 4-arylflavans were assigned with the aid of CD spectra and the strong bands in the 220–240 nm region and ^{1}H-NMR.

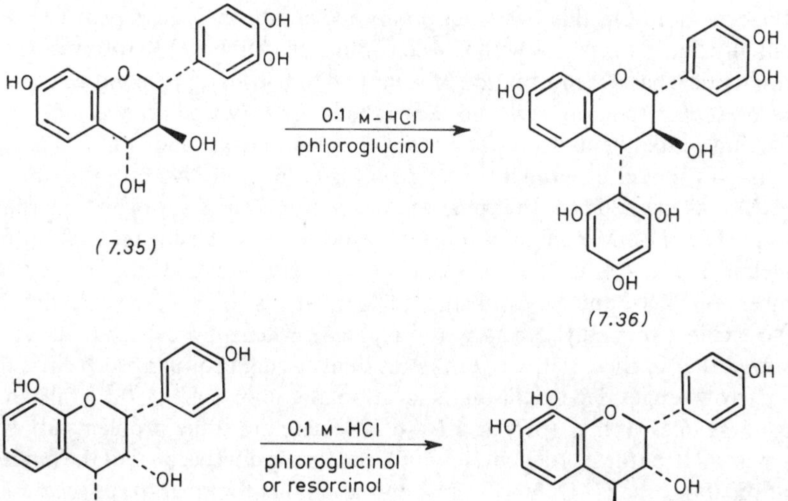

(7.35) —0·1 M-HCl, phloroglucinol→ (7.36)

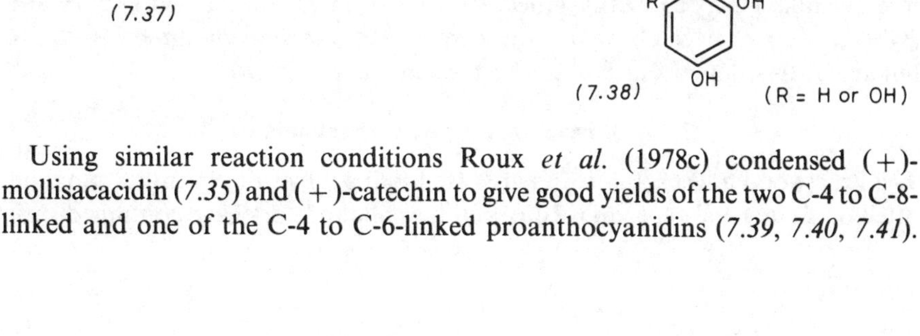

(7.37) —0·1 M-HCl, phloroglucinol or resorcinol→ (7.38) (R = H or OH)

Using similar reaction conditions Roux *et al.* (1978c) condensed (+)-mollisacacidin (*7.35*) and (+)-catechin to give good yields of the two C-4 to C-8-linked and one of the C-4 to C-6-linked proanthocyanidins (*7.39*, *7.40*, *7.41*).

Structures and stereochemistry were assigned using ^{1}H-NMR and CD spectra and both of the C-4 to C-8-linked profisetinidins (*7.39*, *7.40*) were identified with compounds isolated from black wattle or 'Mimosa' extract. These observations support the long-held view (Geissman and Yoshimura, 1966; Creasey and Swain, 1965) that flavan-3,4-diols *via* the reaction of their derived 4-carbocations and nucleophilic phenolic flavan-3-ols represents the mode of biogenesis of the proanthocyanidins found in the woods and bark of many trees (Drewes *et al.*, 1967).

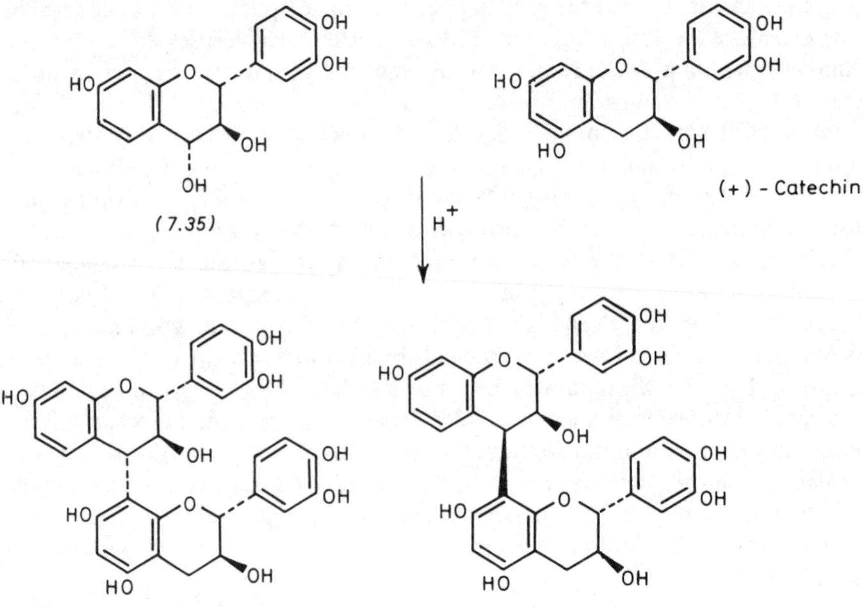

(7.35)

(+) - Catechin

H^+

(7.39)

(7.40)

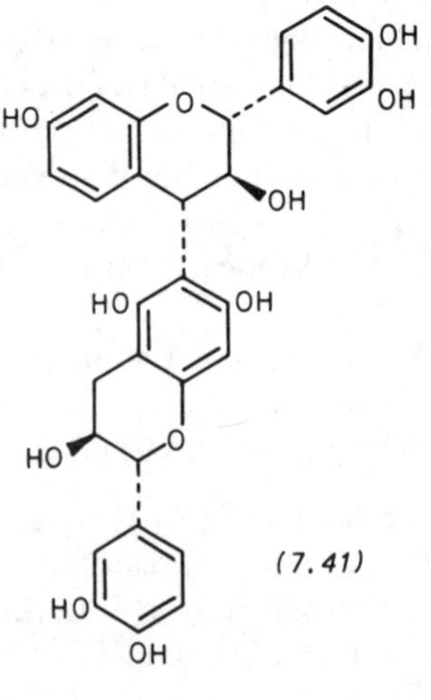

(7.41)

APPENDIX: PROANTHOCYANIDIN NOMENCLATURE

Procyanidins are currently named according to a trivial system introduced by Weinges (Weinges *et al.*, 1968) and later perpetuated and extended by Haslam and his collaborators (Thompson *et al.*, 1972). However, as investigations of the chemistry and biochemistry of the proanthocyanidins have developed over the past decade the isolation of further proanthocyanidin oligomers (some with mixed oxidation patterns in the flavan-3-ol B ring) and of proanthocyanidin polymers has underlined the growing need for an acceptable and usable form of nomenclature in this field. The IUPAC system which uses flavan as the fundamental ring system and the *R* and *S* symbols to signify absolute stereochemistry is fully descriptive but exceedingly cumbersome to use. To the uninitiated it may also be misleading in its description of related centres in a molecule with the same relative configuration at C-4 of the flavan units.

Very recently Porter and his colleagues (Porter *et al.*, 1982) have proposed a form of nomenclature which appears neat and exceedingly attractive. It offers a precise alternative to the fully systematic and more rigorous IUPAC method. Proanthocyanidins are named in an analogous manner to oligo- and polysaccharides, and in the Porter system C-4 of the flavan-3-ol building unit of proanthocyanidins is (in the nomenclature sense) equivalent to the anomeric centre C-1 of the sugar unit in oligosaccharides. The interflavonoid bond is indicated in the same way as the linkage between sugars in oligosaccharides; the bond and its direction are contained in brackets ($4 \rightarrow$).

The system then uses commonly accepted and modified trivial names for the building units of the proanthocyanidins and the $\alpha\beta$ system to denote the configuration of the interflavonoid bond at C-4. These conventions are outlined as follows:

(a) The fundamental flavan structural units of proanthocyanidin oligomers are defined in terms of the familiar monomeric flavan-3-ols. The names catechin, epicatechin, gallocatechin, epigallocatechin, afzelechin, epiafzelechin, fisetinidol, robinetinidol etc., are thus reserved for units with the most commonly encountered 2*R* absolute configuration. Flavan-3-ol units with the 2*S* configuration, whose occurrence in Nature is sporadic, are distinguished by the *enantio* prefix. Thus (+)-epicatechin becomes *ent* epicatechin. The absolute configuration at C-3 for each flavan-3-ol unit is thus automatically defined.

(b) Flavan-3-ol structures are drawn and numbered according to present conventions. The configuration of the interflavonoid bond at C-4 is indicated by the appropriate $\alpha\beta$ nomenclature (IUPAC rules).

The major dimeric procyanidins (Fig. 7.1) thus become:
procyanidin B1: epicatechin-($4\beta \rightarrow 8$)-catechin
procyanidin B2: epicatechin-($4\beta \rightarrow 8$)-epicatechin
procyanidin B3: catechin-($4\alpha \rightarrow 8$)-catechin

procyanidin B4: catechin–(4α → 8)–epicatechin

Structures (*7.25*) and (*7.26*) now are named respectively as:

(*7.25*): gallocatechin–(4α → 8)–epigallocatechin

(*7.26*): gallocatechin–(4α → 8)–catechin

REFERENCES

Bate-Smith, E. C. (1962), *J. Linn. Soc.* (*Bot.*) **58**, 95.

Bate-Smith, E. C. (1972), *Phytochemistry* **11**, 1153, 1755.

Bate-Smith, E. C. (1973), *Phytochemistry* **12**, 907.

Bate-Smith, E. C. (1975), *Phytochemistry* **14**, 1107.

Bate-Smith, E. C. (1977), *Phytochemistry* **16**, 1421.

Bate-Smith, E. C. and Swain, T. (1953), *Chem. Ind.*, 377.

Brown, B. R. and Shaw, M. R. (1974), *J. Chem. Soc.* (*Perkin 1*) 2036.

Clark-Lewis, J. L. (1968), *Aust. J. Chem.* **21**, 2059, 3025.

Clark-Lewis, J. L. and Dainis, I. (1964), *Aust. J. Chem.* **17**, 1170.

Clark-Lewis, J. L. and Dainis, I. (1967), *Aust. J. Chem.* **20**, 2191.

Clark-Lewis, J. L. and Mitsuno, M. (1958), *J. Chem. Soc.* 1724.

Clark-Lewis, J. L. and Mortimer, P. I. (1960), *J. Chem. Soc.* 4106.

Clark-Lewis, J. L. and Porter, J. L. (1972), *Aust. J. Chem.* **25**, 1943.

Clark-Lewis, J. L., Katekar, G. F. and Mortimer, P. I. (1961), *J. Chem. Soc.* 499.

Clark-Lewis, J. L., Baig, M. I. and Thompson, M. J. (1969), *Aust. J. Chem.* **22**, 2645.

Creasey, L. L. and Swain, T. (1965), *Nature* (*London*) **208**, 151.

Drewes, S. E. (1968), *J. Chem. Soc.* (*C*) 1140.

Drewes, S. E. and Roux, D. G. (1964), *Biochem. J.* **90**, 343.

Drewes, S. E. and Roux, D. G. (1966), *Biochem. J.* **98**, 493.

Drewes, S. E., Roux, D. G., Saayman, H. M., Eggers, S. H. and Feeney, J. (1967), *J. Chem. Soc.* (*C*) 1302.

du Preez, I. C. and Roux, D. G. (1970), *J. Chem. Soc.* (*C*), 1800.

du Preez, Rowan, A. C., Roux, D. G. and Feeney, J. (1971), *J. Chem. Soc. Chem. Commun.*, 215.

Fourie, T. G., Roux, D. G. and Ferreira, D. (1974), *Phytochemistry* **13**, 2573.

Freudenberg, K. and Weinges, K. (1960), *Tetrahedron* **8**, 336.

Geissman, T. A. and Yoshimura, N. N. (1966), *Tetrahedron Lett.* 2669.

Grisebach, H. (1968). In *Recent Advances in Phytochemistry, Vol. 1* (eds. T. Mabry, V. C. Runeckles and R. E. Alston), Appleton Century Crofts, New York, p. 379.

Gupta, R. K. and Haslam, E. (1978), *J. Chem. Soc.* (*Perkin 1*) 892.

Haslam, E. (1974a), *Biochem. J.* **139**, 285.

Haslam, E. (1974b), *J. Chem. Soc. Chem. Commun.* 594.

Haslam, E. (1975). In *The Flavonoids* (eds J. B. Harborne, T. Mabry and H. Mabry), Chapman and Hall, London, p. 505.

Haslam, E. (1977), *Phytochemistry* **16**, 1625.

Haslam, E., Fletcher, A. C., Porter, L. J. and Gupta, R. K. (1977a), *J. Chem. Soc.* (*Perkin 1*) 1628.

Haslam, E., Opie, C. T., Porter, L. J. and Jacques, D. (1977b), *J. Chem. Soc.* (*Perkin 1*) 1637.

IUPAC (1979), *Nomenclature of Organic Chemistry* Section F. Pergamon, Oxford.

Jones, W. T., Broadhurst, R. B. and Lyttleton, J. W. (1976), *Phytochemistry* **15**, 1407.

Jurd, L. and Lundin, R. (1968), *Tetrahedron* **24**, 2653.

Keppler, H. H. (1957), *J. Chem. Soc.* 2721.

King, F. E. and Bottomley, W. (1954), *J. Chem. Soc.* 1399.

Mitsuno M. and Yoshizaki, M. (1958), *Chem. Abstr.* **52**, 4105 d.

Oberholzer, M. E., Rall, G. J. H. and Roux, D. G. (1980), *Phytochemistry* **19**, 2503.

Opie, C. T., Porter, L. J. and Haslam, E. (1977), *Phytochemistry* **16**, 99.
Porter, L. J. and Foo, L. Y. (1978), *J. Chem. Soc. Perkin 1*, 1186.
Porter, L. J. and Foo, L. Y. (1980), *Phytochemistry* **19**, 1747.
Porter, L. J., Czochanska, Z., Foo, L. Y., Newman, R. H., Thomas, W. A. and Jones, W. T. (1979), *J. Chem. Soc. Chem. Commun.*, 375.
Porter, L. J., Foo, L. Y. and Hemingway, R. W. (1982), *J. Chem. Soc. (Perkin 1)*, in press.
Porter, L. J., Foo, L. Y., Newman, R. H. and Czochanska, Z. (1980) *J. Chem. Soc.* (*Perkin I*) 2278.
Roberts, J. D., Weigert, F. J., Kroschawitz, J. and Reich, H. J. (1970), *J. Am. Chem. Soc.* **92**, 1338.
Robertson, A. V. (1959), *Can. J. Chem.* **37**, 1946.
Robinson, R. and Robinson, G. M. (1935), *J. Chem. Soc.* 744.
Roux, D. G. (1972), *Phytochemistry* **11**, 1219.
Roux, D. G. and Hundt, H. K. L. (1978), *J. Chem. Soc. Chem. Commun.* 696.
Roux, D. G., Engel, D. W., Hattingh, M. and Hundt, H. K. L. (1978a), *J. Chem. Soc. Chem. Commun.* 695.
Roux, D. G., Botha, J. J. and Ferreira, D. (1978b), *J. Chem. Soc. Chem. Commun.* 698.
Roux, D. G., Botha, J. J. and Ferreira, D. (1978c), *J. Chem. Soc. Chem. Commun.*, 700.
Saayman, H. M. and Roux, D. G. (1965), *Biochem. J.* **96**, 36.
Scopes, P. M., Barrett, M. W., Klyne, W., Fletcher, A. C., Porter, L. J. and Haslam, E. (1979), *J. Chem. Soc.* (*Perkin I*) 2375.
Thompson, R. S., Jacques, D., Tanner, R. J.N. and Haslam, E. (1972), *J. Chem Soc.* (*Perkin I*), 1387.
Tschesche, R., Braun, T. M. and Von Sassen, W. (1980), *Phytochemistry* **19**, 1825.
Weinges, K. (1958), *Liebigs Ann.* **615**, 203.
Weinges, K., Kaltenhauser, W., Marx, H.-D., Nader, E., Nader, F., Perner, J. and Seiler, D. (1968a), *Liebigs Ann.* **711**, 184.
Weinges, K., Goritz, K. and Nader, F. (1968b), *Liebigs Ann.* **715**, 164.
Weinges, K., Bahr, W., Ebert, W., Goritz, K. and Marx, H.-D. (1969), *Fortschr. Chem. Org. Naturst.* **27**, 158.
Weinges, K., Marx H.-D. and Goritz, K. (1970), *Chem. Ber.* **103**, 2336.
Weinges, K., Schilling, G., Mayer, W. and Müller, O. (1973), *Liebigs Ann.* 1471.
Whalley, W. B. (1962). In *The Chemistry of Flavonoid Compounds* (ed. T. A. Geissman), Pergamon Press, Oxford p. 441.

CHECK LIST OF FLAVAN-3-OLS, FLAVAN 3,4-DIOLS AND PROANTHOCYANIDINS

Flavan-3-ols

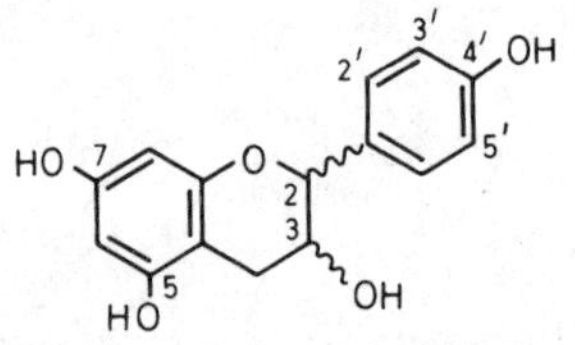

(+)-Afzelechin (2*R*:3*S*, 3,5,7,4′-tetra-OH)
(−)-Afzelechin (2*R*:3*R*, 3,5,7,4′-tetra-OH)
(+)-Epiafzelechin (2*S*:3*S*, 3,5,7,4′-tetra-OH)
(−)-Fisetinidol (2*R*:3*S*, 3,7,3′,4′-tetra-OH)
(−)-Robinetinidol (2*R*:3*S*, 3,7,3′,4′,5′-penta-OH)
(+)-Catechin (2*R*:3*S*, 3,5,7,3′,4′-penta-OH)
(−)-Epicatechin (2*R*:3*R*, 3,5,7,3′,4′-penta-OH)
Symplocoside (epicatechin 3′-methyl ether 7-*O*-glucoside)
(+)-Epicatechin (2*S*:3*S*, 3,5,7,3′,4′-penta-OH)
(+)-Gallocatechin (2*R*:3*S*, 3,5,7,3′,4′,5′-hexa-OH)
(−)-Epigallocatechin (2*R*:3*R*), 3, 5, 7, 3′,4′,5′-hexa-OH)
(−)-Epigallocatechin 4′-methyl ether

(References to flavan-3-ols, see Haslam, 1975; for symplocoside, see Tschesche *et al.*, 1980).

Flavan-3,4-diols

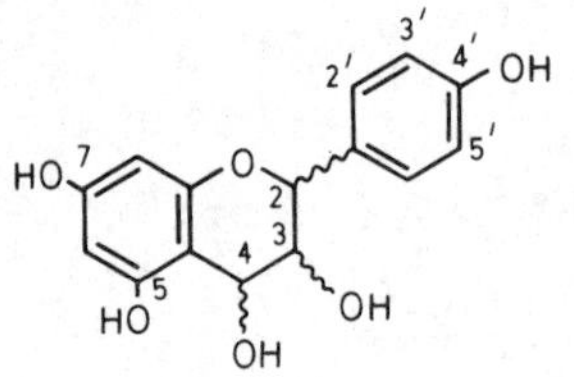

(+)-Guibourtacacidin (2*R*:3*S*:4*S*, 3,4,7,4′-tetra-OH)
isomer (2*R*:3*S*:4*R*)
(+)-Mollisacacidin (2*R*:3*S*:4*R*, 3,4,7,3′,4′-penta-OH)
(+)-Gleditsin (2*R*:3*S*:4*S*, 3,4,7,3′,4′-penta-OH)
7,3,4′-Trimethyl ether
3,4,7,3′,4′-Pentamethyl ether
(−)-Leucofisetinidin (2*S*:3*R*:4*S*, 3,4,7,3′,4′-penta-OH)
(−)-Teracacidin (2*R*:3*R*:4*R*, 3,4,7,8,4′-penta-OH)
(−)-Isoteracacidin (2*R*:3*R*:4*S*, 3,4,7,8,4′-penta-OH)
7,8,4′-Trimethoxyflavan-3,4-diol (2*R*:3*S*:4*R*)
(−)-Melacacidin (2*R*:3*R*:4*R*, 3,4,7,8,3′,4′-hexa-OH)
3,8-Dimethyl ether
(−)-Isomelacacidin (2*R*:3*R*:4*S*, 3,4,7,8,3′4′-hexa-OH)
7,8,3′,4′-Tetramethoxyflavan-3,4-diol (2*R*:3*S*:4*S*)
(+)-Leucorobinetinidin (2*R*:3*S*:4*R*, 3,4,7,3′,4′,5′-hexa-OH)

Proanthocyanidins

Procyanidins B-1, B-2, B-3, B-4 (dimers)
Prodelphinidin dimers
Procyanidin, prodelphinidin polymers
Profisetinidins
Prorobinetinidins
Proguibourtinidins

CHAPTER EIGHT

C-Glycosylflavonoids

J. CHOPIN, M. L. BOUILLANT and E. BESSON

8.1 INTRODUCTION

In line with the general rapid expansion of phytochemistry, the field of *C*-glycosylflavonoids has grown rapidly over the last five years. Naturally occurring *C*-arabinosyl-, *C*-galactosyl- and *C*-rhamnosyl-flavonoids have been added to the already known *C*-glucosyl- and *C*-xylosyl-flavonoids, and a large number of *O*-glycosides derived from these *C*-glycosides have been found. Moreover an unusual carbon-linked sugar residue (6-deoxy-*xylo*-hexos-4-ulosyl) has been identified in *C*-glycosylflavones of *Zea mays*, which inhibit insect development in this plant (Elliger *et al.*, 1980). The natural occurrence of several new types of *C*-glycosylflavonoids (*C*-glycosylchalcones, *C*-glycosylflavanonols, *C*-glycosylisoflavanones) and of tricin-derived *C*-glycosylflavones has been fully established during the period under review.

Besides the well-known Wessely–Moser acid isomerization of 5-hydroxy-*C*-glycosylflavones, which involves hydrolytic opening of the flavone heterocycle and cyclodehydration of the intermediary β-diketone, a further acid isomerization of the sugar heterocycle leading to a mixture of anomeric *C*-pyranosyl- and *C*-furanosyl-flavones has now been definitely proved to occur in *C*-arabinosylflavones (Besson and Chopin, 1980). This explains why the symmetrical 6,8-di-*C*-arabinopyranosyl-5,7-dihydroxyflavones unexpectedly undergo isomerization on acid treatment.

The natural sources of *C*-glycosylflavonoids reported in the last six years are listed in Table 8.1. The number of species are given for each genus, except in the case of the large family surveys that have been carried out in the Cyperaceae, Gramineae, Orchidaceae and Palmae (Harborne, Williams and co-workers). Liverworts have been mainly studied by Markham and Zinsmeister's groups, mosses by Österdahl, pteridophytes by Hiraoka and Wallace, gymnosperms by the Lebreton, Niemann and Wallace groups, monocotyledons by Harborne and Williams, Caryophyllales by Richardson, Labiatae by Weightman, and Polemoniaceae by Smith and co-workers. Some genera have been intensively studied, e.g. *Achillea* (Valant), *Carlina* (Raynaud and co-workers), *Gentiana* (Jacot-Guillarmod, Hostettmann and co-workers), *Larix* (Niemann and co-workers) and *Silene* (Zemtsova, Darmograi). The same is true for some species, e.g. *Avena sativa* (Weissenböck and co-workers), *Crataegus monogyna*

Table 8.1 Natural sources of *C*-glycosylflavonoids

BRYOPHYTA
Apometzgeria (3) Theodor *et al.* (1980, 1981a,b)
Conocephalum (1) Markham *et al.* (1976a)
Hedwigia (1) Österdahl (1978, 1979a,b)
Hymenophyllum (1) Markham and Wallace (1980)
Hymenophyton (2) Markham *et al.* (1976b)
Mnium (1) Vandekerkhove (1978); Österdahl (1979a,c)
Plagiochila (1) Mues and Zinsmeister (1975, 1976)
Takakia (2) Markham and Porter (1979)
Trichomanes (2) Markham and Wallace (1980)

PTERIDOPHYTA
Angiopteridaceae
Angiopteris (4) Wallace and Story (1978); Wallace *et al.* (1979, 1981)
Athyriaceae
Lunathrium (7) Hiraoka (1978)
Matteucia (2) Hiraoka (1978)
Onoclea (1) Hiraoka (1978)
Woodsia (2) Hiraoka (1978)
Dryopteridaceae
Arachnoides (4) Hiraoka (1978)
Cyrtomium (3) Hiraoka (1978)
Dryopteris (13) Hiraoka (1978)
Polystichum (5) Hiraoka (1978)
Lycopodiaceae
Lycopodium (1) Markham and Moore (1980)
Psilotaceae
Psilotum (1) Wallace and Markham (1978a,b)
Tmesipteridaceae
Tmesipteris (2) Wallace and Markham (1978a,b)

SPERMAPHYTA
GYMNOSPERMAE
Cycadaceae
Cycas (3) Lebreton (1980); Niemann and Miller (1975)
Ephedraceae
Ephedra (4) Castledine and Harborne (1976); Wallace (1978, 1979); Thivend *et al.* (1979); Niemann and Miller (1975)
Gnetaceae
Gnetum (3) Wallace (1978); Wallace and Morris (1978); Thivend *et al.* (1979)
Pinaceae
Abies (11) Lebreton *et al.* (1978); Parker *et al.* (1979)
Keteleeria (1) Lebreton *et al.* (1978)
Larix (8) Niemann (1973, 1974, 1975a,b 1979, 1980a,b); Niemann and Miller (1975); Niemann and Baas (1978); Medvedeva *et al.* (1973, 1974); Tjukavkina *et al.* (1975)
Pseudotsuga (1) Niemann and Van Genderen (1980)
Tsuga (4) Lebreton *et al.* (1978)
Podocarpaceae
Podocarpus (2) Lebreton *et al.* (1978); Niemann and Miller (1975)
Stangeriaceae
Stangeria (1) Lebreton (1980)

Taxodiaceae
Sequoia (1) Lebreton *et al.* (1978); Niemann and Miller (1975)
Welwitschiaceae
Welwitschia (1) Wallace (1978); Thivend *et al.* (1979)
ANGIOSPERMAE
MONOCOTYLEDONEAE
Araceae
Acorus (1) El'Yashevich *et al.* (1974a,b)
Asterostigma (1) Markham and Williams (1980)
Bromeliaceae
Pitcairnia (1) Williams (1978)
Tillandsia (5) Williams (1978)
Cyperaceae Williams and Harborne (1977a)
Gramineae
(93% of 274 species from 121 genera) Harborne and Williams (1976); Murray and Williams (1976); (18 sp.) Kaneta and Sugiyama (1973)
Avena sativa Chopin *et al.* (1977a); Effertz and Weissenböck (1980); Nabeta *et al.* (1977); Plesser and Weissenböck (1977); Popovici and Weissenböck (1977); Popovici *et al.* (1977); Weissenböck and Schneider (1974); Weissenböck and Effertz (1974); Weissenböck *et al.* (1976); Weissenböck and Sachs (1977); Weissenböck *et al.* (1978)
Briza (media) Casteldine and Harborne (1976); Chari *et al.* (1980)
Eleusine (9) Hilu *et al.* (1978)
Hordeum vulgare Blume and McClure (1978); Carlin and McClure (1973); Saunders and McClure (1976)
Phragmites (1) Nawwar *et al.* (1980); Jay and Viricel (1980)
Saccharum (120) Williams *et al.* (1974)
Zea mays Saunders and McClure (1976); Elliger *et al.* (1980)
Iridaceae
Crocus (1) Sergeyeva (1977)
Fosteria (1) Ballard and Cruden (1978)
Iris (2) Arisawa *et al.* (1973); Blinova *et al.* (1977)
Rigidella (4) Ballard and Cruden (1978)
Sessilanthera (3) Ballard and Cruden (1978)
Juncaceae
Prionium (1) Williams and Harborne (1975)
Lemnaceae
Spirodela (2) Wallace (1975); Saunders and McClure (1976)
Liliaceae
Asphodelus (1) Abdel-Gawad and Raynaud (1974)
Narthecium (1) Williams (1975)
Paradisia (1) Williams (1975)
Polygonatum (2) Chopin *et al.* (1977b); Najar *et al.* (1978); Morita *et al.* (1976)
Ruscus (1) El Alfy and Paris (1975)
Tulipa (1) Budzianowski and Skrzypczakowa (1978)
Urginea (1) Fernandez *et al.* (1975); Vega (1976); Williams (1975)
Marantaceae
Calathea (1) Williams and Harborne (1977b)
Ctenanthe (2) Williams and Harborne (1977b)
Monotagma (1) Williams and Harborne (1977b)
Sarcophrynium (1) Williams and Harborne (1977b)
Stromanthe (1) Williams and Harborne (1977b)
Orchidaceae (52 genera) Williams (1979)
Palmae Harborne *et al.* (1974)

Table 8.1 (*Contd.*)

Piperaceae
Piper (1) Tillequin *et al.* (1978)
Xyridaceae
Xyris (1) Linard *et al.* (1976)
Zingiberaceae
Costus (7) Williams and Harborne (1977b)

DICOTYLEDONEAE
Acanthaceae
Ecbolium (1) Nair *et al.* (1975a)
Amaranthaceae
Achyranthes (1) Burret (1980)
Aerva (1) Garg *et al.* (1980)
Alternanthera (1) Richardson (1978)
Amaranthus (2) Richardson (1978)
Deeringia (1) Richardson (1978)
Asclepiadaceae
Hoya (4) Niemann *et al.* (1979); Niemann (1980a,b)
Basellaceae
Basella (2) Richardson (1978); Burret (1980)
Boussingaustia (1) Burret (1980)
Cannabinaceae
Cannabis (1) Paris and Paris (1973); Paris *et al.* (1976); Segelman *et al.* (1978)
Caprifoliaceae
Viburnum (1) Saunders and McClure (1976)
Caryophyllaceae
Arenaria (6) Darmograi (1979)
Cerastium (12) Richardson (1978); Darmograi (1979); Dubois (1980); Dubois *et al.* (1980)
Coronaria (1) Darmograi (1976)
Dianthus (3) Boguslavskaya and Beletskii (1978); Seraya *et al.* (1978); Richardson (1978)
Lychnis (5) Richardson (1978); Darmograi (1976)
Melandrium album Brederode and Nigtevecht (1972a,b, 1973, 1974a,b,c)
= *Silene alba, Silene dioica* Zykova and Pivnenko (1975); Zykova *et al.* (1976); Kamsteeg *et al.* (1976); Besson *et al.* (1979a); Wagner *et al.* (1979); Heinsbroek *et al.* (1979, 1980); Brederode *et al.* (1979, 1980)
Silene spp. (47) Richardson (1978); Zemtsova *et al.* (1975); Zemtsova and Dzhumyrko (1976); Darmograi (1976, 1977); Chopin *et al.* (1974); Galle (1974)
Stellaria (3) Richardson (1978); Zoll and Nouvel (1974); Ferreres de Arce (1977)
Spergularia (1) Richardson (1978); Bouillant *et al.* (1979)
Vaccaria (1) Baeva *et al.* (1974)
other genera (10) Richardson (1978)
Chenopodiaceae
Beta (2) Richardson (1978); Burret (1980)
Chenopodium (2) Richardson (1978)
Commelinaceae
Gibasis schiedeana (2) Del Pero de Martinez and Swain (1976, 1977)
Compositae
Achillea (38) Valant (1978); Valant *et al.* (1978, 1980)
Ajania sp. Chumbalov and Zhubaeva (1976)
Atractylis (1) Raynaud and Dombris (1978)

Carlina (4) Raynaud and Rasolojaona (1976, 1978, 1979); Dombris and Raynaud (1977); Besson *et al.* (1979b); Rasolojaona and Raynaud (1979)
Catananche (1) Proliac and Raynaud (1977)
Centaurea (1) Kamanzi and Raynaud (1976)
Flourensia (1) Dillon *et al.* (1976)
Hazardia (12) Clark and Mabry (1978)
Helenium sp. Bierner (1973)
Liatris (5) Wagner *et al.* (1973)
Otospermum sp. Greger (1977)
Coridaceae
Coris (1) Frezet *et al.* (1975)
Cucurbitaceae
Cucumis (1) Monties *et al.* (1976)
Cucurbita (1) Saunders and McClure (1976)
Didiereaceae
Alluaudia (4) Richardson (1978); Rabesa (1980)
Didierea (2) Rabesa (1980)
Dipsacaceae
Cephalaria (1) Galle (1974)
Knautia (1) Zemtsova and Bandyukova (1974)
Gentianaceae
Gentiana (31) Lebreton and Dangy-Caye (1973); Bellmann and Jacot-Guillarmod (1973); Hostettmann *et al.* (1973, 1975); Kaldas *et al.* (1974); Hostettmann and Jacot-Guillarmod (1974a,b, 1975, 1977); Kaldas *et al.* (1975); Jacot-Guillarmod *et al.* (1975); Luong Minh Duc *et al.* (1976, 1980); Marston *et al.* (1976); Goetz *et al.* (1976a,b); Luong Minh Duc and Jacot-Guillarmod (1977); Goetz and Jacot-Guillarmod (1977a,b, 1978); Chulia and Debelmas (1977); Hostettmann-Kaldas and Jacot-Guillarmod (1978); Chulia *et al.* (1978); Burret *et al.* (1978, 1979); Zoring *et al.* (1980)
Hoppea (1) Ghosal *et al.* (1978)
Swertia (2) Miana (1973); Hostettmann and Jacot-Guillarmod (1976)
Geraniaceae
Geranium (2) Boutard and Lebreton (1975); Gonnet (1980) *(G. sylvaticum)*
Labiatae (28 genera) Weightman (1979)
Ajuga (4) Salehian *et al.* (1973); Weightman (1979)
Phlomis (1) Vavilova and Gella (1973a,b)
Teucrium (1) Raynaud and Chouikha (1976)
Leguminosae
Acacia Thieme and Khogali (1974, 1975)
Cladrastis (2) Ohashi *et al.* (1976, 1977a, 1980)
Dalbergia (2) Chawla *et al.* (1974, 1975, 1976); Parthasarathy *et al.* (1974, 1976)
Desmodium (1) Chernobrovaya (1973)
Eysenhardtia (1) Vita-Finzi *et al.* (1980)
Lespedeza (2) Linard *et al.* (1978); Numata *et al.* (1979, 1980)
Lupinus (2) Laman and Volynets (1974); Nicholls and Bohm (1979); Zapesochnaya and Laman (1977)
Melilotus (1) Specht *et al.* (1976)
Onobrychis Moniava and Kemertelidze (1976)
Pterocarpus (1) Gupta *et al.* (1981)
Pueraria (1) Inoue and Fujita (1974, 1977)
Rynchosia (2) Besson *et al.* (1977); Adinarayana *et al.* (1980)
Sarothamnus (1) Brum-Bousquet *et al.* (1977)
Securigera (1) *Coronilla* (1) Jay *et al.* (1980)
Trigonella (1) Sood *et al.* (1976)

Table 8.1 (*Contd.*)

Molluginaceae
Mollugo (3) Chopin *et al.* (1978, 1979); Richardson (1978)
Nyctaginaceae
Mirabilis (3) Richardson (1978)
Ochnaceae
Brackenridgea (1) Bombardelli *et al.* (1974)
Ochna (1) Nair *et al.* (1975b)
Onagraceae
Circaea (3) Boufford *et al.* (1978)
Oxalidaceae
Oxalis (1) Tschesche and Struckmeyer (1976)
Passifloraceae
Passiflora Loehdefink (1976); Quercia *et al.* (1978); Ulubelen and Mabry (1980)
Phytolaccaceae
Trichostigma (1) Richardson (1978)
Plumbaginaceae
Limonium Asen *et al.* (1973)
Polemoniaceae
Cantua (1) Smith *et al.* (1977)
Gilia (3) Smith *et al.* (1977)
Langloisia (1) Smith *et al.* (1977)
Leptodactylon (3) Smith *et al.* (1977)
Linanthus (16) Smith *et al.* (1977)
Loselia (3) Smith *et al.* (1977)
Microsteris (1) Smith *et al.* (1977)
Phlox (19) Levy and Levin (1974); Levy (1976); Smith *et al.* (1977); Levy and Fujii (1978)
Polygonaceae
Fagopyrum (1) Saunders and McClure (1976)
Polygonum (2) Krause (1976a,b); Gonnet (1980) (*P. bistorta*)
Portulacaceae
Calandrinia (2) Richardson (1978)
Montia (1) Richardson (1978)
Ranunculaceae
Adonis (4) Komissarenko *et al.* (1973a,b); Ponomarenko *et al.* (1974); Galle (1974); Wagner *et al.* (1975)
Clematis (6) Dennis and Bierner (1980)
Rhamnaceae
Zizyphus (1) Woo *et al.* (1979)
Rosaceae
Crataegus (2) Nikolov and Litvinenko (1973); Nikolov (1973, 1974, 1975); Nikolov *et al.* (1973, 1976); Nikolov and Vodenicharov (1975)
Rubiaceae
Canephora (1) Paris and Debray (1973)
Rutaceae
Almeidea (1) Jay *et al.* (1979)
Citrus (2) Montoro *et al.* (1974); Tomas *et al.* (1975)
Fortunella (1) Horowitz *et al.* (1974)
Sterculiaceae
Sterculia (1) Nair *et al.* (1976)
Theobroma (1) Jalal and Collin (1977)

Thymelaeaceae
Thymelaea (1) Nawwar *et al.* (1977)
Ulmaceae
Zelkowa (1) Hayashi (1980)
Umbelliferae
Astrantia (1) Hiller *et al.* (1975)
Chaerophyllum aureum Gonnet (1980)
Myrris odorata Gonnet (1980)
Verbenaceae
Vitex (1) Subramanian and Misra (1979)
Zygophyllaceae
Larrea (1) Sakakibara *et al.* (1977)

(Nikolov), *Melandrium album* (= *Silene alba*) (Brederode and Nigtevecht), for physiological, pharmacological and genetical purposes respectively. The occurrence of *C*-glycosylflavonoids in chloroplasts of various plants has been unambiguously demonstrated by Monties (1969), Weissenböck and Schneider (1974) and Saunders and McClure (1976). The evolutionary significance of distribution patterns of *C*-glycosylflavones in the angiosperms has been discussed by Harborne (1977).

Major advances in the most difficult problem in *C*-glycosylflavonoid chemistry, i.e. identification of the glycosyl residue, have involved the use of permethyl ethers of synthetic *C*-glycosylflavonoids in mass spectrometry (Bouillant *et al.*, 1975, 1978) and the application of ^{13}C-NMR spectrometry to natural *C*-glycosylflavonoids of known structure (Chari *et al.*, 1978). Because it can be used with very small amounts of substance, mass spectrometry of permethyl ethers is the method of choice in the study of the nature (hexose, pentose or deoxyhexose) of the *C*-linked sugars in *C*-glycosylflavones which remain by far the most widespread group of *C*-glycosylflavonoids. It is the only method which identifies the nature of the 6-*C*-linked sugar in 6,8-di-*C*-glycosylflavones, owing to the striking fragmentation pattern difference between permethylated 6-*C*-glycosylflavones and 8-*C*-glycosylflavones and to the fact that the ^{13}C-NMR signals of a carbon-linked sugar residue are not significantly affected by the 6 or 8 position of the latter in *C*-glycosylflavones. Its systematic use has disclosed that difficultly separated mixtures of di-*C*-hexosyl- and *C*-hexosyl-*C*-pentosyl-flavones often co-occur in plant extracts. In such cases the characteristic R_F differences observed on TLC between isomeric permethylated *C*-glucosyl- and *C*-galactosyl-flavones on the one hand, permethylated *C*-xylosyl- and *C*-arabinosyl-flavones on the other, has often provided useful separations on the micro-scale.

Moreover, the mass spectrometry of permethyl ethers has solved the difficult problem of locating the position of the *O*-linked sugar in natural 6-*C*-glucosylflavone x″-*O*-glycosides (Bouillant *et al.*, 1979a), following the synthesis of such compounds by *C*-substitution of acacetin with the acetobromo derivatives of neohesperidose, rungiose, cellobiose and rutinose (Besset, 1977;

Bouillant *et al.*, 1980). However, ^{13}C-NMR spectrometry is of more general applicability, since the same problem can be solved in 8- as well as in 6-*C*-glycosylflavone x″-*O*-glycosides. Acyl groups are generally lost during permethylation and their presence must be detected beforehand by alkaline hydrolysis of the natural product under study. Again, with ^{13}C-NMR spectrometry it is possible to identify the acyl groups and their position in the intact molecule. The differentiation by circular dichroism between 6- and 8-*C*-glycosylflavones has been extended to *C*-arabinosyl-, *C*-xylosyl- and *C*-galactosylflavones, whereas *C*-rhamnosylflavones show a different behaviour (Gaffield *et al.*, 1978).

The introduction of the *C*-glucosyl group at the chalcone stage in the biosynthesis of the *C*-glycosylisoflavone puerarin in *Pueraria* roots was suggested by Inoue and Fujita (1974, 1977). The same authors have shown that *C*-glucosylflavones are biosynthesized in *Swertia japonica* via *C*-glucosylflavanones, and that 3′-hydroxylation occurs prior to *C*-glycosylation (Fujita and Inoue, 1979).

Total syntheses of 7,4′-di-*O*-methylbayin (1975), 7,4′-di-*O*-methylpuerarin (1978) and 5,7,4′-tri-*O*-methylvitexin (1979) have been described by Eade and co-workers. The high-yield *C*-glucosylation of 1,3,5-trimethoxybenzene opens the way to the synthesis of almost all types of 8-*C*-glycosyl-5,7-di-*O*-methylflavonoids. The low-yield *C*-glycosylation of 5,7-dihydroxyflavones has still been used in the authors' laboratory for specific purposes (*C*-arabinofuranosylation and *C*-diglycosylation of acacetin, di-*C*-glycosylation of apigenin). HPLC has been recently applied to *C*-glycosylflavonoid analysis (Niemann and Brederode, 1978; Strack *et al.*, 1979). Flavonoid *O*- and *C*-glycosides have been reviewed by Wagner (1974, 1977).

8.2 NATURALLY OCCURRING *C*-GLYCOSYLFLAVONOIDS

Natural *C*-glycosylflavonoids can be divided in two groups: the unhydrolysable 'aglycones' (mono- and di-*C*-glycosylflavonoids) and their hydrolysable derivatives (*O*-glycosides and *O*-acyl derivatives). Mono-*C*-glycosylflavonoids are listed in Table 8.2, di-*C*-glycosylflavonoids in Table 8.3, *C*-glycosylflavonoid *O*-glycosides in Table 8.4 and *O*-acyl-*C*-glycosylflavonoids in Table 8.5. New structures are indicated with an asterisk.

8.2.1 Mono-*C*-glycosylflavonoids

Eight types of mono-*C*-glycosylflavonoids are now well defined: *C*-glycosylflavones, *C*-glycosylflavonols, *C*-glycosylflavanones, *C*-glycosylflavanonols, *C*-glycosylchalcones, *C*-glycosyldihydrochalcones, *C*-glycosylisoflavones and *C*-glycosylisoflavanones. *C*-Glycosylflavones remain by far the most important group. All these flavonoids have a phloroglucinol A ring with the following exceptions: bayin (a flavone); aervanone and palodulcin B (two flavanones);

Table 8.2 Naturally occurring mono-*C*-glycosylflavonoids

Compounds	*Sources*	*References*
C-GLYCOSYLFLAVONES		
Bayin	*Cladrastis platycarpa* (Leguminosae)	Ohashi *et al.* (1976)
(7,4′-di-OH 8-Glc)		
Isovitexin (saponaretin)	Many sources	Horowitz and Gentili (1964) (structure)
(5,7,4′-tri-OH 6-Glc)		
Vitexin	Many sources	Horowitz and Gentili (1964) (structure)
(5,7,4′-tri-OH 8-Glc)		
*8-*C*-β-D-Galactopyranosylapigenin	*Briza media* (Gramineae)	Castledine and Harborne (1976)
(5,7,4′-tri-OH 8-Gal)	*Polygonatum multiflorum* (Liliaceae)	Chopin *et al.* (1977b)
6-*C*-β-D-Xylopyranosylapigenin	*Cerastium arvense* (Caryophyllaceae)	Dubois (1980); Dubois *et al.* (1980)
(5,7,4′-tri-OH 6-Xyl) (Cerarvensin)		
*Isomollupentin	*Spergularia rubra* (Caryophyllaceae)	Bouillant *et al.* (1979b)
(5,7,4′-tri-OH 6-Ara)		
*Mollupentin	*Mollugo pentaphylla* (Molluginaceae)	Chopin *et al.* (1979)
(5,7,4′-tri-OH 8-Ara)		
*3′-Deoxyderhamnosylmaysin	*Zea mays* (Gramineae)	Elliger *et al.* (1980)
[5,7,4′-tri-OH 6-(6-deoxy-*xylo*-hexos-4-ulosyl)]		
Swertisin	Many sources	Komatsu and Tomimori (1966) (structure)
(5,4′-di-OH 7-OMe 6-Glc)		
Isoswertisin	*Mollugo distica* (Molluginaceae)	Chopin *et al.* (1978)
(5,4′-di-OH 7-OMe 8-Glc)	*Gnetum gnemon* (Gnetales)	Wallace and Morris (1978d)
	Rigidella, Sessilanthera, Fosteria (Iridaceae)	Ballard and Cruden (1978)
*Isomolludistin	*Asterostigma reidelianum* (Araceae) (as 2″-*O*-Glc)	Markham and Williams (1980)
(5,4′-di-OH 7-OMe 6-Ara)		
*Molludistin	*Mollugo distica* (Molluginaceae)	Chopin *et al.* (1978)
(5,4′-di-OH 7-OMe 8-Ara)		
*Isocytisoside	*Fortunella margarita* (Rutaceae) (as 2″-*O*-Rha)	Horowitz *et al.* (1974)
(5,7-di-OH 4′-OMe 6-Glc)	*Gentiana pyrenaica* (Gentianaceae) (as 7-*O*-Glc)	Marston *et al.* (1976)
	Securigera coronilla (Leguminosae) (as 2″-*O*-Glc)	Jay *et al.* (1980)

Table 8.2 (*Contd.*)

Compounds	*Sources*	*References*
Cytisoside (5,7-di-OH 4′-OMe 8-Glc)	*Lupinus arboreus* (Leguminosae)	Nicholls and Bohm (1979)
Embigenin (5-OH 7,4′-di-OMe 6-Glc)	*Iris japonica* (Iridaceae) (as 2″-*O*-Rha)	Arisawa *et al.* (1973)
*7,4′-di-*O*-methylisomollupentin (5-OH 7,4′-di-OMe 6-Ara)	*Asterostigma reidelianum* (Araceae)	Markham and Williams (1980)
Iso-orientin (5,7,3′,4′-tetra-OH 6-Glc)	Many sources	Koeppen (1964) (structure)
Orientin (5,7,3′,4′-tetra-OH 8-Glc)	Many sources	Koeppen (1964) (structure)
6-*C*-*β*-D-Xylopyranosyl-luteolin (5,7,3′,4′-tetra-OH 6-Xyl)	*Phlox drummondii* (Polemoniaceae) (as 2″-*O*-Rha)	Mabry *et al.* (1971)
*Derhamnosylmaysin [5,7,3′,4′-tetra-OH 6-(6-deoxy-*xylo*-hexos-4-ulosyl)]	*Zea mays* (Gramineae) (as 2″-*O*-Rha)	Elliger *et al.* (1980)
*Isoswertiajaponin (5,3′,4′-tri-OH 7-OMe 8-Glc)	*Gnetum gnemon* (Gnetaceae) *Phragmites australis* (Gramineae)	Wallace and Morris (1978) Nawwar *et al.* (1980)
Swertiajaponin (5,3′,4′-tri-OH 7-OMe 6-Glc)	Many sources	Komatsu *et al.* (1967) (structure)
Isoscoparin (5,7,4′-tri-OH 3′-OMe 6-Glc)	Many sources	Chopin *et al.* (1968) (structure)
Scoparin (5,7,4′-tri-OH 3′-OMe 8-Glc)	Many sources	Chopin *et al.* (1968) (structure)
*3′-*O*-Methylderhamnosylmaysin [5,7,4′-tri-OH 3′-OMe 6-(6-deoxy-*xylo*-hexos-4-ulosyl)]	*Zea mays* (Gramineae) (as 2″-*O*-Rha)	Elliger *et al.* (1980)
6-*C*-*β*-D-Glucopyranosyldiosmetin (5,7,3′-tri-OH 4′-OMe 6-Glc)	*Citrus limon* (Rutaceae)	Gentili and Horowitz (1968)
8-*C*-*β*-D-Glucopyranosyldiosmetin (5,7,3′-tri-OH 4′-OMe 8-Glc)	*Lupinus arboreus* (Leguminosae)	Nicholls and Bohm (1979)
*7,3′-di-*O*-Methyliso-orientin	*Achillea cretica* (Compositae)	Valant *et al.* (1980)

(5,4′-di-OH 7,3′-di-OMe 6-Glc)	*Phragmites australis* (Gramineae)	Jay and Viricel (1980)
7,3′,4′-Tri-*O*-methyliso-orientin	*Linum maritimum* (Linaceae) (as 2″-*O*-Rha)	Wagner *et al.* (1972)
(5-OH 7,3′,4′-tri-OMe 6-Glc)		
*Isoaffinetin	*Polygonum affine* (Polygonaceae)	Krause (1976a,b)
(5,7,3′,4′,5′-penta-OH 6-Glc)	*Trichomanes venosum* (Hymenophyllaceae)	Markham and Wallace (1980)
*Affinetin	*Trichomanes venosum* (Hymenophyllaceae)	Markham and Wallace (1980)
(5,7,3′,4′,5′-penta-OH 8-Glc)		
*Isopyrenin	*Gentiana pyrenaica* (Gentianaceae)	Marston *et al.* (1976)
(5,7,4′-tri-OH 3′,5′-di-OMe 6-Glc)		
*6-*C*-Glucosyl-5,7-dihydroxy-8,3′,4′,5′-tetramethoxyflavone	*Vitex negundo* (Verbenaceae) (as 5-*O*-Rha)	Subramanian and Misra (1979)
C-GLYCOSYLFLAVONOLS		
*6-*C*-Glucosylkaempferol	*Zelkowa serrata* (Ulmaceae)	Hayashi (1980)
(3,5,7,4′-tetra-OH 6-Glc)		
Keyakinin	*Zelkowa serrata* (Ulmaceae)	Funaoka (1956)
(3,5,4′-tri-OH 7-OMe 6-Glc)		
*6-*C*-Glucosylquercetin	*Zelkowa serrata* (Ulmaceae)	Hayashi (1980)
(3,5,7,3′,4′-penta-OH 6-Glc)		
Keyakinin B	*Zelkowa serrata* (Ulmaceae)	Hillis and Horn (1966)
(3,5,3′,4′-tetra-OH 7-OMe 6-Glc)		
C-GLYCOSYLFLAVANONES		
*Aervanone	*Aerva persica* (Amaranthaceae)	Garg *et al.* (1980)
(7,4′-di-OH 8-Gal)	*Eysenhardtia polystachya* (Leguminosae)	Vita-Finzi *et al.* (1980)
*Palodulcine B	*Eysenhardtia polystachya* (Leguminosae)	Vita-Finzi *et al.* (1980)
(7,3′,4′-tri-OH 8-Gal)		
Hemiphloin	*Tulipa gesneriana* (Liliaceae)	Budzianowski and Skrzypczakowa (1978)
(5,7,4′-tri-OH 6-Glc)		
Isohemiphloin	*Eucalyptus hemiphloia* (Myrtaceae)	Hillis and Horn (1965) (structure)
(5,7,4′-tri-OH 8-Glc)		
C-GLYCOSYLFLAVANONOLS		
*6-*C*-Glucosyldihydrokaempferol	*Zelkowa serrata* (Ulmaceae)	Hayashi (1980)
(3,5,7,4′-tetra-OH 6-Glc)		

Table 8.2 (*Contd.*)

Compounds	*Sources*	*References*
Keyakinol (3,5,4′-tri-OH 7-OMe 6-Glc)	*Zelkowa serrata* (Ulmaceae)	Funaoka (1956)
*6-*C*-Glucosyldihydroquercetin (3,5,7,3′,4′-penta-OH 6-Glc)	*Zelkowa serrata* (Ulmaceae)	Hayashi (1980)
C-GLYCOSYLCHALCONES		
*3′-*C*-Glucosylisoliquiritigenin (2′,4′,4-tri-OH 3′-Glc)	*Cladrastis platycarpa* (Leguminosae)	Ohashi *et al.* (1977a)
C-GLYCOSYLDIHYDROCHALCONES		
Nothofagin (2′,4′,6′4-tetra-OH *C*-Gly)	*Nothofagus fusca* (Fagaceae)	Hillis and Inoue (1967)
Konnanin (2′,4′,6′,3,4-penta-OH *C*-Gly)	*Nothofagus fusca* (Fagaceae)	Hillis and Inoue (1967)
Aspalathin (2′,4′,6′,3,4-penta-OH 3′-Glc)	*Aspalathus linearis* (*acuminatus*) (Leguminosae)	Koeppen and Roux (1965)
C-GLYCOSYLISOFLAVONES		
Puerarin (7,4′-di-OH 8-Glc)	*Pueraria sp.* (Leguminosae)	Murakami *et al.* (1960) (structure)
*8-*C*-Glucosylgenistein (5,7,4′-tri-OH 8-Glc)	*Lupinus luteus* (Leguminosae)	Zapesochnaya and Laman (1977)
*8-*C*-Glucosylprunetin (5,4′-di-OH 7-OMe 8-Glc)	*Dalbergia paniculata* (Leguminosae)	Parthasarathy *et al.* (1974, 1976)
*Volubiline (5-OH 7,4′-di-OMe 8-Rha)	*Dalbergia volubilis* (Leguminosae)	Chawla *et al.* (1974)
*Isovolubiline (5-OH 7,4′-di-OMe 6-Rha)	*Dalbergia volubilis* (Leguminosae)	Chawla *et al.* (1975)
*8-*C*-Glucosylorobol (5,7,3′,4′-tetra-OH 8-Glc)	*Lupinus luteus* (Leguminosae)	Zapesochnaya and Laman (1977)
*Volubilinine (5,7-di-OH 6,4′-di-OMe 8-Glc)	*Dalbergia volubilis* (Leguminosae)	Chawla *et al.* (1976)

Dalpanitin (5,7,4′-tri-OH 3′-OMe 8-Glc)	*Dalbergia paniculata* (Leguminosae)	Adinarayana and Rao (1972)
C-GLYCOSYLISOFLAVANONES		
*Dalpanin 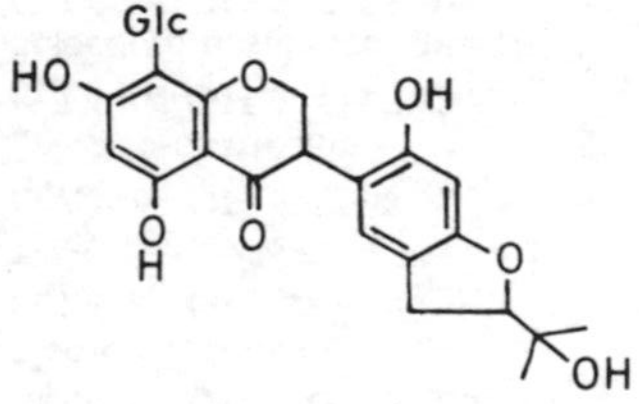	*Dalbergia paniculata* (Leguminosae)	Adinarayana and Rao (1975)
*6-*C*-Glucosyldalbergioidin (5,7,2′,4′-tetra-OH 6-Glc)	*Pterocarpus macrocarpus* (Leguminosae)	Gupta (1981)

* Indicates their discovery as new compounds during the period 1975–1980.

Table 8.3 Naturally occurring di-*C*-glycosylflavonoids

Compounds	*Sources*	*References*
DI-*C*-GLYCOSYLFLAVONES		
Vicenin-2	Many sources	Chopin *et al.* (1969) (structure)
(5,7,4′-tri-OH 6,8-di Glc)		
6,8-Di-*C*-hexosylapigenin	*Plagiochila asplenioides* (Hepaticae)	Mues and Zinsmeister (1976)
(5,7,4′-tri-OH 6,8-di-Hex)	*Hedwigia ciliata* (Bryophyta)	Österdahl (1979a,b)
Vicenin-3	Many sources	Bouillant and Chopin (1971) (structure)
(5,7,4′-tri-OH 6-Glc-8-Xyl)		
Vicenin-1	*Cladrastis platycarpa* (Leguminosae)	Ohashi *et al.* (1977) (NMR)
(5,7,4′-tri-OH 6-Xyl-8-Glc)		
Violanthin	Gramineae spp.	Kaneta and Sugiyama (1973)
5,7,4′-tri-OH 6-Glc-8-Rha)	*Eleusine* sp. (Gramineae)	Hilu *et al.* (1978)
	Angiopteris evecta (Marattiales)	Wallace *et al.* (1979b)
*Isoviolanthin	*Angiopteris evecta* (Marattiales)	Wallace *et al.* (1979b)
(5,7,4′-tri-OH 6-Rha-8-Glc)		Biol and Chopin (1972) (structure)
*Schaftoside	*Silene schafta* (Caryophyllaceae)	Chopin *et al.* (1974) (structure)
(5,7,4′-tri-OH 6-Glc-8-Ara)	Many sources	
*Isoschaftoside	*Flourensia cernua* (Compositae)	Dillon *et al.* (1976)
(5,7,4′-tri-OH 6-Ara-8-Glc)	Many sources	
*Neoschaftoside	*Catananche cerulea* (Compositae)	Proliac *et al.* (1973)
(5,7,4′-tri-OH 6-Glc-8-*Ara*)	Many sources	
*Neoisoschaftoside	*Mnium undulatum* (Bryophyta)	Österdahl (1979a,c)
(5,7,4′-tri-OH 6-*Ara*-8-Glc)		
*Isocorymboside (compound S)	*Polygonatum multiflorum* (Liliaceae)	Chopin *et al.* (1977b)
(5,7,4′-tri-OH 6-Gal-8-Ara)		
*Corymboside	*Carlina corymbosa* (Compositae)	Besson *et al.* (1979b)
(5,7,4′-tri-OH 6-Ara-8-Gal)		
*6,8-Di-*C*-α-L-arabinopyranosylapigenin	*Melilotus alba* (Leguminosae)	Specht *et al.* (1976)
(5,7,4′-tri-OH 6,8-di-Ara)	*Hymenophyton leptopodum* (Hepaticae)	Markham *et al.* (1976b)
	Angiopteris sp. (Marattiales)	Wallace *et al.* (1981)

*6-*C*-Arabinosyl-8-*C*-pentosylapigenins	*Hymenophyton leptopodum* (Hepaticae)	Markham *et al.* (1976b)
(5,7,4′-tri-OH 6-Ara-8-Pen)	*Takakia lepidozioides* (Hepaticae)	Markham and Porter (1979)
	Mollugo pentaphylla (Molluginaceae)	Chopin *et al.* (1979)
	Lespedeza cuneata (Leguminosae)	Numata *et al.* (1979)
	Apometzgeria pubescens (Hepaticae)	Theodor *et al.* (1980)
*Almeidein	*Almeidea guyanensis* (Rutaceae)	Jay *et al.* (1979)
(5,4′-di-OH 7-OMe 6,8-di-Ara)		
*6-*C*-Pentosyl-8-*C*-hexosylacacetin	*Trigonella corniculata* (Leguminosae)	Bouillant *et al.* (1975) (MS)
(5,7-di-OH 4′-OMe 6-Pen-8-Hex)		
*7,4′-Di-*O*-methyl-6,8-di-*C*-α-L-	*Asterostigma reidelianum* (Araceae)	Markham and Williams (1980)
arabinopyranosylapigenin		
(5-OH 7,4′-di-OMe 6,8-di-Ara)		
Lucenin-2	*Hedwigia ciliata* (Bryophyta)	Österdahl (1978) (structure)
(5,7,3′,4′-tetra-OH 6,8-di Glc)	*Spergularia rubra* (Caryophyllaceae)	Bouillant *et al.* (1979) (structure)
	Many sources	
Lucenin-3	Many sources	Bouillant and Chopin (1972) (structure)
(5,7,3′,4′-tetra-OH 6-Glc-8-Xyl)		
Lucenin-1	Many sources	Bouillant and Chopin (1972) (structure)
(5,7,3′,4′-tetra-OH 6-Xyl-8-Glc)		
*Carlinoside	*Carlina vulgaris* (Compositae)	Raynaud and Rasolojaona (1976)
(5,7,3′,4′-tetra-OH 6-Glc-8-Ara)	*Catananche cerulea* (Compositae)	Proliac *et al.* (1973, 1977)
	Takakia lepidozioides (Hepaticae)	Markham and Porter (1979)
	Gibasis schiedeana (Commelinaceae)	Del Pero de Martinez and Swain (1976, 1977)
	Oryza sativa (Gramineae)	Besson and Fukami (unpublished)
*Neocarlinoside	*Oryza sativa* (Gramineae)	Besson and Fukami (unpublished)
(5,7,3′,4′-tetra-OH 6-Glc-8-*Ara*)		
*6-*C*-Arabinosyl-8-*C*-pentosyl-luteolin	*Takakia lepidozioides* (Hepaticae)	Markham and Porter (1979)
(5,7,3′,4′-tetra-OH 6-Ara-8-Pen)		
*6,8-Di-*C*-β-D-glucopyranosylchrysoeriol	*Stellaria holostea* (Caryophyllaceae)	Zoll and Nouvel (1974)
(5,7,4′-tri-OH 3′-OMe 6,8-di-Glc)	*Larrea tridentata* (Zygophyllaceae)	Sakakibara *et al.* (1977)
	Spergularia rubra (Caryophyllaceae)	Bouillant *et al.* (1979)
*6-*C*-Arabinosyl-8-*C*-hexosylchrysoeriol	*Mnium undulatum* (Bryophyta)	Österdahl (1979a,c)
(5,7,4′-tri-OH 3′-OMe 6-Ara-8-Hex)		

Table 8.3 (*Contd.*)

Compounds	*Sources*	*References*
6,8-Di-*C*-glycosyldiosmetin (5,7,3′-tri-OH 4′-OMe 6,8-di-Gly)	*Citrus limon* (Rutaceae)	Chopin *et al.* (1964) Gentili and Horowitz (1968)
*6,8-Di-*C*-β-D-glucopyranosyltricetin (5,7,3′,4′,5′-penta-OH 6,8-di-Glc)	*Plagiochila asplenioides* (Hepaticae) *Apometzgeria pubescens* (Hepaticae)	Mues and Zinsmeister (1976) Theodor *et al.* (1980)
*6,8-Di-*C*-hexopyranosyltricetin (compound 4)	*Plagiochila asplenioides* (Hepaticae)	Mues and Zinsmeister (1976)
*6-*C*-Hexopyranosyl-8-*C*-pentopyranosyl tricetins (compounds 2 and 3)	*Plagiochila asplenioides* (Hepaticae) *Takakia lepidozioides* (Hepaticae)	Mues and Zinsmeister (1976) Markham and Porter (1979)
*6-*C*-Arabinopyranosyl-8-*C*-glucopyranosyl tricetin (5,7,3′,4′,5′-penta-OH 6-Ara 8-Glc)	*Apometzgeria pubescens* (Hepaticae)	Theodor *et al.* (1980, 1981a)
*6,8-Di-*C*-pentosyltricetin (5,7,3′,4′,5′-penta-OH 6,8-di-Pen)	*Takakia lepidozioides* (Hepaticae)	Markham and Porter (1979)
*6-*C*-Hexopyranosyl-8-*C*-pentopyranosyl 3′-*O*-methyltricetin (compound 7) (5,7,4′,5′-tetra-OH 3′-OMe 6-Hex 8-Pen)	*Plagiochila asplenioides* (Hepaticae)	Mues and Zinsmeister (1976)
*6,8-Di-*C*-glucopyranosyltricin (5,7,4′-tri-OH 3′,5′-di-OMe 6,8-di-Glc)	*Apometzgeria pubescens* (Hepaticae)	Theodor *et al.* (1980, 1981a)
*6-*C*-glucopyranosyl 8-*C*-arabinopyranosyl tricin (5,7,4′-tri-OH 3′,5′-di-OMe 6-Glc 8-Ara)	*Apometzgeria pubescens* (Hepaticae)	Theodor *et al.* (1980, 1981a)
*6-*C*-arabinopyranosyl-8-*C*-glucopyrano syltricin (5,7,4′-tri-OH 3′,5′-di-OMe 6-Ara 8-Glc)	*Apometzgeria pubescens* (Hepaticae)	Theodor *et al.* (1980, 1981a)
*6-*C*-Xylosyl-8-*C*-hexosyltricin (5,7,4′-tri-OH 3′,5′-di-OMe 6-Xyl 8-Hex)	*Metzgeria leptoneura* (Hepaticae)	Theodor *et al.* (1981b)
*6,8-Di-*C*-arabinopyranosyltricin (5,7,4′-tri-OH 3′,5′-di-OMe 6,8-di-Ara)	*Apometzgeria pubescens* (Hepaticae)	Theodor *et al.* (1980, 1981a)
*6-*C*-Hexosyl-8-*C*-pentosylapometzgerin (5,7,5′-tri-OH 3′,4′-di-OMe 6-Hex-8-Pen)	*Apometzgeria pubescens* (Hepaticae)	Theodor *et al.* (1980, 1981a)

*6,8-Di-*C*-arabinopyranosylapometzgerin (5,7,5′-tri-OH 3′,4′-diOMe 6,8-di-Ara)	*Apometzgeria pubescens* (Hepaticae)	Theodor *et al.* (1980, 1981a)
DI-*C*-GLYCOSYLDIHYDROCHALCONES		
*3′,5′-Di-*C*-glycosylphloretin (2′,4′,6′,4-tetra-OH 3′,5′-di-Gly)	*Hymenophyton leptopodum* (Hepaticae)	Markham *et al.* (1976b)
DI-*C*-GLYCOSYLISOFLAVONES		
Paniculatin (5,7,4′-tri-OH 6,8-di-Glc)	*Dalbergia paniculata* (Leguminosae)	Narayanan and Seshadri (1971)

* Indicates their discovery as new compounds during the period 1975–1980.

Table 8.4 Naturally occurring *C*-glycosylflavonoid *O*-glycosides

Compounds	*Sources*	*References*
ISOVITEXIN		
7-*O*-Glucoside (saponarin)	Many sources	
*7-*O*-Galactoside (neosaponarin)	*Melandrium album* (Caryophyllaceae)	Wagner *et al.* (1979)
*7-*O*-Xyloside	*Melandrium album* (Caryophyllaceae)	Brederode and Nigtevecht (1972a)
4′-*O*-Glucoside (isosaponarin)	*Gentiana* sp. (Gentianaceae)	Hostettmann *et al.* (1975b)
*2″-*O*-Glucoside	*Oxalis acetosella* (Oxalidaceae)	Tschesche and Struckmeyer (1976)
	Gentiana asclepiadea (Gentianaceae)	Goetz *et al.* (1976a)
	Cucumis melo (Cucurbitaceae)	Monties *et al.* (1976)
	Melandrium album (Caryophyllaceae)	Besson *et al.* (1979a)
*2″-*O*-Xyloside	*Desmodium canadense* (Leguminosae)	Chernobrovaya (1973)
	Passiflora serratifolia (Passifloraceae)	Ulubelen and Mabry (1980)
*2″-*O*-Arabinoside	*Avena sativa* (Gramineae)	Chopin *et al.* (1977a) (structure)
	Melandrium album (Caryophyllaceae)	Besson *et al.* (1979a)
*2″-*O*-Rhamnoside	*Crataegus monogyna* (Rosaceae)	Nikolov *et al.* (1976)
	Melandrium album (Caryophyllaceae)	Besson *et al.* (1979a); Wagner *et al.* (1979)
	Securigera coronilla (Leguminosae)	Jay *et al.* (1980)
*6″-*O*-Arabinoside	*Swertia perennis* (Gentianaceae)	Hostettmann and Jacot-Guillarmod (1976)
*7,2″-Di-*O*-glucoside	*Melandrium album* (Caryophyllaceae)	Brederode and Nigtevecht (1974b)
	Cerastium arvense (Caryophyllaceae)	Dubois (1980); Dubois *et al.* (1980)
*7-*O*-Glucoside 2″-*O*-arabinoside	*Melandrium album* (Caryophyllaceae)	Brederode and Nigtevecht (1972a)
*7-*O*-Glucoside 2″-*O*-rhamnoside	*Melandrium album* (Caryophyllaceae)	Nigtevecht and Brederode (1972)
*7-*O*-Galactoside 2″-*O*-glucoside	*Melandrium album* (Caryophyllaceae)	Wagner *et al.* (1979)
*7-*O*-Galactoside 2″-*O*-rhamnoside	*Melandrium album* (Caryophyllaceae)	Wagner *et al.* (1979)
*7-*O*-Xyloside 2″-*O*-glucoside	*Melandrium album* (Caryophyllaceae)	Brederode and Nigtevecht (1974b)
*7-*O*-Xyloside 2″*O*-arabinoside	*Melandrium album* (Caryophyllaceae)	Brederode and Nigtevecht (1972a)
*7-*O*-Xyloside 2″-*O*-rhamnoside	*Melandrium album* (Caryophyllaceae)	Brederode and Nigtevecht (1972a)
*4′,2″-Di-*O*-glucoside	*Gentiana asclepiadea* (Gentianaceae)	Goetz and Jacot-Guillarmod (1977a)
*4′-*O*-Glucoside 2″-*O*-arabinoside	*Vaccaria segetalis* (Caryophyllaceae)	Baeva *et al.* (1974)

VITEXIN		
7-*O*-Glucoside	*Achillea fragrantissima* (Compositae)	Shalaby *et al.* (1965)
	Cymbidium sp. (Orchidaceae)	Williams (1979)
4′-*O*-Glucoside	*Briza media* (Gramineae)	Williams and Murray (1972)
*2″-*O*-Glucoside	*Polygonatum odoratum* (Liliaceae)	Morita *et al.* (1976)
	Cannabis sativa (Cannabinaceae)	Segelman *et al* (1978)
2″-*O*-Xyloside	*Citrus limonum* (Rutaceae)	Tomas *et al.* (1975)
6″-*O*-Xyloside	*Gypsophila paniculata* (Caryophyllaceae)	Darmograi *et al.* (1968)
*X″-*O*-Arabinoside	*Trichomanes venosum* (Hymenophyllaceae)	Markham and Wallace (1980)
*2″-*O*-Rhamnoside	*Crataegus oxyacantha* (Rosaceae)	Chopin *et al.* (1977a) (structure)
	Crataegus monogyna (Rosaceae)	Nikolov *et al.* (1976)
	Avena sativa (Gramineae)	Chopin *et al.* (1977a) (structure)
*6″-*O*-Rhamnoside	*Larix sibirica* (Pinaceae)	Medvedeva *et al.* (1974)
*7-*O*-Rutinoside	*Phoenix* sp. (Palmae)	Williams *et al.* (1973)
*2″-*O*-Sophoroside	*Polygonatum odoratum* (Liliaceae)	Morita *et al.* (1976)
*6″-*O*-Gentiobioside (marginatoside)	*Piper marginatum* (Piperaceae)	Tillequin *et al.* (1978)
6-*C*-XYLOSYLAPIGENIN (cerarvensin)		
*7-*O*-Glucoside	*Cerastium arvense* (Caryophyllaceae)	Dubois, (1980); Dubois *et al.* (1980)
2″-*O*-Rhamnoside	*Phlox drummondii* (Polemoniaceae)	Bouillant *et al.* (1978) (structure)
6-*C*-ARABINOSYLAPIGENIN (isomollupentin)		
*7,2″-Di-*O*-glucoside	*Spergularia rubra* (Caryophyllaceae)	Bouillant *et al.* (1979b)
3′-DEOXYDERHAMNOSYLMAYSIN		
*2″-*O*-Rhamnoside (3′-deoxymaysin)	*Zea mays* (Gramineae)	Elliger *et al* (1980)
SWERTISIN		
4′-*O*-Glucoside	*Commelina* sp. (Commelinaceae)	Komatsu *et al.* (1968)
*2″-*O*-Glucoside (spinosin)	*Zizyphus vulgaris* (Rhamnaceae)	Woo *et al.* (1979)
X″-*O*-Xyloside	*Iris tingitana* (Iridaceae)	Asen *et al.* (1970)
ISOSWERTISIN		
4′-*O*-Glucoside	*Triticum aestivum* (Gramineae)	Julian *et al.* (1971)
*2″-*O*-Rhamnoside	*Avena sativa* (Gramineae)	Chopin *et al.* (1977a) (structure)
	Mollugo distica (Molluginaceae)	Chopin *et al.* (1978)
MOLLUDISTIN		
*2″-*O*-Glucoside	*Almeidea guyanensis* (Rutaceae)	Jay *et al.* (1979)
*2″-*O*-Rhamnoside	*Mollugo distica* (Molluginaceae)	Chopin *et al.* (1978)

Table 8.4 (*Contd.*)

Compounds	*Sources*	*References*
ISOCYTISOSIDE		
*7-*O*-Glucoside	*Gentiana pyrenaica* (Gentianaceae)	Marston *et al.* (1976)
*2″-*O*-Glucoside	*Securigera coronilla* (Leguminosae)	Jay *et al.* (1980)
*2″-*O*-Rhamnoside (isomargariten)	*Fortunella margarita* (Rutaceae)	Horowitz *et al.* (1974)
CYTISOSIDE		
7-*O*-Glucoside	*Trema aspera* (Ulmaceae)	Oelrichs *et al.* (1968)
*2″-*O*-Rhamnoside (margariten)	*Fortunella margarita* (Rutaceae)	Horowitz *et al.* (1974)
EMBIGENIN		
2″-*O*-Rhamnoside (embinin)	*Iris japonica* (Iridaceae)	Arisawa *et al.* (1973)
7,4′-Di-*O*-METHYLISOMOLLUPENTIN		
*2″-*O*-Glucoside	*Asterostigma reidelianum* (Araceae)	Markham and Williams (1980)
ISO-ORIENTIN		
7-*O*-Glucoside (lutonarin)	Many sources	
*3′-*O*-Glucoside	*Gentiana nivalis* (Gentianaceae)	Hostettmann and Jacot-Guillarmod (1974a)
*3′-*O*-Glucuronide	*Rynchospora eximia* (Cyperaceae)	Williams and Harborne (1977a)
4′-*O*-Glucoside	*Gentiana* sp. (Gentianaceae)	Hostettmann *et al.* (1975b)
*2″-*O*-Glucoside	*Gentiana verna* (Gentianaceae)	Hostettmann and Jacot-Guillarmod (1975)
	Cucumis melo (Cucurbitaceae)	Monties *et al.* (1976)
*2″-*O*-Xyloside	*Desmodium canadense* (Leguminosae)	Chernobrovaya (1973)
*2″-*O*-Arabinoside	*Avena sativa* (Gramineae)	Chopin *et al.* (1977a) (structure)
*2″-*O*-β-L-Arabinofuranoside	*Trichomanes venosum* (Hymenophyllaceae)	Markham and Wallace (1980)
*2″-*O*-Rhamnoside	*Crataegus monogyna* (Rosaceae)	Nikolov *et al.* (1976)
	Securigera coronilla (Leguminosae)	Jay *et al.* (1980)
*6″-*O*-Arabinoside	*Swertia perennis* (Gentianaceae)	Hostettmann and Jacot-Guillarmod (1976)
*4′,2″-Di-*O*-glucoside	*Gentiana asclepiadea* (Gentianaceae)	Goetz and Jacot-Guillarmod (1977a)
7-*O*-Rutinoside	*Triticum aestivum* (Gramineae)	Julian *et al.* (1971)
ORIENTIN		
*7-*O*-Glucoside bisulphate	*Phoenix canariensis* (Palmae)	Harborne *et al.* (1974)

4′-*O*-Glucoside	*Briza media* (Gramineae)	Williams and Murray (1972)
*2″-*O*-Glucoside	*Cannabis sativa* (Cannabinaceae)	Segelman *et al.* (1978)
2″-*O*-Xyloside (adonivernith)	*Adonis vernalis* (Ranunculaceae)	Wagner *et al.* (1975) (structure)
*2″-*O*-β-L-Arabinofuranoside	*Trichomanes venosum* (Hymenophyllaceae)	Markham and Wallace (1980)
*2″-*O*-Rhamnoside	*Crataegus monogyna* (Rosaceae)	Nikolov *et al.* (1976)
7-*O*-Rhamnoside	*Linum usitatissimum* (Linaceae)	Ibrahim and Shaw (1970)
6-*C*-XYLOSYL-LUTEOLIN		
2″-*O*-Rhamnoside	*Phlox drummondii* (Polemoniaceae)	Bouillant *et al.* (1978) (structure)
SWERTIAJAPONIN		
*3′-*O*-Glucoside	*Phragmites australis* (Gramineae)	Nawwar *et al.* (1980)
*2″-*O*-Rhamnoside	*Securigera coronilla* (Leguminosae)	Jay *et al.* (1980)
*3′-*O*-Gentiobioside	*Phragmites australis* (Gramineae)	Nawwar *et al.* (1980)
ISOSCOPARIN		
7-*O*-Glucoside	*Gentiana pyrenaica* (Gentianaceae)	Marston *et al.* (1976)
SCOPARIN		
*X″-*O*-Rhamnosylglucoside	*Crocus reticulatus* (Iridaceae)	Sergeyeva (1977)
*Episcoparin-7-*O*-glucoside (Knautinoside)	*Knautia montana* (Dipsacaceae)	Zemtsova and Bandyukova (1974)
7,3′,4′-Tri-*O*-METHYLISO-ORIENTIN		
2″-*O*-Rhamnoside (linoside B)	*Linum maritimum* (Linaceae)	Chari *et al.* (1978)
ISOPYRENIN		
*7-*O*-Glucoside	*Gentiana pyrenaica* (Gentianaceae)	Marston *et al.* (1976)
6-*C*-GLUCOSYL-5,7-DIHYDROXY-8,3′,4′,5′-TETRAMETHOXYFLAVONE		
5-O-Rhamnoside	*Vitex negundo* (Verbenaceae)	Subramanian and Misra (1979)
PUERARIN		
O-Xyloside	*Pueraria thunbergiana* (Leguminosae)	Murakami *et al.* (1960)

* Indicates their discovery as new compounds during the period 1975–1980.

Table 8.5 Naturally occurring acyl-*C*-glycosylflavonoids

Compounds	*Sources*	*References*
ISOVITEXIN		
*2″-*O*-*trans*-Feruloyl		
*2″-*O*-*trans*-Feruloyl 4′-*O*-glucoside	*Gentiana punctata* (Gentianaceae)	Luong and Jacot-Guillarmod (1977)
*X″-*O*-*trans*-Caffeoyl 2″-*O*-glucoside	*Cucumis melo* (Cucurbitaceae)	Monties *et al.* (1976)
*7-Bisulphate	Palmae	Williams *et al.* (1973); Harborne *et al.* (1974)
VITEXIN		
2″-*O*-*p*-Hydroxybenzoyl	*Vitex lucens* (Verbenaceae)	Horowitz and Gentili (1966)
*2″-*O*-*p*-Coumaroyl	*Trigonella foenum-graecum* (Leguminosae)	Sood *et al.* (1976)
*4‴-*O*-acetyl 2″-*O*-rhamnoside	*Crataegus monogyna* (Rosaceae)	Nikolov *et al.* (1976)
*7-Bisulphate		
*7-*O*-Rutinoside bisulphate	Palmae	Williams *et al.* (1973); Harborne *et al.* (1974)
8-*C*-GALACTOSYLAPIGENIN		
*6″-*O*-Acetyl	*Briza media* (Gramineae)	Chari *et al.* (1980) (structure)
VICENIN-2		
*Feruloyl 6″-*O*-glucoside	*Stellaria holostea* (Caryophyllaceae)	Ferreres de Arce (1977)
*X‴-*O*-Diferuloylhexoside	*Spergularia rubra* (Caryophyllaceae)	Ferreres de Arce (1977)
VICENIN-1		
*Sinapoyl	*Triticum* (wheat germ) (Gramineae)	Galle (1974)
ISOSCHAFTOSIDE		
*Sinapoyl	*Triticum* (wheat germ) (Gramineae)	Galle (1974)
*Feruloyl	*Apometzgeria pubescens* (Hepaticae)	Theodor *et al.* (1980)
SCHAFTOSIDE		
*Feruloyl 6″-*O*-glucoside	*Stellaria holostea* (Caryophyllaceae)	Ferreres de Arce (1977)
ISOSWERTISIN		
*2″-*O*-Acetyl	*Brackenridgea zanguebarica* (Ochnaceae)	Bombardelli *et al.* (1974)
CYTISOSIDE		
O-Acetyl 7-glucoside	*Trema aspera* (Ulmaceae)	Oelrichs *et al.* (1968)
7,4′-DI-*O*-METHYLISOMOLLUPENTIN		
*2″-*O*-Caffeoylglucoside	*Asterostigma reidelianum* (Araceae)	Markham and Williams (1980)

ISO-ORIENTIN		
X″-*O*-Acetyl	*Crataegus monogyna* (Rosaceae)	Nikolov *et al.* (1976)
*2″-*O*-*p*-Hydroxybenzoyl	*Gentiana asclepiadea* (Gentianaceae)	Goetz and Jacot-Guillarmod (1978)
*2″-*O*-*trans*-*p*-Coumaroyl	*Gentiana X marcailhouana*	Luong *et al.* (1980)
*2″-*O*-*trans*-Caffeoyl	*Gentiana burseri*	Jacot-Guillarmod *et al.* (1975)
	Gentiana punctata	Luong and Jacot-Guillarmod (1977)
	Gentiana cruciata	Goetz *et al.* (1976)
	Gentiana X marcailhouana	Luong *et al.* (1980)
*2″-*O*-*trans*-Feruloyl	*Gentiana burseri*	Jacot-Guillarmod *et al.* (1975)
	Gentiana punctata	Luong and Jacot-Guillarmod (1977)
	Gentiana X marcailhouana	Luong *et al.* (1980)
*2″-*O*-*p*-Hydroxybenzoyl 4′-*O*-glucoside	*Gentiana asclepiadea*	Goetz and Jacot-Guillarmod (1978)
*2″-*O*-*trans*-Caffeoyl 4′-*O*-glucoside	*Gentiana punctata*	Luong and Jacot-Guillarmod (1977)
*2″-(*p*-*O*-Glucosyl)*trans*-caffeoyl 4′-*O*-Glucoside	*Gentiana burseri*	Jacot-Guillarmod *et al.* (1975)
*2″-*trans*-Feruloyl 4′-*O*-glucoside	*Gentiana burseri*	Jacot-Guillarmod *et al.* (1975)
	Gentiana punctata	Luong and Jacot-Guillarmod (1977)
*2″-*O*-*trans*-Caffeoylglucoside	*Cucumis melo* (Cucurbitaceae)	Monties *et al.* (1976)
*7-Bisulphate	Palmae	Williams *et al.* (1973); Harborne *et al.* (1974)
ORIENTIN		
*7-Bisulphate		
*7-*O*-Glucoside bisulphate	Palmae	Williams *et al.* (1973); Harborne *et al.* (1974)
SCOPARIN		
*6″-*O*-Acetyl	*Sarothamnus scoparius* (Leguminosae)	Brum-Bousquet *et al.* (1977)
7,3′,4′-TRI-*O*-METHYLISO-ORIENTIN		
6″-*O*-Acetyl 2″-*O*-rhamnoside (linoside A)	*Linum maritimum* (Linaceae)	Chari *et al.* (1978) (structure)
PUERARIN		
4′,6″-di-*O*-Acetyl	*Pueraria tuberosa* (Leguminosae)	Bhutani *et al.* (1969)

3′-*C*-glycosylisoliquiritigenin (a chalcone) and puerarin (an isoflavone) in which the A ring derives from resorcinol; 6-*C*-glucosyl-5,7-dihydroxy-8,3′,4′,5′-tetramethoxyflavanone and volubilinine (an isoflavone) in which the A ring derives from iretol (methoxyphloroglucinol). Most are *C*-β-D-glucopyranosylflavonoids, but *C*-β-D-xylopyranosyl, *C*-β-D-galactopyranosyl, *C*-α-L-arabinopyranosyl, *C*-α-L-rhamnopyranosyl and even *C*-6-deoxy-*xylo*-hexos-4-ulosylflavonoids are also known. These sugar residues are illustrated in Fig. 8.1

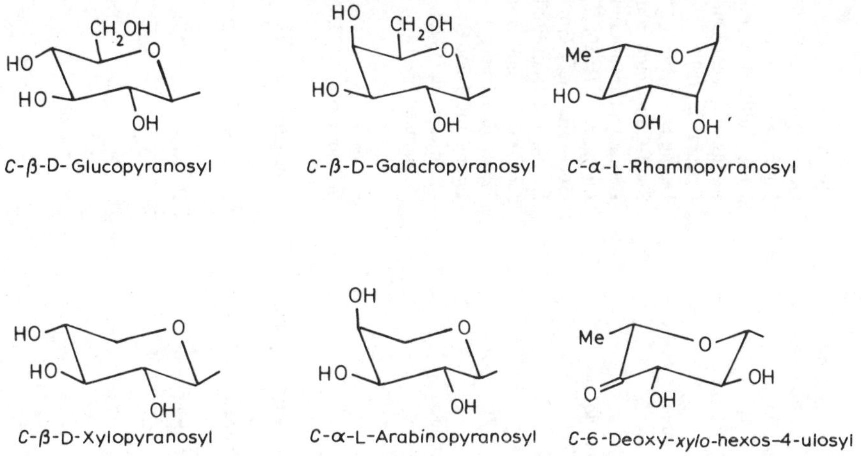

C-β-D-Glucopyranosyl *C*-β-D-Galactopyranosyl *C*-α-L-Rhamnopyranosyl

C-β-D-Xylopyranosyl *C*-α-L-Arabinopyranosyl *C*-6-Deoxy-*xylo*-hexos-4-ulosyl

Fig. 8.1 *C*-Glycosyl residues in natural *C*-glycosylflavonoids.

(a) Mono-*C*-glycosylflavones

Some Wessely–Moser isomers of already known natural *C*-glucosylflavones have now been found in plant extracts: isocytisoside (as *O*-glycosides), isoswertiajaponin, as well as a new di-*O*-methyl ether of iso-orientin, but the major advances in the last five years have been in the identification of new *C*-glycosyl residues and of new substitution types. New natural *C*-glycosyl residues have been found in 8-*C*-β-D-galactopyranosylapigenin from *Briza media* by Castledine and Harborne (1976), in 8-*C*-α-L-arabinopyranosylgenkwanin (molludistine) from *Mollugo distica* by Chopin *et al.* (1978) and in derhamnosylmaysin from *Zea mays* by Elliger *et al.* (1980). The structure of the sugar residue was assigned on the basis of ^{1}H-NMR spectra in the latter case, but on the basis of comparison with synthetic compounds in the former cases. New substitution types in the flavone moiety that have been reported are: 6-*C*-glucosyltricin (isopyrenin) from *Gentiana pyrenaica* by Marston *et al.* (1976), 6-*C*-glycosyltricetin (isoaffinetin) from *Polygonum affine* by Krause (1976a,b), and 8-*C*-glucosyltricetin (affinetin) from *Trichomanes venosum* by Markham and Wallace (1980). In addition 6-*C*-glucosyl-5,7-dihydroxy-8,3′,4′,5′-tetramethoxyflavone has been reported from *Vitex negundo* by Subramanian and Misra (1979), but in this case confirmation of the latter unusual structure by physical methods is still awaited.

Epiorientin and parkinsonins A and B have not been listed in Table 8.2 since a reinvestigation of *Parkinsonia aculeata* failed to reveal any such substances: instead orientin, vitexin, iso-orientin and isovitexin were isolated (Besson *et al.*, 1980). 6-*C*-Xylosylapigenin, already known as the *O*-rhamnoside, has been found in the free state in *Cerastium arvense* and named cerarvensin by Dubois (1980). *C*-Arabinosylflavones appear more widespread than *C*-xylosylflavones, since five examples of the former have already been found against two already known of the latter.

(b) Mono-*C*-glycosylflavonols and flavanonols

Reinvestigation of the extracts from *Zelkowa serrata* led Hayashi (1980) to isolate, beside the already known keyakinin and keyakinol, four new compounds which are considered to be 6-*C*-glucosylkaempferol, 6-*C*-glucosylquercetin, 6-*C*-glucosyldihydrokaempferol and 6-*C*-glucosyldihydroquercetin from their UV and ^{1}H-NMR spectra.

(c) Mono-*C*-glycosylflavanones

Two new compounds of this type have recently been identified, both with a resorcinol-derived A ring and with the *C*-galactosyl residue, in *Eysenhardtia polystachya* by Vita-Finzi *et al.* (1980) and named palodulcins A and B. Indeed the structure assigned to palodulcin A has also been published by Garg *et al.* (1980) for aervanone, isolated from *Aerva persica*.

(d) Mono-*C*-glycosylchalcones and dihydrochalcones

The first example of mono-*C*-glycosylchalcone, again with a resorcinol-derived A ring, was isolated by Ohashi *et al.* (1977a) from *Cladrastis platycarpa* where it co-occurs with the corresponding *C*-glucosylflavone bayin and the apigenin-derived di-*C*-glycosylflavone vicenin-1.

(e) Mono-*C*-glycosylisoflavones and isoflavanones

Six new *C*-glycosylisoflavones have been identified. Two of them, volubiline and isovolubiline from *Dalbergia volubilis* (Chawla *et al.*, 1974, 1975, 1976), are the first examples of mono-*C*-rhamnosylflavonoids. They co-occur with volubilinine, a *C*-glucosylisoflavone in which the A ring is derived from iretol. The three others are 8-*C*-glucosylisoflavones with a phloroglucinol-derived A ring, isolated from *Dalbergia paniculata* (Parthasarathy *et al.*, 1974, 1976) and *Lupinus luteus* (Zapesochnaya and Laman, 1977). Two *C*-glucosylisoflavanones have been described: dalpanin, an 8-*C*-glucoside isolated from *Dalbergia paniculata* by Adinarayana and Rao (1975) and 6-*C*-glucosyldalbergioidin isolated from *Pterocarpus macrocarpus* by Gupta *et al.* (1981), both with a phloroglucinol-derived A ring and a 2′,4′-di-*O*-substitution.

8.2.2 Di-*C*-glycosylflavonoids

Except for one di-*C*-glucosylisoflavone and one poorly characterized di-*C*-glycosyldihydrochalcone, all members of this group are di-*C*-glycosylflavones and the most impressive advances in the last five years have been in their recognition and identification. This has been mainly due to the development of mass spectrometry and TLC of permethyl ethers by Bouillant *et al.* (1975), as already mentioned. By such procedures, it has been shown that mixtures of di-*C*-glucosyl, *C*-xylosyl-*C*-glucosyl and *C*-arabinosyl-*C*-glucosylapigenins are very commonly encountered in plant extracts.

The four isomeric *C*-arabinosyl-*C*-glucosylapigenins which are produced from schaftoside by acid isomerization have all been isolated as natural products. Schaftoside, isolated from *Silene schafta*, was shown to be 6-*C*-glucopyranosyl-8-*C*-arabinopyranosylapigenin by Chopin and Wagner's groups (Chopin *et al.*, 1974) and its acid isomerization product, isoschaftoside, to be identical with 6-*C*-α-L-arabinopyranosyl-8-*C*-β-D-glucopyranosylapigenin synthesized by *C*-arabinopyranosylation of natural vitexin (Biol *et al.*, 1974). Isoschaftoside was later isolated from *Flourensia cernua* by Dillon *et al.* (1976). Another isomer, co-occurring with schaftoside in *Catananche cerulea* had previously been named isoschaftoside (Proliac *et al.*, 1973) but was renamed neoschaftoside after mass spectrometry of the permethyl ether showed it to be a 6-*C*-hexosyl-8-*C*-pentosylapigenin (Bouillant *et al.*, 1975). Its structure as 6-*C*-β-D-glucopyranosyl-8-*C*-β-L-arabinopyranosylapigenin was deduced from the study of synthetic 6-*C*-α and β-L-arabinopyranosyl and furanosylacacetins (Besson, unpublished). Neoisoschaftoside has recently been isolated from *Mnium undulatum* by Österdahl (1979c) and considered to be the Wessely–Moser isomer of neoschaftoside.

Compound S, isolated from *Polygonatum multiflorum*, was shown to be 6-*C*-β-D-galactopyranosyl-8-*C*-α-L-arabinopyranosylapigenin by Chopin *et al.* (1977b) and renamed isocorymboside after identification of corymboside, from *Carlina corymbosa*, as the Wessely–Moser isomer of compound S (Besson *et al.*, 1979b). Isoviolanthin, the Wessely–Moser isomer of natural violanthin, has been recently identified in *Angiopteris evecta* by Wallace *et al.* (1979). A new type of 6,8-di-*C*-hexosylapigenin different from vicenin-2 has been found in *Plagiochila asplenioides* and *Hedwigia ciliata*, but the sugar residues could not be identified. Synthetic 6,8-di-*C*-α-L-arabinopyranosylapigenin has been identified with natural products from several sources. The 7-*O*-methyl ether (almeidein) and the 7,4′-di-*O*-methyl ether were also found as natural compounds and their structure proved by permethylation.

One of the 6-*C*-arabinosyl-8-*C*-pentosylapigenins isolated from *Mollugo pentaphylla* (Chopin *et al.*, 1979), identical with the 6,8-di-*C*-pentosylapigenin from *Lespedeza cuneata* (Numata *et al.*, 1979), is likely 6-*C*-α-L-arabinopyranosyl-8-*C*-β-D-xylopyranosylapigenin (Besson, unpublished). The compound isolated from *Trigonella corniculata* and considered to be 6,8-di-*C*-

glucosylacacetin by Seshadri *et al.* (1972) has been shown to be a 6-*C*-pentosyl-8-*C*-hexosylacacetin by Bouillant *et al.* (1975).

6,8-Di-*C*-β-D-glucopyranosylluteolin (lucenin-2) has been unambiguously characterized in *Hedwigia ciliata* (Österdahl, 1978) and in *Spergularia rubra* (Bouillant *et al.*, 1979b), but lucenins have been found in many plants. Carlinoside, isolated from *Carlina vulgaris*, has been considered as 6-*C*-β-D-glucopyranosyl-8-*C*-α-L-arabinopyranosylluteolin by Raynaud and Rasolojaona (1976) mainly from the MS of its permethyl ether. This was indirectly confirmed by the co-occurrence of carlinoside and neocarlinoside in *Oryza sativa*, the sugar part of the ^{1}H-NMR spectrum of perdeuteriomethylneocarlinoside being identical with that of perdeuteriomethylneoschaftoside (Besson and Fukami, unpublished). One di-*C*-pentosylluteolin co-occurs with the corresponding di-*C*-pentosylapigenin and di-*C*-pentosyltricetin in *Takakia lepidozioides* (Markham and Porter, 1979). 6,8-Di-*C*-β-D-glucopyranosylchrysoeriol synthesized by *C*-glucosylation of natural scoparin (Bouillant, 1976) has been identified in several sources and a 6-*C*-arabinosyl-8-*C*-hexosylchrysoeriol found in *Mnium undulatum* (Österdahl, 1979c). The Hepaticae have been shown by Markham and Zinsmeister's groups to be a very rich source of di-*C*-glycosyltricetin derivatives, since not less than 14 compounds of this type were isolated from *Plagiochila asplenioides*, *Takakia lepidozioides*, *Apometzgeria pubescens*, *Metzgeria conjugata* and *Metzgeria leptoneura*. The flavone moieties were tricetin, 3′-*O*-methyltricetin, tricin and 3′,4′-di-*O*-methyltricetin (apometzgerin) and the identified sugar moieties glucose, arabinose and xylose.

8.2.3 *C*-Glycosylflavonoid *O*-glycosides

C-Glycosylflavones are often found in nature as *O*-glycosides in which the sugar is linked either to a phenolic hydroxyl group (X or X′-*O*-glycosides) or to an hydroxyl group of the glycosyl side chain (X″-*O*-glycosides or *C*-diglycosylflavones), but natural compounds are now known in which both linkages co-occur. Glycosylation of a phenolic hydroxyl group is usually detected and located by changes in UV spectral shifts after hydrolysis. No change occurs with X″-*O*-glycosides which show the same UV spectrum as the corresponding *C*-glycosylflavone, and identification of the glycosidic linkage raises difficult problems which have not found any general solution before the advent of ^{13}C-NMR spectroscopy (see Chapter 2), because mass spectrometry of permethyl ethers only solves the problem as far as 6-*C*-glycosylflavones are concerned (Bouillant *et al.*, 1979a). Moreover, it has been observed that the presence of a highfield acetyl signal in the ^{1}H-NMR spectrum of acetylated 6-*C*-glucosylflavone X″-*O*-glycosides is not diagnostic for the presence of a 2″-*O*-acetyl group (Woo *et al.*, 1979).

Except for the previously reported puerarin *O*-xyloside, all compounds listed in Table 8.4 are *C*-glycosylflavone *O*-glycosides, mainly derived from isovitexin,

vitexin, iso-orientin and orientin. Most of the X″-*O*-glycosides are 2″-and only a few are 6″-*O*-glycosides. No natural 3″- or 4″-*O*-glycoside has yet been described.

Establishment of the structure of vitexin *O*-rhamnoside isolated from various *Crataegus* species has been solved in different ways. The vitexin 2″-*O*-rhamnoside structure was independently determined by mass spectrometry of the permethyl ether (Chopin *et al.*, 1977a), by ^{1}H-NMR (Nikolov *et al.*, 1976) and ^{13}C-NMR (Weinges, 1977). Various isovitexin 7,2″-di-*O*-glycosides have been found in *Melandrium album* (= *Silene alba*) by van Brederode and van Nigtevecht who showed the glycosylation of isovitexin to be governed by a series of distinctive enzymes. Two of them respectively transfer glucose and xylose to the 7-hydroxyl group, three others transfer arabinose, rhamnose and glucose to the 2″-hydroxyl group of isovitexin. That galactose also can be transferred to the 7-hydroxyl group was shown later by Wagner *et al.* (1979). 4′,2″-Di-*O*-glucosides of isovitexin and iso-orientin have been isolated from *Gentiana asclepiadea* (Goetz and Jacot-Guillarmod, 1977a).

Two 8-*C*-triglycosylapigenins are known: polygonatiine (vitexin 2″-*O*-sophoroside) from *Polygonatum odoratum* (Morita *et al.*, 1976) and marginatoside (vitexin 6″-*O*-gentiobioside) from *Piper marginatum* (Tillequin *et al.*, 1978). The presence of 2″-*O*-β-L-arabinofuranosides of iso-orientin and orientin in *Trichomanes venosum* has been suggested by Markham and Wallace (1980) on the basis of ^{13}C-NMR data.

8.2.4 Acyl *C*-glycosylflavonoids

Many new *O*-acylated derivatives of mono- and di-*C*-glycosylflavones, *C*-glycosylflavone *O*-glycosides and *C*-diglycosylflavones have been reported during the last five years. *O*-Acetyl, *O*-*p*-hydroxybenzoyl, *O*-*p*-coumaroyl, *O*-feruloyl, *O*-caffeoyl, *O*-sinapoyl and bisulphate groups are linked either to phenolic hydroxyl groups or to sugar hydroxyl groups. In the latter case, location of the acyl group raises the same problem as that of glycosyl groups, and ^{13}C-NMR spectroscopy again affords a general solution which is only just beginning to be used in this field. In most of the compounds listed in Table 8.5, the highfield shift of 2″-*O*-acetyl in 6- and 8-*C*-glucosylflavones and of 6″-*O*-acetyl in 8-*C*-glucosylflavones was used to locate the acyl group in the 2″ or 6″ position. In some cases, the nature of acyl groups can be deduced from the molecular ion and fragmentation pattern when the acyl group survives permethylation, and the presence of a free hydroxyl group deduced from the molecular ion observed can give information about the position of the lost acyl group from the fragmentation pattern when saponification only occurs after permethylation. In 4‴-*O*-acetylvitexin 2″-*O*-rhamnoside isolated from various *Crataegus* species, the acetyl group is linked to the rhamnose as shown by ^{1}H-NMR (Nikolov *et al.*, 1976) and ^{13}C-NMR (Weinges, 1977). However, in linoside A, the acetyl group is linked to the *C*-glucosyl moiety as shown by ^{13}C-NMR (Chari *et al.*, 1978). Various 2″-*O*-acyl iso-orientins and their 4′-*O*-glucosides have been identified in *Gentiana* by Jacot-Guillarmod and co-

workers. The acyl group of feruloylisoschaftoside from *Apometzgeria pubescens* was first shown to be linked to the 8-*C*-glucosyl residue by mass spectrometry of the permethyl ether and later located at the 2″-position by ^{13}C-NMR (Theodor *et al.*, 1980, 1981b). The existence of feruloyl di-*C*-glycosylflavone *O*-glycosides has been demonstrated in *Stellaria holostea* and *Spergularia rubra* by mass spectrometry of permethyl ethers (Ferreres de Arce, 1977).

8.3 SYNTHESIS OF *C*-GLYCOSYLFLAVONOIDS

Following their successful total synthesis of 7,4′-di-*O*-methylbayin (8-*C*-β-D-glucopyranosyl-7-4′-dimethoxyflavone) (Eade *et al.*, 1975), Eade *et al.* (1978) used the intermediate 3′-β-D-glucopyranosyl-2′-hydroxy-4,4′-dimethoxychalcone for the first synthesis of a *C*-glucosylisoflavone: 7,4′-di-*O*-methylpuerarin (8-*C*-β-D-glucopyranosyl-7,4′-dimethoxyisoflavone). The product of the reaction of 2′-acetoxy-4,4′-dimethoxy-3′-(tetra-acetyl-β-D-glucopyranosyl)chalcone(7)with thallium(III) nitrate in methanol/trimethyl orthoformate solution gave, after acid hydrolysis, a high yield of 7,4′-di-*O*-methylpuerarin (10). From the reaction of the latter chalcone with thallium(III) nitrate in methanol, following the method of Farkas *et al.* (1974), the intermediates dimethyl acetal (8) and enol ether (9) have been isolated (Fig. 8.2).

A more general synthesis of *C*-glycosylflavonoids, exemplified by the synthesis of 5,7,4′-tri-*O*-methylvitexin (8-*C*-β-D-glucopyranosyl-5,7,4′-trimethoxyflavone) has been later provided by Eade and Pham (1979) (Fig. 8.2). The reaction between 1,3,5-trimethoxybenzene (1) and tetra-acetyl-α-D-glucosyl bromide (2) in the presence of zinc oxide gave β-D-glucopyranosyl-2,4,6-trimethoxybenzene tetra-acetate (3) in 70% yield. This tetra-acetate was converted by acetic anhydride and anhydrous aluminium chloride into 3-β-D-glucopyranosyl-2-hydroxy-4,6-dimethoxyacetophenone tetra-acetate (4). Condensation of the latter with 4-methoxybenzaldehyde in the presence of sodium hydroxide gave 3′-β-D-glucopyranosyl-2′-hydroxy-4,4′,6′-trimethoxychalcone (5) from which reaction with selenium dioxide gave 5,7,4′-tri-*O*-methylvitexin (6). A new era is thus opened up, since, from now on, synthesis in good yield of almost every type of 8-*C*-glycosylflavonoid methyl ether may be expected from the *C*-glycosylacetophenones derived from the corresponding glycosyl-2,4,6-trimethoxybenzenes.

In spite of the low yields of 6-*C*-glycosylflavones obtained in the reaction of acetobromosugars with 5,7-dihydroxyflavones, this method has been used in the authors' laboratory to prepare some standard compounds for mass spectral studies or to solve some specific problems. For example, 6-*C*-α-L-arabinopyranosylacacetin and 6-*C*-α-L-arabinofuranosylacacetin were prepared from acacetin and the corresponding acetobromo-L-arabinoses in order to study their acid isomerization products and to compare their chromatographic and spectral properties with the possibly naturally occurring *C*-arabinofuranosylflavones

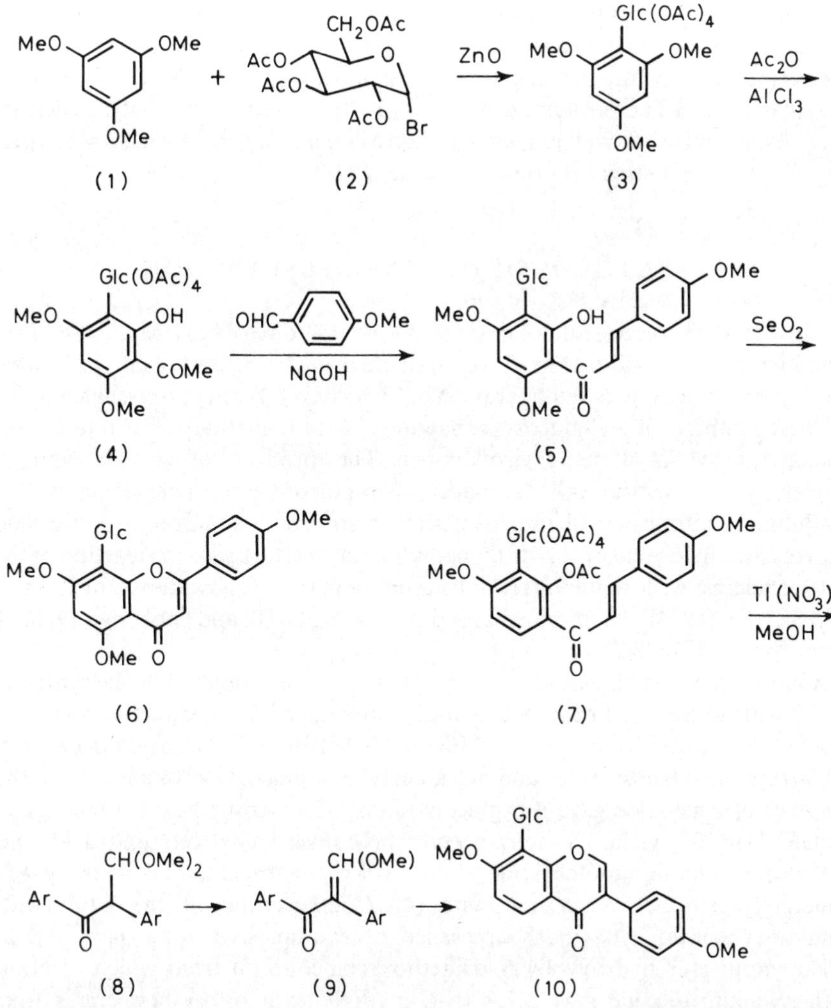

MeO OMe OMe (1) + CH_2OAc AcO AcO O AcO Br (2) $\xrightarrow{ZnO}$ $Glc(OAc)_4$ MeO OMe OMe (3) $\xrightarrow[AlCl_3]{Ac_2O}$

$Glc(OAc)_4$ MeO OH COMe OMe (4) $\xrightarrow[NaOH]{OHC-C_6H_4-OMe}$ Glc MeO OH OMe C O OMe (5) $\xrightarrow{SeO_2}$

Glc MeO O OMe OMe O (6)

$Glc(OAc)_4$ MeO OAc C O OMe (7) $\xrightarrow[MeOH]{Tl(NO_3)_3}$

$CH(OMe)_2$ Ar Ar O (8) → CH(OMe) Ar Ar O (9) → Glc MeO O O OMe (10)

Fig. 8.2 Synthesis of 5,7,4′-tri-*O*-methylvitexin and 7,4′-di-*O*-methylpuerarin.

(Besson, 1977). *C*-Glycosylation of 5,7-dihydroxy-3′,4′,5′-trimethoxyflavone has been achieved with acetobromoglucose and acetobromoarabinose following the discovery of natural *C*-glycosyltricetins and *C*-glycosyltricins, and 6,8-di-*C*-glycosylation was observed in both cases (Lardy, 1981). Similarly 6,8-di-*C*-α-L-arabinopyranosylacacetin was obtained as a byproduct in the previously mentioned *C*-arabinosylation of acacetin, 6,8-di-*C*-β-D-xylopyranosylacacetin and 6,8-di-*C*-β-D-xylopyranosylapigenin in the *C*-xylosylation of acacetin and

apigenin respectively, 6,8-di-*C*-α-L-rhamnopyranosylapigenin in the *C*-rhamnosylation of apigenin (Bouillant, 1976; Besson, 1977). 6,8-Di-*C*-β-D-glucopyranosylchrysoeriol was prepared by reaction of acetobromoglucose with natural scoparin (8-*C*-β-D-glucopyranosylchrysoeriol) for comparison with the natural compound (Bouillant, 1976).

Following the isolation of 7-*O*-, 6-*C*- and 8-*C*-β-neohesperidosylacacetins from *Fortunella margarita* (Horowitz *et al.*, 1974), *C*-neohesperidosylation of acacetin was successfully accomplished with the acetobromo derivative of β-neohesperidose (α-L-rhamnopyranosyl-1 → 2-β-D-glucopyranose) which provided confirmation of the proposed structure (Chopin *et al.*, 1977c; Besset, 1977). Then, because of the difficulty of locating the position of the *O*-linked sugar in 6-*C*-glycosylflavone X″-*O*-glycosides and with the hope of solving this problem by mass spectrometry of permethyl ethers, synthesis of 3″-, 4″- and 6″-*O*-glycosides was undertaken. Reaction of acacetin with the acetobromo derivatives of rungiose (α-L-rhamnopyranosyl-1 → 3-D-glucopyranose), cellobiose (β-D-glucopyranosyl-1 → 4-D-glucopyranose) and rutinose (α-L-rhamnopyranosyl-1 → 6-D-glucopyranose) respectively led to the corresponding 6-*C*-diglycosylacacetins, i.e. the 3″-*O*-rhamnoside, 4″-*O*-glucoside and 6″-*O*-rhamnoside of isocytisoside (6-*C*-glucosylacacetin) (Besset, 1977; Bouillant *et al.*, 1980). 6-*C*-Cellobiosyl 3′,4′-di-*O*-methyl-luteolin was synthesized in the same way (Besset, 1977). These new compounds are listed in Table 8.6.

Table 8.6 New synthetic *C*-glycosylflavones from flavone *C*-glycosylation

Compound	Reference
6-C-Glycosylflavones	
6-*C*-α-L-Arabinopyranosylacacetin	Besson (1977)
6-*C*-α-L-Arabinofuranosylacacetin	Besson (1977)
6-*C*-β-D-Glucopyranosyl-4′-*O*-methyltricin	Lardy (1981)
6-*C*-α-L-Arabinopyranosyl-4′-*O*-methyltricin	Lardy (1981)
6,8-Di-C-glycosylflavones	
6,8-Di-*C*-β-D-glucopyranosylchrysoeriol	Bouillant (1976)
6,8-Di-*C*-α-L-arabinopyranosylacacetin	Besson (1977)
6,8-Di-*C*-β-D-xylopyranosylapigenin	Bouillant (1976)
6,8-Di-*C*-β-D-xylopyranosylacacetin	Besson (1977)
6,8-Di-*C*-α-L-rhamnopyranosylapigenin	Bouillant (1976)
6,8-Di-*C*-β-D-glucopyranosyl-4′-*O*-methyltricin	Lardy (1981)
6,8-Di-*C*-α-L-arabinopyranosyl-4′-*O*-methyltricin	Lardy (1981)
6-C-Diglycosylflavones	
6-*C*-Neohesperidosylacacetin	Besset (1977); Chopin *et al.* (1977c)
6-*C*-Rungiosylacacetin	Bouillant *et al.* (1980)
6-*C*-Cellobiosylacacetin	Besset (1977); Bouillant *et al.* (1980)
6-*C*-Rutinosylacacetin	Besset (1977); Bouillant *et al.* (1980)
6-*C*-Cellobiosyl-3′,4′-di-*O*-methyl-luteolin	Besset (1977)

8.4 IDENTIFICATION OF *C*-GLYCOSYLFLAVONOIDS

8.4.1 Chemical methods

Acid treatment of *C*-glycosylflavonoids is always used for their characterization. Isomerization cannot be used unequivocally for distinguishing asymmetrical from symmetrical (no isomerization) 6,8-di-*C*-glycosylflavones, since 6,8-di-*C*-α-L-arabinopyranosylapigenin gives rise to several isomers under these conditions. Sugar ring isomerization in *C*-arabinosylflavones has been demonstrated by Besson and Chopin (1980) with synthetic 6-*C*-α-L-arabinopyranosyl- and 6-*C*-α-L-arabinofuranosyl-acacetins and with natural 8-*C*-α-L-arabinopyranosyl-genkwanin (molludistine). Sugar ring isomerization of the *C*-arabinosyl moiety is faster than Wessely–Moser isomerization of the flavone moiety when *O*-methyl groups are present in the 7 or 4′ positions. All four 6-*C*-α- and β-L-arabino-pyranosyl and furanosylacacetins can be isolated from the acid isomerization of 6-*C*-α-L-arabinopyranosylacacetin, whereas 8-*C*-α-L-arabinopyranosylgenk-wanin mainly gives a mixture of the two 8-*C*-α isomers under similar conditions. Even after methylation of the phenolic hydroxyl groups, periodate oxidation of spinosin (swertisin 2″-*O*-glucoside) led to abnormal results, 1.9 mol of periodate being consumed and di-*O*-methylswertisin being identified after hydrolysis of the oxidation product erroneously suggesting 3″-*O*-glycosylation (Woo *et al.*, 1979). Ferric chloride oxidation of about 25 mg of *C*-glycosyl-flavonoid is still widely used to identify the carbon-linked sugar residue, e.g. rhamnose in volubilin and isovolubilin (Chawla *et al.*, 1974, 1975). However, glucose and arabinose, but no trace of xylose were formed from the $FeCl_3$ oxidation of vicenin-1 (6-*C*-xylosyl-8-*C*-glucosylapigenin) by Ohashi *et al.* (1977b). In parallel experiments, glucose and arabinose were obtained from schaftoside (6-*C*-glucosyl-8-*C*-arabinosylapigenin), but no sugar could be detected from compound S (6-*C*-galactosyl-8-*C*-arabinosylapigenin) by Chopin *et al.* (1977b). However, $FeCl_3$ oxidation of 3 g of aervanone (8-*C*-galactosyl 7,4′-dihydroxyflavanone) led Garg *et al.* (1980) to 7,8,4′-trihydroxyflavanone in good yield and PC-detectable amounts of galactose and lyxose. This destructive method is now obsolete, since the same amount of sample can be more profitably employed for non-destructive ^{13}C-NMR spectroscopy. Isopropylidenation of *C*-glycosylflavone *O*-glycosides was used by Wagner's group for structural studies. Refluxing with acetone in the presence of dry $CuSO_4$ afforded mono- and di-isopropylidene derivatives from spinosin (a 6-*C*-glucosylflavone 2″-*O*-glucoside) (Woo *et al.*, 1979). Isopropylidenation of glucose residues was similarly observed on prolonged stirring of 2″-*O*-glucosylisovitexin-7-*O*-galactoside with acetone at room temperature in the presence of *p*-toluene sulphonic acid (TsOH) (Wagner *et al.*, 1979). However, isopropylidenation with 2,2-dimethoxypropane and 6M-HCl–dioxan in anhydrous dimethylformamide (DMF) according to Jarman and Ross (1969) was found to be specific for *C*-galactopyranosyl and *C*-arabinopyranosyl residues in mono- and di-*C*-glycosylflavones (Besson, 1977; Chopin *et al.*, 1977b).

8.4.2 Physical methods

(a) ^{1}H-NMR spectrometry

Proton-NMR spectrometry has been widely used to assign the 6 or 8 position to the glycosyl residue in mono-*C*-glycosylflavonoids, from the chemical shift difference exhibited either by H-6 and H-8 protons in 5,7-dihydroxyflavones or by the 2″-acetyl protons in 6- and 8-*C*-glucosylflavone acetates (Gentili and Horowitz, 1968). Absence of a highfield acetyl signal ($\delta \leqslant 1.83$) in the spectrum of the acetylation products of natural acyl *C*-glycosylflavonoids and *O*-glycosides was for a long time the only available test for 2″-substitution. However, the presence of such a signal cannot be used as diagnostic for a 2″-*O*-acetyl, since it was observed in the spectrum of peracetyl 6-*C*-glucosylflavone 2″-*O*-glycosides. In such cases, 2″-*O*-glycosylation was established by mass spectrometry of permethyl ethers or by permethylation followed by hydrolysis and acetylation of the resulting monohydroxypermethyl 6-*C*-glucosylflavone; in the ^{1}H-NMR spectrum of the latter, the acetyl signal was found at high field and shown to be 2″-acetyl by spin decoupling (Tschesche and Struckmeyer, 1976; Woo *et al.*, 1979).

The C-1″ configuration of the glycosyl residue was usually deduced from the coupling constant of H-1″ in the NMR spectra of trimethylsilyl or acetyl derivatives. In the authors' laboratory, perdeuteriomethyl ethers have been used for this purpose, anomeric protons being well separated from other sugar protons, so that one could observe the splitting of the anomeric protons in perdeuteriomethyl (PDM) 6,8-di-*C*-glycosylflavones (Besson *et al.*, 1979b). The structure and configuration of the unusual *C*-glycosyl residue of maysin could be deduced from ^{1}H-NMR spectra of derhamnosylmaysin (Elliger *et al.*, 1980).

(b) ^{13}C-NMR spectrometry

The first published application of ^{13}C-NMR spectrometry in the field of *C*-glycosylflavonoids was the determination of the position of arabinose in iso-orientin 6″-*O*-arabinoside isolated from *Swertia perennis* (Hostettmann and Jacot-Guillarmod, 1976), but the correct assignments for the *C*-glucosyl moiety were given by Chari *et al.* (1978) and used for the structure determination of linosides A and B, spinosin and *Melandrium album* glycosides, all of which are 6-*C*-glucosylflavone 2″-*O*-glycosides. It soon became evident from the comparison of ^{13}C-NMR spectra of natural 6- and 8-*C*-glucosylflavones that the glucose signals remained the same in all cases. This was used by Chopin *et al.* (1977b) to confirm the 6-*C*-galactosyl-8-*C*-arabinosylapigenin structure of compound S isolated from *Polygonatum multiflorum*, by comparing the sugar signals of schaftoside (6-*C*-glucosyl-8-*C*-arabinosylapigenin), iso-orientin (6-*C*-glucosyl-luteolin), molludistin (8-*C*-arabinosylgenkwanin) and 8-*C*-galactosylapigenin with those of compound S, without any assignment of the individual signals. The same technique was used in the structure determination of corymboside (6-*C*-arabinosyl-8-*C*-galactosylapigenin) by Besson *et al.* (1979b). ^{13}C-NMR was also used by

Österdahl (1978) to identify vicenin-2 and lucenin-2 in *Hedwigia ciliata* by comparison of their spectra with those of vitexin, isovitexin, saponarin, iso-orientin and apigenin. The signals of the *C*-galactosyl residue have been obtained from 6″-*O*-acetyl-8-*C*-β-D-galactopyranosylapigenin isolated from *Briza media* (Chari *et al.*, 1980), those of the *C*-xylosyl residue from vicenin-1 (6-*C*-xylosyl-8-*C*-glucosylapigenin) isolated from *Cladrastis platycarpa* (Chopin, unpublished) and those of the *C*-rhamnosyl residue from violanthin (6-*C*-glucosyl-8-*C*-rhamnosylapigenin) isolated from *Viola tricolor* (Chari, 1980). Each *C*-glycosyl residue shows a characteristic spectrum, so that their identification in any given *C*-glycosylflavonoid is now very easy (see Chapter 2).

(c) **Mass spectrometry**

The main advance in this field has been the use of permethyl (PM) or perdeuteriomethyl (PDM) ethers which provide: (1) molecular peaks from mono- and di-*C*-glycosylflavones and their mono- and di-*O*-glycosides; (2) an unambiguous distinction between 6- and 8-*C*-glycosylflavones from the striking difference in their fragmentation patterns; (3) identification of the 6-*C*-glycosyl residue in asymmetrical 6,8-di-*C*-glycosylflavones resulting from the preferential fragmentation of the 6-residue; (4) determination of the position of *O*-glycosidic bonds in 6-*C*-glycosylflavone *O*-glycosides; and (5) characterization of 6-*C*-glycosylflavone 7,2″-di-*O*-glycosides as shown by Bouillant *et al.* (1975, 1978, 1979a,b). Permethylation is carried out on less than 1 mg of *C*-glycosylflavone using MeI and NaH in DMF under anhydrous conditions in a N_2 atmosphere and the crude product purified by TLC on silica gel in $CHCl_3$—EtOAc—Me_2CO(5:4:1) or (5:1:4). In this way, an excellent reproducibility of mass spectra (MS) can be obtained and conclusions can safely be drawn from the relative intensities of ions. Finally the fragmentation pattern (Fig. 8.3) of PM 6-*C*-glycosylflavones can be deduced from the MS of PM derivatives mono-*O*-deuteriomethylated in the 2″-, 3″-, 4″- or 6″- positions (Bouillant *et al.*, 1980), prepared from the corresponding synthetic *O*″-glycosyl-6-*C*-glucosylflavones (Besset, 1977). Application of MS of the PM ethers to the structural study of *C*-glycosylflavones is described below for each type of compound. Main fragments in MS of PM *C*-glycosylflavones are listed in Table 8.7.

(*i*) *PM mono-C-glycosylflavones*

PM 8-C-glycosylflavones At m/z values higher than 150, the MS of PM 5,7-dihydroxy-8-*C*-glycosylflavones, as shown by the spectrum of PM vitexin (Fig. 8.4), is dominated by two major peaks: the molecular ion and the ion *i* in which only a CHOMe group remains of the *C*-glycosyl moiety, with transfer of a sugar methoxyl group to C-1″. The nature of the sugar moiety can be deduced from the difference between the m/z values of *M* and *i* (hexose: 175, deoxyhexose: 145, pentose: 131) and the nature of the flavone moiety from *i* itself (apigenin: 355,

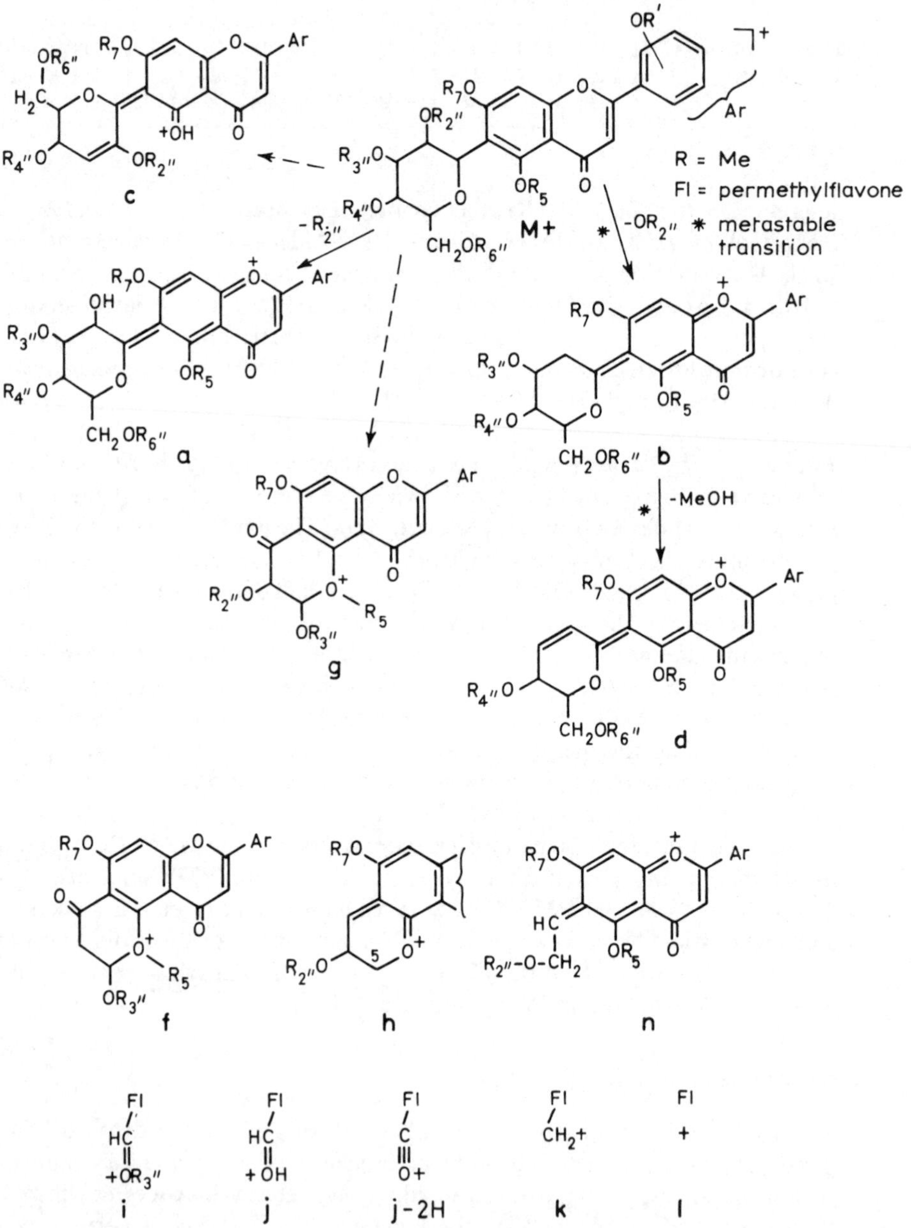

R = Me
Fl = permethylflavone
* metastable transition

M+ → a (−R2″); M+ → b (* −OR2″); b → d (* −MeOH); M+ ⇢ c; M+ ⇢ g

c a b g d f h n i j j-2H k l

Fig. 8.3 Fragmentation pattern of 6-C-glycosylflavones.

Table 8.7 Main fragments in MS of permethyl *C*-glycosylflavones

Common fragments:	$a(M-15)$		$b(M-31)$		$c(M-47)$		$d(M-63)$	
Sugar specific fragments:								
	f	g	h	i	j	k	l	n
Hexose	$M-133$	$M-103$	$M-163$	$M-175$	$M-189$	$M-205$	$M-219$	$M-161$
Deoxyhexose	$M-103$	$M-73$	$M-133$	$M-145$	$M-159$	$M-175$	$M-189$	$M-131$
Pentose	$M-89$	$M-59$	$M-119$	$M-131$	$M-145$	$M-161$	$M-175$	$M-117$

luteolin: 385, tricetin: 415). The other characteristic peaks are $M-14$ and $M-29$ (very weak), *n*, *j*, *k* and *l* (Fig. 8.3). Retro-Diels-Alder (RDA) fragments are very weak. The transition $M \rightarrow i$ is indicated by a metastable ion. The influence of the 5-hydroxyl group is shown by the MS of PM bayin (PM 5-deoxyvitexin) in which the importance of *i*, *j*, *k*, *l*, *n* remains the same, whereas the molecular peak is considerably lowered and new peaks ($M-74$, $M-130$, $M-145$) appear between *M* and *i*, $M-15$ replacing $M-14$ and $M-29$.

PM 6-C-glycosylflavones Again at m/z values higher than 150, the MS of PM 5,7-dihydroxy-6-*C*-glycosylflavones, as shown by the spectrum of PM isovitexin (Fig. 8.4), is clearly different from the preceding ones. The ion *i* remains a major peak, but the molecular ion is weaker (20–40)%) and is followed by important new peaks: *a* ($M-15$), *b* ($M-31$), *c* ($M-47$) independent of the sugar moiety and *g*, *f*, *h* characteristic of the sugar moiety. The relative importance of *i*, *j*, *k*, *l* and *n* remains the same as in 8-*C*-isomers. The transitions $M \rightarrow b$ and $b \rightarrow b$-MeOH are indicated by metastable ions. *b* is usually lower than *i*, except for 6-*C*-xylosylflavones. Another difference between *C*-xylosyl- and *C*-arabinosyl-flavones is that the permethylation of 6-*C*-arabinosylflavones gives, as a byproduct, a compound of which the molecular peak appears at 28 m.u. above that of the PM flavone (6-formyl PM flavone?).

The same fragmentation pattern has been observed in the MS of PM 6-*C*-α and β-L-arabinopyranosyl- and furanosyl-acacetins (Besson, 1977). Some peaks (*M*, $M-47$, $M-79$, $M-117$, $M-119$) are much higher in the furanosyl isomers and a characteristic difference was found in the metastable ions which showed the transitions $M \rightarrow M-15 \rightarrow M-47 \rightarrow M-79$ in the furanosyl isomers instead of $M \rightarrow M-31 \rightarrow M-63$ in the pyranosyl isomers.

(*ii*) *PM 6,8-di-C-glycosylflavones*

Symmetrical 6,8-di-C-glycosylflavones The MS of PM 5,7-dihydroxy-6,8-di-*C*-glycosylflavones bearing the same sugar in positions 6 and 8 is closely similar to that of the corresponding PM 5,7-dihydroxy-6-*C*-glycosylflavones, as shown by the spectrum of PM 6,8-di-*C*-glucosylapigenin (Fig. 8.4). The important molecular ion is followed by all 6-*C*-glycosylflavone peaks and only relative intensities

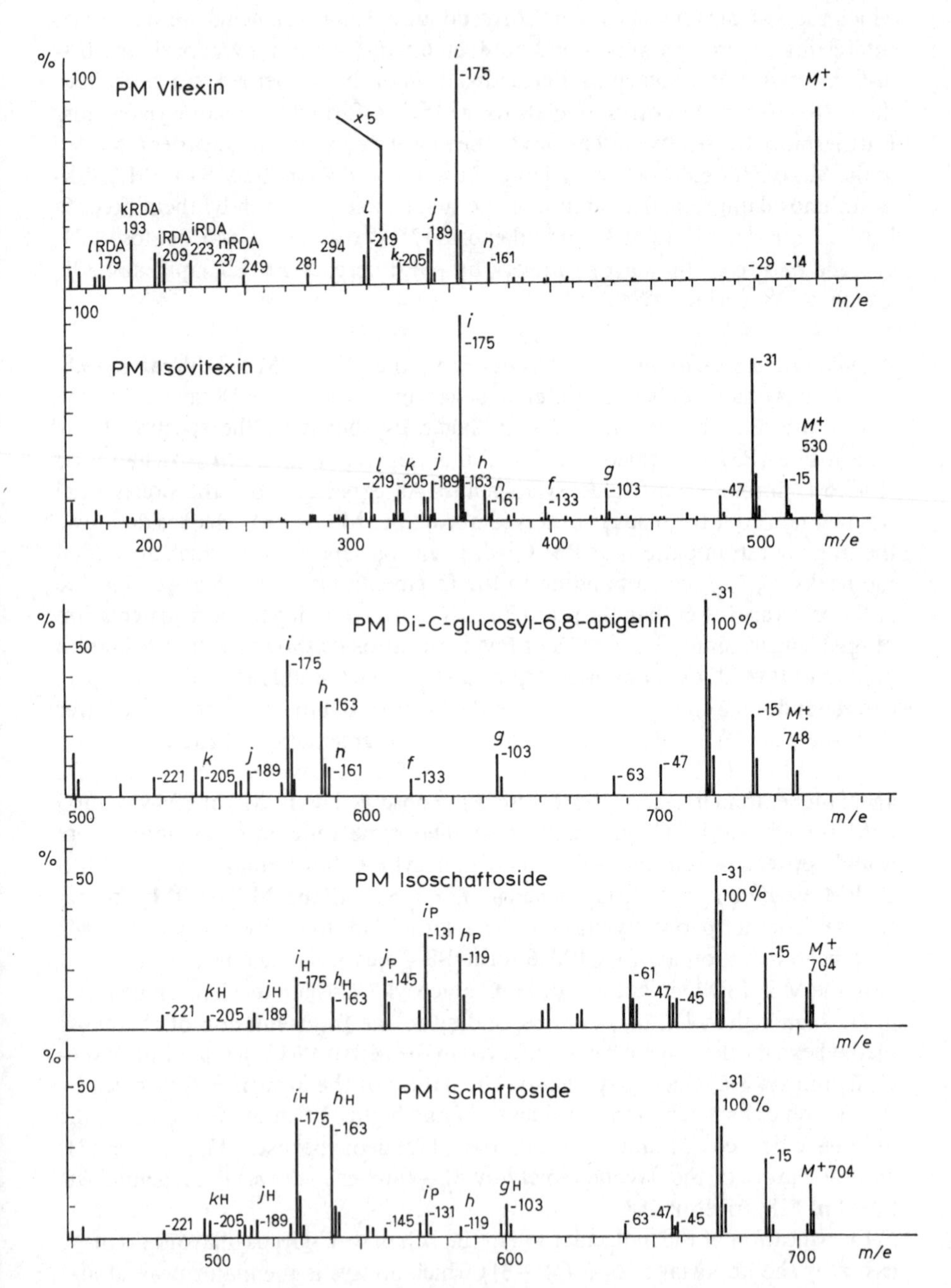

Fig. 8.4 MS of permethyl mono- and di-*C*-glycosylflavones.

are modified, b [$(M-31)$, now the major peak] and h being increased and i decreased. This implies the same fragmentation of the 6-sugar giving ions in which the 8-sugar remains intact. Corresponding RDA fragments are very weak. Interesting differences were found between 6,8-di-*C*-arabinopyranosyl- and 6,8-di-*C*-xylopyranosyl-apigenins. Permethylation of the former led to a byproduct, the MS of which corresponded to a PM 6-formyl-8-*C*-pentosylapigenin. Furthermore the relative intensities of the ions h, i and j are in the order $h > i > j$ in the MS of PM 6,8-di-*C*-xylosylapigenin and $i > h > j$ in the MS of PM 6,8-di-*C*-arabinosylapigenin. The nature of the sugar moiety is given by the difference between the m/z values of M and i (hexose: 175, deoxyhexose: 145, pentose: 131) and the nature of the flavone moiety by the difference $M-2i$ (apigenin: 398, luteolin: 428, tricetin: 458).

Asymmetrical 6,8-di-C-glycosylflavones From the MS of PM 5,7-dihydroxy-6,8-di-*C*-glycosylflavones bearing different sugars in positions 6 and 8 can be deduced the nature and the position of each sugar, as shown by the spectra of PM schaftoside (PM 6-*C*-glucosyl-8-*C*-arabinosylapigenin) and PM isoschaftoside (PM 6-*C*-arabinosyl-8-*C*-glucosylapigenin). As expected from the preferential fragmentation of the 6-sugar observed above, the MS of PM schaftoside shows the fragmentation pattern of PM 6,8-di-*C*-glucosylapigenin to which are added the peaks h_p, i_p, j_p corresponding to the fragmentation of the 8-sugar, but the latter peaks are lower than the peaks h_H, i_H, j_H corresponding to the fragmentation of the 6-sugar. Similarly, the MS of PM isoschaftoside shows the fragmentation pattern of PM 6,8-di-*C*-arabinosylapigenin to which are added the peaks h_H, i_H, j_H corresponding to the fragmentation of the 8-sugar, but the latter peaks are lower than the peaks h_p, i_p, j_p corresponding to the fragmentation of the 6-sugar.

In the MS of PM vicenin-1 (PM 6-*C*-xylosyl-8-*C*-glucosylapigenin), h_p, i_p, j_p are again higher than h_H, i_H, j_H, the reverse being observed in the MS of PM vicenin-3 (PM 6-*C*-glucosyl-8-*C*-xylosylapigenin). The same differences as above were found between 6-*C*-arabinosyl- and 6-*C*-xylosyl-8-*C*-hexosylflavones. In the MS of PM vicenin-1, $h_p > i_p > j_p$, whereas $i_p > h_p > j_p$ in the MS of PM isoschaftoside. Again, permethylation of the latter led to a byproduct, the MS of which corresponded to a PM 6-formyl-8-*C*-hexosylapigenin.

In the MS of PM violanthin (PM 6-*C*-glucosyl-8-*C*-rhamnosylapigenin) h_H, i_H, j_H are higher than h_D, i_D, j_D corresponding to the fragmentation of rhamnose (deoxyhexose), the reverse being observed in the MS of PM isoviolanthin (PM 6-*C*-rhamnosyl-8-*C*-glucosylapigenin). The nature of the 6-sugar is thus given by the difference between the m/z values of M and higher i, that of the 8-sugar by the difference between M and lower i (hexose: 175, deoxyhexose: 145, pentose: 131) and the nature of the flavone moiety by $M-(\text{higher } i + \text{lower } i)$ (apigenin: 398, luteolin: 428, tricetin: 458).

Identification of the molecular ion in the MS of di-*C*-glycosylflavones is made easier by the importance of b $(M-31)$ which always is the major peak at m/z values higher than 150 and by the characteristic peak system M, $M-15$, $M-31$, $M-47$.

(iii) PM isopropylidene C-glycosylflavones

When isopropylidenation was carried out using the conditions described by Jarman and Ross (1969), only *C*-arabinopyranosyl- and *C*-galactopyranosylflavones led to isopropylidene derivatives (Besson, 1977; Chopin *et al.*, 1977b). The reaction being usually incomplete, partial isopropylidenation may take place. For example, isopropylidenation of a natural 6-*C*-galactopyranosyl-8-*C*-arabinopyranosylflavone followed by permethylation of the reaction mixture led to four products separated by silica-gel TLC, which could be identified by their MS as the PM derivatives of the starting compound, two monoisopropylidene and one di-isopropylidene derivative. All of them showed the $M-15$ and $M-31$ peaks originating from the 2-methoxyl group of the permethyl 6-*C*-glycosyl residue and the fragmentation pattern, governed by the 6-*C*-glycosyl residue, remained the same as that of the PM starting compound when the isopropylidene group was attached to the 8-*C*-glycosyl residue only.

(iv) PM 6-C-glycosylflavone O-glycosides

PM 5,7-dihydroxy-6-C-diglycosylflavones As shown by the spectrum of PM 6-*C*-neohesperidosylacacetin (Fig. 8.5) the characteristic feature of the MS of PM 2″-*O*-glycosyl-6-*C*-glycosylflavones compared with those of the corresponding PM

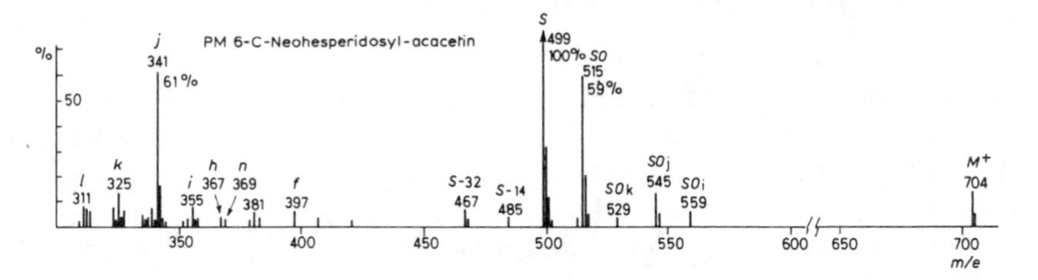

Fig. 8.5 MS of PM 2″-*O*-glycosyl-6-*C*-glycosylflavones.

6-*C*-glycosylflavones is the absence of the $M-15$ and $M-31$ peaks, which are replaced by the ions *SO* (40–60%) and *S* (100%) (Fig. 8.6), derived from the elimination of the permethylated 2″-*O*-glycosyl residue without and with the oxygen atom of the glycosidic bond respectively. Between the rather weak molecular peak (5–15%) and the ion *SO* are found some ions SO_i, SO_j,SO_k (Fig. 8.6) in which fragments of the *O*-glycosyl residue (similar to the fragments *i*, *j*, *k* of PM 6-*C*-glycosylflavones) remain bound to *SO*. At lower m/z values, the most important peak is the ion *j* of the corresponding PM 6-*C*-glycosylflavone. The nature of the flavone moiety is thus given by the m/z value of *j* (apigenin: 341, luteolin: 371, tricetin: 401), that of the 6-*C*-glycosyl residue by $S-j$ (hexose: 158, deoxyhexose: 128, pentose: 114) and that of the 2″-*O*-glycosyl residue by $M-S$ (hexose: 235, deoxyhexose: 205, pentose: 191).

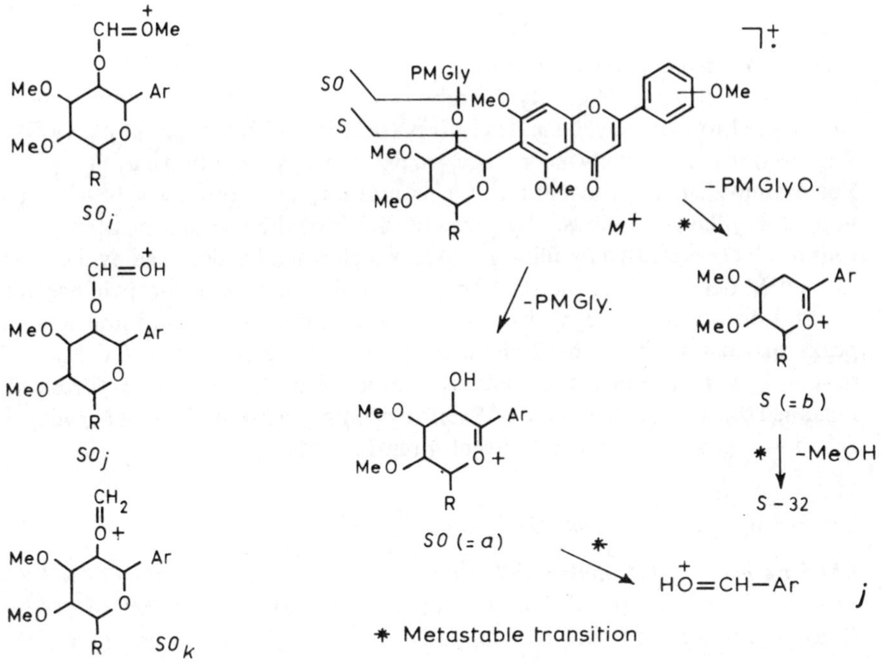

Fig. 8.6 Fragmentation pattern of PM 2″-*O*-glycosyl-6-*C*-glycosylflavones.

As expected from the MS of PM 8-*C*-glycosylflavones, the $M - 15$ and $M - 31$ peaks are also absent from the MS of PM 2″-*O*-glycosyl-8-*C*-glycosylflavones but the ion *S* becomes very weak and the molecular peak stronger, allowing a clear distinction from the 6-isomers (Bouillant *et al.*, 1978).

In the MS of PM 3″-, 4″- or 6″-*O*-glycosyl-6-*C*-glycosylflavones, the presence of $M - 15$ and $M - 31$ peaks proves their derivation from the 2″-methyl and methoxyl groups respectively. The ions *SO* and *S* are much lower than in the MS of PM 2″-*O*-glycosides. The base peak is the ion *j* in 3″-*O*-glycosides and the ion *i* in 4″- or 6″-*O*-glycosides (Besson *et al.*, 1979a). The presence of a peak $S - 14 > S$ in the MS of 6″-*O*-glycosides may be ascribed to the elimination of the C-5″ PM glycosyloxymethyl side chain. Mass spectrometry of the hydrolysis products of PM X″-*O*-glycosyl-6-*C*-glucosylflavones has been shown to provide the complete determination of the position of the *O*″-glycosyl group, characteristic differences being observed in the relative intensities of M, $M - 31$, $M - 89$, $M - 103$, $M - 161$ and $M - 175$ peaks of the four possible monohydroxy PM 6-*C*-glucosylflavones (Bouillant *et al.*, 1979a).

PM 5,7-dihydroxy-6-C-glycosylflavones 7-O-glycosides As shown by the spectrum of PM saponarin (Fig. 8.7), the characteristic features of their MS are the importance of the ion *A* and the presence of two homologous series of peaks, both

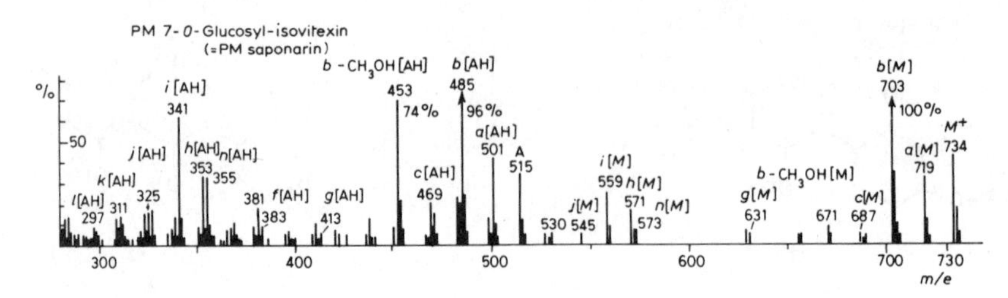

Fig. 8.7 MS of PM 7-*O*-glycosyl-6-*C*-glycosylflavones.

corresponding to the fragmentation of PM 6-*C*-glycosylflavones, the first series related to the molecular peak, the second to the aglycone peak AH (Fig. 8.8). However, the AH – 47 and AH – 63 peaks exhibit a much higher intensity than the corresponding *c* and *b*-MeOH peaks in the MS of PM 6-*C*-glycosylflavones. The nature of the 7-*O*-bound sugar is thus given by the difference *M* – *A* (hexose: 219, deoxyhexose: 189, pentose: 175), the nature of the C-bound sugar by the difference AH – *i* (AH) (hexose: 175, deoxyhexose: 145, pentose: 131) and the nature of the flavone moiety by *i*(AH) itself (apigenin: 341, luteolin: 371, tricetin: 401).

The transfer of one hydrogen atom from the *O*-glycosyl residue to the oxygen atom on rupture of the *O*-glycosidic bond is always observed with PM flavone *O*-glycosides, giving rise to a strong aglycone peak AH. This indeed is the case with PM *C*-glycosylflavone *O*-glycosides when the *O*-bound sugar is linked to the B ring, the ion *A* being absent or very weak. The unusual importance of the ion *A* in the MS of 7-*O*-glycosides may be ascribed to the possible hydrogen transfer from the vicinal *C*-glycosyl instead of the *O*-glycosyl residue.

PM 5,7-dihydroxy-6-C-glycosylflavone 7,2″-di-O-glycosides Owing to the different stability of the 7- and 2″-*O*-glycosidic bonds, the MS of these compounds show the characteristic features observed above in 2″-*O*-glycosides and 7-*O*-glycosides (Fig. 8.9). The molecular peak is easily detected and the absence of *M* – 15 and *M* – 31 peaks replaced by *SO* and *S* shows the loss of the PM 2″-*O*-glycosyl group as a first step in the fragmentation. Then each of the *SO* and *S* peaks loses in a second step the PM 7-*O*-glycosyl group with hydrogen transfer, giving the homologous peaks *SO* (AH) and *S* (AH). Finally *SO* (AH) leads to *j* (AH), the most important peak at lower *m*/*z* values, which indicates the nature of the flavone moiety (apigenin: 327, luteolin: 357, tricetin: 387), that of the 6-*C*-glycosyl residue being given by *S* (AH) – *j* (AH) (hexose: 158, deoxyhexose: 128, pentose: 114), that of the 7-*O*-glycosyl residue by *S* – *S* (AH) (hexose: 218, deoxyhexose: 188, pentose: 174) and that of the 2″-*O*-glycosyl residue by *M* – *S* (hexose: 235, deoxyhexose: 205, pentose: 191).

Confirmation of the 7,2″-*O*-glycosylation can be found in the MS of the hydrolysis product of the PM derivative showing a characteristic fragmentation pattern initiated by the loss of one molecule of water from the 7- and 2″-hydroxyl

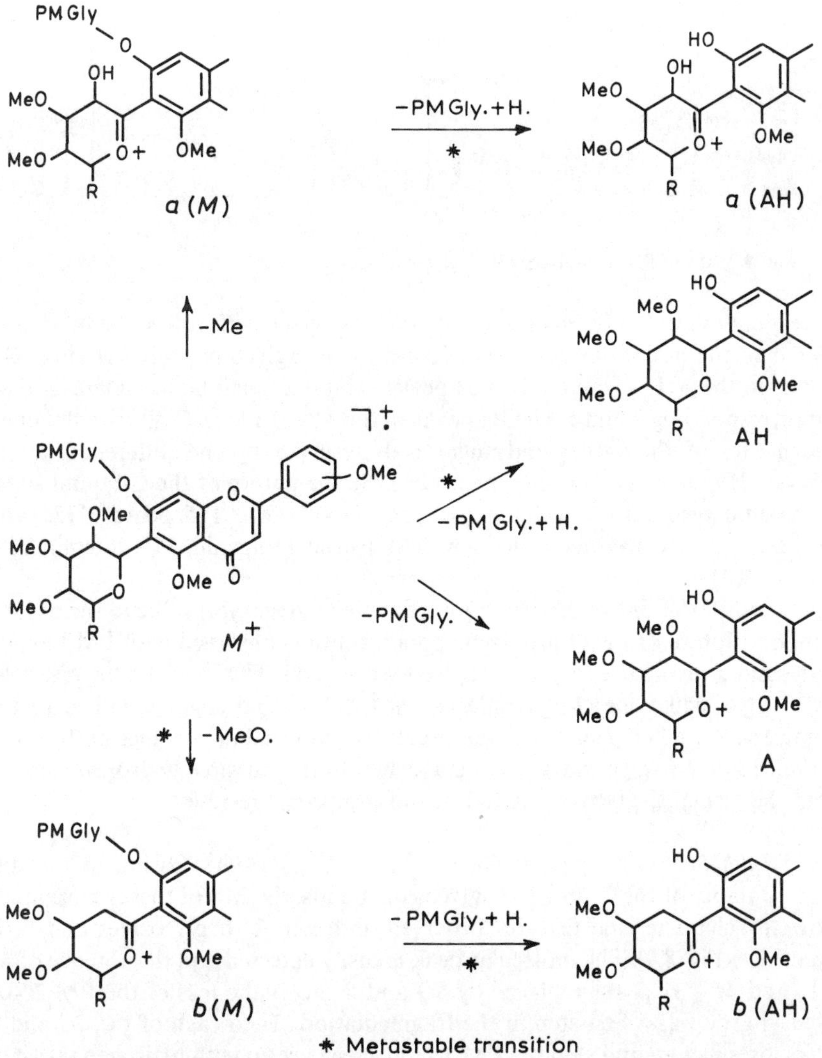

* Metastable transition

Fig. 8.8 Fragmentation pattern of PM 7-*O*-glycosyl-6-*C*-glycosylflavones.

groups. Moreover selective methylation with diazomethane of the phenolic hydroxyl group of this hydrolysis product affords a 2″-monohydroxy derivative showing a characteristic MS (Bouillant *et al.*, 1979b).

(d) Circular dichroism

The chiroptical differentiation of 6- and 8-*C*-glycosylflavones deduced from circular dichroism (CD) studies which had shown that a positive Cotton effect at

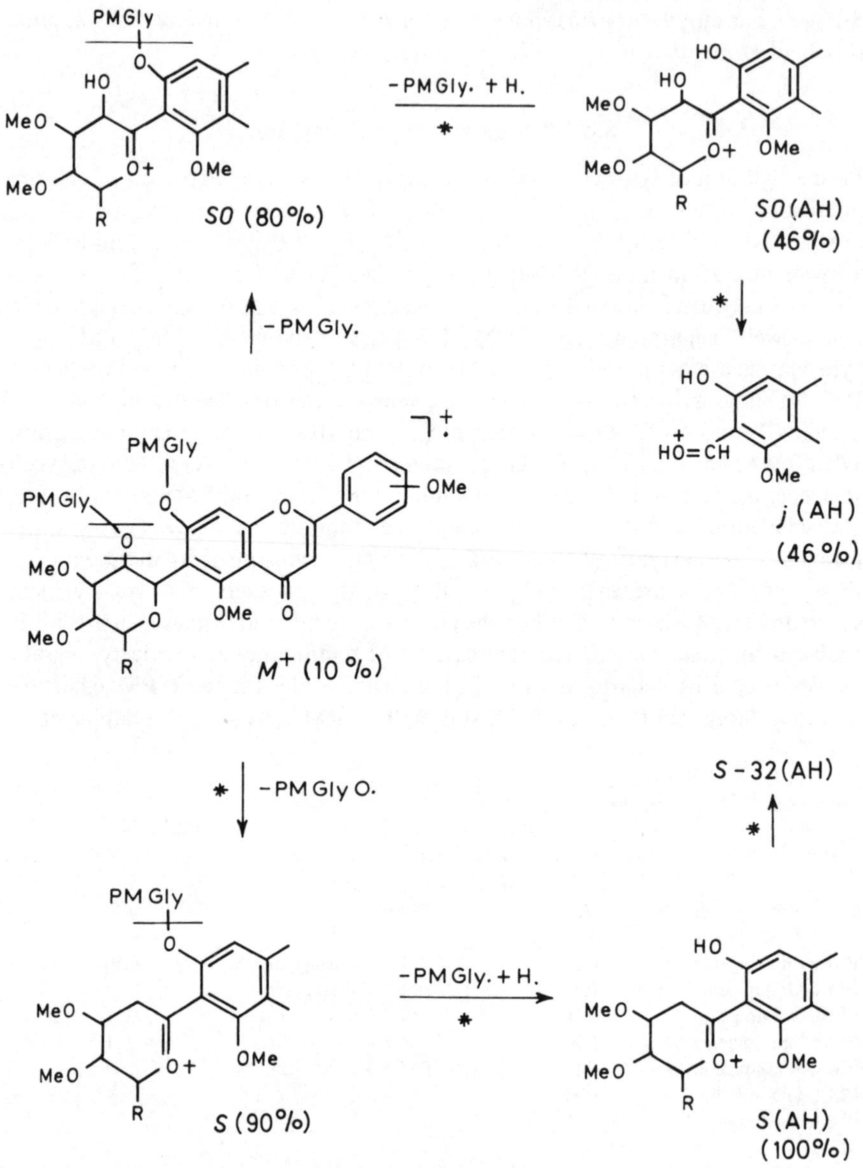

Fig. 8.9 Fragmentation pattern of PM 7,2″-di-*O*-glycosyl-6-*C*-glycosylflavones.

250–275 nm indicates a 6-*C*-linkage whereas a negative one indicates a 8-*C*-linkage has now been extended to *C*-β-D-galactopyranosyl-, *C*-β-D-xylopyranosyl- and *C*-α-L-arabinopyranosyl-flavones, and a quadrant rule has been suggested to explain these chiroptical properties. Both 6-*C*- and

8-*C*-α-L-rhamnopyranosylflavones show a negative CD band at 250–275 nm (Gaffield *et al.*, 1978).

8.4.3 Chromatographic methods

Permethylation of synthetic *C*-glycosylflavones for mass spectrometry disclosed interesting differences in the behaviour of PM ethers on silica-gel TLC (Table 8.8). In $CHCl_3$–EtOAc–Me_2CO(5:4:1)PM 8-*C*-glycosylapigenins show a lower migration than PM 6-*C*-glycosylapigenins, and a large difference was observed between isomeric PM 6-*C*-glycosylapigenins, PM 6-*C*-glucosylapigenin being well separated from PM 6-*C*-galactosylapigenin and PM 6-*C*-xylosylapigenin being well separated from PM 6-*C*-arabinosylapigenin, whereas PM 6-*C*-glucosyl- and 6-*C*-xylosyl-apigenin on the one hand and PM 6-*C*-galactosyl- and 6-*C*-arabinosyl-apigenin on the other show similar migrations. With PM 6,8-di-*C*-glycosylapigenins, interesting differences were found between PM vicenins-1,-2, and -3, which show similar migrations, and PM neoschaftoside, PM schaftoside and PM isoschaftoside, which migrate well below. Owing to the frequent co-occurrence of these di-*C*-glycosylapigenins in many plant extracts, these differences are extremely useful in analysing such mixtures by mass spectrometry of PM ethers. When the glucose of schaftoside and isoschaftoside is replaced by galactose, the corresponding PM compounds respectively migrate slower than PM schaftoside and PM isoschaftoside. Likewise PM 6,8-di-*C*-arabinosylapigenin migrates much slower than PM 6,8-di-*C*-xylosylapigenin.

Table 8.8 TLC of permethyl *C*-glycosylflavones

	R_F value on silica gel in $CHCl_3$–EtO Ac–MeOH (5:4:1)		
	PM Mono-C-glycosylflavones		
Compound	*R_F*	*Compound*	*R_F*
PM 6-Glc-apigenin	0.36	PM 8-Glc-apigenin	0.16
PM 6-Gal-apigenin	0.24	PM 8-Xyl-apigenin	0.14
PM 6-Xyl-apigenin	0.38	PM 8-Rha-apigenin	0.13
PM 6-Ara-apigenin	0.20	PM 8-Glc-luteolin	0.11
PM 6-Rha-apigenin	0.17	PM 8-Xyl-luteolin	0.07
PM 6-Glc-luteolin	0.31		
PM 6-Xyl-luteolin	0.30		
	PM Di-C-glycosylflavones		
Compound	*R_F*	*Compound*	*R_F*
PM vicenin 1	0.43	PM 6,8-di-Xyl-apigenin	0.40
PM vicenin 2	0.40	PM 6,8-di-Rha-apigenin	0.25
PM vicenin 3	0.38	PM 6,8-di-Ara-apigenin	0.15
PM neoschaftoside	0.32	PM lucenin 1	0.35
PM schaftoside	0.28	PM lucenin 2	0.32
PM isoschaftoside	0.23	PM lucenin 3	0.24

Under UV light, PM 6-*C*-glycosylapigenins give blue spots, turning to turquoise on spraying with 50% H_2SO_4, whereas PM 8-*C*-glycosylapigenins give pale-blue spots, turning to yellow–green with H_2SO_4; PM 6-*C*- and 8-*C*-glycosyl-luteolins give pale-blue spots turning to yellow with H_2SO_4; PM 6,8-di-*C*-glycosylapigenins give blue spots turning to yellow–green with H_2SO_4 and PM 6,8-di-*C*-glycosyl-luteolins give pale-blue spots turning to orange–yellow with H_2SO_4.

PM 6-*C*-α-L-arabinopyranosyl (R_F 0.20) and furanosyl (R_F 0.39) acacetins are well separated, whereas the free compounds cannot be separated on PC in 15% HOAc or n-butanol–acetic acid–water (BAW) (4:1:5) and on silica-gel TLC in ethyl acetate–pyridine–water–methanol (80:12:10:5) (APWM). This means that *C*-arabinofuranosylflavones cannot be easily distinguished from their pyranosyl isomers when the usual solvent systems are used. A good separation can be obtained on silica-gel TLC in $CHCl_3$–MeOH (5:1) (Besson, 1977).

The efficiency of chromatographic methods in the separation of *C*-glycosylflavones from complex mixtures has been recently demonstrated by the isolation and identification of 11 di-*C*-glycosides of apigenin, tricin, tricetin and tricetin 3′,4′-dimethyl ether from 80 g of *Apometzgeria pubescens* (Theodor *et al.*, 1980). HPLC has been recently applied to *C*-glycosylflavonoid analysis by Niemann and Brederode (1978) and by Strack *et al.* (1979) (see also Chapter 1).

REFERENCES

Abdel-Gawad, M. and Raynaud, J. (1974), *Plant. Med. Phytother.* **8**, 79.

Adinarayana, D. and Rajasekhara Rao, J. (1972), *Tetrahedron* **28**, 5377.

Adinarayana, D. and Rajasekhara Rao, J. (1975), *Proc. Indian Acad. Sci.* **81A**, 23.

Adinarayana, D., Gunasedar, D., Seligmann, O. and Wagner, H. (1980), *Phytochemistry* **19**, 480.

Arisawa, M., Morita, N., Kondo, Y. and Takemoto, T. (1973), *Yakugaku Zasshi*, **93**, 1655.

Asen, S., Stewart, R. N., Norris, K. H. and Massie, D. R. (1970), *Phytochemistry*, **9**, 619.

Asen, S., Norris, K. H., Stewart, R. N. and Semeniuk, P. (1973), *J. Am. Soc. Hort. Sci.* **98**, 174.

Baeva, R. T., Karryev, M. O., Litvinenko, V. I. and Abubakirov, N.K. (1974), *Khim. Prir. Soedin*, 171.

Ballard, R. E. and Cruden, R. W. (1978), *Biochem. Syst. Ecol.* **6**, 139.

Bellmann, G. and Jacot-Guillarmod, A. (1973), *Helv. Chim. Acta* **56**, 284.

Besset, A. (1977), Thèse de Doctorat de 3e cycle, Université Claude Bernard, Lyon, No. 659.

Besson, E. (1977), Thèse de Doctorat de 3e cycle, Université Claude Bernard, Lyon, No. 655.

Besson, E. and Chopin, J. (1980), *12th IUPAC International Symposium on the Chemistry of Natural Products*, Abstracts, p. 171.

Besson, E., Chopin, J., Krishnaswami, L. and Krishnamurty, H. G. (1977), *Phytochemistry* **16**, 498.

Besson, E., Besset, A., Bouillant, M. L., Chopin, J., Van Brederode, J. and Van Nigtevecht, G. (1979a), *Phytochemistry* **18**, 657.

Besson, E., Dombris, A., Raynaud, J. and Chopin, J. (1979b), *Photochemistry* **18**, 1899.

Besson, E., Chopin, J., Gunasegaran, R. and Ramachandran Nair, A. G. (1980), *Phytochemistry*, **19**, 2787.

Bhutani, S. P., Chibber, S. S. and Seshadri, T. R. (1969), *Indian J. Chem.* **7**, 210.

Bierner, M. W. (1973), *Biochem. Syst. Ecol.* **1**, 55.

Biol, M. C. and Chopin, J. (1972), *C. R. Acad. Sci., Ser. C* **275**, 1523.

Biol, M. C., Bouillant, M. L., Planche, G. and Chopin, J. (1974), *C. R. Acad. Sci., Ser. C* **279**, 409.
Blinova, K. F., Glyzin, V. I. and Prjakhina, N. I. (1977), *Khim. Prir. Soedin*, 116.
Blume, D. E. and McClure, J. W. (1978), *Lloydia* **41**, 651.
Boguslavskaya, L. I. and Beletskii, Y. N. (1978), *Khim. Prir. Soedin*, 801.
Bombardelli, E., Bonati, A., Gabetta, B. and Mustich, G. (1974), *Phytochemistry* **13**, 295.
Boufford, D. E., Raven, P. H. and Averett, J. E. (1978), *Biochem. Syst. Ecol.* **6**, 59.
Bouillant, M. L. (1976), Thèse de Doctorat d'Etat, Université Claude Bernard, Lyon, No. 7652.
Bouillant, M. L. and Chopin, J. (1971), *C. R. Acad. Sci., Ser. C* **273**, 1759.
Bouillant, M. L. and Chopin, J. (1972), *C. R. Acad. Sci., Ser. C* **274**, 193.
Bouillant, M. L., Favre-Bonvin, J. and Chopin, J. (1975), *Phytochemistry* **14**, 2267.
Bouillant, M. L., Besset, A., Favre-Bonvin, J. and Chopin, J. (1978), *Phytochemistry* **17**, 527.
Bouillant, M. L., Besset, A., Favre-Bonvin, J. and Chopin, J. (1979a), *Phytochemistry* **18**, 690.
Bouillant, M. L., Ferreres de Arce, F., Favre-Bonvin, J., Chopin, J., Zoll, A. and Mathieu, G. (1979b), *Phytochemistry* **18**, 1043.
Bouillant, M. L., Besset, A., Favre-Bonvin, J. and Chopin, J. (1980), *Phytochemistry* **19**, 1755.
Boutard, B. and Lebreton, P. (1975), *Plant. Med. Phytother.* **9**, 289.
Brederode, J. van and Nigtevecht, G. van (1972a), *Mol. Gen. Genet.* **118**, 247.
Brederode, J. van and Nigtevecht, G. van (1972b), *Genen. Phaenen.* **15**, 3.
Brederode, J. van and Nigtevecht, G. van (1973), *Mol. Gen. Genet.* **122**, 215.
Brederode, J. van and Nigtevecht, G. van (1974a), *Phytochemistry* **13**, 2763.
Brederode, J. van and Nigtevecht, G. van (1974b), *Biochem. Genet.* **11**, 65.
Brederode, J. van and Nigtevecht, G. van (1974c), *Genetics* **77**, 507.
Brederode, J. van, Chopin, J., Kamsteg, J., Nigtevecht, G. van and Heinsbroek, R. (1979), *Phytochemistry* **18**, 655.
Brederode, J. van, Niemann, G. J. O. and Nigtevecht, G. van (1980), *Planta Med.* **39**, 221.
Brum-Bousquet, M., Tillequin, F. and Paris, R. R. (1977), *Lloydia* **40**, 591.
Budzianowski, J. and Skrzypczakowa, L. (1978), *Phytochemistry* **17**, 2044.
Burret, F. (1980), Thèse de Doctorat de 3e cycle, Lyon, 1.
Burret, F., Chulia, A. J., Debelmas, A. and Hostettmann, K. (1978), *Planta Med.* **34**, 176.
Burret, F., Chulia, A. J. and Debelmas, A. (1979), *Planta Med.* **36**, 178.
Carlin, R. M. and McClure, J. W. (1973), *Phytochemistry* **12**, 1009.
Castledine, R. M. and Harborne, J. B. (1976), *Phytochemistry* **15**, 803.
Chari, V. M. (1980), Private communication.
Chari, V. M., Wagner, H., Schilling, G. and Nesmelyi, A. (1978), *11th IUPAC International Symposium on the Chemistry of Natural Products, Symposium papers*, Vol. 2, p. 279.
Chari, V. M., Harborne, J. B. and Williams, C. A. (1980), *Phytochemistry* **19**, 983.
Chawla, H., Chibber, S. S. and Seshadri, T. R. (1974), *Phytochemistry* **13**, 2301.
Chawla, H., Chibber, S. S. and Seshadri, T. R. (1975), *Indian J. Chem.* **13**, 444.
Chawla, H., Chibber, S. S. and Seshadri, T. R. (1976), *Phytochemistry* **15**, 235.
Chernobrovaya, N. V. (1973), *Khim. Prir. Soedin*, 801.
Chopin, J., Roux, B. and Durix, A. (1964) *C. R. Acad. Sci., Ser. C* **259**, 3111.
Chopin, J., Durix, A. and Bouillant, M. L. (1968), *C. R. Acad. Sci., Ser. C* **266**, 1334.
Chopin, J., Roux, B., Bouillant, M. L., Durix, A., D'Arcy, A., Mabry, T. J. and Yoshioka, Y. H. (1969), *C. R. Acad. Sci., Ser. C* **268**, 980.
Chopin, J., Bouillant, M. L., Wagner, H. and Galle, K. (1974), *Phytochemistry* **13**, 2583.
Chopin, J., Dellamonica, G., Bouillant, M. L., Besset, A., Popovici, G. and Weissenböck, G. (1977a), *Phytochemistry* **16**, 2041.
Chopin, J., Dellamonica, G., Besson, E., Skrzypczakowa, L., Budzianowski, J. and Mabry, T. J. (1977b), *Phytochemistry* **16**, 1999.
Chopin, J., Besset, A. and Bouillant, M. L. (1977c), *C. R. Acad. Sci., Ser. C* **284**, 1007.
Chopin, J., Bouillant, M. L., Ramachandran Nair, A. G., Ramesh, P. and Mabry, T. J. (1978), *Phytochemistry* **17**, 299.
Chopin, J., Besson, E. and Ramachandran Nair, A. G. (1979), *Phytochemistry* **18**, 2059.

Chulia, A. J. and Debelmas, A. (1977), *Plant. Med. Phytother.* **11**, 112.
Chulia, A. J., Hostettmann, K., Bouillant, M. L. and Mariotte, A. M. (1978), *Planta Med.* **34**, 442.
Chumbalov, T. K. and Zhubaeva, R. A. (1976), *Khim. Prir. Soedin*, 661.
Clark, W. D. and Mabry, T. J. (1978), *Biochem. Syst. Ecol.*, **6**, 19.
Darmograi, V. N. (1976), *Khim. Prir. Soedin*, 540.
Darmograi, V. N. (1977), *Khim. Prir. Soedin*, 114.
Darmograi, V. N. (1979), *Khim. Prir. Soedin*, 93.
Darmograi, V. N., Litvinenko, V. I. and Krivenchuk, P. E. (1968), *Khim. Prir. Soedin*, 248.
Del Pero de Martinez, M. and Swain, T. (1976), *Phytochemistry* **15**, 834.
Del Pero de Martinez, M. and Swain, T. (1977), *Biochem. Syst. Ecol.* **5**, 37.
Dennis, W. M. and Bierner, M. W. (1980), *Biochem. Syst. Ecol.* **8**, 65.
Dillon, M. O., Mabry, T. J., Besson, E., Bouillant, M. L. and Chopin, J. (1976), *Phytochemistry* **15**, 1085.
Dombris, A. and Raynaud, J. (1977), *Plant. Med. Phytother.* **12**, 109.
Dubois, M. A. (1980), Thèse de Doctorat de 3e cycle en Pharmacie, Paris.
Dubois, M. A., Zoll, A., Bouillant, M. L. and Favre-Bonvin, J. (1980), *Groupe Polyphenols Annual Meeting*, Neuchâtel, Switzerland.
Eade, R. A. and Pham, H. P. (1979), *Aust. J. Chem.* **32**, 2483.
Eade, R. A., McDonald, F. J. and Simes, J. J. H. (1975), *Aust. J. Chem.* **28**, 2011.
Eade, R. A., McDonald, F. J. and Pham, H. P. (1978), *Aust. J. Chem.* **31**, 2699.
Effertz, B. and Weissenböck, G. (1980), *Phytochemistry* **19**, 1669.
El Alfy, T. S. and Paris, R. R. (1975), *Plant. Med. Phytother.* **9**, 308.
Elliger, C. A., Chan, B. G., Waiss, A. C., Lundin, R. E. and Haddon, W. F. (1980), *Phytochemistry* **19**, 293.
El'Yashevich, O. G., Koreshchuk, K. E., Bezuglaya, N. I. and Drozd, G. A. (1974a), *Farm. Zh. (Kiev)* **29**, 92.
El'Yashevich, O. G., Drozd, G. A., Koreshchuk, K. E., Bezugloya, N. I., Manych, V. I. and Shestak, L. I. (1974b), *Khim. Prir. Soedin*, 94.
Farkas, L., Gottsegen, A., Nogradi, M. and Antus, S. (1974), *J. Chem. Soc. Perkin Trans.* **1**, 305.
Fernandez, M., Renedo, J., Aruppa, T. and Vega, F. (1975), *Phytochemistry* **14**, 586.
Ferreres de Arce, F. (1977), Diplôme d'Etudes Supérieures, Université de Lyon, 1.
Frezet, C., Raynaud, J. and Bouillant, M. L. (1975), *C. R. Acad. Sci., Ser. C.* **280**, 1079.
Fujita, M. and Inoue, T. (1979), *Yakugaku Zasshi* **99**, 165.
Funaoka, K. (1956), *Chem. Abstr.* **50**, 14729.
Gaffield W., Horowitz, R. M., Gentili, B., Chopin, J. and Bouillant, M. L. (1978), *Tetrahedron* **34**, 3089.
Galle, K. (1974), Dissertation, Universität München.
Garg, S. P., Bhushan, R. and Kapoor, R. C. (1980), *Phytochemistry* **19**, 1265.
Gentili, B. and Horowitz, R. M. (1968), *J. Org. Chem.* **33**, 1571.
Ghosal, S., Jaiswal, D. K. and Biswas, K. (1978), *Phytochemistry* **17**, 2119.
Goetz, M. and Jacot-Guillarmod, A. (1977a), *Helv. Chim. Acta* **60**, 1322.
Goetz, M. and Jacot-Guillarmod, A. (1977b), *Helv. Chim. Acta* **60**, 2104.
Goetz, M. and Jacot-Guillarmod, A. (1978), *Helv. Chim. Acta* **61**, 1373.
Goetz, M., Hostettmann, K. and Jacot-Guillarmod, A. (1976a), *Phytochemistry* **15**, 2014.
Goetz, M., Hostettmann, K. and Jacot-Guillarmod, A. (1976b), *Phytochemistry* **15**, 2015.
Gonnet, (1980), Private communication.
Greger, H. (1977). In *The Biology and Chemistry of the Compositae* (eds. V. H. Heywood, J. B. Harborne, and B. L. Turner), Academic Press, London, pp. 899–941.
Gupta, S. R. (1981), *Phytochemistry*, in press.
Harborne, J. B. (1977), *Biochem. Syst. Ecol.* **5**, 7.
Harborne, J. B. and Williams, C. A. (1976), *Biochem. Syst. Ecol.* **4**, 267.
Harborne, J. B., Williams, C. A., Greenham, J. and Moyna, P. (1974), *Phytochemistry* **13**, 1557.
Hayashi, Y. (1980), Private communication.

Heinsbroek, R., Van Brederode, J., Van Nigtevecht, G. and Kamsteeg, J. (1979), *Phytochemistry* **18**, 935.

Heinsbroek, R., Van Brederode, J., Besson, E., Chopin, J., Van Nigtevecht, G., Maas, J. and Kamsteeg, J. (1980), *Phytochemistry* **19**, 1935.

Hiller, K., Gruendemann, E. and Habisch, D. (1975), *Pharmazie* **30**, 809.

Hillis, W. E. and Horn, D. H. S. (1965), *Aust. J. Chem.* **18**, 531.

Hillis, W. E. and Horn, D. H. S. (1966), *Aust. J. Chem.* **19**, 705.

Hillis, W. E. and Inoue, T. (1967), *Phytochemistry* **6**, 59.

Hilu, K. W., de Wet, J. M. J. and Seigler, D. (1978), *Biochem. Syst. Ecol.* **6**, 247.

Hiraoka, A. (1978), *Biochem. Syst. Ecol.* **6**, 171.

Horowitz, R. M. and Gentili, B. (1964), *Chem. Ind.* 498.

Horowitz, R. M. and Gentili, B. (1966), *Chem. Ind.*, 625.

Horowitz, R. M., Gentili, B. and Gaffield, W. (1974), *ACS meeting*, Los Angeles, USA.

Hostettmann, K. and Jacot-Guillarmod, A. (1974a), *Helv. Chim. Acta* **57**, 204.

Hostettmann, K. and Jacot-Guillarmod, A. (1974b), *Helv. Chim. Acta* **57**, 1155.

Hostettmann, K. and Jacot-Guillarmod, A. (1975), *Helv. Chim. Acta* **58**, 130.

Hostettmann, K. and Jacot-Guillarmod, A. (1976), *Helv. Chim. Acta* **59**, 1584.

Hostettmann, K. and Jacot-Guillarmod, A. (1977), *Phytochemistry* **16**, 481.

Hostettmann, K., Bellmann, G., Tabacchi, R. and Jacot-Guillarmod, A. (1973), *Helv. Chim. Acta* **56**, 3050.

Hostettmann, K., Luong Minh Duc, Goetz, M. and Jacot-Guillarmod, A. (1975), *Phytochemistry* **14**, 499.

Hostettmann-Kaldas M. and Jacot-Guillarmod, A. (1978), *Phytochemistry* **17**, 2083.

Ibrahim, R. K. and Shaw, M. (1970), *Phytochemistry* **9**, 1855.

Inoue, T. and Fujita, M. (1974), *Chem. Pharm. Bull.* **22**, 1422.

Inoue, T. and Fujita, M. (1977), *Chem. Pharm. Bull.* **25**, 3226.

Jacot-Guillarmod, A., Luong Minh Duc and Hostettmann, K. (1975), *Helv. Chim. Acta* **58**, 1477.

Jalal, M. A. F. and Collin, H. A. (1977), *Phytochemistry* **16**, 1377.

Jarman, M. and Ross, W. C. J. (1969), *J. Chem. Soc.*, 199.

Jay, M. and Viricel, M. R. (1980), *Phytochemistry*, **19**, 2627.

Jay, M., Gleye, J., Bouillant, M. L., Stanislas, E. and Moretti, C. (1979), *Phytochemistry* **18**, 184.

Jay, M., Voirin, B., Hasan, A., Gonnet, J. F. and Viricel, M. R. (1980), *Biochem. Syst. Ecol.* **8**, 127.

Julian, E. A., Johnson, G., Johnson, D. K. and Donnelly, B. J. (1971), *Phytochemistry* **10**, 3185.

Kaldas, M., Hostettmann, K. and Jacot-Guillarmod, A. (1974), *Helv. Chim. Acta* **57**, 2557.

Kaldas, M., Hostettmann, K. and Jacot-Guillarmod, A. (1975), *Helv. Chim. Acta*, **58**, 2188.

Kamanzi, H. and Raynaud, J. (1976), *Plant. Med. Phytother.* **10**, 78.

Kamsteeg, J., Van Brederode, J. and Van Nigtevecht, G. (1976), *Phytochemistry* **15**, 1917.

Kaneta, M. and Sugiyama, N. (1973), *Agric. Biol. Chem.* **37**, 2663.

Koeppen, B. H. (1964), *Z. Naturforsch.* **19B**, 173.

Koeppen, B. H. and Roux, D. G. (1965), *Tetrahedron Lett.*, 3497.

Komatsu, M. and Tomimori, T. (1966), *Tetrahedron Lett.*, 1611.

Komatsu, M., Tomimori, T. and Makiguchi, Y. (1967), *Chem. Pharm. Bull.* **15**, 1567.

Komatsu, M., Tomimori, T., Takeda, K. and Hayashi, K. (1968), *Chem. Pharm. Bull.* **16**, 1413.

Komissarenko, N. F., Korzennikova, E. P., Angarskaya, M. A. and Koesnikov, D. G. (1973a), *Rast. Res.* **9**, 532.

Komissarenko, N. F., Yatsyuk, V. Ya. and Kirzennikova, E. P. (1973b), *Khim. Prir. Soedin*, 439.

Krause, J. (1976a), *Z. Pflanzenphysiol.* **79**, 372.

Krause, J. (1976b), *Z. Pflanzenphysiol.* **79**, 465.

Laman, N. A. and Volynets, A. P. (1974), *Fiziol. Rast.* **21**, 737.

Lardy, C. (1981), Thèse de Doctorat de 3e cycle, Université Claude Bernard, Lyon, 1.

Lebreton, P. (1980), *Rev. Gen. Bot.* **87**, 133.

Lebreton, P. and Dangy-Caye, M. P. (1973), *Plant. Med. Phytother.* **7**, 87.

Lebreton, P., Boutard, B. and Thivend, S. (1978), *C. R. Acad. Sci., Ser. D*, **287**, 1255.

Levy, M. (1976), *Biochem. Syst. Ecol.* **4**, 249.
Levy, M. and Fujii, K. (1978), *Biochem. Syst. Ecol.* **6**, 117.
Levy, M. and Levin, D. A. (1974), *Am. J. Bot.* **61**, 156.
Linard, A., Jacquemin, H. and Paris, R. (1976), *Plant. Med. Phytother.* **10**, 267.
Linard, A., Delaveau, P. and Paris, R. (1978), *Plant. Med. Phytother.* **12**, 144.
Loehdefink, J. (1976), *Dtsch. Apoth. Ztg.* **116**, 557.
Luong Minh Duc and Jacot-Guillarmod, A. (1977), *Helv. Chim. Acta* **60**, 2099.
Luong Minh Duc, Hostettmann, K. and Jacot-Guillarmod, A. (1976), *Helv. Chim. Acta* **59**, 1294.
Luong Minh Duc, Fombasso, P. and Jacot-Guillarmod, A. (1980), *Helv. Chim. Acta* **63**, 244.
Mabry, T. J., Yoshioka, H., Sutherland, S., Woodland, S., Rahman, W., Ilyas, M., Usmani, J. N., Hameed, N., Chopin, J. and Bouillant, M. L. (1971), *Phytochemistry* **10**, 677.
Markham, K. R. and Porter, L. J. (1979), *Phytochemistry* **18**, 611.
Markham, K. R. and Moore, N. A. (1980), *Biochem. Syst. Ecol.* **8**, 17.
Markham, K. R. and Wallace, J. W. (1980), *Phytochemistry* **19**, 415.
Markham, K. R. and Williams, C. A. (1980), *Phytochemistry*, **19**, 2789.
Markham, K. R., Porter, L. J., Mues, R., Zinsmeister, H. D. and Brehm, B. G. (1976a), *Phytochemistry* **15**, 147.
Markham, K. R., Porter, L. J., Campbell, E. O., Chopin, J. and Bouillant, M. L. (1976b), *Phytochemistry* **15**, 1517.
Marston, A., Hostettmann, K. and Jacot-Guillarmod, A. (1976), *Helv. Chim. Acta* **59**, 2596.
Medvedeva, S. A., Tjukavkina, N. A. and Ivanova, S. Z. (1973), *Khim. Prir. Soedin*, 119.
Medvedeva, S. A., Tjukavkina, N. A. and Ivanova, S. Z. (1974), *Khim. Drev.* **15**, 144.
Miana, G. A. (1973), *Phytochemistry* **12**, 728.
Moniava, I. I. and Kemertelidze, E. P. (1976), *Izv. Akad. Nauk. Gruz. SSR Ser. Chim.* **2**, 119.
Monties, B. (1969), *Bull. Soc. Fr. Physiol. Veg.* **15**, 29.
Monties, B., Bouillant, M. L. and Chopin, J. (1976), *Phytochemistry* **15**, 1053.
Montoro, R., Casas, A. and Primo, E. (1974), *Rev. Agroquim. Tecnol. Aliment.* **14**, 271.
Morita, N., Arisawa, M. and Yoshikawa, A. (1976), *J. Pharm. Soc. J.* **96**, 1180.
Mues, R. and Zinsmeister, H. D. (1975), *Phytochemistry* **14**, 577.
Mues, R. and Zinsmeister, H. D. (1976), *Phytochemistry* **15**, 1757.
Murakami, T., Nishikawa, Y. and Ando, T. (1960), *Chem. Pharm. Bull.* **8**, 688.
Murray, B. G. and Williams, C. A. (1976), *Biochem. Genet.* **14**, 897.
Nabeta, K., Kadota, G. and Tani, T. (1977), *Phytochemistry* **16**, 1112.
Nair, A. G. R., Ramesh, P. and Subramanian, S. S. (1975a), *Phytochemistry* **14**, 1644.
Nair, A. G. R., Ramesh, P. and Subramanian, S. S. (1975b), *Curr. Sci.* **44**, 551.
Nair, A. G. R., Ramesh, P. and Subramanian, S. S. (1976), *Phytochemistry* **15**, 839.
Najar, A., Gujral, V. K. and Gupta, S. R. (1978), *Tetrahedron Lett.*, 2031.
Narayanan, V. and Seshadri, T. R. (1971), *Indian J. Chem.* **9**, 14.
Nawwar, M. A. M., Ishak, M. S., El Sherbieny, A. E. D. A. and Meshaal, S. A. (1977), *Phytochemistry* **16**, 1319.
Nawwar, M. A. M., El Sissi, H. I. and Baracat, H. H. (1980), *Phytochemistry* **19**, 1854.
Nichols, K. W. and Bohm, B. A. (1979), *Phytochemistry* **18**, 1078
Niemann, G. J. (1973), *Phytochemistry* **12**, 2056.
Niemann, G. J. (1974), *Planta Med.* **26**, 101.
Niemann, G. J. (1975a), *Acta Bot. Neerl.* **24**, 65.
Niemann, G. J. (1975b), *Phytochemistry* **14**, 1436.
Niemann, G. J. (1979), *Acta Bot. Neerl.* **28**, 73.
Niemann, G. J. (1980a), *Z. Naturforsch.* **35C**, 514.
Niemann, G. J. (1980b), *Planta Med.* **39**, 221.
Niemann, G. J. and Baas, W. J. (1978), *Z. Naturforsch.* **33C**, 780.
Niemann, G. J. and Brederode, J. van (1978), *J. Chromatogr.* **152**, 523.
Niemann, G. J. and Miller, H. J. (1975), *Biochem. Syst. Ecol.* **2**, 169.
Niemann, G. J. and Van Genderen, H. H. (1980), *Biochem. Syst. Ecol.* **8**, 237.

Niemann, G. J., Baas, W. J., Besson, E. and Chopin, J. (1979), *Z. Naturforsch.* **34C**, 1125.
Nigtevecht, G. van and Brederode, J. van (1972), *Genen. Phaenen.* **15**, 9.
Nikolov, N. (1973), *Farmatsiya (Sofia)*, **23**, 3.
Nikolov, N. (1974), *Farmatziya (Sofia)* **24**, 25.
Nikolov, N. (1975), *Khim. Prir. Soedin*, **11**, 422.
Nikolov, N. and Litvinenko, V. I. (1973), *Farm. Zh. (Kiev)* **28**, 78.
Nikolov, N. and Vodenicharov, R. I. (1975), *Khim. Prir. Soedin*, 423.
Nikolov, N., Batyuk, V. S., Kovalev, I. P. and Ivanov, V. (1973), *Khim. Prir. Soedin*, 116.
Nikolov, N., Horowitz, R. M. and Gentili, B. (1976), *International Congress for Research on Medicinal Plants*, Münich, Abstracts of papers, Section A.
Numata, A., Hokimoto, K., Shimada, A., Yamaguchi, H. and Takaishi, K. (1979), *Chem. Pharm. Bull.* **27**, 602.
Numata, A., Hokimoto, K. and Yamaguchi, H. (1980), *Chem. Pharm. Bull.* **28**, 964.
Oelrichs, P., Marshall, J. T. B. and Williams, D. H. (1968), *J. Chem. Soc.* 941.
Ohashi, H., Goto, M. and Imamura, H. (1976), *Phytochemistry* **15**, 354.
Ohashi, H., Goto, M. and Imamura, H. (1977a), *Phytochemistry* **16**, 1106.
Ohashi, H., Goto, M. and Imamura, H. (1977b), Private communication.
Ohashi, H., Goto, M., Fukuda, J. and Imamura, H. (1980), *Bull. Fac. Agric. Gifu Univ.*, in press.
Österdahl, B. G. (1978), *Acta Chem. Scand.* **32**, 93.
Österdahl, B. G. (1979a), Doctoral dissertation, University of Uppsala, No. 524.
Österdahl, B. G. (1979b), *Acta Chem. Scand.* **33B**, 119.
Österdahl, B. G. (1979c), *Acta Chem. Scand.* **33B**, 400.
Paris, R. R. and Debray, M. (1973), *Plant. Med. Phytother* **7**, 135.
Paris, R. R. and Paris, M. R. (1973), *C. R. Acad. Sci., Ser.* D **277**, 2369.
Paris, R. R., Henri, E. and Paris, M. (1976), *Plant. Med. Phytother.* **10**, 144.
Parker, W. H., Maze, J. and McLachlan, D. G. (1979), *Phytochemistry* **18**, 508.
Parthasarathy, M. R., Seshadri, T. R. and Varma, R. S. (1974), *Curr. Sci.* **43**, 74.
Parthasarathy, M. R., Seshadri, T. R. and Varma, R. S. (1976), *Phytochemistry* **15**, 1025.
Plesser, A. and Weissenböck, G. (1977), *Z. Pflanzenphysiol.* **81**, 425.
Popovici, G. and Weissenböck, G. (1977), *Z. Pflanzenphysiol.* **82**, 450.
Popovici, G., Weissenböck, G., Bouillant, M. L., Dellamonica, G. and Chopin, J. (1977), *Z. Pflanzenphysiol.* **85**, 103.
Ponomarenko, A. A., Komissarenko, N. F., Stukkei, K. L. and Korzennikova, E. P. (1974), *Rast. Resur.* **10**, 63.
Proliac, A. and Raynaud, J. (1977), *Planta Med.* **32**, 68.
Proliac, A., Raynaud, J., Combier, H., Bouillant, M. L. and Chopin, J. (1973), *C. R. Acad. Sci., Ser. D* **277**, 2813.
Quercia, V., Turchetto, L., Pierini, N., Cuozzo, V. and Percaccio, G. (1978), *J. Chromatogr.*, **161**, 396.
Rabesa, Z. A. (1980), Thèse de Doctorat d'Etat, Lyon, No: 8029.
Rasolojaona, L. and Raynaud, J. (1979), *Ann. Pharm. Fr.* **37**, 469.
Raynaud, J. and Chouikha, M. (1976), *Plant. Med. Phytother.* **10**, 199.
Raynaud, J. and Dombris, A. (1978), *Plant. Med. Phytother.* **12**, 220.
Raynaud, J. and Rasolojaona, L. (1976), *C. R. Acad. Sci., Ser. D.* **282**, 1059.
Raynaud, J. and Rasolojaona, L. (1978), *Plant. Med. Phytother.* **12**, 281.
Raynaud, J. and Rasolojaona, L. (1979), *Planta Med.* **37**, 168.
Richardson, M. (1978), *Biochem. Syst. Ecol.* **6**, 283.
Sakakibara, M., Mabry, T. J., Bouillant, M. L. and Chopin, J. (1977), *Phytochemistry,* **16,** 1113.
Salehian, A., Pichon, P. and Prum, N. (1973), *Plant. Med. Phytother.* **7**, 255.
Saunders, J. A. and McClure, J. W. (1976), *Phytochemistry* **15**, 809.
Segelman, A. G., Segelman, F. P., Varma, S. D., Wagner, H. and Seligmann, O. (1977), *J. Pharm. Sci.* **66**, 1358.

Segelman, A. B., Segelman, F. P., Star, A. E., Wagner, H. and Seligman, O. (1978), *Phytochemistry* **17**, 824.
Seraya, L. M., Birke, K., Khimenko, S. V. and Bogulavskaya, L. I. (1978), *Khim. Prir. Soedin*, 802.
Sergeyeva, N. V. (1977), *Khim. Prir. Soedin*, 124.
Seshadri, T. R., Sood, A. R. and Varshney, I. P. (1972), *Indian J. Chem.* **10**, 26.
Shalaby, A. F., Tsingaridas, K. and Steinegger, E. (1965), *Helv. Pharm. Acta*, **40**, 19.
Smith, D. M., Glennie, C. W., Harborne, J. B. and Williams, C. A. (1977), *Biochem. Syst. Ecol.* **5,** 107.
Sood, A. R., Boutard, B., Chadenson, M., Chopin, J. and Lebreton, P. (1976), *Phytochemistry* **15**, 351.
Specht, J. E., Gorz, H. J. and Haskins, F. A. (1976), *Phytochemistry*, **15**, 133.
Strack, D., Fuisting, K. and Popovici, G. (1979), *J. Chromatogr* **176**, 270.
Subramanian, P. M. and Misra, G. S. (1979), *J. Nat. Prod.* **42**, 540.
Theodor, R., Zinsmeister, H. D., Mues, R. and Markham, K. R. (1980), *Phytochemistry* **19**, 1695.
Theodor, R., Markham, K. R., Mues R. and Zinsmeister, H. D. (1981a), *Phytochemistry* **20**, 1457.
Theodor, R., Zinsmeister, H. D., Mues, R. and Markham, K. R. (1981b), *Phytochemistry*, **20**, 1851.
Thieme, H. and Khogali, A. (1974), *Pharmazie* **29**, 352.
Thieme, H. and Khogali, A. (1975), *Pharmazie* **30**, 736.
Thivend, S., Lebreton, P., Ouabonzi, A. and Bouillant, M. L. (1979), *C. R. Acad. Sci., Ser. D* **289**, 465.
Tillequin, F., Paris, M., Jacquemin, H. and Paris, R. R. (1978), *Planta Med.* **33**, 46.
Tjukavkina, N. A., Medvedeva, S. A. and Ivanova, S. V. (1975), *Khim. Drev. (Riga)*, 93.
Tomas, F., Pastor, R. and Carpena, O. (1975), *Rev. Agroquim. Tecnol. Aliment* **15**, 581.
Tschesche, R. and Struckmeyer, K. (1976), *Chem. Ber.* **109**, 2901.
Ulubelen, A. and Mabry, T. J. (1980), *J. Nat. Prod.* **43**, 162.
Valant, K. (1978), Naturwissenschaften **65**, 437.
Valant, K., Besson, E. and Chopin, J. (1978), *Phytochemistry* **17**, 2136.
Valant, K., Besson, E. and Chopin, J. (1980), *Phytochemistry* **19**, 156.
Vandekerkhove, O. (1978), *Z. Pflanzenphysiol.* **86**, 135.
Vavilova, N. K. and Gella, E. H. V. (1973a), *Khim. Prir. Soedin*, 151.
Vavilova, N. K. and Gella, E. H. V. (1973b), *Khim. Prir. Soedin*, 285.
Vega, F. A. (1976), *An. R. Acad. Farm.* **42**, 81.
Vita-Finzi, P., De Bernardi, M., Fronza, G., Mellerio, G., Servettaz Grünanger, O., Vidari, G. and Beltrami, E. (1980), *12th IUPAC International Symposium on the Chemistry of Natural Products*, Abstracts, p. 167.
Wagner, H. (1974), *Fortschr. Chem. Org. Naturst.* **31**, 153.
Wagner, H. (1977). *In Flavonoids and Bioflavonoids, Current Research Trends.* (eds. L. Farkas, M. Gabor and F. Kallay), Akademiai Kiado, Budapest, p. 42.
Wagner, H., Budweg, W., Iyengar, M. A., Volk, O. and Sinn, M. (1972), *Z. Naturforsch.* **27B**, 809.
Wagner, H., Iyengar, A. and Herz, W. (1973), *Phytochemistry* **12**, 2063.
Wagner, H., Rosprim, L. and Galle, K. (1975), *Phytochemistry* **14**, 1089.
Wagner, H., Obermeier, G., Seligmann, O. and Chari, V. M. (1979), *Phytochemistry* **18**, 907.
Wallace, J. W.(1975), *Phytochemistry* **14**, 1765.
Wallace, J. W. (1978), *J. Nat. Prod.* **41**, 650.
Wallace, J. W. (1979), *Am. J. Bot.* **66**, 343.
Wallace, J. W. and Markham, K. R. (1978a), *J. Nat. Prod.* **41**, 651.
Wallace, J. W. and Markham, K. R. (1978b), *Phytochemistry* **17**, 1313.
Wallace, J. W. and Morris, G. (1978), *Phytochemistry* **17**, 1809.
Wallace, J. W. and Story, D. T. (1978), *J. Nat. Prod.* **41**, 651.
Wallace, J. W., Story, D. T., Besson, E. and Chopin, J. (1979), *Phytochemistry* **18**, 1077.
Wallace, J. W., Yoppe, D. L., Besson, E. and Chopin, J. (1981), *Phytochemistry* **20**, 2701.
Weightman, C. (1979), D.E.S.S., Université Lyon 1, N: 78.
Weinges, K. (1977), Private communication.
Weissenböck, G. and Effertz, B. (1974), *Z. Pflanzenphysiol.* **74**, 298.

Weissenböck, G. and Sachs, G. (1977), *Planta* **137**, 49.
Weissenböck, G. and Schneider, V. (1974), *Z. Pflanzenphysiol.* **72**, 23.
Weissenböck, G., Plesser, A. and Trinks, K. (1976), *Ber. Dtsch. Bot. Ges.* **89**, 457.
Weissenböck, G., Effertz, B. and Popovici, G. (1978), *J. Nat. Prod.* **41**, 652.
Williams, C. A. (1975), *Biochem. Syst. Ecol.* **3**, 229.
Willams, C. A. (1978), *Phytochemistry* **17**, 729.
Williams, C. A. (1979), *Phytochemistry* **18**, 803.
Williams, C. A. and Harborne, J. B. (1975), *Biochem. Syst. Ecol.* **3**, 181.
Williams, C. A. and Harborne, J. B. (1977a), *Biochem. Syst. Ecol.* **5**, 45.
Williams, C. A. and Harborne, J. B. (1977b), *Biochem. Syst. Ecol.* **5**, 221.
Williams, C. A. and Murray, B. G. (1972), *Phytochemistry* **11**, 2507.
Williams, C. A., Harborne, J. B. and Clifford, H. T. (1973), *Phytochemistry* **12**, 2417.
Williams, C. A., Harborne, J. B. and Smith, P. (1974), *Phytochemistry* **13**, 1141.
Woo, W. S., Kang, S. S., Shim, S. H., Wagner, H., Mohan, Chari, V., Seligmann, O. and Obermeier, G. (1979), *Phytochemistry* **18**, 353.
Zapesochnaya, G. G. and Laman, N. A. (1977), *Khim. Prir. Soedin*, 862.
Zemtsova, G. N. and Bandyukova, V. A. (1974), *Khim. Prir. Soedin*, 107.
Zemtsova, G. N. and Dzhumyrko, S. F. (1976), *Farmacija, SSSR* **25**, 26.
Zemtsova, G. N., Glyzin, V. Ya and Dzhumyrko, S. F. (1975). *Khim. Prir. Soedin*, 516.
Zoll, A. and Nouvel, G. (1974), *Plant. Med. Phytother.* **8**, 134.
Zykova, N. Y. and Pivnenko, G. P. (1975), *Khim. Prir. Soedin*, 253.
Zykova, N. Y., Madio, M. and Kovu, M. (1976), *Farm. Zh. (Kiev)* **4**, 61.
Zoring, T., Oyungerel, Z. and Laslo, T. (1980), *Khim. Prir. Soedin*, 253.

CHECK LIST OF *C*-GLYCOSYLFLAVONOIDS

Mono-C-glycosylflavones

1 Bayin (7,4′-di-OH 8-Glc)
2 Isovitexin (5,7,4′-tri-OH 6-Glc)
3 7-*O*-Glucoside
4 7-*O*-Galactoside*
5 7-*O*-Xyloside*
6 4′-*O*-Glucoside
7 2″-*O*-Glucoside*
8 2″-*O*-Xyloside*
9 2″-*O*-Arabinoside*
10 2″-*O*-Rhamnoside*
11 6″-*O*-Arabinoside*
12 7,2″-Di-*O*-glucoside*
13 7-*O*-Glucoside 2″-*O*-arabinoside*
14 7-*O*-Glucoside 2″-*O*-rhamnoside*
15 7-*O*-Galactoside 2″-*O*-glucoside*
16 7-*O*-Galactoside 2″-*O*-rhamnoside*
17 7-*O*-Xyloside 2″-*O*-glucoside*
18 7-*O*-Xyloside 2″-*O*-arabinoside*
19 7-*O*-Xyloside 2″-*O*-rhamnoside*
20 4′,2″-Di-*O*-glucoside*
21 4′-*O*-Glucoside 2″-*O*-arabinoside
22 2″-*O*-*trans*-Feruloyl*
23 2″-*O*-*trans*-Feruloyl 4′-*O*-glucoside*
24 X″-*O*-*trans*-Caffeoyl 2″-*O*-glucoside*
25 7-Bisulphate*
26 Vitexin (5,7,4′-tri-OH 8-Glc)
27 7-*O*-Glucoside
28 4′-*O*-Glucoside
29 2″-*O*-Glucoside*
30 2″-*O*-Xyloside
31 6″-*O*-Xyloside
32 X″-*O*-Arabinoside*
33 2″-*O*-Rhamnoside*
34 6″-*O*-Rhamnoside*
35 7-*O*-Rutinoside*
36 2″-*O*-Sophoroside*
37 6″-*O*-Gentiobioside (marginatoside)*
38 2″-*O*-*p*-Hydroxybenzoyl
39 2″-*O*-*p*-Coumaroyl*
40 4‴-*O*-Acetyl 2″-*O*-rhamnoside*
41 7-Bisulphate*
42 7-*O*-Rutinoside bisulphate*
43 8-*C*-*β*-D-Galactopyranosylapigenin (5,7,4′-tri-OH 8-Gal)*
44 6″-*O*-Acetyl*
45 6-*C*-*β*-D-Xylopyranosylapigenin (5,7,4′-tri-OH 6-Xyl)

46 7-*O*-Glucoside*
47 2″-*O*-Rhamnoside
48 Isomollupentin (5,7,4′-tri-OH 6-Ara)*
49 7,2″-Di-*O*-glucoside*
50 Mollupentin (5,7,4′-tri-OH 8-Ara)*
51 3′-Deoxyderhamnosylmaysin [5,7,4′-tri-OH 6-(6-deoxyxylohexos-4-ulosyl)]*
52 2″-*O*-Rhamnoside (3′-deoxymaysin)*
53 Swertisin (5,4′-di-OH 7-OMe 6-Glc)
54 4′-*O*-Glucoside
55 2″-*O*-Glucoside (spinosin)*
56 X″-*O*-Xyloside
57 Isoswertisin (5,4′-di-OH 7-OMe 8 Glc)
58 4′-*O*-Glucoside
59 2″-*O*-Rhamnoside*
60 2″-*O*-Acetyl*
61 Isomolludistin (5,4′-di-OH 7-OMe 6-Ara)*
62 2″-*O*-Glucoside*
63 Molludistin (5,4′-di-OH 7-OMe 8-Ara)*
64 2″-*O*-Glucoside*
65 2″-*O*-Rhamnoside*
66 Isocytisoside (5,7-di-OH 4′-OMe 6-Glc)*
67 7-*O*-Glucoside*
68 2″-*O*-Glucoside*
69 2″-*O*-Rhamnoside (isomargariten)*
70 Cytisoside (5,7-di-OH 4′-OMe 8-Glc)
71 7-*O*-Glucoside
72 2″-*O*-Rhamnoside (margariten)*
73 *O*-Acetyl 7-*O*-glucoside
74 Embigenin (5-OH 7,4′-di-OMe 6-Glc)
75 2″-*O*-Rhamnoside (embinin)
76 7,4′-Di-*O*-methylisomollupentin (5-OH 7,4′-di-OMe 6-Ara)*
77 2″-*O*-Glucoside*
78 2″-*O*-Caffeoylglucoside*
79 Iso-orientin (5,7,3′,4′-tetra-OH 6-Glc)
80 7-*O*-Glucoside
81 3′-*O*-glucoside*
82 3′-*O*-Glucuronide*
83 4′-*O*-Glucoside
84 2″-*O*-Glucoside*
85 2″-*O*-Xyloside*
86 2″-*O*-Arabinoside*
87 2″-*O*-*β*-L-Arabinofuranoside*
88 2″-*O*-Rhamnoside*
89 6″-*O*-Arabinoside*
90 4′,2″-Di-*O*-glucoside*
91 7-*O*-Rutinoside
92 X″-*O*-Acetyl
93 2″-*O*-*p*-Hydroxybenzoyl*
94 2″-*O*-*trans*-*p*-Coumaroyl*
95 2″-*O*-*trans*-Caffeoyl*
96 2″-*O*-*trans*-Feruloyl*
97 2″-*O*-*p*-Hydroxybenzoyl 4′-*O*-Glucoside*
98 2″-*O*-*trans*-Caffeoyl 4′-*O*-Glucoside*
99 2″-(*p*-*O*-Glucosyl)*trans*-caffeoyl 4′-*O*-glucoside*
100 2″-*trans*-Feruloyl 4′-*O*-glucoside*
101 2″-*O*-*trans*-Caffeoylglucoside*
102 7-Bisulphate*
103 Orientin (5,7,3′,4′-tetra-OH 8-Glc)
104 7-*O*-Glucoside bisulphate*
105 4′-*O*-Glucoside
106 2″-*O*-Glucoside*
107 2″-*O*-Xyloside (adonivernith)
108 2″-*O*-*β*-L-Arabinofuranoside*
109 2″-*O*-Rhamnoside*
110 7-*O*-Rhamnoside
111 7-Bisulphate*
112 7-*O*-Glucoside bisulphate*
113 6-*C*-*β*-D-Xylopyranosyl-luteolin (5,7,3′,4′-tetra-OH 6-Xyl)
114 2″-*O*-Rhamnoside
115 Derhamnosylmaysin [5,7,3′,4′-tetra-OH 6-(6-deoxyxylohexos-4-ulosyl)]*
116 2″-*O*-Rhamnoside (maysin)*
117 Swertiajaponin (5,3′,4′-tri-OH 7-OMe 6-Glc)
118 3′-*O*-Glucoside*
119 2″-*O*-Rhamnoside*
120 3′-*O*-Gentiobioside*
121 Isoswertiajaponin (5,3′,4′-tri-OH 7-OMe 8-Glc)*
122 Isoscoparin (5,7,4′-tri-OH 3′-OMe 6-Glc)
123 7-*O*-Glucoside*
124 Scoparin (5,7,4′-tri-OH 3′-OMe 8-Glc)
125 X″-*O*-Rhamnosylglucoside*
126 epi-7-*O*-Glucoside (knautinoside)*
127 3′-*O*-Methylderhamnosylmaysin [5,7,4′-tri-OH 3′-OMe 6-(6-deoxyxylohexos-4-ulosyl)]*
128 2″-*O*-Rhamnoside (3′-*O*-methylmaysin)*
129 6-*C*-Glucosyldiosmetin (5,7,3′-tri-OH 4′-OMe 6-Glc)
130 8-*C*-Glucosyldiosmetin (5,7,3′-tri-OH 4′-OMe 8-Glc)
131 7,3′-Di-*O*-methyliso-orientin (5,4′-di-OH 7,3′-di-OMe 6-Glc)*
132 7,3′,4′-Tri-*O*-methyliso-orientin (5-OH 7,3′,4′-tri-OMe 6-Glc)
133 2″-*O*-Rhamnoside (linoside B)
134 6″-*O*-Acetyl 2″-*O*-rhamnoside (linoside A)
135 Isoaffinetin (5,7,3′,4′,5′-penta-OH 6-Glc)*

136 Affinetin (5,7,3′,4′,5′-penta-OH 8-Glc)*
137 Isopyrenin (5,7,4′-tri-OH 3′,5′-di-OMe 6-Glc)*
138 7-*O*-Glucoside*
139 6-*C*-Glucosyl-5,7-dihydroxy-8,3′,4′,5′-tetramethoxyflavone*
140 5-*O*-Rhamnoside*

Di-C-glycosylflavones

141 Vicenin-2 (5,7,4′-tri-OH 6,8-di-Glc)
142 Feruloyl 6″-*O*-glucoside*
143 X‴-*O*-Diferuloylhexoside*
144 6,8-Di-*C*-hexosylapigenin (5,7,4′-tri-OH 6,8-di-Hex)*
145 Vicenin-3 (5,7,4′-tri-OH 6-Glc-8-Xyl)
146 Vicenin-1 (5,7,4′-tri-OH 6-Xyl-8-Glc)
147 Sinapoyl*
148 Violanthin (5,7,4′-tri-OH 6-Glc-8-Rha)
149 Isoviolanthin (5,7,4′-tri-OH 6-Rha-8-Glc)*
150 Schaftoside (5,7,4′-tri-OH 6-Glc-8-Ara)*
151 Feruloyl 6″-*O*-glucoside*
152 Isoschaftoside (5,7,4′-tri-OH 6-Ara-8-Glc)*
153 Sinapoyl*
154 Feruloyl*
155 Neoschaftoside (5,7,4′-tri-OH 6-Glc-8-Ara)*
156 Neoisoschaftoside (5,7,4′-tri-OH 6-Ara-8-Glc)*
157 Isocorymboside (5,7,4′-tri-OH 6-Gal-8-Ara)*
158 Corymboside (5,7,4′-tri-OH 6-Ara-8-Gal)*
159 6,8-Di-*C*-α-L-arabinopyranosylapigenin (5,7,4′-tri-OH 6,8-di-Ara)*
160 6-*C*-Arabinosyl-8-*C*-pentosylapigenins (5,7,4′-tri-OH 6-Ara-8-Pen)*
161 Almeidein (5,4′-di-OH 7-OMe 6,8-di-Ara)*
162 6-*C*-Pentosyl-8-*C*-hexosylacacetin (5,7-di-OH 4′-OMe 6-Pen-8-Hex)*
163 7,4′-Di-*O*-methyl 6,8-di-*C*-α-L-arabinopyranosylapigenin (5-OH 7, 4′-di-OMe 6,8-di-Ara)
164 Lucenin-2 (5,7,3′,4′-tetra-OH 6,8-di-Glc)
165 Lucenin-3 (5,7,3′,4′-tetra-OH 6-Glc-8-Xyl)
166 Lucenin-1 (5,7,3′,4′-tetra-OH 6-Xyl-8-Glc)
167 Carlinoside (5,7,3′,4′-tetra-OH 6-Glc-8-Ara)*
168 Neocarlinoside (5,7,3′,4′-tetra-OH 6-Glc-8-Ara)*
169 6-*C*-Arabinosyl-8-*C*-pentosyl-luteolin (5,7,3′,4′-tetra-OH 6-Ara-8-Pen)*
170 6,8-Di-*C*-glucosylchrysoeriol (5,7,4′-tri-OH 3′-OMe 6,8-di-Glc)*
171 6-*C*-Arabinosyl-8-*C*-hexosylchrysoeriol (5,7,4′-tri-OH 3′-OMe 6-Ara-8-Hex)*
172 6,8-Di-*C*-glycosyldiosmetin (5,7,3′-tri-OH 4′-OMe 6,8-di-Gly)
173 6,8-Di-*C*-glucosyltricetin (5,7,3′,4′,5′-penta-OH 6,8-di-Glc)*
174 6,8-Di-*C*-hexosyltricetin*
175 6-*C*-Hexosyl-8-*C*-pentosyltricetins*
176 6-*C*-Arabinosyl-8-*C*-glucosyltricetin*
177 6,8-Di-*C*-pentosyltricetin*
178 6-*C*-Hexosyl-8-*C*-pentosyl 3′-*O*-methyltricetin*
179 6,8-Di-*C*-glucopyranosyltricin (5,7,4′-tri-OH 3′,5′-di-OMe 6,8-di-Glc)*
180 6-*C*-Glucosyl-8-*C*-arabinosyltricin*
181 6-*C*-Arabinosyl-8-*C*-glucosyltricin*
182 6-*C*-Xylosyl-8-*C*-hexosyltricin*
183 6,8-Di-*C*-arabinosyltricin*
184 6-*C*-Hexosyl-8-*C*-pentosylapometzgerin (5,7,5′-tri-OH 3′,4′-di-OMe 6-Hex-8-Pen)*
185 6,8-Di-*C*-arabinosylapometzgerin*

C-Glycosylflavonols

186 6-*C*-Glucosylkaempferol (3,5,7,4′-tetra-OH 6-Glc)*
187 Keyakinin (3,5,4′-tri-OH 7-OMe 6-Glc)
188 6-*C*-Glucosylquercetin (3,5,7,3′,4′-penta-OH 6-Glc)*
189 Keyakinin B (3,5,3′,4′-tetra-OH 7-OMe 6-Glc)

C-Glycosylflavanones

190 Aervanone (7,4′-di-OH 8-Gal)*
191 Palodulcine B (7,3′,4′-tri-OH 8-Gal)*
192 Hemiphloin (5,7,4′-tri-OH 6-Glc)
193 Isohemiphloin (5,7,4′-tri-OH 8-Glc)

C-Glycosylflavanonols

194 6-*C*-Glucosyldihydrokaempferol (3,5,7, 4′-tetra-OH 6-Glc)*
195 Keyakinol (3,5,4′-tri-OH 7-OMe 6-Glc)
196 6-*C*-Glucosyldihydroquercetin (3,5,7,3′,4′-penta-OH 6-Glc)*

C-Glycosylchalcone

197 3′-*C*-Glucosylisoliquiritigenin (2′,4′,4-tri-OH 3′-Glc)*

C-Glycosyldihydrochalcones

198 Nothofagin (2′,4′,6′,4-tetra-OH *C*-Gly)
199 Konnanin (2′,4′,6′,3,4-penta-OH *C*-Gly)
200 Aspalathin (2′,4′,6′,3,4-penta-OH 3′-Glc

Di-C-Glycosyldihydrochalcone

201 3′,5′-Di-*C*-glycosylphloretin (2′,4′,6′, 4-tetra-OH 3′,5′-di-Gly)*

C-Glycosylisoflavones

202 Puerarin (7,4′-di-OH 8-Glc)

203 *O*-Xyloside

204 4′,6″-Di-*O*-acetyl

205 8-*C*-Glucosylgenistein (5,7,4′-tri-OH 8-Glc)*

206 8-*C*-Glucosylprunetin (5,4′-di-OH 7-OMe 8-Glc)*

207 Volubilin (5-OH 7,4′-di-OMe 8-Rha)*

208 Isovolubilin (5-OH 7,4′-di-OMe 6-Rha)*

209 8-*C*-Glucosylorobol (5,7,3′,4′-tetra-OH 8-Glc)*

210 Volubilinin (5,7-di-OH 6,4′-di-OMe 8-Glc)*

211 Dalpanitin (5,7,4′-tri-OH 3′-OMe 8-Glc)

Di-C-Glycosylisoflavone

212 Paniculatin (5,7,4′-tri-OH 6,8-di-Glc)

C-Glycosylisoflavanones

213 Dalpanin (cf. Table 8.2)*

214 6-*C*-Glucosyldalbergioidin (5,7,2′,4′-tetra-OH 6-Glc)*

This list includes all known naturally occurring *C*-glycosylflavonoids. Compounds reported as new during the period 1975 to 1980 are asterisked.

CHAPTER NINE

Biflavonoids

HANS GEIGER and CHRISTOPHER QUINN

9.1 COMPLETE LIST OF KNOWN BIFLAVONOIDS

It is only five years since the present authors reviewed the biflavonoids (Geiger and Quinn, 1975). The number of new compounds found during the period 1975 to 1980 is of course limited, but nevertheless there are some very interesting new types, indicating that many more biflavonoids are likely to be uncovered in plants in the future. Three new types of interflavonoid link – 3,3″(1),3,3‴(3), and 6,6″(4) – have been found (see Table 9.1); the formation of all three types can be explained by the mechanism of phenol oxidation discussed earlier (Geiger and Quinn, 1975). Two compounds with a new substitution pattern have been reported (8A.19), which contains both an apigenin and a luteolin moiety, and (8A.21), a *C*-methylated biflavonoid. Until recently glycosides had been found only among the group of *Garcinia* biflavonoids; the recent detection of amentoflavone glycosides (8A.2) in the Psilotales demonstrates that glycosylation is by no means restricted to the former group.

In the list of known biflavonoids (Table 9.1), all new compounds are marked by an asterisk; those which are known only synthetically, but, by analogy, might occur naturally, are given in parentheses. The numbering system of the compounds in this list is as follows: the first numeral refers to the type of interflavonyl link; the letter refers to the type of flavonoid moieties present – A = biflavones, B = flavanone–flavone, C = biflavanones and D = others; and the last numeral refers to the individual compound in each group. The names of the compounds are again those given to them by the authors, since no generally accepted system for naming biflavonoids yet exists, and the full systematic names as used in Chemical Abstracts are too tongue-twisting for practical use.

9.2 ISOLATION, IDENTIFICATION AND STRUCTURE DETERMINATION

No basically new methods of separation and isolation of biflavonoids have been reported during the past five years. For identification purposes, lists of R_F values (Lin and Chen, 1974a, 1975; Dossaji *et al.*, 1975b) and of UV-visible spectra using the standard shift reagents (Dossaji *et al.*, 1975b) have been reported.

Table 9.1 Complete list of known biflavonoids

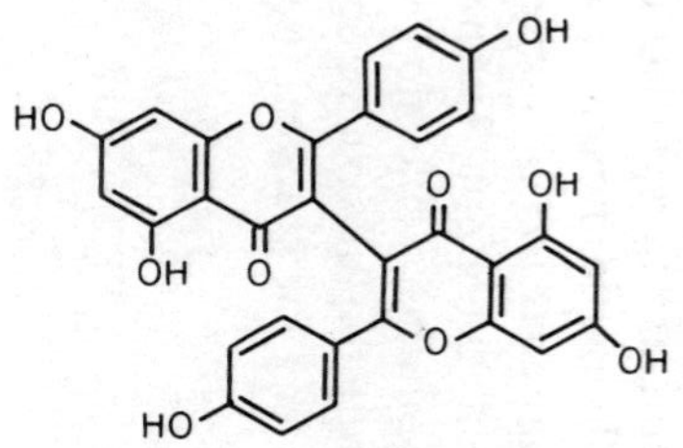

(1A.1 3,3''-Biapigenin)
Geiger and Quinn (1975)

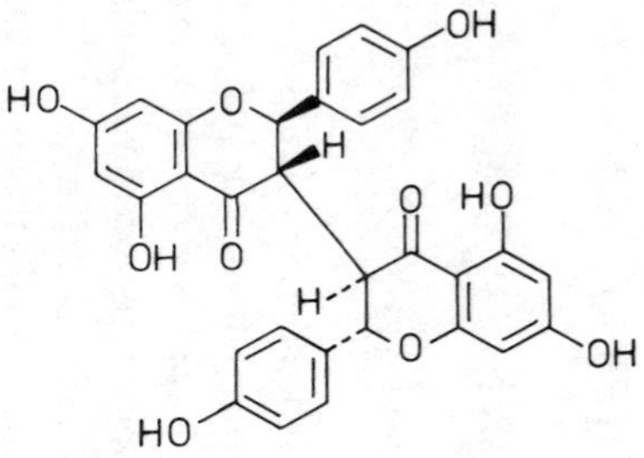

*1C.1 Chamaejasmin
Hwang and Chang (1979)

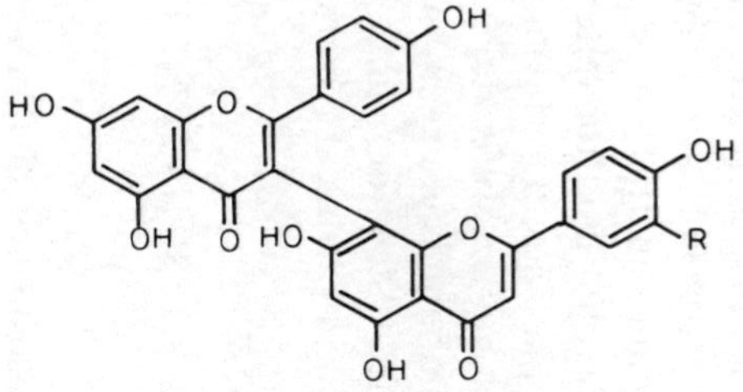

2A

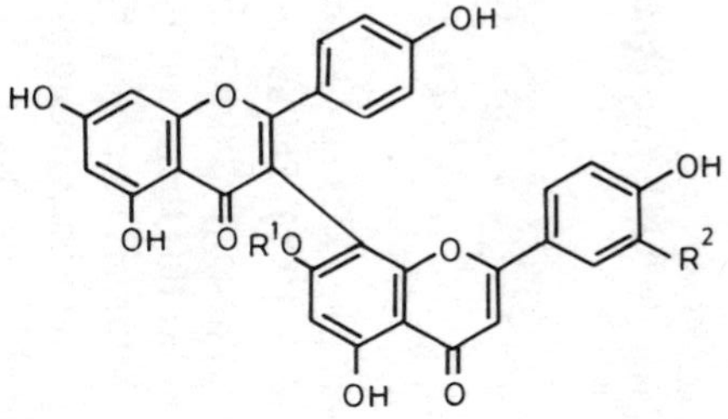

2B

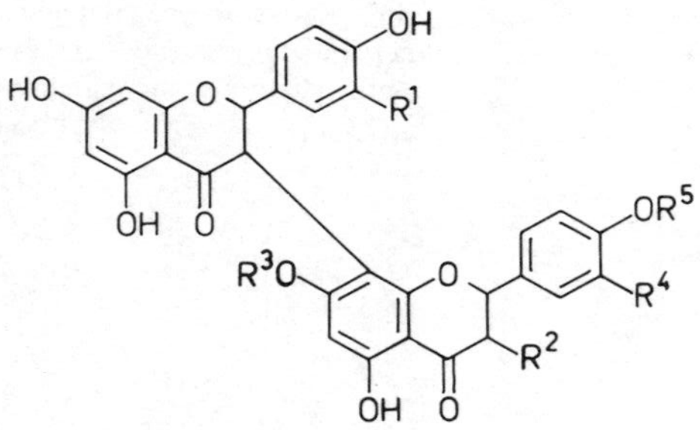

*2C

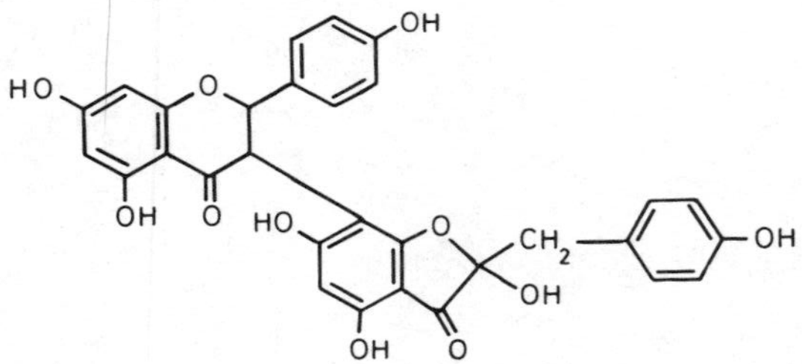

2D

	R					Leading references
2A.1 Bisdehydro-GB1a	H					Geiger and Quinn (1975)
2A.2 Sahranflavone	OH					

	R^1	R^2	Leading references
2B.1 Volkensiflavone (=BGH-III=talbotiflavone)	H	H	Geiger and Quinn (1975)
2B.2 Spicatiside	Glc	H	
2B.3 Fukugetin (=BGH-II=morelloflavone)	H	OH	
2B.4 3-*O*-Methylfukugetin	H	OMe	
2B.5 Fukugiside	Glc	OH	

	R^1	R^2	R^3	R^4	R^5	Leading references
2C.1 GB1a	H	H	H	H	H	Geiger and Quinn (1975)
*2C.2 GB1a-7″-*O*-glucoside	H	H	Glc	H	H	Chen *et al.* (1975c)
2C.3 GB1	H	OH	H	H	H	Geiger and Quinn (1975)
2C.4 GB2a	H	H	H	OH	H	
2C.5 GB2	H	OH	H	OH	H	

Table 9.1 (*Contd.*)

*2C.6 Kolaflavanone	H	OH	H	OH	Me	Cotterill *et al.* (1978)
2C.7 Xanthochymusside	H	OH	Glc	OH	H	Geiger and Quinn (1975)
*2C.8 Manniflavanone	OH	OH	H	OH	H	Crichton and Waterman (1979)
2D.1 Zeyherin						Geiger and Quinn (1975)

*3A

	R^1	R^2	R^3	R^4	R^5	R^6	*Leading references*
*3A.1 Taiwaniaflavone	H	H	H	H	H	H	Kamil *et al.* (1977)
*3A.2 Taiwaniaflavone 7″-methyl ether	H	H	H	H	Me	H	Kawano (1980)
*3A.3 Taiwaniaflavone 4′,7″-dimethyl ether	H	H	Me	H	Me	H	cf also Geiger and Quinn (1975)
(3A.4 Taiwaniaflavone hexamethyl ether)	Me	Me	Me	Me	Me	Me	

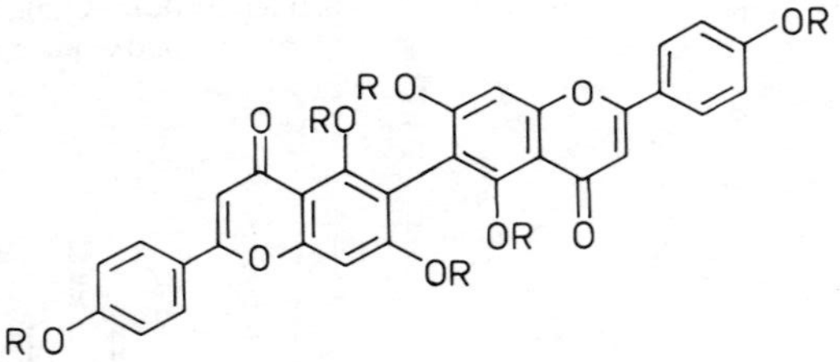

4A

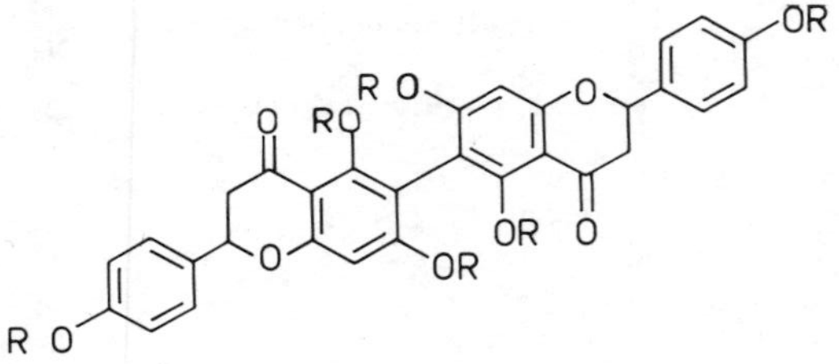

*4C

	R	Leading references
(4A.1 Didehydrosuccedaneaflavone)	H	Lin and Chen (1974b)
(4A.2 Didehydrosuccedaneaflavone hexamethyl ether)	Me	Chen *et al.* (1975d, e)
*4C.1 Succedaneaflavone	H	Chen and Lin (1975)
(4C.2 Succedaneaflavone hexamethyl ether)	Me	Chen *et al.* (1975d,e)

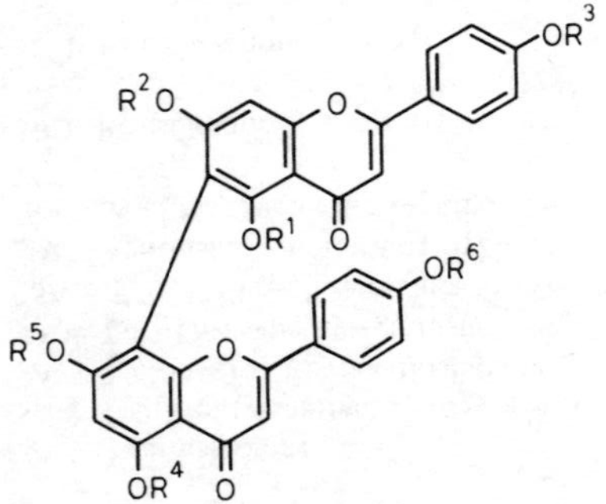

5A

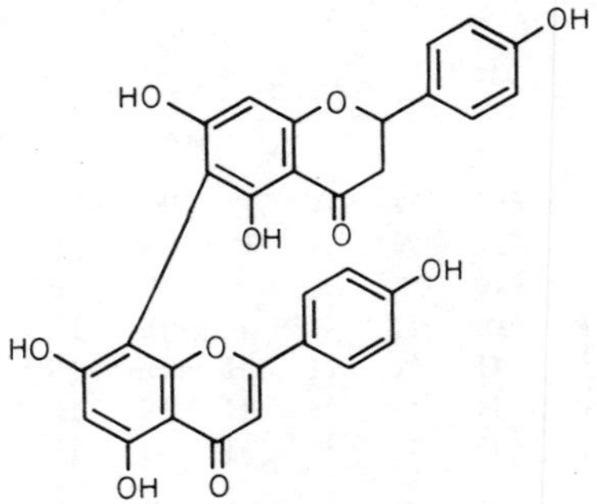

5B

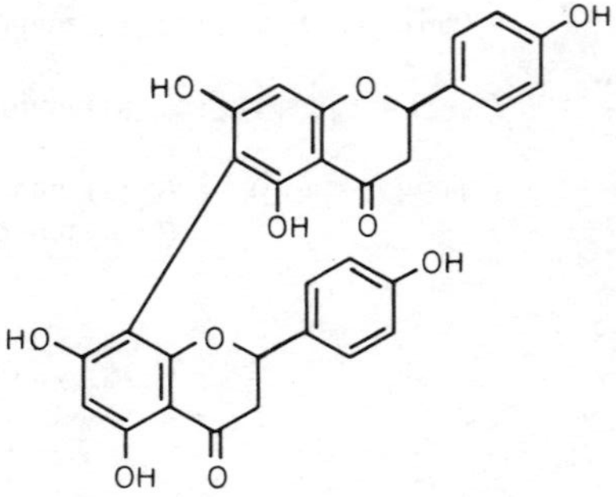

5C

Table 9.1 (*Contd.*)

	R^1	R^2	R^3	R^4	R^5	R^6	*Leading references*
5A.1 Agathisflavone	H	H	H	H	H	H	Geiger and Quinn (1975)
5A.2 7-*O*-Methylagathisflavone (=WA-I)	H	Me	H	H	H	H	
5A.3 7,7″-Di-*O*-methylagathisflavone	H	Me	H	H	Me	H	
5A.4 7,4‴-Di-*O*-methylagathisflavone (=WA-VII)	H	Me	H	H	H	Me	
*5A.5 7,7″,4‴-Tri-*O*-methylagathisflavone	H	Me	H	H	Me	Me	Ilyas *et al.* (1978)
5A.6 Agathisflavone tetramethyl ether	H	Me	Me	H	Me	Me	Geiger and Quinn (1975)
(5A.7 Agathisflavone hexamethyl ether)	Me	Me	Me	Me	Me	Me	Geiger and Quinn (1975); Moriyama *et al.* (1974)
5B.1 Rhusflavone							Geiger and Quinn (1975); Chen and Lin (1976)
5C.1 Rhusflavanone							Geiger and Quinn (1975); Chen *et al.* (1976a)

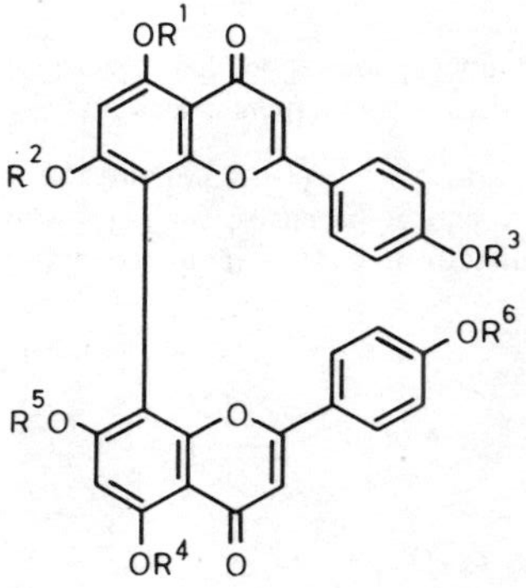

OR1 O R2O OR3 OR6 R5O OR4 O

6A

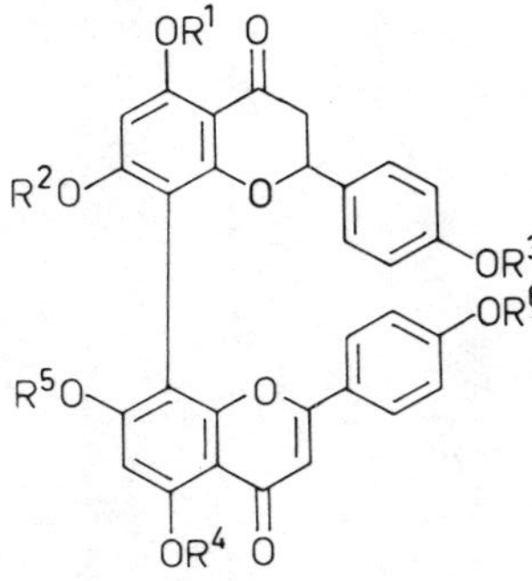

OR1 O R2O OR3 OR6 R5O OR4 O

*6B

OR1 O R2O OR3 OR6 R5O OR4 O

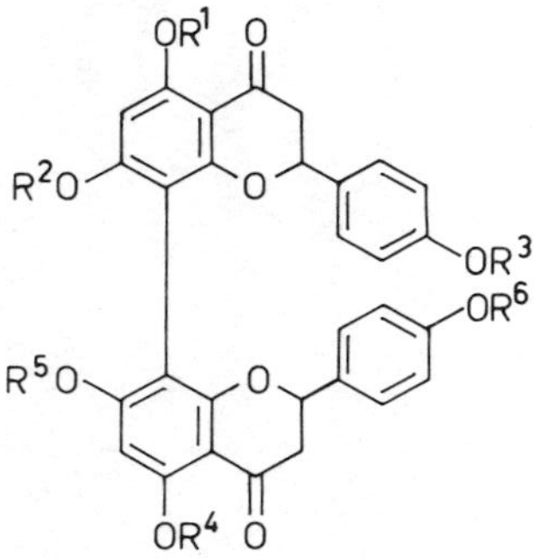

*6C

	R^1	R^2	R^3	R^4	R^5	R^6	*Leading references*
6A.1 Cupressuflavone	H	H	H	H	H	H	Geiger and Quinn (1975)
6A.2 7-*O*-Methylcupressuflavone	H	Me	H	H	H	H	

*6A.3 4′-*O*-Methylcupressuflavone	H	H	Me	H	H	H	Taufeeq *et al.* (1978)
6A.4 7,7″-Di-*O*-methylcupressuflavone	H	Me	H	H	Me	H	Geiger and Quinn (1975); Ahmad and Razaq (1976)
*6A.5 7,4′or 4‴-Di-*O*-methyl cupressuflavone	H	Me	Me	H	H	H	Ilyas *et al.* (1978)
			or				
	H	Me	H	H	H	Me	
6A.6 7,4′,7″-Cupressuflavone trimethyl ether	H	Me	Me	H	Me	H	Geiger and Quinn (1975)
6A.7 Cupressuflavone tetramethyl ether (=WB1=AC3)	H	Me	Me	H	Me	Me	Geiger and Quinn (1975); Ilyas *et al.* (1977b); Ahmad and Razaq (1976)
(6A.8 Cupressuflavone pentamethyl ether)	H	Me	Me	Me	Me	Me	Geiger and Quinn (1975)
(6A.9 Cupressuflavone hexamethyl ether)	Me	Me	Me	Me	Me	Me	Geiger and Quinn (1975); Moriyama *et al.* (1974)
*6B.1 Mesuaferrone B	H	H	H	H	H	H	Raju *et al.* (1976)
(6B.2 Mesuaferrone B hexamethyl ether)	Me	Me	Me	Me	Me	Me	
*6C.1 Neorhusflavanone	H	H	H	H	H	H	Chen *et al.* (1975a); Chen *et al.* (1978)
(6C.2 Neorhusflavanone hexamethyl ether)	Me	Me	Me	Me	Me	Me	

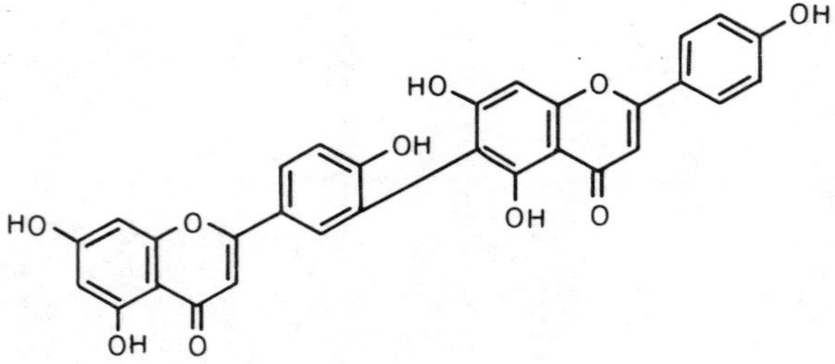

7A.1
Robustaflavone Geiger and Quinn (1975); Lin and Chen (1974d) (isoln. and ppn. of hexamethyl ether).

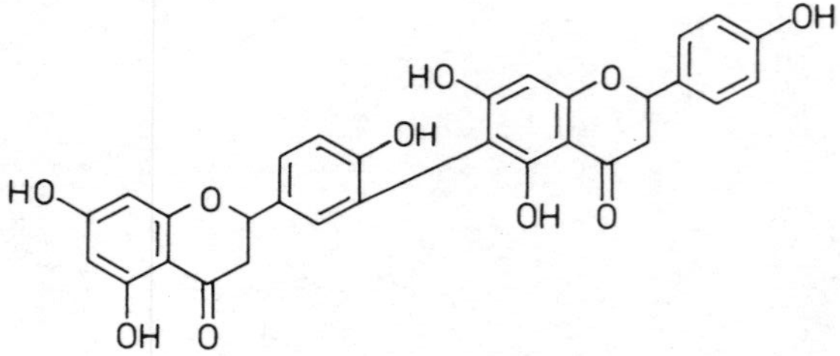

*7C.1
Tetrahydrorobustaflavone Ishratullah *et al.* (1977) (isoln. and ppn. of several methyl ethers)

Table 9.1 (*Contd.*)

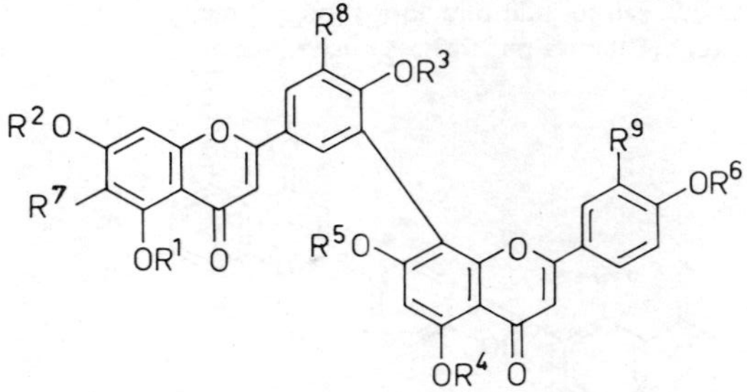

*8A

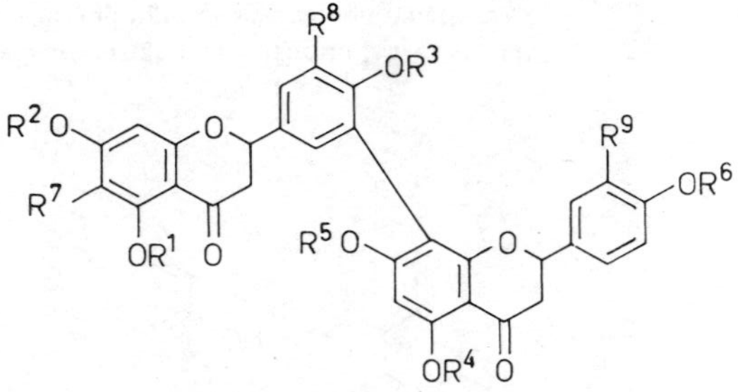

*8B

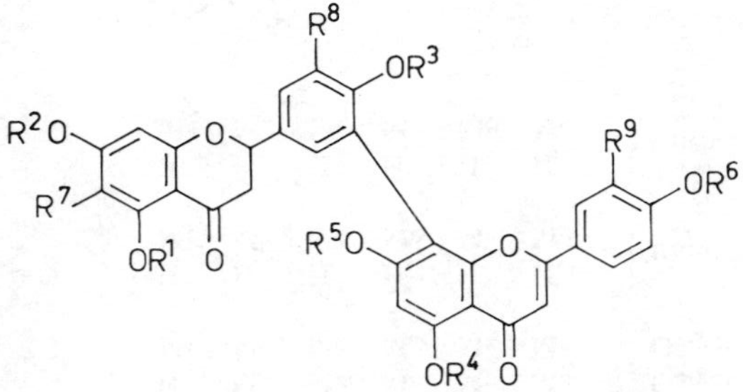

*8C

	R^1	R^2	R^3	R^4	R^5	R^6	R^7	R^8	R^9	*Leading references*
8A.1 Amentoflavone	H	H	H	H	H	H	H	H	H	Geiger and Quinn (1975)
*8A.2 Amentoflavone glucosides	H	{ Glc or H	Glc or H }	H	{ Glc or H	Glc or H }	H	H	H	Wallace and Markham (1978) (detectn. of several glucosides)

8A.3 Sequoiaflavone	H	Me	H	H	H	H	H	H	H	Geiger and Quinn (1975)
8A.4 Bilobetin	H	H	Me	H	H	H	H	H	H	
8A.5 Sotetsuflavone	H	H	H	H	Me	H	H	H	H	Geiger and Quinn (1975); Ilyas *et al.* (1978)
8A.6 Podocarpusflavone A	H	H	H	H	H	Me	H	H	H	Geiger and Quinn (1975)
8A.7 Ginkgetin	H	Me	Me	H	H	H	H	H	H	
8A.8 7,7″-Di-*O*-methylamentoflavone	H	Me	H	H	Me	H	H	H	H	Ilyas *et al.* (1978)
8A.9 Podocarpusflavone B(=putraflavone)	H	Me	H	H	H	Me	H	H	H	
8A.10 Amentoflavone 4′,7″-dimethyl ether	H	H	Me	H	Me	H	H	H	H	Geiger and Quinn (1975)
8A.11 Isoginkgetin	H	H	Me	H	H	Me	H	H	H	
8A.12 Amentoflavone 7″,4‴-dimethyl ether	H	H	H	H	Me	Me	H	H	H	
*8A.13 7,4′,7″-Tri-*O*-methylamentoflavone	H	Me	Me	H	Me	H	H	H	H	Ilyas *et al.* (1978)
8A.14 Sciadopitysin	H	Me	Me	H	H	Me	H	H	H	
8A.15 Heveaflavone	H	Me	H	H	Me	Me	H	H	H	
8A.16 Kayaflavone	H	H	Me	H	Me	Me	H	H	H	Geiger and Quinn (1975)
8A.17 Amentoflavone 7,7″,4′,4‴-tetramethyl ether(=WI3)	H	Me	Me	H	Me	Me	H	H	H	
8A.18 Amentoflavone hexamethyl ether (=Dioonflavone)	Me	Me	Me	Me	Me	Me	H	H	H	
*8A.19 5′-Methoxybilobetin	H	H	Me	H	H	H	H	OMe	H	Joly *et al.* (1980b)
8A.20 5′,8″-Biluteolin	H	H	H	H	H	H	H	OH	OH	Geiger and Quinn (1975); Lindberg *et al.* (1974)
*8A.21 7-*O*-Methyl-6-methylamentoflavone	H	Me	H	H	H	H	Me	H	H	Aqil *et al.* (1976)
8B.1 2,3-Dihydroamentoflavone	H	H	H	H	H	H	H	H	H	
8B.2 2,3-Dihydroamentoflavone 7″,4‴-dimethyl ether	H	H	H	H	Me	Me	H	H	H	Geiger and Quinn (1975)
8B.3 2,3-Dihydrosciadipitysin	H	Me	Me	H	H	Me	H	H	H	
(8B.4 2,3-Dihydroamentoflavonehexamethyl ether)	Me	Me	Me	Me	Me	Me	H	H	H	
8C.1 Tetrahydroamentoflavone	H	H	H	H	H	H	H	H	H	Ishratullah *et al.* (1977); El-Sohly *et al.* (1978); Ahmad *et al.* (1981)

Table 9.1 (*Contd.*)

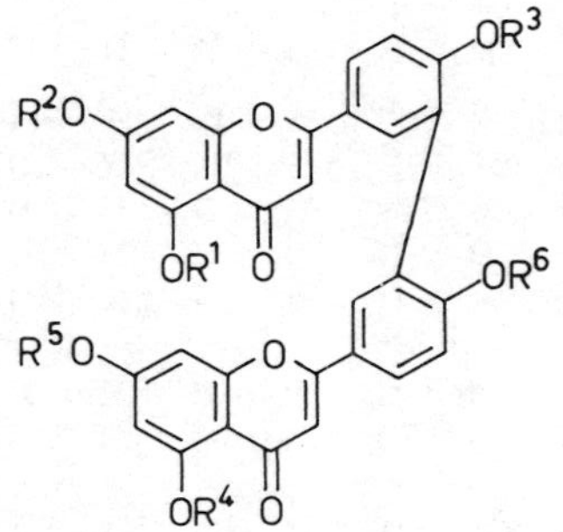

R2O OR1 O OR3 R5O OR4 O OR6

[9A.1($R^1 = R^2 = R^3 = R^4 = R^5 = R^6 =$ Me)3′,3‴-biapigenin-5,5″,7,7″,4′,4‴-hexamethyl ether] Geiger and Quinn (1975)

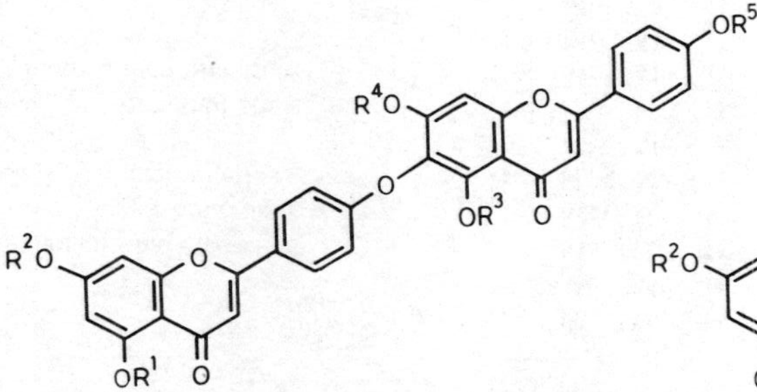

R2O OR1 O R4O OR3 O OR5

10A

R2O OR1 O R4O OR3 O OR5

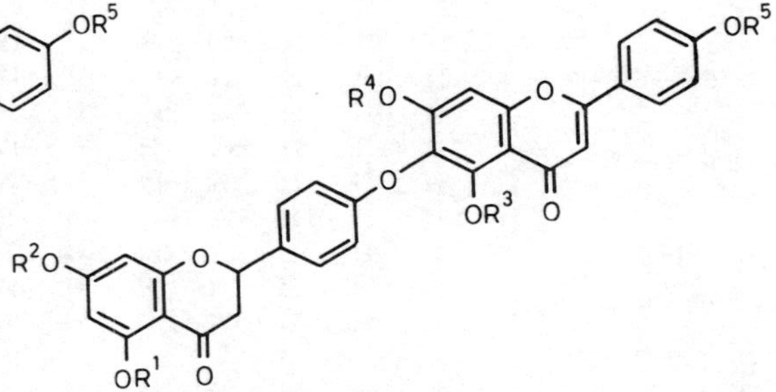

10B

	R^1	R^2	R^3	R^4	R^5	*Leading references*
10A.1 Hinokiflavone	H	H	H	H	H	Geiger and Quinn (1975)
10A.2 Neocryptomerin	H	Me	H	H	H	
10A.3 Isocryptomerin	H	H	H	Me	H	
10A.4 Cryptomerin A	H	H	H	H	Me	
10A.5 Chamaecyparin	H	Me	H	Me	H	
10A.6 Hinokiflavone 7,4‴-dimethyl ether	H	Me	H	H	Me	
10A.7 Cryptomerin B	H	H	H	Me	Me	
10A.8 Hinokiflavone 7,7″,4‴-trimethyl ether	H	Me	H	Me	Me	
(10A.9 Hinokiflavone pentamethyl ether)	Me	Me	Me	Me	Me	
10B.1 2,3-Dihydrohinokiflavone	H	H	H	H	H	Geiger and Quinn (1975)
(10B.2 2,3-Dihydrohinokiflavonepentamethyl ether)	Me	Me	Me	Me	Me	

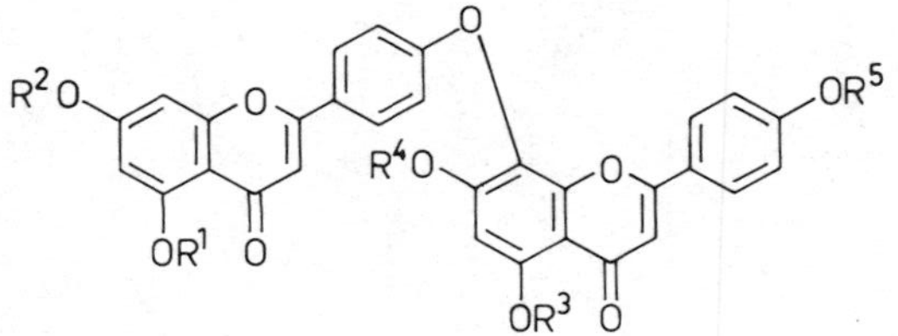

[11A.1($R^1=R^2=R^3=R^4=R^5=Me$)4‴-5,5″,7,7″-pentamethoxy-4′,8″-biflavonyl ether] Geiger and Quinn (1975)

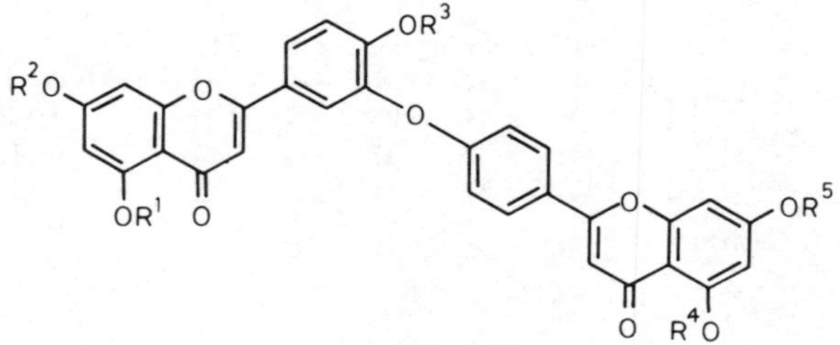

12A

Table 9.1 (*Contd.*)

	R^1	R^2	R^3	R^4	R^5	*Leading references*
12A.1 Ochnaflavone(=OSI)	H	H	H	H	H	Geiger and Quinn (1975); Okigawa *et al.* (1976)
12A.2 Ochnaflavone 4′-methyl ether (=OSII)	H	Me	H	H	H	
12A.3 Ochnaflavone 7,4′-dimethyl ether(=OSII)	H	Me	Me	H	H	
(12A.4 Ochnaflavone 4′,7,7″-trimethyl ether [=OSIV])	H	Me	Me	H	Me	
12A.5 Ochnaflavone pentamethyl ether	Me	Me	Me	Me	Me	

For the determination of biflavonoid structure, ^{13}C-NMR spectroscopy is a new and useful method (Chari *et al.*, 1977) and is described in more detail in Chapter 2. With ^{1}H-NMR spectroscopy, the shift reagent Eu(fod)$_3$ has been used successfully to establish the location of substituents at the 6 and 8 positions of the A rings (Okigawa *et al.*, 1975).

A new mass spectrometric technique is field desorption (FD), which yields only molecular ions, and sometimes doubly charged ions, but no fragments (Geiger, unpublished); it is therefore the best method for determining the molecular weight unequivocally. Its accuracy is, however, unfortunately not sufficient for providing exact elemental composition data.

9.3 SYNTHESIS AND TRANSFORMATIONS

In certain cases (e.g. the partial methyl ethers of agathisflavone of cupressuflavone) synthesis is still the best method to prove the structure. The earlier synthetic work has been reviewed quite comprehensively by Locksley (1973). Recent synthetic work falls into two categories. Some authors are concerned with the synthesis of naturally occurring biflavonoids and their methyl ethers (Ahmad and Razag, 1972, 1976; Chen *et al.*, 1975a,d, 1976a,b, 1978; Gandhi and Tiwari, 1975; Khan *et al.*, 1978; Lindberg *et al.*, 1974; Nakazawa and Wada, 1974; Okigawa *et al.*, 1976). Other authors are carrying out 'pure synthetic work', since the biflavonoids which they are preparing, have not yet been found to be naturally occurring. When new biflavonoids are found or when new syntheses are planned, it would be worthwhile consulting these papers (Berge *et al.*, 1979; Gandhi, 1976, 1977, 1978; Gandhi and Tiwari, 1975, 1976, 1977, 1978; Mathai *et al.*, 1967; Saharia and Tiwari, 1976).

The work on transformation of biflavonoids has been mainly concerned with the dehydrogenation of flavanone to flavone moieties. Lin and Chen (1974b) and Chen and Lin (1976) reported that dehydrogenation with iodine/potassium acetate might lead to isomeric flavone moieties, due to the formation of chalcone intermediates and they demonstrated that dehydrogenation with *N*-bromosuccinimide proceeds without rearrangement. A new reagent for the same reaction –iodine–dimethyl sulphoxide–sulphuric acid – has been described by Fatma *et al.* (1979a).

9.4 PHARMACOLOGY

Since several biflavone-containing plants (e.g. *Viburnum prunifolium* L., *Juniperus communis* L. and *Ginkgo biloba* L.) are used medicinally, some pharmacological and biochemical tests have been carried out on a few biflavonoids. These tests have included spasmolysis (Hörhammer *et al.*, 1967), peripheral vasodilatation, antibradykinin acitivity and antispasmogenic action against prostaglandin PGE$_1$ (Natarajan *et al.*, 1970), inhibition of cyclic GMP

and cyclic AMP phosphodiesterase (Ruckstuhl *et al.*, 1979; Beretz *et al.*, 1979) and inhibition of hepatoma cells (Joly *et al.*, 1980a). In most tests, biflavonoids proved to be more active than their monomeric counterparts.

9.5 SOME UNRELATED BI- AND TRI-FLAVONOID STRUCTURES

The compounds treated in this section are biogenetically unrelated to the other biflavonoids discussed in this chapter. Most of them (13–17; Table 9.2) are indeed closer related to the proanthocyanidins (see Chapter 8) because the 4 position is involved in the interflavonyl link, but they are not proanthocyanidins in the strict sense, since they do not yield anthocyanidins upon acid treatment. Since their occurrence is to our present knowledge quite erratic, they are not dealt with in the next section of this chapter, so that plant sources are mentioned along with their structures (Table 9.2).

Table 9.2 List of miscellaneous di- and trimeric flavonoids

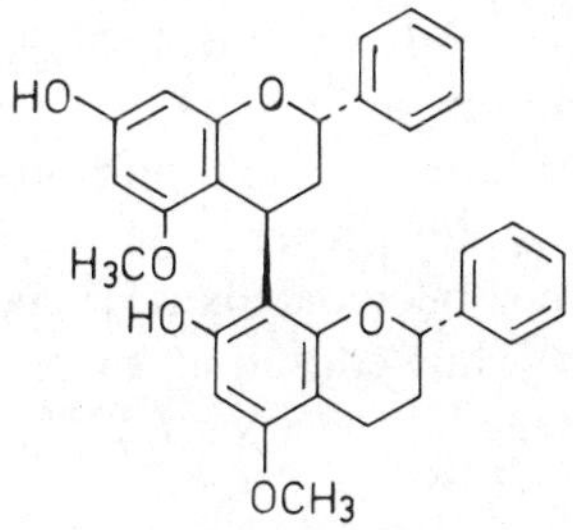

HO, O, H_3CO, HO, O, OCH_3

13.1 'Biflavon 12'; Camarda *et al.* (1981)
occurrence: Dragon's blood [resin of *Daemonorops draco* (Palmae)]

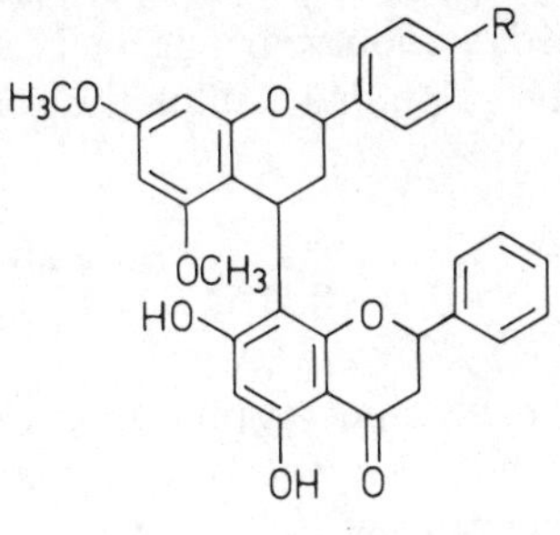

R, H_3CO, O, OCH_3, HO, O, OH, O

14.1 Xanthorrone } Birch *et al.* (1967) and (1969);
14.2 Hydroxyxanthorrone } Birch and Dahl (1974)
occurrence: resin of *Xanthorrhoea* sp. (Liliaceae).

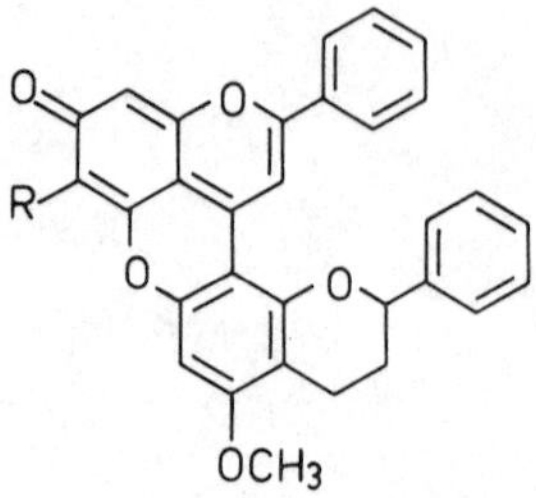

15.1 Nor-dracorubin (R = H); Cardillo *et al.* (1971)
15.2 Dracorubin (R = CH_3); Robertson *et al.* (1950) and references quoted therein.
occurrence: Dragon's blood [resin of *Daemonorops draco* (Palmae)]

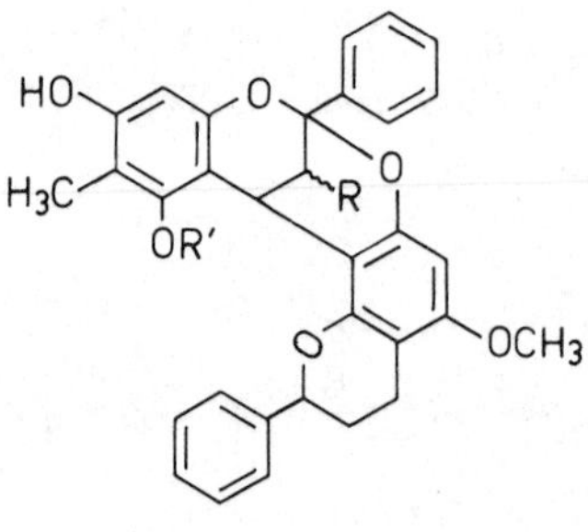

	R	R′	
16.1 'Prodeoxyanthocyanidin 13'	OH	CH_3	Camarda *et al.* (1981)
16.2 'Prodeoxyanthocyanidin 14'	H	CH_3	
16.3 'Prodeoxyanthocyanidin 15'	H	H	
16.4 'Trimeric compound 31'	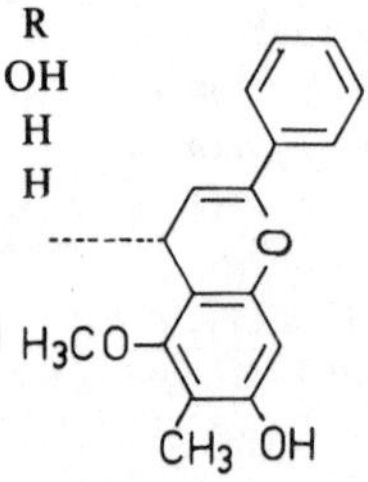	CH_3	

occurrence of 16.1–4: Dragon's blood [resin of *Daemonorops draco* (Palmae)]

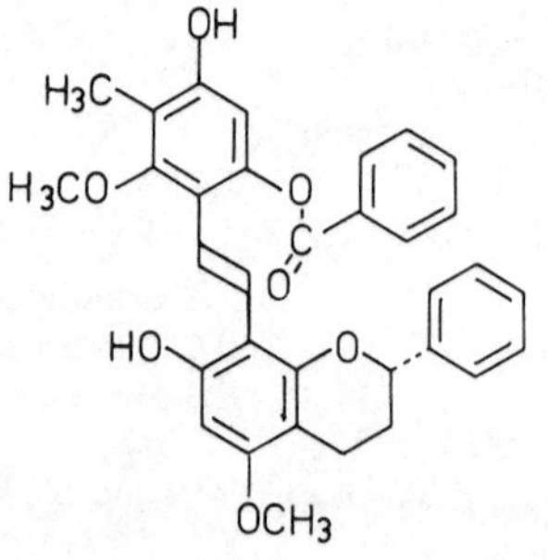

17.1 (2,*S*)-8-*trans*-[2-(6-Benzoyloxy-4-hydroxy-2-methoxy-3-methylphenyl) ethenyl]-5-methoxyflavon-7-ol; Merlini and Nasini (1976).
occurrence: Dragon's blood [resin of *Daemonorops draco* (Palmae)]

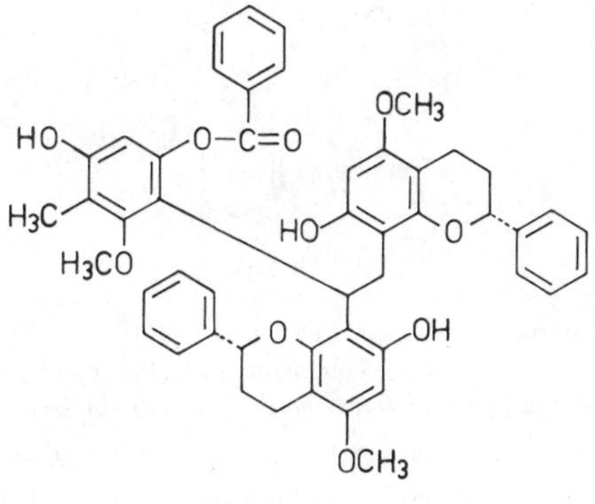

18.1 'Trimeric compound 29'; Camarda *et al.* (1981).
occurrence: Dragons blood [resin of Daemonorops draco (Palmae)]

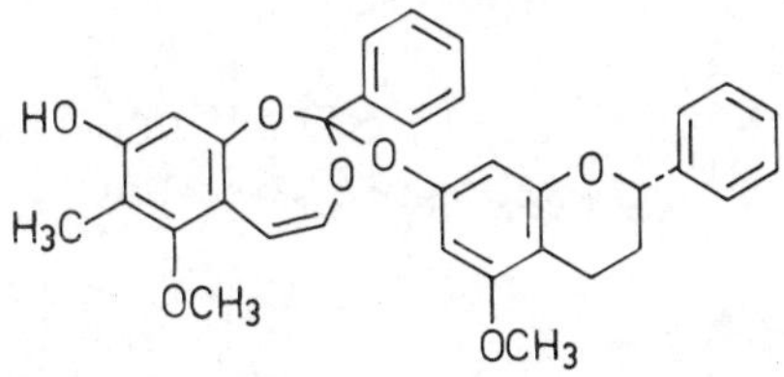

19.1 'Benzodioxepin 21'; Camarda *et al.* (1981)
occurrence: Dragon's blood [resin of *Daemonorops draco* (Palmae)]

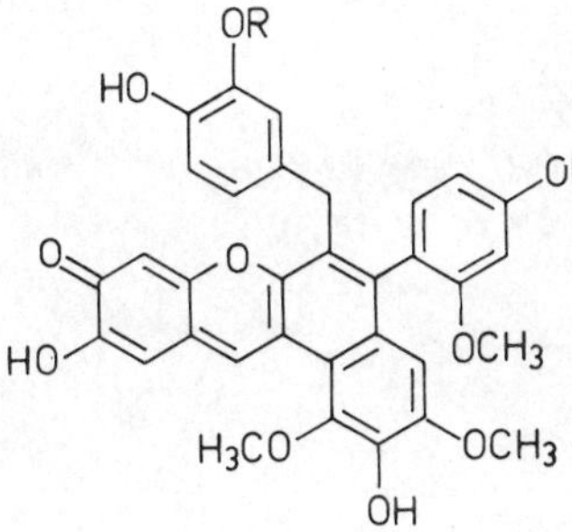

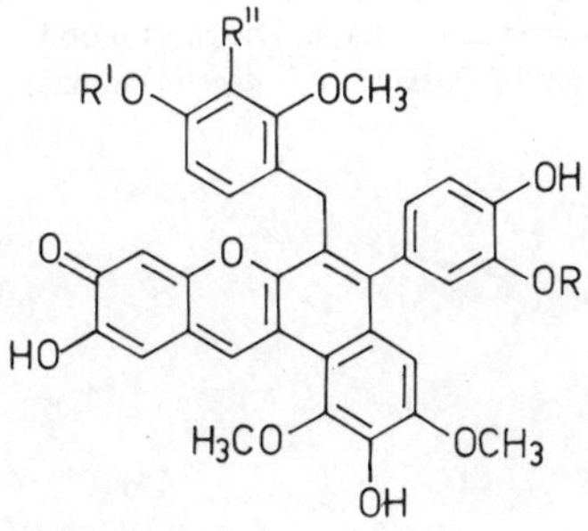

20.1 Santalin A (R = H)
20.2 Santalin B (R = CH_3)

21.1 Santarubin A (R = R″ = H; R′ = CH_3)
21.2 Santarubin B (R = R′ = CH_3; R″ = H)
21.3 Santarubin C (R = R′ = H; R″ = OH)

Arnone *et al.* (1975, 1977a, b, 1981).
occurrence: Red Woods from several *Pterocarpus* spp. and *Baphia nitida* (all Fabaceae).

9.6 NATURAL OCCURRENCE

The natural occurrence of biflavonoids is presented in Table 9.3 and 9.4. For reasons of brevity all compounds isolated from any member of a genus are listed under that genus. Intrageneric variations should be sought in Geiger and Quinn (1975; Tables 13.2–13.5) and in the more recent references listed at the right of the Tables. The number of species listed for each genus is the number reported to contain biflavonoids. The classification of angiosperm taxa is according to Dahlgren (1980).

Biflavonoids are now known to occur in a wide range of vascular plants, but for non-vascular plants there is still only the single record of a rather unusual substance (*viz.*, 5′,8″-biluteolin – 8A.20) in the moss *Dicranum scoparium* Hedw. (Lindberg *et al.*, 1974). A detailed discussion of the range and distribution of biflavonoids may be found in Geiger and Quinn (1975, p. 727 *et seq.*). Although the general picture has changed only slightly since the publication of that account, there have been some significant advances in our knowledge in particular groups, and these will be dealt with here.

9.6.1 Psilotales

Amentoflavone has now been reported in three species of *Psilotum* and two species of *Tmesipteris* (Table 9.3), and so appears to be a characteristic constituent of this very primitive order of living vascular plants. Wallace and Markham (1978) also reported the presence of amentoflavone glycosides in *Psilotum nudum* (L.) Beauv. and *Tmesipteris tannensis* (Spreng.) Bernh.; these are the first reports of a biflavone glycoside in nature. An earlier report that hinokiflavone (in the free state) occurs in *Psilotum* has not been borne out (Wallace & Markham, 1978).

9.6.2 Cycadales

The initial work by Dossaji *et al.* (1973) and others reported in Geiger and Quinn (1975, p. 728) has been extended to a systematic survey of the order encompassing 82 species drawn from all but one of the known genera (Dossaji *et al.*, 1975a). Their work showed a high degree of uniformity in biflavonoid pattern between species of the same genus, but showed significant differences between genera (Table 9.3), so providing support for the existing generic boundaries. In addition, the three families recognized by Johnson (1959) to comprise this order may be characterized by their biflavonoid profiles. Thus, all genera in the Zamiaceae contain amentoflavone and its methyl ethers only, while the monogeneric Cycadaceae possesses both amentoflavone and hinokiflavone derivatives, including 2,3-dihydro derivatives of both (8B and 10B); by contrast, the monotypic Stangeriaceae is devoid of biflavonoids.

Table 9.3 Natural occurrence of biflavonoids in non-flowering plants

	No. species	*Biflavonoid series* 6A	8A	10A	Other	References not in Geiger and Quinn (1975)
BRYALES						
Dicranum	1		20			Lindberg *et al.* (1974)
PSILOTALES						
Psilotum	3		1,2			Cooper-Driver (1977); Wallace and Markham (1978)
Tmesipteris	2		1,2			Cooper-Driver (1977); Wallace and Markham (1978); Voirin and Jay (1977)
SELAGINELLALES						
Selaginella	19		1	1,3		Voirin (1972)
CYCADALES						
Cycadaceae						
Cycas	9		1,5,7	1	8B.1 10B.1	
Zamiaceae						
Lepidozamia	2		1,6,7,11,14,17			
Macrozamia	14		1,4,7,14			
Encephalartos	28		1,4,7,11,14,17			Dossaji *et al.* (1975a) (applies to *Cycas* through *Stangeria*)
Dioon	6		1,3,4,7,14,17,18			
Ceratozamia	5		1,7,14			
Zamia	14		1,3,4,7,14,17			
Bowenia	2		1,4,7,11,14,17			
Microcycas	1		1,4,5,7			
Stangeriaceae						
Stangeria	1		—	—	—	
GINKGOALES						
Ginkgo	1		4,7,11,14,19			Joly *et al.* (1980b)
TAXALES						
Amentotaxus	1		1			
Taxus	3		3,5,7,14			Khan *et al.* (1976)
Torreya	1		16			

CONIFERALES						
Araucariaceae						
Agathis	2	1,2,4	1	1	5A.1,2,4	
Araucaria	5	1,2,4,5,6,7	1,4,5,8,10,13,14,16,17	1	5A.1–5 7A.1	Ilyas *et al.* (1977b)
Cephalotaxaceae						
Cephalotaxus	1		1,3,7,14,21			Aqil *et al.* (1976)
Cupressaceae						
Cupressoideae						
Cupresseae						
Cupressus	6	1,3	1,6	1,3		Taufeeq *et al.* (1978)
Chamaecyparis	2			1,3,4,5		
Thujopsideae						
Thuja	4		1	1		
Thujopsis	1		5,14	1,4,5		
Calocedrus	2			1		
Junipereae						
Juniperus	8	1,4	1,6,14*	1,3,4	7A.1	Mashima *et al.* (1970); Lamer-Zarawska (1975); Fatma *et al.* (1979b)
Callitroideae						
Actinostrobeae						
Callitris	1		1			Prasad and Krishnamurty (1977)
Podocarpaceae						
Podocarpus s.s.	4		1,4,6,9,11,14,16	1,2,4		Bhakuni *et al.* (1974); Rizvi *et al.* (1974); Prasad and Krishnamurty (1976)
Prumnopitys Phil.	1		1,3,4,6	1		
Decussocarpus de Laub.	2		1,6,11,16			
Falcatifolium de Laub.	1		1,5,6,7,*		}	Quinn and Gadek (1981)
Dacrydium s.s.	3		4,5,6,7,11,12,17,*	1	}	Quinn and Gadek (1981)
Lagarostrobos Quinn	2		1,3,5,6,12,17,*	1	}	Quinn and Gadek (1981)
Halocarpus Quinn	3		1,3,5,6,7,11,*	1	}	Quinn and Gadek (1981)
Lepidothamnus Phil.	3	*,7	3,4,5,*	1	}	Quinn and Gadek (1981)

Table 9.3 (*Contd.*)

	No.	Biflavonoid series				References
	species	6A	8A	10A	Other	not in Geiger and Quinn (1975)
Taxodiaceae						
*Cryptomeria**	1			1,4,7		
Cunninghamia	2		3,5,16	1		
Glyptostrobus	1			1,3,4		
Metasequoia	1		1,12,14	1,3	8B.2,3	
Sequoia	1		3, 16	1		
Sequoiadendron	1		1,6	1		
Sciadopitys	1		14			
Taiwania	1		1,3,8	1,3	3A.1–3	Kamil *et al.* (1977), Kawano (1980)
Taxodium	2		1,4,6,12,14	1,3,4,7		Ishratullah *et al.* (1978)

* Indicates the presence of incompletely identified derivatives.

When Johnson (1959) removed *Stangeria eriopus* (Kunze) Nash from the Zamiaceae to a family of its own, he did so on the basis of its frond morphology. While recognizing that there was 'no apparent character in the reproductive structures inconsistent with its inclusion in the Zamiaceae', he concluded that the profound difference between the systems of growth processes involved in the development of the pinnately veined leaflets of *Stangeria* and the longitudinally veined leaflets found elsewhere in the order, neither of which was easily derivable from the other, indicated an early separation in the evolution of the two lines from an ancient procycadalean ancestor with a more generalized vascularization of the frond. The chemical discontinuity between these two morphological types provides strong independent evidence to support the isolated taxonomic position assigned to *Stangeria eriopus* by Johnson.

There is a striking resemblance in the morphology of the frond of *Stangeria* to those of the eusporangiate ferns belonging to the family Marattiaceae, and the absence of biflavonoids from *Stangeria*, as well as from all the ferns (Cooper-Driver, 1977), might be seen as evidence that this genus is one of the more primitive and fern-like cycads. However, as pointed out by Dossaji *et al.* (1975a), this absence may not be a primitive condition, but might have been derived by the secondary loss of the ability to form biflavonoids during the evolution of this line of cycads. Indeed there is a large weight of evidence from internal anatomy and reproductive structures to show that *Stangeria* is a natural member of the Cycadales, and that the morphological resemblance to marattialean ferns has resulted from convergence in one set of characters by members of two quite distant groups. Further, the occurrence of the same type of biflavonoid (*viz.*, amentoflavones) in three other orders of the gymnosperms (Ginkgoales, Coniferales and Taxales) as well as in two groups of non-seed plants (psilophytes and lycopods) is evidence that the ability to form these biflavones was probably evolved early in the development of vascular plants (Geiger and Quinn, 1975; pp. 735–736), and so is almost certainly primitive in the gymnosperms as a whole. Thus it seems most likely that Johnson's common procycadalean ancestor already had the ability to form these biflavones, and that while this has been retained and even elaborated in the rest of the living members of the order, it has been lost in *Stangeria*, just as has been postulated for the conifer family Pinaceae (Geiger and Quinn, 1975; p. 735).

9.6.3 Coniferales

Continuing investigations in the Araucariaceae (e.g. Ilyas *et al.*, 1977b) have confirmed the distinctive complexity of the biflavonoid pattern in this family (Geiger and Quinn, 1975), and have added yet another series of biflavones (*viz.*, robustaflavone – 7A.1), to the four series previously recorded (5A, 6A, 8A and 10A).

Quinn and Gadek (1981) have recently completed an extensive survey of the biflavones of the genus *Dacrydium sensu lato* (family Podocarpaceae). For the

most part, the species were found to contain a complex mixture of amentoflavone derivatives*, together with trace amounts of hinokiflavone, in line with what has been reported elsewhere in the family (Table 9.3). Three species, however, were found to contain cupressuflavone derivatives as their major biflavonoid constituents and only minor amounts of amentoflavones and hinokiflavone. This striking biochemical discontinuity was used, along with other available data, to remove these three species to a new genus, *Lepidothamnus* Phil. (Quinn, 1982), which is considered to have no close affinity with any of the other species in *Dacrydium*. Even within the remainder of the genus, discontinuities were evident in the distribution of various partial methyl ethers of amentoflavone, which supported the previously expressed views of several workers that the whole concept of the genus *Dacrydium* was artificial. Thus these biochemical data were applied, with all other taxonomic evidence, to the task of redefining the generic boundaries, which has led to the recognition of another two new genera, *Halocarpus* Quinn and *Lagarostrobos* Quinn, in addition to a smaller and more appropriately circumscribed *Dacrydium* Sol. ex Lamb. emend. Quinn (Quinn, 1982).

The presence of these biflavonoids in *Lepidothamnus* is only the second record of cupressuflavone derivatives outside the Cupressaceae and Araucariaceae, the other being from a single species of *Casuarina* (Natarajan *et al.*, 1971). There is good evidence from reproductive morphology, karyotype and embryogeny to suggest that the genus *Lepidothamnus* is a natural member of the Podocarpaceae with no special relationship with either of the other two families. Thus it would seem that this is yet another example of the independent evolution of the ability to form cupressuflavone (Geiger and Quinn, 1975; p. 737).

Within the Taxodiaceae, two recent reports are worth noting. First, Ishratullah *et al.* (1978) have isolated several partial methyl ethers of both amentoflavone and hinokiflavone (8A.6,14; 10A.1,3,4,7) from the leaves of *Taxodium mucronatum* Tenore, which were previously reported to contain hinokiflavone alone (Pelter *et al.*, 1969). These results bring this species into line with the biflavonoid pattern found elsewhere in that genus and the family (Geiger and de Groot-Pfleiderer, 1973). Secondly, Kamil *et al.* (1977) report a similar mixture of amentoflavones and hinokiflavones (8A.1,3,8; 10A.1,3) from the monotypic genus *Taiwania*, but also isolated derivatives of the 3,3′-biapigenin, taiwaniaflavone (3A.1,2,3), previously known only as a synthetic compound.

Although there have been several important contributions to our knowledge of the biflavonoids of subfamily Cupressoideae of the Cupressaceae, this subfamily still presents a complex picture. Taufeeq *et al.* (1978) have recorded amentoflavones (8A.1,6), cupressuflavones (6A.1,3) and a monomethyl ether of hinokiflavone from *Cupressus lusitanica* var. *benthami* Carr., in line with earlier reports of members of all three series of biflavones in most members of this

* Subsequent work has identified band 2B as sequoiaflavone, 3C as podocarpusflavone A and 4A as sotetsuflavone.

genus (Geiger and Quinn, 1975; Table 13.3). Recent investigations on a wide range of *Juniperus* species indicate that a similar pattern based on the same three series of biflavones is widespread. Such a pattern may be shown to be a characteristic feature of both genera, although preliminary work by one of us (C.J.Q.) on cultivated material obtained from the Royal Botanic Gardens, Sydney, bears out the absence of cupressuflavone derivatives from the leaves of both *J. chinensis* L. and *J. conferta* Parl. as reported by Pelter *et al.* (1971). The occurrence of cupressuflavone derivatives in at least the majority of species of both *Cupressus* and *Juniperus*, and their absence from *Chamaecyparis* (Table 9.3), calls into question the present taxonomy which separates *Juniperus* and *Cupressus* into different tribes, but considers the latter to be very closely related to *Chamaecyparis*. An extensive chemotaxonomic survey should produce data that would be very valuable in reassessing affinities within the subfamily Cupressoideae.

The only recent report for a member of the subfamily Callitroideae is of amentoflavone in the leaves of *Callitris rhomboidea* R.Br. ex A. and L.C. Rich. (Prasad and Krishnamurty, 1977). Preliminary results of a survey of several species of *Callitris* now underway (C. J. Quinn, unpublished results) indicate the biflavonoid pattern in this genus is both highly uniform and relatively simple, including amentoflavone and its mono- and di-methyl ethers only. This contrasts with the early record of hinokiflavone in one species of *Callitris* by Sawada (1958), and the apparent universal occurrence of hinokiflavone in members of the subfamily Cupressoideae, and appears to support the present arrangement of subfamilies, although investigations of other genera in both subfamilies are obviously needed to confirm any such conclusion.

9.6.4 Angiospermae

The discovery of the occurrence of biflavonoids in the angiosperms has now been extended to a total of 11 families, with the report of amentoflavone in the leaves of *Garuga pinnata* Roxb. (Burseraceae; Ansari *et al.*, 1978) and *Nandina domestica* Thunb. (Nandinaceae; Morita *et al.*, 1974), a 3,8-linked biflavanone (2C.1) from the stem of *Lophiola americana* Wood (Amaryllidaceae; Xue and Edwards, 1980) and chamaejasmine (1C.1) in *Stellera chamaejasmine* L. (Thymeliaceae; Hwang and Chang, 1979). This includes the first report of a biflavonoid in a monocotyledon. Biflavonoids of the 2B and 2C series have now been isolated from the leaves, fruits, bark, wood or roots of 16 species of *Garcinia*, as well as *Allanblackia floribunda* Oliv. and *Pentaphalangium solomonse* Warb., and it would seem that these substances may well be a characteristic feature of the Clusiaceae (Guttiferae). In addition, amentoflavone and its derivatives have been isolated from the leaves of *Calophyllum inophyllum* L. and two species of *Garcinia*, and mesuaferrone A and B (6C.1; 6B.1) from the stamens of *Mesua ferrea* L. (Raju *et al.*, 1976, 1978). Amentoflavone derivatives (8A.1, 6) have also been recorded from a third member of the Euphorbiaceae, *viz.*, *Manihot utilissima* Pohl, by Kamil *et al.* (1974).

Table 9.4 Natural occurrence of biflavonoids in the angiosperms

MAGNOLIIDAE (DICOTYLEDONAE)

	No. species	*Biflavonoid series*				*References not in Geiger and Quinn (1975)*
		2B	*2C*	*8A*	*Other*	
RANUNCULIFLORAE						
Nandinaceae						
Nandina	1			1		Morita *et al.* (1974)
MALVIFLORAE						
Rhamnaceae						
Phyllogeiton	1				2D.1	
Euphorbiaceae						
Hevea	1			15		
Manihot	1			1,6		Kamil *et al.* (1974)
Putranjiva	1			1,6,9		
Thymeliaceae						
Stellera	1				1C.1	Hwang and Chang (1979)
THEIFLORAE						
Ochnaceae						
Ochna	1				12A.1,2,3	Okigawa *et al.* (1976)
Clusiaceae (Guttiferae)						
Allanblackia	1	1,3				
Calophyllum	1			1		Ansari *et al.* (1976); Bandaranayake *et al.* (1975);
Garcinia	16	1–5	1,3–8	1,6		Chen *et al.* (1975b, c); Crichton and Waterman (1979); Cotterill *et al.* (1977, 1978)
Mesua	1				6B.1; 6C.1	Raju *et al.* (1976, 1978)
Pentaphalangium	1		1			Owen and Scheinmann (1974)
ROSIFLORAE						
Casuarinaceae						
Casuarina	4				6A.1; 10A.1	

RUTIFLORAE					
Anacardiaceae					
Rhus	1		1	4C.1; 5A.1; 5B.1; 5C.1; 6C.1; 7A.1; 10A.1	Lin and Chen (1974c, d); Chen and Lin (1975, 1976); Chen *et al.* (1978)
Semecarpus	2		1	7C.1; 8C.1	Ahmad *et al.* (1981); Ishratullah *et al.* (1977)
Toxicodendron	1		1	8C.1	El Sohly *et al.* (1978)
Burseraceae					
Garuga	1		1		Ansari *et al.* (1978)
CORNIFLORAE					
Caprifoliaceae					
Viburnum	1		1		
LILIIDAE (MONOCOTYLEDONAE)					
LILIIFLORAE					
Amaryllidaceae					
Lophiola	1	1			Xue and Edwards (1980)

9.7 EVOLUTIONARY ASPECTS

The extent of the natural occurrence of biflavonoids in angiosperm taxa is still very imperfectly known, so any conclusions on the taxonomic implications or evolutionary status of these biochemical markers must be speculative. There is evidence, however, that while biflavonoids may prove to be a characteristic feature of some families (e.g. the Clusiaceae) or groups of families, in other cases their distribution is isolated or sporadic (e.g. the isolation of amentoflavone from only one of 22 species of *Viburnum* examined by Hörhammer *et al.*, 1966). Further, these occurrences are mainly in families that are regarded neither as being relatively primitive nor as showing any close affinity with one another. Thus, according to a recently published classification of the angiosperms (Dahlgren, 1980), the ten dicotyledonous families with biflavonoids are assigned to six different superorders as set out in Table 9.4. Of these, only the Ranunculiflorae can be said to be relatively primitive among the angiosperms, and the single occurrence of amentoflavone is recorded in one of the more specialized members of the superorder (see Metcalfe and Chalk, 1957, p. 62). A survey of the more primitive angiosperms is obviously needed, but the apparent absence of these dimers from the Magnolialean and most Ranalean families, and their sporadic occurrence in more specialized families including one monocotyledonous family, does not appear to support the view that the latter have simply retained a primitive pre-angiosperm set of metabolic pathways that have been lost in the majority of angiosperms. It seems more likely that the ability to form biflavonoids was lost early in the evolution of the angiosperms, but that it has arisen again as a more recent specialization in several families quite independently.

Members of one of the two series of biflavonoids most common in non-flowering plants (8A or 10A) have been found in some members of all six dicotyledonous superorders, and amentoflavone or its partial methyl ethers have been found in all but one (the Rosiflorae). This homology of biflavones with the non-flowering plants is difficult to reconcile with the conclusion reached above of their independent evolution. In fact, it would appear that amentoflavones are the generalized and therefore primitive form of biflavonoid in the angiosperms as well as in the vascular plants as a whole (see Geiger and Quinn, 1975; p. 736). In order to reconcile these two lines of evidence, it is suggested that, although the ability to form biflavonoids was lost early in the evolution of the angiosperms, much of the genetic information may have been retained, so that only a few reverse mutations were required to restore full function to the pathway. This hypothesis of parallel evolution from a common gene pool would explain the curious distribution of compounds of the same series in so few extremely diverse angiosperm taxa. Subsequent evolution could easily have modified the pathway in order to form the various specialized biflavonoid series (e.g., 2B, 2C and 6B) found in particular families. Hence, while the presence of biflavonoids appears to be primitive in the non-flowering plants, and the angiosperms must have had

their origin amongst them, within the angiosperms these substances would be seen as a secondary specialization that has arisen independently in several families by parallel evolution.

9.8 CONCLUSION

The problems of applying chemical data accumulated piecemeal by many different workers have been discussed by Geiger and Quinn (1975; p. 726), and these problems are still evident in the recent data. For example, although cupressuflavone had already been reported in two species of *Juniperus* by Hameed *et al.* (1973), Ilyas *et al.* (1977a) reported only amentoflavone and hinokiflavone and its monomethyl ether in *J. macropoda* Boiss. Fatma *et al.* (1979b) re-examined the same species and reported, in addition to the above, podocarpusflavone A (8A.6) and cupressuflavone. The problem would be eased if authors made it clear whether they were reporting only some or perhaps only the major biflavonoid fractions present in the species. The presence of further unidentified fractions should always be indicated.

Techniques for the isolation and identification of biflavonoids, as well as the range of commonly occurring structures, are now well established, so that the stage is set for more wide-ranging surveys employing strictly comparable techniques. The studies of Dossaji *et al.* (1975a) and Quinn and Gadek (1981) referred to above are examples of what can be achieved. Such studies, with a stronger taxonomic flavour, as distinct from the strictly chemical approach that has been most common to date, are likely to produce data that will make significant contributions to our understanding of the affinities within several areas of the plant kingdom. In particular, it is desirable that the distribution of biflavonoids among the angiosperms should be investigated, starting with those families where these compounds have already been reported or with families that are closely allied to them. Biflavonoids appear to have come of age as chemotaxonomic characters.

REFERENCES

Ahmad, I., Ishratullah, Kh., Ilyas, M., Rahman, W., Seligmann, O. and Wagner, H. (1981), *Phytochemistry* **20**, 1169.

Ahmad, S. and Razag, S. (1972), *Pak. J. Sci. Ind. Res.* **15**, 361.

Ahmad, S. and Razaq, S. (1976), *Tetrahedron* **32**, 503.

Ansari, W. H., Rahman, W., Barraclough, D., Maynard, R. and Scheinmann, F. (1976), *J. Chem. Soc. Perkin I*, 1458.

Ansari, F. R., Ansari, W. H. and Rahman, W. (1978), *Indian J. Chem.* **16B**, 846.

Aqil, M., Rahman, W., Okigawa, M. and Kawano, N. (1976), *Chem. Ind.*, 567.

Arnone, A., Camarda, L., Merlini, L. and Nasini, G. (1975), *J. Chem. Soc. Perkin I*, 186.

Arnone, A., Camarda, L., Merlini, L., Nasini, G. and Taylor, D. A. H. (1977a), *J. Chem. Soc. Perkin I*, 2116.

Arnone, A., Camarda, L., Merlini, L. and Nasini, G. (1977b), *J. Chem. Soc. Perkin I*, 2118.
Arnone, A., Camarda, L., Merlini, L., Nasini, G. and Taylor, D. A. H. (1981), *Phytochemistry* **20**, 799.
Bandaranayake, M. W., Selliah, S. S., Sultanbawa, M. U. S. and Ollis, D. (1975), *Phytochemistry* **14**, 1878.
Beretz, A., Joly, M., Stoclet, J. C. and Anton, R. (1979), *Planta Med.* **36**, 193.
Berge, D. D., Kale, A. V. and Sharma, T. C. (1979), *Chem. Ind.*, 282.
Bhakuni, D. S., Bittner, M., Sammes, P. G. and Silva, M. (1974), *Rev. Latinoam. Quim.* **5**, 163.
Birch, A. J. and Dahl, C. J. (1974), *Austr. J. Chem.* **27**, 331.
Birch, A. J., Dahl. C. J. and Pelter, A. (1967), *Tetrahedron Lett.* 481.
Birch, A. J., Dahl, C. J. and Pelter, A. (1969), *Austr. J. Chem.* **22**, 423.
Camarda, L., Merlini, L. and Nasini, G. (1981), presented at the International (VI Hungarian) Bioflavonoid Symposium, Munich.
Cardillo, G., Merlini, L. and Nasini, G. (1971), *J. Chem. Soc. (C)*, 3967.
Chari, V. M., Ilyas, M., Wagner, H., Neszmélyi, A., Chen, F.-C., Chen, L.-K., Lin, Y.-C and Lin, Y.-M. (1977), *Phytochemistry* **16**, 1273.
Chen, F.-C. and Lin, Y.-M. (1975), *Phytochemistry* **14**, 1644.
Chen, F.-C. and Lin, Y.-M. (1976), *J. Chem. Soc. Perkin I*, 98.
Chen, F.-C., Lin, Y.-M. and Liang, C.-M. (1974), *Phytochemistry* **13**, 276.
Chen, F.-C., Ho, T.-I., Ueng, T. and Lin, Y.-M. (1975a), *Heterocycles* **3**, 833.
Chen, F.-C., Lin, Y.-M. and Hung, J.-C. (1975b), *Phytochemistry* **14**, 300.
Chen, F.-C., Lin, Y.-M. and Hung, J.-C. (1975c), *Phytochemistry* **14**, 818.
Chen, F.-C., Lin, Y.-M., Shue, Y.-K. and Ueng, T. (1975d), *Heterocycles* **3**, 529.
Chen, F.-C., Ueng, T., Shue, Y.-K. and Lin, Y.-M., (1975e), *Chemistry* (*Chinese Chem. Soc., Taipei*), 91.
Chen, F.-C., Huang, S.-K., Ueng, T. and Lin, Y.-M. (1976a), *Heterocycles* **4**, 1913.
Chen, F.-C., Ueng, T., Ho, T.-I. and Lin, Y.-M. (1976b), *J. Chinese Chem. Soc. Ser. II*, **23**, 115.
Chen, F.-C., Lin, Y.-M. and Lin, Y.-C. (1978), *Heterocycles* **9**, 663.
Cooper-Driver, G. (1977), *Science* **198**, 1260.
Cotterill, P. J., Scheinmann, F. and Puranik, G. S. (1977), *Phytochemistry* **16**, 148.
Cotterill, P. J., Scheinmann, F. and Stenhouse, J. A. (1978), *J. Chem. Soc. Perkin I*, 532.
Crichton, E. G. and Waterman, P. G. (1979), *Phytochemistry* **18**, 1553.
Dahlgren, R. M. T. (1980), *Bot. J. Linn. Soc.* **80**, 91.
Dossaji, S. F., Bell, E. A. and Wallace, J. W. (1973), *Phytochemistry* **12**, 371.
Dossaji, S. F., Mabry, T. J. and Bell, E. A. (1975a), *Biochem. Syst. Ecol.* **2**, 171.
Dossaji, S. F., Mabry, T. J. and Wallace, J. W. (1975b), *Rev. Latinoamer. Quim.* **6**, 37.
El Sohly, M. A., Craig, J. C., Waller, C. W. and Turner, C. E. (1978), *Phytochemistry* **17**, 2140.
Fatma, W., Iqbal, J., Ismail, H., Ishratullah, Kh., Shaida, W. A. and Rahman, W. (1979a), *Chem. Ind.*, 315.
Fatma, W., Taufeeq, H. M., Shaida, W. A. and Rahman, W. (1979b), *Indian J. Chem.* **17B**, 193.
Gandhi, P. (1976), *Indian J. Chem.* **14B**, 1009.
Gandhi, P. (1977), *Curr. Sci.* **46**, 668.
Gandhi, P. (1978), *Experientia* **34**, 15.
Gandhi, P. and Tiwari, R. D. (1975), *J. Indian Chem. Soc.* **52**, 1111.
Gandhi, P. and Tiwari, R. D. (1976), *Indian J. Chem.* **14B**, 632.
Gandhi, P. and Tiwari, R. D. (1977), *J. Indian Chem. Soc.* **54**, 643.
Gandhi, P. and Tiwari, R. D. (1978), *Curr. Sci.* **47**, 576.
Geiger, H. and de Groot-Pfleiderer, W. (1973), *Phytochemistry* **12**, 465.
Geiger, H. and Quinn, C. (1975). In *The Flavonoids* (eds. J. B. Harborne, T. J. Mabry and H. Mabry), Chapman and Hall, London.
Hameed, N., Ilyas, M., Rahman, W., Okigawa, M. and Kawano, N. (1973), *Phytochemistry* **12**, 1494.
Hörhammer, L., Wagner, H. and Reinhardt, H. (1966), *Bot. Mag. Tokyo* **79**, 510.
Hörhammer, L., Wagner, H. and Reinhardt, H. (1967), *Z. Naturforsch.* **22B**, 768.

Hwang, W.-K. and Chang, C.-C. (1979), *K'o Hsueh T'ung Pao* **24**, 24.
Ilyas, M., Ilyas, N. and Wagner, H. (1977a), *Phytochemistry* **16**, 1456.
Ilyas, M., Seligmann, O. and Wagner, H. (1977b), *Z. Naturforsch.* **32C**, 206.
Ilyas, N., Ilyas, M., Rahman, W., Okigawa, M. and Kawano, N. (1978), *Phytochemistry* **17**, 987.
Ishratullah, Kh., Ansari, W. H., Rahman, W., Okigawa, M. and Kawano, N. (1977), *Indian J. Chem.* **15B**, 615.
Ishratullah, Kh., Rahman, W., Okigawa, M. and Kawano, N. (1978), *Phytochemistry* **17**, 335.
Johnson, L. A. S. (1959), *Proc. Linn. Soc. N. S. W.* **84**, 64.
Joly, M., Beck, J. P., Haag-Berruier, M. and Anton, R. (1980a), *Planta Med.* **39**, 230.
Joly, M., Haag-Berrurier, M. and Anton, R. (1980b), *Phytochemistry* **19**, 1999.
Kamil, M., Ilyas, M., Rahman, W., Okigawa, M. and Kawano, N. (1974), *Phytochemistry* **13**, 2619.
Kamil, M., Ilyas, M., Rahman, W., Hasaka, N., Okigawa, M. and Kawano, N. (1977), *Chem. Ind.*, 160.
Kawano, N. (1980), Private communication.
Khan, H. A., Ahmad, F. and Rahman, W. (1978), *Indian J. Chem.* **16B**, 845.
Khan, M. S. Y., Kumar, I., Prasad, J. S., Nagarajan, G. R., Parthasarathy, M. R. and Krishnamurthy, H. G. (1976), *Planta Med.* **30**, 82.
Lamer-Zarawska, E. (1975), *Pol. J. Pharmacol. Pharm.* **27**, 81.
Lin, Y.-M. and Chen, F.-C. (1974a), *Chemistry (Chinese Chem. Soc., Taipei)*, 14.
Lin, Y.-M. and Chen, F.-C. (1974b), *Chemistry (Chinese Chem. Soc., Taipei)*, 67.
Lin, Y.-M. and Chen, F.-C. (1974c), *Phytochemistry* **13**, 657.
Lin, Y.-M. and Chen, F.-C. (1974d), *Phytochemistry* **13**, 1617.
Lin, Y.-M. and Chen, F.-C. (1975), *J. Chromatogr.* **104**, D33.
Lindberg, G., Österdahl, B.-G. and Nilsson, E. (1974), *Chem. Scr.* **5**, 140.
Locksley, H. D. (1973), *Fortschr. Chem. Org. Naturstoffe* **30**, 208.
Mashima, T., Okigawa, M. and Kawano, N. (1979), *J. Pharm. Soc. Jpn.* **90**, 512.
Mathai, K. P., Kanakalakshmi, B. and Sethna, S. (1967), *J. Indian Chem. Soc.* **44**, 148.
Merlini, L. and Nasini, G. (1976), *J. Chem. Soc. Perkin I*, 1570.
Metcalfe, C. R. and Chalk, L. (1957), *Anatomy of the Dicotyledons*, Oxford University Press.
Morita, N., Shimizu, M., Arisawa, M. and Shirataki, Y. (1974), *Chem. Pharm. Bull.* **22**, 2750.
Moriyama, S., Okigawa, M. and Kawano, N. (1974), *J. Chem. Soc. Perkin I*, 2132.
Nakazawa, K. and Wada, K. (1974), *Chem. Pharm. Bull.* **22**, 1326.
Natarajan, S., Murti, V. V. S., Seshadri, T. R. and Ramaswami, A. S. (1970), *Curr. Sci.* **39**, 533.
Natarajan, S., Murti, V. V. S. and Seshadri, T. R. (1971), *Phytochemistry* **10**, 1083.
Okigawa, M., Khan, N. U., Kawano, N. and Rahman, W. (1975), *J. Chem. Soc. Perkin I*, 1563.
Okigawa, M., Kawano, N., Aqil, M. and Rahman, W. (1976), *J. Chem. Soc. Perkin I*, 580.
Owen, P. J. and Scheinmann, F. (1974), *J. Chem. Soc. Perkin I*, 1018.
Pelter, A., Warren, R., Usmani, J. N., Ilyas, M. and Rahman, W. (1969), *Tetrahedron Lett.*, 4259.
Pelter, A., Warren, R., Hameed, N., Ilyas, M. and Rahman, W. (1971), *J. Indian Chem. Soc.* **48**, 204.
Prasad, J. S. and Krishnamurty, H. G. (1976), *Indian J. Chem.* **14B**, 727.
Prasad, J. S. and Krishnamurty, H. G. (1977), *Phytochemistry* **16**, 801.
Quinn, C. J. (1982), *Aust. J. Bot.* **30**(3).
Quinn, C. J. and Gadek, P. (1981), *Phytochemistry* **20**, 677.
Raju, M. S., Srimannarayana, G., Rao, N. V. S., Bala, K. R. and Seshadri, T. R. (1976), *Tetrahedron Lett.*, 4509.
Raju, M. S., Srimannarayana, G. and Rao, N. V. S. (1978), *Indian J. Chem.* **16B**, 167.
Rizvi, S. H. M., Rahman, W., Okigawa, M. and Kawano, N. (1974), *Phytochemistry* **13**, 1990.
Robertson, A., Whalley, W. B. and Yates, J. (1950), *J. Chem. Soc.*, 3117.
Ruckstuhl, M., Beretz, A., Anton, R. and Landry Y. (1979), *Biochem. Pharmacol.* **28**, 535.
Saharia, H. S. and Tiwari, R. D. (1976), *J. Indian Chem. Soc.* **53**, 530.
Sawada, T. (1958), *J. Pharm. Soc. Jpn.* **78**, 1023.

Taufeeq, H. M., Fatma, W., Ilyas, M., Rahman, W. and Kawano, N. (1978), *Indian J. Chem.* **16B**, 655.
Voirin, B. (1972), *Phytochemistry* **11**, 257.
Voirin, B. and Jay, M. (1977), *Phytochemistry* **16**, 2043.
Wallace, J. W. and Markham, K. R. (1978), *Phytochemistry* **17**, 1313.
Xue, L.-X. and Edwards, J. M. (1980), *Planta Med.* **39**, 220.

CHECK LIST OF ALL KNOWN BIFLAVONOIDS

1 3,3″-Biapigenin
2 Chamaejasmin*
3 Bisdehydro Garcinia biflavonoid la
4 Sahranflavone
5 Volkensiflavone
6 Spicatoside
7 Fukugetin
8 3-Methyl ether
9 Fukugiside
10 Garcinia biflavonoid la
11 7″-Glucoside*
12 Garcinia biflavonoid 1
13 Garcinia biflavonoid 2a
14 Garcinia biflavonoid 2
15 Kolaflavanone*
16 Xanthochymusside
17 Manniflavanone*
18 Zeyherin
19 Taiwaniaflavone*
20 7″-Methyl ether*
21 4′,7″-Dimethyl ether*
22 Succedaneaflavone*
23 Agathisflavone
24 7-Methyl ether
25 7,7″-Dimethyl ether
26 7,4‴-Dimethyl ether
27 7,7″,4‴-Trimethyl ether*
28 Tetramethyl ether
29 Rhusflavone
30 Rhusflavanone
31 Cupressuflavone
32 7-Methyl ether
33 4′-Methyl ether*
34 7,7″-Dimethyl ether
35 7,4′-(or 7,4‴)-Dimethyl ether*
36 7,4′,7″-Trimethyl ether
37 Tetramethyl ether
38 Mesuaferrone B*
39 Neorhusflavanone*
40 Robustaflavone
41 Tetrahydrorobustaflavone*
42 Amentoflavone
43 Glucoside*
44 7-Methyl ether (sequoiflavone)
45 4′-Methyl ether (bilobetin)
46 7″-Methyl ether (sotetsuflavone)
47 4‴-Methyl ether (podocarpus flavone A)
48 7,4′-Dimethyl ether (gingketin)
49 7,7″-Dimethyl ether
50 7,4‴-Dimethyl ether (podocarpus flavone B)
51 4′, 7″-Dimethyl ether
52 4′,4‴-Dimethyl ether (isoginkgetin)
53 7″,4‴-Dimethyl ether
54 7,4′,7″-Trimethyl ether
55 7,4′,4‴-Trimethyl ether (sciadopitysin)
56 7,7″,4‴-Trimethyl ether (heveaflavone)
57 4′,7″,4‴-Trimethyl ether (kayaflavone)
58 7,7″,4′,4‴-Tetramethyl ether
59 Hexamethyl ether (dioonflavone)
60 5′-Methoxybilobetin*
61 5′,8″-Biluteolin
62 6-Methylamentoflavone 7-methyl ether*
63 2,3-Dihydroamentoflavone
64 7″,4‴-Dimethyl ether
65 2,3-Dihydrosciadopitysin
66 Tetrahydroamentoflavone
67 Hinokiflavone
68 7-Methyl ether (neocryptomerin)
69 7″-Methyl ether (isocryptomerin)
70 4‴-Methyl ether (cryptomerin A)
71 7,7″-Dimethyl ether (chamaecyparin)
72 7,4‴-Dimethyl ether
73 7″,4‴-Dimethyl ether (cryptomerin B.)
74 7,7″,4‴-Trimethyl ether
75 2,3-Dihydrohinokiflavone
76 Ochnaflavone
77 4′-Methyl ether
78 7,4′-Dimethyl ether
79 Pentamethyl ether

* Newly reported during 1975—1980.

CHAPTER TEN

Isoflavonoids

P. M. DEWICK

10.1 INTRODUCTION

The isoflavonoids are biogenetically related to the flavonoids but constitute a distinctly separate class in that they contain a rearranged C_{15} skeleton and may be regarded as derivatives of 3-phenylchroman. The enzyme(s) responsible for this biochemical rearrangement would appear to be rather specialized, since isoflavonoids have a very limited distribution, being confined essentially to the subfamily Papilionoideae (Lotoideae) of the Leguminosae. There are, however, occasional examples of their occurrence in the subfamily Caesalpinioidae, and in other families (Rosaceae, Moraceae, Amaranthaceae, Podocarpaceae, Chenopodiaceae, Cupressaceae, Iridaceae, Myristicaceae, Stemonaceae) together with recent reports of their isolation from microbial sources. Unless otherwise stated, all plants mentioned in this chapter are members of the Leguminosae.

Structurally, the isoflavonoids may be subdivided into several classes according to oxidation levels in the skeleton, and the complexity of the skeleton, for instance, formation of further heterocyclic rings. The structural variety encountered is surprisingly large, and several new classes have been identified during the period of review.

This chapter covers the important developments in isoflavonoid chemistry and biochemistry since the previous review of Wong (1975) and includes the data reported briefly in the addendum to that chapter. The last few years have seen considerable activity in the isoflavonoid field, and the number of known naturally occurring isoflavonoid aglycones has more than doubled during this period. Reviews covering the occurrence of isoflavonoids (Bandyukova and Kazakov, 1978; Torck, 1976), and the chemistry of coumestans (Darbarwar *et al.*, 1976), pterocarpanoids (pterocarpans, pterocarpenes, coumestans) (Jain and Tuli, 1978) and rotenoids (Jain *et al.*, 1978b) have been published.

10.2 SPECTROSCOPY

Although the established spectral techniques (UV, IR, ^{1}H-NMR, MS etc.) are still essential for structure determination of isoflavonoids, ^{13}C-NMR is a potentially useful tool yet to be generally applied in this area. A number of papers

containing ^{13}C-NMR spectral data for isoflavonoids have been published and are covered in detail in Chapter 2. MS and NMR data, where relevant, will be included in the appropriate class section. One important advance in identification techniques for isoflavonoids is in the use of combined GLC–MS with selected ion monitoring (Woodward, 1980b). Each compound in a mixture is monitored by means of two or three prominent ions chosen because of high relative intensities and because they were not present in the mass spectrum of other compounds of the mixture. It was possible by this technique to detect and measure a wide range of isoflavonoids produced in inoculation droplets from *Monilinia fructicola*–infected *Phaseolus vulgaris*. Detection limits were of the order of 0.1–3 pmol.

As in other flavonoid classes, UV spectroscopy is of prime importance in structure determination, particularly when only small amounts of sample are available. UV spectral data for isoflavones are well documented (Mabry *et al.*, 1970), but no such comprehensive information and diagrams are available for the other classes of isoflavonoids. Because the presence of UV absorbances at particular wavelengths together with the shape of the UV spectrum is so characteristic for some of the isoflavonoid classes, some typical neutral UV spectra (EtOH) are illustrated in Fig. 10.1. The precise absorption wavelength observed will naturally depend on the substitution pattern of the compound, and in some cases, this may result in the presence of an additional band, cf. maackiain and medicarpin.

The use of shift reagents is of less value than in the UV spectroscopy of flavonoids because of the lack of conjugation between the two aromatic nuclei in many classes of isoflavonoids. Most investigators record only shifts with NaOH to establish the presence/absence of phenolic groupings.

Isoflavones exhibit an intense absorption at ~245–275 nm and generally a less intense band or inflection at ~300–340 nm.

Isoflavanones show absorption at ~270 nm and ~310 nm, and the spectrum shape resembles that of a simple aryl aldehyde or ketone.

Isoflavans and *Pterocarpans*, lacking any additional conjugation with the aromatic rings, typically absorb at ~281 and 287 nm, but the two peaks may not always be resolved. With 8,9-disubstituted pterocarpans, an intense band at 300–310 nm is also observed – this is more marked with methylenedioxy rather than dimethoxy substituents and does not occur with 9,10-disubstituted pterocarpans (Ingham and Dewick, 1980b). This extra band also occurs in the corresponding isoflavan spectra (~300 nm), but overlaps the other bands.

6a-Hydroxypterocarpans exhibit spectra identical with the corresponding pterocarpans, but may easily be differentiated since the compounds readily dehydrate in acid solution to yield pterocarpenes. After the addition of one or two drops of

HCl to the cuvette, the characteristic stilbene-like spectrum of the *pterocarpene* rapidly appears. This shows twin peaks at ~335 nm and 353 nm.

Coumestans show intense absorption at about 340–350 nm, dependent on substitution pattern, together with a sharp, low intensity band at 300–310 nm. Their intense blue to violet fluorescence under UV light is, however, their most characteristic property.

Rotenoids usually show absorption in the region 280–300 nm with often a shoulder at ~315 nm, and spectra of *12a-hydroxy-* and *12a-methoxy-rotenoids* are virtually identical.

Dehydrorotenoids typically exhibit two bands at 275–280 nm and 300–320 nm, and further hydroxyl substitution at C-6 has little effect on the spectrum. *6-Oxodehydrorotenoids* absorb similarly, but at the slightly lower wavelengths 260–270 nm and 290–300 nm, and the increased conjugation may result in a low intensity band at higher wavelength, typically 320–350 nm.

Isoflav-3-enes show a broad band at ~320–325 nm in the UV spectrum, and *3-arylcoumarins* a similarly shaped spectrum with absorption in the region 340–360 nm. *3-Aryl-4-hydroxycoumarin* spectra are essentially the same.

2-Arylbenzofuran spectra are stilbene-like and resemble those of pterocarpenes, although usually at rather lower wavelengths, with the twin peaks at ~320 nm and ~335 nm.

Isoflavanquinones exhibit a strong absorption at 265–270 nm, with a weaker band at 360–400 nm, depending on substitution patterns.

Other classes of isoflavonoid are relatively rare, and it is not possible to generalize about typical spectra.

10.3 ISOFLAVONES

Isoflavones (*10.1*) continue to be the most abundant of the natural isoflavonoid derivatives, and over 160 isoflavone aglycones are now recognized. Table 10.1 lists the structures and sources of new isoflavones, and also newer sources of known compounds. The four simple isoflavones daidzein, formononetin, genistein and biochanin A are extremely common, and mention of new sources in the Table is restricted to genus only.

In the last review, only two isoflavones, tlatlancuayin and podospicatin, lacking 4′-oxygenation were known. Two compounds with no B-ring substitution at all have now been isolated, 5-hydroxy-7-methoxyisoflavone from

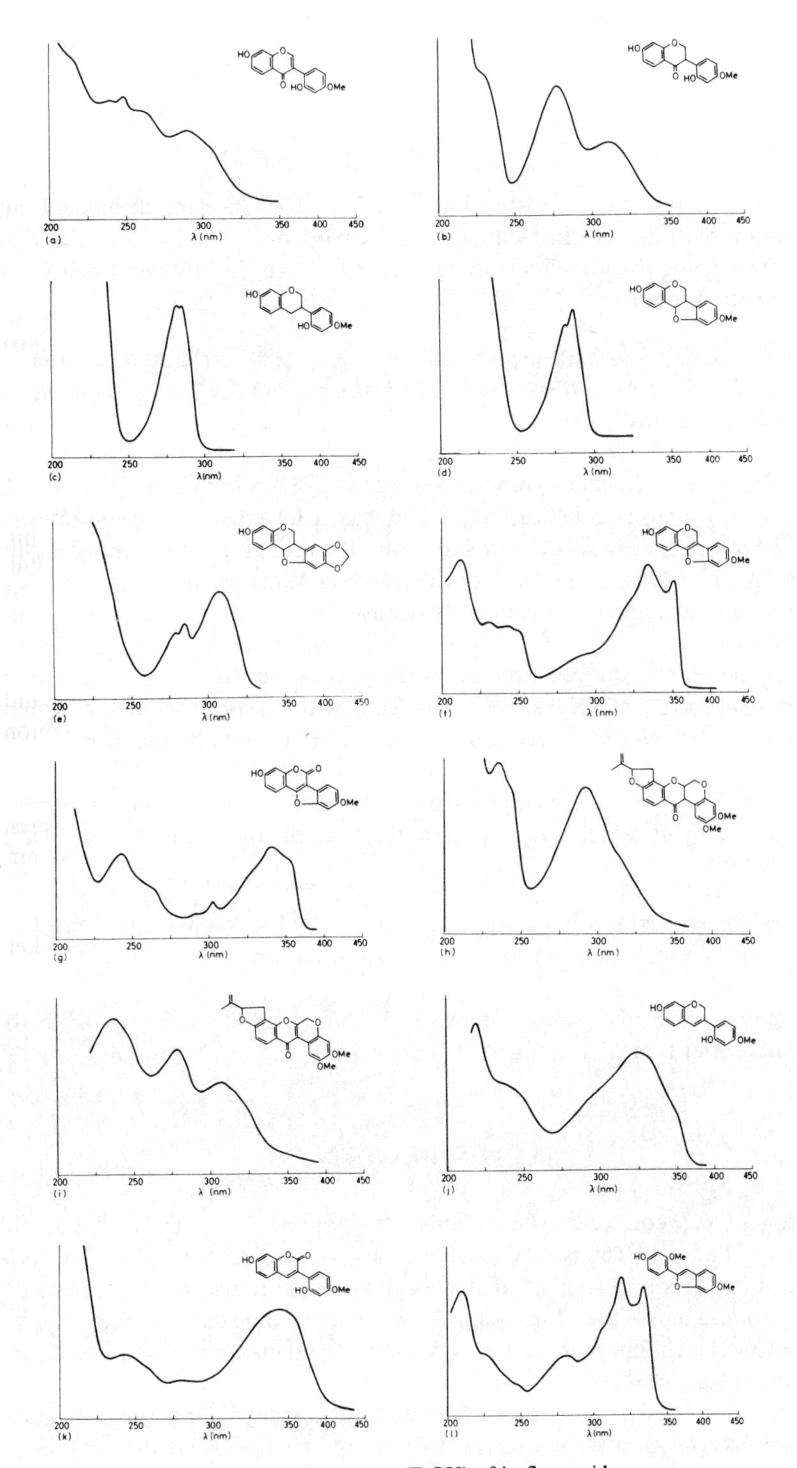

Fig. 10.1 UV spectra (EtOH) of isoflavonoids.

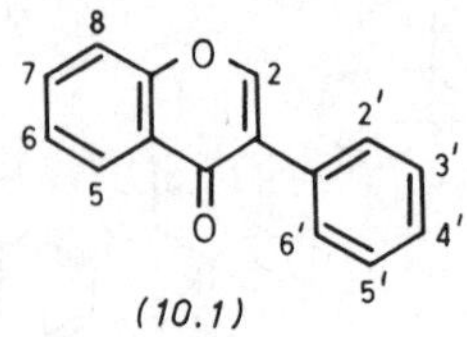

(10.1)

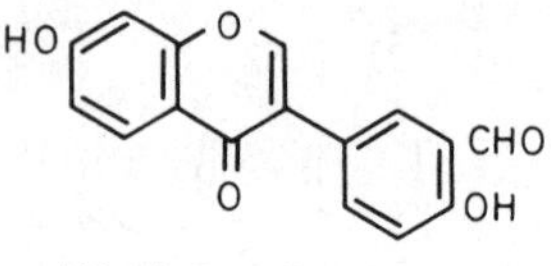

(10.2) Corylinal

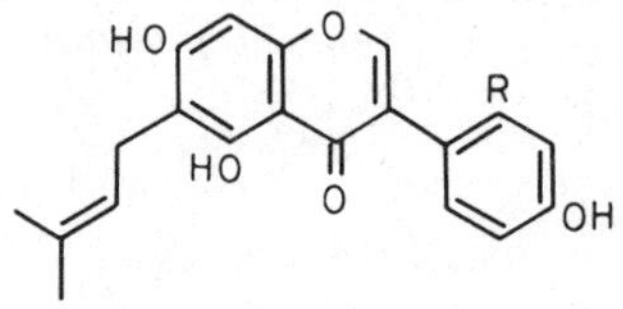

(10.3) R = H, Wighteone

(10.4) R = OH, Luteone

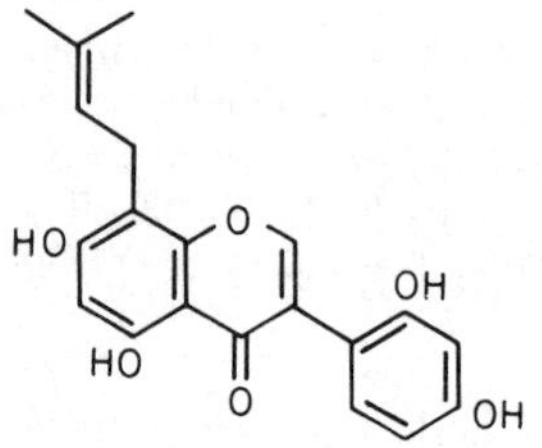

(10.5) 2,3-Dehydrokievitone

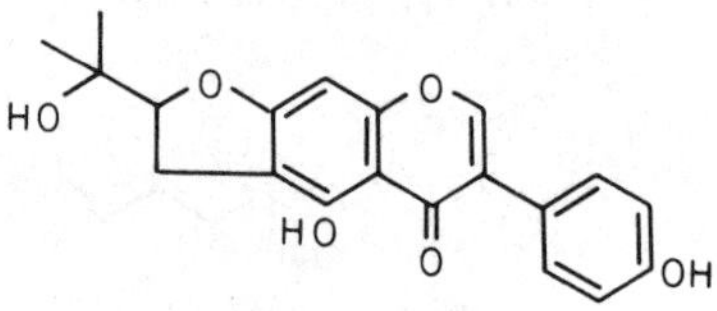

(10.6) Erythrinin C

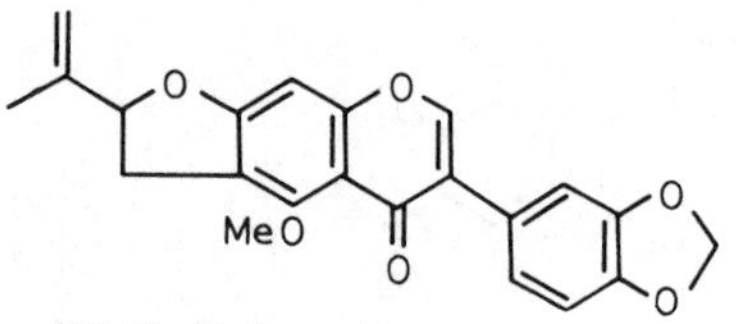

(10.7) Glabrescione A

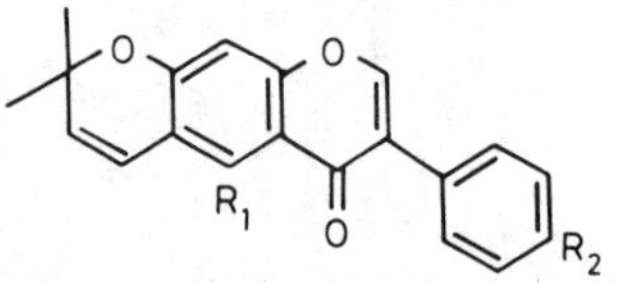

(10.8) R_1 = H, R_2 = OH, Erythrinin A

(10.9) R_1 = R_2 = OH, Alpinumisoflavone

(10.10) R_1 = OH, R_2 = OMe, Alpinumisoflavone-4′-methyl ether

(10.11) R_1 = R_2 = OMe, Alpinumisoflavone dimethyl ether

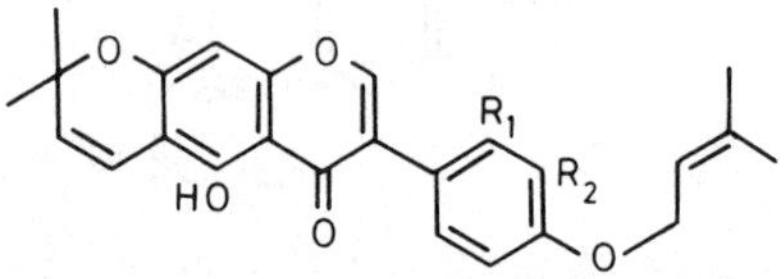

(10.12) R_1 = OH, R_2 = H, Isoauriculatin

(10.13) R_1 = H, R_2 = OH, Isoauriculasin

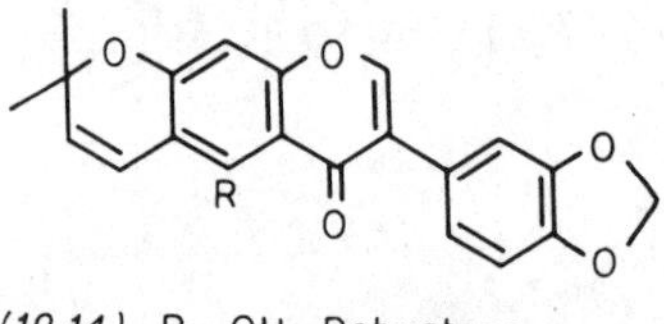

(10.14) R = OH, Robustone

(10.15) R = OMe, Robustone methyl ether

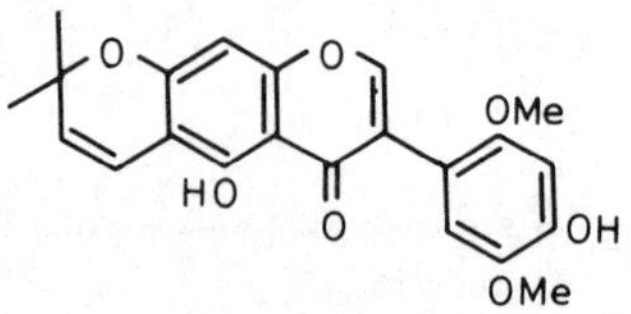

(10.16) Elongatin

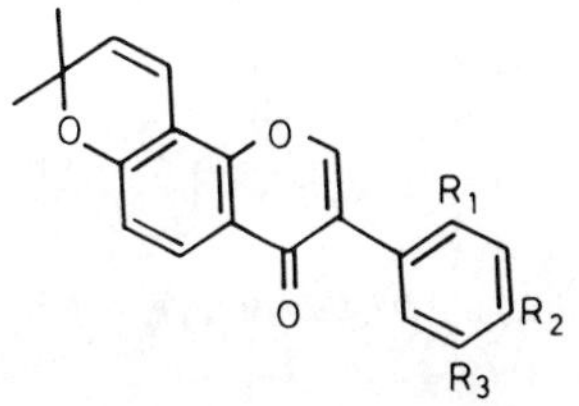

(10.17) $R_1 = R_3 = H$, $R_2 = OMe$, Calopogonium isoflavone A

(10.18) $R_1 = H$, $R_2R_3 = OCH_2O$, Calopogonium isoflavone B

(10.19) $R_1 = R_2 = R_3 = OMe$, Barbigerone

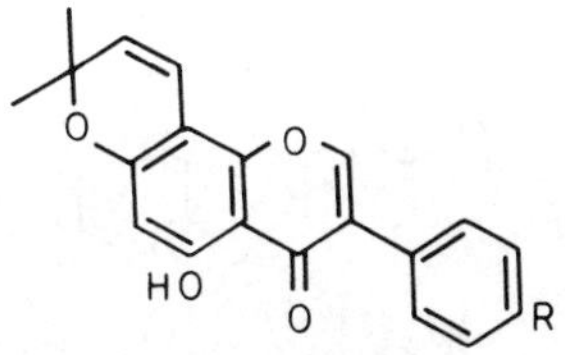

(10.20) R = OH, Derrone

(10.21) R = OMe, Derrone-4′-methyl ether

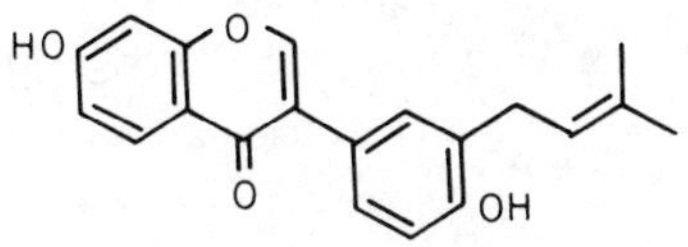

(10.22) Neobavaisoflavone

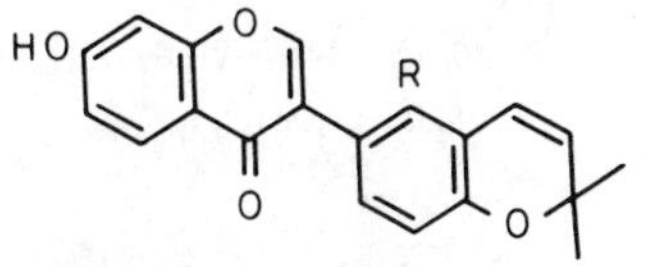

(10.23) R = H, Corylin

(10.25) R = OH, Glabrone

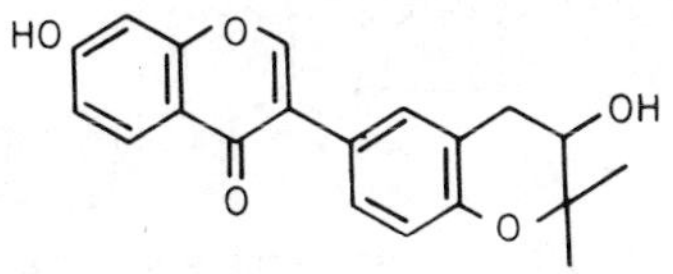

(10.24) Psoralenol

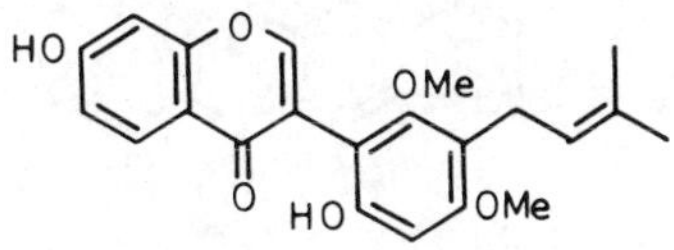

(10.26) Licoricone

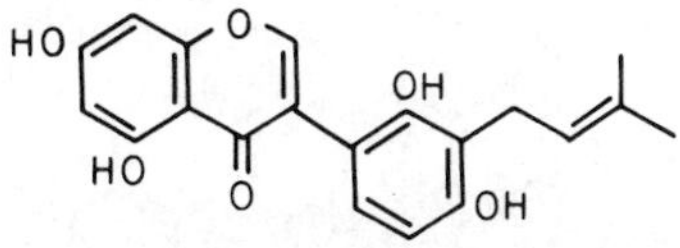

(10.27) Licoisoflavone A

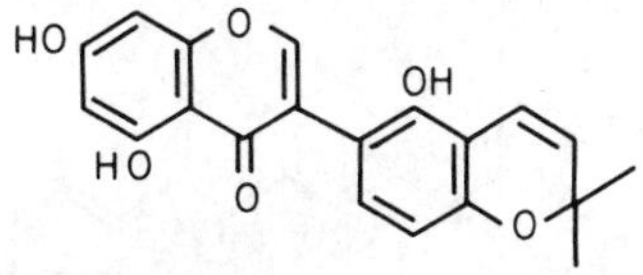

(10.28) Licoisoflavone B

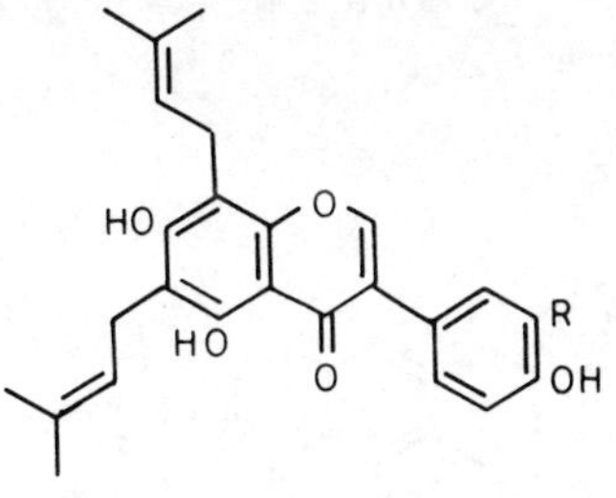

(10.29) R = H, 6,8-Di-(dimethylallyl)-genistein

(10.30) R = OH, 6,8-Di-(dimethylallyl)-orobol

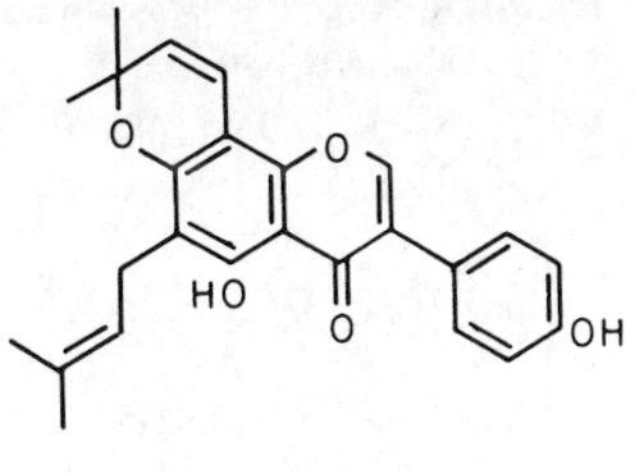

(10.31) Osajin

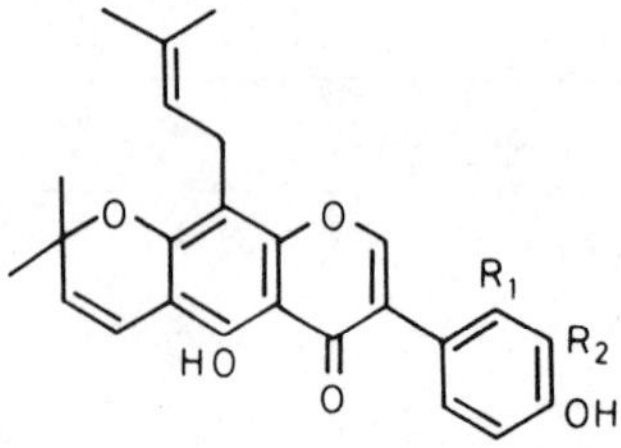

(10.32) R_1 = OH, R_2 = H, Auriculatin
(10.33) R_1 = H, R_2 = OH, Auriculasin

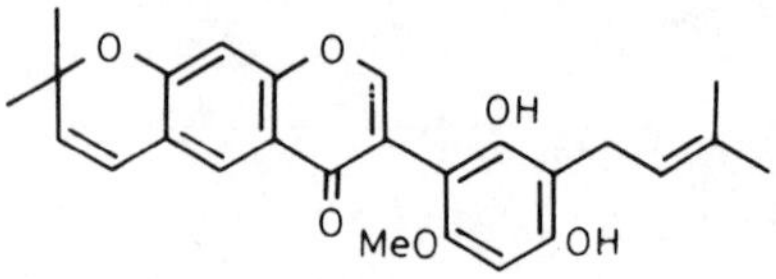

(10.34) Cajaisoflavone

Derris robusta seed (Chibber and Sharma, 1979b), and 5,7-dimethoxyisoflavone from immature fruit of groundnut (*Arachis hypogaea*) infected with *Aspergillus flavus* (Turner *et al.*, 1975). These compounds are remarkable because current theories for the aryl migration during the biosynthesis of isoflavones (see Section 10.18) require the presence of a 4′-(or 2′-) oxygen function. Both tlatlancuayin and podospicatin contain 2′-oxygen functions. Two other isoflavones, 5,7,2′-trihydroxyisoflavone (as its 7-glucoside) from *Cajanus cajan* (Bhanumati *et al.*, 1979b) and betavulgarin (2′-hydroxy-5-methoxy-6,7-methylenedioxyisoflavone or 2′-demethyltlatlancuayin), a phytoalexin from the *Cercospora beticola*-infected leaves of *Beta vulgaris* (Chenopodiaceae) (Geigert *et al.*, 1973) are further examples of 4′-deoxyisoflavones.

Some novel 2-methylisoflavones, also lacking any B-ring substitution, have been isolated from *Glycyrrhiza glabra* root (Bhardwaj *et al.*, 1976, 1977). These are 7-hydroxy-2-methylisoflavone (*10.36*) and its 7-acetyl (*10.38*), 7-methyl (*10.37*) and 8-acetyl (glyzarin) (*10.39*) derivatives. The structures have been

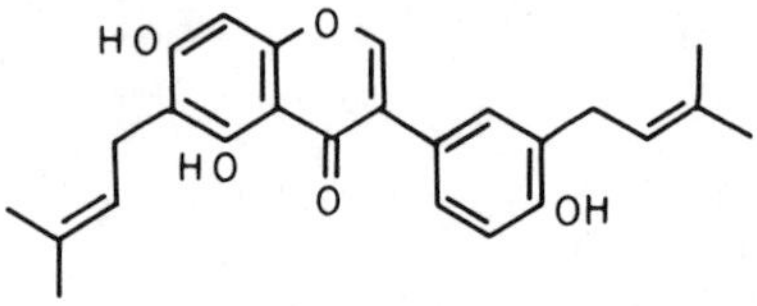

(10.35) 6,3′-Di-(dimethylallyl)-genistein

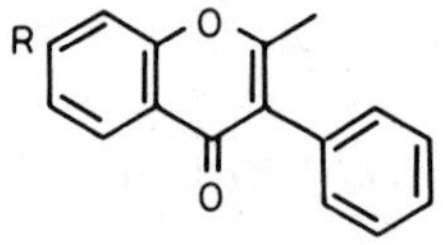

(10.36) R = OH, 7-Hydroxy-2-methylisoflavone
(10.37) R = OMe, 7-Methoxy-2-methylisoflavone
(10.38) R = OCOMe, 7-Acetoxy-2-methylisoflavone

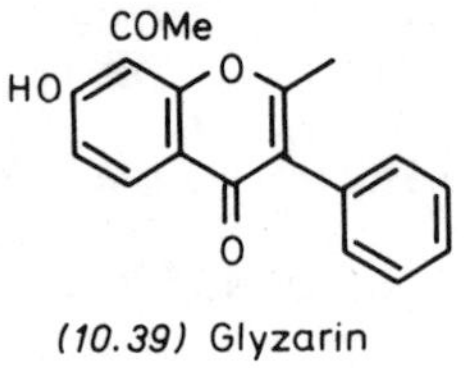

(10.39) Glyzarin

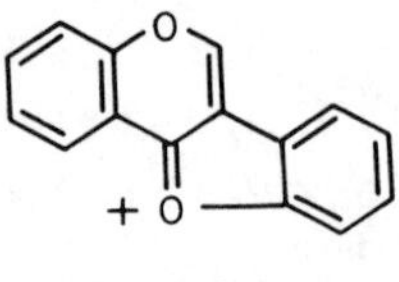

(10.42)

confirmed by Perkin condensation of a deoxybenzoin with acetic anhydride–sodium acetate. Whether or not these compounds are true isoflavonoids, i.e.

Table 10.1 Isoflavones

Isoflavone	*Substituents*								
	5	*6*	*7*	*8*	*2′*	*3′*	*4′*	*5′*	*6′*
Daidzein			OH				OH		
Formononetin			OH				OMe		
Isoformononetin			OMe				OH		
Dimethyldaidzein			OMe				OMe		
7,2′,4′-Trihydroxy-			OH		OH		OH		
2′-Hydroxyformononetin			OH		OH		OMe		
Theralin			OH		OMe		OH		
3′-Hydroxydaidzein			OH			OH	OH		
3′-Hydroxyformononetin (Calycosin)			OH			OH	OMe		
Sayanedin			OMe			OMe	OH		
Cabreuvin			OMe			OMe	OMe		
Pseudobaptigenin (ψ-Baptigenin)			OH			OCH_2O			
7-Methoxy-3′,4′-methylenedioxy-			OMe			OCH_2O			
Koparin			OH		OH	OH	OMe		
2′-Hydroxy-7,3′,4′-trimethoxy-			OMe		OH	OMe	OMe		
Glyzaglabrin			OH		OH	OCH_2O			
7,2′,4′,5′-Tetramethoxy-			OMe		OMe		OMe	OMe	
7,2′-Dimethoxy-4′,5′-methylenedioxy-			OMe		OMe	OCH_2O			
Gliricidin			OH			OH	OMe	OH	
5-Hydroxy-7-methoxy-	OH		OMe						
5,7-Dimethoxy-	OMe		OMe						
5,7,2′-Trihydroxy-(as glucoside)	OH		OH		OH				

Plant sources	*References*
Albizia, Dalbergia, Echinospartum, Glycine, Lespedeza, Machaerium, Phaseolus, Piptanthus, Pterodon, Pueraria, Aspergillus niger	Umezawa *et al.* (1975)
Streptomyces xanthophaeus	Hazato *et al.* (1979)
Albizia, Amorpha, Cajanus, Cicer, Cladrastis, Dalbergia, Echinospartum, Ferreira, Glycyrrhiza, Machaerium, Myroxylon, Onobrychis, Ononis, Pericopsis, Piptanthus, Pueraria, Sophora, Spartium, Trifolium, Vatairea, Virola (Myristicaceae), *Wistaria.*	
Machaerium villosum	Kurosawa *et al.* (1978c)
Dalbergia miscolobium	Gregson *et al.* (1978b)
Pterodon apparicioi	Almeida and Gottlieb (1975)
Phaseolus vulgaris	Woodward (1980a)
Virola caducifolia (Myristicaceae) *V. multinervia*	Braz Filho *et al.* (1977)
Thermopsis alterniflora	Kattaev and Nikonov (1975)
Machaerium villosum	Kurosawa *et al.* (1978c)
Bowdichia nitida	Brown *et al.* (1974)
Cyclobium clausseni	Gottlieb *et al.* (1975)
Machaerium mucronulatum, M. villosum	Kurosawa *et al.* (1978c)
Myroxylon balsamum	Oliveira *et al.* (1978)
M. peruiferum	Maranduba *et al.* (1979)
Sophora secondiflora	Minhaj *et al.* (1976b)
Trifolium pratense	Biggs and Lane (1978)
Pisum sativum	Isogai *et al.* (1970)
Calopogonium mucunoides	Vilain and Jadot (1976)
Myroxylon peruiferum	Maranduba *et al.* (1979)
Cladrastis platycarpa	Ohashi *et al.* (1976)
Dalbergia assamica	Chibber and Sharma (1978)
D. sericea	Parthasarathy *et al.* (1976a)
D. spruceana	Cook *et al.* (1978)
D. stevensonii	Donnelly *et al.* (1973b)
Trifolium pratense	Biggs and Lane (1978)
Calopogonium mucunoides	Vilain and Jadot (1976)
Castanospermum australe	Berry *et al.* (1977)
Myroxylon peruiferum	Maranduba *et al.* (1979)
Glycyrrhiza glabra	Bhardwaj and Singh (1977)
Amorpha fruticosa	Ognyanov and Somleva (1980)
Calopogonium mucunoides	Vilain and Jadot (1976)
Pterodon apparicioi	Galina and Gottlieb (1974)
Gliricidia sepium	Manners and Jurd (1979)
Derris robusta	Chibber and Sharma (1979b)
Arachis hypogaea	Turner *et al.* (1975)
Cajanus cajan	Bhanumati *et al.* (1979b)

Table 10.1 (*Contd.*)

Isoflavone	*Substituents*								
	5	*6*	*7*	*8*	*2′*	*3′*	*4′*	*5′*	*6′*
Genistein	OH		OH				OH		
Biochanin A	OH		OH				OMe		
5-Methylgenistein	OMe		OH				OH		
Prunetin	OH		OMe				OH		
5-Methylbiochanin A	OMe		OH				OMe		
7-Methylbiochanin A (as *C*-rhamnosides)	OH		OMe				OMe		
2′-Hydroxygenistein	OH		OH		OH		OH		
2′-Hydroxybiochanin A	OH		OH		OH		OMe		
Cajanin	OH		OMe		OH		OH		
5,2′-Dihydroxy-7,4′-dimethoxy-	OH		OMe		OH		OMe		
2′-Methoxybiochanin A	OH		OH		OMe		OMe		
Orobol	OH		OH			OH	OH		
3′-Methylorobol	OH		OH			OMe	OH		
Glabrescione B	OMe		OMe			OR	OR*		
5,7-Dihydroxy-3′,4′-methylenedioxy-	OH		OH			OCH_2O			
Derrugenin	OH		OMe		OMe		OH	OMe	
Robustigenin	OH		OMe		OMe		OMe	OMe	
5,7,3′,5′-Tetrahydroxy-4′-methoxy-	OH		OH			OH	OMe	OH	
Texasin		OH	OH				OMe		
Glycitein		OMe	OH				OH		
Kakkatin		OH	OMe				OH		
Afrormosin		OMe	OH				OMe		

Plant sources	*References*
Albizia, Bowdichia, Cajanus, Echinospartum, Glycine, Lespedeza, Pericopsis, Phaseolus, Piptanthus, Pueraria, Spartium, Trifolium,	
Streptomyces griseus	Konig *et al.* (1977)
S. xanthophaeus	Hazato *et al.* (1979)
Albizia, Cicer, Cotoneaster (Rosaceae), *Dalbergia, Echinospartum, Myroxylon, Pericopsis, Pueraria, Sophora, Trifolium, Virola* (Myristicaceae)	
Echinospartum horridum	Grayer-Barkmeijer *et al.* (1978)
Spartium junceum	Ozimina *et al.* (1979)
Dalbergia miscolobium	Formiga *et al.* (1975)
Genista carinalis	Nakov *et al.* (1978)
Echinospartum horridum	Grayer-Barkmeijer *et al.* (1978)
Dalbergia volubilis	Chawla *et al.* (1974, 1975)
Cajanus cajan	Ingham (1976c)
Glycine wightii	Ingham *et al.* (1977)
Lablab niger	Ingham (1977b)
Moghania macrophylla	Prasad and Varma (1977)
Phaseolus vulgaris	Biggs (1975)
Spartium junceum	Ingham and Dewick (1980a)
Virola caducifolia (Myristicaceae)	Braz Filho *et al.* (1976)
V. multinervia	Braz Filho *et al.* (1977)
Cajanus cajan	Ingham (1976c)
Cajanus cajan	Ingham (1976c)
Virola caducifolia (Myristicaceae)	Braz Filho *et al.* (1976)
V. multinervia	Braz Filho *et al.* (1977)
Aspergillus niger, Stemphylium spp.	
Streptomyces neyagawaensis var. *orobolere*	Umezawa *et al.* (1975)
Dalbergia inundata	Almeida and Gottlieb (1974)
Derris glabrescens	Delle Monache *et al.* (1977b)
Lupinus luteus	Laman and Volynets (1974)
Sophora japonica	Komatsu *et al.* (1976b)
*Derris robusta***	Chibber and Sharma (1979d)
Derris robusta	Chibber and Sharma (1979c)
Juniperus macropoda (Cupressaceae)	Sethi *et al.* (1980)
Myroxylon peruiferum	Maranduba *et al.* (1979)
Mildbraedeodendron excelsa	Meegan and Donnelly (1975)
Pueraria spp.	Kubo *el. al.* (1975)
Cladrastis platycarpa	Imamura *et al.* (1974)
Myroxylon peruiferum	Maranduba *et al.* (1979)
Onobrychis viciifolia	
Onobrychis spp.	Ingham (1978b)
Pericopsis elata	
P. mooniana	Fitzgerald *et al.* (1976)

Table 10.1 (*Contd.*)

Isoflavone	*Substituents*								
	5	*6*	*7*	*8*	*2'*	*3'*	*4'*	*5'*	*6'*
Odoratin		OMe	OH			OH	OMe		
Cladrastin		OMe	OH			OMe	OMe		
6,7,3',4'-Tetramethoxy-		OMe	OMe			OMe	OMe		
Fujikinetin		OMe	OH			OCH_2O			
3',4'-Dimethoxy-6,7-methylenedioxy-		OCH_2O				OMe	OMe		
7-Hydroxy-6,2',4',5'-tetramethoxy-		OMe	OH		OMe		OMe	OMe	
6,7,2',4',5'-Pentamethoxy-		OMe	OMe		OMe		OMe	OMe	
Dalpatein		OMe	OH		OMe		OCH_2O		
6-Hydroxy-7,2'-dimethoxy-4',5'-methylenedioxy-		OH	OMe		OMe		OCH_2O		
Milldurone		OMe	OMe		OMe		OCH_2O		
6,7,3',4',5'-Pentamethoxy-		OMe	OMe			OMe	OMe	OMe	
6,7,3'-Trimethoxy-4',5'-methylenedioxy-		OMe	OMe			OMe	OCH_2O		
Retusin			OH	OH			OMe		
8-Methylretusin			OH	OMe			OMe		
8-Methyl-3'-hydroxyretusin			OH	OMe		OH	OMe		
7-Hydroxy-8,3',4'-trimethoxy-			OH	OMe		OMe	OMe		
7,8,2'-Trimethoxy-4',5'-methylenedioxy-			OMe	OMe	OMe		OCH_2O		
Betavulgarin	OMe	OCH_2O			OH				
Tectorigenin	OH	OMe	OH				OH		
7-Methyltectorigenin	OH	OMe	OMe				OH		
Irisolidone	OH	OMe	OH				OMe		

Plant sources	References
Pterodon apparicioi	Galina and Gottlieb (1974)
Wistaria floribunda	Ohashi *et al.* (1979)
Dipteryx odorata	Hayashi and Thomson (1974)
Pterodon apparicioi	Galina and Gottieb (1974)
Cladrastis platycarpa	Imamura *et al.* (1974)
Pterodon apparicioi	Almeida and Gottlieb (1975)
Cladrastis platycarpa	Imamura *et al.* (1974)
Dalbergia sericea	Parthasarathy *et al.* (1976a)
Pterodon apparicioi	Galina and Gottlieb (1974)
Xanthocercis zambesiaca	Harper *et al.* (1976)
Pterodon apparicioi	Galina and Gottlieb (1974)
Mildbraedeodendron excelsa	Meegan and Donnelly (1975)
Pterodon apparicioi	Galina and Gottlieb (1974)
Dalbergia paniculata	Radhakrishniah (1973)
Corydala africana	Campbell and Tannock (1973)
Dalbergia assamica	Chibber and Sharma (1978)
Mildbraedeodendron excelsa	Meegan and Donnelly (1975)
Dalbergia paniculata	Adinarayana and Rao (1975b)
Mildbraedeodendron excelsa	Meegan and Donnelly (1975)
Pterodon apparicioi	Galina and Gottlieb (1974)
Pterodon apparicioi	Galina and Gottlieb (1974)
Mildbraedeodendron excelsa	Meegan and Donnelly (1975)
Dipteryx odorata	Hayashi and Thomson (1974)
Dalbergia retusa	Gregson *et al.* (1978a)
D. variabilis	Kurosawa *et al.* (1978d)
Dipteryx odorata	Hayashi and Thomson (1974)
Pericopsis schliebenii	Fitzgerald *et al.* (1976)
Xanthocercis zambesiaca	Harper *et al.* (1976)
Dipteryx odorata	Hayashi and Thomson (1974)
Myroxylon balsamum	Oliveira *et al.* (1978)
Xanthocercis zambesiaca	Harper *et al.* (1976)
Xanthocercis zambesiaca	Harper *et al.* (1976)
Pterodon apparicioi	Galina and Gottlieb (1974)
Beta vulgaris (Chenopodiaceae)	Geigert *et al.* (1973)
Dalbergia sissoo	Sharma *et al.* (1979b)
D. stevensonii	Donnelly *et al.* (1973b)
D. volubilis	Chawla and Chibber (1977)
Iris germanica (Iridaceae)	Pailer and Franke (1973)
Dalbergia volubilis	Khera and Chibber (1978a)
Iris germanica (Iridaceae)	Pailer and Franke (1973)
I. kashmiriana	Dhar and Kalla (1975)
I. tingitana	El-Emary *et al.* (1980)
Pericopsis mooniana	Fitzgerald *et al.* (1976)

Table 10.1 (*Contd.*)

Isoflavone	*Substituents*								
	5	*6*	*7*	*8*	*2′*	*3′*	*4′*	*5′*	*6′*
5-Methoxyafrormosin	OMe	OMe	OH				OMe		
Irilone	OH	OCH_2O					OH		
Irisolone	OMe	OCH_2O					OH		
Irisolone methyl ether	OMe	OCH_2O					OMe		
Iristectorigenin A	OH	OMe	OH			OH	OMe		
Iristectorigenin B	OH	OMe	OH			OMe	OH		
7-Hydroxy-5,6-dimethoxy-3′,4′-methylenedioxy- [Name dipteryxine pre-empted]	OMe	OMe	OH			OCH_2O			
5,6,7-Trimethoxy-3′,4′-methylenedioxy- [Name odoratine pre-empted]	OMe	OMe	OMe			OCH_2O			
Iriflogenin (as irifloside)	OH	OCH_2O				OMe	OH		
Iriskumaonin	OMe	OCH_2O				OH	OMe		
5,3′,4′-Trimethoxy-6,7-methylenedioxy-	OMe	OCH_2O				OMe	OMe		
Caviunin	OH	OMe	OH		OMe		OMe	OMe	
Irigenin	OH	OMe	OH			OMe	OMe	OH	
Irisflorentin	OMe	OCH_2O				OMe	OMe	OMe	
8-Hydroxygenistein	OH		OH	OH			OH		
Isotectorigenin (pseudotectorigenin)	OH		OH	OMe			OH		
Aurmillone	OH		OH	OMe			OR*		
5,7,3′,4′-Tetrahydroxy-8-methoxy-	OH		OH	OMe		OH	OH		
5,7,3′-Trihydroxy-8,4′-dimethoxy-	OH		OH	OMe		OH	OMe		
5,7,4′-Trihydroxy-8,3′-dimethoxy-(as glycoside)	OH		OH	OMe		OMe	OH		

Plant sources	References
Pueraria thunbergiana	Kurihari and Kikuchi (1973)
Sophora japonica	Komatsu *et al.* (1976b)
Cladrastis platycarpa	Imamura *et al.* (1974)
Wistaria floribunda	Ohashi *et al.* (1979)
Iris germanica (Iridaceae)	Dhar and Kalla (1973)
	Pailer and Franke (1973)
Trifolium pratense	Fraishtat *et al.* (1979a)
Iris florentina (Iridaceae)	Morita *et al.* (1973)
I. germanica	Pailer and Franke (1973)
Iris tingitana (Iridaceae)	El-Emary *et al.* (1980)
Iris germanica (Iridaceae)	Pailer and Franke (1973)
I. unguicularis	Arisawa and Morita (1976)
Streptomyces roseolus†	Chimura *et al.* (1975)
Iris florentina (Iridaceae)	Morita *et al* (1973)
I. germanica	Pailer and Franke (1973)
Dipteryx odorata	Nakano *et al.* (1979)
Corydala africana	Campbell and Tannock (1973)
Dipteryx odorata‡	Nakano *et al.* (1979)
Iris florentina (Iridaceae)	Arisawa *et al.* (1973)
Iris kumaonensis (Iridaceae)	Kalla *et al.* (1978)
I. tingitana	El-Emary *et al.* (1980)
Iris germanica (Iridaceae)	Pailer and Franke (1973)
I. tingitana	El-Emary *et al.* (1980)
Dalbergia inundata	Almeida and Gottlieb (1974)
D. paniculata	Radhakrishniah (1973)
D. spruceana	Cook *et al.* (1978)
Iris florentina (Iridaceae)	Morita *et al.* (1973)
I. germanica	Pailer and Franke (1973)
I. tingitana	El-Emary *et al.* (1980)
I. unguicularis	Arisawa and Morita (1976)
Juniperus macropoda (Cupressaceae)	Sethi *et al.* (1980)
Iris florentina (Iridaceae)	Morita *et al* (1973)
I. germanica	Pailer and Franke (1973)
I. tingitana	El-Emary *et al.* (1980)
Aspergillus niger	Umezawa *et al.* (1975)
Dalbergia sissoo	Dhingra *et al.* (1974)
Aspergillus niger	Umezawa *et al.* (1975)
Millettia auriculata	Raju and Srimannarayana (1978)
Aspergillus niger	Umezawa *et al.* (1975)
Streptomyces roseolus	Chimura *et al.* (1975)
Iris germanica (Iridaceae)	Kawase *et al.* (1973)

Table 10.1 (*Contd.*)

Isoflavone	*Substituents*								
	5	*6*	*7*	*8*	*2′*	*3′*	*4′*	*5′*	*6′*
Platycarpanetin	OMe		OH	OMe		OCH_2O			
Isocaviunin	OH		OH	OMe	OMe		OMe	OMe	
Dipteryxin		OMe	OH	OH			OMe		
8,3′-Dihydroxy-6,7,4′-trimethoxy-		OMe	OMe	OH		OH	OMe		
Petalostetin		OMe	OMe	OMe		OCH_2O			
8,3′,4′-Trimethoxy-6,7-methylenedioxy-		OCH_2O		OMe		OMe	OMe		
5,6,7,8-Tetramethoxy-3′,4′-methylenedioxy-	OMe	OMe	OMe	OMe		OCH_2O			
6-Chlorogenistein	OH	Cl	OH				OH		
6,3′-Dichlorogenistein	OH	Cl	OH			Cl	OH		
Corylinal (*10.2*)									
Wighteone (*10.3*) (erythrinin B)									
Luteone (*10.4*)									
2,3-Dehydrokievitone (*10.5*)									
Erythrinin C (*10.6*)									
Glabrescione A (*10.7*)									
Erythrinin A (*10.8*)									
Alpinumisoflavone (*10.9*)									
Alpinumisoflavone-4′-methyl ether (*10.10*)									
Alpinumisoflavone dimethyl ether (*10.11*)									
Isoauriculatin (*10.12*) (revised structure)									
Isoauriculasin (*10.13*)									
Robustone (*10.14*)									
Robustone methyl ether (*10.15*)									
Elongatin (*10.16*)									
Calopogonium isoflavone A (*10.17*)									
Calopogonium isoflavone B (*10.18*)									
Barbigerone (*10.19*)									
Derrone (*10.20*)									
Derrone-4′-methyl ether (*10.21*)									
Neobavaisoflavone (*10.22*)									
Corylin (*10.23*)									
Psoralenol (*10.24*)									
Glabrone (*10.25*)									
Licoricone (*10.26*)									

Plant sources	References
Cladrastis platycarpa§	Imamura *et al.* (1974)
Dalbergia sissoo	Sharma *et al.* (1979b)
Dipteryx odorata	Hayashi and Thomson (1974)
Streptomyces roseolus	Chimura *et al.* (1975)
Petalostemon candidum	Torrance *et al.* (1979)
Xanthocercis zambesiaca	Harper *et al.* (1976)
Corydala africana	Campbell and Tannock (1973)
Streptomyces griseus	Konig *et al.* (1977)
Streptomyces griseus	Konig *et al.* (1977)
Psoralea corylifolia	Gupta *et al.* (1978)
Erythrina variegata	Deshpande *et al.* (1977)
Glycine wightii	
Lupinus albus	Ingham *et al.* (1977)
Lupinus luteus	Fukui *et al.* (1973)
L. spp.	Harborne *et al.* (1976)
Phaseolus vulgaris	Woodward (1979b)
Erythrina variegata	Deshpande *et al.* (1977)
Derris glabrescens	Delle Monache *et al.* (1977b)
Erythrina variegata	Deshpande *et al.* (1977)
Calopogonium mucunoides	Vilain and Jadot (1976)
Erythrina variegata	Deshpande *et al.* (1977)
Calopogonium mucunoides	Vilain and Jadot (1975)
Derris robusta	Chibber and Sharma (1979a)
Millettia auriculata	Minhaj *et al.* (1976a)
Millettia auriculata	Minhaj *et al.* (1976a)
Derris robusta	Chibber and Sharma (1979a)
Derris robusta	Chibber and Sharma (1979a)
Tephrosia elongata	Smalberger *et al.* (1975)
Calopogonium mucunoides	Vilain and Jadot (1975)
Calopogonium mucunoides	Vilain and Jadot (1976)
Tephrosia barbigera	Vilain (1980)
Derris robusta	Chibber and Sharma (1980)
Calopogonium mucunoides	Vilain and Jadot (1975)
Psoralea corylifolia	Bajwa *et al.* (1974)
Psoralea corylifolia	Jain *et al.* (1974)
Psoralea corylifolia	Suri *et al.* (1978)
Glycyrrhiza glabra	Kinoshita *et al.* (1976)
Glycyrrhiza uralensis	Kaneda *et al.* (1973)

Table 10.1 (*Contd.*)

Isoflavone	*Substituents*								
	5	*6*	*7*	*8*	*2′*	*3′*	*4′*	*5′*	*6′*
Licoisoflavone A (*10.27*) [Name phaseoluteone pre-empted]									
Licoisoflavone B (*10.28*)									
6,8-Di(dimethylallyl)genistein (*10.29*)									
6,8-Di(dimethylallyl)orobol (*10.30*)									
Osajin (*10.31*)									
Auriculatin (*10.32*)									
Auriculasin (*10.33*)									
Cajaisoflavone (*10.34*)									
6,3′-Di(dimethylallyl)genistein (*10.35*)									
7-Hydroxy-2-methyl- (*10.36*)									
7-Methoxy-2-methyl- (*10.37*)									
7-Acetoxy-2-methyl- (*10.38*)									
Glyzarin (*10.39*)									

* $R = -CH_2CH{=}CMe_2$.
† The *Streptomyces* metabolite is reported to be different from that from *Iris* spp.
‡ The *Dipteryx* and *Corydala* metabolites are not identical.
§ This structure for platycarpanetin has been disputed.
** The structure was revised after synthetic studies.

derived from chalcone–flavanone precursors and involving an aryl migration, remains to be established.

Corylinal (*10.2*) is a rather unusual isoflavone from seed of *Psoralea corylifolia* and has been identified as 3′-formyldaidzein (Gupta *et al.*, 1978). It co-occurs with neobavaisoflavone (*10.22*), and may thus arise by oxidation of the dimethylallyl substituent.

Synthesis of natural isoflavones remains an important adjunct to structural elucidation, and the long-established methods of ring closure of a C_1 unit on to an appropriate deoxybenzoin are still widely employed. The use of ethoxalyl chloride, triethyl orthoformate, dimethylformamide–phosphorus oxychloride and ethyl formate as C_1 sources continues (Wagner and Farkas, 1975). Improved yields in the Baker synthesis (ethoxalyl chloride) can be obtained by pyridine hydrochloride decarboxylation (Szabo *et al.*, 1975a), but aromatic methoxyls may also be demethylated, and in the Claisen condensation (ethyl formate), sodium t-butoxide is found to be a better and safer alternative to sodium (Szabo *et al.*, 1975b). Newer routes employing deoxybenzoins have also been reported. Thus Bass (1976) has utilized methanesulphonyl chloride with boron trifluoride–etherate in dimethylformamide as the C_1 unit, and excellent yields of isoflavones

Plant sources	References
Glycyrrhiza spp.	Kinoshita *et al.* (1978b)
Phaseolus vulgaris	Woodward (1979a)
Glycyrrhiza spp.	Saitoh *et al.* (1978)
Millettia pachycarpa	Singhal *et al.* (1980)
Millettia pachycarpa	Singhal *et al.* (1980)
Erythrina variegata	Deshpande *et al.* (1977)
Millettia auriculata	Raju and Srimannarayana (1978)
Millettia auriculata	Minhaj *et al.* (1976a)
Cajanus cajan	Bhanumati *et al.* (1979a)
Millettia pachycarpa	Singhal *et al.* (1980)
Glycyrrhiza glabra	Bhardwaj *et al.* (1976)
Glycyrrhiza glabra	Bhardwaj *et al.* (1976)
Glycyrrhiza glabra	Bhardwaj *et al.* (1976)
Glycyrrhiza glabra	Bhardwaj *et al.* (1977)

(or other chromones) may be obtained. Pelter and Foot (1976) report dimethylformamide dimethylacetal in benzene to be another source of the one-carbon unit with good yields of isoflavones, except for 5,7-dihydroxyisoflavones where protection of these two hydroxyl functions is necessary. By the addition of BF_3–Et_2O as in Bass's method, such protection is no longer essential (Pelter *et al.*, 1978). A further source of the additional carbon atom is provided by *N*-formylimidazole, generated by reaction of formic acid with *NN*-carbonyldi-imidazole (Krishnamurthy and Prasad, 1977). Excellent yields of isoflavone are also reported.

Whilst addition of a one-carbon unit on to a deoxybenzoin can be accomplished efficiently, synthesis of the deoxybenzoin itself may proceed in rather poor yields, and starting materials are usually not readily available. Much more appropriate to isoflavonoid syntheses are routes utilizing chalcones obtained by condensation of the more readily available aromatic acetophenones and aldehydes. The boron trifluoride-catalysed rearrangement of chalcone epoxides still finds occasional use, but yields are never good. However, the thallium nitrate oxidation of 2′-hydroxychalcones (Farkas *et al.*, 1974) has become established as the most popular route to isoflavones. The intermediate acetal is usually

transformed to the isoflavone by treatment with acid, but base-catalysed cyclization is also effective, and thus isoflavones containing acid-labile groupings such as dimethylchromene rings can be synthesized by this route. In such cases, the highly acid-sensitive methoxymethyl groups may be used to protect hydroxyls from oxidation, and the synthesis of isoflavone (*10.40*) (Antus *et al.*, 1976) exemplifies the procedure (Fig. 10.2). Unfortunately, dimethylchromene rings may be oxidized by thallium nitrate to give ring-contracted furano derivatives, and such compounds may be formed as unwanted byproducts, as in the synthesis of corylin methyl ether (*10.41*) (Antus *et al.*, 1979b) (Fig. 10.3). Whilst the thallium nitrate oxidative rearrangement of chalcones seems to be of general application, it has been criticized because nitric acid released during the reaction may lead to the production of nitroisoflavones (Oliveira *et al.*, 1979). Under standardized conditions, however, no nitration is observed (Antus *et al.*, 1979a). The method has also been used to synthesize *O*-and *C*-glycosides (Antus and Nogradi, 1979a,b;. Eade *et al.*, 1978). The structure of 7-methoxy-2-methylisoflavone (10.37) has also been confirmed by its synthesis from α-methyl-2′-hydroxy-4′-methoxychalcone using the thallium nitrate oxidation reaction (Jain *et al.*, 1978a).

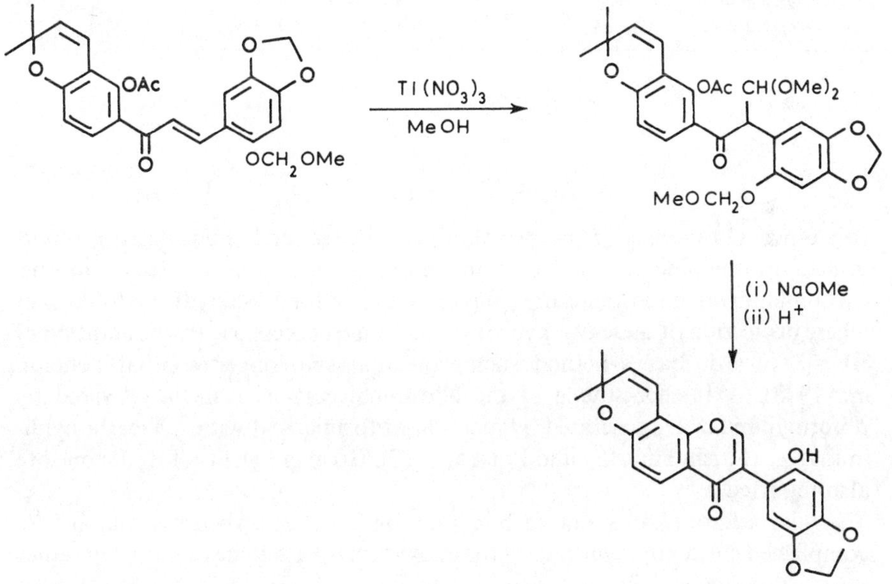

(10.40)

Fig. 10.2

A further route to isoflavones, though of less general synthetic value, is via the rearrangement of 3-bromoflavanones by treating with $AgSbF_6$ in dichloromethane (Pelter *et al.*, 1976).

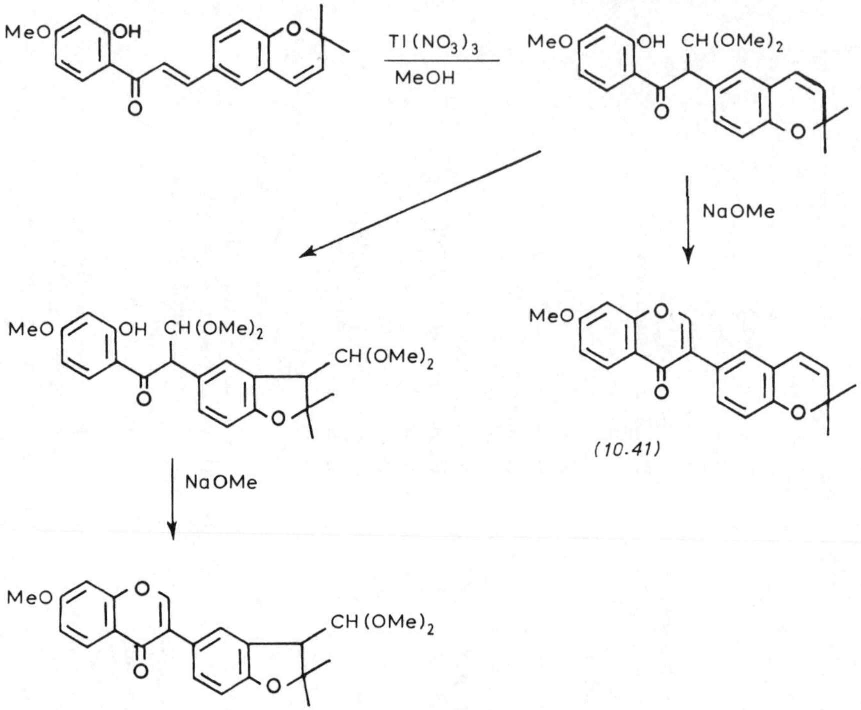

MeO OH Tl(NO$_3$)$_3$ MeOH MeO OH CH(OMe)$_2$ NaOMe MeO OH CH(OMe)$_2$ CH(OMe)$_2$ MeO (10.41) NaOMe MeO CH(OMe)$_2$

Fig. 10.3

The presence of a 2′-substituent in isoflavones can be determined by MS analysis: elimination of a 2′-hydrogen or 2′-substituent gives rise to a resonance-stabilized oxonium ion (*10.42*) (Eguchi *et al.*, 1977). Thus, 2′-unsubstituted isoflavones show a base peak at $M-1$, 2′-methoxyisoflavones a base peak at $M-31$ (cf. Campbell *et al.*, 1969), and 2′-hydroxyisoflavones a prominent $M-17$ peak (cf. Ingham, 1976c). To aid identification of the hydroxylation pattern of isoflavones, Harper *et al.* (1976) have hydrolysed them in base–D_2O, thus producing deuteriated deoxybenzoins. Under these conditions, deuteriation occurs at the methylene and also at positions in the aromatic rings *ortho* or *para* to a free hydroxyl, including that produced by hydrolysis of the pyrone ring. MS analysis of the deoxybenzoin indicates how many deuterium atoms have been incorporated. This method thus differentiates between, say, a 3′-hydroxy-4′-methoxyisoflavone and a 4′-hydroxy-3′-methoxyisoflavone.

A number of isoflavones and glycosides have been isolated from microbial cultures (Chimura *et al.*, 1975; Hazato *et al.*, 1979; Konig *et al.*, 1977; Umezawa *et al.*, 1975), and the question of whether isoflavonoids are restricted to the higher plants is again raised. In all of the above cases, the cultures were grown on a soybean meal medium, and it is quite likely that these isoflavonoids are either soybean-derived or microbial metabolites of soybean isoflavones.

Table 10.2 Isoflavone glycosides

Isoflavone glycoside	*Plant sources*	*References*
Daidzein 7-glucoside (daidzin)	*Glycine max*	Naim *et al.* (1974)
Daidzin-6″-acetate	*Glycine max*	Ohta *et al.* (1979)
Daidzein 7-rhamnoside	*Streptomyces xanthophaeus*	Hazato *et al.* (1979)
Daidzein 7, 4′-dirhamnoside	*Streptomyces xanthophaeus*	Hazato *et al.* (1979)
Formononetin 7-glucoside (ononin)	*Amorpha fruticosa*	Shibata and Shimizu (1978)
	Cladrastis platycarpa	Imamura *et al.* (1974)
	Dalbergia paniculata	Parthasarathy *et al.* (1974)
	Ononis spinosa	Haznagy *et al.* (1978)
	Spartium junceum	Ozimina *et al.* (1979)
	Trifolium spp.	Kazakov *et al.* (1972)
	Wistaria floribunda	Ohashi *et al.* (1979)
Formononetin 7-glucosylglucoside	*Cladrastis platycarpa*	Imamura *et al.* (1974)
Formononetin 7-rutinoside	*Dalbergia paniculata*	Parthasarathy *et al.* (1976b)
Pseudobaptigenin 7-glucoside (rothindin)	*Lupinus luteus*	Laman (1974)
	Ononis spinosa	Haznagy *et al.* (1978)
	Rothia indica	Nair and Subramanian (1976)
5,7,2′-Trihydroxyisoflavone 7-glucoside	*Cajanus cajan*	Bhanumati *et al.* (1979b)
Genistein 7-glucoside (genistin)	*Glycine max*	Naim *et al.* (1974)
	Lupinus luteus	Laman (1974)
	Spartium junceum	Ozimina *et al.* (1979)
	Thermopsis alterniflora	Kattaev and Nikonov (1972)
Genistin-6″acetate	*Glycine max*	Ohta *et al.* (1980)
Genistein 7-glucosylglucoside	*Lupinus luteus*	Laman and Volynets (1974)
Genistein 7-rhamnoside	*Streptomyces xanthophaeus*	Hazato *et al.* (1979)
Genistein 7,4′-dirhamnoside	*Streptomyces xanthophaeus*	Hazato *et al* (1979)
Biochanin A 7-glucoside (sissotrin)	*Cotoneaster serotina* (Rosaceae)	
	C. pannosa	Cooke and Fletcher (1974)
	Dalbergia paniculata	Parthasarathy *et al.* (1974)
Biochanin A 7-rutinoside	*Dalbergia paniculata*	Parthasarathy *et al.* (1976b)
Biochanin A 7-gentiobioside	*Sophora japonica*	Takeda *et al.* (1977)

Biochanin A 7-xylosylglucoside	*Sophora japonica*	Takeda *et al.* (1977)
Prunetin 4′-glucoside (trifoside)	*Genista carinalis*	Nakov *et al.* (1978)
	Trifolium pratense	Kattaev *et al.* (1972)
Prunetin 8-*C*-glucoside	*Dalbergia paniculata*	Parthasarathy *et al.* (1976b)
7-Methylbiochanin A 8-*C*-rhamnoside (volubilin)	*Dalbergia volubilis*	Chawla *et al.* (1974)
7-Methylbiochanin A 6-*C*-rhamnoside (isovolubilin)	*Dalbergia volubilis*	Chawla *et al.* (1975)
5,7-Dihydroxy-3′,4′-methylenedioxyisoflavone 7-glucoside	*Lupinus luteus*	Laman and Volynets (1974)
5,7-Dihydroxy-3′,4′-methylenedioxyisoflavone 7-glucosylglucoside	*Lupinus luteus*	Laman and Volynets (1974)
Glycitein 7-glucoside	*Glycine max*	Naim *et al.* (1974)
Afrormosin 7-glucoside (wistin)	*Amorpha fruticosa*	Shibata and Shimizu (1978)
	Cladrastis platycarpa	Imamura *et al.* (1974)
	Wistaria floribunda	Ohashi *et al.* (1979)
Cladrastin 7-glucoside	*Cladrastis platycarpa*	Ohashi *et al.* (1976)
Fujikinetin 7-glucoside (fujikinin)	*Cladrastis platycarpa*	Imamura *et al.* (1974)
		Ohashi *et al.* (1976)
8-Methylretusin 7-glucoside	*Wistaria floribunda*	Ohashi *et al.* (1979)
8-Methylretusin 7-glucosylglucoside	*Cladrastis platycarpa*	Imamura *et al.* (1974)
Tectorigenin 7-glucoside (tectoridin)	*Dalbergia sissoo*	Sharma *et al.* (1980b)
	D. volubilis	Khera and Chibber (1978a)
	Iris germanica (Iridaceae)	Kawase *et al.* (1973)
	Pueraria montana	Kubo *et al.* (1973)
Tectorigenin 7-gentiobioside	*Dalbergia volubilis*	Khera and Chibber (1978c)
Irisolidone 7-glucoside	*Pueraria lobata*	Kubo *et al.* (1973)
Irisolidone 8-*C*-glucoside (volubilinin)	*Dalbergia volubilis*	Chawla *et al.* (1976)
Irisolidone 7-xylosylglucoside (kakkalide)	*Pueraria thunbergiana*	Kurihari and Kikuchi (1975)
Caviunin 7-glucoside	*Dalbergia sissoo*	Sharma *et al.* (1980b)
	D. paniculata	Radhakrishniah (1979)
Caviunin 7-rhamnosylglucoside	*Dalbergia paniculata*	Rajulu and Rao (1980)
Caviunin 7-gentiobioside	*Dalbergia sissoo*	Sharma *et al.* (1979a)
Irigenin 7-glucoside (iridin)	*Iris florentina* (Iridaceae)	Arisawa *et al.* (1973)
	I. kumaonensis	Kalla *et al.* (1978)
	I. unguicularis	Arisawa *et al.* (1976)
	Juniperus macropoda (Cupressaceae)	Sethi *et al.* (1980)

Table 10.2 (*Contd.*)

Isoflavone glycoside	*Plant sources*	*References*
7-Methyltectorigenin 4′-glucoside	*Dalbergia volubilis*	Khera and Chibber (1978b)
7-Methyltectorigenin 4′-gentiobioside	*Dalbergia volubilis*	Khera and Chibber (1978b)
Irilone 4′-glucoside	*Iris florentina* (Iridaceae)	Tsukida *et al.* (1973)
	Trifolium pratense	Fraishtat *et al.* (1979a)
Irisolone 4′-bioside	*Iris florentina* (Iridaceae)	Tsukida *et al.* (1973)
Iristectorigenin A 7-glucoside	*Iris tectorum* (Iridaceae)	Morita *et al.* (1972)
Iriflogenin 3′-glucoside (irifloside)	*Iris florentina* (Iridaceae)	Arisawa *et al.* (1973)
5,7,4′-Trihydroxy-8,3′-dimethoxyisoflavone 7-glucoside (homotectoridin)	*Iris germanica* (Iridaceae)	Kawase *et al.* (1973)
Isocaviunin 7-glucoside	*Dalbergia sissoo*	Sharma *et al.* (1980b)
(isocaviudin)	*D. paniculata*	Parthasarathy *et al.* (1980)
Isocaviunin 7-gentiobioside	*Dalbergia sissoo*	Sharma *et al.* (1980a)
Platycarpanetin 7-glucoside	*Cladrastis platycarpa*	Imamura *et al.* (1974)
		Ohashi *et al.* (1976)

Isoflavone glycosides (Table 10.2) are isolated much less frequently than aglycones, but a considerable number have now been described. The previously unreported isoflavone 7-methylbiochanin A is known only as its 8-and 6-*C*-rhamnosides volubilin and isovolubilin in *Dalbergia volubilis* (Chawla *et al.*, 1974, 1975). Similarly, 5,7,2′-trihydroxyisoflavone (Bhanumati *et al.*, 1979b), iriflogenin (Arisawa *et al.*, 1973) and 5,7,4′-trihydroxy-8,3′-dimethoxyisoflavone (Kawase *et al.*, 1973) have been isolated only as *O*-glycosides.

10.4 ISOFLAVANONES

Isoflavanones (*10.43*) are much rarer than isoflavones, though the number of known naturally occurring examples continues to grow (Table 10.3). Two

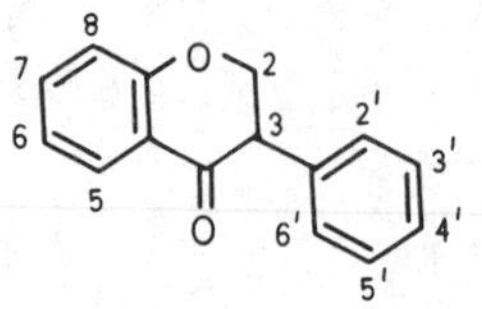

(10.43)

unusual ones isolated from *Lespedeza homoloba* (Ueno *et al.*, 1973a,c) contain geranyl substituents. Lespedeol A (*10.44*) is 6-geranyldalbergioidin and

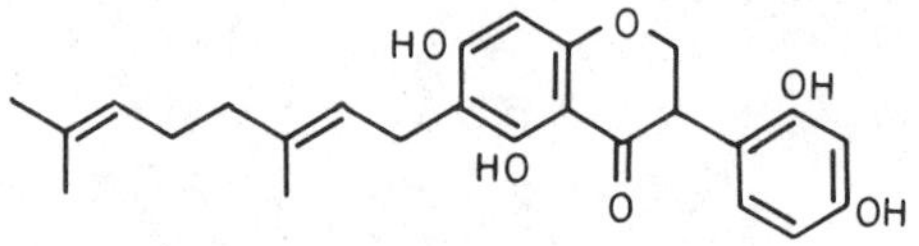

(10.44) Lespedeol A

lespedeol B (*10.45*) is the cyclized chromeno analogue, which may also be obtained by dichlorodicyanobenzoquinone (DDQ) treatment of lespedeol A.

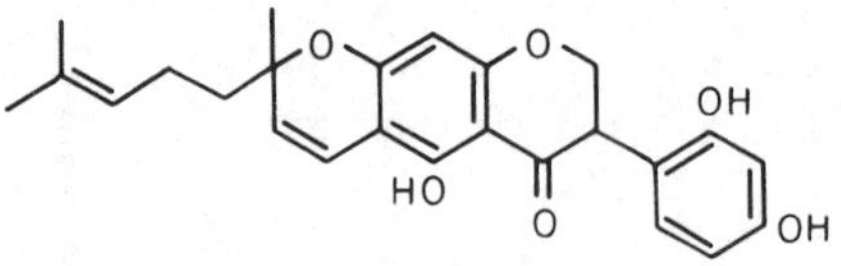

(10.45) Lespedeol B

For many years, the only natural optically active isoflavanone known was 3*R*-sophorol. Since racemization may occur under relatively mild conditions, it was assumed that extraction and purification techniques were responsible for the observed isolation of optically inactive forms. Recently, 3*S*-7,4′-di-

Table 10.3 Isoflavanones

Isoflavanone	*Substituents*									*Plant sources*	*Chirality Optical activity*	*References*
	5	*6*	*7*	*8*	*2′*	*3′*	*4′*	*5′*	*6′*			
7,4′-Dihydroxy-			OH				OH			*Pericopsis mooniana*	*S**	Fitzgerald *et al.* (1976)
7-Hydroxy-4′-methoxy-			OH				OMe			*Myroxylon balsamum*	±	Oliveira *et al.* (1978)
2,3-Dihydroononin			Oglu				OMe			*Ononis spinosa*		Haznagy *et al.* (1978)
7,2′,4′-Trihydroxy-			OH		OH		OH			*Phaseolus vulgaris*	±	Woodward (1980a)
Vestitone			OH		OH		OMe			*Onobrychis viciifolia*		Ingham (1978b)
Sativanone			OH		OMe		OMe			*Dalbergia stevensonii*	±	Donnelly *et al.* (1973b)
										Medicago sativa		Ingham (1979a)
7,3′-Dihydroxy-4′-methoxy-			OH			OH	OMe			*Myroxylon balsamum*	±	Oliveira *et al.* (1978)
Violanone			OH		OMe	OH	OMe			*Dalbergia oliveri*	±	Donnelly and Kavanagh (1974)
3′-Methylviolanone			OH		OMe	OMe	OMe			*Dalbergia cearensis*		Guimaraes *et al.* (1975)
2′-Hydroxy-7,3′,4′-trimethoxy-			OMe		OH	OMe	OMe			*Myroxylon peruiferum*	±	Maranduba *et al.* (1979)
Onogenin			OH		OMe		OCH_2O			*Dalbergia stevensonii*	±	Donnelly *et al.* (1973b)
										Ononis arvensis		Kovalev *et al.* (1975)
Dalbergioidin	OH		OH		OH		OH			*Lablab niger*		Ingham (1977b)
										Lespedeza cyrtobotrya		Miyase *et al.* (1980)
										Phaseolus vulgaris		Woodward (1979a)
Cajanol	OH		OMe		OMe		OH			*Cajanus cajan*		Ingham (1979b)
6,7-Dimethoxy-3′,4′-methylenedioxy-		OMe	OMe				OCH_2O			*Mildbraedeodendron excelsa*	±	Meegan and Donnelly (1975)
Lespedeol A (*10.44*)										*Lespedeza homoloba*		Ueno *et al.* (1973c)
Lespedeol B (*10.45*)										*Lespedeza homoloba*		Ueno *et al.* (1973a).
Neoraunone (*10.46*)										*Neorautanenia amboensis*	±	Breytenbach and Rall (1980)
Ambonone (*10.47*)										*Neorautanenia amboensis*	±	Breytenbach and Rall (1980)
Erosenone (*10.48*)										*Pachyrrhizus erosus*	–	Kalra *et al.* (1977)
5-Deoxykievitone (*10.49*)										*Phaseolus vulgaris*		Woodward (1979b)
Kievitone (*10.50*)										*Lablab niger*		Ingham (1977b)
1″,2″-Dehydrocyclokievitone (*10.51*)										*Phaseolus vulgaris*		Woodward (1979b)
Sophoraisoflavanone A (*10.52*)										*Sophora tomentosa*	–	Komatsu *et al.* (1978b)
Licoisoflavanone (*10.53*)										*Glycyrrhiza*, spp.	±	Saitoh *et al.* (1978)
Isosophoronol (*10.54*)										*Sophora tomentosa*		Delle Monache *et al.* (1977a)

Isosophoranone (*10.55*)	*Sophora tomentosa*	–	Delle Monache *et al.* (1977a)
Cajanone (*10.56*)	*Cajanus cajan*		Preston (1977a)
2′-Methylcajanone (*10.57*)	*Cajanus cajan*	±	Bhanumati *et al.* (1979c)
Dalpanin (*10.58*)	*Dalbergia paniculata*		Adinarayana and Rao (1975a)
Secondifloran (*10.59*)	*Sophora secondiflora*	±	Minhaj *et al.* (1977)

* The chirality was wrongly assigned in this publication.

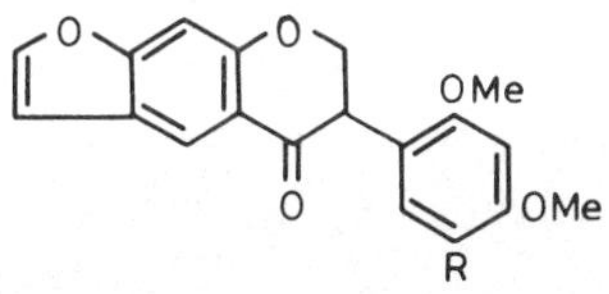

(10.46) R = H, Neoraunone
(10.47) R = OMe, Ambonone

hydroxyisoflavanone {[α]$_D$ as acetate −20.7° ($CHCl_3$)} from *Pericopsis mooniana* (Fitzgerald *et al.*, 1976), sophoraisoflavanone A (*10.52*) {[α]$_D$ −17.3° (EtOH)} (Komatsu *et al.*, 1978b) and isosophoranone (*10.55*) {[α]$_D$ −26° ($CHCl_3$)} (Delle Monache *et al.*, 1977a; Komatsu *et al.*, 1978b) from *Sophora tomentosa* have been isolated in optically active forms. Erosenone {[α]$_D$ −37.7° ($CHCl_3$)} from *Pachyrrhizus erosus* is also an optically active isoflavanone and has been assigned the tentative structure (*10.48*) (Kalra *et al.*, 1977). As with isoflavans,

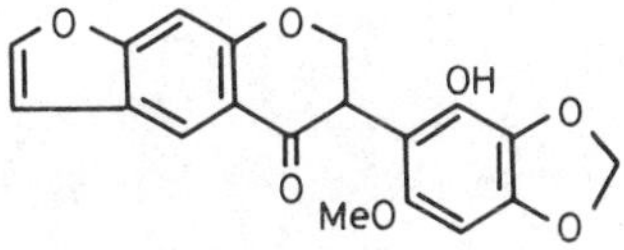

(10.48) Erosenone

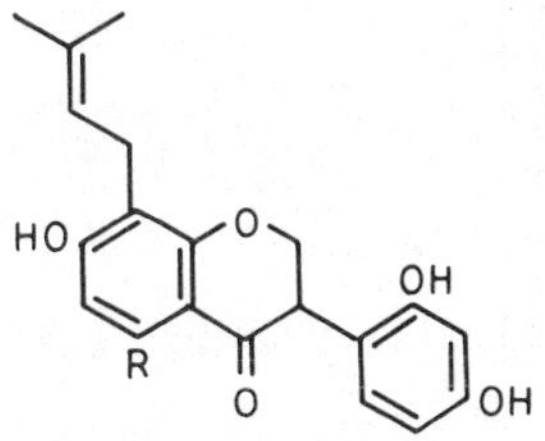

(10.49) R = H, 5-Deoxykievitone
(10.50) R = OH, Kievitone

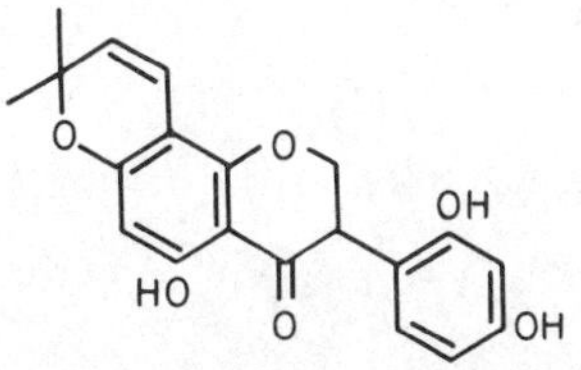

(10.51) 1″, 2″-Dehydrocyclo-kievitone

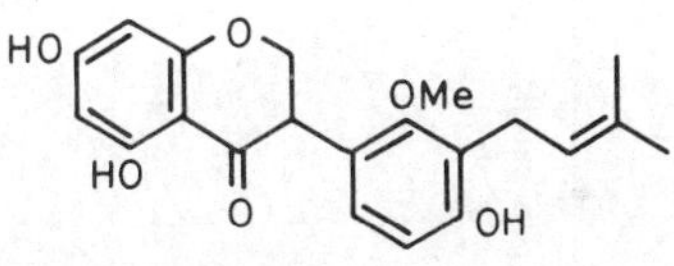

(10.52) Sophoraisoflavanone A

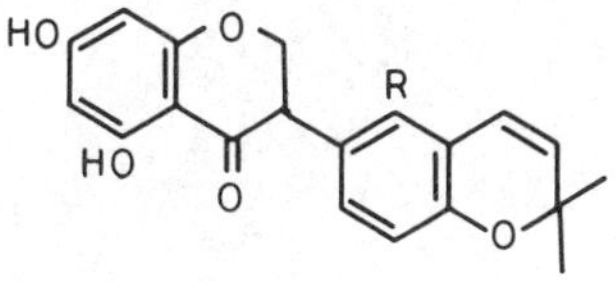

(10.53) R = OH, Licoisoflavanone
(10.54) R = OMe, Isosophoronol

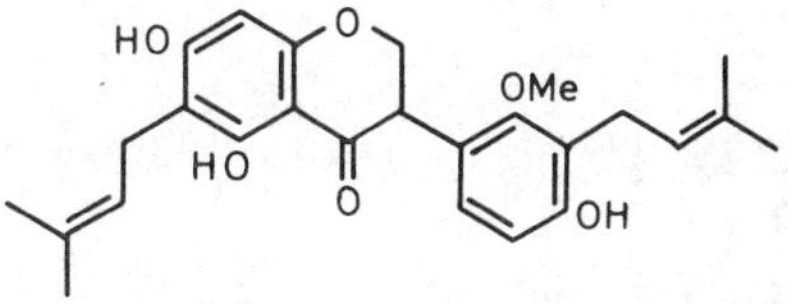

(10.55) Isosophoranone

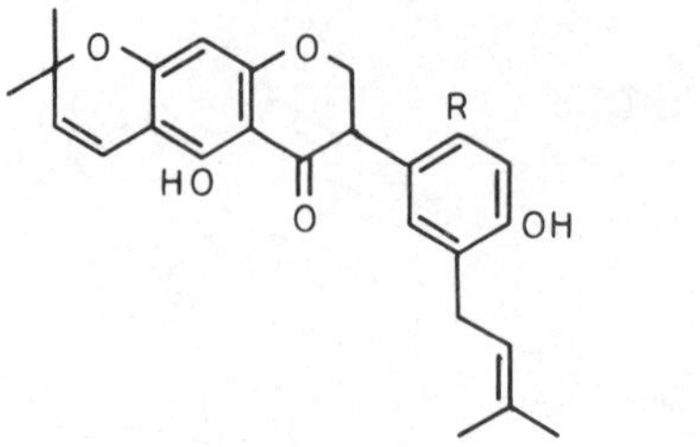

(10.56) R = OH, Cajanone
(10.57) R = OMe, 2'-Methylcajanone

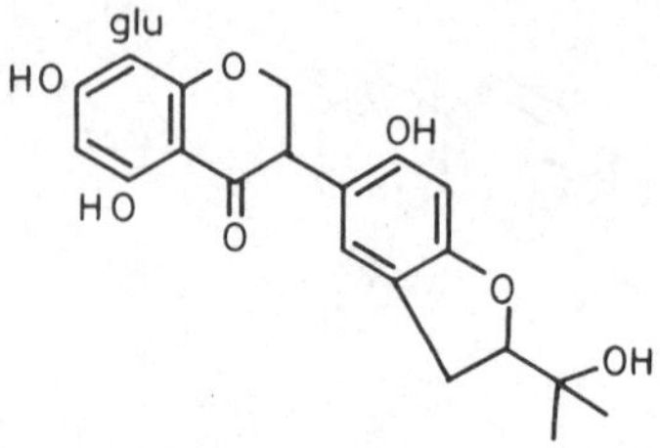

(10.58) Dalpanin

the sign of the rotation is not indicative of the chirality at C-3, but in the ORD curves, 3*S*-isoflavanones show intense negative Cotton effects in the 330–350 nm range (Kurosawa *et al.*, 1978b). Thus, (+)-7,2',4'-trimethoxy-isoflavanone derived from (+)-homopterocarpin via hydrogenolysis to the isoflavan, methylation and $KMnO_4$ oxidation has the same chirality 3*S* as (−)-7,4-diacetoxyisoflavanone (Kurosawa *et al.*, 1978b; Fitzgerald *et al.*, 1976).

In the MS of a number of 2'-hydroxy- and 2'-methoxy-isoflavanones, the intensity of the ring A fragment formed from retro-Diels-Alder (RDA) fragmentation with hydrogen transfer appears to be influenced by the 2'-substituent. Thus with 2'-methoxyisoflavanones, a major ring A fragment is either not observed or is of very low intensity, whereas this fragment is of very high intensity or even base peak with 2'-hydroxyisoflavanones (Ingham, 1976c). However, this cannot be regarded as completely general, since the structure of cajanol was revised after degradative studies (Ingham, 1979b).

Synthetic routes to isoflavanones are in general not efficient, and new approaches are thus welcome. Palladium-catalysed Heck arylation of chroman-4-one enol esters gives very satisfactory yields of isoflavanones (Saito *et al.*, 1973), and uses as starting materials chroman-4-ones and aryl mercuric halides (Fig. 10.4). Isoflavanones may also be obtained from 3-arylcoumarins or 3-aryl-4-hydroxycoumarins via hydroboration followed by chromic acid oxidation (Kirkiacharian, 1975), but this method can result in production of the isoflavan as well as, or instead of the required isoflavanone (Kirkiacharian and Chidiac, 1975; Kirkiacharian and Ravise, 1976) and is thus less attractive.

Although catalytic hydrogenation of isoflavanones normally gives isoflavans, in basic medium ring opening can occur to yield α-methyldeoxybenzoins and the corresponding alcohols (Szabo and Antal, 1976). Similar ring-opening reactions may occur when 2'-methoxyisoflavanones are treated with aluminium chloride in acetonitrile, rather than selective demethylation as occurs with 2'-methoxyisoflavones. Thus, 7,2',4'-trimethoxyisoflavanone yields an α-methylenedeoxybenzoin (Malik and Grover, 1978), but the reaction is by no means general. 5-Methoxyisoflavanones are demethylated at position 5, even if a 2'-methoxy substituent is present.

A novel 3-hydroxyisoflavanone, secondifloran (*10.59*), has been isolated from *Sophora secondiflora* (Minhaj *et al.*, 1977), and although such isoflavonoid

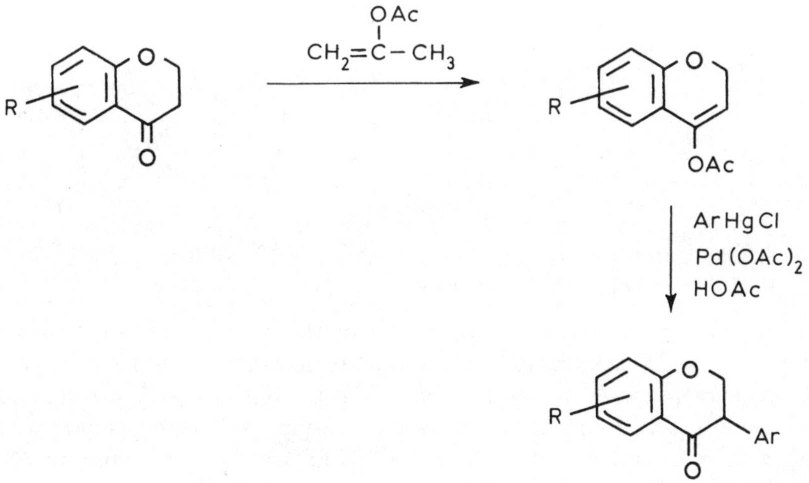

Fig. 10.4

derivatives are known chemically, this is the only example reported in nature. Since the corresponding isoflavanone trimethyl ether is resistant to aerial oxidation, secondifloran is not considered to be an artefact. Secondifloran also contains the uncommon 1,1-dimethylallyl substituent (see Section 10.7).

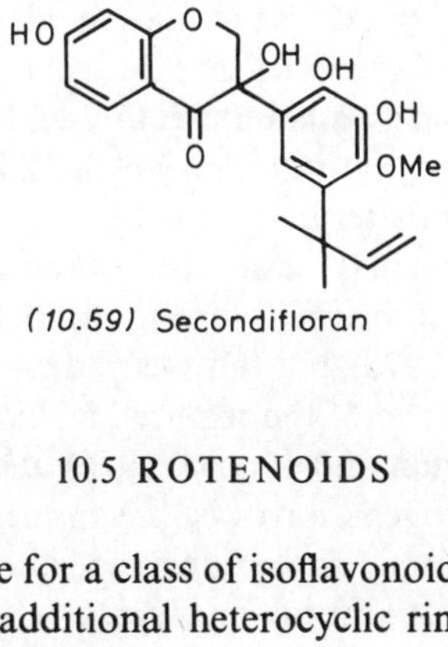

(10.59) Secondifloran

10.5 ROTENOIDS

Rotenoid is a general name for a class of isoflavonoid compound containing an extra carbon atom in an additional heterocyclic ring (*10.60*). A considerable amount of variation in the oxidation levels of this ring system is encountered in nature, and it now becomes necessary to subdivide the natural rotenoids to aid classification. Systematic nomenclature of the rotenoids based on the parent structure rotoxen has never been generally adopted, and a more recent suggestion is that new names based on derran, the reduced structure (*10.60*) should be employed (Jain *et al.*, 1978b). This has merits in that the scheme is analogous to flavan/isoflavan nomenclature, and simple rotenoids would become derranones, dehydrorotenoids derrones etc. However, since most

rotenoids are currently known by trivial names, and these names form the basis for indicating different oxidation levels, it is unlikely that this nomenclature will be widely used. In this review, the group of compounds will be treated in three subdivisions: rotenoids (*10.61*), dehydrorotenoids (*10.62*) and 12a-hydroxy-rotenoids (*10.63*), thus continuing nomenclature already established and in current use. Other variations in oxidation level will be regarded as derivatives of one of these subdivisions.

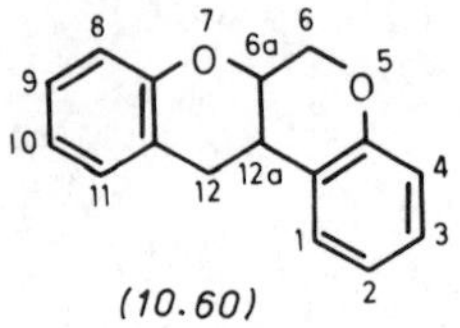

(10.60)

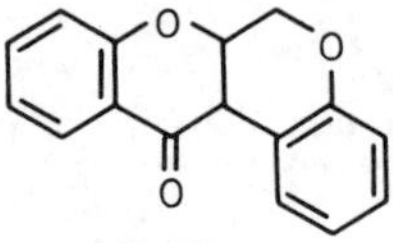

(10.61)

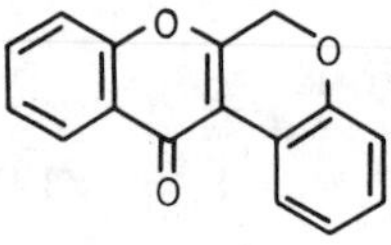

(10.62)

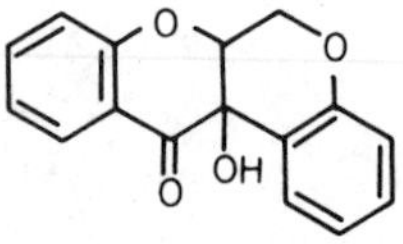

(10.63)

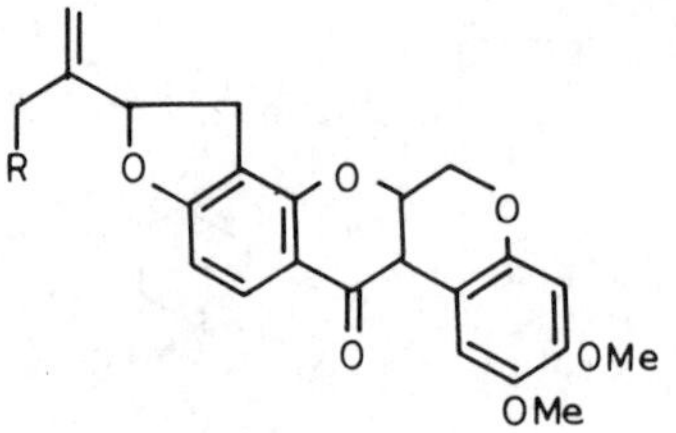

(10.64) R = H, Rotenone
(10.66) R = Oglu, Amorphigenin glucoside

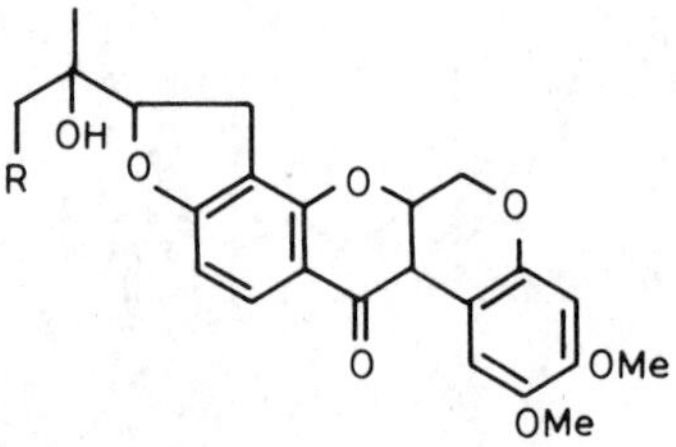

(10.65) R = H, Dalpanol
(10.67) R = OH, Amorphigenol

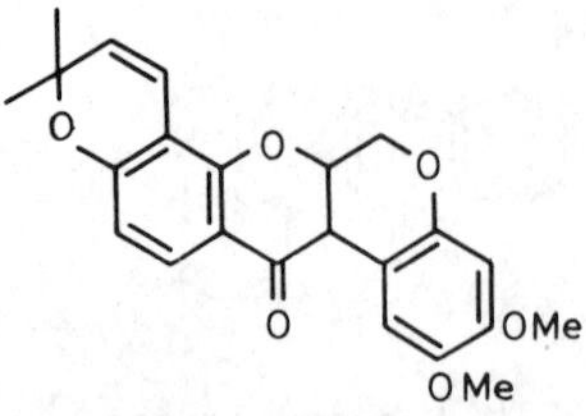

(10.68) Deguelin

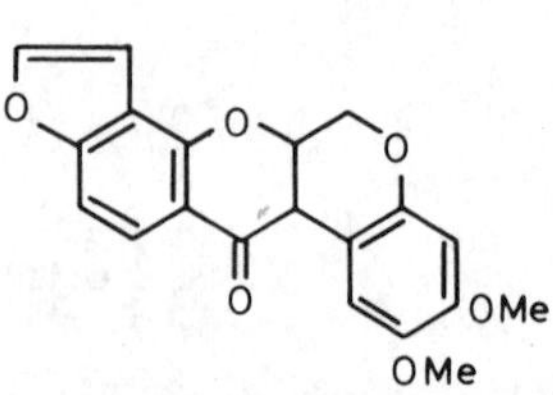

(10.69) Elliptone

(10.70) Sumatrol

(10.71) Toxicarol

(10.72) Villosin

(10.73) R = H, Dehydrorotenone
(10.77) R = OH, Villosol

(10.74) Dehydrodalpanol

(10.75) R = H, Dehydrodeguelin
(10.78) R = OH, Dehydrotoxicarol

(10.76) R = H, Dehydrodolineone
(10.79) R = OMe, Dehydropachyrrhizone

(10.80) R = OH, Stemonal
(10.81) R = OEt, Stemonacetal

(10.82) R_1 = H, R_2 = OH, Amorpholone
(10.83) R_1 = OH, R_2 = OMe, Villinol

(10.84) Stemonone

(10.85) R = H, Rotenonone
(10.86) R = OH, Villosone

(10.87) 12a-Hydroxymunduserone

(10.88) $R_1 = R_2$ = OMe, 12a-Hydroxyrotenone
(10.89) $R_1 R_2 = OCH_2O$, 12a-Hydroxyisomillettone

(10.90) Dalbinol

(10.91) Tephrosin

(10.92) $R_1 = R_2 = OMe$, 12a-Hydroxyerosone
(10.93) $R_1R_2 = OCH_2O$, 12a-Hydroxydolineone

(10.94) Villosinol

(10.95) 11-Hydroxytephrosin

(10.96) 12a-Hydroxypachyrrhizone

(10.97) 12a-Methoxyrotenone

(10.98) $R_1 = R_2 = OMe$, Neobanone
(10.99) $R_1R_2 = OCH_2O$, 12a-Methoxydolineone

(10.100) Clitoriacetal

(10.101) Villol

10.5.1 Rotenoids

Surprisingly, the number of known simple rotenoids has not increased during the review period, although new sources have been reported and a number of glycosides have been isolated (Table 10.4). The presence of rotenoids (and 12a-hydroxyrotenoids) in tissue cultures of *Tephrosia* spp. (Sharma and Khanna, 1975) and *Crotalaria burhia* (Uddin and Khanna, 1979) is worthy of note. Villosin (*10.72*) is 6-hydroxysumatrol isolated from *Tephrosia villosa* pods (Krupadanam *et al.*, 1977) along with other derivatives of rotenone and sumatrol.

The absolute configuration of (−)-rotenone has been established as 6*aS*, 12*aS*, 5′*R* by X-ray analysis of the 8′-bromo derivative (Begley *et al.*, 1975), thus confirming earlier chemical and spectroscopic studies. Rotenone reacts with one molecular equivalent of boron tribromide at low temperature to give the bromide of rotenonic acid (*10.102*) (Carson *et al.*, 1975b) in a reappraisal of earlier results (Unai and Yamamoto, 1973) (Fig. 10.5). The bromide (*10.102*) on treatment with sodium cyanoborohydride in hexamethylphosphorotriamide (HMPT) yields rotenonic acid (*10.103*) and offers a route for the synthesis of specifically labelled derivatives (Carson *et al.*, 1975b). The same overall transformation can be achieved by hydrogenolysis using Pd−$BaSO_4$ in pyridine with substantial, though not complete, stereoselectivity for introduction of hydrogen (Crombie *et al.*, 1979a). The former route proceeds in quite high yields and has been included in the chemical transformation of rotenone to deguelin (*10.68*) (Fig. 10.5) (Anzeveno, 1979a). Rotenonic acid is also a key intermediate in the conversion of rotenone into elliptone (*10.69*) (Fig. 10.5) (Anzeveno, 1979b). These side-chain modifications could find application in the other classes of isoflavonoids.

A number of new rotenoid syntheses have been published. As in many earlier routes, Carson *et al.* (1975a) have exploited the addition of a C_2-unit on to a deoxybenzoin, but using allyl bromide as the source of this unit, cleaving off one carbon via OsO_4-KIO_4 prior to cyclization (Fig. 10.6). This route circumvents dehydrorotenoids as intermediates, but requires protection of the deoxybenzoin from *C*-allylation via an isoflavone structure, and the synthesis of (±)-isorotenone is described. The carbon skeleton of rotenoids can be provided by reacting an aryloxyacetylide with an aromatic aldehyde to give an acetylenic intermediate (Omokawa and Yamashita, 1973). The alcohol produced, after MnO_2 oxidation to the ketone, is thermally cyclized then treated with mild base to yield the rotenoid (Fig. 10.7). This method has been applied in the synthesis of (+)-munduserone (Omokawa and Yamashita, 1973), (±)-9-demethylmunduserone (Omokawa *et al.*, 1975), (±)-deguelin (Omokawa and Yamashita, 1974) and (−)-rotenone (Sasaki and Yamashita, 1979). In the last synthesis, (−)-tuba-aldehyde was used as starting material, and with correct choice of the mild base (pyridine–ethanol), cyclization occurred to yield (−)-rotenone, with no trace of epi-isomer.

Table 10.4 Rotenoids

Rotenoid	*Plant sources*	*References*
(i) *Rotenoids*		
Rotenone (*10.64*)	*Crotalaria burhia*	Uddin and Khanna (1979)
	C. medicaginea	Jain (1979)
	Derris negrensis	Vasconcelos and Maia (1976)
	D. urucu	Braz Filho *et al.* (1975)
	Tephrosia falciformis	Kamal and Jain (1978)
	T. purpurea	
	T. vogelii	Sharma and Khanna (1975)
	Neorautanenia amboensis	Oberholzer *et al.* (1974)
Dalpanol (*10.65*)	*Dalbergia paniculata*	Radhakrishniah (1973)
Dalpanol glucoside	*Dalbergia paniculata*	Radhakrishniah (1973)
Amorphigenin glucoside (*10.66*)	*Amorpha* spp.	Kasymov *et al.* (1968)
Amorphigenol (*10.67*) (as glucoside)	*Amorpha fruticosa*	Kasymov *et al.* (1970)
(as bioside-Amorphol)	*Amorpha* spp.	Kasymov *et al.* (1974)
Deguelin (*10.68*)	*Crotalaria burhia*	Uddin and Khanna (1979)
	Lonchocarpus unifoliatus	Delle Monache *et al.* (1978)
	Tephrosia falciformis	Kamal and Jain (1978)
	T. purpurea	
	T. vogelii	Sharma and Khanna (1975)
Elliptone (*10.69*)	*Crotalaria burhia*	Uddin and Khanna (1979)
	C. medicaginea	Jain (1979)
	Tephrosia falciformis	Kamal and Jain (1978)
	T. purpurea	
	T. vogelii	Sharma and Khanna (1975)
Sumatrol (*10.70*)	*Crotalaria burhia*	Uddin and Khanna (1979)
	Millettia auriculata	Raju and Srimannarayana (1978)
Toxicarol (*10.71*)	*Crotalaria burhia*	Uddin and Khanna (1979)
Villosin (*10.72*)	*Tephrosia villosa*	Krupadanam *et al.* (1977)

(ii) *Dehydrorotenoids*		
Dehydrorotenone (*10.73*)	*Derris negrensis*	Vasconcelos and Maia (1976)
	D. uliginosa	Bose *et al.* (1976)
	D. urucu	Braz Filho *et al.* (1975)
	Neorautanenia amboensis	Oberholzer *et al.* (1974)
	Tephrosia falciformis	Ghanim and Jayaraman (1979)
Dehydrodalpanol (*10.74*)	*Dalbergia paniculata*	Adinarayana and Rao (1975b)
Dehydrodeguelin (*10.75*)	*Amorpha fruticosa*	Ognyanov and Somleva (1980)
Dehydrodolineone (*10.76*)	*Neorautanenia amboensis*	Oberholzer *et al.* (1974)
Villosol (*10.77*) (dehydrosumatrol)	*Tephrosia villosa*	Sarma *et al.* (1976)
Dehydrotoxicarol (*10.78*)	*Amorpha fruticosa*	Reisch *et al.* (1976)
Dehydropachyrrhizone (*10.79*)	*Pachyrrhizus erosus*	Krishnamurti *et al.* (1970)
Stemonal (*10.80*)	*Stemona collinsae* (Stemonaceae)	Shiengthong *et al.* (1974)
Stemonacetal (*10.81*)	*Clitoria macrophylla*	Taguchi *et al.* (1977)
	Stemona collinsae (Stemonaceae)	Shiengthong *et al.* (1974)
Amorpholone (*10.82*)	*Amorpha canescens*	Piatak *et al.* (1975)
Villinol (*10.83*)	*Tephrosia villosa*	Krupadanam *et al.* (1977)
Stemonone (*10.84*)	*Stemona collinsae* (Stemonaceae)	Shiengthong *et al.* (1974)
Rotenonone (*10.85*)	*Neorautanenia amboensis*	Oberholzer *et al.* (1976)
Villosone (*10.86*)	*Tephrosia villosa*	Krupadanam *et al.* (1977)
(iii) *12a-Hydroxyrotenoids*		
12a-Hydroxymunduserone (*10.87*)	*Pachyrrhizus erosus*	Kalra *et al.* (1977)
12a-Hydroxyrotenone (*10.88*)	*Derris urucu*	Braz Filho *et al.* (1975)
	Neorautanenia amboensis	Oberholzer *et al.* (1974)
	Pachyrrhizus erosus	Kalra *et al.* (1977)
	Tephrosia praecans	Camele *et al.* (1980)
12a-Hydroxyisomilletone (*10.89*)	*Neorautanenia amboensis*	Oberholzer *et al.* (1976)
Dalbinol (*10.90*) (12a-hydroxyamorphigenin)	*Dalbergia assamica*	Chibber and Sharma (1978)
	D. latifolia	Chibber and Khera (1978)
Dalbin (dalbinol glucoside)	*Dalbergia assamica*	Chibber and Sharma (1978)
	D. latifolia	Chibber and Khera (1979)
Tephrosin (*10.91*)	*Crotalaria burhia*	Uddin and Khanna (1979)
	C. medicaginea	Jain (1979)
	Derris urucu	Braz Filho *et al.* (1975)

Table 10.4 (*Contd.*)

Rotenoid	*Plant sources*	*References*
	Lonchocarpus unifoliatus	Delle Monache *et al.* (1978)
	Tephrosia falciformis	Kamal and Jain (1978)
	T. praecans	Camele *et al.* (1980)
	T. purpurea	
	T. vogelii	Sharma and Khanna (1975)
12a-Hydroxyerosone (*10.92*)	*Neorautanenia amboensis*	Breytenbach and Rall (1980)
	Pachyrrhizus erosus	Kalra *et al.* (1977)
12a-Hydroxydolineone (*10.93*)	*Neorautanenia amboensis*	Oberholzer *et al.* (1974)
	Pachyrrhizus erosus	Krishnamurti *et al.* (1970)
Villosinol (*10.94*) (12a-hydroxysumatrol)	*Tephrosia villosa*	Sarma *et al.* (1976)
11-Hydroxytephrosin (*10.95*)	*Amorpha fruticosa*	Mitscher *et al.* (1979)
12a-Hydroxypachyrrhizone (*10.96*)	*Pachyrrhizus erosus*	Krishnamurti *et al.* (1970)
12a-Methoxyrotenone (*10.97*)	*Neorautanenia amboensis*	Oberholzer *et al.* (1974)
Neobanone (*10.98*)	*Neorautanenia amboensis*	Oberholzer *et al.* (1976)
12a-Methoxydolineone (*10.99*)	*Neorautanenia amboensis*	Oberholzer *et al.* (1974)
Clitoriacetal (*10.100*)	*Clitoria macrophylla*	Taguchi *et al.* (1977)
Villol (*10.101*)	*Tephrosia villosa*	Krupadanam *et al.* (1977)

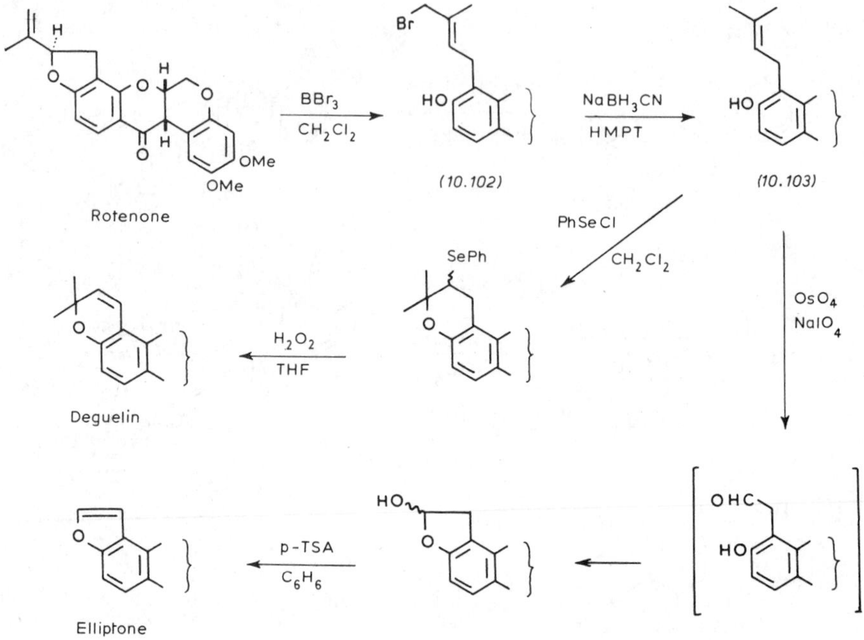

Fig. 10.5

In a further new route to rotenoids via dehydrorotenoids (Fig. 10.8), the β-keto ester (*10.104*) reacts thermally with the appropriate phenol, and rotenoids can be obtained by hydrogenation (Verhe and Schamp, 1975). Because the A ring of natural rotenoids contains, without exception, 2,3-dimethoxy or 2,3-methylenedioxy substituents, this route may prove more useful than the unusual starting materials might indicate.

10.5.2 Dehydrorotenoids

Because 6a,12a-dehydrorotenoids may be formed by the ready dehydration of 12a-hydroxyrotenoids, themselves formed by aeration of rotenoid solutions, the natural occurrence of both of these classes has been disputed in the past. It is now accepted that dehydrorotenoids are plant products and not artefacts, and dehydro derivatives of eight rotenoids are recognized. Those recently reported are listed in Table 10.4.

6-Oxy derivatives of dehydrorotenoids have also been isolated. Stemonal (*10.80*) from roots of the Thai medicinal plant *Stemona collinsae* (Stemonaceae) (Shiengthong *et al.*, 1974) and stemonacetal (*10.81*) from *S. collinsae* and also *Clitoria macrophylla* (Taguchi *et al.*, 1977) are 6-hydroxy and 6-ethoxy derivatives respectively of dehydrosermundone, amorpholone (*10.82*) from root

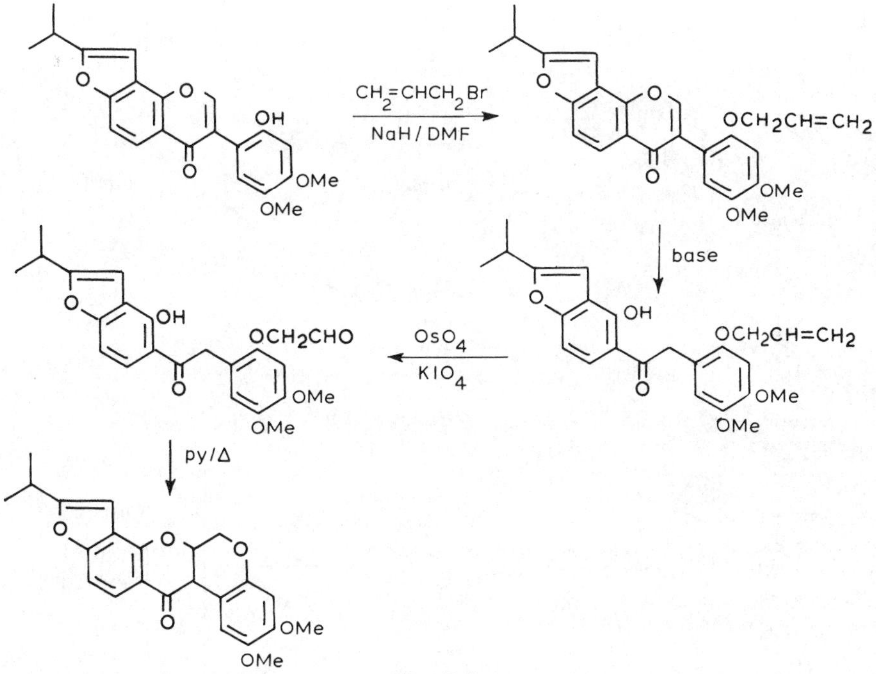

$CH_2=CHCH_2Br$ / NaH / DMF

base

OsO_4 / KIO_4

py/Δ

Fig. 10.6

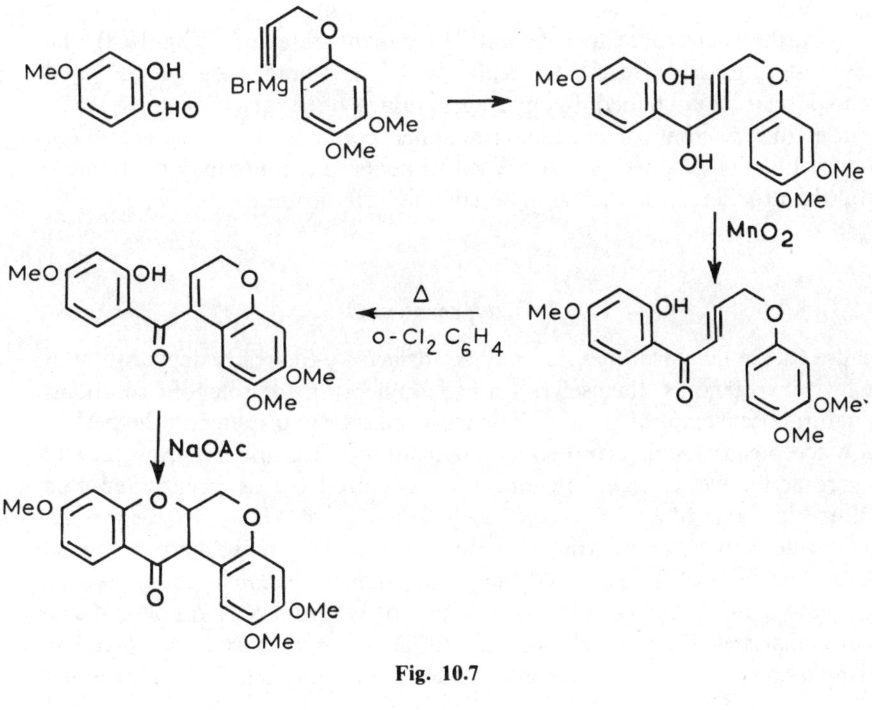

MnO_2

Δ / o-$Cl_2C_6H_4$

NaOAc

Fig. 10.7

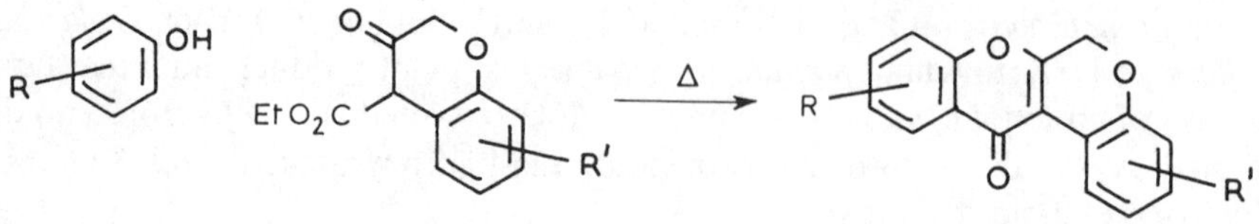

Fig. 10.8

of *Amorpha canescens* (Piatak *et al.*, 1975) is 6-hydroxydehydrorotenone, and villinol (*10.83*) from *Tephrosia villosa* pods (Krupadanam *et al.*, 1977) is 6-methoxydehydrosumatrol. Stemonone (*10.84*) and villosone (*10.86*) are 6-keto dehydrorotenoids co-occurring with stemonal and villinol respectively, and rotenonone (*10.85*) has been isolated from *Neorautanenia amboensis* (Oberholzer *et al.*, 1976).

As with the dehydrorotenoids, it is important to establish that 6-oxygenated compounds do not arise as artefacts, since photochemical studies (Chubachi and Hamada, 1974; Krupadanam *et al.*, 1978) have shown that these types of compounds can be produced by UV irradiation of rotenoid and dehydrorotenoid solutions. The conversion into 6-oxorotenoids can be almost total in the case of some 6a,12a-dehydrorotenoids (Krupadanam *et al.*, 1978).

10.5.3 12a-Hydroxyrotenoids

(±)-Tephrosin (*10.91*) has been known for many years but was often regarded as an artefact of deguelin via aerial oxidation (Wong, 1975). With the isolation of optically active tephrosin and millettosin, these compounds were accepted to be natural products. 12a-Hydroxy derivatives of most of the known rotenoids have now been reported (Table 10.4), and many of these have also been isolated in optically active forms. Where absolute configurations have been reported, these compounds have the 6a*S*, 12a*S* configuration, as do the natural rotenoids. Further evidence that these materials must be regarded as natural products has been obtained during biosynthetic studies in *Amorpha fruticosa* seedlings (Crombie *et al.*, 1979b). 12a-Hydroxyamorphigenin (*10.90*) with a specific activity some six times greater than that of the co-occurring amorphigenin was isolated, and it could thus not have arisen as an artefact of isolation.

Optically active 12a-methoxy derivatives of rotenone, erosone (neobanone) and dolineone have also been reported in *Neorautanenia amboensis* (Oberholzer *et al.*, 1974, 1976). The structure of neobanone has been confirmed by total synthesis, via the rotenoid erosone, aerial oxidation to 12a-hydroxyerosone, then methylation (Oberholzer *et al.*, 1977a).

6,12a-Dihydroxyrotenoids are a further variation in rotenoid structure, and two examples of these have been isolated from plants, namely clitoriacetal (*10.100*) which co-occurs with the 6-ethoxydehydrorotenoid stemonacetal in

Clitoria macrophylla (Taguchi *et al.*, 1977) and villol (*10.101*) from *Tephrosia villosa* pods (Krupadanam *et al.*, 1977). This last plant produces an interesting array of rotenoid variants, namely 6-hydroxyrotenoid, dehydrorotenoid, 6-methoxy- and 6-keto-dehydrorotenoid, 12a-hydroxyrotenoid and 6,12a-dihydroxyrotenoid derivatives.

10.6 PTEROCARPANS

Pterocarpans (*10.105*) are widely distributed in leguminous plants both as heartwood and bark constituents, and also in young tissue challenged by micro-organisms, since many function as phytoalexins (see Section 10.20). It is convenient to subdivide the group into pterocarpans, 6a-hydroxypterocarpans (*10.106*) and pterocarpenes (*10.107*), in general reserving the nomenclature pterocarpan for the fully reduced ring system (*10.105*).

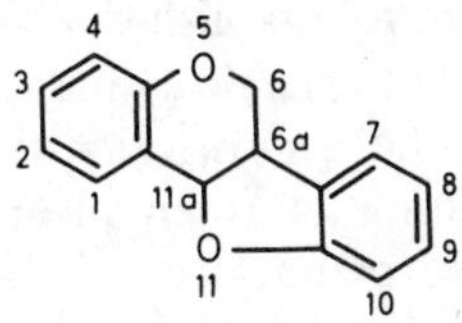

(*10.105*)

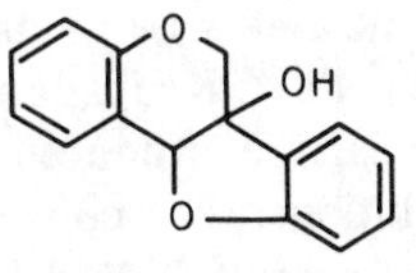

(*10.106*)

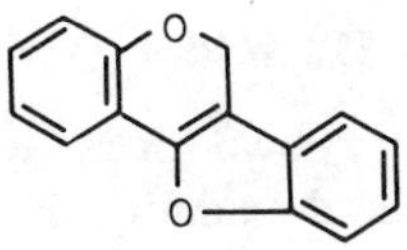

(*10.107*)

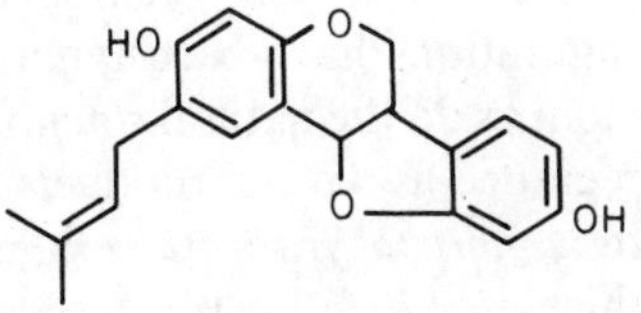

(*10.108*) Homoedudiol

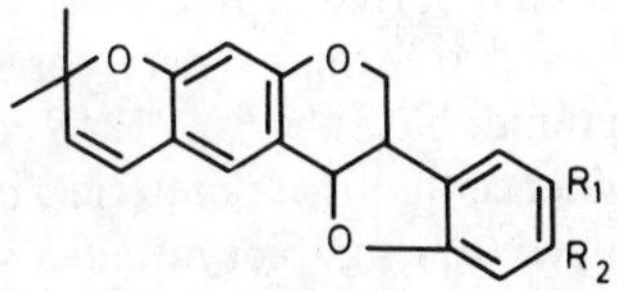

(*10.109*) R_1 = H, R_2 = OH, Neorautenol

(*10.110*) R_1R_2 = OCH_2O, Neorautenane

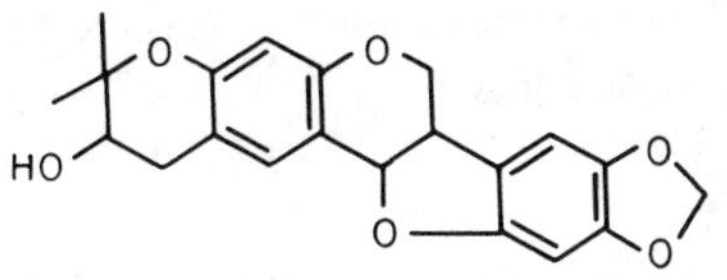

(*10.111*) Neorautanol

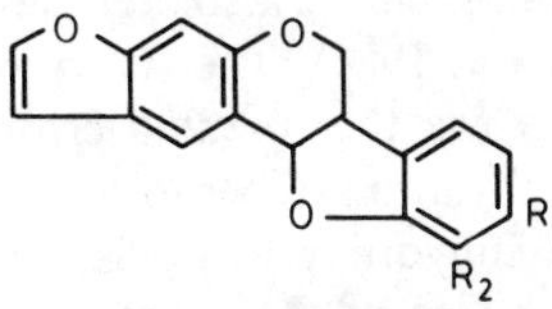

(*10.112*) R_1 = OH, R_2 = H, Neodunol

(*10.113*) R_1 = R_2 = OMe, Ambonane

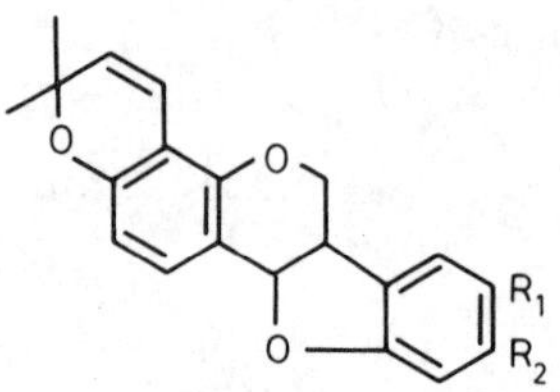

O O O R_1 R_2

(10.114) $R_1 = H$, $R_2 = OMe$, Hemileiocarpin
(10.115) $R_1 R_2 = OCH_2O$, Leiocarpin

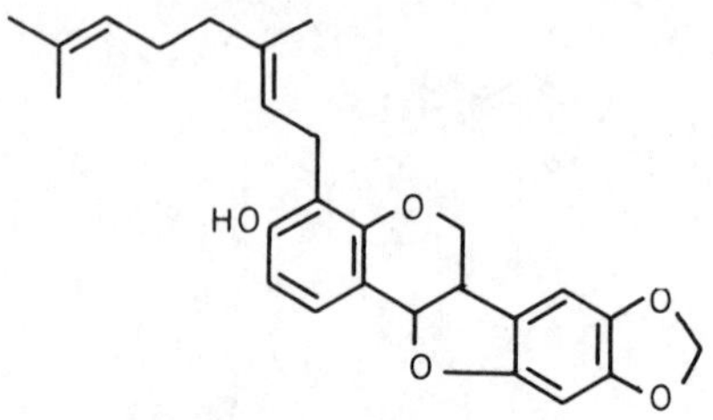

HO O O O O

(10.116) Nitiducol

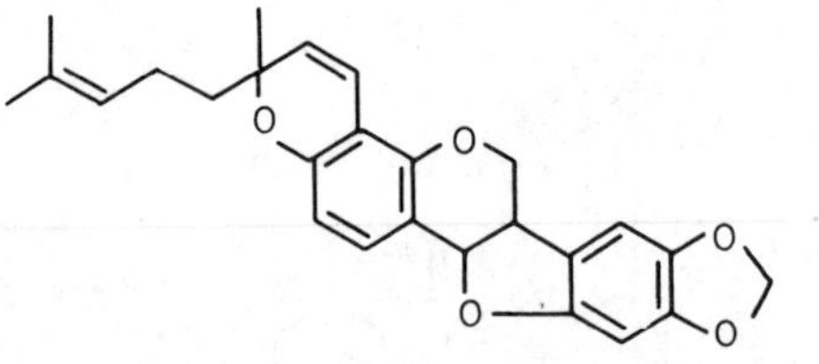

O O O O O

(10.117) Nitiducarpin

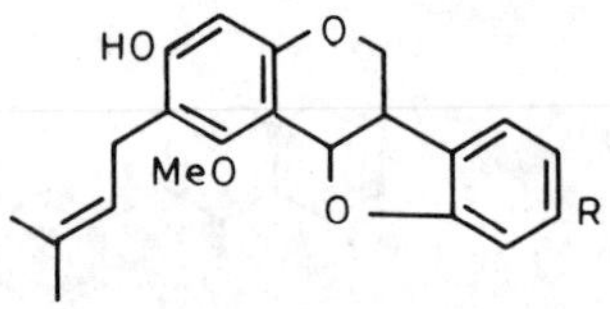

HO O MeO O R

(10.118) R = OH, Edudiol
(10.119) R = OMe, Edulenol

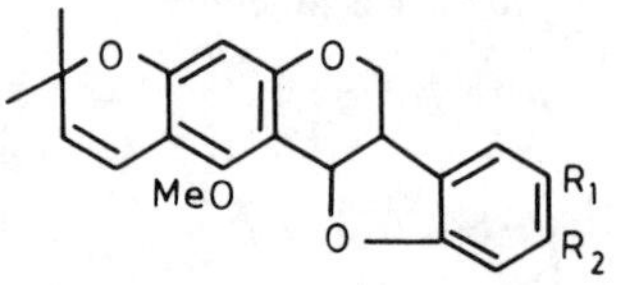

O O MeO O R_1 R_2

(10.120) $R_1 = H$, $R_2 = OH$, Edulenanol
(10.121) $R_1 = H$, $R_2 = OMe$, Edulenane
(10.122) $R_1 = OMe$, $R_2 = OH$, Desmodin

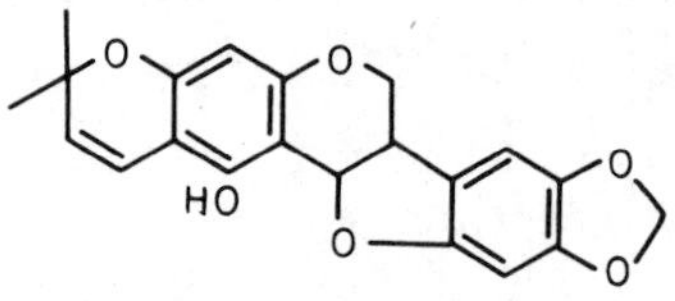

O O HO O O O

(10.123) Neorautenanol

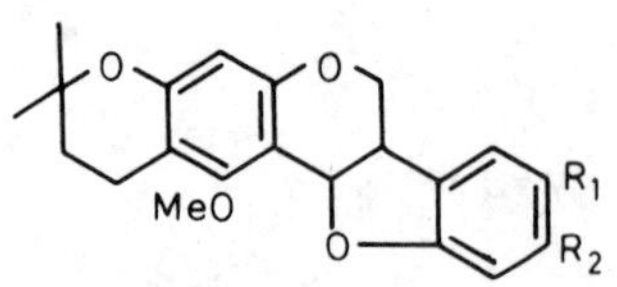

O O MeO O R_1 R_2

(10.124) $R_1 = H$, $R_2 = OMe$, Edulane
(10.125) $R_1R_2 = OCH_2O$, Neorautanin

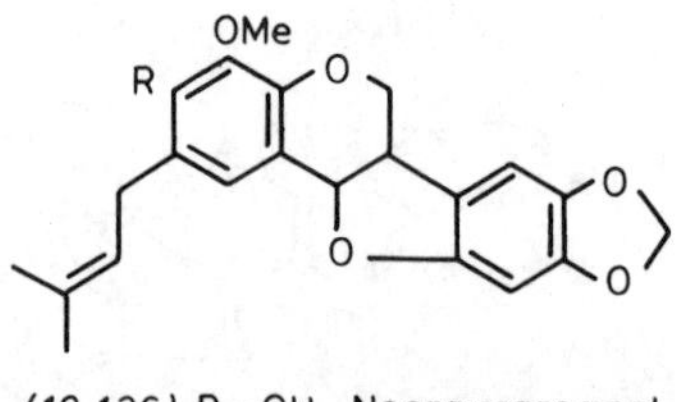

OMe R O O O O

(10.126) R = OH, Neoraucarpanol
(10.127) R = OMe, Neoraucarpan

(10.128) R=OH, Phaseollidin

(10.129) R=OMe, Sandwicensin

(10.130) Lespedezin

(10.131) 1-Methoxyphaseollidin

(10.132) Gangetinin

(10.133) Lespein

(10.134) Acanthocarpan

(10.135) Glyceollin IV

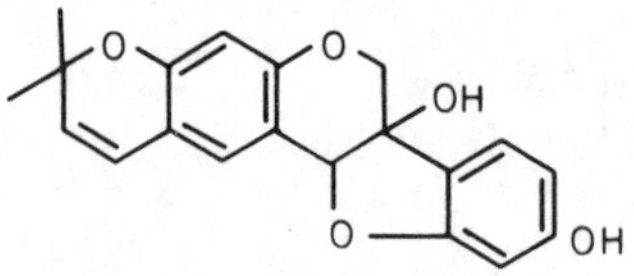

O O OH O OH

(10.136) Glyceollin II

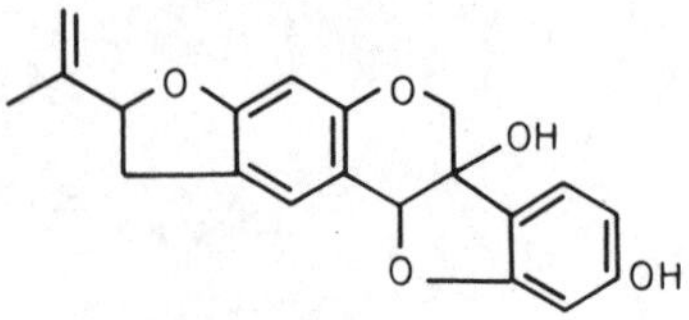

O O OH O OH

(10.137) Glyceollin III

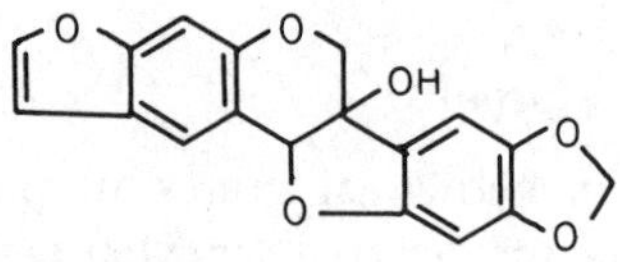

O O OH O O O

(10.138) Neobanol

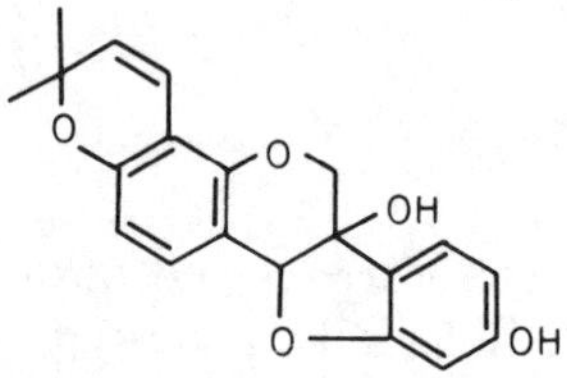

O O OH O OH

(10.139) Glyceollin III

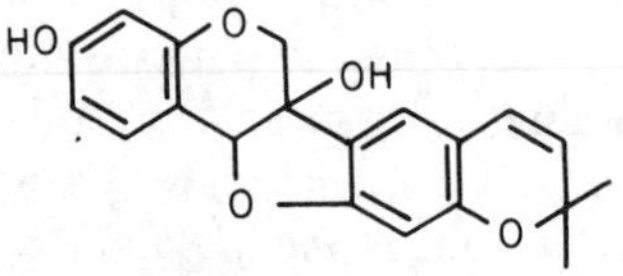

HO O OH O O

(10.140) Tuberosin

HO O OH O R

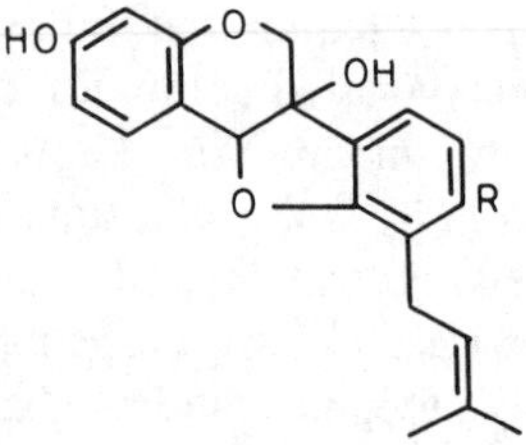

(10.141) R = OH, Sandwicarpin
(10.142) R = OMe, Cristacarpin

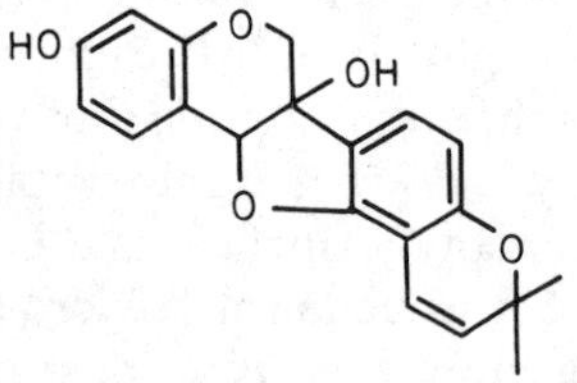

HO O OH O O

(10.143) 6a-Hydroxyphaseollin

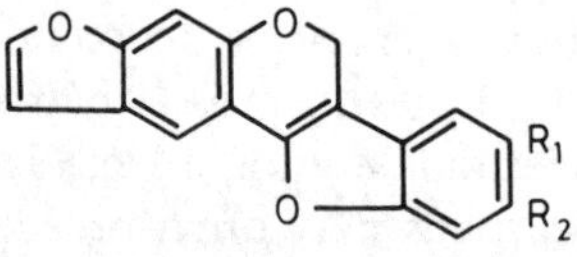

O O R_1 O R_2

(10.144) R_1 = H, R_2 = OH, Neorauteen
(10.145) R_1R_2 = OCH_2O, Neoduleen

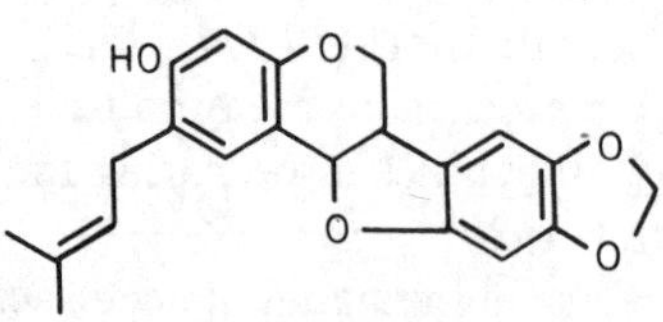

HO O O O O

(10.146) Edunol

HO O O O

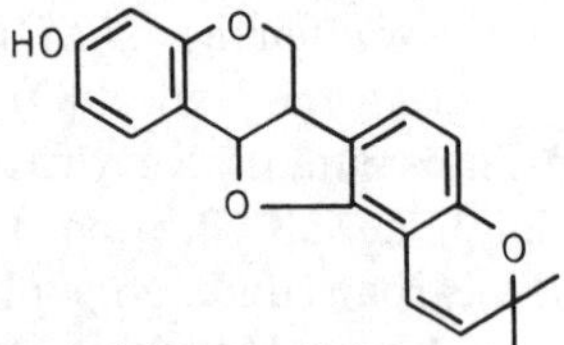

(10.147) Phaseollin

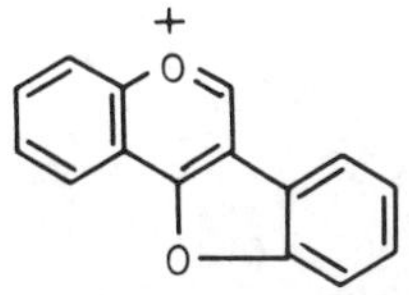

(10.148)

10.6.1 Pterocarpans

New pterocarpans and further sources of known structures are presented in Table 10.5. It is interesting to note that medicarpin (previously known as demethylhomopterocarpin, but this name is too cumbersome for general usage) is now undoubtedly the most common of natural pterocarpans since it functions as a phytoalexin in tissues of many legumes. A wide variety of substitution patterns is encountered, but 3,9- and 3,8,9-oxygenation patterns predominate, and 1-oxygenated compounds (corresponding to position 5 in isoflavones) are relatively uncommon. Unusual geranyl substitutents have been found in nitiducol (*10.116*) and nitiducarpin (*10.117*) from *Dalbergia nitidula* bark (Van Heerden *et al.*, 1978) and in lespedezin (*10.130*) from *Lespedeza homoloba* bark (Ueno *et al.*, 1973b). Lespein (*10.133*), also isolated from the latter plant is 6a-dimethylallylphaseollidin and represents the first example of a 6a-alkylated pterocarpan.

Pterocarpans are easily recognized from their characteristic UV and NMR spectra, but assignment of substitution may be complicated, since MS fragmentation may give peaks which could arise from either ring (Wong, 1975). Hydrogenolysis to 2′-hydroxyisoflavans is the usual means of overcoming this difficulty. The reverse reaction can be brought about by treatment with DDQ (provided that there is a hydroxyl or methoxyl at position 7 of the isoflavan) (Cornia and Merlini, 1975; Gunning *et al.*, 1977), and this synthetic route to pterocarpans may often proceed more smoothly (Ingham and Dewick, 1979) than the established method via borohydride reduction of 2′-hydroxyisoflavones (Wagner and Farkas, 1975). Excess reagent must be avoided, since DDQ also converts the pterocarpan into a coumestan in excellent yield (Ferreira *et al.*, 1974; Gunning *et al.*, 1977). A further new approach to pterocarpans is by the Li_2PdCl_4–catalysed Heck arylation of chromenes (Fig. 10.9), and this has been used to synthesize the natural pterocarpans pterocarpin (Horino and Inoue, 1976), neorautenane (*10.110*) and neorautanol (*10.111*) (via maackiain) (Breytenbach and Rall, 1980). Although pterocarpans are produced in excellent yields, the route is somewhat limited by difficulties in preparing the chromene starting materials (Breytenbach and Rall, 1980).

Pterocarpans undergo photolytic cleavage in methanol or acetic acid solution to give 3,4-*trans*-2′-hydroxy-4-methoxy (or acetoxy)-isoflavans (Fig. 10.10). The

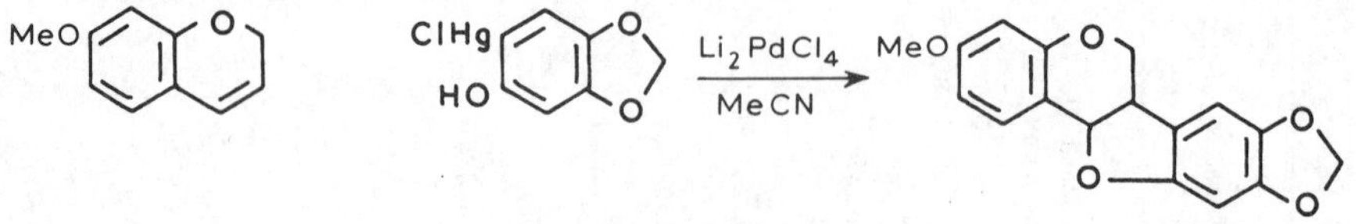

Fig. 10.9

methoxy compounds slowly revert to pterocarpans, rapidly on treatment with dilute HCl, similar results being obtained after hydrolysis of the acetoxy compounds. Cyclization is dependent on an effective leaving group at C-4, and the evidence favours a carbonium ion mechanism (Van der Meerwe *et al.*, 1978).

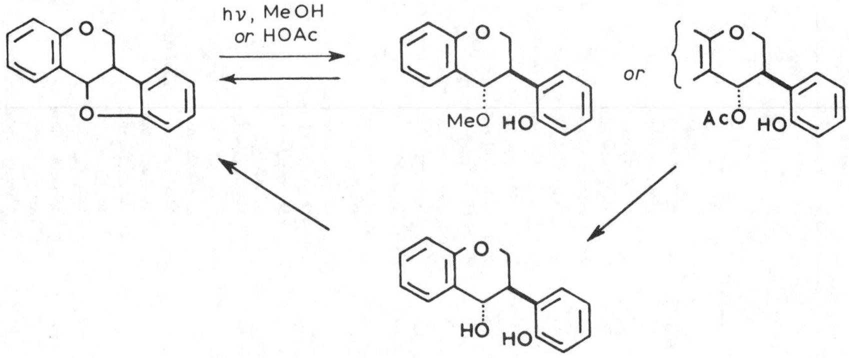

Fig. 10.10

The reactions of pterocarpans and their interrelationship with other classes of isoflavonoid have been summarized in flow-chart form (Gunning *et al.*, 1977).

Few pterocarpan glycosides have been reported in nature, but these derivatives do offer a means of resolving synthetic pterocarpan racemates, and the 3-glucoside has been used to resolve (±)-medicarpin (Antus and Nogradi, 1979a). Though pterocarpans contain two chiral centres, only *R,R* and *S,S* configurations are stereochemically possible. Laevorotatory, dextrorotatory and racemic forms are known in nature, and as a result of chemical degradation, NMR and ORD studies, it has been generally accepted that all laevorotatory pterocarpans have the 6a*R*, 11a*R* absolute configuration, and dextrorotatory ones the 6a*S*, 11a*S* configuration (Wong, 1975). Pterocarpans display multiple Cotton effects in the 250–350 nm region of their ORD spectra, and these may also be used to aid chirality assignments (Cook *et al.*, 1978). X-ray crystallographic study of (+)-edunol (*10.146*) as its 4-bromo-3-methyl derivative confirmed this (Leipoldt *et al.*, 1977), though in a recent communication (Breytenbach and Rall, 1980), the dextrorotatory nature of the pterocarpan has been admitted to be erroneous, and

Table 10.5 Pterocarpans

Pterocarpan	*Substituents*								*Plant sources*	*Optical activity*	*References*
	1	*2*	*3*	*4*	*7*	*8*	*9*	*10*			
(i) *Pterocarpans*											
Demethylmedicarpin			OH				OH		*Albizia procera*	+	Deshpande and Shastri (1977)
									Erythrina crista-galli		Ingham and Markham (1980)
									E. sandwicensis		Ingham (1980)
									Melilotus alba		Ingham (1976a)
									Pachyrrhizus erosus		Ingham (1979g)
									Phaseolus vulgaris	–	Woodward (1980a)
									Psophocarpus tetragonolobus		Ingham and Markham (1980)
Medicarpin (Demethylhomopterocarpin)			OH				OMe		*Canavalia ensiformis*	–	Lampard (1974)
									Caragana spp.		Ingham (1979d)
									Cicer arietinum		Ingham (1976d)
									Dalbergia cearensis	±	Guimaraes *et al.* (1975)
									D. ecastophyllum	±, +	Matos *et al.* (1975)
									D. nitidula	+	Letcher and Shirley (1976)
									D. oliveri	+	Donnelly and Kavanagh (1974)
									D. sericea		Ingham (1979f)
									D. spruceana	±	Cook *et al.* (1978)
									D. stevensonii	±, –	Donnelly *et al.* (1973b)
									D. variabilis	±, +	Kurosawa *et al.* (1978d)
									D. volubilis	+	Khera and Chibber (1978a)
									Derris amazonica	+	Braz Filho *et al.* (1975)
									Factorovskya aschersoniana		Ingham (1979c)
									Lathyrus nissolia		Robeson and Ingham (1979)
									Machaerium acutifolium	+	Ollis *et al.* (1978b)
									M. kuhlmannii	+	
									M. nictitans	+	Ollis *et al.* (1978a)
									M. vestitum	+	Kurosawa *et al.* (1978a)
									Medicago spp.		Ingham (1979a)
									Melilotus alba		Ingham (1976)
									M. spp.		Ingham (1977a)
									Myroxylon balsamum	+	Oliveira *et al.* (1978)
									Neorautanenia amboensis	–	Breytenbach and Rall (1980)

				Onobrychis spp.		Ingham (1978b)
				Osteophleum platyspermum (Myristicaceae)	±	Braz Filho *et al.* (1977)
				Parochetus communis		Ingham (1979c)
				Pericopsis schliebenii	−	Fitzgerald *et al.* (1976)
				Platymiscium trinitatis	+	Craveiro and Gottlieb (1974)
				Trifolium arvense		Ingham and Dewick (1977)
				T. hybridum		Ingham (1976b)
				T. spp.		Ingham (1978c)
				Trigonella spp.		Ingham and Harborne (1976)
				Vicia faba		Hargreaves *et al.* (1976)
				Vigna unguiculata	−	Lampard (1974)
Medicarpin glucoside	Oglu		OMe	*Medicago sativa*	−	Sakagami *et al.* (1974)
				Ononis spinosa		Haznagy *et al.* (1978)
Isomedicarpin	OMe		OH	*Psophocarpus tetragonolobus*	−	Preston (1977b)
Homopterocarpin	OMe		OMe	*Bowdichia nitida*		Brown *et al.* (1974)
				Machaerium villosum	+	Kurosawa *et al.* (1978c)
				Pericopsis angolensis	−	Fitzgerald *et al.* (1976)
				P. schliebenii	−	Fitzgerald *et al.* (1976)
Maackiain	OH	OCH_2O		*Caragana* spp.		Ingham (1979d)
				Cicer arietinum		Ingham (1976d)
				Dalbergia oliveri	+	Donnelly and Kavanagh (1974)
				D. sericea		Ingham (1979f)
				D. spruceana		Cook *et al.* (1978)
				D. stevensonii	±, −	Donnelly *et al.* (1973b)
				Osteophleum platyspermum (Myristicaceae)	±	Braz Filho *et al.* (1977)
				Pericopsis schliebenii	−	Fitzgerald *et al.* (1976)
				Pisum sativum		Ingham (1979e)
				Sophora japonica	+	Komatsu *et al.* (1976b)
				S. microphylla		Briggs *et al.* (1975)
				S. subprostrata	−	Komatsu *et al.* (1976a)
				S. tetraptera		Briggs *et al.* (1975)
				S. tomentosa	−	Komatsu *et al.* (1978a)
				Trifolium arvense		Ingham and Dewick (1977)
				T. hybridum		Ingham (1976b)
				T. spp.		Ingham (1978c)
				Trigonella spp.		Ingham and Harborne (1976)

Table 10.5 (*Contd.*)

Pterocarpan	Substituents 1	2	3	4	7	8	9	10	Plant sources	Optical activity	References
Trifolirhizin [(−)-Maackiain glucoside]			Oglu			OCH_2O			*Ononis arvensis*	−	Kovalev *et al.* (1976)
									O. spinosa	−	Haznagy *et al.* (1978)
									Sophora subprostrata	−	Komatsu *et al.* (1976a)
Trifolirhizin-6′-monoacetate									*Sophora subprostrata*	−	Komatsu *et al.* (1976a)
Pterocarpin			OMe			OCH_2O			*Sophora subprostrata*	−	Komatsu *et al.* (1976a)
Vesticarpan			OH				OMe	OH	*Machaerium vestitum*	+	Kurosawa *et al.* (1978a)
									Platymiscium trinitatis	+	Craveiro and Gottlieb (1974)
Nissolin			OH				OH	OMe	*Lathyrus nissolia*		Robeson and Ingham (1979)
Methylnissolin			OH				OMe	OMe	*Lathyrus nissolia*		Robeson and Ingham (1979)
3-Hydroxy-2,9-dimethoxy-		OMe	OH				OMe		*Pisum sativum*	−	Pueppke and Van Etten (1975)
Sparticarpin		OMe	OMe				OH		*Spartium junceum*	−	Ingham and Dewick (1980a)
2,3,9-Trimethoxy-		OMe	OMe				OMe		*Pisum sativum*	−	Pueppke and Van Etten (1975)
2,8-Dihydroxy-3,9-dimethoxy-		OH	OMe			OH	OMe		*Swartzia laevicarpa*	−	Braz Filho *et al.* (1980)
Mucronucarpan		OH	OMe				OMe	OH	*Machaerium mucronulatum*	+	Kurosawa *et al.* (1978c)
2,8-Dihydroxy-3,9,10-trimethoxy-		OH	OMe			OH	OMe	OMe	*Swartzia laevicarpa*	−	Braz Filho *et al.* (1980)
3,4,9-Trihydroxy-			OH	OH			OH		*Melilotus alba*		Ingham (1976a)
3,4-Dihydroxy-9-methoxy-			OH	OH			OMe		*Melilotus alba*		Ingham (1976a)
4-Hydroxy-3,9,-dimethoxy-			OMe	OH			OMe		*Melilotus alba*		Ingham (1976a)
3,4,9-Trimethoxy			OMe	OMe			OMe		*Myroxylon peruiferum*	−	Maranduba *et al.* (1979)
3,4-Dihydroxy-8,9-methylene-dioxy-			OH	OH		OCH_2O			*Dalbergia spruceana*	−	Cook *et al.* (1978)
3-Hydroxy-4-methoxy-8,9-methylenedioxy-			OH	OMe		OCH_2O			*Dalbergia spruceana*	−	Cook *et al.* (1978)
									Trifolium hybridum		Ingham (1976b)
4-Hydroxy-3-methoxy-8,9-methylenedioxy-			OMe	OH		OCH_2O			*Dalbergia spruceana*	−	Cook *et al.* (1978)
8-Hydroxy-3,4,9,10-tetramethoxy-			OMe	OMe		OH	OMe	OMe	*Swartzia laevicarpa*	−	Braz Filho *et al.* (1980)
Trifolian	OH	OMe	OH			OCH_2O			*Trifolium pratense*	−	Fraishtat *et al.* (1979b)
4-Hydroxy-2,3,9-trimethoxy-		OMe	OMe	OH			OMe		*Pisum sativum*	−	Pueppke and Van Etten (1975)

2,8-Dihydroxy-3,4,9,10-tetra-methoxy-	OH	OMe	OMe	OH	OMe	OMe	*Swartzia laevicarpa*	–	Braz Filho *et al.* (1980)
Homoedudiol (*10.108*)							*Neorautanenia edulis*	–	Brink *et al.* (1974b)
							Pachyrrhizus erosus		Ingham (1979g)
Neorautenol (*10.109*)							*Neorautanenia edulis*	–	Brink *et al.* (1974b)
Neorautenane (*10.110*)							*Neorautanenia amboensis*	–	Breytenbach and Rall (1980)
Neorautanol (*10.111*)							*Neorautanenia amboensis*	–	Breytenbach and Rall (1980)
Neodunol (*10.112*)							*Neorautanenia amboensis*	–	Breytenbach and Rall (1980)
							N. edulis	–	Brink *et al.* (1974b)
							Pachyrrhizus erosus		Ingham (1979g)
Ambonane (*10.113*)							*Neorautanenia amboensis*	–	Breytenbach and Rall (1980)
Hemileiocarpin (*10.114*)							*Dalbergia nitidula*	+	Van Heerden *et al.* (1978)
Leiocarpin (*10.115*)							*Dalbergia nitidula*	+	Van Heerden *et al.* (1978)
Nitiducol (*10.116*)							*Dalbergia nitidula*	+	Van Heerden *et al.* (1978)
Nitiducarpin (*10.117*)							*Dalbergia nitidula*	+	Van Heerden *et al.* (1978)
Edudiol (*10.118*)							*Neorautanenia edulis*	–	Brink *et al.* (1977)
Edulenol (*10.119*)							*Neorautanenia amboensis*	–	Brink *et al.* (1977)
							N. edulis		Rall *et al.* (1972)
Edulenanol (*10.120*)							*Neorautanenia amboensis*	–	Breytenbach and Rall (1980)
Edulenane (*10.121*)							*Neorautanenia amboensis*	–	Breytenbach and Rall (1980)
Desmodin (*10.122*)							*Desmodium gangeticum*	–	Purushothaman *et al.* (1975)
Neorautenanol (*10.123*)							*Neorautanenia amboensis*	–	Breytenbach and Rall (1980)
Edulane (*10.124*)							*Neorautanenia edulis*	–	Brink *et al.* (1977)
Neorautanin (*10.125*)							*Neorautanenia edulis*	–	Brink *et al.* (1977)
Neoraucarpanol (*10.126*)							*Neorautanenia amboensis*	–	Brink *et al.* (1977)
Neoraucarpan (*10.127*)							*Neorautanenia edulis*	–	Brink *et al.* (1977)
Phaseollidin (*10.128*)							*Erythrina corallodendron*		Ingham (1980)
							E. crista-galli	–	Ingham and Markham (1980)
							E. lysistemon		Ingham (1980)
							E. sandwicensis	–	Ingham (1980)
							Lablab niger		Ingham (1977b)
							Psophocarpus tetragonolobus		Preston (1977b)
									Ingham (1978a)

Table 10.5 (*Contd.*)

Pterocarpan	*Substituents*								*Plant sources*	*Optical activity*	*References*
	1	*2*	*3*	*4*	*7*	*8*	*9*	*10*			
Sandwicensin (*10.129*)									*Erythrina sandwicensis*	–	Ingham (1980)
Lespedezin (*10.130*)									*Lespedeza homoloba*	±	Ueno *et al.* (1973b)
1-Methoxyphaseollidin (*10.131*)									*Psophocarpus tetragonolobus*	–	Preston (1977b)
Gangetinin (*10.132*)									*Desmodium gangeticum*	–	Puroshothaman *et al.* (1975)
Lespein (*10.133*)									*Lespedeza homoloba*	–	Ueno *et al.* (1973b)
(ii) *6a-Hydroxypterocarpan*											
3,6a,9-Trihydroxypterocarpan			OH				OH		*Erythrina sandwicensis*		Ingham (1980)
									Glycine max	–	Lyne and Mulheirn (1978)
6a-Hydroxymedicarpin			OH				OMe		*Melilotus alba*		Ingham (1976a)
									Trifolium pratense		Bilton *et al.* (1976)
6a-Hydroxyisomedicarpin			OMe				OH		*Melilotus alba*	–	Ingham (1976a)
Variabilin (homopisatin)			OMe				OMe		*Caragana* spp.		Ingham (1979d)
									Dalbergia variabilis	+	Kurosawa *et al.* (1978d)
									Lathyrus odoratus		Robeson *et al.* (1980)
									Lens culinaris	–	Robeson (1978)
									Trifolium pratense		Bilton *et al.* (1976)
6a-Hydroxymaackiain			OH			OCH_2O			*Trifolium pratense*		Bilton *et al.* (1976)
Pisatin			OMe			OCH_2O			*Caragana* spp.		Ingham (1979d)
									Lathyrus odoratus		Robeson *et al.* (1980)
									L. spp.		Robeson and Harborne (1977)
									Trifolium pratense		Bilton *et al.* (1976)
3,6a,7-Trihydroxy-9-methoxy-pterocarpan			OH		OH		OMe		*Trifolium pratense*		Ingham (1976a)
3,6a,7-Trihydroxy-8,9-methylenedioxypterocarpan			OH		OH	OCH_2O			*Trifolium pratense*		Ingham (1976a)
Acanthocarpan (*10.134*)									*Caragana acanthophyllum*		Ingham (1979d)
Glyceollin IV (*10.135*)									*Glycine max*	–	Lyne and Mulheirn (1978)
Glyceollin II (*10.136*)									*Glycine max*	–	Lyne *et al.* (1976)
Glyceollin III (*10.137*)									*Glycine max*	–	Lyne *et al.* (1976)

Neobanol (*10.138*)						*Neorautanenia amboensis*	–	Oberholzer *et al.* (1976)
Glyceollin I (*10.139*)						*Glycine max*	–	Burden and Bailey (1975)
							–	Lyne *et al.* (1976)
Tuberosin (*10.140*)						*Pueraria tuberosa*	+	Joshi and Kamat (1973)
Sandwicarpin (*10.141*) (6a-hydroxyphaseollidin)						*Erythrina sandwicensis*	–	Ingham (1980)
Cristacarpin (*10.142*)						*Erythrina crista-galli*	–	Ingham and Markham (1980)
						E. sandwicensis	–	Ingham (1980)
						Psophocarpus tetragonolobus		Ingham and Markham (1980)
6a-Hydroxyphaseollin (*10.143*)						*Phaseolus vulgaris*		Van den Heuvel (1976)
							–	Woodward (1980a)
(iii) *Pterocarpenes*								
3,9-Dihydroxy-	OH			OH		*Lespedeza cyrtobotrya*		Miyase *et al.* (1980)
						Tetragonolobus maritimus		Ingham and Dewick (1978)
Flemichapparin-B (anhydropisatin)	OMe		OCH_2O			*Derris urucu*		Braz Filho *et al.* (1975)
						Sophora japonica		Komatsu *et al.* (1976b)
Bryacarpene-5	OMe			OMe	OMe	*Brya ebenus*		Ferreira *et al.* (1974)
Bryacarpene-2	OMe		OMe	OMe	OH	*Brya ebenus*		Ferreira *et al.* (1974)
Bryacarpene-3	OMe		OMe	OMe	OMe	*Brya ebenus*		Ferreira *et al.* (1974)
3-Hydroxy-4-methoxy-8,9-methylenedioxy-	OH	OMe	OCH_2O			*Swartzia ulei*		Formiga *et al.* (1975)
Bryacarpene-4	OMe	OH		OMe	OMe	*Brya ebenus*		Ferreira *et al.* (1974)
Bryacarpene-1	OMe	OH	OMe	OMe	OH	*Brya ebenus*		Ferreira *et al.* (1974)
Neorauteen (*10.144*)						*Neorautanenia edulis*		Brink *et al.* (1974b)
Neoduleen (*10.145*)						*Neorautanenia edulis*		Brink *et al.* (1974b)

the absolute configuration has been reversed. The same conclusion was reached in an X-ray crystallographic study of (−)-phaseollin (*10.147*) (DeMartinis *et al.*, 1978), but this structure was assigned without the aid of a heavy atom, and these authors have given no indication why the *S,S* configuration should be preferred over *R,R*. At this moment, the picture is very confusing, and definitive conclusions about absolute configurations are awaited. Thus, in Table 10.5, optical rotations rather than absolute configurations are expressed.

10.6.2 6a-Hydroxypterocarpans

Pisatin (6a-hydroxypterocarpin), a phytoalexin produced by *Pisum* spp. on fungal infection, was the only member of this class to be recognized for some years. A related phytoalexin from soya bean (*Glycine max*) was erroneously assigned the structure 6a-hydroxyphaseollin (*10.143*) (Sims *et al.*, 1972), but since 6a-hydroxyphaseollin was subsequently known as a fungal metabolite of phaseollin (Burden *et al.*, 1974), the soya phytoalexin structure was reassigned as (*10.139*) (Burden and Bailey, 1975). The material isolated, however, after HPLC purification, separated into three 6a-hydroxypterocarpan components which are now named glyceollins I (*10.139*), II (*10.136*) and III (*10.137*) (Lyne *et al.*, 1976). Further minor 6a-hydroxypterocarpan phytoalexins of *Glycine max* have since been identified (Lyne and Mulheirn, 1978). These and other 6a-hydroxypterocarpans are listed in Table 10.5.

Acanthocarpan (*10.134*), a phytoalexin from the *Helminthosporium carbonum*-infected leaflets of *Caragana acanthophylla*, has been characterized as a *C*-methyl derivative of pisatin (Ingham, 1979d), although the position of substitution (2 or 4) could not be established with certainty from MS data, and insufficient material was available for a NMR spectrum. The only other natural *C*-methylated isoflavonoid known is the isoflavanone ougenin.

Chirality assignments in 6a-hydroxypterocarpans are perhaps more confused now than with pterocarpans. It has generally been assumed that replacement of a hydrogen at 6a with a hydroxyl will have no effect on optical rotation, and thus (−)-6a-hydroxypterocarpans should have the same absolute configuration as (−)-pterocarpans. A number of researchers have, however, misinterpreted the rules for *R,S* nomenclature and assigned the wrong chirality. Thus, although substitution of the 6a-hydrogen with a hydroxyl in a 6a*R*,11a*R*-pterocarpan does not change the absolute configuration, it does affect the nomenclature, and the product is now 6a*S*,11a*S* (Fig. 10.11). Literature reports include several errors, and it has been necessary for this discrepancy to be brought to the general attention of other workers (Sicherer and Sicherer-Roetman, 1980; Ingham and Markham, 1980).

6a-Hydroxypterocarpans are characterized by their rapid dehydration to pterocarpenes in the presence of acids, and this is conveniently noted by UV spectroscopy (see Section 10.2). A similar process is encountered on MS analysis, and $M-18$ is a major fragment. Unless the compound contains isopentenyl or

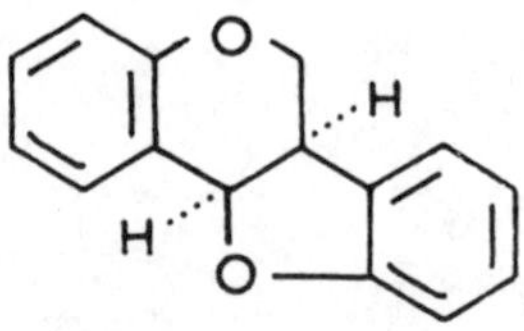

6a*R*, 11a*R* 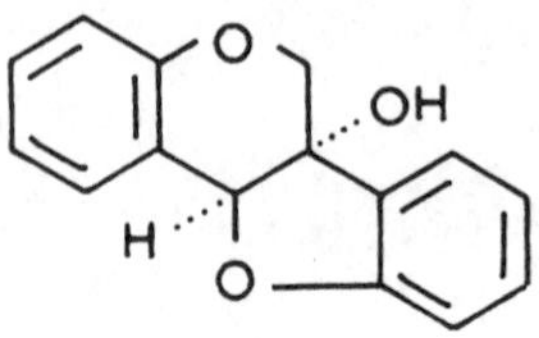   6a*S*, 11a*S*

Fig. 10.11

chromene ring substituents for example, further fragmentation may be limited to loss of hydrogen from this intermediate pterocarpene (see Section 10.6.3). NMR spectra are much simpler than with pterocarpans, and in some cases the two C-6 protons are magnetically equivalent, thus showing a singlet instead of the usual double doublet (Ingham and Markham, 1980). This equivalence is presumably accidental and cannot be predicted.

10.6.3 Pterocarpenes

Pterocarpenes could, in some cases, arise as artefacts of 6a-hydroxy-pterocarpans, but there is now a considerable selection of naturally occurring examples (Table 10.5). The heartwood of *Brya ebenus* has proved a very rich source, five bryacarpenes having been isolated from this material (Ferreira *et al.*, 1974). As mentioned above, MS fragmentation of pterocarpenes is frequently limited to loss of hydrogen to yield the very stable ion (*10.148*, cf. *10.42*). By adjusting the spectrometer settings, it is possible to produce doubly charged ion mass spectra, and under such conditions, more meaningful fragmentation results (Eguchi *et al.*, 1976).

Generally, pterocarpenes are synthesized by acid-catalysed cyclization of 2′-hydroxyisoflavanones (Farkas *et al.*, 1974). They are somewhat labile in solution, and are readily oxidized to coumestans, even by atmospheric oxygen. This conversion may be brought about particularly efficiently by treatment with DDQ (Ferreira *et al.*, 1974).

10.7 ISOFLAVANS

The simplest example of the isoflavans (*10.149*) is 7,4′-dihydroxyisoflavan (equol), an animal metabolite undoubtedly produced by degradation of simple isoflavones such as formononetin and daidzein obtained in the diet (Wong, 1975). Every plant-derived isoflavan, however, contains a 2′-oxygen substituent, as well as 7,4′-oxygenation. This feature has been associated with possible biosynthetic derivation by reduction of pterocarpans (Wong, 1975), and

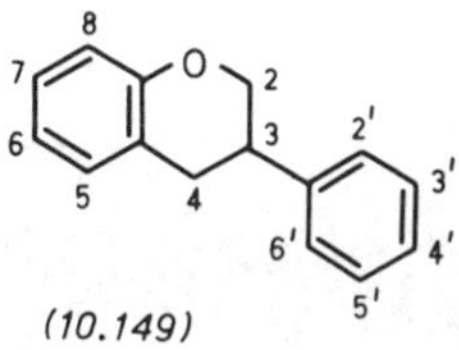

(10.149)

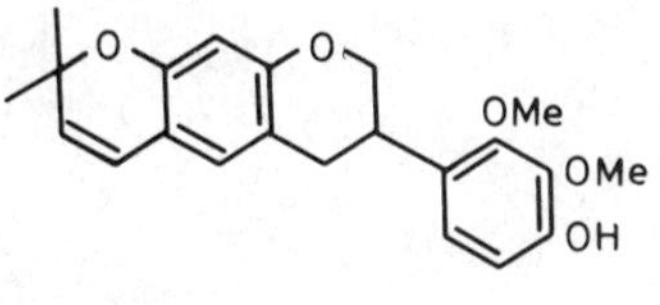

(10.150) Spherosinin

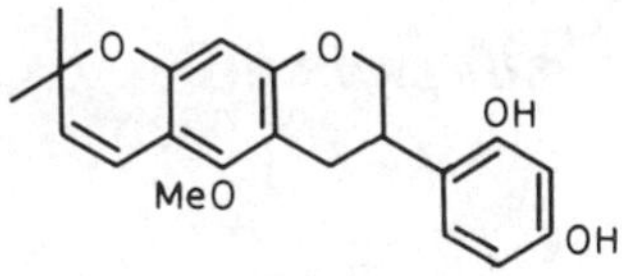

(10.151) Neorauflavane

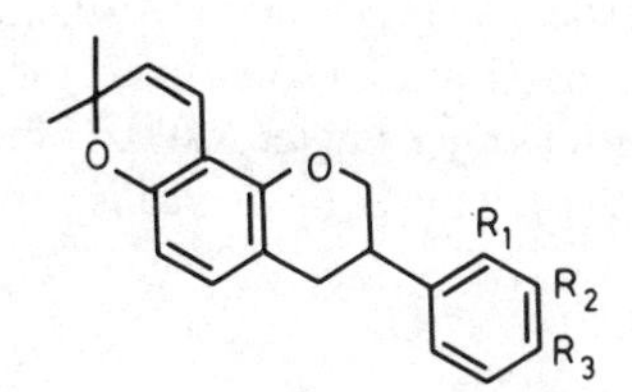

(10.152) $R_1 = R_3 = OH$, $R_2 = H$, Glabridin

(10.153) $R_1 = OH$, $R_2 = H$, $R_3 = OMe$, 4'-Methylglabridin

(10.154) $R_1 = R_3 = OH$, $R_2 = OMe$, 3'-Methoxyglabridin

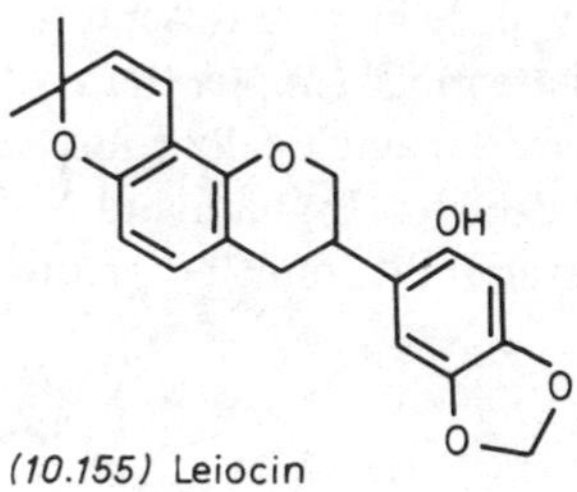

(10.155) Leiocin

isoflavan biosynthesis is indeed closely related to that of pterocarpans, but not necessarily in this manner (see Section 10.18). Isoflavans often function as phytoalexins, and many of the new structures reported (Table 10.6) have been isolated during antifungal screens. Typically, the isoflavan phytoalexins are simple hydroxy/methoxy-substituted structures. 5-Oxygenation is a rare feature in isoflavan structures (cf. Section 10.6). Complex geranyl-substituted isoflavans, namely heminitidulan (*10.156*), nitidulin (*10.157*) and nitidulan (*10.158*)

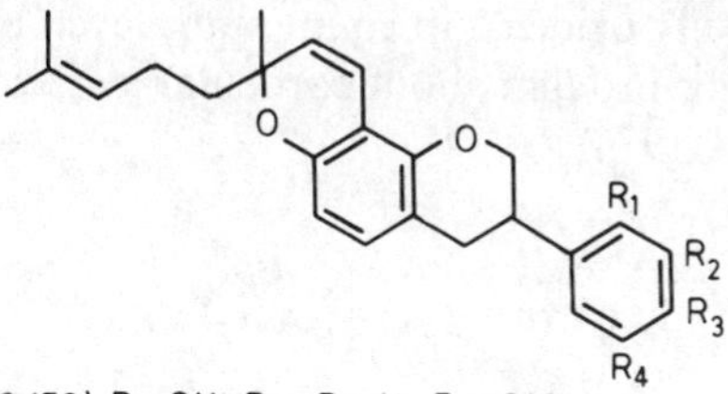

(10.156) $R_1 = OH$, $R_2 = R_4 = H$, $R_3 = OMe$ Heminitidulan

(10.157) $R_1 = R_2 = OH$, $R_3 = OMe$, $R_4 = H$ Nitidulin

(10.158) $R_1 = OH$, $R_2 = H$, $R_3R_4 = OCH_2O$ Nitidulan

have been isolated from the bark of *Dalbergia nitidula* along with pterocarpan analogues (Van Heerden *et al.*, 1978). Although 3,3-dimethylallyl substituents

Table 10.6 Isoflavans

Isoflavan	*Substituents*									*Plant sources*	*Chirality/ optical activity*	*References*
	5	*6*	*7*	*8*	*2′*	*3′*	*4′*	*5′*	*6′*			
Demethylvestitol			OH		OH		OH			*Anthyllis vulneraria*		Ingham (1977c)
										Erythrina sandwicensis		Ingham (1980)
										Lablab niger		Ingham (1977b)
										Lotus corniculatus		Ingham (1977c)
										L. hispidus		Ingham and Dewick (1979)
										L. uliginosus		Ingham (1977c)
										Phaseolus vulgaris	*R*	Woodward (1980a)
										Tetragonolobus spp.		Ingham (1977c)
Vestitol			OH		OH		OMe			*Cyclobium clausseni*	*R*	Gottlieb *et at.* (1975)
										Dalbergia ecastophyllum	*S*	Matos *et al.* (1975)
										D. sericea		Ingham (1979f)
										D. variabilis	*S*	Kurosawa *et al.* (1978d)
										Factorovskya aschersoniana		Ingham (1979c)
										Lotus corniculatus	*R*	Bonde *et al.* (1973)
										L. hispidus		Ingham and Dewick (1979)
										L. uliginosus		Ingham (1977c)
										Machaerium vestitum	*S*	Kurosawa *et al.* (1978a)
										Medicago spp.		Ingham (1979a)
										Melilotus alba		Ingham (1976a)
										Onobrychis spp.		Ingham (1978b)
										Trifolium arvense		Ingham and Dewick (1977)
										T. hybridum		Ingham (1976b)
										T. spp.		Ingham (1978c)
										Trigonella spp.		Ingham and Harborne (1976)
Isovestitol			OH		OMe		OH			*Anthyllis vulneraria*		Ingham (1977c)
										Erythrina sandwicensis		Ingham (1980)
										Lablab niger		Ingham (1977b)
										Tetragonolobus spp.		Ingham (1977c)
										Trifolium arvense		Ingham and Dewick (1977)
										T. spp.		Ingham (1978c)
Neovestitol			OMe		OH		OH			*Dalbergia sericea*		Ingham (1979f)

Table 10.6 (*Contd.*)

Isoflavan	*Substituents* 5	6	7	8	2′	3′	4′	5′	6′	*Plant sources*	*Chirality/optical activity*	*References*
Sativan			OH		OMe		OMe			*Derris amazonica*	*S*	Braz Filho *et al.* (1975)
										Lotus corniculatus	*R*	Bonde *et al.* (1973)
										L. hispidus		Ingham and Dewick (1979)
										Medicago sativa	*R*	Ingham and Millar (1973)
										M. spp.		Ingham (1979a)
										Trifolium arvense		Ingham and Dewick (1977)
										T. hybridum		Ingham (1976b)
										T. spp.		Ingham (1978c)
										Trigonella spp.		Ingham and Harborne (1976)
Isosativan			OMe		OH		OMe			*Dalbergia ecastophyllum*	*S*	Matos *et al.* (1975)
										Medicago scutellata		Ingham (1979a)
										Trifolium hybridum		Ingham (1976b)
										T. spp.		Ingham (1978c)
Arvensan			OMe		OMe		OH			*Trifolium arvense*		Ingham and Dewick (1977)
										T. stellatum		Ingham (1978c)
Laxifloran (spherosin)			OH		OMe	OMe	OH			*Lablab niger*		Ingham (1977b)
										Sphaerophysa salsula	+	Kattaev *et al.* (1975)
Mucronulatol			OH		OMe	OH	OMe			*Astragalus cicer*		
										A. gummifer		Ingham and Dewick (1980b)
										A. lusitanicus	*R*	Teresa *et al.* (1979)
										A. pyrenaicus		Ingham and Dewick (1980b)
										Dalbergia cearensis	*R*	Guimaraes *et al.* (1975)
										D. ecastophyllum	*RS*	Donnelly *et al.* (1973a)
										D. oliveri	*RS*	Donnelly and Kavanagh (1974)
										D. variabilis	*RS*	Kurosawa *et al.* (1978d)
										Machaerium mucronulatum	*S, RS*	Kurosawa *et al.* (1978c)
										M. opacum	*S*	Ollis *et al.* (1978c)
										M. vestitum	*S*	Kurosawa *et al.* (1978a)
										M. villosum	*S*	Kurosawa *et al.* (1978c)
Isomucronulatol			OH		OH	OMe	OMe			*Gliricidia sepium*	*R*	Manners and Jurd (1979)
										Glycyrrhiza glabra		Ingham (1977d)

Compound								Source		Reference
Astraciceran			OH		OMe		OCH_2O	*Astragalus cicer*		
								A. pyrenaicus		Ingham and Dewick (1980b)
5-Methoxyvestitol	OMe		OH		OH		OMe	*Lotus hispidus*		Ingham and Dewick (1979)
Bryaflavan		OH	OH		OMe	OH	OMe	*Brya ebenus*	*S*	Ferreira *et al.* (1974)
8-Demethylduartin			OH	OH	OMe	OH	OMe	*Dalbergia ecastophyllum*	*R*	Donnelly *et al.* (1973a)
Duartin			OH	OMe	OMe	OH	OMe	*Machaerium mucronulatum*	*S*	Kurosawa *et al.* (1978c)
								M. opacum	*S*	Ollis *et al.* (1978c)
								M. villosum	*S*	Kurosawa *et al.* (1978c)
Spherosinin (*10.150*)								*Sphaerophysa salsula*	+	Kattaev *et al.* (1975)
Neorauflavane (*10.151*)								*Neorautanenia edulis*	–	Brink *et al.* (1974a)
Glabridin (*10.152*)								*Glycyrrhiza glabra*	*R*	Saitoh *et al.* (1976)
4′-Methylglabridin (*10.153*)								*Glycyrrhiza glabra*	*R*	Mitscher *et al.* (1980)
3′-Methoxyglabridin (*10.154*)								*Glycyrrhiza glabra*	*R*	Mitscher *et al.* (1980)
Leiocin (*10.155*)								*Dalbergia nitidula*	*S*	Van Heerden *et al.* (1978)
Heminitidulan (*10.156*)								*Dalbergia nitidula*	*S*	Van Heerden *et al.* (1978)
Nitidulin (*10.157*)								*Dalbergia nitidula*	*S*	Van Heerden *et al.* (1978)
Nitidulan (*10.158*)								*Dalbergia nitidula*	*S*	Van Heerden *et al.* (1978)
Leiocinol (*10.159*)								*Dalbergia nitidula*	*S*	Van Heerden *et al.* (1978)
2′-Methylphaseollidin-isoflavan (*10.160*)								*Vigna unguiculata*		Preston (1975)
Phaseollinisoflavan (*10.161*)								*Glycyrrhiza glabra*	*R*	Mitscher *et al.* (1980)
2′-Methylphaseollin-isoflavan (*10.162*)								*Phaseolus vulgaris*	+	Van Etten (1973)
α,α-Dimethylallylcyclolobin (*10.163*)								*Cyclobium clausseni*	*R*	Gottlieb *et al.* (1975)
Unanisoflavan (*10.164*)								*Sophora secondiflora*	–	Minhaj *et al.* (1976b)
Hispaglabridin A (*10.165*)								*Glycyrrhiza glabra*	*R*	Mitscher *et al.* (1980)
Hispaglabridin B (*10.166*)								*Glycyrrhiza glabra*	*R*	Mitscher *et al.* (1980)

are common in isoflavonoid structures, related 1,1-dimethylallyl substituents are rare. Two isoflavans with this unusual grouping, probably formed by Claisen-type rearrangement of an *O*-3,3-dimethylallyl grouping, have been reported, α,α-dimethylallylcyclolobin (*10.163*) from *Cyclobium clausseni* (Gottlieb *et al*., 1975) and its 4′-methyl ether unanisoflavan (*10.164*) from *Sophora secondiflora*

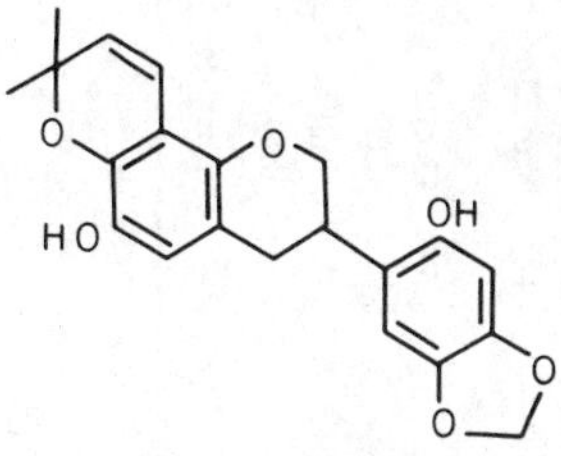

(10.159) Leiocinol

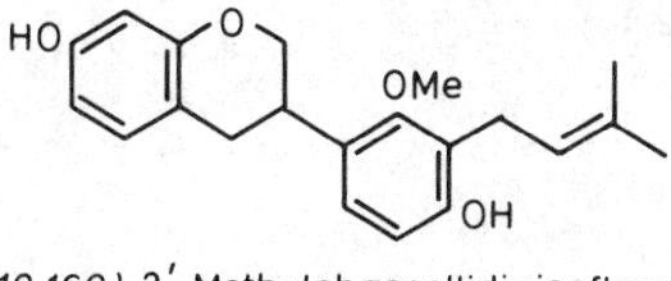

(10.160) 2′-Methylphaseollidinisoflavan

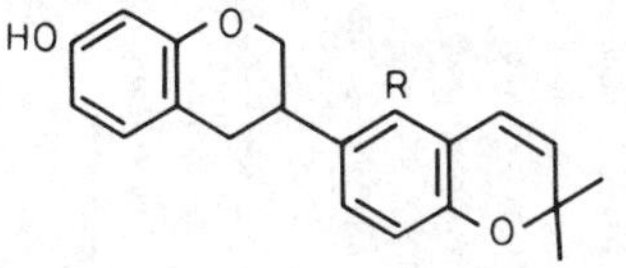

(10.161) R = OH, Phaseollinisoflavan
(10.162) R = OMe, 2′-Methylphaseollin-isoflavan

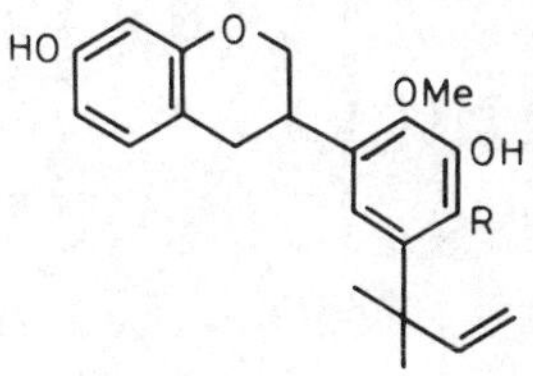

(10.163) R = OH, α,α-Dimethylallyl-cyclolobin
(10.164) R = OMe, Unanisoflavan

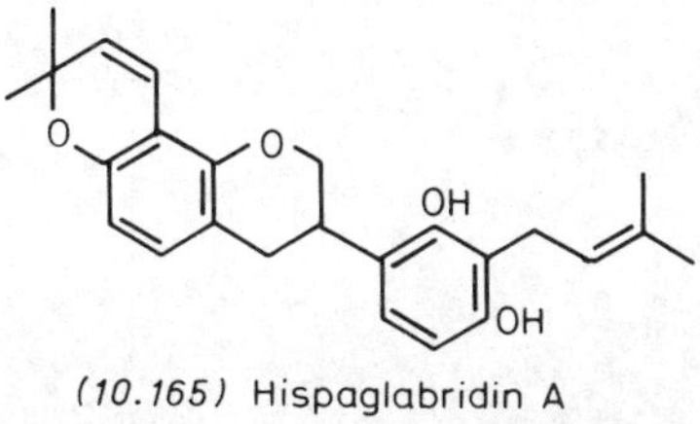

(10.165) Hispaglabridin A

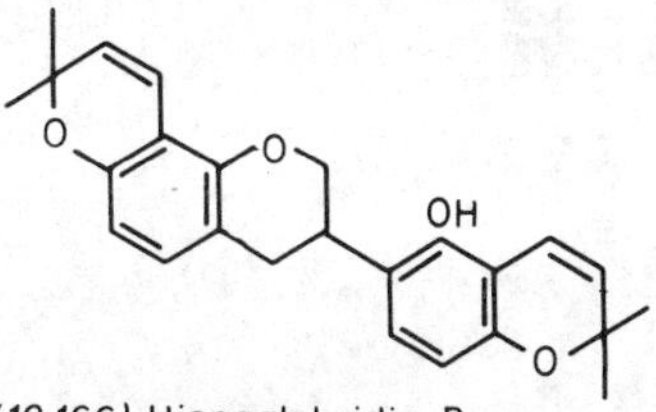

(10.166) Hispaglabridin B

(Minhaj *et al*., 1976b). Unanisoflavan could not be found in a second sample of *S. secondiflora*, but the related 3-hydroxyisoflavanone secondifloran was isolated (Minhaj *et al*., 1977). *Cyclobium clausseni* also yielded a material biscyclolobin, formulated as a dehydrative dimer based on cyclolobin, and a tentative structure such as (*10.167*) is suggested (Gottlieb *et al*., 1975).

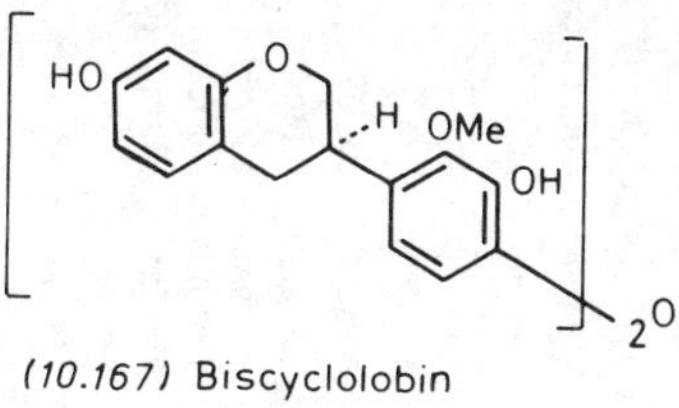

(10.167) Biscyclolobin

The absolute configuration of isoflavans cannot be assigned merely by optical rotation measurements, but 3*S*-isoflavans exhibit a negative Cotton effect in the 260–300 nm region of their ORD curves (Kurosawa *et al.*, 1978b). Thus, (+)-vestitol, (−)-mucronulatol and (−)-duartin all have the 3*S*- configuration. This assignment is, however, based partly on knowledge of the absolute configuration of pterocarpans, and any reappraisal there could naturally change current assignments. It is important to realize that both ORD and NMR characteristics are markedly influenced by the preferred conformation of the molecule in solution, and it is essential not to restrict ORD measurements to long-wavelength regions (Kurosawa *et al.*, 1978b).

Isoflavans are usually synthesized by catalytic hydrogenation of isoflavones (Wagner and Farkas, 1975). The particular condition of the catalyst is an important consideration, and to avoid hydrogenation of aromatic rings, and to optimize isoflavan yield, pretreatment of Pd–C catalysts in acetic acid with oxygen is regarded as desirable (Lamberton *et al.*, 1978).

10.8 QUINONE DERIVATIVES

10.8.1 Isoflavonequinone

Bowdichione (*10.168*) is the only example of this class, and was isolated from the heartwood of *Bowdichia nitida* (Brown *et al.*, 1974), along with 3′-hydroxyformononetin, genistein and homopterocarpin. This coloured compound displays both isoflavone and quinone characteristics in its MS fragmentation, showing RDA fragmentation in the isoflavone portion and loss of CO + Me in the quinone. Its constitution was confirmed by oxidation of 3′-hydroxyformononetin with Fremy's salt.

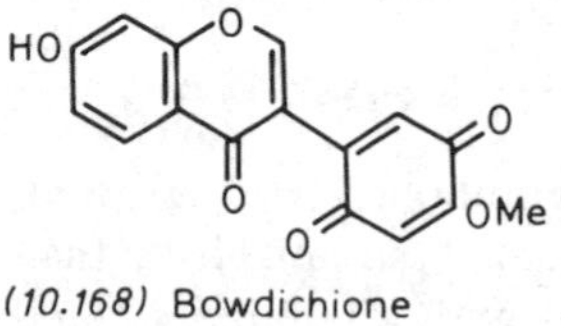

(*10.168*) Bowdichione

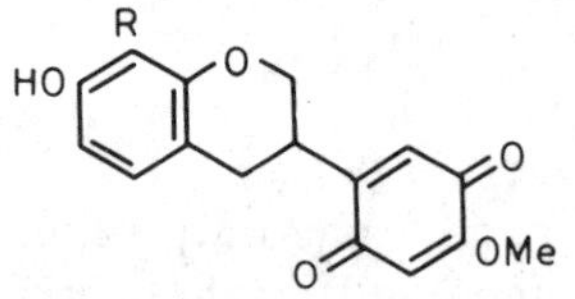

(*10.169*) R = OH, Claussequinone
(*10.171*) R = OMe, Mucroquinone

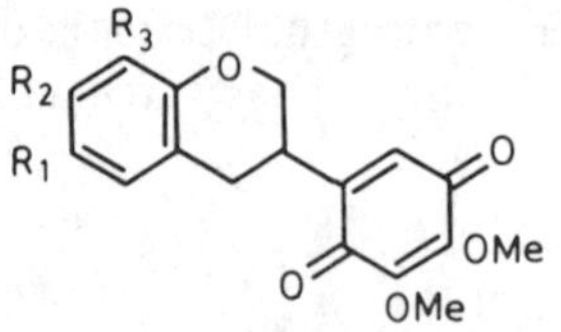

(*10.170*) $R_1 = R_2 =$ OMe, $R_3 =$ H, Abruquinone A
(*10.172*) $R_1 =$ H, $R_2 =$ OH, $R_3 =$ OMe, Amorphaquinone
(*10.173*) $R_1 =$ OH, $R_2 = R_3 =$ OMe, Abruquinone C
(*10.174*) $R_1 = R_2 = R_3 =$ OMe, Abruquinone B

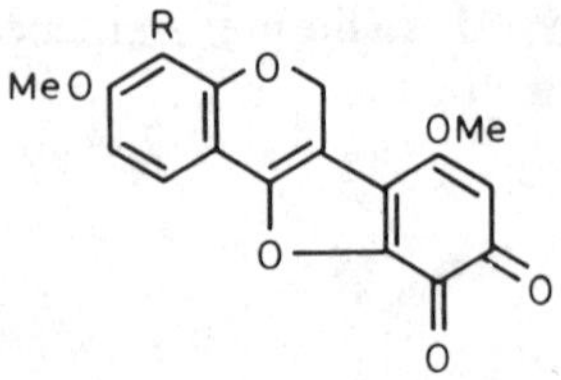

(*10.175*) R = H, 4-Deoxybryaquinone
(*10.176*) R = OH, Bryaquinone

10.8.2 Isoflavanquinones

Three isoflavanquinones abruquinones-A -B and -C (*10.170*, *10.174*, *10.173*) from *Abrus precatorius* root (Lupi *et al.*, 1979) and amorphaquinone (*10.172*) from *Amorpha fruticosa* root (Shibata and Shimizu, 1978) are recently reported examples of this isoflavonoid class, to be added to the previously known claussequinone and mucroquinone (Table 10.7). Mucroquinone (*10.171*) is known in both enantiomeric forms. Absolute configurations have been assigned by ORD in essentially the same manner as with isoflavans, 3*S*-isoflavanquinones exhibiting negative Cotton effects in the 260–300 nm region, although materials with the same absolute configuration can produce ORD curves almost enantiomeric in the 400–500 nm region (Kurosawa *et al.*, 1978b). Confirmation has been achieved by the oxidation of isoflavans of known configuration. As with isoflavans, NMR characteristics are influenced by preferred conformations in solution. Isoflavanquinones are readily reduced to the corresponding quinol, but these derivatives are unstable and reoxidize in the presence of atmospheric oxygen. RDA fragmentation with hydrogen transfer to ring A predominates in the MS.

Table 10.7 Isoflavanquinones

Isoflavanquinone	*Plant sources*	*Chirality*	*References*
Claussequinone (*10.169*)	*Cyclobium clausseni*	*R*	
	C. vecchi	*R*	Gottlieb *et al.* (1975)
Abruquinone A (*10.170*)	*Abrus precatorius*	*S*	Lupi *et al.* (1979)
Mucroquinone (*10.171*)	*Cyclobium clausseni*	*R*	Gottlieb *et al.* (1975)
	Machaerium mucronulatum	*S*	Kurosawa *et al.* (1978c)
Amorphaquinone (*10.172*)	*Amorpha fruticosa*		Shibata and Shimizu (1978)
Abruquinone C (*10.173*)	*Abrus precatorius*	*S*	Lupi *et al.* (1979)
Abruquinone B (*10.174*)	*Abrus precatorius*	*S*	Lupi *et al.* (1979)

10.8.3 Pterocarpenequinones

Two purple pterocarpenequinones, bryaquinone (*10.176*) and 4-deoxybryaquinone (*10.175*) have been isolated in small amounts from the heartwood of *Brya ebenus* (Ferreira *et al.*, 1975) along with a number of pterocarpenes. Confirmation of structure was made with the aid of reductive acetylation (Zn–Ac_2O–NaOAc) to the pterocarpene acetate, then hydrogenolysis to the isoflavan derivative. Whether or not these compounds are artefacts remains to be established.

10.9 3-ARYL-4-HYDROXYCOUMARINS

One new example of this class, glabrescin (*10.177*) from the seed of *Derris glabrescens* (Delle Monache *et al.*, 1977b) has been reported. This is a 4-methoxy

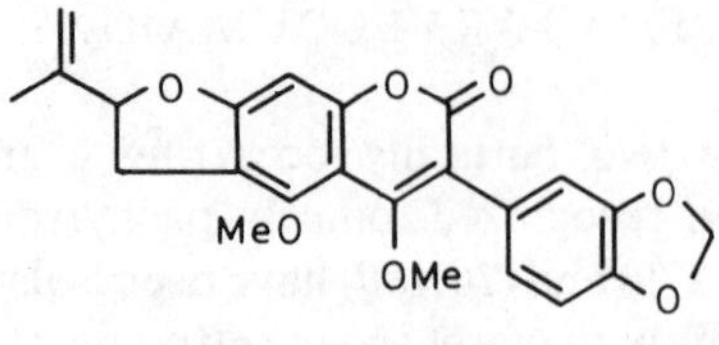

(10.177) Glabrescin

derivative, and co-occurs in this plant with a further example derrusnin (*10.178*). So far, these isoflavonoids have not been found outside the genus *Derris* and all compounds known have 5-methoxy substituents thus favouring the coumarin

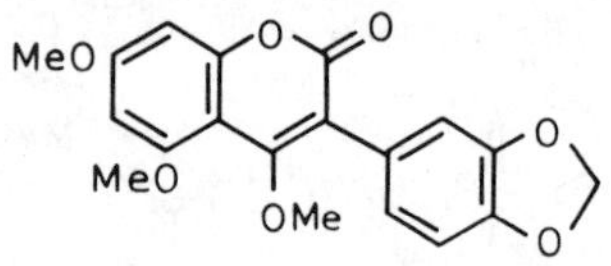

(10.178) Derrusnin

tautomer rather than the 2-hydroxyisoflavone form (Wong, 1975). A new approach to the synthesis of 3-aryl-4-hydroxycoumarins via Meerwein arylation of 4-hydroxycoumarins with diazonium salts in the presence of a cupric catalyst has been described (Sawhney and Mathur, 1976). A useful conversion of isoflavones into 3-aryl-4-hydroxycoumarins may be achieved by treatment of the isoflavone with aqueous hydroxylamine to yield an intermediate oxime and isoxazole (Fig. 10.12). The isoxazole under basic conditions gives the coumarin imine from which the arylhydroxycoumarin may be obtained by hydrolysis (Szabo *et al.*, 1980).

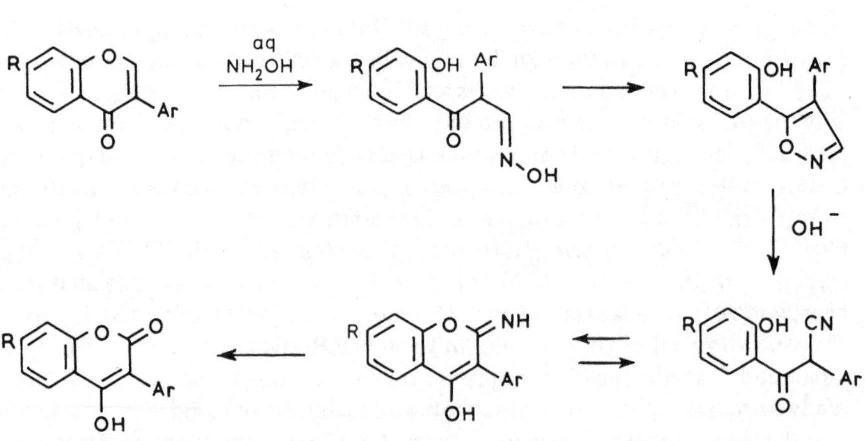

Fig. 10.12

10.10 3-ARYLCOUMARINS

For many years, only two naturally occurring examples of this class of isoflavonoids have been recognized, namely pachyrrhizin and neofolin. Two further compounds (*10.179*) and (*10.180*) have been isolated from the heartwood of *Dalbergia oliveri*, along with pterocarpan and coumestan analogues (Donnelly and Kavanagh, 1974). A fifth example, glycyrin (*10.181*) is reported in the root of *Glycyrrhiza* spp. (Kinoshita *et al.*, 1978a). These compounds characteristically show a low-field ethylenic singlet for H-4 in their NMR spectra, and in their MS behave as typical coumarins losing 28 m.u. (−CO).

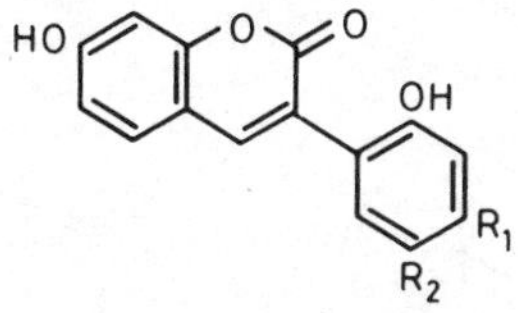

(*10.179*) $R_1 = OMe$, $R_2 = H$
(*10.180*) $R_1R_2 = OCH_2O$

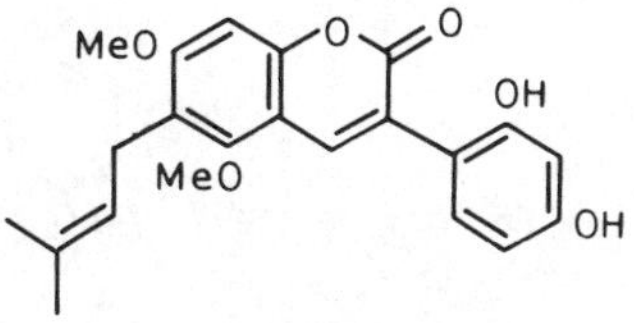

(*10.181*) Glycyrin

Synthesis of 3-arylcoumarins may be effected in several ways. Perkin condensation of the appropriate phenylacetic acid and aldehyde in acetic anhydride is the standard preparation (Donnelly and Kavanagh, 1974; Jurd and Manners, 1977), but other classes of isoflavonoid may be modified, e.g. oxidation of isoflav-3-enes with CrO_3 (Jurd and Manners, 1977), oxidation of isoflavans (not 2′-hydroxy) with DDQ (Ferreira *et al.*, 1974) and reduction of 3-aryl-4-hydroxycoumarin tosylates with Zn–HCl (Ahluwalia and Mehta, 1978).

10.11 ISOFLAV-3-ENES

Isoflav-3-enes have been known chemically for many years, being synthesized by dehydration of isoflavanols or by controlled acid treatment of pterocarpans. They are, however, very reactive, especially in solution, and this is presumably why their existence as natural products has not been established until recently. Neorauflavene (*10.186*) from *Neorautanenia edulis* (Brink *et al.*, 1974a) was the first example reported, followed by sepiol (*10.183*) and 2′-methylsepiol (*10.185*) from *Gliricidia sepium* (Jurd, 1976; Jurd and Manners, 1977) and glabrene (*10.187*) from *Glycyrrhiza glabra* root (Kinoshita *et al.*, 1976). Two further examples, haginin A and B (*10.184*, *10.182*) have recently been isolated from heartwood of *Lespedeza cyrtobotrya* (Miyase *et al.*, 1980) (Table 10.8). Isoflav-3-enes show typical allylic coupling in their NMR spectrum, and in the MS of sepiol and 2′-methylsepiol, very prominent ions at $M-1$ were observed due to ready formation of the isoflavylium ion. Characterization is aided by conversion into the isoflavan by hydrogenation, or by CrO_3 oxidation to the corresponding 3-arylcoumarin.

Table 10.8 Isoflav-3-enes

Isoflav-3-ene	*Plant sources*	*References*
Haginin B (*10.182*)	*Lespedeza cyrtobotrya*	Miyase *et al.* (1980)
Sepiol (*10.183*)	*Gliricidia sepium*	Jurd and Manners (1977)
Haginin A (*10.184*)	*Lespedeza cyrtobotrya*	Miyase *et al.* (1980)
2′-Methylsepiol (*10.185*)	*Gliricidia sepium*	Jurd and Manners (1977)
Neorauflavene (*10.186*)	*Neorautanenia edulis*	Brink *et al.* (1974a)
Glabrene (*10.187*)	*Glycyrrhiza glabra*	Kinoshita *et al.* (1976)

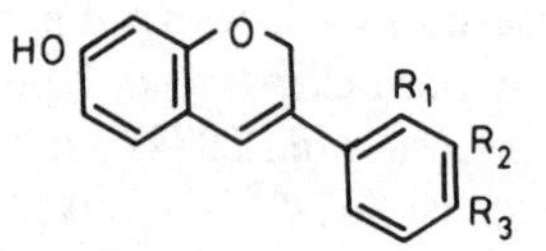

(10.182) R_1=OMe, R_2=H, R_3=OH, Haginin B
(10.183) R_1=R_2=OH, R_3=OMe, Sepiol
(10.184) R_1=R_2=OMe, R_3=OH, Haginin A
(10.185) R_1=R_3=OMe, R_2=OH, 2′-Methylsepiol

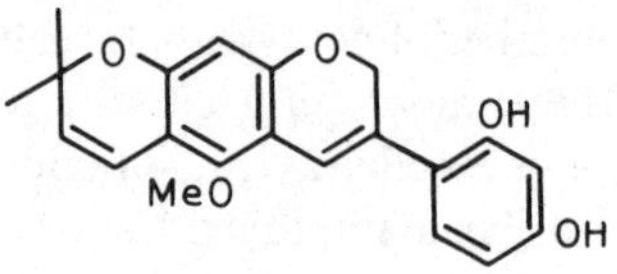

(10.186) Neorauflavene

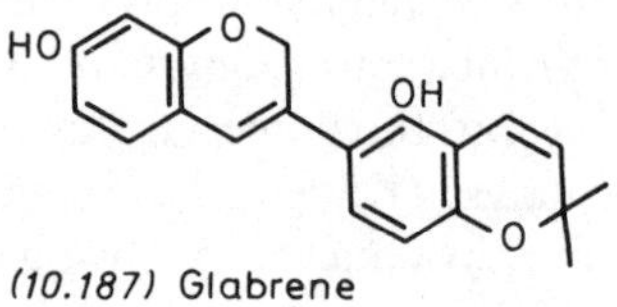

(10.187) Glabrene

These compounds appear to play an important role in the biosynthesis of coumestans (see Section 10.18), and it is anticipated that further naturally occurring examples will be reported in due course.

10.12 COUMESTANS

Coumestans (*10.188*) are widely distributed and are easily recognized in solution or on chromatograms from their intense bright-blue or violet fluorescence under UV light. Only three new coumestans have been reported during the period of review, tephrosol from *Tephrosia villosa* (Rao and Srimannarayana, 1980), and corylidin (*10.191*) (Gupta *et al.*, 1977) and psoralidin oxide (*10.190*) (Gupta *et al.*,

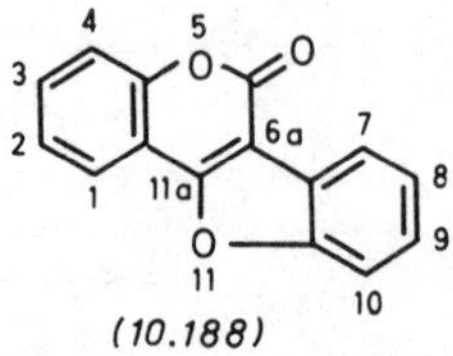

(10.188)

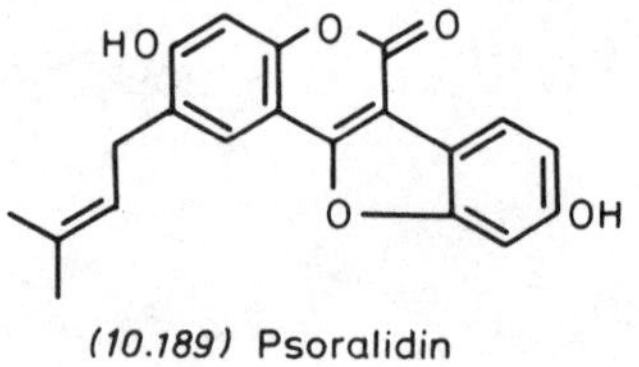

(10.189) Psoralidin

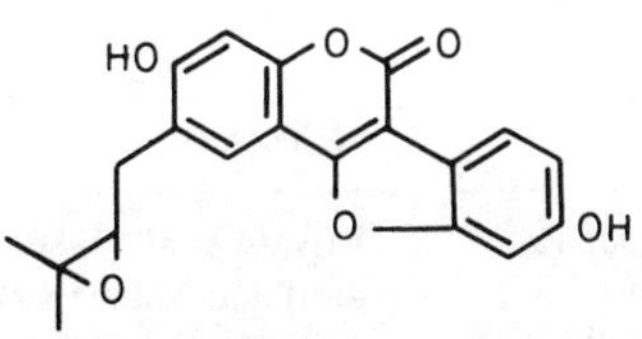

(10.190) Psoralidin oxide

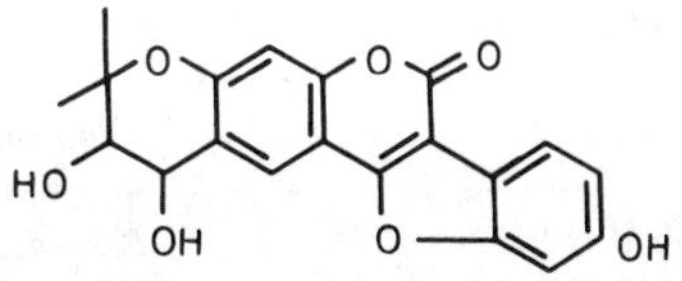

(10.191) Corylidin

1980) from *Psoralea corylifolia* (Table 10.9). Several numbering systems have been employed in the past for coumestan nomenclature (Wong, 1975) and despite the occasional appearance of the old ones, most authors have adopted the systematic numbering also applied to pterocarpans.

A number of new synthetic routes to coumestans have been published. The use of DDQ under mild conditions to oxidize pterocarpans and pterocarpenes results in excellent yields of coumestans (Ferreira *et al.*, 1974), and this reagent has also been employed to convert 2′-hydroxyisoflav-3-enes and 2′-hydroxy-3-arylcoumarins into coumestans, by analogy with the biosynthetic process (Martin and Dewick, 1978). DDQ oxidation of 2′-hydroxy-3-arylcoumarins is a much less efficient route than the others mentioned, and although lead tetra-acetate is an alternative oxidizing agent (Kurosawa and Nogami, 1976), yields are still relatively low. However, these methods do allow isoflavones to be modified into coumestans. 3-Aryl-4-hydroxycoumarins may also be oxidatively cyclized to coumestans by refluxing in diphenyl ether in the presence of Pd–C catalyst (Kappe and Schmidt, 1972), and the syntheses of coumestrol and dimethylcoumestrol have been achieved by this method (Kappe and Brandner, 1974). The electrochemical synthesis of coumestans (Grujic *et al.*, 1976) from a 4-hydroxycoumarin and a catechol is reminiscent of the Wanzlick synthesis (Wagner and Farkas, 1975).

10.13 α-METHYLDEOXYBENZOINS

Angolensin remained the sole example of this class of natural product for a quarter of a century, but the 2-*O*-methyl derivative (*10.192*) has since been extracted from heartwood of *Pericopsis elata*, where it co-occurs with angolensin (Fitzgerald *et al.*, 1976). Although both compounds isolated from this source have been assigned the *R*-configuration, they exhibit different specific rotations, angolensin being laevorotatory $\{[\alpha]_D - 114^\circ\ (CHCl_3)\}$ and 2-methylangolensin dextrorotatory $\{[\alpha]_D + 5.3^\circ\ (CHCl_3)\}$. Biochanin A and afrormosin were also

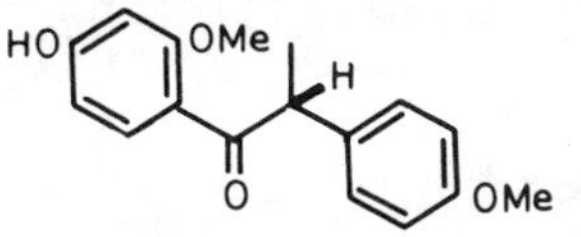

(10.192) *R*-2-Methylangolensin

Table 10.9 Coumestans

Coumestan	*Substituents*								*Plant sources*	*References*
	1	*2*	*3*	*4*	*7*	*8*	*9*	*10*		
Coumestrol			OH				OH		*Phaseolus lunatus*	Rich *et al.* (1977)
									P. vulgaris	Gnanamanickan and Patil (1977)
									Trifolium alexandrinum	Shemesh *et al.* (1976)
9-Methylcoumestrol			OH				OMe		*Dalbergia oliveri*	Donnelly and Kavanagh (1974)
									D. stevensonii	Donnelly *et al.* (1973b)
									Myroxylon balsamum	Oliveira *et al.* (1978)
									Trifolium pratense	Dewick (1977)
3-Hydroxy-8,9-dimethoxy-			OH			OMe	OMe		*Myroxylon balsamum*	Oliveria *et al.* (1978)
Medicagol			OH			OCH_2O			*Dalbergia oliveri*	Donnelly and Kavanagh (1974)
									D. stevensonii	Donnelly *et al.* (1973b)
									Sophora tomentosa	Komatsu *et al.* (1978a)
									Trifolium pratense	Dewick (1977)
Tephrosol		OMe	OH			OCH_2O			*Tephrosia villosa*	Rao and Srimannarayana (1980)
Psoralidin (*10.189*)									*Phaseolus lunatus*	Rich *et al.* (1977)
Psoralidin oxide (*10.190*)									*Psoralea corylifolia*	Gupta *et al.* (1980)
Corylidin (*10.191*)									*Psoralea corylifolia*	Gupta *et al.* (1977)

isolated from *P. elata*, and it is generally assumed that α-methyldeoxybenzoins also share an isoflavonoid origin. It is thus interesting to note that catalytic hydrogenation of isoflavanone in basic media can lead to 2-hydroxy-α-methyldeoxybenzoin and the corresponding alcohol (Szabo and Antal, 1976),

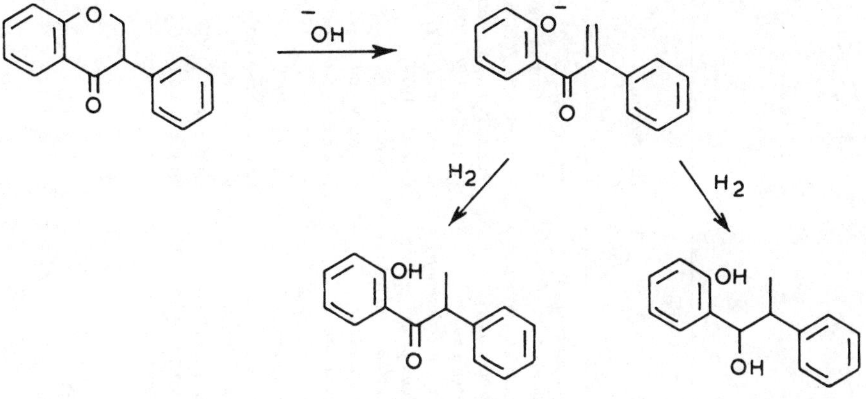

Fig. 10.13

derived presumably via the ring-opened α-methylene derivative (Fig. 10.13). Isoflavones may also be converted into α-methyldeoxybenzoins by Birch reduction using lithium in liquid ammonia (Major *et al.*, 1980). A re-examination of *Pterocarpus angolensis* heartwood has led to the isolation of *S*-4-methylangolensin (*10.193*), together with *R*-angolensin and an unusual pair of epimeric sesquiterpene derivatives, cadinyl ethers of *R*-angolensin (*10.194*) (Bezuidenhoudt *et al.*, 1980).

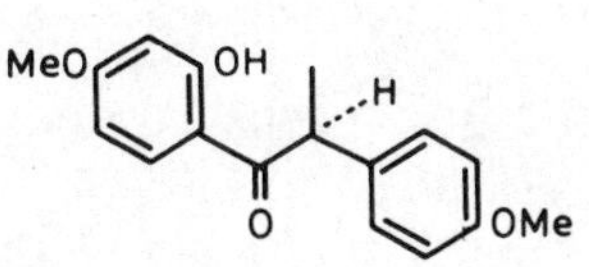

(*10.193*) S-4-Methylangolensin

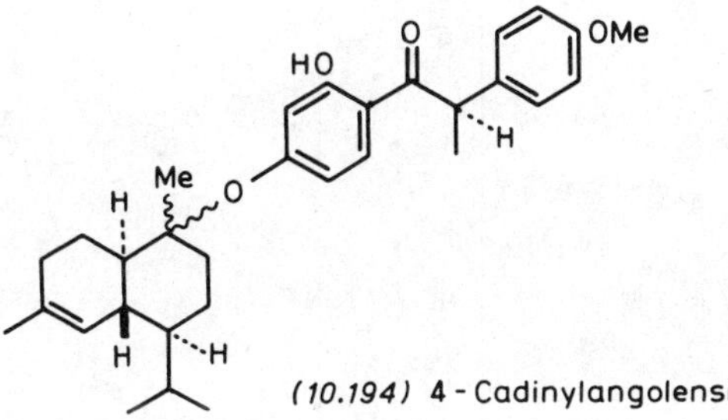

(*10.194*) 4-Cadinylangolensin

10.14 2-ARYLBENZOFURANS

The variety of structures encountered in the known naturally occurring 2-arylbenzofurans suggests that several different biosynthetic origins are involved. Thus, egonol (*10.195*) appears to be of lignan/neolignan origin and may

be derived by loss of one carbon atom from a bisarylpropanoid. Benzofuran (*10.196*) co-occurs with oxyresveratrol and probably arises by oxidative cyclization of this stilbene in a process analogous to laboratory conversions.

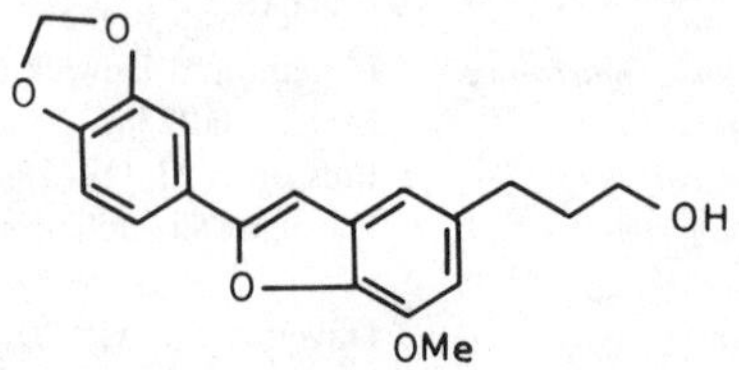

(*10.195*) Egonol

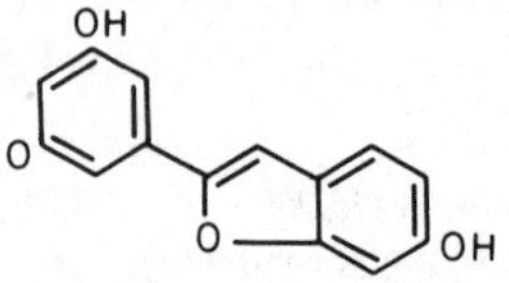

(*10.196*)

However, a third group of 2-arylbenzofurans (Table 10.10) co-occur in leguminous plants with isoflavonoids having related substitution patterns, and a biosynthetic origin by loss of one carbon from an isoflavonoid is likely. Pterofuran (*10.199*) from *Pterocarpus indicus* was first to be isolated in 1964 and was found to co-occur with formononetin, pterocarpin and angolensin (Cooke and Rae, 1964). A biosynthetic sequence involving loss of C-6 from a coumestan, by analogy with alkaline degradation of coumestans (Wong, 1975), is generally

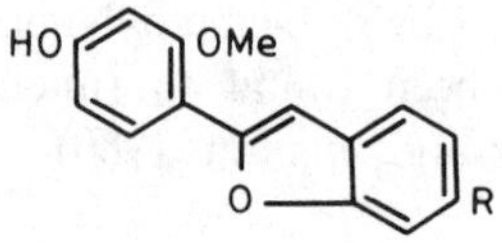

(*10.197*) R = OH, 6-Demethylvignafuran
(*10.198*) R = OMe, Vignafuran

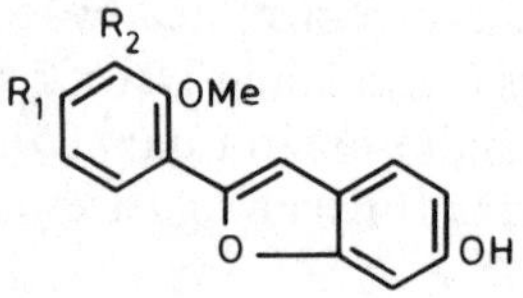

(*10.199*) R_1 = OMe, R_2 = OH, Pterofuran
(*10.200*) R_1 = OH, R_2 = OMe, Isopterofuran

postulated. The co-occurrence of (*10.201*) with 3-hydroxy-8,9-dimethoxycoumestan in *Myroxylon balsamum* (Oliveira *et al.*, 1978) and (*10.202*) and (*10.203*) with medicagol in *Sophora tomentosa* (Komatsu *et al.*, 1978a) lend support for this view. Neoraufurane (*10.204*), though, co-occurs with similarly substituted pterocarpan, isoflav-3-ene and isoflavan derivatives, e.g. edudiol, neorauflavene and neorauflavane respectively (Brink *et al.*, 1974a, 1977), and in this case, the benzofuran moiety is presumably related to the acetate-derived ring of the isoflavonoid, not the shikimate-derived ring as above. This latter relationship may well operate in the case of vignafuran, a phytoalexin from

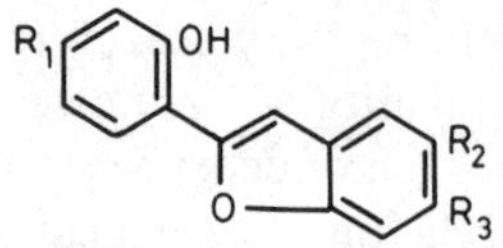

(*10.201*) R_1 = OH, R_2 = R_3 = OMe
(*10.202*) R_1 = OH, R_2R_3 = OCH_2O
(*10.203*) R_1 = OMe, R_2R_3 = OCH_2O

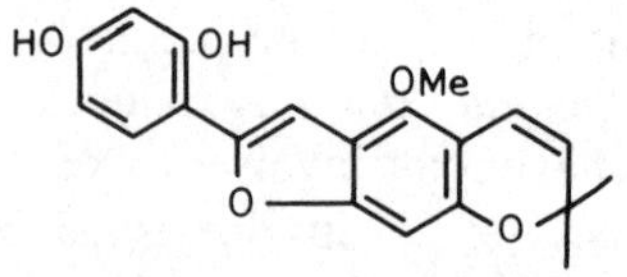

(*10.204*) Neoraufurane

Table 10.10 2-Arylbenzofurans

2-Arylbenzofuran	*Plant sources*	*References*
6-Demethylvignafuran (*10.197*)	*Coronilla emerus*	Dewick and Ingham (1980)
	Tetragonolobus maritimus	Ingham and Dewick (1978)
Vignafuran (*10.198*)	*Lablab niger*	Ingham (1977b)
	Vigna unguiculata	Preston *et al.* (1975)
Isopterofuran (*10.200*)	*Coronilla emerus*	Dewick and Ingham (1980)
(2,4-Dihydroxyphenyl)-5,6-dimethoxybenzofuran (*10.201*)	*Myroxylon balsamum*	Oliveira *et al.* (1978)
(2,4-Dihydroxyphenyl)-5,6-methylenedioxybenzofuran (*10.202*)	*Sophora tomentosa*	Komatsu *et al.* (1978a)
(2-Hydroxy-4-methoxyphenyl)-5,6-methylenedioxybenzofuran (*10.203*)	*Sophora tomentosa*	Komatsu *et al.* (1978a)
Neoraufurane (*10.204*)	*Neorautanenia edulis*	Brink *et al.* (1974a)

Colletotrichum lindemuthianum-infected cowpea (*Vigna unguiculata*) (Preston *et al.*, 1975), as demonstrated by feeding experiments with labelled phenylalanine precursors (Martin and Dewick, 1979; see Section 10.18). Demethylvignafuran (*10.197*) and isopterofuran (*10.200*) have also been found to function as phytoalexins (Ingham and Dewick, 1978; Dewick and Ingham, 1980).

OH MeO OMe CHO OH OMe O OMe

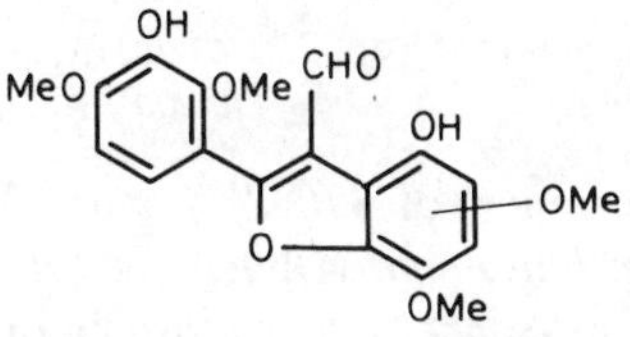

(*10.205*) Bryebinal

The UV spectra of 2-arylbenzofurans are stilbene-like and closely resemble those of pterocarpenes. Little structural information is usually obtained from the MS where the only major fragments result from loss of methyls and CO. Several synthetic approaches have been employed to confirm structures. Hoesch synthesis from a phenylacetonitrile and a phenol is limited because the position of coupling cannot be predicted and a mixture of 2-arylbenzofurans results (Preston *et al.*, 1975). A route exploiting the similarity of 2-arylbenzofurans to the isoflavonoids involves as key step the formation of deoxybenzoins by base hydrolysis of 2′-benzyloxyisoflavones. Debenzylation and acid-catalysed ring closure produced 2-arylbenzofurans in reasonable yields (Ingham and Dewick, 1978; Dewick and Ingham, 1980). Essentially the same transformation, but in a different sequence, was achieved by debenzylation, hydrolysis, then HBr–Ac_2O cyclization of the acetal derived by thallium nitrate oxidation of

2-benzyloxychalcones (Meyer-Dayan *et al.*, 1978). Vignafuran and pterofuran have also been synthesized by reaction of a copper(I) arylacetylide with an *o*-halogenophenol (Duffley and Stevenson, 1977), a method which has wide applicability to all types of 2-arylbenzofuran derivatives.

A further route by treatment of isoflavylium perchlorates with H_2O_2 in acidified propan-2-ol gives good yields of 2-arylbenzofurans, and is of particular interest in that it mimics the possible biosynthetic process because the 2-aryl substituent is provided by the B ring of the isoflavonoid (Deschamps-Vallet *et al.*, 1979) (Fig. 10.14). This biogenetic analogy is also exemplified in two other conversions, though of less synthetic value (Fig. 10.14). Treatment of a chalcone with excess thallium acetate gave a coumaranone instead of the required isoflavone, and the coumaranone, after reduction with sodium borohydride, eliminated formic acid to give a 2-arylbenzofuran (Ollis *et al.*, 1970). A similar ring-contraction process leading to a coumaranone has been observed on treatment of the isoflavone elongatin with alkaline H_2O_2 (Smalberger *et al.*, 1975). The second example involves reduction of a 3-arylcoumarin with AlH_3 to the corresponding alcohol which yields a 2-arylbenzofuran on treatment with DDQ (Kinoshita *et al.*, 1978a).

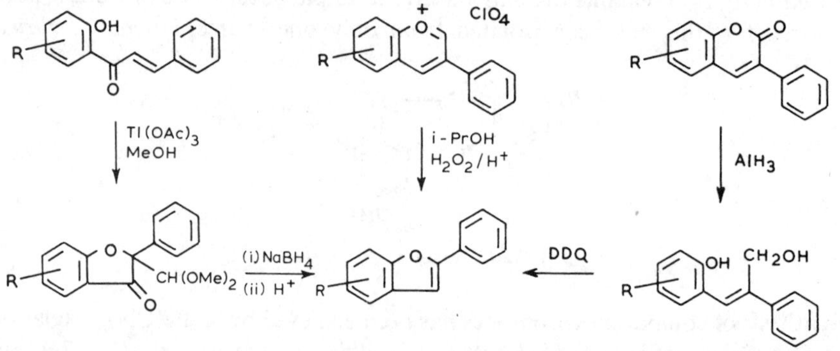

Fig. 10.14

Bryebinal, a minor yellow pigment from *Brya ebenus*, has been identified as a 3-formyl-2-arylbenzofuran, and has been assigned the tentative structure (*10.205*) (Ferreira *et al.*, 1975). Since it co-occurs with a number of pterocarpenes, bryebinal could represent an intermediate between these compounds and 2-arylbenzofurans if such a relationship exists.

10.15 ISOFLAVANOL

(+)-Ambanol (*10.206*) is the only known naturally occurring isoflavanol and was isolated from the root of *Neorautanenia amboensis* (Oberholzer *et al.*,

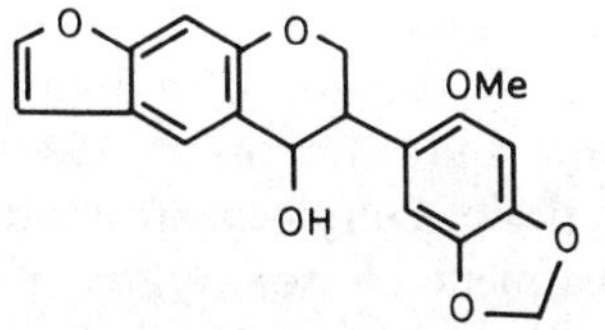

(10.206) Ambanol

1977b). On treatment with acid, it readily eliminates water to yield the corresponding isoflav-3-ene. Synthesis of (±)-ambanol was achieved by potassium borohydride reduction of the isoflavone dehydroneotenone. 2′-Hydroxyisoflavanols are regarded as important intermediates in the biosynthesis of pterocarpans, isoflavans and other classes of isoflavonoid (see Section 10.18), but it is unlikely that such isoflavanols will ever be isolated because of their ready cyclization to pterocarpans. Such cyclization is prevented by the 2′-methoxy group in ambanol.

10.16 COUMARONOCHROMONE

Lisetin (*10.207*) remains the only natural example of this class of isoflavonoid, and so far, this has been isolated from only one source *Piscidia erythrina.*

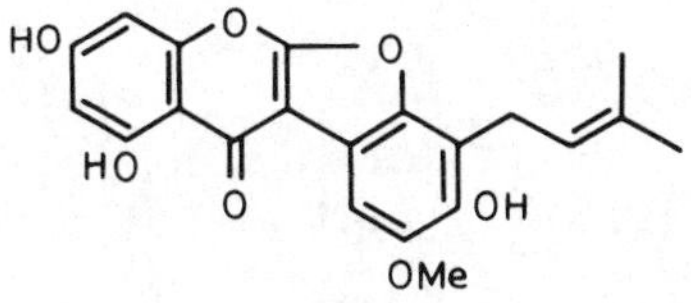

(10.207) Lisetin

Synthesis of coumaronochromones has been achieved by $K_3Fe(CN)_6$ oxidation of 2′-hydroxyisoflavones (Falshaw *et al.*, 1966; Tsukayama *et al.*, 1979), but recent studies have shown that other oxidizing agents may be employed. Ag_2CO_3 in quinoline (Antus and Nogradi, 1979b), and Pb(IV) or Mn(III) acetates (Kurosawa and Araki, 1979) are all suitable reagents.

10.17 MICROBIAL METABOLITES OF ISOFLAVONOIDS

A number of modified isoflavonoid structures have been identified when specific isoflavonoids have been introduced into microbial cultures. These probably represent intermediates in the metabolic pathway employed by the microorganism to detoxify the isoflavonoid. Because these intermediates are not yet recognized as normal plant products, it is appropriate to consider them separately. It may be, though, that some of the isoflavonoids already dealt with in

the other sections are in fact microbial metabolites themselves, since it is not usual for researchers to employ aseptically grown plants. Particularly in the case of phytoalexin studies, it is almost impossible in microbially infected tissues to differentiate between plant and microbial metabolism, and a multiple phytoalexin response could well be a mixture of induced plant products and their microbial metabolites. Rapid detoxification of phytoalexins to reduce their antifungal properties may be an essential step for fungi to become pathogenic in a particular plant; in most cases investigated, the metabolites identified tend to be less fungitoxic than the substrate. The data in Table 10.11 have been subdivided according to the mechanism of initial metabolism, i.e. hydrogenolysis-type reactions, hydroxylation, demethylation and others.

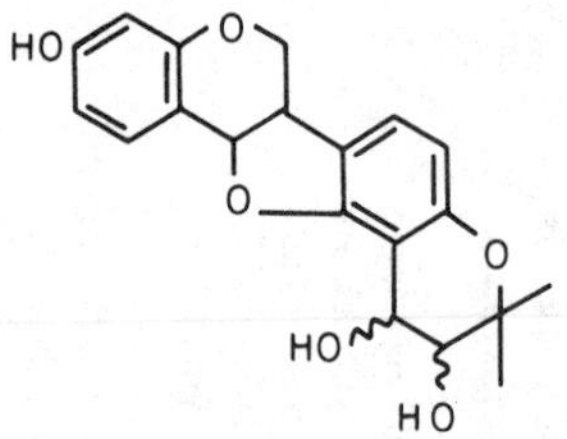

(10.208)

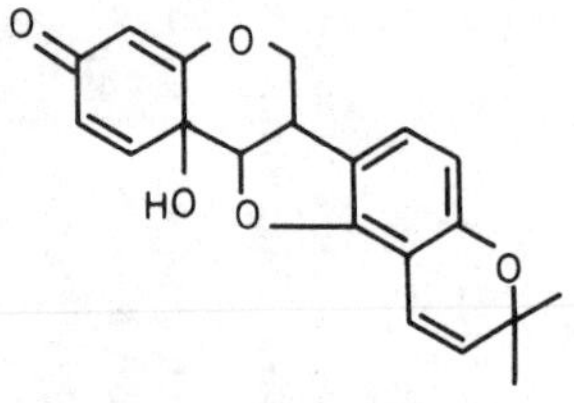

(10.209) 1a-Hydroxyphaseollone

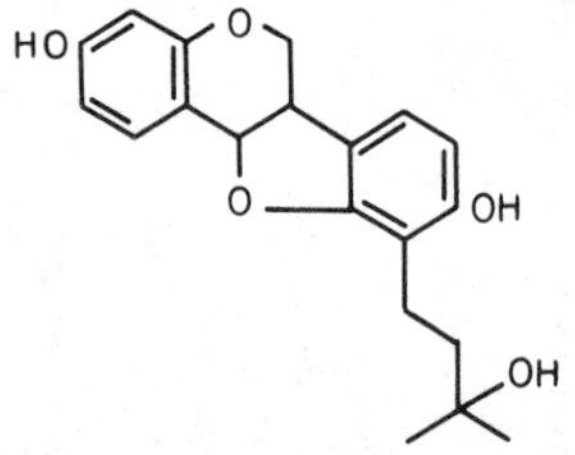

(10.210) Phaseollidin hydrate

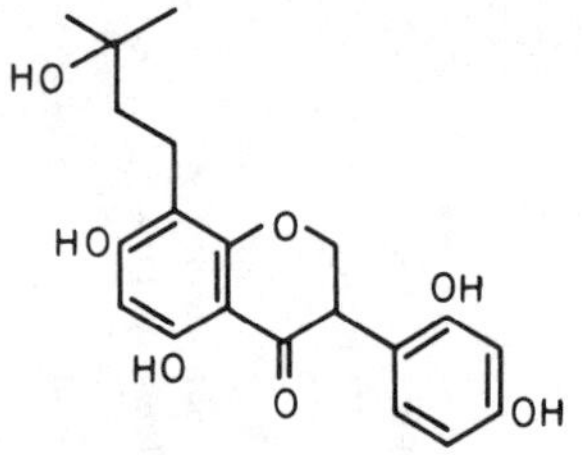

(10.211) Kievitone hydrate

The ability of fungi to convert pterocarpans into 2′-hydroxyisoflavans, a reaction analogous to hydrogenolysis, has been recognized for some time. *Stemphylium botryosum* converts several pterocarpan phytoalexins in this manner. Consequently, the common co-occurrence of isoflavans with pterocarpans in phytoalexin samples may be the result of fungal metabolism, although some isoflavan phytoalexins can be induced by abiotic treatments (Dewick and Martin, 1979), and isoflavans in many cases are more fungitoxic than the corresponding pterocarpans (Harborne and Ingham, 1978). 3-Hydroxy-2, 9-dimethoxypterocarpan is similarly degraded by *Fusarium solani* f.sp. *pisi* cultures to 7,2′-dihydroxy-6,4′-dimethoxyisoflavan (Pueppke and Van Etten, 1976). The pterocarpan, isolated from *F. solani* f.sp. *pisi*-infected pea (*Pisum sativum*) is itself believed to be a fungal metabolite of a pea isoflavonoid, since its synthesis could not be induced abiotically.

Table 10.11 Microbial metabolites of isoflavonoids

Substrate	*Organism*	*Product(s)*	*Reference*
(i) *Hydrogenolysis reactions*			
Medicarpin	*Stemphylium botryosum*	Vestitol	Steiner and Millar (1974)
Maackiain	*Stemphylium botryosum*	7,2′-Dihydroxy-4′,5′-methylenedioxyisoflavan (Maackiain-isoflavan)	Higgins (1975)
Phaseollin	*Stemphylium botryosum*	Phaseollinisoflavan	Higgins *et al.* (1974)
3-Hydroxy-2,9-dimethoxy-pterocarpan	*Fusarium solani* f.sp. *pisi*	7,2′-Dihydroxy-6,4′-dimethoxyisoflavan	Pueppke and Van Etten (1976)
(ii) *Hydroxylation reactions*			
Medicarpin	*Botrytis cinerea*	6a-Hydroxymedicarpin	Ingham (1976a)
Medicarpin	*Sclerotinia trifoliorum*	6a-Hydroxymedicarpin	Bilton *et al.* (1976)
Maackiain	*Sclerotinia trifoliorum*	6a-Hydroxymaackiain	Bilton *et al.* (1976)
Phaseollin	*Botrytis cinerea*	6a-Hydroxyphaseollin	Van den Heuvel and Glazener (1975)
Phaseollin	*Colletotrichum lindemuthianum*	6a-Hydroxyphaseollin 6a,7-Dihydroxyphaseollin	Burden *et al.* (1974)
(iii) *Demethylation reactions*			
Pisatin	*Ascochyta pisi*	6a-Hydroxymaackiain	Van't Land *et al.* (1975)
	Nectria haematococca	6a-Hydroxymaackiain	Van Etten *et al.* (1980)
	Fusarium solani f.sp. *pisi*	6a-Hydroxymaackiain	Van Etten *et al.* (1975)
	F. anguioides	6a-Hydroxymaackiain	Lappe and Barz (1978)
	F. avenaceum	6a-Hydroxymaackiain	Lappe and Barz (1978)
	F. oxysporum f. sp. *pisi*	6a-Hydroxymaackiain 3-Hydroxymaackiainisoflavan	Fuchs *et al.* (1980)
Medicarpin	*Fusarium proliferatum*	Demethylmedicarpin 7,2′,4′-Trihydroxyisoflavan	Weltring and Barz (1980)
Homopterocarpin	*Fusarium poliferatum*	Isomedicarpin Demethylmedicarpin 7,2′,4′-Trihydroxyisoflavan	Weltring and Barz (1980)

5,7,4′-Trimethoxyisoflavone	*Fusarium* spp.	5-Methylbiochanin A Biochanin A Genistein Orobol	Barz *et al.* (1976)
(iv) *Other reactions*			
Phaseollin	*Septoria nodorum*	*cis*- and *trans*-3′,4′-Dihydroxy-3′,4′-dihydrophaseollin (*10. 208*)	Bailey *et al.* (1977)
	Fusarium solani f.sp. *phaseoli*	la-Hydroxyphaseollone (*10.209*)	Van den Heuvel *et al.* (1974)
	Cladosporium herbarum	la-Hydroxyphaseollone (*10.209*)	Van den Heuvel and Glazener (1975)
Phaseollidin	*Fusarium solani* f.sp. *phaseoli*	Phaseollidin hydrate (*10.210*)	Smith *et al.* (1980)
Kievitone	*Fusarium solani* f.sp. *phaseoli*	Kievitone hydrate (*10.211*)	Kuhn *et al.* (1977)

6a-Hydroxylation of pterocarpans is also a common fungal modification. Thus, cultures of *Sclerotinia trifoliorum* modify medicarpin and maackiain into their 6a-hydroxy derivatives (Bilton *et al.*, 1976). In the presence of plant tissue, methylation at the 7-hydroxyl may also occur and variabilin and pisatin have been isolated from *S. trifoliorum* - or *Botrytis cinerea*-infected *Trifolium pratense* (Bilton *et al.*, 1976). Phaseollin may be 6a-hydroxylated also, and in *Colletotrichum lindemuthianum*, further aromatic hydroxylation takes place, since 6a,7-dihydroxyphaseollin has been identified as a second metabolite (Burden *et al.*, 1974). Similar 6a,7-dihydroxy derivatives of medicarpin and maackiain have been isolated from the *Trifolium pratense–Colletotrichum coffeanum* combination (Ingham, 1976a).

The 6a-hydroxypterocarpan pisatin is typically metabolized by demethylation at position 3 yielding 6a-hydroxymaackiain, and this transformation has been observed in cultures of several organisms. In *Fusarium oxysporum* f.sp. *pisi*, further metabolism occurs and a hydrogenolysis-type reaction yields 3-hydroxymaackiainisoflavan (Fuchs *et al.*, 1980). The metabolism of medicarpin and homopterocarpin by *Fusarium proliferatum* proceeds via demethylmedicarpin and hydrogenolysis to 7,2′,4′-trihydroxyisoflavan (Weltring and Barz, 1980). However, homopterocarpin is demethylated initially at position 9, and the metabolic pathway proceeds through isomedicarpin rather than medicarpin. A sequence of demethylations followed by aromatic hydroxylation, then ring cleavage is believed to operate in the metabolism of isoflavones by a number of *Fusarium* species (Barz *et al* ., 1976). Incubation of 5,7,4′-trimethoxyisoflavone resulted in the production of 5-methylbiochanin A, biochanin A, genistein and orobol.

A combination of plant and fungal metabolism is probably responsible for the production of 4-hydroxy derivatives of medicarpin, demethylmedicarpin and homopterocarpin in *Botrytis cinerea*-infected *Melilotus alba* (Ingham, 1976a), along with vestitol, demethylmedicarpin, 6a-hydroxymedicarpin and 6a-hydroxyisomedicarpin. All are regarded as metabolites of medicarpin and are illustrative of the processes described above.

The metabolic pathway employed to detoxify an isoflavonoid is markedly dependent on the organism concerned. Thus, phaseollin, in addition to the routes already outlined, has been observed to be converted into two other metabolites. *Septoria nodorum* attacks the dimethylchromene system and *cis*- and *trans*-diols result (Bailey *et al.*, 1977), whereas *Fusarium solani* f.sp. *phaseoli* produces the metabolite la-hydroxyphaseollone (Van den Heuvel *et al.*, 1974). When phaseollidin is the substrate, however, this latter organism hydrates the dimethylallyl substituent producing phaseollidin hydrate (Smith *et al.*, 1980), an analogous modification of the isoflavanone kievitone yielding kievitone hydrate (Kuhn *et al.*, 1977). An enzymic system from *F. solani* f.sp. *phaseoli* catalyses the same conversions (Kuhn and Smith, 1979).

10.18 BIOSYNTHESIS OF ISOFLAVONOIDS

The isoflavonoids share a common biosynthetic pathway with the flavonoids as far as chalcone intermediates, but then a 1,2-aryl migration occurs to give the characteristic rearranged skeleton. The mechanism of this rearrangement step is still completely unknown despite the amount of biosynthetic research undertaken, but from the information available, it is possible that only two chalcones, 2′,4′,4-trihydroxychalcone and 2′,4′,6′,4-tetrahydroxychalcone, normally act as substrates for aryl migration (Dewick, 1978). An oxidative reaction is undoubtedly involved and the site of oxidation would appear to be the 4-hydroxyl, since 4-methoxychalcones are accepted for isoflavonoid biosynthesis only after demethylation (Hahlbrock and Grisebach, 1975). Of the several hundred known naturally occurring isoflavonoids, very few indeed lack the corresponding 4′-oxygen substituent. An attractive mechanism is the hypothesis of Pelter *et al.* (1971) in which phenolic oxidation of the chalcone leads to a spirodienone intermediate (*10.212*) (Fig. 10.15). Decomposition of this intermediate via a proton would give daidzein or genistein, and if methylation via *S*-adenosylmethionine is associated with the aryl migration, formononetin or biochanin A would be formed. These four isoflavones could then, by further substitution, act as precursors of virtually all other natural isoflavonoids. (Similar oxidation of 2-hydroxychalcones could lead to 2′-hydroxy/methoxyisoflavones and would account for a number of the known 4′-deoxyisoflavonoids.)

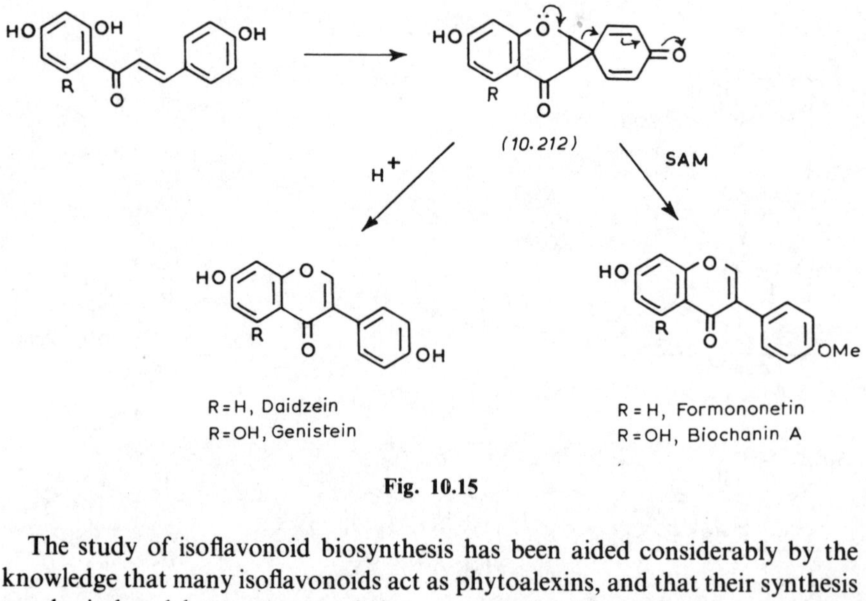

Fig. 10.15

The study of isoflavonoid biosynthesis has been aided considerably by the knowledge that many isoflavonoids act as phytoalexins, and that their synthesis can be induced by treatment of the appropriate plant with fungi, or, more

appropriate to biosynthetic experiments, by UV light or heavy-metal salts. Thus in feeding experiments, the precursor can be added at precisely the time of maximum synthesis of the isoflavonoid, and extremely high incorporations of precursor (e.g. 1–10 %) are achieved.

In $CuCl_2$-treated seedlings of red clover (*Trifolium pratense*), labelled 2′,4′,4-trihydroxychalcone and formononetin were good precursors of the pterocarpan phytoalexins (−)-medicarpin and (−)-maackiain, whereas 2′,4′-dihydroxy-4-methoxychalcone and daidzein were poor precursors (Dewick, 1975a), illustrating that methylation is probably associated with the aryl-migration step. Such results do not contradict the findings of Wengenmayer *et al.* (1974), who have isolated an enzyme preparation catalysing the 4′-methylation of daidzein, since a small, but significant incorporation of daidzein into formononetin was also observed. However, a further, major route from 2′,4′,4-trihydroxychalcone to formononetin, not via daidzein is implicated by the incorporation data.

The pathway from formononetin to medicarpin probably proceeds via 2′-hydroxylation to (*10.213*), reduction to the isoflavanol (*10.214*), then further reduction to the isoflavanone (*10.215*) followed by cyclization via the carbonium ion (*10.216*) (Fig. 10.16). The conversion of (*10.213*) into medicarpin is thus analogous to the borohydride reduction sequence used for chemical synthesis of

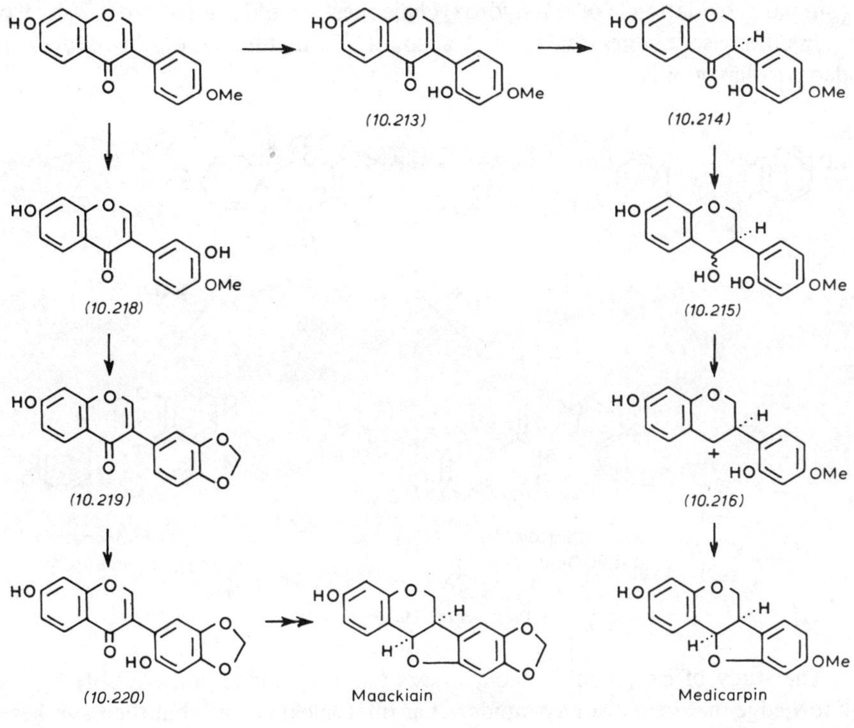

Fig. 10.16

pterocarpans. Isoflavone (*10.213*) and (±)-isoflavanone (*10.214*) were extremely good precursors of medicarpin, but not maackiain in $CuCl_2$-treated red clover, and the pathway to maackiain presumably branches at formononetin (Dewick, 1977). Pterocarpene (*10.217*) was a rather poor precursor of medicarpin, and (±)-medicarpin was not incorporated into maackiain. In further experiments (Dewick and Ward, 1978), the isoflavones formononetin, calycosin (*10.218*), pseudobaptigenin (*10.219*) and 7,2′-dihydroxy-4′,5′-methylenedioxyisoflavone (*10.220*) were all excellent precursors of maackiain, and form a logical sequence for elaboration of the substitution pattern of this pterocarpan (Fig. 10.16). The reduction sequence from isoflavone (*10.213*) to (−)-medicarpin is undoubtedly completely stereospecific, and this was demonstrated in UV- and $CuCl_2$-treated fenugreek (*Trigonella foenum-graecum*) seedlings where [2D]-isoflavone (*10.221*) gave rise to (−)-medicarpin with deuterium located exclusively in the 6-pro-*R* position (*10.222*), and thus indicated an overall *trans* addition of hydrogen to the double bond (Dewick and Ward, 1977). Location of the label was accomplished by the use of deuterium NMR.

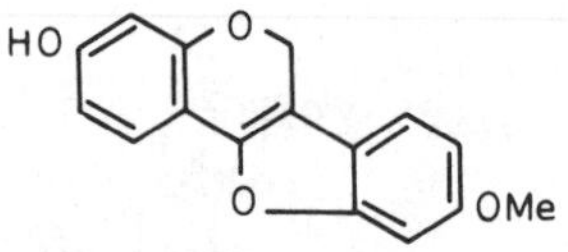

HO O O OMe

(*10.217*)

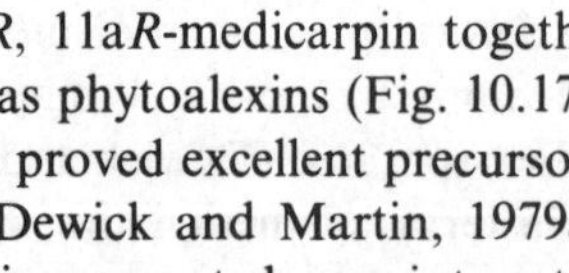

HO O D O HO OMe

(*10.221*)

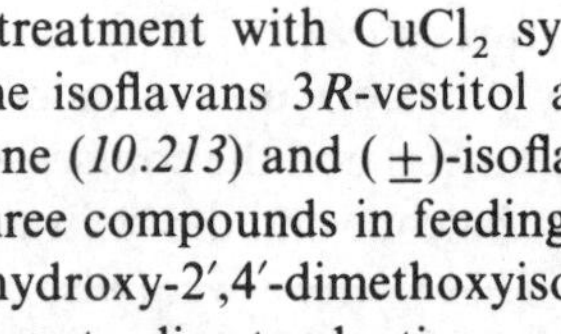

H HO O D H H O OMe

(*10.222*)

Lucerne (*Medicago sativa*) seedlings on treatment with $CuCl_2$ synthesize (−)-6a*R*, 11a*R*-medicarpin together with the isoflavans 3*R*-vestitol and 3*R*-sativan as phytoalexins (Fig. 10.17). Isoflavone (*10.213*) and (±)-isoflavanone (*10.214*) proved excellent precursors of all three compounds in feeding experiments (Dewick and Martin, 1979a), but 7-hydroxy-2′,4′-dimethoxyisoflavone was not incorporated, even into sativan, ruling out a direct reductive sequence to this isoflavan. To ascertain whether isoflavans are biosynthesized via pterocarpans (cf. hydrogenolysis reactions) or whether the reverse sequence occurs (cf. DDQ oxidations), labelled (±)-medicarpin and (±)-vestitol were tested as precursors. Medicarpin was incorporated into vestitol and sativan, and vestitol into medicarpin and sativan, thus demonstrating the interconversion of these two compounds. However, incorporation data, and the results of kinetic feeding experiments with phenylalanine suggested that medicarpin and vestitol were

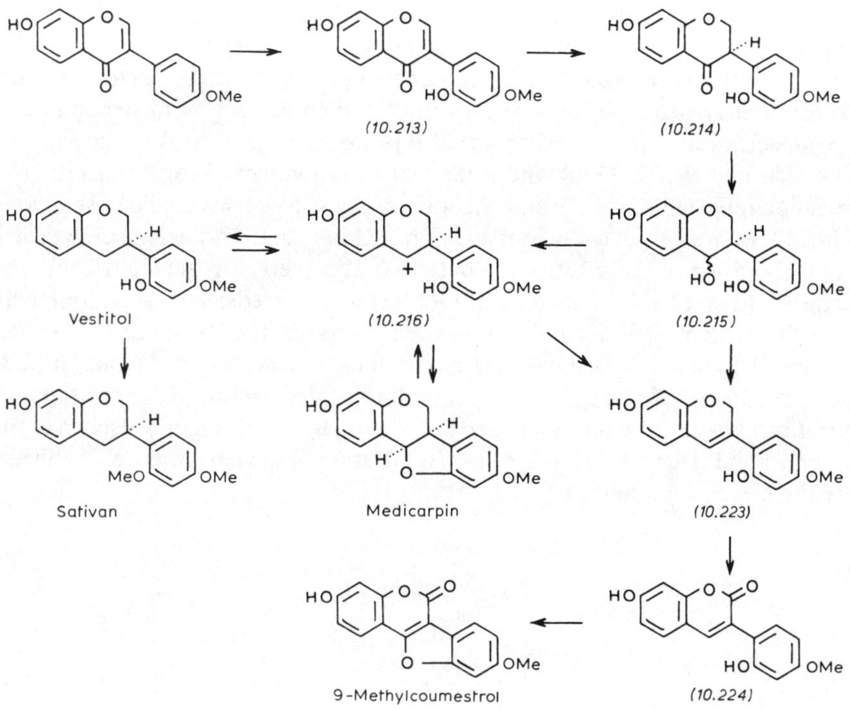

Fig. 10.17

synthesized simultaneously from a common intermediate, probably the carbonium ion (*10.216*). Sativan was synthesized later than the other two phytoalexins and is most likely derived by methylation of vestitol. Red clover seedlings also had the ability to transform vestitol into medicarpin, but isoflavanone (*10.214*) was a much better precursor.

In a further series of feeding experiments in lucerne seedlings, 2′,4′,4-trihydroxychalcone and formononetin proved good precursors of medicarpin, vestitol, sativan and the coumestan 9-methylcoumestrol, but 2′,4′-dihydroxy-4-methoxychalcone and daidzein were poorly incorporated (Dewick and Martin, 1979b). These results were also interpreted in terms of methylation occurring during the aryl migration, and further evidence was obtained by also studying incorporations into coumestrol. Thus daidzein was a precursor of coumestrol but not of 9-methylcoumestrol, despite the fact that synthesis of the latter compound is greatly stimulated by the induction process. Since the isoflavones formononetin and (*10.213*), and the corresponding isoflavanones were all good precursors of the phytoalexins and 9-methylcoumestrol, a metabolic grid may be involved in their biosynthetic origin, although the major route appeared to be via isoflavone (*10.213*).

The interconversion of medicarpin and vestitol via a carbonium ion is a chemically viable process (Van der Meerwe *et al.*, 1978), but in nature would probably require suitable stabilization of such an intermediate. When (±)-[6a-^{3}H,9-*O*-Me-^{14}C]medicarpin and (±)-[3-^{3}H,4′-*O*-Me-^{14}C]vestitol were fed to induced lucerne seedlings (Martin and Dewick, 1978), they were both incorporated into the other phytoalexins with no change in ^{3}H:^{14}C ratio, thus excluding the involvement of an alternative Δ^3 intermediate such as an isoflav-3-ene. The isoflav-3-ene (*10.223*) was not incorporated into the phytoalexins, but it did prove to be a most efficient precursor of the coumestan, as was the arylcoumarin (*10.224*). The biosynthesis of coumestans had been considered to proceed via allylic oxidation of pterocarpenes, derived by cyclization of 2′-hydroxyisoflavanones (Hahlbrock and Grisebach, 1975), but this had never been proven in feeding experiments because of the ready aerial oxidation of pterocarpenes to coumestans. These more recent results suggest that a different route operates, namely allylic oxidation of a 2′-hydroxyisoflav-3-ene to a 2′-hydroxy-3-arylcoumarin, followed by cyclization and further oxidation to the coumestan. Isoflav-3-ene (*10.223*) may be produced by dehydration of isoflavanol (*10.215*), or by loss of a proton from carbonium ion (*10.216*), the latter route explaining the observed significant incorporation of medicarpin into 9-methylcoumestrol (Martin and Dewick, 1978) (Fig. 10.17).

These phytoalexin studies have indicated that the B-ring oxygenation pattern appears to be built up, generally at the isoflavone level, by the sequences 4′→2′,4′ and 4′→4′,5′(≡ 3′,4′)→2′,4′,5′. The A-ring pattern of 7 or 5,7 is determined at the chalcone level, but this can be modified further (i.e. 7→6,7) at the isoflavone level, as is demonstrated by the incorporation of formononetin (Dewick, 1978) and texasin (6,7-dihydroxy-4′-methoxyisoflavone) (Al-Ani and Dewick, 1980) into afrormosin (7-hydroxy-6,4′-dimethoxyisoflavone) in sainfoin (*Onobrychis viciifolia*) seedlings. Once again, 4′-methylation appears to be effected during aryl migration. The origin of the acetate-derived ring of the 6a-hydroxypterocarpan phytoalexin pisatin has been studied by feeding sodium [1,2-$^{13}C_2$]acetate to $CuCl_2$-treated pea (*Pisum sativum*) pods (Stoessl and Stothers, 1979). By analysing ^{13}C−^{13}C coupling in the ^{13}C-NMR spectrum of the pisatin produced, it was possible to show that carbons 1 and 1a, 2 and 3, and 4 and 4a had been incorporated as intact C_2 units, that cyclization of the polyketide must have occurred as in *a* rather than *c*, and that reduction of the 'lost' oxygen function must have occurred before cyclization (Fig. 10.18). By contrast, a similar experiment in parsley cell suspension cultures produced the flavonoids apigenin and kaempferol in which randomization of label occurred, demonstrating that free rotation of the acetate-derived ring had occurred and thus proving the intermediacy of 2′,4′,6′,4-tetrahydroxychalcone (Light and Hahlbrock, 1980). It is remarkable that this amount of information can be inferred from a single feeding experiment and illustrates the potential of ^{13}C-NMR studies in biosynthesis. More conventional precursor studies in $CuCl_2$-treated pea pods and seedlings (Dewick and Banks, 1980) have shown that formononetin,

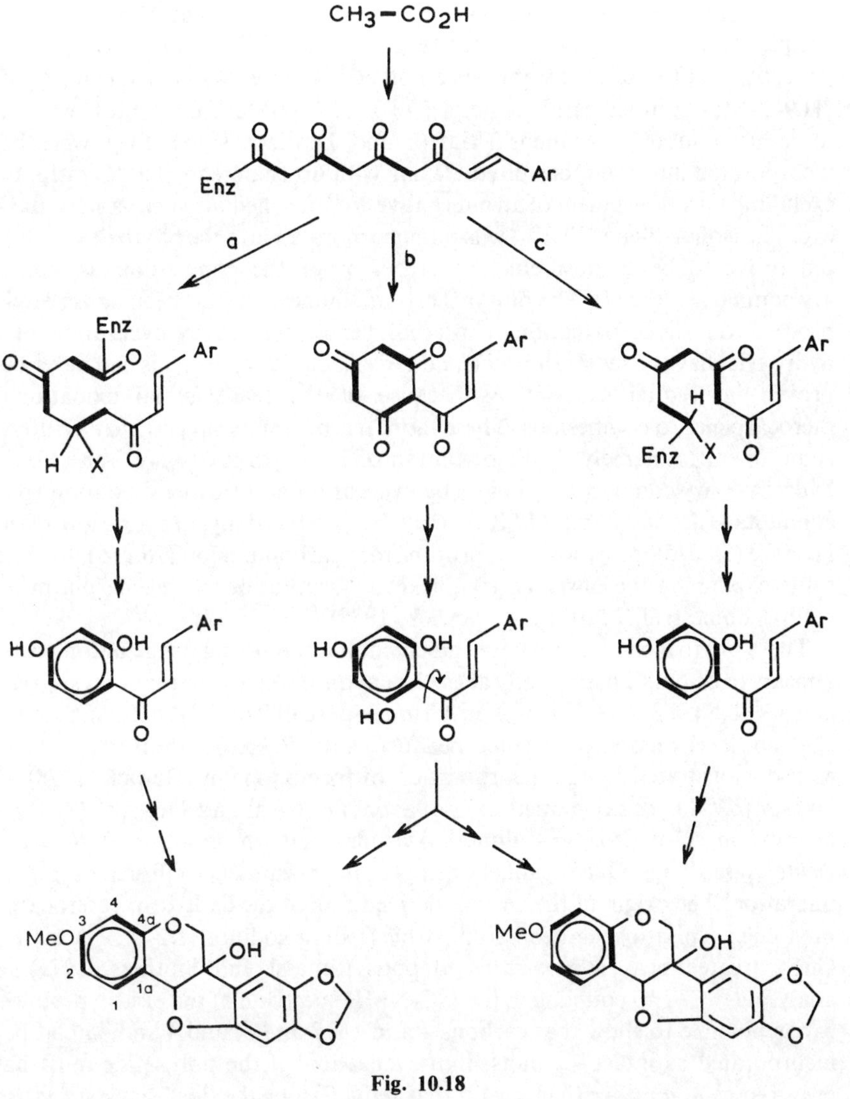

Fig. 10.18

calycosin, pseudobaptigenin, 7,2′-dihydroxy-4′,5′-methylenedioxyisoflavone, maackiain and 6a-hydroxymaackiain all function as efficient precursors of pisatin, and these represent a logical sequence for the biosynthetic pathway. Some incorporations of the pterocarpene anhydropisatin were also recorded and the precise role, if any, of this compound awaits clarification. This 6a-hydroxylation of pterocarpans is paralleled by fungal modification of some phytoalexins (see Section 10.17).

On the basis of feeding experiments with labelled chalcones, isoflavones and rotenoids (Crombie *et al.*, 1973), a tentative pathway to the rotenoid amorphigenin (*10.230*) via 9-demethylmunduserone (*10.225*), rotenonic acid (*10.226*) and rotenone (*10.229*) had been established. A route involving the epoxide of rotenonic acid (*10.227*), cyclization to dalpanol (*10.228*), then dehydration to rotenone (*10.229*) and finally hydroxylation was postulated (Fig. 10.19). In further feeding experiments in seedlings of *Amorpha fruticosa* (Crombie *et al.*, 1979b), the presence of rotenone, dalpanol and rot-2′-enonic acid (*10.226*) has been established by isotope dilution, although labelled rotenonic acid could be isolated only after short feeding times, in keeping with its role as a biosynthetic intermediate. Rot-3′-enonic acid was not incorporated into amorphigenin, although it had earlier been established that isopentenyl alcohol was a better precursor than dimethylallyl alcohol (Crombie *et al.*, 1979a). Rot-2′-enonic acid, labelled mainly in the 4′ position, was a good precursor of amorphigenin, dalpanol and 12a-hydroxyamorphigenin, the specific activities of the latter two compounds being some six times higher than that of amorphigenin. This fits with the involvement of dalpanol as a biosynthetic intermediate, and also establishes that the 12a-hydroxyrotenoid is not an artefact, but a true natural product. Degradation of labelled rotenone from a further experiment indicated that, after correction for the labelling pattern of the precursor, cyclization of the dimethylallyl side chain was accomplished with virtually complete stereoselectivity. From this, the configuration of the postulated epoxide intermediate can be assigned as in (*10.227*). Neither of the 6′-epimers of amorphigenol was

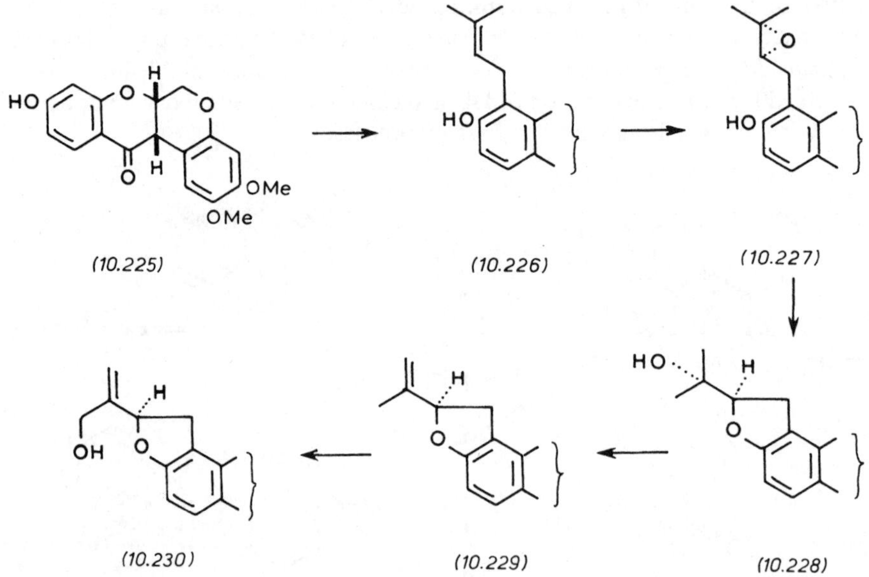

Fig. 10.19

incorporated into amorphigenin, and thus allylic hydroxylation of rotenone must be involved rather than hydroxylation of dalpanol followed by dehydration (Crombie *et al.*, 1979c). Of interest is that the label from [7′-^{14}C]rotenone becomes distributed equally between 7′ and 8′ of amorphigenin, but the reason is unknown.

The glucosylation of daidzein to the 7-*O*-glucoside daidzin and the 8-*C*-glucoside puerarin has been investigated in *Pueraria lobata* root (Inoue and Fujita, 1977). By a series of competition feeding experiments with 2′,4′,4-trihydroxychalcone and 7,4′-dihydroxyflavanone mixtures, it was demonstrated that the chalcone was the more immediate precursor of both glucosides. Daidzin was a poor precursor of puerarin, and although daidzein was well incorporated into daidzin, it proved a poor precursor of puerarin. Thus, since 2′,4′,4-trihydroxychalcone was well incorporated into both glycosides, it was suggested that daidzin was formed by glucosylation of daidzein, but puerarin was perhaps produced via the chalcone-*C*-glucoside. This conclusion has not been confirmed by testing the appropriate chalcone-*C*-glucoside as a precursor.

The isoflavonoid nature of the 2-arylbenzofuran phytoalexin vignafuran has been demonstrated in UV-induced cowpea (*Vigna unguiculata*) seedlings using [U-^{14}C]-, [1-^{14}C]- and [2-^{14}C]-phenylalanine precursors (Martin and Dewick, 1979). During the biosynthetic process, C-3 of phenylalanine appears to be lost, and the resulting labelling pattern shows that the phenylalanine-derived aromatic ring becomes the 2-aryl substituent, and not part of the benzofuran system (Fig. 10.20). Thus, a pathway to 2-arylbenzofurans by loss of C-6 from a coumestan is excluded. Chemical analogies for possible alternative routes are noted in Section 10.14. Incorporation levels were very small, most of the administered label being found in kievitone, an isoflavanone phytoalexin of *V. unguiculata*. The low rate of biosynthesis was probably responsible for the observed lack of incorporation of labelled chalcone and isoflavone compounds administered which may nevertheless be precursors.

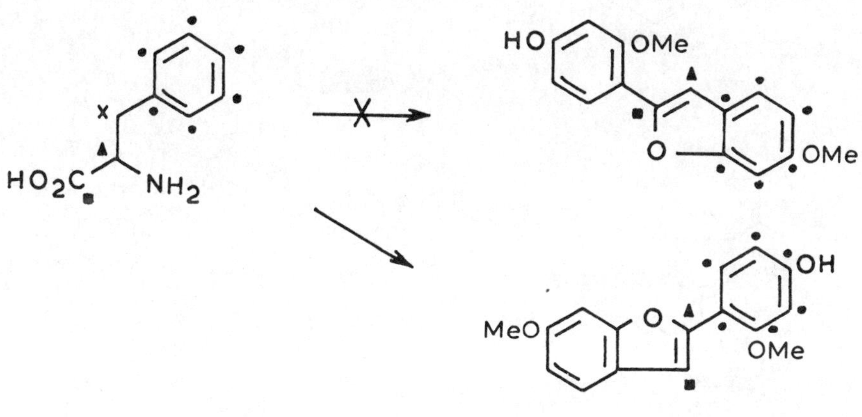

X
HO_2C
NH_2
HO
OMe
O
OMe
OH
MeO
O
OMe

Fig. 10.20

Biosynthetic experiments have established that the homoisoflavonoid eucomin is formed in bulbs of *Eucomis bicolor* (Liliaceae) by addition of a carbon atom from methionine on to a C_{15} chalcone-type skeleton (Dewick, 1975b), and 2′-methoxychalcones appear to be biosynthetic precursors (Fig. 10.21). Thus the homoisoflavonoids are formed by a pathway not involving the 1,2-aryl migration step characteristic of isoflavonoids and it is now inappropriate to consider these compounds along with the isoflavonoids. The 'homoisoflavonoids' merely represent a further modification of the unrearranged flavonoid-type skeleton.

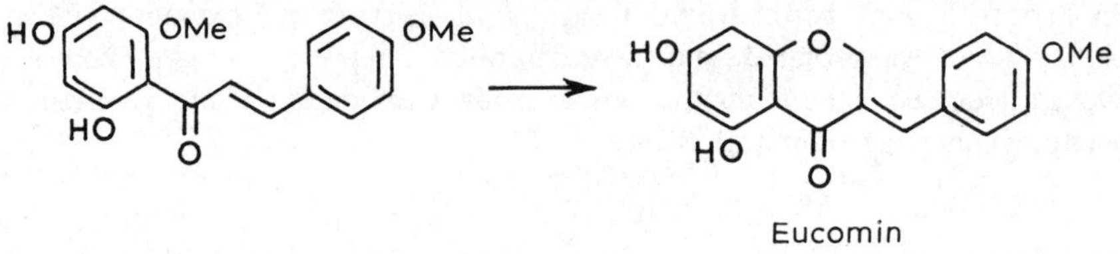

HO OMe OMe HO O → HO O OMe HO O

Eucomin

Fig. 10.21

Enzymic studies of isoflavonoid biosynthesis are much less advanced than those on flavonoids, but several enzyme preparations have been purified and investigated. An *S*-adenosylmethionine:isoflavone-4′-*O*-methyltransferase from cell suspension cultures of *Cicer arietinum* catalyses the 4′-methylation of daidzein and genistein (Wengenmayer *et al.*, 1974), and a dimethylallylpyrophosphate:trihydroxypterocarpan dimethylallyltransferase has been reported in a fraction from elicitor-induced soybean (*Glycine max*) cotyledons (Zähringer *et al.*, 1979). The latter enzyme catalyses the dimethylallylation of 3,6a,9-trihydroxypterocarpan and is probably a key enzyme in the biosynthesis of the glyceollins, the soybean phytoalexins. Another prenyltransferase from *Lupinus albus* catalyses the 6-dimethylallylation of the isoflavones genistein and 2′-hydroxygenistein to give the antifungal isoflavones wighteone and luteone respectively (Schröder *et al.*, 1979). Detailed discussion of these enzymes is included in Chapter 11.

10.19 BIOGENETIC RELATIONSHIPS AMONG THE ISOFLAVONOIDS

Because of the lack of experimental data, discussions of biogenetic relationships (e.g. Wong, 1975) have had to rely heavily on chemical interconversions of various isoflavonoid classes and information about the natural co-occurrence of various compounds. Metabolic schemes presented previously have thus been largely speculative. As biosynthetic evidence increases, it is now possible to

interrelate a large number of isoflavonoid classes, and to produce a scheme based entirely on direct experimental data (Fig. 10.22). This scheme is, of course, far from complete, but differs considerably from earlier ones, since chemically based hypotheses have not always been proved correct. In particular, no central role is ascribed to pterocarpenes. The inclusion of other classes of isoflavonoids must await further experimentation.

Knowledge of the co-occurrence of various isoflavonoids is nevertheless an essential foundation for biosynthetic research. The studies of Woodward (1979a,b,1980a) on *Monilinia fructicola*-infected *Phaseolus vulgaris* illustrate the value of such work. From drop diffusates, a whole range of 5-oxy- and 5-deoxy-isoflavonoids was isolated, and these compounds represent a nearly complete biosynthetic sequence from the simple isoflavones daidzein and genistein to phytoalexins phaseollin and kievitone.

10.20 BIOLOGICAL PROPERTIES OF ISOFLAVONOIDS

The oestrogenic activity of simple isoflavones and coumestans, and the insecticidal properties of rotenoids have been extensively discussed earlier (Wong, 1975; McClure, 1975). At that time, the antifungal properties of isoflavonoids and their function as phytoalexins were only just being recognized. Phytoalexins are antifungal compounds induced by a plant in response to infection or attempted invasion by fungi or other micro-organisms, but may also be induced by treatment with abiotic agents such as toxic chemicals or UV light (Harborne and Ingham, 1978). During the period of review, a considerable amount of data about new isoflavonoid phytoalexin structures and the mode of action of these materials has been published, and a number of important reviews have appeared (e.g. Van Etten and Pueppke, 1976; Gross, 1977; Grisebach and Ebel, 1978; Harborne and Ingham, 1978; Mahadevan, 1979).

Pterocarpans and isoflavans account for the bulk of isoflavonoid phytoalexins, although 6a-hydroxypterocarpans, isoflavanones, isoflavones and 2-arylbenzofurans may also function in this role. Medicarpin is by far the most common of pterocarpan phytoalexins and is sometimes accompanied by the methylenedioxy analogue maackiain. Vestitol and sativan are the most commonly encountered isoflavans. In the Phaseoleae and related tribes, more complex prenylated isoflavonoid derivatives, e.g. phaseollin and phaseollidin, are produced (Ingham and Markham, 1980). *Vicia faba* (broad bean) and *Lens esculenta* (lentil) are atypical amongst the Leguminosae in producing mainly furanoacetylene structures, e.g. wyerone rather than isoflavonoids, although traces of medicarpin in *V. faba* (Hargreaves *et al.*, 1976) and variabilin in *L. esculenta* (Robeson, 1978) have been isolated. Stilbene structures rather than isoflavonoids have been reported in *Arachis hypogaea* (Ingham, 1976e; Keen and Ingham, 1976) and *Trifolium dubium* and *T. campestre* (Ingham, 1978c), and *Lathyrus odoratus* produces unusual chromone derivatives as well as

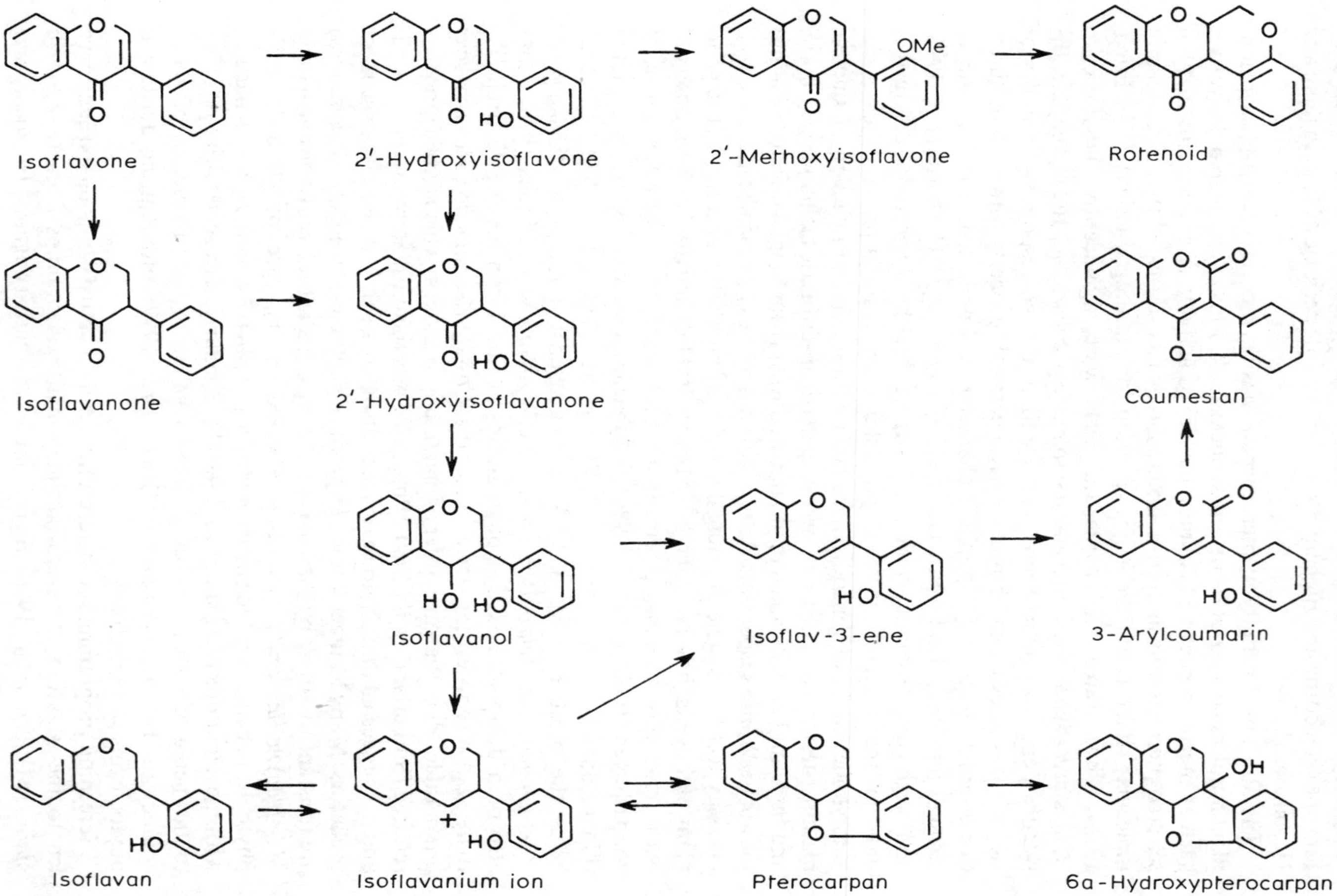

Isoflavone

2′-Hydroxyisoflavone

2′-Methoxyisoflavone

Rotenoid

Isoflavanone

2′-Hydroxyisoflavanone

Coumestan

Isoflavanol

Isoflav-3-ene

3-Arylcoumarin

Isoflavan

Isoflavanium ion

Pterocarpan

6a-Hydroxypterocarpan

Fig. 10.22

6a-hydroxypterocarpans (Robeson *et al.*, 1980). The pattern of phytoalexins formed by a particular plant appears to be a useful taxonomic marker, not only for tribe (Harborne and Ingham, 1978), but also at species level (e.g. Ingham and Harborne, 1976).

The antifungal activity of pterocarpans has been correlated with their non-planar molecular shape (Perrin and Cruickshank, 1969), since planar derivatives such as pterocarpenes and coumestans lacked activity. A more recent study of selected pterocarpans and other isoflavonoids has suggested that a non-planar molecular shape is not the major criterion for antifungal activity (Van Etten, 1976). Thus, three planar pterocarpenes were significantly fungitoxic to *Aphanomyces euteiches* and *Fusarium solani* f.sp. *cucurbitae*, and in one case, the pterocarpene was much more active than the corresponding non-planar 6a-hydroxypterocarpan. Some physicochemical property other than three-dimensional shape, e.g. lipophilic character, is implicated. The isolation of strongly antifungal isoflavones luteone (Fukui *et al.*, 1973; Harborne *et al.*, 1976) and wighteone (Ingham *et al.*, 1977) from healthy *Lupinus* tissues further indicates that three-dimensional molecular shape is not a prerequisite.

Phytoalexins have generally been isolated after fungal infection of suitable plant tissue, in most cases, by employing the drop-diffusate technique (Harborne and Ingham, 1978), in which phytoalexins diffuse from fungus-infected cells into the applied spore suspension. Antifungal compounds can then be identified by a bioassay technique. The antifungal activity of phytoalexins is not restricted to plant pathogens, however, and a significant level of activity has been observed with some phytoalexins, especially phaseollinisoflavan, against a range of zoopathogens (Gordon *et al.*, 1980). Antibacterial activity has also been noted (Gnanamanickam and Patil, 1977; Wyman and Van Etten, 1978), kievitone being the most effective of those tested against a range of *Pseudomonas*, *Xanthomonas* and *Achromobacter* species (Wyman and Van Etten, 1978). It has also been demonstrated that phytoalexins exhibit insect-feeding-deterrent activity against root-feeding larvae of scarabs (*Costelytra zealandica*, *Heteronychus arator*) at levels of the same order of magnitude as those at which fungal growth is reduced (Sutherland *et al.*, 1980). Phaseollin was the most active of those tested, and it is suggested that phytoalexins may also play a role in defence against insect predators. Many legumes known to produce phytoalexins (e.g. *Medicago sativa*) are resistant to attack by these larvae. Activity against the root-knot nematode *Meloidogyne incognita* has been observed in the case of glyceollin, this phytoalexin being induced when roots of soya bean (*Glycine max*) are challenged with the nematode (Kaplan *et al.*, 1980a,b). Similarly, a hypersensitive response to the nematode *Pratylenchus scribneri* resulted in the production of active coumestans by *Phaseolus lunatus* (Rich *et al.*, 1977), although no antifungal phytoalexins were detected.

Screening programmes to detect antimicrobial activity in plant extracts have led to the isolation of several antimicrobial isoflavonoids from *Glycyrrhiza glabra* (Mitscher *et al.*, 1980). Significant *in vitro* activity against *Haphylococcus*

aureus and *Mycobacterium swegmatis* were observed for the isoflavans hispaglabridin-A and -B, glabridin and 4′-methylglabridin, and two flavanones. In contrast to phytoalexins, these are not induced metabolites, though are closely related structurally to phytoalexins such as phaseollinisoflavan, which was also identified in the extract. A similar screening of *Amorpha fruticosa* demonstrated that the rotenoid 11-hydroxytephrosin possessed antimicrobial activity (Mitscher *et al.*, 1979), and suggests that a systematic study of isoflavonoids as potential antimicrobial agents may well be worth while.

Several simple isoflavonoids, pseudobaptigenin, calycosin, maackiain and irilone, present in red clover (*Trifolium pratense*) root suppress the growth of wheat coleoptiles *in vitro* (Popravko *et al.*, 1979), and glycosides of genistein and pseudobaptigenin from *Lupinus luteus* have similar properties (Laman, 1974). Though not confirmed, these compounds could act as endogenous growth regulators. The isoflavonoid content of red clover decreased during the autumn following hardening and vernalization.

The isolation of a number of isoflavones and their glycosides from microbial sources has been commented upon in Section 10.3. Whether or not these are true microbial metabolites remains to be established, but they were found to inhibit specific enzyme systems, including β-galactosidase (Hazato *et al.*, 1979), catechol-*O*-methyltransferase (Chimura *et al.*, 1975) and dopa decarboxylase (Umezawa *et al.*, 1975).

REFERENCES

Adinarayana, D. and Rao, J. R. (1975a), *Proc. Indian Acad. Sci. Sect. A* **81**, 23 (*Chem. Abstr.* **82**, 108844).

Adinarayana, D. and Rao, J. R. (1975b), *Indian J. Chem.* **13**, 425.

Ahluwalia, V. K. and Mehta, S. (1978), *Indian J. Chem.* **16B**, 977.

Al-Ani, H. A. M. and Dewick, P. M. (1980), *Phytochemistry,* **19**, 2337.

Almeida M. E. L de and Gottlieb, O. R. (1974), *Phytochemistry* **13**, 751.

Almeida, M. E. L. de and Gottlieb, O. R. (1975), *Phytochemistry* **14**, 2716.

Antus, S. and Nogradi, M. 1979a), *Chem. Ber.* **112**, 480.

Antus, S. and Nogradi, M. (1979b), *Acta Chim. Acad. Sci. Hung.* **100**, 179 (*Chem. Abstr.* **92**, 59169).

Antus, S., Farkas, L., Gottsegen, A., Kardos-Balogh, Z. and Nogradi, M. (1976), *Chem. Ber.* **109**, 3811.

Antus, S., Farkas, L. and Gottsegen, A. (1979a), *Acta Chim. Acad. Sci. Hung.* **102**, 205 (*Chem. Abstr.* **93**, 70357).

Antus, S., Gottsegen, A., Nogradi, M. and Gergely, A. (1979b), *Chem. Ber.* **112**, 3879.

Anzeveno, P. B. (1979a), *J. Org. Chem.* **44**, 2578.

Anzeveno P. B. (1979b), *J. Heterocyclic Chem.* **16**, 1643.

Arisawa, M. and Morita, N. (1976), *Chem. Pharm. Bull.* **24**, 815

Arisawa, M., Morita, N., Kondo, Y. and Takemoto, T. (1973), *Chem. Pharm. Bull.* **21**, 2323.

Arisawa, M., Kizu, H. and Morita, N. (1976), *Chem. Pharm. Bull.* **24**, 1609.

Bailey, J. A., Burden, R. S., Mynett, A. and Brown, C. (1977), *Phytochemistry* **16**, 1541.

Bajwa, B. S., Khanna, P. L. and Seshadri, T. R. (1974), *Indian J. Chem.* **12**, 15.

Bandyukova, V. A. and Kazakov, A. L. (1978), *Khim. Prir. Soedin*, 669.

Barz, W., Schlepphorst, R. and Laimer, J. (1976), *Phytochemistry* **15**, 87.

Bass, R. J. (1976), *J. Chem. Soc. Chem. Commun.*, 78.
Begley, M. J., Crombie, L. and Whiting, D. A. (1975), *J. Chem. Soc. Chem. Commun.*, 850.
Berry, R. C., Eade, R. A. and Simes, J. J. H. (1977), *Aust. J. Chem.* **30**, 1827.
Bezuidenhoudt, B. C. B., Brandt, E. V., Roux, D. G. and Van Rooyen, P. H. (1980), *J. Chem. Soc. Perkin I*, 2179.
Bhanumati, S., Chhabra, S. C. and Gupta, S. R. (1979a), *Phytochemistry* **18**, 1254.
Bhanumati, S., Chhabra, S. C., Gupta, S. R. and Krishnamoorthy, V. (1979b), *Phytochemistry* **18**, 365.
Bhanumati, S., Chhabra, S. C., Gupta, S. R. and Krishnamoorthy, V. (1979c), *Phytochemistry* **18**, 693.
Bhardwaj, D. K. and Singh, R. (1977), *Curr. Sci.* **46**, 753.
Bhardwaj, D. K., Murari, R., Seshadri, T. R. and Singh, R. (1976), *Phytochemistry* **15**, 352.
Bhardwaj, D. K., Seshadri, T. R. and Singh, R. (1977). *Phytochemistry* **16**, 402.
Biggs, R. (1975), *Aust. J. Chem.* **28**, 1389.
Biggs, D. R. and Lane, G. A. (1978), *Phytochemistry* **17**, 1683.
Bilton, J. N., Debnam, J. R. and Smith, I. M. (1976), *Phytochemistry* **15**, 1411.
Bonde, M. R., Millar, R. L. and Ingham, J. L. (1973), *Phytochemistry* **12**, 2957.
Bose, P. C., Kirtaniya, C. L. and Adityachaudhury, N. (1976) *Indian J. Chem.* **14B**, 1012.
Braz Filho, R., Gottlieb, O. R., Mourao, A. P., Rocha, A.I.da and Oliveira, F. S. (1975), *Phytochemistry* **14**, 1454.
Braz Filho, R., Pedreira. G., Gottlieb, O. R. and Maia, J. G. S. (1976), *Phytochemistry* **15**, 1029.
Braz Filho, R., Gottlieb, O. R., Moraes, A.A.de, Pedreira, G., Pinho, S.L.V., Magalhaes, M. T. and Ribeiro, M.N.De S. (1977), *Lloydia* **40**, 236.
Braz Filho, R., Moraes, M.P.L.de and Gottleib, O. R. (1980), *Phytochemistry* **19**, 2003.
Breytenbach, J. C. and Rall, G. J. H. (1980), *J. Chem. Soc., Perkin I*, 1804.
Briggs, L. H., Cambie, R. C. and Montgomery, R. K. (1975), N.Z.J.Sci. **18**, 555 (*Chem. Abstr.* **84**, 132818).
Brink, A. J., Rall, G. J. H. and Engelbrecht, J. P. (1974a), *Tetrahedron* **30**, 311.
Brink, A. J., Rall, G. J. H. and Engelbrecht, J. P. (1974b), *Phytochemistry* **13**, 1581.
Brink, A. J., Rall, G. J. H. and Breytenbach, J. C. (1977), *Phytochemistry* **16**, 273.
Brown, P. M., Thomson, R. H., Hausen, B. M. and Simatupang, M. H. (1974), *Liebigs. Ann. Chem.*, 1295.
Burden, R. S. and Bailey, J. A. (1975), *Phytochemistry* **14**, 1389.
Burden, R. S., Bailey, J. A. and Vincent, G. G. (1974), *Phytochemistry* **13**, 1789.
Camele, G., Delle Monache, F., Delle Monache, G. and Marini-Bettolo, G. B. (1980), *Phytochemistry* **19**, 707.
Campbell, R. V. M. and Tannock, J. (1973), *J. Chem. Soc. Perkin 1*, 2222.
Campbell, R. V. M., Harper, S. H. and Kemp, A. D. (1969), *J. Chem. Soc. C.* 1787.
Carson, D., Cass, M. W., Crombie, L. and Whiting, D. A. (1975a), *J. Chem. Soc. Chem. Commun.* 802.
Carson, D., Crombie, L. and Whiting, D. A. (1975b), *J. Chem. Soc. Chem. Commun.* 851.
Chawla, H. M. and Chibber, S. S. (1977), *Indian J. Chem.* **15B**, 492.
Chawla, H., Chibber, S. C. and Seshadri, T. R. (1974), *Phytochemistry* **13**, 2301.
Chawla, H. M., Chibber, S. S. and Seshadri, T. R. (1975), *Indian J. Chem.* **13**, 444.
Chawla, H. M., Chibber, S. S. and Seshadri, T. R. (1976), *Phytochemistry* **15**, 235.
Chibber, S. S. and Khera, U. (1978), *Phytochemistry* **17**, 1442.
Chibber, S. S. and Khera, U. (1979), *Phytochemistry* **18**, 188.
Chibber, S. S. and Sharma, R. P. (1978), *Natl. Acad. Sci. Lett. (India)* **1**, 253 (*Chem. Abstr.* **90**, 3117).
Chibber, S. S. and Sharma, R. P. (1979a), *Indian J. Chem.* **18B**, 471.
Chibber, S. S. and Sharma, R. P. (1979b), *Planta Med.* **36**, 379.
Chibber, S.S. and Sharma, R. P. (1979c), *Phytochemistry* **18**, 1082.
Chibber, S. S. and Sharma, R. P. (1979d), *Phytochemistry* **18**, 1583.
Chibber, S. S. and Sharma, R. P. (1980), *Phytochemistry* **19**, 1857.

Chimura, H., Sawa, T., Kumada, Y., Naganawa, H., Matsuzaki, M., Takita, T., Hamada, M., Takeuchi, T. and Umezawa, H. (1975) *J. Antibiot. (Tokyo)* **28**, 619.
Chubachi, M. and Hamada, M. (1974), *Chem. Lett.*, 397.
Cook, J. T., Ollis, W. D., Sutherland, I. O. and Gottleib, O. R. (1978), *Phytochemistry* **17**, 1419.
Cooke, R. G. and Fletcher, R. A. H. (1974), *Aust. J. Chem.* **27**, 1377.
Cooke, R. G. and Rae, I. D. (1964), *Aust. J. Chem.* **17**, 379.
Cornia, M. and Merlini, L. (1975), *J. Chem. Soc. Chem. Commun.* 428.
Craveiro, A. A. and Gottlieb, O. R. (1974), *Phytochemistry* **13**, 1629.
Crombie, L., Dewick, P. M. and Whiting, D. A. (1973), *J. Chem. Soc. Perkin I*, 1285.
Crombie, L., Holden, I., Kilbee, G. W. and Whiting, D. A. (1979a), *J. Chem. Soc. Chem. Commun.*, 1142.
Crombie, L., Holden, I., Kilbee, G. W. and Whiting, D. A. (1979b), *J. Chem. Soc. Chem. Commun.*, 1143.
Crombie, L., Holden, I., Kilbee, G. W. and Whiting, D. A. (1979c), *J. Chem. Soc. Chem. Commun.*, 1144.
Darbarwar, M., Sundaramurthy, V. and Subba Rao, N. V. (1976), *J. Sci. Ind. Res.* **35**, 297.
Delle Monache, G., Delle Monache, F., Marini-Bettolo, G. B., Albuquerque, M.M.de, Mello, J. F. de and Lima, O. G. de (1977a), *Gazz. Chim. Ital.* **107**, 189.
Delle Monache, F., Valera, G. C., Zapata, D. S. de and Marini-Bettolo, G. B. (1977b), *Gazz. Chim. Ital.* **107**, 403.
Delle Monache, F., Suarez, L. E. C. and Marini-Bettolo, G. B. (1978), *Phytochemistry* **17**, 1812.
DeMartinis, C., Mackay, M. F. and Poppleton, B. J. (1978), *Tetrahedron* **34**, 1849.
Deschamps-Vallet, C., Ilotse, J. B., Meyer-Dayan, M. and Molho, D. (1979), *Tetrahedron Lett.*, 1109.
Deshpande, V. H. and Shastri, R. K. (1977), *Indian J. Chem.* **15B**, 201.
Deshpande, V. H., Pendse, A. D. and Pendse, R. (1977), *Indian J. Chem.* **15B**, 205.
Dewick, P. M. (1975a), *Phytochemistry* **14**, 979.
Dewick, P. M. (1975b), *Phytochemistry* **14**, 983.
Dewick, P. M. (1977), *Phytochemistry* **16**, 93.
Dewick, P. M. (1978), *Phytochemistry* **17**, 249.
Dewick, P. M. and Banks, S. W. (1980), *Planta Med.* **39**, 287.
Dewick, P. M. and Ingham, J. L. (1980), *Phytochemistry* **19**, 289.
Dewick, P. M. and Martin, M. (1979a), *Phytochemistry* **18**, 591.
Dewick, P. M. and Martin, M. (1979b), *Phytochemistry* **18**, 597.
Dewick, P. M. and Ward, D. (1977), *J. Chem. Soc. Chem. Commun.* 338.
Dewick, P. M. and Ward, D. (1978), *Phytochemistry* **17**, 1751.
Dhar, K. L. and Kalla, A. K. (1973), *Phytochemistry* **12**, 734.
Dhar, K. L. and Kalla, A. K. (1975), *J. Indian Chem. Soc.* **52**, 784.
Dhingra, V. K., Seshadri, T. R. and Mukerjee, S. K. (1974), *Indian J. Chem.* **12**, 1118.
Donnelly, D. M. X. and Kavanagh, P. J. (1974), *Phytochemistry* **13**, 2587.
Donnelly, D. M. X., Keegan, P. J. and Prendergast, J. P. (1973a), *Phytochemistry* **12**, 1157.
Donnelly, D. M. X., Thompson, J. C., Whalley, W. B. and Ahmad, S. (1973b), *J. Chem. Soc. Perkin I*, 1737.
Duffley, R. P. and Stevenson, R. (1977), *J. Chem. Soc. Perkin I*, 802.
Eade, R. A., McDonald, F. J. and Pham, H-P. (1978), *Aust. J. Chem.* **31**, 2699.
Eguchi, S., Nakayama, M. and Hayashi, S. (1976), *Org. Mass Spectrom.* **11**, 574.
Eguchi, S., Haze, M., Nakayama, M. and Hayashi, S. (1977), *Org. Mass Spectrom.* **12**, 51.
El-Emary, N. A., Kobayashi, Y. and Ogihara, Y. (1980), *Phytochemistry* **19**, 1878.
Falshaw, C. P., Ollis, W., Moore, J. A. and Magnus, K. (1966), *Tetrahedron, Suppl. No.* **7**, 333.
Farkas, L., Gottsegen, A., Nogradi, M. and Antus, S. (1974), *J. Chem. Soc. Perkin I*, 305.
Ferreira, M. A., Moir, M. and Thomson, R. H. (1974), *J. Chem. Soc. Perkin I*, 2429.
Ferreira, M. A., Moir, M. and Thomson, R. H. (1975), *J. Chem. Soc. Perkin I*, 1113.
Fitzgerald, M. A., Gunning, P. J. M. and Donnelly, D. M. X. (1976), *J. Chem. Soc. Perkin I*, 186.

Formiga, M. D., Gottlieb, O. R., Mendes, P. H., Koketsu, M., Almeida, M. E. L. de, Pereira, M. O. da S. and Magalhaes, M. T. (1975), *Phytochemistry* **14**, 828.

Fraishtat, P. D., Popravko, S. A. and Vul'fson, N. S. (1979a), *Bioorg. Khim.* **5**, 228 (*Chem. Abstr.* **90**, 148457).

Fraishtat, P. D., Popravko, S. A. and Vul'fson, N. S. (1979b); *Bioorg. Khim.* **5**, 1879 (*Chem. Abstr.* **92**, 177392).

Fuchs, A., Vries, F. W. de, Landheer, C. A. and Van Veldhuirzen, A. (1980), *Phytochemistry* **19**, 917.

Fukui, H., Egawa, H., Koshimizu, K. and Mitsui, T. (1973), *Agric. Biol. Chem.* **37**, 417.

Galina, E, and Gottlieb, O. R. (1974), *Phytochemistry* **13**, 2593.

Geigert, J., Stermitz, F. R., Johnson, G., Maag, D. D. and Johnson, D. K. (1973), *Tetrahedron* **29**, 2703.

Ghanim, A. and Jayaraman, I. (1979), *Indian J. Chem.* **17B**, 648.

Gnanamanickam, S. S. and Patil, S. S. (1977), *Physiol. Plant Pathol.* **10**, 159.

Gordon, M. A., Lapa, E. W., Fitter, M. S. and Lindsay, M. (1980), *Antimicrob. Agents Chemother.* **17**, 120.

Gottlieb, O. R., Oliveira, A. B. de, Goncalves, T. M. M., Oliveira, G. G. de and Pereira, S. A. (1975), *Phytochemistry* **14**, 2495.

Grayer-Barkmeijer, R. J., Ingham, J. L. and Dewick, P. M. (1978), *Phytochemistry* **17**, 829.

Gregson, M., Ollis, W. D., Redman, B. T., Sutherland, I. O., Dietrichs, H. H. and Gottlieb, O. R. (1978a), *Phytochemistry* **17**, 1395.

Gregson, M., Ollis, W. D., Sutherland, I. O., Gottlieb, O. R. and Magalhaes, M. T. (1978b), *Phytochemistry* **17**, 1375.

Grisebach, H. and Ebel, J. (1978), *Angew. Chem. Int. Ed.* **17**, 635.

Gross, D. (1977), *Fortschr. Chem. Org. Naturst.* **34**, 187.

Grujic, Z., Tabakovic, I. and Trkovnik, M. (1976), *Tetrahedron Lett.* 4823.

Guimaraes, I. S. de S., Gottlieb, O. R., Andrade, C. H. S. and Magalhaes, M. T. (1975), *Phytochemistry* **14**, 1452.

Gunning, P. J. M., Kavanagh, P. J., Meegan, M. J. and Donnelly, D. M. X. (1977), *J. Chem. Soc. Perkin I*, 691.

Gupta, B. K., Gupta, G. K., Dhar, K. L. and Atal, C. K. (1980), *Phytochemistry* **19**, 2232.

Gupta, G. K., Dhar, K. L. and Atal, C. K. (1977), *Phytochemistry* **16**, 403.

Gupta, G. K., Dhar, K. L. and Atal, C. K. (1978), *Phytochemistry* **17**, 164.

Hahlbrock, K. and Grisebach, H. (1975). In *The Flavonoids* (eds. J. B. Harborne, T. J. Mabry and H. Mabry), Chapman and Hall, London, p. 866.

Harborne, J. B. and Ingham, J. L. (1978). In *Biochemical Aspects of Plant and Animal Coevolution* (ed. J. B. Harborne), Academic Press, London and New York, p. 343.

Harborne, J. B., Ingham, J. L., King, L. and Payne, M. (1976), *Phytochemistry* **15**, 1485.

Hargreaves, J. A., Mansfield, J. W. and Coxon, D. T. (1976), *Nature (London)* **262**, 318.

Harper, S. H., Shirley, D. B. and Taylor, D. A. (1976), *Phytochemistry* **15**, 1019.

Hayashi, T. and Thomson, R. H. (1974), *Phytochemistry* **13**, 1943.

Hazato, T., Naganawa, H., Kumagai, M., Aoyagi, T. and Umezawa, H. (1979), *J. Antibiot. Tokyo* **32**, 217.

Haznagy, A., Toth, G. and Tamas, J. (1978), *Arch. Pharm.* **311**, 318.

Higgins, V. J. (1975), *Physiol. Plant Pathol.* **6**, 5.

Higgins, V. J., Stoessl, A. and Heath, M. C. (1974), *Phytopathology* **64**, 105.

Horino, H. and Inoue, N. (1976), *J. Chem. Soc. Chem. Commun.*, 500.

Imamura, H., Hibino, Y. and Ohashi, H. (1974), *Phytochemistry* **13**, 757.

Ingham, J. L. (1976a), *Phytochemistry* **15**, 1489.

Ingham, J. L. (1976b), *Z. Naturforsch.* **31c**, 331.

Ingham, J. L. (1976c), *Z. Naturforsch.* **31c**, 504.

Ingham, J. L. (1976d), *Phytopathol. Z.* **87**, 353.

Ingham, J. L. (1976e), *Phytochemistry* **15**, 1791.

Ingham, J. L. (1977a), *Z. Naturforsch.* **32c**, 449.

Ingham, J. L. (1977b), *Z. Naturforsch.* **32c**, 1018.
Ingham, J. L. (1977c), *Phytochemistry* **16**, 1279.
Ingham, J. L. (1977d), *Phytochemistry* **16**, 1457.
Ingham, J. L. (1978a), *Phytochemistry* **17**, 165.
Ingham, J. L. (1978b), *Z. Naturforsch.* **33c**, 146.
Ingham, J. L. (1978c), *Biochem. Syst. Ecol.* **6**, 217.
Ingham, J. L. (1979a), *Biochem. Syst. Ecol.* **7**, 29.
Ingham, J. L. (1979b), *Z. Naturforsch.* **34c**, 159.
Ingham, J. L. (1979c), *Z. Naturforsch.* **34c**, 290.
Ingham, J. L. (1979d), *Z. Naturforsch.* **34c**, 293.
Ingham, J. L. (1979e), *Z. Naturforsch.* **34c**, 296.
Ingham, J. L. (1979f), *Z. Naturforsch.* **34c**, 630.
Ingham, J. L. (1979g), *Z. Naturforsch.* **34c**, 683.
Ingham, J. L. (1980), *Z. Naturforsch.* **35c**, 384.
Ingham, J. L. and Dewick, P. M. (1977), *Z. Naturforsch* **32c**, 446.
Ingham, J. L. and Dewick, P. M. (1978), *Phytochemistry* **17**, 535.
Ingham, J. L. and Dewick, P. M. (1979), *Phytochemistry* **18**, 1711.
Ingham, J. L. and Dewick, P. M. (1980a), *Z. Naturforsch.* **35c**, 197.
Ingham, J. L. and Dewick, P. M. (1980b), *Phytochemistry* **19**, 1767.
Ingham, J. L. and Harborne, J. B. (1976), *Nature (London)* **260**, 241.
Ingham, J. L. and Markham, K. R. (1980), *Phytochemistry* **19**, 1203.
Ingham, J. L. and Millar, R. L. (1973) *Nature (London)* **242**, 125.
Ingham, J. L., Keen, N. T. and Hymowitz, T. (1977), *Phytochemistry* **16**, 1943.
Inoue, T. and Fujita, M. (1977), *Chem. Pharm. Bull.* **25**, 3226.
Isogai, Y., Komoda, Y. and Okamoto, T. (1970), *Chem. Pharm. Bull.* **18**, 1872.
Jain, A. C. and Tuli, D. K. (1978), *J. Sci. Indian Res.* **37**, 287.
Jain, A. C., Gupta, G. K. and Rao, P. R. (1974), *Ind. J. Chem.* **12**, 659.
Jain, A. C., Khazanchi, R. and Gupta, R. C. (1978a), *Bioorg. Chem.* **7**, 493.
Jain, A. C., Kumar, A. and Kohli, A. K. (1978b), *J. Sci. Ind. Res.* **37**, 606.
Jain, S. C. (1979), *J. Nat. Prod.* **42**, 689.
Joshi, B. S. and Kamat, V. N. (1973), *J. Chem. Soc. Perkin I*, 907.
Jurd, L. (1976), *Tetrahedron Lett.* 1741.
Jurd, L. and Manners, G. D. (1977), *J. Agric. Food Chem.* **25**, 723.
Kalla, A. K., Bhan, M. K. and Dhar, K. L. (1978), *Phytochemistry* **17**, 1441.
Kalra, A. J., Krishnamurti, M. and Nath, M. (1977), *Indian J. Chem.* **15B**, 1084.
Kamal, R. and Jain, S. C. (1978), *Planta Med.* **33**, 418.
Kaneda, M., Saitoh, T., Iitaka, Y. and Shibata, S. (1973), *Chem. Pharm. Bull.* **21**, 1338.
Kaplan, D. T., Keen, N. T. and Thomason, I. J. (1980a), *Physiol. Plant Pathol.* **16**, 309.
Kaplan, D. T., Keen, N. T. and Thomason, I. J. (1980b), *Physiol. Plant Pathol.* **16**, 319.
Kappe, T. and Brandner, A. (1974), *Z. Naturforsch.* **29b**, 292.
Kappe, T. and Schmidt, H. (1972), *Org. Prep. Proced. Int.* **4**, 233 (*Chem. Abstr.* **78**, 71950).
Kasymov, A. U., Kondratenko, E. S. and Abubakirov, N. K. (1968), *Khim. Prir. Soedin*, 326. (*Chem. Abstr.* **70**, 88190).
Kasymov, A. U., Kondratenko, E. S., Rashkes, Y. V. and Abubakirov, N. K. (1970), *Khim. Prir. Soedin*, 197 (*Chem. Abstr.* **73**, 77501).
Kasymov, A. U., Kondratenko, E. S. and Abubakirov, N. K. (1974), *Khim. Prir. Soedin*, 464 (*Chem. Abstr.* **82**, 82953).
Kattaev, N. S. and Nikonov, G. K. (1972), *Khim. Prir. Soedin*, 648 (*Chem. Abstr.* **78**, 108181).
Kattaev, N. S. and Nikonov, G. K. (1975), *Khim. Prir. Soedin*, 140 (*Chem. Abstr.* **83**, 111105).
Kattaev, N. S., Kharlamov, I. A., Akhmedkhodzhaeva, N. M., Nikonov, G. K. and Khalmaton, K. K. (1972), *Khim. Prir. Soedin*, 806 (*Chem. Abstr.* **78**, 94807).
Kattaev, N. S., Nikonov, G. K. and Rashkes, Y. V. (1975), *Khim. Prir. Soedin*, 147 (*Chem. Abstr.* **83**, 128646).

Kawase, A., Ohta, N. and Yagishita, K. (1973), *Agric. Biol. Chem.* **37**, 145.
Kazakov, A. L., Shinkarenko, A. L. and Oganesyan, E. T. (1972), *Khim. Prir. Soedin*, 804 (*Chem. Abstr.* **78**, 108243).
Keen, N. T. and Ingham, J. L. (1976), *Phytochemistry* **15**, 1794.
Khera, U. and Chibber, S. S. (1978a), *Indian J. Chem.* **16B**, 78.
Khera, U. and Chibber, S. S. (1978b), *Indian J. Chem.* **16B**, 641.
Khera, U. and Chibber, S. S. (1978c), *Phytochemistry* **17**, 596.
Kinoshita, T., Saitoh, T. and Shibata, S. (1976), *Chem. Pharm. Bull.* **24**, 991.
Kinoshita, T., Saitoh, T. and Shibata, S. (1978a), *Chem. Pharm. Bull.* **26**, 135.
Kinoshita, T., Saitoh, T. and Shibata, S. (1978b), *Chem. Pharm. Bull.* **26**, 141.
Kirkiacharian, B. S. (1975), *J. Chem. Soc. Chem. Commun.* 162.
Kirkiacharian, B. S. and Chidiac, H. (1975), *C. R. Acad. Sci. Ser. C* **280**, 775.
Kirkiacharian, B. S. and Ravise, A. (1976), *Phytochemistry* **15**, 907.
Komatsu, M., Yokoe, I., Shirataki, Y. and Chen, J. (1976a), *Phytochemistry* **15**, 1089.
Komatsu, M., Yokoe, I. and Shirataki, Y. (1976b), *J. Pharm. Soc. Jpn.* **96**, 254.
Komatsu, M., Yokoe, I. and Shirataki, Y. (1978a), *Chem. Pharm. Bull.* **26**, 1274.
Komatsu, M., Yokoe, I. and Shirataki, Y. (1978b), *Chem. Pharm. Bull.* **26**, 3863.
Konig, W. A., Krauss, C. and Zähner, H. (1977), *Helv. Chim. Acta* **60**, 2071.
Kovalev, V. N., Spirodonov, V. N., Borisov, M. I., Kovalev, I. P., Gordienko, V. G. and Kolesnikov, D. D. (1975), *Khim. Prir. Soedin*, 354. (*Chem. Abstr.* **84**, 40717.)
Kovalev, V. N., Borisov, M. I., Spiridonov, V. N., Kovalev, I. and Gordienko, V. G. (1976), *Khim. Prir. Soedin*, **104**, (*Chem. Abstr.* **85**, 59566).
Krishnamurti, M., Sambhy, Y. R. and Seshadri, T. R. (1970), *Tetrahedron* **26**, 3023.
Krishnamurty, H. G. and Prasad, J. S. (1977), *Tetrahedron Lett.* 3071.
Krupadanam, G. L. D., Sarma, P. N., Srimannarayana, G. and Subba Rao, N. V. (1977), *Tetrahedron Lett.* 2125.
Krupadanam, G. L. D., Srimannarayana, G. and Subba Rao, N. V. (1978), *Indian J. Chem.* **16B**, 770.
Kubo, M., Fujita, K., Nishimura, H., Naruto, S. and Namba, K. (1973), *Phytochemistry* **12**, 2547.
Kubo, M., Sasaki, M., Namba, K., Naruto, S. and Nishimura, H. (1975), *Chem. Pharm. Bull.* **23**, 2449.
Kuhn, P. J. and Smith, D. A. (1979), *Physiol. Plant Pathol.* **14**, 179.
Kuhn, P. J., Smith, D. A. and Ewing, D. F. (1977), *Phytochemistry* **16**, 296.
Kurihari, T. and Kikuchi, M. (1973), *J. Pharm. Soc. Jpn.* **93**, 1201.
Kurihara, T. and Kikuchi, M. (1975), *J. Pharm. Soc. Jpn.* **95**, 1283.
Kurosawa, K. and Araki, F. (1979), *Bull. Chem. Soc. Jpn.* **52**, 529.
Kurosawa, K. and Nogami, K. (1976), *Bull. Chem. Soc. Jpn.* **49**, 1955.
Kurosawa, K., Ollis, W. D., Redman, B. T., Sutherland, I. O. and Gottlieb, O. R. (1978a), *Phytochemistry* **17**, 1413.
Kurosawa, K., Ollis, W. D., Redman, B. T., Sutherland, I. O., Alves, H. M. and Gottlieb, O. R. (1978b), *Phytochemistry* **17**, 1423.
Kurosawa, K., Ollis, W. D., Sutherland, I. O., Gottlieb, O. R. and Oliveira, A. B. de, (1978c), *Phytochemistry* **17**, 1405.
Kurosawa, K., Ollis, W. D., Sutherland, I. O. and Gottlieb, O. R. (1978d), *Phytochemistry* **17**, 1417.
Lamen, N. A. (1974), *Fiziol. Rast.* **21**, 301 (*Chem. Abstr.* **82**, 82511).
Laman, N. A. and Volynets, A. P. (1974), *Khim. Prir. Soedin*, 162 (*Chem. Abstr.* **81**, 132774).
Lamberton, J. A., Suares, H. and Watson, K. G. (1978), *Aust. J. Chem.* **31**, 455.
Lampard, J. F. (1974), *Phytochemistry* **13**, 291.
Lappe, U. and Barz, W. (1978), *Z. Naturforsch.* **33c**, 301.
Leipoldt, J. G., Rall, G. J. H., Roux, D. G. and Breytenbach, J. C. (1977), *J. Chem. Soc. Chem. Commun.* 349.
Letcher, R. M. and Shirley, I. M. (1976), *Phytochemistry* **15**, 353.
Light, R. J. and Hahlbrock, K. (1980), *Z. Naturforsch.* **35c**, 717.

Lupi, A., Delle Monache, F., Marini-Bettolo, G. B., Costa, D. L. B. and Albuquerque, I. L. D'. (1979), *Gazz. Chim. Ital.* **109**, 9.

Lyne, R. L. and Mulheirn, L. J. (1978), *Tetrahedron Lett.* 3127.

Lyne, R. L., Mulheirn, L. J. and Leworthy, D. P. (1976), *J. Chem. Soc. Chem. Commun.* 497.

Mabry, T. J., Markham, K. R. and Thomas, M. B. (1970), *The Systematic Identification of Flavonoids*, Springer-Verlag, New York, Heidelberg and Berlin.

Mahadevan, A. (1979), *J. Sci. Ind. Res.* **38**, 156.

Major, A., Nagy, Z. and Nogradi, M. (1980), *Acta Chim. Acad. Sci. Hung.* **104**, 85 (*Curr. Abstr. Chem. Index Chemicus*, **78**, 301542).

Malik, M. L. and Grover, S. K. (1978), *Tetrahedron Lett.* 1859.

Manners, G. D. and Jurd, L. (1979), *Phytochemistry* **18**, 1037.

Maranduba, A., Oliveira, A. B. de, Oliveira, G. G. de, Reis, J. E. de P. and Gottlieb, O. R. (1979), *Phytochemistry* **18**, 815.

Martin, M. and Dewick, P. M. (1978), *Tetrahedron Lett.* 2341.

Martin, M. and Dewick, P. M. (1979), *Phytochemistry* **18**, 1309.

Matos, F. J. de A., Gottlieb, O. R. and Andrade, C. H. S. (1975), *Phytochemistry* **14**, 825.

McClure, J. W. (1975) in *The Flavonoids* (eds. J. B. Harborne, T. J. Mabry and H. Mabry), Chapman and Hall, London, p. 970.

Meegan, M. J. and Donnelly, D. M. X. (1975), *Phytochemistry* **14**, 2283.

Meyer-Dayan, M., Bodo, B., Deschamps-Vallet, C. and Molho, D. (1978), *Tetrahedron Lett.* 3359.

Minhaj, N., Khan, H., Kapoor, S. K. and Zaman, A. (1976a), *Tetrahedron* **32**, 749.

Minhaj, N., Khan, H., Zaman, A. and Dean, F. M. (1976b), *Tetrahedron Lett.* 2391.

Minhaj, N., Tasneem, K., Khan, K. Z. and Zaman, A. (1977), *Tetrahedron Lett.* 1145.

Mitscher, L. A., Al-Shamma, A., Haas, T., Hudson, P. B. and Park, Y. H. (1979), *Heterocycles* **12**, 1033.

Mitscher, L. A., Park, Y. H., Clark, D. and Beal, J. L. (1980), *J. Nat. Prod.* **43**, 259.

Miyase, T., Ueno, A., Noro, T. and Fukushima, S. (1980), *Chem. Pharm. Bull.* **28**, 1172.

Morita, N., Shimokoriyama, M., Shimizu, M. and Arisawa, M. (1972), *Chem. Pharm. Bull.* **20**, 730.

Morita, N., Arisawa, M., Kondo, Y. and Takemoto, T. (1973), *Chem. Pharm. Bull.* **21**, 600.

Naim, M., Gestetner, B., Zilkah, S., Birk, Y. and Bondi, A. (1974), *J. Agric. Food Chem.* **22**, 806.

Nair, A. G. R. and Subramanian, S. S. (1976), *Indian J. Chem.* **14B**, 801.

Nakano, T., Alonso, J., Grillet, R. and Martin, A. (1979), *J. Chem. Soc. Perkin I*, 2107.

Nakov, N., Tolkatschev, O. N. and Achtardjiev, C. (1978), *Pharmazie* **33**, 463.

Oberholzer, M. E., Rall, G. J. H. and Roux, D. G. (1974), *Tetrahedron Lett.* 2211.

Oberholzer, M. E., Rall, G. J. H. and Roux, D. G. (1976), *Phytochemistry* **15**, 1283.

Oberholzer, M. E., Rall, G. J. H. and Roux, D. G. (1977a), *J. Chem. Soc. Perkin I*, 423.

Oberholzer, M. E., Rall, G. J. H. and Roux, D. G. (1977b), *Tetrahedron Lett.* 1165.

Ognyanov, S. and Somleva, T. (1980), *Planta Med.* **38**, 279.

Ohashi, H., Goto, M. and Imamura, H. (1976), *Phytochemistry* **15**, 354.

Ohashi, H., Fujiyama, T. and Imamura, H. (1979), *Gifu Daigaku Nogakubu Kenkyu Hokoku*, 123 (*Chem. Abstr.* **93**, 4053).

Ohta, N., Kuwata, G., Akahori, H. and Watanabe, T. (1979), *Agric. Biol. Chem.* **43**, 1415.

Ohta, N., Kuwata, G., Akahori, H. and Watanable, T. (1980), *Agric. Biol. Chem.* **44**, 469.

Oliveira, A. B. de, Madruga, M. I. L. M. and Gottlieb, O. R. (1978), *Phytochemistry* **17**, 593.

Oliveira A. B. de, Oliveira, G. G. de, Pimenta, L. de O., Madruga, M. I. L. M., Reis, J. E. de P. and Gottlieb, O. R. (1979), *Rev. Latinoam. Quim.* **10**, 122 (*Chem. Abstr.* **92**, 94184).

Ollis, W. D., Ormand, K. L. and Sutherland, I. O. (1970), *J. Chem. Soc. C*, 119.

Ollis, W. D., Redman, B. T., Roberts, R. J., Sutherland, I. O., Gottlieb, O. R. and Magalhaes, M. T. (1978a), *Phytochemistry* **17**, 1383.

Ollis, W. D., Redman, B. T., Sutherland, I. O. and Gottlieb, O. R. (1978b), *Phytochemistry* **17**, 1379.

Ollis, W. D., Sutherland, I. O., Alves, H. M. and Gottlieb, O. R. (1978c), *Phytochemistry* **17**, 1401.

Omokawa, H. and Yamashita, K. (1973), *Agric. Biol. Chem.* **37**, 1717.

Omokawa, H. and Yamashita, K. (1974), *Agric. Biol. Chem.* **38**, 1731.

Omokawa, H., Kouya, S. and Yamashita, K. (1975), *Agric. Biol. Chem.* **39**, 393.
Ozimina, I. I., Bandyukova, V. A. and Kazakov, A. L. (1979), *Khim. Prir. Soedin*, 858 (*Chem. Abstr.* **93**, 22588).
Pailer, M. and Franke, F. (1973), *Monatsh. Chem.* **104**, 1394.
Parthasarathy, M. R., Seshadri, T. R. and Varma, R. S. (1974), *Curr. Sci.* **43**, 74.
Parthasarathy, M. R., Seshadri, T. R. and Varma, R. S. (1976a), *Phytochemistry* **15**, 226.
Parthasarathy, M. R., Seshadri, T. R. and Varma, R. S. (1976b), *Phytochemistry* **15**, 1025.
Parthasarathy, M. R., Sharma, P. and Kalidhar, S. B. (1980), *Indian J. Chem.* **19B**, 429.
Pelter, A. and Foot, S. (1976), *Synthesis*, 326.
Pelter, A., Bradshaw, J. and Warren, R. F. (1971), *Phytochemistry* **10**, 835.
Pelter, A., Ward, R. S. and Balasubramanian, M. (1976), *J. Chem. Soc. Chem. Commun.*, 151.
Pelter, A., Ward, R. S. and Ashdown, D. H. J. (1968), *Synthesis*, 843.
Perrin, D. R. and Cruickshank, I. A. M. (1969), *Phytochemistry* **8**, 971.
Piatak, D. M., Flynn, G. A. and Sorensen, P. D. (1975), *Phytochemistry* **14**, 1391.
Popravko, S. A., Kokolova, S. A., Fraishtat, P. D. and Kononenko, G. P. (1979), *Bioorg. Khim.* **5**, 1654 (*Chem. Abstr.* **92**, 37791).
Prasad, J. S. and Varma, R. S. (1977), *Phytochemistry* **16**, 1120.
Preston, N. W. (1975), *Phytochemistry* **14**, 1131.
Preston, N. W. (1977a), *Phytochemistry* **16**, 143.
Preston, N. W. (1977b), *Phytochemistry* **16**, 2044.
Preston, N. W., Chamberlain, K. and Skipp, R. A. (1975), *Phytochemistry* **14**, 1843.
Pueppke, S. G. and Van Etten, H. D. (1975), *J. Chem. Soc. Perkin I*, 946.
Pueppke, S. G. and Van Etten, H. D. (1976), *Physiol. Plant Pathol.* **8**, 51.
Purushothaman, K. K., Chandrasekharan, S., Balakrishna, K. and Connolly, J. D. (1975), *Phytochemistry* **14**, 1129.
Radhakrishniah, M. (1973), *Phytochemistry* **12**, 3003.
Radhakrishniah, M. (1979), *J. Indian Chem. Soc.* **56**, 81.
Raju, K. V. S. and Srimannarayana, G. (1978), *Phytochemistry* **17**, 1065.
Rajulu, K. G. and Rao, J. R. (1980), *Phytochemistry* **19**, 1563.
Rall, G. J. H., Brink, A. J. and Engelbrecht, J. P. (1972), *J. S. Afr. Chem. Inst.* **25**, 131 (*Chem. Abstr.* **77**, 139846).
Rao, P. P. and Srimannarayana, G. (1980), *Phytochemistry* **19**, 1272.
Reisch, J., Gombos, M., Szendrei, K. and Novak, I. (1976), *Phytochemistry* **15**, 234.
Rich, J. R., Keen, N. T. and Thomason, I. J. (1977), *Physiol. Plant Pathol.* **10**, 105.
Robeson, D. J. (1978), *Phytochemistry* **17**, 807.
Robeson, D. J. and Harborne, J. B. (1977), *Z. Naturforsch.* **32c**, 289.
Robeson, D. J. and Ingham, J. L. (1979), *Phytochemistry* **18**, 1715.
Robeson, D. J., Ingham, J. L. and Harborne, J. B. (1980), *Phytochemistry* **19**, 2171.
Saito, R., Izumi, T. and Kasahara, A. (1973), *Bull Chem. Soc. Jpn.* **46**, 1776.
Saitoh, T., Kinoshita, T. and Shibata, S. (1976), *Chem. Pharm. Bull.* **24**, 752.
Saitoh, T., Noguchi, H. and Shibata, S. (1978), *Chem. Pharm. Bull.* **26**, 144.
Sakagami, Y., Kumai, S. and Suzuki, A. (1974), *Agric. Biol. Chem.* **38**, 1031.
Sarma, P. N., Srimannarayana, G. and Subba Rao, N. V. (1976), *Indian J. Chem.* **14B**, 152.
Sasaki, I. and Yamashita, K. (1979), *Agric. Biol. Chem.* **43**, 137.
Sawhney, K. N. and Mathur, K. B. L. (1976), *Indian J. Chem.* **14B**, 518.
Schröder, G., Zähringer, U., Heller, W., Ebel, J. and Grisebach, H. (1979), *Arch. Biochem. Biophys.* **194**, 635.
Sethi, M. L., Taneja, S. C., Agarwal, S. G., Dhar, K. L. and Atal, C. K. (1980), *Phytochemistry* **19**, 1831.
Sharma, A., Chibber, S. S. and Chawla, H. M. (1979a), *Phytochemistry* **18**, 1253.
Sharma, A., Chibber, S. S. and Chawla, H. M. (1980a), *Phytochemistry* **19**, 715.
Sharma, A., Chibber, S. S. and Chawla, H. M. (1980b), *Indian J. Chem.* **19B**, 237.
Sharma, R. and Khanna, P. (1975), *Indian J. Exp. Biol.* **13**, 84 (*Chem. Abstr.* **83**, 5041).

Sharma, S., Chibber, S. S. and Chawla, H. M. (1979b), *Indian J. Chem.* **18B**, 472.
Shemesh, M., Ayalon, N. and Lindner, H. R. (1976), *J. Agric. Sci.* **87**, 467.
Shibata, H. and Shimizu, S. (1978), *Heterocycles* **10**, 85.
Shiengthong, D., Donavanik, T., Uaprasert, V., Roengsumran, S. and Massey-Westropp, R. A. (1974), *Tetrahedron Lett.*, 2015.
Sicherer, C. A. X. G. F. and Sicherer-Roetman, A. (1980), *Phytochemistry* **19**, 485.
Sims, J. J., Keen, N. T. and Honwad, V. K. (1972), *Phytochemistry* **11**, 827.
Singhal, A. K., Sharma, R. P., Thyagarajan, G., Herz, W. and Govindan, S. V. (1980), *Phytochemistry* **19**, 929.
Smalberger, T. M., Vleggaar, R. and Weber, J. C. (1975), *Tetrahedron* **31**, 2297.
Smith, D. A., Kuhn, P. J., Bailey, J. A. and Burden, R. S. (1980), *Phytochemistry* **19**, 1673.
Steiner, P. W. and Millar, R. L. (1974), *Phytopathology* **64**, 586.
Stoessl, A. and Stothers, J. B. (1979), *Z. Naturforsch.* **34c**, 87.
Suri, J. L., Gupta, G. K., Dhar, K. L. and Atal, C. K. (1978), *Phytochemistry* **17**, 2046.
Sutherland, O. R. W., Russell, G. R., Biggs, D. R. and Lane, G. A. (1980), *Biochem. Syst. Ecol.* **8**, 73.
Szabo, V. and Antal, E. (1976), *Magy. Kem. Foly.* **82**, 474 (*Chem. Abstr.* **86**, 55118).
Szabo, V., Borbely, S. and Darbai, M. (1975a), *Magy. Kem. Foly.* **81**, 311 (*Chem. Abstr.* **83**, 178724).
Szabo, V., Borbely, S., Farkas, E. and Tolnai, S. (1975b), *Magy. Kem. Foly.* **81**, 220 (*Chem. Abstr.* **83**, 79033).
Szabo, V., Borda, J. and Theisz, E. (1980), *Acta Chim. Acad. Sci. Hung.* **103**, 271 (*Curr. Abstr. Chem. Index Chemicus* **78**, 301085).
Taguchi, H., Kanchanapee, P. and Amatayakul, T. (1977), *Chem. Pharm. Bull.* **25**, 1026.
Takeda, T., Ishiguro, I., Masegi, M. and Ogihara, Y. (1977), *Phytochemistry* **16**, 619.
Teresa, J. De P., Aubanell, J. C. H. and Grande, M. (1979), *An. Quin.* **75**, 1005 (*Chem. Abstr.* **92**, 160594).
Torck, M. (1976), *Fitoterapia* **47**, 195.
Torrance, S. J., Wiedhopf, R. M., Hoffman, J. J. and Cole, J. R. (1979), *Phytochemistry* **18**, 366.
Tsukayama, M., Horie, T., Masumura, M., Nakayama, M. and Hayashi, S. (1979) *Heterocycles* **12**, 1539.
Tsukida, K., Saiki, K. and Ito, M. (1973), *Phytochemistry* **12**, 2318.
Turner, R. B., Lindsey, D. L., Davis, D. D. and Bishop, R. D. (1975), *Mycopathologia* **57**, 39.
Uddin, A. and Khanna, P. (1979), *Planta Med.* **36**, 181.
Ueno, A., Ichikawa, M., Fukushima, S., Saiki, Y. and Morinaga, K. (1973a), *Chem. Pharm. Bull.* **21**, 2712.
Ueno, A., Ichikawa, M., Fukushima, S., Saiki, Y., Noro, T., Morinaga, K. and Kuwano, H. (1973b), *Chem. Pharm. Bull.* **21**, 2715.
Ueno, A., Ichikawa, M., Miyase, T., Fukushima, S., Saiki, Y. and Morinaga, K. (1973c), *Chem. Pharm. Bull.* **21**, 1734.
Umezawa, H., Tobe, H., Shibamoto, N., Nakamura, F., Nakamura, K., Matsuzaki, M. and Takeuchi, T. (1975), *J. Antibiot. (Tokyo)* **28**, 947.
Unai, T. and Yamamoto, I. (1973), *Agric. Biol. Chem.* **37**, 897.
Van den Heuvel, J. (1976), *Neth. J. Plant Pathol.* **82**, 153.
Van den Heuvel, J. and Glazener, J. A. (1975), *Neth. J. Plant. Pathol.* **81**, 125.
Van den Heuvel, J., Van Etten, H. D., Serum, J. W., Coffen, D. L. and Williams, T. H. (1974), *Phytochemistry* **13**, 1129.
Van der Meerwe, P. J., Rall, G. J. H. and Roux, D. G. (1978), *J. Chem. Soc. Chem. Commun.*, 224.
Van Etten, H. D. (1973), *Phytochemistry* **12**, 1791.
Van Etten, H. D. (1976), *Phytochemistry* **15**, 655.
Van Etten, H. D. and Pueppke, S. G. (1976). In *Biochemical Aspects of Plant–Parasite Relationships* (eds. J. Friend, and D. R. Threlfall), Academic Press, London and New York, p. 239.
Van Etten, H. D., Pueppke, S. G. and Kelsey, T. C. (1975), *Phytochemistry* **14**, 1103.
Van Etten, H. D., Matthews, P. S., Tegtmeier, K. J., Dietert, M. F. and Stein, J. I. (1980), *Physiol. Plant Pathol.* **16**, 257.

Van Heerden, F. R., Brandt, E. V. and Roux, D. G. (1978), *J. Chem. Soc. Perkin I*, 137.

Van't Land, B. G., Wiersma-Van Duin, E. D. and Fuchs, A. (1975), *Acta Bot. Neerl.* **24**, 251.

Vasconcelos, M. N. L. and Maia, J. G. S. (1976), *Acta Amazonica* **6**, 59. (*Chem. Abstr.* **85**, 119558).

Verhe, R. and Schamp, N. (1975), *Experientia* **31**, 759.

Vilain, C. (1980), *Phytochemistry* **19**, 988.

Vilain, C. and Jadot, J. (1975), *Bull. Soc. R. Sci. Liege* **44**, 306 (*Chem. Abstr.* **84**, 74136).

Vilain, C. and Jadot, J. (1976), *Bull. Soc. R. Sci. Liege* **45**, 468 (*Chem. Abstr.* **86**, 152677).

Wagner, H. and Farkas, L. (1975). In *The Flavonoids* (eds. J. B. Harborne, T. J. Mabry and H. Mabry), Chapman and Hall, London, p. 127.

Weltring, K. and Barz, W. (1980), *Z. Naturforsch.* **35c**, 399.

Wengenmayer, H., Ebel, J. and Grisebach, H. (1974), *Eur. J. Biochem.* **50**, 135.

Wong, E. (1975). In *The Flavonoids* (eds. J. B. Harborne, T. J. Mabry and H. Mabry), Chapman and Hall, London, p. 743.

Woodward, M. D. (1979a), *Phytochemistry* **18**, 363.

Woodward, M. D. (1979b), *Phytochemistry* **18**, 2007.

Woodward, M. D. (1980a), *Phytochemistry* **19**, 921.

Woodward, M. D. (1980b), *Physiol. Plant Pathol.* **17**, 17.

Wyman, J. G. and Van Etten, H. D. (1978), *Phytopathology* **68**, 583.

Zähringer, U., Ebel, J., Mulheirn, L. J., Lyne, R. L. and Grisebach, H. (1979), *FEBS Lett.* **101**, 90.

CHECK LIST OF KNOWN NATURAL ISOFLAVONOID AGLYCONES

Isoflavones

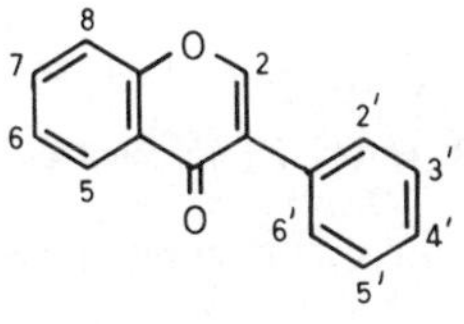

1 Daidzein, 7,4′-$(OH)_2$
2 Formononetin, 7-OH,4′-OMe
3 Isoformononetin, 4′-OH, 7-OMe
4 Dimethyldaidzein, 7,4′-$(OMe)_2$
5 Durlettone, 7-OMe,4′-$OCH_2CH=CMe_2$
*6 7,2′,4′-$(OH)_3$
*7 2′-Hydroxyformononetin, 7,2′-$(OH)_2$, 4′-OMe
*8 Theralin, 7,4′-$(OH)_2$,2′-OMe
9 3′-Hydroxydaidzein, 7,3′,4′-$(OH)_3$
10 7,4′-$(OH)_2$,3′-OMe
11. 3′-Hydroxyformononetin, calycosin, 7,3′-$(OH)_2$,4′-OMe
12 Sayanedin, 4′-OH,7,3′-$(OMe)_2$
13 Cladrin, 7-OH,3′,4′-$(OMe)_2$
14 Cabreuvin, 7,3′,4′-$(OMe)_3$
15 Pseudobaptigenin, 7-OH,3′,4′-OCH_2O-
*16 7-OMe, 3′,4′-OCH_2O-
17 Maximaisoflavone B, 7-$OCH_2CH=CMe_2$, 3′,4′-OCH_2O-
*18 Koparin, 7,2′,3′-$(OH)_3$, 4′-OMe
*19 2′-OH, 7,3′,4′-$(OMe)_3$
*20 Glyzaglabrin, 7,2′-$(OH)_2$, 3′,4′-OCH_2O-
21 Maximaisoflavone C, 7-$OCH_2CH=CMe_2$, 2′-OMe, 4′,5′-OCH_2O-
*22 7,2′,4′,5-$(OMe)_4$
*23 7,2′-$(OMe)_2$, 4′,5′-OCH_2O-
24 Baptigenin, 7,3′,4′,5′-$(OH)_4$
*25 Gliricidin, 7,3′,5′-$(OH)_3$, 4′-OMe
*26 5-OH,7-OMe
*27 5,7-$(OMe)_2$
*28 5,7,2′-$(OH)_3$
29 Genistein, 5,7,4′-$(OH)_3$
30 Biochanin A, 5,7-$(OH)_2$, 4′-OMe
31 5-Methylgenistein, 7,4′-$(OH)_2$, 5-OMe
32 Prunetin, 5,4′-$(OH)_2$, 7-OMe
*33 5-Methylbiochanin A, 7-OH, 5,4′-$(OMe)_2$
*34 7-Methylbiochanin A, 5-OH, 7,4′-$(OMe)_2$
*35 2′-Hydroxygenistein, 5,7,2′,4′-$(OH)_4$
*36 2′-Hydroxybiochanin A, 5,7,2′-$(OH)_3$, 4′-OMe
*37 Cajanin, 5,2′,4′-$(OH)_3$, 7-OMe
*38 5,2′-$(OH)_2$, 7,4′-$(OMe)_2$ (?)

*39 2′-Methoxybiochanin A, 5,7-$(OH)_2$, 2′,4′-$(OMe)_2$
40 Orobol, 5,7,3′,4′-$(OH)_4$
41 3′-Methylorobol, 5,7,4′-$(OH)_3$, 3′-OMe
42 Pratensein, 5,7,3′-$(OH)_3$, 4′-OMe
43 Santal, 5,3′,4′-$(OH)_3$, 7-OMe
*44 Glabrescione B, 5,7-$(OMe)_2$, 3′,4′-$(OCH_2CH=CMe_2)_2$
*45 5,7-$(OH)_2$, 3′,4′-OCH_2O-
46 Derrustone, 5,7-$(OMe)_2$, 3′,4′-OCH_2O-
*47 Derrugenin, 5,4′-$(OH)_2$, 7,2′,5′-$(OMe)_3$
*48 Robustigenin, 5-OH, 7,2′,4′,5′-$(OMe)_4$
*49 5,7,3′,5′-$(OH)_4$, 4′-OMe
50 6,7,4′-$(OH)_3$
51 Texasin, 6,7-$(OH)_2$, 4′-OMe
52 Glycitein, 7,4′-$(OH)_2$, 6-OMe
*53 Kakkatin, 6,4′-$(OH)_2$, 7-OMe
54 Afrormosin, 7-OH, 6,4′-$(OMe)_2$
*55 Odoratin, 7,3′-$(OH)_2$,6,4′-$(OMe)_2$
56 Cladrastin, 7-OH, 6,3′,4′-$(OMe)_3$
57 6,7,3′,4′-$(OMe)_4$
58 Fujikinetin, 7-OH, 6-OMe, 3′,4′-OCH_2O-
59 6,7-$(OMe)_2$, 3′,4′-OCH_2O-
*60 6,7-OCH_2O-, 3′,4′-$(OMe)_2$
61 6,7,2′,3′,4′-$(OMe)_5$
*62 7-OH, 6,2′,4′,5′-$(OMe)_4$
63 6,7,2′,4′,5′-$(OMe)_5$
64 Dalpatein, 7-OH, 6,2′-$(OMe)_2$, 4′,5′-OCH_2O-
65 6-OH, 7,2′-$(OMe)_2$, 4′,5′-OCH_2O-
66 Milldurone, 6,7,2′-$(OMe)_3$, 4′,5′-OCH_2O-
*67 6,7,3′,4′,5′-$(OMe)_5$
68 6,7,3′-$(OMe)_3$, 4′,5′-OCH_2O-
69 Retusin, 7,8-$(OH)_2$, 4′-OMe
70 8-Methylretusin, 7-OH, 8,4′-$(OMe)_2$
*71 8-Methyl-3′-hydroxyretusin, 7,3′-$(OH)_2$, 8,4′-$(OMe)_2$
*72 7-OH, 8,3′,4′-$(OMe)_3$
73 Maximaisoflavone A, 7,8; 3′,4′-$(-OCH_2O-)_2$
*74 7,8,2′-$(OMe)_3$, 4′,5′-OCH_2O-
75 Betavulgarin, 2′-OH, 5-OMe, 6,7-OCH_2O-
76 Tlatlancuayin, 5,2′-$(OMe)_2$, 6,7-OCH_2O-
77 6-Hydroxygenistein, 5,6,7,4′-$(OH)_4$
78 Tectorigenin, 5,7,4′-$(OH)_3$, 6-OMe
79 7-Methyltectorigenin, 5,4′-$(OH)_2$, 6,7-$(OMe)_2$
80 Muningin, 6,4′-$(OH)_2$, 5,7-$(OMe)_2$
81 Irisolidone, 5,7-$(OH)_2$, 6,4′-$(OMe)_2$
*82 5-Methoxyafrormosin, 7-OH, 5,6,4′-$(OMe)_3$
83 7,4′-Dimethyltectorigenin, 5-OH, 6,7,4′-$(OMe)_3$
84 Irilone, 5,4′-$(OH)_2$, 6,7-OCH_2O-
85 Irisolone, 4′-OH, 5-OMe, 6,7-OCH_2O-
*86 Irisolone methyl ether, 5,4′-$(OMe)_2$, 6,7-OCH_2O-
87 Podospicatin, 5,7,2′-$(OH)_3$, 6,5′-$(OMe)_2$
88 Iristectorigenin A, 5,7,3′-$(OH)_3$, 6,4′-$(OMe)_2$
*89 Iristectorigenin B, 5,7,4′-$(OH)_3$, 6,3′-$(OMe)_2$
*90 7-OH, 5,6-$(OMe)_2$, 3′,4′-OCH_2O-
91 5,6,7-$(OMe)_3$, 3′,4′-OCH_2O-
92 Iriflogenin, 5,4′-$(OH)_2$, 3′-OMe, 6,7-OCH_2O-
*93 Iriskumaonin, 3′-OH, 5,4′-$(OMe)_2$, 6,7-OCH_2O-
94 5,3′,4′-$(OMe)_3$, 6,7-OCH_2O-
95 Caviunin, 5,7-$(OH)_2$, 6,2′,4′,5′-$(OMe)_4$
96 Irigenin, 5,7,5′-$(OH)_3$, 6,3′,4′-$(OMe)_3$
97 Irisflorentin, 5,3′,4′,5′-$(OMe)_4$, 6,7-OCH_2O-
*98 8-Hydroxygenistein, 5,7,8,4′-$(OH)_4$
*99 Isotectorigenin, pseudotectorigenin, 5,7,4′-$(OH)_3$, 8-OMe
*100 Aurmillone, 5,7-$(OH)_2$, 8-OMe, 4′-$OCH_2CH=CMe_2$
*101 5,7,3′,4′-$(OH)_4$, 8-OMe
*102 5,7,3′-$(OH)_3$, 8,4′-$(OMe)_2$
*103 5,7,4′-$(OH)_3$, 8,3′-$(OMe)_2$
*104 Platycarpanetin, 7-OH, 5,8-$(OMe)_2$, 3′,4′-OCH_2O-(?)
*105 Isocaviunin, 5,7-$(OH)_2$, 8,2′,4′,5′-$(OMe)_4$
*106 Dipteryxin, 7,8-$(OH)_2$, 6,4′-$(OMe)_2$
*107 8,3′-$(OH)_2$, 6,7,4′-$(OMe)_3$
*108 Petalostetin, 6,7,8-$(OMe)_3$, 3′,4′-OCH_2O-
*109 8,3′,4′-$(OMe)_3$, 6,7- OCH_2O-
110 5,6,7,8-$(OMe)_4$, 3′,4′-OCH_2O-
*111 6-Chlorogenistein, 5,7,4′-$(OH)_3$, 6-Cl
*112 6,3′-Dichlorogenistein, 5,7,4′-$(OH)_3$, 6,3′-Cl_2
*113 Corylinal, 7,4′-$(OH)_2$, 3′-CHO
*114 Wighteone, erythrinin B, 5,7,4′-$(OH)_3$, 6-$CH_2CH=CMe_2$
*115 Luteone, 5,7,2′,4′-$(OH)_4$, 6-$CH_2CH=CMe_2$
116 Derrubone, 5,7-$(OH)_2$, 3′,4′-OCH_2O-, 6-$CH_2CH=CMe_2$
*117 2,3-Dehydrokievitone, 5,7,2′,4′-$(OH)_4$, 8-$CH_2CH=CMe_2$

*118 Erythrinin C, 5,4′-$(OH)_2$, 6,7-$CH_2CH(COHMe_2)O$-
*119 Glabrescione A, 5-OMe, 3′,4′-OCH_2O-, 6,7-$CH_2CH(CMe=CH_2)O$-
120 Dehydroneotenone, 2′-OMe, 4′,5′-OCH_2O-, 6,7-CH=CH-O-
*121 Erythrinin A, 4′-OH, 6,7-CH=CH-CMe_2O-
122 Alpinumisoflavone, 5,4′-$(OH)_2$, 6,7-CH=$CHCMe_2O$-
*123 Alpinumisoflavone-4′-methyl ether, 5-OH, 4′-OMe, 6,7-CH=$CHCMe_2O$-
*124 Alpinumisoflavone dimethyl ether, 5,4′-$(OMe)_2$, 6,7-CH=$CHCMe_2O$-
125 Parvisoflavone B, 5,2′,4′-$(OH)_3$, 6,7-CH=$CHCMe_2O$-
126 Isoauriculatin, 5,2′-$(OH)_2$, 4′-OCH_2CH=CMe_2, 6,7-CH=$CHCMe_2O$-
*127 Isoauriculasin, 5,3′-$(OH)_2$, 4′-OCH_2CH=CMe_2 6,7-CH=$CHCMe_2O$-
128 Robustone, 5-OH, 3′,4′-OCH_2O-, 6,7-CH=$CHCMe_2O$-
129 Robustone methyl ether, 5-OMe, 3′,4′-OCH_2O-, 6,7-CH=$CHCMe_2O$-
*130 Elongatin, 5,4′-$(OH)_2$,2′,5′-$(OMe)_2$,6,7-CH=$CHCMe_2O$-
*131 Calopogonium isoflavone A, 4′-OMe, 7,8-$OCMe_2CH$=CH-
*132 Calopogonium isoflavone B, 3′,4′-OCH_2O-, 7,8-$OCMe_2CH$=CH-
*133 Barbigerone, 2′,4′,5′-$(OMe)_3$, 7,8-$OCMe_2CH$=CH-
134 Jamaicin, 2′-OMe, 4′,5′-OCH_2O-, 7,8-$OCMe_2CH$=CH-
135 Ferrugone, 2′,5′-$(OMe)_2$, 3′,4′-OCH_2O-, 7,8-$OCMe_2CH$=CH-
*136 Derrone, 5,4′-$(OH)_2$, 7,8-$OCMe_2CH$=CH-
*137 Derrone-4′-methyl ether, 5-OH, 4′-OMe, 7,8-$OCMe_2CH$=CH-
138 Parvisoflavone A, 5,2′,4′-$(OH)_3$, 7,8-$OCMe_2CH$=CH-
139 Toxicarol isoflavone, 5-OH, 2′,4′,5′-$(OMe)_3$, 7,8-$OCMe_2CH$=CH-
140 Durmillone, 6-OMe, 3′,4′-OCH_2O-, 7,8-$OCMe_2CH$=CH-
141 Ichthynone, 6,2′-$(OMe)_2$, 4′5′-OCH_2O-, 7,8-$OCMe_2CH$=CH-
142 Neobavaisoflavone, 7,4′-$(OH)_2$, 3′-CH_2CH=CMe_2
*143 Corylin, 7-OH, 3′,4′-CH=$CHCMe_2O$-
*144 Psoralenol, 7-OH, 3′,4′-$CH_2CHOHCMe_2O$-
*145 Glabrone, 7,2′-$(OH)_2$, 3′,4′-CH=$CHCMe_2O$-
146 Licoricone, 7,6′-$(OH)_2$, 2′,4′-$(OMe)_2$, 3′-CH_2CH=CMe_2
*147 Licoisoflavone A, 5,7,2′,4′-$(OH)_4$, 3′-CH_2CH=CMe_2
*148 Licoisoflavone B, 5,7,2′-$(OH)_3$, 3′,4′-CH=$CHCMe_2O$-
149 Piscerythrone, 5,7,2′,4′-$(OH)_4$, 5′-OMe, 3′-CH_2CH=CMe_2
150 Piscidone, 5,7,3′,4′-$(OH)_4$, 6′-OMe, 2′-CH_2CH=CMe_2 *or* 5,7,3′,6′-$(OH)_4$, 4′-OMe, 2′CH_2CH=CMe_2(?)
*151 6,8-Di(dimethylallyl)genistein, 5,7,4′-$(OH)_3$, 6,8-$(CH_2CH=CMe_2)_2$
*152 6,8-Di(dimethylallyl)orobol, 5,7,3′,4′-$(OH)_4$, 6,8-$(CH_2CH=CMe_2)_2$
153 Osajin, 5,4′-$(OH)_2$, 6-CH_2CH=CMe_2, 7,8-$OCMe_2CH$=CH-
154 Scandinone, nallanin, 4′-OH, 5-OMe, 6-CH_2CH=CMe_2, 7,8-$OCMe_2CH$=CH-
155 Pomiferin, 5,3′,4′-$(OH)_3$, 6-CH_2CH=CMe_2, 7,8-$OCMe_2CH$=CH-
156 Auriculin, 5,2′-$(OH)_2$, 4′-OMe, 8-CH_2CH=CMe_2, 6,7-CH=$CHCMe_2O$-
157 Scandenone, warangalone, 5,4′-$(OH)_2$, 8-CH_2CH=CMe_2, 6,7-CH=$CHCMe_2O$-
158 Auriculatin, 5,2′,4′-$(OH)_3$, 8-CH_2CH=CMe_2, 6,7-CH=$CHCMe_2O$-
*159 Auriculasin, 5,3′,4′-$(OH)_3$, 8-CH_2CH=CMe_2, 6,7-CH=$CHCMe_2O$-
*160 Cajaisoflavone, 2′,4′-$(OH)_2$, 6′-OMe, 3′-CH_2CH=CMe_2, 6,7-CH=$CHCMe_2O$-
161 Munetone, 2′-OMe, 6,7; 3′,4′-$(-CH=CHCMe_2O-)_2$
162 Mundulone, 2′-OMe, 6,7-CH_2-$CHOHCMe_2O$-, 3′,4′-CH=$CHCMe_2O$-
*163 6,3′-Di(dimethylallyl)genistein, 5,7,4′-$(OH)_3$, 6,3′-$(CH_2CH=CMe_2)_2$
164 Chandalone, 5,4′-$(OH)_2$, 3′-CH_2-CH=CMe_2, 6,7-CH=$CHCMe_2O$-
*165 7-OH, 2-Me
*166 7-OMe, 2-Me
*167 7-OCOMe, 2-Me
*168 Glyzarin, 7-OH, 8-COMe, 2-Me

Isoflavanones

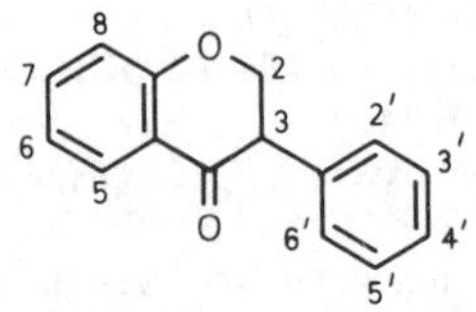

*1 7,4'-$(OH)_2$
*2 7-OH,4'-OMe
*3 7,2',4'-$(OH)_3$
*4 Vestitone, 7,2'-$(OH)_2$, 4'-OMe
5 Sativanone, 7-OH, 2',4'-$(OMe)_2$
*6 7,3'-$(OH)_2$, 4'-OMe
7 Violanone, 7,3'-$(OH)_2$, 2',4'-$(OMe)_2$
*8 3'-Methylviolanone, 7-OH, 2',3',4'-$(OMe)_3$
*9 2'-OH, 7,3',4'-$(OMe)_3$
10 Sophorol, 7,2'-$(OH)_2$, 4',5'-OCH_2O-
11 Onogenin, 7-OH, 2'-OMe, 4',5'-OCH_2O-
12 5,7-$(OH)_2$, 4'-OMe
13 Padmakastein, 5,4'-$(OH)_2$, 7-OMe
14 Dalbergioidin, 5,7,2',4'-$(OH)_4$
15 Ferreirin, 5,7,2'-$(OH)_3$, 4'-OMe
16 Homoferreirin, 5,7-$(OH)_2$, 2',4'-$(OMe)_2$
*17 Cajanol, 5,4'-$(OH)_2$, 7,2'-$(OMe)_2$
18 Parvisoflavanone, 5,7,4'-$(OH)_3$,2',3'-$(OMe)_2$
19 6,7-$(OMe)_2$, 3',4'-OCH_2O-
20 Ougenin, 5,2',4'-$(OH)_3$, 7,3'-$(OMe)_2$, 6-Me
*21 Lespedeol A, 5,7,2',4'-$(OH)_4$, 6-CH_2CH $=CMeCH_2CH_2CH=CMe_2$
*22 Lespedeol B, 5,2',4'-$(OH)_3$, 6,7-CH $=CHCMe(CH_2CH_2CH=CMe_2)O$-
*23 Neoraunone, 2',4'-$(OMe)_2$, 6,7-CH $=CHO$-
24 Nepseudin, 2',3',4'-$(OMe)_3$ 6,7-CH $=CHO$-
*25 Ambonone, 2',4',5'-$(OMe)_3$, 6,7-CH $=CHO$-
26 Neotenone, 2'-OMe, 4',5'-OCH_2O-, 6,7-$CH=CHO$-
*27 Erosenone, 2'-OH, 6'-OMe, 3',4'-OCH_2O-, 6,7-$CH=CHO$-(?)
*28 5-Deoxykievitone, 7,2',4'-$(OH)_3$, 8-$CH_2CH=CMe_2$
29 Kievitone, 5,7,2',4'-$(OH)_4$, 8-CH_2 $CH=CMe_2$
*30 Kievitone hydrate, 5,7,2',4'-$(OH)_4$, 8-$CH_2CH_2CMe_2OH$ (Metabolite)
*31 1'',2''-Dehydrocyclokievitone, 5,2',4'-$(OH)_3$, 7,8-$OCMe_2CH=CH$-
*32 Sophoraisoflavanone A, 5,7,4'-$(OH)_3$, 2'-OMe, 3'-$CH_2CH=CMe_2$
*33 Licoisoflavanone, 5,7,2'-$(OH)_3$, 3',4'-$CH=CHCMe_2O$-
*34 Isosophoronol, 5,7-$(OH)_2$, 2'-OMe, 3',4'-$CH=CHCMe_2O$-
35 5,7,2'-$(OH)_3$, 4',5'-$OCH(CMe_2OH)CH_2$-
*36 Isosophoranone, 5,7,4'-$(OH)_3$, 2'OMe, 6,3'-$(CH_2CH=CMe_2)_2$
*37 Cajanone, 5,2',4'-$(OH)_3$, 5'-CH_2CH $=CMe_2$, 6,7-$CH=CHCMe_2O$-
*38 2'-Methylcajanone, 5,4'-$(OH)_2$, 2'-OMe, 5'-$CH_2CH=CMe_2$, 6,7-CH $=CHCMe_2O$-
*39 Secondifloran, 7,3,2',3'-$(OH)_4$, 4'-OMe, 5'-$CMe_2CH=CH_2$

Rotenoids

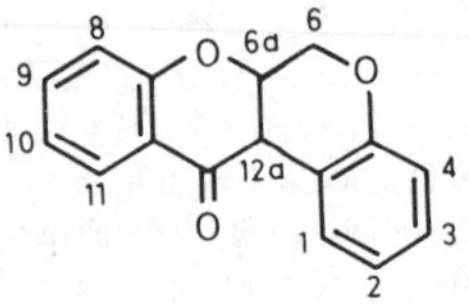

(i) *Rotenoids*
1 Munduserone, 2,3,9-$(OMe)_3$
2 Sermundone, 11-OH, 2,3,9-$(OMe)_3$
3 Rotenone, 2,3-$(OMe)_2$, 8,9-$CH_2CH(CMe$ $=CH_2)O$-
4 Isomillettone, 2,3-OCH_2O-, 8,9-$CH_2CH(CMe=CH_2)O$-
5 Dalpanol, 2,3-$(OMe)_2$, 8,9-$CH_2CH(CMe_2OH)O$-
6 Amorphigenin, 2,3-$(OMe)_2$, 8,9-$CH_2CH[C(CH_2OH)=CH_2]O$-
7 Dihydroamorphigenin, 2,3-$(OMe)_2$, 8,9-$CH_2CH(CMeCH_2OH)O$-
8 Amorphigenol, 2,3-$(OMe)_2$, 8,9-$CH_2CH[CMe(OH)CH_2OH]O$-
9 Deguelin, 2,3-$(OMe)_2$, 8,9-CH $=CHCMe_2O$-
10 Millettone, 2,3-OCH_2O-, 8,9-$CH=CHCMe_2O$-
11 Myriconol, 9-OH, 2,3-$(OMe)_2$, 10-$CH=CHCH_2CH=CH_2$(?)
12 Elliptone, 2,3-$(OMe)_2$, 8,9-$CH=CHO$-
13 Erosone, 2,3-$(OMe)_2$, 9,10-$OCH=CH$-
14 Dolineone, 2,3-OCH_2O, 9,10-$OCH=CH$-
15 Sumatrol, 11-OH, 2,3-$(OMe)_2$, 8,9-$CH_2CH(CMe=CH_2)O$-

16 Toxicarol, 11-OH, 2,3-$(OMe)_2$, 8,9-CH $=CHCMe_2O$-
17 Malaccol, 11-OH, 2,3-$(OMe)_2$, 8,9-CH =CHO-
18 Pachyrrhizone, 8-OMe, 2,3-OCH_2O-, 9,10-OCH=CH-
19 Villosin, 11,6-$(OH)_2$, 2,3-$(OMe)_2$, 8,9-$CH_2CH(CMe=CH_2)O$-

(ii) *Dehydrorotenoids*

1 Dehydrorotenone, 6a,12a-dehydro, 2,3-$(OMe)_2$, 8,9-$CH_2CH(CMe=CH_2)O$-
*2 Dehydrodalpanol, 6a,12a-dehydro, 2,3-$(OMe)_2$, 8,9-$CH_2CH(CMe_2OH)O$-
3 Dehydrodeguelin, 6a,12a-dehydro, 2,3-$(OMe)_2$, 8,9-$CH=CHCMe_2O$-
4 Dehydromillettone, 6a,12a-dehydro, 2,3-OCH_2O-, 8,9-$CH=CHCMe_2O$-
5 Dehydrodolineone, 6a,12a-dehydro, 2,3-OCH_2O-, 9,10-OCH=CH-
*6 Villosol, dehydrosumatrol, 6a,12a-dehydro, 11-OH, 2,3-$(OMe)_2$, 8,9-$CH_2CH(CMe=CH_2)O$-
*7 Dehydrotoxicarol, 6a,12a-dehydro, 11-OH, 2,3-$(OMe)_2$, 8,9-CH $=CHCMe_2O$-
*8 Dehydropachyrrhizone, 6a,12a-dehydro, 8-OMe, 2,3-OCH_2O-, 9,10-OCH=CH-
9 Stemonal, 6a,12a-dehydro, 11,6-$(OH)_2$ 2,3,9-$(OMe)_3$
10 Stemonacetal, 6a,12a-dehydro, 11-OH, 2,3,9-$(OMe)_3$, 6-OEt
*11 Amorpholone, 6a,12a-dehydro, 6-OH, 2,3-$(OMe)_2$, 8,9-$CH_2CH(CMe=CH_2)O$-
*12 Villinol, 6a,12a-dehydro, 11-OH, 2,3,6-$(OMe)_3$, 8,9-$CH_2CH(CMe=CH_2)O$-
13 Stemonone, 6a,12a-dehydro, 6-oxo, 11-OH, 2,3,9-$(OMe)_3$
*14 Rotenonone, 6a,12a-dehydro, 6-oxo, 2,3-$(OMe)_2$, 8,9-$CH_2CH(CMe=CH_2)O$-
*15 Villosone, 6a,12a-dehydro, 6-oxo, 11-OH, 2,3-$(OMe)_2$, 8,9-CH_2CH(CMe $=CH_2)O$-

(iii) *12a-Hydroxyrotenoids*

*1 12a-Hydroxymunduserone, 12a-OH, 2,3,9-$(OMe)_3$
2 12a-Hydroxyrotenone, 12a-OH, 2,3-$(OMe)_2$, 8,9-$CH_2CH(CMe=CH_2)O$-
*3 12a-Hydroxyisomillettone, 12a-OH, 2,3-OCH_2O-, 8,9-$CH_2CH(CMe=CH_2)O$-
*4 Dalbinol, 12a-hydroxyamorphigenin, 12a-OH, 2,3-$(OMe)_2$ 8,9-$CH_2CH[C(CH_2OH)=CH_2]O$-
5 Tephrosin, 12a-OH, 2,3-$(OMe)_2$, 8,9-$CH=CHCMe_2O$-
6 Millettosin, 12a-OH, 2,3-OCH_2O-,8,9-CH $=CHCMe_2O$-
*7 12a-Hydroxyerosone, 12a-OH, 2,3-$(OMe)_2$, 9,10-OCH=CH-
*8 12a-Hydroxydolineone, 12a-OH, 2,3-OCH_2O-, 9,10-OCH=CH-
*9 Villosinol, 12a-hydroxysumatrol, 11,12a-$(OH)_2$, 2,3-$(OMe)_2$, 8,9-CH_2CH(CMe $=CH_2)O$-
*10 11-Hydroxytephrosin, 11,12a-$(OH)_2$, 2,3-$(OMe)_2$, 8,9-$CH=CHCMe_2O$-
*11 12a-Hydroxypachyrrhizone, 12a-OH, 8-OMe, 2,3-OCH_2O-, 9,10-OCH=CH-
12 12a-Methoxyrotenone, 2,3,12a-$(OMe)_3$, 8,9-$CH_2CH(CMe=CH_2)O$-
*13 Neobanone, 2,3,12a-$(OMe)_3$, 9,10-OCH =CH-
14 12a-Methoxydolineone, 12a-OMe, 2,3-OCH_2O-, 9,10-OCH=CH-
*15 Clitoriacetal, 6,11,12a-$(OH)_3$, 2,3,9-$(OMe)_3$
*16 Villol, 6,11,12a-$(OH)_3$, 2,3-$(OMe)_2$, 8,9-$CH_2CH(CMe=CH_2)O$-

Pterocarpans

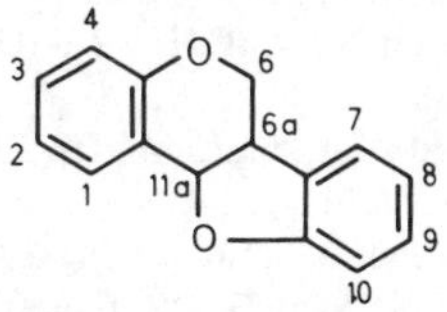

(i) *Pterocarpans*

*1 Demethylmedicarpin, 3,9-$(OH)_2$
2 Medicarpin, demethylhomopterocarpin, 3-OH, 9-OMe
*3 Isomedicarpin, 9-OH, 3-OMe
4 Homopterocarpin, 3,9-$(OMe)_2$
5 Maackiain, 3-OH, 8,9-OCH_2O-
6 Pterocarpin, 3-OMe, 8,9-OCH_2O-
*7 Vesticarpan, 3,10-$(OH)_2$, 9-OMe
*8 Nissolin, 3,9-$(OH)_2$, 10-OMe
*9 Methylnissolin, 3-OH, 9,10-$(OMe)_2$
10 Philenopteran, 3,9-$(OH)_2$, 7,10-$(OMe)_2$
11 9-Methylphilenopteran, 3-OH, 7,9,10-$(OMe)_3$
*12 3-OH, 2,9-$(OMe)_2$
*13 Sparticarpin, 9-OH, 2,3-$(OMe)_2$
*14 2,3,9-$(OMe)_3$
*15 2,8-$(OH)_2$, 3,9-$(OMe)_2$

16 2-Hydroxypterocarpin, 2-OH, 3-OMe, 8,9-OCH_2O-
17 2-Methoxypterocarpin, 2,3-$(OMe)_2$, 8,9-OCH_2O-
*18 Mucronucarpan, 2,10-$(OH)_2$, 3,9-$(OMe)_2$
*19 2,8-$(OH)_2$, 3,9,10-$(OMe)_3$
*20 3,4,9-$(OH)_3$
*21 3,4-$(OH)_2$, 9-OMe
*22 4-OH, 3,9-$(OMe)_2$
23 3-OH, 4,9-$(OMe)_2$
24 3,4,9-$(OMe)_3$
25 3,4-$(OH)_2$, 8,9-OCH_2O-
26 3-OH, 4-OMe, 8,9-OCH_2O-
27 4-OH, 3-OMe, 8,9-OCH_2O-
28 3,4-$(OMe)_2$, 8,9-OCH_2O-
*29 8-OH, 3,4,9,10-$(OMe)_4$
*30 Trifolian, 1,3-$(OH)_2$, 2-OMe, 8,9-OCH_2O-
*31 4-OH, 2,3,9-$(OMe)_3$
*32 2,8-$(OH)_2$, 3,4,9,10-$(OMe)_4$
*33 Homoedudiol, 3,9-$(OH)_2$, 2-$CH_2CH=CMe_2$
34 Edunol, 3-OH, 8,9-OCH_2O-, 2-CH_2 $CH=CMe_2$
*35 Neorautenol, 9-OH, 2,3-CH $=CHCMe_2O$-
*36 Neorautenane, 8,9-OCH_2O-, 2,3-$CH=CHCMe_2O$
37 Neorautane, 8,9-OCH_2O-, 2,3-$CH_2CH_2CMe_2O$-
*38 Neorautanol, 8,9-OCH_2O-, 2,3-$CH_2CHOHCMe_2O$-
*39 Neodunol, 9-OH, 2,3-$CH=CHO$-
40 Neodulin, 8,9-OCH_2O, 2,3-$CH=CHO$-
*41 Ambonane, 9,10-$(OMe)_2$, 2,3-CH $=CHO$-
*42 Hemileiocarpin, 9-OMe, 3,4-$OCMe_2CH$ $=CH$-
43 Leiocarpin, 8,9-OCH_2O, 3,4-$OCMe_2CH$ $=CH$-
*44 Nitiducol, 3-OH, 8,9-OCH_2O-, 4-CH_2CH $=CMeCH_2CH_2CH=CMe_2$
*45 Nitiducarpin, 8,9-OCH_2O-, 3,4-$OCMe(CH_2CH_2CH=CMe_2)CH$ $=CH$-
*46 Edudiol, 3,9-$(OH)_2$, 1-OMe, 2-CH_2CH $=CMe_2$
47 Edulenol, 3-OH, 1,9-$(OMe)_2$, 2-CH_2CH $=CMe_2$
*48 Edulenanol, 9-OH, 1-OMe, 2,3-CH $=CHCMe_2O$-
*49 Edulenane, 1,9-$(OMe)_2$, 2,3-CH $=CHCMe_2O$-
*50 Desmodin, 9-OH, 1,8-$(OMe)_2$, 2,3-CH $=CHCMe_2O$-
*51 Neorautenanol, 1-OH, 8,9-OCH_2O-, 2,3-$CH=CHCMe_2O$-
*52 Edulane, 1,9-$(OMe)_2$, 2,3-$CH_2CH_2CMe_2O$-
*53 Neorautanin, 1-OMe, 8,9-OCH_2O-, 2,3-$CH_2CH_2CMe_2O$-
*54 Neoraucarpanol, 3-OH, 4-OMe, 8,9-OCH_2O-, 2-$CH_2CH=CMe_2$
*55 Neoraucarpan, 3,4-$(OMe)_2$, 8,9-OCH_2O-, 2-$CH_2CH=CMe_2$
56 Ficinin, 4-OMe, 8,9-OCH_2O-, 2,3-CH $=CHO$-
57 Phaseollidin, 3,9-$(OH)_2$, 10-CH_2CH $=CMe_2$
*58 Sandwicensin, 9-methylphaseollidin, 3-OH, 9-OMe, 10-$CH_2CH=CMe_2$
*59 Phaseollidin hydrate, 3,9-$(OH)_2$, 10-$CH_2CH_2CMe_2OH$ (Metabolite)
60 Phaseollin, 3-OH, 9,10-$OCMe_2CH=CH$-
*61 3-OH, 9,10-$OCMe_2CHOH$ CHOH- (Metabolite)
*62 Lespedezin, 3,9-$(OH)_2$, 10-CH_2CH $=CMeCH_2CH_2CH=CMe_2$
*63 1-Methoxyphaseollidin, 3,9-$(OH)_2$, 1-OMe, 10-$CH_2CH=CMe_2$
64 Ficifolinol, 3,9-$(OH)_2$, 2,8-$(CH_2CH$ $=CMe_2)_2$
65 Folitenol, 3-OH, 2-$CH_2CH=CMe_2$, 9,10-$OCMe_2CH=CH$-
66 Folinin, 2,3-$CH_2CH_2CMe_2O$-, 9,10-$OCMe_2CH=CH$-
67 Gangetin, 9-OH, 1-OMe, 10-CH_2CH $=CMe_2$, 2,3-$CH=CHCMe_2O$-
*68 Gangetinin, 1-OMe, 2,3; 10,9-(-CH $=CHCMe_2O$-$)_2$
*69 Lespein, 3,9-$(OH)_2$, 6a, 10-$(CH_2CH$ $=CMe_2)_2$
*70 la-Hydroxyphaseollone (Metabolite)

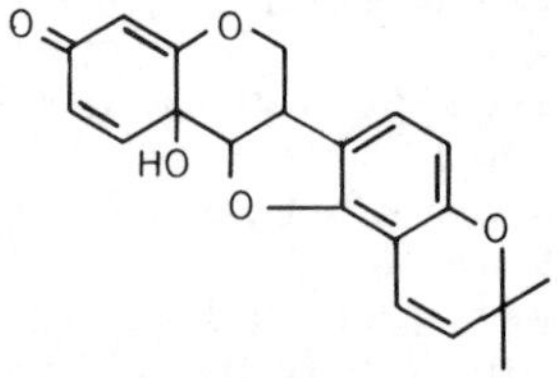

(ii) *6a-Hydroxypterocarpans*
*1 3,9,6a-$(OH)_3$
*2 6a-Hydroxymedicarpin, 3,6a-$(OH)_2$, 9-OMe

*3 6a-Hydroxyisomedicarpin, 9,6a-$(OH)_2$, 3-OMe
4 Variabilin, homopisatin, 6a-OH, 3,9-$(OMe)_2$
*5 6a-Hydroxymaackiain, 3,6a-$(OH)_2$, 8,9-OCH_2O-
6 Pisatin, 6a-OH, 3-OMe, 8,9-OCH_2O-
*7 3,7,6a-$(OH)_3$, 9-OMe
*8 3,7,6a-$(OH)_3$, 8,9-OCH_2O-
*9 Acanthocarpan, 6a-OH, 3-OMe, 8,9-OCH_2O-, 2 or 4-Me(?)
*10 Glyceollin IV, 9,6a-$(OH)_2$, 3-OMe, 2-$CH_2CH = CMe_2$
*11 Glyceollin II, 9,6a-$(OH)_2$, 2,3-CH $= CHCMe_2O$-
*12 Glyceollin III, 9,6a-$(OH)_2$, 2,3-$CH_2CH(CMe = CH_2)O$-
*13 Neobanol, 6a-OH, 8,9-OCH_2O-, 2,3-CH = CHO-
*14 Glyceollin I, 9,6a-$(OH)_2$, 3,4-$OCMe_2CH$ = CH-
15 Tuberosin, 3,6a-(OH_2), 8,9-CH $= CHCMe_2O$-
*16 Sandwicarpin, 6a-hydroxyphaseollidin, 3,9,6a-$(OH)_3$, 10-$CH_2CH = CMe_2$
*17 Cristacarpin, 3,6a-$(OH)_2$, 9-OMe, 10-$CH_2CH = CMe_2$
*18 6a-Hydroxyphaseollin, 3,6a-$(OH)_2$, 9,10-$OCMe_2CH = CH$-
*19 3,7,6a-$(OH)_3$, 9,10-$OCMe_2CH = CH$- (Metabolite)

(*iii*) *Pterocarpenes*
*1 6a, 11a-dehydro, 3,9-$(OH)_2$
2 6a, 11a-dehydro, 3,9-$(OMe)_2$
3 Flemichapparin B, anhydropisatin, 6a,11a-dehydro, 3-OMe,8,9-OCH_2O-
*4 Bryacarpene-5, 6a,11a-dehydro, 3,9,10-$(OMe)_3$
*5 Bryacarpene-2, 6a,11a-dehydro, 10-OH, 3,8,9-$(OMe)_3$
*6 Bryacarpene-3, 6a,11a-dehydro, 3,8,9,10-$(OMe)_4$
*7 6a,11a-dehydro, 3-OH, 4-OMe, 8,9-OCH_2O-
*8 Bryacarpene-4, 6a,11a-dehydro, 4-OH, 3,9,10-$(OMe)_3$
*9 Bryacarpene-1, 6a,11a-dehydro,4,10-$(OH)_2$, 3,8,9-$(OMe)_3$
10 Leiocalycin, 6a,11a-dehydro, 2-OH, 1,3-$(OMe)_2$, 8,9-OCH_2O-
*11 Neorauteen, 6a,11a-dehydro, 9-OH, 2,3-CH = CHO-
*12 Neoduleen, 6a, 11a-dehydro, 8,9-OCH_2O-, 2,3-CH = CHO-

Isoflavans

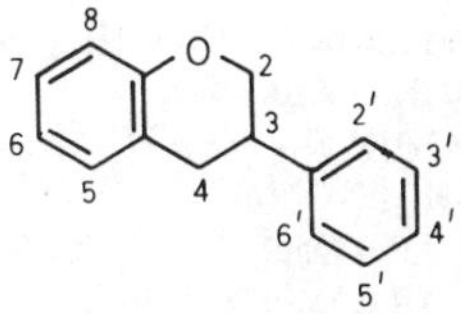

1 Equol, 7,4′-$(OH)_2$ (Metabolite)
*2 Demethylvestitol, 7,2′,4′-$(OH)_3$
3 Vestitol, 7,2′-$(OH)_2$, 4′-OMe
*4 Isovestitol, 7,4′-$(OH)_2$, 2′-OMe
*5 Neovestitol, 2′,4′-$(OH)_2$, 7-OMe
6 Sativan, 7-OH, 2′,4′-$(OMe)_2$
7 Isosativan, 2′-OH, 7,4′-$(OMe)_2$
*8 Arvensan, 4′-OH, 7,2′-$(OMe)_2$
9 Laxifloran, spherosin, 7,4′-$(OH)_2$, 2′,3′-$(OMe)_2$
10 Mucronulatol, 7,3′-$(OH)_2$, 2′,4′-$(OMe)_2$
*11 Isomucronulatol, 7,2′-$(OH)_2$, 3′,4′-$(OMe)_2$
*12 Maackiainisoflavan, 7,2′-$(OH)_2$, 4′,5′-OCH_2O-(Metabolite)
*13 Astraciceran, 7-OH, 2′-OMe, 4′,5′-OCH_2O-
14 Lonchocarpan, 7,4′-$(OH)_2$, 2′,3′,6′-$(OMe)_2$
*15 5-Methoxyvestitol, 7,2′-$(OH)_2$, 5,4′-$(OMe)_2$
*16 7,2′-$(OH)_2$, 6,4′-$(OMe)_2$ (Metabolite)
*17 Bryaflavan, 6,7,3′-$(OH)_3$, 2′,4′-$(OMe)_2$
18 8-Demethylduartin, 7,8,3′-$(OH)_3$, 2′,4′-$(OMe)_2$
19 Duartin, 7,3′-$(OH)_2$, 8,2′,4′-$(OMe)_3$
*20 Spherosinin, 4′-OH, 2′,3′-$(OMe)_2$, 6,7-CH $= CHCMe_2O$-
21 Neorauflavane, 2′,4′-$(OH)_2$, 5-OMe, 6,7-$CH = CHCMe_2O$-
*22 Glabridin, 2′,4′-$(OH)_2$, 7,8-$OCMe_2CH$ = CH-
*23 4′-Methylglabridin, 2′-OH, 4′-OMe, 7,8-$OCMe_2CH = CH$-
*24 3′-Methoxyglabridin, 2′,4′-$(OH)_2$, 3′-OMe, 7,8-$OCMe_2CH = CH$-
*25 Leiocin, 2′-OH, 4′,5′-OCH_2O-, 7,8-$OCMe_2CH = CH$-
*26 Heminitidulan, 2′-OH, 4′-OMe, 7,8-$OCMe(CH_2CH_2CH = CMe_2)CH$ = CH-
*27 Nitidulin, 2′,3′-$(OH)_2$, 4′-OMe, 7,8-$OCMe(CH_2CH_2CH = CMe_2)$ CH = CH-

*28 Nitidulan, 2′-OH, 4′,5′-OCH_2O-, 7,8-$OCMe(CH_2CH_2CH=CMe_2)$ $CH=CH$-
*29 Leiocinol, 6,2′-$(OH)_2$, 4′,5′-OCH_2O-, 7,8-$OCMe_2CH=CH$-
*30 2′-Methylphaseollidinisoflavan, 7,4′-$(OH)_2$, 2′-OMe, 3′-$CH_2CH=CMe_2$
31 Phaseollinisoflavan, 7,2′-$(OH)_2$, 3′,4′-$CH=CHCMe_2O$-
32 2′-Methylphaseollinisoflavan, 7-OH, 2′-OMe, 3′,4′-$CH=CHCMe_2O$-
*33 α,α-Dimethylallylcyclolobin, 7,3′,4′-$(OH)_3$, 2′-OMe, 5′-$CMe_2CH=CH_2$
*34 Unanisoflavan, 7,3′-$(OH)_2$, 2′,4′-$(OMe)_2$, 5′-$CMe_2CH=CH_2$
*35 Hispaglabridin A, 2′,4′-$(OH)_2$, 3′-$CH_2CH=CMe_2$, 7,8-$OCMe_2CH=CH$-
*36 Hispaglabridin B, 2′-OH, 7,8; 4′, 3′-(-$OCMe_2CH=CH$-$)_2$
37 Licoricidin, 5,2′,4′-$(OH)_3$, 6,3′-$(CH_2CH=CMe_2)_2$
*38 3-Hydroxymaackiainisoflavan, 7,3,2′-$(OH)_3$, 4′,5′-OCH_2O- (Metabolite)
*39 Biscyclolobin, (7,3′-$(OH)_2$, 2′-OMe, 4′-$)_2O$ (?)

Isoflavonequinone

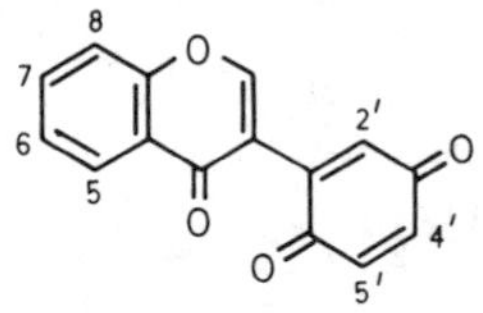

*1 Bowdichione, 7-OH, 4′-OMe

Isoflavanquinones

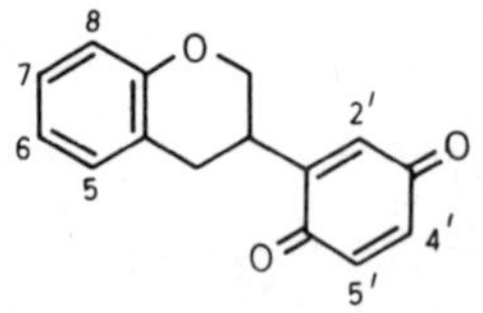

1 Claussequinone, 7-OH, 4′-OMe
*2 Abruquinone A, 6,7,4′,5′-$(OMe)_4$
3 Mucroquinone, 7-OH, 8,4′-$(OMe)_2$
*4 Amorphaquinone, 7-OH, 8,4′,5′-$(OMe)_3$
*5 Abruquinone C, 6-OH, 7,8,4′,5′-$(OMe)_4$
*6 Abruquinone B, 6,7,8,4′,5′-$(OMe)_5$

Pterocarpenequinones

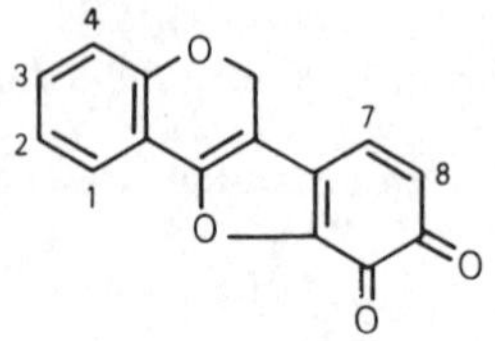

*1 4-Deoxybryaquinone, 3,7-$(OMe)_2$
*2 Bryaquinone, 4-OH, 3,7-$(OMe)_2$

3-*Arylcoumarins*

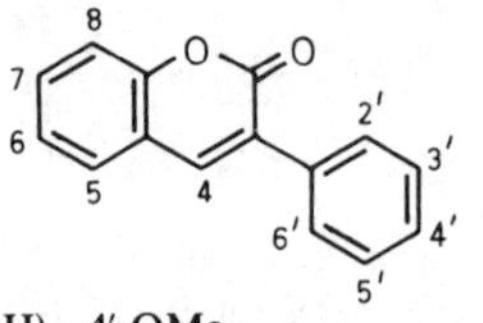

*1 7,2′-$(OH)_2$, 4′-OMe
*2 7,2′-$(OH)_2$, 4′,5′-OCH_2O-
3 Pachyrrhizin, 2′-OMe, 4′,5′-OCH_2O-, 6,7-$CH=CHO$-
*4 Glycyrin, 2′,4′-$(OH)_2$, 5,7-$(OMe)_2$, 6-$CH_2CH=CMe_2$
5 Neofolin, 8,2′-$(OMe)_2$, 4′,5′-OCH_2O-, 6,7-$CH=CHO$-

3-*Aryl-4-hydroxycoumarins*

1 Derrusnin, 4,5,7-$(OMe)_3$, 3′,4′-OCH_2O-
2 Robustic acid, 4-OH, 5,4′-$(OMe)_2$, 6,7-$CH=CHCMe_2O$-
3 Robustic acid methyl ether, 4,5,4′-$(OMe)_3$, 6,7-$CH=CHCMe_2O$-
4 Robustin, 4-OH, 5-OMe, 3′,4′-OCH_2O-, 6,7-$CH=CHCMe_2O$-
5 Robustin methyl ether, 4,5-$(OMe)_2$, 3′,4′-OCH_2O-, 6,7-$CH=CHCMe_2O$-
*6 Glabrescin, 4,5-$(OMe)_2$, 3′,4′-OCH_2O-, 6,7-$CH_2CH(CMe=CH_2)O$-
7 Scandenin, 4,4′-$(OH)_2$, 5-OMe, 6-$CH_2CH=CMe_2$, 7,8-$OCMe_2CH=CH$-
8 Lonchocarpic acid, 4,4′-$(OH)_2$, 5-OMe, 8-$CH_2CH=CMe_2$, 6,7-$CH=CHCMe_2O$-
9 Lonchocarpin, 4-OH, 5,4′-$(OMe)_2$, 8-$CH_2CH=CMe_2$, 6,7-$CH=CHCMe_2O$-

Isoflav-3-enes

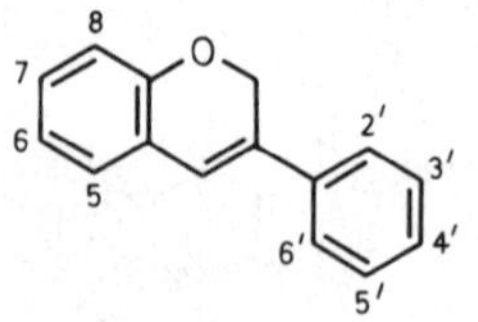

*1 Haginin B, 7,4′-$(OH)_2$, 2′-OMe
*2 Sepiol, 7,2′,3′-$(OH)_3$, 4′-OMe
*3 Haginin A, 7,4′-$(OH)_2$, 2′,3′-$(OMe)_2$
*4 2′-Methylsepiol, 7,3′-$(OH)_2$, 2′,4′-$(OMe)_2$
5 Neorauflavene, 2′,4′-$(OH)_2$, 5-OMe, 6,7-$CH=CHCMe_2O$-
*6 Glabrene, 7,2′-$(OH)_2$, 3′,4′-$CH=CHCMe_2O$-

Coumestans

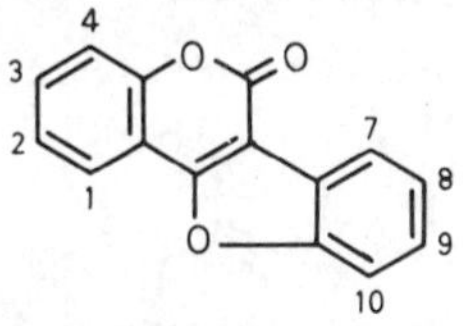

1 Coumestrol, 3,9-$(OH)_2$
2 9-Methylcoumestrol, 3-OH, 9-OMe
3 Dimethylcoumestrol, 3,9-$(OMe)_2$
4 3,9-$(OH)_2$, 8-OMe
5 3-OH, 8,9-$(OMe)_2$
6 Medicagol, 3-OH, 8,9-OCH_2O-
7 Flemichapparin C, 3-OMe, 8,9-OCH_2O-
8 Repensol, 3,7,9-$(OH)_3$
9 Trifoliol, 3,7-$(OH)_2$, 9-OMe
10 Demethylwedelolactone, 1,3,8,9-$(OH)_4$
11 Wedelolactone, 1,8,9-$(OH)_3$,3-OMe
12 Lucernol, 2,3,9-$(OH)_3$
*13 Tephrosol, 3-OH, 2-OMe, 8,9-OCH_2O-
14 2-OH, 3-OMe, 8,9-OCH_2O-
15 Sativol, 4,9-$(OH)_2$, 3-OMe
16 2-OH, 1,3-$(OMe)_2$, 8,9-OCH_2O-
17 Psoralidin, 3,9-$(OH)_2$, 2-$CH_2CH=CMe_2$
*18 Psoralidin oxide, 3,9-$(OH)_2$, 2-CH_2 CH–O–CMe_2
*19 Corylidin, 9-OH, 2,3-$CHOHCHOHCMe_2O$-
20 Erosnin, 8,9-OCH_2O-, 2,3-$CH=CHO$-
21 Glycyrol, 1,9-$(OH)_2$, 3-OMe, 2-$CH_2CH=CMe_2$
22 1-Methylglycyrol, 9-OH, 1,3-$(OMe)_2$, 2-$CH_2CH=CMe_2$
23 Isoglycyrol, 9-OH, 3-OMe, 1,2-$OCMe_2CH_2CH_2$-
24 Sojagol, 3-OH, 9,10-$OCMe_2CH_2CH_2$-

α-Methyldeoxybenzoins

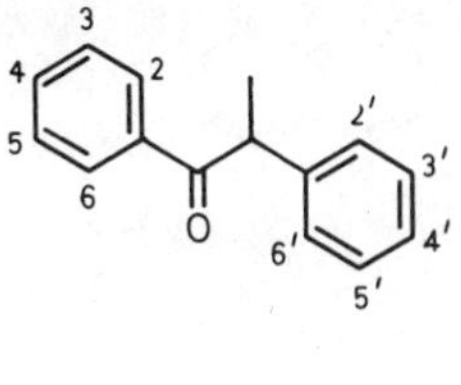

1 Angolensin, 2,4-$(OH)_2$, 4′-OMe
*2 2-Methylangolensin, 4-OH, 2,4′-$(OMe)_2$
*3 4-Methylangolensin, 2-OH, 4,4′-$(OMe)_2$
*4 4-Cadinylangolensin, 2-OH, 4′-OMe, 4-O-$C_{15}H_{25}$

2-Arylbenzofurans

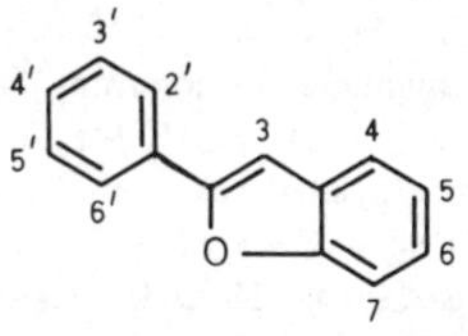

*1 6-Demethylvignafuran, 4′,6-$(OH)_2$, 2′-OMe
*2 Vignafuran, 4′-OH, 2′,6-$(OMe)_2$
3 Pterofuran, 3′,6-$(OH)_2$, 2′,4′-$(OMe)_2$
*4 Isopterofuran, 4′,6-$(OH)_2$, 2′,3′-$(OMe)_2$
*5 2′,4′-$(OH)_2$, 5,6-$(OMe)_2$
*6 2′,4′-$(OH)_2$, 5,6-OCH_2O-
*7 2′-OH, 4′-OMe, 5,6-OCH_2O-
*8 Neoraufurane, 2′,4′-$(OH)_2$, 4-OMe, 5,6-$CH=CHCMe_2O$-
*9 Bryebinal, 3′,4-$(OH)_2$, 2′,4′,7,5 or 6-$(OMe)_4$, 3-CHO(?)

Isoflavanol

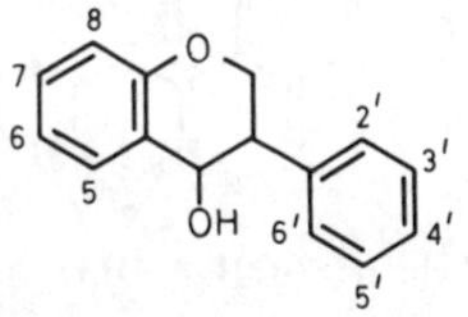

*1 Ambanol, 2′-OMe, 4′,5′-OCH_2O-, 6,7-$CH=CHO$-

Coumaronochromone

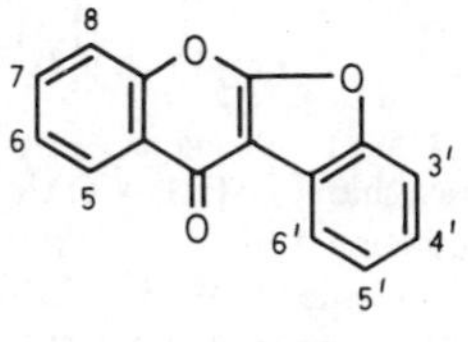

1. Lisetin, 5,7,4′-$(OH)_3$, 5′-OMe, 3′-$CH_2CH=CMe_2$

CHAPTER ELEVEN

Biosynthesis

J. EBEL and K. HAHLBROCK

11.1 INTRODUCTION

Our present knowledge in the field of flavonoid biosynthesis is based on a combination of earlier results from radioactive tracer studies *in vivo* and more recent data obtained at the enzyme level *in vitro*. In the past few years, the enzymology of flavonoid biosynthesis has made particularly rapid progress. It became firmly established that all classes of flavonoids derive their carbon skeleton from compounds of intermediary cell metabolism through the action of two consecutive (general phenylpropanoid and flavonoid glycoside) pathways. Although the first (general phenylpropanoid) pathway is shared with some other biosynthetic routes, it is discussed in some detail in the first part of this chapter, because it may in some cases determine the substitution pattern of the flavonoid end products through structural modifications introduced at the level of the substrates supplied for their synthesis. The second, major part then deals with the reactions of the various flavonoid pathways proper.

11.2 GENERAL PHENYLPROPANOID METABOLISM

Phenylpropanoid units derived from the shikimate pathway are common structural elements of all flavonoid compounds and of various other classes of phenylpropanoids, such as lignin, stilbenes and cinnamate esters. The sequence of reactions converting phenylalanine into the CoA ester derivatives of substituted cinnamic acids was therefore termed 'general phenylpropanoid metabolism' (Ebel *et al.*, 1974). The enzymes catalysing the individual steps are phenylalanine ammonia-lyase, cinnamate 4-hydroxylase and 4-coumarate:CoA ligase (Fig. 11.1, Table 11.1). A close metabolic relationship between these three enzymes is apparent from their interdependent regulation in a number of plant tissues either during developmental changes or in response to changes in the environmental conditions.

Phenylalanine ammonia-lyase catalyses the *anti* elimination of ammonia and the (*pro*-3*S*)-proton from L-phenylalanine to yield *trans*-cinnamic acid. The enzymes from many plants also convert L-tyrosine into *trans*-4-coumarate, albeit with much lower efficiency. There is much evidence that both substrates are of

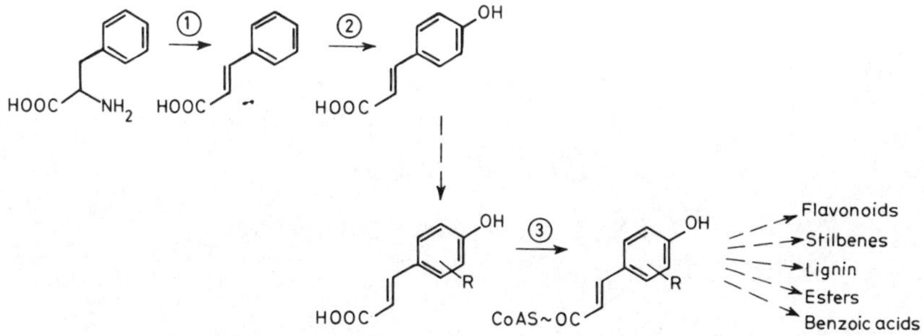

Fig. 11.1 Scheme illustrating the reactions of general phenylpropanoid metabolism. The enzymes marked by numbers are listed in Table 11.1. The CoA esters of 4-coumarate and other substituted cinnamic acids are central intermediates at the branch point between the general metabolism and the various subsequent, specific pathways. The dashed arrow indicates (a) hypothetical 3- and 5-hydroxylase(s) as well as a methyltransferase which is not a typical enzyme of this sequence. R = H, OH, or OCH_3.

physiological importance. Various properties of phenylalanine ammonia-lyases from a large number of sources have been studied, including their widely differing reactivities towards tyrosine. No tyrosine ammonia-lyase has so far

Table 11.1 List of enzymes mentioned in Figures

Enzyme	*EC number*	*Key to Figure*
General phenylpropanoid metabolism		
Phenylalanine ammonia-lyase	4.3.1.5	Fig. 11.1, No. 1
Cinnamate 4-hydroxylase	1.14.13.11	Fig. 11.1, No. 2
4-Coumarate:CoA ligase	6.2.1.12	Fig. 11.1, No. 3
Additional reaction preceding the CoA ligase step		
S-Adenosyl-L-methionine: caffeate 3-*O*-methyltransferase	2.1.1.-	Fig. 11.1
Flavone and flavonol glycoside pathways		
Acetyl-CoA carboxylase	6.4.1.2	Fig. 11.3, No. 1
Chalcone synthase	—	Fig. 11.3, No. 2
Chalcone isomerase	5.5.1.6	Fig. 11.3, No. 3
'Flavonoid 3-hydroxylase'	—	Fig. 11.3, No. 4
'Flavonoid oxidase'	—	Fig. 11.3, No. 5
'Flavonoid 3′-hydroxylase'	—	Fig. 11.3, No. 6
S-Adenosyl-L-methionine:flavonoid 3′-*O*-methyltransferase	2.1.1.42	Fig. 11.3, No. 7
UDP-Glucose:flavonoid 7-*O*-glucosyltransferase	2.4.1.81	Fig. 11.3, No. 8
UDP-Glucose:flavonol 3-*O*-glucosyltransferase	2.4.1.-	Fig. 11.3, No. 9
UDP-Apiose/UDP-xylose synthase	—	Fig. 11.3, No. 10
UDP-Apiose:flavone 7-*O*-glucoside 2″-*O*-apiosyltransferase	2.4.2.25	Fig. 11.3, No. 11
Malonyl-CoA:flavonoid 7-*O*-glycoside malonyltransferase	2.3.1.-	Fig. 11.3, No. 12
Malonyl-CoA:flavonol 3-*O*-glucoside malonyltransferase	2.3.1.-	Fig. 11.3, No. 13

been found which does not possess phenylalanine ammonia-lyase activity. Most of this work has been reviewed by Hanson and Havir (1972a,b) and Camm and Towers (1973).

Phenylalanine ammonia-lyase has attracted considerable attention because it provides the link between primary metabolism and the phenylpropanoid pathways (Fig. 11.1). Regulation of phenylalanine ammonia-lyase activity *in vivo* may be exerted through both allosteric effects and modulation of the enzyme level by various stimuli. A strong correlation between flavonoid production and the expression of phenylalanine ammonia-lyase activity in developing plant tissues or after treatment of plant tissues or plant cell cultures with light, micro-organisms, microbial products (elicitors) or hormones and other chemicals has been observed (see Section 11.4.10). It should be noted, however, that phenylalanine ammonia-lyase cannot be regarded as the 'key enzyme' of flavonoid biosynthesis, because of its involvement in other phenylpropanoid pathways.

The second enzyme of general phenylpropanoid metabolism, cinnamate 4-hydroxylase (*trans*-cinnamate 4-mono-oxygenase), catalyses the 4-hydroxylation of cinnamate to yield 4-coumarate. The enzyme has been isolated from a variety of plants and its properties have been studied.

Cinnamate 4-hydroxylase was first detected in microsomal membranes of pea seedlings by Russell and Conn (1967). Microsomal systems have since provided convenient sources of the enzyme (Russell, 1971; Potts *et al.*, 1974; Tanaka *et al.*, 1974; Tanaka and Uritani, 1974; Young and Beevers, 1976; Rich and Lamb, 1977; Pfändler *et al.*, 1977). Tanaka *et al.* (1974) concluded from their studies of microsomal fractions of *Brassica* that cinnamate 4-hydroxylase activity is not associated with the endoplasmic reticulum (ER). By contrast, later studies by others have convincingly demonstrated that the enzyme is associated with the ER in *Ricinus* endosperm tissues (Young and Beevers, 1976), Jerusalem artichoke (*Helianthus tuberosus*) tuber (Benveniste *et al.*, 1977) and *Sorghum* seedlings (Saunders *et al.*, 1977).

Cinnamate 4-hydroxylase is a mono-oxygenase which catalyses the NADPH- and O_2-dependent hydroxylation of *trans*-cinnamate to give *trans*-4-coumarate (Russell, 1971; Potts *et al.*, 1974; Pfändler *et al.*, 1977). The enzyme is highly substrate-specific and is inhibited by *cis*-cinnamate and by the reaction product, *trans*-4-coumarate (Russell, 1971; Pfändler *et al.*, 1977). Further investigations have demonstrated the involvement of cytochrome *P*-450 in the enzyme-mediated reaction (Potts *et al.*, 1974; Rich and Lamb, 1977).

The involvement of cytochrome *P*-450, the occurrence of a 'NIH-shift' during the hydroxylation reaction (Reed *et al.*, 1973) and the requirement of a lipid component for enzyme activity (Büche and Sandermann, 1973) provide some evidence for a basic similarity between plant and animal microsomal hydroxylation. Differences appear to exist, however, with regard to the mechanism of induction of these hydroxylases and their substrate specificities (Büche and Sandermann, 1973; Pfändler *et al.*, 1977).

The CoA esters of 4-coumarate and of other substituted cinnamic acids play a central role as intermediates at the branch point between general phenylpropanoid metabolism and the various subsequent specific pathways (Fig. 11.1). The function of CoA esters of substituted cinnamic acids as substrates in the biosynthesis of flavonoids (Hahlbrock, 1977, 1981; Hahlbrock and Grisebach, 1979), lignin (Hahlbrock, 1977; Gross, 1977; Hahlbrock and Grisebach, 1979), stilbenes (Rupprich and Kindl, 1978; Schöppner and Kindl, 1979; Rupprich *et al.*, 1980; Fritzemeier and Kindl, 1981) and cinnamate esters (Rhodes and Wooltorton, 1976; Ulbrich *et al.*, 1976; Ulbrich and Zenk, 1980) has been demonstrated in recent years.

The enzyme responsible for CoA ester synthesis from various cinnamic acids is 4-coumarate:CoA ligase* (Fig. 11.1, Table 11.1). A cinnamate-activating activity was first reported for extracts from leaves of spinach beet (*Beta vulgaris*) (Walton and Butt, 1970, 1971). 4-Coumarate:CoA ligases were isolated from various plant tissues and partially purified (Gross and Zenk, 1974; Knobloch and Hahlbrock, 1975, 1977; Ranjeva *et al.*, 1975 a,b, 1976; Wallis and Rhodes, 1977). The existence of isoenzymes has been demonstrated in soybean (*Glycine max*) cell cultures, pea (*Pisum sativum*) shoots and *Petunia hybrida* leaves (Table 11.2).

The basic properties of 4-coumarate:CoA ligase from various sources are fairly uniform. A molecular weight of 55 000 has been reported for both isoenzymes from soybean (Lindl *et al.*, 1973; Knobloch and Hahlbrock, 1975) and for the *Forsythia* enzyme (Gross and Zenk, 1974). A similar molecular weight (approximately 60 000) has been estimated for the parsley enzyme (Knobloch and Hahlbrock, 1977; Ragg *et al.*, 1981). 4-Coumarate:CoA ligase is specific for the activation of substituted cinnamic acids in the presence of ATP, Mg^{2+} and CoASH, with the respective cinnamoyl-AMP being an intermediate in this reaction (Rhodes and Wooltorton, 1973; Gross and Zenk, 1974; Knobloch and Hahlbrock, 1975, 1977). The substrate specificities of the ligases from several plants have been studied (Table 11.2). No significant activation of 4-hydroxybenzoate, differently substituted phenylacetic acids, phenylalanine, tyrosine and of several aliphatic mono- and di-carboxylic acids was observed with the enzymes from *Forsythia*, *Glycine max* and *Petroselinum hortense* (Gross and Zenk, 1974; Knobloch and Hahlbrock, 1975, 1977). The enzyme acts only on the *trans* form of 4-coumarate (Knobloch and Hahlbrock, 1975, 1977).

Much interest has been focused on the question as to whether the activities of distinct 4-coumarate:CoA ligases are related to specific phenylpropanoid pathways. Most of the enzymes studied exhibit a preference for only a few substituted cinnamic acids, such as 4-coumarate, caffeate and ferulate, and do

* Some authors have designated the ligase as 'hydroxycinnamate:CoA ligase'. Because 4-coumarate was one of the most efficient substrates for nearly all ligases reported so far to be involved in phenylpropanoid metabolism (Gross, 1977; Knobloch and Hahlbrock, 1977), and the EC number was assigned to 4-coumarate:CoA ligase, this latter term is used here throughout.

not activate sinapate or other highly methoxylated cinnamic acids (Table 11.2). For example, 4-coumarate:CoA ligase from irradiated parsley cells, which is thought to be involved in the biosynthesis of both flavonoids (Knobloch and Hahlbrock, 1977) and cinnamate esters (Kühnl, unpublished results), shows relatively low K_m values and high rates of conversion for 4-coumarate, caffeate and ferulate. Their substitution patterns correspond to those of the flavones and flavonols identified as aglycones in parsley cells (Kreuzaler and Hahlbrock, 1973). However, a comparatively narrow substrate specificity of the chalcone synthase from these cells, which utilizes only 4-coumaroyl-CoA with high efficiency (Section 11.4.2) and the occurrence of a flavonoid-specific methyltransferase (Section 11.4.6) leave some doubt as to whether caffeoyl-CoA and feruloyl-CoA are natural substrates of the flavonoid glycoside pathway in parsley. Ligases with substrate specificities similar to that of the parsley enzyme have been isolated from various sources (Table 11.2). Their possible functions in flavonoid, lignin and cinnamate ester biosynthesis have been discussed by several authors (Gross *et al.*, 1975; Rhodes and Wooltorton, 1973, 1975a,b, 1976; Rhodes *et al.*, 1976; Gross, 1977).

An important contribution to the elucidation of the metabolic function of 4-coumarate:CoA ligase was the discovery of isoenzymes (Table 11.2). The two isoenzymes of soybean (*Glycine max*) cell cultures differ markedly in their substrate specificities and other characteristic properties (Knobloch and Hahlbrock, 1975). Isoenzyme 1 shows relatively low K_m values and high rates of conversion for the three typical substrates of the lignin pathway, 4-coumarate, ferulate and sinapate, indicating that these acids are substrates of the enzyme *in vivo*. Isoenzyme 2 has relatively high affinities for 4-coumarate and caffeate, but does not activate sinapate. Therefore the assignment of a function in lignin biosynthesis to isoenzyme 1 is obvious from its substrate specificity. A similar clear-cut relation between isoenzyme 2 and a specific phenylpropanoid pathway cannot be deduced with certainty merely from its substrate specificity. Both isoenzymes could provide the substrate(s) for the flavonoid pathway, 4-coumaroyl-CoA and perhaps caffeoyl-CoA (Hrazdina *et al.*, 1976; Saleh *et al.*, 1978), with similar efficiency (Knobloch and Hahlbrock, 1975). However, the involvement of isoenzyme 2 in the flavonoid pathway was indirectly concluded from the great similarity of its substrate specificity with that of 4-coumarate:CoA ligase from parsley cells (Knobloch and Hahlbrock, 1977), whose involvement in flavonoid biosynthesis seems obvious. It cannot be excluded, however, that isoenzyme 2 is also involved in lignin biosynthesis in soybean cells (Hahlbrock and Grisebach, 1979).

Two ligase isoenzymes from extracts of pea (*Pisum sativum*) shoots with possible metabolic functions similar to those of the soybean isoenzymes have been separated (Wallis and Rhodes, 1977; Butt and Wilkinson, 1979). The two enzymes not only showed different substrate specificities (Table 11.2) but also responded differently to illumination of etiolated pea shoots (Butt and Wilkinson, 1979). Both forms were present in buds and stems, but while the

Table 11.2 4-Coumarate:CoA ligases from various sources

Source of enzyme	*Number of ligases reported*	*Basis for distinction of isoenzymes*	*Type of assay**	*Substrate specificity: Relative rate of conversion (4-coumarate = 100 %) and K_m value (in parenthesis, $\times 10^5$ mol/l)*			
				4-Coumarate	Caffeate	Ferulate	Sinapate
Acer saccharinum	?†		A	100	59	50	0
Brassica napo-brassica	1		A	100(1.4)	68	73(3.1)	2
Forsythia suspensa	1		A	100(2.5)	62(1.9)	96(2.0)	0
Glycine max	2	DEAE-cellulose chromatography, disc-gel electrophoresis					
Isoenzyme 1			B	100(3.2)	56(4.0)	56(0.9)	46(1.1)
Isoenzyme 2			B	100(1.7)	87(1.4)	96(13)	0
Lycopersicon esculentum	1		A	100(0.14)	77(0.89)	89	0
Petroselinum hortense	1		B	100(1.4)	85(2.0)	90(3.0)	0
Petunia hybrida	3	DEAE-cellulose and hydroxyapatite chromatography	A				
Isoenzyme 1a				100(6)	300(10)	0	0
Isoenzyme 1b				100(10)	0	0	80(17)
Isoenzyme 2				100(5)	0.2(480)	162(7)	0
Pisum sativum	2	DEAE-cellulose chromatography					
'Peak' 1				100	38	78	56
'Peak' 2				100	27	63	2
Solanum tuberosum	?†		A	100(0.62)	51(1.6)	77(1.4)	0
Taxus baccata	?†		A	100	80	92	‡

* Data were determined either (A) with substrate concentrations of the order of 0.2–0.7 mmol/l or (B) as relative *V* values.

† Not determined.

‡ Not reported.

activities of both forms increased in stems upon illumination, only ligase 2 activity increased in buds. Butt and Wilkinson (1979) concluded that ligase 1 is involved in lignin synthesis in stems and ligase 2 may be involved in both lignin and flavonoid biosynthesis in stems and leaves.

An even more complex system is present in *Petunia hybrida* leaves which produce flavonoids, cinnamate esters (caffeoyl- and feruloylquinic acids) and lignin (Ranjeva *et al.*, 1975a, 1976). After chromatography of cell free extracts from *Petunia* leaves on DEAE-cellulose and hydroxyapatite, three forms of

Remarks	Reference
	Gross *et al.* (1975)
Isolated from roots, activity increased by aging or ethylene treatment	Rhodes and Wooltorton (1973) Rhodes and Wooltorton (1975a) Rhodes *et al.* (1976)
Isolated from young xylem	Gross *et al.* (1973) Gross *et al.* (1974) Gross *et al.* (1975)
Isolated from cell suspension cultures, activity dependent on growth stage of culture	Knobloch and Hahlbrock (1975)
Isolated from fruits	Rhodes and Wooltorton (1976)
Isolated from cell suspension cultures, activity increased by irradiation, dilution, elicitor treatment of cells	Knobloch and Hahlbrock (1977)
Isolated from leaves	Ranjeva *et al.* (1975a) Ranjeva *et al.* (1976)
Isolated from shoots	Wallis and Rhodes (1977) Butt and Wilkinson (1979)
Isolated from tubers, activity increased upon aging of potato tuber disks in the light	Rhodes and Wooltorton (1975b)
	Gross *et al.* (1975)

4-coumarate:CoA ligase were isolated (Table 11.2). All three isoenzymes activated 4-coumarate but differed significantly with respect to their specificities for caffeate, ferulate and sinapate. The involvement of the three ligases in the formation of flavonoids, cinnamate esters and lignin respectively was concluded from their differential sensitivity to feedback inhibition by naringenin and cinnamate esters (Ranjeva *et al.*, 1976, 1979).

Other interesting properties of 4-coumarate:CoA ligases from parsley and soybean are their inhibition by the substrates, 4-coumarate and CoA, and the

product, AMP (Knobloch and Hahlbrock, 1975, 1977). Thus, a combination of pronounced substrate specificities, the occurrence of isoenzymes and complex regulatory effects of various metabolites indicate an important role of this enzyme in controlling the distribution of cinnamoyl-CoA esters among different pathways of phenylpropanoid metabolism (Hahlbrock and Grisebach, 1979).

Some of the specific pathways into which general phenylpropanoid metabolism branches require, in addition to 4-hydroxylation, further substitution of the aromatic ring. The most prominent additional reactions are hydroxylation and subsequent methylation in the 3 and 5 positions. Of the enzymes involved, only methyltransferases have so far been characterized to any considerable extent (Poulton, 1981). Some of these transferases have been assigned to lignin biosynthesis. Since flavonoid-specific methyltransferases have been found in various plants, too, it appears unlikely that general phenylpropanoid metabolism includes a 'general' methyltransferase (Hahlbrock and Grisebach, 1979). Phenolases catalyse the hydroxylation of 4-coumarate to caffeate. However, the significance of these enzymes in general phenylpropanoid metabolism and flavonoid biosynthesis is not established (Hahlbrock and Grisebach, 1975). The results of genetic and biochemical investigations with *Silene dioica* petals were interpreted as indicating that in this plant the 3′-hydroxylation pattern of the B ring of anthocyanidins is determined at the 4-coumaroyl-CoA stage (Kamsteeg *et al.*, 1980d, 1981).

The question of whether the activated products of 4-coumarate substitution, caffeoyl-CoA and feruloyl-CoA, are intermediates in flavonoid biosynthesis is discussed below in connection with the substrate specificity of chalcone synthase (Section 11.4.2).

11.3 INDIVIDUAL PATHWAYS OF FLAVONOID BIOSYNTHESIS

All classes of flavonoids are biosynthetically closely related, with a chalcone being the first common intermediate. Earlier feeding experiments with radioactively labelled precursors have established that the carbon skeleton is derived from acetate and phenylalanine; ring A is formed by a head-to-tail condensation of three acetate units and ring B as well as carbon atoms 2,3 and 4 of the heterocyclic ring C arise from phenylalanine. Most of the results of this work have been reviewed previously by Hahlbrock and Grisebach (1975). The biosynthetic relationships of the flavonoids as concluded mainly from labelling experiments *in vivo* are illustrated in Fig. 11.2. More recent investigations at the enzymic level have largely confirmed the hypothetic steps which had been deduced from incorporation experiments. In particular, detailed knowledge of the central reaction of flavonoid biosynthesis, the condensation of the acyl residues from one molecule of 4-coumaroyl-CoA and three molecules of malonyl-CoA (Fig. 11.3), was obtained by several authors (see below). Extensive studies of the enzymology and regulation of flavone and flavonol glycoside

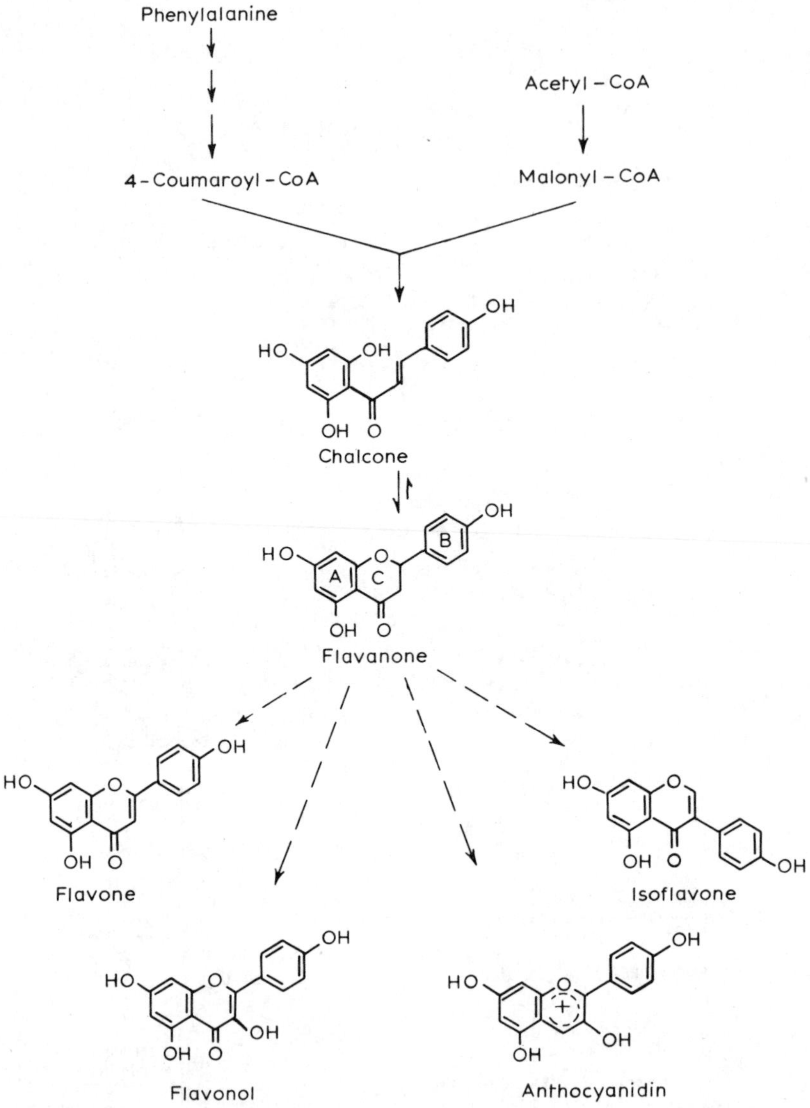

Fig. 11.2 Scheme illustrating the position of the chalcone as the first common intermediate in the biosynthesis of all classes of flavonoid.

biosynthesis revealed many details of the reactions involved in the formation of these two important classes of flavonoid (Section 11.4). On the other hand, relatively little beyond the results of tracer studies is known about the synthesis of anthocyanins (Section 11.5), isoflavonoids (Section 11.6) and other flavonoids (Section 11.7).

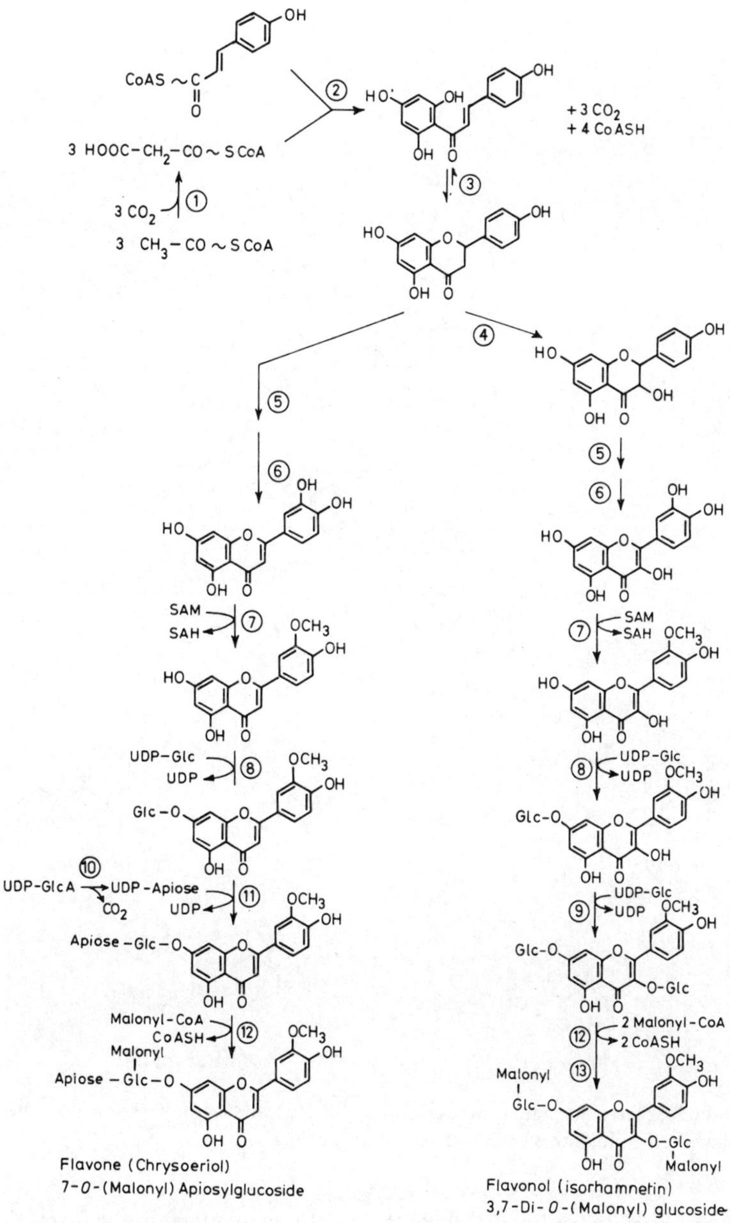
CoAS ~ C
3 HOOC–CH2–CO ~ SCoA
3 CO2
3 CH3 – CO ~ SCoA
+ 3 CO2
+ 4 CoASH
SAM
SAH
UDP–Glc
UDP
Glc–O
UDP-GlcA
UDP-Apiose
CO2
Apiose – Glc – O
Malonyl-CoA
CoASH
Malonyl
Apiose – Glc – O
2 Malonyl-CoA
2 CoASH
O-Glc
Flavone (Chrysoeriol)
7-O-(Malonyl) Apiosylglucoside
Flavonol (isorhamnetin)
3,7-Di-O-(Malonyl) glucoside

Fig. 11.3 Scheme illustrating the sequence of reactions of the flavone and flavonol glycoside pathways. The enzymes marked by numbers are listed in Table 11.1. SAM, *S*-adenosyl-L-methionine; SAH, *S*-adenosyl-L-homocysteine.

11.4 CHALCONES, FLAVANONES, FLAVONES AND FLAVONOLS

The chalcone/flavanone isomers are the central intermediates in the synthesis of all flavonoids (Hahlbrock and Grisebach, 1975; Hahlbrock, 1981). Compelling evidence has now been presented that the chalcone is the immediate product of the synthase reaction (see below). In a second step, chalcone isomerase catalyses the stereospecific formation of the (−)(2*S*)-flavanone from the corresponding chalcone.

The frequent co-occurrence of chalcones, flavanones, flavones and flavonols suggests that their biosynthetic pathways are closely related. *In vivo*, chalcones with a phloroglucinol-type substitution in ring A are exclusive intermediates in the formation of 5, 7-dihydroxyflavonoids, while chalcones with a resorcinol-type substitution in ring A are selectively converted into 7-hydroxyflavonoids (cf. Hahlbrock and Grisebach, 1975). However, an enzyme catalysing the synthesis of a chalcone with a resorcinol structure in ring A has not been reported. Further, the actual mechanisms through which flavanones are converted into flavones, flavonols and other flavonoids are in most cases not well understood. Only a few studies with cell-free extracts have demonstrated the oxidation of flavanones to the corresponding flavones and the hydroxylation of flavanones in the 3 position to yield the corresponding dihydroflavonols. Several properties of the enzymes catalysing flavanone oxidation and 3-hydroxylation indicate that they are not peroxidases (Section 11.4.4). This is in contrast to earlier reports which demonstrated the conversion of 4,2′,4′-trihydroxychalcone to the corresponding dihydroflavonol, flavonol and aurone either by plant extracts or by horseradish peroxidase (Wong, 1965; Wong and Wilson, 1972; Rathmell and Bendall, 1972).

The most frequently occurring substitution patterns of the aromatic ring B require further hydroxylation and methylation reactions at positions 3′ and 5′. The existence of flavonoid-specific 3′-hydroxylases and 3′-*O*- and 4′-*O*-methyltransferases in several plants suggests that at least in some tissues the introduction of 3′-hydroxyl and 3′*O*- and 4′-*O*-methyl groups into these compounds takes place after formation of the flavonoid ring structure (Sections 11.4.5 and 11.4.6).

The biosynthesis of flavones and flavonols has been studied most extensively in cultured parsley (*Petroselinum hortense*) cells. Leaves and seeds from parsley plants contain large amounts of flavone and flavonol glycosides with a phloroglucinol-type substitution in ring A. A high concentration of these compounds occurs also in irradiated suspension cultures of parsley cells derived from leaf petioles. More than 20 different flavonoid glycosides were found. These are derived from only three flavone and three flavonol aglycones, all of which have very similar substitution patterns (Kreuzaler and Hahlbrock, 1973). The chemical structures of the aglycones and their proposed biosynthetic relations are shown in Fig. 11.4. Except for the characteristic 3-hydroxyl group of the flavonols, the six aglycones differ only with respect to their substitution patterns in the 3′ position.

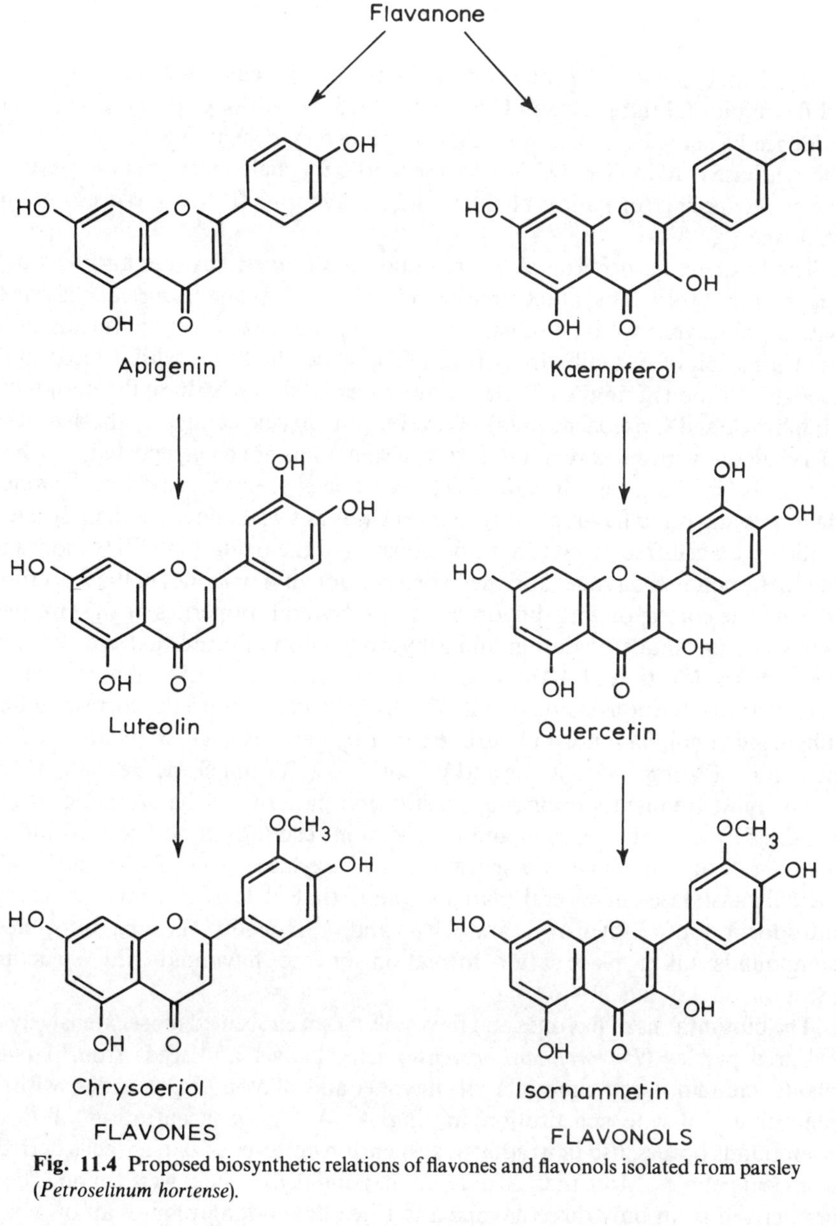

Flavanone

Apigenin

Luteolin

Chrysoeriol

FLAVONES

Kaempferol

Quercetin

Isorhamnetin

FLAVONOLS

Fig. 11.4 Proposed biosynthetic relations of flavones and flavonols isolated from parsley (*Petroselinum hortense*).

The flavonoid glycosides from parsley were isolated either as malonylated or as non-acylated flavone and flavonol 7-*O*-glucosides, flavone 7-*O*-apiosyl (1→ 2)-glucosides and flavonol 3,7-*O*-diglucosides (Kreuzaler and Hahlbrock, 1973). Examples are given in Fig. 11.3. The site of malonylation of flavone and flavonol glycosides is not known with certainty, but is deduced from indirect

evidence to be the 6-*O*- position of the glucose moiety (Hahlbrock, 1972). Many tissues possess high esterase activity which might impair the isolation of acylated compounds. It is therefore possible that flavonoid glycosides occur more frequently in the acylated form than normally found in plant extracts.

The flavonoid glycoside pathway in parsley consists of about 13 enzymes which catalyse several consecutive steps of flavone and flavonol biosynthesis (Fig. 11.3). The first step is the conversion of acetyl-CoA to malonyl-CoA, which serves as substrate for three enzymes, chalcone synthase and two malonyltransferases. In the second step, the chalcone is formed from 4-coumaroyl-CoA (and possibly from other substituted cinnamoyl-CoA esters) and malonyl-CoA. The chalcone is isomerized to the corresponding flavanone, which is further converted into the basic flavone and flavonol structures. The subsequent steps are substitutions of ring B of the aglycones by hydroxylation and *O*-methylation. The final steps are glycosylation of the aglycones and acylation of the resulting glycosides. The enzymes involved in the individual steps, as elucidated with parsley cells, are listed in Table 11.1.

11.4.1 Formation of malonyl-CoA

In parsley, malonyl-CoA is required for two types of reaction in flavonoid biosynthesis, the formation of the aromatic ring A and the malonylation of flavone and flavonol glycosides. Malonyl-CoA can be supplied in higher plants by the ATP-dependent carboxylation of acetyl-CoA catalysed by acetyl-CoA carboxylase:

$$CH_3CO \sim SCoA + HCO_3^- + ATP \rightleftharpoons HOOC\text{-}CH_2\text{-}CO \sim SCoA + ADP + P_i$$

Plant acetyl-CoA carboxylase has been studied in different tissues and is usually regarded as an enzyme of fatty acid biosynthesis (Heinstein and Stumpf, 1969; Kannangara and Stumpf, 1972; Brock and Kannangara, 1976; Nielsen *et al.*, 1979; Mohan and Kekwick, 1980; Stumpf, 1980). However, it was deduced from the regulation of acetyl-CoA carboxylase activity in irradiated parsley cells (Section 11.4.10) that this enzyme is also an integral member of the flavonoid glycoside pathway (Ebel and Hahlbrock, 1977).

The acetyl-CoA carboxylase which is involved in flavonoid biosynthesis in parsley has been purified from irradiated cells and characterized to some extent (Egin-Bühler *et al.*, 1980). The biotin-containing enzyme has a molecular weight of about 840 000 and is probably composed of four subunits of $M_r = 210\,000$ each. Similar structural properties have been found in comparative studies for the enzyme of wheat germ (Egin-Bühler *et al.*, 1980). These results suggest that the three catalytic subsites of the two carboxylases from these phylogenetically divergent plants, the biotin-carboxylation site, the transcarboxylation site and the biotin-carrying site reside in the same enzyme subunit. This type of structural organization has been found also for animal acetyl-CoA carboxylases (Tanabe *et al.*, 1975; Mackall and Lane, 1977; Ahmad *et al.*, 1978) which catalyse the first

committed step in fatty acid biosynthesis. Parsley acetyl-CoA carboxylase catalyses the carboxylation of the CoA esters of acetate, propionate and butyrate at decreasing rates in this order and does not accept hexanoyl-CoA and methylcrotonoyl-CoA as substrates (Egin-Bühler and Ebel, unpublished). The apparent K_m values for the substrates of the carboxylation reaction are: acetyl-CoA, 0.15 mmol/1; HCO_3^-, 1.0 mmol/1; $MgATP^{2-}$, 0.07 mmol/1. Recent successful application of the affinity adsorbent avidin-monomer Sepharose **4B** to the isolation of the parsley carboxylase substantially improves and simplifies the purification of this enzyme (Egin-Bühler and Ebel, unpublished).

11.4.2 Formation of chalcone

The formation of a chalcone is the exceptional step in flavonoid biosynthesis which is both specific for this pathway and common to all flavonoids. The enzyme catalysing this step, chalcone synthase, can therefore be regarded as the key enzyme of flavonoid biosynthesis. The product of general phenylpropanoid metabolism, 4-coumaroyl-CoA, serves as one of two substrates for the synthase reaction. The overall reaction consists of three successive condensation steps with 'acetate units' derived from malonyl-CoA. The result is the elongation of the aliphatic side chain of 4-coumarate by six carbon atoms which then cyclize to give the aromatic ring A (Fig. 11.3). The putative formation of a poly-β-ketoacyl intermediate according to the 'acetate rule' (Birch and Donovan, 1953; Lynen and Tada, 1961) was initially postulated on the basis of tracer experiments *in vivo* (Grisebach, 1959).

Chalcone synthesis *in vitro* was first demonstrated with a crude extract from irradiated parsley cells which contained all enzymes of the flavonoid glycoside pathway (Kreuzaler and Hahlbrock, 1972). Owing to the presence of chalcone isomerase in the extract, the synthase reaction led to the formation of the flavanone, naringenin. More recently, however, it was shown that the immediate product is the corresponding chalcone and not the flavanone (Heller and Hahlbrock, 1980; Light and Hahlbrock, 1980). Essentially the same result was obtained by Sütfeld and Wiermann (1980a, 1981) working with anthers from tulip (*Tulipa* cv. 'Apeldoorn') and petals from *Cosmos sulphureus*. Further support for the notion that the enzyme is a chalcone synthase – and not a flavanone synthase, as presumed earlier (Kreuzaler and Hahlbrock, 1975a) – comes from genetic studies on the control of chalcone isomerase activity (see below).

Chalcone synthase has been purified from cell cultures of parsley (Kreuzaler and Hahlbrock, 1975a; Kreuzaler *et al.*, 1979) and *Haplopappus gracilis* (Saleh *et al.*, 1978) and from parts of tulip and *Cosmos* plants (Sütfeld and Wiermann, 1981). The enzymes from parsley and tulip have apparent molecular weights in the range from 55 000 to 80 000 (Kreuzaler *et al.*, 1979; Sütfeld *et al.*, 1978). Chalcone synthase has no cofactor requirements. Substrates are malonyl-CoA and the CoA ester of a substituted cinnamic acid. 4-Coumaroyl-CoA is in all

known cases more efficient than caffeoyl-CoA and feruloyl-CoA. In tulip anthers, caffeoyl-CoA and feruloyl-CoA, in addition to 4-coumaroyl-CoA, were proposed to function as substrates of chalcone synthase *in vivo* (Sütfeld *et al.*, 1978). The substitution pattern of the starter molecule(s) for chalcone synthesis in parsley is not known for certain, but the possible occurrence of a flavonoid-specific 3′-hydroxylase and the unequivocal demonstration of a 3′-*O*-methyltransferase might rule out caffeoyl-CoA and feruloyl-CoA.

The sequence of reactions shown in Fig. 11.5 was deduced from the occurrence of several side products of the chalcone synthase reaction with the partially purified enzymes from parsley and *Haplopappus* (Kreuzaler and Hahlbrock, 1975b; Hrazdina *et al.*, 1976; Saleh *et al.*, 1978). Under certain assay conditions, chalcone synthase catalyses only one or two (instead of three) condensation steps. This leads to the formation of short-chain release products containing either one 'acetate unit' (benzalacetones or dihydropyrones) or two 'acetate units' (styrylpyrones). The respective release products were formed with all three substrates tested *in vitro*, 4-coumaroyl-CoA, caffeoyl-CoA and feruloyl-CoA with both the parsley and *Haplopappus* enzymes (Hrazdina *et al.*, 1976; Saleh *et al.*, 1978). The formation of release products can be greatly reduced by using proper assay conditions (Schröder *et al.*, 1979a). No such products were observed with the enzyme from tulip anthers (Sütfeld *et al.*, 1978).

Other artificial side reactions in the chalcone synthase assay, which have been very useful for the elucidation of the reaction mechanism, are CO_2 exchange and decarboxylation of malonyl-CoA (Kreuzaler *et al.*, 1978). Their occurrence can be explained by assuming the intermediate formation of an acetyl-CoA carbanion from malonyl-CoA during the condensation reaction (Fig. 11.6). Both these side reactions are unlikely to occur *in vivo*. The apparent K_m values for malonyl-CoA in the side reactions are two to three orders of magnitude higher, and the pH optimum is much lower, than the respective values for malonyl-CoA in the complete chalcone synthase reaction (Heller *et al.*, 1980). With chalcone synthase preparations from parsley and tulip, apparent K_m values for all possible substrates leading to chalcone formation at appreciable rates *in vitro* were around 2 μmol/l (Hrazdina *et al.*, 1976; Sütfeld *et al.*, 1978), or somewhat higher (18 μmol/l) in the case of malonyl-CoA with the parsley enzyme (Heller *et al.*, 1980), indicating a high affinity of these enzymes for their substrates in the complete synthase reaction. Both enzymes are strongly inhibited by the reaction products, chalcone (or flavanone) and coenzyme A.

The activity of chalcone synthase has been related to the biosynthesis of all major classes of flavonoids in a variety of plants: flavones and flavonols in parsley (Hahlbrock *et al.*, 1976), chalcones and flavonols in tulip (Herdt *et al.*, 1978; Sütfeld *et al.*, 1978), anthocyanins in *Haplopappus gracilis* (Saleh *et al.*, 1978), in *Hippeastrum* (Hrazdina *et al.*, 1978) and in red cabbage (*Brassica oleracea*) (Hrazdina and Creasy, 1979), isoflavonoids in soybean (*Glycine max*) (Zähringer *et al.*, 1978) and *C*-glycosylflavones in oat (*Avena sativa*) (Fuisting and Weissenböck, 1980).

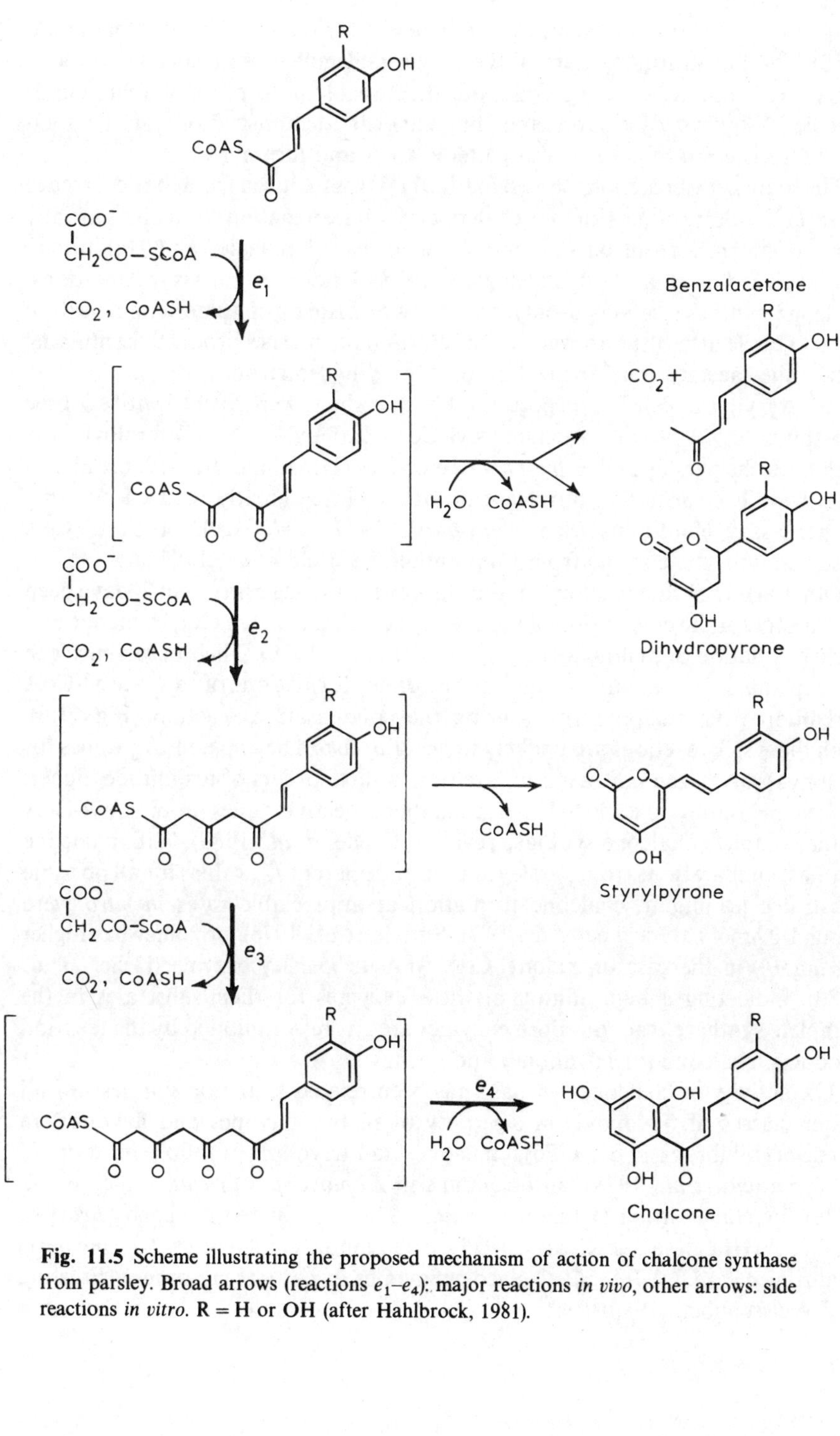

Fig. 11.5 Scheme illustrating the proposed mechanism of action of chalcone synthase from parsley. Broad arrows (reactions e_1–e_4): major reactions *in vivo*, other arrows: side reactions *in vitro*. R = H or OH (after Hahlbrock, 1981).

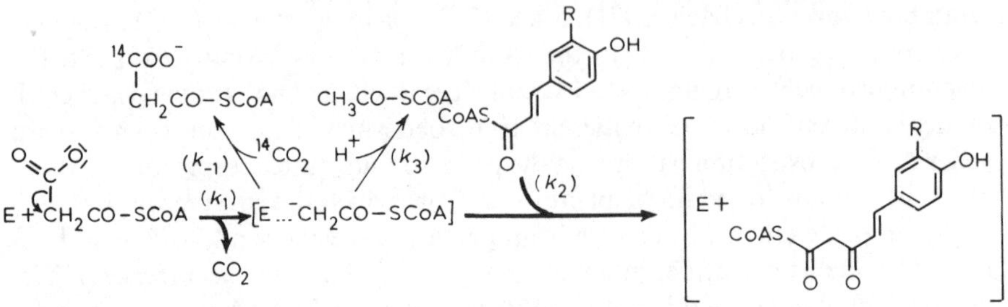

Fig. 11.6 Proposed mechanism of reaction e_1 (Fig. 11.5) of chalcone synthase from parsley. Broad arrows (rate constants k_1 and k_2): putative reactions *in vivo*, other arrows (rate constants k_{-1} and k_3): side reactions *in* vitro. R = H or OH, E = chalcone synthase (after Hahlbrock, 1981).

11.4.3 Isomerization of chalcone

From the results described above it is clear that the product of the chalcone synthase reaction is the chalcone and not the isomeric flavanone. Thus, the function of chalcone isomerase appears to be the provision of the flavanone as the substrate for conversion into other flavonoids (Fig. 11.2).

Chalcone isomerase was the first isolated enzyme of flavonoid metabolism (Moustafa and Wong, 1967). The enzyme has been partially purified from a variety of flavonoid-producing plants and plant tissues and extensively studied (cf. Hahlbrock and Grisebach, 1975). In summary, the enzyme shows the following characteristics. It catalyses the formation of the 6-membered heterocyclic ring C of flavanones from the corresponding chalcones and has no cofactor requirements. The kinetics of isomerization of 2′,4,4′-trihydroxychalcone (isoliquiritigenin) to 7,4′-dihydroxyflavanone (liquiritigenin) by the soybean (*Glycine max*) enzyme have been examined and have indicated that at pH 7.6 the reaction was reversible with an equilibrium constant of 37 in favour of the flavanone (Boland and Wong, 1975). The product was shown to have the (-) (2*S*) configuration (Moustafa and Wong, 1967). The stereochemistry of the isomerase reaction at position C-3 was elucidated by NMR studies with deuterated flavanones (Hahlbrock *et al.*, 1970a). When [3-^{2}H]isoliquiritigenin was used as substrate for either one of two chalcone isomerase isoenzymes from mung bean (*Phaseolus aureus*) seedlings, deuterium was located preferentially in the equatorial position at C-3 of liquiritigenin. Conversely, when the enzymic cyclization was carried out in deuterium oxide with unlabelled chalcone, deuterium was preferentially introduced in the axial position at C-3 of the flavanone.

Chalcone isomerases from seedlings or leaves of mung bean, chick pea (*Cicer arietinum*), parsley or *Datisca cannabina* have been separated by polyacrylamide gel electrophoresis into a number of activities as measured by the disappearance of the chalcone in a spectrophotometric assay (Hahlbrock *et al.*, 1970b;

Grambow and Grisebach, 1971). Since KCN was not added to avoid possible interference of peroxidase with the assay (Boland and Wong, 1975) and the reaction products were not unequivocally identified, it is not known whether all of these different enzyme activities are true isoenzymes of chalcone isomerase. In general, the substitution patterns of the preferred substrates correspond to those of the flavonoids found in the plant tissue from which the enzyme was isolated. For example, chalcone isomerase from parsley is specific for 4,2′,4′6′-tetrahydroxychalcone as substrate, in agreement with the exclusive occurrence of 5,7-dihydroxyflavonoids in this plant (Hahlbrock *et al.*, 1970b). Chalcone glucosides will not serve as substrates of isomerases, in all cases so far studied.

A molecular weight of 15 600 has been determined for the enzyme from soybean seeds (Boland and Wong, 1975). Effects of iodoacetamide and sodium tetrathionate on the soybean enzyme, and a strong dependence of the activity on pH, were interpreted as indicating that a sulfhydryl group is not involved in the reaction mechanism and that the catalytic site is probably an imidazole group which assists in ring closure at C-2 of the chalcone (Boland and Wong, 1975, 1979).

Kuhn *et al.* (1978) investigated the genetic control of chalcone isomerase activity in 18 lines of *Callistephus chinensis*. The absence of the enzyme from four mutants (recessive genotypes) was correlated with an abnormally high level of chalcone glucoside accumulation. In genotypes with the wild-type allele, anthocyanin glycosides were formed without intermediate chalcone accumulation. Similar results have been found in studies with different genotypes of *Dianthus caryophyllus* (Forkmann and Dangelmayr, 1980) and *Petunia hybrida* (Forkmann and Kuhn, 1979). The importance of chalcone isomerase in the formation of various basic flavonoid structures is also strongly indicated by the regulatory behaviour of chalcone isomerase activity during the light-induced flavonoid glycoside formation in parsley cells (Section 11.4.10) and during developmentally regulated phenylpropanoid metabolism in *Tulipa* anthers (Wiermann, 1972, 1979). When chalcone isomerase activity was measured at different stages of pollen development in *Tulipa* anthers, highest enzyme activity occurred at a period when the concentration of chalcones in the tissue decreased rapidly and that of flavonols increased.

11.4.4 Modifications at ring C

The flavanone, naringenin, is the putative substrate for flavone and dihydroflavonol formation in flavone and flavonol glycoside biosynthesis (Fig. 11.3). Studies by Sutter *et al.* (1975) with protein extracts from primary leaves of parsley which catalysed the oxidation of naringenin to the corresponding flavone, apigenin, in the presence of Fe^{2+} ions and (an) other heat-stable non-proteinaceous cofactor(s) have now been extended to parsley cell cultures as source of the enzyme. Britsch *et al.* (1981) demonstrated that the enzyme from parsley cells required 2-oxoglutarate, Fe^{2+} and possibly ascorbate as cofactors.

The same protein extract, in the presence of these cofactors, also catalysed the 3-hydroxylation of naringenin to the dihydroflavonol, dihydrokaempferol and the oxidation of dihydrokaempferol to kaempferol. The proposed sequence of reactions which would be in agreement with the experimental data thus far obtained is shown in Fig. 11.7. The number of enzymes involved in the formation of apigenin and kaempferol and further details of these complex reactions are under investigation. According to the cofactor requirement, the soluble enzyme(s) belong(s) to the class of 2-oxoglutarate-dependent dioxygenases (Abbott and Udenfriend, 1974).

In contrast to the studies with the enzyme from parsley cells, oxidation of flavanones to flavones in flower extracts from *Antirrhinum majus* was catalysed by an enzyme activity in the microsomal fraction which required NADPH as co-substrate (Stotz and Forkmann, 1981).

3-Hydroxylation of the flavanone intermediate is also an obligatory step in anthocyanin biosynthesis (Grisebach, 1979). Formation of dihydrokaempferol from naringenin was found with dimethyl sulphoxide (DMSO)-permeabilized cells of *Haplopappus gracilis* cell cultures which produced cyanidin 3-*O*-glucoside in response to irradiation with UV light (Fritsch and Grisebach, 1975). Recent results have proved that flowers of *Matthiola incana* and *Antirrhinum majus* are valuable systems for studies of 3-hydroxylation in relation to anthocyanin biosynthesis (Forkmann *et al.*, 1980; Forkmann and Stotz, 1981). Enzyme preparations from flowers of defined genotypes of both plants contain 3-hydroxylase activity which catalyses the formation of dihydrokaempferol and dihydroquercetin from naringenin and eriodictyol respectively. Again, 2-oxoglutarate, Fe^{2+} and ascorbate are required as cofactors. The gene controlling 3-hydroxylation in *M. incana* is not yet known, but in *Antirrhinum majus* the gene *inc* seems to control the conversion of flavanones into dihydroflavonols (Harrison and Stickland, 1974). Investigations of different genotypes of *A. majus* demonstrate a correlation between a block of the anthocyanin pathway by recessive alleles of the gene *inc* and the absence of 3-hydroxylase activity (Forkmann and Stotz, 1981).

11.4.5 Hydroxylation of ring B

The most frequently occurring structural modifications of flavonoid aglycones are hydroxylation and subsequent methylation in the 3′ and 5′ positions of ring B. The synthesis of 3′-hydroxynaringenin (eriodictyol) from caffeoyl-CoA can be catalysed under certain assay conditions *in vitro* by chalcone synthase. If this compound were synthesized in sufficient amounts *in vivo* and accepted as substrate for flavone and flavonol formation, the 3′-hydroxyl group would not need to be introduced at a later stage. However, investigations with cell-free extracts from cell cultures of *H. gracilis* and parsley and flowers of *M. incana* and *A. majus* demonstrated the presence of a 3′-hydroxylase in microsomal fractions which required NADPH as a cofactor (Fritsch and Grisebach, 1975; Forkmann

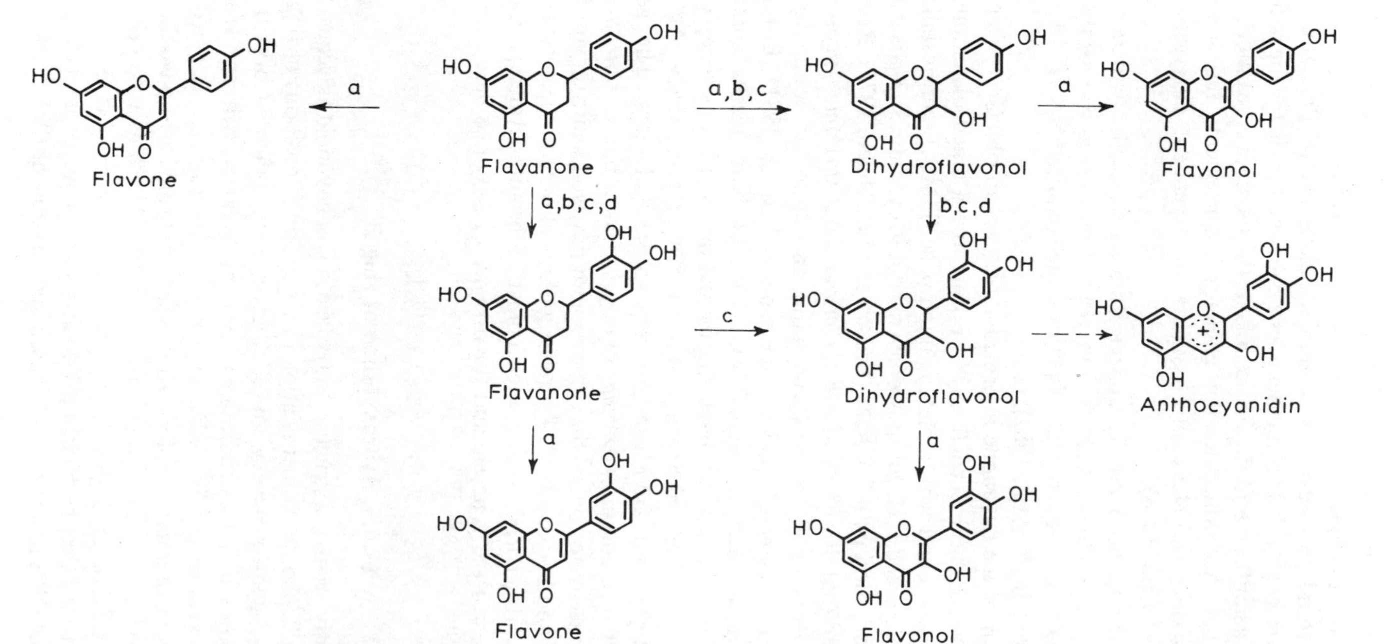

Fig. 11.7 Conversion by 3-hydroxylation, 3′-hydroxylation and oxidation of naringenin to various substituted flavones, dihydroflavonols, flavonols and anthocyanidins by cell-free extracts from *Petroselinum hortense* cell cultures (a), *Haplopappus gracilis* cell cultures (b), flowers of *Matthiola incana* (c) and flowers of *Antirrhinum majus* (d).

et al., 1980; Forkmann and Stotz, 1981; Britsch *et al.*, 1981). It can be assumed that these enzymes are haem-containing mono-oxygenases (Ullrich and Duppel, 1975), like cinnamate 4-hydroxylase (Section 11.2). In extracts from *H. gracilis* and *M. incana*, 3′-hydroxylation yielded eriodictyol and dihydroquercetin from naringenin and dihydrokaempferol respectively (Fig. 11.7). A strict correlation between 3′-hydroxylase activity of *M. incana* and the wild-type allele b^+ controlling 3′-hydroxylation in this plant was found. Basically similar results were obtained from experiments with flower extracts of *A. majus*. Chemogenetic and enzymic studies on different genotypes suggest that the 3′-hydroxyl group of flavonoids in *A. majus* is introduced at the stage of a C_{15} intermediate. 3′-Hydroxylation in microsomal fractions from *A. majus* was readily demonstrated with naringenin as substrate, but the formation of dihydroquercetin from dihydrokaempferol could not be shown.

11.4.6 Methylation

Methylation of 3′,4′-dihydroxy-substituted flavonoids has been extensively studied in a number of plants (Poulton, 1981). Methylation in the 3′ position of luteolin and quercetin is catalysed by a specific *S*-adenosyl-L-methionine: flavonoid 3′-*O*-methyltransferase from parsley cell cultures (Ebel *et al.*, 1972; Ebel and Hahlbrock, 1977). Luteolin and its 7-*O*-glucoside served as substrates with the lowest K_m and highest V/K_m values (Table 11.3). The enzyme methylated eriodictyol at a low rate and showed no activity towards mono-phenols such as 4-coumarate or apigenin. Under appropiate conditions (e.g. pH 9.3), the enzyme also methylated caffeate, albeit with a K_m which was about 35-fold higher and a V/K_m which was about 20-fold lower than the respective values for luteolin. This low specificity of the methyltransferase for caffeate, together

Table 11.3 Substrate specificities of parsley (*Petroselinum hortense*) and soybean (*Glycine max*) flavonoid *O*-methyltransferases

Source of enzyme	*Petroselinum hortense**			*Glycine max†*		
Substrate	K_m ($\times 10^6$ *mol/l*)	*Relative* V_{max}	*Relative* V_{max}/K_m	K_m ($\times 10^6$ *mol/l*)	*Relative* V_{max}	*Relative* V_{max}/K_m
Eriodictyol	1200	17	0.01	75	35	0.47
Luteolin	46	100	2.2	16 (16.7)	100 (138)	6.2 (8.3)
Luteolin 7-*O*-glucoside	31	114	3.7	28	40	1.4
Quercetin	—‡	—‡	—‡	35	210	6.0
Caffeate	1600	168	0.11	(770)	(92)	(0.12)

* Rates measured at pH 9.3 (Ebel *et al.*, 1972).
† Rates measured at pH 7.8 or pH 8.7 (in parenthesis) (Poulton *et al.*, 1977).
‡ Not measured.

with the fact that 3-methylcaffeoyl-CoA (feruloyl-CoA) is a poor substrate of chalcone synthase from parsley (Section 11.4.2), indicates that methylation of flavonoids in this plant occurs predominantly or exclusively after completion of the C_{15} skeleton. In parsley, the same methyltransferase is probably associated with both the flavone and flavonol glycoside pathways (Ebel and Hahlbrock, 1977). In general, flavonoid *O*-methyltransferases differ considerably from caffeate *O*-methyltransferases with respect to both substrate specificity and various other properties (Table 11.4). A flavonoid *O*-methyltransferase which was remarkably similar to the enzyme from parsley has been isolated from soybean (*Glycine max*) cell cultures and separated from caffeate *O*-methyltransferase (Poulton *et al.*, 1976, 1977). The flavonoid-specific enzyme closely resembled parsley flavonoid *O*-methyltransferase in several respects. *Ortho*-dihydroxy-substituted phenols were methylated specifically at the *meta* position. A pronounced preference was shown towards flavonoid substrates, as demonstrated by their low K_m and high V/K_m values (Table 11.3). Flavones and flavonols were methylated more efficiently than flavanones and dihydroflavonols. The substrate specificities indicate that in soybean, as in parsley, methyl groups are introduced into flavonoid compounds after formation of the flavonoid ring structure.

The question of whether methylation precedes glucosylation in the biosynthesis of methylated flavonoid glycosides cannot be answered unambiguously. As with the parsley 3′-*O*-methyltransferase, both luteolin and its 7-*O*-glucoside were efficient substrates of the soybean 3′-*O*-methyl transferase. On the other hand, quercetin 3-*O*- glucoside and 3-*O*-rutinoside (rutin) were methylated at lower rates than quercetin. These results suggest that, at least in the biosynthesis of methylated flavonol 3-*O*-glucosides, methylation precedes glucosylation (Poulton *et al.*, 1977). Some other properties of flavonoid *O*-methyltransferases are listed in Table 11.4.

During pollen development in *Tulipa* anthers, these organs successively accumulate simple cinnamic acid derivatives, chalcones, flavonols and finally anthocyanins (Wiermann, 1970; Quast and Wiermann, 1973). Methylated derivatives, such as ferulate and 3′-*O*-methylquercetin (isorhamnetin), constitute a considerable portion of the phenylpropanoids produced. Two distinct *S*-adenosyl-L-methionine:*ortho*-dihydroxyphenol *meta*-*O*-methyltransferases were detected in protein extracts of tulip anthers and could be partially separated upon protein fractionation (Sütfeld and Wiermann, 1978). The properties of the two enzymes correlated closely with those of caffeate and flavonoid *O*-methyltransferases from soybean cell cultures. Analysis of the distribution of these enzymes between the pollen and tapetum fractions of the anthers revealed that both methyltransferases were predominantly present in the tapetum fraction (Herdt *et al.*, 1978). However, the activity changes of the two enzymes during various stages of pollen development were remarkably different (Sütfeld and Wiermann, 1980b). While a close correlation between high activity of caffeate *O*-methyltransferase and the accumulation of large amounts of ferulic acid

derivatives was observed, no such relationship between the level of flavonoid *O*-methyltransferase activity and the formation of methylated flavonol glycosides was found. Since chalcone synthase from tulip anthers possesses a relatively broad substrate specificity *in vitro*, utilizing 4-coumaroyl-CoA, caffeoyl-CoA and feruloyl-CoA with similar efficiencies (see above), it seems possible that in tulip anthers methylation occurs, at least in part, at the C_9 stage (Sütfeld and Wiermann, 1980b).

Three distinct *ortho*-dihydroxyphenol *O*-methyltransferases were identified in leaves of *Nicotiana tabacum* var. Samsun NN (Legrand *et al.*, 1976a,b, 1978). They could be separated by ion-exchange chromatography on DEAE-cellulose and distinguished both by differences in specificity towards a variety of *ortho*-diphenolic substrates and by the degree of enhancement of their activities upon virus infection of tobacco leaves. Caffeate was most efficiently methylated by transferase I, and quercetin, the major *ortho*-dihydroxyphenol of tobacco leaves, by transferases II and III. Rutin (quercetin 3-*O*-rutinoside) was not a substrate for either one of the enzymes, again suggesting that methylation precedes glycosylation.

Although flavonoids *O*-methylated in ring A frequently occur in nature (Harborne *et al.*, 1975), *in vitro* *O*-methylation of ring A is not well documented. Citrus tissues, especially the fruit flavedo, produce a complement of polymethylated flavones characteristic of the genus (Swift, 1967; Kefford and Chandler, 1970). Recently, investigations of a stepwise methylation of quercetin and other flavonoids by cell-free extracts of peel, root, or callus tissues of calamondin (*Citrus mitis*) have been reported (Brunet *et al.*, 1978; Brunet and Ibrahim, 1980). Product identification by thin-layer chromatography indicated that methylation occurred at almost all hydroxyl groups of the various substrates tested. The authors concluded that citrus tissues contain several transferases methylating hydroxyl groups at different positions of the A, B and C ring of the flavonoid skeleton.

Texasin (6,7-dihydroxy-4′-methoxyisoflavone) was efficiently methylated by the flavonoid *O*-methyltransferase from soybean cell cultures, but the exact position of methylation remains unknown (Poulton *et al.*, 1977). Methylation of quercetin and luteolin at both 7 and 3′ positions has been observed with a partially purified methyltransferase of tobacco (*Nicotiana tabacum*) cell cultures (Tsang and Ibrahim, 1979a,b). Two forms of the enzyme were resolved by DEAE-cellulose chromatography. One enzyme catalysed the methylation of quercetin almost exclusively at the 7 position to yield rhamnetin, whereas the other methylated caffeate predominantly at the 3 position to give ferulate. It has not been reported whether one of the transferases isolated from tobacco leaves (Legrand *et al.*, 1978) is also capable of methylating quercetin or luteolin in the 7 position. One important aspect of further investigations should concern the question of whether A-ring methylation is catalysed by transferase activities with broad substrate specificity or by distinct enzymes with high preference for specific hydroxyl groups.

Table 11.4 Methyltransferases from various sources

Source of enzyme	*Number of transferases reported*	*Basis for distinction between enzymes*	*Substrate specificity: compounds converted and position of methylation (in parenthesis)*	*pH optimum*	*Effects of Mg^{2+} ions and EDTA*	*Remarks*	*Reference*
Cicer arietinum cell cultures	1		Daidzein and genistein (4′ position)	9.0	None	Inhibition by SAH*	Wengenmayer *et al.* (1974)
Glycine max cell cultures	2	Sephadex G-200 chromatography, stability upon storage					
1. Caffeate 3-*O*-methyltransferase			1. Caffeate, 5-hydroxy-ferulate, 3,4,5-trihydroxy cinnamate (3 position)	6.5–7.0	None	Inhibition by SAH, luteolin, and quercetin	Poulton *et al.* (1976)
2. Flavonoid 3′-*O*-methyltransferase			2. Caffeate, 5-hydroxy-ferulate (3 position), luteolin and 7-*O*-gluco-side, eriodictiol, quer-cetin (3′ position)	8.6–8.9	Stimulation by Mg^{2+}, inhibition by EDTA	Inhibition by SAH	Poulton *et al.* (1977)
Nicotiana tabacum leaves	3	DEAE-cellulose chromatography	Caffeate, 5-hydroxyferulate (3 position), protocate-chuate (3 and 4 positions), esculetin (6,7-dihydroxy-coumarin, 6 and 7 positions), quercetin (3′ position)	?†	?†	Different specificities of the three transferases toward phenolic substrates	Legrand *et al.* (1976b, 1978)
cell cultures	2	DEAE-cellulose chromatography	Caffeate, 5-hydroxyferulate, 3,4,5-trihydroxycinnamate (3 and 4 positions), daphnetin and esculetin (6 and 7 positions), luteolin and quercetin (7 and 3′ positions)	7.3 and 8.3	None		Tsang and Ibrahim (1979a,b)
Petroselinum hortense cell cultures	1		Caffeate (3 position) luteolin and 7-*O*-glucoside, eriodictyol, quercetin (3′ position)	9.7	Stimulation by Mg^{2+}, inhibition by EDTA		Ebel *et al.* (1972)

Tulipa anthers	2	Sephadex G-200 chromatography, isoelectric focusing, disc-gel electrophoresis					Sütfeld and Wiermann (1978)
1. Caffeate 3-*O*-methyltransferase			1. Caffeate, 5-hydroxyferulate (3 position)	7.6	None	Inhibition by SAH	
2. Flavonoid 3′-*O*-methyltransferase			2. Luteolin, quercetin (3′ position)	7.6	Mg^{2+} required for full activity	Inhibition by SAH	

* *S*-Adenosyl-L-homocysteine.

† Data not reported.

11.4.7 Glycosylation

It is most likely that glycosylation occurs subsequent to all other substitutions and modifications of the flavonoid ring structure (Hahlbrock and Grisebach, 1975). The various 7-*O*-glycosides and 3,7-di-*O*-glycosides in parsley are formed by sequential glycosylation steps as indicated in Fig. 11.3. The first reaction of this sequence (No. 8) is identical for flavones and flavonols and is catalysed by UDP-glucose:flavone/flavonol 7-*O*-glucosyltransferase (Table 11.5). The enzyme has a broad specificity for several flavone, flavonol and flavanone aglycones, but does not glucosylate various other phenolic compounds, including isoflavones and cyanidin (Sutter *et al.*, 1972). Quercetin 3-*O*-glucoside is not further glucosylated in the 7-*O*-position (Sutter and Grisebach, 1973). This suggests that the reaction sequence for the 7-*O*- and 3-*O*-glucosylation of flavonols in parsley is as indicated in Fig. 11.3. UDP-glucose or TDP-glucose can serve as glucosyl donors *in vitro*.

The second glucosylation of flavonols is catalysed by UDP-glucose:flavonol 3-*O*-glucosyltransferase (Fig. 11.3, No. 9) which was completely separated from 7-*O*-glucosyltransferase on the basis of different net charges of the two enzymes (Sutter and Grisebach, 1973). Flavonol 3-*O*-glucosyltransferase exhibits a strict positional specificity and catalyses the 3-*O*-glucosylation of a number of flavonols, including quercetin 7-*O*-glucoside, *in vitro* (Table 11.5). It is not known whether both aglycones and 7-*O*-glucosides are substrates *in vivo*, but this is likely because, at least in parsley, 3-*O*-glucosides do not serve as substrates for 7-*O*-glucosylation (see above). Glucosyl transfer from UDP-glucose to the 3-*O* position of kaempferol or quercetin was freely reversible in the presence of 3-*O*-glucosyltranferase (Sutter and Grisebach, 1975). It is not known at present whether or not this interesting feature is of significance *in vivo*.

Further glycosylation of the sugar moiety of flavonoid glucosides occurs in the biosynthesis of apiosylglucosides (Fig. 11.3). The branched-chain sugar, apiose, is found in parsley only in flavone glycosides and not in other glycosides (Sandermann, 1969). The synthesis of the sugar nucleotide, UDP-apiose (reaction No. 10), and the apiosyltransfer from UDP-apiose (No. 11) are therefore specifically related to flavonoid biosynthesis in this plant. UDP-apiose synthase (complete name: UDP-D-apiose/UDP-D-xylose synthase) catalyses the NAD^+-dependent conversion of UDP-D-glucuronate to give two products *in vitro*, UDP-apiose and UDP-xylose. The significance of UDP-xylose synthesis is not known, but might be an artificial side reaction occurring only under the assay conditions *in vitro*. The ratio of UDP-apiose/UDP-xylose formation remained constant during a 1400-fold purification of the synthase from irradiated parsley cells (Matern and Grisebach, 1977). Gel electrophoresis under denaturing conditions and sedimentation-equilibrium measurements showed that the highly purified enzyme was composed of two proteins with molecular weights of about 65 000 and 86 000 which were present in approximately equimolar amounts. Each protein contained two apparently identical subunits. Only the larger protein showed synthase activity. Studies on the reaction meachanism indicated

that an L-*threo*-4-pentosulose was an intermediate in the synthesis of both sugar nucleotides. The reaction appears not to involve an intermediate bound to the enzyme as a Schiff base (Baron and Grisebach, 1973; Matern and Grisebach, 1977) in contrast to the conversion of fructose 1,6-bisphosphate by class I aldolase (Horecker *et al.*, 1972).

Transfer of the apiosyl residue from UDP-apiose to flavone 7-*O*-glucosides is catalysed in parsley by UDP-D-apiose:flavone 7-*O*-glucoside 2″-*O*-apiosyltransferase (Fig. 11.3, No. 11). The reaction products are flavone 7-*O*-[β-D] apiofuranosyl(1→2)[β-D]glucosides, such as apiin (apigenin 7-*O*-apiosylglucoside) or graveobioside B (chrysoeriol 7-*O*-apiosylglucoside) (Ortmann *et al.*, 1972). The enzyme is specific for UDP-apiose as glycosyl donor and acts on 7-*O*-glucosides of a variety of flavones, flavanones and isoflavones (Table 11.5), but does not accept flavonol 7-*O*- or 3-*O*-glucosides, flavonoid aglycones or free glucose as substrates. The substrate specificity of apiosyltransferase *in vitro* is compatible with the observation that only flavone, but no flavonol apiosylglucosides, occur in parsley.

Several enzymic reactions in crude extracts catalysing the formation of flavonol glycosides from the corresponding aglycones and sugar nucleoside diphosphates have been reported (for an earlier review, see Hahlbrock and Grisebach, 1975). Here, only the more recent work will be discussed. UDP-glucose:flavonol and/or UDP-glucose:anthocyanidin 3-*O*-glucosyltransferases have been demonstrated in protein extracts from various sources (Table 11.5). In those cases where flavonols and anthocyanidins were tested, most of the 3-*O*-glucosyltransferase activities glucosylated both types of substrate, albeit with different efficiencies. The 3-*O*-glucosyltransferases had a pronounced positional specificity. In the cases investigated, glucosylation could not be detected at the 5 position (Kamsteeg *et al.*, 1978a) or 7 position (Saleh *et al.*, 1976a,b). Anthocyanidin 3-*O*-glucosyltransferase from *Silene dioica* specifically glucosylated anthocyanidins, but did not convert flavonols into the respective 3-*O*-glucosides (Kamsteeg *et al.*, 1978a). It is not known whether 3-*O*-glucosylation of both flavonols and anthocyanidins by other enzyme preparations reflects a broader substrate specificity or the presence of more than one enzyme exhibiting different specificities. All 3-*O*-glucosyltransferases used UDP-glucose as glucosyl donor. Optimal activities between pH 7.5 and pH 8.5 have been reported for the various 3-*O*-glucosyltransferase reactions.

One important aspect of research concerning flavonol/anthocyanidin 3-*O*-glucosyltransferase is the use of genetically defined mutants. In maize (*Zea mays*) the amount of 3-*O*-glucosyltransferase activity in different plant tissues and the dosage of the wild-type *Bz* allele are directly correlated (Larson and Coe, 1977). Dooner and Nelson (1977a, 1979a) reported a more complex control of this enzyme by the action of four genes in maize endosperm. *Bz* appeared to be the structural gene, whereas the nature of the genes *C*, *R* and *Vp* is not known and their effect seemed to be indirect. The *Bz* locus in maize is one of many loci at which controlling element-induced phenotypic changes are known (Dooner and Nelson, 1977b, 1979b).

Table 11.5 Glycosyltransferases from various sources

Source of enzyme	Flavonoid compounds tested as possible substrates	Position of glycosylation	Glycosyl donor	pH Optimum	Reference
Brassica oleracea seedlings	Various anthocyanidins, flavonols, flavones, dihydroquercetin, naringenin	3	UDP-glucose	8	Saleh *et al.* (1976b)
	Malvidin	3	UDP-glucose		Hrazdina *et al.* (1980)
Glycine max cell cultures	Various flavonols, cyanidin, apigenin, luteolin	3	UDP-glucose	8.5	Poulton and Kauer (1977)
Haplopappus gracilis cell cultures	Various anthocyanidins, flavonols, flavones, dihydroquercetin, naringenin	3	UDP-glucose	8	Saleh *et al.* (1976a)
Hippeastrum petals	Malvidin	3	UDP-glucose		Hrazdina *et al.* (1978)
Petroselinum hortense cell cultures					
1. Flavone/flavonol 7-*O*-glucosyl-transferase	Various flavones, flavonols, flavanones, quercetin 3-*O*-glucoside, cyanidin, daidzein, biochanin A	7	UDP-glucose, TDP-glucose	7.5	Sutter *et al.* (1972)
2. Flavonoid 7-*O*-glucoside apiosyltrans-ferase	7-Glucosides of various flavones, flavanones, isoflavones, flavonol 7-*O*-glucosides	2″	UDP-apiose	7.0	Ortmann *et al.* (1972)
3. Flavonol 3-*O*-glucosyl-transferase	Various flavonols, quercetin 7-*O*-glucoside, dihydroquercetin	3	UDP-glucose	8.0	Sutter and Grisebach (1973, 1975)

Petunia hybrida corollas	Cyanidin, delphinidin	3	UDP-glucose	7.5–8.0	Kho *et al.* (1978)
Phaseolus vulgaris seedlings	Malvidin	3	UDP-glucose		Hrazdina *et al.* (1980)
Pisum sativum seedlings	Malvidin	3	UDP-glucose		Hrazdina *et al.* (1980)
Silene alba petals, green parts	Isovitexin, iso-orientin vitexin	2″	UDP-glucose	8.5 (isovitexin) 7.5 (iso-orientin)	van Brederode *et al.* (1979)
Silene dioica petals					
1. Anthocyanidin 3-*O*-glucosyl-transferase	Several anthocyanidins, cyanidin 3-*O*-glucoside cyanidin 3-*O*-rhamnosyl-glucoside, kaempferol, quercetin	3	UDP-glucose	7.5	Kamsteeg *et al.* (1978a)
2. Anthocyanidin 3-*O*-glucoside rhamnosyltrans-ferase	3-*O*-Glucosides of cyanidin, pelargonidin, delphinidin	6″	UDP-rhamnose	8.1	Kamsteeg *et al.* (1980a)
3. Anthocyanidin 3-rhamnosyl-glucoside 5-*O*-glucosyl-transferase	Anthocyanidin 3-*O*-glucosides, anthocyanidin 3-*O*-rhamnosyl-glucosides	5	UDP-glucose	6.5 (3-*O*-glucosides) 7.4 (3-*O*-rhamnosyl-glucosides)	Kamsteeg *et al.* (1978b) Kamsteeg *et al.* (1980b)
Spinacea oleracea seedlings	Malvidin	3			Hrazdina *et al.* (1980)
Tulipa leaves, petals, pollen	Quercetin, malvidin	3	UDP-glucose		Hrazdina *et al.* (1978) Wiermann and Buth-Weber (1980)
Zea mays seeds, seedlings, pollen	Quercetin anthocyanidins	3	UDP-glucose	8.2	Larson (1971) Larson and Lonergan (1973)

Other glucosyltransferases (Table 11.5) whose activities have been related specifically to anthocyanin biosynthesis are discussed in Section 11.5.

11.4.8 Acylation

Malonylation (Fig. 11.3, Nos. 12 and 13) of the sugar moieties is the last step in the biosynthesis of flavonoid glycosides in cultured parsley cells (Hahlbrock, 1972; Ebel and Hahlbrock, 1977). Several malonylated flavone and flavonol glycosides have been isolated from irradiated cell suspension cultures of parsley (Kreuzaler and Hahlbrock, 1973), and it is possible that the corresponding non-malonylated compounds were formed artificially by deacylation during the isolation procedure (Matern *et al.*, 1981). Two malonyltransferases have been isolated from irradiated parsley cells and extensively purified (Matern *et al.*, 1981). Both transferases have an apparent molecular weight of approximately 50 000 and are probably glycoproteins. One enzyme (Fig. 11.3, No. 12) is most active with flavone and flavonol 7-*O*-glycosides as substrates, such as apigenin 7-*O*-glucoside and the 7-*O*-apiosylglucosides of apigenin and chrysoeriol. The other enzyme (Fig. 11.3, No. 13) preferentially malonylates the 3-*O*-glucosides of kaempferol, quercetin and isorhamnetin. All flavone and flavanone 7-*O*-glucosides tested were relatively poor substrates of the second enzyme. The substrate specificities of the two enzymes *in vitro* are in good agreement with the pattern of malonylated flavonoid glycosides occurring in the cell cultures *in vivo*. Further acylations will be discussed below in Section 11.5.

11.4.9 Proposed sequence of reactions

Our present knowledge of the biosynthetic pathways of flavone and flavonol glycosides is illustrated in Fig. 11.3. The sequence of the individual reactions has been derived mainly from enzyme studies in parsley but is likely to be the same in other plants. The scheme indicates that some of the reactions of both flavone and flavonol glycoside formation (e.g. reactions 7,8 and 12) are probably catalysed by the same enzyme. Other reactions (e.g. 9,10,11 and 13) are specific for either flavone or flavonol glycoside formation. As discussed above, little is known at present about reactions 4–6, and their exact sequence is unknown. Some of the early steps in flavone and flavonol glycoside synthesis are probably common to the formation of most, or all, flavonoid classes, including anthocyanins and isoflavonoids.

11.4.10 Regulation of enzyme activities

Regulation of flavonoid biosynthesis is closely correlated with differentiation and development in the plant tissue producing the pigments. Various types of endogenous and exogenous stimuli, including hormones, nutrients, light and temperature, are known to exert regulatory influences on the enzymes of

flavonoid biosynthesis. Even a brief summary of this complex subject would be far beyond the scope of this chapter. For recent reviews, the reader is referred to articles by Hahlbrock and Grisebach (1979) and Wiermann (1981). To give here only one illustrative example, Fig. 11.8 shows the time courses of large light-

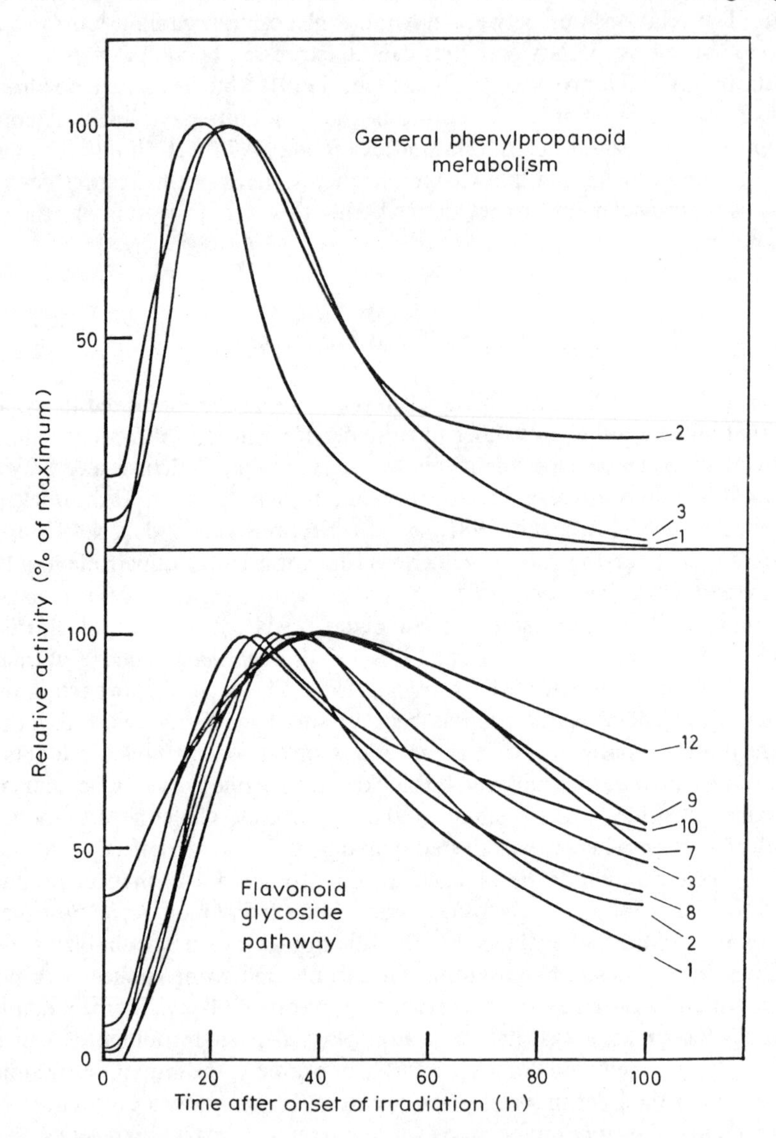

Fig. 11.8 Activity changes for the three enzymes of general phenylpropanoid metabolism and for eight enzymes of the flavonoid glycoside pathway during continuous irradiation of a previously dark-grown parsley cell culture. The numbering of enzymes is the same as in Table 11.1 (after Hahlbrock *et al.*, 1976; Ebel and Hahlbrock, 1977; Heller and Hahlbrock, 1980).

induced changes in the activities of the enzymes related to flavone and flavonol glycoside biosynthesis in cultured parsley cells. In this particular case, the changes in enzyme activities were shown to be caused by corresponding changes in the activities of the respective mRNAs (Schröder *et al.*, 1979b; Gardiner *et al.*, 1980). The relationships between flavonoid glycoside accumulation rates and enzyme as well as mRNA activities can be expressed by simple mathematical equations (cf. Hahlbrock and Grisebach, 1979). Under certain conditions, phenylalanine ammonia-lyase seems to be the rate-limiting enzyme of flavonoid biosynthesis in parsley cells (Hahlbrock *et al.*, 1976). It should be noted, however, that our limited knowledge of other systems in this respect does not allow us to draw conclusions yet as to whether this is the exception or rather the rule.

11.5 ANTHOCYANINS

Tracer studies with ^{3}H- and ^{14}C-labelled precursors have confirmed the general rule that anthocyanins, like other flavonoids, are synthesized from acetate and phenylalanine. These experiments also established that chalcones and dihydroflavonols are intermediates in anthocyanin biosynthesis (cf. Hahlbrock and Grisebach, 1975; Grisebach, 1981; see also Sections 11.3 and 11.4.4). Further strong support for the role of chalcones, flavanones and dihydroflavonols as intermediates has been derived from studies with genetically defined lines of *Antirrhinum majus* (Harrison and Stickland, 1974; Stickland and Harrison, 1974), *Petunia hybrida* (Kho *et al.*, 1975, 1977), *Matthiola incana* (Forkmann, 1977, 1980) and *Zea mays* (McCormick, 1978). The general approach in these studies with defined genotypes has been to supplement putative biosynthetic intermediates subsequent to the genetic block. For example, mutants of *M. incana*, which were assumed to be blocked in chalcone synthesis, accumulated cyanidin glycosides upon feeding of the appropriate chalcone, flavanone or dihydroflavonol, whereas substituted cinnamic acids were ineffective. Mutants of *P. hybrida* thought to be blocked in dihydroflavonol formation produced cyanidin and delphinidin glycosides when dihydroflavonols were supplied as precursors. Additional evidence for the function of dihydroflavonols as intermediates in anthocyanin biosynthesis was obtained from studies with plant tissues which had been exposed to potent inhibitors of phenylalanine ammonia-lyase, such as amino-oxyacetate or α-amino-oxy-β-phenylpropionate (Amrhein *et al.*, 1976; Amrhein and Gödeke, 1977). For example, the inhibition by amino-oxyacetate of the light-induced formation of cyanidin glycosides in buckwheat (*Fagopyrum esculentum*) hypocotyls was completely or partially reversed in the presence of dihydrokaempferol or dihydroquercetin (Amrhein, 1979).

The co-occurrence of anthocyanins and flavonols in a large variety of plants suggests that anthocyanin biosynthesis may be closely related to flavonol biosynthesis (Fig. 11.3 and Fig. 11.7). In agreement with this assumption, the

enzymes chalcone synthase, chalcone isomerase and flavanone 3-hydroxylase have been demonstrated in a number of anthocyanidin-forming plants and plant tissues. In suspension-cultured cells of *Haplopappus gracilis*, irradiation with UV light greatly stimulated the activities of phenylalanine ammonia-lyase, chalcone synthase and chalcone isomerase concomitant with cyanidin glycoside accumulation (Wellmann *et al.*, 1976). Flavanone 3-hydroxylase activity was demonstrated in cell-free extracts from flowers of *Matthiola incana* and *Antirrhinum majus* which synthesize large amounts of anthocyanins (Forkmann *et al.*, 1980; Forkmann and Stotz, 1981). The important function of chalcone synthase, chalcone isomerase and flavanone 3-hydroxylase in anthocyanin biosynthesis has also been documented by studies with defined genotypes of several plants in which certain genetic blocks in the anthocyanin pathway were correlated with the absence of the respective enzyme activity (Kuhn *et al.*, 1978; Forkmann and Kuhn, 1979; Forkmann and Dangelmayr, 1980; Forkmann and Stotz, 1981; Spribille and Forkmann, 1981, see also above). The conversion of dihydroflavonols into anthocyanidins has yet to be demonstrated in cell-free extracts.

As in the cases of flavones and flavonols (Sections 11.4.7 and 11.4.8), glycosylation and acylation are probably the last steps in anthocyanin biosynthesis. It can be assumed that the first glycosylation occurs at the 3-*O*-position of the anthocyanidin. Because of the instability of the flavylium cation, 3-*O*-glycosylation is very likely a prerequisite for anthocyanin accumulation *in vivo*. Further glycosyl moieties may then be transferred to the 3-*O*-glycosyl residue and/or to the 5 position of anthocyanidin 3-*O*-glycosides. Several 3-*O*-glucosyltransferases from various sources are listed in Table 11.5; some of their properties have been discussed in Section 11.4.7. Most of these enzymes efficiently glucosylate both anthocyanidins and flavonols, whereas a 3-*O*-glucosyltransferase from *Silene dioica* was reported to be specific for anthocyanidins (Kamsteeg *et al.*, 1978a).

In petals of *S. dioica*, additional glycosylations occur in the biosynthesis of anthocyanidin 3,5-di-*O*-glycosides. According to Kamsteeg *et al.* (1980a), glycosylation of 3-*O*-glucosides is catalysed by UDP-rhamnose:anthocyanidin 3-*O*-glucoside 6″-*O*-rhamnosyltransferase which transfers the rhamnosyl moiety to the 6-hydroxyl group of the 3-*O*-linked glucose. The 3-*O*-glucosides of pelargonidin, cyanidin and delphinidin are more efficient substrates for rhamnosylation than the 3,5-di-*O*-glucosides. A third transferase from *S. dioica* catalyses the transfer of glucose from UDP-glucose to the 5-hydroxyl group of anthocyanins (Kamsteeg *et al.*, 1978b). The substrate specificity of the 5-*O*-glucosyltransferase was reported to be pH-dependent (Kamsteeg *et al.*, 1980b).

The anthocyanidin glycosides from petals of *S. dioica* occur mostly in the acylated form, with 4-coumarate or caffeate being the acyl residue. An enzyme catalysing the transfer of the 4-coumaroyl or caffeoyl moiety from the corresponding CoA ester to the 4-hydroxyl group of the rhamnosyl moiety of anthocyanidin 3-*O*-rhamnosylglucosides or 3-*O*-rhamnosylglucoside-5-*O*-glucosides has been demonstrated in this plant tissue (Kamsteeg *et al.*, 1980c).

11.6 ISOFLAVONOIDS

As in the case of anthocyanins, most of our knowledge about the biosynthesis of isoflavonoids originates from studies with radioactive isotopes. Again, it was established that acetate gives rise to ring A and that phenylalanine or cinnamate (or ring-substituted cinnamic acids) are incorporated into ring B and carbon atoms 2,3 and 4 of the heterocyclic ring C (cf. Hahlbrock and Grisebach, 1975). The mode of incorporation of phenylpropanoid compounds into isoflavonoids is depicted in Fig. 11.9. The central reaction where the isoflavonoid pathway branches off from the common biosynthetic pathway of all flavonoids is a 1→2 aryl migraton. Since chalcones and flavanones are efficient precursors of isoflavonoids, the migration of ring B from the former β to the former α carbon atom of the phenylpropanoid precursor must take place after formation of the basic C_{15} skeleton. However, the actual mechanism of aryl migration is not known (see Chapter 10). The chemical structures of isoflavonoids exhibit a wide range of modifications. The conversion of basic into structurally modified isoflavonoids is well documented (Chapter 10).

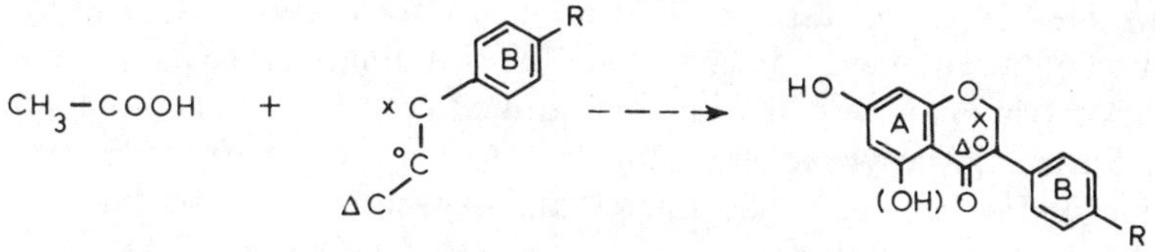

CH_3-COOH + [phenylpropanoid precursor] ---→ [isoflavonoid]

Fig. 11.9 Mode of incorporation of phenylpropanoid compounds into isoflavonoids.

Recent studies on the induced synthesis of isoflavonoid phytoalexins in legumes have contributed to some extent to the still limited knowledge about the enzymology of isoflavonoid biosynthesis. Phytoalexins constitute a chemically heterogeneous class of naturally occurring antimicrobial agents, including isoflavonoid as well as non-flavonoid compounds. They are produced in various plants in response to microbial attack (Grisebach and Ebel, 1978). For example, various tissues of soybean (*Glycine max*) respond to the fungal pathogen *Phytophthora megasperma* f. sp. *glycinea* and to an elicitor isolated from this fungus by synthesizing large amounts of glyceollin, which was identified as a mixture of structurally related pterocarpanoid, i.e. isoflavonoid, compounds (Burden and Bailey, 1975; Lyne *et al.*, 1976; Lyne and Mulheirn, 1978; Grisebach and Ebel, 1978). The accumulation of glyceollin in challenged cotyledons and cell cultures of soybean coincides with large increases in the activities of several enzymes of both general phenylpropanoid metabolism (phenylalanine ammonia-lyase, cinnamate 4-hydroxylase and 4-coumarate:CoA ligase) and flavonoid biosynthesis (chalcone synthase and chalcone isomerase) (Ebel *et al.*, 1976; Zähringer *et al.*, 1978; Ebel, 1979; Hille *et al.*, 1982). This coincidence confirms the conclusions drawn from tracer studies that the same

enzymes are involved in the initial steps of isoflavonoid biosynthesis as in other flavonoid pathways. Further investigations demonstrated a dimethylallyl pyrophosphate:3,6a,9-trihydroxypterocarpan dimethylallyltransferase in a particulate fraction from elicitor-treated cotyledons and cell suspension cultures of soybean (Zähringer *et al.*, 1979, 1981). The major product of dimethylallyl transfer *in vitro* was 2-dimethylallyl-trihydroxypterocarpan. A minor product was the 4-substituted pterocarpan (Zähringer *et al.*, 1979, 1981). Both products are likely intermediates in the biosynthesis of the various structurally related soybean phytoalexins. Introduction of a prenyl substituent into genistein (5,7,4′-trihydroxyisoflavone) and 2′-hydroxygenistein is catalysed by a particulate fraction from white lupin (*Lupinus albus*) (Schröder *et al.*, 1979c). Another isoflavonoid phytoalexin, phaseollin, accumulates in cell suspension cultures of french bean (*Phaseolus vulgaris*) in response to treatment with an elicitor from the bean pathogen, *Colletotrichum lindemuthianum*, or with an artificial elicitor, autoclaved ribonuclease A. This accumulation is accompanied by large increases in the activities of enzymes of both general phenylpropanoid metabolism and flavonoid biosynthesis (Dixon and Bendall, 1978a,b; Dixon and Lamb, 1979).

4′-*O*-Methylation of ring B of isoflavonoids is catalysed by a 4′-*O*-methyltransferase from seedlings or cell suspension cultures of chick pea (*Cicer arietinum*) (Wengenmayer *et al.*, 1974). The enzyme is specific for 4′-hydroxyisoflavones, such as daidzein (7,4′-dihydroxyisoflavone) or genistein, and does not methylate 4-coumarate, naringenin, apigenin nor luteolin (Table 11.4). The high specificity of this transferase for isoflavones suggests that 4′-*O*-methylation of isoflavonoids takes place after aryl migration.

11.7 OTHER FLAVONOIDS

Minor groups of flavonoids are the biflavonoids, proanthocyanidins, *C*-glycosylflavonoids and neoflavonoids. Feeding experiments with radioactively labelled precursors have demonstrated that they are synthesized according to the general scheme for flavonoid formation (Fig. 11.2), but as yet no enzymic studies have been reported with these compounds. For the chemistry, biochemistry and taxonomic distribution of these compounds see Chapters 7, 8 and 9.

REFERENCES

Abbott, M. T. and Udenfriend, S. (1974). In *Molecular Mechanism of Oxygen Activation* (ed. O. Hayaishi), Academic Press, New York, pp. 167–214.

Ahmad, F., Ahmad, P. M., Pieretti, L. and Watters, G. T. (1978), *J. Biol. Chem.* **253**, 1733.

Amrhein, N. (1979), *Phytochemistry* **18**, 585.

Amrhein, N. and Gödeke, K. H. (1977), *Plant Sci. Lett.* **8**, 313.

Amrhein, N., Gödeke, K. H. and Kefeli, V. I. (1976), *Ber. Dtsch. Bot. Ges.* **89**, 247.

Baron, D. and Grisebach, H. (1973), *Eur. J. Biochem.* **38**, 153.

Benveniste, I., Salaun, J. P. and Durst, F. (1977), *Phytochemistry* **16**, 69.
Birch, A. J. and Donovan, F. W. (1953), *Aust. J. Chem.* **6**, 360.
Boland, M. J. and Wong, E. (1975), *Eur. J. Biochem.* **50**, 383.
Boland, M. J. and Wong, E. (1979), *Bioorg. Chem.* **8**, 1.
Britsch, L., Hellen, W. and Grisebach, H. (1981), *Z. Naturforsch* **36c**, 742.
Brock, K. and Kannangara, C. G. (1976), *Carlsberg Res. Commun.* **41**, 121.
Brunet, G. and Ibrahim, R. K. (1980), *Phytochemistry* **19**, 741.
Brunet, G., Saleh, N. A. M. and Ibrahim, R. K. (1978), *Z. Naturforsch.* **33C**, 786.
Burden, R. S. and Bailey, J. A. (1975), *Phytochemistry* **14**, 1389.
Butt, V. S. and Wilkinson, E. M. (1979). In *Regulation of Secondary Product and Plant Hormone Metabolism* (eds. M. Luckner and K. Schreiber), Pergamon Press, Oxford, Vol. 55, pp. 147–154.
Büche, T. and Sandermann, H., Jr. (1973), *Arch. Biochem. Biophys.* **158**, 445.
Camm, E. L. and Towers, G. H. N. (1973), *Phytochemistry* **12**, 961.
Dixon, R. A. and Bendall, D. S. (1978a), *Physiol. Plant Pathol.* **13**, 283.
Dixon, R. A. and Bendall, D. S. (1978b), *Physiol. Plant Pathol.* **13**, 295.
Dixon, R. A. and Lamb, C. J. (1979), *Biochim. Biophys. Acta* **586**, 453.
Dooner, H. K. and Nelson, O. E. (1977a), *Biochem. Genet.* **15**, 509.
Dooner, H. K. and Nelson, O. E., Jr. (1977b), *Proc. Natl. Acad. Sci. USA* **74**, 5623.
Dooner, H. K. and Nelson, O. E. (1979a), *Genetics* **91**, 309.
Dooner, H. K. and Nelson, O. E., Jr. (1979b), *Proc. Natl. Acad. Sci. USA* **76**, 2369.
Ebel, J. (1979). In *Regulation of Secondary Product and Plant Hormone Metabolism* (eds. M. Luckner and K. Schreiber), Pergamon Press, Oxford, Vol. 55, pp. 155–162.
Ebel, J. and Hahlbrock, K. (1977), *Eur. J. Biochem.* **75**, 201.
Ebel, J., Hahlbrock, K. and Grisebach, H. (1972), *Biochim. Biophys. Acta* **268**, 313.
Ebel, J., Schaller-Hekeler, B., Knobloch, K.-H., Wellmann, E., Grisebach, H. and Hahlbrock, K. (1974), *Biochim. Biophys. Acta* **362**, 417.
Ebel, J., Ayers, A. R. and Albersheim, P. (1976), *Plant Physiol.* **57**, 775.
Egin-Bühler, B., Loyal, R. and Ebel, J. (1980), *Arch. Biochem. Biophys.* **203**, 90.
Forkmann, G. (1977), *Planta* **137**, 159.
Forkmann, G. (1980), *Planta* **148**, 157.
Forkmann, G. and Dangelmayr, B. (1980), *Biochem. Genet.* **18**, 519.
Forkmann, G. and Kuhn, B. (1979), *Planta* **144**, 189.
Forkmann, G. and Stotz, G. (1981), *Z. Naturforsch.* **36C**, 411.
Forkmann, G., Heller, W. and Grisebach, H. (1980), *Z. Naturforsch.* **35C**, 691.
Fritsch, H. and Grisebach, H. (1975), *Phytochemistry* **14**, 2437.
Fritzemeier, K.-H. and Kindl, H. (1981), *Planta* **151**, 48.
Fuisting, K. and Weissenböck, G. (1980), *Z. Naturforsch.* **35C**, 973.
Gardiner, S. E., Schröder, J., Matern, U., Hammer, D. and Hahlbrock, K. (1980), *J. Biol. Chem.* **255**, 10752.
Grambow, H. J. and Grisebach, H. (1971), *Phytochemistry* **10**, 789.
Grisebach, H. (1959). In *Proceedings of the IVth International Congress of Biochemistry, Wien, 1958*, Pergamon Press, London, Vol. 2, pp. 56–70.
Grisebach, H. (1979). In *Biochemistry of Plant Phenolics* (eds. T. Swain, J. B. Harborne and Ch. F. van Sumere), Plenum Press, New York, pp. 221–248.
Grisebach, H. (1981). In *Anthocyanins as Food Colours* (ed. P. Markakis), Academic Press, New York, pp. 69–92.
Grisebach, H. and Ebel, J. (1978), *Angew. Chem.* **90**, 668; *Angew. Chem. Int. Ed. Engl.* **17**, 635.
Gross, G. G. (1977), *Recent Adv. Phytochem.* **11**, 141.
Gross, G. G. and Zenk, M. H. (1974), *Eur. J. Biochem.* **42**, 453.
Gross, G. G., Stöckigt, J., Mansell, R. L. and Zenk, M. H. (1973), *FEBS Lett.* **31**, 283.
Gross, G. G., Mansell, R. L. and Zenk, M. H. (1975), *Biochem. Physiol. Pflanzen* **168**, 41.
Hahlbrock, K. (1972), *FEBS Lett.* **28**, 65.
Hahlbrock, K. (1977). In *Plant Tissue Culture and Its Biotechnological Application* (eds. W. Barz, E. Reinhard and M. H. Zenk), pp. 95–111.

Hahlbrock, K. (1981). In *The Biochemistry of Plants* (eds. E. E. Conn and P. K. Stumpf), Vol. 7, Academic Press, New York, Vol. 7, pp. 425–456.

Hahlbrock, K. and Grisebach, H. (1975). In *The Flavonoids* (eds. J. B. Harborne, T. J. Mabry and H. Mabry), Chapman and Hall, London, pp. 866–915.

Hahlbrock, K. and Grisebach, H. (1979), *Annu. Rev. Plant Physiol.* **30**, 105.

Hahlbrock, K., Wong, E., Schill, L. and Grisebach, H. (1970a), *Phytochemistry* **9**, 949.

Hahlbrock, K., Zilg, H. and Grisebach, H. (1970b), *Eur. J. Biochem.* **15**, 13.

Hahlbrock, K., Knobloch, K.-H., Kreuzaler, F., Potts, J. R. M. and Wellmann, E. (1976), *Eur. J. Biochem.* **61**, 199.

Hanson, K. R. and Havir, E. A. (1972a), *The Enzymes* **7**, 75.

Hanson, K. R. and Havir, E. A. (1972b), *Recent Adv. Phytochem.* **4**, 45.

Harborne, J. B., Mabry, T. J. and Mabry, H. (eds.) (1975), *The Flavonoids*, Chapman and Hall, London.

Harrison, B. J. and Stickland, R. G. (1974), *Heredity* **33**, 112.

Heinstein, P. F. and Stumpf, P. K. (1969), *J. Biol. Chem.* **214**, 5374.

Heller, W. and Hahlbrock, K. (1980), *Arch. Biochem. Biophys.* **200**, 617.

Heller, W., Gardiner, S. E. and Hahlbrock, K. (1980), *Hoppe-Seyler's Z. Physiol. Chem.* **361**, 265.

Herdt, E., Sütfeld, R. and Wiermann, R. (1978), *Cytobiologie* **17**, 433.

Hille, A., Purwin, C. and Ebel, J. (1982), *Plant Cell Reports*, in press.

Horecker, B. L., Tsolas, O. and Lai, C. Y. (1972), *The Enzymes* **7**, 213.

Hrazdina, G. and Creasy, L. L. (1979), *Phytochemistry* **18**, 581.

Hrazdina, G., Kreuzaler, F., Hahlbrock, K. and Grisebach, H. (1976), *Arch. Biochem. Biophys.* **175**, 392.

Hrazdina, G., Wagner, G. J. and Siegelmann, H. W. (1978), *Phytochemistry* **17**, 53.

Hrazdina, G., Alscher-Herman, R. and Kish, V. M. (1980), *Phytochemistry* **19**, 1355.

Kamsteeg, J., van Brederode, J. and van Nigtevecht, G. (1978a), *Biochem. Genet.* **16**, 1045.

Kamsteeg, J., van Brederode, J. and van Nigtevecht, G. (1978b), *Biochem. Genet.* **16**, 1059.

Kamsteeg, J., van Brederode, J. and van Nigtevecht, G. (1980a), *Z. Naturforsch.* **35C**, 249.

Kamsteeg, J., van Brederode, J. and van Nigtevecht, G. (1980b), *Z. Pflanzenphysiol.* **96**, 87.

Kamsteeg, J., van Brederode, J., Hommels, C. H. and van Nigtevecht, G. (1980c), *Biochem. Physiol. Pflanzen* **175**, 403.

Kamsteeg, J., van Brederode, J. and van Nigtevecht, G. (1980d), *Phytochemistry* **19**, 1459.

Kamsteeg, J., van Brederode, J., Verschuren, P. M. and van Nigtevecht, G. (1981), *Z. Pflanzenphysiol.* **102**, 435.

Kannangara, C. G. and Stumpf, P. K. (1972), *Arch. Biochem. Biophys.* **152**, 83.

Kefford, J. F. and Chandler, B. V. (1970). In *The Chemical Constituents of Citrus Fruits*, Academic Press, New York.

Kho, K. F. F., Bennink, G. J. H. and Wiering, H. (1975), *Planta* **127**, 271.

Kho, K. F. F., Bolsman-Louwen, A. C., Vuik, J. C. and Bennink, G. J. H. (1977), *Planta* **135**, 109.

Kho, K. F. F., Kamsteeg, J. and van Brederode, J. (1978), *Z. Pflanzenphysiol.* **88**, 449.

Knobloch, K.-H. and Hahlbrock, K. (1975), *Eur. J. Biochem.* **52**, 311.

Knobloch, K.-H. and Hahlbrock, K. (1977), *Arch. Biochem. Biophys.* **184**, 237.

Kreuzaler, F. and Hahlbrock, K. (1972), *FEBS Lett.* **28**, 69.

Kreuzaler, F. and Hahlbrock, K. (1973), *Phytochemistry* **12**, 1149.

Kreuzaler, F. and Hahlbrock, K. (1975a), *Eur. J. Biochem.* **56**, 205.

Kreuzaler, F. and Hahlbrock, K. (1975b), *Arch. Biochem. Biophys.* **169**, 84.

Kreuzaler, F., Light, R. J. and Hahlbrock, K. (1978), *FEBS Lett.* **94**, 175.

Kreuzaler, F., Ragg, H., Heller, W., Tesch, R., Witt, I., Hammer, D. and Hahlbrock, K. (1979), *Eur. J. Biochem.* **99**, 89.

Kuhn, B., Forkmann, G. and Seyffert, W. (1978), *Planta* **138**, 199.

Larson, R. L. (1971), *Phytochemistry* **10**, 3073.

Larson, R. L. and Coe, E. H., Jr. (1977), *Biochem. Genet.* **15**, 153.

Larson, R. L. and Lonergan, C. M. (1973), *Cereal Res. Commun.* **1**, 13.

Legrand, M., Fritig, B. and Hirth, L. (1976a), *Phytochemistry* **15**, 1353.

Legrand, M., Fritig, B. and Hirth, L. (1976b), *FEBS Lett.* **70**, 131.
Legrand, M., Fritig, B. and Hirth, L. (1978), *Planta* **144**, 101.
Light, R. J. and Hahlbrock, K. (1980), *Z. Naturforsch.* **35C**, 717.
Lindl, T., Kreuzaler, F. and Hahlbrock, K. (1973), *Biochim. Biophys. Acta* **302**, 457.
Lyne, R. L. and Mulheirn, L. J. (1978), *Tetrahedron Lett.*, 3127.
Lyne, R. L., Mulheirn, L. J. and Leworthy, D. P. (1976), *J. Chem. Soc. Chem. Commun.*, 497.
Lynen, F. and Tada, M. (1961), *Angew. Chem.* **73**, 513.
Mackall, J. C. and Lane, M. D. (1977), *Biochem. J.* **162**, 635.
Matern, U. and Grisebach, H. (1977), *Eur. J. Biochem.* **74**, 303.
Matern, U., Potts, J. R. M. and Hahlbrock, K. (1981), *Arch. Biochem. Biophys.*, **208**, 233.
McCormick, S. (1978), *Biochem. Genet.* **16**, 777.
Mohan, S. B. and Kekwick, R. G. O. (1980), *Biochem. J.* **187**, 667.
Moustafa, E. and Wong, E. (1967), *Phytochemistry* **6**, 625.
Nielsen, N. C., Adee, A. and Stumpf, P. K. (1979), *Arch. Biochem. Biophys.* **192**, 446.
Ortmann, R., Sutter, A. and Grisebach, H. (1972), *Biochim. Biophys. Acta* **289**, 293.
Pfändler, R., Scheel, D., Sandermann, H., Jr. and Grisebach, H. (1977), *Arch. Biochem. Biophys.* **178**, 315.
Potts, J. R. M., Weklych, R. and Conn, E. E. (1974), *J. Biol. Chem.* **249**, 5019.
Poulton, J. (1981). In *The Biochemistry of Plants* (eds. E. E. Conn and P. K. Stumpf), Academic Press, New York, Vol. 7, pp. 667–723.
Poulton, J. E. and Kauer, M. (1977), *Planta* **136**, 53.
Poulton, J., Hahlbrock, K. and Grisebach, H. (1976), *Arch. Biochem. Biophys.* **176**, 449.
Poulton, J. E., Hahlbrock, K. and Grisebach, H. (1977), *Arch. Biochem. Biophys.* **180**, 543.
Quast, L. and Wiermann, R. (1973), *Experientia* **29**, 1165.
Ragg, H., Kuhn, D. N. and Hahlbrock, K. (1981), *J. Biol. Chem.* **256**, 10061.
Ranjeva, R., Faggion, R. and Boudet, A. (1975a), *Physiol. Veg.* **13**, 725.
Ranjeva, R., Boudet, A., Harada, H. and Marigo, G. (1975b), *Biochim. Biophys. Acta* **399**, 23.
Ranjeva, R., Boudet, A. M. and Faggion, R. (1976), *Biochemie* **58**, 1255.
Ranjeva, R., Boudet, A. M. and Alibert, G. (1979). In *Regulation of Secondary Product and Plant Hormone Metabolism* (eds. M. Luckner and K. Schreiber), Pergamon Press, Oxford, Vol. 55, pp. 91–100.
Rathmell, W. G. and Bendall, D. S. (1972), *Biochem. J.* **127**, 125.
Reed, D. J., Vimmerstedt, J., Jerina, D. M. and Daly, J. W. (1973), *Arch. Biochem. Biophys.* **154**, 642.
Rhodes, M. J. C. and Wooltorton, L. S. C. (1973), *Phytochemistry* **12**, 2381.
Rhodes, M. J. C. and Wooltorton, L. S. C. (1975a), *Phytochemistry* **14**, 1235.
Rhodes, M. J. C. and Wooltorton, L. S. C. (1975b), *Phytochemistry* **14**, 2161.
Rhodes, M. J. C. and Wooltorton, L. S. C. (1976), *Phytochemistry* **15**, 947.
Rhodes, M. J. C., Hill, A. C. R. and Wooltorton, L. S. C. (1976), *Phytochemistry* **15**, 707.
Rich, P. R. and Lamb, C. J. (1977), *Eur. J. Biochem.* **72**, 353.
Rupprich, N. and Kindl, H. (1978), *Hoppe Seyler's Z. Physiol. Chem.* **359**, 165.
Rupprich, N., Hildebrand, H. and Kindl, H. (1980), *Arch. Biochem. Biophys.* **200**, 72.
Russell, D. W. (1971), *J. Biol. Chem.* **246**, 3870.
Russell, D. W. and Conn, E. E. (1967), *Arch. Biochem. Biophys.* **122**, 256.
Saleh, N. A. M., Fritsch, H., Witkop, P. and Grisebach, H. (1976a), *Planta* **133**, 41.
Saleh, N. A. M., Poulton, J. E. and Grisebach, H. (1976b), *Phytochemistry* **15**, 1865.
Saleh, N. A. M., Fritsch, H., Kreuzaler, F. and Grisebach, H. (1978), *Phytochemistry* **17**, 183.
Sandermann, H., Jr. (1969), *Phytochemistry* **8**, 1571.
Saunders, J. A., Conn, E. E., Lin, C. H. and Shimada, M. (1977), *Plant Physiol.* **60**, 629.
Schöppner, A. and Kindl, H. (1979), *FEBS Lett.* **108**, 349.
Schröder, J., Heller W. and Hahlbrock, K. (1979a), *Plant Sci. Lett.* **14**, 281.
Schröder, J., Kreuzaler, F., Schäfer, E. and Hahlbrock, K. (1979b), *J. Biol. Chem.* **254**, 57.
Schröder, G., Zähringer, U., Heller, W., Ebel, J. and Grisebach, H. (1979c), *Arch. Biochem. Biophys.* **194**, 635.

Spribille, R. and Forkmann, G. (1981) *Z. Naturforsch.* **36C**, 619.
Stickland, R. G. and Harrison, B. J. (1974), *Heredity* **33**, 108.
Stotz, G. and Forkmann, G. (1981) *Z. Naturforsch.* **36c**, 737.
Stumpf, P. K. (1980). In *The Biochemistry of Plants* (ed. P. K. Stumpf), Academic Press, New York, Vol. 4, pp. 177–204.
Sütfeld, R. and Wiermann, R. (1978), *Biochem. Physiol. Pflanzen* **172**, 111.
Sütfeld, R. and Wiermann, R. (1980a), *Arch. Biochem. Biophys.* **201**, 64.
Sütfeld, R. and Wiermann, R. (1980b), *Z. Pflanzenphysiol.* **97**, 283.
Sütfeld, R. and Wiermann, R. (1981), *Z. Naturforsch.* **36C**, 30.
Sütfeld, R., Kehrel, B. and Wiermann, R. (1978), *Z. Naturforsch.* **33C**, 841.
Sutter, A. and Grisebach, H. (1973), *Biochim. Biophys. Acta* **309**, 289.
Sutter, A. and Grisebach, H. (1975), *Arch. Biochem. Biophys.* **167**, 444.
Sutter, A., Ortmann, R. and Grisebach, H. (1972), *Biochim. Biophys. Acta* **258**, 71.
Sutter, A., Poulton, J. and Grisebach, H. (1975), *Arch. Biochem. Biophys.* **170**, 547.
Swift, J. (1967), *J. Agric. Food Chem.* **15**, 99.
Tanabe, T., Wada, K., Okazaki, T. and Numa, S. (1975), *Eur. J. Biochem.* **57**, 15.
Tanaka, Y. and Uritani, I. (1974), *Agric. Biol. Chem.* **38**, 1547.
Tanaka, Y., Kojima, M. and Uritani, I. (1974), *Plant Cell Physiol.* **15**, 843.
Tsang, Y.-F. and Ibrahim, R. K. (1979a), *Z. Naturforsch.* **34C**, 46.
Tsang, Y.-F. and Ibrahim, R. K. (1979b), *Phytochemistry* **18**, 1131.
Ulbrich, B. and Zenk, M. H. (1980), *Phytochemistry* **19**, 1625.
Ulbrich, B., Stöckigt, J. and Zenk, M. H. (1976), *Naturwissenschaften* **63**, 484.
Ullrich, V. and Duppel, W. (1975), *The Enzymes* **12**, 253.
Van Brederode, J., Chopin, J., Kamsteeg, J., van Nigtevecht, G. and Heinsbroek, R. (1979), *Phytochemistry* **18**, 655.
Wallis, P. J. and Rhodes, M. J. C. (1977), *Phytochemistry* **16**, 1891.
Walton, E. and Butt, V. S. (1970), *J. Exp. Bot.* **21**, 887.
Walton, E. and Butt, V. S. (1971), *Phytochemistry* **10**, 295.
Wellmann, E., Hrazdina, G. and Grisebach, H. (1976), *Phytochemistry* **15**, 913.
Wengenmayer, H., Ebel, J. and Grisebach, H. (1974), *Eur. J. Biochem.* **50**, 135.
Wiermann, R. (1970), *Planta* **95**, 133.
Wiermann, R. (1972), *Planta* **102**, 55.
Wiermann, R. (1979). In *Regulation of Secondary Product and Plant Hormone Metabolism* (eds. M. Luckner and K. Schreiber), Pergamon Press, Oxford, Vol. 55, pp. 231–239.
Wiermann, R. (1981). In *The Biochemistry of Plants* (eds. E. E. Conn and P. K. Stumpf), Academic Press, New York, Vol. 7, pp. 85–116.
Wiermann, R. and Buth-Weber, M. (1980), *Protoplasma* **104**, 307.
Wong, E. (1965), *Biochim. Biophys. Acta* **111**, 358.
Wong, E. and Wilson, J. M. (1972), *Phytochemistry* **11**, 875.
Young, O. and Beevers, H. (1976), *Phytochemistry* **15**, 379.
Zähringer, U., Ebel, J. and Grisebach, H. (1978), *Arch. Biochem. Biophys.* **188**, 450.
Zähringer, U., Ebel, J., Mulheirn, L. J., Lyne, R. L. and Grisebach, H. (1979), *FEBS Lett.* **101**, 90.
Zähringer, U., Schaller, E. and Grisebach, H. (1981), *Z. Naturforsch.* **36C**, 234.

CHAPTER TWELVE

Mammalian Metabolism of Flavonoids

LESLIE A. GRIFFITHS

12.1 INTRODUCTION

Although earlier investigations of the mammalian metabolism of flavonoid compounds were largely concerned with the identification of their urinary metabolites (cf. De Eds, 1968), more recent studies have centred on the role of the intestinal microflora in the catabolism of flavonoid molecules, on the significance of the biliary-enteric route of excretion and on the disposition of flavonoids in mammalian tissues following oral and parenteral administration. Additionally the introduction of chemically modified flavonoids and of synthetic phenylchromones as therapeutic agents has promoted metabolic and pharmacokinetic investigations on compounds showing close structural relationships with the naturally occurring flavonoids.

The reported biological activity of certain flavonoids in specific mammalian systems has also resulted in the initiation of studies on their interaction with isolated enzymes, cell constituents and membranes, which may be of importance in the mediation of pharmacological effects. Interest in the metabolism of other flavonoids has been stimulated by the finding that the aglycones of certain naturally occurring flavonol glycosides are mutagenic (Bjeldanes and Chang, 1977; Sugimura *et al.*, 1977; MacGregor and Jurd, 1978) and that the aglycone quercetin may, following oral ingestion, give rise to neoplasms of the gastrointestinal tract (Pamukcu *et al.*, 1980a,b) (see Sections 12.2.5 and 12.2.11).

Shorter specialized reviews of mammalian flavonoid biochemistry in relation to (a) nutrition (Kühnau, 1976), (b) the intestinal microflora (Griffiths, 1975), (c) biotransformation (Wurm, 1975; Scheline, 1978), (d) biliary excretion (Griffiths and Hackett, 1977) and (e) pharmacological activity (Gabor, 1975) have been published over the period 1975–1980. Since the last comprehensive review of mammalian flavonoid metabolism was that of De Eds (1968), an attempt has been made in the current review to provide coverage from 1968 to 1980.

12.2 BIOTRANSFORMATION

12.2.1 Ring fission

Three main types of ring scission have been postulated for the cleavage of flavonoid molecules *in vivo* (Fig. 12.1).

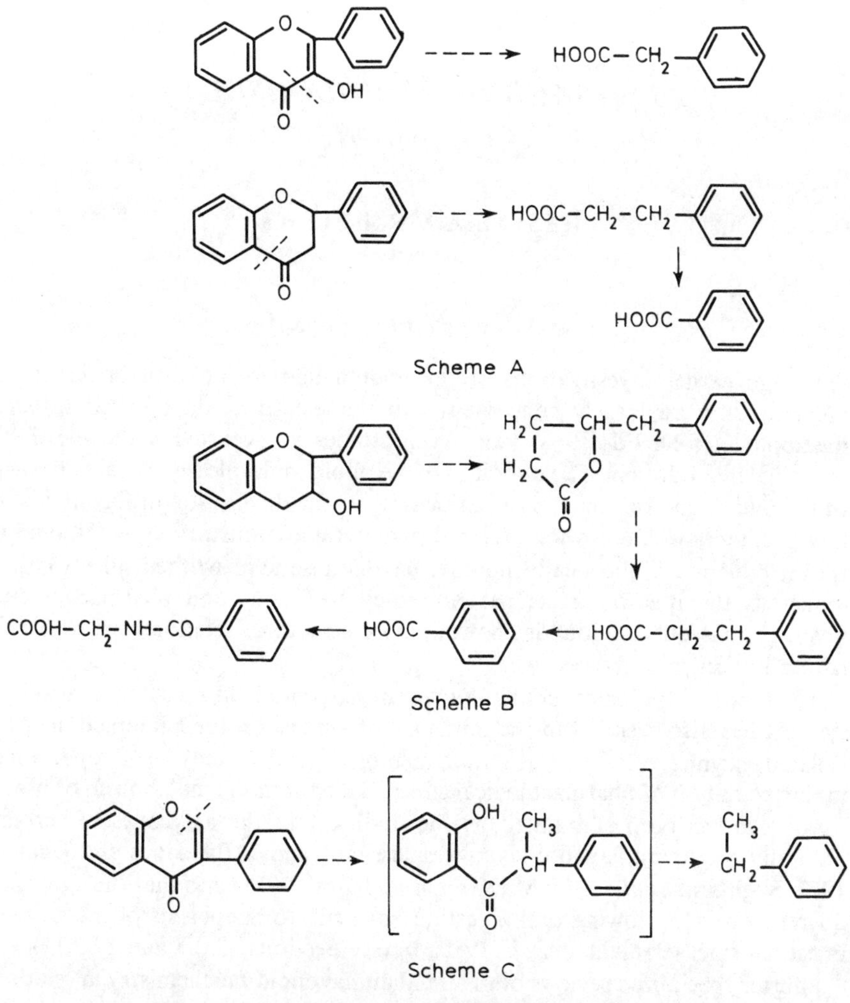

O OH O ----→ HOOC–CH2–

O O ---→ HOOC–CH2–CH2– ↓ HOOC–

Scheme A

O OH ---→ H2C—CH–CH2– H2C O C O

COOH–CH2–NH–CO– ← HOOC– ← HOOC–CH2–CH2–

Scheme B

O O ---→ [OH CH3 CH C O] ---→ CH3 CH2–

Scheme C

Fig. 12.1 Postulated scission of flavonoid molecules *in vivo*.

Scheme A scission was originally advanced by Booth *et al.* (1956) based on their investigations on the metabolism of the flavonols rutin and quercetin in which the formation of phenolic acids of the C_6–C_2 type were observed. A modified form of scheme A, where cleavage occurs between C_4 and C_5, was later advanced by Booth *et al.* (1958a,b) to account for the observed formation of C_6–C_3 phenolic acids from the flavanones, naringenin, hesperetin and eriodictyol.

Scheme B scission was proposed by Oshima and Watanabe (1958) and extended by Griffiths (1964) and Das and Griffiths (1969) for the catabolism of (+)-catechin and related flavanols via the obligate intermediates δ-phenyl-γ-valerolactones.

Scheme C for isoflavones is suggested by the findings of Batterham *et al.* (1965) and Griffiths and Smith (1972a), who reported the formation of *p*-ethylphenol from genistein by sheep and rat microflora respectively and by the experiments of Shutt and Braden (1968) and Batterham *et al.* (1971), who identified a phenyl-α-methylbenzyl ketone intermediate, namely 2,4-dihydroxyphenyl-α-4-hydroxyphenylethyl ketone (*O*-desmethylangolensin) as a urinary metabolite of formononetin in sheep. Formation of *p*-ethylphenol was not observed, however, following, the administration of formononetin, indicating that the absence of the 5-hydroxyl group from the isoflavone molecule restricts cleavage beyond the phenyl-α-methylbenzyl ketone intermediate.

Other classes of flavonoid may also undergo ring fission. The anthocyanidin, delphinidin and the anthocyanins, malvin and pelargonin, were shown by Griffiths and Smith (1972a,b) to be degraded *in vivo* and *in vitro* to phenolic metabolites which were, however, dissimilar to those reported to be formed from the corresponding flavanols and flavonols, indicating a different pathway of degradation for compounds possessing a flavylium structure. Studies on the proanthocyanidin, leucocyanidin by Masquelier *et al.* (1965) and Claveau and Masquelier (1966) have shown that this compound undergoes ring fission in the rat to phenolic metabolites, including homovanillic acid and *m*-hydroxyphenylacetic acid. Although the dihydrochalcones, phlorrhizin (phloridzin) and phloretin are known to undergo ring fission following oral administration to the rat (Booth *et al.*, 1958a,b; Griffiths and Smith, 1972a), the naturally occurring chalcones, butein and isoliquiritigenin do not give rise to detectable urinary ring fission products (Brown and Griffiths, unpublished observation). This may be attributable to the A ring of these chalcones showing resorcinol-type substitution as opposed to the phloroglucinol substitution of flavonoids known to undergo ring fission (see below).

The early studies on the metabolism of flavonols, flavones and flavanones by Booth and his associates (Booth *et al.*, 1956, 1958a,b) which led to recognition that ring fission to phenolic acids was a major route of degradation of flavonoid molecules have been confirmed and extended by a number of subsequent investigations. Two groups of flavonoid the former showing 4′-monohydroxylation (Griffiths and Smith, 1972a) and the latter 3′,4′5′-trihydroxylation (Griffiths and Smith, 1972b) were the subject of comparative studies. It was shown that only those flavonoids, which possessed free 5,7-dihydroxylation in the A ring underwent cleavage to ring fission products analogous to those reported by Booth and his associates. The absence of either 5- or 7-hydroxylation or the presence of *O*-methyl substitution in those positions reduced susceptibility to cleavage. Additionally it has been shown that hydroxyethylation on the free ring hydroxyls of rutin to give the pharmacologically active hydroxyethylrutosides (Courbat *et al.*, 1966a,b) results in compounds [*viz.* 7,3′,4′-trihydroxyethylrutoside (*12.1*) and 5,7,3′,4′-tetrahydroxyethylrutoside (*12.2*)] which are resistant to ring scission, when administered orally or parenterally to the rat (Barrow and Griffiths, 1972, 1974a,b). The highly methoxylated flavonols nobiletin and

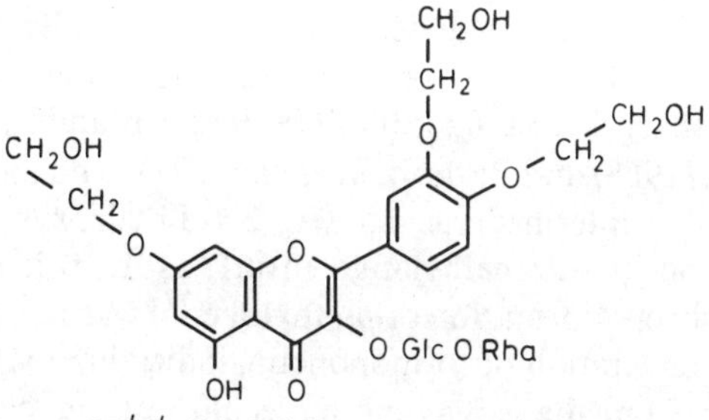

(12.1) 7, 3′,4′-Tri-O-(β-hydroxyethyl)rutoside
(rutoside ≡ rutin)

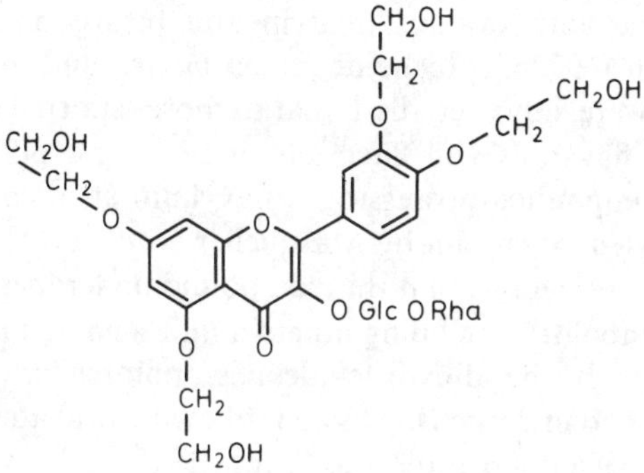

(12.2) 5,7,3′,4′-Tetra-O-(β-hydroxyethyl)rutoside

tangeretin have also been reported to be resistant to ring scission and likewise the 5- and 7-*O*-methylquercetins (De Eds, 1968). Methylation of the 3-hydroxyl group of (+)-catechin also results in a derivative [3-*O*-methyl-(+)-catechin (*12.3*)] which is resistant to ring scission in the rat and mouse (Griffiths and Hackett, 1978). Oral administration of the non-hydroxylated compounds flavone and flavanone resulted in no significant ring cleavage of the administrated compounds (Das and Griffiths, 1966; Das *et al.*, 1973; Buset and Scheline, 1980; Svardal *et al.*, 1980).

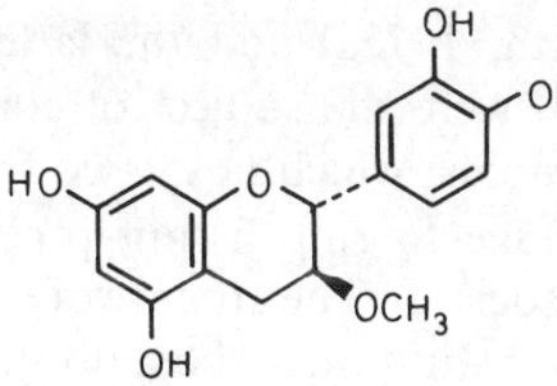

(12.3) 3-O-Methyl-(+)-catechin

12.2.2 Site of ring fission

The early demonstration that excretion of the urinary ring fission products of (+)-catechin (Griffiths, 1964) and of quercetin (Nakagawa *et al.*, 1965) was

markedly reduced by co-administration of oral antibiotics suggested that certain flavonoids were significantly metabolized by the intestinal microflora. Subsequently, investigations on a range of flavonoids, including apigenin, apiin, kaempferol, robinin, myricetin, myricitrin and tricetin (Griffiths and Smith, 1972a,b), provided evidence not only that the metabolism of these compounds was dependent *in vivo* upon the participation of the intestinal flora but also showed that many were degraded by intestinal micro-organisms *in vitro* during anaerobic incubation to metabolites identical with those isolated from the urine of dosed animals. The degradation of other flavonoids by intestinal micro-organisms under conditions of anaerobic incubation *in vitro* has been reported by Scheline (1968b), Das (1969), Simpson *et al.* (1969), Krishnamurty *et al.* (1970), Honohan *et al.* (1976) and Harmand and Blanquet (1978).

Although these observations indicated that the intestinal microflora was capable of effecting degradation of these flavonoids to the ring fission products detected in mammalian urine, it remained to be determined whether the ring fission was wholly mediated by intestinal bacteria or whether additionally, mammalian tissue enzymes also participated in cleavage of these molecules. Investigations employing germ-free animals maintained in a sterile Trexler isolator showed, however, that such animals were wholly unable to effect ring fission of a group of orally administrated flavonoids, which included (+)-catechin, rutin, apigenin, myricetin, hesperidin and naringin (Griffiths and Barrow, 1972). Conventionalization (i.e. deliberate reinfection of the gastrointestinal tract with the microflora derived from normal animals) conferred on the ex-germ-free animals the ability to effect ring fission to the urinary metabolites normally found. Accordingly it was concluded that ring cleavage of flavonoid compounds is normally wholly mediated by the gastrointestinal microflora and not by enzyme systems present in mammalian tissues.

As several authors have suggested that ring fission of flavonoid molecules may occur in the liver (Takacs and Gabor, 1975; Förster, 1975; Harmand and Blanquet, 1978), the ability of rat liver to degrade (+)-[U-^{14}C]catechin was recently investigated using the liver perfusion technique of Hems *et al.* (1966). Although large amounts of labelled flavanol conjugates were detected in both the perfusate and the formed bile, no labelled phenolic acids or phenyl-γ-valerolactones known to be formed *in vivo* from (+)-catechin by rats with a viable microflora were detected (Griffiths *et al.*, 1980). The possibility that some flavonoids may undergo ring cleavage in liver is, however, suggested by the liver perfusion experiments of Takacs and Gabor (1975), who reported that the flavonol glycoside, rutin, underwent ring fission during perfusion of the rat liver to give 3,4-dihydroxyphenyl-γ-valerolactone and several phenolic acids. Although 3,4-dihydroxyphenyl-γ-valerolactone is well established as a metabolite of (+)-catechin (Oshima and Watanabe, 1958; Das and Griffiths, 1968, 1969), it has, however, not previously been reported as metabolite of a flavonol. Since a non-labelled rutin was used in these studies, the possibility that the

phenolic metabolites detected may have arisen from dietary precursors or endogenous liver metabolism cannot be excluded.

Although Förster (1975, 1978) has also claimed that ring fission of a flavonoid, namely 7-mono-*O*-(β-hydroxyethyl)rutoside (*12.4*), was observed in the course of his liver perfusion experiments in the rat, the only evidence advanced was the apparent disappearance of the perfused fluorescent flavonoid from the perfusate as no ring fission products were identified.

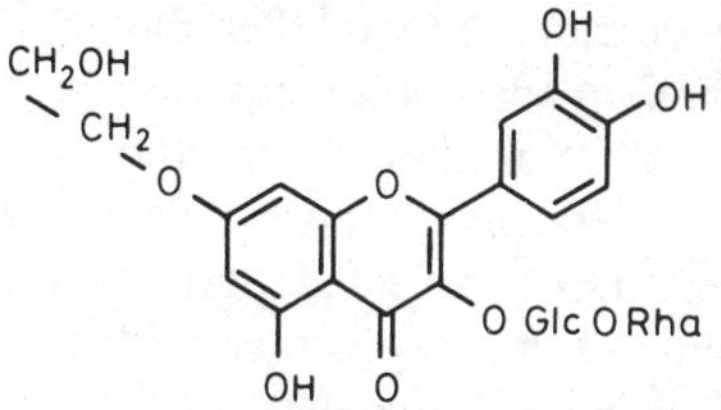

(*12.4*) 7-*O*-(β-hydroxyethyl) rutoside

12.2.3 Intermediates in flavonoid ring fission

Although investigation has revealed the identity of the phenolic metabolites derived from the B ring of many flavonoids (cf. De Eds, 1968), limited information is available concerning both the early intermediates of ring scission in the mammalian intestine and the subsequent fate of the A ring fragment. Metabolites derived from the A ring have seldom been detected as urinary metabolites in the rat, but evidence has been presented that phloroglucinol intermediates may occur in the course of degradation of flavonoids possessing 5,7-dihydroxylation. An early report by Kallianos *et al.* (1959) claimed that phloroglucinol and phloroglucinolcarboxylic acid were detectable in the gastro-intestinal tract of the rat following quercetin administration and it has been reported that excretion of phloroglucinol glucuronides occur following the administration of leucocyanidin to the rat (Masquelier *et al.*, 1965; Claveau and Masquelier, 1966). Since, however, acid treatment at elevated temperatures was used by both groups of workers to recover phloroglucinol from complex mixtures of metabolites, the possibility of direct chemical degradation cannot be excluded, especially as Masri *et al.* (1959) have demonstrated that quercetin is degraded to phloroglucinol and other products under similar conditions of acid treatment.

Investigations on flavonoid catabolism by rat microflora *in vitro* have likewise failed to provide definitive information on the fate of the A ring (Scheline, 1968a; Griffiths and Smith, 1972a,b). The detection of phloroglucinol as an intermediate in the microfloral degradation of the dihydrochalcone, phlorrhzin (Griffiths and Smith, 1972a) is of limited importance, as it appears probable that the absence of an oxygen bridge from compounds of this type may result in different cleavage reactions from those found in *O*-heterocycles.

Studies by Simpson *et al.* (1969) and Cheng *et al.* (1971) using isolated rumen micro-organisms under conditions of anaerobic incubation have, however, demonstrated that phloroglucinol is, under these conditions, a metabolite of rutin, quercetin, naringin and naringenin. The phloroglucinol detected was, however, further metabolized on extended incubation. The ability to degrade flavonoids via phloroglucinol was shown to be effected by the rumen micro-organism, *Butyrivibrio* sp. (Cheng *et al.*, 1971) which is probably absent from the gastrointestinal tracts of non-ruminant species.

De Eds (1968) has also made reference to unpublished experiments in which phloroglucinol was detected as a transient intermediate of quercetin following incubation with rat faeces for shorter periods of time but no quantitative data were reported. The available evidence that phloroglucinol intermediates are obligate intermediates in non-ruminant species is therefore very slender and clearly requires re-examination by modern techniques.

Although phloroglucinol has been postulated as an intermediate in the degradation of flavones, flavonols and flavanones (cf. De Eds, 1968), studies have established that the flavanols including (+)-catechin (Oshima and Watanabe, 1958; Griffiths, 1964; Das and Griffiths, 1969; Das, 1971, 1974), (−)-epicatechin (Oshima *et al.*, 1960) and epi-afzelechin (Griffiths and Smith, 1972a) are degraded via phenyl-γ-valerolactones which are derived from both A and B ring carbon atoms (Fig. 12.1, Scheme B).

Furthermore, evidence has been presented based on experiments using a (+)-[^{14}C]catechin preparation, predominantly labelled in the A ring, that the carbons of this ring are largely metabolized *in vivo* to $^{14}CO_2$, although radioactivity was also associated with the phenyl-γ-valerolactones excreted (Das and Griffiths, 1969) which, as noted above, contain carbon atoms derived from the A ring. [^{14}C]Quercetin has also been shown to give rise to respired $^{14}CO_2$ (15% of dose) on administration to the rat (Kallianos *et al.*, 1959), but as this early study employed a randomly labelled quercetin, the carbon atom precursors of the $^{14}CO_2$ could not be identified.

The possibility that aliphatic intermediates might be formed in the course of metabolism of the A ring carbons to $^{14}CO_2$ was suggested on the basis of data obtained in an investigation *in vivo* of (+)-[^{14}C]catechin metabolism in the rat (Das and Griffiths, 1969), but it was not until 1972 that evidence of the identity of such an aliphatic intermediate was obtained. Using cell-free extracts of a *Pseudomonoas* sp. isolated from rat faeces, Jeffrey *et al.* (1972b) obtained evidence for the metabolism of the A ring of dihydrogossypetin via the four-carbon intermediate, oxaloacetic acid. Since it had previously been reported that dihydrogossypetin and taxifolin were intermediates in the metabolism of (+)-catechin (Fig. 12.2) by this pseudomonad (Jeffrey *et al.*, 1969, 1972a), these findings may also be of considerable significance in relation to flavanol metabolism. It is, however, important to note that to date *none* of the intermediates formed by the cell-free pseudomonad extracts have been reported as *in vivo* metabolites of (+)-catechin in the rat and secondly that formation of

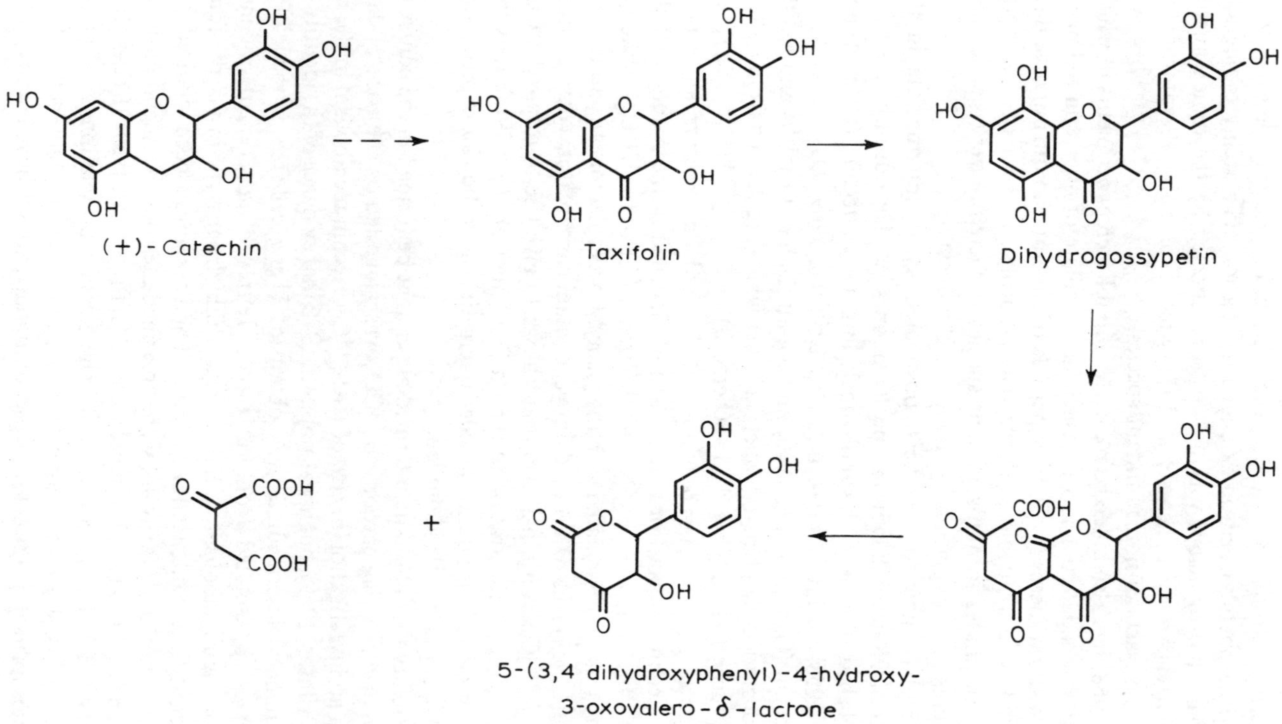

(+)-Catechin

Taxifolin

Dihydrogossypetin

5-(3,4 dihydroxyphenyl)-4-hydroxy-3-oxovalero-δ-lactone

Fig. 12.2 Metabolism of (+)-catechin by a pseudomonad from rat faeces (Jeffrey *et al.*, 1972a,b).

the early intermediates, taxifolin and dihydrogossypetin, required highly aerobic conditions which would not normally obtain in the mammalian intestine which favours the activities of anaerobic or micro-aerophilic organisms.

A soil pseudomonad, *P. putida*, has also been isolated by elective culture on quercetin by Schultz *et al.* (1974), which has been shown to possess enzymes able to effect 8-hydroxylation of certain flavones and flavanones followed by ring fission to give a pyrone intermediate and ultimately oxaloacetic acid, and in the case of quercetin, protocatechuic acid (Fig. 12.3). Since quercetin degradation *in vivo* results largely in the urinary excretion of phenylacetic acid derivatives (Booth *et al.*, 1956), the products of the pseudomonad pathway are again seen to be dissimilar to those observed in mammalian studies. Although there is therefore no reason to suppose that the isolated pseudomonads are responsible for flavonoid degradation in the mammalian intestine, the metabolic data reported in these studies indicate new mechanisms of ring fission, which in a modified form may be of significance in intestinal metabolism.

It is of interest to note that the structural requirements for flavonoid ring fission *in vivo*, namely the possession of free hydroxyls at C-5 and C-7, are also important for ring cleavage by the pseudomonad pathway of Schultz *et al.* (1974), suggesting dependence upon similar enzymic mechanisms. However, additional structural requirements were noted by Schultz *et al.* for cleavage by the pseudomonad pathway, indicating greater specificity of the oxygenases involved. It was noted that the second dioxygenase (Fig. 12.3) required a hydroxyl at C-3 which must be at an unsaturated site. In consequence, taxifolin did not serve as a substrate despite the fact that it has been found to be a substrate for the pseudomonad isolated by Jeffrey *et al.* (1972). Conversely, quercetin which is readily degraded by preparations of the pseudomonad of Schultz *et al.* was not metabolized by the pseudomonad employed by Jeffrey *et al.* These apparent anomalies may be attributable to the fact that the enzyme preparations were initially obtained by induction with different substrates, namely quercetin and (+)-catechin. The wider specificity observed in relation to degradation of flavonoids by the intestinal microflora may point to participation of a number of induced enzymes. Evidence has been presented that the administration of certain flavonoids, namely robinin and 7-*O*-hydroxyethylrutoside, over an extended period of time may result in elevated excretion of the microfloral ring-fission products (Griffiths and Smith, 1972; Barrow and Griffiths, 1974b).

12.2.4 Species variation in flavonoid metabolism

Since many of the earlier investigations on flavonoid metabolism were carried out using a limited number of animal species, relatively few differences in the metabolic fate of flavonoids were noted, although Booth *et al.* (1958) reported that whilst hesperetin and hesperidin gave rise to large amounts of *m*-hydroxyphenylpropionic acid in rat urine, the major metabolite in man was 3-hydroxy-4-methoxyphenylhydracrylic acid.

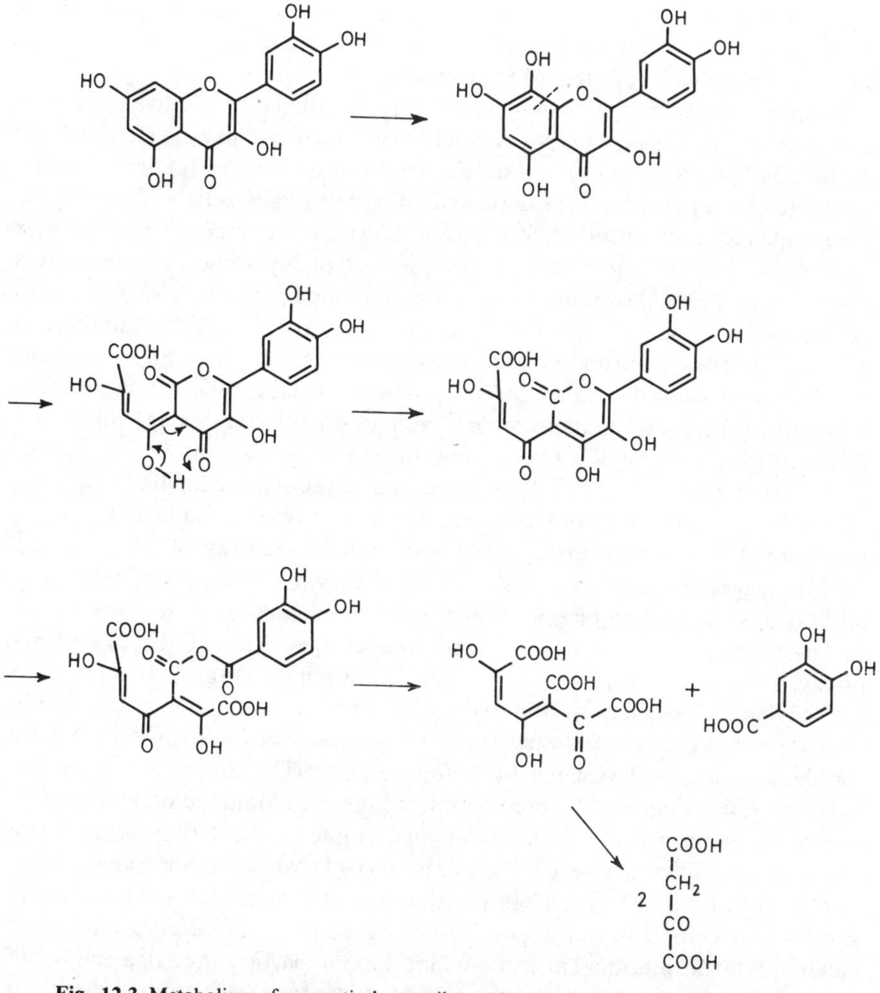

Fig. 12.3 Metabolism of quercetin by a soil pseudomonad (Schultz *et al.*, 1974).

More recent investigations on the metabolism of the flavonol, (+)-catechin in a wide range of mammalian species has shown considerable species variation in relation to the urinary ring fission products. Whereas the early study of Oshima *et al.* (1958) indicated that in the rabbit, the major phenolic acid metabolites were vanillic acid, protocatechuic acid and *m*-hydroxybenzoic acid, the studies of Griffiths (1962, 1964) and Das and Griffiths (1969) on the rat indicated that *m*-hydroxyphenylpropionic acid and *m*-hydroxyhippuric acid were the major metabolites in this species. In the guinea pig, however, it was found that *m*-hydroxybenzoic acid was the major metabolite and that *m*-hydroxyphenylpropionic acid was present in trace amounts only (Das and Griffiths,

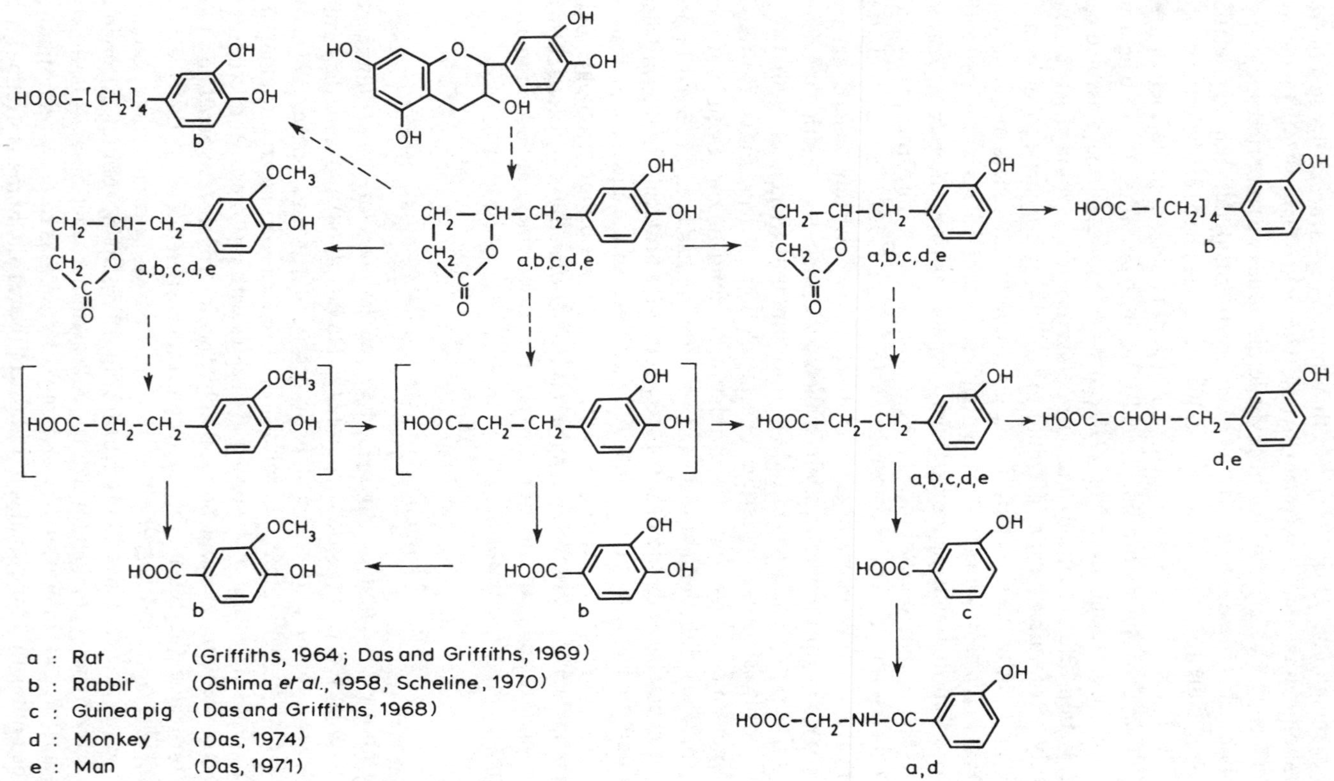

a : Rat (Griffiths, 1964; Das and Griffiths, 1969)
b : Rabbit (Oshima *et al.*, 1958, Scheline, 1970)
c : Guinea pig (Das and Griffiths, 1968)
d : Monkey (Das, 1974)
e : Man (Das, 1971)

Fig. 12.4 Species variation in (+)-catechin catabolism.

1968). Investigations in primate species indicated that in man the metabolic pattern resembled that of the rat in that *m*-hydroxyphenylpropionic acid was the major metabolite (Das, 1971), but that in monkey (*Macaca iris* sp.) the major metabolite was *m*-hydroxyphenylhydracrylic acid (Das, 1974). In all species investigated, the formation of phenylvalerolactones was reported and it appears probable that these are important obligatory intermediates in the degradation of (+)-catechin to phenolic acids, as was demonstrated in the rat employing *m*-[^{14}C]hydroxyphenylvalerolactone which was shown to be metabolized to *m*-[^{14}C]hydroxyphenylpropionic acid (Das and Griffiths, 1969). These interrelationships are shown in Fig. 12.4. Since evidence has been presented earlier (see Section 12.2.2) that the catabolism of flavonoids, including (+)-catechin, is largely attributable to the intestinal microflora, it appears probable that the divergent pathways shown in Fig. 12.4 are due to differences in the microflora present in the gastrointestinal tracts of different species. This is in accord with the observation that a suspension of micro-organisms from the *rabbit* gastrointestinal tract is able to metabolize (+)-catechin to 5-(3,4-dihydroxyphenyl)valeric acid and 5-(3-hydroxyphenyl)valeric acid (Scheline, 1970), whereas incubation of (+)-catechin with micro-organisms from the *rat* intestine yielded mainly *m*-hydroxyphenylpropionic acid, whilst phenylvaleric acids were either not detected (Das, 1969) or were detected only in trace amounts (Scheline, 1970).

Not only is species variation seen in relation to degradative pathways of flavonoid metabolism but also in relation to conjugation reactions. Recently 3,3′ (or 4′)-*O*-dimethyl-(+)-catechin glucuronide has been observed to be the major urinary metabolite of 3-*O*-[^{14}C]methyl-(+)-catechin in the rat where it amounts to 64–92% of the total radioactivity. In the mouse and marmoset monkey, however, this conjugate was found to be a minor metabolite and the major metabolite was the free aglycone 3,3′-*O*-dimethyl-(+)-catechin, which accounted for 68% of the total activity in mouse urine and 78% in marmoset urine (Hackett and Griffiths, 1981).

12.2.5 Metabolic hydrolysis of flavonoid glycosides

The ability of mammals to hydrolyse the linkages of orally administered flavonoid glycosides is widely reported (cf. De Eds, 1968) and since the resultant aglycones are in part excreted in faeces (Griffiths and Smith, 1972a,b), it appeared probable that hydrolysis occurred in the gastrointestinal tract. Anaerobic incubation of flavonoid glycosides with the intestinal microflora has also resulted in the detection of aglycone intermediates. Scheline (1968b) demonstrated the release of quercetin and hesperetin when their respective glycosides rutin and hesperidin were incubated with rat microflora. Similarly, release of the aglycones of myricitrin (Fig. 12.5), apiin, robinin, naringin and phlorrhizin has been demonstrated under similar conditions (Griffiths and Smith, 1972a,b). Interest in flavonoid glycosidases has recently been stimulated by the finding that microbial glycosidases of the rat and human intestine are able

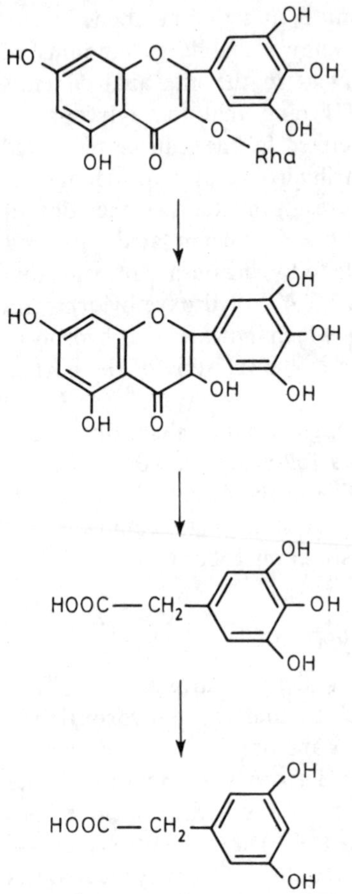

Fig. 12.5 Degradation of myricitrin by the intestinal microflora of the rat (Griffiths and Smith, 1972b).

to release mutagenic aglycones from certain flavonoid glycosides, which are known to be of widespread occurrence in vegetable food stuffs (Brown and Dietrich, 1979; Tamura *et al.*, 1980). Since, moreover, parenteral administration of the flavonoid glycosides naringin and hesperidin to biliary cannulated rats, in which contact between the administered flavonoid and the gut microflora is precluded, resulted in the excretion of conjugates of the intact glycosides but not of the corresponding aglycones (Hackett *et al.*, 1979), it appears probable that glycosidases able to hydrolyse these compounds are absent from mammalian tissues. This conclusion is supported by earlier observations on the metabolism of the flavonoid glycosides rutin, naringin and hesperidin in germ-free rats, when the unchanged glycosides but not the free aglycones were detected on examination of the faeces (Griffiths and Barrow, 1972).

Limited information is available on the fate of the flavonoid conjugates which are known to be extensively excreted in bile (see Section 12.3.3). It appears

probable, however, that the flavonoid glucuronides undergo metabolic hydrolysis in the lumen of the intestine, as it is known that large amounts of β-glucuronidase of microfloral origin are present in the intestinal contents (Marsh *et al.*, 1952; Hawksworth *et al.*, 1971) and that the aglycones of certain flavonoid conjugates, known to be excreted in bile, can be recovered from the faeces. Administration of 7,3′,4′-trihydroxyethylrutoside and 7-monohydroxyethylrutoside has been shown to result in extensive excretion of their glucuronide conjugates in the bile of bile-duct-cannulated rats, but examination of the faeces of non-cannulated rats following dosage revealed that the major faecal metabolites were the aglycones, 7,3′,4′-trihydroxyethylquercetin and 7-monohydroxyethylquercetin respectively, which indicates that hydrolysis of both sugar and glucuronide linkages occurs in the intestine of the normal animal (Barrow and Griffiths, 1974a,b).

It is also of interest here to note that, although flavanol glucuronides can be recovered from the faeces of germ-free rats following the administration (+)-catechin, these conjugates are not detectable in the faeces of normal rats (Griffiths and Barrow, 1972), again indicating that these biliary conjugates are under normal conditions degraded by the intestinal microflora.

12.2.6 Hydroxylation

A number of flavonoids have been reported to undergo hydroxylation *in vivo*. Both chrysin and tectochrysin were reported to undergo 4′-hydroxylation following oral administration to the rat (Griffiths and Smith, 1972a). Following administration of flavone to the guinea pig, several hydroxylation products were detected in urine including 4′-hydroxyflavone and 3′,4′-dihydroxyflavone (Das and Griffiths, 1966). Recently Buset and Scheline (1979) have reported hydroxylation of flavanone following oral administration to the rat. Urinary metabolites isolated and identified by GC–MS included 6-hydroxyflavanone, 6-hydroxyflavan-4-β-ol and *trans*-3-hydroxyflavan-4-β-ol (Fig. 12.6). Little hydroxylation was reported in ring B.

In a parallel investigation on the biliary and urinary metabolites of a group of naturally occurring polyhydroxylated flavanones following parenteral and oral administration to the rat, metabolic hydroxylation was detected only as a minor pathway of a single flavanone, *viz*. naringenin which gave rise to trace amounts of eriodictyol (Hackett *et al.*, 1979). Evidence has been presented by Jeffrey *et al.* (1972) that the dihydroflavonol, taxifolin, was hydroxylated to dihydrogossypetin by a pseudomonad present in rat faeces, but no evidence is available whether this reaction occurs *in vivo* in the rat intestine.

12.2.7 Reductive reactions

Although the excretion of *m*-hydroxyphenacyl derivatives following the administration of 3′,4′-dihydroxylated flavonoids is widely reported the available

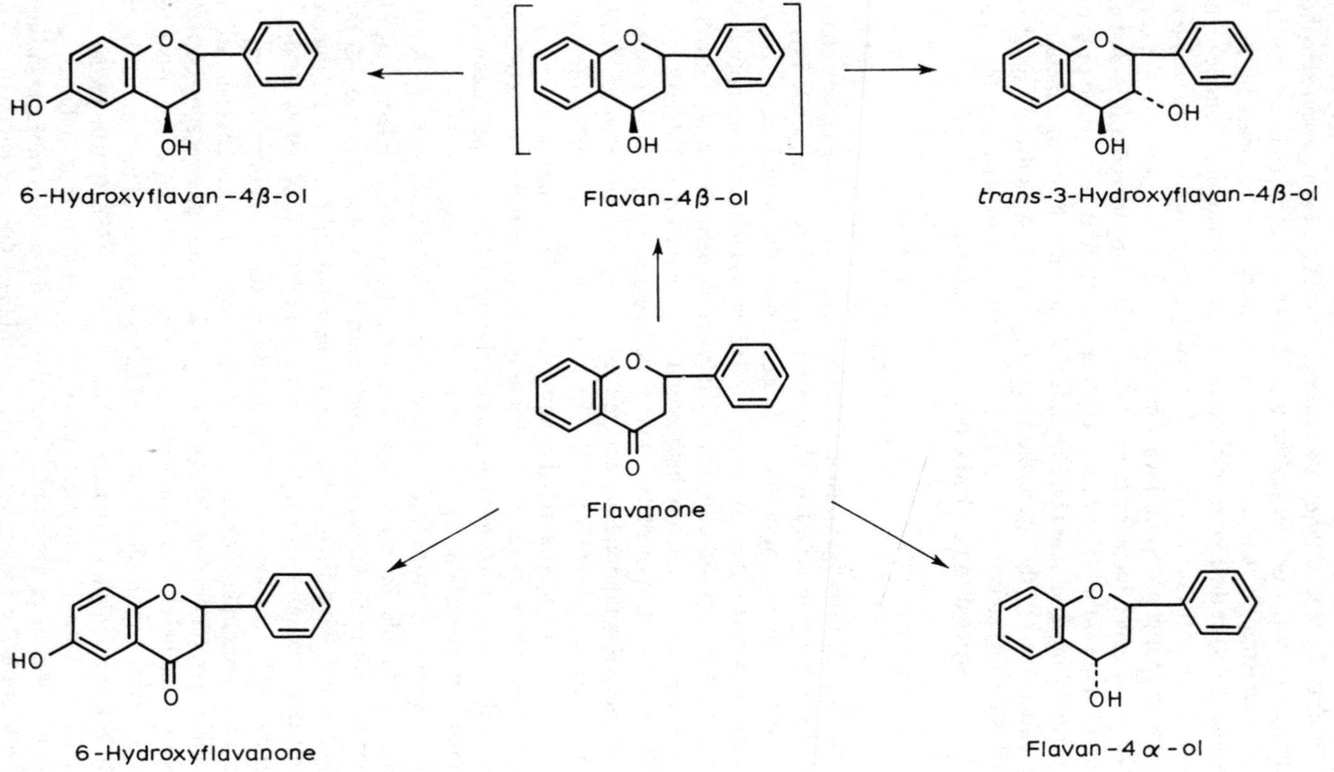

6-Hydroxyflavan-4β-ol

Flavan-4β-ol

trans-3-Hydroxyflavan-4β-ol

Flavanone

6-Hydroxyflavanone

Flavan-4α-ol

Fig. 12.6 Major metabolites of flavonone (Buset and Scheline, 1979).

evidence indicates that dehydroxylation occurs *following* ring fission to *o*-dihydric phenacyl products, which are then dehydroxylated by the intestinal microflora (cf. De Eds, 1968). No evidence has to date been presented to indicate that direct reductive dehydroxylation of *intact* flavonoids occurs in the mammal.

An interesting reductive reaction has recently been reported by Buset and Scheline (1979) in their investigations on the metabolism of flavanone in the rat. Examination of the urinary metabolites using a GC–MS technique has revealed that the carbonyl group of flavanone is reduced to give a number of carbinol metabolites, which included flavan-4-α-ol, *trans*-3-hydroxyflavan-4-β-ol and 6-hydroxyflavan-4-β-ol (Fig. 12.6). Another recent study on a group of naturally occurring plant flavanones has, however, shown that their major metabolites in bile and urine are the glucuronide conjugates of the aglycone or glycoside administered (Hackett *et al.*, 1979), indicating that carbinol formation is probably restricted to non-hydroxylated compounds.

12.2.8 Conjugation

Although early investigations on flavonoid metabolism were largely concerned with the ring fission products which were excreted in urine (cf. De Eds, 1968), more recent investigations have focused upon the nature of the flavonoid conjugates excreted predominantly in bile but also in urine after both oral and parenteral administration, which it has recently been shown (see also Section 12.3.3) may represent a very high proportion of the dose (Table 12.1).

Whilst administration of flavonoid *glycosides* may result in biliary excretion of the *unchanged* compound as reported for naringin (Hackett *et al.*, 1979) and 5,7,3′,4′-tetrahydroxyethylrutoside (Barrow and Griffiths, 1972), more usually excretion of conjugates of the administered flavonoid is observed. Thus glucuronides of the flavanones, hesperetin and naringenin, are excreted in bile after both oral or parenteral administration and likewise the glucuronides of naringin and hesperidin (Hackett *et al.*, 1979).

It is important to note that whilst flavonoid *glycosides* may be excreted unchanged in bile, flavonoid *aglycones* are not excreted in bile to a significant extent without prior conjugation. The flavanols, (+)-catechin and 3-*O*-methyl-(+)-catechin, have been shown to be excreted in rat bile as glucuronide conjugates (Das and Sothy, 1971; Barrow and Griffiths, 1971; Griffiths and Hackett, 1978; Hackett and Griffiths, 1981), the flavanones, naringenin and hesperetin, are similarly excreted (Honohan *et al.*, 1976; Hackett *et al.*, 1979), but the chalcones, butein and 2′,4′,4-trihydroxychalcone, have recently been shown to be excreted in bile mainly as sulphate conjugates (Griffiths *et al.*, 1980; Brown and Griffiths, unpublished). Silybin (*12.5*) is reported to be excreted in bile as both glucuronide and sulphate (Bülles *et al.*, 1975). It will be noted that conjugation with glucuronic acid or with sulphate not only elevates the molecular weight of the administered flavonoid but also confers considerable polarity upon the molecule and in consequence favours elimination via the biliary route (see Section 12.3.3).

Table 12.1 Excretion of flavonoids and their conjugates in bile and urine of the rat

Compound	*Route*	*Dose*	*Excretion in bile (% of dose)*	*Excretion urine (% of dose)*	*Period of collection (h)*	*References*
(+)-Catechin	i.v.	25 mg*	42.7	32.6	24	Das and Sothy (1971)
	oral	10 mg*	34.0	19.6	24	Shaw and Griffiths (1980)
3-*O*-Methyl-(+)-catechin	i.v.	25 mg/kg	57.4	17.5	24	Hackett and Griffiths (1980)
	oral	25–30 mg/kg	48.3	16.4	48	
Hesperetin	i.p.	0.15 mg*	100.7	ND	24	Honohan *et al.* (1976)
	oral	0.29 mg*	56.5	ND	24	
Naringin	i.p.	5 mg*	89.7	0		Hackett *et al.* (1979)
	oral	50 mg*	11.4	2.3	48	
7-Monohydroxyethylrutoside†	i.v.	5 mg/kg	78.1	18.4	24	Barrow and Griffiths (1974a)
	oral	7.4 mg/kg	22.6	4.0	72	Barrow and Griffiths (1974b)
7,3′,4′-Tri-*O*-(β-hydroxy-ethyl) rutoside	i.v.	1.1 mg/kg	65.3	14.3	24	Barrow and Griffiths (1974a)
	oral	1.4 mg/kg	13.2	1.3	72	Barrow and Griffiths (1974b)
Silybin	i.v.	20 mg/kg	76.0	8.0	48	Bülles *et al.* (1975)
	oral	2–20 mg/kg	20.0	ND	48	
Flavoxate	i.v.	5 mg/kg	22.1	17.6	24	Setnikar *et al.* (1975)
	oral	30 mg/kg	58.2	20.0	24	

ND Not determined.
* Dose per rat.
† Rutoside ≡ rutin.

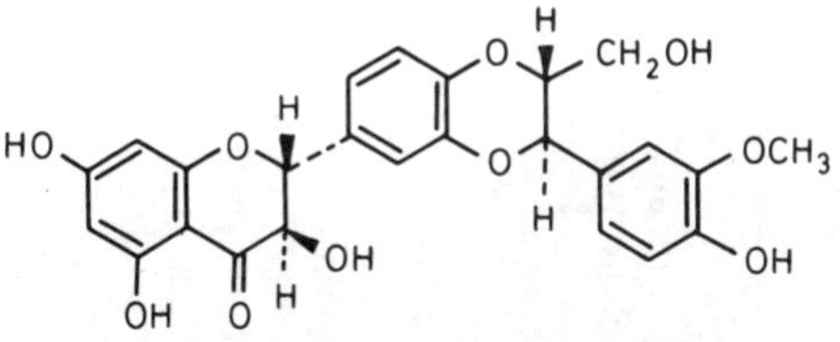

(12.5) Silybin

12.2.9 Site(s) of formation of flavonoid conjugates

Although studies on flavonoid metabolism *in vivo* are numerous, a very limited amount of research upon the metabolism of flavonoids *in vitro* by mammalian tissues has been reported and information on the site(s) of formation of flavonoid glucuronides is sparse. Evidence has been presented, however, that rabbit liver microsomal fractions are able to effect the transfer of glucuronic acid from UDP-glucuronic acid to the isoflavones, biochanin A, formononetin, daidzein, genistein and equol to give the corresponding monoglucuronides (Labow and Layne, 1972). Using a rat liver perfusion technique, Förster *et al.* (1972) demonstrated the formation of a blue fluorescent metabolite from perfused 7,3′,4′-trihydroxyethylrutoside which was subsequently identified as a glucuronide (Förster, 1977) and which appears to be identical with 7,3′,4′-trihydroxyethylrutoside 5-glucuronide detected earlier in the bile and urine of the intact rat (Barrow and Griffiths, 1974a). The formation of flavanol glucuronides in the perfused liver has also been demonstrated using [U-^{14}C] (+)-catechin. Labelled flavanol glucuronides were detected in both the perfusate and in the formed bile (Griffiths *et al.*, 1980).

The limited information available on the origin of flavonoid glucuronides in the mammal thus clearly indicates that the liver is able to form conjugates of this type. However, the relative importance of this organ in relation to other possible sites, such as the intestinal mucosa which has been implicated as an important site of phenol conjugation by Powell *et al.* (1974), remains to be evaluated by further research.

12.2.10 Methylation

Although many reports have been published indicating that methoxylated phenolic acids are excreted by the mammal following the administration of certain flavonoids especially those possessing 3′,4′-dihydroxylation of the B ring, the view has usually been taken that these metabolites arise largely if not wholly by methylation of phenolic acid intermediates formed as the direct products of ring fission (cf. De Eds, 1968). Such a view is supported by evidence that *o*-dihydric C_6–C_1, C_6–C_2 and C_6–C_3 acids readily undergo methylation *in vivo* (Booth *et al.*, 1955, 1957) and *in vitro* (Masri *et al.*, 1962) and by failure in earlier

investigations to detect intact methylated flavonoids among the products of excretion.

Recently, however, evidence has been obtained that the methyl ethers of certain administered flavonoids are excreted as their glucuronide and sulphate conjugates in the bile of rats (Griffiths and Hackett, 1978; Griffiths *et al.*, 1980; Shaw and Griffiths, 1980). Studies on the flavanol, (+)-catechin indicated that following both oral (Barrow and Griffiths, 1971) and parenteral administration (Das and Sothy, 1971) flavanol conjugates were excreted in bile. The major biliary metabolite was not identified in the earlier studies, but it has now been identified by mass spectrometry and other techniques as 3′-*O*-methyl-(+)-catechin glucuronide (Griffiths *et al.*, 1980). The metabolite was shown, following oral administration of [U-^{14}C] (+)-catechin to the rat, to account for approx. 50 % of the total biliary radioactivity which corresponded to 34 % of the administered dose. Perfusion of rat liver with [U-^{14}C](+)-catechin was also shown to result in rapid transformation of the flavanol to the methylcatechin glucuronide.

Incubation of non-labelled (+)-catechin with a homogenate of rat liver containing added *S*-[*methyl*-^{14}C] adenosylmethionine resulted in the formation of a radioactive product which was shown by HPLC and paper chromatography to be identical with the aglycone of the biliary metabolite (Shaw and Griffiths, 1980).

The synthetic derivative, 3-*O*-methyl-(+)-catechin has also been reported to undergo methylation in the rat giving rise to large amounts of 3,3′(or 4′)-*O*-dimethyl-(+)-catechin glucuronide which was excreted in bile and urine. Mass spectrometric examination of the aglycone obtained by β-glucuronidase hydrolysis of the major biliary metabolite indicated that the second methyl substituent had been introduced into a hydroxyl of ring B (Griffiths and Hackett, 1978; Hackett and Griffiths, 1981). The free aglycone was also detected in rat urine as a minor metabolite.

Comparative studies in the mouse and marmoset showed that methylation of 3-*O*-methyl-(+)-catechin occurred to a similar extent as in the rat but that the major urinary metabolite in both species was the free aglycone, 3,3′(or 4′)-*O*-dimethyl-(+)-catechin whilst the corresponding glucuronide was a minor metabolite (Hackett and Griffiths, 1981).

The chalcone, butein (2′,4′,3,4-tetrahydroxychalcone), has also been shown to undergo methylation *in vivo* in the rat. Following parenteral administration, the major biliary metabolite of butein was shown to be a sulphate conjugate. The aglycone, released by treatment with arylsulphatase, was shown by mass spectrometry to be a methylbutein, which possessed UV absorption characteristics and fluorescence identical with 3-*O*-methylbutein (Griffiths *et al.*, 1980). Similarly 2′,3,4-trihydroxychalcone has been shown to undergo 3-*O*-methylation and conjugation with sulphate in the rat. Parenteral administration of the flavonol glycoside, rutin, resulted in the biliary excretion of small amounts of glucuronide conjugates, the aglycone of which was shown by UV spectral

analysis to be the 3′-methyl ether of quercetin, namely isorhamnetin (Brown and Griffiths, unpublished observations). Trace amounts of isorhamnetin have previously been detected in rat urine following the administration of quercetin (Peter *et al.*, 1966). It is important to note that methyl ethers have been detected only in respect of flavonoids possessing *o*-dihydric substitution in ring B of the molecule.

The subsequent metabolic fate of methylated flavonoids secreted in bile is determined by their susceptibility to ring fission in the intestine. An intact aglycone of the metabolite 3-*O*-methylbutein sulphate was, after metabolic hydrolytic cleavage, detectable in rat faeces (Brown and Griffiths, unpublished observation). The aglycone of the biliary metabolite 3′-*O*-methyl-(+)-catechin glucuronide was not, however, detectable in faeces of the rat, indicating that the 3′-*O*-methyl-(+)-catechin, like (+)-catechin, is readily degraded by the intestinal microflora (Shaw and Griffiths, unpublished observations). It appears possible therefore that the earlier reported urinary metabolites of (+)-catechin, 3-methoxy-4-hydroxyphenyl-γ-valerolactone (Oshima and Watanabe, 1958; Das and Griffiths, 1969) and vanillic acid (Oshima *et al.*, 1958) may in part be derived from the methylated biliary metabolite. Since it is also well established that methoxylated aromatic acids are readily demethylated by gut microorganisms (Scheline, 1966, 1968a; Griffiths, 1970), a significant proportion of the phenolic acids detected in urine after the administration of (+)-catechin and other flavonoids may be derived from methylated intermediates.

The possibility that methylated flavonoids secreted in bile may undergo *direct* dealkylation also requires consideration in view of the reports, that the isoflavones, biochanin A (4′-*O*-methylgenistein) and formononetin (4′-*O*-methyldaidzein), are demethylated to genistein and daidzein respectively when incubated with rumen fluid (Nilsson *et al.*, 1967) and that small amounts of apigenin are detectable in urine following the administration of acacetin (4′-methylapigenin) to the rat (Griffiths and Smith, 1972a).

12.2.11 Interactions of flavonoids with enzymes and other cell constituents

A number of flavonoids have been reported to have specific effects upon the activity of enzyme systems of mammalian cells, and certain of these may be of importance in mediating pharmacological action. Schwabe and Flohe (1972) reported inhibition of rat liver catechol-*O*-methyltransferase by certain flavonoids and have shown that the structural requirements for inhibitory activity are a flavonol ring system, a keto group at C-4 and hydroxylation at C-5. Gugler and Dengler (1973) have also demonstrated inhibition of *human* liver catechol-*O*-methyltransferase by flavonoids. They noted that flavonoid glycosides were less active than the corresponding aglycones and that reduction of the double bond between C-2 and C-3 diminished the inhibitory effect. Both groups reported that the presence of a keto group at C-4 and *o*-dihydroxylation in the B ring enhanced inhibition, and Schwabe and Flohe noted that compounds

possessing *o*-dihydric substitution could themselves serve as substrates for methylation *in vitro*. Recent investigations (Griffiths *et al.*, 1980) have shown that this reaction may be of major importance in the metabolism of flavonoids *in vivo*.

Certain flavones and flavonols have been reported to inhibit bovine pancreatic ribonuclease at concentrations of 10^{-3}–10^{-4} M. The effect is dependent on the presence of a keto group at C-4 and of certain hydroxyl groups. Flavanones and flavanonols lacked activity as did all flavonoid glycosides examined (Mori and Noguchi, 1970). The ability to inhibit lens aldose reductase is shown by a large number of flavonoids (Varma *et al.*, 1975; Varma and Kinoshita, 1976). At concentrations of 10^{-5} M, rutin and quercitrin showed inhibition of 95%.

The flavanol (+)-catechin has been shown to produce a half-maximal inhibition of the prostaglandin synthetase of the renal medulla at a concentration of 5×10^{-5} M (Baumann and von Bruchhausen, 1979). (+)-Catechin has also recently been shown to be among a group of flavonoids able to inhibit cyclic GMP phosphodiesterase selectively (Ruckstuhl *et al.*, 1979). Certain ATPases were at an earlier date shown to be inhibited by the flavonol quercetin (Carpenedo *et al.*, 1969), and the significance for membrane-linked activity was discussed.

Certain flavones including 5,6- and 7,8-benzoflavones have been shown to exert an inhibitory effect on mono-oxygenases *in vitro* (Diamond and Gelboin, 1969; Wiebel and Gelboin, 1975) but to induce microsomal mono-oxygenase activity *in vivo* (Boobis *et al.*, 1977; Fang and Strobel, 1978; Wattenberg *et al.*, 1968). The latter authors also studied the structural requirements for induction and noted that increasing hydroxylation in the flavone series reduced inducing capacity but that the corresponding methoxy compounds were active. Active compounds included the naturally occurring flavones tangeretin and nobiletin. Several flavones have recently been shown to stimulate mammalian epoxide hydrase activity (Alworth *et al.*, 1980).

Since both arylhydrocarbon mono-oxygenase and epoxide hydrase are known to be important in the formation of carcinogenic metabolites, these findings are of considerable medical interest and further research on the ability of naturally occurring flavonoids to induce the formation of these enzymes *in vivo* appears to be an area of priority.

The interaction of flavonoids with the genetic material of the cell is indicated by numerous reports that certain flavonoids display mutagenicity in the Ames test (cf. Brown, 1980). Several studies have been carried out to determine the structural features necessary for mutagenic activity (McGregor and Jurd, 1978; Hardigree and Epler, 1978) and McGregor and Jurd concluded that for mutagenic activity in *Salmonella typhimurium* strain TA 98, a flavonoid should possess a free 3-hydroxyl substituent, a double bond in the 2,3 position, a keto group at the 4 position and a structure which permits the proton of the 3-hydroxyl group to tautomerize to a 3-keto compound. A high level of mutagenic activity was shown by quercetin, and Bjeldanes and Chang (1977) have shown that the mutagenic activity of quercetin is tripled by incubation with

the microsomal fraction of liver. Although several reported metabolites of quercetin (*viz.* phenolic acids arising by ring fission) were examined for activity, all were found to be non-mutagenic and it was concluded that the proximate carcinogen was an unidentified metabolite.

Quercetin has been shown to be carcinogenic also in the rat following oral dosage, as neoplasms were subsequently observed in the gastrointestinal tract and bladder (Pamukcu *et al.*, 1980a,b). As quercetin is poorly absorbed (less than 1 % of dose) from the gastrointestinal tract (Gugler *et al.*, 1975), the development of neoplasms in the bladder epithelium appears somewhat surprising, but it is perhaps important to bear in mind that activation of the small amounts of absorbed quercetin may occur during transit through the liver as the microsomes of rat liver have been shown at least *in vitro* to increase the mutagenicity of quercetin (Bjeldanes and Chang, 1977).

12.2.12 Metabolism of synthetic and chemically modified flavonoids

In addition to the investigations reported on the metabolism of naturally occurring flavonoids which are reviewed in preceding sections, the introduction of a number of chemically modified and synthetic compounds as pharmacological agents or food additives has resulted in metabolic study of compounds showing structural features absent from the natural plant flavonoids.

The synthetic flavone, flavoxate, (*12.6*), an antispasmodic drug was, following oral and intravenous administration to the rat, reported to be metabolized to a conjugate of 3-methylflavone-8-carboxylic acid (*12.7*) which was excreted in bile

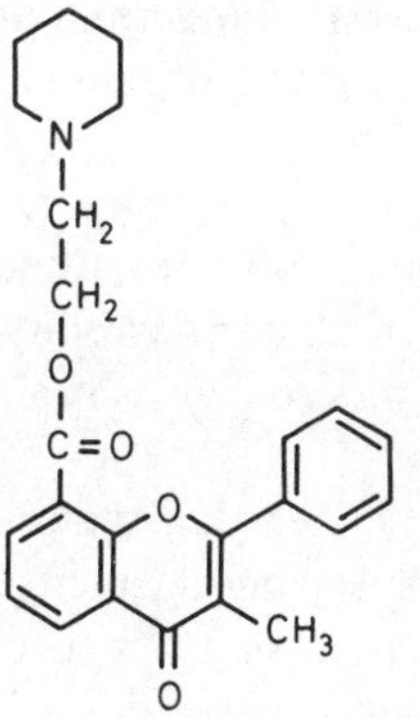

(*12.6*) Flavoxate

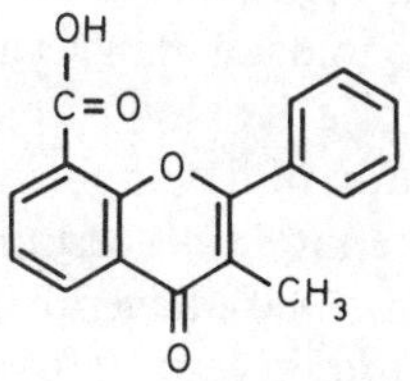

(*12.7*) 3-Methylflavone-8-carboxylic acid

and urine. Up to 58 % of the dose was recovered from the bile of orally dosed biliary-cannulated animals within 24 h of administration. Total recovery from bile and urine under these conditions was 78 % (Setnikar *et al.*, 1975). Although the authors suggest the remainder of the dose is probably metabolized by ring

fission, the pattern of substitution of the molecule, especially the absence of A-ring hydroxylation, makes this improbable (see Section 12.2.1).

Metabolic study of the vaso-active hydroxyethylrutosides obtained by the hydroxyethylation of rutin has shown that the hydroxyethyl groups are not only resistant to metabolic change but confer greater metabolic stability upon the *O*-heterocyclic ring system, which as a consequence of hydroxyethylation becomes resistant to microfloral ring fission (Barrow and Griffiths, 1972, 1974a,b), whereas rutin is readily degraded to phenolic acids and CO_2 following oral administration to the rat (Booth *et al.*, 1956; Petrakis *et al.*, 1959). 7,4′[^{14}C] Di-*O*-(β-hydroxyethyl)rutoside, 7,3′,4′[^{14}C]-tri-*O*-(β-hydroxyethyl)rutoside (Fig. 12.7) and 5,7,3′,4′[^{14}C]-tetra-*O*-(β-hydroxyethyl)rutoside are degraded *in vivo* only to the aglycone level, as the corresponding [^{14}C]hydroxy-ethylquercetins can be recovered from the faeces (Barrow and Griffiths, 1972, 1974a,b). No ^{14}C labelled ring-fission products were detected in urine which contained small amounts of unchanged hydroxyethylrutosides and their glucu-ronide conjugates. Large amounts of the unchanged hydroxyethylrutosides and their glucuronides have been recovered from bile (see Section 12.3.3). No $^{14}CO_2$ was detected in the respired air.

A new therapeutic agent, 3-*O*-methyl-(+)-catechin, has also been shown to be resistant to microfloral ring fission (Griffiths and Hackett, 1978), unlike the parent compound, (+)-catechin, which readily undergoes ring fission in the rat (Griffiths, 1964). The major metabolite of 3-*O*-methyl-(+)-catechin was shown to be 3,3′(or 4′)-dimethyl-*O*-(+)-catechin glucuronide which was secreted in both bile and urine in the rat, whereas in the mouse and marmoset, the aglycone was the major urinary metabolite (Hackett and Griffiths, 1981). Metabolic studies utilizing 3-*O*-[*methyl*-^{14}C] methyl-(+)-catechin showed that the 3-*O*-methyl substituent was resistant to metabolic attack and no $^{14}CO_2$ was detected in respired air following oral administration to the rat (Hackett and Griffiths, 1981).

Although the possible use of chemically modified flavonoid derivatives, especially substituted dihydrochalcones as sweetening agents, has been explored in a number of laboratories, only limited data have been published on the metabolism of these compounds. Recently, however, investigations have been reported upon the fate of a group of sulphopropylated [^{14}C]hesperetin derivatives (Brown *et al.*, 1978). Although [^{14}C]hesperetin is known to be well absorbed (84% of dose) in the bile-duct-ligated rat (Honohan *et al.*, 1976), absorption of orally administered 7-*O*-[^{14}C]sulphopropylhesperetin, 4-*O*-[^{14}C] sulphopropylhesperetin dihydrochalcone and 2,4-[^{14}C]di-*O*-sulphopropyl-hesperetin dihydrochalcone by these rats was found to decrease sharply with both sulphopropylation and reductive opening of the dihydropyrone ring. The sulphopropyl derivatives were also reported to be largely resistant to microfloral degradation.

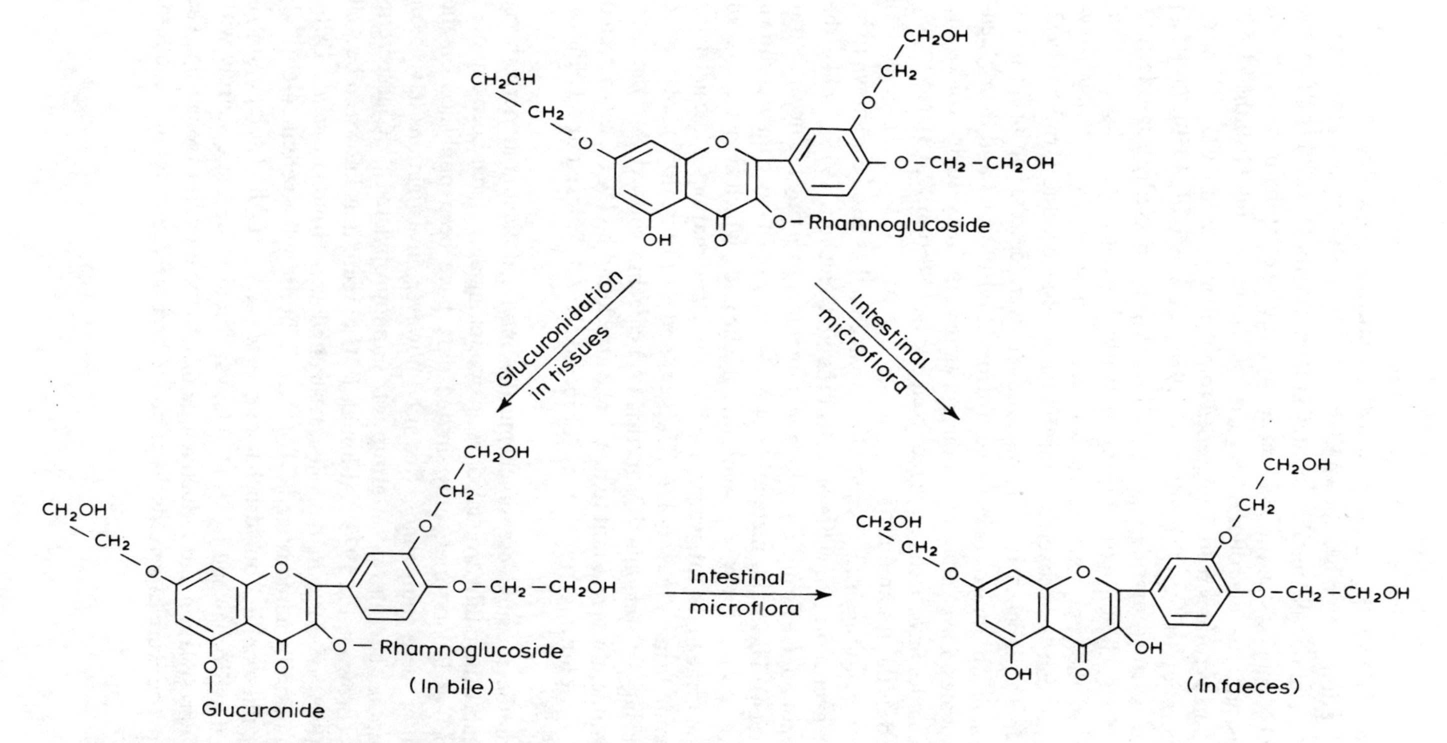

Fig. 12.7 Mammalian metabolism of 7,3′4′-tri-*O*-(*β*-hydroxyethyl)rutoside.

12.3 METABOLIC DISPOSITION

Studies on the distribution of radiochemically labelled compounds and their metabolites in the organs and tissues of animals can provide useful information, not only on the fate of the compound, but also on their sites of metabolism, the routes of excretion and the sequence of metabolite formation.

12.3.1 Distribution in organs, tissues and biological fluids

The earliest detailed study of the metabolic disposition of a flavonoid was that of Petrakis *et al.* (1959), who orally administered randomly labelled [^{14}C]quercetin to the rat and reported on the distribution of radioactivity in a range of organs, tissues and biological fluids. They showed that 12 h after administration, radioactivity was largely associated with gastrointestinal contents (44.2%), although significant levels were present in lungs (11.5%), respired CO_2 (15%), kidney (0.3%), urine (3.9%) and blood (0.2%). Rather surprisingly, no significant radioactivity was detected in liver. The urinary metabolites detected included the C_6–C_2 phenolic acids reported by Booth *et al.* (1956). Following intraperitoneal administration of [^{14}C]quercetin the formation of ^{14}C-labelled C_6–C_2 phenolic acids was reduced but [^{14}C]vanillic acid was detected in urine. No $^{14}CO_2$ was respired. This indicates not only that the metabolism of quercetin is determined by the route of administration but also suggests that the $^{14}CO_2$ production observed after oral administration was dependent upon contact with the intestinal microflora, which is largely diminished by the *intraperitoneal* administration of quercetin. It is of interest here to note that Pfeifer *et al.* (1970) have shown that parenteral administration of rutin (quercetin rhamnoglucoside) results in deposition of the compound at the injection site and in the bile ducts, and in consequence effective biliary excretion is minimal (Barrow and Griffiths, 1971, 1974a).

Following the oral administration of [^{3}H] rutin to rats, Tamemasa *et al.* (1976) showed that trace amounts of radioactivity only were detectable in bile, whereas 66% of the administered radioactivity was excreted in urine. The labelled metabolites of rat urine were not identified but their chromatographic properties suggest that they are largely ring fission products. The levels of radioactivity in a wide range of tissues was reported.

Early investigation of the disposition of flavanols showed that following oral administration of a [^{14}C]catechin preparation containing non-labelled ascorbic acid, radioactivity was detectable in the muscles, liver, spleen, adrenals and kidneys of scorbutic guinea pigs (Shamrai and Zaprometov, 1964). The [^{14}C]catechin preparation used was stated to contain a mixture of labelled catechins including (−)-epigallocatechin gallate, epicatechin gallate, (−)-epigallocatechin and (−)-epicatechin. The authors noted that the highest levels of radioactivity were observed in the adrenals. This was confirmed in a later publication (Shamrai and Fedurov, 1967), when additional data were presented

which indicated that 30 % of the administered ^{14}C was eliminated as $^{14}CO_2$, 30 % in the faeces and 10 % in the urine.

The distribution of the leucoanthocyanidin, 5,7,3′,4′-[^{14}C]tetrahydroxyflavan-3,4-diol, in the tissues of the mouse has been studied by Laparra and Michaud (1973) using a radioautographic technique. Following intravenous administration, radioactivity disappeared rapidly from the systemic circulation, accumulating initially in liver and kidney and subsequently in skin, the periosteum of bone and in cartilaginous structures. Later, large amounts of radioactivity were observed in the gastrointestinal tract. Although low levels of activity were detected in the heart, high activity was found to be associated with aorta. Later, high activity was also observed in the intervertebral cartilages of the spinal column. The authors suggest that localization of the radioactive tetrahydroxyflavandiol in these cartilaginous structures may be attributable to the availability of reactive anionic sites in the mucopolysaccharides.

Recently Harmand and Blanquet (1977) have investigated the distribution of radioactivity in the rat following the intravenous administration of a mixed [^{14}C]flavanol preparation containing (+)-catechin, (−)-epicatechin and several proanthocyanidin dimers. Radioactivity in blood was reported to reach a maximum value at 9 h, whilst radioactivity in connective tissues including cartilage, trachea, aorta and skin peaked at 15 h. The data obtained showed a specific accumulation of radioactivity in the latter tissues, resembling the pattern of distribution of [^{14}C]tetrahydroxyflavandiol reported by Laparra and Michaud (1973). A specific accumulation was also shown in respect of the thyroid, pituitary and adrenal glands. Radioactivity in liver was also shown to attain maximal values at 15 h. Subsequently, radioactivity equivalent to 49 % of the intravenous dose was excreted in urine, 3 % in faeces and 1 % in expired air within 24 h.

The disposition of the chemically modified flavonoids, 7,3′,4′-tri-*O*-(*β*-hydroxyethyl)rutoside and 7-mono-*O*-(*β*-hydroxyethyl)rutoside, has recently been investigated following intravenous administration to the mouse (Hackett and Griffiths, 1977). Blood concentrations of both compounds fell rapidly, whilst accumulation of radioactivity in the liver and kidneys was observed. The period of greatest decline in liver ^{14}C (0–4 h) corresponded to the period of maximal biliary excretion. ^{14}C was also readily detected in the more vascularized tissues, namely spleen and lung. In none of the experiments was radioactivity detected in brain, indicating that neither the administered compounds nor their metabolites were able to cross the blood–brain barrier. At no stage was ^{14}C detected in peritoneal fat or in bone samples. The ^{14}C in carcasses of animals killed between 3 and 72 h after intravenous dosage was associated mainly with gastrointestinal contents, indicating the importance of the biliary route of excretion.

Employing radioautography, the disposition of an intravenous dose of [^{14}C]trihydroxyethylrutoside was investigated in mice sacrificed at 15 min, 30 min and 4 h after administration (Fig. 12.8; Hackett and Griffiths, unpublished observation). Following initial concentration in the liver (l), radio-

activity was transferred via the gall bladder (gb) and bile duct to the upper and lower intestine (ui, li). Radioactivity was also observed by this technique in kidney (k) and in the urinary bladder (ub). The major metabolites present in mouse faeces have been shown to be the corresponding aglycones of the rutosides administered (Hackett and Griffiths, 1977).

Radioautographic examination of [^{14}C]hydroxyethylrutoside-dosed mice did not reveal localization of label in the connective tissue sites, including cartilage of the intervertebral discs, trachea and aorta, which were identified by Laparra and Michaud (1973) and Harmand and Blanquet (1977) as specific sites of accumulation of radioactivity following [^{14}C]flavan-3,4-diol and/or [^{14}C]flavan-3-ol administration.

Following the injection of [^{14}C]trihydroxyethylrutoside and [^{14}C]monohydroxyethylrutoside into the tail vein of pregnant mice on the 18th day of pregnancy, it was shown that 5 min after injection, the time corresponding to peak tissue levels in male mice, ^{14}C in all foetuses was less than 0.1% of the administered dose. Foetuses from mice killed 3 h after injection of either compound contained no detectable ^{14}C (Hackett and Griffiths, 1977a,b).

12.3.2 Blood levels

Although the plasma levels attained in several mammalian species following the administration of certain flavonoids employed in clinical practice has been reported, blood pharmacokinetic data are available for relatively few flavonoids.

Plasma levels of the flavonol, quercetin, following intravenous administration to man were monitored by Gugler *et al.* (1975) using a specific and highly sensitive assay method. Disappearance of quercetin from plasma after a single dose was apparently bi-exponential with a mean terminal half life of 2.4 h. The decline of the plasma level curve after the distribution phase was attributed by the authors mainly to metabolism of the drug. Following oral administration, however, no measurable plasma concentration was observed and the authors concluded that unchanged quercetin did not reach the systemic circulation.

Following the oral administration of a [^{14}C]hydroxyethylrutoside preparation to a group of human volunteers, peak plasma levels were observed from 2 to 9 h (Fig. 12.9). These levels then showed a progressive decline until at 40 h, a low ^{14}C level was established which then declined slowly over the following 120 h when the experiment was terminated (Hackett *et al.*, 1976). The maximal plasma levels observed were approx. 0.3 μg/ml. Insufficient ^{14}C-labelled metabolite(s) were present in the blood plasma to permit identification, but since the hydroxyethylrutosides present in this preparation are known to be resistant to ring fission (Barrow and Griffiths, 1972, 1974a,b), it is probable that the radioactivity detected in plasma was associated with unchanged hydroxyethylrutosides or their glucuronide conjugates.

Similar quantitative data have subsequently been presented by Förster (1977) following oral administration of labelled hydroxyethyl rutosides to human

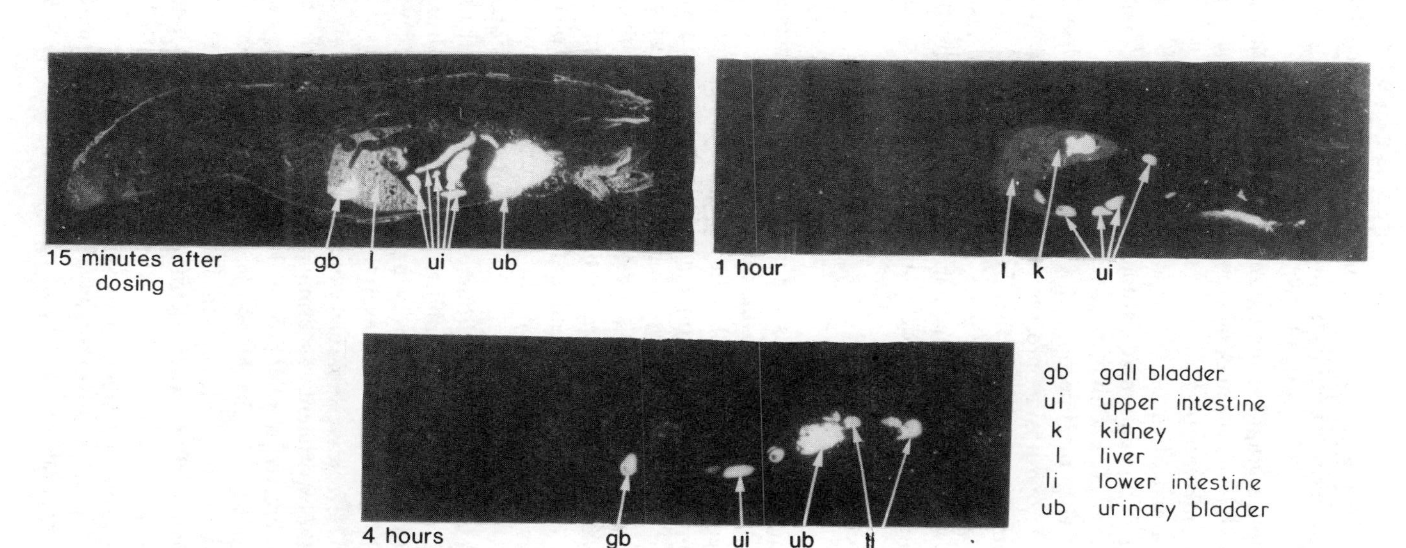

Fig. 12.8 Distribution of ^{14}C in the mouse following intravenous administration of 7,3′4,′-[^{14}C] tri-*O*-(*β*-hydroxyethyl) rutoside.

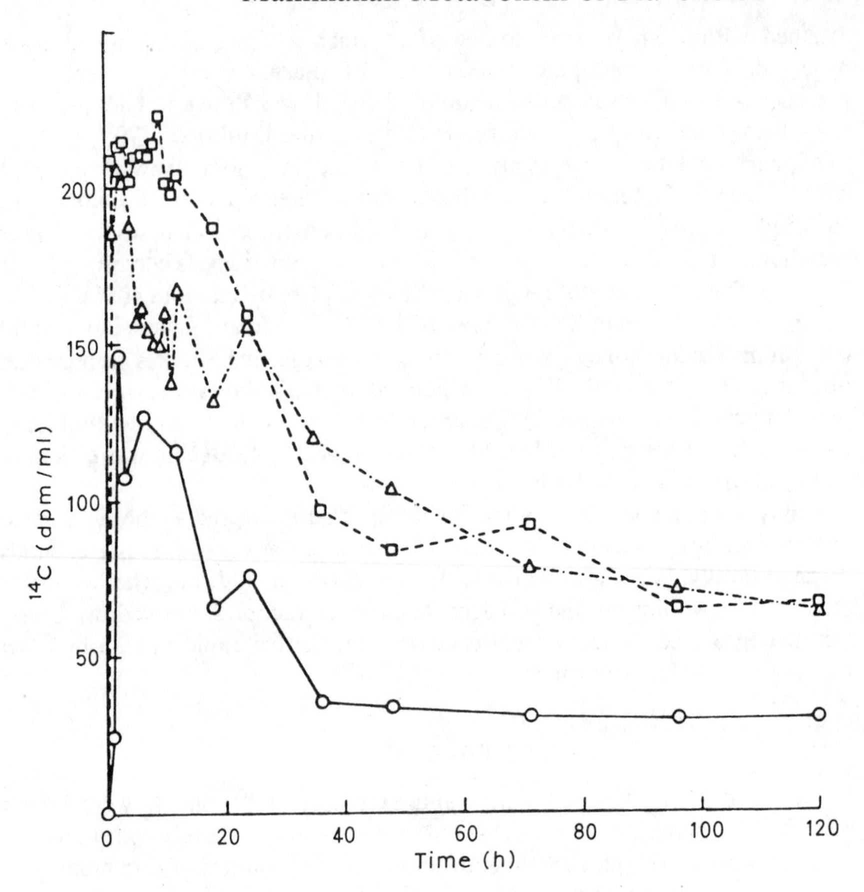

Fig. 12.9 ^{14}C levels in the plasma of three human subjects following a single oral dose of [^{14}C]hydroxyethylrutosides.

volunteers. Additionally, data obtained by intravenous infusion were reported which showed that the hydroxyethylrutosides are under these conditions rapidly removed from the systemic circulation.

The plasma levels of hydroxyethylrutoside in man have also been determined by circular dichroism measurements (Jung *et al.*, 1977), following the administration of a non-labelled hydroxyethylrutoside preparation. After intravenous administration, it was observed that the major part of the hydroxyethylrutosides disappeared from the blood within 1–2 h. Following oral administration maximal concentrations were found over the 6–10 h period.

Investigation on plasma levels of dogs following the oral administration of a ^{14}C-labelled mixed hydroxyethylrutoside preparation showed that peak levels were attained between 5 and 6 h (Mirkovitch *et al.*, 1973). Studies employing 7-[^{14}C]mono-*O*-(β-hydroxyethyl)rutoside also showed that ^{14}C peak levels were

attained within 6 h of oral dosage after which a slow decline to 72 h was observed. Chromatographic examination of plasma samples revealed the presence of a single radioactive metabolite, which was shown to be 7-mono-*O*-(β-hydroxyethyl)rutoside glucuronide (Hackett and Griffiths, 1979).

A pharmacokinetic investigation of (+)-catechin in man showed that peak plasma levels of metabolites of this flavanol determined by a colorimetric technique occurred within 1–2 h after oral administration when up to 17 μg of metabolites/ml serum were observed. These peak blood levels corresponded to 2.5–3 % of the ingested dose (Giles and Gumma, 1973). Subsequently, however, Balant *et al.* (1979) using a highly specific HPLC technique concluded that at the maximum serum concentrations, 10 % of serum metabolites represented unchanged (+)-catechin, 40 % corresponded to metabolites possessing an intact phloroglucinol ring, whilst 50 % represented other unidentified metabolites of (+)-catechin. The apparent half life of unchanged (+)-catechin was estimated from sera data to be 1–1.3 h.

It may be concluded that high serum levels of administered flavonoids are not readily sustained, whether single doses of the compounds are administered orally or parenterally. The data available do not, however, indicate whether this is attributable wholly to the efficient hepatic extraction observed in biliary cannulated animals and to renal excretion or whether rapid uptake by other tissues plays a significant part.

12.3.3 Routes of excretion

Earlier studies on the metabolism and excretion of flavonoids were largely restricted to metabolites eliminated in urine which were mainly compounds of small molecular weight, subsequently shown to be products of ring fission (see Section 12.2.1). These metabolites were sometimes observed to be accompanied by small or trace amounts of the flavonoid administered or its glucuronide conjugates. In 1971, evidence was presented that the biliary-enteric route was of major importance for the excretion of intact flavonoid molecules and their conjugates (Barrow and Griffiths, 1971; Das and Sothy, 1971).

(a) Biliary excretion

Following parenteral administration of [^{14}C](+)-catechin to the bile-duct-cannulated rat, Das and Sothy showed that 33–44 % of the administered radioactivity was excreted in bile largely as flavanol conjugates. Parallel investigations showed that following *oral* as well as parenteral administration, conjugates of (+)-catechin, naringin and 7,3′,4′-tri-*O*-(β-hydroxyethyl)rutoside were excreted in rat bile (Barrow and Griffiths, 1971). Following intraperitoneal administration of the latter hydroxyethylrutoside to intact rats, large amounts of the aglycone were detected in faeces. The corresponding aglycones of naringin, naringenin and (+)-catechin were not detected under similar experimental

conditions in faeces, but this can be attributed to the known susceptibility of the latter to degradation by the intestinal microflora (Griffiths and Smith, 1972a; Griffiths, 1964; Das, 1969).

Further investigations have established that biliary excretion is an important factor in the metabolism and pharmacokinetics of many flavonoid compounds (Table 12.1). Flavonoids now known to be excreted in bile include: flavanols (Barrow and Griffiths, 1971; Das and Sothy, 1971; Griffiths and Hackett, 1978; Hackett and Griffiths, 1980; Harmand *et al.*, 1978; Shaw and Griffiths, 1980), flavanones (Honohan *et al.*, 1976; Hackett *et al.*, 1979; Buset and Scheline, 1980), flavolignans (Bülles *et al.*, 1975; Mennicke *et al.*, 1979), anthocyanins (Lietti *et al.*, 1976), flavones (Oustrin *et al.*, 1977; Setnikar *et al.*, 1975; Svardal *et al.*, 1980), flavonols (Barrow and Griffiths, 1971, 1972, 1974a,b), chalcones (Griffiths *et al.*, 1980), dihydrochalcones (Barrow and Griffiths, 1971) and the *C*-glycosylflavone, orientin (luteolin-8-*C*-glucopyranoside) (Paris and Delaveau, 1977).

The major factors leading to the excretion of organic compounds in significant quantity in rat bile have been identified by Smith (1973) as (a) a molecular weight exceeding 325 ($\pm$50) and (b) the degree of polarity of the compound excreted. Since most *non-glycosidic* flavonoids show low polarity and possess a molecular weight below the described threshold, they would not be expected to be readily excreted in bile. However, recent studies have shown that the major products of hepatic metabolism of many flavonoids are the corresponding glucuronides (Barrow and Griffiths, 1974a,b; Bülles *et al.*, 1975; Hackett and Griffiths, 1979; Griffiths *et al.*, 1980; Shaw and Griffiths, 1980), which not only possess considerable polarity but also show molecular weights in excess of the 325 threshold.

Flavonoid glycosides, although of low polarity, do possess molecular weights above this described threshold and in consequence may be excreted to an appreciable extent in bile without glucuronidation. This has been noted in relation to 5,7,3′,4′-tetra-*O*-(*β*-hydroxyethyl)rutoside (Barrow and Griffiths, 1972) and naringin (Hackett *et al.*, 1979).

The substitution of certain flavonoids has been shown to be an important factor in determining the extent of biliary excretion. In studies on the hydroxyethylrutoside series (Barrow and Griffiths, 1972, 1974a,b), it was shown that progressive hydroxyethylation of the parent rutin molecule resulted in a reduction in the proportion of the dose excreted in bile. The highest levels of biliary excretion were observed in respect of the mono- and di-hydroxyethylrutosides and the lowest shown by the fully hydroxyethylated 5,7,3′,4′-*O*-(*β*-tetrahydroxyethyl)rutoside. The latter rutoside alone was excreted in larger amounts in urine than in bile. Progressive hydroxyethylation not only diminished biliary excretion but also reduced the percentage of the flavonoid excreted as glucuronide, which accords well with the view that glucuronide formation facilitates excretion of flavonoids by the biliary route.

Dosage levels may also be an important factor in determining the extent of biliary excretion. Hackett *et al.* (1979) have shown that progressive reduction of

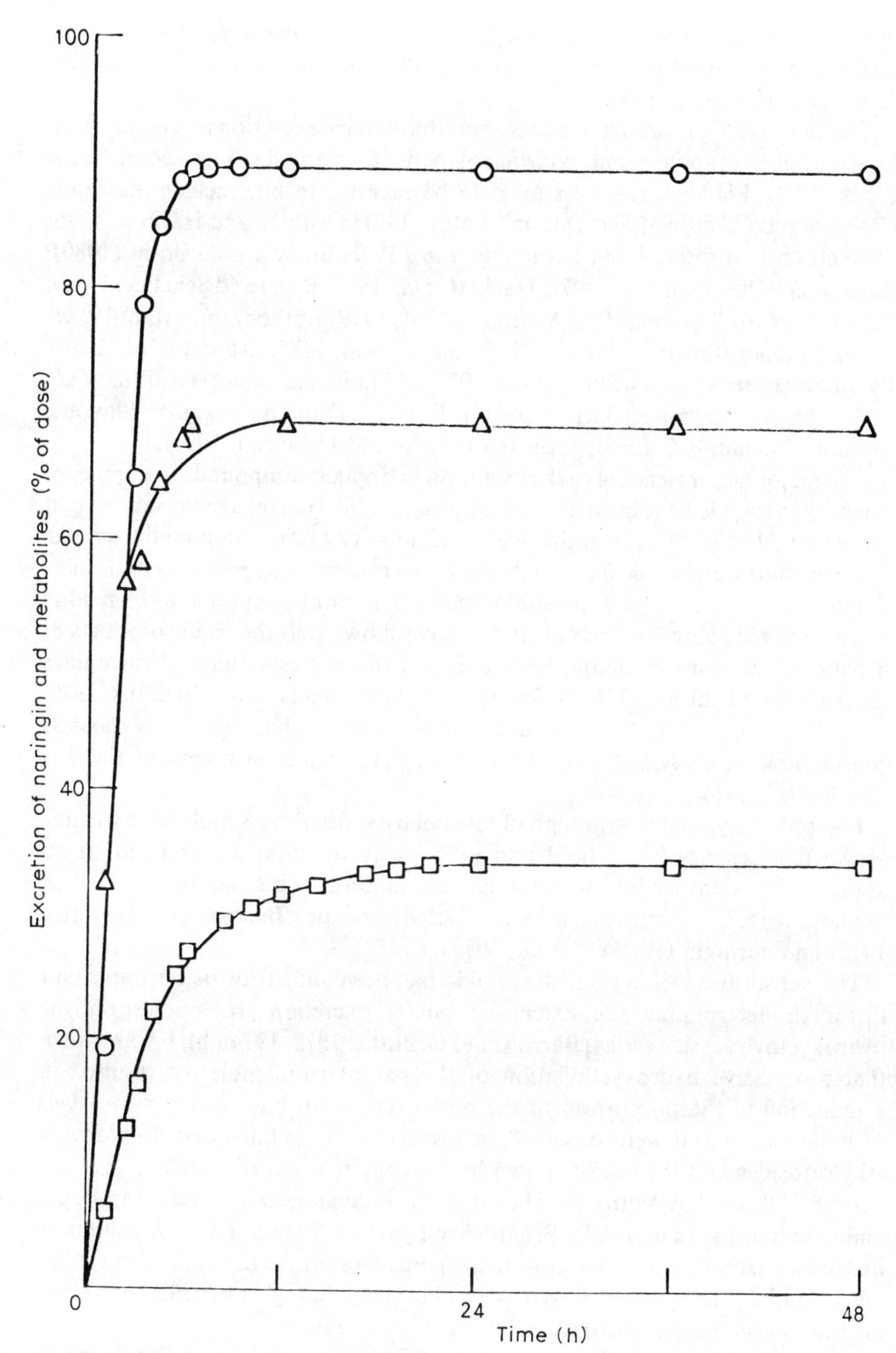

Fig. 12.10 The excretion of naringin and its metabolites in bile as a percentage of dose following intraperitoneal administration to the rat. Dosage levels per rat: 5 mg: ○ 10 mg: △ 50 mg: □

the intraperitoneal dose of certain flavanones results in elevation of the percentage of the dose excreted in bile. The inverse relationship between dose and the magnitude of biliary excretion in respect of naringin is shown in Fig. 12.10.

Although the extent of biliary excretion of flavonoid is usually greater after parenteral administration than after oral administration, this is usually seen to be a consequence of slow absorption from the gastrointestinal tract rather than attributable to any change in the partitioning of metabolites between the biliary and the renal route, as the total excretion by both these routes is usually significantly less after oral administration (Table 12.1)

(b) Enterohepatic cycling

Although biliary excretion has been established as a major route of excretion of many flavonoids limited information is available concerning the subsequent fate of the conjugates entering the intestine. The presence of large amounts of β-glucuronidase in intestinal contents (Marsh *et al.*, 1952; Hawkesworth *et al.*, 1971) and of micro-organisms able to effect ring scission of the released aglycones (see Section 12.2.2) provides the means by which the biliary metabolites of many naturally occurring flavonoids may be degraded to compounds of low molecular weight. Additionally, however, the possibility of reabsorption of the biliary metabolites with the establishment of an enterohepatic cycle requires consideration. Evidence that the hydroxyethylrutosides undergo enterohepatic cycling in the rat has recently been presented (Hackett and Griffiths, 1977b). Using intercannulated rat preparations, the biliary metabolites, formed following the parenteral dosage of [^{14}C]monohydroxyethylrutoside and [^{14}C]trihydroxyethylrutoside to the first animal of each pair, were transferred by a bile-duct cannula into the gut lumen of a second animal and the levels of radioactivity determined subsequently in the bile and urine of the second animal. The results obtained showed that a significant uptake of the biliary metabolites from the gut of the second animal had occurred. The radioactive constituents of the bile and urine were shown to consist largely of the glucuronide conjugates of the administered hydroxyethylrutoside. The evidence obtained does not, however, reveal whether deconjugation of the biliary metabolites precedes absorption followed by resynthesis, but this appears probable in view of the known polarity of the glucuronide conjugates. Since enterohepatic cycling extends the period during which pharmacologically active compounds persist in the systemic circulation, these findings may be of significance in relation to the maintenance of therapeutic levels in man but the possibility of species differences should not be excluded.

(a) Urinary excretion

Although as noted above, the flavonoid metabolites excreted in urine are largely ring fission products of low molecular weight, small amounts of flavonoids and

their conjugates have not been infrequently observed. Following the oral administration of naturally occurring flavonols, urinary excretion of flavonols and their conjugates is minimal. In a definitive study on the disposition of quercetin in man, Gugler *et al.* (1975) reported that following oral administration to man, neither quercetin nor quercetin conjugates were detectable in urine at any time following ingestion. Following the oral administration of rutin to the rat, Barrow and Griffiths (1974a) were likewise unable to detect the administered compound or its conjugates in urine. Although an early investigation failed to detect the excretion of the rutin derivative, 7,3′,4′-tri-*O*-(β-hydroxyethyl)rutoside in human urine following the oral administration of the unlabelled compound (Wienert and Gahlen, 1970), subsequent investigation revealed that following oral administration to the rat low but significant urinary excretion (1.7–2.5%) of the administered compound and of its glucuronide(s) occurred (Barrow and Griffiths, 1974b). In man following the oral administration of ^{14}C-labelled hydroxyethylrutosides, 1.6–2% of the administered radioactivity was excreted in urine as unchanged hydroxyethylrutoside. Total urinary radioactivity corresponding to 3.05–5.97% of the dose was observed (Hackett *et al.*, 1976). Subsequent investigations on the fate of orally administered [^{14}C]7,3′,4′-tri-*O*-(β-hydroxyethyl)rutoside in man (Tan *et al.*, 1978) showed that following β-glucuronidase treatment of the urine 5.3% of the administered ^{14}C could be recovered from the urine.

In a recent pharmacokinetic study on the absorption and elimination of the flavone, [^{3}H]diosmin, it has been claimed that 25% of the dose is eliminated in urine (Oustrin *et al.*, 1977). Much of this radioactivity may, however, be associated with ring fission products, earlier reported in rat urine after diosmin administration (Booth *et al.*, 1958).

Following the oral administration of [U-^{14}C] (+)-catechin to the rat, it was shown that 50–63% of the administered radioactivity was excreted in urine but only 0.12% was attributable to unchanged (+)-catechin (Das and Griffiths, 1969). In man, Balant *et al.* (1979) have shown by means of an HPLC technique that only 0.5% of an oral dose of (+)-catechin appears in human urine as unchanged (+)-catechin. Although unchanged (+)-catechin is thus a minor urinary metabolite, recent investigation has shown that following the oral dosage of [^{14}C] (+)-catechin to the rat, a high proportion of the urinary radioactivity (approx. 30%) is associated with the flavonoid conjugate, 3′-*O*-methyl-(+)-catechin glucuronide (Shaw and Griffiths, unpublished observation) which is also known to be the major biliary metabolite of (+)-catechin in the rat (Griffiths *et al.*, 1980).

The large amounts of radioactivity (*ca.* 50% of dose) excreted in the urine of the rat following oral dosage with the related flavanol, 3-*O*-methyl-(+)-catechin have been shown to be largely associated with the metabolite, 3,3′ (or 4′)-dimethyl-(+)-catechin glucuronide which accounts for approx. 80% of the urinary radioactivity in this species (Hackett and Griffiths, 1981). In the mouse and marmoset, however, the major urinary metabolite is the free aglycone,

3,3′-dimethyl-(+)-catechin. The high level of urinary excretion of metabolites containing an intact flavonoid ring system is probably attributable to high absorption of orally administered 3-*O*-methyl-(+)-catechin associated with resistance of the molecule to microfloral ring fission.

Although the extent of urinary excretion of orally administered flavonoids has been employed by a number of investigators as a measure of absorption, it is evident that any assessment of flavonoid absorption based on urinary excretion data alone may lead to erroneous conclusions on two counts. Firstly, it has been shown that many flavonoids are excreted to a significant extent in bile; indeed this may represent the major proportion of the absorbed dose. Secondly, many flavonoids undergo degradation in the intestine to compounds of low molecular weight which are readily absorbed and subsequently excreted in urine. Much of the urinary radioactivity measured, following the administration of a labelled flavonoid, may be attributable to intestinal metabolic products of this type.

It is, therefore, important that in assessing flavonoid absorption by excretion data that both biliary and urinary metabolites should be fully characterized and a measure of absorption obtained by summation of recoveries of intact flavonoids and flavonoid conjugates from both bile and urine. Although the excretion of ring fission products provides no evidence of absorption of the parent flavonoid molecules, it is, however, important to give consideration in specific cases to the possibility that the fission products may be derived from biliary metabolites rather than from the unabsorbed flavonoid.

REFERENCES

Alworth, W. L., Dang, C. C., Ching, L. M. and Viswanathan, T. (1980), *Xenobiotica* **10**, 395.

Balant, L., Burki, B., Wermeille, M. and Golden, G. (1979), *Arzneimittel-Forsch.* **29**, 1758.

Barrow, A. and Griffiths, L. A. (1971), *Biochem. J.* **125**, 24P.

Barrow, A. and Griffiths, L. A. (1972), *Xenobiotica* **2**, 575.

Barrow, A. and Griffiths, L. A. (1974a), *Xenobiotica* **4**, 1.

Barrow, A. and Griffiths, L. A. (1974b), *Xenobiotica* **4**, 743.

Batterham, T. J., Hart, N. K., Lamberton, J. A. and Braden, A. W. H. (1965), *Nature (London)* **206**, 509

Batterham, T. J., Shutt, D. A., Hart, N. K., Braden, A. W. H. and Tweeddale, H. J. (1971), *Aust. J. Agric. Res.* **22**, 131.

Baumann, J. and Von Bruchhausen, F. (1979), *Naunyn-Schmiedeberg's Arch. Pharmacol.* **306**, 85.

Bjeldanes, L. F. and Chang, G. W. (1977), *Science* **197**, 577.

Boobis, A. R., Nebert, D. W. and Felton, J. S. (1977), *Mol. Pharmacol.* **13**, 259.

Booth, A. N., Murray, C. W., De Eds, F. and Jones, F. T. (1955), *Fed. Proc. Fed. Am. Soc. Exp. Biol.* **14**, 321.

Booth, A. N., Murray, C. W., Jones, F. T. and De Eds, F. (1956), *J. Biol. Chem.* **223**, 251.

Booth, A. N., Emerson, A. H., Jones, F. T. and De Eds, F. (1957), *J. Biol. Chem.* **229**, 51.

Booth, A. N., Jones, F. T. and De Eds, F. (1958a) *J. Biol. Chem.* **230**, 661.

Booth, A. N., Jones, F. T. and De Eds, F. (1958b) *J. Biol. Chem.* **233**, 280.

Brown, J. P. (1980), *Mutat. Res.* **75**, 243.

Brown, J. P. and Dietrich, P. S. (1979), *Mutat. Res.* **66**, 223.

Brown, J. P., Crosby, G. A., Du Bois, G. E., Enderlin, F. E., Hale, R. L. and Wingard, R. E. (1978), *J. Agric. Food Chem.* **26**, 1418.

Bülles, H., Bülles, J., Krumbiegel, G., Mennicke, W. H. and Nitz, D. (1975), *Arzneimittel-Forsch.* **25**, 902.
Buset, H. and Scheline, R. R. (1979), *Biomed. Mass Spectrom.* **6**, 212.
Buset, H. and Scheline, R. R. (1980), *Acta Pharm. Suec.* **17**, 157.
Carpenedo, F., Bortignon, D., Bruni, A. and Santi, R. (1969), *Biochem. Pharmacol.* **18**, 1495.
Cheng, K. J., Krishnamurty, H. G., Jones, G. A. and Simpson, F. J. (1971), *Can. J. Microbiol.* **17**, 129.
Claveau, P. and Masquelier, J. (1966), *Can. J. Pharm. Sci.* **1**, 74.
Courbat, P., Favre, J., Guerne, R. and Uhlmann, G. (1966a), *Helv. Chim. Acta* **49**, 1202.
Courbat P., Uhlmann, G. and Guerne, R. (1966b) *Helv. Chim. Acta* **49**, 1420.
Das, N. P. (1969), *Biochim. Biophys. Acta* **177**, 668.
Das, N. P. (1971), *Biochem. Pharmacol.* **20**, 3435.
Das, N. P. (1974), *Drug Metab. Dispos.* **2**, 209.
Das, N. P. and Griffiths, L. A. (1966), *Biochem. J.* **98**, 488.
Das, N. P. and Griffiths, L. A. (1968), *Biochem. J.* **110**, 449.
Das, N. P. and Griffiths, L. A. (1969), *Biochem. J.* **115**, 831.
Das, N. P. and Sothy, S. P. (1971), *Biochem. J.* **125**, 417.
Das, N. P., Scott, K. N. and Duncan, J. H. (1973), *Biochem. J.* **136**, 903.
De Eds, F. (1968). In *Comprehensive Biochemistry* (ed. Florkin), Vol. 20, Elsevier, London.
Diamond, L. and Gelboin, H. V. (1969), *Science* **166**, 1023.
Fang, W-F. and Strobel, H. W. (1978), *Cancer Res.* **38**, 2939.
Förster, H. (1975). In *Topics in Flavonoid Chemistry and Biochemistry* (eds. L. Farkas, M. Garbor and F. Kallay), Elsevier, Amsterdam, p. 257.
Förster, H. (1977). In *Flavonoids and Bioflavonoids* (eds. L. Farkas, M. Gabor and F. Kallay), Elsevier, Amsterdam, p. 333.
Förster, H. (1978). In *O-(β-Hydroxyethyl) rutoside. Experimentelle und Klinische Ergebnisse* (eds W. Voelter and G. Jung), Springer-Verlag, Berlin, p. 43.
Förster, H., Bruhn, V. and Hoos I. (1972), *Arzneimittel-Forsch.* **22**, 1312.
Gabor, M. (1975), *Abriss der Pharmakologie von Flavonoiden*, Akademiai Kiado, Budapest, p. 24.
Giles, A. R. and Gumma, A. (1973), *Arzneimittel-Forsch.* **23**, 98.
Griffiths, L. A. (1962), *Nature (London)* **194**, 869.
Griffiths, L. A. (1964), *Biochem. J.* **92**, 173.
Griffiths, L. A. (1970), *Experientia* **26**, 723.
Griffiths, L. A. (1975). In *Topics in Flavonoid Chemistry and Biochemistry* (eds L. Farkas, M. Garbor and F. Kallay), Elsevier, Amsterdam, p. 201.
Griffiths, L. A. and Barrow, A. (1972), *Biochem. J.* **130**, 1161.
Griffiths, L. A. and Hackett, A. M. (1977). In *Flavonoids and Bioflavonoids* (eds. L. Farkas, M. Gabor and F. Kallay), Elsevier, Amsterdam, p. 325.
Griffiths, L. A. and Hackett, A. M. (1978), *Abstr. 7th Int. Congr. Pharmacol., 833.*
Griffiths, L. A. and Smith, G. E. (1972a), *Biochem. J.* **128**, 901.
Griffiths, L. A. and Smith, G. E. (1972b), *Biochem. J.* **130**, 141.
Griffiths, L. A., Hackett, A. M., Shaw, I. C. and Brown, S. (1980). In *Recent Developments in Mass Spectrometry in Biochemistry, Medicine and Environmental Research* (ed. A. Frigerio), Elsevier, Amsterdam, p. 91.
Gugler, R. and Dengler, H. J. (1973), *Naunyn-Schmiedeberg's Arch. Pharmacol.* **276**, 223.
Gugler, R., Leschnik, M. and Dengler, H. J. (1975), *Eur. J. Clin. Pharmacol.* **9**, 229.
Hackett, A. M. and Griffiths, L. A. (1977a), *Xenobiotica* **7**, 641.
Hackett, A. M. and Griffiths, L. A. (1977b), *Experientia* **33**, 161.
Hackett, A. M. and Griffiths, L. A. (1979), *Eur. J. Drug Metab.* **4**, 207.
Hackett, A. M. and Griffiths, L. A. (1981), *Drug. Metab. Dispos.* **9**, 54.
Hackett, A. M., Marsh, I., Barrow, A. and Griffiths, L. A. (1979), *Xenobiotica* **9**, 491.
Hackett, A. M., Griffiths, L. A., Luyckx, A. S. and van Cauwenberge, H. (1976), *Arzneimittel-Forsch.* **26**, 925.

Hardigree, A. A. and Epler, J. L. (1978), *Mutat. Res.* **58**, 231.
Harmand, M. F. and Blanquet, P. (1978), *Eur. J. Drug Metab.* **3**, 31.
Hawksworth, G., Drasar, B. S. and Hill, M. J. (1971), *J. Med. Microbiol.* **4**, 451.
Hems, R., Ross, B. D., Berry, M. N. and Krebs, H. A. (1966), *Biochem. J.* **101**, 284.
Honohan, T., Hale, R. L., Brown, J. P. and Wingard, R. E. (1976), *J. Agric. Food Chem.* **24**, 906.
Jeffrey, A. M., Knight, M. and Evans, W. C. (1969), *J. Gen. Microbiol.* **56**, 1.
Jeffrey, A. M., Knight, M. and Evans, W. C. (1972a), *Biochem. J.* **130**, 373.
Jeffrey, A. M., Jerina, D. M., Self, R. and Evans, W. C. (1972b), *Biochem. J.* **130**, 383.
Jung, G., Ottnad, M. and Voelter, W. (1977), *Eur. J. Drug Metab.* **2**, 131.
Kallianos, A. G., Petrakis, P. S. and Shetlar, M. R. (1959), *Arch. Biochem. Biophys.* **81**, 430.
Krishnamurty, H. G., Cheng, K-J, Jones, G. A., Simpson, F. J. and Watkin, J. E. (1970), *Can. J. Microbiol.* **16**, 759.
Kühnau, J. (1976), *Wld. Rev. Nutr. Diet* **24**, 117.
Labow, R. S. and Layne, D. S. (1972), *Biochem. J.* **128**, 491.
Laparra, J. and Michaud, J. (1973), *C. R. Acad. Sci. Paris* **276**, 2847.
Lietti, A., Cristoni, A. and Picci, M. (1976), *Arzneimittel-Forsch.* **26**, 829.
MacGregor, J. T. and Jurd, L. (1978), *Mutat. Res.* **54**, 297.
Marsh, C. A., Alexander, F. and Levvy, G. A. (1952), *Nature (London)* **170**, 163.
Masquelier, J., Claveau, P. and Golse, J. (1965), *Bull. Soc. Pharm. Bordeaux* **104**, 193.
Masri, M. S., Booth, A. N. and De Eds, F. (1959), *Arch. Biochem. Biophys.* **85**, 284.
Masri, M. S., Booth, A. N. and De Eds, F. (1962), *Biochim. Biophys. Acta* **65**, 495.
Mennicke, W. H., Lang, W. and Lorenz, D. (1979), *Hepatology (Falk Foundation)*, **79**, 411.
Mirkovitch, V., Robinson, J. W. L., Bel, F. and Gumma, A. (1973), *Arzneimittel-Forsch.* **23**, 967.
Mori, S. and Noguchi, I. (1970), *Arch. Biochem. Biophys.* **139**, 444.
Nakagawa, Y., Shetlar, M. R. and Wender, S. H. (1965), *Biochim. Biophys. Acta* **97**, 233.
Nilsson, A., Hill, J. L. and Davies, H. L. (1967), *Biochim. Biophys. Acta* **148**, 92.
Oshima, Y. and Watanabe, H. (1958), *J. Biochem. (Tokyo)* **45**, 973.
Oshima, Y., Watanabe, H. and Isakari, S. (1958), *J. Biochem. (Tokyo)* **45**, 861.
Oshima, Y., Watanabe, H. and Kuwazuka, S. (1960), *Bull. Agric. Chem. Soc. (Jpn.)* **24**, 497.
Oustrin, J., Fauran, M. J. and Commanay, L. (1977), *Arzneimittel-Forsch.* **27**, 1688.
Pamukcu, A. M., Hatcher, J., Taguchi, H. and Bryan, G. T. (1980a), *Proc. Am. Assoc. Cancer Res.* **21**, 74.
Pamukcu, A. M., Hatcher, J., Taguchi, H. and Bryan, G. T. (1980b), *Cancer Res.* **40**, 3468.
Paris, R. R. and Delaveau, P. (1977), *Plant. Med. Phytother.* **11**, No. spec., 198.
Peter, H., Fisel, J. and Weisser, W. (1966), *Arzneimittel-Forsch.* **16**, 719.
Petrakis, P. L., Kallianos, A. G., Wender, S. H. and Shetlar, M. R. (1959), *Arch. Biochem. Biophys.* **85**, 264.
Pfeifer, K., Mehnert, N. and Hulsmann, N. (1970), *Dte Gesundh-Wesen* **25**, 386.
Powell, G. M., Miller, J. J., Olavesen, A. H. and Curtis, C. G. (1974), *Nature (London)* **252**, 234.
Ruckstuhl, M., Beretz, A., Anton, R. and Landry, Y. (1979), *Biochem. Pharmacol.* **28**, 535.
Scheline, R. R. (1966), *J. Pharm. Pharmacol.* **18**, 664.
Scheline, R. R. (1968a), *Acta Pharmacol. Toxicol.* **26**, 189.
Scheline, R. R. (1968b), *Acta Pharmacol. Toxicol.* **26**, 332.
Scheline, R. R. (1970), *Biochim. Biophys. Acta* **222**, 228.
Scheline, R. R. (1978), *Mammalian Metabolism of Plant Xenobiotics*, Academic Press, London, p. 295.
Schultz, E., Engle, F. E. and Wood, J. M. (1974), *Biochemistry* **13**, 1768.
Schwabe, K-P. and Flohe, L. (1972), *Hoppe-Seyler's Z. Physiol. Chem.* **353**, 476.
Setnikar, I., Cova, A. and Magistretti, M. J. (1975), *Arzneimittel-Forsch.* **25**, 1916.
Shamrai, E. F. and Fedurov, V. V. (1967), *Ukrayins'ki Biokhimichnyi Zhurnal* **39**, 83.
Shamrai, E. F. and Zaprometov, M. N. (1964), *Biokhimiya* **29**, 595.
Shaw, I. C. and Griffiths, L. A. (1980), *Xenobiotica* **10**, 905.
Shutt, D. A. and Braden, A. W. H. (1968), *Aust. J. Agric. Res.* **19**, 545.

Simpson, F. J., Jones, G. A. and Wolin, E. A. (1969), *Can. J. Microbiol.* **15**, 972.
Smith, R. L. (1973), *The Excretory Function of Bile*. Chapman and Hall, London.
Sugimura, T., Nagao, M., Matsushima, T., Yahagi, T., Seino, Y., Shirai, A., Sawamura, M., Natori, S., Yoshihira, K., Fukuoka, M. and Kuroyanagi, M. (1977), *Proc. Jpn. Acad. Ser. B* **53**, 194.
Svardal, A., Buset, H. and Scheline, R. R. (1980), *Acta Pharm. Suec.* **18**, 55.
Takacs, O. and Gabor, M. (1975). In *Topics in Flavonoid Chemistry and Biochemistry* (eds. L. Farkas, M. Gabor and F. Kallay), Elsevier, Amsterdam, p. 227.
Tamura, G., Gold, C., Ferro-Luzzi, A. and Ames, B. N. (1980), *Proc. Natl. Acad. Sci. U.S.A.* **77**, 4961.
Tamemasa, O., Goto, R. and Ogura, S. (1976), *Pharmacometrics (Japan)*, **12**, 193.
Tan, H. S. I., Mowery, P. J., Ritschel, W. A. and Neu, C. (1978), *J. Pharm. Sci.* **67**, 1142.
Varma, S. D. and Kinoshita, J. H. (1976), *Biochem. Pharmacol.* **25**, 2505.
Varma, S. D., Mikuri, I. and Kinoshita, J. H. (1975), *Science* **188**, 1215.
Wattenberg, L. W., Page, M. A. and Leong, J. L. (1968), *Cancer Res.* **28**, 934.
Wiebel, F. J. and Gelboin, H. V. (1975), *Biochem. Pharmacol.* **24**, 1511.
Wienert, V. and Gahlen, W. (1970), *Hautarzt* **21**, 278.
Wurm, G. (1975), *Dtsch. Apothek. Z.* **115**, 355.

Plant Species Index

Subject Index

Individual flavonoids are listed in the various check lists of known compounds (see under check lists) and can be traced back in the preceding chapter. A list of all ^{13}C NMR spectra illustrated appears on pp. 51–56.